GROUP REPRESENTATION THEORY FOR PHYSICISTS

GROUP REPRESENTATION THEORY FOR PHYSICISTS

JIN-QUAN CHEN

World Scientific
Singapore • New Jersey • London • Hong Kong

Published by

World Scientific Publishing Co. Pte. Ltd.
P O Box 128, Farrer Road, Singapore 9128

USA office: World Scientific Publishing Co., Inc.
687 Hartwell Street, Teaneck, NJ 07666, USA

UK office: World Scientific Publishing Co. Pte. Ltd.
73 Lynton Mead, Totteridge, London N20 8DH, England

GROUP REPRESENTATION THEORY FOR PHYSICISTS

ISBN 9971-50-105-8
9971-50-099-X (pbk)

Printed in Singapore by JBW Printers & Binders Pte. Ltd.

Foreword

A new generation of books on group theory for physicists has appeared over the last ten years. Many of them deal only with elementary particle physics or with condensed matter physics. This volume by Prof. Jin-Quan Chen is a serious attempt to cover a broad range of applications of group theory to physics. It begins with an introduction to the elements of group theory and the theory of representations. Representations of finite groups and character theory are carefully treated and applied in later chapters to point groups and space groups, where thorough and practical information is given about molecular and crystal groups.

The permutation group is discussed in detail, and its use in finding the irreps of unitary groups is carefully and completely covered. Lie groups and Lie algebras are treated sufficiently to enable their use in elementary particle physics, with a thorough presentation of Dynkin diagrams and the reduction of products.

For spectroscopy and, in particular, nuclear spectroscopy, this is the first up-to-date and thorough treatment of the calculation of isoscalar factors and coefficients of fractional parentage. There are extensive tables and indications of computational methods.

With his collaborators in Nanjing and Philadelphia, Chen has contributed extensively to new developments over the past decade. These researches are incorporated into the present book, of which a preliminary version was published a few years ago in the People's Republic of China.

It has been a pleasure for me to talk with Chen many times about group theory and a special pleasure to welcome the appearance of this book.

Morton Hamermesh
University of Minnesota

Preface

Conscious of the frustration we experienced as we tried to learn group theory, and apply it to problems in physics, F. Wang, M. J. Gao and I endeavoured in 1974 to carry out a systematic reform of traditional group representation theory. Our aim was to establish a new approach to group representations in accordance with the concept and method used in quantum mechanics, so that it would be much more accessible to physicists and chemists.

The breakthrough came in late 1974. A new approach to group representation gradually came into being and with it a new method, the eigenfunction method (EFM), emerged for finding primitive characters, irreducible bases, Clebsch-Gordan (CG) coefficients, etc. From 1976–1978, the EFM was translated into codes and for the first time the CG coefficients and the outer-product reduction coefficients for the permutation group $S_2 - S_6$ were computed by the EFM. The EFM was extended in 1979 to the computation of the CG coefficients, Racah coefficients, isoscalar factors, subduction coefficients of unitary groups. From 1981–1987, the new approach was applied to space groups and graded unitary groups and established several codes for space groups, SU(n) CG coefficients, $\mathrm{SU}(mn) \supset \mathrm{SU}(m) \times \mathrm{SU}(n)$ and $\mathrm{SU}(m+n) \supset \mathrm{SU}(m)\times \mathrm{SU}(n)$ coefficients of fractional parentage, etc.

The new approach has been introduced at various universities and institutions, such as the City College of the City University of New York, the University of Chicago, Clarkson, Dalhousie, Drexel, Duke, Hong Kong, Minnesota, Maryland, Montreal, Singapore, Yale, and Los Alamos National Laboratory, as well as at the IXth and XIVth International Colloquium on Group Theoretical Method in Physics held in Mexico 1980, and Seoul 1985, respectively.

The Chinese book *A New Approach to Group Representation Theory* was written in 1981 and published in 1985 and has since been used as a textbook in Nanjing University for undergraduate and graduate students. Prior to the publication of the Chinese book, the manuscript had been used as a textbook in Nanjing University and Drexel University in Philadelphia.

The present book was based on the Chinese book and the new developments during the period of 1981–1987. All chapters except chapters 4 and 7 have been re-written.

I am indebted to my collaborators Mei-Juan Gao who took part in the writing of Chapter 8 and performed most of the practical computations, Guang-Qun Ma who was involved in the writing of Chapter 10, and it is he who first attempted to use the EFM to treat the space group representation, Fan Wang and Xuan-Gen Chen for their invaluable suggestions and criticism. I am grateful to Professors Xiao-Qian Zhou, Shi-Shu Wu and Duan Feng for their constant encouragement and patient reading of the manuscript. Thanks are due also to Professors Tan Lu, Fu-Cho Pu, Qin-Yue Qu, Hong-Zhou Sun, Rui-Bao Tao, Cheng-Li Wu, Rong-Jue Wei, Xi-De Xie and Guang-Xiang Xu for their enthusiasm in this work and support, and to Xue-Chu He and Jia-Song Wang for their help in the programming. I am grateful to many students who have offered helpful criticism and raised challenging questions.

Mr. Pei-Ning Wang typed the initial draft of Chapters 4–9, and my son, Bing-Qing wrote several software codes for facilitating the typing and proof-reading of the manuscript. Without their help it would be inconceivable that the re-writing could have been accomplished within half a year.

Special mention must be made of Professors B. Bayman, M. Hamermesh and K.T. Hecht for their constant interest and many illuminating discussions, and D.H. Feng, who gave me the opportunity and encouragement to teach a course on group theory at Drexel University.

Last but not least I would like to thank Professors L. C. Biedenharn, J. Birman, U. Fano, R. Gilmore, J. Ginocchio, J. J. Griffin, F. Iachello, J. Paldus, J. Patera and K. K. Phua for their stimulating discussion and the hospitality extended to me during my visits to their institutions.

Nanjing University Jin-Quan Chen
March, 1987

Contents

6.5. The CSCO-II and irreducible bases of SO_3 . . . 291
6.6. The intrinsic state of SO_3 . . . 292
6.7. The projection state of SO_3 . . . 293
6.8. Irreducible tensors of SO_3 and $\overline{SO}_3$ . . . 295
6.8.1. The irreducible tensor of the adjoint rep of SO_3 and $\overline{SO}_3$ . . . 295
6.8.2. Irreducible tensor of SO_3 and $\overline{SO}_3$ in general cases . . . 295

Chapter 7 The Unitary Groups

7.1. Unitary groups in coordinate space and state space . . . 298
7.2. Relations between the CSCO-I and generators of unitary groups and permutation groups . . . 301
7.2.1. The Gel'fand invariants . . . 301
7.2.2. The relation between the CSCO-I of permutation groups and unitary groups . . . 302
7.2.3. Relations between the generators of unitary groups and permutation groups . . . 305
7.3. The CSCO-II and CSCO-III of U_n and SU_n . . . 306
7.4. The Gel'fand basis and Gel'fand matrix elements . . . 308
7.5. The Gel'fand basis of unitary groups and quasi-standard basis of permutation groups . . . 311
7.5.1. The CSCO-II of unitary groups and CSCO of the broken chains of permutation groups . . . 311
7.5.2. The labeling and finding of the Gel'fand basis . . . 313
7.6. The contragredient representation . . . 320
7.7. The CG coefficients of SU_n group . . . 322
7.7.1. The CG coefficients of U_n and the IDC of the permutation group . . . 322
7.7.2. The procedure for evaluating the SU_n CG coefficients . . . 325
7.7.3. Phase convention . . . 327
7.8. The CG coefficients of SU_n and the $S_f \supset S_{f_1} \otimes S_{f_2}$ irreducible basis . . . 327
7.9. The $SU_{mn} \supset SU_m \times SU_n$ irreducible basis . . . 328
7.9.1. The CG coefficients of S_f and the $SU_{mn} \supset SU_m \times SU_n$ irreducible basis . . . 328
7.9.2. The irreps $([\nu_1], [\nu_2])$ of the groups SU_m and SU_n contained in the irrep $[\nu]$ of SU_{mn} . . . 332
7.9.3. Representation transformation between the $SU_{mn} \supset SU_m \times SU_n$ irreducible basis and the SU_{mn} Gel'fand basis . . . 332
7.10. The $SU_{n_1n_2n_3} \supset SU_{n_1} \times SU_{n_2} \times SU_{n_3}$ irreducible bases and the Racah coefficients of permutation groups* . . . 333
7.11. $SU_{n_1n_2n_3n_4} \supset SU_{n_1} \times SU_{n_2} \times SU_{n_3} \times SU_{n_4}$ irreducible basis and the 9ν coefficients of the permutation group* . . . 335
7.12. $SU_{m+n} \supset SU_m \otimes SU_n$ irreducible basis . . . 337
7.12.1. The IDC of permutation groups and $SU_{m+n} \supset SU_m \otimes SU_n$ irreducible bases . . . 337
7.12.2. The content of irreps $([\nu_1], [\nu_2])$ of $SU_m \otimes SU_n$ in the irrep of SU_{m+n} . . . 338

7.12.3. The representation transformation between the $SU_{m+n} \supset SU_m \otimes SU_n$ irreducible basis and the Gel'fand basis of SU_{m+n} . . . 338
7.13. The isoscalar factors and the fractional parentage coefficients . . . 340
7.13.1. Isoscalar factors . . . 340
7.13.2. The orbital fractional parentage coefficients (CFP) . . . 342
7.13.3. The spin-isospin CFP . . . 345
7.13.4. The total CFP . . . 346
7.13.5. Eigenfunction method for evaluating the CFP . . . 347
7.14. $S_f \supset S_{f_1} \otimes S_{f_2} \otimes S_{f_3}$ irreducible basis and SU_n Racah coefficients* . . . 353
7.15. $S_f \supset S_{f_1} \otimes S_{f_2} \otimes S_{f_3} \otimes S_{f_4}$ irreducible basis and the 9ν coefficients of SU_n* . . . 355
7.15.1. The 9ν coefficients of SU_n . . . 355
7.15.2. Evaluation of the Racah coefficients and 9ν coefficients of SU_n . . . 356
7.16. $SU_{mn} \supset SU_m \times SU_n$ CFP . . . 358
7.16.1. $SU_{mn} \supset SU_m \times SU_n$ CFP and $S_{f_1+f_2} \supset S_{f_1} \otimes S_{f_2}$ ISF . . . 358
7.16.2. The evaluation of the $SU_{mn} \supset SU_m \times SU_n$ many-particle CFP . . . 360
7.16.3. The symmetries of the $SU_{mn} \supset SU_m \times SU_n$ ISF . . . 362
7.16.4. More examples . . . 364
7.16.5. $SU_{4(2l+1)} \supset (SU_{2l+1} \supset SO_3) \times (SU_4 \supset SU_2 \times SU_2)$ ISF and total CFP . . . 365
7.17. The $SU_{m+n} \supset SU_m \otimes SU_n$ CFP* . . . 366
7.17.1. The $S_f \supset S_{f-1}$ outer-product ISF (The $SU_f \supset SU_{f-1} \otimes U_1$ ISF) . . . 366
7.17.2. The $S_f \supset S_{f_{12}} \otimes S_{f_{34}}$ outer-product ISF ($SU_f \supset SU_{f_{12}} \otimes SU_{34}$ ISF) . . . 369
7.17.3. The $\mathcal{SU}_{m+n} \supset \mathcal{SU}_n$ ISF and $S_f \supset S_{f_{12}} \otimes S_{f_{34}}$ outer-product ISF . . . 371
7.17.4. The evaluation of $\mathcal{SU}_{m+n} \supset \mathcal{SU}_m \otimes \mathcal{SU}_n$ ISF . . . 371
7.17.5. Symmetries of the $\mathcal{SU}_{m+n} \supset \mathcal{SU}_m \otimes \mathcal{SU}_n$ ISF . . . 372
7.18. The SU_n singlet factor . . . 373
7.19. Second quantized expression for the CFP . . . 374
7.19.1. One-particle CFP . . . 374
7.19.2. Two-particle CFP . . . 377
7.19.3. The CFP in the interacting boson model . . . 378

Chapter 8 The Point Groups

8.1. Basic operations of point groups . . . 380
8.1.1. Basic operations and their faithful reps . . . 380
8.2. Some commonly used point groups . . . 385
8.3. The CSCO-I and CSCO-II of point groups . . . 389
8.3.1. The conventional labeling for point group irreps (Mullikan notation) . . . 389
8.3.2. The CSCO-II for the commonly used point groups . . . 395
8.4. Irreducible matrix elements and irreducible basis of point groups G . . . 395
8.4.1. Irreducible matrices . . . 395
8.4.2. Irreducible basis . . . 398
8.4.3. The splitting of atomic levels and the $O_3 \supset G \supset G(s)$ basis . . . 399
8.5. The CG coefficients of point groups . . . 402
8.5.1. The CG series of point groups . . . 402
8.5.2. The CG coefficients of point groups . . . 403

Appendix

Contents of Some Important Tables

Glossary

1. Group

E	the Euclidean group
G	a group or a representation group
$\overline{G}$	the intrinsic group of G
$\mathbf{G}$	a space group
$\mathbf{G}^{\dagger}$	double group or double space group
$\mathbf{g}$	a point group
$\mathbf{G}_0$	isogonal point group of the space group $\mathbf{G}$
$\mathbf{G}_0(\mathbf{k})$	little co-group
$G(\mathbf{k})$	little group
$G_{\mathbf{k}}, G'_{\mathbf{k}}$	representation groups
$\|G\|, \|\mathbf{G}\|$	the order of G or $\mathbf{G}$
G_s	a subgroup of G
$G(s) = G(1) \supset G(2) \supset \ldots$	a canonical subgroup chain of G
$\overline{G}(s) = \overline{G}(1) \supset \overline{G}(2) \supset \ldots$	a canonical subgroup chain of $\overline{G}$
$G(s')$	a non-canonical subgroup chain
$\mathbf{P}$	holosymmetric point group of the crystal system
$\mathbf{P}(\mathbf{k})$	symmetric group of $\mathbf{k}$
$S_n(\mathcal{S}_n)$	permutation group in coordinate (state) space
$S_n(\omega)$	a permutation group in the indices (ω)
$\mathbf{T}$	translation group, or lattice group
$U_n, SU_n(\mathcal{U}_n, S\mathcal{U}_n)$	unitary group in coordinate (state) space

2. Group elements

$a, R_a, R(a), R, S, T, \ldots$	group elements
α, β, γ	group elements of a point group
$p(\wp)$	permutation operator in coordinate (state) space
$Q = \binom{\omega_0}{\omega}$	order-preserving permutation
$R_i = \{\gamma_i \| \mathbf{V}(\gamma_i)\}'$ $= \exp[i\mathbf{k}\cdot\mathbf{V}(\gamma_i)]\{\gamma_i \| \mathbf{V}(\gamma_i)\}$	group element of the rep group $G'_{\mathbf{k}}$

3. Spaces and basis vectors

$L, \mathcal{L}$	any rep space
L_g	group space of a group G, or a rep group G
$L_c(L_n)$	class space of a group G (a rep group $G'_{\mathbf{k}}$)
L_ν	eigenspace of the CSCO-I of $G, L \subset L_g$
$\mathcal{L}_\nu$	eigenspace of the CSCO-I of $G, \mathcal{L}_\nu \subset \mathcal{L}$
$L_{(\nu)k}$	the k-th irreducible space of G
$L(\mathbf{k})$	group space of the rep group $G'_{\mathbf{k}}$
$\mathcal{L}(\mathbf{k})$	eigenspace of the translation operator $\{\varepsilon \vert \mathbf{R}_n\}$
$\mathcal{L}_{(^*\mathbf{k})}$	representation space of a space group $\mathbf{G}$
$\vert\omega_0\rangle = \vert i_1 i_2 \ldots i_n\rangle$	normal order state (without repeated state labels)
$\vert\omega\rangle = \vert i_1 i_2 \ldots i_n\rangle$	normal order state (with repeated state labels)

4. Complete set of commuting operators

$C(C(i))$	CSCO-I, or CSCO of $G(G(i))$
$C(s) = (C(1), C(2), \ldots)$	complete set of commuting operators of $G(s)$
$\overline{C}(s) = (\overline{C}(1), \overline{C}(2), \ldots)$	complete set of commuting operators of $\overline{G}(s)$
$C(s')$	CSCO of a non-canonical subgroup chain $G(s')$
$M = (C, C(s))$	CSCO-II of G
$K = (C, C(s), \overline{C}(s))$	CSCO-III of G
$C(n)$	CSCO-I of S_n, or the two-cycle class operator of S_n
$\mathcal{C}(n)$	CSCO-I of the state permutation group $\mathcal{S}_n$

5. Irreducible basis

$\psi_m^{(\nu)}$	$G \supset G(s)$ irreducible basis
$\psi_m^{(\nu)k}, P_m^{(\nu)k}$	$G \supset G(s)$ and $\overline{G} \supset \overline{G}(s)$ irreducible basis
$\vert Y_m^{(\nu)}\rangle$	Yamanouchi basis
$\vert Y_m^{(\nu)}(\omega)\rangle$	Yamanouchi basis of $S_n(\omega)$
$\left\vert \begin{smallmatrix}[\nu]\\ W\end{smallmatrix}\right\rangle, \left\vert \left(\begin{smallmatrix}[\nu]\\ (m)\end{smallmatrix}\right)\right\rangle$	Gel'fand basis for a unitary group
$\left\vert [\nu], {}^{\tau[\nu_1][\nu_2]}_{m_1 m_2}\right\rangle$	$S_n \supset S_{n_1} \otimes S_{n_2}$ basis
$\left\vert [\nu], {}^{\tau[\nu_1][\nu_2]}_{W_1 W_2}\right\rangle$	$SU_{mn} \supset SU_m \times SU_n$ or $SU_{m+n} \supset SU_m \otimes SU_n$ basis
$\psi_{\mathbf{k}} = \exp[i(\mathbf{k} + \mathbf{K}_m)\cdot\mathbf{r}]$	irreducible basis of the lattice group

6. Coefficients

$C^{(\nu)\tau, m}_{\nu_1 m_1, \nu_2 m_2}$	CG coefficients

$C^{(\nu)\tau,m}_{\nu_1 m_1,\nu_2 m_2,\omega}$	induction coefficients (IDC) of the permutation group
$\begin{bmatrix} \nu\mathbf{k}\nu'\mathbf{k}' & \nu''\mathbf{k}''\theta \\ \sigma a\sigma' a' & \sigma'' a'' \end{bmatrix}$	CG coefficients for space groups
$C^{(\nu)\tau,\beta\Lambda}_{\nu_1\beta_1\Lambda_1,\nu_2\beta_2\Lambda_2}$	$G \supset G_1$ isoscalar factor (ISF)
$C^{[\nu]\beta,[\nu']\beta'}_{\sigma\sigma',\mu\mu'}$	$S_n \supset S_{n-1}$ ISF or $S_n \supset S_{n-1}$ outer-product ISF
$C^{[\nu]\beta,\tau[\nu']\beta'[\nu'']\beta''}_{[\sigma]\theta\sigma'\sigma'',[\mu]\varphi\mu'\mu''}$	$S_n \supset S_{n_1} \otimes S_{n_2}$ ISF, or $S_n \supset S_{n_1} \otimes S_{n_2}$ outer-product ISF
$C^{[\nu]\tau,\beta[\sigma]\theta[\mu]\varphi}_{[\nu']\beta'\sigma'\mu',[\nu'']\beta''\sigma''\mu''}$	$SU_{mn} \supset SU_m \times SU_n$ or $SU_{m+n} \supset SU_m \otimes SU_n$ ISF
$C^{[\nu]\tau,\alpha L}_{[\nu_1]\alpha_1 L_1,[\nu_2]\alpha_2 L_2}$	$SU_{2l+1} \supset SO_3$ ISF (or CFP)
$C^{[\nu]\tau,\beta ST}_{[\nu_1]\beta_1 S_1 T_1,[\nu_2]\beta_2 S_2 T_2}$	$SU_4 \supset SU_2 \times SU_2$ ISF (or CFP)
$\left\langle \begin{smallmatrix}[\nu]\\ m\end{smallmatrix} \middle\vert [\nu], \begin{smallmatrix}\tau[\nu_1][\nu_2]\\ m_1 m_2\end{smallmatrix} \right\rangle$, or $\langle[\nu]m\vert\tau[\nu_2]m_2\rangle$,	subduction coefficients (SDC) of permutation groups
$\left\langle \begin{smallmatrix}[\nu]\\ W\end{smallmatrix} \middle\vert [\nu], \begin{smallmatrix}\tau[\nu_1][\nu_2]\\ W_1 W_2\end{smallmatrix} \right\rangle$,	$SU_{m+n} \downarrow (SU_m \otimes SU_n)$ SDC, or $SU_{mn} \downarrow (SU_m \times SU_n)$ SDC
$R^{[\nu]k}(\omega)$	normalization coefficients

7. Miscellaneous

$C_i, C(\varphi)$	class operators
CFP	fractional parentage coefficients
$D^{(\nu)}(R_i)$	the irreducible matrix of the group element R_i
$\tilde{D}$	the transpose of the matrix D
EFM	eigenfunction method
$\varepsilon_i(\nu_1\nu_2\nu)$	phase factor
$h_\nu, \vert\nu\vert$	dimension of the irrep
irrep	irreducible representation
ISF	isoscalar factor
N	number of classes
n	number of linearly independent class operators of a rep group G
ν_0	the adjoint rep
$P^{(\nu)k}_m$	generalized projection operator
$P^{(\nu)}$	projection operator
rep	representation

The main notations for the space group are listed below:

Genealogical relation for the space group and its point group

$$\mathbf{G} \supset \mathbf{G}(\mathbf{k}) \supset \mathbf{G}(s) \supset \mathbf{T} \,,$$
$$\mathbf{P} \supset \mathbf{G}_0 \supset \mathbf{G}_0(\mathbf{k}) \,,$$
$$\mathbf{P} \supset \mathbf{P}(\mathbf{k}) \,,$$
$$\mathbf{G}_0(\mathbf{k}) = \mathbf{P}(\mathbf{k}) \cap \mathbf{G}_0$$

Coset decomposition

$$\mathbf{G} = \sum_{\sigma} {}_1^q \oplus \{\beta_\sigma|\mathbf{V}_\sigma\}\mathbf{G}(\mathbf{k}) \,,$$

$$\mathbf{G}_0 = \sum_{\sigma} {}_1^q \oplus \beta_\sigma \mathbf{G}_0(\mathbf{k})$$

Table 1. Space-group elements, IRB and irrep.

	$\mathbf{G}$	$\mathbf{G}(\mathbf{k})^{a}$	$\mathbf{T}$	$\mathbf{G}_{\mathbf{k}}'^{b}$	$\mathbf{G}_0(\mathbf{k})^{c}$
group element	$\{\alpha_i|\mathbf{V}\ (\alpha_i)+\mathbf{R}_n\}$	$\{\gamma_i|\ \mathbf{V}(\gamma_i)+\mathbf{R}_n\}$	$\{\varepsilon|\mathbf{R}_n\}$	$\{\gamma_i|\mathbf{V}\ (\gamma_i)\}'$	γ
basis vector	$\psi^{(\nu)}_{\mathbf{k}_\sigma a} = \{\beta_\sigma|\mathbf{V}_\sigma\}\psi^{(\nu)}_{\mathbf{k},a}$	$\psi^{(\nu)}_{\mathbf{k},a}$	$\psi_{\mathbf{k}} = e^{i(\mathbf{k}+\mathbf{K}_m)\cdot r}$	$\psi^{(\nu)}_{\mathbf{k},a}$	
irrep	$D^{(*\mathbf{k})(\nu)}_{\tau b,\sigma a}$	$D^{(\mathbf{k})(\nu)}_{ba}$	$\exp(-i\mathbf{k}\cdot\mathbf{R}_n)$	$D^{(\mathbf{k})(\nu)}_{ba}$	$\Delta^{(\nu)}_{ba}$
dimension	qh_ν	h_ν	1	h_ν	h_ν

[a]The irrep of $\mathbf{G}(\mathbf{k})$ is $D^{(\mathbf{k})(\nu)}(\{\gamma|\mathbf{c}\}) = \exp(-i\mathbf{k}\cdot\mathbf{c})\Delta^{(\nu)}(\gamma)$.

[b]$\{\gamma_j|\mathbf{V}\ (\gamma_j)\}' = R_j = \exp[i\mathbf{k}\cdot\mathbf{V}\ (\gamma_j)]\ \{\gamma_i|\mathbf{V}(\gamma_i)\}$, $j = 1,2,\ldots,|\ \mathbf{G}_0\ (\mathbf{k})\ |$, are the active elements of the rep group $\mathbf{G}'_{\mathbf{k}}$.

[c]$\Delta^{(\nu)}$ is the projective irrep of the little co-group $\mathbf{G}_0(\mathbf{k})$ and $\Delta^{(\nu)}(\gamma) = D^{(\mathbf{k})(\nu)}(\{\gamma|\mathbf{V}(\gamma)\}')$.

Introduction

Group theory plays a very important role in physics and chemistry, and its importance continues to grow seemingly endlessly. The representation theory for both finite and compact Lie groups is treated extensively in numerous books and articles. However, they are basically of the same category. This theory, which we call the traditional group rep theory, seems to be perfect from the mathematical point of view. Nevertheless, it is not totally satisfying from a practical or physical point of view. Many sophisticated physicists, who were quite at home in their own fields, seemed to be afraid of group theory and expressed their dissatisfaction with traditional rep theory (Sokolov 1956, Salam 1963, Lipkin 1966). In his opening speech for "Seminars on Theoretical Physics" held in Trieste in 1962, Salam said "In 1951, I had the good fortune of listening to Professor Racah's lecture on Lie groups at Princeton. After attending these lectures, I thought, "This is really too hard. I cannot learn all this ... All this is too damned hard and unphysical."

Therefore, the first serious drawback of the traditional group rep theory is that it is unphysical. Group theory was introduced into mathematics as early as 1810, and the theory of group rep was developed mainly by mathematicians during the 1920's, before quantum mechanics was formulated; in this respect it is unlike calculus, which was invented about the same time as Newton's laws were discovered. Second, there is no general method for treating various kinds of group representation problems. Any given technique applies only to a particular problem and for a particular group. Not only do the methods for dealing with point groups, permutation groups, space groups, and Lie groups all differ drastically from each other, but the methods for finding the characters, irreducible basis (IRB), irreducible matrices and Clebsch-Gordan (CG) coefficients also vary from one to the other. Therefore, in many cases, these methods are more of an art than a science. Third, in physical applications, we often need to construct an IRB $\psi_m^{(\nu)}$ symmetry adapted to a given group chain $G \supset G(s)$. The standard method is to use the generalized projection operator

$$P_{mk}^{(\nu)} = \frac{h_\nu}{g} \sum_R D_{mk}^{(\nu)}(R)^* R,$$

where $D^{(\nu)}(R)$ are irreducible matrices in the $G \supset G(s)$ IRB, which in turn depend on the $G \supset G(s)$ IRB $\psi_m^{(\nu)}$ through the relation

$$D_{mk}^{(\nu)}(R) = \langle \psi_m^{(\nu)} | R | \psi_k^{(\nu)} \rangle.$$

Now the trouble is that the $G \supset G(s)$ IRB is not known yet. Thus we are at an impasse when both the matrices and IRB are unknown.

As pointed out by Salam (1963), a battle has raged between the amateurs and professional group theorists; the amateurs have maintained that everything one needs from the theory of groups can be discovered by the light of nature provided one knows how to multiply two matrices. As an amateur myself, in this book I have introduced a new (certainly, non-professional) approach to group rep theory and it is quite interesting to note that the foundation of the new approach is precisely the theory of the complete set of commuting operators (CSCO) initiated by Dirac, the prince of amateurs in the field of group theory.

The special features of the approach are as follows.

1. *Accessibility*: The new representation theory for groups is essentially an extension of representation theory in quantum mechanics. Group rep theory is intimately related to quantum mechanics just as calculus is to classical mechanics. Thus it is easily acceptable to physicists. For a group G, three kinds of CSCO, the CSCO-I, -II, and -III are introduced, which are, roughly speaking, the CSCO for the class space, the irreducible space, and the group space, respectively. They are the analogous of $\mathbf{J}^2$, $(\mathbf{J}^2,\ J_z)$, $(\mathbf{J}^2,\ J_z, J_3)$, respectively, for the rotation group in three-dimensional space.

2. *Versatility*: It is constructive in nature, leading to a new method, the so-called eigenfunction method (EFM) for determining group rep. The problems of determining (1) the primitive characters and isoscalar factors, (2) the $G \supset G(s)$ IRB and CG coefficients, and (3) the irreducible matrices are all reduced to a single recipe: seeking the eigenvectors (or eigenfunctions) of the CSCO-I, -II, and -III of G, respectively. The EFM proves to be very powerful and versatile in treating point groups, permutation groups, unitary groups, graded unitary groups and space groups, and for both vector and projective reps. The EFM for the primitive characters, IRB, irreducible matrices, and CG coefficients of a discrete group (both for finite and infinite types) is simpler than conventional methods and is flexible enough to obtain the irreducible basis adapted to any given group chain $G \supset G(s)$ without need of any knowledge of the irreducible matrices, or conversely, to obtain all the irreducible matrices in any given $G \supset G(s)$ classification without any knowledge of the irreducible basis.

3. *Applicability*: Since the ultimate step of the method is the diagonalization of the representative matrices of a certain type of CSCO, the procedure can be easily translated into a computer program. Several standard codes are already available.

The book is self-contained and suitable for self-study, and is a combination of a textbook and a monograph. By ignoring some proofs and some paragraphs or passages marked with asterisks, the book becomes an easily readable textbook. The theory is developed starting with concrete examples and leading up to more abstract conclusions, as well as from the special to the general, supplemented with abundant illustrative examples. The stress is on the EFM technique rather than on strict rigor. Some theorems are cited without proof, since the theorems are easily understandable and their proofs can be found in almost every textbook. A knowledge of elementary group theory is not necessary for reading the book, but a knowledge of the linear algebra and representation theory in quantum mechanics was assumed.

The various important coefficients, such as the Clebsch-Gordan, Racah, subduction and induction coefficients, the isoscalar factors and fractional parentage coefficients are discussed in detail for point groups, permutation groups, unitary groups and space groups. Tables for several useful coefficients are given. Some new dualities between the permutation group and unitary groups are disclosed and are fully exploited for computing many coefficients of unitary groups in terms of those of the permutation groups. The theory on roots and weights in Lie groups is also reformulated in the spirit of representation theory of quantum mechanics. The applications of group theory in quantum mechanics are discussed with emphasis on application to many-body systems. The connection between the new and traditional approaches is discussed. There should be no difficulty for a reader of the present book to understand the conclusions derived in other textbooks or in the literature, although the derivations given here may be totally different.

Tables and figures are indexed according to their section numbers. For example, Table(Fig.) x.y-n denotes the n-th table (figure) in Sec. x.y. If there is only a single table or figure in the section, then the suffix "-1" will be omitted. References are indicated by the names of the first authors followed by the year of publication; if this is still not sufficient, then an index $[x]$ will precede the year.

Chapter 1

Elements of Group Theory

1.1. Definition of Group

A set of elements (or operators) $\{a, b, c, \ldots\}$ or $\{R_a : a = 1, 2, 3, \ldots\}$, or $\{R, S, T, \ldots\}$ is called a group G, if a multiplication rule is defined for any two elements so that the product ab has a definite meaning and the following four postulates are satisfied:

1. *Closure*: If a and b belong to the set, then ab also belongs to the set.
2. *Associativity*: $a(bc) = (ab)c$.
3. There exists the identity element e such that $ae = ea = a$ for any a belonging to the set.
4. There exists the inverse element, i.e., for each element a, there is a corresponding element b such that $ab = ba = e$. b is called the inverse element of a and denoted by $b = a^{-1}$.

Notice that, in general $ab \neq ba$. Therefore the order of multiplication is important.

An *Abelian group* is one whose elements are commuting with one another, i.e., $[a, b] = ab - ba = 0$.

A *finite group* is one having a finite number of the elements.

For a *continuous group*, the group elements may be labeled by parameters which (or some of which) vary continuously. A continuous group is an infinite group.

In the following we use $G = \{a\}$ to denote a group, and use $a \in G$ to denote that a is an element of G (read: a belongs to G).

The *order* of a finite group G is defined as the total number of its elements and will be denoted by g or $|G|$.

For a finite group, we must have $R^n = e$ for some positive integer n. The smallest possible positive n for which $R^n = e$ is called the *order* of the element R, and denoted as $|R|$.

The set of elements $R, R^2, R^3, \ldots, R^n (= e)$ forms a group, called the cyclic group of order $n = |R|$, and is often denoted by C_n.

The *generators* of a group G are the elements $a, b, c \ldots$ of G, if every element of G is expressible as a finite product of powers (including negative powers) of $a, b, c, \ldots$.

From mathematical point of view, the product ab of two elements a and b can be defined arbitrarily. In physics, we are mainly interested in the group of transformations, or the group of operators. In such cases, the group elements $R_a, R_b \ldots$ represent a set of transformations or operators, and the product $R_a R_b = R_c$ is defined as the resultant operation of first operating with R_b and then with R_a.

Definition 1.1: Two sets S and S' are said to be commutative and denoted by $[S, S']=0$, if each element of S commutes with each element of S'.

All the operations which leaves a system (or a geometry object) unchanged (i.e. it appears not to have changed after the operation) form a group called the *symmetry group* of the system (or the object).

Obviously, the Hamiltonian H of a microscopic system commutes with the symmetry group G of the system, i.e.,

$$[H, G] = 0 . \tag{1-1a}$$

According to Definition 1.1, this implies that

$$[H, R_a] = 0 \quad \text{for } R_a \in G . \tag{1-1b}$$

Examples of groups:

1. All the integers under addition form an infinite discrete group.
2. The integers modulo n (i.e., $a = b$ if $a - b = mn, m$ being an integer) under addition form a group called the *group of integers modulo n* and denoted as Z_n.
3. The n complex numbers $\exp(2\pi mi/n), m = 0, 1, \ldots, n-1$ form the cyclic group under multiplication.

   All the previous three groups are Abelian.
4. Rotation group R_3 in three dimensions: A system with spherical symmetry is invariant under rotations through any angle φ about any axis $\mathbf{n}(\theta', \varphi')$ passing through its center. All these operations

$$R_{\mathbf{n}(\theta',\varphi')}(\varphi), \quad 0 \le \theta' \le \pi, \quad 0 \le \varphi' \le 2\pi, \quad 0 \le \varphi \le \pi \tag{1-2a}$$

   form the three-dimensional rotation group R_3. The identity is $R_{\mathbf{n}}(0)$, and the inverse of $R_{\mathbf{n}(\theta',\varphi')}(\varphi)$ is

$$R^{-1}_{\mathbf{n}(\theta',\varphi')}(\varphi) = R_{\mathbf{n}(\pi-\theta',\pi+\varphi')}(\varphi) . \tag{1-2b}$$

   Since θ', φ' and φ are continuous variables, and two rotations are in general not commuting, R_3 is a continuous non-Abelian group.
5. Rotation group R_2 in two dimensions: A linear molecule such as CO is invariant under rotations through any angle about z-axis passing through the line connecting the centers of the atoms (Fig. 1.1-1).

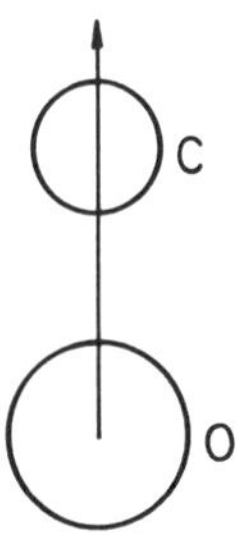

Fig. 1.1-1. The molecule CO.

The symmetry operations are

$$R_z(\varphi), \quad 0 \le \varphi \le 2\pi . \tag{1-2c}$$

They form the two-dimensional rotation group R_2. It is clear that

$$R_z(\varphi_1)R_z(\varphi_2) = R_z(\varphi_2)R_z(\varphi_1) = R_z(\varphi_1 + \varphi_2) ;$$

thus R_2 is a continuous Abelian group.

6. Space inversion group G_I or C_i consists of two elements: the identity e and the space inversion element I which brings the point $P(x,y,z)$ to $P'(-x,-y,-z)$.

7. The group C_{3v}: The ammonia molecule NH_3 (Figs. 1.1-2a and 1.1-2b) has six symmetry operations,

$$e, C_3, C_3^2, \sigma_1, \sigma_2, \sigma_3 \; , \tag{1-3}$$

where $C_3 = R_z(120°)$, $C_3^2 = R_z(240°)$, and σ_i are reflection planes containing the z-axis and the vertices i, as shown in Fig. 1.1-2b. The six operations in Eq. (1-3) form the symmetry group of NH_3, denoted as C_{3v}.

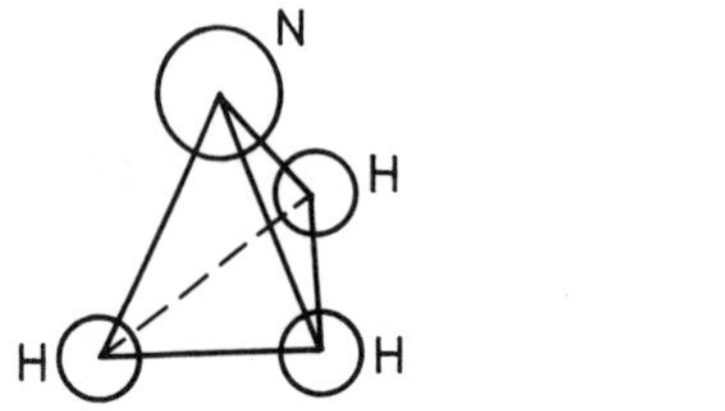

Fig. 1.1-2a. The ammonia molecule NH_3 with the symmetry C_{3v}.

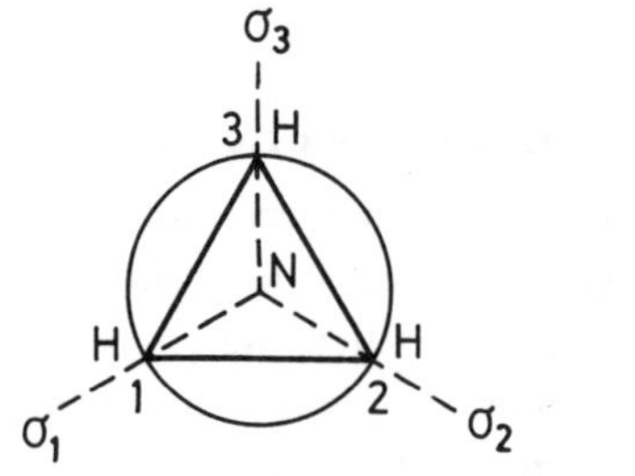

Fig. 1.1-2b. The ammonia molecule NH_3 seen from the top.

According to the interchange of the vertices 1,2,3 under the operations (1-3), we can obtain the multiplication table for C_{3v}. Notice that the anticlockwise rotation is taken to be positive, and the reflection planes σ_i are fixed in space (i.e., they do not change with the vertices). For example we have the sequence shown in Fig. 1.1-2c.

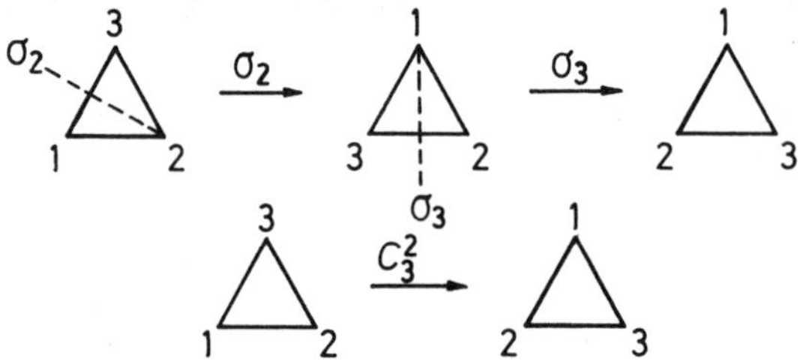

Fig. 1.1-2c.

Hence we see that $\sigma_2\sigma_3 = C_3^2$. The multiplication relation for the elements of C_{3v} is listed in Table 1.1. Such a table is called a group table.

Table 1.1. The group table of C_{3v}.

a \ b (ab)	e	σ_3	σ_2	σ_1	C_3	C_3^2
e	e	σ_3	σ_2	σ_1	C_3	C_3^2
σ_3	σ_3	e	C_3^2	C_3	σ_1	σ_2
σ_2	σ_2	C_3	e	C_3^2	σ_3	σ_1
σ_1	σ_1	C_3^2	C_3	e	σ_2	σ_3
C_3	C_3	σ_2	σ_1	σ_3	C_3^2	e
C_3^2	C_3^2	σ_1	σ_3	σ_2	e	C_3

8. If a set of matrices constitutes a group under matrix multiplication, then it is called a *matrix group*. For example the eight matrices $\{\pm R_1, \pm R_2, \pm R_3, \pm R_4\}$ form a matrix group, where

$$R_1 = \begin{pmatrix} 1 & 0 \\ 0 & 1 \end{pmatrix}, R_2 = \begin{pmatrix} 1 & 0 \\ 0 & -1 \end{pmatrix}, R_3 = \begin{pmatrix} 0 & 1 \\ 1 & 0 \end{pmatrix}, R_4 = \begin{pmatrix} 0 & -1 \\ 1 & 0 \end{pmatrix}. \qquad (1\text{-}4)$$

Ex. 1.1. Check the multiplication Table 1.1.

Ex. 1.2. Construct the group table for the group C_{4v} consisting of the following eight elements: $(e, C_4, C_4^2, C_4^3, \sigma_1, \sigma_2, \sigma_3, \sigma_4)$, where $C_4 = R_z(90°)$, and σ_i are reflection planes shown in Fig. 1.1-3.

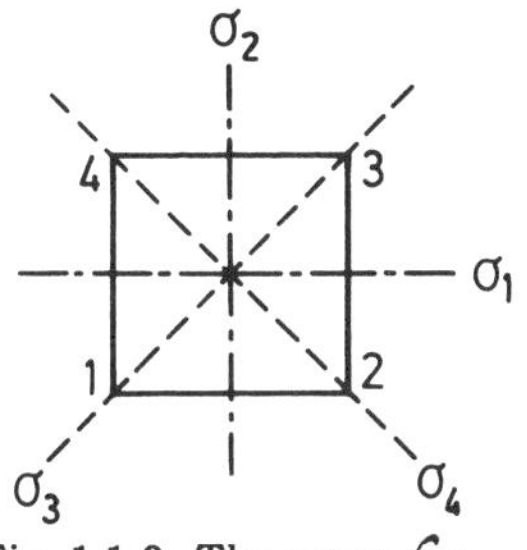

Fig. 1.1-3. The group C_{4v}.

Ex. 1.3. Construct the group table for the matrix group (1-4). (Hint: a 4×4 table is sufficient.)

1.2. Permutation Group S_n

1.2.1 Definition of S_n

All the $n!$ permutations

$$\begin{pmatrix} 1, & 2, & \dots & n \\ p_1, & p_2, & \dots & p_n \end{pmatrix} \equiv \begin{pmatrix} i \\ p_i \end{pmatrix} \qquad (1\text{-}5)$$

form a group $S_n(1, 2, \dots, n) \equiv S_n$, called the *permutation group* or *symmetric group*. Equation (1-5) denotes a permutation of the indices i to p_i. The product of $R_1 R_2$ of two permutations is defined as the resultant permutation of first permuting with R_2 and then with R_1. For example,

$$R_1 = \begin{pmatrix} 1 & 2 & 3 \\ 2 & 1 & 3 \end{pmatrix}, \quad R_2 = \begin{pmatrix} 1 & 2 & 3 \\ 3 & 2 & 1 \end{pmatrix},$$

$$R_1 R_2 = \begin{pmatrix} 1 & 2 & 3 \\ 2 & 1 & 3 \end{pmatrix} \begin{pmatrix} 1 & 2 & 3 \\ 3 & 2 & 1 \end{pmatrix} = \begin{pmatrix} 1 & 2 & 3 \\ 3 & 1 & 2 \end{pmatrix},$$

$$R_2 R_1 = \begin{pmatrix} 1 & 2 & 3 \\ 3 & 2 & 1 \end{pmatrix} \begin{pmatrix} 1 & 2 & 3 \\ 2 & 1 & 3 \end{pmatrix} = \begin{pmatrix} 1 & 2 & 3 \\ 2 & 3 & 1 \end{pmatrix}.$$

One easily sees that
$[R_1, R_2] = 0$, if R_1 and R_2 do not involve the permutation of the same index,
$[R_1, R_2] \neq 0$, if R_1 and R_2 involve the permutation of the same index. Obviously we have

$$\begin{pmatrix} 1 & 2 & 3 \\ p_1 & p_2 & p_3 \end{pmatrix} = \begin{pmatrix} 2 & 1 & 3 \\ p_2 & p_1 & p_3 \end{pmatrix} = \begin{pmatrix} 3 & 2 & 1 \\ p_3 & p_2 & p_1 \end{pmatrix} = \dots$$

and the inverse of $\begin{pmatrix} i \\ p_i \end{pmatrix}$ is $\begin{pmatrix} p_i \\ i \end{pmatrix}$. For example, the permutation group S_3 has the following 3!=6 permutations:

$$R_1 = e = \begin{pmatrix} 1 & 2 & 3 \\ 1 & 2 & 3 \end{pmatrix}, \quad R_2 = \begin{pmatrix} 1 & 2 & 3 \\ 2 & 1 & 3 \end{pmatrix}, \quad R_3 = \begin{pmatrix} 1 & 2 & 3 \\ 3 & 2 & 1 \end{pmatrix},$$

$$R_4 = \begin{pmatrix} 1 & 2 & 3 \\ 1 & 3 & 2 \end{pmatrix}, \quad R_5 = \begin{pmatrix} 1 & 2 & 3 \\ 2 & 3 & 1 \end{pmatrix}, \quad R_6 = \begin{pmatrix} 1 & 2 & 3 \\ 3 & 1 & 2 \end{pmatrix}. \tag{1-6}$$

1.2.2. Permutation expressed in terms of cycles and transpositions

Define:

$$e = (i)\ , \qquad \text{one-cycle} \quad (i \text{ remains to be } i),$$

$$p_{ij} = (ij) = \begin{pmatrix} ij \\ ji \end{pmatrix}, \qquad \text{two-cycle, or transposition,}$$

$$p_{ijk} = (ijk) = \begin{pmatrix} ijk \\ jki \end{pmatrix}, \qquad \text{three-cycle.}$$

Similarly, $(p_1, p_2 \ldots p_k)$ is termed as a k-cycle and k is called the *length* of the cycle. $(p_1, p_2 \ldots p_k)$ is the generator of the cyclic group C_k.

Any permutation can be expressed as a product of cycles without common indices

$$\begin{pmatrix} 1 & 2 & 3 & 4 & 5 & 6 & 7 & 8 \\ 4 & 3 & 7 & 1 & 5 & 8 & 2 & 6 \end{pmatrix} = (14)(237)(5)(68)\ . \tag{1-7}$$

Remarks:

1. In expressing a permutation as a product of cycles, the order with which we write the cycles is irrelevant, since there are no common indices among the different cycles.

2. One-cycles can be omitted. Hence the permutation (1-7) can be written as (14) (237) (68).

3. A cycle can be written in several forms, such as

$$(\alpha\beta\gamma\delta) = (\beta\gamma\delta\alpha) = (\gamma\delta\alpha\beta) = (\delta\alpha\beta\gamma)\ .$$

Any transposition can be expressed as a product of adjacent transpositions by the following recursive formula:

$$(i, i+v) = (i+1, i+v)(i, i+1)(i+1, i+v). \tag{1-8a}$$

Example 1: The group S_4. Letting $v = 2$ in (1-8a),

$$(13) = (23)(12)(23), \quad (24) = (34)(23)(34). \tag{1-8b}$$

Letting $v = 3$ in (1-8a) and using (1-8b),

$$(14) = (24)(12)(24) = (34)(23)(34)(12)(34)(23)(34). \tag{1-8c}$$

The following important relations are easily verified:

$$1.\ (\alpha\beta\gamma\delta\varepsilon) = (\alpha\beta)(\beta\gamma\delta\varepsilon) = (\alpha\beta\gamma)(\gamma\delta\varepsilon) = (\alpha\beta\gamma\delta)(\delta\varepsilon)$$
$$= (\alpha\beta)(\beta\gamma)(\gamma\delta)(\delta\varepsilon)\ . \qquad (1\text{-}9a)$$
$$2.\ (\alpha\beta\gamma\delta\varepsilon) = (\alpha\varepsilon)(\alpha\beta\gamma\delta) = (\alpha\delta\varepsilon)(\alpha\beta\gamma) = (\alpha\gamma\delta\varepsilon)(\alpha\beta)$$
$$= (\alpha\varepsilon)(\alpha\delta)(\alpha\gamma)(\alpha\beta). \qquad (1\text{-}9b)$$
$$3.\ (\alpha\beta\gamma\delta\varepsilon)^{-1} = (\alpha\varepsilon\delta\gamma\beta) = (\varepsilon\delta\gamma\beta\alpha). \qquad (1\text{-}10)$$

Note: The order of the cycles in the above equations is crucial, since they involve common indices.

The generators of S_n are the $n-1$ adjacent transpositions $(i, i+1), i = 1, 2, \ldots, n-1$.

A permutation can be written as a product of transpositions in many ways but the number N of factors is always even or odd. Therefore for any permutation p, we can define a *permutation parity* δ_p by

$$\delta_p = (-1)^N, \qquad (1\text{-}11)$$

and the permutation with δ_p=1(-1) is called an even (odd) permutation. Obviously, if $p = p_1 p_2$, then $\delta_p = \delta_{p_1}\delta_{p_2}$. The k-cycle $(i_1 i_2 \ldots i_k)$ has the parity $(-1)^{k-1}$.

Example 2: The six elements (1-6) of S_3 can be written in terms of cycles as

$$R_1 = e, \qquad R_2 = \begin{pmatrix} 1 & 2 & 3 \\ 2 & 1 & 3 \end{pmatrix} = (12), \quad R_3 = \begin{pmatrix} 1 & 2 & 3 \\ 3 & 2 & 1 \end{pmatrix} = (13)$$
$$R_4 = \begin{pmatrix} 1 & 2 & 3 \\ 1 & 3 & 2 \end{pmatrix} = (23), \quad R_5 = \begin{pmatrix} 1 & 2 & 3 \\ 2 & 3 & 1 \end{pmatrix} = (123), \quad R_6 = \begin{pmatrix} 1 & 2 & 3 \\ 3 & 1 & 2 \end{pmatrix} = (132)\ , \qquad (1\text{-}12)$$

where e, (123) and (132) are even, while (12), (13) and (23) are odd. The group table of S_3 is shown in Table 1.2.

Table 1.2 Group of table of S_3.

$a = cb$ (b across, c down)	e_1	$(12)_2$	$(13)_3$	$(23)_4$	$(123)_5$	$(132)_6$
e_1	e_1	$(12)_2$	$(13)_3$	$(23)_4$	$(123)_5$	$(132)_6$
$(12)_2$	$(12)_2$	e_1	$(132)_6$	$(123)_5$	$(23)_4$	$(13)_3$
$(13)_3$	$(13)_3$	$(123)_5$	e_1	$(132)_6$	$(12)_2$	$(23)_4$
$(23)_4$	$(23)_4$	$(132)_6$	$(123)_5$	e_1	$(13)_3$	$(12)_2$
$(123)_5$	$(123)_5$	$(13)_3$	$(23)_4$	$(12)_2$	$(132)_6$	e_1
$(132)_6$	$(132)_6$	$(23)_4$	$(12)_2$	$(13)_3$	e_1	$(123)_5$

1.3. Subgroup

If there is a subset of elements in a group G, which by itself forms a group G_s under the same multiplication rule as that of G, then G_s is said to be a *subgroup* of G, denoted as $G \supset G_s$.

The subgroup G_s itself may contain a subgroup G'_s etc.; they form what is called a group (or subgroup) chain

$$G \supset G_s \supset G'_s \supset \ldots$$

Examples of subgroups:

1. $R_3 \supset R_2$. The two-dimensional rotation group is a subgroup of the three-dimensional rotation group.

2. The group $\mathcal{C}_{3v}$ has four subgroup chains, namely

$$\mathcal{C}_{3v} \supset \mathcal{C}_3, \quad \mathcal{C}_{3v} \supset \mathcal{C}_{s_i}, \quad i = 1, 2, 3$$

$$\mathcal{C}_3 = (e, C_3, C_3^2), \quad \mathcal{C}_{s_i} = (e, \sigma_i).$$

3. $S_4 \supset S_3 \supset S_2$.

Besides this, the group S_4 has a variety of subgroups, e.g., $S_3(134)$, $S_3(234), \ldots$ as well as the following two Abelian subgroups:

$$\text{the four-group:} \quad F = \{e, (12)(34), (13)(24), (23)(14)\}\ ; \tag{1-13}$$

$$\text{the cyclic group:} \quad \mathcal{C}_4 = \{e, (1234), (1234)^2 (= (13)(24)), (1234)^3\}\ . \tag{1-14}$$

Similarly the group S_n has the group chain $S_n \supset S_{n-1} \supset S_{n-2} \supset \ldots \supset S_2$.

The *alternative group* $\mathcal{A}_n$ is the group consisting of all the even permutations of S_n.

Ex. 1.4. Using (1-9) check Table 1.2.

Ex. 1.5. Construct the group table for the four-group.

1.4. Isomorphism and Homomorphism

Two groups G and G' are said to be *isomorphic*($G \approx G'$) if their elements can be put into a 1-1 correspondence which preserves the multiplication rule, i.e., corresponding to $ab = c$, we have $a'b' = c'$.

Two groups are said to be *anti-isomorphic* if their elements have a 1-1 correspondence and corresponding to $ab = c$ we have $b'a' = c'$ (instead of $a'b' = c'$).

If $ab = c$, then $b^{-1}a^{-1} = c^{-1}$. Letting $R_a = a^{-1}, R_b = b^{-1}, \ldots$ then the set $\{R_a\}$ forms a group $\tilde{G}$ which is isomorphic with G'. The difference between the group $\tilde{G}$ and G is merely a matter of nomenclature for the elements. Therefore, if G is anti-isomorphic to G', then essentially G is isomorphic to G'.

Two groups which are isomorphic are the same group from the standpoint of abstract groups, though they may be totally different realizations. Therefore, sometimes we simply use $G = G'$ to indicate that G is isomorphic to G'.

Example 1: The group of integers modulo n, Z_n, is isomorphic to the cyclic group C_n.

Example 2: From Tables 1.1 and 1.2 one sees that $\mathcal{C}_{3v}$ is isomorphic to S_3:

$$\begin{array}{cccccc} e, & C_3, & C_3^2, & \sigma_1, & \sigma_2, & \sigma_3. \\ e, & (123), & (132), & (23), & (13), & (12) \end{array} \ . \tag{1-15}$$

Actually, the isomorphism (1-15) can also be established in the following way without using the group table.

First label the three vertices of the triangle in a definite way, such as ${}_1\overset{3}{\triangle}_2$, which is called the original triangle. Under the group operations, the vertices interchange among themselves. The group operator is represented by the permutation of the vertices of the original triangle. For example, under the rotation C_3.

$$ {}_1\overset{3}{\triangle}_2 \xrightarrow{C_3} {}_3\overset{2}{\triangle}_1 \ , \quad C_3 \leftrightarrow (123) \ . $$

Thus C_3 corresponds to the permutation (123), which signifies that after the operation C_3, the vertex 1 goes to where 2 was, 2 to where 3 was, and 3 to where 1 was.

Using the above method, we can easily obtain the isomorphism (1-15). This method applies also to other point groups [see Eq. (3-21), Table 8.2-2 and 8.2-3 etc.].

Note that the correspondence $C_3 \leftrightarrow (123)$ only refers to the original triangle. For example, from

$$ {}_1\overset{3}{\triangle}_2 \xrightarrow{\sigma_1} {}_1\overset{2}{\triangle}_3 \xrightarrow{C_3} {}_2\overset{3}{\triangle}_1 $$

one sees that the effect of the operation C_3 on the second triangle is not to move the vertex 1 to where 2 was, etc. Summarizing, in discussing the isomorphism between C_{3v} and S_3, C_3 always corresponds to (123), while in discussing the effect of C_3 on a triangle, it always rotates the triangle through 120°.

Example 3: The subgroups $S_3(124), S_3(134)$ and $S_3(234)$ are all isomorphic to $S_3 \equiv S_3(123)$.

Cayley's Theorem: Every finite group G is isomorphic to a subgroup of the permutation group $S_{|G|}$.

Suppose that $R_a(R_1, R_2, \dots, R_{|G|}) = (R_{a_1}, R_{a_2}, \dots R_{a_{|G|}})$, which shows that the effect of multiplying the group elements of G from the left by an element R_a is a permutation of the group elements. Therefore the group element R_a corresponds to the permutation

$$ R_a \longleftrightarrow p_a = \begin{pmatrix} 1 & 2 & \dots & |G| \\ a_1 & a_2 & \dots & a_{|G|} \end{pmatrix} . \qquad (1\text{-}16) $$

p_a is an element of the permutation group $S_{|G|}$. It is easy to show that $p_1, p_2, \dots, p_{|G|}$ form a group isomorphic to G.

Reading the group table horizontally, we can easily establish the isomorphism (1-16). For example from Table 1.2 we get the isomorphism of S_3 to a subgroup of S_6, i.e.,

$$ \begin{aligned} e = e &= (1)(2)(3)(4)(5)(6), & p_{12} &= \begin{pmatrix} 123456 \\ 216543 \end{pmatrix} = (12)(36)(45), \\ p_{13} &= \begin{pmatrix} 123456 \\ 351624 \end{pmatrix} = (13)(25)(46), & p_{23} &= \begin{pmatrix} 123456 \\ 465132 \end{pmatrix} = (14)(26)(35), \\ p_{123} &= \begin{pmatrix} 123456 \\ 534261 \end{pmatrix} = (156)(234), & p_{132} &= \begin{pmatrix} 123456 \\ 642315 \end{pmatrix} = (165)(243). \end{aligned} \qquad (1\text{-}16') $$

Suppose that to each element a of a group G, there is an element a' in a group G'; however, there may be several elements in G mapped to the same a' in G', and if corresponding to $ab = c$, we have $a'b' = c'$, then we say that G is *homomorphic* to G', denoted as $G \rightarrow G'$.

If G is homomorphic to $G', G \rightarrow G'$, the set of elements of G which is mapped onto the identity of G' is called the *kernel of the homomorphism* (Bacry 1977).

Obviously, the isomorphism is a special case of the homomorphism.

Every group has a simplest homomorphic mapping, i.e. let each element of the group correspond to the identity element.

For example, C_{3v} is homomorphic to S_2. The mapping is

$$e, \quad C_3, \quad C_3^2 \to e, \qquad \sigma_1, \sigma_2, \sigma_3 \to (12) \,, \tag{1-16''}$$

and the kernel is (e, C_3, C_3^2).

An isomorphism of a group with itself, i.e. a a 1-1 correspondence between elements of the group preserving multiplication is called an *automorphism*.

An automorphism of G can be regarded as a linear transformation in the $|G|$-dimensional vector space. All these transformations form the group of automorphisms.

If this correspondence is brought about by conjugation (see Sec. 1.5),

$$G \to R_a G R_a^{-1} ,$$

then it is called an *inner automorphism*. The inner automorphism forms a subgroup of the automorphism.

Excluding the inner automorphism, the remaining automorphisms are called the *outer automorphisms*.

For example, suppose there is an automorphism,

$$(e, a, a^2) \xrightarrow{\phi} (e, a^2, a),$$

then

$$\phi = \begin{pmatrix} 1 & 0 & 0 \\ 0 & 0 & 1 \\ 0 & 1 & 0 \end{pmatrix} .$$

1.5. Conjugate Classes

An element b of a group G is said to be *conjugate* to an element a if we can find an element u in G such that

$$b = uau^{-1} . \tag{1-17}$$

We refer to uau^{-1} as a *conjugate operation* on a with u. From (1-17) we have $a = u^{-1}bu$, i.e., if b is conjugate to a, then a is conjugate to b. It is easy to show that if a is conjugate to b, and b is conjugate to c, then a is conjugate to c. Any element is conjugate to itself.

The elements conjugate to one another from a *conjugate class*, or simply a *class*. Each element of G belongs to one of the classes. The number of classes, denoted by N, is an important characteristic of the group. Designate the number of elements in the class i by g_i, then we have $g = \sum_{i=1}^{N} g_i$.

How do we find the classes?

Method 1: By multiplying the b-th column of the group table, such as Table 1.2, from the left with the element R_b^{-1}, we obtain a new table, where the elements in each row belong to the same class. This new table is referred to as the *class structure table*.

Method 2: It is easy to show that the two elements which lie symmetrically with respect to the diagonal line in the group table belong to the same class.

An *ambivalent class* is a class of G which contains element as well as their inverses.

Remarks :

1. In every group, the identity e forms a class by itself.
2. In an Abelian group, each element forms a class by itself:
$a = uau^{-1} = uu^{-1}a = a.$

Example 1: Group R_3. Suppose that there are two rotations through the same angle φ but about different axes, say z and n axes (see Fig. 1.5). These two rotations are related by

$$R_n(\varphi) = R(z \to \mathbf{n})R_z(\varphi)R^{-1}(z \to \mathbf{n}) \,. \tag{1-18}$$

where $R(z \to \mathbf{n})$ is a rotation which takes the z-axis into the $\mathbf{n}$-axis. Therefore, all rotations through the same angle φ belong to the same class.

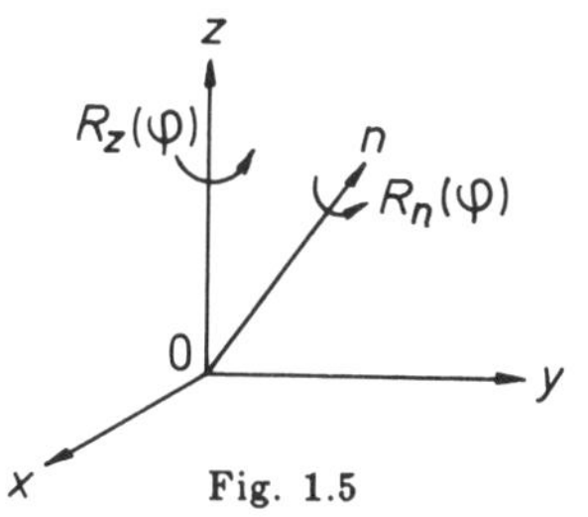

Fig. 1.5

The geometric meaning of (1-18) is that a rotation through angle φ about the axis $\mathbf{n}$ can be thought of as the net result of three rotations: first the rotation $R^{-1}(z \to \mathbf{n})$ takes the axis $\mathbf{n}$ to z, then $R_z(\varphi)$ performs a rotation through angle φ about the z-axis, and finally $R(z \to \mathbf{n})$ takes z back to $\mathbf{n}$.

All the classes of R_3 are ambivalent on account of (1-2b). The number of classes of R_3 is infinite.

Example 2: The group C_{3v} has three ambivalent classes, namely,

$$e, \quad (C_3, C_3^2), \quad (\sigma_1, \sigma_2, \sigma_3).$$

Example 3: Classes of the permutation group. Let $a = (ij), p = \begin{pmatrix} i \\ p_i \end{pmatrix}$ thus $p^{-1} = \begin{pmatrix} p_i \\ i \end{pmatrix}$. It is easy to see that

$$b = pap^{-1} = p(ij)p^{-1} = (p_i p_j),$$

since pap^{-1} represents the permutation $p_i \to i \to j \to p_j$ and $p_j \to j \to i \to p_i$, which is equivalent to $p_i \leftrightarrow p_j$. Similarly we have

$$p(ij)(klm\ldots)p^{-1} = p(ij)p^{-1}p(klm\ldots)p^{-1} = (p_i p_j)(p_k p_l p_m \cdots) \,. \tag{1-19}$$

For example, $a = (12)(345), p = (135)$; permuting the indices in $(12)(345)$ according to what is specified by p, i.e., $1 \to 3 \to 5 \to 1$, we get

$$pap^{-1} = (32)(541) = (23)(154).$$

As mentioned before, in expressing a permutation by a product of independent cycles, the order of the cycle factors is irrelevant. We can always write the cycles in the order of decreasing length of the cycles. Suppose that the longest one is an i_1-cycle, the next one is an i_2-cycle $(i_1 > i_2 > \ldots)$. We can use $(i_1 i_2 \ldots)$ to represent the *cycle structure* of a permutation. It is clear that $n = i_1 + i_2 + \ldots + i_n$. For example, the cycle structure of the permutation (237)(14)(58)(5) is (3221) = (32^21).

From (1-19) we know that conjugate elements of a permutation group have the same cycle structure, and vice versa. Thus the elements belonging to the same class have the same permutation parity.

According to the cycle structure, it is easy to write down the conjugate classes of a permutation group. For instance

Group S_3 has N=3 classes

$$\begin{aligned}
&1.\ (111) = (1^3), && g_1 = 1;\ e\ . \\
&2.\ (21), && g_2 = 3;\ (12),\ (23),\ (13)\ . \\
&3.\ (3), && g_3 = 2;\ (123),\ (132)\ .
\end{aligned} \tag{1-20}$$

Group S_4 has $N = 4$ classes

$$\begin{aligned}
&1.\ (1^4), && g_1 = 1;\ e. \\
&2.\ (21^2), && g_2 = 6;\ (12),\ (13),\ (14),\ (23),\ (24),\ (34). \\
&3.\ (31), && g_3 = 8;\ (123),\ (132),\ (124),\ (142),\ (134),\ (143),\ (234),\ (243). \\
&4.\ (4), && g_4 = 6;\ (1234),\ (1243),\ (1324),\ (1342),\ (1423),\ (1432). \\
&5.\ (2^2), && g_5 = 3;\ (12)(34),\ (13)(24),\ (14)(23)\ .
\end{aligned} \tag{1-21}$$

From Eq. (1-10) one sees that the permutation a and a^{-1} have the same cycle structure, and thus belong to the same class. In other words, all classes of S_n are ambivalent classes.

Note: Two elements belonging to the same class of group G, may belong to different classes of a subgroup of G. For example, (12)(34), (13)(24) and (14)(23) belong to the same class of S_4, but each forms a class by itself for the four-group (1-13), which is a subgroup of S_4, since in the four-group we cannot find an element u such that (12)(34) $=u(13)(24)u^{-1}$.

Let us put the cycle structure in the following way

$$\underbrace{(\cdot)\cdots(\cdot)}_{f_1}\ \underbrace{(\cdot\cdot)\cdots(\cdot\cdot)}_{f_2}\ \underbrace{(\cdots)\cdots(\cdots)}_{f_3}\cdots,$$

i.e., there are f_1 1-cycles, f_2 2-cycles,..., and f_n n-cycles, $n = \sum_{k=1}^{n} kf_k$. It can be shown (Hamermesh, 1962) that the number of the permutations of S_n which have this cycle structure is

$$g_{(f_1f_2\ldots)} = \frac{n!}{f_1!f_2!\ldots f_n!2^{f_2}3^{f_3}\cdots n^{f_n}}. \tag{1-22}$$

Letting $f_1 = n - k$, $f_k = 1$, and all other $f_i = 0$, we get the number of elements belonging to the k-cycle class

$$g_{(k)} = \frac{n!}{(n-k)!k} = \binom{n}{k}(k-1)! \tag{1-23}$$

where $\binom{n}{k}$ is the binary combination coefficient.

Ex. 1.6. From group Table 1.2 construct the class structure table of S_3.

Ex. 1.7. Calculate the number of elements contained in each class of S_4 and S_5.

Ex. 1.8. Find out the classes of the group C_{4v}.

1.6. Cosets, Lagrange Theorem

Let $G_s = (H_1, H_2, \dots H_h)$ be a subgroup of $G, |G_s| (= h) < |G|, H_1 = e$. Suppose that a is an element of G which is not contained in G_s; we form the products $aH_1, aH_2, \dots, aH_h$, and denote the set of these h elements symbolically by aG_s. The products aH_i are all different, for if $aH_i = aH_j$, we would have $H_i = H_j$. Also none of them is contained in G_s, for if $aH_i = H_j$, then $a = H_j H_i^{-1}$, and a would belong to G_s, contrary to our assumption. If G_s and aG_s have not exhausted the group G, then we proceed as before; pick some element b of G which belongs neither to G_s nor to aG_s, and form the set bG_s. The set bG_s will again yield $|G_s|$ new elements of G. Continuing this process, the group G is decomposed into several disjoint sets, each containing $|G_s|$ elements, denoted symbolically as

$$\begin{aligned} G &= G_s + aG_s + bG_s + \dots + dG_s \\ &= (e + a + b + \dots + d)G_s \end{aligned} \tag{1-24}$$

where the plus sign "+" should be understood as the union "∪" for sets. The sets $aG_s, bG_s, \dots$ are called the *left cosets* of G_s in G, and $a, b, \dots$ are called the *representatives* of the cosets.

Suppose that $aH_i = a_i$ for $i = 1, 2, \dots h$, i.e.

$$aG_s = (a_1, a_2, \dots, a_h) \ , \quad a_1 = a \ .$$

It is easy to see that the cosets aG_s and $a_k G_s$ coincide, i.e.,

$$aG_s = a_k G_s, \quad \text{for any } a_k \in aG_s,$$

since

$$a_k H_i = aH_k H_i = aH_{ki}.$$

Therefore, any element in the coset aG_s can be chosen as the coset representative, and for given G_s, the left coset decomposition of G is unique.

Analogously, we have the right coset decomposition

$$G = G_s + G_s a' + G_s b' + \dots + G_s d'. \tag{1-25}$$

This leads to

Lagrange Theorem: If G_s is a subgroup of G, then $|G_s|$ divides $|G|$, or

$$|G|/|G_s| = m(\text{integer}), \tag{1-26}$$

m is called the *index* of the subgroup G_s.

From Eq. (1-26) we immediately know that the order $|R|$ of any element R of G divides $|G|$, i.e., $|G|/|R|$ =integer. If $|G|$ is a prime number, then $|R|$ is necessarily either equal to one (for $R = e$) or $|G|$ (for $R \neq e$). Hence we have

Theorem 1.1: Let G be a finite group of order p where p is a prime number, then G is a cyclic group.

The left coset decomposition of S_3 is

$$\begin{aligned} S_3 &= (e + (13) + (23))[e, (12)] = (e + (123) + (132))[e, (12)] \\ &= [e, (12)] + [(13), (123)] + [(23), (132)]. \end{aligned}$$

Its right coset decomposition is

$$S_3 = [e,(12)](e+(13)+(23)) = [e,(12)](e+(132)+(123))$$
$$= [e,(12)] + [(13),(132)] + [(23),(123)].$$

As is seen from the above, the left cosets a_iG_s in general do not coincide with the right cosets G_sa_i.

1.7. Invariant Subgroup

If the left coset and right coset of G_s with any element a of G are the same,

$$aG_s = G_sa, \quad a \in G \ , \tag{1-27}$$

then we say G_s is an *invariant* (or *normal*) *subgroup* of G.

The invariant subgroup can also be defined in the following way. If $G_s = (H_1, H_2, \ldots H_h)$ and

$$aH_ia^{-1} \in G_s \ , \quad i = 1,2,\ldots,h \ , \quad a \in G \ , \tag{1-28}$$

i.e., if G_s is invariant under the group of inner automorphism, then G_s is an invariant subgroup of G.

Equation (1-28) tells us that an invariant subgroup G_s contains either all or none of the elements in a class of G. The reverse is also true, i.e., if a subgroup G_s of G consists of entire classes of G, then G_s is an invariant subgroup of G. Therefore, any subgroup of an Abelian group is an invariant subgroup. Thus we have the definition

Simple group G :	if $G \not\supset$ an invariant subgroup.
Semi-simple group G :	if $G \not\supset$ an invariant Abelian subgroup.

where $\not\supset$ reads as "does not contain."

Example:

1. The group R_3 is simple, but R_2 is not.
2. The group S_3 is not semi-simple, since it contains the Abelian invariant subgroup $\mathcal{A}_3 = (e,(123),(132))$.

The *center* of a group G is the set of elements which commute with every element of G. The center of a group G is an invariant subgroup of G.

1.8. Factor Group*

Let G_s be an invariant subgroup of G. We use G/G_s to denote the set of cosets of G_s in G:

$$\frac{G}{G_s} = \{G_s, a_2G_s, \ldots, a_mG_s\}. \tag{1-29}$$

Define a product on the set G/G_s according to the rule

$$aG_s \times bG_s = abG_s. \tag{1-30}$$

Then the set G/G_s is a group under the multiplication rule (1-30), called the *factor* or *quotient group* of G relative to the invariant subgroup G_s. The identity of the factor group G/G_s is the subgroup G_s.

Theorem 1.2: If $G \to G'$ is a homomorphic mapping with kernel G_s, then G_s is an invariant subgroup of G and the factor group G/G_s is isomorphic with G'.

Proof: If $a \to e', b \to e'$, then $ab \to e'$. Besides, $aa^{-1} = e \to e'$, hence if $a \to e'$, then $a^{-1} \to e'$. Therefore, G_s is a subgroup.

Since $G_s \to e', aG_s \to a'e' = a', G_s a \to e'a' = a'$. Thus aG_s coincides with $G_s a$, namely G_s is an invariant subgroup of G. Obviously, G/G_s is isomorphic to G'.

Example: $S_3/\mathcal{A}_3 = \{\mathcal{A}_3, \mathcal{B}\}$ is a factor group of order 2, where

$$\mathcal{A}_3 = ((e, (123), (132)), \quad \mathcal{B} = ((12), (13), (23)).$$

The multiplication relations are

$$\mathcal{A}_3\mathcal{A}_3 = \mathcal{A}_3, \quad \mathcal{B}\mathcal{B} = \mathcal{A}_3, \quad \mathcal{B}\mathcal{A}_3 = \mathcal{A}_3\mathcal{B} = \mathcal{B}\ . \tag{1-31}$$

Therefore $S_3/\mathcal{A}_3 \approx S_2$,

$$\mathcal{A}_3 \to e, \quad \mathcal{B} \to (12). \tag{1-32}$$

In the homomorphic mapping (1-32) of $S_3 \to S_2$, the kernel is $\mathcal{A}_3$, which is an invariant subgroup of S_3.

Ex. 1.9. Show that the factor group S_4/F is isomorphic to S_3.

1.9. Direct Product and Semi-Direct Product

Let there be two independent (thus commuting with one another) groups $G = \{a\}$ and $G' = \{a'\}$, which may be associated with different multiplication rules. Form the $|G||G'|$ pairs (a, a') (or abbreviated as aa') and define the product of pairs by

$$(a, a')(b, b') = (ab, a'b'),$$

then these $|G||G'|$ pairs form a group, called the *direct product* of G and G' and denoted as $G \times G'$. Obviously, $G \times G'$ is identical with $G' \times G$.

As a special case of the above, the groups G and G' can be subgroups of a larger group. For example, let G_1 and G_2 be two subgroups of G and commuting with one another

$$[H_i^{(1)}, H_j^{(2)}] = 0, \quad i = 1, \dots, |G_1|, \quad j = 1, \dots, |G_2|, \quad H_i^{(1)} \in G_1, \quad H_j^{(2)} \in G_2. \tag{1-33}$$

The $|G_1||G_2|$ products $\{H_i^{(1)}H_j^{(2)}\}$ form the direct product group $G_1 \times G_2$. $G_1 \times G_2$ is a subgroup of G, denoted as $G \supset G_1 \times G_2$. Since $[G_1, G_2] = 0$, both G_1 and G_2 are invariant subgroups of $G_1 \times G_2$.

Let G be a group with subgroups G_1 and G_2 such that

1. The coset $H_i^{(2)}G_1 = G_1H_i^{(2)}$, for any $H_i^{(2)} \in G_2$,
2. Any element of G can be expressed as $R = H_i^{(1)}H_j^{(2)}$,
3. The intersection of G_1 and G_2 is the identity, then G is called the *semi-direct product* of G_1 and G_2, denoted by

$$G = G_1 \wedge G_2.$$

Notice that G_1 is an invariant subgroup of G, but G_2 is not necessarily invariant. In the expression $G_1 \wedge G_2$ we always write the invariant subgroup first.

Example 1: The cyclic $\mathcal{C}_6$ of order 6.

$$\mathcal{C}_6 = G_1 \times G_2, \quad G_1 = (e, a^2, a^4), \quad G_2 = (e, a^3).$$

Example 2: $S_n(n = n_1 + n_2)$ has two commuting subgroups $S_{n_1}(12\dots n_1)$ and $S_{n_2}(n_1 + 1, \dots n)$; we have $S_n \supset S_{n_1} \times S_{n_2}$.

Example 3: The rotation group R_3 commutes with the space inversion group G_I. The direct product of R_3 and G_I is called the *orthogonal group* in three-dimensional space, denoted by O_3,

$$O_3 = R_3 \times G_I.$$

Example 4: $\mathcal{A}_3 = (e, (123), (132))$ is an invariant subgroup of S_3. The permutation group S_3 is a semi-direct product of $\mathcal{A}_3$ and S_2,

$$S_3 = \mathcal{A}_3 \wedge S_2. \tag{1-34}$$

Example 5: $S_4 = F \wedge S_3$, F being the four-group.

The notion of direct or semi-direct product seems to be the inverse of the notion of factor group. Their exact relations are as follows.

1. If $G = G_1 \times G_2$, then $G/G_1 = G_2$, and $G/G_2 = G_1$.
2. If $G = G_1 \wedge G_2$, then $G/G_1 = G_2$.
3. If $G/G_1 = G_2$, then

$$G = G_1 \wedge G_2, \quad \text{if } G_2 \text{ is a subgroup of } G,$$
$$G = G_1 \times G_2, \quad \text{if } G_2 \text{ is an invariance subgroup of } G.$$

4. If $G/G_1 = G_2, G/G_2 = G_1$, then $G = G_1 \times G_2$.
   For example, from $S_3 = \mathcal{A}_3 \wedge S_2$ and $S_4 = F \wedge S_3$, we have $S_3/\mathcal{A}_3 = S_2$, and $S_4/F = S_3$ respectively.

Ex. 1.10. Show that $O_3/R_3 = Z_2$.

Chapter 2
Group Representation Theory

2.1. Linear Vector Space

2.1.1. Definition of linear vector space

We are familiar with the usual three-dimensional vector space. We now need to generalize this concept. Suppose that a set L consists of elements $\mathbf{x}, \mathbf{y}, \ldots$. The set will be called a *linear vector space* L if the element of which can be multiplied by a complex α or added to one another and satisfy the following conditions:

1. If $\mathbf{x}$, $\mathbf{y} \in L$, then $\mathbf{x} + \mathbf{y} \in L$.
2. $(\alpha + \beta)\mathbf{x} = \alpha\mathbf{x} + \beta\mathbf{x}$.
3. $(\alpha\beta)\mathbf{x} = \alpha(\beta\mathbf{x})$. (2-1)
4. $1\,\mathbf{x} = \mathbf{x}$.
5. $\alpha(\mathbf{x} + \mathbf{y}) = \alpha\mathbf{x} + \alpha\mathbf{y}$.
6. The set L contains a zero element (null vector), $\mathbf{0}$, such that $\mathbf{x} + \mathbf{0} = \mathbf{x}$, for all $\mathbf{x} \in L$.

$\mathbf{x}, \mathbf{y}, \ldots$ are called *vectors* of the space L.

For example, the set of all $n \times n$ matrices forms a linear vector space of dimension n^2; now $\mathbf{x}$, $\mathbf{y}\ldots$ represent matrices and the components of $\mathbf{x}$, $\mathbf{y}, \ldots$ are the matrix elements $x_{ij}, y_{ij}, \ldots$. The components of $\alpha\,\mathbf{x}$ and $\mathbf{x}+\mathbf{y}$ are αx_{ij} and $x_{ij} + y_{ij}$, respectively; while the null vector is the null matrix. Later, we shall see that $\mathbf{x}$, $\mathbf{y}, \ldots$, may represent the group elements $R_a, R_b, \ldots$

The definition of linear dependence of vectors is just the usual one.

In the n-dimensional space L_n, any n linearly independent vectors $\mathbf{u}_1$, $\mathbf{u}_2, \ldots, \mathbf{u}_n$ form a set of basis vectors, or provide a basis (coordinate system) in L_n. Any vector $\mathbf{x}$ in L_n can be expressed as a linear combination of the basis vectors

$$\mathbf{x} = \sum_i x_i \mathbf{u}_i \; . \tag{2-2}$$

The coefficients x_i are called the coordinates of the vector $\mathbf{x}$ in the basis $\mathbf{u}_1, \ldots, \mathbf{u}_n$. $\mathbf{u}_1, \ldots, \mathbf{u}_n$ can be the n state vectors $|\varphi_1\rangle, \ldots, |\varphi_n\rangle$ in quantum mechanics, which form a complete set in L_n. Any state vector $|\psi\rangle$ can be expanded as

$$|\psi\rangle = \sum_i a_i |\varphi_i\rangle \; . \tag{2-3}$$

In the space of $n \times n$ matrices, the basis vectors $\mathbf{u}_i$ can be chosen as the matrices $\mathbf{e}_{jk}$,

$$\mathbf{e}_{jk} = \overset{k\text{-th column}}{\begin{pmatrix} & \vdots & \\ & \ldots 1 \ldots & \\ & \vdots & \end{pmatrix}} j\text{-th row} \tag{2-4}$$

All the elements of $\mathbf{e}_{jk}$ are zeroes except the element at the j-th row and k-th column, which is equal to 1. Any matrix $\mathbf{x}$ is expressible in terms of the basis vectors $\mathbf{e}_{jk}$,

$$\mathbf{x} = \sum_{jk} x_{jk}\mathbf{e}_{jk} . \tag{2-5}$$

2.1.2 Covariant and contravariant

Suppose that the coordinate system $\mathbf{u}_i$ is changed into a new coordinate system $\mathbf{u}'_i$ through a transformation $B = (b_{ij})$,

$$\mathbf{u}'_i = \sum_j b_{ij}\mathbf{u}_j . \tag{2-6a}$$

Under the transformation, the coordinates of a vector $\mathbf{x}$ changes accordingly, while the abstract vector $\mathbf{x}$ is kept invariant:

$$\mathbf{x} = \sum_i x_i\mathbf{u}_i = \sum_i x'_i\mathbf{u}'_i . \tag{2-7}$$

From (2-6a) we have

$$\mathbf{u}_i = \sum_j (B^{-1})_{ij}\mathbf{u}'_j . \tag{2-8}$$

Substituting (2-8) into (2-7), and noting that $\mathbf{u}'_i$ are linearly independent, we obtain

$$x'_j = \sum_i x_i(B^{-1})_{ij} .$$

Let

$$A = \tilde{B}^{-1} \tag{2-9}$$

where $\sim$ represents the transposition of the matrix. We have

$$x'_i = \sum_j a_{ij}x_j . \tag{2-6b}$$

Let $\mathbf{u}(\mathbf{u}')$ be the column vector formed out of $\mathbf{u}_i$ $(\mathbf{u}'_i)$, and $\mathbf{x}(\mathbf{x}')$ be the column vector formed out of $x_i(x'_i)$. Equation (2-6) can be recast in the following form,

$$\mathbf{u}' = B\mathbf{u} , \tag{2-10a}$$

$$\mathbf{x}' = A\mathbf{x} = \tilde{B}^{-1}\mathbf{x} . \tag{2-10b}$$

Equations (2-10) show that the basis $\mathbf{u}$ and the coordinate vector $\mathbf{x}$ transform in different ways. We call a vector *covariant* (*contravariant*) if it transforms as the basis $\mathbf{u}$ (the coordinate vector $\mathbf{x}$).

For orthogonal transformation $\tilde{B} = B^{-1}$, we have $A = \tilde{B}^{-1} = B$, and the covariant and contravariant are identical.

2.1.3. Metric tensor

In quantum mechanics, the scalar product of two basis vectors $|\varphi_i\rangle$ and $|\varphi_j\rangle$ is defined as

$$(\varphi_i, \varphi_j) \equiv \langle\varphi_i|\varphi_j\rangle = \int \varphi_i^*(\mathbf{x})\varphi_j(\mathbf{x})d\mathbf{x} = g_{ij} = g_{ji}^* = \langle\varphi_j|\varphi_i\rangle^* . \tag{2-11}$$

(g_{ij}) is a hermitian matrix and is called the *metric tensor.* In quantum mechanics, we generally use an orthonormal basis for which the metric tensor $g_{ij} = \delta_{ij}$, i.e. a unit matrix.

In an n-dimensional space L_n, n orthonormal vectors form a complete set. The orthonormality is expressed as

$$\langle \varphi_i | \varphi_j \rangle = \delta_{ij} \ , \tag{2-12a}$$

$$\sum_{i=1}^{n} |\varphi_i\rangle \ \langle \varphi_i| = 1 \ . \tag{2-12b}$$

n linearly indepenent but not orthogonal vectors still form a complete set in the space L_n. However, (2-12) is replaced by the more general expression

$$\langle \varphi_i | \varphi_j \rangle = g_{ij} \ , \tag{2-13a}$$

$$\sum_{i,j=1}^{n} |\varphi_i\rangle (g^{-1})_{ij} \langle \varphi_j| = 1 \tag{2-13b}$$

where g^{-1} is the inverse of the matrix (g_{ij}).

If $g_{ij} = g_i \delta_{ij}$, then $(g^{-1})_{ij} = g_i^{-1}\delta_{ij}$. Therefore, for an orthogonal but not normalized basis, we have

$$g_i^{-1} \langle \varphi_i | \varphi_j \rangle = \delta_{ij} \ , \tag{2-14a}$$

$$\sum_{i=1}^{n} g_i^{-1} |\varphi_i\rangle \langle \varphi_i| = 1 \ . \tag{2-14b}$$

An alternative way of treating the non-orthonormal basis $\{|\varphi_i\rangle\}$ is by introducing the so-called *dual basis* $\{|\overline{\varphi}_i\rangle\}$, which is orthogonal to $\{|\varphi_i\rangle\}$,

$$\langle \overline{\varphi}_i | \varphi_j \rangle = \delta_{ij} \ . \tag{2-15a}$$

The completeness relation (2-13b) becomes

$$\sum_{i=1}^{n} |\varphi_i\rangle \langle \overline{\varphi}_i| = 1 \ . \tag{2-15b}$$

The above definition can be generalized to any linear vector space L. The scalar product is defined as

$$(\mathbf{u}_i, \mathbf{u}_j) = g_{ij} = g_{ji}^* \ , \tag{2-16}$$

where g_{ij} are specified complex numbers. The different specifications of the values g_{ij} correspond to different definitions of the scalar product. Once the metric tensor g_{ij} is given, the scalar product of any two vectors in the space L is specified

$$(\mathbf{x}, \mathbf{y}) = \left(\sum_i x_i \mathbf{u}_i, \sum_j y_j \mathbf{u}_j \right) = \sum_{ij} x_i^* g_{ij} y_j \ . \tag{2-17}$$

The scalar product is an invariant, since both $\mathbf{x}$ and $\mathbf{y}$ are invariants (see (2-7)).

A space L on which a scalar product is defined is called a *unitary space.* The different definitions of the scalar product in the same space L gives different unitary spaces. From now on, we only discuss unitary spaces, and whenever we talk about basis vectors, we always assume that they are orthonormal unless otherwise stated. Section 2.12 is devoted to non-orthonormal bases.

2.2 Linear Operator and Its Representation

If under the action of an operator R, any vector $\mathbf{x}$ in a space L changes to another vector $\mathbf{y}$ in L,

$$\mathbf{y} = R\mathbf{x} \,, \tag{2-18}$$

then the space L is said to be *closed* under the operation R, or to be an *invariant space* of the operator R. Equation (2-18) defines a mapping of the vector space onto itself. The operator R is said to be linear if

$$R(\mathbf{x}+\mathbf{z}) = R\mathbf{x} + R\mathbf{z}, \quad R(\alpha\mathbf{x}) = \alpha R\mathbf{x} \,, \tag{2-19}$$

where the action of R on $\mathbf{x}$ and $\mathbf{y}$ can be defined arbitrarily.

It is to be noted that no coordinate system is specified in the definition of the operator, so that the operator has an intrinsic significance.

For practical application, we always introduce a coordinate system. We employ the notation used in quantum mechanics. Let $\{|\varphi_i\rangle : i = 1, 2, \ldots, n\}$, be an orthonormal basis in a space L. Suppose that the space L is an invariant space of R; thus $R|\varphi_i\rangle$ can be expressed as a linear combination of the n basis vectors;

$$R|\varphi_i\rangle = \sum_j D_{ji}(R)|\varphi_j\rangle \,. \tag{2-20a}$$

Using the orthonormality of the basis vectors, one has

$$D_{ji}(R) = \langle\varphi_j|R|\varphi_i\rangle \,. \tag{2-21}$$

The matrix $D(R)$ is called, in the language of quantum mechanics, the *representative* of the operator R in the representation $\{\varphi_i\}$. Given $D(R)$, the action of the operator R on the basis, and in turn on any state vector $|\psi\rangle = \sum_i a_i|\varphi_i\rangle$, in L, is determined:

$$R|\psi\rangle = R\sum_i a_i|\varphi_i\rangle = \sum_i a_i R|\varphi_i\rangle = \sum_{ij} D_{ji}(R)a_i|\varphi_j\rangle \,. \tag{2-22a}$$

Notice that here the coefficients a_i are regarded as constants and the operator R acts only on the basis vectors, $|\varphi_i\rangle \xrightarrow{R} |\varphi_i'\rangle$.

However, the action of R on $|\psi\rangle$ can also be viewed in the following way: The basis is kept unchanged, while the coordinate of $|\psi\rangle$ (i.e., in the language of quantum mechanics, the wave function of $|\psi\rangle$ in the representation $\{\varphi_i\}$) changes from $\{a_i\}$ to $\{a_i'\}$,

$$R|\psi\rangle = |\psi'\rangle = \sum_j a_j'|\varphi_j\rangle \,. \tag{2-22b}$$

Comparing (2-22a) with (2-22b), one has

$$a_j' = \sum_i D_{ji}(R)a_i \,. \tag{2-20b}$$

Equations (2-20) can be rewritten as

$$R|\boldsymbol{\varphi}\rangle = \tilde{D}(R)|\boldsymbol{\varphi}\rangle \,, \tag{2-23a}$$

$$R\mathbf{a} = \mathbf{a}' = D(R)\mathbf{a} \tag{2-23b}$$

where $|\boldsymbol{\varphi}\rangle$ and $\mathbf{a}$ are column vectors. Notice the different transformation properties of $|\boldsymbol{\varphi}\rangle$ and $\mathbf{a}$.

From RR^{-1} and Eq. (2-12b),

$$\sum_j \langle\varphi_i|R|\varphi_j\rangle\langle\varphi_j|R^{-1}|\varphi_k\rangle = \delta_{ik}. \tag{2-24a}$$

Therefore

$$D(R^{-1}) = D^{-1}(R) . \tag{2-24b}$$

That is, the *representation of the inverse operator* R^{-1} is equal to the inverse of the representation matrix of R.

For any operator R in a space L, we can define another operator $R^\dagger$ by

$$\langle\varphi_j|R^\dagger|\varphi_i\rangle = \langle R\varphi_j|\varphi_i\rangle, \quad i,j = 1,2,\ldots n . \tag{2-25a}$$

$R^\dagger$ is called the *adjoint*, or *hermitian conjugate*, operator of R. From (2-11) and (2-25a) we have

$$D(R^\dagger) = \tilde{D}^*(R) \equiv D^\dagger(R). \tag{2-25b}$$

i.e., in an orthonormal basis, the representation of the adjoint operator $R^\dagger$ is the hermitian conjugate of the matrix representing the operator R.

An operator is *self-adjoint*, or hermitian, if it is identical with its adjoint:

$$R^\dagger = R , \tag{2-26a}$$

$$\langle\varphi_j|R|\varphi_i\rangle = \langle R\varphi_j|\varphi_i\rangle, \quad i,j = 1,2,\ldots n . \tag{2-26b}$$

From (2-25b), it is known that in an orthonormal basis, a hermitian operator is represented by a hermitian matrix,

$$D(R) = D^\dagger(R) . \tag{2-26c}$$

An operator R is said to be *unitary* if

$$\langle R\psi|R\psi'\rangle = \langle\psi|\psi'\rangle . \tag{2-27a}$$

By using Eq. (2-25a), one has

$$\langle R\psi|R\psi'\rangle = \langle\psi|R^\dagger R|\psi'\rangle = \langle\psi|\psi'\rangle . \tag{2-27b}$$

Therefore, a unitary operator satisfies

$$R^\dagger R = RR^\dagger = 1, \quad R^\dagger = R^{-1} . \tag{2-28a, b}$$

Introducing a basis, and using Eqs. (2-28b), (2-25b) and (2-24b), one has

$$D(R^\dagger) = D(R^{-1}) , \tag{2-29a}$$

$$D^\dagger(R) = D^{-1}(R) . \tag{2-29b}$$

i.e., in an orthonormal basis, the representation of a unitary operator is a unitary matrix.

It should be emphasized that the question of whether an operator is self-adjoint or unitary, i.e., whether Eq. (2-26a) or (2-28) holds or not, solely depends on the definition of the action of the operator in the given space and the definition of the scalar product, but is independent of the choice of the basis. However, the correctness of Eqs. (2-26c) and (2-29) does depend on the choice of the basis. In fact they are correct only for the orthonormal basis. The generalization to the non-orthonormal basis is given by Eqs. (2-118) and (2-121a) in Sec. 2.12.

Suppose that we change the basis $|\varphi_i\rangle$ to a new one $|\varphi'_i\rangle$ through a unitary transformation $B = (b_{ik})$:

$$|\varphi'_i\rangle = \sum_k b_{ik}|\varphi_k\rangle \ . \tag{2-30}$$

The representation of the operator R in the new basis is

$$D'_{ji}(R) = \langle\varphi'_j|R|\varphi'_i\rangle \ . \tag{2-31}$$

Substituting (2-30) into (2-31), we obtain a relation between the new and old representatives of R,

$$\begin{aligned} D'(R) &= AD(R)A^{-1}, \\ A &= B^* \ . \end{aligned} \tag{2-32}$$

The above discussion applies to any linear vector space with the following modifications in notations,

$$\begin{gathered} |\varphi_i\rangle \to \mathbf{u}_i, \quad |\psi\rangle \to \mathbf{x}, \quad a_i \to x_i, \\ \langle\varphi_i|R|\varphi_j\rangle \to (\mathbf{u}_i, R\mathbf{u}_j), \quad \langle R\varphi_i|R\varphi_j\rangle \to (R\mathbf{u}_i, R\mathbf{u}_j). \end{gathered} \tag{2-33}$$

2.3. Complete Set of Commuting Operators

2.3.1. *Eigenspace of self-adjoint operators*

Suppose that C is a self-adjoint operator in thespace L_n, and

$$C|\varphi_b\rangle = \sum_a C_{ab}|\varphi_a\rangle \ . \tag{2-34a}$$

$$C_{ab} = D_{ab}(C) = \langle\varphi_a|C|\varphi_b\rangle \ . \tag{2-34b}$$

Now let us find the eigenvectors of C

$$|\psi^\nu\rangle = \sum_a u_{\nu a}|\varphi_a\rangle \ , \tag{2-35a}$$

$$C|\psi^\nu\rangle = \nu|\psi^\nu\rangle \ . \tag{2-35b}$$

Substituting Eq. (2-35a) into (2-35b) and multiplying from the left by $\langle\varphi_a|$, we obtain

$$\sum_{b=1}^{n}(C_{ab} - \nu\delta_{ab})u_{\nu b} = 0 \ , \quad a = 1, 2, \ldots n \ . \tag{2-36}$$

The condition for the existence of non-trivial solutions to (2-36) is

$$\det|C_{ab} - \nu\delta_{ab}|_1^n = \prod_{i=1}^{k}(\nu - \nu_i)^{M_{\nu_i}} = 0 \ . \tag{2-37}$$

Equation (2-37) is called the *characteristic equation* or *expectation equation*, from which the eigenvalues ν_i along with their degeneracies M_{ν_i} can be determined. Here we assume that there are k distinct eigenvalues $\nu_1, \nu_2, \ldots, \nu_k$.

$$\sum_{i=1}^{k} M_{\nu_i} = n \ . \tag{2-38}$$

Substituting an eigenvalue, say ν, into Eq. (2-36), we obtain M_ν linearly independent eigenvectors,

$$|\psi_j^{(\nu)}\rangle, \quad j = 1, 2, \ldots, M_\nu . \tag{2-39}$$

They form an M_ν-dimensional subspace L_ν of the space L. L_ν is called the *eigenspace* of the operator C. Any vector in L_ν,

$$|\phi^{(\nu)}\rangle = \sum_{i=1}^{M_\nu} c_i |\psi_i^{(\nu)}\rangle , \tag{2-40}$$

is an eigenvector of C with the eigenvalue ν. The eigenspace can be defined symbolically by

$$CL_\nu = \nu L_\nu , \tag{2-41a}$$

or

$$L_\nu = \{|\psi^\nu\rangle : \quad C|\psi^\nu\rangle = \nu|\psi^\nu\rangle\} . \tag{2-41b}$$

Since C is assumed to be self-adjoint, the eigenvectors with different eigenvalues ν are orthogonal, i.e., the eigenspaces L_ν are orthogonal to one another. Thus the space L of dimension n is decomposed into a direct sum of k orthogonal eigenspaces of C, denoted by

$$L_n = \sum_{\nu=1}^{k} \oplus L_\nu . \tag{2-42}$$

2.3.2 Complete set of commuting operators (CSCO)

A set of commuting operators $C = (C_1, C_2, \ldots, C_l)$ is said to be a CSCO in a space L, if in L all the eigenvalues of C are non-degenerate. Let $|\psi_\lambda\rangle$ be a simultaneous eigenvector of C. It satisfies simultaneously the eigenequations

$$\begin{pmatrix} C_1 \\ \vdots \\ C_l \end{pmatrix} |\psi_{\lambda_1\ldots\lambda_l}\rangle = \begin{pmatrix} \lambda_1 \\ \vdots \\ \lambda_l \end{pmatrix} |\psi_{\lambda_1\ldots\lambda_l}\rangle , \tag{2-43a}$$

or written more compactly as

$$C|\psi_\lambda\rangle = \lambda|\psi_\lambda\rangle, \quad \lambda = (\lambda_1 \ldots \lambda_l) , \tag{2-43b}$$

$$|\psi_\lambda\rangle = \sum_a u_{\lambda a}|\varphi_a\rangle . \tag{2-43c}$$

In the space L_n, C has and only has n distinct sets of eigenvalues. One λ corresponds to only one eigenvector $|\psi_\lambda\rangle$. The n eigenvectors $|\psi_\lambda\rangle$ form a complete set in L_n. If all C_i are self-adjoint, then the eigenvectors with different λ are orthogonal, i.e., the eigenvectors of a CSCO form a orthonormal and complete set in L_n

$$\langle\psi_\lambda|\psi_{\lambda'}\rangle = \delta_{\lambda,\lambda'}, \quad \sum_\lambda |\psi_\lambda\rangle\langle\psi_\lambda| = 1 . \tag{2-44a, b}$$

Using Eq. (2-43c), Eq. (2-44a, b) can be expressed as

$$\sum_a u^*_{\lambda a} u_{\lambda' a} = \delta_{\lambda\lambda'}, \quad \sum_\lambda u^*_{\lambda a} u_{\lambda b} = \delta_{ab} . \tag{2-44c, d}$$

In other words, $\| u_{\lambda a} \|$ is a unitary matrix.

It must be stressed that a CSCO is related to a particular space. A CSCO in a space L_n is in general no longer a CSCO in another space L_m.

Theorem 2.1: A CSCO for a vector space L of finite dimensionality can always be chosen to consist of only a single operator.

Proof: We know that the basis vector $|\psi_\lambda\rangle$ for an n-dimensional vector space can always be labeled uniquely by a single discrete parameter λ, $\lambda = 1, 2, \ldots n$. Without loss of generality, $|\psi_\lambda\rangle$ can be assumed to be orthonormal. Let us construct a linear operator C by

$$C = \sum_{\lambda=1}^{n} \lambda |\psi_\lambda\rangle\langle\psi_\lambda| \,. \tag{2-44e}$$

Obviously we have

$$C|\psi_\lambda\rangle = \lambda|\psi_\lambda\rangle \,. \tag{2-44f}$$

Thus each basis vector is an eigenvector of the linear operator C with the parameter λ as the eigenvalue. According to the hypothesis of the uniqueness of the label λ, the single operator C is evidently a CSCO of the space L.

From Dirac (1958, Sec. 19) we have

Theorem 2.2: If L is an invariant subspace of a linear operator A, which commutes with each operator of the CSCO of the space L, then A is a function of the CSCO.

2.4. Group Representations

A representation (rep) of a group G is a homomorphism of G onto a group $D(G)$, the elements of which are operators in a space L, or matrices if a basis is chosen for L. L is called the representation space, and its dimension is called the *dimension* or *degree* of the rep $D(G)$.

If the operators are linear, $D(G)$ is said to be linear. As most reps used in physics are linear, in this book we only deal with linear reps.

A rep is said to be *faithful* if the mapping $G \to D(G)$ is an isomorphism.

From the homomorphism we have

$$D(RS) = D(R)D(S), \quad R, S \in G \,, \tag{2-45a}$$

$$D(e) = I \text{ (unit matrix)} \,. \tag{2-45b}$$

Hence

$$D(R^{-1}) = [D(R)]^{-1} \,. \tag{2-45c}$$

Ex. 2.1. If G_s is the kernel of the homomorphism $G \to D(G)$, then D(G) is a faithful representation of the factor group G/G_s.

Since state vectors in quantum mechanics are defined only up to a phase, we might deal with a group rep up to a phase. The conditions (2-45a) and (2-45c) are replaced by

$$D(RS) = \eta(R, S)D(R)D(S) \,, \tag{2-45a$'$}$$

$$D(R^{-1}) = \eta^*(R, R^{-1})[D(R)]^{-1} \,, \tag{2-45b$'$}$$

where $|\eta(R, S)| = 1$, and the phase factor $\eta(R, S)$ form what is called a *factor system.* The rep defined by (2-45′) is termed as the *projective* (or *ray*) rep, while the rep defined by (2-45) is termed as the *vector* rep.

It is easy to show that if a space L_n is an invariant subspace of G, then L_n form a representation space of G. Furthermore, if a basis $\{\varphi_i\}$ is chosen for L_n, the matrix representatives (in the usual sense used in quantum mechanics) of all the group elements in the rep space $L_n = \{\varphi_i\}$ constitute a rep D of the group G. $|\varphi_i\rangle$ are called the basis vectors *carrying* the rep D.

Proof: Since L_n is invariant under G,

$$|\varphi_i'\rangle = R_a|\varphi_i\rangle = \sum_{j=1}^{n} D_{ji}(R_a)|\varphi_j\rangle, \quad i\ 1\ldots n, \quad a = 1\ldots g\ . \tag{2-46}$$

$$D_{ji}(R_a) = \langle\varphi_j|R_a|\varphi_i\rangle\ . \tag{2-47}$$

Equation (2-46) can be put into matrix form

$$R_a\boldsymbol{\varphi} = \tilde{D}(R_a)\boldsymbol{\varphi}\ . \tag{2-48}$$

It is straightforward to show that the matrices defined by (2-47) satisfy the requirement of (2-45) (for non-orthonormal basis, see Sec. 2.12).

The rep $D(R) = 1$ for all $R \in G$, is called the *identity representation.* Any abstract group has the identity rep.

The rep $D^*(G)$ which is obtained from the rep $D(G)$ by complex conjugation, is called the *conjugate representation.*

If D^* is equivalent to D, then D is said to be a self-conjugate rep.

The rep $\tilde{D}^{-1}(G)$ of G is called the *contragredient rep.* If $\{\mathbf{u}_i\}$ carry the rep $D(G)$ and $\{x_i\}$ carry the rep $\tilde{D}^{-1}(G)$, from Eqs. (2-7) and (2-10) we know that

$$\mathbf{x} = \sum_i x_i\mathbf{u}_i$$

is invariant under the group G.

For unitary reps (see Sec. 2.5),

$$\tilde{D}^{-1}(G) = D^*(G)\ , \tag{2-49}$$

namely, the contragredient rep is just the conjugate rep.

Theorem 2.3: A rep D is real if and only if it conincides with its conjugate ($D = D^*$).
Theorem 2.4: If the character of a rep D is real, then D is self-conjugate.
Theorem 2.5: The character of a rep D is complex if and only if D is not self-conjugate.

Example 1: The group R_2. The two unit vectors $\mathbf{u}_1$ and $\mathbf{u}_2$ form a basis in a 2-dimensional space. The action of a rotation operator $R(\varphi)$ on the basis vectors $\mathbf{u}_i$ is defined as a rotation of $\mathbf{u}_1$ and $\mathbf{u}_2$ through angle φ

$$\begin{aligned} R(\varphi)\mathbf{u}_1 = \mathbf{u}_1' &= \cos\varphi\mathbf{u}_1 + \sin\varphi\mathbf{u}_2, \\ R(\varphi)\mathbf{u}_2 = \mathbf{u}_2' &= -\sin\varphi\mathbf{u}_1 + \cos\varphi\mathbf{u}_2\ . \end{aligned} \tag{2-50}$$

From the definition (2-46) or (2-47), we get a 2-dimensional rep of R_2

$$D(\varphi) = D(R(\varphi)) = \begin{pmatrix} \cos\varphi & -\sin\varphi \\ \sin\varphi & \cos\varphi \end{pmatrix}\ , \quad 0 \le \varphi \le 2\pi\ . \tag{2-51}$$

The infinite number of matrices (2-51) form a matrix group, which is a faithful rep of the group R_2.

Example 2: Representations of S_3. Suppose that there are three electrons, two are in the state α (spin up) and one in the state β (spin down). The possible states in the spin space for the three electrons are

$$\begin{aligned} \varphi_1 &= x_\alpha(\xi_1)x_\alpha(\xi_2)x_\beta(\xi_3) = |\alpha\alpha\beta\rangle\ , \\ \varphi_2 &= x_\alpha(\xi_1)x_\beta(\xi_2)x_\alpha(\xi_3) = |\alpha\beta\alpha\rangle\ , \\ \varphi_3 &= x_\beta(\xi_1)x_\alpha(\xi_2)x_\alpha(\xi_3) = |\beta\alpha\alpha\rangle\ , \end{aligned} \tag{2-52}$$

where ξ_i is the spin coordinate of the i-th electron. $\{\varphi_i\}$ forms a 3-dimensional space. We define the action of an element p of the permutation group S_3 upon φ_i as a permutation of the subscripts of the spin coordinates ξ_i. For example

$$(123)|\alpha\alpha\beta\rangle = (123)\big(x_\alpha(\xi_1)x_\alpha(\xi_2)x_\beta(\xi_3)\big) = x_\alpha(\xi_2)x_\alpha(\xi_3)x_\beta(\xi_1) = |\beta\alpha\alpha\rangle \ . \qquad (2\text{-}53)$$

In general, the operation of a cycle permutation $(jkm\ldots)$ on an n-particle product state $|i_1 i_2 \ldots i_n\rangle$ is to move the state index i_j from its original j-th place to the k-th place, and move i_k to the m-th place, etc. For example

$$(356)|i_1 i_2 i_3 i_4 i_5 i_6\rangle = |i_1 i_2 i_6 i_4 i_3 i_5\rangle \ .$$

Obviously, the space $\{\varphi_i\}$ is closed under the group S_3, and is a rep space of S_3. From Eq. (2-46), it is easy to find the representative matrices:

$$\begin{matrix} D(e) & D(12) & D(13) & D(23) & D(123) & D(132) \\ \begin{pmatrix} 1&0&0\\0&1&0\\0&0&1 \end{pmatrix} & \begin{pmatrix} 1&0&0\\0&0&1\\0&1&0 \end{pmatrix} & \begin{pmatrix} 0&0&1\\0&1&0\\1&0&0 \end{pmatrix} & \begin{pmatrix} 0&1&0\\1&0&0\\0&0&1 \end{pmatrix} & \begin{pmatrix} 0&1&0\\0&0&1\\1&0&0 \end{pmatrix} & \begin{pmatrix} 0&0&1\\1&0&0\\0&1&0 \end{pmatrix} \end{matrix} \ . \qquad (2\text{-}54)$$

It should be stressed that the operation p on φ_i is defined as permuting the *subscripts of the coordinate indices* rather than the subscripts of the basis vectors. Therefore, we have $(12)\ \varphi_1 = \varphi_1$ rather than $(12)\ \varphi_1 = \varphi_2$.

Ex. 2.2. Verify that Eq. (2-54) is a faithful rep of S_3.

Ex. 2.3. Show that the four matrices in Eq. (1-4) form a projective rep of the point group $C_{2v} = (e, C_2^z, \sigma^x, \sigma^y)$, where σ^x and σ^y are reflection plans with normals in the x and y direction, respectively.

From the above discussion we know that the reps considered in group theory and in quantum mechanics are almost the same thing. The only differences are (1) the rep space in quantum mechanics is always a physical space, while the rep space in group theory can be any linear vector space; (2) the operators in quantum mechanics correspond to observables and are thus always self-adjoint, while the group operators may be non self-adjoint; (3) for a non-orthonormal basis, the definitions of the representative matrix of an operator are different (see Sec. 2.12).

2.5. Unitary Representation

If all the matrices of a rep are unitary,

$$D^\dagger(R_a) = D^{-1}(R_a) \ , \quad a = 1, 2, \ldots g \ ,$$

then the rep is called a *unitary* rep. According to Sec. 2.3, the rep of a unitary operator in an orthonormal basis must be a unitary matrix, while in a non-orthonormal basis, it can be unitary or non-unitary (see Sec. 8.7).

If both the bases $\{\varphi_i\}$ and $\{\varphi'_i\}$ are orthonormal, then the rep $D(G)$ defined by Eq. (2-46) is unitary. For example, the reps given by Eqs. (2-51) and (2-54) are unitary.

In Sec. 2.4, we used unit vectors as basis for the group R_2. It is more common to use wave functions $\varphi(\mathbf{x})$ as the basis for a rep of R_2 or R_3. Now let us define the action of a rotation $R_z(\varphi)$ on the wave function (or a field) $\varphi(\mathbf{x})$. We define $\varphi'(\mathbf{x}) = R_z(\varphi)\varphi(\mathbf{x})$ as the new field resulting from rotating the old field $\varphi(\mathbf{x})$ through angle φ about the z-axis. Here the coordinate axes are kept unchanged. Under the rotation $R_z(\varphi)$, a point $P(\mathbf{x}_0)$ in the space moves to the point $P'(\mathbf{x}'_0)$ (see Fig. 2.5)

$$\mathbf{x}'_0 = R\mathbf{x}_0 \ . \qquad (2\text{-}55)$$

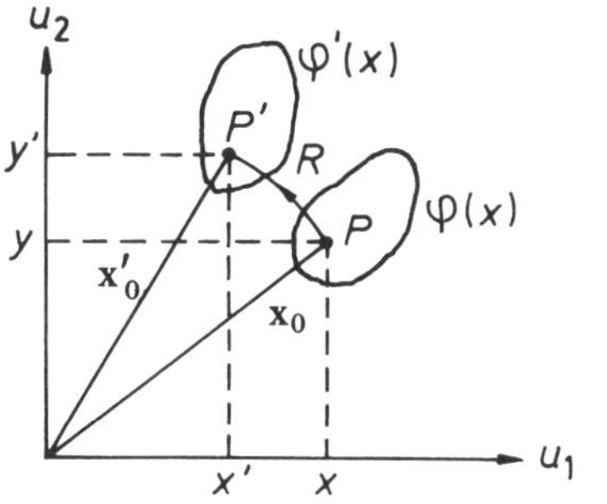

Fig. 2.5

Evidently, the value of the new field $\varphi'(\mathbf{x})$ at the new point P' is equal to the value of the old field $\varphi(\mathbf{x})$ at the original point P

$$\varphi'(\mathbf{x}'_0) = \varphi(\mathbf{x}_0) \ . \tag{2-56}$$

From Eq. (2-55), one has $\mathbf{x}_0 = R^{-1}\mathbf{x}'_0$, therefore

$$\varphi'(\mathbf{x}'_0) = \varphi(R^{-1}\mathbf{x}'_0) \ . \tag{2-57}$$

Letting $\mathbf{x}'_0 \to \mathbf{x}$ (since $\mathbf{x}'_0$ is an arbitrary point in space), one has

$$R\varphi(\mathbf{x}) = \varphi'(\mathbf{x}) = \varphi(R^{-1}\mathbf{x}) \ . \tag{2-58}$$

In the following we shall adhere to Eq. (2-58) as the definition for the operation of the rotation R upon a wave function $\varphi(\mathbf{x})$, i.e., rotating the field while keeping the coordinate system unchanged.

Example 1: The group R_2. Suppose that there are two points $P(x,y)$ and $P'(x',y')$ (see Fig. 2.5), and under the rotation $R(\varphi)$, the point P goes over to P',

$$\begin{pmatrix} x' \\ y' \end{pmatrix} = R(\varphi)\begin{pmatrix} x \\ y \end{pmatrix} = \begin{pmatrix} \cos\varphi & -\sin\varphi \\ \sin\varphi & \cos\varphi \end{pmatrix}\begin{pmatrix} x \\ y \end{pmatrix} \ . \tag{2-59}$$

Assume that $\varphi_1(\mathbf{x})$ and $\varphi_2(\mathbf{x})$ are wave functions of a p_x electron and a p_y electron, respectively. Ignoring all irrelevant factors, $\varphi_1(\mathbf{x}) = x, \varphi_2(\mathbf{x}) = y$.

From (2-58) we have

$$\begin{pmatrix} \varphi_1 \\ \varphi_2 \end{pmatrix} \xrightarrow{R} \begin{pmatrix} \varphi'_1 \\ \varphi'_2 \end{pmatrix} = \begin{pmatrix} \cos\varphi & \sin\varphi \\ -\sin\varphi & \cos\varphi \end{pmatrix}\begin{pmatrix} x \\ y \end{pmatrix} = \tilde{D}(\varphi)\begin{pmatrix} \varphi_1 \\ \varphi_2 \end{pmatrix} \ . \tag{2-60a}$$

Therefore, with $\varphi_1 = x$, and $\varphi_2 = y$ as our basis, the representation of R_2 is

$$D(\varphi) = \begin{pmatrix} \cos\varphi & -sin\varphi \\ \sin\varphi & \cos\varphi \end{pmatrix} \ , \tag{2-60b}$$

which is identical with (2-51). The difference between Eq. (2-59) and Eq. (2-60a) merits special attention. When x and y are regarded as coordinates, they transform according to (2-59), and if regarded as wave functions, they transform according to (2-60a). Henceforth, whenever we talk about using x, y and z as our basis, we always understand that they are wave functions instead of coordinates.

Having defined the scalar product (2-11) and the action (2-58) of group elements on $\varphi(\mathbf{x})$, we can check whether the group operator R is unitary. It is clear that

$$\langle R\varphi(\mathbf{x})|R\psi(\mathbf{x})\rangle = \int [R\varphi(\mathbf{x})]^* R\psi(\mathbf{x})d\mathbf{x} = \int \varphi^*(\mathbf{x}')\psi(\mathbf{x}')d\mathbf{x} \ , \tag{2-61}$$

where

$$\mathbf{x}' = R^{-1}\mathbf{x} .$$

Furthermore, the volume element is unchanged under the rotation,

$$d\mathbf{x}' = d\mathbf{x} . \tag{2-62}$$

Substituting this into (2-61), we have

$$\langle R\varphi(\mathbf{x})|R\psi(\mathbf{x})\rangle = \langle\varphi(\mathbf{x})|\psi(\mathbf{x})\rangle .$$

Thus the rotation operator is unitary.

In configuration space, the action of other operators such as space inversion, plane reflection, etc., are also defined by Eq. (2-58). Under these operations, we also have $d\mathbf{x}' = d\mathbf{x}$. Therefore, they are also unitary.

The above discussion can be easily generalized to many-particle wave functions,

$$\begin{aligned} \Psi(\mathbf{X}) &= \psi_1(\mathbf{x}_1)\psi_2(\mathbf{x}_2)\ldots\psi_n(\mathbf{x}_n), \\ \Phi(\mathbf{X}) &= \varphi_1(\mathbf{x}_1)\varphi_2(\mathbf{x}_2)\ldots\varphi_n(\mathbf{x}_n) , \\ d\mathbf{X} &= d\mathbf{x}_1 d\mathbf{x}_2\ldots d\mathbf{x}_n = d\mathbf{X}' . \end{aligned} \tag{2-63a}$$

$$R\Psi(\mathbf{X}) = \Psi(R^{-1}\mathbf{X}) \equiv \psi_1(R^{-1}\mathbf{x}_1)\psi_2(R^{-1}\mathbf{x}_2)\ldots\psi_n(R^{-1}\mathbf{x}_n) . \tag{2-63b}$$

Consequently, these operators are still unitary in the rep space spanned by many-particle wave functions.

Example 2: Representations of C_{3v}. Obviously (x, y, z) is a rep space of C_{3v}. Let us find the rep of C_{3v} in this basis. We use $\sigma^{(\varphi)}$ to designate a reflection plane, whose intersection with the xy plane has the orientation angle φ with respect to the x-axis (see Fig. 8.1-2). The three reflection planes of C_{3v} as shown in the figure below Table 8.3-5 are thus represented by

$$\sigma^{(1)} = \sigma^{0^\circ} , \quad \sigma^{(2)} = \sigma^{120^0}, \quad \sigma^{(3)} = \sigma^{60^0} . \tag{2-64a}$$

For the problem here it is convenient to use cylindrical coordinates (ρ, φ, z). A function $\psi(\rho, \varphi, z)$ under σ^θ goes over to

$$\varphi'(\rho, \varphi, z) = \varphi(\rho, 2\theta - \varphi, z) . \tag{2-64b}$$

Using Eqs. (2-60b) and (2-64b) we can obtain the rep of C_{3v} which is left to the reader as an exercise.

Ex. 2.4. Construct the rep of C_{3v} in the basis (x, y, z).

Ex. 2.5. Construct the rep of C_{4v} in the basis (x, y).

2.6. Regular Rep and Group Algebra

2.6.1. Definition of regular representation

A function Φ is said to possess *no* symmetry whatsoever with respect to a group G, if under the action of the g elements of G, it generates g linearly independent functions. Φ is said to be *totally symmetric* if it is invariant under G. Functions whose symmetry behavior lies between these two are said to have *partial symmetry*. For example, with respect to S_3, $|\alpha\beta\gamma\rangle$, $|\alpha\alpha\beta\rangle$ and $|\alpha\alpha\alpha\rangle$ are functions without symmetry, with partial symmetry and totally symmetric, respectively.

We shall always use Φ_0 to represent a function without any symmetry. For simplicity, we assume that under the action of the g elements of G, it generates g orthonormal functions

$$\varphi_a = R_a\Phi_0 \quad a = 1, 2, \ldots, g . \tag{2-65a}$$

$$\langle \varphi_a | \varphi_b \rangle = \delta_{ab} \ . \tag{2-65b}$$

The g-dimensional rep carried by $\varphi_1 \ldots , \varphi_g$ is called the *regular rep* of G

$$R_c \varphi_b = \sum_{a=1}^{g} D_{ab}(R_c) \varphi_a \ . \tag{2-66}$$

Using Eq. (2-65a), it can be rewritten as

$$R_c R_b \Phi_0 = \sum_{a=1}^{g} D_{ab}(R_c) R_a \Phi_0 \ . \tag{2-67}$$

Therefore, the matrix elements of the regular rep are given by

$$D_{ab}(R_c) = \begin{cases} 1, & \text{when} \quad R_c R_b = R_a, \\ 0, & \text{otherwise.} \end{cases} \tag{2-68}$$

The condition $R_c R_b = R_a$ is equivalent to

$$R_c R_b \Phi_0 = R_a \Phi_0, \quad \text{i.e.,} \ R_c \varphi_b = \varphi_a \ . \tag{2-69}$$

From Eq (2-68) it is seen that in any row or any column of a matrix $D(R)$ of the regular rep, there is only one nonvanishing element (being equal to unity), and that the diagonal elements of all matrices are zeroes except for the matrix representing the identity element.

Example: Regular rep of S_3. Applying the six permutations (1-12) of S_3 to $\Phi_0 = |\alpha\beta\gamma\rangle$, we get the following six functions

$$\begin{matrix} \varphi_1 = e\Phi_0 & \varphi_2 = (12)\Phi_0 & \varphi_3 = (13)\Phi_0 & \varphi_4 = (23)\Phi)_0 & \varphi_4 = (123)\Phi_0 & \varphi_6 = (132)\Phi_0 \\ |\alpha\beta\gamma\rangle \ , & |\beta\alpha\gamma\rangle \ , & |\gamma\beta\alpha\rangle \ , & |\alpha\gamma\beta\rangle \ , & |\gamma\alpha\beta\rangle \ , & |\beta\gamma\alpha\rangle \ . \end{matrix} \tag{2-70}$$

We can find the regular rep according to definition (2-66). However the simplest way of obtaining the regular rep is by using the group table. In fact the c-th row of Table 1.2 gives the representation matrix of the c-th element R_c of S_3,

$$\begin{aligned} D(e) &= \{123456\}, \quad & D(12) &= \{216543\}, \quad & D(13) &= \{351624\}, \\ D(23) &= \{465132\}, \quad & D(123) &= \{534261\}, \quad & D(132) &= \{642315\} \ , \end{aligned} \tag{2-71a}$$

where $\{a_1 a_2 a_3 \ldots\}$ denotes a matrix with the matrix elements all equal to one at the row a_i and column i, and zeroes elsewhere, e.g.,

$$D(123) = \{534261\} = \begin{pmatrix} 0&0&0&0&0&1 \\ 0&0&0&1&0&0 \\ 0&1&0&0&0&0 \\ 0&0&1&0&0&0 \\ 1&0&0&0&0&0 \\ 0&0&0&0&1&0 \end{pmatrix} \ . \tag{2-71b}$$

As we shall see, the regular rep plays a crucial role in the group rep theory.

2.6.2. Group space

Equations (2-66) and (2-67) are equivalent. Dividing both sides of Eq. (2-67) by Φ_0, we have

$$R_c R_b = \sum_{a=1}^{a} D_{ab}(R_c) R_a \ . \tag{2-72}$$

Thus we can regard the g elements $\{R_a\}$ of the group G, instead of the g functions $\{\varphi_a = R_a\Phi_0\}$, as the basis vectors for the regular rep. Now that R_a are basis vectors, we can define the addition of R_a and the multiplication of R_a with complex numbers. (Notice that there are no such operations in the definition of the group). Thus the g elements $\{R_a\}$ form a linear vector space of dimension g, which is called the ***group space***. The mapping of the group space to the space $\{\varphi_a\}$ is a 1-1 mapping. The metric tensor in the space $\{\varphi_a\}$ is defined by (2-65b), while the metric tensor for the group space can be defined in the same way:

$$\langle R_a|R_b\rangle \equiv (R_a, R_b) = g_{ab} = \delta_{ab} \ . \tag{2-73}$$

Any vector P in the group space is expressible in terms of the g basis vectors

$$P = \sum_{a=1}^{g} u_a R_a = \sum_{a=1}^{g} u(R_a) R_a \ , \tag{2-74}$$

where u_a are complex numbers. From (2-73) and (2-74) we have the scalar product of two vectors $P^{(i)}$ and $P^{(j)}$:

$$\langle P^{(i)}|P^{(j)}\rangle \equiv (P^{(i)}, P^{(j)}) = \sum_{a=1}^{g} u_a^{(i)*} u_a^{(j)} \ . \tag{2-75}$$

2.6.3. Group algebra

In Eq. (2-66), R_c is an operator and φ_b is a basis vector, while in the corresponding Eq. (2-72), R_c is an operator and R_b is a basis vector. In order to distinguish operators from basis vectors, we might put a caret ^ above the operator, thus (2-72) becomes

$$\hat{R}_c R_b = R_c R_b = \sum_{a=1}^{g} D_{ab}(R_c) R_a \ . \tag{2-76}$$

i.e., the action of $\hat{R}_c$ upon R_b is defined as the group product R_cR_b. Therefore, in fact the symbol ^ is redundant and will be omitted henceforth. However, we should keep in mind that a group element now can play double roles; *it can serve either as an operator or as a basis vector.* Thus in the group space, in addition to the operation of linear combination.

$$P^{(i)} + P^{(j)} = \sum_a (u_a^{(i)} + u_a^{(j)}) R_a \in L_g \ , \tag{2-77}$$

we have another operation, i.e., multiplication according to the multiplication rule of the group. Clearly, the product of any vectors in L_g still belongs to L_g

$$P^{(i)} P^{(j)} = \sum_{ab} u_a^{(i)} u_b^{(j)} R_a R_b \in L_g \ . \tag{2-78}$$

Definition 2.1: A linear space is called an *algebra* if the linear space is closed under a multiplication rule.

An equivalent definition is as follows.

Definition 2.2: A set of elements is said to form an *algebra* if the set is closed under linear combinations and multiplications.

The group space is closed under group multiplication and thus form an algebra, called the ***group algebra***, and the vector P in Eq. (2-74) is called an *element* of the group algebra. If B is a subspace of A and is closed under the same multiplication rule as that of the algebra A, then B is an algebra by itself, and is called a *subalgebra* of the algebra A.

In analogy with the definition of the group rep $D(G)$, we can define the rep of an algebra. The representative of an element $P = \sum_a u_a R_a$ of the group algebra is defined by

$$D(P) = \sum_a u_a D(R_a) . \qquad (2\text{-}79)$$

Thus a rep of a group gives a rep of the group algebra, and vice versa.

2.7. Space of Functions on the Group

A vector P in the group space L_g can be expressed as (2-74), where the g complex numbers u_a, the coordinates of the vector P in the basis $\{R_a\}$, can be regarded as the components of a g-dimensional vector $\mathbf{u}$

$$\mathbf{u} = (u_1, u_2, \ldots, u_g) = (u(R_1), u(R_2), \ldots, u(R_g)) . \qquad (2\text{-}80)$$

On the other hand, Eq. (2-80) can also be regarded as a function $u(R)$ defined on the group manifold. For discrete groups, $u(R)$ is a discrete function, i.e., it is defined only at the g "points" $R = R_1, R_2, \ldots, R_g$. $u(R)$ or $u(R_a)$ is called a *function on the group* (manifold). Here the argument R or R_a should be regarded as a variable. The totality of $\mathbf{u}$ forms a space L_F, called the *space of functions on the group*. For a finite group, there are only g independent vectors in L_F. We may choose the following g vectors as a basis of L_F

$$\mathbf{e}_1 = (1, 0, \ldots, 0), \quad \mathbf{e}_2 = (0, 1, 0, \ldots, 0), \ldots, \quad \mathbf{e}_g = (0, 0, \ldots, 1) . \qquad (2\text{-}81)$$

The metric tensor is defined by

$$(\mathbf{e}_a, \mathbf{e}_b) = g_{ab} = \delta_{ab} . \qquad (2\text{-}82)$$

Any vector in L_F can be expressed as

$$\mathbf{u} = \sum_{a=1}^{g} u(R_a)\mathbf{e}_a , \qquad (2\text{-}83)$$

The scalar product of two functions on the group is thus equal to

$$\langle \mathbf{u}^{(i)} | \mathbf{u}^{(j)} \rangle = \sum_{a=1}^{g} u^{(i)}(R_a)^* u^{(j)}(R_a) . \qquad (2\text{-}84)$$

We have already defined the operation of a group element R on a vector P in group space by (cf. Eq. (2-22a))

$$RP = R \sum_{a=1}^{g} u(R_a) R_a = \sum_{a=1}^{g} u(R_a) R R_a . \qquad (2\text{-}85)$$

Let

$$R R_a = R_b, \quad R_a = R^{-1} R_b ;$$

substituting them into (2-85) and changing the summation index from a to b (cf. Eq. (2-22b))

$$RP = P' = \sum_{b=1}^{g} u(R^{-1} R_b) R_b = \sum_{b=1}^{g} u'(R_b) R_b , \qquad (2\text{-}86)$$

i.e., the operation of R on the vector P can be looked upon as changing the coordinates of P from $u(R_a)$ to $u'(R_a)$ while keeping the basis $\{R_a\}$ unchanged. We thus define the operation of a group element on functions on the group as

$$u'(R_a) = Ru(R_a) = u(R^{-1}R_a) \ . \tag{2-87a}$$

It has the same form as the operation of a rotation operator on wave functions.

Notice that

$$RSu(R_a) = Ru(S^{-1}R_a) \neq u(R^{-1}S^{-1}R_a) \ ,$$

but instead we have

$$RSu(R_a) = u((RS)^{-1}R_a) = u(S^{-1}R^{-1}R_a) \ . \tag{2-87b}$$

From Eqs. (2-84) and (2-87) it is easy to prove that

$$\langle R\mathbf{u}^{(i)}|R\mathbf{u}^{(j)}\rangle = \sum_{a=1}^{g}[u^{(i)}(R^{-1}R_a)]^* u^{(j)}(R^{-1}R_a) = \sum_{b=1}^{g}[u^{(j)}(R_b)]^* u(R_b) = \langle \mathbf{u}^{(i)}|\mathbf{u}^{(j)}\rangle \ .$$

Therefore, group elements are unitary operators in the space L_F.

Summarizing, there is a 1-1 correspondence between the configuration space $\{\varphi_a\}$ of (2-65a), the group space $\{R_a\}$ and the space of function on the group, $\{u(R_a)\}$. Each of them carries the regular representation of the group G.

2.8. Equivalent Representations and Characters

If $D(G)$ and $D'(G)$ are two reps of a group G in the space L and L', respectively, and they are related by

$$D'(R) = AD(R)A^{-1}, \quad R \in G \ , \tag{2-88}$$

where A is a square matrix, then we say that the reps D and D' are equivalent, and L and L' are equivalent rep spaces.

Evidently, two equivalent reps must have the same dimension, and two reps with different dimensions cannot be equivalent.

If we make the space L and L' identical, then the relation (2-88) merely represents a change of basis [cf. Eq. (2-32)].

Since the trace of a matrix is invariant under a similarity transformation

$$\begin{aligned} \mathrm{Tr}\, D(R)' &= \sum_i D'_{ii}(R) = \sum_{ikl} A_{ik} D_{kl}(R) A^{-1}_{li} = \sum_{kl} \delta_{kl} D_{kl}(R) \\ &= \sum_k D_{kk}(R) = \mathrm{Tr}\ D(R) \ , \end{aligned} \tag{2-89}$$

we call the trace of $D(R)$ the *character* of the element R in the rep D, and is denoted by

$$\chi(R) = \mathrm{Tr}\ D(R) \ . \tag{2-90}$$

Therefore, equivalent reps have the same set of characters.

From Eq. (2-90) it follows that $\chi(e)$ is equal to the dimension of the representation.

As with Eq. (2-89), from (1-17) we can show that all elements of the same class have the same character. Thus the character is a function of classes. We use χ_i^μ to designate the character of the i-th class in the μ-th rep.

As we said that there are an infinite number of reps of a group G which are equivalent to a given rep $D(G)$. Due to the simplicity of the unitary rep, it is most desirable to work with the unitary rep. However, not every rep of any group is equivalent to a unitary rep, except those of finite groups and the so-called compact Lie groups (see Sec. 5.5). We have

Theorem 2.6: Any representation of a finite group is equivalent to a unitary representation.

In the following we only deal with the unitary rep.

Ex. 2.6. Prove that the character is only a function of the class.

2.9. Reducible and Irreducible Representations

From the two reps of $G, D^{(1)}(G)$ and $D^{(2)}(G)$, with dimensions h_1 and h_2, it is trivial to obtain a new rep D with dimension $h_1 + h_2$ by[1)]

$$D(R) = \left(\begin{array}{c|c} D^{(1)}(R) & 0 \\ \hline 0 & D^{(2)}(R), \end{array}\right) . \tag{2-91}$$

The rep D is called the direct sum of the reps $D^{(1)}$ and $D^{(2)}$, and is denoted by

$$D = D^{(1)} \oplus D^{(2)} . \tag{2-92}$$

Evidently, the character of the rep D is

$$\chi(R) = \chi^{(1)}(R) + \chi^{(2)}(R) . \tag{2-93}$$

The converse to the above is: Given a rep $D(G)$, can we find a similarity transformation T to bring all the matrices $D(R)$ to the same block-diagonal form? If we can, then the rep D is said to be *reducible*;[2)] otherwise, D is *irreducible*. For example, if

$$TD(R_a)T^{-1} = \left(\begin{array}{c|c} D^{(1)}(R_a) & 0 \\ \hline 0 & D^{(2)}(R_a) \end{array}\right) , \quad a = 1, 2 \dots g , \tag{2-94}$$

then we say that the rep $D(G)$ is reduced into a direct sum of the two reps $D^{(1)}$ and $D^{(2)}$, denoted by

$$D = D^{(1)} \oplus D^{(2)}.$$

An alternative definition of a reducible rep space is:

Definition 2.3: If in a rep space L we can find two subspaces $L^{(1)}$ and $L^{(2)}$, which are invariant[3)] under the group G, then we say that the rep space L is *reducible* and can be reduced into a direct sum of two rep spaces

$$L = L^{(1)} \oplus L^{(2)} , \tag{2-95}$$

otherwise, we say that the space L is *irreducible*.

[1)] Using the following multiplication rule for the block-diagonal matrices,

$$D(R)D(S) = \left(\begin{array}{c|c} D^{(1)}(R)D^{(1}(S) & 0 \\ \hline 0 & D^{(2)}(R)D^{(2)}(S) \end{array}\right) ,$$

it is easy to show that $D(R)$ form a rep of G.

[2)] To be exact, what we call reducible here refers to a fully reducible rep (see Sec. 5.1).

[3)] For a fully reducible rep, we only need to find one invariant subspace $L^{(1)}$ of G, since for such a case, the remaining subspace $L^{(2)}$ is necessarily invariant under G.

A basis of the irreducible space is called an *irreducible* basis, denoted by φ_m^ν or $|\nu m\rangle$, where ν is the label for the irreducible rep, and m enumerates the basis vectors of the irreducible space.

Definition 2.4: An invariant subspace is said to be minimal if it does not contain any non-trivial invariant subspace (i.e., the null space and the subspace itself) with respect to G.

Now we can have a more succint definition for the irreducible space.

Definition 2.5: A minimal invariant subspace of a group G is called an irreducible space of G.

Similarly, we can define a minimal invariant subspace under an algebra A as the irreducible space of A and the rep generated by the irreducible space of A as the irrep *of the algebra.*

Example 1: The group R_2. Equation (2-60b) gives a 2-dimensional rep of R_2 in the basis $\varphi_1 = x$ and $\varphi_2 = y$. If we make the following transformation (the task of finding such a transformation is a fundamental problem in the group rep theory and will be discussed in great detail in the subsequent chapters) on the basis

$$\varphi_1' = \frac{-1}{\sqrt{2}}(x+iy) = rY_{11}(\theta,\varphi), \quad \varphi_2' = \frac{1}{\sqrt{2}}(x-iy) = rY_{1-1}(\theta,\varphi) \ , \tag{2-96}$$

we shall see that this 2-dimensional rep will be reduced into two one- dimensional reps. In order to show this, we compare Eq. (2-96) with Eq. (2-30), and obtain

$$B = \begin{pmatrix} -\frac{1}{\sqrt{2}} & -\frac{i}{\sqrt{2}} \\ \frac{1}{\sqrt{2}} & -\frac{i}{\sqrt{2}} \end{pmatrix} . \tag{2-97}$$

Using Eqs. (2-88), (2-60b) and (2-97), we obtain the rep of R_2 in the new basis (2-96) as

$$D'(\varphi) = B^* D(\varphi)\tilde{B} = \begin{pmatrix} e^{-i\varphi} & 0 \\ 0 & e^{i\varphi} \end{pmatrix} . \tag{2-98}$$

Ex. 2.7. Show that if the basis in (2-52) undergoes the following transformation

$$\begin{pmatrix} \psi_1 \\ \psi_2 \\ \psi_3 \end{pmatrix} = \begin{pmatrix} \sqrt{\frac{1}{3}} & \sqrt{\frac{1}{3}} & \sqrt{\frac{1}{3}} \\ \sqrt{\frac{2}{3}} & -\sqrt{\frac{1}{6}} & -\sqrt{\frac{1}{6}} \\ 0 & \sqrt{\frac{1}{2}} & -\sqrt{\frac{1}{2}} \end{pmatrix} \begin{pmatrix} \varphi_1 \\ \varphi_2 \\ \varphi_3 \end{pmatrix} , \tag{2-99}$$

the 3-dimensional rep (2-54) will be reduced.

If a rep $D(G)$ has the block-diagonal form (2-94), we shall say that $D(G)$ is in a *reduced form.* We inquire whether $D^{(1)}$ and $D^{(2)}$ can be further reduced. Since each subspace can be treated independently, we discuss $D^{(1)}$ and $D^{(2)}$ separately. For example, the following transformation

$$\left(\begin{array}{c|c} T_1' & 0 \\ \hline 0 & 1 \end{array}\right) \begin{array}{l} \}h_1 \\ \}h_2 \end{array} \tag{2-100}$$

only changes $D^{(1)}$ without affecting $D^{(2)}$. We may choose T_1' appropriately such that $D^{(1)}$ becomes block-diagonal. We continue this process until $D(G)$ becomes a direct sum of irreducible representations,

$$D(R) = \begin{pmatrix} D^{(\nu_1)}(R) & & \\ & D^{(\nu_2)}(R) & \\ & & \ddots \end{pmatrix} . \tag{2-101}$$

or

$$D(R) = \sum_i \oplus D^{(\nu_i)} = D^{(\nu_1)} \oplus D^{(\nu_2)} \oplus \dots \tag{2-102}$$

Suppose that the dimension of the irrep (ν_i) is $|\nu_i|$, then $h = \sum_i |\nu_i|$. The reducible representation space L is correspondingly decomposed into a direct sum of a number of irreducible spaces

$$L = \sum_i \oplus L_{\nu_i} = L_{\nu_1} \oplus L_{\nu_2} \oplus \dots \tag{2-103a}$$

Among these irreps, some of them may be equivalent. Equivalent irreps are considered to be the same irrep, therefore, in the reducible rep D, an irrep $D^{(\nu)}$ may occur more than once, say τ_ν times. Thus, Eq. (2-102) can be rewritten as

$$D = \sum_\nu \oplus \tau_\nu D^{(\nu)} , \tag{2-103b}$$

where the sum runs only over the inequivalent irreps.

The character of a reducible rep is called the *compound character*, while the character of an irreducible representation is called the *simple*, or *primitive character.*

Using Eq. (2-103b), the compound character is expressible in terms of the simple characters

$$\chi_i = \sum_\nu \tau_\nu \chi_i^{(\nu)} . \tag{2-103c}$$

If a rep $D(G)$ has been reduced to a form such as Eq. (2-101) where all $D^{(\nu_i)}$ are irreducible, then we say $D(G)$ is in *totally reduced form.*

Since a fully reducible rep can always be reduced to a sum of irreps of lower dimensions, the problem of finding all reps of a finite group or compact Lie group is thus simplified to the problem of finding all the inequivalent irreps of G.

Before discussing how to reduce a rep, we introduce Theorem 2.7.

Theorem 2.7: If an operator C commutes with a group G, then the eigenspace L_λ of C is a representation space of G.

Proof: According to the assumption

$$[C, R_a] = 0, \quad a = 1, \dots, g . \tag{2-104}$$

Let $\psi^{(\lambda)}$ be a vector in the eigenspace L_λ of C.

$$C\psi^{(\lambda)} = \lambda\psi^{(\lambda)} , \tag{2-105a}$$

$$CR_a\psi^{(\lambda)} = R_aC\psi^{(\lambda)} = \lambda R_a\psi^{(\lambda)} , \quad a = 1, \dots, g . \tag{2-105b}$$

Therefore, $R_a\psi^{(\lambda)}$ is still an eigenvector of C with the same eigenvalue λ. It means that $R_a\psi^\lambda \in L_\lambda$, i.e., L_λ is an invariant subspace of G, and thus is a rep space of G.

If a rep space of G is decomposed to a direct sum of k eigenspaces of C, $L = \sum_{i=1}^{k} \oplus L_{\lambda_i}$, then the corresponding rep D is reduced to a direct sum of k reps

$$D = \sum_{i=1}^{k} \oplus D^{(\lambda_i)}.$$

2.10. Subduced and Induced Representations

A *subduced representation* is one obtained by restricting an irrep $D^{(\nu)}$ of a group G to elements of a subgroup G_s of G, denoted by $D^{(\nu)} \downarrow G_s$.

The subduced rep is in general a reducible rep of G_s, which can be reduced to a direct sum of the irreps of G_s,

$$D^{(\nu)} \downarrow G_s = \sum_{\mu} \oplus \tau_{\mu}^{(\nu)} D^{(\mu)}(G_s), \tag{2-106a}$$

where $\tau_{\mu}^{(\nu)}$ is the number of times that the irrep (μ) of G_s occurs in the subduced rep $D^{(\nu)} \downarrow G_s$.

Definition 2.6: If $\tau_{\mu}^{(\nu)} \leq 1$ for all possible ν and μ, then G_s is said to be a *canonical subgroup* of G.

Definition 2.7: A group chain

$$G \supset G_1 \supset G_2 \supset \ldots \supset G_n$$

is said to be a *canonical subgroup chain* if G_{i+1} is a canonical subgroup of G_i for $i = 0, 1, 2, \ldots, n-1$ with $G_0 = G$, and G_n being an Abelian group.

The above process for obtaining a rep of a subgroup G_s from an irrep of the larger group G is called *subduction*.

The reverse process is called *induction*, i.e., the construction a rep of a larger group G from an irrep of its subgroup G_s. The left coset decomposition of G with respect to G_s is written as

$$G = \sum_{i=1}^{q} a_i G_s \ , \tag{2-106b}$$

where a_i are coset representatives, $a_1 = e$ and $q = |G|/|G_s|$.

Let $|\mu m\rangle$ be the basis vector of an irrep (μ) of G_s, with dimension h_μ. Let

$$|\mu m i\rangle \equiv a_i |\mu m\rangle \ , \quad i = 1, 2, \ldots, q \ , \quad m = 1, 2, \ldots, h_\mu \ . \tag{2-106c}$$

For any $R \in G, Ra_i$ can be expressed as

$$Ra_i = a_j H \ , \quad H \in G_s \ . \tag{2-106d}$$

Thus

$$\begin{aligned} R|\mu m i\rangle &= Ra_i|\mu m\rangle = a_j H|\mu m\rangle \\ &= a_j \sum_{m'} D^{(\mu)}_{m'm}(H)|\mu m'\rangle = \sum_{m'} D^{(\mu)}_{m'm}(H)|\mu m' j\rangle \ , \end{aligned} \tag{2-106e}$$

where $D^{(\mu)}$ is the irrep of G_s. Therefore, the qh_μ vectors $|\mu m i\rangle$ are closed under the group G and thus carry a rep of G, called the *induced representation* and denoted by $D^{(\mu)} \uparrow G$, or $(\mu) \uparrow G$. The induced rep is in general reducible, i.e.,

$$D^{(\mu)} \uparrow G = \sum_{\nu} \oplus \tau_{\nu}^{(\mu)} D^{(\nu)}(G) \tag{2-106f}$$

where $\tau_{\nu}^{(\mu)}$ is the number of occurrences of the irrep (ν) of G in the induced rep $(\mu) \uparrow G$.

Frobenious reciprocity theorem: The number of times that $D^{(\nu)}(G)$ occurs in $D^{(\mu)} \uparrow G$ is equal to the number of times that the irrep $D^{(\mu)}(G_s)$ appears in $D^{(\nu)} \downarrow G_s$, i.e.,

$$\tau_{\nu}^{(\mu)} = \tau_{\mu}^{(\nu)} \ . \tag{2-106g}$$

From (2-106e) we know that the representation matrix of any element R of G in the induced rep $(\mu) \uparrow G$ is block diagonalized,

$$D^{(\mu)\uparrow G}_{kj,mi}(R) = \mathcal{G}_{ji}(R) D^{(\mu)}_{km}(a_j^{-1} R a_i) , \tag{2-106h}$$

where $\mathcal{G}(R)$ is called the ground rep of the group G with respect to the subgroup G_s,

$$\mathcal{G}_{ji}(R) = \begin{cases} 1 & \text{if } a_j^{-1} R a_i \in G_s , \\ 0 & \text{otherwise} . \end{cases} \tag{2-106i}$$

It is seen that the ground rep looks like the regular rep in that there is only one nonvanishing matrix element in each row or column. It is convenient to introduce the modified ground rep $\mathcal{D}(R)$ as

$$\mathcal{D}_{ji}(R) = \begin{cases} H , & \text{if } a_j^{-1} R a_i = H \in G_s , \\ 0 , & \text{otherwise} . \end{cases} \tag{2-106j}$$

For example, the modified ground rep of the permutation (14) of S_4 with respect to its subgroup $S_3 \times S_1$ is

$$\mathcal{D}(14) = \begin{bmatrix} 0 & 0 & 0 & (123) \\ 0 & (13) & 0 & 0 \\ 0 & 0 & (13) & 0 \\ (132) & 0 & 0 & 0 \end{bmatrix} \tag{2-106k}$$

with four coset representatives ordered as $e, (34), (234)$, and (1234).

Ex. 2.8. Find the modified ground reps of all the transpositions of S_4 with respect to its subgroup $S_3 \times S_1$.

2.11. Schur's Lemma

Schur's lemma 1: If C is an operator commuting with a group G, and L_ν is an invariant subspace of C and an irreducible space of G, then L_ν is necessarily an eigenspace of C.

Proof: We use the method of *reduction ad absurdum.* Since L_ν is an invariant subspace of C, we can find eigenvectors of C in L_ν and group them into eigenspaces of C. Suppose that L_ν can be decomposed into a direct sum of two eigenspaces of C, $L_\nu = L_1 \oplus L_2, CL_i = \lambda^i L_i, i = 1, 2$. Then according to Theorem 2.2, the rep space L_ν would be decomposed into a direct sum of two rep spaces of G, in contradiction with the assumption of the irreducibility of L_ν. Therefore, the only possibility is that $CL_\nu = \lambda^\nu L_\nu$.

Assume that $\psi_1^{(\nu)}, \dots, \psi_{h_\nu}^{(\nu)}$ are the basis vectors of the irreducible space L_ν. Schur's lemma 1 tells us

$$C\psi_i^{(\nu)} = \lambda^{(\nu)} \psi_i^{(\nu)}, \quad i = 1, 2, \dots, h_\nu , \tag{2-107}$$

$$\langle \psi_i^{(\nu')} | C | \psi_j^{(\nu)} \rangle = \delta_{\nu\nu'} D_{ij}^{(\nu)}(C) = \delta_{\nu\nu'} \delta_{ij} \lambda^{(\nu)} . \tag{2-108}$$

Therefore, in an irreducible space L_ν of a group G, the representative of an operator which commutes with the group G and has L_ν as its invariant subspace is a multiple of the unit matrix.

Another form of Schur's lemma 1 is

Schur's lemma 1′: If the matrices $D(R)$ form an irrep of a group G, and if

$$AD(R) = D(R)A, \quad \text{for all } R \in G , \tag{2-109}$$

then A=const. I.

In other words, if a matrix commutes with all the matrices of an irrep, then the matrix is necessarily a multiple of the unit matrix.

Proof: In the irreducible space we solve the eigenequation

$$A\mathbf{x} = \lambda \mathbf{x}.$$

By Eq. (2-109), $D(R)\mathbf{x}$ is also an eigenvector of A with the same eigenvalue λ. Therefore the eigenspace of A is an invariant subspace of the group G. But this would mean that D is reducible, unless this subspace is the whole space, or the null space. The first possibility implies that A has only one eigenvalue, i.e., $A = \lambda I$, the second that $A = 0$. QED.

The Schur's lemma plays a central role in group rep theory, which gives the criterion for irreducibility.

Criterion for irreducibility: A rep is irreducible if and only if the only matrices which commute with all matrices of the rep are multiples of the unit matrix.

Consequently, to determine whether a rep D is irreducible, it suffices to find the matrix A from the matrix equation (2-109) and to see whether A is a multiple of the unit matrix.

It is to be noted that the inverse of Schur's lemma is not true. All we can say is that the eigenspace of an operator C which commutes with the group G is a representation space of G, which is in general reducible. What we are going to do in the next chapter is to enlarge the single operator C in Theorem 2.7 to a complete set of commuting operators (CSCO), all of which are commuting with G, so that the common eigenspaces of the CSCO are irreducible spaces of G. Thus the problem of finding irreps of a group G is converted into that of diagonalizing a complete set of commuting operators in the reducible representation space.

Ex. 2.9. The group $SL(2,C)$ is represented by the set of regular 2×2 complex matrices. Show that this rep is irreducible. By applying Schur's lemma, find the center of this group. Generalize to the case of $GL(n,C)$.

2.12. Appendix: Non-Orthonormal Basis

The following points should be especially taken care of for a non-orthonormal basis $\{|\varphi_i\rangle\}$.

2.12.1 Two definitions of the representation of an operator

(1) Definition in group representation theory: Suppose that under a group G the non-orthonormal basis $|\varphi_i\rangle$ transforms as

$$R|\varphi_i\rangle = \sum_j D_{ji}(R)|\varphi_j\rangle \; . \tag{2-110a}$$

Then the matrix in (2-110a), $D(R)$, is still defined as the representative of R. However, (2-47) is no longer true. Instead we have

$$D_{ji}(R) = \langle\overline{\varphi}_j|R|\varphi_i\rangle \; , \tag{2-110b}$$

where $|\overline{\varphi}_i\rangle$ is the dual basis defined in (2-15). From Eq. (2-15) one gets

$$|\overline{\varphi}_j\rangle = \sum_k (g^{-1})_{kj}|\varphi_k\rangle \; , \tag{2-111a}$$

$$|\varphi_k\rangle = \sum_i g_{ik}|\overline{\varphi}_i\rangle \; . \tag{2-111b}$$

Once the operation of R on $|\varphi_i\rangle$ is known, from Eq. (2-110a) we immediately get the representative $D(R)$ without having to find the dual basis first, and then using (2-110b).

We now use Eq. (2-15) to prove that the rep defined by (2-110b) obeys the requirement (2-45). From $RR^{-1} = 1$, one has

$$\sum_j \langle\overline{\varphi}_i|R|\varphi_j\rangle\langle\overline{\varphi}_j|R^{-1}|\varphi_k\rangle = \langle\overline{\varphi}_i|\varphi_k\rangle = \delta_{ik} \; ,$$

$$\therefore \quad D(R^{-1}) = D^{-1}(R) \; . \tag{2-112}$$

Furthermore,

$$\langle\overline{\varphi}_i|RS|\varphi_k\rangle = \sum_j \langle\overline{\varphi}_i|R|\varphi_j\rangle\langle\overline{\varphi}_j|S|\varphi_k\rangle \ , \tag{2-113a}$$

$$\langle\overline{\varphi}_i|e|\varphi_j\rangle = \langle\overline{\varphi}_i|\varphi_j\rangle = \delta_{ij} \ . \tag{2-113b}$$

Hence Eq. (2-45) holds.

(2) Definition in quantum mechanics. The representative $\mathcal{D}(R)$ of an operator R is defined by

$$\mathcal{D}_{ji}(R) = \langle\varphi_j|R|\varphi_i\rangle \ . \tag{2-114}$$

Using Eq. (2-13b), one gets

$$\langle\varphi_i|RS|\varphi_l\rangle = \sum_{jk}\langle\varphi_i|R|\varphi_j\rangle(g^{-1})_{jk}\langle\varphi_k|S|\varphi_l\rangle \ ,$$

i.e.

$$\mathcal{D}(RS) = \mathcal{D}(R)g^{-1}\mathcal{D}(S) \ . \tag{2-115a}$$

From Eq. (2-13a) it follows that

$$\langle\varphi_i|e|\varphi_j\rangle = \langle\varphi_i|\varphi_j\rangle = g_{ij} \ ,$$

$$\mathcal{D}(e) = g \ . \tag{2-115b}$$

With $S = R^{-1}$ in (2-115a) and using (2-115b) one obtains

$$\mathcal{D}(R^{-1}) = g\mathcal{D}^{-1}(R)g \ . \tag{2-115c}$$

One sees clearly that the rep defined in quantum mechanics by Eq. (2-114) does not satisfy the group representation requirement of (2-45).

(3) The relation between the two kinds of reps: Multiplying Eq. (2-110a) from the left by $\langle\varphi_k|$, we obtain a relation between the two:

$$\mathcal{D}(R) = gD(R) \ . \tag{2-116}$$

2.12.2. Representation of an adjoint operator

(1) Using Eq. (2-110) one has

$$\begin{aligned}\langle\varphi_i|R^\dagger|\varphi_k\rangle &= \langle\varphi_i|\sum_j D_{jk}(R^\dagger)\varphi_j\rangle = \sum_j g_{ij}D_{jk}(R^\dagger) \\ &= \langle R\varphi_i|\varphi_k\rangle = \langle\sum_j D_{ji}(R)\varphi_j|\varphi_k\rangle = \sum_j D^*_{ji}(R)g_{jk} \ .\end{aligned}$$

$$D(R^\dagger) = g^{-1}\tilde{D}^*(R)g \ . \tag{2-117}$$

Hence $D(R^\dagger) \neq D^\dagger(R) = \tilde{D}^*(R)$. Therefore, under the definition (2-110b), the representative matrix of a self-adjoint operator is no longer hermitian. Instead, its matrix satisfies the condition

$$D(A) = g^{-1}\tilde{D}^*(A)g \ . \tag{2-118}$$

(2) On the contrary, under the definition (2-114) one has

$$\mathcal{D}(R^\dagger) = \mathcal{D}^\dagger(R) \equiv \tilde{\mathcal{D}}^*(R) \ . \tag{2-119}$$

Therefore, under the definition (2-114) the representative matrix of a self-adjoint operator remains hermitian for non-orthonormal basis.

2.12.3. Representation of a unitary operator

(1) From $R^\dagger = R^{-1}$, one has

$$D(R^\dagger) = D(R^{-1}) \, , \tag{2-120a}$$

while from (2-112) and (2-117),

$$D(R^\dagger) = D(R^{-1}) = D^{-1}(R) = g^{-1}\tilde{D}^*(R)g \, . \tag{2-121a}$$

Consequently, $D^{-1}(R) \neq \tilde{D}^*(R)$, i.e., $D(R)$ is in general no longer unitary, although in special cases it may still be unitary (see Sec. 8.7).

(2) From $R^\dagger = R^{-1}$ one gets

$$\mathcal{D}(R^\dagger) = \mathcal{D}(R^{-1}) \, . \tag{2-120b}$$

Using Eqs. (2-119) and (2-115c), one has

$$\mathcal{D}(R^\dagger) = \mathcal{D}^\dagger(R) = \mathcal{D}(R^{-1}) = g\mathcal{D}^{-1}(R)g \, . \tag{2-121b}$$

Therefore $\mathcal{D}^\dagger(R) \neq \mathcal{D}^{-1}(R)$, i.e., $\mathcal{D}(R)$ is, in general, no longer unitary.

2.12.4. Representation transformation

(1) If we switch to the new basis $|\varphi_i'\rangle$ given in Eq. (2-30), the corresponding new dual basis becomes

$$|\bar{\varphi}_i'\rangle = \sum_j a_{ij}|\bar{\varphi}_j\rangle \, . \tag{2-122}$$

It is easy to show that for preserving the orthonormality $\langle\bar{\varphi}_j'|\varphi_i'\rangle = \delta_{ij}$, the matrix $A = |a_{ij}|$ has to satisfy the condition

$$A = (B^\dagger)^{-1} \, . \tag{2-123a}$$

For a real B, it becomes

$$A = \tilde{B}^{-1} \, , \tag{2-123b}$$

i.e., in such a case, the basis $|\varphi_i\rangle$ is covariant, while the dual basis $|\bar{\varphi}_i\rangle$ is contravariant. (Examples are given in Sec. 10.7).

The new representative of the operator R is

$$D_{ji}'(R) = \langle\bar{\varphi}_j'|R|\varphi_i'\rangle \, . \tag{2-124}$$

Substituting Eqs. (2-30) and (2-122) into (2-124), and using (2-123a), we get the relation

$$D'(R) = A^*D(R)\tilde{B} = \tilde{B}^{-1}D(R)\tilde{B} \, . \tag{2-125}$$

Setting $B = T^{-1}$, we obtain the familiar formula

$$D'(R) = TD(R)T^{-1} \, . \tag{2-126}$$

It is easy to show that: (1) under a unitary transformation, an orthonormal basis goes over to another orthonormal basis; (2) the transformation between two orthonormal bases is necessarily unitary; (3) the transformation between an orthonormal basis and an non-orthornomal basis is necessarily non-unitary.

(2) The relation between the new and old metric tensors is

$$g_{il}' = \langle\varphi_i'|\varphi_l'\rangle = \sum_{jk} b_{ij}^* g_{jk} b_{lk} \, ,$$

i.e.,

$$g' = B^*g\tilde{B} \, . \tag{2-127}$$

One can also show that

$$\mathcal{D}'(R) = B^*\mathcal{D}(R)\tilde{B} \, . \tag{2-128}$$

2.12.5. Eigenvectors of a self-adjoint operator

The non-orthonormal basis vectors $|\varphi_b\rangle$ can be linearly combined into eigenvectors of a self-adjoint operator C,

$$C|\psi_\lambda\rangle = \lambda|\psi_\lambda\rangle \,, \tag{2-129}$$

$$|\psi_\lambda\rangle = \sum_b u_{\lambda b}|\varphi_b\rangle \,. \tag{2-130}$$

This can be done under both definitions (2-110b) and (2-114) for the representation matrices.

(1) Substituting (2-130) into (2-129), then multiplying from the left by $\langle\overline{\varphi}_a|$, and making use of (2-15a) and (2-110b), one gets

$$\sum_b (D_{ab}(C) - \lambda\delta_{ab})u_{\lambda b} = 0 \,. \tag{2-131}$$

We see that it has exactly the same form as Eq. (2-36) for the orthonormal basis.

Assuming that C is self-adjoint, then

$$\langle\psi_{\lambda'}|C|\psi_\lambda\rangle = \lambda\langle\psi_{\lambda'}|\psi_\lambda\rangle = \langle C\psi_{\lambda'}|\psi_\lambda\rangle = \lambda'\langle\psi_{\lambda'}|\psi_\lambda\rangle \,,$$

i.e.,

$$(\lambda - \lambda')\langle\psi_{\lambda'}|\psi_\lambda\rangle = 0 \,. \tag{2-132}$$

Therefore, the eigenvectors $|\psi_\lambda\rangle$ of C are orthogonal in λ, and can be normalized.

If C is a CSCO for the space L, then each eigenvalue λ is unique, and the orthonormality and completeness (2-44a, b) remain true; however Eqs. (2-44c, d) are not true. Substituting (2-130) into (2-44a), we get the orthonormality relation

$$\sum_{ab} u^*_{\lambda a} g_{ab} u_{\lambda' b} = \delta_{\lambda\lambda'} \,. \tag{2-133a}$$

Define

$$v_{\lambda a} = \sum_b g_{ab} u_{\lambda b} \,. \tag{2-134}$$

Then (2-133a) can be rewritten as

$$\sum_a v^*_{\lambda a} u_{\lambda' a} = \delta_{\lambda\lambda'} \,. \tag{2-133b}$$

On multiplying (2-130) from the left by $\langle\varphi_a|$, one gets

$$\langle\varphi_a|\psi_\lambda\rangle = \sum_b g_{ab} u_{\lambda b} = v_{\lambda a} \,. \tag{2-135}$$

On multiplying (2-44b) from the left by $\langle\overline{\varphi}_c|$, and from the right by $|\varphi_b\rangle$, as well as using (2-130) and (2-135), one obtains the completeness,

$$\sum_\lambda v^*_{\lambda b} u_{\lambda c} = \delta_{bc} \,. \tag{2-136a}$$

Using (2-134), we may cast it in the form

$$\sum_{\lambda a} u^*_{\lambda a} g_{ab} u_{\lambda c} = \delta_{bc} \,. \tag{2-136b}$$

Finally, multiplying $|\psi_\lambda\rangle = \sum_c u_{\lambda c}|\varphi_c\rangle$ by $v^*_{\lambda b}$, summing over λ, and using (2-136a), we get the inverse of (2-130):

$$|\varphi_b\rangle = \sum_\lambda v^*_{\lambda b}|\psi_\lambda\rangle = \sum_{\lambda a} u^*_{\lambda a} g_{ab}|\psi_\lambda\rangle \ . \tag{2-137a}$$

Equation (2-136) shows that the transformation $\|u_{\lambda b}\|$ from the non-orthonormal basis $|\varphi_b\rangle$ to the orthonormal basis $|\psi_\lambda\rangle$ is not unitary.

For an orthogonal but unnormalized basis with the metric tensor $g_{ab} = g_a\delta_{ab}$, (2-136) reduces to

$$\begin{aligned} \sum_a g_a u^*_{\lambda a} u_{\lambda' a} &= \delta_{\lambda\lambda'} \ , \\ \sum_\lambda g_a u^*_{\lambda a} u_{\lambda b} &= \delta_{ab} \ , \end{aligned} \tag{2-138}$$

and its dual basis is

$$|\overline{\varphi}_a\rangle = (g_a)^{-1}|\varphi_a\rangle = (\langle\varphi_a|\varphi_a\rangle)^{-1}|\varphi_a\rangle \ . \tag{2-139}$$

(2) Applying the operator C to (2-130), we have

$$\sum_b (C - \lambda)|\varphi_b\rangle u_{\lambda b} = 0 \ . \tag{2-140a}$$

Multiplying (2-140a) from the left by $\langle\varphi_a|$ gives

$$\sum_b (\mathcal{D}_{ab}(C) - \lambda g_{ab}) u_{\lambda b} = 0 \ , \tag{2-140b}$$

or

$$(\mathcal{D}(C) - \lambda g)\mathbf{u}_\lambda = 0 \ . \tag{2-140c}$$

It is equivalent to the eigenequations

$$[g^{-1}\mathcal{D}(C) - \lambda \cdot I]\mathbf{u}_\lambda = 0 \quad \text{or} \quad [g^{-\frac{1}{2}}\mathcal{D}(C)g^{-\frac{1}{2}} - \lambda \cdot I]\mathbf{u}_\lambda = 0 \ . \tag{2-141}$$

According to (2-116), $g^{-1}\mathcal{D}(C) = D(C)$. Therefore, (2-141) and (2-131) are exactly the same. It means that the eigenvectors of C are the same regardless of the definition for the representation matrix of C. This is what one would expect.

For an orthornomal basis, $|\overline{\varphi}_a\rangle = |\varphi_a\rangle, \mathcal{D} = D$, all the differences discussed above between the cases (1) and (2) disappear, and we are back to our previous discussion.

Henceforth we shall use definition (2-110b) for the representative of an operator.

Finally, one interesting point is worth mentioning. From (2-137a), i.e.,

$$|\varphi_a\rangle = \sum_\lambda v^*_{\lambda a}|\psi_\lambda\rangle$$

as well as (2-111) and (2-134), one has

$$|\overline{\varphi}_a\rangle = \sum_\lambda u^*_{\lambda a}|\psi_\lambda\rangle \ . \tag{2-137b}$$

If we regard $|\varphi_a\rangle$ as an eigenvector of a non-self-adjoint operator $\mathcal{O}$ with the eigenvalue γ_a,

$$\mathcal{O}|\varphi_a\rangle = \gamma_a|\varphi_a\rangle \ , \tag{2-142a}$$

then the dual basis $|\overline{\varphi}_a\rangle$ is an eigenvector of the adjoint operator $\mathcal{O}^\dagger$ with the eigenvalue complex conjugate to γ_a,

$$\mathcal{O}^\dagger|\overline{\varphi}_a\rangle = \gamma^*_a|\overline{\varphi}_a\rangle \ . \tag{2-142b}$$

To show this, it suffices to prove that the solutions to (2-142a, b) obey the condition $\langle\overline{\varphi}_a|\varphi_b\rangle = \delta_{ab}$,

$$\langle\overline{\varphi}_a|\mathcal{O}|\varphi_b\rangle = \gamma_b\langle\overline{\varphi}_a|\varphi_b\rangle = \langle\mathcal{O}^\dagger\overline{\varphi}_a|\varphi_b\rangle = \gamma_a\langle\overline{\varphi}_a|\varphi_b\rangle \ .$$
$$(\gamma_a - \gamma_b)\langle\overline{\varphi}_a|\varphi_b\rangle = 0.$$

Therefore, after normalization we have $\langle\overline{\varphi}_a|\varphi_b\rangle = \delta_{ab}$.

Define the column vectors

$$\mathbf{u}_\lambda = \mathrm{col}(u_{\lambda a}, u_{\lambda b}, \dots) \ , \quad \mathbf{v}_a = \mathrm{col}(v^*_{\lambda a}, v^*_{\lambda' a}, \dots)$$
$$\overline{\mathbf{v}}_a = \mathrm{col}(u^*_{\lambda a}, u^*_{\lambda' a}, \dots) \ .$$

Evidently, $\mathbf{u}_\lambda$ is the representative of the eigenvector $|\psi_\lambda\rangle$ of the self-adjoint operator C in the non-orthonormal basis $\{\varphi_a\}$, while $\mathbf{v}_a(\overline{\mathbf{v}}_a)$ is the representative of the eigenvector $|\varphi_a\rangle(|\overline{\varphi_a}\rangle)$ of the non-self-adjoint operator $\mathcal{O}(\mathcal{O}^\dagger)$ in the orthonormal basis $\psi_\lambda\}$. Therefore, (2-133a) and (2-136b) are the orthonormality and completeness relations of $\mathbf{u}_\lambda$, while (2-133b) and (2-136a) are the "orthonormality and completeness" relations of $\mathbf{v}_a(\overline{\mathbf{v}}_a)$. In other words, (2-133) and (2-136) can be regarded either as the orthonormality and completeness relations of the eigenvectors of the self-adjoint operator C, or those of the non-self-adjoint operators $\mathcal{O}$ and $\mathcal{O}^\dagger$.

Chapter 3

Representation Theory for Finite Groups

This chapter is the kernel of a new approach to group representation theory. Two routes are available for presenting the material. The first one simply follows the path traced out according to the sequence of topics of the book, i.e., following the original steps by which the theory was developed. We use the permutation group S_3, the simplest non-Abelian finite group, as a guide for establishing the general theory and show how the question of reducing a group rep is intimately related to the elimination of degeneracy in quantum mechanics. The second route is as follows: after reading Theorem 3.12, jump to Secs 3.7, 3.8 and 3.9, then turn back to Theorem 3.14 in Sec. 3.3. The first route is more instructive and easily mastered by the novice, while the second one is more elegant and rigorous from the mathematical point of view. Therefore, these two routes are complementary in their advantages and disadvantages. The reader may choose one according to his own preferences. The extensive review article (Chen, [15] 1985) was written following the second route.

3.1. Class Space and Class Operators

3.1.1. Class operators

A sum of all elements belonging to the same class is called *class operator*. Supposing that the i-th class has g_i elements $R_1, R_2, \ldots, R_{g_i}$, the class operator C_i is

$$C_i = \sum_{l=1}^{g_i} R_l^{(i)} , \quad i = 1, 2, \ldots, N . \tag{3-1}$$

Sometimes we will use the so called average class operator C^i,

$$C^i = C_i / g_i . \tag{3-2}$$

Remark: The group operators R are unitary, but the class operators may not.

The class operators have the following three important properties.

(1) They commute with any element of the group G.

$$[C_i, R_a] = 0 , \quad a = 1, 2, \ldots, g . \tag{3-3}$$

Proof: According to (1-17), a conjugate operation on an element $R_l^{(i)}$ only changes it to another element $R_{l'}^{(i)}$ of the same class:

$$R_a C_i R_a^{-1} = \sum_{l=1}^{g_i} R_a R_l^{(i)} R_a^{-1} = \sum_{l'=1}^{g_i} R_{l'}^{(i)} = C_i . \tag{3-4}$$

Thus Eq. (3-3) holds.

(2) With the aid of Eqs. (3-1) and (3-3), we see that they commute with each other:

$$[C_i, C_j] = 0 . \tag{3-5}$$

(3) They are closed under group mutliplication:

$$C_i C_j = \sum_{k=1}^{N} C_{ij}^{k} C_k . \tag{3-6a}$$

where C_{ij}^{k} are integers.

Proof: According to

$$\sum_{a=1}^{g} R_a R_l^{(i)} R_a^{-1} = \sum_{a=1}^{g} R_a (R_b R_l^{(i)} R_b^{-1}) R_a^{-1} = \sum_{a=1}^{g} R_a R_{l'}^{(i)} R_a^{-1} ,$$

we know that in the above equation, all the elements of the i-th class are on the same footing. Therefore

$$\sum_{a=1}^{g} R_a C_i R_a^{-1} = g_i \sum_{a=1}^{g} R_a R_l^{(i)} R_a^{-1} .$$

From (3-3) and the above equation, we get

$$C_i = \frac{g_i}{g} \sum_{a=1}^{g} R_a R_l^{(i)} R_a^{-1} , \tag{3-6b}$$

or

$$\sum_{a=1}^{g} R_a R_l^{(i)} R_a^{-1} = \frac{g}{g_i} C_i . \tag{3-6c}$$

On the other hand, from (3-3) we also have

$$C_i C_j = \frac{1}{g} \sum_{a=1}^{g} R_a (C_i C_j) R_a^{-1} . \tag{3-6d}$$

According to (3-6c), the right-hand side of Eq. (3-6d) must consist of complete classes. Thus Eq. (3-6a) holds.

From the requirement that both sides of (3-6a) should have the same number of elements, one has

$$g_i g_j = \sum_{k=1}^{N} C_{ij}^{k} g_k . \tag{3-6e}$$

In Sec. 3.11 it will be shown that the coefficients C_{ij}^{k} determine the simple characters of the group G, and are called the *structure constants* of a finite group. By Eq. (3-5), we infer that C_{ij}^{k} is symmetric in i and j, i.e.,

$$C_{ij}^{k} = C_{ji}^{k} . \tag{3-7}$$

Because of the above three properties, the class operator plays a decisive role in the new approach to group representation theory.

Theorem 3.1: Any operator constructed out of group operators and commuting with the group is necessarily a class operator or a linear combination of the class operators of the group.

Proof: Assume that the operator

$$A = \sum_{a=1}^{g} x_a R_a$$

commutes with the group G, where x_a are complex numbers. This means that

$$A = R_a A R_a^{-1}, \quad a = 1, 2, \dots, g .$$

Therefore,

$$A = \frac{1}{g}\sum_{a=1}^{g} R_a A R_a^{-1} = \sum_{b=1}^{g} x_b \left[\frac{1}{g}\sum_{a=1}^{g} R_a R_b R_a^{-1}\right] = \sum_{b=1}^{g} \frac{x_b}{g_b} C_b ,$$

where C_b is the class operator containing the element R_b.

3.1.2. Class algebra

The N-dimensional vector space spanned by the N class operators C_i of a group G is called the *class space* L_c, which is a subspace of the g-dimensional group space. Any vector Q in L_c can be expressed as a linear combination of the basis vectors C_i,

$$Q = \sum_{i=1}^{n} q_i C_i = \sum_{i=1}^{n} q(C_i) C_i , \tag{3-8}$$

where $q_i = q(C_i)$ are complex numbers. The sum of two vectors in L_c is defined by

$$Q^{(\nu)} + Q^{(\mu)} = \sum_{i=1}^{N} (q_i^{(\nu)} + q_i^{(\mu)}) C_i \in L_c . \tag{3-9}$$

Since $L_c \subset L_g$, the metric tensor g_{ij} in L_c is determined by the metric tensor in L_g. From (2-73) and (3-1a) we have

$$\langle C_i | C_j \rangle = g_{ij} = g_i \delta_{ij} . \tag{3-10}$$

Therefore $\{C_i\}$ is an orthogonal but not normalized basis. The scalar product of two vectors in L_c is thus equal to

$$\langle Q^{(\nu)} | Q^{(\mu)} \rangle = \sum_{i} g_i q_i^{(\nu)*} q_i^{(\mu)} . \tag{3-11}$$

Furthermore, we can use (3-6a) to define the product of two vectors in L_c, i.e.,

$$Q^{(\nu)} Q^{(\mu)} = \sum_{i} q_i^{(\nu)} C_i \sum_{j} q_j^{(\mu)} C_j = \sum_{ijk} (q_i^{(\nu)} q_j^{(\mu)} C_{ij}^{k}) C_k \in L_c . \tag{3-12}$$

From (3-9) and (3-12) it is seen that the set $\{Q^{(\nu)}\}$ is closed under linear combination and group multiplication; therefore the class operators C_i form an algebra, called the *class algebra*, which is a subalgebra of the group algebra.

3.1.3. Space of functions on classes

The N complex numbers in (3-8) can be regarded either as coordinates of the vector Q in the basis $\{Q_i\}$,

$$\mathbf{q} = (q_1, q_2, \dots, q_N) = (q(C_1), q(C_2), \dots, q(C_N)) , \tag{3-13}$$

or as a *function defined on classes*. For discrete groups, it is a discrete function and only has definite values at the N "points". The totality of $q(C_i)$ forms a space, called the *space of functions on classes*. There is a 1-1 mapping between the class space and the space of functions on classes. In analogy with Eq. (3-11), the scalar product of two vectors $\mathbf{q}^{(\nu)}$ and $\mathbf{q}^{(\mu)}$ in the space of functions on classes is defined by

$$\langle \mathbf{q}^{(\nu)} | \mathbf{q}^{(\mu)} \rangle = \sum_{i=1}^{N} g_i q^{(\nu)*}(C_i) q^{(\mu)}(C_i) . \tag{3-14}$$

3.1.4. The natural representation of a class algebra

In the group algebra, group elements are both operators and basis vectors, and the g group elements form the basis of the regular rep of the group. Similarly, in the class algebra, class operators are both operators and basis vectors, and the N class operators form a basis of a rep, called the *natural* rep, of the class algebra. Analogous to (2-76), we have

$$\hat{C}_i C_j = C_i C_j = \sum_k D_{kj}(C_i) C_k \ , \tag{3-15a}$$

which may be written as

$$C_i \begin{pmatrix} C_1 \\ C_2 \\ \vdots \\ C_N \end{pmatrix} = \tilde{D}(C_i) \begin{pmatrix} C_1 \\ C_2 \\ \vdots \\ C_N \end{pmatrix} \ . \tag{3-15b}$$

Comparing (3-15a) with (3-6a), the matrix representative of C_i in the natural rep is

$$D_{kj}(C_i) = C_{ij}^k \ . \tag{3-16}$$

Just as there is no necessity for distinguishing between R and $\hat{R}$, it is not necessary to distinguish between C_i and $\hat{C}_i$.

Example 1: The group S_3. The three class operators of S_3 are

$$C_1 = e, \quad C_2 = (12) + (13) + (23), \quad C_3 = (123) + (132) \ . \tag{3-17}$$

From the group table of S_3, we obtain the multiplication table for the class operators of S_3.

Table 3.1. Multiplication table of the class operators of S_3.

$C_i \backslash C_j$ ($C_i C_j$)	C_1	C_2	C_3
C_1	C_1	C_2	C_3
C_2	C_2	$3(C_1 + C_3)$	$2C_2$
C_3	C_3	$2C_2$	$2C_1 + C_3$

According to (3-7), Table 3.1 is symmetric with respect to the diagonal. From Eq. (3-16) and Table 3.1, we obtain the natural rep of the class operators of S_3, i.e.,

$$D(C_1) = \begin{pmatrix} 1 & 0 & 0 \\ 0 & 1 & 0 \\ 0 & 0 & 1 \end{pmatrix} , \quad D(C_2) = \begin{pmatrix} 0 & 3 & 0 \\ 1 & 0 & 2 \\ 0 & 3 & 0 \end{pmatrix} , \quad D(C_3) = \begin{pmatrix} 0 & 0 & 2 \\ 0 & 2 & 0 \\ 1 & 0 & 1 \end{pmatrix} \ . \tag{3-18}$$

Remark: Since the basis $\{C_i\}$ is not normalized, $D(C_2)$ and $D(C_3)$ are not hermitian, though the operators C_2 and C_3 are self-adjoint.

Example 2: The group S_4. The group S_4 has five class operators,

$$\begin{aligned} &C_1 = e, \quad C_2 = \sum_{j>i=1}^{4} (ij) \ , \quad C_3 = \sum_{k>j>i=1}^{4} [(ijk) + (ikj)] \ , \\ &C_4 = (1234) + (1243) + (1324) + (1342) + (1423) + (1432), \\ &C_5 = (12)(34) + (13)(24) + (14)(23) \ . \end{aligned} \tag{3-19}$$

The products of C_2 with the five class operators are

$$C_2 \begin{pmatrix} C_1 \\ C_2 \\ C_3 \\ C_4 \\ C_5 \end{pmatrix} = \begin{pmatrix} 0 & 1 & 0 & 0 & 0 \\ 6 & 0 & 3 & 0 & 2 \\ 0 & 4 & 0 & 4 & 0 \\ 0 & 0 & 3 & 0 & 4 \\ 0 & 1 & 0 & 2 & 0 \end{pmatrix} \begin{pmatrix} C_1 \\ C_2 \\ C_3 \\ C_4 \\ C_5 \end{pmatrix} . \tag{3-20a}$$

The natural representation of C_2 is then given by

$$D(C_2) = \begin{pmatrix} 0 & 6 & 0 & 0 & 0 \\ 1 & 0 & 4 & 0 & 1 \\ 0 & 3 & 0 & 3 & 0 \\ 0 & 0 & 4 & 0 & 2 \\ 0 & 2 & 0 & 4 & 0 \end{pmatrix} . \tag{3-20b}$$

Example 3: The group C_{6v}. The group C_{6v} is the symmetry group of a hexagon (see Fig. 3.1). It has 12 elements: six rotations $C_6^n = R(2n\pi/6), n = 1, 2, \ldots, 6$, and six reflection

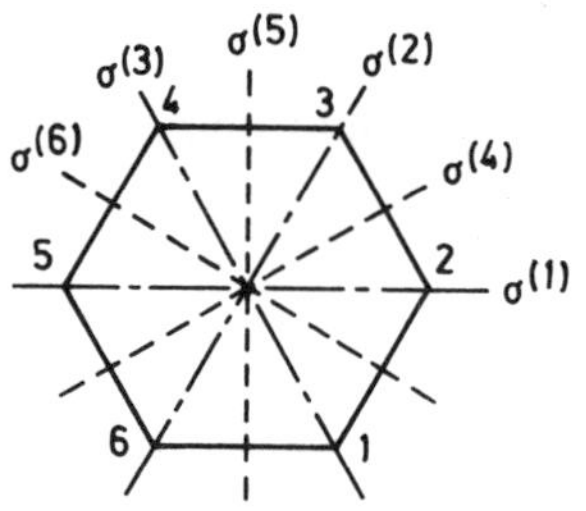

Fig. 3.1. The group C_{6v}.

planes $\sigma^{(1)}, \ldots, \sigma^{(6)}$. The group elements can be represented in terms of the permutations of the six vertices of the hexagon, namely,

$$\begin{array}{cccccc} e & C_6 & C_6^2 & C_6^3 & C_6^4 & C_6^5 \\ e\,, & (123456), & (135)(246)\,, & (14)(25)(36)\,, & (153)(264)\,, & (165432)\,, \\ \\ \sigma^{(1)} & \sigma^{(2)} & \sigma^{(3)} & \sigma^{(4)} & \sigma^{(5)} & \sigma^{(6)} \\ (13)(46)\,, & (15)(24)\,, & (26)(35)\,, & (14)(23)(56)\,, & (16)(25)(34)\,, & (12)(36)(45)\,. \end{array} \tag{3-21}$$

Therefore, C_{6v} is isomorphic to a subgroup of S_6. C_{6v} has six class operators,

$$\begin{aligned} &C_1 = e\,, \quad C_2 = C_6^3\,, \qquad C_3 = C_6^2 + C_6^4\,, \quad C_4 = C_6^1 + C_6^5\,, \\ &C_5 = \sigma^{(1)} + \sigma^{(2)} + \sigma^{(3)}\,, \qquad\qquad C_6 = \sigma^{(4)} + \sigma^{(5)} + \sigma^{(6)}\,. \end{aligned} \tag{3-22}$$

From the isomorphism (3-21), the product of group elements of C_{6v} can be found from the multiplication rule of the permutation group. For example, from (123456)(13)(46)=(14)(23)(56), we infer that $C_6\sigma^{(1)} = \sigma^{(4)}$. In this way we can easily write the multiplication relations for the class operators of C_{6v}. For example,

$$C_4 \begin{pmatrix} C_1 \\ C_2 \\ C_3 \\ C_4 \\ C_5 \\ C_6 \end{pmatrix} = \begin{pmatrix} C_4 \\ C_3 \\ 2C_2 + C_4 \\ 2C_1 + C_3 \\ 2C_6 \\ 2C_5 \end{pmatrix} , \quad C_5 \begin{pmatrix} C_1 \\ C_2 \\ C_3 \\ C_4 \\ C_5 \\ C_6 \end{pmatrix} = \begin{pmatrix} C_5 \\ C_6 \\ 2C_5 \\ 2C_6 \\ 3C_1 + 3C_3 \\ 3C_2 + 3C_4 \end{pmatrix} . \tag{3-23}$$

3.2. The First Kind of CSCO of G (CSCO-I)

As we had indicated Theorem 2.7 is the basis of the new approach for reducing a group rep. Since each class operator commutes with the group G, it is natural to choose one of them, say C_i, as the operator C in Theorem 2.7, and decompose the representation space L of G into a direct sum of the eigenspaces L_{λ_i} of C_i, i.e.,

$$C_i L_{\lambda_i} \equiv \lambda_i L_{\lambda_i}, \\ L = \sum_{\lambda_i} \oplus L_{\lambda_i} . \tag{3-24}$$

However, the subspace L_{λ_i} may be still reducible. We will then need to choose another class operator C_j, and decompose L_{λ_i} into a direct sum of the eigenspace of C_j, i.e.,

$$C_j L_{\lambda_i \lambda_j} = \lambda_j L_{\lambda_i \lambda_j} , \tag{3-25}$$

$$L_{\lambda_i} = \sum_{\lambda_j} \oplus L_{\lambda_i \lambda_j} , \quad L = \sum_{\lambda_i \lambda_j} \oplus L_{\lambda_i \lambda_j} . \tag{3-26}$$

Obviously, each subspace $L_{\lambda_i \lambda_j}$ is a simultaneous eigenspace of the operators C_i and C_j, and gives a representation space for G. We continue this process until each subspace cannot be further reduced by adding extra class operators of G. Now the crucial question is how to choose the class operators $C_i, C_j, \dots$ so that their simultaneous eigenspaces cannot be further decomposed by adding extra class operators? Moreover, are these eigenspaces irreducible spaces of G? To solve this problem, let us start with the decomposition of the class space into a direct sum of the irreducible spaces of the class algebra.

3.2.1. Reduction of the natural representation of the class algebra

The natural rep $D_{kj}(C_i) = C_{ij}^k$ is an N-dimensional reducible representation of the class algebra. Since the N class operators commute with each other, the N matrices $D(C_i), i = 1, 2, \dots N$, can be diagonalized simultaneously. In the language of group representation theory, we have

Theorem 3.2: The natural rep D of the class algebra can be reduced to a direct sum of N one-dimensional (and thus necessarily irreducible) reps of the class algebra.

For simplicity in exposition, we first assume that all the class operators of G are self-adjoint. For reducing the natural rep, we choose one class operator, say C_{i_1}, and find its eigenvectors Q in the class space;

$$C_{i_1} Q = \lambda_{i_1} Q , \tag{3-27a}$$

$$Q = \sum_{j=1}^{N} q_j C_j . \tag{3-27b}$$

This amounts to diagonalizing the matrix $D(C_{i_1})$. From (2-36) and (3-16) we have

$$\sum_{j=1}^{N} (C_{i_1 j}^k - \lambda_{i_1} \delta_{jk}) q_j = 0, \quad k = 1, 2, \dots , N . \tag{3-28}$$

$$\det|C_{i_1 j}^k - \lambda_{i_1} \delta_{jk}| = \prod_{\nu=1}^{n} (\lambda_{i_1} - \lambda_{i_1}^{\nu})^{M_\nu} = 0 \tag{3-29}$$

where the integer M_ν is the degeneracy of the eigenvalue $\lambda_{i_1}^{\nu}$. If there are N distinct eigenvalues, i.e., if all the degeneracies $M_\nu = 1$, then C_{i_1} is a CSCO of the class space. Corresponding to each eigenvalue $\lambda_{i_1}^{\nu}$, there is a unique (up to a normalization factor) solution $\mathbf{q}^{(\nu)} = (q_1^{(\nu)}, \ldots, q_N^{(\nu)})$, which in turn gives an eigenvector $Q^{(\nu)}$ of C_{i_1},

$$Q^{(\nu)} = \sum_j q_j^{(\nu)} C_j \; . \tag{3-27c}$$

From Eqs. (3-5) and (3-27a) one has

$$C_{i_1}(C_j Q^{(\nu)}) = C_j C_{i_1} Q^{(\nu)} = \lambda_{i_1}^{\nu}(C_j Q^{(\nu)}) \; .$$

It shows that $C_j Q^{(\nu)}$ is still an eigenvector of C_{i_1} with the eigenvalue $\lambda_{i_1}^{\nu}$. Since each $\lambda_{i_1}^{\nu}$ corresponds to a unique eigenvector $Q^{(\nu)}$, we have

$$C_j Q^{(\nu)} = \text{const} Q^{(\nu)} = \lambda_j^{\nu} Q^{(\nu)} \; . \tag{3-30}$$

Hence we know that the eigenvectors of the CSCO C_{i_1} are necesarily eigenvectors of all other class operators. This implies that with $Q^{(\nu)}$ as the new basis, the N class operators are all diagonalized. In the language of quantum mechanics, the non-diagonal rep of the N class operators has been transformed into the diagonal rep i.e.,

$$D(C_i) = \begin{pmatrix} C_{i1}^1 & \cdots & C_{iN}^1 \\ \cdots & \cdots & \cdots \\ C_{i1}^N & \cdots & C_{iN}^N \end{pmatrix} \to \begin{pmatrix} \lambda_i^{(\nu_1)} & & & \\ & \lambda_i^{(\nu_2)} & & \\ & & \ddots & \\ & & & \lambda_i^{(\nu_N)} \end{pmatrix} . \tag{3-31}$$

If a certain eigenvalue, say $\lambda_{i_1}^{\mu}$, is $M_\mu(>1)$-fold degenerate, then C_{i_1} is no longer a CSCO of the class space. Corresponding to $\lambda_{i_1}^{\mu}$, from Eq. (3-28) we can obtain M_μ linearly independent solutions $Q_1^{(\mu)}, \ldots, Q_{M_\mu}^{(\mu)}$. They form a M_μ-dimensional eigenspace L_μ of C_{i_1}. In such a case, we can pick another class operator, say C_{i_2}, and make linear combinations of $Q_1^{(\mu)}, \ldots Q_{M_\mu}^{(\mu)}$, so that they are eigenvectors of C_{i_2} as well. We continue this process until we find a set of class operators

$$C = (C_{i_1}, C_{i_2} \ldots, C_{i_l}) \; , \tag{3-32}$$

whose simultaneous eigenspaces are all one-dimensional. C is then a CSCO of the class space. The simultaneous eigenvectors of C satisfy the following l eigenequations,

$$\begin{pmatrix} C_{i_1} \\ \vdots \\ C_{i_l} \end{pmatrix} Q^{(\nu)} = \begin{pmatrix} \lambda_{i_1}^{\nu} \\ \vdots \\ \lambda_{i_l}^{\nu} \end{pmatrix} Q^{(\nu)} \; , \tag{3-33a}$$

which can be written more succintly as

$$C Q^{(\nu)} = \lambda^{(\nu)} Q^{(\nu)} \; , \tag{3-33b}$$

$$\lambda^{(\nu)} = (\lambda_{i_1}^{\nu}, \lambda_{i_2}^{\nu}, \ldots, \lambda_{i_l}^{\nu}) \; . \tag{3-33c}$$

3.2.2. The CSCO-I of G

Definition 3.1: If a set of operators $C = (C_i, C_{i_2}, \ldots, C_{i_l})$ selected out of the class operators is a CSCO of the class space, then C is called a CSCO of the first kind of the group G, or simply a CSCO of G, designated as CSCO-I.

Theorem 3.3: The set of N class operators $(C_1, C_2, \ldots, C_N)$ is necessarily a CSCO of G.

Proof: We regard the diagonal matrix elements $(\lambda_i^{(\nu_1)}, \lambda_i^{(\nu_2)}, \ldots, \lambda_i^{(\nu_N)})$ in Eq. (3-31) as a column vector, and place the N column vectors together to form a matrix M,

$$M = \begin{pmatrix} \lambda_1^{(\nu_1)} & \lambda_2^{(\nu_1)} & \cdots & \lambda_N^{(\nu_1)} \\ \lambda_1^{(\nu_2)} & \lambda_2^{(\nu_2)} & \cdots & \lambda_N^{(\nu_2)} \\ \vdots & \vdots & & \vdots \\ \lambda_1^{(\nu_N)} & \lambda_2^{(\nu_N)} & \cdots & \lambda_N^{(\nu_N)} \end{pmatrix} . \tag{3-34}$$

The i-th column vector $(\lambda_i^{(\nu_1)}, \lambda_i^{(\nu_2)}, \ldots, \lambda_i^{(\nu_N)})$ is a representative of the class operator C_i. Since the N class operators are linearly independent, the N column vectors in the matrix M are necessarily also linearly independent. Therefore, the rank of the matrix M is equal to N, which in turn implies that the N row vectors $(\lambda_1^{(\nu_i)}, \lambda_2^{(\nu_i)}, \ldots, \lambda_N^{(\nu_i)})$, i=1,2,.. . , N, are also linearly independent. Consequently, no two row vectors can be identical. Stated differently, the set of operators $(C_1, \ldots, C_N)$ has N distinct sets of eigenvalues, and thus is a CSCO of G.

From Theorem 2.2, we have

Theorem 3.4: Any class operator C_i of a group G is a function of the CSCO of G.

In turn, we have also Theorem 3.5.

Theorem 3.5: Any eigenvector of the CSCO of G is necessarily a simultaneous eigenvector of all the N class operators of G,

$$C_i Q^{(\nu)} = \lambda_i^{\nu} Q \ , \quad i = 1, 2, \ldots, N \ . \tag{3-33d}$$

The choice of the CSCO of G is not unique. In applications, we always want the number l of the operators which form C to be as small as possible. Therefore, we should choose such operators which have as many as possible distinct eigenvalues as the members of the CSCO. However, different CSCO's of G are equivalent in the sense that they have, according to Theorem 3.5, identical eigenvectors $Q^{(\nu)}, \nu = 1, 2, \ldots, N$. Thus we have Theorem 3.6.

Theorem 3.6: Different CSCO's of G are equivalent.

For groups whose simple characters are unknown, we can use the previous method to find the CSCO-I. As to groups with known simple characters (as is the case for most finite groups), it is trivial to find their CSCO by using the method of Sec. 3.12. For permutation groups, we can also use formula (4-3a) to obtain their CSCO. The CSCO of the permutation groups $S_2 - S_{10}$ and the point groups are listed in Table 3.2-1 and Table 8.3, respectively. From Table 3.2-1 one sees that for the permutation groups $S_2 - S_5$ and S_7, only a single operator, the 2-cycle class operator, is sufficient to form a CSCO, while for S_6, and $S_8 - S_{14}$ (the results for $S_{11} - S_{14}$ are not shown in Table 3.2-1), two class operators are needed to form a CSCO, i.e., $C = (C_{(2)}, C_{(3)})$.

From Table 8.3, one sees that for all point groups commonly used in physics we need at most three class operators to form a CSCO. Therefore, the number l of the class operators contained in a CSCO of G, is much less than the number of classes, and it is precisely this fact that makes the new approach to group rep theory powerful in practical calculation.

Table 3.2-1. The new and old labelling schemes for irreps of permutation groups.

S_2	[2]
$\lambda_{(2)}$	1
$\lambda_{(3)}$	

S_3	[3]	[21]
$\lambda_{(2)}$	3	0
$\lambda_{(3)}$		

S_4	[4]	[31]	[22]
$\lambda_{(2)}$	6	2	0
$\lambda_{(3)}$			

S_5	[5]	[41]	[32]	[31²]
$\lambda_{(2)}$	10	5	2	0
$\lambda_{(3)}$				

S_6	[6]	[51]	[42]	[41²]	[33]	[321]
$\lambda_{(2)}$	15	9	5	3	3	0
$\lambda_{(3)}$				4	−8	

S_7	[7]	[61]	[52]	[51²]	[43]	[421]	[3²1]	[41³]
$\lambda_{(2)}$	21	14	9	7	6	3	1	0
$\lambda_{(3)}$								

S_8	[8]	[71]	[62]	[61²]	[53]	[521]	[51³]	[4²]	[431]	[42²]	[4211]	[3²2]
$\lambda_{(2)}$	28	20	14	12	10	7	4	8	4	2	0	0
$\lambda_{(3)}$							16		−8		0	−16

S_9	[9]	[81]	[72]	[71²]	[63]	[621]	[61³]	[54]	[531]	[52²]	[521²]	[4²1]	[432]	[431²]	[51⁴]	[3³]
$\lambda_{(2)}$	36	27	20	18	15	12	9	12	8	6	4	6	3	1	0	0
$\lambda_{(3)}$						24		0							24	−24

S_{10}	[10]	[91]	[82]	[81²]	[73]	[721]	[71³]	[64]	[631]	[62²]	[621²]	[61⁴]	[5²]	[541]	[532]	[531²]	[52²1]	[4²2]	[4²1²]	[43²]	[4321]	[521³]
$\lambda_{(2)}$	45	35	27	25	21	18	15	17	13	11	9	5	15	10	7	5	3	5	3	3	0	0
$\lambda_{(3)}$							60					40	0			0	0	−20	−12	−24	−15	15

1. The first row is the partition $[\nu]=[\nu_1,\nu_2 \ldots]$ (see Sec. 4.1). The second and third rows are eigenvalues of the 2- and 3-cycle class operators of S_n, which are related to the partitions as [see (4-25) and (7-38d)]

$$\lambda_{(2)}^{[\nu]}=\frac{n}{2}+\frac{1}{2}\sum_l \nu_l(\nu_l-2l)\ , \qquad \lambda_{(3)}^{[\nu]}=\frac{1}{3}\left\{2n-\frac{3}{2}n^2+\sum_l \nu_l\left[\nu_l^2-\left(3l-\frac{3}{2}\right)\nu_l+3l(l-1)\right]\right\}.$$

2. $\lambda_{(3)}^{[\nu]}$ are listed only when $\lambda_{(2)}^{[\nu]}$ are degenerate.

3. The eigenvalues of the 2- and 3-cycle class operators for the conjugate $[\tilde{\nu}]$ are $\lambda_{(2)}^{[\tilde{\nu}]}=-\lambda_{(2)}^{[\nu]}$, and $\lambda_{(3)}^{[\tilde{\nu}]}=\lambda_{(3)}^{[\nu]}$. Only the positive $\lambda_{(2)}^{[\nu]}$ are tabulated.

According to Theorem 2.1, the CSCO of a finite group G can always be chosen so as to consist of a single operator. If $C = (C_1, \ldots, C_l)$ is a CSCO of G, we can always find a linear combination of these l class operators so that

$$C = k_1 C_1 + \ldots + k_l C_l \tag{3-35}$$

is a CSCO of G. With known eigenvalues λ_i^ν of C_i, which are simply related to the simple characters as $\lambda_i^\nu = (g_i/h_\nu)\chi_i^\nu$ [see (3-67)], it is easy to choose the coefficients $k_1, k_2, \ldots, k_l$, so that the single operator C has N distinct eigenvalues $\lambda^\nu = \sum_{i=1}^{l} k_i \lambda_i^\nu$, and is thus a CSCO of G.

For example, for the permutation group S_6, we can choose

$$C = C_{(2)} + 3C_{(3)} , \tag{3-36}$$

as a CSCO, which has $N = 11$ (i.e., the number of classes of S_6) distinct eigenvalues, as listed in Table 3.2-2.

Table 3.2-2. Eigenvalues of the two kinds of CSCO-I of S_6, $(C_{(2)}, C_{(3)})$ and $(C_{(2)} + 3C_{(3)})$.

λ \ partitions	[6]	[51]	[42]	[411]	[33]	[321]	[222]	$[31^3]$	$[2^2 1^2]$	$[21^4]$	$[1^6]$
$\lambda_{(2)}$	15	9	5	3	3	0	−3	−3	−5	−9	−15
$\lambda_{(3)}$	40	16	0	4	−8	−5	−8	4	0	16	40
$3\lambda_{(3)}+\lambda_{(2)}$	135	57	5	15	−21	−15	−27	9	−5	39	105

In the following, the CSCO of a group G can be understood either as a set of l class operators of (3-32), or as a single operator (3-35). Likewise, the eigenvalue $\lambda^{(\nu)}$ of C can be understood either as set of eigenvalues (3-33c), or as a single eigenvalue. However, for brevity in exposition, in proving theorems the CSCO of G is always assumed to consist of only a single operator C.

Theorem 3.7: An Abelian finite group G has $|G|$ one-dimensional irreps.

Proof: For an Abelian group, the group space coincides with the class space, and the irreps of the group algebra coincides with the irreps of the class algebra. Therefore Theorem 3.7 follows Theorem 3.2.

3.2.3. The CSCO of a direct product group $G_1 \times G_2$

Suppose that $C^{(i)}$ are the CSCO of the groups G_i with N_i classes, $i = 1, 2$. The direct product group $G_1 \times G_2$ has $N_1 N_2$ classes. One can easily see that the set of operators

$$C = (C^{(1)}, C^{(2)}) \tag{3-37}$$

has $N_1 N_2$ distinct sets of eigenvalues $(\lambda^{(1)}, \lambda^{(2)})$ in the class space of $G_1 \times G_2$; therefore $C = (C^{(1)}, C^{(2)})$ is the CSCO of $G_1 \times G_2$. (See Sec. 8.3 for examples.)

3.2.4. The case of non-self-adjoint class operators

Theorem 3.8: The CSCO of any finite group is equivalent to a self-adjoint CSCO.

Proof: Suppose that among the N class operators, n_1 are ambivalent and $2n_2$ are nonambivalent. The class operators are

$$C_i = \sum_{l=1}^{g_i} R_l^{(i)} , \quad C_{i'} = \sum_{l=1}^{g_i} (R_l^{(i)})^{-1} , \quad i = n_1 + 1, \ldots, n_1 + n_2 , \tag{3-38}$$

where $N = n_1 + 2n_2$. Using the unitarity of the group operators, we immediately see that the ambivalent class operators are self-adjoint, while the nonambivalent class operators satisfy

$$C_{i'} = C_i^\dagger \ , \quad i = n_1 + 1, \dots n_1 + n_2 \ . \tag{3-39}$$

Of the $2n_2$ non-self-adjoint operators $(C_i, C_{i'})$ we can construct another $2n_2$ self-adjoint operators

$$K_j = C_l + C_{l'}, \quad K_j' = i(C_l - C_{l'}) \ , \tag{3-40}$$

$$j = l - n_1 \ , \quad l = n_1 + 1, \dots, n_1 + n_2 \ . \tag{3-41}$$

Thus we have a set of N self-adjoint operators

$$C' = (C_1, \dots, C_{n_1}, \quad K_1, \dots, K_{n_2}, \quad K_1', \dots, K_{n_2}') \ .$$

By using the procedure for proving Theorem 3.3, C' is seen to be a CSCO of G, while according to Theorem 3.6 any CSCO of G is necessarily equivalent to C'.

The significance of Theorem 3.8 is that from now on in proving theorems we can always assume that the CSCO of G is self-adjoint and utilize all the results obtained in quantum mechanics related to the self-adjoint CSCO.

For example, for the cyclic group $\{a, a^2, a^3, a^4 = e\}$, the element a has four eigenvalues, 1, $i, -i$ and -1 and can be chosen as the CSCO of the group. If we choose $[(a + a^{-1}), -i(a - a^{-1})]$ as the CSCO, then all the eigenvalues are real [(2,0), (0,2), (0,−2), (0,0)]. We may also choose $2(a + a^{-1}) - i(a - a^{-1})$ as the CSCO with the eigenvalues 4, 2, −2 and 0.

Since C is equivalent to a self-adjoint CSCO, the eigenvectors of C obey the orthonormality and completeness relations

$$\langle Q^{(\nu)} | Q^{(\mu)} \rangle = \delta_{\nu\mu}, \quad \sum_\nu | Q^{(\nu)} \rangle \langle Q^{(\nu)} | = 1 \ , \tag{3-42}$$

which can be rewritten as

$$\sum_{i=1}^{N} g_i q_i^{(\nu)*} q_i^{(\mu)} = \delta_{\nu\mu} \ , \tag{3-43a}$$

$$\sum_{\nu=1}^{N} g_i q_i^{(\nu)*} q_j^{(\nu)} = \delta_{ij} \ . \tag{3-43b}$$

Notice that Eq. (3-43) has the same form as Eq. (2-138).

3.2.5. The groups S_3 and C_{6v}

Example 1: Reduction of the natural rep of the class algebra of S_3. The natural rep of the S_3 class algebra is given in (3-18). Let us first diagonalize the matrix $D(C_2)$. We will find three distinct eigenvalues $\lambda = 3, 0, -3$. Consequently, C_2 is a CSCO of S_3. Its three eigenvectors are

$$\lambda = 3, \qquad Q^{(3)} = \sqrt{\frac{1}{6}}(C_1 + C_2 + C_3) \ ,$$

$$\lambda = 0, \qquad Q^{(0)} = \sqrt{\frac{1}{6}}(2C_1 - C_3) \ ,$$

$$\lambda = -3, \quad Q^{(-3)} = \sqrt{\frac{1}{6}}(C_1 - C_2 + C_3) \tag{3-44a}$$

where the normalization is determined by (3-43a) with $g_1 = 1, g_2 = 3, g_3 = 2$ and the phase is decided by the requirement that $q_e^{(\nu)} > 0$ for reasons to be explained in Sec. 3.11. In the new basis $\{Q^{(3)}, Q^{(0)}, Q^{(-3)}\}$, the rep matrices of all the class operators are diagonal, i.e.,

$$\mathcal{D}(C_1) = \begin{pmatrix} 1 & 0 & 0 \\ 0 & 1 & 0 \\ 0 & 0 & 1 \end{pmatrix} , \mathcal{D}(C_2) = \begin{pmatrix} 3 & 0 & 0 \\ 0 & 0 & 0 \\ 0 & 0 & -3 \end{pmatrix} , \mathcal{D}(C_3) = \begin{pmatrix} 2 & 0 & 0 \\ 0 & -1 & 0 \\ 0 & 0 & 2 \end{pmatrix} . \tag{3-44b}$$

Example 2: The group $\mathcal{C}_{6v}$. From (3-23) we obtain the representative of the class operator C_4 in the natural representation,

$$D(C_4) = \left(\begin{array}{cccc|cc} 0 & 0 & 0 & 2 & & \\ 0 & 0 & 2 & 0 & & \mathbf{0} \\ 0 & 1 & 0 & 1 & & \\ 1 & 0 & 1 & 0 & & \\ \hline & & \mathbf{0} & & 0 & 2 \\ & & & & 2 & 0 \end{array}\right) . \tag{3-45}$$

It is block diagonal and the two submatrices can be diagonalised individually with the results

$$\begin{aligned} \lambda_4 &= -1 \,, \quad Q^{(-1)} = \sqrt{\frac{1}{12}}(2C_1 + 2C_2 - C_3 - C_4), \\ \lambda_4 &= 1 \,, \quad Q^{(1)} = \sqrt{\frac{1}{12}}(2C_1 - 2C_2 - C_3 + C_4), \end{aligned} \tag{3-46}$$

$$\begin{aligned} \lambda_4 &= 2 \quad \text{(double root)} \,, \quad Q_1 = C_1 + C_2 + C_3 + C_4 \,, \quad Q_2 = C_5 + C_6, \\ \lambda_4 &= -2 \quad \text{(double root)} \,, \quad Q_3 = C_1 - C_2 + C_3 - C_4 \,, \quad Q_4 = C_5 - C_6 \,. \end{aligned} \tag{3-47}$$

Due to the two-fold degeneracy of $\lambda_4 = \pm 2, C_4$ is not a CSCO of $\mathcal{C}_{6v}$. We have to choose another class operator, say C_5, and find the simultaneous eigenvectors of C_4 and C_5. For the single roots, it is easy to check that $Q^{(\pm 1)}$ are also eigenvectors of C_5 with eigenvalues zero. For the double roots, we have to solve the following eigenequations to get the eigenvectors of C_5:

$$C_5(b_1Q_1 + b_2Q_2) = \lambda_5(b_1Q_1 + b_2Q_2) \,, \quad C_5(b_3Q_3 + b_4Q_4) = \lambda_5(b_3Q_3 + b_4Q_4) \,. \tag{3-48}$$

With the help of (3-23) and (3-47), the solutions can be found as follows:

$$\begin{aligned} (\lambda_4, \lambda_5) = \quad (2,3), \quad & Q^{(2,3)} = \sqrt{\frac{1}{12}}(Q_1 + Q_2) = \sqrt{\frac{1}{12}}(C_1 + C_2 + C_3 + C_4 + C_5 + C_6), \\ (2,-3), \quad & Q^{(2,-3)} = \sqrt{\frac{1}{12}}(Q_1 - Q_2) = \sqrt{\frac{1}{12}}(C_1 + C_2 + C_3 + C_4 - C_5 - C_6), \\ (-2,3), \quad & Q^{(-2,3)} = \sqrt{\frac{1}{12}}(Q_3 + Q_4) = \sqrt{\frac{1}{12}}(C_1 - C_2 + C_3 - C_4 + C_5 - C_6), \\ (-2,-3), \quad & Q^{(-2,-3)} = \sqrt{\frac{1}{12}}(Q_3 - Q_4) = \sqrt{\frac{1}{12}}(C_1 - C_2 + C_3 - C_4 - C_5 + C_6). \end{aligned} \tag{3-49}$$

Now all degeneracies are lifted. Therefore $C = (C_4, C_5)$ is a CSCO of $\mathcal{C}_{6v}$. If we let

$$C = 2C_4 + C_5$$

then it has six distinct eigenvalues $\lambda = 2\lambda_4 + \lambda_5 = 7, 1, -1, -7, -2, 2$, and thus is also a CSCO of $\mathcal{C}_{6v}$.

Ex. 3.1. Diagonalize the matrix (3-20b) and find the irreducible basis vectors $Q^{(\nu)}$ for the S_4 class algebra.

Ex. 3.2. Find the CSCO of the group $\mathcal{C}_{4v}$ and its eigenvectors in the class space.

3.3. Projection Operator $P^{(\nu)}$

3.3.1. *Decomposition of the regular rep into inequivalent reps of G*

From Eqs. (3-5) and (3-27b) it follows that C and $Q^{(\nu)}$ are commutative. By (3-33b) we have

$$C(Q^{(\nu)}Q^{(\mu)}) = (CQ^{(\nu)})Q^{(\mu)} = \lambda^{(\nu)}(Q^{(\nu)}Q^{(\mu)}) = Q^{(\nu)}(CQ^{(\mu)}) = \lambda^{(\mu)}(Q^{(\nu)}Q^{(\mu)}) \ .$$

Therefore

$$Q^{(\nu)}Q^{(\mu)} = \delta_{\nu\mu}\eta_\nu Q^{(\nu)} \ , \tag{3-50}$$

where η_ν is a constant depending only on ν. Letting

$$P^{(\nu)} = \eta_\nu^{-1}Q^\nu \ , \tag{3-51}$$

we have

$$P^{(\nu)}P^{(\mu)} = \delta_{\nu\mu}P^{(\nu)} \ . \tag{3-52}$$

In mathematics, the operators obeying Eq. (3-52) are called *idempotents*.

From (3-43b), we obtain the inverse expansion of (3-27b),

$$C_i = \sum_{\nu=1}^{N} g_i q_i^{(\nu)*}Q^{(\nu)}. \tag{3-53}$$

Multiplying this result from the right by $Q^{(\mu)}$ and using (3-50), we have

$$\lambda_i^{(\nu)} = \eta_\nu g_i q_i^{(\nu)*} \ . \tag{3-54}$$

Combining (3-51), (3-53) and (3-54), we obtain

$$C_i = \sum_{\nu=1}^{N} \lambda_i^{(\nu)}P^{(\nu)} \ . \tag{3-55}$$

On putting $C_1 = e$, and noting that its eigenvalue is $\lambda_e^\nu = 1$, we finally get

$$e = \sum_{\nu=1}^{N} P^{(\nu)} \ . \tag{3-56}$$

This is the decomposition formula for the identity element of G. The implication of this formula will be illucidated in Eq. (3-60).

Now let us turn to the group space.

Theorem 3.9: The eigenvectors $P^{(\nu)}$ of the CSCO C of G are projection operators onto the eigenspaces L_ν of C, and the group space can be decomposed into a direct sum of the N mutually orthogonal rep spaces L_ν.

Proof: Clearly, $P^{(\nu)}$ is an eigenvector of the CSCO of G, i.e.,

$$CP^{(\nu)} = \lambda^{(\nu)}P^{(\nu)} \ . \tag{3-57}$$

From (3-57) we have

$$C(P^{(\nu)}R_a) = (CP^{(\nu)})R_a = \lambda^{(\nu)}(P^{(\nu)}R_a) \ . \tag{3-58}$$

Therefore the space

$$L_\nu \equiv P^{(\nu)}L_g \equiv \{P^{(\nu)}R_a : a = 1, 2, \ldots, g\} \tag{3-59}$$

is an eigenspace of C with the eigenvalue λ^ν. Furthermore, L_ν and L_μ are orthogonal for $\lambda^\nu \neq \lambda^\mu$. Using (3-56) we obtain

$$L_g = eL_g = \sum_{\nu=1}^{N} P^{(\nu)} L_g = \sum_{\nu=1}^{N} \oplus L_\nu . \tag{3-60a}$$

Thus the group space L_g is decomposed into a direct sum of N mutually orthogonal eigenspace of G, each giving a rep space of G, and $P^{(\nu)}$ is the projection operator onto the rep space L_ν.

Example: The eigenvectors $Q^{(3)}, Q^{(0)}$ and $Q^{(-3)}$ (which differ from $P^{(\nu)}$ by constant factors) project the group space of S_3 onto three eigenspaces of C:

One-dimensional: $L_{(3)} = e + (12) + (13) + (23) + (123) + (132)$;

One-dimensional: $L_{(-3)} = e - (12) - (13) - (23) + (123) + (132)$;

Four-dimensional: $L_{(0)}$, the basis of which can be chosen as $\{Q^{(0)}e, Q^{(0)}(12), Q^{(0)}(13), Q^{(0)}(123)\}$, i.e.,

$$\begin{aligned} L_{(0)} = \{&2e - (123) - (132),\ 2(12) - (13) - (23)\ , \\ &2(13) - (23) - (12),\ 2(123) - (132) - e\}\ . \end{aligned}$$

Theorem 3.10: In any rep space of G, the possible eigenvalues of the CSCO of G cannot go beyond the N sets determined in the class space of G.

Proof: According to (3-56), for any rep space $\mathcal{L}$ we have

$$\mathcal{L} = e\mathcal{L} = \sum_{\nu=1}^{N} P^{(\nu)} \mathcal{L} = \sum_{\nu=1}^{N} \oplus \mathcal{L}_\nu\ . \tag{3-60b}$$

If $\mathcal{L}_\mu$ is an eigenspace of C belonging to an eigenvalue λ^μ other than those determined in the class space of G, then we must have

$$\langle \mathcal{L}_\nu | \mathcal{L}_\mu \rangle = 0\ , \quad \text{for} \quad \nu = 1, 2, \dots, N\ .$$

By virtue of (3-60b), this implies that

$$\langle \mathcal{L} | \mathcal{L}_\mu \rangle = 0\ .$$

Since $\mathcal{L}$ is an arbitrary space, $\mathcal{L}_\mu$ is necessarily a null space.

Theorem 3.11: In the group space of G, the CSCO of G has N and only N distinct eigenvalues $\lambda^\nu, \nu = 1, 2, \dots, N$, determined from the class space of G.

Proof: Here the first part is trivial, since the group space contains the class space as its subspace, while the second part follows from Theorem 3.10.

Theorem 3.12: The rep spaces which are eigenspaces of the CSCO of G belonging to different eigenvalues are inequivalent.

Proof: If the two representation spaces $\mathcal{L}_\nu$ and $\mathcal{L}_{\nu'}$ with different eigenvalues λ^ν and $\lambda^{\nu'}$ were equivalent, than the matrices of the CSCO of G in the two rep spaces must relate to each other by

$$D^{(\nu')}(C_i) = TD^{(\nu)}(C_i)T^{-1}\ . \tag{3-61}$$

On the other hand, since $\mathcal{L}_\nu$ and $\mathcal{L}_{\nu'}$ are eigenspaces of C, the rep of C in $\mathcal{L}_\nu(\mathcal{L}_{\nu'})$ must be equal to the unit matrix multiplied by the eigenvalue $\lambda^\nu(\lambda^{\nu'})$:

$$D^{(\nu)}(C) = \lambda^\nu \bullet I, \quad D^{(\nu')}(C) = \lambda^{\nu'} \bullet I\ . \tag{3-62}$$

Substituting Eq. (3-62) into Eq. (3-61), we get $\lambda^\nu = \lambda^{\nu'}$, which contradicts the hypothesis. Thus the theorem is proved.

Combining Theorems 3.9 and 3.12 we have

Theorem 3.9′: The group space can be decomposed into a direct sum of N inequivalent rep spaces L_ν, each being an eigenspace of the CSCO of G.

It must be emphasized that the rep space $\mathcal{L}_\nu$ is in general still *reducible*. Suppose that it can be decomposed into a direct sum of τ_ν irreducible spaces

$$\mathcal{L}_\nu = \mathcal{L}_{(\nu)1} \oplus \mathcal{L}_{(\nu)2} \oplus \ldots \oplus \mathcal{L}_{(\nu)\tau_\nu} . \tag{3-63}$$

Then we can show that

Theorem 3.13: The irreducible spaces $\mathcal{L}_{(\nu)k}$ with different eigenvalues λ^ν are inequivalent.

The reader who prefers mathematical rigorous to the analysis of concrete problems may jump to Secs. 3.7, 3.8 and 3.9, and then turn back here.

Theorem 3.14: Irreducible reps with the same eigenvalue $\lambda^{(\nu)}$ are equivalent.

The proof of this theorem is given in Sec. 3.9. There we shall also prove that a finite group G with N classes has N and only N inequivalent irreps.

Therefore, if in a space we have found an h_ν-dimensional irrep of G with the eigenvalue λ^ν, then an irrep of G in any space with the same eigenvalue λ^ν must be also of dimension h_ν.

3.3.2. Label for irreps

From Theorems 3.13 and 3.14, we conclude that an inequivalent irrep of G can be labeled uniquely by the eigenvalue λ^ν of the CSCO of G. In the following, we use ν to represent the eigenvalue λ^ν as well as the irrep label.

In the traditional theory on finite groups, inequivalent irreps are labeled by the simple characters. What is the relation between these two labeling schemes? According to Theorem 3.4, any class operator of G is a function of the CSCO of G, i.e.,

$$C_i = F_i(C) . \tag{3-64}$$

Therefore, the eigenvalue of C_i is a function of the eigenvalue $\lambda^{(\nu)}$ of C, i.e.,

$$\lambda_i^{(\nu)} = F_i(\lambda^{(\nu)}) . \tag{3-65}$$

The relation between the eigenvalue $\lambda_i^{(\nu)}$ and the simple character of $\chi_i^{(\nu)}$ can be found as follows. From (3-1) we have

$$D^{(\nu)}(C_i) = \sum_{l=1}^{g_i} D^{(\nu)}(R_l^{(i)}) . \tag{3-66}$$

Making use of the fact that in the irrep $(\nu), C_i$ is a multiple of the unit matrix and that the elements of the same class have the same character, and taking the trace of Eq. (3-66) we have

$$h_\nu \lambda_i^{(\nu)} = g_i \chi_i^{(\nu)} , \tag{3-67}$$

where h_ν is the dimension of the irrep (ν). Thus

$$\chi_i^{(\nu)} = \frac{h_\nu}{g_i} \lambda_i^{(\nu)} . \tag{3-68}$$

From (3-65) and (3-68) we get

$$\chi_i^{(\nu)} = \frac{h_\nu}{g_i} F_i(\lambda^{(\nu)}) . \tag{3-69}$$

This shows that the N simple characters are functions of the eigenvalue λ^ν of the CSCO of G. Therefore, the N simple characters are not functionally independent.

We use two examples to illustrate Eq. (3-64).

Example 1: The group S_3.

$$C = C_2 = \sum_{i>j=1}^{3} (ij) , \quad C_3 = \frac{1}{3}(C)^3 - e . \tag{3-70}$$

Example 2: The group S_4.

$$C = C_2 = \sum_{i>j=1}^{4} (ij) , \quad C_3 = -\frac{1}{48}(C)^4 + \frac{13}{12}(C)^2 - 4e . \tag{3-71}$$

$$C_4 = \frac{1}{16}(C)^3 - \frac{5}{4}(C)^2 , \quad C_5 = \frac{1}{32}(C)^4 - \frac{9}{8}(C)^2 + 3e . \tag{3-72}$$

3.3.3. Decomposition of an arbitrary rep space

According to Theorem 3.14, reps generated by the irreducible spaces $\mathcal{L}_{(\nu)k}$ in (3-63) are all equivalent and labeled by the same irrep label (ν). Therefore, if a function $\psi^{(\nu)}$ belongs to the eigenspace $\mathcal{L}_\nu$ of C, then $\psi^{(\nu)}$ is necessarily a basis vector for one of those equivalent irreps, or a linear combination of their basis vectors. This leads to Definition 3.2.

Definition 3.2: If a vector $\psi^{(\nu)}$ belongs to the eigenspace $\mathcal{L}_\nu$ of the CSCO of G, the $\psi^{(\nu)}$ is said to belong to the irrep (ν) of G.

Remark: Since the eigenspace $\mathcal{L}_\nu$ is in general reducible, if both $\psi_1^{(\nu)}$ and $\psi_2^{(\nu)}$ belong to the irrep (ν), it does not necessarily mean that they belong to the same irreducible space. For example, $\psi_1^{(\nu)}$ may belong to the first irreducible space $\mathcal{L}_{(\nu)1}$, while $\psi_2^{(\nu)}$ may belong to the second irreducible space $\mathcal{L}_{(\nu)2}$.

Theorem 3.15: A sufficient and necessary condition for a function $\psi^{(\nu)}$ to belong to the irrep (ν) of a group G is that $\psi^{(\nu)}$ be an eigenfunction of the CSCO of G, i.e.,

$$C\psi^{(\nu)} = \nu\psi^{(\nu)} . \tag{3-73}$$

Proof: The necessity follows from the following consideration. Suppose $\psi^{(\nu)}$ is a vector of an irreducible space L of G. Obviously L is a rep space for any class operator of G and thus is an invariant subspace of C. According to Schur's lemma 1, $\psi^{(\nu)}$ is necessarily an eigenfunction of C.

The sufficiency follows trivially from Definition 3.2.

Theorem 3.15 is the corner stone of the eigenfunction method (EFM) for group representations. It converts the problem of finding irreps of a group G into that of finding the eigenfunctions of the CSCO of G, i.e., diagonalizing the representative of the CSCO of G in the reducible basis $\varphi_1, \ldots, \varphi_n$. Let

$$\psi^{(\nu)} = \sum_{i=1}^{n} a_i \varphi_i \tag{3-74}$$

be the sought-for eigenfunctions of C. The eigenvalue of ν and coefficients a_i are to be determined by

$$\begin{pmatrix} C_{11} & C_{12} & \dots & C_{1n} \\ C_{21} & C_{22} & \dots & C_{2n} \\ \dots & \dots & \dots & \dots \\ C_{n1} & C_{n2} & \dots & C_{nn} \end{pmatrix} \begin{pmatrix} a_1 \\ a_2 \\ \vdots \\ a_n \end{pmatrix} = \nu \begin{pmatrix} a_1 \\ a_2 \\ \vdots \\ a_n \end{pmatrix}, \tag{3-75a}$$

where

$$C_{ij} = \langle \varphi_i | C | \varphi_j \rangle \ . \tag{3-76a}$$

For the majority of point groups and for all permutation groups, C is self-adjoint and the C_{ij} are real. In such cases, $\| C_{ij} \|$ is a real and symmetric matrix,

$$C_{ij} = C_{ji} \ . \tag{3-76b}$$

From (3-75a) and (3-76b) we obtain

$$\sum_j \langle \varphi_j | C | \varphi_i \rangle a_j = \nu a_i. \tag{3-75b}$$

The last result is very useful. It shows that in order to obtain an equation for the variable a_i, one only needs to know the action of C on the state φ_i.

A natural extension of Theorem 3.15 is Theorem 3.16.

Theorem 3.16: A necessary and sufficient condition for $\psi^{(\nu)}_{\lambda(s_1),\lambda(s_2),\dots}$ to belong to the irreps $\nu, \lambda(s_1), \lambda(s_2), \dots$ of a subgroup chain $G \supset G(s_1) \supset G(s_2) \supset \dots$ is that it satisfies the following eigenequations:

$$\begin{pmatrix} C \\ C(s_1) \\ C(s_2) \\ \vdots \end{pmatrix} \psi^{(\nu)}_{\lambda(s_1),\lambda(s_2),\dots} = \begin{pmatrix} \nu \\ \lambda(s_1) \\ \lambda(s_2) \\ \vdots \end{pmatrix} \psi^{(\nu)}_{\lambda(s_1),\lambda(s_2),\dots}\,, \tag{3-77a}$$

where $C(s_i)$ is the CSCO of $G(s_i)$.

Usually there is more than one value of $\lambda(s_1)$ that goes with the same quantum number ν, since the irrep of G is reducible in restricting to its subgroup $G(s_1)$,

$$D^{(\nu)} \downarrow G(s_1) = \sum_i \oplus \tau_i^{(\nu)} D^{\lambda^i(s_1)}(G(s_1)) \ .$$

If $G(s_1)$ is a canonical subgroup of G, then $\tau_i^{(\nu)} \leq 1$. The same applies to the relation between $\lambda(s_1)$ and $\lambda(s_2)$, etc.

For simplicity, we use $G(s)$ to denote the subgroup chain $G(s_1) \supset G(s_2) \supset \dots$, $C(s)$ the set of operators $(C(s_1), C(s_2), \dots,)$ and m the set of eigenvalues $(\lambda(s_1), \lambda(s_2), \dots)$. Theorem 3.16 can be reformulated as

Theorem 3.16: A necessary and sufficient condition for $\psi^{(\nu)}_m$ to belong to the irrep (ν, m) of the group chain $G \supset G(s)$ is that $\psi^{(\nu)}_m$ satisfies the eigenequations

$$\begin{pmatrix} C \\ C(s) \end{pmatrix} \psi^{(\nu)}_m = \begin{pmatrix} \nu \\ m \end{pmatrix} \psi^{(\nu)}_m \ . \tag{3-77b}$$

$C(s)$ will be referred to as the CSCO of the subgroup chain $G(s)$.

Suppose that the eigenspace $\mathcal{L}_\nu = \{\psi^{(\nu)} : C\psi^{(\nu)} = \nu\psi^{(\nu)}\}$ is an irreducible space of G; then the degeneracy M_ν of the eigenvalue ν in Eq. (3-75a) is equal to the dimension of the irrep $(\nu), M_\nu = h_\nu$. If $G \supset G(s)$ is a canonical subgroup chain, from the discussion following (3-77a) we know that this degeneracy can be totally lifted by the eigenequations of $C(s)$, i.e., the degeneracies for the sets of eigenvalues (ν, m_i) are all equal to unity for $i = 1, 2, \dots h_\nu$. (Noting that the degeneracies of (ν, m_i) must be independent of i, since if $\mathcal{L}_\nu$ contains the irrep (ν) τ_ν times, then each component m has to occur τ_ν times.)

Conversely, if we can find a subgroup chain $G \supset G(s)$, such that the set of the eigenvalues (ν, m) of $(C, C(s))$ is non-degenerate, then $G \supset G(s)$ is a canonical subgroup chain, and $\mathcal{L}_\nu$ is irreducible. Corresponding to each $(\nu, m), m = m_1, \dots, m_{h_\nu}$, there is only one eigenvector. These h_ν eigenvectors $\psi_m^{(\nu)}$ carry an irrep of G. Therefore, for such cases the irreducible basis for $G \supset G(s)$ can be found from Eq. (3-77b) without any knowledge of the characters or irreducible matrices.

With the known irreducible basis, we can use (2-20a) or (2-21) to find the irreducible matrices.

Suppose now that $\mathcal{L}_\nu$ is reducible as shown in (3-63), and $G \supset G(s)$ is a canonical subgroup chain. Then the degeneracy M_ν of ν in Eq. (3-75a) is equal to $M_\nu = \tau_\nu h_\nu$, while the degeneracy of (ν, m) is equal to the multiplicity τ_ν. For a given (ν, m), from Eq. (3-77b) we can find τ_ν linearly independent eigenvectors $\psi_m^{(\nu)\tau}, \tau = 1, 2, \dots, \tau_\nu$, which can be chosen as orthogonal in the multiplicity index τ.

Remark: Since here the multiplicity indices τ are chosen independently for each component index m, in general $\psi_m^{(\nu)\tau}$ and $\psi_{m'}^{(\nu)\tau}$ do not belong to the same irreducible space. The problem of how to find the irreducible basis for such cases is relegated to Sec. 3.13.

3.4. Reduction of Representations of $\mathcal{C}_{3v}$, S_2 and S_3

3.4.1. The group $\mathcal{C}_{3v}$

We begin with the simple problem of finding an irreducible basis for the point group $\mathcal{C}_{3v}$ out of the polynomials of x, y, z of degree one and two. Using the notation of Eq. (2-64a), the CSCO of $\mathcal{C}_{3v}$ is $C = \sigma^{0^\circ} + \sigma^{120^\circ} + \sigma^{60^\circ}$.

To find the irreducible bases, we solve the eigenequation (3-73). Using (2-64b) we immediately have

$$\begin{aligned} &Cz = 3z, \\ &C \exp(i\varphi) = \exp(-i\varphi) + \exp[i(240^\circ - \varphi)] + \exp[i(120^\circ - \varphi)] = 0\ , \\ &C \exp(-i\varphi) = 0\ . \end{aligned}$$

Here the eigenvalue 3 is a single root, and the corresponding eigenspace $\mathcal{L}_{(3)}, \{z\}$, is one-dimensional and thus irreducible. The irrep (3) of $\mathcal{C}_{3v}$ is the identity rep. On the other hand the eigenvalue 0 is a double root, and $\mathcal{L}_{(0)} = (e^{i\varphi}, e^{-i\varphi})$ carries a two-dimensional rep of $\mathcal{C}_{3v}$. Let us try to lift the degeneracy by finding the eigenvectors of the CSCO of a subgroup, say $\mathcal{C}_s = (e, \sigma^{0^\circ})$, of $\mathcal{C}_{3v}$. Now $C(s) = \sigma^{0^\circ}$, and its eigenvectors are x and y corresponding to the eigenvalues 1 and -1. Therefore $(\nu, m) = (0, \pm 1)$, has no degeneracy. Accordingly, $\mathcal{C}_{3v} \supset \mathcal{C}_s$ is a canonical subgroup chain, and the space $\mathcal{L}_{(0)}$ is irreducible. Therefore, out of x, y and z, we can construct the bases for the one- and two-dimensional irreps (3) and (0), respectively:

$$\psi_1^{(3)} = z\ , \quad \psi_1^{(0)} = x\ , \quad \psi_{-1}^{(0)} = y\ . \tag{3-78a}$$

Similarly, out of the six reducible basis vectors x^2, y^2, z^2, xy, xz and yz, we can obtain the following $\mathcal{C}_{3v} \supset \mathcal{C}_s$ irreducible bases;

$$\begin{aligned} &\psi_1^{(3)1} = x^2 + y^2\ , && \psi_1^{(3)2} = z^2, \\ &\begin{cases} \psi_1^{(0)1} = \rho^2 \cos 2\varphi = x^2 - y^2\ , \\ \psi_{-1}^{(0)1} = \rho^2 \sin 2\varphi = -2xy\ , \end{cases} && \begin{cases} \psi_1^{(0)2} = xz\ , \\ \psi_{-1}^{(0)2} = yz\ . \end{cases} \end{aligned} \tag{3-78b}$$

This shows that both irreps (3) and (0) occur twice.

Ex. 3.3. Construct the $C_{3v} \supset C_3$ irreducible basis which is linear in x and y, and find the corresponding irreducible matrices.

Ex. 3.4. Construct the irreducible basis of C_{4v} from the polynomials of x, y and z of degree one and two, and find the irreducible matrices for the following two subgroup chains (using Fig. 1.1-3 and the result of Ex. 3.2):

$$(a) \quad C_{4v} \supset (e, \sigma_4) \, , \qquad (b) \quad C_{4v} \supset C_4 \, .$$

3.4.2. The group S_2

Suppose that there are two electrons, one with spin up (α) and the other with spin down (β). There are two possible states in the spin space,

$$\psi_1 = |\alpha\beta\rangle = \chi_\alpha(1)\chi_\beta(2) \, , \quad \varphi_2 = |\beta\alpha\rangle = \chi_\beta(1)\chi_\alpha(2)$$

which form a reducible (the regular) rep of S_2. The CSCO of S_2 is $C = (12)$. The eigenfunctions of C are easily found, i.e.,

$$\psi^{(1)} = \sqrt{\frac{1}{2}}(\varphi_1 + \varphi_2) \, , \quad \psi^{(-1)} = \sqrt{\frac{1}{2}}(\varphi_1 - \varphi_2) \, . \tag{3-79a}$$

They give the basis vectors for the two one-dimensional irreps (1) and (−1). $\psi^{(1)}(\psi^{(-1)})$ is symmetric (antisymmetric) under the interchange of the spin coordinate indices 1 and 2, and is labelled conventionally by the so-called Young tableau $\boxed{1}\boxed{2}$ $\left(\begin{matrix}\boxed{1}\\\boxed{2}\end{matrix}\right)$ [to be discussed extensively in Sec. 4.3]. Thus Eq. (3-79a) can be written as

$$\left| \boxed{1}\boxed{2} \right\rangle = \sqrt{\frac{1}{2}}(\varphi_1 + \varphi_2) \, , \quad \left| \begin{matrix}\boxed{1}\\\boxed{2}\end{matrix} \right\rangle = \sqrt{\frac{1}{2}}(\varphi_1 - \varphi_2) \, . \tag{3-79b}$$

3.4.3. The group S_3 in the configuration $\alpha^2\beta$

Now we have three electrons in the configuration $\alpha^2\beta$. The three possible states

$$\varphi_1 = |\alpha\alpha\beta\rangle \, , \quad \varphi_2 = |\alpha\beta\alpha\rangle \, , \quad \varphi_3 = |\beta\alpha\alpha\rangle \, , \tag{3-80a}$$

carry a 3-dimensional rep of S_3. There are two ways of reducing this rep: (1) using the EFM, (2) using the projection operator method.

(1) The EFM. The CSCO of S_3 is

$$C = C(3) = (12) + (23) + (13) \, . \tag{3-80b}$$

The effect of the permutations (ij) on φ_l is given in Table 3.4.

Table 3.4.

φ_l / $(ij)\varphi_l$ / (ij)	$\varphi_1 = \lvert\alpha\alpha\beta\rangle$	$\varphi_2 = \lvert\alpha\beta\alpha\rangle$	$\varphi_3 = \lvert\beta\alpha\alpha\rangle$
(12)	φ_1	φ_3	φ_2
(23)	φ_2	φ_1	φ_3
(13)	φ_3	φ_2	φ_1

Let $\varphi^{(\nu)} = a_1\varphi_1 + a_2\varphi_2 + a_3\varphi_3$. From Eqs. (3-75a), (3-80) and Table 3.4, we have

$$\begin{pmatrix} 1 & 1 & 1 \\ 1 & 1 & 1 \\ 1 & 1 & 1 \end{pmatrix} \begin{pmatrix} a_1 \\ a_2 \\ a_3 \end{pmatrix} = \nu \begin{pmatrix} a_1 \\ a_2 \\ a_3 \end{pmatrix} . \tag{3-81}$$

This result gives a unique eigenvector for the single root $\nu = 3$, namely,

$$\psi^{(3)} = |\, \boxed{1}\boxed{2}\boxed{3} \,\rangle = \sqrt{\frac{1}{3}}(\varphi_1 + \varphi_2 + \varphi_3) . \tag{3-82}$$

It gives the totally symmetric, i.e., the identity rep of S_3, $D^{(3)}(R_a) = 1$. The irreducible basis is labelled traditionally by the Young tableau $\boxed{1}\boxed{2}\boxed{3}$.

From (3-81) we have a double root $\nu = 0$. Since from the above example we already know that $\nu = 0$ is a two-dimensional irrep of C_{3v}, and S_3 is isomorphic to C_{3v}, the double root $\nu = 0$ of (3-81) must correspond to a two-dimensional irrep of S_3. Substituting $\nu = 0$ into (3-81), we only get one independent equation

$$a_1 + a_2 + a_3 = 0 . \tag{3-83}$$

Clearly the solutions to (3-83) are not unique. Any two independent solutions yields a two-dimensional irrep of S_3. For example, we may have the following three sets of solutions giving rise to three reps $D^{(0)}, D^{(0)'}$ and $D^{(0)''}$,

$$\begin{cases} \psi_1^{(0)} = \sqrt{\frac{1}{6}}(2\varphi_1 - \varphi_2 - \varphi_3), \\ \psi_{-1}^{(0)} = \sqrt{\frac{1}{2}}(\varphi_2 - \varphi_3). \end{cases} \begin{cases} \psi_1^{'(0)} = \sqrt{\frac{1}{6}}(\varphi_1 + \varphi_2 - 2\varphi_3), \\ \psi_{-1}^{'(0)} = \sqrt{\frac{1}{2}}(\varphi_1 - \varphi_2). \end{cases}$$
$$\begin{cases} \psi_1^{''(0)} = \sqrt{\frac{1}{6}}(\varphi_1 - 2\varphi_2 + \varphi_3), \\ \psi_{-1}^{''(0)} = \sqrt{\frac{1}{2}}(\varphi_1 - \varphi_3). \end{cases} \tag{3-84}$$

The three sets of basis here are related to one another by similarity transformations; therefore the irreps $D^{(0)}$, $D^{(0)'}$ and $D^{(0)''}$ are equivalent. This shows that the CSCO of G only determine the irreps up to an equivalence.

In order to fix the irreducible basis $\psi_m^{(\nu)}$ completely, we usually require that it also belongs to a definite irrep of the subgroup S_2, i.e., we need the $S_3 \supset S_2$ basis. According to Theorem 3.16, it is necessarily a simultaneous eigenfunction of $C(3)$ and $C(2), C_2 = (12)$ being the CSCO of S_2:

$$\begin{pmatrix} C(3) \\ C(2) \end{pmatrix} \psi_m^{(\nu)} = \begin{pmatrix} \nu \\ m \end{pmatrix} \psi_m^{(\nu)} . \tag{3-85}$$

Using Table 3.4, the solutions to the eigenequation of $C(2)$ are

$$m = 1, \quad \text{double root} , \quad a_2 = a_3 , \tag{3-86}$$

$$m = -1, \quad \text{single root} , \quad a_1 = 0, \quad a_2 = -a_3 . \tag{3-87}$$

Combining these with Eq. (3-83), we obtain two simultaneous eigenfunctions which are precisely $\psi_1^{(0)}$ and $\psi_{-1}^{(0)}$ of Eq. (3-84). The totally symmetric basis $\psi^{(3)}$ is of course an eigenvector of $C(2)$ with eigenvalue +1; thus it can be rewritten as $\psi_1^{(3)}$.

Summarizing, the CSCO of $S_3, C(3)$, is a CSCO in the class space of S_3, but it is no longer a CSCO in the configuration space $\{\varphi_i\}$. $(C(3), C(2))$ is the CSCO for the canonical subgroup chain $S_3 \supset S_2$, which is called the CSCO-II of S_3 (for a general definition, see Sec. 3.8), and is a CSCO in each irreducible space of S_3.

In solving eigenequations (3-85), we just adopted the procedure of first solving that for $C(3)$, and then those for $C(2)$. In fact, for hand calculation, it is more convenient to do the opposite. For example, from Eq. (3-86) we obtain the solution $a_2 = a_3$ for $m = 1$. Let us choose a_1 and a_2 as the independent variables. From (3-81) pick out two equations involving a_2 and a_3, and use $a_2 = a_3$ to eliminate the variable a_3. As a result, we obtain the following two equations

$$\begin{cases} a_1 + 2a_2 = \nu a_1, \\ a_1 + 2a_2 = \nu a_2, \end{cases} \qquad \begin{vmatrix} 1-\nu & 2 \\ 1 & 2-\nu \end{vmatrix} = \nu(\nu - 3) = 0 . \tag{3-88}$$

The solutions are immediately found to be

$$\begin{aligned} (\nu, m) &= (3, 1) , \quad a_1 = a_2 = a_3 , \\ (\nu, m) &= (0, 1) , \quad a_1 = -2a_2 = -2a_3 . \end{aligned} \tag{3-89}$$

Traditionally, the two basis vectors of the two-dimensional irrep of S_3 are labelled by the following two Young tableaux,

$$\begin{aligned} \psi_1^{(0)} &= \left| \begin{array}{|c|c|} \hline 1 & 2 \\ \hline 3 \\ \cline{1-1} \end{array} \right\rangle = \sqrt{\frac{1}{6}}(2\varphi_1 - \varphi_2 - \varphi_3) , \\ \psi_{-1}^{(0)} &= \left| \begin{array}{|c|c|} \hline 1 & 3 \\ \hline 2 \\ \cline{1-1} \end{array} \right\rangle = \sqrt{\frac{1}{2}}(\varphi_2 - \varphi_3) . \end{aligned} \tag{3-90}$$

$\psi_1^{(0)}(\psi_{-1}^{(0)})$ is symmetric (anti-symmetric) in the indices 1 and 2. Note that from (3-82) and (3-90) we can get the transformation matrix from the reducible basis $\{\varphi_i\}$ to the irreducible bases $(\psi_1^{(3)}, \psi_1^{(0)}, \psi_{-1}^{(0)})$ and the matrix is precisely the one given by Eq. (2-99).

With the help of (2-47) and Table 3.4, we can get the rep matrices for the irrep (0) of S_3 with the basis $(\psi_1^{(0)}, \psi_{-1}^{(0)})$, i.e.,

$$D^{(0)}(e) = \begin{pmatrix} 1 & 0 \\ 0 & 1 \end{pmatrix}, \quad D^{(0)}(12) = \begin{pmatrix} 1 & 0 \\ 0 & -1 \end{pmatrix}, \quad D^{(0)}(23) = \begin{pmatrix} -1/2 & \sqrt{3}/2 \\ \sqrt{3}/2 & 1/2 \end{pmatrix}. \tag{3-91a}$$

Using (1-8b) and (1-9), we can write all the remaining matrices

$$\begin{aligned} D^{(0)}(13) &= \begin{pmatrix} -1/2 & -\sqrt{3}/2 \\ -\sqrt{3}/2 & 1/2 \end{pmatrix}, \quad D^{(0)}(123) = \begin{pmatrix} -1/2 & \sqrt{3}/2 \\ -\sqrt{3}/2 & -1/2 \end{pmatrix}, \\ D^{(0)}(132) &= \tilde{D}^{(0)}(123) = \begin{pmatrix} -1/2 & -\sqrt{3}/2 \\ \sqrt{3}/2 & -1/2 \end{pmatrix}. \end{aligned} \tag{3-91b}$$

Notice that the irreducible basis in (3-90) does not have a definite symmetry with respect to the interchange of the coordinate indices 1 and 3, or 2 and 3. It is due to the fact that

$$[(12), (23)] \neq 0 , \quad [(12) , (13)] \neq 0 , \tag{3-92}$$

and thus the three transpositions (12), (23) and (13) cannot be diagonalized simultaneously.

It is easy to see that $(\psi_1^{'(0)}, \psi_{-1}^{'(0)})$ and $(\psi_1^{''(0)}, \psi_{-1}^{''(0)})$ in Eq. (3-84) are the irreducible bases of $S_3 \supset S_2(23)$ and $S_3 \supset S_2(13)$, respectively,

Ex. 3.5. Show that in the state $\psi_1^{(0)} = \left| \begin{array}{|c|c|} \hline 1 & 2 \\ \hline 3 \\ \cline{1-1} \end{array} \right\rangle$ the probabilities for the particles 1 and 3 being symmetric, and for the particles 2 and 3 being symmetric, are both equal to $\frac{1}{4}$, while in the state $\psi_{-1}^{(0)} = \left| \begin{array}{|c|c|} \hline 1 & 3 \\ \hline 2 \\ \cline{1-1} \end{array} \right\rangle$, the same probabilities are equal to $\frac{3}{4}$.

Ex. 3.6. Use the EFM to reduce the three-dimensional rep carried by $\{|\alpha\beta\beta\rangle, |\beta\alpha\beta\rangle, |\beta\beta\alpha\rangle\}$.

(2) The projection operator method. Applying the operator $Q^{(3)}$ of (3-44a) to φ_1, we have

$$\psi^{(3)} = aQ^{(3)}\varphi_1 = \sqrt{\frac{1}{3}}(\varphi_1 + \varphi_2 + \varphi_3), \tag{3-93}$$

where a is a constant. Applying $Q^{(3)}$ to φ_2 and φ_3 gives the same result. Therefore, $Q^{(3)}$ projects out a one-dimensional rep space. Applying $Q^{(0)}$ to φ_1, φ_2 and φ_3 we get three functions

$$\begin{aligned}
\psi_1^{(0)} &= bQ^{(0)}\varphi_1 = \sqrt{\frac{1}{6}}(2\varphi_1 - \varphi_2 - \varphi_3)\ , \\
\psi_2^{(0)} &= bQ^{(0)}\varphi_2 = \sqrt{\frac{1}{6}}(2\varphi_2 - \varphi_1 - \varphi_3)\ , \\
\psi_3^{(0)} &= bQ^{(0)}\varphi_3 = \sqrt{\frac{1}{6}}(2\varphi_3 - \varphi_1 - \varphi_2)\ ,
\end{aligned} \tag{3-94}$$

where only two of them are linearly independent. Hence $Q^{(0)}$ projects out a two-dimensional rep space. Equation (3-93) is identical to (3-82), however (3-94) differs from (3-90) in that (a) Eq. (3-90) is the $S_3 \supset S_2$ basis while (3-94) is not; (b) the basis vectors in (3-90) are orthogonal, while those of Eq. (3-94) are not.

The advantage of the EFM is that once the CSCO of G and $G(s)$ are known, we can solve the eigen equation (3-77b) directly to obtain the $G \supset G(s)$ irreducible basis without first finding $Q^{(\nu)}$. Furthermore, the EFM is easily programmable.

More will be said about the projection operator method in Sec. 3.13.

3.4.4. The group S_3 in the configuration $\alpha\beta\gamma$

Suppose that we have three particles in the configuration $\alpha\beta\gamma$. The system has six possible states $\varphi_1, \ldots, \varphi_6$ (see Eq. (2-70)), which carry the regular rep of S_3. The irreducible basis $\psi_m^{(\nu)} = \sum_i u_i\varphi_i$ satisfies the eigenequations (3-85). To find the coefficients u_i, we have to diagonalize the regular representation matrices $D(C(3)) = D(12) + D(23) + D(13)$, and $D(C(2)) = D(12)$, simultaneously. With the help of (2-71), we find the eigenequations of $C(3)$ and $C(2)$ as follows.

$$\begin{pmatrix} -\nu & 1 & 1 & 1 & 0 & 0 \\ 1 & -\nu & 0 & 0 & 1 & 1 \\ 1 & 0 & -\nu & 0 & 1 & 1 \\ 1 & 0 & 0 & -\nu & 1 & 1 \\ 0 & 1 & 1 & 1 & -\nu & 0 \\ 0 & 1 & 1 & 1 & 0 & -\nu \end{pmatrix} \begin{pmatrix} u_1 \\ u_2 \\ u_3 \\ u_4 \\ u_5 \\ u_6 \end{pmatrix} = 0, \quad \begin{pmatrix} -m & 1 & 0 & 0 & 0 & 0 \\ 1 & -m & 0 & 0 & 0 & 0 \\ 0 & 0 & -m & 0 & 0 & 1 \\ 0 & 0 & 0 & -m & 1 & 0 \\ 0 & 0 & 0 & 1 & -m & 0 \\ 0 & 0 & 1 & 0 & 0 & -m \end{pmatrix} \begin{pmatrix} u_1 \\ u_2 \\ u_3 \\ u_4 \\ u_5 \\ u_6 \end{pmatrix} = 0. \tag{3-95a,b}$$

The eigenvalues ν can be obtained by inspection. For $\nu = 0$, the rank of the matrix in (3-95a) is two; therefore $\nu = 0$ is a four-fold root. According to Theorem 3.11, in the group space $C(3)$ has the same eigenvalues 3,0, and -3 as those of the class space. Thus the matrix in (3-95a) has another two single roots $\nu = \pm 3$ with the eigenvectors

$$\psi^{(3)} = \psi_1^{(3)} = \left|\ \boxed{1}\boxed{2}\boxed{3}\ \right\rangle = \sqrt{\frac{1}{6}}(\varphi_1 + \varphi_2 + \varphi_3 + \varphi_4 + \varphi_5 + \varphi_6)\ , \tag{3-96}$$

$$\psi^{(-3)} = \psi_{-1}^{(-3)} = \left|\ \begin{array}{c}\boxed{1}\\\boxed{2}\\\boxed{3}\end{array}\ \right\rangle = \sqrt{\frac{1}{6}}(\varphi_1 - \varphi_2 - \varphi_3 - \varphi_4 + \varphi_5 + \varphi_6)\ . \tag{3-97}$$

It is easy to verify that $\psi^{(3)}(\psi^{(-3)})$ is totally symmetric (antisymmetric) and therefore belongs to the irrep $m = 1\ (-1)$ of S_2. The totally antisymmetric rep is labeled by the Young tableau $\begin{array}{|c|}\hline 1\\ \hline 2\\ \hline 3\\ \hline\end{array}$ and is called the *alternative representation* of S_3, in which all odd (even) permutations are represented by $-1(+1)$,

$$D(12) = D(23) = D(13) = -1, \quad D(e) = D(123) = D(132) = 1\ . \tag{3-98}$$

For the four-fold root $\nu = 0$, from (3-95a) we only get two independent equations, namely,

$$\nu = 0: \quad u_1 + u_5 + u_6 = 0\ , \quad u_2 + u_3 + u_4 = 0\ . \tag{3-99}$$

Furthermore, from (3-95b) we have three equations:

$$m = \pm 1: \ u_1 = \pm u_2, \ u_3 = \pm u_6, \ u_4 = \pm u_5\ . \tag{3-100}$$

Combining the last two equations, for both $(\nu, m) = (0, 1)$ and $(0, -1)$ we obtain only one independent equation

$$(\nu, m) = (0, 1): \ u_1 + u_3 + u_4 = 0\ , \tag{3-101}$$

$$(\nu, m) = (0, -1): \ u_1 - u_3 - u_4 = 0\ . \tag{3-102}$$

Therefore, the solutions are not unique. Why is this so? From Eq. (3-91) we know that $\nu = 0$ is a two-dimensional irrep, $h_\nu = 2$. Here $\nu = 0$ is a four-fold root i.e., $M_\nu = 4$. Since $M_\nu = \tau_\nu h_\nu$, the irrep (0) must occur twice ($\tau_\nu = 2$) in the regular rep of S_3. In other words in the decomposition of the group space of S_3

$$L_g \equiv L_{(3)} + L_{(0)} + L_{(-3)}\ , \tag{3-103}$$

the four-dimensional eigenspace of $C(3)$ can be reduced into a direct sum of two equivalent irreducible spaces of dimension 2:

$$L_{(0)} = L_{(0)1} \oplus L_{(0)2}\ . \tag{3-104}$$

How does one decompose the eigenspace $L_{(0)}$, or more generally, how does one decompose the eigenspace $L_{(\nu)}$ of the CSCO of G? According to Theorem 2.7, if we can find a new operator $\overline{C}_1$ which commutes with G, then the eigenspace $L_{(\nu)}$ can be further decomposed into a direct sum of the eigenspaces of $\overline{C}_1$,

$$L_{(\nu)} = \sum_{k_1} \oplus L_{(\nu)k_1}, \quad \overline{C}_1 L_{(\nu)k_1} = k_1 L_{(\nu)k_1}\ . \tag{3-105}$$

If the subspace $L_{(\nu)k_1}$ is still reducible, then we need to find another operator $\overline{C}_2$ which commutes with both G, and $\overline{C}_1$, and decompose $L_{(\nu)k_1}$ into a direct sum of the eigenspaces of $\overline{C}_2$, etc. until we find a set of operators $\overline{C}(s) = (\overline{C}_1\overline{C}_2\ldots)$ such that the simultaneous eigenspaces of $\overline{C}(s)$ are irreducible. The eigenvalue $k = (k_1, k_2, \ldots)$ of $\overline{C}(s)$ thus provides new label to distinguish the equivalent irreps contained in $L_{(\nu)}$.

By Theorems 3.1 and 3.4, we cannot find such operators from the group G. In other words, it is impossible to use the remaining class operators of G which are not contained in the CSCO of G to further decompose the eigenspace $L_{(\nu)}$. Therefore, we have to search for another group $\overline{G}$ which commutes with G and thus will provide some new operators with which to decompose the eigenspace $L_{(\nu)}$. In the following section, we will first solve this problem for the permutation group.

3.5. State Permutation Group

Suppose that there are n particles occupying n distinct single particle states, $i_1, i_2, \ldots, i_n$. The state

$$\begin{aligned}\Psi(X) \equiv \Psi(x_1, \ldots, x_n) &= \varphi_{i_1}(x_1)\varphi_{i_2}(x_2)\ldots\varphi_{i_n}(x_n) \\ &= |\omega_0\rangle = \prod_{a=1}^{n} \varphi_{i_a}(x_a) = |i_1 i_2, \ldots, i_n\rangle\end{aligned} \tag{3-106}$$

is referred to as the normal order state, in which the kth particle is in the kth single particle state. The ordering of the single particle states is specified right at the beginning. For example, for the orbital angular momentum $l = 1$, one may specify $m = 1, 0, -1$ as the states i_1, i_2, i_3, respectively. However, once the ordering is specified, we must abide by it through the whole analysis. Under the above specification, $|10-1\rangle$ is a normal order state, and all the other states, such as $|01-1\rangle, |-101\rangle$ are not.

Up to now, the element p of a permutation group S_n is defined as a permutation of the subscripts of the coordinates $\{x_a\}$. We now introduce another kind of permutation, designated by the script letter $\wp$, which permutes the subscripts of the single particle states $\{i_a\}$, and is called the state permutation (Bohr 1969). All the permutations of the n state indices form a group, called the *state permutation group* $\mathcal{S}_n$. For example

$$\wp_{123}|i_1 i_2 i_3\rangle = |i_2 i_3 i_1\rangle\ , \quad \wp_{23}|i_2 i_3 i_1\rangle = |i_3 i_2 i_1\rangle\ , \tag{3-107a}$$

in contrast to the coordinate permutations

$$p_{123}|i_1 i_2 i_3\rangle = |i_3 i_1 i_2\rangle\ , \quad p_{23}|i_2 i_3 i_1\rangle = |i_2 i_1 i_3\rangle\ . \tag{3-108}$$

For simplicity in notation, we often use $\alpha, \beta, \gamma, \delta, \ldots$ in place of $i_1, i_2, i_3, i_4, \ldots$ Thus $\wp_{12} = (\alpha\beta)$ denotes an interchange of α and β, while $\wp_{123} = (\alpha\beta\gamma)$ denotes a cycle permutation $\alpha \to \beta \to \gamma \to \alpha$. With this notation, Eq. (3-107a) becomes

$$\wp_{123}|\alpha\beta\gamma\rangle = |\beta\gamma\alpha\rangle\ , \quad \wp_{23}|\beta\gamma\alpha\rangle = |\gamma\beta\alpha\rangle\ . \tag{3-107b}$$

The permutation group $\mathcal{S}_n$ has the following properties.

1. S_n and $\mathcal{S}_n$ are isomorphic.
2. S_n and $\mathcal{S}_n$ commute, i.e.,

$$[p_a, \wp_b] = 0\ . \tag{3-109}$$

3. For the normal order state we have

$$p|\omega_0\rangle = \wp^{-1}|\omega_0\rangle\ . \tag{3-110}$$

The first and second points are self evident, and the third point is proved as follows. Let $p = \binom{a}{a'}$, i.e., p does the coordinate transformation $x_a \to x_{a'}$ and let $\wp = \binom{i_a}{i_{a'}}$, i.e., it performs the change $i_a \to i_{a'}$. We have, then,

$$p\wp|\omega_0\rangle = p\wp \prod_{a=1}^{n} \varphi_{i_a}(x_a) = \prod_{a'=1}^{n} \varphi_{i_{a'}}(x_{a'}) = |\omega_0\rangle\ . \tag{3-111}$$

This proves Eq. (3-110).

Remark 1: Equation (3-110) does not imply that $p = \wp^{-1}$. It only shows that when acting on the normal order state, $p = \wp^{-1}$, while for other states, $p \neq \wp^{-1}$ in general.

Remark 2: The term state permutation was used also in Weyl (1946) and Hamermesh (1962). Although in appearance their definition of the state permutation is same as that given here, i.e., it is a permutation of the state indices, in fact the two definitions are entirely different. In our case each single particle state is assigned a definite index from the beginning regardless of it being occupied by which particle. In Weyl (1946) or Hamermesh (1962), the state index k is assigned to the single particle state occupied by the k-th particle regardless of which single particle state it is in. For example, in the product state $|10-1\rangle$, they took $i_1=1, i_2=0, i_3=-1$, while in the product state $|-101\rangle$, they took $i_1=-1, i_2=0, i_3=1$. According to their definition, any n-particle product state is always of the form $\prod_{a=1}^{n}\varphi_{i_a}(x_a)$. Therefore what they called a state permutation $\mathbf{p}$ is always equal to the inverse of the coordinate permutation p, i.e.,

$$\mathbf{p}=p^{-1} \,. \tag{3-112}$$

Thus the group $\{\mathbf{p}\}$ is nothing new but essentially the coordinate permutation group, and is useless for our purpose. In this book we always use Eq. (3-107) as the definition for state permutations.

Applying the $n!$ elements of the group S_n or $\mathcal{S}_n$ to the normal order state $\Psi_0=|\omega_o\rangle$, we obtain the $n!$ states

$$\varphi_a=p_a|\omega_0\rangle=\wp_a^{-1}|\omega_0\rangle=p_a|i_1,i_2,\ldots,i_n\rangle \,. \tag{3-113}$$

They form a basis for the regular rep of S_n as well as of $\mathcal{S}_n$.

Using (3-109) and the fact that the classes of the permutation group are all ambivalent, we have

$$\mathcal{C}_j\varphi_a=\mathcal{C}_j p_a|\omega_0\rangle=p_a\mathcal{C}_j|\omega_0\rangle=p_aC_{j'}|\omega_0\rangle=p_aC_j|\omega_0\rangle=C_jp_a|\omega_0\rangle=C_j\varphi_a \,, \tag{3-114}$$

where $C_{j'}$ is defined by Eq. (3-38). Since φ_a in the above is an arbitrary function,

$$\mathcal{C}_j=C_j \,, \quad j=1,2,\ldots,N \,. \tag{3-115a}$$

Therefore the CSCO of $\mathcal{S}_n$ and S_n are equal, i.e.,

$$\mathcal{C}(n)=C(n) \,. \tag{3-115b}$$

According to Theorem 3.15, this implies that if $\psi^{(\nu)}$ belongs to the irrep (ν) of S_n, it must also belong to the irrep (ν) of $\mathcal{S}_n$.

From the last two results, the class operators of $\mathcal{S}_n$ do not give any new operators. Fortunately, the class operators $\mathcal{C}_k(i)$ of the subgroup $\mathcal{S}_i$ are not equal to the class operators $C_k(i)$ of $S_i, i=n-1,\ldots,2$. This can be seen from the fact that $[\mathcal{C}_k(i),S_n]=0$, while $[C_k(i),S_n]\neq 0$. We use $\mathcal{C}(i)$ to denote the CSCO of the subgroup $\mathcal{S}_i$ of the state permutation group $\mathcal{S}_n$. Then the set of operators $\mathcal{C}(s)=(\mathcal{C}(n-1),\mathcal{C}(n-2),\ldots,\mathcal{C}(2))$ provides a set of new operators, which can be used to decompose the eigenspace of the CSCO of S_n. Before going to discuss the general case, we return to the suspended problem in Sec. 3.4 on the reduction of the rep space $L_{(0)}$ of S_3.

3.6. Reduction of the Regular Rep of S$_3$

1. With the new operator $\mathcal{C}(2)=\wp_{12}$, the eigenequation (3-85) is extended to

$$\begin{pmatrix} C(3)\\ C(2)\\ \mathcal{C}(2)\end{pmatrix}\psi_m^{(\nu)k}=\begin{pmatrix}\nu\\ m\\ k\end{pmatrix}\psi_m^{(\nu)k} \,, \quad \psi_m^{(\nu)k}=\sum_{i=1}^{6}u_i\varphi_i \,, \tag{3-116}$$

where φ_i were defined in (2-70). From the definition $\wp_{12}=(\alpha\beta)$, we have

$$\wp_{12}|\alpha\beta\gamma\rangle=|\beta\alpha\gamma\rangle \,, \quad \wp_{12}|\gamma\beta\alpha\rangle=|\gamma\alpha\beta\rangle \,, \quad \wp_{12}|\alpha\gamma\beta\rangle=|\beta\gamma\alpha\rangle \,,$$

i.e.,

$$\wp_{12}\varphi_1=\varphi_2, \quad \wp_{12}\varphi_3=\varphi_5 \quad ,\wp_{12}\varphi_4=\varphi_6 \,. \tag{3-117}$$

Since $(\wp_{12})^2 = 1$, the eigenvalues are $k = \pm 1$. From (3-116) and (3-117) we obtain

$$k = \pm 1: \ u_1 = \pm u_2, \ u_3 = \pm u_5, \ u_4 = \pm u_6 \ . \tag{3-118}$$

Combining Eqs. (3-110), (3-101), (3-102) and (3-118), we obtain the following four eigenvectors

$$(\nu, m, k) = (0,1,1) \ , \psi_1^{(0)1}(\alpha\beta\gamma) = \left| \ \begin{array}{|c|c|}\hline 1 & 2 \\ \hline 3 \\ \cline{1-1}\end{array} \ \begin{array}{|c|c|}\hline \alpha & \beta \\ \hline \gamma \\ \cline{1-1}\end{array} \right\rangle$$

$$= \sqrt{\frac{1}{12}}[2(\varphi_1 + \varphi_2) - (\varphi_3 + \varphi_4 + \varphi_5 + \varphi_6)] \ .$$

$$(0,-1,1) \ , \psi_{-1}^{(0)1}(\alpha\beta\gamma) = \left| \ \begin{array}{|c|c|}\hline 1 & 2 \\ \hline 3 \\ \cline{1-1}\end{array} \ \begin{array}{|c|c|}\hline \alpha & \gamma \\ \hline \beta \\ \cline{1-1}\end{array} \right\rangle = \frac{1}{2}[-\varphi_3 + \varphi_4 - \varphi_5 + \varphi_6] \ ,$$

$$(0,1,-1), \psi_1^{(0)-1}(\alpha\beta\gamma) = \left| \ \begin{array}{|c|c|}\hline 1 & 2 \\ \hline 3 \\ \cline{1-1}\end{array} \ \begin{array}{|c|c|}\hline \alpha & \gamma \\ \hline \beta \\ \cline{1-1}\end{array} \right\rangle = \frac{1}{2}[-\varphi_3 + \varphi_4 + \varphi_5 - \varphi_6]$$

$$(0,-1,-1) \ , \psi_{-1}^{(0)-1}(\alpha\beta\gamma) = \left| \ \begin{array}{|c|c|}\hline 1 & 3 \\ \hline 2 \\ \cline{1-1}\end{array} \ \begin{array}{|c|c|}\hline \alpha & \gamma \\ \hline \beta \\ \cline{1-1}\end{array} \right\rangle = \sqrt{\frac{1}{12}}[2(\varphi_1 - \varphi_2) + \varphi_3 + \varphi_4 - \varphi_5 - \varphi_6] \ . \tag{3-119a}$$

Because of the fact that $C(3) = \mathcal{C}(3)$ and Theorem 3.16, the eigenvector $\psi_m^{(\nu)k}$ of (3-116) is the irreducible basis (ν, m) of the group chain $S_3 \supset S_2$, and the irreducible basis (ν, k), of the group chain $\mathcal{S}_3 \supset \mathcal{S}_2$.

In Sec. 3.4, we used tableaux filled with numbers, the Young tableaux, to designate a $S_3 \supset S_2$ irreducible basis. Analogously, we use tableaux filled with single particle states α, β, γ, referred to as the Weyl tableaux (see Sec. 4.8 for a general definition), to designate a $\mathcal{S}_3 \supset \mathcal{S}_2$ irreducible basis. Therefore, instead of the quantum number ν, m and k, we can use two tableaux, one Young tableau Y_m^ν and one Weyl tableau W_k^ν, to label a $S_3 \supset S_2$ and $\mathcal{S}_3 \supset \mathcal{S}_2$ irreducible basis. The totally symmetric (antisymmetric) state in (3-96) is also totally symmetric (antisymmetric) in the state indices α, β and γ and can be labeled by two tableaux as

$$\psi_1^{(3)1}(\alpha\beta\gamma) = \left| \ \begin{array}{|c|c|c|}\hline 1 & 2 & 3 \\ \hline\end{array} \ \begin{array}{|c|c|c|}\hline \alpha & \alpha & \beta \\ \hline\end{array} \right\rangle = \sqrt{\frac{1}{6}}[\varphi_1 + \varphi_2 + \varphi_3 + \varphi_4 + \varphi_5 + \varphi_6]$$

$$\psi_{-1}^{(-3)-1}(\alpha\beta\gamma) = \left| \ \begin{array}{|c|}\hline 1 \\ \hline 2 \\ \hline 3 \\ \hline\end{array} \ \begin{array}{|c|}\hline \alpha \\ \hline \beta \\ \hline \gamma \\ \hline\end{array} \right\rangle = \sqrt{\frac{1}{6}}[\varphi_1 - \varphi_2 - \varphi_3 - \varphi_4 + \varphi_5 + \varphi_6] \ , \tag{3-119b}$$

$(\psi_1^{(0)1}, \psi_{-1}^{(0)1})$ and $(\psi_1^{(0)-1}, \psi_{-1}^{(0)-1})$ form the two irreducible spaces $L_{(0)1}$ and $L_{(0)-1}$, respectively.

It is easy to check that the irreducible matrices of S_3 in these two spaces are the same, and are identical with (3-91). Therefore, these two irreps, which are distinguished by the quantum number k of $\mathcal{C}(2)$, the CSCO of the state permutation group $\mathcal{S}_2$, are not only equivalent but also identical (due to our special choice for the phases of the solutions). The four-dimensional eigenspace $L_{(0)}$ of $C(3)$ is thus decomposed into a direct sum of the two eigenspaces of $\mathcal{C}(2)$,

$$L_{(0)} = L_{(0)1} \oplus L_{(0)-1} \ , \quad \mathcal{C}(2) L_{(0)k} = k L_{(0)k} \ . \tag{3-120}$$

The set of operators, $K = (C(3), C(2), \mathcal{C}(2))$, forms a CSCO in the regular rep space of S_3, and is called the CSCO-III of S_3 (for a general definition, see Sec. 3.8). The simultaneous eigenspaces of K are all one-dimensional. Thus the six-dimensional group space S_3 is fully decomposed into a direct sum of six one-dimensional spaces

$$L_g = \sum_{\nu m k} \oplus L_{(\nu)mk} \ . \tag{3-121}$$

Similarly, $(\psi_1^{(0)1}, \psi_1^{(0)-1})$ and $(\psi_{-1}^{(0)1}, \psi_{-1}^{(0)-1})$ form two irreducible spaces of $\mathcal{S}_3$. The irreducible matrices of $\mathcal{S}_3$ in these two spaces are also identical with (3-91). These two equivalent irreps of the state permutation group are distinguished by the quantum number of the operator $C(2)$, the CSCO of the coordinate permutation group S_2.

2. In the above discussion, we first solve the eigenequation of $C(3)$, and then those of $C(2)$ and $\mathcal{C}(2)$. For actual calculations, it is more convenient to first use the eigenequations of $C(2)$ and $\mathcal{C}(2)$ to eliminate the non-independent variables u_i and thus to reduce the order of the eigen equation of $C(3)$. The concrete steps are as follows.

(a) Applying the operators $C(2) = (12)$ and $\mathcal{C}(2) = (\alpha\beta)$ to φ_i, we obtain Table 3.6-1, which in turn gives (3-122).

Table 3.6-1

φ_i / $R\varphi_i$ / R	φ_1	φ_2	φ_3	φ_4	φ_5	φ_6
	$\|\alpha\beta\gamma\rangle$	$\|\beta\alpha\gamma\rangle$	$\|\gamma\beta\alpha\rangle$	$\|\alpha\gamma\beta\rangle$	$\|\gamma\alpha\beta\rangle$	$\|\beta\gamma\alpha\rangle$
(12)	φ_2	φ_1	φ_6	φ_5	φ_4	φ_3
$\wp_{12}$	φ_2	φ_1	φ_5	φ_6	φ_3	φ_4

$$\begin{aligned} m = \pm 1: \ u_1 &= \pm u_2, \ u_3 = \pm u_6, \ u_4 = \pm u_5 , \\ k = \pm 1: \ u_1 &= \pm u_2, \ u_3 = \pm u_5, \ u_4 = \pm u_6 . \end{aligned} \tag{3-122}$$

Combining these two sets of equations, we get

$$(m,k) = (1,1): \ u_1 = u_2, \qquad u_3 = u_4 = u_5 = u_6, \tag{3-123a}$$

$$(-1,-1): \ u_1 = -u_2, \qquad u_3 = u_4 = -u_5 = -u_6, \tag{3-123b}$$

$$(1,-1): \ u_1 = u_2 = 0, \quad u_3 = -u_4 = -u_5 = u_6 , \tag{3-123c}$$

$$(-1,1): \ u_1 = u_2 = 0, \quad u_3 = -u_4 = u_5 = -u_6. \tag{3-123d}$$

(b) For $(m,k) = (1,1)$ and (-1, -1), only two independent variables are left, which may be chosen as u_1 and u_3. Two more equations involving u_1 and u_3 are needed. They can be found from the eigenequation of $C(3)$. To this end, according to (3-75b), we only need to know the effect of $C(3)$ on the states φ_1 and φ_3, which are listed in Table 3.6-2.

Table 3.6-2

p / $p\varphi_i$ / φ_i	(12)	(13)	(23)
φ_1	φ_2	φ_3	φ_4
φ_3	φ_6	φ_1	φ_5

Using Eq. (3-75b) and Table 3.6-2, we obtain

$$u_2 + u_3 + u_4 = \nu u_1 , \quad u_1 + u_5 + u_6 = \nu u_3 . \tag{3-124}$$

Combining (3-123a) and (3-124) leads to

$$\begin{aligned} u_1 + 2u_3 &= \nu u_1 , \\ u_1 + 2u_3 &= \nu u_3 , \end{aligned} \quad \begin{vmatrix} 1-\nu & 2 \\ 1 & 2-\nu \end{vmatrix} = \nu(\nu - 3) = 0 . \tag{3-125}$$

From (3-125) and (3-123a), we obtain the eigenvectors $\psi_1^{(3)1}$ and $\psi_1^{(0)1}$, which are just the result (3-119).

Combining (3-124) and (3-123b) leads to

$$\begin{aligned} -u_1 + 2u_3 &= \nu u_1 \ , \\ u_1 - 2u_3 &= \nu u_3 \ , \end{aligned} \qquad \begin{vmatrix} -1-\nu & 2 \\ 1 & -2-\nu \end{vmatrix} = \nu(\nu+3) = 0 \ . \tag{3-126}$$

From (3-126) and (3-123b), we obtain the eigenvectors $\psi_{-1}^{(-3)-1}$ and $\psi_{-1}^{(0)-1}$, which are again just Eq. (3-119).

For $(m, k) = (1, -1)$ and $(-1, 1)$, Eqs. (3-123c) and (3-123d) alone give two unique eigenvectors, which are easily recognized as $\psi_{1}^{(0)-1}$ and $\psi_{-1}^{(0)1}$. They are identical to the result shown in (3-119a).

3.7. Intrinsic Group

In Secs. 3.4-3.6, we discussed the reduction of the rep of S_3 carried by the product states. Now we proceed to generalize it to a general theory. For the coordinate permutation group S_n, we have found the state permutation group $\mathcal{S}_n$ which is isomorphic to and commuting with S_n. Now for any group G we wish to find a group $\overline{G}$ which is anti-isomorphic (or isomorphic, since there is no essential difference between isomorphism and anti-isomorphism) to G and commuting with G.

3.7.1. Definition of the intrinsic group

Definition 3.3: For each element R of a group G, we can define a corresponding operator $\overline{R}$ in the group space L_g through the following *defining equation:*

$$\overline{R}S = SR, \quad \text{for all } S \in L_g \ . \tag{3-127}$$

The group formed by the totality of the operators $\overline{R}$ is called the *intrinsic group* of G, or simply the intrinsic group $\overline{G}$ if no confusion will arise.

Let us first show that the operators $\overline{R}$ do form a group. According to (3-127), the operation of $\overline{R}$ on a vector S in L_g is to change it into another vector SR. It should be emphasized that (3-127) is the defining equation for the operator $\overline{R}$ rather than an identity relation. Hence it is not permissible to multiply (3-127) from the right by another vector T of L_g, i.e.,

$$\overline{R}ST \neq SRT \ . \tag{3-128}$$

Instead, we should regard ST as a new vector in L_g and then use (3-127) to obtain

$$\overline{R}ST = \overline{R}(ST) = STR \ . \tag{3-129}$$

Suppose that the multiplication relation for the group G is

$$RS = U \ . \tag{3-130a}$$

From Eqs. (3-127) and (3-129) we have

$$\overline{S}\overline{R}T = \overline{S}TR = TRS = TU = \overline{U}T \ . \tag{3-131}$$

Since T is an arbitrary vector in L_g, one has

$$\overline{S}\overline{R} = \overline{U} \ . \tag{3-130b}$$

Therefore, there is a 1-1 correspondence between the elements $\overline{R}$ and R, and (3-130) shows that the totality of $\overline{R}$ forms a group $\overline{G}$ which is anti-isomorphic to the group G.

From (3-127) one has

$$S\overline{R}T = STR \quad \text{for all } T \in L_g \ . \tag{3-132}$$

Comparing (3-129) with the last result, and noting that T is an arbitrary vector in L_g, one has

$$\overline{R}S = S\overline{R} \tag{3-133a}$$

or

$$[\overline{R}, S] = 0 \,. \tag{3-133b}$$

Therefore, $\overline{G}$ commutes with G.

Note the essential difference between (3-127) and (3-133a). The latter is an identity relation, while the former is not. The difference comes from the fact that in (3-127), S is regarded as a vector in L_g, while in (3-133a), S is an operator in L_g. The rule for determing whether a group element S is to be regarded as a basis vector or as an operator is very simple. If S is the last element behind an intrinsic group operator, then S should be looked upon as a basis vector; if S is followed by another group elements of G, then S should be regarded as an operator.

3.7.2. Regular representation of intrinsic group

Equation (3-127) shows that the group space L_g forms a representation space for the intrinsic group G:

$$\overline{R}_b R_c = \sum_{a=1}^{g} D_{ac}(\overline{R}_b) R_a = R_c R_b \,, \tag{3-134}$$

Therefore

$$D_{ac}(\overline{R}_b) = \begin{cases} 1 \,, & \text{for} \quad R_c R_b = R_a \\ 0 \,, & \text{otherwise} \end{cases} \,. \tag{3-135}$$

The rep $D(\overline{G})$ will be called the regular rep of the intrinsic group $\overline{G}$. $D(\overline{G})$ was referred to as the inverted rep of G in Boerner (1963), since $D(\overline{G})$ is anti-isomorphic to G.

Comparing (3-135) with (2-68), one sees that in obtaining the regular reps of G and $\overline{G}$ from the group table of G, the indices c and b interchange their roles. Thus reading Table 1.2 vertically instead of horizontally, one immediately obtains the regular rep of the intrinsic permutation group $\overline{S}_3$,

$$\begin{aligned} D(\overline{e}) &= \{123456\}, \quad D(\overline{12}) = \{215634\}, \quad D(\overline{13}) = \{361542\}, \\ D(\overline{23}) &= \{456123\}, \quad D(\overline{123}) = \{542361\}, \quad D(\overline{132}) = \{634215\} \,. \end{aligned} \tag{3-136}$$

Let A be a vector in the group space L_g,

$$A = \sum_a u_a R_a \,. \tag{3-137a}$$

Its representation matrix $D_{ab}(A) = \langle a|A|b\rangle$ is determined by

$$A R_b = \sum_a \langle a|A|b\rangle R_a \,. \tag{3-137b}$$

The corresponding intrinsic element is

$$\overline{A} = \sum_a u_a \overline{R}_a \,. \tag{3-137c}$$

Its representation matrix $D_{ab}(\overline{A})$ is decided upon by

$$\overline{A} R_b = R_b A = \sum_a \langle a|\overline{A}|b\rangle R_a \,. \tag{3-137d}$$

Theorem 3.17: The matrices $D(\overline{A})$ and $\tilde{D}(A)$ are similar matrices, $\tilde{D}(A)$ being the transpose of $D(A)$.

Proof: Taking the Hermitian conjugate of (3-137b), letting $A \leftrightarrow A^\dagger$, and using the unitarity of the group operators,

$$R_b^{-1}A = \sum_a \langle b|A|a\rangle R_a^{-1} . \tag{3-137e}$$

Comparing (3-137d) with (3-137e), we obtain

$$\langle a|\overline{A}|b\rangle = \langle b^{-1}|A|a^{-1}\rangle , \tag{3-137f}$$

where

$$\langle b^{-1}|A|a^{-1}\rangle = \langle R_b^{-1}|A|R_a^{-1}\rangle .$$

Since R_a^{-1} and R_b^{-1} are still vectors of the group space, the matrices $\langle b^{-1}|A|a^{-1}\rangle$ and $\langle b|A|a\rangle$ are similar. Thus (3-137f) shows that $D(\overline{A})$ and $\tilde{D}(A)$ are similar matrices, i.e.,

$$D(\overline{A}) = T\tilde{D}(A)T^{-1} . \tag{3-137g}$$

Ex. 3.7. Using group Table 1.2, find the subgroup of S_6 which is isomorphic to the intrinsic group $\overline{S}_3$, and verify that this subgroup commutes with the subgroup of S_6, Eq. (1-16), which is isomorphic to S_3.

3.7.3. Action of intrinsic group elements on functions on the group

In analogy with (2-85) and (2-86), we can deduce from (3-127) the operation of the intrinsic group elements on functions on the group, $u(S)$. Applying $\overline{R}$ to any vector $P = \sum_S u(S)$ in L_g gives

$$\overline{R}P = \overline{R}\sum_S u(S)S = PR = \sum_S u(S)SR = \sum_S u(SR^{-1})S .$$

Therefore we have

$$\overline{R}u(S) = u(SR^{-1}) . \tag{3-138}$$

3.7.4. Properties of the intrinsic group

1. $\overline{G}$ commutes with G, i.e.,

$$[\overline{R}, S] = 0 . \tag{3-139}$$

2. $\overline{G}$ and G are anti-isomorphic.

Therefore, all the conclusions about the group G apply to the intrinsic group $\overline{G}$ as well. For example, if $C = k_1C_{i_1} + \ldots + k_lC_{i_l}$ is a CSCO of G, then $\overline{C} = k_1\overline{C}_{i_1} + \ldots + k_l\overline{C}_{i_l}$ is a CSCO of $\overline{G}, \overline{C}_i$ being the class operators of $\overline{G}$,

$$\overline{C}_i = \sum_{l=1}^{g_i} \overline{R}_l^{(i)} . \tag{3-140}$$

If G has N inequivalent irreps (ν), so as $\overline{G}$. If G has a subgroup chain $G \supset G_1 \supset G_2 \supset \ldots$, then $\overline{G}$ has also a subgroup chain $\overline{G} \supset \overline{G}_1 \supset \overline{G}_2 \supset \ldots$, etc.

Theorem 3.18: The CSCO of $\overline{G}$ and G are equal:

$$\overline{C} = C . \tag{3-141}$$

Proof: From definition (3-127),

$$\overline{C}_i R = \left(\sum_l \overline{R}_l^{(i)}\right) R = R \sum_l R_l^{(i)} = RC_i \ . \tag{3-142}$$

On the other hand, the class operators of G commute with any elements of $G, RC_i = C_i R$. Hence

$$\overline{C}_i R = C_i R \ . \tag{3-143}$$

Since R is an arbitrary element, we have $\overline{C}_i = C_i$. QED.

Because of (3-141), if $P^{(\nu)}$ is an eigenvector of C, then it is necessarily an eigenvector of $\overline{C}$, and if $\psi^{(\nu)}$ belongs to the irrep (ν) of G, then it must also belong to an irrep of the intrinsic group $\overline{G}$, which can be labeled by the same quantum number ν.

From (3-134) and (3-138) one sees that the operators of the intrinsic group are unitary both in group space and in the space of functions on the group. For example, the regular rep of $\overline{S}_3$ in Eq. (3-136) is unitary.

3.7.5. Some remarks

The following points are worth noting.

1. From (3-127) it is seen that, for an Abelian group, the intrinsic group $\overline{G}$ coincides with the group G itself.

2. The intrinsic group element $\overline{R}$ defined by (3-127) is not a conjugate element of the group element R. From (3-127) a formal relation between $\overline{R}$ and R can be written as

$$\overline{R} = SRS^{-1} \ , \quad \text{when } R \text{ acts on } S \ . \tag{3-144}$$

It should be stressed that the last result is not an identity either. It only shows that when acting on $S, \overline{R}$ is equivalent to the operator SRS^{-1}, and that while acting on another vector $T, \overline{R}$ will be equivalent to TRT^{-1}. In other words, the element S in Eq. (3-144) is not fixed; it changes according to which vector the operator T in acting upon. This is a most important point, albeit a little bit tricky, for the understanding of the intrinsic group.

In contrast, a conjugate element T of the group element R is

$$T = S_0 R S_0^{-1} \ , \tag{3-145}$$

where S_0 is a fixed element of G. Equation (3-145) is an identity relation and T is an element of G.

3. One should not confuse the intrinsic group $\overline{G}$ with the group G' brought about by an inner automorphism (Sec. 1.4):

$$G' = R_a G R_a^{-1} \ . \tag{3-146}$$

4. It is important to distinguish between the subgroup $\overline{G}(1)$ of the intrinsic group $\overline{G}$ of G and the intrinsic group $\overline{G}'(1)$ of the subgroup $G(1)$ of G. $\overline{G}(1)$ is defined in the whole group space of G, while $\overline{G}'(1)$ is defined in the group space of $G(1)$. Let $R, R(1)$, $\overline{R}(1)$, and $\overline{R}'(1)$ be the group elements of the groups $G, G(1), \overline{G}(1)$ and $\overline{G}'(1)$, respectively. The definition for the groups $\overline{G}(1)$ and $\overline{G}'(1)$ are, respectively,

$$\overline{R}(1)R = RR(1) \quad \text{for all } R \in G \ , \tag{3-147}$$

$$\overline{R}'(1)S(1) = S(1)R(1), \quad \text{for all } S(1) \in G(1) \ . \tag{3-148}$$

Obviously, $\overline{G}(1)$ commutes with the whole group G, while $\overline{G}'(1)$ commutes only with the subgroup $G(1)$, i.e.,

$$[\overline{G}(1), G] = 0 \ , \quad [\overline{G}'(1), G] \neq 0 \quad [\overline{G}'(1), G(1)] = 0 \ .$$

3.7.6. Intrinsic state (regular rep case)

Equations (3-127) and (3-138) define the action of the intrinsic group elements in group space and the space of functions on the group. We now address ourselves to the problem of defining the action of the intrinsic group elements in configuration space. The problem is rather involved. Let us first discuss the regular representation case. Suppose that $\Psi_0(X)$ is a function in the configuration space without any symmetry. Under the action of the g elements of G, it gives g orthogonal functions

$$\varphi_a(X) = R_a \Psi_0(X) \ , \quad a = 1, 2, \ldots g \ , \tag{3-149}$$

which carry the regular rep of G. As mentioned before, $\overline{R}S = S\overline{R}$ is not an identity; therefore, it does not hold when acting on all these g functions. However, we may require it to hold when acting on one of them, say $\Psi_0(X)$,

$$\overline{R}S\Psi_0(X) = SR\Psi_0(X) \ . \tag{3-150}$$

Letting S be the identity element we have

$$\overline{R}\Psi_0(X) = R\Psi_0(X) = \Psi_0(R^{-1}X) \ . \tag{3-151}$$

In this way, we have defined the action of $\overline{R}$ on the chosen state $\Psi_0(X)$. We call the state $\Psi_0(X)$ which satisfies (3-151) the *intrinsic state* of the group G (the physical interpretation of the intrinsic state for the rotation group will be given in Chapter 6). Equation (3-151) shows that when acting on the intrinsic state, $\overline{R}$ is equivalent to R.

From (3-127) and (3-151), we obtain the action of $\overline{R}$ on any function $\varphi_a(X) = R_a\Psi_0(X)$:

$$\overline{R}\varphi_a(X) = \overline{R}R_a\Psi_0(X) = R_a R\Psi_0(X) \ . \tag{3-152}$$

Notice that

$$\overline{R}\varphi_a(X) \neq R\varphi_a(X) = RR_a\Psi_0(X) \ . \tag{3-153}$$

The remaining question is which state should be chosen as the intrinsic state? In principle, any state from $\varphi_1, \ldots, \varphi_g$ can be chosen as the intrinsic state. However once chosen, it should be kept fixed throughout the whole analysis. In some cases, the choice of the intrinsic state is related to the choice of representation. For example, in the SU_3 quark model of particle physics, there are three representations, the *I-spin*, *U-spin* and *V-spin* representations (Feld 1969). The intrinsic states for these three reps are (Chen [18] 1979)

$$\begin{aligned} &\text{(a) } I-\text{spin} \ , \quad \Psi_0(I) = |uds\rangle \ , \\ &\text{(b) } U-\text{spin} \ , \quad \Psi_0(U) = |dsu\rangle \ , \\ &\text{(c) } V-\text{spin} \ , \quad \Psi_0(V) = |sud\rangle \ , \end{aligned} \tag{3-154}$$

where u, d and s are the up, down and strange quarks, respectively. The transformations between these three reps are given in (3-226c).

Example: Let us find the effect of the intrinsic permutation $(\overline{12})$ of $\overline{S}_3$ on the six states in Table 3.6-1. The normal-ordered state $|\alpha\beta\gamma\rangle$ is chosen as the intrinsic state and the result is shown in Table 3.7.

Table 3.7

φ_a	$\varphi_1 = \Phi_0$ $\|\alpha\beta\gamma\rangle$	$\varphi_2 = (12)\Phi_0$ $\|\beta\alpha\gamma\rangle$	$\varphi_3 = (13)\Phi_0$ $\|\gamma\beta\alpha\rangle$	$\varphi_4 = (23)\Phi_0$ $\|\alpha\gamma\beta\rangle$	$\varphi_5 = (123)\Phi_0$ $\|\gamma\alpha\beta\rangle$	$\varphi_6 = (132)\Phi_0$ $\|\beta\gamma\alpha\rangle$
$(\overline{12})\varphi_a$	$(12)\Phi_0$ $\|\beta\alpha\gamma\rangle$	Φ_0 $\|\alpha\beta\gamma\rangle$	$(13)(12)\Phi_0$ $\|\gamma\alpha\beta\rangle$	$(23)(12)\Phi_0$ $\|\beta\gamma\alpha\rangle$	$(123)(12)\Phi_0$ $\|\gamma\beta\alpha\rangle$	$(132)(12)\Phi_0$ $\|\alpha\gamma\beta\rangle$

One sees that acting on the product state, the intrinisc permutation $(\overline{12})$ is equivalent to the state permutation $\wp_{12} = (\alpha\beta)$. This does not occur fortuitously. We are going to show that the state-permutation group $\mathcal{S}_n$ is a realization of the intrinsic permutation group $\overline{S}_n$ in the space of product states.

3.7.7. Intrinsic permutation group and state permutation group

Theorem 3.19: With the normal order state $|\omega_0\rangle = |i_1 i_2 \ldots i_n\rangle$ as the intrinsic state, the intrinsic permutation is the inverse of the state permutation, i.e.,

$$\overline{p} = \wp^{-1} . \tag{3-155}$$

Proof: From the definition (3-151) one has

$$\overline{p}|\omega_0\rangle = p|\omega_0\rangle . \tag{3-156}$$

Using the fact that S_n commutes with both $\overline{S}_n$ and $\mathcal{S}_n$, and using (3-110), one obtains

$$\overline{p}\varphi_a = \overline{p}p_a|\omega_0\rangle = p_a p|\omega_0\rangle = p_a\wp^{-1}|\omega_0\rangle = \wp^{-1}p_a|\omega_0\rangle = \wp^{-1}\varphi_a . \tag{3-157}$$

Since φ_a is an arbitrary function, (3-155) is true.

Henceforth, for the treatment of product states, we always use the state-permutation group in place of the more abstract intrinsic permutation group.

The advantage of choosing the normal-order state as the intrinsic state is that the intrinisc quantum numbers, which are relative to the intrinsic state, of the permutation group will be directly related to the quantum numbers of the unitary groups (see Sec. 7.5).

Since the classes of the permutation group are all ambivalent from (3-155) we know that the class operators of $\overline{S}_n$ and $\mathcal{S}_n$, as well as their respective subgroups, $\overline{S}_i$ and $\mathcal{S}_i$, are equal. Therefore, the CSCO of $\overline{S}_i$ and $\mathcal{S}_i$ are equal,

$$\overline{C}(i) = \mathcal{C}(i) , \quad i = n, n-1, \ldots, 2 . \tag{3-158}$$

3.8. CSCO-II and CSCO-III of G

In this section the discussion on the group S_3 is to be extended to any finite group G. Let us seek the eigenvectors of the CSCO of G in the group space L_g,

$$P^{(\nu)} = \sum_{a=1}^{g} u_{\nu a} R_a , \tag{3-159}$$

$$C P^{(\nu)} = \nu P^{(\nu)} . \tag{3-160}$$

Written in matrix form (3-160) becomes

$$\sum_{b=1}^{g} (C_{ab} - \nu\delta_{ab}) u_{\nu b} = 0 . \tag{3-161}$$

The matrix $\|C_{ab}\|$ is the representative of the operator C in the regular rep:

$$C R_b = \sum_{a=1}^{a} C_{ab} R_a, \quad C_{ab} = \langle R_a|C|R_b\rangle \equiv (R_a, C R_b) . \tag{3-162}$$

C is a CSCO for the N-dimensional class space but is no longer a CSCO for the g-dimensional group space L_g. According to Theorem 3.11, the eigenvalues of C in L_g remain the N distinct values $\nu_1, \nu_2, \ldots, \nu_N$, determined in the class space, but each eigenvalue ν has the degeneracy $M_\nu \geq 1$. From Eqs. (3-160) and (2-37) one has

$$\det|C_{ab} - \nu\delta_{ab}|_1^g = \prod_{i=1}^{n}(\nu - \nu_i)^{M_{\nu_i}} = 0 , \tag{3-163}$$

$$\sum_{i=1}^{N} M_{\nu_i} = g . \tag{3-164a}$$

All the eigenvectors $P^{(\nu_i)}$ of C with the same eigenvalue ν_i form the eigenspace L_{ν_i} of C, whose dimension is the degeneracy M_{ν_i}. The group space is decomposed into a direct sum of the N eigenspaces,

$$L_g = \sum_{\nu=1}^{N} \oplus L_\nu . \tag{3-164b}$$

To lift the degeneracy of ν entirely, we have to add extra operators to C, so that it becomes a CSCO of the g-dimensional group space L_g. These extra operators must commute with one another and with C. By virtue of Theorem 3.4, the class operators that are not included in C are useless for lifting the degeneracy. The possible candidates for these extra operators are the CSCO of subgroups of G. Suppose that G has a subgroup chain

$$G \supset G(s) , \quad G(s) = G(s_1) \supset G(s_2) \supset \ldots \tag{3-164c}$$

We use $C(s_i)$ to designate the CSCO of the subgroup $G(s_i)$ and $C(s)$ the set of operators $(C(s_1), C(s_2), \ldots)$. Obviously, the operators $C(s_i)$ commute with C and with one another.

By the anti-isomorphism between G and $\overline{G}$, we have the intrinsic subgroup chain

$$\overline{G} \supset \overline{G}(s) , \quad \overline{G}(s) = \overline{G}(s_1) \supset \overline{G}(s_2) \supset \ldots \tag{3-164d}$$

and the CSCO $\overline{C}(s) = (\overline{C}(s_1), \overline{C}(s_2), \ldots)$. $\overline{C}(s)$ commutes with both C and $C(s)$, i.e.,

$$[C(s) , \overline{C}(s)] = 0 . \tag{3-165}$$

$\overline{C}(s)$ are also candidates for these extra operators to be added to the CSCO of G.

Definition 3.4: If, starting from the group G, we can find a group chain (3-164c) and the corresponding operator set K,

$$K = (C, C(s), \overline{C}(s)) , \tag{3-166a}$$

$$C(s) = (C(s_1), C(s_2), \ldots) , \quad \overline{C}(s) = (\overline{C}(s_1), \overline{C}(s_2), \ldots) , \tag{3-166b}$$

such that K is a CSCO for the g-dimensional group space L_g, then K is called a CSCO-III of G, while

$$M = (C, C(s)) , \tag{3-166c}$$

is called a CSCO-II of G.

According to the definition, in group space K has g distinct sets of eigenvalues

$$\begin{aligned} &\lambda = (\nu, m, k) , \\ &m = (\lambda(s_1), \lambda(s_2), \ldots), \quad k = (\overline{\lambda}(s_1) , \overline{\lambda}(s_2), \ldots) . \end{aligned} \tag{3-166d}$$

Theorem 3.20: If $G \supset G(s)$ is a canonical subgroup chain, then $M = (C, C(s))$ is a CSCO-II of G, and $K = (C, C(s), \overline{C}(s))$ is a CSCO-III of G.

The proof of this theorem is left as an exercise to the reader.

In many cases, the operator set K in (3-166a) is over determined in the sense that the following K may already be a CSCO in L_g:

$$K = (C', C'(s), \overline{C}'(s)) , \quad C'(s) = (C'(s_1), C'(s_2)\ldots), \quad \overline{C}'(s) = (\overline{C}'(s_1), \overline{C}'(s_2)\ldots) , \tag{3-167a}$$

where C' and $C'(s_i)$ involve only some of the class operators contained in the CSCO of G and $G(s_i)$, respectively. For examples, see Exercises 3.13 and 3.14 and Sec. 4.5.

Furthermore, in analogy with (3-35), each operator set $C'(s_i)$ can be assumed to consist of only a single operator, and we can use a single operator A and $\overline{A}$,

$$A = \sum_i d_i C'(s_i) , \quad \overline{A} = \sum_i d_i \overline{C}'(s_i) \tag{3-167b}$$

to replace $C'(s)$ and $\overline{C}'(s)$, where the coefficients d_i are properly chosen so that

$$K = (C', A, \overline{A}) \tag{3-167c}$$

is a CSCO in L_g. The above consideration leads to a more general definition for the CSCO-II and CSCO-III of G.

Definition 3.4′: A set of commuting operators $K = (C', A, \overline{A})$, with C' commuting with G, and A an operator set or an operator in L_g, is called a CSCO-III of G if K is a CSCO of L_g, and

$$M = (C', A) , \tag{3-167d}$$

is called a CSCO-II of G.

Therefore, each of the operator sets K of (3-166a), (3-167a) and (3-167c) is a possible form of the CSCO-III of G. In the following we still use $K = (C, C(s), \overline{C}(s))$ to denote a CSCO-III of G, keeping in mind that actually K could take other forms as well.

Although for an arbitary group G, we cannot yet prove the existence of a canonical subgroup chain, for the finite groups commonly used in physics, it is easy to find their canonical subgroup chains.

In practical applications, the subgroup chain $G \supset G(s)$ is often specified by the physical problem being studied (see Sec. 3.19). Therefore, it may happen that the physically required subgroup chain $G \supset G(s)$ is not a canonical one. In such a case, $(C, C(s), \overline{C}(s))$ is not a CSCO in L_g, and thus is not a CSCO-III of G.

According to Dirac (1958, Sec. 14), any set of commuting operators can be made into a complete set of commuting operators by adding certain operators to it. Let us first add one operator say ξ, which can always be written as a linear combination of the g group operators, $\xi = \sum_a \xi_a R_a$, ξ_a being complex numbers, since we are working in the group space L_g. Corresponding to the operator ξ, there is the intrinsic operator $\overline{\xi} = \sum_a \xi_a \overline{R}_a$. Thus the additional operators always occur in pairs. If $(C, C(s), \xi, \overline{C}(s), \overline{\xi})$ is not yet a CSCO of L_g, we can add another pair of operators η and $\overline{\eta}$, etc., until

$$(C, C''(s), \overline{C}''(s)) \tag{3-168a}$$

is a CSCO of L_g, where

$$C''(s) = (C(s), \xi, \eta, \ldots) , \quad \overline{C}''(s) = (\overline{C}(s), \overline{\xi}, \overline{\eta}, \ldots) . \tag{3-168b}$$

Hence we can always find a CSCO-III of G regardless of whether $G \supset G(s)$ is canonical or not. To be specific, in the following we shall assume that $G \supset G(s)$ is a canonical subgroup chain unless otherwise stated. For the noncanonical case, all we need to do is to reinterpret the meaning of $C(s)$ and $\overline{C}(s)$ in accordance with (3-168a).

We can also choose a single operator

$$K = C + \sum_i [\alpha_i C(s_i) + \beta_i \overline{C}(s_i)] , \tag{3-168c}$$

as a CSCO-III of G, where the coefficients α_i and β_i are properly chosen so that K has g distinct eigenvalues.

3.9. Full Reduction of the Regular Representation

3.9.1. Eigenvectors of the CSCO-III *of G*

Let $P_m^{(\nu)k}$ be the eigenvectors of the CSCO-III of G in the group space L_g,

$$\begin{pmatrix} C \\ C(s) \\ \overline{C}(s) \end{pmatrix} P_m^{(\nu)k} = \begin{pmatrix} \nu \\ m \\ k \end{pmatrix} P_m^{(\nu)k}, \quad P_m^{(\nu)k} = \sum_{a=1}^{g} u_{\nu mk,a} R_a . \tag{3-169a}$$

Using $\overline{R}S = SR$, we can rewrite the third equation in (3-169a) as

$$\overline{C}(s) P_m^{(\nu)k} = P_m^{(\nu)k} C(s) = k P_m^{(\nu)k} , \tag{3-169b}$$

where $m(k)$ denotes the set of eigenvalues (3-166d) if $C(s)[\overline{C}(s)]$ is interpreted as in (3-166b), or simply the eigenvalue of $A(\overline{A})$, if $C(s)[\overline{C}(s)]$ is interpreted as $A(\overline{A})$. To be specific, in the following we assume that

$$C(s) = A , \quad \overline{C}(s) = \overline{A} . \tag{3-169c}$$

We first introduce Theorem 3.21.

Theorem 3.21: The representatives of the operators A and $\overline{A}$ in the eigenspace L_ν of C are similar matrices.

Proof: Let the M_ν basis vectors in L_ν be

$$\varphi_i = \sum_a v_{ia} R_a , \quad i = 1, 2, \ldots , M_\nu , \tag{3-170a}$$

$$C\varphi_i = \nu \varphi_i . \tag{3-170b}$$

Using Eq. (3-137f), one obtains

$$\langle \varphi_i | \overline{A} | \varphi_j \rangle = \sum_{ab} \langle v_{ia} R_a | \overline{A} | v_{jb} R_b \rangle = \sum_{ab} v_{ia}^* v_{jb} \langle b^{-1} | A | a^{-1} \rangle = \langle \varphi_j' | A | \varphi_i' \rangle , \tag{3-170c}$$

where

$$\varphi_i' = \sum_a v_{ia}^* R_a^{-1} .$$

Taking the Hermitian conjugate of (3-170b), and using the fact that C is self-adjoint and R_a is unitary, one has

$$C\varphi_i' = \nu \varphi_i' . \tag{3-171}$$

This shows that $\varphi_i' \in L_\nu$, thus φ_i' is necessarily a linear combination of the M_ν basis vectors φ_i. Therefore (3-170c) tells us that the representative matrices of A and $\overline{A}$ in L_ν are similar, i.e., $D^{(\nu)}(A) \sim D^{(\nu)}(\overline{A})$.

Suppose that $D^{(\nu)}(A)$ has $|\nu|$ distinct eigenvalues $m_1, \ldots, m_{|\nu|}$, while $D^{(\nu)}(\overline{A})$ has $|\overline{\nu}|$ distinct eigenvalues $k_1, \ldots, k_{|\overline{\nu}|}$. Since $(C, \overline{A})$ commutes the group G and thus its eigenspace $L_{(\nu)k} = \{P_{m_i}^{(\nu)k} : i = 1, 2, \ldots, |\nu|\}$ is necessarily a rep space of G, the degeneracy of m_i is necessarily independent of i, as mentioned in Sec. 3.4.

Furthermore, according to the hypothesis that $(C, A, \overline{A})$ is a CSCO of L_g, $(A, \overline{A})$ is necessarily a CSCO in each eigenspace L_ν, $\nu = 1, 2, \ldots, N$. Therefore the degeneracy of m_i in L_ν is totally eliminated by the eigenvalues of $\overline{A}$. In other words the degeneracy of m_i must be equal to the number $|\overline{\nu}|$ of the distinct eigenvalues of $\overline{A}$. Therefore, the characteristic equation for $D^{(\nu)}(A)$ has the form

$$\det|D^{(\nu)}(A) - m \bullet I| = \prod_{i=1}^{|\nu|} (m - m_i)^{|\overline{\nu}|} . \tag{3-172a}$$

Similarly, interchanging G and $\overline{G}$, we have

$$\det|D^{(\nu)}(\overline{A}) - k \bullet I| = \prod_{i=1}^{|\overline{\nu}|} (k - k_i)^{|\nu|} . \tag{3-172b}$$

Thanks to Theorem 3.21, $D^{(\nu)}(A)$ and $D^{(\nu)}(\overline{A})$ have identical characteristic equations; therefore we have

$$|\nu| = |\overline{\nu}| = h_\nu , \quad m_i = k_i, \quad i = 1, 2, \ldots, h_\nu . \tag{3-172c}$$

In the following, $m(k)$ is also used as an index to enumerating the eigenvalues, and thus we can write $m, k = 1, 2, \ldots, h_\nu$.

The conservation of dimension in (3-172a) or (3-172b) gives

$$h_\nu^2 = M_\nu . \tag{3-172d}$$

This leads to Theorem 3.22.

Theorem 3.22: The dimension of the eigenspace L_ν of C is necessarily a square of an integer.

Stated differently, the number of distinct sets of eigenvalues (m, k) of $(A, \overline{A})$ in L_ν has to be equal to the dimension of L_ν.

From Eqs. (3-164a) and (3-172d) we get

$$g = \sum_{i=1}^{N} h_{\nu_i}^2 . \tag{3-173}$$

Now the eigenspace L_ν of C is further decomposed into h_ν rep spaces of G,

$$L_\nu = \sum_{k=1}^{h_\nu} \oplus L_{(\nu)k} , \quad L_{(\nu)k} = \{P_m^{(\nu)k} : m = 1, 2, \ldots, h_\nu\} . \tag{3-174}$$

Analogously, since $(C, C(s))$ commutes with $\overline{G}$, the eigenspace of $(C, C(s))$,

$$\overline{L}_{(\nu)m} = \{P_m^{(\nu)k} : k = 1, 2, \ldots, h_\nu\} , \tag{3-174'}$$

is a rep space of the intrinsic group $\overline{G}$. Hence, the space L_ν can also be decomposed into h_ν rep spaces $\overline{L}_{(\nu)m}$ of $\overline{G}$,

$$L_\nu = \sum_{m=1}^{h_\nu} \oplus \overline{L}_{(\nu)m} . \tag{3-175}$$

The eigenequation (3-169a) can be written in matrix form,

$$\sum_{b=1}^{g}\left[\left\langle a\left|\begin{matrix}C\\C(s)\\\overline{C}(s)\end{matrix}\right|b\right\rangle-\begin{pmatrix}\nu\\m\\k\end{pmatrix}\delta_{ab}\right]u_{\nu mk,b}=0\ , \tag{3-176}$$

where $\langle a|C|b\rangle = (R_a, CR_b) = C_{ab}$, and

$$C(s)R_b = \sum_a \langle a|C(s)|b\rangle R_a\ , \tag{3-177}$$

$$\overline{C}(s)R_b = R_bC(s) = \sum_a \langle a|\overline{C}(s)|b\rangle R_a\ . \tag{3-178}$$

The eigenvectors of the CSCO-III form an orthonormal and complete set in the group space L_g,

$$\langle P_m^{(\nu)k}|P_{m'}^{(\nu')k'}\rangle = \delta_{\nu\nu'}\delta_{mm'}\delta_{kk'}\ , \tag{3-179a}$$

$$\sum_{\nu mk} |P_m^{(\nu)k}\rangle\langle P_m^{(\nu)k}| = 1\ . \tag{3-179b}$$

Using (3-169a), (3-179) becomes

$$\sum_a u^*_{\nu mk,a}u_{\nu'm'k',a} = \delta_{\nu\nu'}\delta_{mm'}\delta_{kk'}\ , \tag{3-180a}$$

$$\sum_{\nu mk} u^*_{\nu mk,a}u_{\nu mk,b} = \delta_{ab}\ . \tag{3-180b}$$

With the help of the last result, the inverse expansion of (3-169a) is

$$R_a = \sum_{\mu ji} u^*_{\mu ji,a}P_j^{(\mu)i}\ . \tag{3-181}$$

3.9.2. The representations $D^{(\nu)k}(G)$ and $D^{(\nu)m}(\overline{G})$

Since C commutes with the group elements and in turn with $P_m^{(\nu)k}$,

$$CP_j^{(\mu)i}P_m^{(\nu)k} = \mu P_j^{(\mu)i}P_m^{(\nu)k} = \nu P_j^{(\mu)i}P_m^{(\nu)k}\ . \tag{3-182}$$

Therefore

$$(\nu-\mu)P_j^{(\mu)i}P_m^{(\nu)k} = 0\ . \tag{3-183}$$

From (3-169), it follows that

$$C(s)(P_j^{(\mu)i}P_m^{(\nu)k}) = j(P_j^{(\mu)i}P_m^{(\nu)k})\ , \tag{3-184}$$

$$\overline{C}(s)(P_j^{(\mu)i}P_m^{(\nu)k}) = P_j^{(\mu)i}\overline{C}(s)P_m^{(\nu)k} = k(P_j^{(\mu)i}P_m^{(\nu)k})\ . \tag{3-185}$$

Furthermore

$$P_j^{(\mu)i}C(s)P_m^{(\nu)k} = P_j^{(\mu)i}(C(s)P_m^{(\nu)k}) = mP_j^{(\mu)i}P_m^{(\nu)k} = (P_j^{(\mu)i}C(s))P_m^{(\nu)k} = iP_j^{(\mu)i}P_m^{(\nu)k}\ .$$

Thus

$$(m-i)P_j^{(\mu)i}P_m^{(\nu)k} = 0\ . \tag{3-186}$$

Due to the nondegeneracy of the eigenvalues of the CSCO-III, the above results imply that

$$P_j^{(\mu)i} P_m^{(\nu)k} = \delta_{\nu\mu}\delta_{im}(\xi^{\nu}_{mjk})^{-1} P_j^{(\nu)k} , \tag{3-187}$$

where ξ^{ν}_{mjk} is a constant to be decided upon.

Applying R_a of (3-181) to $P_m^{(\nu)k}$ and using Eq. (3-187), one gets

$$R_a P_m^{(\nu)k} = \sum_j D_{jm}^{(\nu)k}(R_a) P_j^{(\nu)k} , \tag{3-188a}$$

$$D_{jm}^{(\nu)k}(R_a) = (\xi^{\nu}_{mjk})^{-1} u^*_{\nu jm,a} . \tag{3-188b}$$

Thus the eigenvectors $P_m^{(\nu)k}(m = 1, 2, \ldots , h_\nu)$ form the basis of the k-th rep (ν) of the group G with the rep matrices $D^{(\nu)k}(R_a), k = 1, 2, \ldots , h_\nu$. Similarly, for the group $\overline{G}$ we have

$$\overline{R}_a P_m^{(\nu)k} = \sum_i D_{ik}^{(\nu)m}(\overline{R}_a) P_m^{(\nu)i} , \tag{3-189a}$$

$$D_{ik}^{(\nu)m}(\overline{R}_a) = (\xi^{\nu}_{kmi})^{-1} u^*_{\nu ki,a} . \tag{3-189b}$$

These show that the eigenvectors $P_m^{(\nu)k}(k = 1, 2, \ldots , h_\nu)$ form the basis of the mth rep (ν) of the intrinsic group $\overline{G}$ with the rep matrices $D^{(\nu)m}(\overline{R}_a)$, $m = 1, 2, \ldots h_\nu$.

Ex. 3.8. Check that in the eigenspace $L_{(0)}$ of the CSCO of S_3, $\mathcal{D}^{(0)}$ (12) and $\mathcal{D}^{(0)}(\overline{12})$ are similar matrices, where $L_{(0)} = \{\mathbf{1-6, 5-6, 2-3, 3-4}\}$, the boldface integers being shorthand notation for the group elements.

3.9.3. The standard phase choice for $P_m^{(\nu)k}$

The eigenvector $P_m^{(\nu)k}$ in (3-169a) can be determined only up to a phase factor. Until now the phase has been assumed to be chosen arbitrarily. From (3-187) one sees that the constant ξ^{ν}_{mjk} depends on the phase choice of $P_m^{(\nu)k}$. Let us make the ansatz

$$P_j^{(\nu)m} P_m^{(\nu)k} = (\xi^{\nu})^{-1} P_j^{(\nu)k} , \tag{3-190}$$

where ξ^ν depends only on ν; later we shall show how this can be achieved by a suitable choice of the phases.

With (3-190), (3-188b) now becomes

$$D_{jm}^{(\nu)}(R_a) = (\xi^{\nu})^{-1} u^*_{\nu jm,a} . \tag{3-191}$$

From (3-180a) and (3-191) we have

$$\sum_{a=1}^{g} |D_{jm}^{(\nu)}(R_a)|^2 = |\xi^{\nu}|^{-2} . \tag{3-192}$$

Summing over the index m from 1 to h_ν, and using the unitarity of the rep $D^{(\nu)}$, we obtain

$$|\xi^{\nu}|^2 = h_\nu / g . \tag{3-193}$$

Choosing ξ^ν to be real and positive, we have

$$\xi^{\nu} = (h_\nu/g)^{\frac{1}{2}} . \tag{3-194}$$

Hence Eq. (3-187) becomes

$$P_j^{(\nu)i} P_m^{(\nu)k} = \sqrt{\frac{g}{h_\nu}} \delta_{\nu\mu} \delta_{im} P_j^{(\nu)k} \ . \tag{3-195}$$

We see that the real, positive choice for ξ^ν corresponds to the requirement that the coefficient $u_{\nu mm,e}$ in front of the identity e in $P_m^{(\nu)m}$ be always real and positive,

$$u_{\nu mm,e} > 0, \quad m = 1, 2, \ldots, h_\nu \ . \tag{3-196}$$

Equations (3-188), (3-189) and (3-169a) reduce to

$$RP_m^{(\nu)k} = \sum_{m'} D_{m'm}^{(\nu)}(R) P_{m'}^{(\nu)k} \quad k = m_1, m_2, \ldots, m_{h_\nu}, \tag{3-197a}$$

$$\overline{R} P_m^{(\nu)k} = P_m^{(\nu)k} R = \sum_{k'} D_{k'k}^{(\nu)}(\bar{R}) P_m^{(\nu)k'}, \quad m = m_1, m_2, \ldots, m_{h_\nu} \ , \tag{3-197b}$$

$$D_{mk}^{(\nu)}(R_a) = D_{km}^{(\nu)}(\overline{R}_a) = \sqrt{\frac{g}{h_\nu}} u_{\nu mk,a}^* \ , \tag{3-198}$$

$$P_m^{(\nu)k} = \sqrt{\frac{h_\nu}{g}} \sum_a D_{mk}^{(\nu)*}(R_a) R_a \ . \tag{3-199}$$

While Eq. (3-180) becomes

$$\frac{h_\nu}{g} \sum_{a=1}^{g} D_{mk}^{(\nu)*}(R_a) D_{m'k'}^{(\nu')}(R_a) = \delta_{\nu\nu'} \delta_{mm'} \delta_{kk'} \ , \tag{3-200a}$$

$$\sum_{\nu=1}^{N} \sum_{m,k=1}^{h_\nu} \frac{h_\nu}{g} D_{mk}^{(\nu)*}(R_a) D_{mk}^{(\nu)}(R_b) = \delta_{ab} \tag{3-200b}$$

The phase choice (3-190) and (3-194) which leads to the results (3-197)-(3-200) is referred to as the *standard phase choice*. From (3-196) and (3-198) we know that the steps for reaching the standard phase choice are as follows.

1. The coefficient in front of the identity element e in the eigenvector $P_m^{(\nu)m}$ should be real and positive.

2. Among the h_ν reps of G, the phases of the basis vectors of one rep, say the first one, $P_m^{(\nu)k=1}, m = 2, 3, \ldots, h_\nu$, can be chosen arbitrarily.

3. The phase of the eigenvector $P_m^{(\nu)k}$ for $k \neq 1$ can be fixed by requiring that the coefficient $u_{\nu mk,a}$ in front of a certain element R_a be equal to

$$(h_\nu/g)^{\frac{1}{2}} \langle P_m^{(\nu)1} | R_a | P_k^{(\nu)1} \rangle^* \ , \tag{3-201}$$

where the element R_a can be chosen arbitrarily so long as $u_{\nu mk,a} \neq 0$. The matrix element (3-201) can be calculated with the help of the group table.

Finally we need to show that the system of eigenvectors $P_m^{(\nu)k}$ of (3-199) satisfy the ansatz (3-190). Using (3-199), (3-197a) and (3-200a), we will verify that the ansatz (3-190) is indeed satisfied.

From (3-197) and (3-198) one sees that, under the standard phase choice, the h_ν reps, $D^{(\nu)k}(G), k = 1, 2, \ldots, h_\nu$, become identical; the h_ν reps $D^{(\nu)m}(\overline{G}), m = 1, 2, \ldots, h_\nu$, also become identical, and $D^{(\nu)}(\overline{G})$ is identical to $\widetilde{D}^{(\nu)}(G)$. Therefore, we have Theorem 3.23.

Theorem 3.23: The reps $D^{(\nu)k}(G)$ $[D^{(\nu)m}(\overline{G})]$ with the same eigenvalue ν of the CSCO of G are equivalent and can be made identical to one another by using the standard phase choice.

3.9.4. The irreducibility of $D^{(\nu)}(G)$

Theorem 3.24: The N inequivalent reps $D^{(\nu)}(G)$ resulting from the decomposition of the group space L_g of G are irreducible.

Proof: Suppose that an $h_\nu \times h_\nu$ matrix A satisfies

$$AD^{(\nu)}(R_a) = D^{(\nu)}(R_a)A \ , \quad a = 1, 2, \ldots , g \ , \tag{3-202}$$

or

$$\sum_{m'} A_{mm'} D^{(\nu)}_{m'k}(R_a) = \sum_{m'} D^{(\nu)}_{mm'}(R_a) A_{m'k} \ . \tag{3-203}$$

Multiplying both sides of (3-203) by $(h_\nu/g)D^{(\nu)*}_{ik}(R_a)$, and summing over a, from (3-200a) we obtain

$$A_{mi} = \delta_{mi} A_{kk} \ . \tag{3-204}$$

This shows that the only matrix which commutes with all the matrices of the rep $D^{(\nu)}$ is a multiple of the unit matrix. According to Schur's lemma 1′, $D^{(\nu)}(G)$ is irreducible. Similarly, the reps $D^{(\nu)}(\overline{G})$ for the intrinsic group are also irreducible.

Now we can give the CSCO-II of G an alternative definition.

Definition 3.5: If all the eigenspaces of $(\overline{C}, \overline{C}(s))$ are irreducible spaces of G, then $(\overline{C}, \overline{C}(s))$ is called a CSCO-II of the intrinsic group $\overline{G}$, while the corresponding operator $(C, C(s))$, the CSCO-II of G.

Therefore, the question raised at the beginning of Chapter 3 is answered. The operator set sought after, whose eigenspaces are irreducible spaces of G, is the CSCO-II, $(\overline{C}, \overline{C}(s))$, of the intrinsic group $\overline{G}$. Evidently, the CSCO-II of a group G is a CSCO for any irreducible space of G.

One of the advantages of the EFM for constructing the irreducible basis or irreps of a group G is that the subgroup chain $G(s)$ used to classify the irreducible basis or irrep can be chosen at will without the restriction that the subgroup has to be an invariant subgroup of G (Dirl 1977).

However, if, for some circumstances, we are only interested in obtaining the irreducible basis without the requirement that it be in a certain classification scheme, then we pay attention only to the operator set $C(s)$, without bothering about its related subgroup chain. In such cases, the eigenvalue of $C(s)$ is merely used to distinguish between the components of an irreducible basis, and $C(s)$ can be chosen differently for different irreps. The choice of $C(s)$ can be arbitrary so long as its eigenvalues can provide enough labels for the basis vectors of the same irrep. For example, for two-dimensional irreps of the point or space groups, the possible choice of $C(s)$ is a (plane) reflection operator, or a two-fold rotation C_2; for irreps with $h_\nu = 3(4)$, it is a three-fold (four-fold) rotation $C_3(C_4)$.

From (3-173), we have Theorem 3.25 (Burnside's Theorem).

Theorem 3.25: The regular rep of G contains N inequivalent irreps $D^{(\nu)}, N$ being the number of classes of G; the number of times each irrep occurs is equal to its dimension.

For example, for the permutation group $S_3, g = 6, N = 3$. There are three inequivalent irreps $\nu = 3$, 0, and -3, with dimensions 1,2, and 1, respectively, $6{=}1{+}2^2{+}1$. For a non-Abelian abstract group of order eight, the only possibility to fulfil (3-173) is $8{=}1{+}1{+}1{+}1{+}2^2$, since it must have the one-dimensional identity rep. Therefore, it has four one-dimensional and one two-dimensional inequivalent irreps.

The eigenvectors $P^{(\nu)k}_m$ are the $G \supset G(s)$ and $\overline{G} \supset \overline{G}(s)$ irreducible basis vectors. If m and k are interpreted as in (3-166d), then $P^{(\nu)k}_m$ belongs to the irreps $\nu, \lambda(s_1), \lambda(s_2), \ldots$ of $G \supset G(s_1) \supset G(s_2) \supset \ldots$, and belongs to the irreps $\nu, \overline{\lambda}(s_1), \overline{\lambda}(s_2), \ldots$ of $\overline{G} \supset \overline{G}(s_1) \supset \overline{G}(s_2) \supset \ldots$

Equation (3-197) shows that under the action of $R(\overline{R})$, $P_m^{(\nu)k}$ only changes its external (intrinsic) quantum number $m(k)$.

3.9.5. The EFM for $G \supset G(s)$ irreducible matrices

According to (3-198), once the eigenvectors of the CSCO-III of G have been computed, the irreducible matrices for the $G \supset G(s)$ basis, referred to as the $G \supset G(s)$ irreducible matrices, can be readily obtained. It should be stressed that (3-198) *holds only for the standard phase choice.*

Theorem 3.26: Regarded as functions on the group, the complex conjugate of the $G \supset G(s)$ irreducible matrix elements are eigenfunctions of the CSCO-III of G,

$$\begin{pmatrix} C \\ C(s) \\ \overline{C}(s) \end{pmatrix} D_{mk}^{(\nu)*}(R_a) = \begin{pmatrix} \nu \\ m \\ k \end{pmatrix} D_{mk}^{(\nu)*}(R_a) . \tag{3-205}$$

Proof: It suffices to prove that, regarded as functions on the group, $D_{mk}^{(\nu)}(R_a)^*$ is the $G \supset G(s)$ irreducible basis vector (ν, m) and the $\overline{G} \supset \overline{G}(s)$ irreducible basis vector (ν, k).

Using (2-87a) and (2-45a), one has

$$RD_{mk}^{(\nu)*}(R_a) = D_{mk}^{(\nu)*}(R^{-1}R_a) = \sum_{m'} D_{mm'}^{(\nu)*}(R^{-1}) D_{m'k}^{(\nu)*}(R_a) . \tag{3-206}$$

Thus

$$RD_{mk}^{(\nu)*}(R_a) = \sum_{m'} D_{m'm}^{(\nu)}(R) D_{m'k}^{(\nu)*}(R_a) . \tag{3-207}$$

Similarly, from (3-138) and (2-45a), we have

$$\overline{R}D_{mk}^{(\nu)*}(R_a) = D_{mk}^{(\nu)*}(R_a R^{-1}) = \sum_{k'} D_{kk'}^{(\nu)}(R) D_{mk'}^{(\nu)*}(R_a) . \tag{3-208}$$

Comparing the above with (3-197), we know that $D_{mk}^{(\nu)}(R_a)^*$ is indeed the $G \supset G(s)$ and $\overline{G} \supset \overline{G}(s)$ irreducible basis.

Therefore, in the space of functions on the group, the eigenfunctions of the CSCO-III of G, the $G \supset G(s)$ and $\overline{G} \supset \overline{G}(s)$ irreducible basis, and the complex conjugate of the $G \supset G(s)$ irreducible matrix elements, are all different names of the same thing.

To simplify our notation, let us assume that $G(s) = G_f \supset G_{f-1} \supset \ldots \supset G_1$, with the CSCO $C(s) = (C(f), C(f-1), \ldots, C(1))$ and the eigenvalue $m = (\lambda_f, \lambda_{f-1}, \ldots \lambda_1)$. From

$$[C(j), R_i] = 0, \quad j = f, f-1, \ldots, i, \quad R_i \in G_i , \tag{3-209}$$

we know that when acting upon $P_m^{(\nu)k}$, the elements R_i of the subgroup G_i cannot change the eigenvalues $\lambda_f, \lambda_{f-1}, \ldots, \lambda_i$. Therefore

$$D_{m'm}^{(\nu)}(R_i) = \langle P_{m'}^{(\nu)k} | R_i | P_m^{(\nu)k} \rangle = \delta_{\lambda'_f \lambda_f} \cdots \delta_{\lambda'_i \lambda_i} D_{\overline{m}'\overline{m}}^{\lambda_i}(R_i) , \tag{3-210}$$

where $D_{\overline{m}'\overline{m}}^{\lambda_i}(R_i)$ is the matrix elements for the irrep (λ_i) of the subgroup G_i,

$$D_{\overline{m}'\overline{m}}^{(\lambda_i)}(R_i) = \langle P_{\overline{m}'}^{(\lambda_i)\overline{k}} | R_i | P_{\overline{m}}^{(\lambda_i)\overline{k}} \rangle . \tag{3-211}$$

Equation (3-210) shows that the $G \supset G(s)$ irreducible matrices have the nice property that they are block-diagonal for the elements belonging to any subgroup contained in the subgroup chain $G(s)$ of G.

3.9.6. Generalized irreducible matrices

Sometimes we need the matrix elements of a group operator R_a between two irreducible basis vectors adapted to different subgroup chains,

$$\mathcal{D}^{(\nu)}_{m\kappa}(R_a) = \langle \psi^{(\nu)}_m | R_a | \phi^{(\nu)}_\kappa \rangle , \tag{3-212}$$

where $\psi^{(\nu)}_m$ and $\varphi^{(\nu)}_\kappa$ are the $G \supset G(s)$ and $G \supset G(s)'$ irreducible bases, respectively. $\mathcal{D}^{(\nu)}_{m\kappa}(R_a)$ is called the generalized irreducible matrices. In analogy with Eq. (3-205), it can be shown that $\mathcal{D}^{(\nu)}_{m\kappa}(R_a)^*$ satisfies the following eigenequations

$$\begin{pmatrix} C \\ C(s) \\ \overline{C}(s)' \end{pmatrix} \mathcal{D}^{(\nu)}_{m\kappa}(R_a)^* = \begin{pmatrix} \nu \\ m \\ \kappa \end{pmatrix} \mathcal{D}^{(\nu)}_{m\kappa}(R_a)^* , \tag{3-213}$$

where $C(s)'$ is the CSCO of the subgroup chain $G(s)'$.

From (3-213) and a normalization condition similar to (3-200a) with D replaced by $\mathcal{D}$, we can evaluate the generalized matrix elements.

3.9.7. Reduction of the regular representation in configuration space

We can also carry out the reduction of the regular rep in configuration space. The procedure is exactly the same as in the group space. We only need to make the following substitutions $R_a \to \varphi_a, P^{(\nu)k}_m \to \psi^{(\nu)k}_m$. Correspondingly, we have the following equations:

$$\begin{pmatrix} C \\ C(s) \\ \overline{C}(s) \end{pmatrix} \psi^{(\nu)k}_m = \begin{pmatrix} \nu \\ m \\ k \end{pmatrix} \psi^{(\nu)k}_m , \tag{3-214}$$

$$\psi^{(\nu)k}_m = \sum_{a=1}^{g} u_{\nu mk,a}\varphi_a(X) = \sqrt{\frac{h_\nu}{g}} \sum_{a=1}^{g} D^{(\nu)*}_{mk}(R_a)\varphi_a(X) . \tag{3-215}$$

$$R\psi^{(\nu)k}_m = \sum_{m'} D^{(\nu)}_{m'm}(R)\psi^{(\nu)k}_{m'} , \tag{3-216a}$$

$$\overline{R}\psi^{(\nu)k}_m = \sum_{k'} D^{(\nu)}_{kk'}(R)\psi^{(\nu)k'}_m \tag{3-216b}$$

$$\langle \psi^{(\nu)k}_m | \psi^{(\nu')k'}_{m'} \rangle = \delta_{\nu\nu'}\delta_{mm'}\delta_{kk'} ,$$

$$\sum_{\nu mk} |\psi^{(\nu)k}_m\rangle\langle\psi^{(\nu)k}_m| = 1 . \tag{3-217}$$

The only case which merits special consideration is that when the g basis functions φ_a are non-orthogonal. We shall in fact meet such a case in Chapter 8.

To further elucidate the transformation property of $\psi^{(\nu)k}_m$ under the group G and $\overline{G}$, we construct the following array $\Psi^{(\nu)}$ out of the h^2_ν functions $\psi^{(\nu)k}_m$ belonging to the irrep (ν)

$$\Psi^{(\nu)} = \begin{pmatrix} \psi^{(\nu)1}_1 & \psi^{(\nu)2}_1 & \cdots & \psi^{(\nu)h_\nu}_1 \\ \psi^{(\nu)1}_2 & \psi^{(\nu)2}_2 & \cdots & \psi^{(\nu)h_\nu}_2 \\ \vdots & \vdots & & \vdots \\ \psi^{(\nu)1}_{h_\nu} & \psi^{(\nu)2}_{h_\nu} & \cdots & \psi^{(\nu)h_\nu}_{h_\nu} \end{pmatrix} . \tag{3-218}$$

Equation (3-216a) [(3-216b)] tells us that under the operation $R_a(\overline{R}_a)$, each column (row) in (3-218) transforms according to the same irrep $D^{(\nu)}(\widetilde{D}^{(\nu)})$.

3.9.8. Example: the group S_3

In Eq. (3-119), we already obtained the eigenvectors $\psi_m^{(\nu)k}$, or equivalently, $P_m^{(\nu)k}$, of the CSCO-III of the permutation group S_3, but we evaded the phase problem. We now follow the three steps of subsection 3.9.4 to determine the phases of the eigenvectors. First we extract the factor $\sqrt{h_\nu/g}$ out of the coefficients $u_{\nu mk,a}$. Then the emerging result of Eq. (3-119) is as shown in Table 3.9. The phases for $\psi_m^{(\nu)m}$, $(\nu, m) = (3,1), (-3,-1), (0,1)$ and $(0,-1)$ are fixed by step

Table 3.9. The standard basis $\psi_m^{(\nu)k}(P_m^{(\nu)k})$ of S_3 and $\overline{S}_3$ and the standard matrix elements.

$[D^{(\nu)}_{mk}(R_a)]$ / $\lambda_3, \lambda_2, \bar{\lambda}_2$; ν, m, k	Y_m	W_k (Φ_a / R_a)	Norm $\sqrt{\frac{h_\nu}{6}}$	$\|\alpha\beta\gamma\rangle$ e	$\|\beta\alpha\gamma\rangle$ (12)	$\|\gamma\beta\alpha\rangle$ (13)	$\|\alpha\gamma\beta\rangle$ (23)	$\|\gamma\alpha\beta\rangle$ (123)	$\|\beta\gamma\alpha\rangle$ (132)	Irreducible matrix $D^{(\nu)}_{ij}(R_a)$
3, 1, 1	[1 2 3]	[α β γ]	$\sqrt{\frac{1}{6}}$	1	1	1	1	1	1	$D^{(3)*}_{11}$
−3, −1, −1	[1 / 2 / 3]	[α / β / γ]	$\sqrt{\frac{1}{6}}$	1	−1	−1	−1	1	1	$D^{(-3)*}_{11}$
0, 1, 1	[1 2 / 3]	[α β / γ]	$\sqrt{\frac{1}{3}}$	1	1	$-\frac{1}{2}$	$-\frac{1}{2}$	$-\frac{1}{2}$	$-\frac{1}{2}$	$D^{(0)*}_{11}$
0, −1, 1	[1 3 / 2]	[α β / γ]	$\sqrt{\frac{1}{3}}$	0	0	$-\frac{\sqrt{3}}{2}$	$\frac{\sqrt{3}}{2}$	$-\frac{\sqrt{3}}{2}$	$\frac{\sqrt{3}}{2}$	$D^{(0)*}_{21}$
0, 1, −1	[1 2 / 3]	[α γ / β]	$\sqrt{\frac{1}{3}}$	0	0	$-\frac{\sqrt{3}}{2}$	$\frac{\sqrt{3}}{2}$	$\frac{\sqrt{3}}{2}$	$-\frac{\sqrt{3}}{2}$	$D^{(0)*}_{12}$
0, −1, −1	[1 3 / 2]	[α γ / β]	$\sqrt{\frac{1}{3}}$	1	−1	$\frac{1}{2}$	$\frac{1}{2}$	$-\frac{1}{2}$	$-\frac{1}{2}$	$D^{(0)*}_{22}$

1. The phase of $\psi_{-1}^{(0)1}$ can be chosen freely. Suppose it has been chosen as shown in the fourth row of Table 3.9. To determine the phase of $\psi_1^{(0)-1}$, i.e., the fifth row vector in Table 3.9, we need to calculate a nonvanishing matrix element $D^{(0)}_{1-1}(R_a)$ from the known irreducible basis $\psi_m^{(0)k=1}$, where R_a can be chosen arbitrarily; for example, we might choose $R_a = R_3 = (13)$. From the third and fourth rows of Table 3.9 and using Table 1.2 we can calculate

$$\begin{aligned} D^{(0)}_{1-1}(R_3)^* &= \langle \psi_1^{(0)1} | (13) | \psi_{-1}^{(0)1} \rangle^* = \sqrt{\frac{1}{48}} \langle 2(\mathbf{1}+\mathbf{2}) - \mathbf{3} - \mathbf{4} - \mathbf{5} - \mathbf{6} | \mathbf{3} | - \mathbf{3} + \mathbf{4} - \mathbf{5} + \mathbf{6} \rangle \\ &= \sqrt{\frac{1}{48}} \langle 2(\mathbf{1}+\mathbf{2}) - \mathbf{3} - \mathbf{4} - \mathbf{5} - \mathbf{6} | - \mathbf{1} + \mathbf{6} - \mathbf{2} + \mathbf{4} \rangle = -\sqrt{3}/2, \end{aligned} \tag{3-219}$$

where the boldfaced integers are shorthand notation for φ_a or R_a. The phase of $\varphi_1^{(0)-1}$ is now determined by requiring that its coefficient u_3 has the same sign as that of $D^{(0)}_{1-1}(R_3)^* = -\sqrt{3}/2$. Having adjusted the phase, thanks to (3-198), we can read off from Table 3.9 all the irreps of S_3. The result is identical with (3-91).

If the phase of $\psi_1^{(0)-1}$ had not been chosen appropriately, say if it was chosen opposite to that given in Table 3.9, and if we still used (3-198) to obtain matrix elements from $u_{\nu mk,a}$, we would have obtained

$$D'(13) = \begin{pmatrix} -\frac{1}{2} & \frac{\sqrt{3}}{2} \\ -\frac{\sqrt{3}}{2} & \frac{1}{2} \end{pmatrix} \tag{3-220}$$

and $D'(13)D'(13)$ would be equal to $-I/2$ (I being the unit matrix) destroying the homomorphism $(13)(13) \to e$. Thus the last result is false.

For the latter phase choice, we have to calculate the matrix elements from (3-188a). For example, we can obtain

$$D^{(0)1}(13) = \begin{pmatrix} -\frac{1}{2} & -\frac{\sqrt{3}}{2} \\ -\frac{\sqrt{3}}{2} & \frac{1}{2} \end{pmatrix}, \quad D^{(0)-1}(13) = \begin{pmatrix} -\frac{1}{2} & \frac{\sqrt{3}}{2} \\ \frac{\sqrt{3}}{2} & \frac{1}{2} \end{pmatrix}. \tag{3-221a}$$

Now the irreps $D^{(0)1}$ and $D^{(0)-1}$ are no longer identical, but are related by a similarity transformation,

$$D^{(0)1}(R) = \begin{pmatrix} 1 & 0 \\ 0 & -1 \end{pmatrix} D^{(0)-1}(R) \begin{pmatrix} 1 & 0 \\ 0 & -1 \end{pmatrix}. \tag{3-221b}$$

Ex. 3.9. From Table 3.9, write out all the irreducible matrices of S_3.

Ex. 3.10. Using S_3 as example, check Eq. (3-216).

Ex. 3.11. Find the $C_{3v} \supset C_3$ irreps by decomposing the regular rep of C_{3v} and compare the result with Ex. 3.3.

Ex. 3.12. Decompose the regular rep of the matrix group given in Ex. 1.3.

Ex. 3.13. Suppose that we have the matrix group

$$M = (I, -I, i\sigma_x, -i\sigma_x, i\sigma_y, -i\sigma_y, i\sigma_z, -i\sigma_z),$$

where I is the unity matrix and $\sigma_{x,y,z}$ are the Pauli matrices.

(a) Find two forms of the CSCO-III, one in the form of (3-166a) and the other in the form of (3-167a).

(b) Decompose the regular rep of the matrix group.

(c) Discuss the relation between the two-dimensional irreps of M and the Pauli matrices.

Ex. 3.14. The same problem as Ex. 3.13(a) and (b), but referred to the group C_{4v} (See Fig. 1.1-3). Find the irreps for the following two subgroup chains: (1) $C_{4v} \supset (e, \sigma_1)$; (2) $C_{4v} \supset C_4$ and compare the results with those of Ex. 3.4.

3.10. Generalized Projection Operator

3.10.1. Properties of $P_{mk}^{(\nu)}$

Define the quantity

$$P_{mk}^{(\nu)} = \sqrt{\frac{h_\nu}{g}} P_m^{(\nu)k} = \frac{h_\nu}{g} \sum_a D_{mk}^{(\nu)*}(R_a) R_a . \tag{3-222}$$

From (3-195) one has

$$P_{mi}^{(\nu)} P_{jk}^{(\mu)} = \delta_{\nu\mu} \delta_{ij} P_{mk}^{(\nu)} . \tag{3-223}$$

By taking the hermitian conjugate of (3-222) and using the unitarity of R_a and $D^{(\nu)}$, we obtain

$$\begin{aligned} (P_{mk}^{(\nu)})^\dagger &= \frac{h_\nu}{g} \sum_a D_{mk}^{(\nu)}(R_a) R_a^\dagger = \frac{h_\nu}{g} \sum_a D_{mk}^{(\nu)}(R_a^{-1}) R_a \\ &= \frac{h_\nu}{g} \sum_a D_{km}^{(\nu)*}(R_a) R_a = P_{km}^{(\nu)} . \end{aligned} \tag{3-224}$$

There are many names for the operator $P^{(\nu)}_{mk}$, such as the normal unit operator (Rutherford 1948), unit operator (Dirl 1977), shift operator (Bohr 1969), irreducible symmetry operator (Folland 1977), etc. We prefer to call it the *generalized projection operator*.

From (3-151) and (3-197) it is seen that, acting upon the intrinsic state $\Phi_0(X)$, $P^{(\nu)k}_m$ yields the $G \supset G(s)$ and $\overline{G} \supset \overline{G}(s)$ irreducible basis,

$$\psi^{(\nu)k}_m(X) = P^{(\nu)k}_m \Phi_0(X) . \tag{3-225}$$

It should be stressed that the meaning of the intrinsic quantum number k in configuration space depends on the choice of the intrinsic state. For example, applying $P^{(\nu)}_{mk}$ to another state $\varphi_a(X) = R_a\Phi_0(X)$, we obtain another state, namely,

$$\phi^{(\nu)k}_m(X) = P^{(\nu)k}_m \varphi_a(X) . \tag{3-226a}$$

which is still the $G \supset G(s)$ irreducible basis (ν, m), but is not the $\overline{G} \supset \overline{G}(s)$ irreducible basis (ν, k) with $\Phi_0(X)$ as the intrinsic state. From (3-225) and (3-197b), $\phi^{(\nu)k}_m$ can be expressed as a linear combination of the $\overline{G} \supset \overline{G}(s)$ irreducible basis $\psi^{(\nu)k}_m$ with $\Phi_0(X)$ as the intrinsic state,

$$\phi^{(\nu)k}_m(X) = \sum_{k'} D^{(\nu)}_{kk'}(R_a)\psi^{(\nu)k'}_m(X) . \tag{3-226b}$$

As the choice of the intrinsic state is arbitrary, we may choose $\varphi_a(X)$ as the intrinsic state $\Phi'_0(X)$. Thus $\phi^{(\nu)k}_m$ is the $\overline{G} \supset \overline{G}(s)$ irreducible basis (νk) with $\varphi_a(X)$ as the intrinsic state.

In Sec. 3.7 we introduced the I-, U-, and V-spin representations. From (3-226b) we can easily obtain the relations between the wave functions in different representations. Using

$$\Phi_0(U) = (132)\Phi_0(I), \quad \Phi_0(V) = (123)\Phi_0(I) ,$$

and letting $R_a = (132)$ and (123) in (3-226b), we have

$$\psi^{(\nu)k}_m(U) = \sum_{k'} D^{(\nu)}_{kk'}(132)\psi^{(\nu)k'}_m(I), \quad \psi^{(\nu)k}_m(V) = \sum_{k'} D^{(\nu)}_{kk'}(123)\psi^{(\nu)k'}_m(I) . \tag{3-226c}$$

More involved relations for the quark model are given in Chen ([18], 1979).

By using (3-222), (3-223) and (3-225), we have

$$P^{(\nu)}_{lm'}\psi^{(\nu)k}_m = \sqrt{\frac{g}{h_\nu}} P^{(\nu)}_{lm'}P^{(\nu)}_{mk}\Phi_0 = \delta_{mm'}\psi^{(\nu)k}_l , \tag{3-227}$$

i.e., when $m = m'$, the operator $P^{(\nu)}_{lm}$ shifts the m-th component to the l-the component of the (ν) irreducible basis.

From (3-200b) we obtain the inverse of (3-222):

$$R_a = \sum_{\nu=1}^{N}\sum_{m=1}^{h_\nu}\sum_{k=1}^{h_\nu} D^{(\nu)}_{mk}(R_a)P^{(\nu)}_{mk} . \tag{3-228}$$

Letting $R_a = e$, we deduce the decomposition of the identity element, i.e.,

$$e = \sum_{\nu=1}^{N}\sum_{m=1}^{h_\nu} P^{(\nu)}_{mm} . \tag{3-229}$$

From (3-224) and (3-223), we find

$$(P^{(\nu)}_{mm})^\dagger = P^{(\nu)}_{mm} , \tag{3-230}$$

$$P_{mm}^{(\nu)} P_{ll}^{(\nu)} = \delta_{ml} P_{mm}^{(\nu)} \ . \tag{3-231}$$

It shows that $P_{mm}^{(\nu)}$ is a self-adjoint idempotent, called the *primitive idempotent.* $P_{mm}^{(\nu)}$ is a projection operator onto the m-th component of the m-th ν irreducible basis, $\psi_m^{(\nu)m} = P_{mm}^{(\nu)}\Phi_0$. It should be emphasized that neither $\{P_{mm}^{(\nu)} : m = 1, 2, \dots, h_\nu\}$ nor $\{P_{mm}^{(\nu)}\Phi(X) : m = 1, 2, \dots, h_\nu\}$ with arbitrary $\Phi(X)$ forms an irreducible basis. Only when the irrep (ν) occurs once in the rep space $\{R_a\Phi(X) : a = 1, 2, \dots, g\}$ does $\{P_{mm}^{(\nu)}\Phi(X) : m = 1, 2, \dots, h_\nu\}$ form an irreducible basis.

From (3-197a), it follows that

$$R_a P_{kk}^{(\nu)} = \sum_{k'} D_{k'k}^{(\nu)}(R_a) P_{k'k}^{(\nu)} \ . \tag{3-232}$$

Thus the set $\{R_a P_{kk}^{(\nu)} : a = 1, 2, \dots, g\}$ belongs to the k-th ν-irreducible space $L_{(\nu)k}$, and $P_{kk}^{(\nu)}$ is the generator of the subspace $L_{(\nu)k}$. Equation (3-229) corresponds to the decomposition of L_g into a direct sum of the irreducible spaces of $G, L_g = \sum_{\nu k} \oplus L_{(\nu)k}$. In a parallel way, $\{P_{mm}^{(\nu)} R_a : a = 1, 2, \dots, g\}$ belongs to the irreducible space $\overline{L}_{(\nu)m}$ of the intrinsic group $\overline{G}$, and Eq. (3-229) also corresponds to the decomposition of L_g into a direct sum of the irreps of $\overline{G}, L_g = \Sigma_{m\nu} \oplus \overline{L}_{(\nu)m}$.

Let us introduce the definition

$$P^{(\nu)} = \sum_{k=1}^{h_\nu} P_{kk}^{(\nu)} \ . \tag{3-233}$$

From (3-231) it is readily seen that

$$P^{(\nu)} P^{(\mu)} = \delta_{\nu\mu} P^{(\nu)} \ . \tag{3-234}$$

The idempotent $P^{(\nu)}$ is the generator of the eigenspace L_ν of the CSCO of G, and (3-233) corresponds to the decomposition (3-174).

From Eqs. (2-90) and (3-222),

$$P^{(\nu)} = \frac{h_\nu}{g} \sum_{i=1}^{N} \chi_i^{(\nu)*} C_i \ . \tag{3-235a}$$

Obviously, $P^{(\nu)}$ is the eigenoperator of the CSCO of G with the eigenvalue ν and (3-234) is precisely (3-52). Therefore, the operator $P^{(\nu)}$ in (3-233) which is obtained from a contraction of $P_{kk}^{(\nu)}$ is identical with $P^{(\nu)} = Q^{(\nu)}/\eta_\nu$ in (3-51). Comparing the last result with Eqs. (3-51) and (3-27b), one has

$$\frac{h_\nu}{g} \chi_i^{(\nu)*} = \frac{1}{\eta_\nu} q_i^{(\nu)} \ . \tag{3-236}$$

Setting $m = k$ and $m' = k'$ in (3-200a) and summing over m and m', one gets

$$\sum_{i=1}^{N} \frac{g_i}{g} \chi_i^{(\nu)*} \chi_i^{(\nu')} = \delta_{\nu\nu'} \tag{3-237a}$$

Hence we see that if $(g_i/g)^{\frac{1}{2}} \chi_i^{(\nu)} (i, \nu = 1, 2, \dots, N)$ are regarded as matrix elements of a $N \times N$ matrix, then any two row vectors of the matrix are orthonormal. According to a general theorem of linear algebra, any two column vectors of such a matrix are necessarily also orthonormal, i.e.,

$$\sum_{\nu=1}^{N} \frac{g_i}{g} \chi_i^{(\nu)*} \chi_j^{(\nu)} = \delta_{ij} \ . \tag{3-237b}$$

Equations (3-237a,b) are the two orthogonality theorems of the character, and either one may be used as a criterion for the irreducibility of a rep.

Using (3-237b) we are able to find the inverse of Eq. (3-235a):

$$C_i = \sum_{\nu=1}^{N} \frac{g_i}{h_\nu} \chi_i^{(\nu)} P^{(\nu)} \,. \tag{3-235b}$$

Since the characters of the inequivalent irreps form an orthogonal vector system, two inequivalent irreps cannot have the same characters and irreps with equal characters are equivalent. From Sec. 2.8 we also know that equivalent reps have equal characters. Therefore we have

Theorem 3.27: The equality of the characters is the necessary and sufficient condition for two irreps to be equivalent.

In addition, (3-69) shows that the simple characters are uniquely decided by the eigenvalues of the CSCO of G. Consequently we have

Theorem 3.28: The equality of the eigenvalue ν of the CSCO of G is the necessary and sufficient condition for two irreps to be equivalent.

According to Theorem 3.10, in any rep space, the eigenvalues of the CSCO of G cannot go beyond the N values determined in the class space. Therefore we have

Theorem 3.29: Any irrep of a group G is equivalent to one of the irreps resulting from the decomposition of the regular rep of G.

Hence the regular rep of a group G contains all the inequivalent irreps of G and we get

Theorem 3.30: A group G with N classes has N and only N inequivalent irreps.

3.10.2. A recursive method for $P_{mk}^{(\nu)}$

When the simple characters of a group G as well as those of its subgroups contained in a canonical subgroup chain $G(s)$ are known, the following recursive method (Chen 1979) can be used for constructing the generalized projection operator $P_{mk}^{(\nu)}$ without solving the eigenequations (3-169a).

Suppose $G(s) = G(1) \supset G(2) \supset, \ldots$ with the CSCO $C(s) = \big(C(1), C(2) \ldots\big)$. Their irreps are labeled by $\nu_1, \nu_2, \ldots$ Then up to a constant factor we have

$$\begin{aligned} P_{mm}^{(\nu)} &= P^{(\nu)} P^{(\nu_1)} P^{(\nu_2)} \ldots , \\ m &= (\nu_1, \nu_2, \nu_3, \ldots) \,. \end{aligned} \tag{3-238}$$

Proof: From $[C, C(i)] = 0, [C(i), C(j)] = 0, i, j = 1, 2, \ldots$ and Eq. (3-235a), we know that $C(i)$ commutes with any factor $P^{(\nu_j)}$ in (3-238). Therefore

$$\begin{aligned} C P_{mm}^{(\nu)} &= \nu P_{mm}^{(\nu)} , \\ C(i) P_{mm}^{(\nu)} &= P^{(\nu)} P^{(\nu_1)} \ldots \big(C(i) P^{(\nu_i)}\big) \ldots = \nu_i P_m^{(\nu)m} , \\ \overline{C}(i) P_{mm}^{(\nu)} &= P_{mm}^{(\nu)} C(i) = \nu_i P_{mm}^{(\nu)} , \end{aligned} \tag{3-239}$$

which is a special case of (3-169a).

Let $P_{m_1 m_1}^{(\nu_1)}$ be the generalized projection operator of G_1 for the subgroup chain $G(1) \supset G(2) \supset G(3) \supset, \ldots$, then we have

$$P^{(m)} \equiv P_{m_1 m_1}^{(\nu_1)} = P^{(\nu_1)} P^{(\nu_2)} P^{(\nu_3)}, \ldots , \quad m = (\nu_1, \nu_2, \nu_3, \ldots), \quad m_1 = (\nu_2, \nu_3, \ldots) \,. \tag{3-240a}$$

From (3-238) and (3-240a) we have

$$P^{(\nu)}_{mm} = P^{(\nu)} P^{(m)} = P^{(\nu)} P^{(\nu_1)}_{m_1 m_1} \ . \tag{3-240b}$$

It shows that the idempotent $P^{(\nu)}_{mm}$ of G is expressible in terms of the projection operator of G and the idempotent $P^{(\nu)}_{m_1 m_1}$ of its subgroup $G(1)$. Thus a recursive procedure is established.

Similarly, up to a constant factor we have

$$\begin{aligned} P^{(\nu)}_{mk} &= P^{(\nu)} P^{(m)} R P^{(k)} \ , \\ P^{(k)} &= P^{(\bar{\nu}_1)} P^{(\bar{\nu}_2)} P^{(\bar{\nu}_3)} \dots \ , \ k = (\bar{\nu}_1, \bar{\nu}_2, \bar{\nu}_3, \dots) \ , \end{aligned} \tag{3-241}$$

where R is an appropriate element of G and can be chosen freely so long as $P^{(m)} R P^{(k)} \neq 0$.

Example: The group S_3. Ignoring the constant factors, from Eq. (3-44) we have the projection operators of S_3 and S_2,

$$\begin{aligned} P^{(\nu)} &= P^{(0)} = (2e - (123) - (132)) \ . \\ P^{(m)} &= P^{(\pm 1)} = (e \pm (12)). \end{aligned}$$

Therefore

$$\begin{aligned} P^{(0)}_{1,1} &= P^{(0)} P^{(1)} = [2(e + (12)) - (13) - (23) - (123) - (132)] \ . \\ P^{(0)}_{-1,1} &= P^{(0)} P^{(-1)} (23) P^{(1)} \\ &= [2e - (123) - (132)][e - (12)](23)[e + (12)] \\ &= 3[-(13) + (23) - (123) + (132)] \ , \end{aligned} \tag{3-242}$$

in agreement with Table 3.9.

Ex. 3.15. Using the result of Ex. 3.1, find the primitive idempotents $P^{(2)}_{mm}$ of S_4 for $m = (3,1), (0,1)$ and $(0,-1)$.

3.11. Eigenfunction Method for Characters

The EFM given below provides a simple method for computing characters of finite groups. From (3-236), (3-54), and (3-67), one has

$$|\eta_\nu|^2 = \frac{g}{h_\nu^2} \ . \tag{3-243}$$

Choosing η_ν to be real and positive, we find

$$\eta_\nu = \sqrt{g}/h_\nu \ . \tag{3-244}$$

Substituting this into (3-236), one obtains the relation between the simple character $\chi^{(\nu)}_i$ and the eigenvector $q^{(\nu)}_i$,

$$\chi^{(\nu)}_i = \sqrt{g} q^{(\nu)*}_i \ . \tag{3-245}$$

Thus the problem of finding characters is converted into that of finding the eigenvectors of the CSCO of G in class space.

If the simple characters $\chi^{(\nu)}_i$ are regarded as a function $\chi^{(\nu)}(C_i)$ on the classes, then from (3-205) we know that the complex conjugate of the simple character is an eigenfunction of the CSCO of G,

$$C\chi^{(\nu)*}(C_i) = \nu \chi^{(\nu)*}(C_i) \ . \tag{3-246}$$

Equation (3-246) remains true for compact Lie groups.

The EFM for characters is summarized as follows:

1. Find the representatives of the l class operators contained in the CSCO of $G, C = (C_{i_1}, C_{i_2}, \ldots, C_{i_l})$,

$$\| C^k_{i_1,j} \|^N_1 , \| C^k_{i_2,j} \|^N_1 , \ldots .$$

2. From

$$\sum_j (C^k_{ij} - \lambda^{(\nu)}_i \delta_{jk}) q^{(\nu)}_j = 0, \quad i = i_1, i_2, \ldots \tag{3-247}$$

find the simultaneous eigenvectors $\mathbf{q}^{(\nu)}$ with the normalization $\sum_{i=1}^N g_i |q^{(\nu)}_i|^2 = 1$ and the phase choice that $q^{(\nu)}_e$ is real and positive.

3. From $\chi^{(\nu)}_i = \sqrt{g} q^{(\nu)*}_i$, we obtain the characters.

For example, according to (3-245), from (3-44a), (3-46) and (3-49) we can easily obtain the simple characters of $S_3 (g = 6)$ and $C_{6v} (g = 12)$, listed in Tables 3.11-1 and 3.11-2.

Table 3.11-1. Character table of S_3.

$\lambda^{(\nu)}$	class / partitions	C_1 (1^3)	C_2 (12)	C_3 (3)
3	[3]	1	1	1
−3	[1^3]	1	−1	1
0	[21]	2	0	−1

Table 3.11-2. Character table of C_{6v}.

$\lambda^{(\nu)}_4$	$\lambda^{(\nu)}_5$	class / old labels	C_1	C_2	C_3	C_4	C_5	C_6
2	3	A_1	1	1	1	1	1	1
2	−3	A_2	1	1	1	1	−1	−1
−2	3	B_1	1	−1	1	−1	1	−1
−2	−3	B_2	1	−1	1	−1	−1	1
−1	0	E_1	2	2	−1	−1	0	0
1	0	E_2	2	−2	−1	1	0	0

Before the EFM was proposed, several conventional methods were available for determining the simple characters. These included (a) Jones' method (1975), (b) Boerner's method (1963), (c) Bradley and Cracknell's method (1972), (d) Burnside's method (1955) and (e) Dixon's method (1967). The EFM differs from these methods in the introduction of the CSCO for the class space and thus greatly simplifies the calculation.

Ex. 3.16. Find the simple character of S_4 (using the results of Ex. 3.1).

Ex. 3.17. Find the simple character of C_{4v}.

3.12. The Applications of Simple Characters

From the previous discussion, we see that in the new approach the character steps down from its dominant role in traditional group theory. We can carry out the reduction of a rep of G without any knowledge of the simple characters. However, if the simple characters are known, as is the case with all commonly used finite groups, then they can be used to simplify calculations.

The simple characters have the following major applications in the new approach.

1. Finding the CSCO of G. From the character table and the relation $\lambda_i^{(\nu)} = (g_i/h_\nu)\chi_i^{(\nu)}$, we can construct the following matrix,

$$\begin{pmatrix} \lambda_1^{(\nu_1)} & \lambda_2^{(\nu_1)} & \cdots & \lambda_N^{(\nu_1)} \\ \lambda_1^{(\nu_2)} & \lambda_2^{(\nu_2)} & \cdots & \lambda_N^{(\nu_2)} \\ \cdots & \cdots & \cdots & \cdots \\ \lambda_1^{(\nu_N)} & \lambda_2^{(\nu_N)} & \cdots & \lambda_N^{(\nu_N)} \end{pmatrix} . \tag{3-248}$$

If we can find a column in the matrix, say column i_1, which has N distinct eigenvalues, then the class operator C_{i_1} is a CSCO of G. Otherwise, we look for two columns, say columns i_1 and i_2. If the N pairs of eigenvalues $(\lambda_{i_1}^{(\nu_1)}, \lambda_{i_2}^{(\nu_1)}), \ldots, (\lambda_{i_1}^{(\nu_N)}, \lambda_{i_2}^{(\nu_N)})$ are all different, then (C_{i_1}, C_{i_2}) is a CSCO of G, etc. Therefore, if the simple characters are known, it is trivial to find the CSCO of G.

2. Determining the multiplicity of an irrep in a reducible rep. In Eq. (2-103b) we introduced the multiplicity τ_ν, i.e., the number of times the irrep (ν) occurs in a rep D with the characters χ_i,

$$D = \sum_\nu \oplus \tau_\nu D^{(\nu)} , \quad \chi_i = \sum_\nu \tau_\nu \chi_i^{(\nu)} . \tag{3-249}$$

From (3-237) and (3-249) we obtain an important expression for the multiplicity τ_ν,

$$\tau_\nu = \sum_i \frac{g_i}{g} \chi_i \chi_i^{(\nu)*} . \tag{3-250}$$

If we are only interested in which irreps a given rep will decompose into, we can get the answer from our last result without solving the eigenfunction of the CSCO of G.

Example: Find the multiplicity τ_ν of the irrep (ν) in the regular rep of G.

The answer $\tau_\nu = h_\nu$ is already known. Here we merely use (3-250) to check the result. From Sec. 2.6 we know that in the regular rep only the identity has a nonvanishing character,

$$\chi_e = g ; \quad \chi_i = 0 , \quad i \neq e .$$

Using $g_e = 1$ and $\chi_e^{(\nu)} = h_\nu$, from (3-250) we get $\tau_\nu = h_\nu$.

Equation (3-250) tells us that if two reps have equal characters, then both will have the same block-diagonal form (2-101) after reduction, and therefore are identical except for the order of appearance of $D^{(\nu_i)}(R)$. Thus they can be transformed into an equivalent reduced form and, therefore, are themselves equivalent. On the other hand, two equivalent reps must have the same characters. Thus we have

Theorem 3.31: The equality of characters is the necessary and sufficient condition for the equivalence of two reps.

Theorem 3.32: Given an irrep D with the characters $\chi(R_a)$, the quantity

$$\frac{1}{g}\sum_{a=1}^{g} \chi(R_a^2)$$

equals $+1$ if D is real, -1 if D is self-conjugate but not real, 0 if D is not equivalent to its conjugate (Bacry 1977).

3.13. Reduction of Non-Regular Reps (EFM for Irreducible Basis)

3.13.1. Canonical subgroup chains with $\tau_\nu = 1$

Suppose that there are n orthonormal wave functions,

$$\varphi_a(X) , \quad a = 1, 2, \ldots n , \tag{3-251a}$$

which carry a reducible rep of G, and we need to find the irreducible basis

$$\psi_m^{(\nu)} = \sum_a u_{\nu m,a}\varphi_a , \tag{3-251b}$$

adapted to a canonical subgroup chain $G \supset G(s)$. $\psi_m^{(\nu)}$ satisfies

$$\begin{pmatrix} C \\ C(s) \end{pmatrix} \psi_m^{(\nu)} = \begin{pmatrix} \nu \\ m \end{pmatrix} \psi_m^{(\nu)} , \tag{3-252a}$$

or

$$\sum_{b=1}^{N} \left[\left\langle \varphi_a \middle| \begin{matrix} C \\ C(s) \end{matrix} \middle| \varphi_b \right\rangle - \begin{pmatrix} \nu \\ m \end{pmatrix} \delta_{ab} \right] u_{\nu m,b} = 0 . \tag{3-252b}$$

If the eigenvalue (ν, m) is a single root, it means that the irrep ν occurs only once, and the h_ν eigenvectors $\psi_m^{(\nu)}$ carry the irrep ν of G.

In the foregoing procedure, a knowledge of the irreducible matrices is not necessary. However, in some cases, certain conventional or standard irreducible matrices for the $G \supset G(s)$ basis are given. In order that the irreducible basis found from the EFM be consistent (including the phase) with the standard matrices, we can use the following technique.

We need only find one component, say $\psi_m^{(\nu)}$, from (3-252). Using the known matrix elements, we can construct an operator $F_{m'm}(R)$, a suitable linear combination of the group elements, by means of which the other m'-th component can be derived from the known component m successively,

$$\psi_{m'}^{(\nu)} = F_{m'm}^{(\nu)}(R)\psi_m^{(\nu)} . \tag{3-253}$$

The form of the operator $F_{m'm}^{(\nu)}(R)$ is very simple for the finite groups commonly used and can be easily found. For example, suppose that $R_a\psi_m^{(\nu)} = c_1\psi_m^{(\nu)} + c_2\psi_{m'}^{(\nu)}$; then

$$F_{mm'}^{(\nu)}(R) = (R_a - c_1)/c_2 . \tag{3-253$'$}$$

3.13.2. Canonical subgroup chain with $\tau_\nu > 1$

Suppose that the eigenvalue (ν, m) is a τ_ν-fold root; then this fact indicates that the irrep ν occurs τ_ν times, and for a given (ν, m), there are τ_ν linearly independent solutions to (3-252),

$$\psi_m^{(\nu)\tau} , \quad \tau = 1, 2, \ldots , \tau_\nu .$$

The eigenvectors $\psi_m^{(\nu)\tau}$ can be chosen to be orthogonal in the multiplicity label τ. However, it should be stressed that the eigenvectors $\psi_m^{(\nu)\tau}, m = 1, 2, \ldots , h_\nu$, chosen arbitrarily except for the requirement of orthogonality with respect to τ, in general do not generate an irrep of G. To obtain the irreducible basis we can use either of the following two methods.

1. Using intrinsic quantum number: In reducing a regular rep, the h_ν distinct eigenvalues k of $C(s)$ provide just enough labels for distinguishing the h_ν equivalent irreps. While in reducing a non-regular rep, an irrep ν may occur only $\tau_\nu < h_\nu$ times. For such cases, there are too many intrinsic quantum numbers k. Now the question is, can we still use the intrinsic quantum number to distinguish the τ_ν equivalent irrep? If the answer is yes, then how?

For non-regular reps, the intrinsic state $\Phi_0(X)$ must have certain symmetries (otherwise applying $|G|$ elements to it will generate $|G|$ basis vectors carrying the regular rep of G). Suppose that it is invariant under a set of elements $\{T_\alpha : \alpha = 1, 2, \ldots |G_{in}|\}$,

$$T_\alpha\Phi_0(X) = \Phi_0(X), \quad T_\alpha \in G_{in} . \tag{3-254}$$

The set forms a subgroup G_{in} of the group G, called the *symmetry group of the intrinsic state.* The left coset decomposition of G with respect to G_{in} is denoted as

$$G = \sum_{i=1}^{q} \oplus a_i G_{in} , \quad a_1 = e . \tag{3-255a}$$

Applying the $|G|$ elements R to $\Phi_0(X)$ we can get only $q = |G|/|G_{in}|$ linear independent states φ_i which carry a non-regular rep of G,

$$\varphi_i = a_i \Phi_0(X) . \tag{3-255b}$$

Now let us inspect the action of an intrinsic group element $\overline{R}$ on the basis vector $\varphi_1 = \Phi_0(X)$:

$$\begin{aligned} \overline{R}\Phi_0(X) &= R\Phi_0(X) \\ \overline{R}\Phi_0(X) &= \overline{R}T_\alpha\Phi_0(X) = T_\alpha R\Phi_0(X) . \end{aligned} \tag{3-256}$$

Since in general $R\Phi_0(X) \neq T_\alpha R\Phi_0(X)$, unless R belongs to the symmetry group G_{in}, a contradiction arises here. Therefore, the intrinsic group elements do not have a definite meaning in the non-regular rep space, except those which belong to the intrinsic subgroup G_{in} and thus are equivalent to the identity.

For example, the symmetry group for the intrinsic state $\Phi_0 = |\alpha\beta\beta\gamma\rangle$ is $G_{in} = \{e, (23)\}$. The intrinsic permutation $(\overline{12})$ is meaningless,

$$\begin{aligned} (\overline{12})\Phi_0 &= (12)\Phi_0 = (12)|\alpha\beta\beta\gamma\rangle = |\beta\alpha\beta\gamma\rangle , \\ (\overline{12})\Phi_0 &= (\overline{12})(23)\Phi_0 = (23)(12)\Phi_0 = |\beta\beta\alpha\gamma\rangle . \end{aligned}$$

Nevertheless, according to the following theorem, we can still extract out of the intrinsic group something which remains meaningful in the non-regular rep space.

Theorem 3.33: If the class operators $C_i(1)$ of a subgroup G_1 of G are commuting with the symmetry group G_{in} of an intrinsic state,

$$[C_i(1), T_\alpha] = 0 , \quad \alpha = 1, 2, \dots , |G_{in}| , \tag{3-257}$$

then the class operators $\overline{C}_i(1)$ of the corresponding intrinsic group $\overline{G}_1$ have a definite meaning.

Proof: According to Eq. (3-257) we have

$$\overline{C}_i(1)\Phi_a(X) = \overline{C}_i(1)R_a\Phi_0(X) = R_a C_i(1)\Phi_0(X) . \tag{3-258}$$

$$\begin{aligned} \overline{C}_i(1)\Phi_a(X) &= \overline{C}_i(1)R_a T_\alpha\Phi_0(X) = R_a T_\alpha C_i(1)\Phi_0(X) \\ &= R_a C_i(1) T_\alpha \Phi_0(X) = R_a C_i(1)\Phi_0(X) . \end{aligned} \tag{3-259}$$

In summary, for the non-regular rep case, the meaningful operator set $\overline{C}(s')$ can be found according to the following steps:

(a) Find the symmetry group G_{in} for the chosen intrinsic state $\Phi_0(X)$.

(b) Find the subgroup chain $G(s')$ whose CSCO, $C(s')$, commutes with G_{in}.

Stated differently, $G(s')$ results from deleting all those subgroups in the canonical subgroup chain $G(s)$ whose CSCO, denoted by $C(s'')$, do not commute with G_{in}.

(c) Then the corresponding intrinsic operator set $\overline{C}(s')$ has a definite meaning.

Obviously, $G \supset G(s')$ is no longer a canonical subgroup chain. For example, for the intrinsic state $|\alpha\beta\beta\gamma\rangle, G_{in} = \{e, (23)\}, G(s') = S_3$, and the meaningful operator set is $\overline{C}(s') = \overline{C}(3)$, in contrast to the regular rep case, for which $G(s) = S_3 \supset S_2$, and $\overline{C}(s) = (\overline{C}(3), \overline{C}(2))$.

In the case when single group elements have no definite meaning in a given space, we can no longer talk about irreducible bases of the group in the space. However, due to the fact that the

CSCO's of the intrinsic group $\overline{G}$ and the subgroup chain $\overline{G}(s')$ do have a definite meaning, for convenience in exposition, we might as well call the simultaneous eigenfunction of $(\overline{C},\overline{C}(s'))$ as a $\overline{G} \supset \overline{G}(s')$ *quasi irreducible basis.*

Having found the meaningful operator set $\overline{C}(s')$, we can use it to lift the degeneracy of (ν, m) in Eq. (3-252).

Theorem 3.34: The set of operators $(C, C(s), \overline{C}(s'))$ is a CSCO for the non-regular rep space $\{a_i\Phi_0(X) : i = 1, 2 \ldots, q\}$.

Although we cannot yet prove this theorem, no counter examples have been found.

Therefore, the intrinsic operator set $\overline{C}(s')$ provides just enough quantum numbers for distinguishing the τ_ν equivalent irreps.

Equation (3-252) is extended to

$$\begin{gathered}
\psi_m^{(\nu)\kappa} = \sum_a u_{\nu m\kappa,a}\varphi_a \, , \\
\begin{pmatrix} C \\ C(s) \\ \overline{C}(s') \end{pmatrix} \psi_m^{(\nu)\kappa} = \begin{pmatrix} \nu \\ m \\ \kappa \end{pmatrix} \psi_m^{(\nu)\kappa} \, . \\
\sum_{b=1}^{N} \left[\left\langle \varphi_a \middle| \begin{matrix} C \\ C(s) \\ \overline{C}(s') \end{matrix} \middle| \varphi_b \right\rangle - \begin{pmatrix} \nu \\ m \\ \kappa \end{pmatrix} \delta_{ab} \right] u_{\nu m\kappa,b} = 0 \, , \\
m = m_1, m_2, \ldots, m_{h_\nu}, \quad \kappa = \kappa_1, \kappa_2, \ldots, \kappa_{\tau_\nu}, \quad N = \sum_\nu \tau_\nu h_\nu \, .
\end{gathered} \tag{3-260}$$

Note the relation between the quantum number sets κ and k: κ results from deleting all the quantum numbers of the now meaningless operators $C(s'')$ in the set k.

The set of eigenvectors $\{\psi_m^{(\nu)\kappa} : m = 1, 2, \ldots, h_\nu\}$ is the $G \supset G(s)$ basis for the κ-th irrep ν. If the phases of the eigenvectors with different κ are chosen arbitrarily, the τ_ν irreps carried by the τ_ν sets of eigenvectors are only equivalent instead of being identical. For obtaining τ_ν sets of the $G \supset G(s)$ irreducible bases which will transform according to the same irrep $D^{(\nu)}$,

$$R\psi_m^{(\nu)\kappa} = \sum_{m'} D_{m'm}^{(\nu)}(R)\psi_{m'}^{(\nu)\kappa} \, , \tag{3-261}$$

we can use the technique introduced in the previous subsection, i.e., we only need to find the τ_ν eigenvectors $\psi_m^{(\nu)\kappa}$, $\kappa = 1, 2, \ldots, \tau_\nu$, for a particular component m; the other components can be obtained by recursively using the formula

$$\psi_{m'}^{(\nu)\kappa} = F_{m'm}^{(\nu)}(R)\psi_m^{(\nu)\kappa} \, . \tag{3-262}$$

Although (3-261) parallels (3-216a), there is no counterpart of (3-216b), since single intrinsic group elements have no meaning.

The q basis vectors $\psi_m^{(\nu)\kappa}$ form an orthonormal complete set in the non-regular rep space. The orthonomality relation is

$$\langle \psi_{m'}^{(\nu')\kappa'} | \psi_m^{(\nu)\kappa} \rangle = \delta_{\nu\nu'}\delta_{mm'}\delta_{\kappa\kappa'} \, . \tag{3-263}$$

2. Without using the intrinsic quantum number: For a given (ν, m) find τ_ν linearly independent solutions to Eq. (3-252). After the orthogonalization procedure, they become $\psi_m^{(\nu)j}$, $j = 1, 2, \ldots, \tau_\nu$. The other components are obtained again by

$$\psi_{m'}^{(\nu)j} = F_{m'm}^{(\nu)}(R)\psi_m^{(\nu)j}. \tag{3-264}$$

Now the index j is merely an additional label rather than an intrinsic quantum number, and the τ_ν sets of irreducible bases are chosen freely apart from the orthogonalization requirement.

3.13.3. Non-canonical subgroup chain

Now let us assume that the physically needed group chain $G \supset G(s)$ is not canonical. Therefore the corresponding operator set $(C, C(s))$ is in general not a CSCO in irreducible spaces of G; in other words, the eigenvalue of $(C, C(s))$ is in general not sufficient to label uniquely an irreducible basis vector. The non-canonical subgroup chains which we often come across in physics is of the form $G \supset G_1 \times G_2 \supset G(s_1) \times G(s_2)$, where $G \supset G(s_i), i = 1, 2$ are canonical. Let $(C_i, C(s_i))$ be the CSCO-II of G_i. Now the operator set

$$(C; C(s)) = (C; C_1, C(s_1), C_2, C(s_2))$$

is not a CSCO-II of G. We use $\psi^{(\nu)\kappa}_{\beta,\mu_1 m_1 \mu_2 m_2}$ to designate an irreducible basis vector adapted to the following group chain,

$$\begin{array}{ccccccc} G \supset & G_1 \times & G_2 \supset & G(s_1) \times & G(s_2), & \overline{G} \supset & \overline{G}(s') \\ \nu & \mu_1 & \mu_2 & m_1 & m_2 & \nu & \kappa \end{array} .$$

While β is an additional quantum number whose range is equal to the number of times, $(\mu_1 \mu_2 \nu)$, that the irrep (μ_1, μ_2) of $G_1 \times G_2$ occurs in the irrep ν of G. The irreducible basis satisfies the eigenequations

$$\begin{pmatrix} C \\ C_1 \\ C(s_1) \\ C_2 \\ C(s_2) \\ \overline{C}(s') \end{pmatrix} \psi^{(\nu)\kappa}_{\beta,\mu_1 m_1 \mu_2 m_2} = \begin{pmatrix} \nu \\ \mu_1 \\ m_1 \\ \mu_2 \\ m_2 \\ \kappa \end{pmatrix} \psi^{(\nu)\kappa}_{\beta,\mu_1 m_1 \mu_2 m_2} \tag{3-265}$$

$$\kappa = \kappa_1, \kappa_2, \ldots, \kappa_{\tau_\nu}, \quad \beta = 1, 2, \ldots, (\mu_1 \mu_2 \nu) .$$

For a given $(\nu, \mu_1, m_1, \mu_2, m_2, \kappa)$, there are $(\mu_1 \mu_2 \nu)$ linearly independent solutions to (3-265), which can be orthogonalized with respect to the index β. We only need to get the solutions for a particular m_1 and m_2. The remaining components of the irreducible basis can be obtained by

$$\psi^{(\nu)\kappa}_{\beta,\mu_1 m_1' \mu_2 m_2} = F^{(\nu_1)}_{m_1' m_1}(R_1)\, \psi^{(\nu)\kappa}_{\beta,\mu_1 m_1 \mu_2 m_2}, \quad \psi^{(\nu)\kappa}_{\beta,\mu_1 m_1 \mu_2 m_2'} = F^{(\nu_2)}_{m_2' m_2}(R_2)\, \psi^{(\nu)\kappa}_{\beta \mu_1 m_1 \mu_2 m_2} . \tag{3-266}$$

If the irrep (ν) of G occurs only once in the reducible space $\{\varphi_a\}$, then the equation for $\overline{C}(s')$ as well as the quantum number κ are redundant.

3.13.4. Projection operator method

In the traditional approach, the principal method for constructing an irreducible basis of a finite group is the projection operator method. The procedure for projecting out an irreducible basis from a reducible basis $\{\varphi_a\}$ can be summarized as follows.

1. Compute the characters χ_i of the reducible rep carried by $\{\varphi_a\}$.

2. Using (3-250), decompose the compound character χ_i into a sum of the simple characters of $G, \chi_i = \sum_\nu \tau_\nu \chi_i^\nu$.

3. Find the irreducible matrices $D^{(\nu)}(R)$ for those irreps for which $\tau_\nu \geq 1$.

4. Apply the generalized projection operator $P^{(\nu)}_{mk}$, (3-222), to one of the reducible basis vectors, say φ_1. If the result is not zero, then by varying m while keeping ν and k fixed, we can find a set of irreducible basis $\psi^{(\nu)}_m$.

5. If $P^{(\nu)}_{mk}\varphi_1 = 0$, then we change k to another quantum number k', and repeat step 4. If $P^{(\nu)}_{mk}\varphi_1 = 0$, for any k, then we change φ_1 to φ_2, and repeat steps 4 and 5, until we find a set of irreducible basis $\psi^{(\nu)}_m$.

6. If $\tau_\nu > 1$, by varying the quantum number k, we can get τ_ν sets of linearly independent but usually not orthogonal irreducible bases.

The projection operator method, though simple in principle, suffers from several drawbacks:

1. There is no rule governing the choice of the quantum number k and the basis vector φ_1 on which the projection operator applies.

2. In the case $\tau_\nu > 1$, the irreducible bases obtained from the projections with different k are in general neither linearly independent, nor orthogonal. If τ_ν is large, it is tedious to pick out the linearly independent ones.

3. The irreducible matrices for all group elements must be known beforehand. If according to a physical problem, we need the irreducible bases adapted to a given subgroup chain $G \supset G(s)$, then we must use the $G \supset G(s)$ irreducible matrices in constructing the projection operator (3-222). Often these are not available in textbooks, and thus have to be computed by hand. However there is no general and yet simple method for obtaining the $G \supset G(s)$ irreps in the traditional group represenation theory.

4. The procedure is laborious and become untractable for groups of larger order.

The EFM is simpler and especially suitable for computer calculation. With the EFM, we can obtain the $G \supset G(s)$ irreducible basis without any prior knowledge of the $G \supset G(s)$ irreps.

3.14. Kronecker Product of Representations

3.14.1 Clebsch-Gordan series

Suppose that $\psi_{m_1}^{(\nu_1)}(x_1)(m_1 = 1,2,\ldots,h_{\nu_1})$ and $\varphi_{m_2}^{(\nu_2)}(x_2)(m_2 = 1,2,\ldots,h_{\nu_2})$ carry the irreps ν_1 and ν_2 of a group G, respectively. The $h_{\nu_1}h_{\nu_2}$ products

$$|m_1m_2\rangle = \psi_{m_1}^{(\nu_1)}(x_1)\varphi_{m_2}^{(\nu_2)}(x_2) \tag{3-267}$$

form a rep space of G,

$$\begin{aligned} R|m_1m_2\rangle &= \psi_{m_1}^{(\nu_1)}(R^{-1}x_1)\varphi_{m_2}^{(\nu_2)}(R^{-1}x_2) = \sum_{m_1'm_2'} D_{m_1'm_1}^{(\nu_1)}(R)D_{m_2'm_2}^{(\nu_2)}(R)|m_1'm_2'\rangle \\ &= \sum_{m_1'm_2'} D_{m_1'm_2',m_1m_2}^{(\nu_1\times\nu_2)}(R)|m_1'm_2'\rangle \ , \end{aligned} \tag{3-268}$$

$$D_{m_1'm_2',m_1m_2}^{(\nu_1\times\nu_2)}(R) = D_{m_1'm_1}^{(\nu_1)}(R)D_{m_2'm_2}^{(\nu_2)}(R) \ . \tag{3-269}$$

It generates a rep of G, called the *direct product rep* or *Kronecker product* of the irreps ν_1 and ν_2, denoted by $(\nu_1) \times (\nu_2)$. Equation (3-269) can be written as

$$D^{(\nu_1\times\nu_2)} = D^{(\nu_1)} \otimes D^{(\nu_2)} \ , \tag{3-270}$$

where the symbol $\otimes$ indicates the direct product of matrices. The product rep in general can be reduced into irreps of G,

$$D^{(\nu_1\times\nu_2)} = D^{(\nu_1)} \otimes D^{(\nu_2)} = \sum_{\nu_3}(\nu_1\nu_2\nu_3)D^{(\nu_3)} \ , \tag{3-271}$$

where $(\nu_1\nu_2\nu_3)$ is the number of times that the irrep ν_3 occurs in the product rep. Eq.(3-271) is referred to as the Clebsch-Gordan (CG) series, and often shortened to

$$(\nu_1) \times (\nu_2) = \sum_{\nu_3}(\nu_1\nu_2\nu_3)(\nu_3) \ . \tag{3-272}$$

From (3-269) it is easy to calculate the character of the product rep.

$$\begin{aligned}\chi^{(\nu_1\times\nu_2)}(R) &= \sum_{m_1m_2} D^{(\nu_1\times\nu_2)}_{m_1m_2,m_1m_2}(R) \\ &= \sum_{m_1m_2} D^{(\nu_1)}_{m_1m_1}(R)D^{(\nu_2)}_{m_2m_2}(R) \\ &= \chi^{(\nu_1)}(R)\chi^{(\nu_2)}(R)\ ,\end{aligned} \tag{3-273a}$$

which can be decomposed into a sum of the primitive characters,

$$\chi^{(\nu_1)}(R)\chi^{(\nu_2)}(R) = \sum_{\nu_3}(\nu_1\nu_2\nu_3)\chi^{(\nu_3)}(R)\ . \tag{3-273b}$$

Using (3-250) we get

$$(\nu_1\nu_2\nu_3) = \sum_i \frac{g_i}{g}\chi_i^{(\nu_1)}\chi_i^{(\nu_2)}\chi_i^{(\nu_3)*}\ . \tag{3-274}$$

It is clear that

$$(\nu_1\nu_2\nu_3) = (\nu_2\nu_1\nu_3)\ . \tag{3-275}$$

For groups with real characters, we have

$$(\nu_1\nu_2\nu_3) = (\nu_i\nu_j\nu_k)\ , \tag{3-276}$$

where (i,j,k) is any permutation of (1,2,3).

We say a group G is *simply reducible* if the Kronecker product of any two irreps of G contains each irrep no more than once. Therefore, for a simple reducible group, all the coefficients $(\nu_1\nu_2\nu_3) \le 1$.

For example, the permutation group S_3 and S_4, and the rotation group R_3 are simply reducible.

3.14.2. Symmetrized and antisymmetrized squares

It is easy to show that the product rep $(\nu_1)\times(\nu_2)$ can be reduced into a symmetric and an antisymmetric product reps

$$(\nu)\times(\nu) = [(\nu)\times(\nu)]_s \oplus [(\nu)\times(\nu)]_a\ , \tag{3-277}$$

where $[(\nu_1)\times(\nu_2)]_s$ is called the *symmetric square* with the basis

$$\Psi_{jl} = \frac{1}{\sqrt{2(1+\delta_{jl})}}(\psi_j^{(\nu)}\varphi_l^{(\nu)} + \psi_l^{(\nu)}\varphi_j^{(\nu)}),\quad j\le l = 1,\dots,h_\nu\ . \tag{3-278}$$

Its dimension is $\binom{h_\nu}{2} + h_\nu = h_\nu(h_\nu+1)/2$. The representation matrices are

$$D^{[(\nu)\times(\nu)]_s}_{ik,jl} = \frac{1}{\sqrt{(1+\delta_{ik})(1+\delta_{jl})}}[D^{(\nu)}_{ij}D^{(\nu)}_{kl} + D^{(\nu)}_{il}D^{(\nu)}_{kj}] \tag{3-279}$$

where $i\le k$, and $j\le l$. $[(\nu_1)\times(\nu_2)]_a$ is called the *antisymmetric square* with the basis

$$\Phi_{jl} = \frac{1}{\sqrt{2}}(\psi_j^{(\nu)}\varphi_l^{(\nu)} - \psi_l^{(\nu)}\varphi_j^{(\nu)}),\quad j< l = 1,\dots,h_\nu\ . \tag{3-280}$$

Its dimension is $\binom{h_\nu}{2} = h_\nu(h_\nu-1)/2$. The representation matrices are

$$D^{[(\nu)\times(\nu)]_a}_{ik,jl} = D^{(\nu)}_{ij}D^{(\nu)}_{kl} - D^{(\nu)}_{il}D^{(\nu)}_{kj}\ , \tag{3-281}$$

where $i < k$ and $j < l$. Both the symmetric and antisymmetric squares are in general reducible reps of G.

Ex. 3.18. Show that $\chi^{[\nu_1\times\nu_2]_s}(R) = \frac{1}{2}[\chi(R)^2 + \chi(R^2)]$ and $\chi^{[\nu_1\times\nu_2]_a}(R) = \frac{1}{2}[\chi(R)^2 - \chi(R^2)]$.

3.15. The CG Coefficients

3.15.1. Definition and properties of the CG coefficients

To effect the reduction of (3-271), the product basis vectors of (3-267) need to be recombined into the $G \supset G(s)$ irreducible basis

$$\Psi_m^{(\nu)\tau}(x_1, x_2) = \sum_{m_1 m_2} C^{(\nu)\tau,m}_{\nu_1 m_1,\nu_2 m_2} \psi^{(\nu_1)}_{m_1}(x_1)\varphi^{(\nu_2)}_{m_2}(x_2) \,, \qquad \tau = 1, 2, \ldots, (\nu_1\nu_2\nu) \,, \tag{3-282}$$

where τ is called the *outer multiplicity label* and $C^{(\nu)\tau,m}_{\nu_1 m_1,\nu_2 m_2}$ is called the *CG coefficients* or *Wigner coefficients.* If $(\nu_1\nu_2\nu) = 1$, then the corresponding CG coefficient is said to be multiplicity free.

Notice that the subgroups $G(s_1), G(s_2)$ and $G(s)$ used to characterize the irreducible bases $\psi^{\nu_1}_{m_1}, \varphi^{\nu_2}_{m_2}$ and $\Psi^{(\nu)\tau}_m$ may be different from one another or may be identical.

Since Eq. (3-282) is a transformation between two sets of orthonormal bases, the CG coefficients satisfy the unitarity relations

$$\sum_{m_1 m_2} (C^{(\nu)\tau,m}_{\nu_1 m_1,\nu_2 m_2})^* C^{(\nu')\tau',m'}_{\nu_1 m_1,\nu_2 m_2} = \delta_{\nu\nu'}\delta_{\tau\tau'}\delta_{mm'} \,, \tag{3-283}$$

$$\sum_{\nu\tau m} (C^{(\nu)\tau m}_{\nu_1 m_1,\nu_2 m_2})^* C^{(\nu)\tau m}_{\nu_1 m_1',\nu_2 m_2'} = \delta_{m_1 m_1'}\delta_{m_2 m_2'} \,. \tag{3-284}$$

Applying the group element R to the left-hand side of Eq. (3-282), we have

$$R\Psi^{(\nu)\tau}_m = \sum_{m'} D^{(\nu)}_{m'm}(R)\Psi^{(\nu)\tau}_{m'} = \sum_{\substack{m' \\ m_1' m_2'}} D^{(\nu)}_{m'm}(R) C^{(\nu)\tau m'}_{\nu_1 m_1',\nu_2 m_2'} \psi^{(\nu_1)}_{m_1'}\varphi^{(\nu_2)}_{m_2'} \,, \tag{3-285}$$

and to the right-hand side of (3-282), we obtain

$$R \sum_{m_1 m_2} C^{(\nu)\tau m}_{\nu_1 m_1,\nu_2 m_2} \psi^{(\nu_1)}_{m_1}\varphi^{(\nu_2)}_{m_2} = \sum_{\substack{m_1 m_2 \\ m_1' m_2'}} C^{(\nu)\tau m}_{\nu_1 m_1,\nu_2 m_2} D^{(\nu_1)}_{m_1' m_1}(R) D^{(\nu_2)}_{m_2' m_2}(R) \psi^{(\nu_1)}_{m_1'}\varphi^{(\nu_2)}_{m_2'} \,. \tag{3-286}$$

Since the product basis vectors are linearly independent, we have

$$\sum_{m'} C^{(\nu)\tau m'}_{\nu_1 m_1',\nu_2 m_2'} D^{(\nu)}_{m'm}(R) = \sum_{m_1 m_2} C^{(\nu)\tau m}_{\nu_1 m_1,\nu_2 m_2} D^{(\nu_1)}_{m_1' m_1}(R) D^{(\nu_2)}_{m_2' m_2}(R) \,. \tag{3-287}$$

Multiplying both sides of (3-287) by $(C^{(\nu)\tau,m}_{\nu_1\overline{m}_1,\nu_2\overline{m}_2})^*$, summing over ν, τ, and m, and using (3-284), one has

$$\sum_{\nu\tau m m'} (C^{(\nu)\tau,m}_{\nu_1 m_1,\nu_2 m_2})^* C^{(\nu)\tau m'}_{\nu_1 m_1',\nu_2 m_2'} D^{(\nu)}_{m'm}(R) = D^{(\nu_1)}_{m_1' m_1}(R) D^{(\nu_2)}_{m_2' m_2}(R) \,. \tag{3-288}$$

Multiplying both sides of (3-288) by $D^{(\overline{\nu})*}_{\overline{m}'\overline{m}}(R)$, summing over R and using (3-200a), gives

$$\frac{1}{g}\sum_R D^{(\nu)*}_{m'm}(R) D^{(\nu_1)}_{m_1' m_1}(R) D^{(\nu_2)}_{m_2' m_2}(R) = \frac{1}{h_\nu}\sum_\tau (C^{(\nu)\tau m}_{\nu_1 m_1,\nu_2 m_2})^* C^{(\nu)\tau m'}_{\nu_1 m_1',\nu_2 m_2'} \,. \tag{3-288'}$$

Under the condition that all the irreducible matrices are real, and on putting $m' = m, m'_1 = m_1$ and $m'_2 = m_2$ in (3-288'), we have

$$\frac{1}{g}\sum_R D^{(\nu)}_{mm}(R) D^{(\nu_1)}_{m_1 m_1}(R) D^{(\nu_2)}_{m_2 m_2}(R) = \frac{1}{h_\nu} \sum_\tau |C^{(\nu)\tau, m}_{\nu_1 m_1, \nu_2 m_2}|^2 \ . \tag{3-289}$$

Therefore we obtain the symmetry relation

$$\sum_\tau |C^{(\nu)\tau, m}_{\nu_1 m_1, \nu_2 m_2}|^2 / h_\nu = \sum_\tau |C^{(\nu_2)\tau, m_2}_{\nu_1 m_1, \nu m}|^2 / h_{\nu_2} = \sum_\tau |C^{(\nu_1)\tau, m_1}_{\nu m, \nu_2 m_2}|^2 / h_{\nu_1} \ . \tag{3-290}$$

When either ν_1 or ν_2 is the identity rep I, the CG coefficients are trivial. From

$$R(\psi^{(\nu_1)}_{m_1} \varphi^{(I)}) = (R\psi^{(\nu_1)}_{m_1})(R\varphi^{(I)}) = \sum_{m'_1} D^{(\nu_1)}_{m'_1 m_1}(R) (\psi^{(\nu_1)}_{m'_1} \varphi^{(I)})$$

one sees that the product basis $\psi^{(\nu_1)}_{m_1} \varphi^{(I)}$ remains the $G \supset G(s_1)$ irreducible basis (ν_1, m_1). Therefore one obtains

$$C^{(\nu)m}_{\nu_1 m_1, I} = C^{(\nu)m}_{I, \nu_1 m_1} = \delta_{\nu \nu_1} \delta_{m m_1} \ . \tag{3-291}$$

3.15.2. The EFM for CG coefficients

The CG coefficients have wide applications in physics and there is abundant literature devoted to this problem. The conventional methods for computing the CG coefficients are essentially the projection operator method and its variations. For an extensive review of the subject the reader is referred to the review article by Chen [15] 1985. In the following we only discuss the EFM.

According to (3-252), $\Psi^{(\nu)\tau}_m$ satisfies the eigenequations

$$\begin{pmatrix} C \\ C(s) \end{pmatrix} \Psi^{(\nu)\tau}_m = \begin{pmatrix} \nu \\ m \end{pmatrix} \Psi^{(\nu)\tau}_m \ . \tag{3-292}$$

Inserting (3-282) into (3-292) and multiplying from the left by $\langle m'_1 m'_2 |$, we have

$$\sum_{m_1 m_2} \left(\left\langle m'_1 m'_2 \left| \begin{matrix} C \\ C(s) \end{matrix} \right| m_1 m_2 \right\rangle - \begin{pmatrix} \nu \\ m \end{pmatrix} \delta_{m_1 m'_1} \delta_{m_2 m'_2} \right) C^{\nu \tau m}_{\nu_1 m_1, \nu_2 m_2} = 0 \ , \tag{3-293}$$

$$\tau = 1, 2, \ldots (\nu_1 \nu_2 \nu) \ .$$

This shows that CG coefficients result from a diagonalization of the representative matrix of the CSCO-II of G in the product basis. The matrix elements of a class operator C_i are given by

$$\langle m'_1 m'_2 | C_i | m_1 m_2 \rangle = \sum_{l=1}^{g_i} D^{(\nu_1)}_{m'_1 m_1}(R^{(i)}_l) D^{(\nu_2)}_{m'_2 m_2}(R^{(i)}_l) \ . \tag{3-294}$$

Notice that

$$\langle m'_1 m'_2 | C_i | m_1 m_2 \rangle \neq D^{(\nu_1)}_{m'_1 m_1}(C_i) D^{(\nu_2)}_{m'_2 m_2}(C_i) \ .$$

Using (3-294) we can calculate the matrix elements of the CSCO-II. From the characteristic equation of (3-293), we can obtain the eigenvalue (ν, m) along with its degeneracy, which gives the multiplicity $(\nu_1 \nu_2 \nu)$ in the CG series (3-272).

When $(\nu_1 \nu_2 \nu) > 1$, for a given (ν, m), there are $(\nu_1 \nu_2 \nu)$ sets of linearly independent solutions to Eq. (3-293). Subject to the orthogonality requirement with respect to the multiplicity label τ, i.e.,

$$\langle \Psi^{(\nu)\tau}_m | \Psi^{(\nu)\tau'}_m \rangle = \delta_{\tau \tau'},$$

$$\sum_{m_1 m_2} (C^{\nu \tau m}_{\nu_1 m_1, \nu_2 m_2})^* C^{\nu \tau' m}_{\nu_1 m_1, \nu_2 m_2} = \delta_{\tau \tau'} \ , \tag{3-295}$$

the $(\nu_1, \nu_2 \nu)$ sets of solutions can be chosen arbitrarily.

For each possible ν we take only the CG coefficients for a particular m from (3-293); the remaining CG coefficients of the irrep (ν) should be evaluated with the help of (3-264), i.e.,

$$C^{(\nu)\tau,m'}_{\nu_1 m'_1,\nu_2 m'_2} = \sum_{m_1 m_2} \langle m'_1 m'_2 | F^{(\nu)}_{m'm}(R) | m_1 m_2 \rangle C^{(\nu)\tau,m}_{\nu_1 m_1,\nu_2 m_2} . \tag{3-296}$$

It is to be noted that the multiplicity separation is arbitrary, and that the following linear combination satisfies all our requirements for the CG coefficient,

$$C^{(\nu)\theta,m}_{\nu_1 m_1,\nu_2 m_2} = \sum_\tau S^{(\nu)}_{\theta\tau} C^{(\nu)\tau,m}_{\nu_1 m_1,\nu_2 m_2} , \tag{3-297}$$

where $S^{(\nu)}$ is a $(\nu_1\nu_2\nu) \times (\nu_1\nu_2\nu)$ unitary matrix. Hence the CG coefficient can be determined only up to a unitary transformation.

The advantage of the EFM for the CG coefficients lies in the fact that here only the irreducible matrices of a few group elements (which are contained in the CSCO-II of G) are required, while in the projection operator method, the irreducible matrices of all the $|G|$ group elements are required. Another feature of the EFM is that the CG series and CG coefficients are obtained simultaneously.

3.16. Isoscalar Factors

The question we address in this section is how to find the CG coefficients for the $G \supset G_1 \supset G_1(s)$ irreducible basis when the CG coefficients for the $G_1 \supset G_1(s)$ irreducible basis are known.

The CG coefficients for the $G_1 \supset G_1(s)$ irreducible basis are defined through

$$|\Lambda_\theta, m\rangle = \sum_{m_1 m_2} C^{\Lambda_\theta,m}_{\Lambda_1 m_1,\Lambda_2 m_2} |\Lambda_1 m_1\rangle |\Lambda_2 m_2\rangle , \quad \theta = 1,2,\ldots(\Lambda_1\Lambda_2\Lambda), \tag{3-298}$$

where θ is the outer multiplicity label for the group G_1.

The $G \supset G_1 \supset G_1(s)$ irreducible basis is designed by $\left|{(\nu) \atop \beta\Lambda m}\right\rangle$, where $\beta(= 1,2,\ldots a_\Lambda)$ is the *inner multiplicity* which takes account of the multiple occurrence of the irrep Λ of G_1 in the irrep ν of G,

$$D^{(\nu)} = \sum_\Lambda \oplus a_\Lambda D^{(\Lambda)} , \tag{3-299a}$$

or

$$(\nu) = \sum_\Lambda a_\Lambda \cdot (\Lambda) . \tag{3-299b}$$

The CG coefficients for the $G \supset G_1 \supset G_1(s)$ irreducible basis are defined through

$$\left|{(\nu)_\tau \atop \beta\Lambda m}\right\rangle = \sum_{\substack{\beta_1\Lambda_1 m_1 \\ \beta_2\Lambda_2 m_2}} C^{(\nu)_\tau,\beta\Lambda m}_{\nu_1\beta_1\Lambda_1 m_1,\nu_2\beta_2\Lambda_2 m_2} \left|{(\nu_1) \atop \beta_1\Lambda_1 m_1}\right\rangle \left|{(\nu_2) \atop \beta_2\Lambda_2 m_2}\right\rangle , \quad \tau = 1,2,\ldots(\nu_1\nu_2\nu) . \tag{3-300}$$

An alternative way for constructing the $G \supset G_1 \supset G_1(s)$ irreducible basis in (3-300) is that of first using the $G_1 \supset G_1(s)$ CG coefficients to couple $\left|{(\nu_1) \atop \beta_1\Lambda_1 m_1}\right\rangle$ and $\left|{(\nu_2) \atop \beta_2\Lambda_2 m_2}\right\rangle$ into the irreducible basis $|\Lambda_\theta, m\rangle$ of G_1, i.e.,

$$\left[\left|{(\nu_1) \atop \beta_1\Lambda_1}\right\rangle \left|{(\nu_2) \atop \beta_2\Lambda_2}\right\rangle\right]^{\Lambda_\theta}_m = \sum_{m_1 m_2} C^{\Lambda_\theta,m}_{\Lambda_1 m_1,\Lambda_2 m_2} \left|{(\nu_1) \atop \beta_1\Lambda_1 m_1}\right\rangle \left|{(\nu_1) \atop \beta_2\Lambda_2 m_2}\right\rangle , \tag{3-301}$$

and then using the so-called $G \supset G_1$ *isoscalar factor* (ISF), or the *reduced Wigner coefficients*, $C^{(\nu)\tau,\beta\Lambda_\theta}_{\nu_1\beta_1\Lambda_1,\nu_2\beta_2\Lambda_2}$, to combine (3-301) into the irreducible basis of G,

$$\left|\begin{matrix}(\nu)_\tau \\ \beta\Lambda m\end{matrix}\right\rangle = \sum_{\beta_1\Lambda_1\beta_2\Lambda_2\theta} C^{(\nu)_\tau,\beta\Lambda_\theta}_{\nu_1\beta_1\Lambda_1,\nu_2\beta_2\Lambda_2}\left[\left|\begin{matrix}(\nu_1) \\ \beta_1\Lambda_1\end{matrix}\right\rangle\left|\begin{matrix}(\nu_2) \\ \beta_2\Lambda_2\end{matrix}\right\rangle\right]^{\Lambda_\theta}_m . \tag{3-302}$$

From Eqs. (3-300)-(3-302) one has

$$C^{(\nu)_\tau,\beta\Lambda m}_{\nu_1\beta_1\Lambda_1 m_1,\nu_2\beta_2\Lambda_2 m_2} = \sum_\theta C^{(\nu)_\tau,\beta\Lambda_\theta}_{\nu_1\beta_1\Lambda_1,\nu_2\beta_2\Lambda_2} C^{\Lambda_\theta,m}_{\Lambda_1 m_1,\Lambda_2 m_2} . \tag{3-303}$$

This is the celebrated *Racah factorization lemma*. It tells us that from the the $G \supset G_1$ ISF and $G_1 \supset G_1(s)$ CG coefficients, we can easily construct the $G \supset G_1 \supset G_1(s)$ CG coefficients. Equation (3-303) is also valid for Lie groups.

Racah's lemma can be applied to any link in the group chain. For example, the CG coefficients for the group chain $G \supset G_1 \supset G_2 \supset G_3$ can be expressed schematically as

$$(G \supset G_1 \supset G_2 \supset G_3)CGC = (G \supset G_1)\text{ISF} \times (G_1 \supset G_2)\text{ISF} \times (G_2 \supset G_3)CGC . \tag{3-304}$$

Therefore, the calculation of the CG coefficients for a large group G is reduced to the computation of the ISF in each link of the group chain starting from the group G.

Equation (3-302) represents a similarity transformation between two sets of orthonormal bases, therefore the ISF satisfy the unitarity relations

$$\sum_{\substack{\beta_1\Lambda_1\theta \\ \beta_2\Lambda_2}} (C^{(\nu)_\tau,\beta\Lambda_\theta}_{\nu_1\beta_1\Lambda_1,\nu_2\beta_2\Lambda_2})^* C^{(\nu')_{\tau'},\beta'\Lambda_\theta}_{\nu_1\beta_1\Lambda_1,\nu_2\beta_2\Lambda_2} = \delta_{\nu\nu'}\delta_{\tau\tau'}\delta_{\beta\beta'} , \tag{3-305}$$

$$\sum_{\nu\tau\beta}(C^{(\nu)_\tau,\beta\Lambda_\theta}_{\nu_1\beta_1\Lambda_1,\nu_2\beta_2\Lambda_2})^* C^{(\nu)_\tau,\beta\Lambda_{\theta'}}_{\nu_1\beta_1'\Lambda_1',\nu_2\beta_2'\Lambda_2'} = \delta_{\beta_1\beta_1'}\delta_{\beta_2\beta_2'}\delta_{\Lambda_1\Lambda_1'}\delta_{\Lambda_2\Lambda_2'}\delta_{\theta\theta'} . \tag{3-306}$$

Note that the quantum number Λ is kept fixed in the above equations.

The inverse of (3-302) is

$$\left[\left|\begin{matrix}(\nu_1) \\ \beta_1\Lambda_1\end{matrix}\right\rangle \left|\begin{matrix}(\nu_2) \\ \beta_2\Lambda_2\end{matrix}\right\rangle\right]^{\Lambda_\theta}_m = \sum_{\nu\tau\beta}(C^{(\nu)_\tau\beta\Lambda_\theta}_{\nu_1\beta_1\Lambda_1,\nu_2\beta_2\Lambda_2})^*\left|\begin{matrix}(\nu)_\tau \\ \beta\Lambda m\end{matrix}\right\rangle . \tag{3-307}$$

The EFM for the ISF will be further discussed in Secs. 4.19, 7.16 and 7.17.

The CG coefficients discussed in this book are all real except those for the space groups. For the real CG coefficients, we omit the complex conjugate symbol *.

3.17. Irreducible Tensor of a Group G

3.17.1. *Definition of irreducible tensor*

We say that a collection of operators

$$T^{(\nu)}_m,\ m = 1, 2, \ldots, h_\nu \tag{3-308a}$$

is a set of irreducible tensors of a group G, if under the action of the group G it transforms as

$$RT^{(\nu)}_m R^{-1} = \sum_{m'} D^{(\nu)}_{m'm}(R)T^{(\nu)}_{m'} . \tag{3-308b}$$

Equivalently we say that $T^{(\nu)}_m$ is the m-th component of the ν-th irreducible tensor of G.

If an operator H is invariant under the group G,

$$RHR^{-1} = H \ ,$$

or written as

$$[H, R] = 0 \ , \quad R \in G \ , \tag{3-309}$$

then H is called a *scalar operator* of the group G, or an *invariant* of G, or an *irreducible tensor* of the identity rep.

The EFM can also be used to construct irreducible tensors. For example, from the one-body operators, $O_i, i = 1, 2, 3$, we can construct the following irreducible tensors of the permutation group S_3,

$$\begin{aligned} O^{(3)} &= \sqrt{\frac{1}{3}}(O_1 + O_2 + O_3) \\ O^{(0)}_1 &= \sqrt{\frac{1}{6}}(O_1 + O_2 - 2O_3) \\ O^{(0)}_{-1} &= \sqrt{\frac{1}{2}}(O_1 - O_2) \ . \end{aligned} \tag{3-310a}$$

Similarly, for the two-body operators O_{ij}, we can obtain the irreducible tensor of S_3,

$$\begin{aligned} O^{(3)} &= \sqrt{\frac{1}{3}}(O_{12} + O_{13} + O_{23}) \ , \\ O^{(0)}_1 &= \sqrt{\frac{1}{6}}(2O_{12} - O_{13} - O_{23}) \ , \\ O^{(0)}_{-1} &= \sqrt{\frac{1}{2}}(O_{13} - O_{23}) \ . \end{aligned} \tag{3-310b}$$

By means of the CG coefficients of the group G, two irreducible tensors $T^{(\nu_1)}$ and $U^{(\nu_2)}$ of the group G can be coupled into another irreducible tensor $V^{(\nu)}$:

$$V^{(\nu)_\tau}_m = \sum_{m_1 m_2} C^{(\nu)_\tau, m}_{\nu_1 m_1, \nu_2 m_2} T^{(\nu_1)}_{m_1} U^{(\nu_2)}_{m_2} \ . \tag{3-311}$$

To see this, note that

$$\begin{aligned} RV^{(\nu)_\tau}_m R^{-1} &= \sum_{m_1 m_2} C^{(\nu)_\tau, m}_{\nu_1 m_1, \nu_2 m_2} RT^{(\nu_1)}_{m_1} R^{-1} RU^{(\nu_2)}_{m_2} R^{-1} \\ &= \sum_{m_1 m_2 m'_1 m'_2} C^{(\nu)_\tau, m}_{\nu_1 m_1, \nu_2 m_2} D^{(\nu_1)}_{m'_1 m_1}(R) D^{(\nu_2)}_{m'_2 m_2}(R) T^{(\nu_1)}_{m'_1} U^{(\nu_2)}_{m'_2} \\ &= \sum_{m' m'_1 m'_2} D^{(\nu)}_{m' m}(R) C^{(\nu)_\tau, m'}_{\nu_1 m'_1, \nu_2 m'_2} T^{(\nu_1)}_{m'_1} U^{(\nu_2)}_{m'_2} = \sum_{m'} D^{(\nu)}_{m' m}(R) V^{(\nu)_\tau}_{m'} \ , \end{aligned} \tag{3-312}$$

where Eq. (3-287) has been used.

The most familiar example of an irreducible tensor is the spherical harmonics Y_{lm}, $m = -l, \ldots, l$, which is an irreducible tensor of the rotation group R_3.

3.17.2. The Wigner-Eckart theorem

In analogy with Eq. (3-312) we can prove that

$$\varphi^{(\nu'_2)_\tau}_{m'_2} = \sum_{m m_1} C^{(\nu'_2)_\tau, m'_2}_{\nu_1 m_1, \nu m} T^{(\nu)}_m \psi^{(\nu_1)}_{m_1} \tag{3-313}$$

is again an irreducible basis of G. The inverse is

$$T_m^{(\nu)}\psi_{m_1}^{(\nu_1)} = \sum_{\nu_2' m_2' \tau} (C_{\nu_1 m_1, \nu m}^{(\nu_2')_\tau, m_2'})^* \varphi_{m_2'}^{(\nu_2')_\tau} . \tag{3-314}$$

With the help of (3-314), the matrix element of $T_m^{(\nu)}$ can be expressed as

$$\langle \psi_{m_2}^{(\nu_2)} | T_m^{(\nu)} | \psi_{m_1}^{(\nu_1)} \rangle = \sum_{\nu_2' m_2' \tau} (C_{\nu_1 m_1, \nu m}^{(\nu_2')_\tau, m_2'})^* \langle \psi_{m_2}^{(\nu_2)} | \varphi_{m_2'}^{(\nu_2')_\tau} \rangle . \tag{3-315}$$

According to Schur's lemma (2-108), we have

$$\langle \psi_{m_2}^{(\nu_2)} | \varphi_{m_2'}^{(\nu_2')_\tau} \rangle = \langle \psi_{m_2}^{(\nu_2)} | 1 | \varphi_{m_2'}^{(\nu_2')_\tau} \rangle = \text{const.}\ \delta_{\nu_2 \nu_2'} \cdot \delta_{m_2 m_2'} , \tag{3-316a}$$

where the constant is independent of m and m' and is a function of ν_1, ν_2, ν and τ, denoted by

$$\text{const.} = \langle \psi^{(\nu_2)} \ \| T^{(\nu)} \| \ \psi^{(\nu_1)} \rangle^{(\tau)} \tag{3-316b}$$

and called the reduced matrix element. We follow the definition in Rose (1957) which is related to the Edmonds' (1957) definition by

$$\langle \psi^{(\nu_2)} \ \| T^{(\nu)} \| \ \psi^{(\nu_1)} \rangle = \sqrt{\frac{1}{h_{\nu_2}}} \langle \psi^{(\nu_2)} \ \| T^{(\nu)} \| \ \psi^{(\nu_1)} \rangle_{\text{Edmonds}} .$$

From (3-315) and (3-316) we obtain the celebrated *Wigner-Eckart theorem*,

$$\langle \psi_{m_2}^{(\nu_2)} | T_m^{(\nu)} | \psi_{m_1}^{(\nu_1)} \rangle = \sum_\tau (C_{\nu_1 m_1, \nu m}^{(\nu_2)_\tau, m_2})^* \langle \psi^{(\nu_2)} \| \ T^{(\nu)} \ \| \psi^{(\nu_1)} \rangle^\tau . \tag{3-317}$$

The inverse of (3-317) gives the irreducible matrix element

$$\langle \psi^{(\nu_2)} \ \| T^{(\nu)} \| \ \psi^{(\nu_1)} \rangle^{(\tau)} = \sum_{m m_1} C_{\nu_1 m_1, \nu m}^{(\nu_2)_\tau, m_2} \langle \psi_{m_2}^{(\nu_2)} | T_m^{(\nu)} | \psi_{m_1}^{(\nu_1)} \rangle . \tag{3-318}$$

According to the Wigner-Eckart theorem, the matrix element of an irreducible tensor is expressible in terms of a sum of products of two factors, one is a symmetry-related geometric factor, the CG coefficient, and the other is a physical factor, the reduced matrix element. The specific properties of the states and the operator enter the physical factor only. It is precisely this fact that makes the Wigner-Eckart theorem invaluable in physics.

For simply reducible groups, the multiplicity index τ is redundant, and (3-317) reduces to

$$\langle \psi_{m_2}^{(\nu_2)} | T_m^{(\nu)} | \psi_{m_1}^{(\nu_1)} \rangle = (C_{\nu_1 m_1, \nu m}^{\nu_2 m_2})^* \langle \psi^{(\nu_2)} \ \| T^{(\nu)} \| \ \psi^{(\nu_1)} \rangle , \tag{3-319}$$

i.e., the matrix element is factorizable.

If $T^{(\nu)}$ is an invariant operator H of the group G, letting $\nu = I$ (identity rep) in (3-319) we have

$$\langle \psi_{m_2}^{(\nu_2)} | H | \psi_{m_1}^{(\nu_1)} \rangle = \delta_{\nu_1 \nu_2} \delta_{m_1 m_2} \langle \psi^{(\nu_1)} \| \ H \| \ \psi^{(\nu_1)} \rangle = E_{\nu_1} \delta_{\nu_1 \nu_2} \delta_{m_1 m_2} . \tag{3-320}$$

Thus we get back to Schur's lemma (2-108).

The irreducible tensor for point groups and permutation groups have been discussed by Griffith (1962) and Vanagas (1972).

The Wigner-Eckart theorem is also applicable to compact Lie groups.

3.18. Symmetries of the CG Coefficients and ISF

The CG coefficients and ISF have many symmetries (Hamermesh 1962, Derome 1966, Wybourne 1974, Butler 1975, Gao 1985). The labor involved in the computation of these coefficients can be greatly reduced by considering their symmetry properties with respect to the permutation of their arguments and their relation to complex conjugate rep.

1. Since the CG series for $(\nu_1)\times(\nu_2)$ and $(\nu_2)\times(\nu_1)$ are identical, we must have

$$C^{(\nu)_\tau,m}_{\nu_1 m_1,\nu_2 m_2} = \varepsilon_1 C^{(\nu)_\tau,m}_{\nu_2 m_2,\nu_1 m_1} , \tag{3-321}$$

where $\varepsilon_1 = \varepsilon_1(\nu_1\nu_2\nu_\tau)$ is a phase factor. For $\nu_1 \neq \nu_2$, it is determined by the phase convention, while for $\nu_1 = \nu_2$, $\varepsilon_1 = 1$ if ν belongs to the symmetric square, and $\varepsilon_1 = -1$ if ν belongs to the antisymmetric square.

2. We use $(\overline{\nu})$ and $(\overline{m})$ to denote the contragredient reps of G and $G(s)$. For unitary rep we have

$$D^{(\overline{\nu})}_{\overline{m}'\overline{m}}(R) = D^{(\nu)*}_{m'm}(R) . \tag{3-322}$$

Taking complex conjugate of (3-287) and making use of Eq. (3-322), we obtain a second symmetry relation

$$C^{(\nu)_\tau,m}_{\nu_1 m_1,\nu_2 m_2} = \varepsilon_2 (C^{(\overline{\nu})_\tau,\overline{m}}_{\overline{\nu}_1\overline{m}_1,\overline{\nu}_2\overline{m}_2})^* , \tag{3-323}$$

where $\varepsilon_2 = \varepsilon_2(\nu_1\nu_2\nu_\tau)$ is a phase factor.

3. From the Hermitian conjugate of (3-308b) one has

$$R T^{(\nu)\dagger}_m R^{-1} = \sum_{m'} D^{(\nu)^*}_{m'm}(R) T^{(\nu)\dagger}_{m'} .$$

It shows that $T^{(\nu)^\dagger}_m$ transforms contragrediently to $T^{(\nu)}_m$, i.e.,

$$T^{(\nu)^\dagger}_m = T^{(\overline{\nu})}_{\overline{m}} . \tag{3-324a}$$

Therefore one obtains

$$\langle \nu_1 m_1 | T^{(\nu)}_m | \nu_2 m_2 \rangle = \langle \nu_1 m_1 | T^{(\overline{\nu})^\dagger}_{\overline{m}} | \nu_2 m_2 \rangle . \tag{3-324b}$$

Let us first assume that the Kronecker products $(\nu)\times(\nu_1)$ and $(\nu)\times(\nu_2)$ are simply reducible. From our last result and the Wigner-Eckart Theorem (3-319) we obtain

$$(C^{\nu_1 m_1}_{\nu m,\nu_2 m_2})^* \langle \nu_1 \| T^{(\nu)} \| \nu_2 \rangle = C^{\nu_2 m_2}_{\overline{\nu}\,\overline{m},\nu_1 m_1} \langle \nu_2 \| T^{(\overline{\nu})} \| \nu_1 \rangle^* . \tag{3-325}$$

Squaring, summing over m_1, m_2 and m, and using the orthogonality of the CG coefficients, we have

$$\sqrt{h_{\nu_1}} |\langle \nu_1 \| T^{(\nu)} \| \nu_2 \rangle| = \sqrt{h_{\nu_2}} |\langle \nu_2 \| T^{(\overline{\nu})} \| \nu_1 \rangle . \tag{3-326a}$$

On substituting (3-326a) into (3-325) and using (3-323) and (3-321) we obtain another symmetry relation

$$\sqrt{\frac{1}{h_\nu}} C^{(\nu)m}_{\nu_1 m_1,\nu_2 m_2} = \varepsilon_3 \sqrt{\frac{1}{h_{\nu_1}}} C^{(\overline{\nu}_1)\overline{m}_1}_{\overline{\nu}\,\overline{m},\nu_2 m_2} . \tag{3-327a}$$

If the multiplicity is larger than 1, then (3-326a) is replaced by

$$h_{\nu_1} \sum_\tau |\langle \nu_1 \| T^{(\nu)} \| \nu_2 \rangle^{(\tau)}|^2 = h_{\nu_2} \sum_\tau |\langle \nu_2 \| T^{(\overline{\nu})} \| \nu_1 \rangle^{(\tau)}|^2 , \tag{3-326b}$$

and (3-327a) is in general no longer true.

In the case when all the irreps of a group G are real, the CG coefficients obey Eq. (3-290). Through suitable linear combinations (Hamermesh 1962), the CG coefficients can be made to satisfy the following symmetry

$$\sqrt{\frac{1}{h_\nu}} C^{(\nu)\tau,m}_{\nu_1 m_1,\nu_2 m_2} = \varepsilon_3' \sqrt{\frac{1}{h_{\nu_1}}} C^{(\nu_1)\tau,m_1}_{\nu m,\nu_2 m_2} . \tag{3-327b}$$

From the symmetries of the CG coefficients, we obtain the symmetries of the ISF:

1.
$$C^{(\nu)\tau,\beta\Lambda_\theta}_{\nu_1\beta_1\Lambda_1,\nu_2\beta_2\Lambda_2} = \eta_1 C^{(\nu)\tau,\beta\Lambda_\theta}_{\nu_2\beta_2\Lambda_2,\nu_1\beta_1\Lambda_1} . \tag{3-328}$$

When $\nu_1 \neq \nu_2$, the phase factor η_1 is determined by the phase convention.

2. If the multiplicity label θ is redundant, from Eqs. (3-323) and (3-303) we have

$$C^{(\nu)\tau,\beta\Lambda}_{\nu_1\beta_1\Lambda_1,\nu_2\beta_2\Lambda_2} = \eta_2 (C^{(\bar{\nu})\tau,\beta\bar{\Lambda}}_{\bar{\nu}_1\beta_1\bar{\Lambda}_1,\bar{\nu}_2\beta_2\bar{\Lambda}_2})^* . \tag{3-329}$$

3. When both the multiplicity labels τ and θ are redundant,

$$\sqrt{\frac{h_\Lambda}{h_\nu}} C^{(\nu),\beta\Lambda}_{\nu_1\beta_1\Lambda_1,\nu_2\beta_2\Lambda_2} = \eta_3 \sqrt{\frac{h_{\Lambda_1}}{h_{\nu_1}}} C^{(\bar{\nu}_1),\beta_1\bar{\Lambda}_1}_{\bar{\nu}\beta\bar{\Lambda},\nu_2\beta_2,\Lambda_2} . \tag{3-330a}$$

4. When all the irreps are real,

$$\sqrt{\frac{h_\Lambda}{h_\nu}} C^{(\nu)\tau,\beta\Lambda\theta}_{\nu_1\beta_1\Lambda_1,\nu_2\beta_2\Lambda_2} = \eta_3' \sqrt{\frac{h_{\Lambda_1}}{h_{\nu_1}}} C^{(\nu_1)\tau,\beta_1\Lambda_1\theta}_{\nu\beta\Lambda,\nu_2\beta_2\Lambda_2} . \tag{3-330b}$$

Remark: For the multiplicity-free case, the symmetries (3-321), (3-323) and (3-327b), or equivalently, (3-328), (3-329) and (3-330b), are satisfied automatically. For non-multiplicity-free cases, the above symmetries can be imposed to reduce the arbitrariness in the multiplicity separation.

3.19. Applications of Group Theory in Quantum Mechanics

Modern developments in all branches of physics are putting more and more emphasis on the role of symmetries of the underlying physical systems and the application of group theory in quantum mechanics becomes more and more diversified, and is a central subject in numerous books (Lobel 1968, 1971, 1975 and references therein). In this section, we merely sketch some key points without going into great detail.

3.19.1. When G is the symmetry group of the Hamiltonian

1. *Simplifying the problem of solving the Schrödinger equation.* Suppose that $\{\varphi_a : a = 1,2,\ldots,n\}$ carries a rep $\mathcal{D}$ for both the Hamiltonian H and its symmetry group G,

$$\mathcal{D}_{ab}(H) = \langle\varphi_a|H|\varphi_b\rangle , \tag{3-331a}$$

$$\mathcal{D}_{ab}(R) = \langle\varphi_a|R|\varphi_b\rangle, \quad \text{for } R \in G . \tag{3-331b}$$

The energies of the system are decided upon by the expectation equation

$$\det|\mathcal{D}(H) - E \cdot I|_1^n = 0 . \tag{3-331c}$$

Instead of directly solving Eq. (3-331c), we first find the irreducible bases of the symmetry group G.

$$\psi^{(\nu)\kappa}_m = \sum_{a=1}^{n} a^{(\nu)\kappa}_{m,a} \varphi_a, \quad m = 1,2,\ldots,h_\nu, \quad \kappa = 1,2,\ldots,\tau_\nu , \tag{3-332a}$$

$$n = \sum_{\nu=1}^{N} \tau_\nu h_\nu \ . \tag{3-332b}$$

In the new bases (3-332a), both the reps $\mathcal{D}(H)$ and $\mathcal{D}(G)$ become block-diagonal, i.e.,

$$\mathcal{D}(G) \to \sum_{\nu} \oplus \tau_\nu D^{(\nu)}(G) \ , \tag{3-333a}$$

where

$$D^{(\nu)}_{m'm}(R) = \langle \psi^{(\nu)\kappa}_{m'} | R | \psi^{(\nu)\kappa}_{m} \rangle \ . \tag{3-333b}$$

According to Schur's lemma (3-320),

$$\mathcal{D}(H) \to \sum_{\nu} \oplus h_\nu D^{(\nu)}(H) \ , \tag{3-334a}$$

$$D^{(\nu)}_{\kappa'\kappa}(H) = \langle \psi^{(\nu)\kappa'}_{m} | H | \psi^{(\nu)\kappa}_{m} \rangle \ . \tag{3-334b}$$

It is interesting to note that (a) the eigenspace of $(C, C(s))$, $\mathcal{L}_{(\nu)m} = \{\psi^{(\nu)\kappa}_m : \kappa = 1, 2, \ldots, \tau_\nu\}$ is an invariant subspace for both the Hamiltonian H and the intrinsic operator set $\overline{C}(s')$ introduced in (3-260). However, H does not commute with $\overline{C}(s')$ which is a CSCO of the subspace $\mathcal{L}_{(\nu)m}$, otherwise H would be a function of $\overline{C}(s')$; (b) the dimension of the matrix $D^{(\nu)}(H)[D^{(\nu)}(G)]$ is equal to the number of times that the matrix $D^{(\nu)}(G)[D^{(\nu)}(H)]$ occurs in the rep $\mathcal{D}(G)$, $[\mathcal{D}(H)]$.

Therefore, the eigenequation (3-331c) is decomposed into N expectation equations with the much lower orders τ_ν,

$$\det|D^{(\nu)}(H) - E \cdot I|_1^{\tau_\nu} = 0 \ . \tag{3-334c}$$

In general, τ_ν is much smaller than n. In this way, the problem of solving the Schrödinger equation is greatly simplified.

Suppose that in (3-334c) there are n_ν distinct eigenvalues E^ν_i, $i = 1, 2, \ldots, n_\nu$, each with the degeneracy θ^ν_i. For a given E^ν_i, from (3-334c) we can get θ^ν_i eigenfunctions of the Hamiltonian,

$$\begin{gathered} \Psi^{(\nu)\theta}_m(E^\nu_i) = \sum_{\kappa=1}^{\tau_\nu} a^\theta_\kappa(E^\nu_i)\psi^{(\nu)\kappa}_m \ , \quad \tau_\nu = \sum_{i=1}^{n_\nu} \theta^\nu_i \ , \\ i = 1, 2, \ldots, n_\nu, \quad \theta = 1, 2, \ldots, \theta^\nu_i \ . \end{gathered} \tag{3-334d}$$

Note that the degeneracy of the energy level E^ν_i is $\theta^\nu_i h_\nu$.

If in (3-334c) there are τ_ν distinct eigenvalues, $E^\nu_i, i = 1, 2, \ldots, \tau_\nu$, then E^ν_i can serve as the multiplicity label κ in (3-332a) for differentiating the τ_ν sets of equivalent bases of the irrep ν. In this case, the degeneracy of the energy E^ν_i is equal to the dimension of the irrep ν.

If a certain irrep ν occurs only once in the rep $\mathcal{D}(G)$, i.e., $\tau_\nu = 1$, then the irreducible basis $\psi^{(\nu)}_m$ is just the eigenfunction of the Hamiltonian, $\Psi^{(\nu)}_m(E) = \psi^{(\nu)}_m$. This means that for this special case, from group theory alone we can obtain the eigenfunction of the Hamiltonian. Of course, this happens only for very special energy levels; otherwise quantum mechanics could be substituted by group theory and this is evidently absurd.

2. *Using the eigenvalue of the CSCO-II of G to label energy levels.* Since $[H, G] = 0$, we have

$$[H, C] = 0, \quad [H, C(s)] = 0 \ , \tag{3-335}$$

i.e., C and $C(s)$ are constants of motion. Therefore the eigenvalues (ν, m) of the CSCO-II of $G, (C, C(s))$, are good quantum numbers and can be used to label (but usually not uniquely) the energy level of the system.

3. According to Theorem 2.7, the degenerate wave functions belonging to the same energy E carry a rep $\mathcal{D}^E$ for the group G, which is in general reducible.

Non-accidental degeneracy: If the rep $\mathcal{D}^E(G)$ is irreducible, say it is the irrep $D^{(\nu)}$, then the degeneracy of the energy level E is equal to the dimension h_ν of the irrep ν, and the eigenvalues of the CSCO-II of G provides enough labels to distinguish the degenerate states with the energy E. This kind of degeneracy is called *non-accidental degeneracy* which is entirely due to the inherent symmetry of the system.

Accidental degeneracy: If the rep $\mathcal{D}^E(G)$ is reducible, $\mathcal{D}^E = \sum_\nu \theta_\nu D^{(\nu)}$, the degeneracy of the energy level is equal to $\sum_\nu \theta_\nu h_\nu$. Sometimes it is called accidental degeneracy. However, one should distinguish the following two cases. (a) The full degeneracy may happen by chance or by design. For example, a numerical parameter in the Hamiltonian may be adjusted until two energy levels cross. This is truely an *accidental degeneracy*. (b) If, however, one finds what appears to be systematical accidental degeneracies, not just a single level, but in many levels, then it is usually a sign that the degeneracies are not accidental, but are due to the existence of some hitherto unsuspected higher symmetry. In such cases, one should not use the word "accidental."

As an example, consider the hydrogen atom for which the potential is $V = -1/r$. The obvious symmetry group is the rotation group R_3. The energy of the hydrogen atom depends only on the principle quantum number n but is independent of the angular momentum l, which is the irrep label for R_3 (Sec. 6.3). Therefore, the rep $\mathcal{D}^{(n)}$ of R_3 carried by the degenerate wave functions with the same energy E_n is reducible: $\mathcal{D}^{(n)} = \sum_{l=0}^{n} \oplus D^{(l)}$. These degeneracies are systematic and are due to the fact that the hydrogen has a higher symmetry group, the four-dimensional rotation group R_4 (Wybourne 1974). The generators of the group R_4 are the three components of the angular momentum, L_x, L_y, and L_z, and the three components of the vector

$$A = -\left[\frac{1}{2Me^2}(\mathbf{p}\times\mathbf{L} - \mathbf{L}\times\mathbf{p}) - \frac{\mathbf{r}}{r}\right] .$$

For the alkali atom, $V = -\frac{1}{r} + \frac{\epsilon}{r^2}$, the degeneracy of the energy in l is lifted, but the degeneracies in the magnetic quantum number m persist. The $(2l+1)$-fold degeneracy is non-accidental which is characteristic of the central force and cannot be reduced without breaking the rotational symmetry of the potential.

3.19.2. Splitting of the energy level due to a perturbation

Let $H = H_0 + H_1$, where H_0 is the principle term and H_1 is a perturbation. We also assume that

$$[H_0, G] = 0, \quad [H_1, G_1] = 0, \quad \text{but } [H_1, G] \neq 0 , \tag{3-336}$$

where G_1 is a subgroup of G. After the onset of the perturbation, the symmetry group of the system is lowered from G to G_1.

Evidently, the best way to treat such a system is to use the $G \supset G_1 \supset G_1(s)$ irreducible basis $\psi_{\Lambda m}^{(\nu)}$, where for simplicity we assume that $G \supset G_1 \supset G_1(s)$ is a canonical subgroup chain, and $G_1(s)$ can be chosen arbitrarily so long as $G_1 \supset G_1(s)$ is canonical. According to (3-320) we have

$$\langle \psi_{\Lambda m}^{(\nu)} | H_0 | \psi_{\Lambda' m'}^{(\nu')} \rangle = E_\nu^{(0)} \delta_{\nu\nu'} \delta_{\Lambda\Lambda'} \delta_{mm'} , \tag{3-337a}$$

$$\langle \psi_{\Lambda m}^{(\nu)} | H_1 | \psi_{\Lambda' m'}^{(\nu')} \rangle = E_{\nu\nu'\Lambda}^{(1)} \delta_{\Lambda\Lambda'} \delta_{mm'} . \tag{3-337b}$$

The diagonal element of H_1 is

$$\langle \psi_{\Lambda m}^{(\nu)} | H_1 | \psi_{\Lambda m}^{(\nu)} \rangle = E_{\nu\Lambda}^{(1)} . \tag{3-337c}$$

Without the perturbation, the symmetry group of the system is G, and ν, Λ, and m are all good quantum numbers, and the degeneracy of each energy level, if there is no accidental degeneracy, is h_ν.

With the perturbation, only Λ and m are good quantum numbers, while ν is only an approximate quantum number, and each level $E_\nu^{(0)}$ is split into several sublevels, as schematically shown in Fig. 3.2. The number of the sublevels is determined by the subduction of the irrep $D^{(\nu)}$ of G with respect to its subgroup G_1.

$$D^{(\nu)} \downarrow G_1 = \sum_\Lambda \oplus \tau_\Lambda D^{(\Lambda)}, \tag{3-337d}$$

where the multiplicity τ_Λ can be decided upon from (3-250). The magnitude of the splitting is determined by (3-337c). Assuming that there is no accidental degeneracy, the degeneracy of each sublevel Λ is equal to the dimension h_Λ of the irrep Λ of G_1.

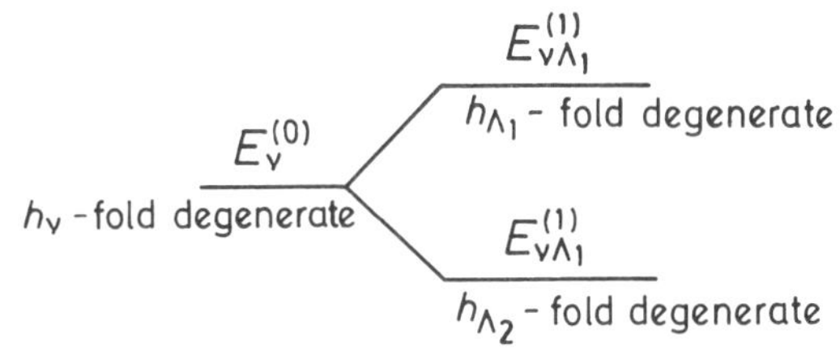

Fig. 3.2. The splitting of energy levels due to symmetry breaking.

We thus see that with the help of group theory, we can easily predict the number of the sublevels into which a given original energy level is split, as well as the magnitude of the splitting. Furthermore, the $G \supset G_1 \supset G_1(s)$ irreducible basis is the approximate (without considering the mixing of the various irreps of G) wave function for the perturbated Hamiltonian.

3.19.3. Dynamical symmetry

If the Hamiltonian of a system is a function of the CSCO of the subgroups contained in a group chain $G \supset G_1 \supset G_2 \supset \ldots$, we say that the system has a dynamical symmetry. Letting C_i be the CSCO of G_i, we have

$$H = F(C, C_1, C_2, \ldots) = F(C, C(s)) , \tag{3-338a}$$

where $C(s) = (C_1, C_2, \ldots)$. In this case, $[H, G] \neq 0$; however

$$[H, C] = 0, \quad [H, C(s)] = 0 . \tag{3-338b}$$

Therefore the eigenvalues (ν, m) of $(C, C(s))$ are good quantum numbers. Naturally, we choose the $G \supset G_1 \supset G_2 \supset \ldots$ irreducible basis as our basis for the system. In this basis, the Hamiltonian is diagonalized. Thus we can obtain an analytic expression for the energy

$$E = F(\nu, \mu_1, \mu_2, \ldots) = F(\nu, m) , \tag{3-338c}$$

without having to solve the Schrödinger equation

It induces splitting of the energy level associated with an irrep of G, but not mixing of the energy levels associated with different irreps of G.

Examples of the dynamical symmetries are legion:

1. The quark model in particle physics (de Swart 1963, Lichtenberg 1978). The group chains are

$$SU_3 \supset SU_2 \times U_1, \quad SU_4 \supset SU_3 \times U_1 \supset SU_2 \times U_1, \text{ etc.}$$

2. The interacting boson model in nuclear physics (Arima, 1979, see Sec. 9.9). There are three dynamical subgroup chains

$$\begin{aligned} &U_6 \supset U_5 \supset SO_5 \supset SO_3, \\ &U_6 \supset SU_3 \supset SO_3, \\ &U_6 \supset SO_6 \supset SO_3 . \end{aligned} \tag{3-338d}$$

Sections 9.5-9.10 are devoted to the various dynamical symmetries of nuclei.

3.19.4. General case

For the general case which does not fall into the previous three cases, group theory still plays an important role in the many-body problem. We may choose an appropriate group chain $G \supset G(s)$ and use the $G \supset G(s)$ irreducible basis to diagonalize the Hamiltonian. The appropriate group chain means that (1) $G(s)$ contains the symmetry group of the system and (2) the matrix elements of the Hamiltonian H in the basis are easily calculable (see Chap. 9).

3.19.5. Selection rule

Suppose that the operators inducing transitions between energy levels form an irreducible tensor of the symmetry group G of a system. According to the Wigner-Eckart theorem (3-317), the transition amplitude between two states belonging to the irreps ν_i and ν_f is proportional to

$$\langle\psi_{m_f}^{(\nu_f)}|T_m^{(\nu)}|\psi_{m_i}^{(\nu_i)}\rangle = \sum_\tau (C_{\nu_i m_i, \nu m}^{(\nu_f)\tau, m_f})^* \langle\psi^{(\nu_f)} \, \|T^{(\nu)}\| \, \psi^{(\nu_i)}\rangle^{(\tau)} \, . \tag{3-339}$$

If the irrep ν_f is not contained in the Kronecker product $(\nu) \times (\nu_i)$, then the CG coefficient in (3-339) is zero, and the transition amplitude vanishes. Therefore, based on the CG series of the symmetry group G, we can predict which transitions are forbidden. For the simply reducible case, (3-339) also gives the relative strengths of transitions from a given initial state $\psi_{m_i}^{(\nu_i)}$ to different final states belonging to the same irrep ν_f,

$$|\langle\psi_{m'_f}^{(\nu_f)}|T_{m'}^{(\nu)}|\psi_{m_i}^{(\nu_i)}\rangle|^2 \, / \, |\langle\psi_{m_f}^{(\nu_f)}|T_m^{(\nu)}|\psi_{m_i}^{(\nu_i)}\rangle|^2 = (C_{\nu_i m_i, \nu m'}^{\nu_f m'_f})^2 / (C_{\nu_i m_i, \nu m}^{\nu_f m_f})^2 \, .$$

Hence from the CG coefficients we can infer the relative strengths of the transitions.

Ex. 3.19. Show that in the Kronecker product space the generalized projection operator is equal to $P_{mk}^{(\nu)} = \sum_{\tau=1}^{(\nu_1\nu_2\nu)} |\nu m\tau\rangle\langle\nu k\tau|$.

Ex. 3.20. Using Theorem 2.2 and the fact that $[C, \overline{C}(s)] = 0$, and C is a CSCO of the class space, is it possible to conclude that the intrinsic operator set $\overline{C}(s)$ is a function of C?

3.20. Summary

Thus far, all the important theorems for finite groups have been re-established through a route quite different from the traditional one. The traditional approach relies heavily on the character theory, whereas the new approach is based on the decomposition of the regular rep space by a set of commuting operators. The former approach may seem to be more elegant from a mathematical point of view. However, it has the fatal drawback that it does not provide us with any practical method for reducing the regular rep. The new approach, though a bit lengthy in proving some theorems, is very instructive in nature. It not only offers more insights into the group structure, thereby revealing the duality between the group G and its intrinsic group $\overline{G}$, but also gives a simple and universal method for decomposing the regular rep into irreps subducted according to any given subgroup chain without any knowledge of the characters. Consequently, the main advantages of the new approach are its applicability and flexibility.

The basic contents of this chapters can be summarized in the following seven theorems.

Theorem I: The eigenoperator $P^{(\nu)}$ of the CSCO-I of G in the class space is the projection operator onto the irrep (ν) of G,

$$CP^{(\nu)} = \nu P^{(\nu)},$$
$$P^{(\nu)} = \frac{h_\nu}{g} \sum_i \chi_i^{(\nu)*} C_i \, . \tag{3-340}$$

Theorem II: The eigenfunctions of the CSCO-I of G in the space of functions on classes are complex conjugates of the simple characters

$$C\chi^{(\nu)*}(C_i) = \nu\chi^{(\nu)*}(C_i) . \tag{3-341}$$

The eigenfunctions of the CSCO-I obey the orthonormality and completeness

$$\sum_{i=1}^{N} \frac{g_i}{g}\chi^{(\nu)*}(C_i)\chi^{(\nu')}(C_i) = \delta_{\nu\nu'} \tag{3-342}$$

$$\sum_{\nu=1}^{N} \frac{g_i}{g}\chi^{(\nu)*}(C_i)\chi^{(\nu)}(C_j) = \delta_{ij} . \tag{3-343}$$

The simple characters can be determined from (3-341) and (3-342).

Theorem III: A necessary and sufficient condition for $\psi^{(\nu)}$ to belong to the irrep (ν) of G is that $\psi^{(\nu)}$ satisfies the eigenequation

$$C\psi^{(\nu)} = \nu\psi^{(\nu)} . \tag{3-344}$$

Theorem IV: A necessary and sufficient condition for $\psi_m^{(\nu)}$ to belong to the irreps (ν, m) of the group chain $G \supset G(s)$ is that $\psi_m^{(\nu)}$ satisfy the eigenequations

$$\begin{pmatrix} C \\ C(s) \end{pmatrix} \psi_m^{(\nu)} = \begin{pmatrix} \nu \\ m \end{pmatrix} \psi_m^{(\nu)} . \tag{3-345}$$

If $G \supset G(s)$ is a canonical subgroup chain, then $(C, C(s))$ is a CSCO-II of G.

Theorem V: A necessary and sufficient condition for $\psi_m^{(\nu)\kappa}$ to belong to the irreps (ν, m) and (ν, κ) of the group chains $G \supset G(s)$ and $\overline{G} \supset \overline{G}(s')$, respectively, is that $\psi_m^{(\nu)\kappa}$ satisfy the eigenequations

$$\begin{pmatrix} C \\ C(s) \\ \overline{C}(s') \end{pmatrix} \psi_m^{(\nu)\kappa} = \begin{pmatrix} \nu \\ m \\ \kappa \end{pmatrix} \psi_m^{(\nu)\kappa} . \tag{3-346}$$

Here the intrinsic subgroup chain $\overline{G}(s')$ does not necessarily correspond to the subgroup chain $G(s)$. If $G \supset G(s)$ is a canonical subgroup chain, then $(C, C(s), \overline{C}(s))$ is a CSCO-III of G.

Theorem VI: The eigenoperator of the CSCO-III of G in the group space is the generalized projection operator $P_{m\kappa}^{(\nu)}$

$$\begin{pmatrix} C \\ C(s) \\ \overline{C}(s) \end{pmatrix} P_{mk}^{(\nu)} = \begin{pmatrix} \nu \\ m \\ k \end{pmatrix} P_{mk}^{(\nu)} . \tag{3-347}$$

$$P_{mk}^{(\nu)} = \frac{h_\nu}{g} \sum_a D_{mk}^{(\nu)*}(R_a)R_a . \tag{3-348}$$

Theorem VII: The eigenfunctions of the CSCO-III of G in the space of functions on the group (manifold) are complex conjugates of the irreducible matrix elements in the $G \supset G(s)$ basis,

$$\begin{pmatrix} C \\ C(s) \\ \overline{C}(s) \end{pmatrix} D_{mk}^{\nu)*}(R_a) = \begin{pmatrix} \nu \\ m \\ k \end{pmatrix} D_{mk}^{(\nu)*}(R_a) . \tag{3-349}$$

The eigenfunctions of the CSCO-III of G satisfy the orthonormality and completeness

$$\sum_{a} \frac{h_\nu}{g} D^{(\nu)*}_{mk}(R_a) D^{(\nu')}_{m'k'}(R_a) = \delta_{\nu\nu'}\delta_{mm'}\delta_{kk'} , \tag{3-350a}$$

$$\sum_{\nu mk} \frac{h_\nu}{g} D^{(\nu)*}_{mk}(R_a) D^{(\nu)}_{mk}(R_b) = \delta_{ab} . \tag{3-350b}$$

From the Eqs. (3-349) and (3-350a) we can determine the irreducible matrix elements.

Chapter 4

Representation Theory of the Permutation Group

The permutation group has important applications in physics for many-particle systems. The importance lies in the fact that a system of identical particles has permutation symmetry and that there exist many deep and delicate inter-relations between the permutation group and the unitary group. The rep theory of the permutation group is well-established through the efforts of Young, Frobenious, Yamanouchi and others. This theory has many advantages. For example, it gives the branching law for reducing the irrep of S_n into those of S_{n-1}, the intuitive and elegant way of labelling the irreps and the irreducible bases by the Young diagrams and the Young tableaux, etc.

However, from the practical point of view, this theory has some serious shortcomings. (1) It is too difficult for physicists to grasp the theory quickly. Soklov(1956) pointed out: "Due to the fact that group theory, especially the theory of representations and characters of the permutation group, is extremely difficult even for the specialists, that there arose the tendency of opposing the so-called 'group-pest' in quantum mechanics." (2) When it comes to actual calculation, it is rather tedious to obtain the characters, the Yamanouchi bases, the Clebsch-Gordan coefficients, etc.

From the general theory for finite groups and the example of the permutation group S_3 in the last chapter, it is seen that the new approach to group rep theory is an independent theory in the sense that we can obtain, through straightforward, standard and easily programmable calculations, all the results without using the traditional Young-Yamanouchi theory. Nevertheless, it has its own drawbacks, namely, it is not possible to obtain general conclusions about the dimensionality of irreps and the branching law prior to concrete calculations.

From the above discussion it is seen that the advantages and disadvantages of the Young-Yamanouchi theory and the new approach of the permutation group are complementary; one's advantage is just the other's disadvantage. Thus, from the practical point of view, we can combine these two to give a powerful method for handling the rep of the permutation group.

The first four sections of this chapter deal mainly with the relations between the new and old theories of the permutation group and some general conclusions obtained in the traditional theory of the permutation group. They are simple in themselves, yet the paths along which these conclusions were reached are by no means straightforward. If one is only interested in the application of the permutation group, it is sufficient to understand these conclusions without the need of tracing them back to their origins. Readers interested in the traditional theory of permutation groups are referred to Rutherford (1948) and Hamermesh (1962). The remaining sections deal with the problem of how to use these general conclusions borrowed from the traditional theory of the permutation group to simplify the calculation based on the EFM.

4.1. Partition, Young Diagrams and Eigenvalues of CSCO-I

In the traditional theory of the permutation group, we use partitions or the Young diagrams to label the irreps of the permutation group S_n.

The splitting-up of n into a sum of n integers as in

$$\begin{aligned} &n = \nu_1 + \nu_2 + \ldots + \nu_n\,, \\ &\nu_1 \geq \nu_2 \geq \ldots \geq \nu_n \geq 0\,, \end{aligned} \tag{4-1}$$

is called a *partition*. For example, 4=4+0+0+0, 4=3+1+0+0, 4=2+2+0+0, 4=2+1+1+0, 4=1+1+1+1. We use the symbol $[\nu] = [\nu_1\nu_2\ldots\nu_n]$ to designate a partition. We usually drop those ν_i's which are zeroes. Thus $n = 4$ has the following partitions [4], [31], [22], [211], and [1111], or in abbreviated form [4], [31], $[2^2]$, $[21^2]$ and $[1^4]$. In the traditional theory of the permutation group it was shown that irreps of the permutation group S_n are labeled by partitions of n. The number of partitions of n is just the number of inequivalent irreps of S_n.

A partition $[\nu_1\nu_2\ldots]$ can be pictured by a *Young diagram* in which boxes are arranged in rows and columns with ν_i boxes occupying the first ν_i positions (counted from the left) of the i-th row. As an example, the Young diagrams corresponding to the partitions of $n = 4$ are

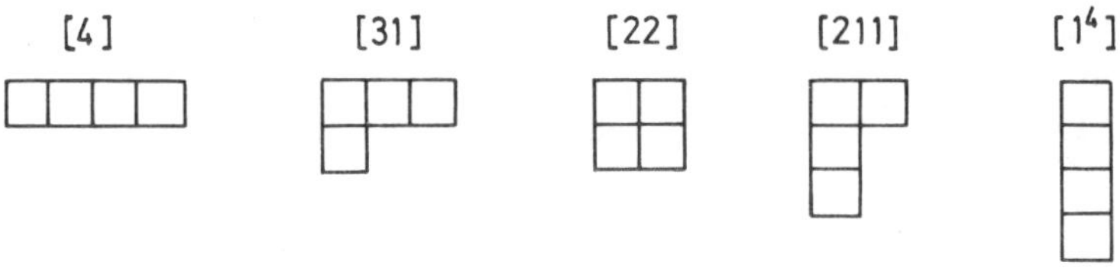

(4-2)

There is a one-to-one correspondence between the eigenvalues of the CSCO-I of S_n and the partitions or Young diagrams. It can be proved (see Eqs. (4-25) and (7-38d)) that the relations between the eigenvalues of the 2-cycle and 3-cycle class operators and the partitions $[\nu] = [\nu_1\nu_2\ldots\nu_k]$ are given by

$$\begin{aligned} \lambda_{(2)}^{(\nu)} &= \frac{n}{2} + \frac{1}{2}\sum_{l=1}^{k} \nu_l(\nu_l - 2l), \\ \lambda_{(3)}^{(\nu)} &= \frac{1}{3}\left\{2n - \frac{3}{2}n^2 + \sum_{l=1}^{k} \nu_l\left[\nu_l^2 - \left(3l - \frac{3}{2}\right)\nu_l + 3l(l-1)\right]\right\}. \end{aligned} \tag{4-3}$$

In Table 3.2-1 the eigenvalues $\lambda_{(2)}^{(\nu)}$ and $\lambda_{(3)}^{(\nu)}$ are listed along with the corresponding partitions.

Using the conclusion that each partition labels uniquely an irrep of the permutation group and the relation (4-3) or the like (such as the relation between the eigenvalues of 4-cycle class operator and the partitions, etc.), we can easily find the CSCO-I of the permutation group. For example, if $\lambda_{(2)}^{(\nu)}$ are distinct for all partitions of n, then the 2-cycle class operator $C_{(2)}$ is a CSCO-I of S_n. Otherwise, we look at the pairs of values $(\lambda_{(2)}^{(\nu)}, \lambda_{(3)}^{(\nu)})$; if they are different for all partitions of n, then $(C_{(2)}, C_{(3)})$ is a CSCO-I of S_n, etc.

The reason why an irrep of S_n is labeled by a partition of n can be understood from the facts that a partition $[\nu] = [\nu_1, \nu_2, \ldots \nu_n]$ corresponds to a class label $(\nu) = (\nu_1\nu_2\ldots\nu_n)$ of S_n, and the number of inequivalent irreps of S_n is equal to the number of classes.

Two Young diagrams or partitions are said to be *conjugate* if they are obtained from each other by an interchange of rows and columns. The partition conjugate to $[\nu]$ is designated as $[\tilde{\nu}]$. For example, in (4-2), [31] and [211] are conjugate partitions.

A Young diagram $[\nu]$ is said to be *self-conjugate* if $[\nu] = [\tilde{\nu}]$. For example, [22] and [311] are self-conjugate.

Two reps labeled by $[\nu]$ and $[\tilde{\nu}]$ are called *conjugate reps* (not to be confused with the complex conjugate rep). A rep labeled by self-conjugate Young diagram is called a *self-conjugate rep*. It is clear that conjugate reps always appear in pairs.

4.2. Characters of Permutation Group

The Frobenius theory of characters (Hamermesh 1962) plays a crucial role in the traditional theory of the permutation group. It is an elegant, nevertheless very difficult theory.

The dimension of the irrep $[\nu] = [\nu_1\nu_2\ldots\nu_k]$ of the permutation group S_n is given by the Frobenius formula (Bohr 1969)

$$h_{[\nu]} = \chi_e^{(\nu)} = n! \prod_{i<j\le k} (\nu_i - \nu_j + j - i) \Big/ \prod_{i=1}^{k} (\nu_i + k - i)! \,. \tag{4-4a}$$

A simpler formula is given by Robinson (1961). With each box of a Young diagram we associate a *hook* consisting of the given box (the head node) together with all these boxes that are in the same row and in the same column as the head node. The number of boxes in the hook is called the *hook length*. The dimension $h_{[\nu]}$ can be expressed as

$$h_{[\nu]} = n! \Big/ \prod_i l_i \,, \tag{4-4b}$$

where $\Pi_i l_i$ is the product of all the hook lengths in the Young diagram $[\nu]$. For example

$$h_{[31]} = \frac{4!}{\begin{array}{|c|c|c|}\hline 4 & 2 & 1 \\ \hline 1 \\ \cline{1-1}\end{array}} = \frac{4!}{4\cdot2\cdot1\cdot1} = 3 \,,$$

$$h_{[321]} = \frac{6}{\begin{array}{|c|c|c|}\hline 5 & 3 & 1 \\ \hline 3 & 1 \\ \cline{1-2} 1 \\ \cline{1-1}\end{array}} = \frac{6!}{5\cdot3\cdot3\cdot1\cdot1\cdot1\cdot1} = 16 \,.$$

The dimensions of the irreps of the permutation groups up to S_6 are given in Table A1 in the Appendix.

For the simple characters of the permutation groups up to S_7, see Hamermesh (1962) p. 276, up to S_{10}, see Littlewood (1958). The EFM can also be used to calculate the simple characters of the permutation group. The characters of S_{11} have been calculated in this way by Gao (1976).

Ex. 4.1. Use Eq. (4-4b) to calculate the dimensions of the irreps of $S_3 - S_6$.

4.2.1. Characters of conjugate representations

In the Frobenius theory, using the so-called "regular application of r dots method," the relation between characters of conjugate representations has been shown to be

$$\chi_i^{(\tilde{\nu})} = \delta_i \chi_i^{(\nu)} \,, \tag{4-5}$$

where δ_i is the parity of the i-th class. This can be proved much more easily by using the EFM.

Suppose the CSCO-I of S_n is $C = (C_1, \ldots, C_l)$. From (3-28) we know that $q_j^{(\nu)}$ satisfy the following sets of eigenequations

$$\sum_{j=1}^{N} C_{ij}^k q_j^{(\nu)} = \lambda_i^{(\nu)} q_k^{(\nu)}, \quad i = 1, 2, \ldots l \,. \tag{4-6}$$

Since the permutation parities at both sides of $C_iC_j = \sum_k C_{ij}^k C_k$ must be equal, we have

$$C_{ij}^k = 0, \quad \text{for} \quad \delta_i\delta_j \neq \delta_k \,. \tag{4-7}$$

From (4-6) and (4-7)

$$\sum_{j=1}^{N} C_{ij}^{k}(\delta_j q_j^{(\nu)}) = (\delta_i \lambda_i^{(\nu)})(\delta_k q_k^{(\nu)}), \quad i = 1, 2, \ldots, l\,. \tag{4-8}$$

Therefore if $\lambda^{(\nu)} = (\lambda_1^{(\nu)}, \ldots, \lambda_l^{(\nu)})$ is an eigenvalue set of C, then

$$\lambda^{(\tilde{\nu})} = (\delta_1 \lambda_1^{(\nu)}, \ldots \delta_l \lambda_l^{(\nu)}) \tag{4-9a}$$

is necessarily an eigenvalue set of C, and the eigenvector of the irrrep $(\tilde{\nu})$ is

$$q_j^{(\tilde{\nu})} = \delta_j q_j^{(\nu)}, \quad j = 1, 2, \ldots, N\,. \tag{4-9b}$$

From this result and the relation $\chi_j^{(\nu)} = \sqrt{g} q_j^{(\nu)*}$ (Eq. (3-245)), we obtain (4-5).

4.3. Branching Law, Young-Yamanouchi Basis, and Young Tableaux

An irreducible representation $[\nu]$ of S_n is in general reducible with respect to its subgroup S_{n-1}, $[\nu] \downarrow S_{n-1} = \sum_{\nu'} \oplus D^{[\nu']}(S_{n-1})$, where the Young diagrams $[\nu']$ result from removing one box in all possible way in the Young diagram $[\nu]$, and each irrep $[\nu']$ occurs only once (Hamermesh 1962).

For example the irrep [431] of S_8 contains each of the irreps [43], [421] and [331] once.

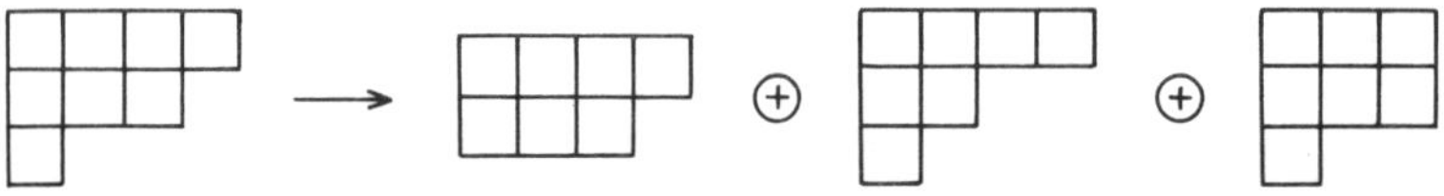

Dimension: 70 = 14 + 35 + 21 (4-10)

In general, we have

$$[\nu_1 \nu_2 \ldots \nu_i \ldots \nu_n] = \sum_{i=n}^{1}{}' [\nu_1, \nu_2, \ldots \nu_i - 1, \ldots \nu_n]. \tag{4-11a}$$

The prime in the summation indicates that we must have $\nu_i - 1 \geq 0$ and $\nu_i - 1 > \nu_{i+1}$ in order that $[\nu']$ is a Young diagram. The dimensions of the both sides of (4-11a) must be equal; therefore

$$h_{[\nu_1 \ldots \nu_i \ldots \nu_n]} = \sum_{i=n}^{1}{}' h_{[\nu_1 \ldots \nu_i - 1 \ldots \nu_n]} \tag{4-11b}$$

Equation (4-11a) is referred to as the *branching law* for the permutation group.

4.3.1. The Young-Yamanouchi basis

The $S_n \supset S_{n-1} \supset \ldots \supset S_2 \supset S_1$ irreducible basis is now accepted as the *standard basis* of the permutation group. It is also called the Young-Yamanouchi basis. From the branching law (4-11a) and the fact that the group S_2 is Abelian, we know that the group chain $S_n \supset S_{n-1} \supset \ldots \supset S_2$ is canonical (the last group S_1 is redundant).

A *Young tableau* is an arrangement of the numbers 1,2..., n in a Young diagram in which numbers increase as one moves to the right and as one goes down. A Young tableau is denoted by $Y_m^{(\nu)}$, where m is the index of the tableau.

From a given Young tableau $Y_m^{[\nu]}$, one obtains another Young tableau $Y_{m'}^{[\nu']}$ involving $n-1$ numbers by removing the box containing the number n; by removing the box with the number

$n-1$ one obtains yet another Young tableau $Y_{m''}^{[\nu'']}$ and so on. We use $|[\nu]m\rangle$ or $|Y_m^{[\nu]}\rangle$ to denote a Yamanouchi basis vector. The symbol $|Y_m^{[\nu]}\rangle$ will stand for an irreducible basis vector belonging to the irrep $[\nu], [\nu'], [\nu''] \ldots, [1]$ of the group $S_n, S_{n-1}, S_{n-2}, \ldots, S_1$. For example

$$\begin{array}{c} \begin{array}{|c|c|c|} \hline 1 & 3 & 4 \\ \hline 2 & 5 & \\ \cline{1-2} \end{array} \\ [32] \end{array} \longrightarrow \begin{array}{c} \begin{array}{|c|c|c|} \hline 1 & 3 & 4 \\ \hline 2 & & \\ \cline{1-1} \end{array} \\ [31] \end{array} \longrightarrow \begin{array}{c} \begin{array}{|c|c|} \hline 1 & 3 \\ \hline 2 & \\ \cline{1-1} \end{array} \\ [21] \end{array} \longrightarrow \begin{array}{c} \begin{array}{|c|} \hline 1 \\ \hline 2 \\ \hline \end{array} \\ [11] \end{array} \longrightarrow \begin{array}{c} \begin{array}{|c|} \hline 1 \\ \hline \end{array} \\ [1] \end{array}$$

The irreducible basis vector $\begin{array}{|c|c|c|} \hline 1 & 3 & 4 \\ \hline 2 & 5 & \\ \cline{1-2} \end{array}$ belongs to the irreps [32], [31], [21], [11] and [1] of the groups S_5, S_4, S_3, S_2 and S_1, respectively. The irrep of the last group S_1 is always [1], and thus can be omitted.

The labeling by the sequence $[\nu], [\nu'], [\nu''], \ldots, [1]$ is complete in the sense that no two basis vectors of S_n will have the same sequence and that, to every sequence, there corresponds a basis vector of S_n. By the completeness of this labelling system, it follows immediately that the dimension of the irrep $[\nu]$ is equal to the number of distinct tableaux obtainable from the Young diagram $[\nu]$.

It is seen that the decomposition of a Young tableau is consistent with the branching law.

Besides the Young tableaux, we can use the *Yamanouchi symbols* $(r_n r_{n-1} \ldots r_2 r_1)$ to label a Yamanouchi basis vector, where r_i is the row number of the letter i in the Young tableau. According to the definition of the Young tableaux, the letter 1 is always in the upper left-hand corner of a Young tableau; therefore $r_1 \equiv 1$.

Once we have Yamanouchi symbols for each Young tableau of the irrep $[\nu]$, we arrange them in decreasing page order and assign an index m to each symbol. For example

Young tableaux	$\begin{array}{\|c\|c\|c\|} \hline 1 & 2 & 3 \\ \hline 4 & & \\ \cline{1-1} \end{array}$	$\begin{array}{\|c\|c\|c\|} \hline 1 & 2 & 4 \\ \hline 3 & & \\ \cline{1-1} \end{array}$	$\begin{array}{\|c\|c\|c\|} \hline 1 & 3 & 4 \\ \hline 2 & & \\ \cline{1-1} \end{array}$
Yamanouchi symbols	(2111)	(1211)	(1121)

It can be seen that the largest Yamanouchi symbol corresponds to the smallest index $m = 1$.

The Young tableaux for permutation groups up to S_6 are listed in Table 4.4-1. The Yamanouchi symbols for $S_3 - S_5$ are listed in Table 4.4-2.

For typographical reason, in the following we often delete the square boxes in Young tableaux.

4.4. Yamanouchi Matrix Elements

From the branching law (4-11a) and the fact that the Yamanouchi basis is an irreducible basis classified according to the irreps of $S_n \supset S_{n-1} \supset \ldots \supset S_2$, we know that in the irrep $[\nu]$ of S_n, the representatives $D^{[\nu]}(R)$ of the elements R belonging to the subgroup S_{n-1} must take the block-diagonal form

$$D^{[\nu_1 \ldots \nu_i \ldots \nu_n]}(R) = \sum_{i=n}^{1}{}' \oplus D^{[\nu_1 \ldots \nu_i - 1 \ldots \nu_n]}(R), \quad R \in S_{n-1} \ . \tag{4-12}$$

Table 4.4-1. The phase factors $\Lambda_m^{[\nu]*}$, Young tableaux $Y_m^{[\nu]}$ and the corresponding eigenvalues $\lambda=\Sigma_{f=3}^{n}(2f-5)\,\lambda_f$.†

$[\nu]$	[21]		[31]			[22]		[211]		
m	1	2	1	2	3	1	2	1	2	3
Λ_m^ν	+	−	+	−	+	+	−	+	−	+
Y_m^ν	12 3	13 2	123 4	124 3	134 2	12 34	13 24	12 3 4	13 2 4	14 2 3
λ			9	6		0		−6		

$[\nu]$	[41]				[32]					[311]					
m	1	2	3	4	1	2	3	4	5	1	2	3	4	5	6
Λ_m^ν	+	−	+	−	+	−	+	+	−	+	−	+	+	−	+
Y_m^ν	1234 5	1235 4	1245 3	1345 2	123 45	124 35	134 25	125 34	135 24	123 4 5	124 3 5	134 2 5	125 3 4	135 2 4	145 2 3
λ	46	34	31		19	16		10		9	6		−6		

$[\nu]$	[221]					[2111]				[51]				
m	1	2	3	4	5	1	2	3	4	1	2	3	4	5
Λ_m^ν	+	−	−	+	−	+	−	+	−	+	−	+	−	+
Y_m^ν	12 34 5	13 24 5	12 35 4	13 25 4	14 25 3	12 3 4 5	13 2 4 5	14 2 3 5	15 2 3 4	12345 6	12346 5	12356 4	12456 3	13456 2
λ	−10		−16			−31				134	109	97	94	

$[\nu]$	[42]									[411]					
m	1	2	3	4	5	6	7	8	9	1	2	3	4	5	6
Λ_m^ν	+	−	+	−	+	−	+	+	−	+	−	+	−	+	−
Y_m^ν	1234 56	1235 46	1245 36	1345 26	1236 45	1246 35	1346 25	1256 34	1356 24	1234 5 6	1235 4 6	1245 3 6	1345 2 6	1236 4 5	1246 3 5
λ	81	69	66		54	51		45		67	55	52		30	27

$[\nu]$	[411]				[33]					[321]					
m	7	8	9	10	1	2	3	4	5	1	2	3	4	5	6
Λ_m^ν	+	+	−	+	+	−	+	+	−	+	−	+	+	−	−
Y_m^ν	1346 2 5	1256 3 4	1356 2 4	1456 2 3	123 456	124 356	134 256	125 346	135 246	123 45 6	124 35 6	134 25 6	125 34 6	135 24 6	123 46 5
λ		15			40	37		31		19	16		10		9

*The definition of $\Lambda_m^{[\nu]}$ is given in Sec. 4.7.

†See Eq. (4-31). The eigenvalues for the totally symmetric bases are $\overset{[3]}{3}$, $\overset{[4]}{21}$, $\overset{[5]}{76}$, $\overset{[6]}{176}$.

cont'd

$[\nu]$	[321]										[222]				
m	7	8	9	10	11	12	13	14	15	16	1	2	3	4	5
Λ_m^ν	+	−	−	+	−	−	+	+	−	+	+	−	−	+	−
Y_m^ν	124 36 5	134 26 5	125 36 4	135 26 4	145 26 3	126 34 5	136 24 5	126 35 4	136 25 4	146 25 3	12 34 56	13 24 56	12 35 46	13 25 46	14 25 36
λ	6		−6			−10		−16			−31		−37		

$[\nu]$	$[31^3]$										[2211]				
m	1	2	3	4	5	6	7	8	9	10	1	2	3	4	5
Λ_m^ν	+	−	+	+	−	+	−	+	−	+	+	−	−	+	−
Y_m^ν	123 4 5 6	124 3 5 6	134 2 5 6	125 3 4 6	135 2 4 6	145 2 3 6	126 3 4 5	136 2 4 5	146 2 3 5	156 2 3 4	12 34 5 6	13 24 5 6	12 35 4 6	13 25 4 6	14 25 3 6
λ	−12	−15		−27			−52				−45		−51		

$[\nu]$	[2211]				$[21^4]$				
m	6	7	8	9	1	2	3	4	5
Λ_m^ν	+	−	+	−	+	−	+	−	+
Y_m^ν	12 36 4 5	13 26 4 5	14 26 3 5	15 26 3 4	12 3 4 5 6	13 2 4 5 6	14 2 3 5 6	15 2 3 4 6	16 2 3 4 5
λ	−66				−94				

For example, corresponding to Eq. (4-10) we have

$$D^{[431]}(R) = \begin{pmatrix} D^{[43]}(R) & & \\ & D^{[421]}(R) & \\ & & D^{[331]}(R) \end{pmatrix}, \quad R \in S_7 . \tag{4-13}$$

More examples can be found in Table 4.4-2.

Once the irreducible matrices of the $n-1$ generators of S_n, (12), (23),..., $(n-1, n)$, are known, the irreducible matrices of all elements of S_n can be found from (1-8a) , (1-9) and the like.

There is a simple rule for finding the irreducible matrix of the adjacent transpositions $(i, i-1), i = 2, 3, \ldots, n$ in the Yamanouchi basis of S_n:

1. If $i-1$ and i are in the same row or column of a Young tableau $Y_m^{[\nu]}$,

$$(i-1, i)|Y_m^{[\nu]}\rangle = \pm|Y_m^{[\nu]}\rangle, \quad \text{when } i-1 \text{ and } i \text{ are in the same} \begin{cases} \text{row} \\ \text{column.} \end{cases} \tag{4-14a}$$

2. If $i-1$ and i are neither in the same row nor the same column of $Y_m^{[\nu]}$,

$$D^{[\nu]}_{m'm}(i-1,i) = \langle Y^{[\nu]}_{m'}|(i-1,i)|Y^{[\nu]}_m\rangle$$
$$= \begin{cases} 1/\sigma\,, & m'=m \\ \sqrt{\sigma^2-1}/|\sigma|\,, & \text{when } Y^{[\nu]}_{m'} = (i-1,i)Y^{[\nu]}_m\,, \\ 0 & \text{otherwise}\,, \end{cases} \tag{4-14b}$$

where σ is called the *axial distance* from $i-1$ to i in the Young tableau $Y_m^{[\nu]}$ and is defined as following: Starting from the position of $i-1$ in $Y_m^{[\nu]}$, we proceed by any rectangular route, one box at a time, until we reach the position of i. Counting plus one for each step made upwards or to the right, and minus one for each step made downwards or to the left, the resulting number of steps made will be the axial distance. For the Young tableaux $\begin{array}{|c|c|c|}\hline 1&2&4\\\hline 3&5\\\cline{1-2}\end{array}$ and $\begin{array}{|c|c|c|}\hline 1&2&3\\\hline 4&5\\\cline{1-2}\end{array}$ the axial distances from 3 to 4 are $+3$ and -3, respectively. Thus

$$\left\langle \begin{array}{|c|c|c|}\hline 1&2&4\\\hline 3&5\\\cline{1-2}\end{array} \right| (34) \left| \begin{array}{|c|c|c|}\hline 1&2&4\\\hline 3&5\\\cline{1-2}\end{array} \right\rangle = +\frac{1}{3} \qquad \left\langle \begin{array}{|c|c|c|}\hline 1&2&3\\\hline 4&5\\\cline{1-2}\end{array} \right| (34) \left| \begin{array}{|c|c|c|}\hline 1&2&3\\\hline 4&5\\\cline{1-2}\end{array} \right\rangle = -\frac{1}{3}$$

$$\left\langle \begin{array}{|c|c|c|}\hline 1&2&4\\\hline 3&5\\\cline{1-2}\end{array} \right| (34) \left| \begin{array}{|c|c|c|}\hline 1&2&3\\\hline 4&5\\\cline{1-2}\end{array} \right\rangle = \sqrt{\frac{8}{3}}\,, \qquad \left\langle \begin{array}{|c|c|c|}\hline 1&2&3\\\hline 4&5\\\cline{1-2}\end{array} \right| (34) \left| \begin{array}{|c|c|c|}\hline 1&2&5\\\hline 3&4\\\cline{1-2}\end{array} \right\rangle = 0 \tag{4-15}$$

The axial distance can be calculated by

$$\sigma = c_i - c_{i-1} - (r_i - r_{i-1}), \tag{4-16}$$

where $r_i, r_{i-1}(c_i, c_{i-1})$ are the row (column) numbers of the letters i and $i-1$ in the Young tableau $Y_m^{[\nu]}$ respectively.

The standard method for deriving the Yamanouchi matrix elements (4-14b) is rather laborious. In Sec. 4.20, it will be derived easily by using the EFM (Zhu 1983).

Care must be exercised to distinguish between the Young tableaux $Y_m^{[\nu]}$ and the basis vectors $|Y_m^{[\nu]}\rangle$. If $Y_{m'}^{[\nu]} = (i-1,i)Y_{m'}^{[\nu]}$ then according to Eq. (4-14b) we have

$$(i-1,i)|Y_m^{[\nu]}\rangle = \frac{1}{\sigma}\left|Y_m^{[\nu]}\right\rangle + \frac{\sqrt{\sigma^2-1}}{|\sigma|}\left|Y_{m'}^{[\nu]}\right\rangle\;; \tag{4-17}$$

however

$$(i-1,i)|Y_m^{[\nu]}\rangle \neq |Y_{m'}^{[\nu]}\rangle = |(i-1,i)Y_m^{[\nu]}\rangle.$$

The off-diagonal matrix elements in Eq. (4-14b) are always chosen to be positive. This is called the *Yamanouchi phase convention*, and the matrix elements are called the *Yamanouchi matrix elements.* In this book we will also adopt this phase convention.

Equation (4-14) hold for any $i = 2, \ldots, n$. Therefore a Yamanouchi basis vector $|Y_m^{[\nu]}\rangle$ is symmetric (antisymmetric) with respect to any adjacent indices which are at the same row (column) in $Y_m^{[\nu]}$, but has no symmetry with respect to other indices (non-adjacent indices on the same row or column, or adjacent indices in different rows or columns). For example, the irreducible basis vector

$$\left| \begin{array}{|c|c|c|c|}\hline 1&2&3&11\\\hline 4&7&8\\\cline{1-3} 5&9&12\\\cline{1-3} 6&10\\\cline{1-2}\end{array} \right\rangle$$

is symmetric in the indices (1,2,3) and (7,8) and antisymmetric in the indices (4,5,6) and (9,10), but has no definite symmetry with respect to other indices.

Ex. 4.2. Find the Yamanouchi matrices of the irreps [42] and [321] for the adjacent transpositions of S_6.

Table 4.4-2 gives the Yamanouchi matrices of the adjacent permutations of S_3 to S_5. Since they are symmetric matrices, only the upper triangles of the matrices are given.

Table 4.4-2. Yamanouchi matrix elements of adjacent transpositions for $S_2 - S_5$.

S_3: 

$$\begin{array}{c} 211 \\ 121 \end{array} \overset{(12)}{\begin{pmatrix} 1 & 0 \\ & -1 \end{pmatrix}} \quad \overset{(23)}{\begin{pmatrix} -\frac{1}{2} & \frac{\sqrt{3}}{2} \\ & \frac{1}{2} \end{pmatrix}}$$

S_4: 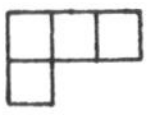

$$\begin{array}{c} 2111 \\ 1211 \\ 1121 \end{array} \overset{(12)}{\left(\begin{array}{c|cc} 1 & 0 & 0 \\ \hline & 1 & 0 \\ & & -1 \end{array}\right)} \quad \overset{(23)}{\left(\begin{array}{c|cc} 1 & 0 & 0 \\ \hline & -\frac{1}{2} & \frac{\sqrt{3}}{2} \\ & & \frac{1}{2} \end{array}\right)} \quad \overset{(34)}{\begin{pmatrix} -\frac{1}{3} & \frac{\sqrt{8}}{3} & 0 \\ & \frac{1}{3} & 0 \\ & & 1 \end{pmatrix}}$$

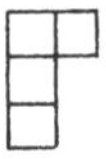

$$\begin{array}{c} 3211 \\ 3121 \\ 1321 \end{array} \overset{(12)}{\left(\begin{array}{cc|c} 1 & 0 & 0 \\ & -1 & 0 \\ \hline & & -1 \end{array}\right)} \quad \overset{(23)}{\left(\begin{array}{cc|c} -\frac{1}{2} & \frac{\sqrt{3}}{2} & 0 \\ & \frac{1}{2} & 0 \\ \hline & & -1 \end{array}\right)} \quad \overset{(34)}{\begin{pmatrix} -1 & 0 & 0 \\ & -\frac{1}{3} & \frac{\sqrt{8}}{3} \\ & & \frac{1}{3} \end{pmatrix}}$$

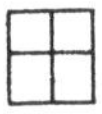

$$\begin{array}{c} 2211 \\ 2121 \end{array} \overset{(12)}{\begin{pmatrix} 1 & 0 \\ & -1 \end{pmatrix}} \quad \overset{(23)}{\begin{pmatrix} -\frac{1}{2} & \frac{\sqrt{3}}{2} \\ & \frac{1}{2} \end{pmatrix}} \quad \overset{(34)}{\begin{pmatrix} 1 & 0 \\ & -1 \end{pmatrix}}$$

S_5: 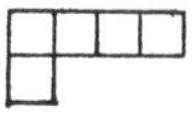

$$\begin{array}{c} 21111 \\ 12111 \\ 11211 \\ 11121 \end{array} \overset{(12)}{\left(\begin{array}{c|ccc} 1 & & & \\ \hline & 1 & 0 & 0 \\ & & 1 & 0 \\ & & & -1 \end{array}\right)} \overset{(23)}{\left(\begin{array}{c|ccc} 1 & & & \\ \hline & 1 & 0 & 0 \\ & & -\frac{1}{2} & \frac{\sqrt{3}}{2} \\ & & & \frac{1}{2} \end{array}\right)} \overset{(34)}{\left(\begin{array}{c|ccc} 1 & & & \\ \hline & -\frac{1}{3} & \frac{\sqrt{8}}{3} & 0 \\ & & \frac{1}{3} & 0 \\ & & & 1 \end{array}\right)} \overset{(45)}{\begin{pmatrix} -\frac{1}{4} & \frac{\sqrt{15}}{4} & 0 & 0 \\ & \frac{1}{4} & 0 & 0 \\ & & 1 & 0 \\ & & & 1 \end{pmatrix}}$$

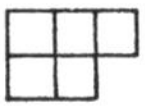

$$\begin{array}{c} 22111 \\ 21211 \\ 21121 \\ 12211 \\ 12121 \end{array} \overset{(12)}{\left(\begin{array}{ccc|cc} 1 & & & & \\ & 1 & & & \\ & & -1 & & \\ \hline & & & 1 & \\ & & & & -1 \end{array}\right)} \quad \overset{(23)}{\left(\begin{array}{ccc|cc} 1 & 0 & 0 & & \\ & -\frac{1}{2} & \frac{\sqrt{3}}{2} & & 0 \\ & & \frac{1}{2} & & \\ \hline & & & -\frac{1}{2} & \frac{\sqrt{3}}{2} \\ & & & & \frac{1}{2} \end{array}\right)}$$

cont'd

$$(34)\quad \left(\begin{array}{ccc|cc} -\frac{1}{3} & \frac{\sqrt{8}}{3} & 0 & & \\ & \frac{1}{3} & 0 & 0 & \\ & & 1 & & \\ \hline & & & 1 & 0 \\ & & & & -1 \end{array}\right) \qquad (45)\quad \begin{pmatrix} 1 & 0 & 0 & 0 & 0 \\ & -\frac{1}{2} & 0 & \frac{\sqrt{3}}{2} & 0 \\ & & -\frac{1}{2} & 0 & \frac{\sqrt{3}}{2} \\ & & & \frac{1}{2} & 0 \\ & & & & \frac{1}{2} \end{pmatrix}^{1)}$$

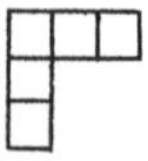

$$\begin{array}{c} 32111 \\ 31211 \\ 31121 \\ 13211 \\ 13121 \\ 11321 \end{array} \quad (12)\quad \left(\begin{array}{ccc|ccc} 1 & & & & & \\ & 1 & & & & \\ & & -1 & & & \\ \hline & & & 1 & & \\ & & & & -1 & \\ & & & & & -1 \end{array}\right) \qquad (23)\quad \left(\begin{array}{ccc|ccc} 1 & 0 & & & & \\ & -\frac{1}{2} & \frac{\sqrt{3}}{2} & & 0 & \\ & & \frac{1}{2} & & & \\ \hline & & & -\frac{1}{2} & \frac{\sqrt{3}}{2} & 0 \\ & & & & \frac{1}{2} & 0 \\ & & & & & -1 \end{array}\right)$$

$$(34)\quad \left(\begin{array}{ccc|ccc} -\frac{1}{3} & \frac{\sqrt{8}}{3} & 0 & & & \\ & \frac{1}{3} & 0 & & & \\ & & 1 & & & \\ \hline & & & -1 & 0 & 0 \\ & & & & -\frac{1}{3} & \frac{\sqrt{8}}{3} \\ & & & & & \frac{1}{3} \end{array}\right) \qquad (45)\quad \begin{pmatrix} -1 & 0 & 0 & 0 & 0 & 0 \\ & -\frac{1}{4} & 0 & \frac{\sqrt{15}}{4} & 0 & 0 \\ & & -\frac{1}{4} & 0 & \frac{\sqrt{15}}{4} & 0 \\ & & & \frac{1}{4} & 0 & 0 \\ & & & & \frac{1}{4} & 0 \\ & & & & & 1 \end{pmatrix}$$

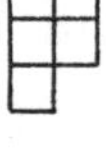

$$\begin{array}{c} 32211 \\ 32121 \\ 23211 \\ 23121 \\ 21321 \end{array} \quad (12)\quad \left(\begin{array}{cc|ccc} 1 & & & & \\ & -1 & & & \\ \hline & & 1 & & \\ & & & -1 & \\ & & & & -1 \end{array}\right) \qquad (23)\quad \left(\begin{array}{cc|ccc} -\frac{1}{2} & \frac{\sqrt{3}}{2} & & & \\ & & & 0 & \\ & \frac{1}{2} & & & \\ \hline & & -\frac{1}{2} & \frac{\sqrt{3}}{2} & 0 \\ & & & \frac{1}{2} & 0 \\ & & & & -1 \end{array}\right)$$

---

1) The matrices (45), (15), (25), (35) of Hamermesh of p. 229 are incorrect.

cont'd

$$
\overset{(34)}{\left(\begin{array}{cc|ccc}
1 & 0 & & 0 & \\
 & -1 & & & \\
\hline
 & & -1 & 0 & 0 \\
 & & & -\frac{1}{2} & \frac{\sqrt{8}}{3} \\
 & & & & \frac{1}{3}
\end{array}\right)}
\qquad
\overset{(45)}{\left(\begin{array}{ccccc}
-\frac{1}{2} & 0 & \frac{\sqrt{3}}{2} & 0 & 0 \\
 & -\frac{1}{2} & 0 & \frac{\sqrt{3}}{2} & 0 \\
 & & \frac{1}{2} & 0 & 0 \\
 & & & \frac{1}{2} & 0 \\
 & & & & -1
\end{array}\right)}
$$

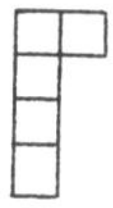

$$
\begin{array}{l}
43211 \\ 43121 \\ 41321 \\ \\ 14321
\end{array}
\overset{(12)}{\left(\begin{array}{ccc|c}
1 & & & \\
 & -1 & & \\
 & & -1 & \\
\hline
 & & & -1
\end{array}\right)}
\overset{(23)}{\left(\begin{array}{ccc|c}
-\frac{1}{2} & \frac{\sqrt{3}}{2} & 0 & \\
 & \frac{1}{2} & 0 & \\
 & & -1 & \\
\hline
 & & & -1
\end{array}\right)}
\overset{(34)}{\left(\begin{array}{ccc|c}
-1 & 0 & 0 & \\
 & -\frac{1}{3} & \frac{\sqrt{8}}{3} & \\
 & & \frac{1}{3} & \\
\hline
 & & & -1
\end{array}\right)}
\overset{(45)}{\left(\begin{array}{cccc}
-1 & 0 & 0 & 0 \\
 & -1 & 0 & 0 \\
 & & -\frac{1}{4} & \frac{\sqrt{15}}{4} \\
 & & & \frac{1}{4}
\end{array}\right)}
$$

4.5. The CSCO-II of Permutation Groups

Since the group chain $S_n \supset S_{n-1} \supset \ldots \supset S_2$ is canonical, the set of operators

$$M = (C(n), \;\; C(n-1), \ldots C(2)) \tag{4-18}$$

is a CSCO-II of S_n, where $C(f), f = n, \ldots 2$, is the CSCO-I of S_f. For $6 \le f \le 14, C(f)$ contains 2- and 3-cycle class operators $C_{(2)}(f)$, $C_{(3)}(f)$; for $f \ge 15, C(f)$ contains more operators. We are now going to prove that the set of operators $(C(n), \ldots, C(2))$ is over-complete.

Theorem 4.1: The $(n-1)$ 2-cycle class operators $(C_{(2)}(n), \ldots, C_{(2)}(2))$ of the group chain $S_n \supset S_{n-1} \supset \ldots \supset S_2$ constitute a CSCO-II of S_n.

Proof: It suffices to prove that a Yamanouchi basis vector of S_n can be labeled uniquely by the eigenvalue $\lambda_n, \lambda_{n-1}, \ldots, \lambda_2$ of the $(n-1)$ 2-cycle class operators. For $n \le 5$, a single class operator $C_{(2)}(n)$ is the CSCO-I of S_n, so the theorem is trivial for $n \le 5$.

Now supposing it holds for S_n, we want to prove that it also holds for S_{n+1}. According to the branching law this in turn amounts to proving that if there are m Young diagrams $[\nu^1] \ldots [\nu^m]$ of S_{n+1}, which correspond to the same eigenvalue λ_{n+1}, and if $[\overline{\nu}^i]$ is a Young diagram resulting from removing one box from the Young diagram $[\nu^i]$ (see Fig. 4.5-1), then $[\overline{\nu}^i] \ne [\overline{\nu}^j]$ for $i \ne j$.

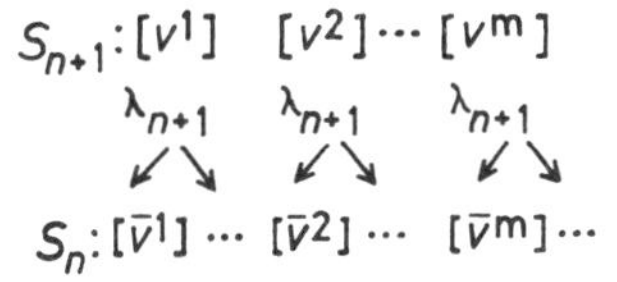

Fig. 4.5-1a.

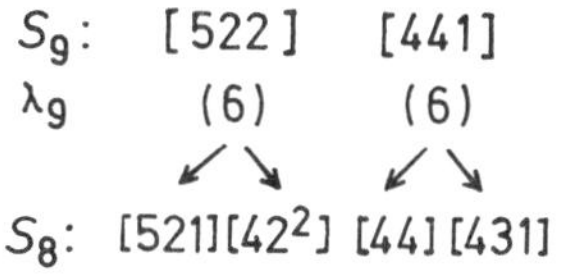

Fig. 4.5-1b.

It again amounts to proving that all the eigenvalues corresponding to the Young diagrams $[\mu^i]$ which result from adding one box to the same Young diagram $[\overline{\nu}]$ of S_n are distinct (see the Fig. 4.5-2). Let $[\mu^i]$ be the Young diagram resulting from adding one box in the i-th row of the Young diagram $[\overline{\nu}] = [\nu_1 \ldots \nu_i \ldots \nu_n]$,

$$[\mu^i] = [\nu_1, \nu_2, \ldots \nu_{i-1}, \;\; \nu_i + 1, \;\; \nu_{i+1}, \ldots] \,, \tag{4-19}$$

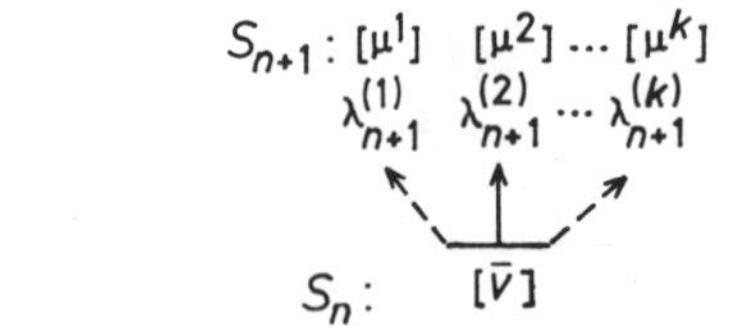

Fig. 4.5-2a.

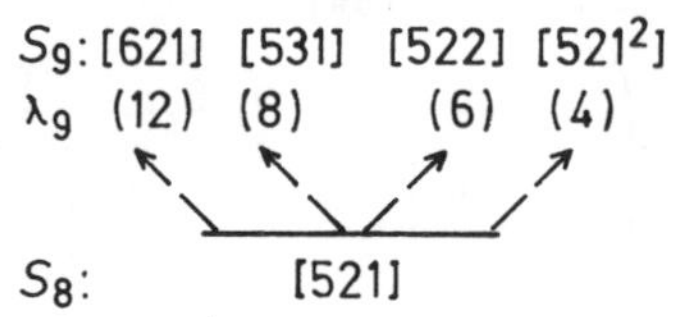

Fig. 4.5-2b.

From Eq. (4-3a) and (4-19) one has

$$\lambda_{n+1}^{(i)} = (n+1)/2 + \frac{1}{2}\sum_l (\nu_l + \delta_{li})(\nu_l + \delta_{li} - 2l) \ . \tag{4-20}$$

If $\lambda_{n+1}^{(i)} = \lambda_{n+1}^{(j)}$, from Eq. (4-20) one has

$$\nu_i - i = \nu_j - j \ . \tag{4-21}$$

Supposing $i < j$, it must have $\nu_i \geq \nu_j$ because of the rule for the Young diagrams, and thus Eq. (4-21) holds only when $i = j$. The theorem is therefore proved. Therefore the basis vectors of different irreps of S_n with the same eigenvalue λ_n must follow different routes in the process of reduction with respect to the subgroups $S_{n-1}, S_{n-2}, \ldots, S_2$. Consequently, even though the eigenvalues λ_n of the 2-cycle class operator $C_{(2)}(n)$ have degeneracies in the class space, the set of eigenvalues $(\lambda_n, \lambda_{n-1}, \ldots \lambda_2)$ still can label a Yamanouchi basis vector uniquely.

Henceforth, to simplify the notation, $C(n)$ is used to designate the 2-cycle class operator instead of $C_{(2)}(n)$, unless otherwise stated.

The above conclusion can be seen more clearly from the branching diagram. Figure 4.5-3 gives the branching law for S_n with $n \leq 6$. The numbers below each partition of S_n are the eigenvalues λ_n. Starting from each partition of S_n, each "flight route" along the arrows corresponds to a Yamanouchi basis vector of S_n. We can use either a sequence of partitions or a sequence of eigenvalues through which the route passes to label a basis vector.

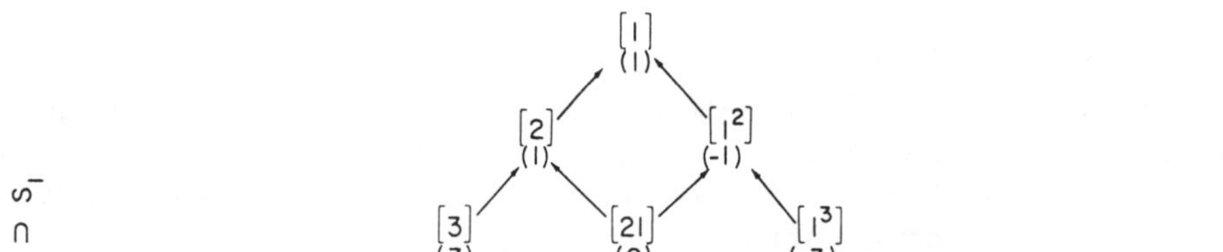

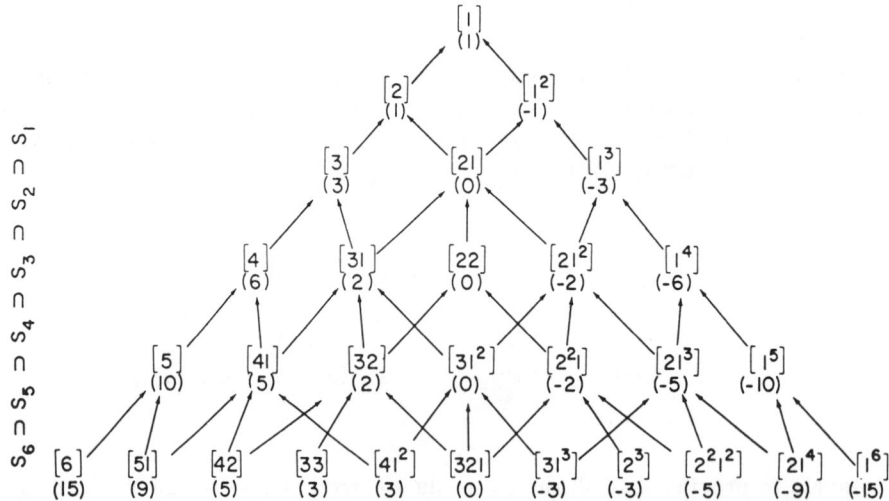

Fig. 4.5-3. The branching diagram for the group S_6.

Now we have three ways to label a Yamanouchi basis vector: the $(n-1)$ eigenvalues $(\lambda_n, \ldots, \lambda_2)$ of the CSCO-II of S_n, the $(n-1)$ Yamanouchi symbol $(r_n, \ldots, r_2, r_1 = 1)$, and the Young tableau. They have a one-to-one correspondence. As an example, in Table 4.5 we list the three labelings for bases of the irreps [411] and [33]. It is seen that the degeneracy of $\lambda_6 = 3$ does not prevent us from labeling the basis vectors uniquely by the set of eigenvalues $(\lambda_6, \lambda_5 \dots, \lambda_2)$.

Table 4.5. Three kinds of labeling of the Yamanouchi basis for the irreps [411] and [33].

ψ_λ	$\psi_{3,5,6,3,1}$	$\psi_{3,5,2,3,1}$	$\psi_{3,5,2,0,1}$	$\psi_{3,5,2,0,-1}$	$\psi_{3,0,2,3,1}$
$\psi(r)$	$\psi(321111)$	$\psi(312111)$	$\psi(311211)$	$\psi(311121)$	$\psi(132111)$
Young tableau	1 2 3 4 / 5 / 6	1 2 3 5 / 4 / 6	1 2 4 5 / 3 / 6	1 3 4 5 / 2 / 6	1 2 3 6 / 4 / 5
ψ_λ	$\psi_{3,2,2,3,1}$	$\psi_{3,2,2,0,1}$	$\psi_{3,2,2,0,-1}$	$\psi_{3,2,0,0,1}$	$\psi_{3,2,0,0,-1}$
$\psi(r)$	$\psi(222111)$	$\psi(221211)$	$\psi(221121)$	$\psi(212211)$	$\psi(212121)$
Young tableau	1 2 3 / 4 5 6	1 2 4 / 3 5 6	1 3 4 / 2 5 6	1 2 5 / 3 4 6	1 3 5 / 2 4 6

The use of the eigenvalues to label a basis vector is easily acceptable to physicists. It has the following special features:

1. The eigenvalues $\lambda_f, f = n, \dots 2$, besides specifying that the basis vector ψ_λ belongs to the irrep (λ_f) of S_f, also gives the difference between the number of the symmetric bonds and that of the antisymmetric bonds for the first f particles in the basis function $\psi_\lambda = \psi_m^{[\nu]}$. This can be shown as follows. In evaluating the expectation value of the transposition (ij), $\psi_m^{[\nu]}$ can be expressed in the following way

$$\psi_m^{[\nu]} = S_{ij}\psi_{ij}^{(s)} + A_{ij}\psi_{ij}^{(a)} ,$$

where $\psi_{ij}^{(s)}$ $(\psi_{ij}^{(a)})$ stands for the wave function which is symmetric (antisymmetric) in the indices i and j. Thus

$$\lambda_n^{(\nu)} = \sum_{i>j=1}^{n} (|S_{ij}|^2 - |A_{ij}|^2) = n_s - n_a . \tag{4-22}$$

For the totally symmetric (antisymmetric) irrep , λ_n reaches the maximum (minimum) $(\lambda_n)_{\max} = \binom{n}{2}[(\lambda_n)_{\min} = -\binom{n}{2}]$

Now let us find the relation between the eigenvalue $\lambda_n^{[\nu]}$ and the partition $[\nu]$. Let $Y_1^{[\nu]}$ be the Yamanouchi basis vector having the maximum Yamanouchi numbers. We split the 2-cycle class operator $C(n)$ into three parts

$$C(n) = \sum_{j\geq i=1}^{n} (ij) = T_r + T_c + T_m , \tag{4-23a}$$

where $T_r(T_c)$ is the sum of the transpositions i and j which are in the same rows (columns) of $Y_1^{[\nu]}$ and T_m is the remaining part of $C(n)$. Let A be the product of the antisymmetrizers for the columns of $Y_1^{[\nu]}$. For instance, if

$$Y_1^{[\nu]} = \begin{array}{|c|c|c|}\hline 1 & 2 & 3 \\ \hline 4 & 5 \\ \cline{1-2} 6 \\ \cline{1-1}\end{array}$$

then

$$A = \mathcal{A}(146)\mathcal{A}(25) ,$$

where $\mathcal{A}$ are antisymmetrizers [see Eq. (4-68b)],

$$\mathcal{A}(i_1 i_2 \ldots i_n) = \frac{1}{n!} \sum_p \delta_p p \,, \quad p \in S_n(i_1 \ldots i_n) \,.$$

From

$$[C(n), A] = 0, \quad [T_c, A] = 0 \,, \tag{4-23b,c}$$

we have

$$[T_r + T_m, \ A] = 0. \tag{4-23d}$$

Using Eq. (4-23d) we get

$$\begin{aligned} C(n) A |Y_1^{[\nu]}\rangle &= A T_r |Y_1^{[\nu]}\rangle + T_c A |Y_1^{[\nu]}\rangle + A T_m |Y_1^{[\nu]}\rangle \\ &= \left[\frac{1}{2} \sum_i \nu_i(\nu_i - 1) - \frac{1}{2} \sum_i \mu_i(\mu_i - 1) \right] A |Y_1^{[\nu]}\rangle + A T_m |Y_1^{[\nu]}\rangle \,, \end{aligned} \tag{4-24a}$$

where $[\mu_1 \mu_2 \ldots] = [\tilde{\nu}]$ is the partition conjugate to $[\nu]$. In the following we are going to show that $A T_m |Y_1^{[\nu]}\rangle$ is identically zero.

Suppose that i and j are the two numbers in the same row of $Y_1^{[\nu]}$, and i and k are the two numbers in the same column of $Y_1^{[\nu]}$. According to

$$\begin{matrix} i \ldots j \\ \vdots \\ k \end{matrix} \xrightarrow{(ij)} \begin{matrix} j \ldots i \\ \vdots \\ k \end{matrix} \xrightarrow{(jk)} \begin{matrix} k \ldots i \\ \vdots \\ j \end{matrix} \xrightarrow{(ik)} \begin{matrix} i \ldots k \\ \vdots \\ j \end{matrix}$$

we know that

$$(jk) = (ik)(jk)(ij) \,,$$

where (ij) belongs to T_r, and (ik) belongs to T_c. Consequently, we have

$$A(jk)|Y_1^{[\nu]}\rangle = A(ik)(jk)(ij)|Y_1^{[\nu]}\rangle = A(ik)(jk)|Y_1^{[\nu]}\rangle = -A(jk)|Y_1^{[\nu]}\rangle \,,$$

where the property $A(ik) = -A$ has been used. Thus we have shown that

$$A(jk)|Y_1^{[\nu]}\rangle = 0 \,,$$

which implies that

$$A T_m |Y_1^{[\nu]}\rangle = 0 \,. \tag{4-24b}$$

From (4-24a) and (4-24b) and the fact that $A|Y_1^{[\nu]}\rangle$ still belongs to the irrep $[\nu]$ (but is not a Yamanouchi basis vector) and that the eigenvalue $\lambda_n^{[\nu]}$ only depends on the irrep $[\nu]$, we get

$$\lambda_n^{[\nu]} = \frac{1}{2} \sum_i \nu_i(\nu_i - 1) - \frac{1}{2} \sum_i \mu_i(\mu_i - 1) \,. \tag{4-24c}$$

Comparing Eq. (4-22) with (4-24c) we infer that

$$n_s = \frac{1}{2} \sum_i \nu_i(\nu_i - 1), \tag{4-24d}$$

$$n_a = \frac{1}{2} \sum_i \mu_i(\mu_i - 1). \tag{4-24e}$$

n_a can be rewritten as

$$n_a = \sum_{l=1}^{n} (\nu_l - \nu_{l+1}) l(l-1)/2 . \tag{4-24f}$$

Substituting it into (4-24c) finally gives

$$\lambda_n^{[\nu]} = \frac{1}{2} \sum_i \nu_l (\nu_l - 2l + 1) . \tag{4-25}$$

Equation (4-24c) provides us with a simple method for calculating the eigenvalue associated with the Young diagram $[\nu]$: all the boxes in the same row (column) are regarded as symmetric (antisymmetric), while the boxes which are neither in the same row nor the same column are counted as half symmetric and half antisymmetric and thus do not contribute to $\lambda_n^{[\nu]}$.

2. From (4-9a) we know that the conjugate Young diagrams have opposite eigenvalues. Consequently self-conjugate Young diagrams correspond to zero eigenvalue:

$$\nu_n^{[\nu]} = 0, \quad \text{if} \quad [\nu] = [\tilde{\nu}] .$$

The counting of n_s and n_a in a Young diagram $[\nu]$ can be further simplified by the fact that the self-conjugate diagram inside the Young diagram $[\nu]$ has no contribution to $\lambda_n^{[\nu]}$. For example

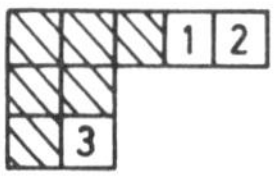

$$\lambda^{[522]} = 0 + 3 + 4 + 1 - 2 = 6.$$

The hatched diagram contributes zero. Adding successively box 1 and box 2 give three and four symmetric bonds, while adding box 3 gives 1 symmetric and 2 antisymmetric bond(s). Therefore $\lambda^{[522]} = 8 - 2 = 6$.

3. From Eq. (4-25) we know that associated to each basis vector ψ_λ, there is a basis vector $\psi_{-\lambda}$ where $-\lambda \equiv (-\lambda_n, -\lambda_{n-1}, \dots, -\lambda_2)$. Since $\pm\lambda_n$ correspond to conjugate representations, ψ_λ and $\psi_{-\lambda}$ must belong to conjugate Young tableaux, i.e.,

$$(\nu m) = (\lambda_n \dots \lambda_2) , \quad (\tilde{\nu}\tilde{m}) = (-\lambda_n \dots - \lambda_2) \equiv (\tilde{\lambda}_n \dots \tilde{\lambda}_2) . \tag{4-26a}$$

For example one has

$$\psi_{3,2,2,3,1} = \begin{array}{|c|c|c|} \hline 1 & 2 & 3 \\ \hline 4 & 5 & 6 \\ \hline \end{array} \qquad \psi_{-3,-2,-2,-3,-1} = \begin{array}{|c|c|} \hline 1 & 4 \\ \hline 2 & 5 \\ \hline 3 & 6 \\ \hline \end{array} \tag{4-26b}$$

4. ψ_ν are orthonormal:

$$\begin{aligned} &\langle \psi_\lambda | \psi_{\lambda'} \rangle = \delta_{\lambda\lambda'} , \\ &\delta_{\lambda\lambda'} = \delta_{\lambda_n \lambda'_n} \delta_{\lambda_{n-1} \lambda'_{n-1}} \cdots \delta_{\lambda_2 \lambda'_2} . \end{aligned} \tag{4-26c}$$

5. The enumeration of the Yamanouchi basis vectors according to decreasing page order of the Yamanouchi symbols is equivalent to enumerating them according to decreasing page order of the eigenvalues $(\lambda_n, \lambda_{n-1}, \dots, \lambda_2)$.

According to the branching diagram, such as Fig. 4.5-3, for example, by taking a suitable linear combination of the $(n-1)$ 2-cycle class operators, one can easily construct a single operator

$$M = \sum_{f=2}^{n} k_f C(f) , \tag{4-27a}$$

such that M is a CSCO-II of S_n. To this end, we only need to choose the coefficients k_f properly so as to make the eigenvalues λ of the operator M all distinct for each Yamanouchi basis vector of S_n,

$$\lambda = \sum_{f=2}^{n} k_f \lambda_f \ . \tag{4-27b}$$

For instance

$$M = \sum_{f=2}^{n} (f+7) C(f) \ , \tag{4-27c}$$

is a CSCO-II of S_n for $n \leq 6$.

Ex. 4.3. Give the three labeling schemes for the Yamanouchi basis vectors of the irrep [311].

4.6. The EFM for Yamanouchi Basis (I)

The Yamanouchi basis can be found by solving the following eigenequation

$$\begin{pmatrix} C(n) \\ C(s) \end{pmatrix} \psi_m^{(\nu)} = \begin{pmatrix} \nu \\ m \end{pmatrix} \psi_m^{(\nu)} \ , \quad C(s) = \big(C(n-1), \dots C(2)\big) \ . \tag{4-28a}$$

which may be written as

$$C(f)\psi_m^{(\nu)} = \lambda_f \psi_m^{(\nu)}, \quad f = n, n-1, \dots, 2 \ . \tag{4-28b}$$

$$C(f) = \sum_{i>j=1}^{f} (ij) \ ,$$

$$\psi_m^{(\nu)} = \psi_\lambda, \quad \lambda = (\nu, m) = (\lambda_n, \lambda_{n-1}, \dots, \lambda_2) \ ,$$

or more concisely as

$$M\psi_\lambda = \lambda\psi_\lambda, \quad M = \big(C(n), C(n-1), \dots C(2)\big) \ . \tag{4-28c}$$

If we take the single operator M in (4-27a) as the CSCO-II of S_n, then the problem of finding the Yamanouchi basis is reduced to that of solving the eigenequation of a single operator M,

$$M\psi_\lambda = \lambda\psi_\lambda \ , \quad M = \sum_{f=2}^{n} k_f C(f) \ . \tag{4-29}$$

For calculation by a computer (4-29) is much simpler than (4-28). Nevertheless, for calculations by hand, (4-28) is preferable. From now on, the eigenequations $M\psi_\lambda = \lambda\psi_\lambda$ can be understood either as (4-28) or (4-29).

In practical calculations, we often solve first the eigenequation of $C(2)$, and then solve the eigenequations of $C(3), \dots, C(n)$ in the two eigenspaces of $C(2)$ with the eigenvalues $\lambda_2 = \pm 1$. Likewise, $C(3), \dots, C(n)$ can be combined linearly into a single operator

$$M' = \sum_{f=3}^{n} a_f C(f) \ , \tag{4-30a}$$

such that all the eigenvalues of M' are different,

$$\lambda = \sum_{f=3}^{n} a_f \lambda_f \ . \tag{4-30b}$$

Solving the eigenequation of M' in the eigenspaces of $C(2)$ gives the Yamanouchi basis. For example, for $n \leq 6$ we can choose

$$M' = \sum_{f=3}^{n}(2f-5)C(f) \ , \quad \lambda = \sum_{f=3}^{n}(2f-5)\lambda_f \ . \tag{4-31}$$

Table 4.4-1 gives the eigenvalues λ of the Yamanouchi basis vectors with $\lambda_2 = 1$.

Example 1. Find the Yamanouchi basis of S_4 in the configuration $\alpha^3\beta$. The reducible basis vectors are

$$\varphi_1 = |\alpha\alpha\alpha\beta\rangle \ , \quad \varphi_2 = |\alpha\alpha\beta\alpha\rangle \ , \quad \varphi_3 = |\alpha\beta\alpha\alpha\rangle \ , \quad \varphi_4 = |\beta\alpha\alpha\alpha\rangle \ . \tag{4-32}$$

$$C(4) = \begin{pmatrix} 3 & 1 & 1 & 1 \\ 1 & 3 & 1 & 1 \\ 1 & 1 & 3 & 1 \\ 1 & 1 & 1 & 3 \end{pmatrix} , \quad C(3) = \begin{pmatrix} 3 & 0 & 0 & 0 \\ 0 & 1 & 1 & 1 \\ 0 & 1 & 1 & 1 \\ 0 & 1 & 1 & 1 \end{pmatrix} , \quad C(2) = \begin{pmatrix} 1 & 0 & 0 & 0 \\ 0 & 1 & 0 & 0 \\ 0 & 0 & 0 & 1 \\ 0 & 0 & 1 & 0 \end{pmatrix} . \tag{4-33}$$

Using the elimination method discussed before, a simultaneous diagonalization of the three matrices of (4-33) gives the Yamanouchi bases of S_4.

$$\begin{aligned}
\psi_{6,3,1} &= \left| \begin{array}{|c|c|c|c|}\hline 1 & 2 & 3 & 4 \\ \hline\end{array} \right\rangle = \frac{1}{2}(\varphi_1 + \varphi_2 + \varphi_3 + \varphi_4) \\
\psi_{2,3,1} &= \left| \begin{array}{|c|c|c|}\hline 1 & 2 & 3 \\ \hline 4 \\ \cline{1-1}\end{array} \right\rangle = \frac{1}{\sqrt{12}}(3\varphi_1 - \varphi_2 - \varphi_3 - \varphi_4) \\
\psi_{2,0,1} &= \left| \begin{array}{|c|c|c|}\hline 1 & 2 & 4 \\ \hline 3 \\ \cline{1-1}\end{array} \right\rangle = \frac{1}{\sqrt{6}}(2\varphi_2 - \varphi_3 - \varphi_4) \\
\psi_{2,0,-1} &= \left| \begin{array}{|c|c|c|}\hline 1 & 3 & 4 \\ \hline 2 \\ \cline{1-1}\end{array} \right\rangle = \frac{1}{\sqrt{2}}(\varphi_3 - \varphi_4)
\end{aligned} \tag{4-34}$$

In Sec. 3.13, it was pointed out that once we know one component of an irreducible basis, we can use (3-253) to obtain the other components. From Eqs. (4-17) and (4-14) one has

$$\begin{aligned}
|Y_{m'}^{[\nu]}\rangle &= [(T - D_{mm}^{[\nu]}(T))/D_{m'm}^{[\nu]}(T)]|Y_m^{[\nu]}\rangle, \\
Y_{m'}^{[\nu]} &= TY_m^{[\nu]}, \quad T = (i-1, i), \quad i = 2, 3, \ldots n \ ,
\end{aligned} \tag{4-35}$$

where $D_{m'm}^{[\nu]}(T)$ is given by (4-14). Comparing (4-35) with (3-253), it is seen that

$$F_{m'm}^{(\nu)} = (T - D_{mm}^{(\nu)}(T))/D_{m'm}^{(\nu)}(T) \ . \tag{4-36a}$$

From the eigenequation (4-29) (or (4-28)) we can find τ_ν eigenvectors $\psi_m^{(\nu)\tau}$, $\tau = 1, \ldots, \tau_\nu$, for a particular (ν, m). Choosing the appropriate adjacent permutations T and using the Yamanouchi matrix elements (4-14) (or Table 4.4-1 for S_n with $n \leq 5$), we find, from

$$\psi_{m'}^{(\nu)} = F_{m'm}^{(\nu)}\psi_m^{(\nu)\tau} \ , \tag{4-36b}$$

all the other components successively. For the irrep [32] of S_5, from the component $\left|{}^{123}_{45}\right\rangle$, we can obtain all the other components by applying the adjacent permutations in the following way

$$\left|\begin{matrix}1\,2\,3\\4\,5\end{matrix}\right\rangle \overset{(34)}{\longrightarrow} \left|\begin{matrix}1\,2\,4\\3\,5\end{matrix}\right\rangle \overset{(23)}{\longrightarrow} \left|\begin{matrix}1\,3\,4\\2\,5\end{matrix}\right\rangle \overset{(45)}{\longrightarrow} \left|\begin{matrix}1\,3\,5\\2\,4\end{matrix}\right\rangle$$

$$\downarrow (45)$$

$$\left|\begin{matrix}1\,2\,5\\3\,4\end{matrix}\right\rangle .$$

The EFM for finding the Yamanouchi basis can be summarized as follows:

1. For a given configuration $(\alpha)^{n_1}(\beta)^{n_2}(\gamma)^{n_3}\ldots$ of an n-particle system, write down all possible n-particle product states φ_a, $a = 1, 2, \ldots, N$, where

$$N = \frac{n!}{n_1!n_2!n_3!\ldots} . \tag{4-37}$$

2. Let $\psi_\lambda = \sum_a u_{\lambda a}\varphi_a$ be the required Yamanouchi basis. $u_{\lambda a}$ are solutions of the following eigenequations

$$\begin{gathered}\sum_b (\langle\varphi_a|C(f)|\varphi_b\rangle - \lambda_f\delta_{ab})u_{\lambda b} = 0 ,\\ a = 1, 2, \ldots, N, \quad f = n, n-1, \ldots, 2 .\end{gathered} \tag{4-38}$$

By means of

$$C(f) = C(f-1) + \sum_{i=1}^{f-1} p_{if} , \tag{4-39}$$

Eq. (4-38) can be simplified to

$$\sum_b \left[\sum_{i=1}^{f-1}\langle\varphi_a|p_{if}|\varphi_b\rangle - (\lambda_f - \lambda_{f-1})\delta_{ab}\right]u_{\lambda b} = 0 , \quad f = 2, 3, \ldots, n . \tag{4-40}$$

with the convention that $\lambda_1 \equiv 0$.

3. According to the branching diagram, choose an appropriate set of eigenvalues $(\lambda_2, \lambda_3, \ldots)$; substituting this into (4-40) and solve (4-40) in the sequence $f = 2, 3, \ldots$

4. If the eigenvalue $\lambda = (\nu, m)$ has no degeneracy, from (4-40) we get a unique solution $\psi_m^{(\nu)}$; if it has τ-fold degeneracy, we can get τ orthornormal solutions $\psi_m^{(\nu)1}, \ldots, \psi_m^{(\nu)\tau}$.

5. Using (4-36), find out all the other components of the Yamanouchi basis.

Example 2: Find the Yamanouchi basis of S_4 for the configuration $\alpha^2\beta\gamma$. In this configuration there are $N = 12$ product states listed in the first column of Table 4.6-1. The indices of these states are given in the second column. The result of applying $C(12) = (12)$ to these 12 states is shown in the third column.

Table 4.6-1.[†*]

φ_a	a \ b (ij)	(12)	(13)	(23)	(14)	(24)	(34)
$\lvert\alpha\alpha\beta\gamma\rangle$	1	1	2	3			6
$\lvert\beta\alpha\alpha\gamma\rangle$	2	3	1	2	7	11	8
$\lvert\alpha\beta\alpha\gamma\rangle$	3	2		1			10
$\lvert\gamma\alpha\beta\alpha\rangle$	4	5	8	12	1	4	7
$\lvert\alpha\gamma\beta\alpha\rangle$	5	4		10			9
$\lvert\alpha\alpha\gamma\beta\rangle$	6	6	7	9			1
$\lvert\gamma\alpha\alpha\beta\rangle$	7	9	6	7	2	12	4
$\lvert\beta\alpha\gamma\alpha\rangle$	8	10		11			2
$\lvert\alpha\gamma\alpha\beta\rangle$	9	7		6			5
$\lvert\alpha\beta\gamma\alpha\rangle$	10	8	12	5	10	6	3
$\lvert\beta\gamma\alpha\alpha\rangle$	11	12	5	8			11
$\lvert\gamma\beta\alpha\alpha\rangle$	12	11		4			12

[†] $(ij)\varphi_a = \varphi_b$.

[*] The blanks indicate that these entries are useless for our purpose.

Since we only need to find the solution of one component for each irrep, and any irrep, except the totally symmetric or antisymmetric irrep, of S_n has the eigenvalues $\lambda_2 = \pm 1$, the value of λ_2 can be chosen arbitrarily, say $\lambda_2 = -1$. From the first and second columns of Table 4.6-1, we immediately have

$$\lambda_2 = -1: \; u_2 = -u_3 \,, \quad u_4 = -u_5, \quad u_7 = -u_9 \,, \quad u_{10} = -u_8, \quad u_{11} = -u_{12} \,. \tag{4-41}$$

From (4-41), the number of independent variables are reduced from 12 to 7. Out of the 12 we can choose 7 as independent variable; for example we may choose $u_1, u_2, u_4, u_6, u_7, u_{10}$ and u_{11}. Next, we solve the eigenequation of $C(3)$. From (3-75b) it is seen that to obtain the matrix elements of $C(3)$, we only need to apply the permutations (13) and (23) to the states $\varphi_1, \varphi_2, \varphi_4, \varphi_6, \varphi_7, \varphi_{10}$ and φ_{11}. The result is given in Table 4.6-1. From (3-75b) and the rows 1,2,4,6 in Table 4.6-1 we get

$$\begin{aligned} &u_1 + u_2 + u_3 = \lambda_3 u_1, \quad u_3 + u_1 + u_2 = \lambda_3 u_2, \\ &u_5 + u_8 + u_{12} = \lambda_3 u_4, \quad u_6 + u_7 + u_9 = \lambda_3 u_6 \,. \end{aligned} \tag{4-42a}$$

According to Fig. 4.5-3, it is seen that apart from the totally symmetric and antisymmetric irreps, any irrep of S_n has components with $\lambda_3 = 0$. From Eq. (4-42a) with $\lambda_3 = 0$, we get the three independent equations

$$u_1 + u_2 + u_3 = 0, \quad u_5 + u_8 + u_{12} = 0 \,, \quad u_6 + u_7 + u_9 = 0 \,. \tag{4-42b}$$

It is easily seen that the remaining rows 7, 10, 11 in Table 4.6-1 do not yield new equations. Combining (4-42b) and (4-41), we have

$$u_1 = u_6 = 0, \quad u_{12} = u_4 + u_{10} \,. \tag{4-43}$$

Therefore the seven independent variables are reduced to four, which can be chosen as u_2, u_4, u_7 and u_{10}.

Similarly applying the permutations (14), (24) and (34) to $\varphi_2, \varphi_4, \varphi_7$ and φ_{10}, we get the result listed in Table 4.6-1. From (4-40) and rows 2,4,7,10 of Table 4.6-1, we get the eigen-equation of $C(4)$ (noting that $\lambda_3 = 0$),

$$\begin{aligned}
u_7 + u_{11} + u_8 &= -u_4 + u_7 - 2u_{10} &&= \lambda_4 u_2 \,, \\
u_1 + u_4 + u_7 &= u_4 + u_7 &&= \lambda_4 u_4 \,, \\
u_2 + u_{12} + u_4 &= u_2 + 2u_4 + u_{10} &&= \lambda_4 u_7 \,, \\
u_{10} + u_6 + u_3 &= -u_2 + u_{10} &&= \lambda_4 u_{10} \,.
\end{aligned} \tag{4-44}$$

$$\begin{vmatrix} -\lambda_4 & -1 & 1 & -2 \\ 0 & 1-\lambda_4 & 1 & 0 \\ 1 & 2 & -\lambda_4 & 1 \\ -1 & 0 & 0 & 1-\lambda_4 \end{vmatrix} = \lambda_4(\lambda_4 - 2)^2(\lambda_4 + 2) = 0 \tag{4-45}$$

Equation (4-45) tells us that the irrep $\lambda_4 = 0$ (corresponding to [22] from Table 3.2-1) and $\lambda_4 = -2$([211]) each occurs once and $\lambda_4 = 2$ ([31]) occurs twice. In the following we will find the solutions for each of the eigenvalues λ_4.

(a) $\lambda_4 = 0$: Substituting $\lambda_4 = 0$ into (4-44), we obtain ratios between u_2, u_4, u_7 and u_{10}. Using (4-41) and (4-43), we have[1)]

$$\psi^{(1)}_{0,-1} = \left| \begin{array}{|c|c|}\hline 1 & 3 \\ \hline 2 & 4 \\ \hline \end{array} \;\; \begin{array}{|c|c|}\hline \alpha & \alpha \\ \hline \beta & \gamma \\ \hline \end{array} \right\rangle = \frac{1}{\sqrt{8}}[\varphi_3 + \varphi_4 + \varphi_8 + \varphi_9 - \varphi_2 - \varphi_5 - \varphi_7 - \varphi_{10}] \,. \tag{4-46a}$$

From (4-36) and the fourth column of Table 4.6-1, as well as the matrix elements of the irrep [22] in Table 4.4-2, we obtain another component of the irrep [22]

$$\begin{aligned}
\psi^{(0)}_{0,1} &= \left| \begin{array}{|c|c|}\hline 1 & 2 \\ \hline 3 & 4 \\ \hline \end{array} \;\; \begin{array}{|c|c|}\hline \alpha & \alpha \\ \hline \beta & \gamma \\ \hline \end{array} \right\rangle = [(23) - D^{[22]}_{22}(23)]\psi^{(0)}_{0,-1}/D^{[22]}_{12}(23) = \left[(23) - \frac{1}{2}\right]\psi^{(0)}_{0,-1}\Big/\frac{\sqrt{3}}{2} \\
&= \frac{1}{\sqrt{24}}[2(\varphi_1 + \varphi_6 + \varphi_{11} + \varphi_{12}) - (\varphi_2 + \varphi_3 + \varphi_4 + \varphi_5 + \varphi_7 + \varphi_8 + \varphi_9 + \varphi_{10}] \,.
\end{aligned} \tag{4-46b}$$

(b) $\lambda_4 = -2$: Similarly we find

$$\begin{aligned}
\psi^{(-2)}_{0,-1} &= \left| \begin{array}{|c|c|}\hline 1 & 3 \\ \hline 2 & \multicolumn{1}{c}{} \\ \cline{1-1} 4 & \multicolumn{1}{c}{} \\ \cline{1-1} \end{array} \;\; \begin{array}{|c|c|}\hline \alpha & \alpha \\ \hline \beta & \multicolumn{1}{c}{} \\ \cline{1-1} \gamma & \multicolumn{1}{c}{} \\ \cline{1-1} \end{array} \right\rangle = \frac{1}{\sqrt{48}}[3(\varphi_3 - \varphi_2 + \varphi_7 - \varphi_9) \\
&\quad + 2(\varphi_{11} - \varphi_{12}) - \varphi_4 + \varphi_5 + \varphi_8 - \varphi_{10}] \,,
\end{aligned}$$

$$\begin{aligned}
\psi^{(-2)}_{0,1} &= \left| \begin{array}{|c|c|}\hline 1 & 2 \\ \hline 3 & \multicolumn{1}{c}{} \\ \cline{1-1} 4 & \multicolumn{1}{c}{} \\ \cline{1-1} \end{array} \;\; \begin{array}{|c|c|}\hline \alpha & \alpha \\ \hline \beta & \multicolumn{1}{c}{} \\ \cline{1-1} \gamma & \multicolumn{1}{c}{} \\ \cline{1-1} \end{array} \right\rangle = [(23) - D^{[211]}_{22}(23)]\psi^{(-2)}_{0,-1}/D^{[211]}_{12}(23) \\
&= \frac{1}{\sqrt{16}}[2(\varphi_1 - \varphi_6) - (\varphi_2 + \varphi_3 + \varphi_4 + \varphi_5) + (\varphi_7 + \varphi_8 + \varphi_9 + \varphi_{10})] \,.
\end{aligned}$$

$$\begin{aligned}
\psi^{(-2)}_{-3-1} &= \left| \begin{array}{|c|c|}\hline 1 & 4 \\ \hline 2 & \multicolumn{1}{c}{} \\ \cline{1-1} 3 & \multicolumn{1}{c}{} \\ \cline{1-1} \end{array} \;\; \begin{array}{|c|c|}\hline \alpha & \alpha \\ \hline \beta & \multicolumn{1}{c}{} \\ \cline{1-1} \gamma & \multicolumn{1}{c}{} \\ \cline{1-1} \end{array} \right\rangle = [(34) - D^{[211]}_{22}(34)]\psi^{(-2)}_{0,-1}/D^{[211]}_{32}(34) \\
&= \sqrt{\frac{1}{6}}(\varphi_4 + \varphi_{10} + \varphi_{11} - \varphi_5 - \varphi_8 - \varphi_{12}) \,.
\end{aligned} \tag{4-47}$$

---

1)The Weyl tableaux $\begin{array}{|c|c|}\hline \alpha & \alpha \\ \hline \beta & \gamma \\ \hline \end{array}$, etc. can be ignore for the moment. This will be discussed later in Sec. 4.8.

(c) $\lambda_4 = 2$: This is a double root. From (4-44), we obtain only two independent equations

$$u_4 = u_7, \quad u_2 = -u_{10} . \tag{4-48}$$

The solutions of (4-48) are not unique. We may for example choose $(u_2, u_4) = (0,1)$ and $(1,0)$, and obtain the following two solutions

$$\psi_{0,-1}^{(2)\tau} = \left| \begin{smallmatrix} 1 & 3 & 4 \\ 2 & & \end{smallmatrix} \quad \begin{smallmatrix} \alpha & \alpha & \beta \\ \gamma & & \end{smallmatrix} \right\rangle = \sqrt{\frac{1}{6}}(-\varphi_4 + \varphi_5 - \varphi_7 + \varphi_9 + \varphi_{11} - \varphi_{12}) . \tag{4-49a}$$

$$\psi_{0,-1}^{\prime(2)\theta} = \sqrt{\frac{1}{6}}(\varphi_2 - \varphi_3 + \varphi_8 - \varphi_{10} + \varphi_{11} - \varphi_{12}). \tag{4-49b}$$

After orthogonalization, the second solution becomes

$$\psi_{0,-1}^{(2)\theta} = \left| \begin{smallmatrix} 1 & 3 & 4 \\ 2 & & \end{smallmatrix} \quad \begin{smallmatrix} \alpha & \alpha & \gamma \\ \beta & & \end{smallmatrix} \right\rangle = N \left[\psi_{0,-1}^{\prime(2)\theta} - \langle \psi_{0,-1}^{(2)\tau} | \psi_{0,-1}^{\prime(2)\theta} \rangle \psi_{0,-1}^{(2)\tau} \right]$$

$$= \sqrt{\frac{1}{48}}[3(-\varphi_2 + \varphi_3 - \varphi_8 + \varphi_{10}) - 2(\varphi_{11} - \varphi_{12}) + (-\varphi_4 + \varphi_5 - \varphi_7 + \varphi_9)] . \tag{4-49c}$$

From (4-36) and the Yamanouchi matrix elements of [31] in Table 4.4-2, we obtain the remaining components

$$\psi_{3,1}^{(2)\tau} = \left| \begin{smallmatrix} 1 & 2 & 3 \\ 4 & & \end{smallmatrix} \quad \begin{smallmatrix} \alpha & \alpha & \beta \\ \gamma & & \end{smallmatrix} \right\rangle = \frac{1}{6}\left[3(\varphi_1 + \varphi_2 + \varphi_3) - \sum_{a=4}^{12} \varphi_a \right] .$$

$$\psi_{3,1}^{(2)\theta} = \left| \begin{smallmatrix} 1 & 2 & 3 \\ 4 & & \end{smallmatrix} \quad \begin{smallmatrix} \alpha & \alpha & \gamma \\ \beta & & \end{smallmatrix} \right\rangle = \frac{1}{\sqrt{18}}[2(\varphi_6 + \varphi_7 + \varphi_9) - \varphi_4 - \varphi_5 - \varphi_8 - \varphi_{10} - \varphi_{11} - \varphi_{12}] .$$

$$\psi_{0,1}^{(2)\tau} = \left| \begin{smallmatrix} 1 & 2 & 4 \\ 3 & & \end{smallmatrix} \quad \begin{smallmatrix} \alpha & \alpha & \beta \\ \gamma & & \end{smallmatrix} \right\rangle = \frac{1}{\sqrt{18}}[2(\varphi_6 + \varphi_8 + \varphi_{10}) - \varphi_4 - \varphi_5 - \varphi_7 - \varphi_9 - \varphi_{11} - \varphi_{12}] .$$

$$\psi_{0,1}^{(2)\theta} = \left| \begin{smallmatrix} 1 & 2 & 4 \\ 3 & & \end{smallmatrix} \quad \begin{smallmatrix} \alpha & \alpha & \gamma \\ \beta & & \end{smallmatrix} \right\rangle = \frac{1}{12}[3(2\varphi_1 - \varphi_2 - \varphi_3) - 4(\varphi_{11} + \varphi_{12})$$

$$+ 5(\varphi_4 + \varphi_5) + 2\varphi_6 - \varphi_7 - \varphi_8 - \varphi_9 - \varphi_{10}] . \tag{4-49d}$$

We had started from the 12 reducible basis vectors, and have already found 11 irreducible basis vectors; therefore there is one left over, which must belong to either the totally symmetric or antisymmetric irrep. Starting from (4-41), we found all solutions corresponding to $\lambda_2 = -1$. Consequently, the one which is left over must belong to $\lambda_2 = 1$, i.e., the totally symmetric irrep. We can write this irreducible basis down immediately, i.e.,

$$\psi_{3,1}^{(6)} = \left| \begin{smallmatrix} 1 & 2 & 3 & 4 \end{smallmatrix} \quad \begin{smallmatrix} \alpha & \alpha & \beta & \gamma \end{smallmatrix} \right\rangle = \frac{1}{\sqrt{12}} \sum_{a=1}^{12} \varphi_a . \tag{4-50}$$

Supposing that the wave functions of the single particle states α, β and γ are proportional to 1, x and y respectively, from (4-46) and (4-47) we have

$$\begin{aligned}
\psi^{(0)}_{0,1} &= \left| \begin{array}{|c|c|}\hline 1&2\\\hline 3&4\\\hline\end{array} \right\rangle = [2x_1y_2 + 2x_3y_4 - (x_1+x_2)(y_3+y_4)] + [x \leftrightarrow y] \,, \\
\psi^{(0)}_{0,-1} &= \left| \begin{array}{|c|c|}\hline 1&3\\\hline 2&4\\\hline\end{array} \right\rangle = [(x_1-x_2)(y_3-y_4)] + [x \leftrightarrow y] \,, \\
\psi^{(-2)}_{0,1} &= \left| \begin{array}{|c|c|}\hline 1&2\\\hline 3\\\cline{1-1} 4\\\cline{1-1}\end{array} \right\rangle = [(x_1+x_2)y_3 - (x_1+x_2-2x_3)y_4] - [x \leftrightarrow y] \,, \\
\psi^{(-2)}_{0,-1} &= \left| \begin{array}{|c|c|}\hline 1&3\\\hline 2\\\cline{1-1} 4\\\cline{1-1}\end{array} \right\rangle = [(x_1-x_2)(y_3-3y_4) + 2x_1y_2] - [x \leftrightarrow y] \,, \\
\psi^{(-2)}_{-3,-1} &= \left| \begin{array}{|c|c|}\hline 1&4\\\hline 2\\\cline{1-1} 3\\\cline{1-1}\end{array} \right\rangle = \begin{vmatrix} x_1 & y_1 & 1 \\ x_2 & y_2 & 1 \\ x_3 & y_3 & 1 \end{vmatrix} \,.
\end{aligned} \tag{4-51}$$

Here $[x \leftrightarrow y]$ denotes an interchange of x and y in the previous term.[2)]

4.7. The CSCO-III of the Permutation Group

4.7.1. CSCO-III

According to Secs. 8 and 4.5, we know that the CSCO-III of S_n consists of $(2n-3)$ 2-cycle class operators

$$K = (C(n);\; C(n-1), C(n-2), \ldots C(2);\; \overline{C}(n-1),\; \overline{C}(n-2)\;, \ldots \overline{C}(2)) \,, \tag{4-52a}$$

where $\overline{C}(f)$ is the 2-cycle class operator of the intrinsic permutation group $\overline{S}_f$.

In the space spanned by n-particle product states, it is more convenient to use the following operator as the CSCO-III,

$$K = (C(n), C(n-1), \ldots, C(2), \mathcal{C}(n-1), \ldots, \mathcal{C}(2)) \,, \tag{4-52b}$$

where $\mathcal{C}(f)$ is the 2-cycle class operator of the state permutation group $\mathcal{S}_f$.

The problem of decomposing the regular representation of S_n is converted into that of solving the eigenequation of the operator K. To see this, let $\psi^{(\nu)k}_m$ be the irreducible basis

$$\psi^{(\nu)k}_m = \sum_{a=1}^{n!} u_{\nu mk,a} \varphi_a \,, \tag{4-53a}$$

which satisfy the eigenequations

$$\begin{pmatrix} C(n) \\ C(s) \\ \overline{C}(s) \end{pmatrix} \psi^{(\nu)k}_m = \begin{pmatrix} \nu \\ m \\ k \end{pmatrix} \psi^{(\nu)k}_m \,. \tag{4-53b}$$

$$C(s) = (C(n-1), \ldots, C(2)), \quad \overline{C}(s) = (\overline{C}(n-1), \ldots, \overline{C}(2)) \,,$$

$$\nu = \lambda_n \,, \quad m = (\lambda_{n-1}, \ldots, \lambda_2) \,, \quad k = (\overline{\lambda}_{n-1}, \ldots, \overline{\lambda}_2) \,.$$

[2)]The result for the first two basis vectors of the irrep [211] given in Bohr (1969) Section IC is incorrect.

From Eqs. (4-53a) and (4-53b), we obtain the eigenequations satisfied by $u_{\nu mk,a}$,

$$\begin{aligned} &\sum_b K_{ab} u_{\lambda b} = \lambda u_{\lambda a} \\ &K_{ab} = \langle \varphi_a | K | \varphi_b \rangle, \quad \lambda = (\nu, m, k) \ . \end{aligned} \tag{4-53c}$$

In analogy with (4-29), the $(2n-3)$ 2-cycle class operators in (4-52a) can be combined into a single operator K such that it is a CSCO-III of S_n,

$$K = \sum_{f=2}^{n} a_f C(f) + \sum_{f=2}^{n-1} \overline{a}_f \overline{C}(f) \ . \tag{4-54}$$

Therefore the problem of diagonalizing simultaneously the $(2n-3)$ operators is reduced to that of diagonalizing a single operator. The operator K in (4-53c) can now be understood either as a set of $2n-3$ operator or just a single operator (4-54).

For example, the CSCO-III of S_3 can be chosen as

$$K = 3C(3) + 2C(2) + \overline{C}(2) \ . \tag{4-55}$$

K has six distinct eigenvalues 12, −12, 3, 1, −1, −3, corresponding to $(\nu, m, k) =$ (3, 1, 1), (−3, −1, −1), (0, 1, 1), (0, 1, −1), (0, −1, 1), (0, −1, −1). From Eqs. (4-55), (3-96) and (3-136), we obtain the representative of the operator K in the regular representation,

$$K = \begin{pmatrix} 0 & 6 & 3 & 3 & 0 & 0 \\ 6 & 0 & 0 & 0 & 3 & 3 \\ 3 & 0 & 0 & 0 & 4 & 5 \\ 3 & 0 & 0 & 0 & 5 & 4 \\ 0 & 3 & 4 & 5 & 0 & 0 \\ 0 & 3 & 5 & 4 & 0 & 0 \end{pmatrix} \ . \tag{4-56}$$

A diagonalization of K leads to (3-119).

4.7.2. *The labeling for Yamanouchi basis of S_n and $\mathcal{S}_n$*

The Yamanouchi basis $\psi_m^{(\nu)k}$ of S_n and $\mathcal{S}_n$ (or $\overline{S}_n$) can be labeled by the $2n-3$ quantum numbers, among which $(\nu, m) = (\lambda_n, \lambda_{n-1}, \ldots, \lambda_2)$ characterizes the transformation property of the basis under S_n and $(\nu, k) = (\lambda_n, \overline{\lambda}_{n-1}, \ldots, \overline{\lambda}_2)$ characterizes the transformation property of the basis under $\mathcal{S}_n$ (or $\overline{S}_n$). As we mentioned in Sec. 4.5, the quantum numbers (ν, m), the Yamanouchi symbol $(r_n r_{n-1} \cdots r_2)$, and the Young tableau $Y_m^{[\nu]}$ are all equivalent to one another. Obviously, we can let the quantum number (ν, k) correspond to another Yamanouchi symbol $(\overline{r}_n \overline{r}_{n-1} \ldots \overline{r}_2)$, or another Young tableau $Y_k^{[\nu]}$. It follows that we have three equivalent ways for labeling the irreducible basis of S_n and $\overline{S}_n$:

1. Two sets of quantum numbers, (ν, m) and (ν, k),
2. Two sets of Yamanouchi symbols $(r_n \ldots r_2)$ and $(\overline{r}_n \ldots \overline{r}_2)$,
3. Two Young tableaux $Y_m^{[\nu]}$ and $Y_k^{[\nu]}$.

Since the Yamanouchi basis of the intrinsic group $\overline{S}_n$ is the Yamanouchi basis of the state permutation group $\mathcal{S}_n$, which can be labeled by the Weyl tableau $W_k^{(\nu)}$. $W_k^{(\nu)}$ is obtained by replacing the letter a in the Young tableau $Y_k^{(\nu)}$ by the state index $i_a, a = 1, 2, \ldots, n$. Clearly the Yamanouchi basis $|W_k^{(\nu)}\rangle$ of $\mathcal{S}_n$ and the Yamanouchi basis $|Y_k^{(\nu)}\rangle$ of $\overline{S}_n$ have the same properties, namely, $|W_k^{(\nu)}\rangle$ is symmetric (antisymmetric) with respect to two adjacent state indices which are in the same row (column) and is neither symmetric nor antisymmetric in all other cases. Thus, if $i_1 = \alpha, i_2 = \beta, i_3 = \gamma$ and $i_4 = \delta$, then $\left|{}^{\alpha\beta\gamma}_{\delta}\right\rangle$ is symmetric in the indices

α, β, γ but has no symmetry with respect to the indices γ and δ. Therefore we have the fourth labeling scheme.

4. A Young tableau $Y_m^{(\nu)}$ and a Weyl tableau $W_k^{(\nu)}$.

4.7.3. Phase convention and the principle term

In order that the eigensolutions $\psi_m^{(\nu)k}$ of (4-53) satisfy the Yamanouchi phase convention, we must give a rule for choosing the phase of $\psi_m^{(\nu)k}$.

Let R_{mk} be a permutation operator which transfer the Young tableau $Y_k^{(\nu)}$ to $Y_m^{(\nu)}$,

$$Y_m^{(\nu)} = R_{mk} Y_k^{(\nu)} \ . \tag{4-57}$$

R_{mk} must be expressible in terms of the adjacent permutations

$$R_{mk} = T_\alpha T_\beta \ldots T_\delta \ . \tag{4-58}$$

From (4-17) and (4-58) we have

$$\begin{aligned} D_{mk}^{(\nu)}(R_{mk}) &= \sum_{st\ldots u} D_{ms}^{(\nu)}(T_\alpha) D_{st}^{(\nu)}(T_\beta) \ldots D_{uk}^{(\nu)}(T_\delta) \\ &= D_{mi}^{(\nu)}(T_\alpha) D_{ij}^{(\nu)}(T_\beta) \ldots D_{lk}^{(\nu)}(T_\delta) > 0^{3)} \ , \end{aligned} \tag{4-59a}$$

where

$$Y_l = T_\delta Y_k, \ldots, \quad Y_i = T_\beta Y_j, \quad Y_m = T_\alpha Y_i \ .$$

According to the Yamanouchi convention (4-14b), the non-diagonal matrix elements of adjacent permutations in (4-59a) are all positive; therefore

$$D_{mk}^{(\nu)}(R_{mk}) > 0 \ . \tag{4-59b}$$

Equation (4-59b) gives the rule for determining the phase of the eigenvector

$$\mathbf{u}_{\nu mk} = (u_{\nu mk,1}, u_{\nu mk,2}, \ldots, u_{\nu mk,g}) \ .$$

According to the quantum numbers (νm) and (νk), we determine the Young tableaux $Y_m^{(\nu)}$ and $Y_k^{(\nu)}$. Clearly there is one and only one operator $R_{a_0} = R_{mk}$ satisfying (4-57). Adjust the overall phase of the vector $\mathbf{u}_{\nu mk}$ so that the term $u_{\nu mk, a_0} > 0$. Such a term will be called the *principle term*. In Table 4.9, the framed terms are the principle terms. With n-particle product states as the reducible basis, it is easy to know which term is the principle term a_0 in $\psi_m^{(\nu)k} = |Y_m^{(\nu)}, W_k^{(\nu)}\rangle = \sum_a u_{\nu mk,a} \varphi_a$. For example, for the irreducible basis vector

$$\left| \begin{array}{|c|c|c|} \hline 1 & 2 & 4 \\ \hline 3 & 5 \\ \cline{1-2} 6 \\ \cline{1-1} \end{array} \quad \begin{array}{|c|c|c|} \hline \alpha & \gamma & \varphi \\ \hline \beta & \delta \\ \cline{1-2} \varepsilon \\ \cline{1-1} \end{array} \right\rangle, \text{ the principle term is } \varphi_{a_0} = |\alpha\gamma\beta\varphi\delta\varepsilon\rangle \ . \tag{4-60}$$

It can be shown that the rule (4-60) for determining the principle term applies also to the case where there are identical single particle states in the Weyl tableau; see Table 4.9. It is easy to check that all the irreducible bases of S_3 and S_4 obtained in the previous sections fulfill this simple phase rule.

4.7.4. The matrix elements of conjugated irreps

Since the CSCO-III consists of 2-cycle class operators, the permutation parity of K is negative. Therefore

$$K_{ab} = 0, \quad \text{when} \quad \delta_a \delta_b \neq -1 \ . \tag{4-61}$$

3) Notice that there is no summation index any more, since from (4-14b) we have that $D_{mj}^{(\nu)}(T_\beta) = 0$ for $Y_m^{(\nu)} \neq T_\beta Y_j^{(\nu)}$.

From (4-53c) and (4-61) one has

$$\sum_{b=1}^{g} K_{ab}(\delta_b u_{\lambda b}) = (-\lambda)(\delta_a u_{\lambda a}) \,. \tag{4-62}$$

This means that if $\{u_{\lambda a}\}$ is a solution corresponding to the eigenvalue λ, then there must exist a solution

$$u_{-\lambda a} = \pm \delta_a u_{\lambda a} \,, \tag{4-63}$$

corresponding to the eigenvalue $-\lambda$. From Sec. 4.5 we know that ψ_λ and $\psi_{-\lambda}$ are conjugate bases, i.e., if

$$\psi_\lambda = \psi_m^{(\nu)k} = |Y_m^\nu, Y_k^\nu\rangle = \sum_a u_{\lambda a}\varphi_a \,, \tag{4-64a}$$

then

$$\psi_{-\lambda} \equiv \psi_{\tilde{m}}^{(\tilde{\nu})\tilde{k}} = |\tilde{Y}_m^\nu, \tilde{Y}_k^\nu\rangle = \sum_a u_{-\lambda a}\varphi_a \,, \tag{4-64b}$$

where $-\lambda = (-\nu, -m, -k) = (\tilde{\nu}, \tilde{m}, \tilde{k})$. $\tilde{Y}_m^{(\nu)} \equiv Y_{\tilde{m}}^{(\tilde{\nu})}$ and $\tilde{Y}_k^{(\nu)} \equiv Y_{\tilde{k}}^{(\tilde{\nu})}$ are the conjugate Young tableaux of $Y_m^{(\nu)}$ and $Y_k^{(\nu)}$, respectively.

The choice of the sign in (4-63) is determined by the phase convention. We use the Young-Yamanouchi phase convention which requires that the principle term $u_{\lambda a_0}$ be positive. If $Y_m = R_{mk}Y_k$, then $\tilde{Y}_m = R_{mk}\tilde{Y}_k$. This means that if $u_{\lambda a_0}$ is the principle term in ψ_λ, then $u_{-\lambda a_0}$ is the principle term in $\psi_{-\lambda}$. In order that $u_{-\lambda a_0} > 0$, (4-63) must take the following form

$$u_{-\lambda a} = \delta_{a_0}\delta_a u_{\lambda a} \,. \tag{4-65}$$

According to $D_{mk}^{(\nu)}(R_a) = \sqrt{g/h_\nu}\, u_{\nu mk,a}^*$, (4-65) can be rewritten as

$$D_{-m-k}^{(-\nu)}(R_a) = \delta_{a_0}\delta_a D_{mk}^{(\nu)}(R_a), \quad R_{a_0} = R_{mk} \,. \tag{4-66a}$$

A Young tableau $Y_m^{(\nu)}$ can be obtained from the Young tableau $Y_1^{(\nu)}$ (corresponding to the maximum Yamanouchi symbol) through a unique permutation p, $Y_m^{(\nu)} = pY_1^{(\nu)}$. To each Young tableau $Y_m^{(\nu)}$, we associate a phase factor $\Lambda_m^\nu = \delta_p$.

From (4-57) we have

$$\delta_{a_0} = \Lambda_m^\nu \Lambda_k^\nu \,. \tag{4-66b}$$

Inserting this into (4-66a) we obtain the relationship between the matrix elements of the irrep (ν) and its conjugate $(\tilde{\nu})$.

$$D_{\tilde{m}\tilde{k}}^{(\tilde{\nu})}(R_a) = \delta_a \Lambda_m^\nu \Lambda_k^\nu D_{mk}^{(\nu)}(R_a)^{4)} \,, \tag{4-67}$$

4.7.5. The symmetrizer and anti-symmetrizer

The eigenvectors $P_{mk}^{[\nu]}$ in the group space of the CSCO-III of S_n are called the *orthogonal unit* $O_{mk}^{[\nu]}$ (Rutherford 1948),

$$P_{mk}^{[\nu]} = O_{mk}^{[\nu]} = \frac{h_\nu}{n!}\sum_a D_{mk}^{[\nu]}(p_a)p_a \,. \tag{4-68a}$$

---

4) Under Jahn's phase convention, $D_{\tilde{m}\tilde{k}}^{(\tilde{\nu})}(R_a) = \delta_a\, D_{mk}^{(\nu)}(R_a)$.

$P^{[n]}$ and $P^{[1^n]}$ are called the symmetrizer and anti-symmetrizer, respectively and denoted by

$$\mathcal{S} = \frac{1}{n!}\sum_{a=1}^{n!} p_a, \quad \mathcal{A} = \frac{1}{n!}\sum_{a=1}^{n!} \delta_a p_a . \tag{4-68b}$$

4.8. Quasi-Standard Basis of the Permutation Group

4.8.1 State permutation group (for the case with repeated state labels)

In Secs. 3.5 and 4.7 we discussed the state permutation group $\mathcal{S}_n$ for the case when there were n particles occupying n distinct single particle states. However, in physical applications, we deal more often with cases corresponding to n particles occupying $l < n$ single particle states. For example, consider an n-particle system in the configuration $(m_1)^{n_1}(m_2)^{n_2}\dots(m_l)^{n_l}, \sum_i n_i = n$. In order to define a state permutation for the cases with repeated state labels, we assign the l state labels $m_1, m_2, \dots, m_l$ to the n state indices in the following way:

$$i_1 = \dots = i_{n_1} = m_1, \quad i_{n_1+1} = \dots = i_{n_1+n_2} = m_2, \dots, \quad i_{n-n_l+1} = \dots = i_n = m_l , \tag{4-69a}$$

i.e., the indices $1\dots, n_1$ correspond to $m_1; n_1+1, \dots, n_1+n_2$ correspond to $m_2, \dots,$ and $n - n_l + 1, \dots, n$ correspond to m_l.

As in the case without repeated state labels, the permutation operator $\wp$ of the state permutation group $\mathcal{S}_n$ is defined by

$$\wp_{ab}|\dots i_a \dots i_b \dots\rangle = |\dots i_b \dots i_a \dots\rangle . \tag{4-69b}$$

When there are no repeated state labels, i.e., for the configuration $(m_1)^1(m_2)^1\dots(m_n)^1$, each single particle state m_a corresponds to a unique state index i_a. Therefore the meaning of state permutations is unambiguous. For the case with repeated state labels, for example, in the configuration $(m_1)^{n_1}(m_2)^{n_2}\dots(m_l)^{n_l}$, a single particle state may correspond to several state indices. At the beginning we can specify the state indices in a normal order state $|\omega\rangle$ in the following way

$$\begin{aligned} &|m_1 \dots m_1, \ m_2 \dots m_2, \ \dots, \ m_l \dots m_l\rangle \\ &\equiv |\omega\rangle = |i_1 \dots i_{n_1}, i_{n_1+1} \dots \ i_{n_1+n_2}, \ \dots i_{n-n_l+1}, \dots i_n\rangle . \end{aligned} \tag{4-70a}$$

Applying the $n!$ state permutations to $|\omega\rangle$, we can get N linearly independent states with

$$N = n!/(n_1!n_2!\dots n_l!) . \tag{4-70b}$$

Among them, one is the normal order state, while the remaining $N-1$ are non-normal order states. For these non-normal order states, we are no longer able to distinguish which m_1 corresponds to i_1, which m_1 corresponds to $i_2, \dots,$ which m_2 corresponds to $i_{n_1+1}, \dots,$ etc. Therefore the meaning of state permutation is ambiguous.

This is not surprising if we remember that the state permutation group is a realization of the intrinsic permutation group $\overline{S}_n$ and that with the normal order state (4-70a) as the intrinsic state, single intrinsic group element loses meaning; only the CSCO of the group $\overline{S}_n$ and the CSCO of some subgroups of $\overline{S}_n$ still retain a definite meaning. In Sec. 3.13 we described a method for finding these meaningful operators.

The symmetry group of the intrinsic state $|\omega\rangle$ of (4-70a) is

$$G_{in} = S_{n_1}(1, \dots, n_1) \times S_{n_2}(n_1+1, \dots, n_1+n_2) \times \dots \times S_{n_l}(n - n_l + 1, \dots n) .$$

The operators which commute with one another and with G_{in} are[5)]

$$(C, C(s')) = (C(n); C(n-n_l), \dots, C(n_1+n_2), C(n_1)) . \tag{4-71a}$$

Therefore the CSCO, $(\overline{C}, \overline{C}(s'))$, and in turn the CSCO

$$(\mathcal{C}; \mathcal{C}(s')) = (\mathcal{C}(n); \quad \mathcal{C}(n-n_l), \dots, \mathcal{C}(n_1+n_2), \mathcal{C}(n_1)) \tag{4-71b}$$

of the group chain $\mathcal{S}_n \supset \mathcal{S}_{n-n_l} \supset \dots \supset \mathcal{S}_{n_1+n_2} \supset \mathcal{S}_{n_1}$ have a definite meaning. In other words, the result of the action of any operator in (4-71b) is independent of which m_1 is regarded as i_1 which m_1 is regarded as $i_2, \dots$, and which m_2 is regarded as i_{n_1+1}, etc. For example, for the configuration $(\alpha)(\beta)^2(\gamma)$ $(i_1 = \alpha, i_2 = i_3 = \beta, i_4 = \gamma)$,

$$\mathcal{C}(3)|\beta\alpha\beta\gamma\rangle = \mathcal{C}(3)|i_2 i_1 i_3 i_4\rangle = \mathcal{C}(3)|i_3 i_1 i_2 i_4\rangle = |\beta\beta\alpha\gamma\rangle + |\beta\alpha\beta\gamma\rangle + |\alpha\beta\beta\gamma\rangle . \tag{4-72}$$

Equation (4-72) shows that the action of $\mathcal{C}(3)$ is independent of which β is assigned as i_2 or i_3.

4.8.2. The quasi-standard basis of the permutation group

The group chain $\mathcal{S}_n \supset \mathcal{S}_{n-n_l} \supset \dots \supset \mathcal{S}_{n_1+n_2} \supset \mathcal{S}_{n_1}$ will be called a broken chain, in contrast to the "perfect chain" $\mathcal{S}_n \supset \mathcal{S}_{n-1} \supset \dots \supset \mathcal{S}_2$. The broken chain is not canonical, since in the irreducible space (ν) with dimension h_ν, the CSCO of the broken chain, $(\mathcal{C}, \mathcal{C}(s'))$, only has $\tau_\nu < h_\nu$ sets of eigenvalues. According to Subsection 2 of Sec. 3.13, we may as well call the simultaneous eigenfunctions $\psi^{(\nu)\kappa}$ of $(\mathcal{C}, \mathcal{C}(s'))$ as the quasi-irreducible basis in the $G \supset G(s')$ classification. $\psi^{(\nu)\kappa}$ satisfy the following eigenequations

$$\begin{pmatrix} \mathcal{C} \\ \mathcal{C}(s') \end{pmatrix} \psi^{(\nu)\kappa} = \begin{pmatrix} \nu \\ \kappa \end{pmatrix} \psi^{(\nu)\kappa}, \quad \kappa = \kappa_1, \dots, \kappa_{\tau_\nu} , \tag{4-73a}$$

$$\mathcal{C}(s') = (\mathcal{C}(n-n_l), \dots, \mathcal{C}(n_1+n_2), \mathcal{C}(n_1)) , \tag{4-73b}$$

$$\kappa = (\overline{\lambda}(n-n_l), \dots, \overline{\lambda}(n_1+n_2), \overline{\lambda}(n_1)) . \tag{4-73c}$$

$\psi^{(\nu)\kappa}$ is referred to as the quasi-standard basis of the state permutation group $\mathcal{S}_n$. The quasi-standard basis $\psi^{(\nu)\kappa}$ is orthogonal in the quantum numbers ν and κ; however

$$\wp\psi^{(\nu)\kappa} \neq \sum_{\kappa'} D^{(\nu)}_{\kappa'\kappa}(p)\psi^{(\nu)\kappa'} , \tag{4-74}$$

since a single group element $\wp$ of $\mathcal{S}_n$ does not have a definite action on $\psi^{(\nu)\kappa}$.

Because the n_1-single particle state $|m_1 \dots m_1\rangle$ must belong to the totally symmetric rep of the state permutation group $\mathcal{S}_{n_1}, \mathcal{C}(n_1)$ can be taken as the 2-cycle class operator $\mathcal{C}_{(2)}(n_1)$. Thus the eigenvalue of $\mathcal{C}_{(2)}(n_1)$ is simply

$$\overline{\lambda}(n_1) = n_1(n_1-1)/2 .$$

The method for decomposing a non-regular representation of $\mathcal{S}_n$ can be summarized as follows:

The standard basis (ν, m) of $\mathcal{S}_n$ and the quasi-standard basis (ν, κ) of $\mathcal{S}_n$ is expanded in terms of the N linearly independent basis vectors φ_a generated from $|\omega\rangle$,

$$\psi^{(\nu)\kappa}_m = \sum_{a=1}^{N} u_{\nu m\kappa, a}\varphi_a , \tag{4-75a}$$

[5)]In this section $C(i)$ stands for the CSCO of $\mathcal{S}_i$ rather than the 2-cycle class operator of $\mathcal{S}_i$.

$$\varphi_a = p_a|\underbrace{m_1 \dots m_1}_{n_1}, \dots, \underbrace{m_l \dots m_l}_{n_l}\rangle = p_a|\omega\rangle \ . \tag{4-75b}$$

$\psi_m^{(\nu)\kappa}$ satisfy the following eigenequation

$$\begin{pmatrix} C_{(2)}(n) \\ C(s) \\ \mathcal{C}(s') \end{pmatrix} \psi_m^{(\nu)\kappa} = \begin{pmatrix} \nu \\ m \\ \kappa \end{pmatrix} \psi_m^{(\nu)\kappa} \ ,$$

$$m = m_1, m_2, \dots, m_{h_\nu}$$
$$\kappa = \kappa_1, \kappa_2, \dots, \kappa_{\tau_\nu} \ , \quad N = \sum_\nu \tau_\nu h_\nu \ , \tag{4-76}$$

where $C(s)$ remains to be the set of $(n-2)$ 2-cycle class operators

$$C(s) = (C_{(2)}(n-1), C_{(2)}(n-2), \dots C_{(2)}(2)) \ .$$

From (4-75) and (4-76) we obtain

$$\sum_{b=1}^{N} \left[\left\langle \varphi_a \middle| \begin{matrix} C_{(2)}(n) \\ C(s) \\ \mathcal{C}(s') \end{matrix} \middle| \varphi_b \right\rangle - \begin{pmatrix} \nu \\ m \\ \kappa \end{pmatrix} \delta_{ab} \right] u_{\nu m \kappa, b} = 0 \ . \tag{4-77}$$

$\psi_m^{(\nu)\kappa}$ satisfy the orthogonality and completeness condition

$$\begin{aligned} &\langle \psi_m^{(\nu)\kappa} | \psi_{m'}^{(\nu')\kappa'} \rangle = \delta_{\nu\nu'} \delta_{mm'} \delta_{\kappa\kappa'} \ , \\ &\sum_{\nu m \kappa} |\psi_m^{(\nu)\kappa}\rangle\langle\psi_m^{(\nu)\kappa}| = 1 \ , \\ &\sum_{a=1}^{N} u^*_{\nu m \kappa, a} u_{\nu' m' \kappa', a} = \delta_{\nu\nu'} \delta_{mm'} \delta_{\kappa\kappa'} \ , \\ &\sum_\nu \sum_{m=1}^{h_\nu} \sum_{\kappa=1}^{\tau_\nu} u^*_{\nu m \kappa, a} u_{\nu m \kappa, a'} = \delta_{aa'} \ . \end{aligned} \tag{4-78}$$

Remark: Although (4-78) is similar in appearance to (3-217) and (3-180), here the coefficients $u_{\nu m \kappa, a}$ are not related to the irreducible matrix elements $D^{(\nu)}_{m\kappa}(R_a)$. An example for finding $\psi_m^{(\nu)\kappa}$ is given in Sec. 4.9.

4.8.3. Projection operator and quasi-standard basis

It is usually taken for granted (Bohr 1969 Appendix IC, Patterson 1976) that for the case of the non-regular rep one can still use the generalized projection operator $P_m^{(\nu)k}$ to project out the standard basis of S_n out of the normal order state (4-70a).

$$\psi_m^{(\nu)k}(\omega) = P_m^{(\nu)k}|\omega\rangle, \quad P_m^{(\nu)k} = \sqrt{\frac{h_\nu}{g}} \sum_{a=1}^{n!} D^{(\nu)}_{mk}(p_a) p_a \ . \tag{4-79}$$

At first sight, it seems that using (3-169) and the relation $\overline{p} = \wp^{-1}$, we can easily "prove" that the basis (4-79) satisfy the following equations

$$\begin{pmatrix} C \\ C(s) \\ \mathcal{C}(s) \end{pmatrix} \psi_m^{(\nu)k}(\omega) = \begin{pmatrix} \nu \\ m \\ k \end{pmatrix} \psi_m^{(\nu)k}(\omega) \ , \tag{4-80a}$$

$$\wp\psi_m^{(\nu)k}(\omega) = \sum_{k'} D_{k'k}^{(\nu)}(p)\psi_m^{(\nu)k'}(\omega) \ , \tag{4-80b}$$

regardless whether the normal order state $|\omega\rangle$ contains identical state labels or not. However according to the discussion in the previous subsection we know that among the eigenequations of $\mathcal{C}(s) = (\mathcal{C}(n-1), \mathcal{C}(n-2), \ldots, \mathcal{C}(2))$, only the eigenequations of $\mathcal{C}(s')$ [(4-73)] are meaningful, and the remaining are meaningless. Thus among the quantum numbers $k = (\overline{\lambda}(n-n_1), \ldots, \overline{\lambda}(2))$, only the quantum numbers $\kappa = (\overline{\lambda}(n-n_l), \ldots, \overline{\lambda}(n_1+n_2), \overline{\lambda}(n_1))$ are true, and the rest are false and should be discarded. We also know that (4-80b) is not true. What then is wrong with the "proof" of (4-80), and in what sense can (4-80) be regarded as correct?

We say that (4-80) holds only formally, namely, if each $\psi_m^{(\nu)k}$ is regarded as an independent basis vector no matter whether it is zero, or $\psi_m^{(\nu)k}$ and $\psi_m^{(\nu)k'}$ are linearly dependent and unnormalized. Therefore it only means that the operators $P_m^{(\nu)k}$ are the standard basis of the intrinsic permutation group $\overline{S}_n$ and does not imply in the least that the $\psi_m^{(\nu)k}$ are the standard basis of the state permutation group $\mathcal{S}_n$.

Let us now pass on to the problem of how to correctly use and interpret the projection operator when it acts on the normal order state $|\omega\rangle$ with repeated state labels. In such cases $\psi_m^{(\nu)k}(\omega)$ of (4-79) is an unnormalized standard basis vector (ν, m) of S_n and a quasi-standard basis vector (ν, κ) of the state permutation group $\mathcal{S}_n$. Dropping the subscript m for clarity one has

$$\psi^{(\nu)k}(\omega) = P^{(\nu)k}|\omega\rangle = R^{(\nu)k}(\omega)\psi^{(\nu)\kappa}(\omega) \ , \tag{4-81a}$$

where we use $\psi^{(\nu)k}(\omega)$(k include the false quantum numbers) and $\psi^{(\nu)\kappa}(\omega)$($\kappa$ are the true quantum numbers) to denote an unnormalized and normalized basis vectors, respectively. $R^{(\nu)k}(\omega)$ is a normalization constant which can be calculated from

$$R^{(\nu)k}(\omega) = \left(\sqrt{\frac{n!}{h_\nu}}\langle\omega|P_k^{(\nu)k}|\omega\rangle\right)^{\frac{1}{2}} = \left[\left\langle\omega\right|\sum_p D_{kk}^{(\nu)}(p)p\left|\omega\right\rangle\right]^{\frac{1}{2}} = \left[{\sum_p}' D_{kk}^{(\nu)}(p)\right]^{\frac{1}{2}} . \tag{4-81b}$$

In deriving (4-81b), Eqs. (3-224) and (3-195) have been used. The prime in the summation over p in (4-81b) indicates that the permutation p has to satisfy the condition $p|\omega\rangle = |\omega\rangle$. For example, one has

$$R^{[\nu]k}(\alpha\alpha\beta\gamma\gamma) = 1 + D_{kk}^{[\nu]}(12) + D_{kk}^{[\nu]}(45) + D_{kk}^{[\nu]}((12)(45)) \ . \tag{4-81c}$$

When all the state labels in $|\omega\rangle$ are distinct, the only operator which satisfies $p|\omega\rangle = |\omega\rangle$ is $p = e$ (identity), and thus $R^{(\nu)k}(\omega) = 1$. In such a case, the quasi-standard basis becomes the standard basis.

A much better method for computing the normalization factor $R^{[\nu]k}(\omega)$ is given in Chen([16], 1986).

Table 4.8 gives the normalization factors $R^{(\nu)k}(\omega)$ for the permutation groups $S_2 - S_4$.

It is possible that several projection operators $P_m^{(\nu)k}$ in (4-81a) with different k give rise to the same quasi-standard basis $\psi^{(\nu)\kappa}$, since different k may lead to the same quantum number κ after deleting the false quantum numbers from k.

For example, for the configuration $\alpha\beta^2\delta$, only $\mathcal{C}(4)$ and $\mathcal{C}(3)$ have definite action while $\mathcal{C}(2)$ is meaningless and the eigenvalue $\overline{\lambda}_2$ of $\mathcal{C}(2)$ should be deleted from k. Deleting the false quantum number $\overline{\lambda}_2 = \pm 1$ from k=(0,1) and (0,-1) leads to the same quantum $\kappa = 0$. From (4-81) and Table 4.8 we have

$$\psi^{(2),0,*}(\alpha\beta\beta\delta) = \sqrt{2}\psi^{(2),0,1}(\alpha\beta\beta\delta) = \sqrt{\frac{2}{3}}\psi^{(2),0,-1}(\alpha\beta\beta\delta) \ , \tag{4-82}$$

where $\psi^{(2),0,*}$ is a quasi-standard basis vector of $\mathcal{S}_4$, the asterisk at the position of $\overline{\lambda}_2$ indicates that it is not an eigenvector of $\mathcal{C}(2)$.

Table 4.8. Normalization factors $R^{(\nu)k}(\omega)$.

(ω) \ $[\nu]$	[3]	[21]	
m	1	1	2
$(\alpha\beta\beta)$	$\sqrt{2}$	$\sqrt{\frac{1}{2}}$	$\sqrt{\frac{3}{2}}$

$(\overline{\omega})$ \ $[\nu]$	[4]	[31]			[22]		[211]		
m	1	1	2	3	1	2	1	2	3
$(\alpha\beta\gamma\gamma)$	$\sqrt{2}$	$\sqrt{\frac{2}{3}}$	$\sqrt{\frac{4}{3}}$	$\sqrt{2}$	$\sqrt{2}$			$\sqrt{\frac{2}{3}}$	$\sqrt{\frac{4}{3}}$
$(\alpha\alpha\beta\beta)$	2	$\sqrt{\frac{4}{3}}$	$\sqrt{\frac{8}{3}}$		2				
$(\alpha\beta\beta\beta)$	$\sqrt{6}$	$\sqrt{\frac{2}{3}}$	$\sqrt{\frac{4}{3}}$	2					

(ω) \ $[\nu]$	[5]	[41]				[32]				
m	1	1	2	3	4	1	2	3	4	5
$(\alpha\beta\gamma\delta\delta)$	$\sqrt{2}$	$\sqrt{\frac{3}{4}}$	$\sqrt{\frac{5}{4}}$	$\sqrt{2}$	$\sqrt{2}$	$\sqrt{2}$	$\sqrt{\frac{1}{2}}$	$\sqrt{\frac{1}{2}}$	$\sqrt{\frac{3}{2}}$	$\sqrt{\frac{3}{2}}$
$(\alpha\alpha\beta\gamma\gamma)$	2	$\sqrt{\frac{3}{2}}$	$\sqrt{\frac{5}{2}}$	2		2	1		$\sqrt{3}$	
$(\alpha\beta\beta\gamma\gamma)$	2	$\sqrt{\frac{3}{2}}$	$\sqrt{\frac{5}{2}}$	1	$\sqrt{3}$	2	$\frac{1}{2}$	$\sqrt{\frac{3}{4}}$	$\sqrt{\frac{3}{4}}$	$\frac{3}{2}$
$(\alpha\beta\gamma\gamma\gamma)$	$\sqrt{6}$	1	$\sqrt{\frac{5}{3}}$	$\sqrt{\frac{10}{3}}$	$\sqrt{6}$	$\sqrt{\frac{2}{3}}$	$\sqrt{\frac{3}{4}}$		2	
$(\alpha\alpha\beta\beta\beta)$	$\sqrt{12}$	$\sqrt{2}$	$\sqrt{\frac{10}{3}}$	$\sqrt{\frac{20}{3}}$		$\sqrt{\frac{4}{3}}$	$\sqrt{\frac{8}{3}}$		$\sqrt{8}$	
$(\alpha\alpha\alpha\beta\beta)$	$\sqrt{12}$	$\sqrt{\frac{9}{2}}$	$\sqrt{\frac{15}{2}}$			$\sqrt{12}$				
$(\alpha\beta\beta\beta\beta)$	$\sqrt{24}$	$\sqrt{\frac{3}{2}}$	$\sqrt{\frac{5}{2}}$	$\sqrt{5}$	$\sqrt{15}$					

(ω) \ $[\nu]$	[311]						[221]					[21³]			
m	1	2	3	4	5	6	1	2	3	4	5	1	2	3	4
$(\alpha\beta\gamma\delta\delta)$		$\sqrt{\frac{3}{4}}$	$\sqrt{\frac{3}{4}}$	$\sqrt{\frac{5}{4}}$	$\sqrt{\frac{5}{4}}$	$\sqrt{2}$	$\sqrt{\frac{1}{2}}$	$\sqrt{\frac{1}{2}}$	$\sqrt{\frac{3}{2}}$	$\sqrt{\frac{3}{2}}$				$\sqrt{\frac{3}{4}}$	$\sqrt{\frac{5}{4}}$
$(\alpha\alpha\beta\gamma\gamma)$		$\sqrt{\frac{3}{2}}$		$\sqrt{\frac{5}{2}}$			1		$\sqrt{3}$						
$(\alpha\beta\beta\gamma\gamma)$		$\sqrt{\frac{3}{8}}$	$\sqrt{\frac{9}{8}}$	$\sqrt{\frac{5}{8}}$	$\sqrt{\frac{15}{8}}$		$\frac{1}{2}$	$\sqrt{\frac{3}{4}}$	$\sqrt{\frac{3}{4}}$	$\frac{3}{2}$					
$(\alpha\beta\gamma\gamma\gamma)$			1		$\sqrt{\frac{5}{3}}$	$\sqrt{\frac{10}{3}}$									
$(\alpha\alpha\beta\beta\beta)$															
$(\alpha\alpha\alpha\beta\beta)$															
$(\alpha\beta\beta\beta\beta)$															

[1] A blank in the table means $R^{(\nu)k}(\omega)=0$.

[2] If all the single particle states in $|\omega\rangle$ are the same, then for the totally symmetric rep $[n]$, we have $R^{[n]}(\omega) = \sqrt{1/n!}$.

4.8.4. *The labeling of the quasi-standard basis*

In Sec. 4.7 we discussed the standard basis $\psi_m^{(\nu)k}(\omega_0)$ of S_n and $\overline{S}_n$, where $|\omega_0\rangle$ represents a normal order state with no repeated state labels. The projected state $\psi_m^{(\nu)k}(\omega)$ of (4-79) out of a normal order state $|\omega\rangle$ with repeated state labels can be obtained from $\psi_m^{(\nu)k}(\omega_0)$ simply by letting $\omega_0 \to \omega$. For example

$$\psi_m^{(\nu)k}(\alpha\beta\beta\delta) = \psi_m^{(\nu)k}(\alpha\beta\gamma\delta)|_{\gamma=\beta} \ . \tag{4-83a}$$

This procedure is called *assimilation.* In Sec. 4.7, we use the Weyl tableau W_k to replace the quantum number k, i.e., use $\psi_m^{(\nu)}(W_k)$ to denote $\psi_m^{(\nu)k}(\omega_0)$. Application of the assimilation procedure to $\psi_m^{(\nu)}(W_k)$ yields a Weyl table with repeated state labels. For instance, choosing $(\nu)k = (2), 0, 1$ *and* $(2), 0, -1$, we can express (4-83a) in terms of the Weyl tableau

$$\psi_m^{(2),0,1}(\alpha\beta\beta\delta) = \psi_m^{[31]}\left(\begin{array}{|c|c|c|}\hline \alpha & \beta & \delta \\ \hline \gamma \\ \cline{1-1}\end{array}\right)\Bigg|_{\gamma=\beta} ,$$

$$\psi_m^{(2),0,-1}(\alpha\beta\beta\delta) = \psi_m^{[31]}\left(\begin{array}{|c|c|c|}\hline \alpha & \gamma & \delta \\ \hline \beta \\ \cline{1-1}\end{array}\right)\Bigg|_{\gamma=\beta} . \tag{4-83b}$$

It is thus seen that we can use Weyl tableaux with repeated state labels to denote the quasi-standard basis of the permutation group. Equation (4-82) can be rewritten as

$$\psi^{(2),0,*}(\alpha\beta\beta\delta) = \left|\begin{array}{|c|c|c|}\hline \alpha & \beta & \delta \\ \hline \beta \\ \cline{1-1}\end{array}\right\rangle = \sqrt{2}\psi\left(\begin{array}{|c|c|c|}\hline \alpha & \beta & \delta \\ \hline \gamma \\ \cline{1-1}\end{array}\right)\Bigg|_{\gamma=\beta} = \sqrt{\frac{2}{3}}\psi\left(\begin{array}{|c|c|c|}\hline \alpha & \gamma & \delta \\ \hline \beta \\ \cline{1-1}\end{array}\right)\Bigg|_{\gamma=\beta} .$$

It is easy to prove that if there are two identical state labels appearing in the same column of a Weyl tableau W_k, then the basis vector $|W_k\rangle = 0$. For example, we have

$$\left|\begin{array}{|c|c|c|}\hline \alpha & \gamma & \delta \\ \hline \alpha \\ \cline{1-1}\end{array}\right\rangle \propto \psi\left(\begin{array}{|c|c|c|}\hline \alpha & \gamma & \delta \\ \hline \beta \\ \cline{1-1}\end{array}\right)\Bigg|_{\beta=\alpha} = 0 \ .$$

Since the basis vector $\psi\left(\begin{array}{|c|c|c|}\hline \alpha & \gamma & \delta \\ \hline \beta \\ \cline{1-1}\end{array}\right)$ is antisymmetric under the interchange of α and β, and therefore is vanishing when $\alpha = \beta$.

Now we give a general definition for the Weyl tableau.

A *Weyl tableau* is a Young diagram $[\nu]$ whose boxes are filled with the single particle states $m_1, m_2, \ldots, m_l$ under the restrictions that

(a) the same m value may not appear twice in any column,

(b) the m values must read in increasing order (first all m_1's, then m_2's, etc.) as we read from left to right in any row as well as from top to bottom in any column.

The labeling of the quasi-standard basis of the state permutation group in terms of the Weyl tableau is realized through the following scheme:

A Weyl tableau W_k filled with $m_1, \ldots, m_l$ is equivalent to l partitions $[\nu], [\nu^{l-1}], [\nu^{l-2}], \ldots, [\nu^2], [\nu^1]$. Here $[\nu^{l-1}], [\nu^{l-2}] \ldots$ are obtained by successively removing the boxes filled with m_l, $m_{l-1}, \ldots$ from the Weyl tableau W_k. The basis vector $|W_k\rangle$ is defined as the one belonging to the irreps $[\nu], [\nu^{l-1}], \ldots, [\nu^2], [\nu^1]$ of the state permutation group $S_n, S_{n-n_l}, \ldots, S_{n_1+n_2}$ and S_{n_1}. For example,

$$\underset{S_9[432]}{\begin{array}{|c|c|c|c|}\hline \alpha & \alpha & \beta & \gamma \\ \hline \beta & \beta & \delta \\ \cline{1-3} \gamma & \delta \\ \cline{1-2}\end{array}} \longrightarrow \underset{S_7[421]}{\begin{array}{|c|c|c|c|}\hline \alpha & \alpha & \beta & \gamma \\ \hline \beta & \beta \\ \cline{1-2} \gamma \\ \cline{1-1}\end{array}} \longrightarrow \underset{S_5[32]}{\begin{array}{|c|c|c|}\hline \alpha & \alpha & \beta \\ \hline \beta & \beta \\ \cline{1-2}\end{array}} \longrightarrow \underset{S_2[2]}{\begin{array}{|c|c|}\hline \alpha & \alpha \\ \hline\end{array}} \tag{4-84}$$

Finally we turn to the labeling question. We have the following four schemes for labeling the quasi-standard basis of S_n.

1. By $\psi^{(\nu)k}(\omega)$: Here $|\omega\rangle$ is a normal order state.

Using Table 3.2-1, the l sets of eigenvalues $(\nu, k) = (\overline{\lambda}(n), \overline{\lambda}(n-n_l), \dots, \overline{\lambda}(n_1))$ can be converted into l partitions, $[\nu] = [m_{1l}m_{2l}\dots m_{ll}]$, $[\nu^{l-1}] = [m_{1l-1}m_{2l-1}\cdots m_{l-1l-1}], \dots,$ $[\nu^2] = [m_{12}, m_{22}]$ and $[\nu^1] = [m_{11}]$. Thus it is equivalent to the second labeling scheme.

2. By the Gel'fand pattern

$$\begin{pmatrix} [\nu] \\ (m) \end{pmatrix} = \begin{pmatrix} m_{1l} \ m_{2l} \dots m_{ll} \\ m_{1l-1} \cdots m_{l-1,l-1} \\ \cdots \\ m_{12}m_{22} \\ m_{11} \end{pmatrix}, \quad [\nu] = [m_{1l} \dots m_{ll}] \,. \tag{4-85}$$

It belongs to the irreps $[\nu] = [m_{1l} \dots m_{ll}], \dots [m_{12}m_{22}]$ and $[m_{11}]$ of $S_n, \dots, S_{n_1+n_2}$ and S_{n_1}, respectively.

3. By a Weyl tableau $|W_k\rangle$.

4. By the Yamanouchi symbol and the normal order state, i.e., by $\psi^{[\nu]r_m}(\omega)$.

Let us take the following three examples to illustrate these labeling schemes

(i) $|\omega\rangle = |\alpha\alpha\beta\gamma\gamma\rangle$

$$\psi^{(5),*,3,1}(\omega) = \left| \begin{pmatrix} 4 & & 1 & & 0 \\ & 3 & & 0 & \\ & & 2 & & \end{pmatrix} \right\rangle = \left| \begin{array}{|c|c|c|c|} \hline \alpha & \alpha & \beta & \gamma \\ \hline \gamma \\ \cline{1-1} \end{array} \right\rangle \propto \psi^{[41]21111}(\omega) \propto \psi^{[41]12111}(\omega) \,.$$

This belongs to the irreps [41], [3] and [2] of S_5, S_3 and S_2.

(ii) $|\omega\rangle = |\alpha\alpha\beta\beta\beta\rangle$

$$\psi^{(2),*,*,1}(\omega) = \left| \begin{pmatrix} 3 & & 2 \\ & 2 & \end{pmatrix} \right\rangle = \left| \begin{array}{|c|c|c|} \hline \alpha & \alpha & \beta \\ \hline \beta & \beta \\ \cline{1-2} \end{array} \right\rangle \propto \psi^{[32]22111}(\omega) \propto \psi^{[32]21211}(\omega) \propto \psi^{[32]12211}(\omega)$$

This belongs to the irreps [32] and [2] of S_5 and S_2.

(iii) $|\omega\rangle = |\alpha\beta\beta\beta\gamma\gamma\rangle$

$$\psi^{(3,4),*,2,*,*}(\omega) = \left| \begin{pmatrix} 4 & & 1 & & 1 \\ & 3 & & 1 & \\ & & 1 & & \end{pmatrix} \right\rangle = \left| \begin{array}{|c|c|c|c|} \hline \alpha & \beta & \beta & \gamma \\ \hline \beta \\ \cline{1-1} \gamma \\ \cline{1-1} \end{array} \right\rangle,$$

$$\psi^{(3,-8),*,2,*,*}(\omega) = \left| \begin{pmatrix} 3 & & 3 & & 0 \\ & 3 & & 1 & \\ & & 1 & & \end{pmatrix} \right\rangle = \left| \begin{array}{|c|c|c|} \hline \alpha & \beta & \beta \\ \hline \beta & \gamma & \gamma \\ \hline \end{array} \right\rangle \,.$$

Here (3,4) and (3,−8) are the eigenvalues of the CSCO-I, $C = (C_{(2)}(6), C_{(3)}(6))$. It is seen that the eigenvalue 3 of $C_{(2)}(6)$ is no longer sufficient to specify the irrep $[\nu]$, since the group chain $S_6 \supset S_4 \supset S_1$ is a broken chain. This is why the operators $C(j)(j = n, n-n_l, \dots, n_1+n_2, n_1)$ in (4-73a) should be the CSCO-I of S_j instead of the 2-cycle class operator of S_j.

It is seen that the first three labeling schemes are in one-to-one correspondence

$$\psi^{(\nu)\kappa}(\omega) = \left| \begin{pmatrix} [\nu] \\ m \end{pmatrix} \right\rangle = |W_\kappa^{[\nu]}\rangle \,, \tag{4-86}$$

while the labeling with the Yamanouchi symbol r_m is not unique, as there may be several r_m's corresponding to the same state. If there are no identical state labels in $|\omega\rangle$, i.e., for the standard basis of S_n, the four labeling schemes will all be in one-to-one correspondence.

We used the flight route of Fig. 4.5-3 to represent a Yamanouchi basis vector of S_n or $\mathcal{S}_n$. The flight route is labeled by the names $\lambda_n, \lambda_{n-1}, \ldots, \lambda_2$ for the $n-1$ 'airports' through which it passes. We can also use a flight route to represent a quasi-standard basis vector of $\mathcal{S}_n$. But now it is an 'express flight route' which only passes $l < n$ 'airports'. The express flight route is labeled by the names $\overline{\lambda}(n), \overline{\lambda}(n-n_l), \ldots, \overline{\lambda}(n_1)$ of the l 'airports.' Obviously, the total number τ_ν of possible express routes is less than the number h_ν of all possible local routes. Below we use the irrep $\nu = 5$ (or $[\nu] = [41]$) of S_5 as an example to illustrate the difference between these two cases.

configuration $\alpha\beta\gamma\delta\varepsilon$

Irrep (5) has $h_\nu = 4$ standard basis vectors

$$\begin{array}{|c|c|c|c|}\hline \alpha&\beta&\gamma&\delta\\\hline \varepsilon\\\cline{1-1}\end{array} = \psi^{(5),6,3,1}(\alpha\beta\gamma\delta\varepsilon)$$

$$\begin{array}{|c|c|c|c|}\hline \alpha&\beta&\gamma&\varepsilon\\\hline \delta\\\cline{1-1}\end{array} = \psi^{(5),2,3,1}(\alpha\beta\gamma\delta\varepsilon)$$

$$\begin{array}{|c|c|c|c|}\hline \alpha&\beta&\delta&\varepsilon\\\hline \gamma\\\cline{1-1}\end{array} = \psi^{(5),2,0,1}(\alpha\beta\gamma\delta\varepsilon)$$

$$\begin{array}{|c|c|c|c|}\hline \alpha&\gamma&\delta&\varepsilon\\\hline \beta\\\cline{1-1}\end{array} = \psi^{(5),2,0,-1}(\alpha\beta\gamma\delta\varepsilon)$$

Starting from the airport (5) there are $h_\nu = 4$ flight routes: (5,6,3,1), (5,2,3,1), (5,2,0,1), and (5,2,0,−1).

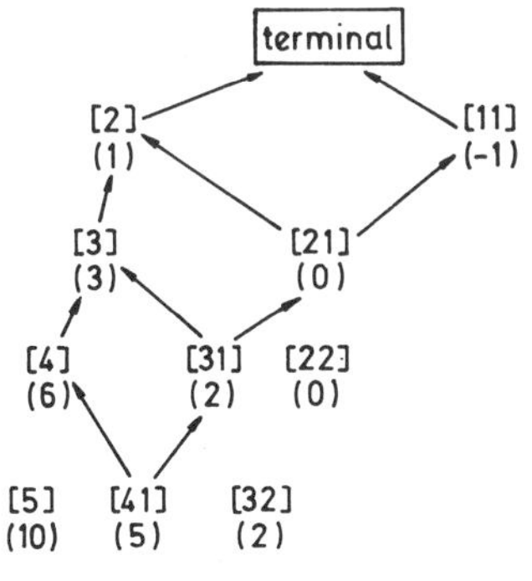

Fig. 4.8-1a. The standard basis of S_5.

configuration $(\alpha)^2\beta(\gamma)^2$

Irrep (5) has $\tau_\nu = 2$ quasi-standard basis vectors

$$\begin{array}{|c|c|c|c|}\hline \alpha&\alpha&\beta&\gamma\\\hline \gamma\\\cline{1-1}\end{array} = \psi^{(5),*,3,1}(\alpha\alpha\beta\gamma\gamma)$$

$$\begin{array}{|c|c|c|c|}\hline \alpha&\alpha&\gamma&\gamma\\\hline \beta\\\cline{1-1}\end{array} = \psi^{(5),*,0,1}(\alpha\alpha\beta\gamma\gamma)$$

Starting from the airport (5), there are $\tau_\nu = 2$ express flight routes: (5, *, 3, 1) and (5, *, 0, 1).

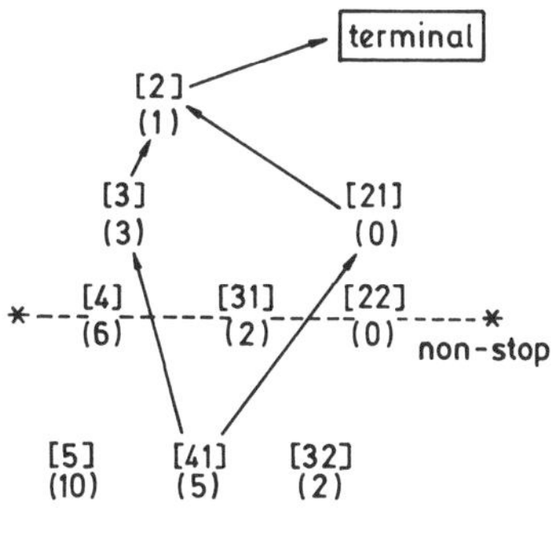

Fig. 4.8-1b. The quasi-standard basis of $\mathcal{S}_5$.

4.9. The EFM for Yamanouchi Basis (II)

In this section we will use the Weyl tableaux and the intrinsic quantum number to simplify the calculation of the standard basis of S_n. The basis calculated in this way is also the quasi-standard basis of S_n. The first three steps and the fifth step are identical to those given in Sec. 4.6. Only the fourth step should be modified into:

4a. According to the configuration $(\alpha)^{n_1}(\beta)^{n_2}\ldots$ of an n-particle system, write out all possible Weyl tableaux. The number of the Weyl tableaux corresponding to a given partition $[\nu]$ gives the multiplicity τ_ν. According to the corresponding relation between the Young diagrams $[\nu]$ and the eigenvalues λ_n, Eq. (4-3a) or Fig. 4.5-3, we then know the eigenvalue λ_n and its degeneracy τ_ν. Thus the problem of finding the roots of the expectation equation is avoided.

Example: The configuration $\alpha^2\beta\gamma$ has the following Weyl tableaux:

$$\begin{array}{|c|c|c|c|}\hline \alpha & \alpha & \beta & \gamma \\ \hline \end{array} \qquad \begin{array}{|c|c|c|}\hline \alpha & \alpha & \beta \\ \hline \gamma \\ \cline{1-1} \end{array} \qquad \begin{array}{|c|c|c|}\hline \alpha & \alpha & \gamma \\ \hline \beta \\ \cline{1-1} \end{array} \qquad \begin{array}{|c|c|}\hline \alpha & \alpha \\ \hline \beta & \gamma \\ \hline \end{array} \qquad \begin{array}{|c|c|}\hline \alpha & \alpha \\ \hline \beta \\ \cline{1-1} \gamma \\ \cline{1-1} \end{array} \tag{4-87a}$$

Therefore $\tau_{[4]} = \tau_{[22]} = \tau_{[211]} = 1$ and $\tau_{[31]} = 2$. According to Fig. 4.5-3, we know that the corresponding eigenvalues are $\lambda_4 = 6, 0, -2$ (single roots) and $\lambda_4 = 2$ (double root).

4b. For the $\tau_\nu = 1$ case, the eigenequations of $(C, C(s))$ are sufficient to fix the solution completely. For the $\tau_\nu > 1$ case, we further need the eigenequations of $\mathcal{C}(s')$ to fix the solutions. From the Weyl tableaux determined in step (4a) above, we can know the corresponding eigenvalues κ of $\mathcal{C}(s')$ without solving the expectation equation. On substituting the eigenvalues into (4-77), we easily find the solutions $u_{\nu m\kappa,a}$.

As an example, we again take the configuration $\alpha^2\beta\gamma$ discussed in Sec. 4.6. For this configuration, the elements of S_2 are equivalent to the identity element, and the CSCO of S_3 is meaningful. In Sec. 4.6, we arbitrarily chose two independent solutions of (4-48a). Now we can employ the eigenequation of $\mathcal{C}(s') = \mathcal{C}(3)$ to fix the solutions completely. In (4-48a) there are two independent variables which can be chosen as u_4 and u_2. Applying $\mathcal{C}(3)$ on $\varphi_4 = |\gamma\alpha\beta\alpha\rangle$, we obtain

$$\begin{aligned} \mathcal{C}(3)\varphi_4 &= \mathcal{C}(3)|\gamma\alpha\beta\alpha\rangle = (\wp_{12} + \wp_{23} + \wp_{13})|i_4 i_1 i_3 i_2\rangle \\ &= |\gamma\alpha\beta\alpha\rangle + |\gamma\alpha\alpha\beta\rangle + |\gamma\beta\alpha\alpha\rangle = \varphi_4 + \varphi_7 + \varphi_{12}\ . \end{aligned} \tag{4-87b}$$

Using (3-75b) we get an equation involving the variable u_4,

$$u_4 + u_7 + u_{12} = \overline{\lambda}_3 u_4 . \tag{4-87$'$}$$

From (4-43) and (4-48a), it reads

$$3u_4 + u_{10} = \overline{\lambda}_3 u_4 . \tag{4-87c}$$

The Weyl tableau $\begin{array}{|c|c|c|}\hline \alpha & \alpha & \beta \\ \hline \gamma \\ \cline{1-1} \end{array}$ in (4-87a) corresponds to $\overline{\lambda}_3 = 3$. Substituting it into (4-87c), we get a solution: $u_{10} = 0, u_4 = 1$. This is just the first solution we chose for (4-48a) in Sec. 4.6. Another Weyl tableau $\begin{array}{|c|c|c|}\hline \alpha & \alpha & \gamma \\ \hline \beta \\ \cline{1-1} \end{array}$ in (4-87a) corresponds to $\overline{\lambda}_3 = 0$. The corresponding solution of (4-87c) is: $u_4 = -1$ and $u_{10} = 3$. It is just the solution (4-49c). The coincidence of the solution obtained here and those in Sec. 4.6 is, of course, fortuitous.

The third and fourth steps in Sec. 4.6 can also be modified into the following steps: Associate each permissible Weyl tableau W_κ to a Young tableau Y_m. From Y_m and W_κ write down the corresponding eigenvalues (ν, m, κ). Substituting the eigenvalues into (4-76), we can find the eigenfunction $\psi_m^{(\nu)\kappa} = |Y_m, W_\kappa\rangle$. Then using (4-36) we will get the other components.

Ex. 4.4. Find the standard basis of S_4 and the quasi-standard basis of S_4 for the configuration $\alpha\beta\gamma^2$ (the answer is given in Table 4.9).

Table 4.9. The standard basis[1)] of S_4 and quasi-standard basis of S_4 for the configuration $(\alpha)(\beta)(\gamma)^2$.

$\lambda_4\lambda_3\lambda_2$	$\bar{\lambda}_2$	Young tableau	Weyl tableau	$\|\alpha\beta\gamma\gamma\rangle$	$\|\beta\alpha\gamma\gamma\rangle$	$\|\gamma\alpha\beta\gamma\rangle$	$\|\gamma\beta\alpha\gamma\rangle$	$\|\beta\gamma\alpha\gamma\rangle$	$\|\alpha\gamma\beta\gamma\rangle$	$\|\gamma\alpha\gamma\beta\rangle$	$\|\gamma\beta\gamma\alpha\rangle$	$\|\gamma\gamma\beta\alpha\rangle$	$\|\gamma\gamma\alpha\beta\rangle$	$\|\beta\gamma\gamma\alpha\rangle$	$\|\alpha\gamma\gamma\beta\rangle$
6, 3, 1	1	1234	$\alpha\beta\gamma\gamma$	$\boxed{\sqrt{\frac{1}{12}}}$	$\sqrt{\frac{1}{12}}$	$\sqrt{\frac{1}{12}}$	$\sqrt{\frac{1}{12}}$	$\sqrt{\frac{1}{12}}$	$\sqrt{\frac{1}{12}}$	$\sqrt{\frac{1}{12}}$	$\sqrt{\frac{1}{12}}$	$\sqrt{\frac{1}{12}}$	$\sqrt{\frac{1}{12}}$	$\sqrt{\frac{1}{12}}$	$\sqrt{\frac{1}{12}}$
2, 3, 1	1	123 4		$\boxed{\sqrt{\frac{1}{12}}}$	$\sqrt{\frac{1}{12}}$	$\sqrt{\frac{1}{12}}$	$\sqrt{\frac{1}{12}}$	$\sqrt{\frac{1}{12}}$	$\sqrt{\frac{1}{12}}$	$-\sqrt{\frac{1}{12}}$	$-\sqrt{\frac{1}{12}}$	$-\sqrt{\frac{1}{12}}$	$-\sqrt{\frac{1}{12}}$	$-\sqrt{\frac{1}{12}}$	$-\sqrt{\frac{1}{12}}$
2, 0, 1	1	124 3	$\alpha\beta\gamma$ γ	$\boxed{\sqrt{\frac{1}{6}}}$	$\sqrt{\frac{1}{6}}$	$-\sqrt{\frac{1}{24}}$	$-\sqrt{\frac{1}{24}}$	$-\sqrt{\frac{1}{24}}$	$-\sqrt{\frac{1}{24}}$	$\sqrt{\frac{1}{24}}$	$\sqrt{\frac{1}{24}}$	$-\sqrt{\frac{1}{6}}$	$-\sqrt{\frac{1}{6}}$	$\sqrt{\frac{1}{24}}$	$\sqrt{\frac{1}{24}}$
2, 0, −1	1	134 2				$-\sqrt{\frac{1}{8}}$	$-\sqrt{\frac{1}{8}}$	$\sqrt{\frac{1}{8}}$	$\boxed{\sqrt{\frac{1}{8}}}$	$-\sqrt{\frac{1}{8}}$	$-\sqrt{\frac{1}{8}}$			$\sqrt{\frac{1}{8}}$	$\sqrt{\frac{1}{8}}$
2, 3, 1	−1	123 4								$\sqrt{\frac{1}{6}}$	$-\sqrt{\frac{1}{6}}$	$-\sqrt{\frac{1}{6}}$	$\sqrt{\frac{1}{6}}$	$-\sqrt{\frac{1}{6}}$	$\boxed{\sqrt{\frac{1}{6}}}$
2, 0, 1	−1	124 3	$\alpha\gamma\gamma$ β			$\sqrt{\frac{3}{16}}$	$-\sqrt{\frac{3}{16}}$	$-\sqrt{\frac{3}{16}}$	$\boxed{\sqrt{\frac{3}{16}}}$	$\sqrt{\frac{1}{48}}$	$-\sqrt{\frac{1}{48}}$	$\sqrt{\frac{1}{12}}$	$-\sqrt{\frac{1}{12}}$	$-\sqrt{\frac{1}{48}}$	$\sqrt{\frac{1}{48}}$
2, 0, −1	−1	134 2		$\boxed{\frac{1}{2}}$	$-\frac{1}{2}$	$-\frac{1}{4}$	$\frac{1}{4}$	$-\frac{1}{4}$	$\frac{1}{4}$	$-\frac{1}{4}$	$\frac{1}{4}$			$-\frac{1}{4}$	$\frac{1}{4}$
−2, 0, 1	−1	12 3 4				$\frac{1}{4}$	$-\frac{1}{4}$	$-\frac{1}{4}$	$\boxed{\frac{1}{4}}$	$-\frac{1}{4}$	$\frac{1}{4}$	$-\frac{1}{2}$	$\frac{1}{2}$	$\frac{1}{4}$	$-\frac{1}{4}$
−2, 0, −1	−1	13 2 4	$\alpha\gamma$ β γ	$\boxed{\sqrt{\frac{1}{12}}}$	$-\sqrt{\frac{1}{12}}$	$-\sqrt{\frac{1}{48}}$	$\sqrt{\frac{1}{48}}$	$-\sqrt{\frac{1}{48}}$	$\sqrt{\frac{1}{48}}$	$\sqrt{\frac{3}{16}}$	$-\sqrt{\frac{3}{16}}$			$\sqrt{\frac{3}{16}}$	$-\sqrt{\frac{3}{16}}$
−2, −3, −1	−1	14 2 3		$\boxed{\sqrt{\frac{1}{6}}}$	$-\sqrt{\frac{1}{6}}$	$\sqrt{\frac{1}{6}}$	$-\sqrt{\frac{1}{6}}$	$\sqrt{\frac{1}{6}}$	$-\sqrt{\frac{1}{6}}$						
0, 0, 1	1	12 34	$\alpha\beta$ $\gamma\gamma$	$\boxed{\sqrt{\frac{1}{6}}}$	$\sqrt{\frac{1}{6}}$	$-\sqrt{\frac{1}{24}}$	$-\sqrt{\frac{1}{24}}$	$-\sqrt{\frac{1}{24}}$	$-\sqrt{\frac{1}{24}}$	$-\sqrt{\frac{1}{24}}$	$-\sqrt{\frac{1}{24}}$	$\sqrt{\frac{1}{6}}$	$\sqrt{\frac{1}{6}}$	$-\sqrt{\frac{1}{24}}$	$-\sqrt{\frac{1}{24}}$
0, 0, −1	1	13 24				$-\sqrt{\frac{1}{8}}$	$-\sqrt{\frac{1}{8}}$	$\sqrt{\frac{1}{8}}$	$\boxed{\sqrt{\frac{1}{8}}}$	$\sqrt{\frac{1}{8}}$	$\sqrt{\frac{1}{8}}$			$-\sqrt{\frac{1}{8}}$	$-\sqrt{\frac{1}{8}}$

1)The framed entries are the principle terms [see (4-60)].

4.10. The Inner Product and the CG Series of Permutation Groups

The state of a microscopic particle is described by a wave function $\psi(q)$, q representing all the coordinates. For example, for an electron, q represents the orbital and spin coordinates; for a nucleon, q includes the orbital, spin and isospin coordinates; for a quark, q denotes the orbital, spin, flavor and color coordinates. In treating the permutation symmetry of an n-particle system, we usually first consider the symmetry for the individual degrees of freedom and then treat the symmetry governing the full problem with all degrees of freedom included. For example, we divide the coordinate q into two parts

$$q = (x, \xi) ,$$

where x and ξ may represent the orbital and spin-isospin coordinates, respectively, or represent spin and isospin coordinates, respectively, etc. We now have three realization of the permutation group S_n, i.e.,

permutation group,	permutation operator,	permutation object,	irreducible basis
$S_n(x)$	${}^x p$	x	$\varphi_{m_1}^{(\nu_1)}(x)$
$S_n(\xi)$	${}^\xi p$	ξ	$\psi_{m_2}^{(\nu_2)}(\xi)$
$S_n(q)$	${}^q p \equiv p$	$q = (x, \xi)$	$\Psi_m^{(\nu)\tau}(q)$

According to the definition, we have

$$[{}^x p, {}^\xi p] = 0 , \quad p = {}^x p \, {}^\xi p .$$

The permutation group $S_n(q)$ is called the *inner-product* of $S_n(x)$ and $S_n(\xi)$. Suppose $\varphi_{m_1}^{\nu_1}(x)$ and $\psi_{m_2}^{\nu_2}(\xi)$ are the irreducible basis of $S_n(x)$ and $S_n(\xi)$, respectively. The reduction of the product (i.e., the Kronecker product) rep with the basis $\varphi_{m_1}^{\nu_1}(x)\psi_{m_2}^{\nu_2}(\xi)$ into the irreps of $S_n(q)$ is called the *inner-product reduction* of the permutation group. In other words, the inner product reduction deals with the problem of how to linearly combine the products of wave functions with definite permutation symmetries in each individual degree of freedom into a wave function with a definite permutation symmetry in the overall degrees of freedom.

The CG series of the permutation group is designated

$$[\nu_1] \times [\nu_2] = \sum_{\nu} (\nu_1 \nu_2 \nu)[\nu] .$$

The multiplicity $(\nu_1\nu_2\nu)$ is determined by (3-274). Noting all primitive characters of the permutation group are real, (3-274) becomes

$$(\nu_1 \nu_2 \nu) = \frac{1}{g} \sum_i g_i \chi_i^{(\nu_1)} \chi_i^{(\nu_2)} \chi_i^{(\nu)} . \tag{4-88}$$

$(\nu_1\nu_2\nu)$ satisfies the relation (3-276).

With the help of $\chi_i^{(\tilde{\nu})} = \delta_i \chi_i^{(\nu)}$ [see (4-5)], from (4-88) one has

$$(\nu_1 \nu_2 \nu) = (\tilde{\nu}_1 \tilde{\nu}_2 \nu) = (\nu_1 \tilde{\nu}_2 \tilde{\nu}) = (\tilde{\nu}_1 \nu_2 \tilde{\nu}) . \tag{4-89}$$

According to $\chi_i^{[n]} = 1$ and $\chi_i^{[1^n]} = \delta_i$, we know immediately that

$$[\nu] \times [n] = [\nu], \quad [\nu] \times [1^n] = [\tilde{\nu}] . \tag{4-90}$$

From the simple characters of the permutation group and (4-88), one can calculate the CG series of the permutation group. Tables of the CG series for the permutation group up to S_8 are given by Itzykson (1966). Table 4-10 gives the CG series of $S_3 - S_5$.

Table 4-10. CG series of $S_3 - S_5$*.

(a) S_3 group

	[3]	[21]	[1^3]
↑[21] × [21]	1	1	1

(b) S_4 group

	[4]	[31]	[22]	[211]	[1^4]	
↑[31] × [31]	1	1	1	1		[31] × [211], [211] × [31]
[31] × [22]		1		1		[31] × [22], [211] × [22]
[22] × [22]	1		1		1	[22] × [22], [22] × [22]↓
	[1^4]	[21^2]	[2^2]	[31]	[4]	

(c) S_5 group

	[5]	[41]	[32]	[311]	[2^21]	[21^3]	[1^5]	
↑[41] × [41]	1	1	1	1				[41] × [21^3], [21^3] × [41]
[41] × [32]		1	1	1	1			[41] × [2^21], [21^3] × [32]
[41] × [311]		1	1	1	1	1		[41] × [311], [21^3] × [311]
[32] × [32]	1	1	1	1	1	1		[32] × [2^21], [221] × [32]
[32] × [311]		1	1	2	1	1		[32] × [311], [221] × [311]
[311] × [311]	1	1	2	1	2	1	1	[311] × [311], [311] × [311]↓
	[1^5]	[21^3]	[2^21]	[311]	[32]	[41]	[5]	

*The left(right) heading matches the top(bottom) heading.

Ex. 4.5. Using (4-88) and the characters of S_4 in Ex. 3.16, construct the CG series of S_4.

4.11. Calculation of the CG Coefficients of the Permutation Group

We use $|m_1 m_2\rangle$ (or $(m_1 m_2)$, as shown in Table 4.13) to denote the product basis

$$|m_1 m_2\rangle = \varphi_{m_1}^{[\nu_1]}(x)\psi_{m_2}^{[\nu_2]}(\xi) = |Y_{m_1}^{[\nu_1]} Y_{m_2}^{[\nu_2]}\rangle \ . \tag{4-91}$$

Employing the CG coefficients $C_{\nu_1 m_1, \nu_2 m_2}^{(\nu)\tau, m}$, the product basis can be linearly combined into the irreducible basis $\Psi_m^{(\nu)\tau}(q)$ of the group $S_n(q)$

$$\Psi_m^{(\nu)\tau}(q) = \sum_{m_1 m_2} C_{\nu_1 m_1, \nu_2 m_2}^{(\nu)\tau, m} \varphi_{m_1}^{(\nu_1)}(x)\psi_{m_2}^{(\nu_2)}(\xi), \quad \tau = 1, 2, \ldots (\nu_1\nu_2\nu) \ , \tag{4-92a}$$

where τ is the multiplicity label. From (4-91) and (4-92a), we obtain

$$C_{\nu_1 m_1, \nu_2 m_2}^{[\nu]\tau, m} = \langle Y_m^{[\nu]\tau}(q) | Y_{m_1}^{[\nu_1]} Y_{m_2}^{[\nu_2]}\rangle \ . \tag{4-92b}$$

The CG coefficients of the permutation group are quite useful (see Secs. 7.9 and 7.16). Hamermesh (1962) has calculated the CG coefficients for the product [311]× [311] by a recursive method. Vanagas (1972) gave some algebraic formulas for the CG coefficients in a few special cases. Schindler (1977) calculated the CG coefficients of S_n for $n \leq 6$ using the tensor product decomposition method.

The problems of the CG coefficients and CG series can be solved simultaneously by the EFM. From (3-293) and (4-28c) we obtained the eigenequation satisfied by the CG coefficients of S_n,

$$\begin{gathered} \sum_{m_1 m_2} \left[\langle m_1' m_2' | C(f) | m_1 m_2\rangle - \lambda_f \delta_{m_1' m_1} \delta_{m_2' m_2}\right] C_{\nu_1 m_1, \nu_2 m_2}^{(\nu)\tau m} = 0 \\ f = n, n-1, \ldots, 2, \quad \tau = 1, 2, \ldots, (\nu_1\nu_2\nu), \\ (\nu, m) = (\lambda_n, \lambda_{n-1}, \cdots, \lambda_2) \ . \end{gathered} \tag{4-93}$$

The matrix elements of the 2-cycle class operator can be calculated recursively by using

$$\langle m_1'm_2'|C(f)|m_1m_2\rangle = \langle m_1'm_2'|C(f-1)|m_1m_2\rangle + \sum_{i=1}^{f-1} D^{(\nu_1)}_{m_1'm_1}(if)D^{(\nu_2)}_{m_2'm_2}(if) \ . \tag{4-94}$$

Let $N = h_1, h_2, h_i$ being the dimension of the irrep $[\nu_i]$ of S_f. The representative of $C(f)$ in the product (or uncoupled) rep is a $N \times N$ matrix given by

$$\mathbf{C}(f) = |\langle m_1'm_2'|C(f)|m_1m_2\rangle|_1^N \ . \tag{4-95}$$

The unitary transformation matrix which brings the $n-1$ matrices $\mathbf{C}(n), \mathbf{C}(n-1), \ldots, \mathbf{C}(2)$ simultaneously into diagonal form gives the CG coefficients.

It is easy to show that the matrix elements of $\mathbf{C}(f)$ have the following symmetries:

1. $\mathbf{C}(f)$ is a real and symmetric matrix, i.e.,

$$\begin{aligned}\langle m_1'm_2'|C(f)|m_1m_2\rangle &= \langle m_1m_2|C(f)|m_1'm_2'\rangle \\ &= \langle m_1m_2'|C(f)|m_1'm_2\rangle = \langle m_1'm_2|C(f)|m_1m_2'\rangle \ .\end{aligned} \tag{4-96}$$

2. From (4-67), we have

$$\langle \tilde{m}_1'\tilde{m}_2'|C(f)|\tilde{m}_1\tilde{m}_2\rangle = \Lambda^{\nu_1}_{m_1'}\Lambda^{\nu_2}_{m_2'}\Lambda^{\nu_1}_{m_1}\Lambda^{\nu_2}_{m_2}\langle m_1'm_2'|C(f)|m_1m_2\rangle \ . \tag{4-97}$$

From (4-97) we can derive a symmetry relation for the CG coefficients. Rewriting (4-93) in the form

$$\sum_{\tilde{m}_1\tilde{m}_2} \langle \tilde{m}_1'\tilde{m}_2'|C(f)|\tilde{m}_1\tilde{m}_2\rangle C^{\nu\tau m}_{\tilde{\nu}_1\tilde{m}_1,\tilde{\nu}_2\tilde{m}_2} = \lambda_f C^{\nu\tau m}_{\tilde{\nu}_1\tilde{m}_1',\tilde{\nu}_2\tilde{m}_2'} \ , \tag{4-98}$$

and inserting (4-97) into (4-98), we have

$$\sum_{m_1m_2} \langle m_1'm_2'|C(f)|m_1m_2\rangle \Lambda^{\nu_1}_{m_1}\Lambda^{\nu_2}_{m_2} C^{\nu\tau m}_{\tilde{\nu}_1\tilde{m}_1,\tilde{\nu}_2\tilde{m}_2} = \lambda_f \Lambda^{\nu_1}_{m_1'}\Lambda^{\nu_2}_{m_2'} C^{\nu\tau m}_{\tilde{\nu}_1\tilde{m}_1',\tilde{\nu}_2\tilde{m}_2'} \ . \tag{4-99}$$

Comparing (4-93) with (4-99), we see that

$$\Lambda^{\nu_1}_{m_1}\Lambda^{\nu_2}_{m_2} C^{\nu\tau m}_{\tilde{\nu}_1\tilde{m}_1,\tilde{\nu}_2\tilde{m}_2} = \eta^{\nu_\tau}_m \sum_{\tau'} a_{\tau\tau'} C^{\nu\tau' m}_{\nu_1 m_1,\nu_2 m_2} \ . \tag{4-100a}$$

Therefore

$$C^{\nu\tau m}_{\tilde{\nu}_1\tilde{m}_1,\tilde{\nu}_2\tilde{m}_2} = \eta^{\nu_\tau}_m \Lambda^{\nu_1}_{m_1}\Lambda^{\nu_2}_{m_2} \sum_{\tau'} a_{\tau\tau'} C^{\nu\tau' m}_{\nu_1 m_1,\nu_2 m_2} \ . \tag{4-100b}$$

If $[\tilde{\nu}_1] \neq [\nu_1]$, or $[\tilde{\nu}_2] \neq [\nu_2]$, we can always choose the CG coefficients so that they satisfy the following symmetry

$$C^{\nu\tau m}_{\tilde{\nu}_1\tilde{m}_1,\tilde{\nu}_2\tilde{m}_2} = \eta^{\nu_\tau}_m \Lambda^{\nu_1}_{m_1}\Lambda^{\nu_2}_{m_2} C^{\nu\tau m}_{\nu_1 m_1,\nu_2 m_2} \ . \tag{4-101}$$

We now proceed to prove that the phase factor $\eta^{\nu_\tau}_m$ is independent of m and depends only on ν_1, ν_2, ν and τ, i.e.,

$$\eta^{\nu_\tau}_m = \varepsilon_2(\nu_1\nu_2\nu_\tau) \equiv \varepsilon_2 = \pm 1 \ . \tag{4-102}$$

Using (4-92a), (4-101), (4-97) and (3-287) we have

$$\begin{aligned}\langle \Psi^{[\nu]\tau}_m|p|\Psi^{[\nu]\tau}_{m'}\rangle = D^{(\nu)}_{mm'}(p) &= \sum_{\tilde{m}_1\tilde{m}_2\tilde{m}_1'\tilde{m}_2'} C^{\nu\tau m}_{\tilde{\nu}_1\tilde{m}_1,\tilde{\nu}_2\tilde{m}_2} C^{\nu\tau m'}_{\tilde{\nu}_1\tilde{m}_1',\tilde{\nu}_2\tilde{m}_2'}\langle \tilde{m}_1\tilde{m}_2|p|\tilde{m}_1'\tilde{m}_2'\rangle \\ &= \eta^{\nu\tau}_m\eta^{\nu\tau}_{m'} \sum_{m_1m_2m_1'm_2'} C^{\nu\tau,m}_{\nu_1m_1,\nu_2m_2} C^{\nu\tau,m'}_{\nu_1m_1',\nu_2m_2'} D^{(\nu_1)}_{m_1m_1'}(p)D^{(\nu_2)}_{m_2m_2'}(p) \\ &= \eta^{\nu\tau}_m\eta^{\nu\tau}_{m'} D^{(\nu)}_{mm'}(p) \ .\end{aligned}$$

Therefore

$$\eta_m^{\nu\tau}\eta_{m'}^{\nu\tau} = 1 , \tag{4-103}$$

for any m and m'. Consequently $\eta_m^{\nu_\tau}$ is independent of m.

From (4-102) and (4-101), we finally arrive at an important relation:

$$C^{\nu\tau m}_{\tilde{\nu}_1\tilde{m}_1,\tilde{\nu}_2\tilde{m}_2} = \varepsilon_4(\nu_1\nu_2\nu_\tau)\Lambda^{\nu_1}_{m_1}\Lambda^{\nu_2}_{m_2}C^{\nu\tau m}_{\nu_1 m_1,\nu_2 m_2} . \tag{4-104}$$

The phase factor $\varepsilon_4 = \pm 1$ is dictated by the overall phase convention (see (4-122a)).

By virtue of relation (4-104), we only need to calculate the CG coefficients for those $[\nu_1]$ and $[\nu_2]$ whose row length is greater than or equal to the column length.

If both $[\nu_1]$ and $[\nu_2]$ are self-conjugate, the following situations should be discussed separately.

(a) $(\nu_1\nu_2\nu) = 1$. In these cases the relation (4-104) holds automatically.

(b) $(\nu_1\nu_2\nu) = 2$. Two situations appear here. (i) Two sets of the CG coefficients $C^{[\nu]\alpha,m}_{\nu_1 m_1,\nu_2 m_2}$ and $C^{[\nu]\beta,m}_{\nu_1 m_1,\nu_2 m_2}$ are both "symmetric" (i.e., the phase factor $\varepsilon_4 = 1$ for $\tau = \alpha,\beta$) or both "antisymmetric" (i.e., $\varepsilon_4 = -1$ for $\tau = \alpha,\beta$). In such an instance, the CG coefficents calculated fulfill (4-104) automatically. (ii) Two sets of the CG coefficients contain both the "symmetric" and "antisymmetric" components. In this case they can be recombined linearly into a "symmetric" and an "antisymmetric" CG coefficient associated with $\varepsilon_4 = 1$ and -1, respectively.

(c) $(\nu_1\nu_2\nu) \geq 3$. See Chen ([10], 1981) and Gao (1985).

3. From $C(2) = {}^{x}C(2){}^{\xi}C(2)$, and

$$^{x}C(2)\varphi^{[\nu_1]}_{m_1}(x) = \lambda_2^{(1)}\varphi^{[\nu_1]}_{m_1}(x), \quad {}^{\xi}C(2)\psi^{[\nu_2]}_{m_2}(\xi) = \lambda_2^{(2)}\psi^{[\nu_2]}_{m_2}(\xi) ,$$
$$\lambda_2^{(i)} = \pm 1, \quad i = 1,2 ,$$

we have

$$\begin{aligned} C(2)|m_1m_2\rangle &= \lambda_2|m_1m_2\rangle , \\ \lambda_2 &= \lambda_2^{(1)}\lambda_2^{(2)} . \end{aligned} \tag{4-105}$$

Thus $\mathbf{C}(2)$ is already diagonalized in the uncoupled representation. We only need to diagonalize the $n-2$ matrices $\mathbf{C}(3)$, $\mathbf{C}(4)$, ... , $\mathbf{C}(n)$.

4. From $[C(f), C(2)] = 0$, we have

$$\langle m_1'm_2'|C(f)C(2)|m_1m_2\rangle = \langle m_1'm_2'|C(2)C(f)|m_1m_2\rangle .$$

Therefore

$$\begin{aligned} \lambda_2\langle m_1'm_2'|C(f)|m_1m_2\rangle &= \lambda_2'\langle m_1'm_2'|C(f)|m_1m_2\rangle , \\ \langle m_1'm_2'|C(f)|m_1m_2\rangle &= 0, \quad \text{for} \quad \lambda_2 \neq \lambda_2' . \end{aligned} \tag{4-106}$$

Equation (4-106) tells us that there is no coupling between the basis vectors with $\lambda_2 = 1$ (denoted as $|m_1m_2\rangle_+$) and those with $\lambda_2 = -1$ (denoted as $|m_1m_2\rangle_-$). By appropriate ordering of $|m_1m_2\rangle$, the matrices $\mathbf{C}(f)$ becomes block-diagonalized, i.e.,

$$\mathbf{C}(f) = \left(\begin{array}{c|c} \mathbf{C}^{(+)}(f) & 0 \\ \hline 0 & \mathbf{C}^{(-)}(f) \end{array}\right) , \quad f = 3,4,\ldots,n, \tag{4-107}$$

$$\mathbf{C}^{(\pm)}(f) = \|\langle m_1'm_2'|C(f)|m_1m_2\rangle_\pm\|_1^{N\pm} , \tag{4-108}$$

where $N_+(N_-)$ is the number of the basis vectors with $\lambda_2 = 1(-1)$. Consequently, the calculation of the CG coefficients is reduced to a diagonalization of the $n-2$ operators in the eigenspaces $\lambda_2 = \pm 1$ of $C(2)$ separately. This fact greatly simplifies both the calculation and the tabulation of the CG coefficients. Each CG coefficient table can be decomposed into two sub-tables, one corresponding to $\lambda_2 = 1$ and the other to $\lambda_2 = -1$. See Sec. 4.13.

Example 1: The CG coefficients of S_3. S_3 has three irreps [3], [21] and $[1^3]$. According to (4-90), we only need to calculate the CG coefficients for $[21]\times[21]$. The standard basis of S_3 is

$$\varphi_1^{[21]} = \left| \begin{array}{|c|c|}\hline 1 & 2 \\ \hline 3 \\ \cline{1-1}\end{array} \right\rangle , \quad \lambda_2^{(1)} = 1 \, ; \quad \varphi_2^{[21]} = \left| \begin{array}{|c|c|}\hline 1 & 3 \\ \hline 2 \\ \cline{1-1}\end{array} \right\rangle , \quad \lambda_2^{(1)} = -1 \, .$$

$$|m_1 m_2\rangle = \varphi_{m_1}^{[21]} \psi_{m_2}^{[21]} \, .$$

The ordering of the product basis vectors $|m_1 m_2\rangle = \varphi_{m_1}^{[21]} \psi_{m_2}^{[21]}$ is taken as $|11\rangle, |22\rangle, |12\rangle, |21\rangle$. The first two belong to $\lambda_2 = 1$, and the other two belong to $\lambda_2 = -1$. From (4-94) and Table 4.4-2, we obtain the matrix

$$\mathbf{C}(3) = \left(\begin{array}{cc:cc} 3/2 & 3/2 & & \\ 3/2 & 3/2 & & \\ \hdashline & & -3/2 & 3/2 \\ & & 3/2 & -3/2 \end{array} \right) . \tag{4-109}$$

The eigenvalues of $\mathbf{C}(3)$ are $\lambda_3 = 3, 0, 0, -3$. Comparing it with Fig. 4.5-3, we obtain the CG series

$$[21] \times [21] = [3] + [21] + [1^3] \, . \tag{4-110}$$

From the eigenequation of $C(3)$, we obtain the CG coefficients, and therefore the coupled basis

$$\Psi_1^{(3)} = \sqrt{\frac{1}{2}} |11\rangle + |22\rangle) \, , \quad \Psi_{-1}^{(-3)} = \sqrt{\frac{1}{2}} (|12\rangle - |21\rangle) \, .$$

$$\Psi_1^{(0)} = \sqrt{\frac{1}{2}} (|11\rangle - |22\rangle) \, , \quad \Psi_{-1}^{(0)} = -\sqrt{\frac{1}{2}} (|12\rangle + |21\rangle) \, ,$$

It is seen that the irreps (3) and (0), or [3] and [21], belong to the symmetric square, and the irrep (-3), or $[1^3]$, belongs to the antisymmetric square. In Table 4.13, we use $[\ ]_s$ and $[\ ]_a$ to represent the symmetric and anti-symmetric squares. Thus Eq. (4-110) is written as

$$[21] \times [21] = [3]_s + [21]_s + [1^3]_a .$$

Example 2: Find the CG coefficients of S_4 for $[31]\times[22]$. The ordering of the product basis vectors $|m_1 m_2\rangle = \varphi_{m_1}^{[31]} \psi_{m_2}^{[22]}$ is taken as

$$|m_1 m_2\rangle = |11\rangle, |21\rangle, |32\rangle; |12\rangle, |22\rangle, |31\rangle .$$

The ordering of the standard basis of S_4 is given in Table 4.4-1. In the above equation, the first three states belong to $\lambda_2 = 1$ and the rest belong to $\lambda_2 = -1$.

Utilizing the matrix elements of [31] and [22] given in Table 4.4-2, it is easy to construct the matrices of $C(3)$ and $C(4)$, namely

$$\mathbf{C}(4) = \left(\begin{array}{ccc|ccc} 0 & \sqrt{2} & -\sqrt{2} & & & \\ \sqrt{2} & 1 & 1 & & & \\ -\sqrt{2} & 1 & 1 & & & \\ \hline & & & 0 & -\sqrt{2} & -\sqrt{2} \\ & & & -\sqrt{2} & -1 & 1 \\ & & & -\sqrt{2} & 1 & -1 \end{array}\right) ,$$

$$\mathbf{C}(3) = \left(\begin{array}{ccc|ccc} 0 & 0 & 0 & & & \\ 0 & 3/2 & 3/2 & & & \\ 0 & 3/2 & 3/2 & & & \\ \hline & & & 0 & 0 & 0 \\ & & & 0 & -3/2 & 3/2 \\ & & & 0 & 3/2 & -3/2 \end{array}\right) . \tag{4-111}$$

From the expectation equation of $\mathbf{C}(4)$ one obtains

$$\det|\mathbf{C}(4) - \lambda_4 I| = (\lambda_4 - 2)^3(\lambda_4 + 2)^3 = 0 .$$

Comparing this with Table 3.2-1, we have

$$[31] \times [22] = [31] + [211] .$$

By diagonalizing $\mathbf{C}(4)$ and $\mathbf{C}(3)$, we obtain the CG coefficients listed in Table 4.13-2b.

Computer calculations of the CG coefficients

The most prominent feature of the EFM for the CG coefficients is that it is easily programmable. The main steps are as follows:

1. For a given $[\nu_1]$ and $[\nu_2]$ note down, from the known CG series $(\nu_1\nu_2\nu)$ [e.g., Table 4.10-1 or tables given by Itzykson (1966)] and from Table 4.4-1, the eigenvalues $\lambda = k_n\lambda_n + \ldots + k_3\lambda_3$ for the first components of each possible irrep $[\nu]$.

2. Divide the product basis vectors into two groups and calculate the matrices $\mathbf{C}^{(+)}(f)$ in (4-108). We only need to solve the eigenequation of the operator $M = \sum_f k_f C(f)$ in the $\lambda_2 = +1$ subspace, since the first component of all irreps, except the totally antisymmetric irrep $[1^n]$, correspond to $\lambda_2 = 1$, and the CG coefficients for $[1^n]$ is trivial (see (4-124)).

3. Form the matrices

$$\mathbf{M}^{(+)} = k_n\mathbf{C}^{(+)}(n) + \ldots + k_2\mathbf{C}^{(+)}(3) . \tag{4-112a}$$

Substitute the values λ found in step 1 into the following equation

$$(\mathbf{M}^{(+)} - \lambda\mathbf{I})\mathbf{U}^{\tau} = 0 , \quad \tau = 1, 2, \ldots , (\nu_1\nu_2\nu) , \tag{4-112b}$$

where $\mathbf{U}^{\tau} = \{C^{[\nu]\tau, m=1}_{\nu_1 m_1, \nu_2 m_2}\}$ is a column vector with m_1, m_2 as the row index. Solving (4-112b), we find the CG coefficients of first component of the τ-th irrep (ν).

4. The overall phase convention. The phase of the eigenvector $\mathbf{U}^{\tau}$ is dictated by the overall phase convention, which is stipulated as following: First arrange the product basis vectors

$|m_1m_2\rangle$ according to the order $|1,1\rangle, |1,2,\rangle, \ldots, |1,h_{\nu_2}\rangle, |2,1\rangle, \ldots, |2,h_{\nu_2}\rangle, \ldots |h_{\nu_1},h_{\nu_2}\rangle$; then demand that the first nonvanishing component of the vector $\mathbf{U}^\tau$ be positive, i.e.,

$$C^{[\nu]\tau,1}_{\nu_1m_1,\nu_2m_2}|_{(m_1,m_2)=\min}\rangle 0 , \tag{4-113}$$

where $(m_1m_2)_{\min}$ means taking the index m_1 as small as possible followed by taking m_2 as small as possible for which the CG coefficient $C^{[\nu]\tau,1}_{\nu_1m_1,\nu_2m_2}$ is nonzero.

5. Substituting $\Psi^{[\nu]\tau}_m = \sum_{cd} C^{[\nu]\tau m}_{\nu_1c,\nu_2d}|cd\rangle$ into (4-35) and multiplying by $\langle ab|$ from the left, we get

$$C^{\nu_\tau m'}_{\nu_1a,\nu_2b} = \frac{1}{D^{(\nu)}_{m'm}(T)} \sum_{cd}[D^{(\nu_1)}_{ac}(T)D^{(\nu_2)}_{bd}(T) - D^{(\nu)}_{mm}(T)\delta_{ac}\delta_{bd}]C^{\nu_\tau m}_{\nu_1c,\nu_2d} . \tag{4-114}$$

Starting from the CG coefficients $C^{[\nu]\tau,1}_{\nu_1m_1,\nu_2m_2}$ found from (4-112b), using (4-114) and choosing appropriate transpositions, we can find all other components of the CG coefficients of the τ-th irrep (ν).

Notice that we cannot use the intrinsic quantum number as the multiplicity label for the CG coefficients and the IDC (induction coefficients) to be discussed in Sec. 4.15, since in the space of the product states we cannot define a meaningful intrinisic state.

Two EFM programs were available for computing the CG coefficients of the permutation group; one uses the non-recursive method just discussed (the program in Algol-60 was published in Chen([10] 1981), and the other employs the recursive method (Gao 1985) to be discussed in Sec. 4.19.

4.12. Properties of the CG Coefficients of Permutation Groups

The CG coefficients of the permutation group have some special properties (Gao 1985) besides the general results given in Sec. 3.15. These properties can be exploited to reduce the number of independent CG coefficients that have to be calculated, as well as to check the correctness of the calculated CG coefficients.

1. Unitarity

$$\begin{aligned} \sum_{m_1m_2} C^{[\nu]\tau,m}_{\nu_1m_1,\nu_2m_2} C^{[\nu']\tau',m'}_{\nu_1m_1,\nu_2m_2} &= \delta_{\nu\nu'}\delta_{\tau\tau'}\delta_{mm'}, \\ \sum_{\nu\tau m} C^{[\nu]\tau,m}_{\nu_1m_1,\nu_2m_2} C^{[\nu]\tau,m}_{\nu_1m_1',\nu_2m_2'} &= \delta_{m_1m_1'}\delta_{m_2m_2'}. \end{aligned} \tag{4-115}$$

2.

$$C^{[\nu]\tau,m}_{\nu_1m_1,\nu_2m_2} = \varepsilon_1(\nu_1\nu_2\nu_\tau)C^{[\nu]\tau,m}_{\nu_2m_2,\nu_1m_1} . \tag{4-116a}$$

From the overall phase convention (4-113), it is seen that

$$\varepsilon_1(\nu_1\nu_2\nu_\tau) = \text{sign}(C^{[\nu]\tau,1}_{\nu_1m_1,\nu_2m_2}|_{(m_2m_1)=\min}) , \tag{4-116b}$$

where $(m_2m_1)_{\min}$ means that we first take the index m_2 as small as possible and then take m_1 as small as possible.

For $\nu_1 = \nu_2$ let

$$\delta_{\nu_\tau} = \varepsilon_1(\nu_1\nu_1\nu_\tau) . \tag{4-116c}$$

If $\delta_{\nu_\tau} = 1$ (-1), then the irrep $[\nu]_\tau$ belongs to the symmetric (antisymmetric) square $[(\nu_1)\times(\nu_2)]_s([(\nu_1)\times(\nu_2)]_a)$.

3. From (3-327b) we have for $\nu_1 \neq \nu_2 \neq \nu \neq \nu_1$,

$$\frac{1}{\sqrt{h_\nu}}C^{[\nu]\tau,m}_{\nu_1m_1,\nu_2m_2} = \varepsilon_2(\nu_1\nu_2\nu_\tau)\frac{1}{\sqrt{h_{\nu_1}}}C^{[\nu_1]\tau,m_1}_{\nu m,\nu_2m_2} = \varepsilon_3(\nu_1\nu_2\nu_\tau)\frac{1}{\sqrt{h_{\nu_2}}}C^{[\nu_2]\tau,m_2}_{\nu_1m_1,\nu m} , \tag{4-117}$$

with the phase factor

$$\varepsilon_2(\nu_1\nu_2\nu_\tau) = \mathrm{sign}(C^{[\nu]\tau,m}_{\nu_1 1,\nu_2 m_2}|_{(mm_2)=\min}) , \tag{4-118}$$

$$\varepsilon_3(\nu_1\nu_2\nu_\tau) = \mathrm{sign}(C^{[\nu]\tau,m}_{\nu_1 m_1,\nu_2 1}|_{(m_1 m)=\min}) . \tag{4-119}$$

4. For $\nu_1 = \nu_2 \neq \nu$, (4-117) still holds. Besides we have

$$C^{[\nu_1]\tau,m_1}_{\nu m,\nu_1 m_2} = \delta_{\nu_\tau} C^{[\nu_1]\tau,m_2}_{\nu m,\nu_1 m_1} , \quad \delta_{\nu_\tau} = \varepsilon_1(\nu_1\nu_1\nu_\tau) , \tag{4-120}$$

which can be easily proved by using (4-116c) and (4-117).

5. The symmetries for $\nu_1 = \nu_2 = \nu$ are more involved (see Gao 1985).

6. The symmetries of the CG coefficients under conjugation are

$$\begin{aligned} C^{[\nu]\tau,m}_{\nu_1 m_1,\nu_2 m_2} &= \varepsilon_4(\nu_1\nu_2\nu_\tau)\Lambda^{\nu_2}_{m_2}\Lambda^{\nu_2}_{m_2} C^{[\nu]\tau,m}_{\nu_1 m_1,\tilde{\nu}_2\tilde{m}_2} \\ &= \varepsilon_5(\nu_1\nu_2\nu_\tau)\Lambda^{\nu_1}_{m_1}\Lambda^{\nu}_{m} C^{[\tilde{\nu}]\tau,\tilde{m}}_{\tilde{\nu}_1\tilde{m}_1,\nu_2 m_2} = \varepsilon_6(\nu_1\nu_2\nu_\tau)\Lambda^{\nu_2}_{m_2}\Lambda^{\nu}_{m} C^{[\tilde{\nu}]\tau,\tilde{m}}_{\nu_1 m_1,\tilde{\nu}_2\tilde{m}_2} , \end{aligned} \tag{4-121}$$

$$\varepsilon_4(\nu_1\nu_2\nu_\tau) = \mathrm{sign}(\Lambda^{\nu_1}_{m_1}\Lambda^{\nu_2}_{m_2} C^{[\nu]\tau,1}_{\nu_1 m_1,\nu_2 m_2}|_{(\overline{m}_1\overline{m}_2)}) , \tag{4-122a}$$

$$\varepsilon_5(\nu_1\nu_2\nu_\tau) = \mathrm{sign}(\Lambda^{\nu_1}_{m_1}\Lambda^{\nu}_{m} C^{[\nu]\tau,h_\nu}_{\nu_1 m_1,\nu_2 m_2}|_{(\overline{m}_1\underline{m}_2)}) , \tag{4-122b}$$

$$\varepsilon_6(\nu_1\nu_2\nu_\tau) = \mathrm{sign}(\Lambda^{\nu_2}_{m_2}\Lambda^{\nu}_{m} C^{[\nu]\tau,h_\nu}_{\nu_1 m_1,\nu_2 m_2}|_{(\underline{m}_1,\overline{m}_2)}) , \tag{4-122c}$$

where $(\overline{m}_1\overline{m}_2)$ means first taking the index m_1 as large as possible and then taking m_2 as large as possible, whereas $(\overline{m}_1\underline{m}_2)$ means taking m_1 as large as possible and then taking m_2 as small as possible.

Notice that for non-multiplicity free cases, (4-121) in general only holds when the partitions $[\nu_1]$ and $[\nu_2]$ (or $[\nu_1]$ and $[\nu]$, or $[\nu_2]$ and $[\nu]$) are not self-conjugate simultaneously. If both are self-conjugate (for S_n with $n \leq 7$, then they must equal), the imposition of the symmetry

$$C^{[\nu]\tau,m}_{\nu_1 m_1,\nu_1 m_2} = \pm\Lambda^{\nu_1}_{m_1}\Lambda^{\nu_1}_{m_2} C^{[\nu]\tau,m}_{\nu_1\tilde{m}_1,\nu_1\tilde{m}_2}, \quad \text{for} \quad [\nu_1] = [\tilde{\nu}_1] \neq [\nu] , \tag{4-123a}$$

or

$$C^{[\nu]\tau,m}_{\nu_1 m_1,\nu m_2} = \pm\Lambda^{\nu}_{m_2}\Lambda^{\nu}_{m} C^{[\nu]\tau,\tilde{m}}_{\nu_1 m_1,\nu\tilde{m}_2} \quad \text{for} \quad [\nu] = [\tilde{\nu}] \neq [\nu_1] , \tag{4-123b}$$

will be of help for the multiplicity separation.

For multiplicity-free cases, the symmetries (4-116)-(4-123) are satisfied automatically; however the phases now enter of necessity instead of being determined by the phase convention. For non-multiplicity-free cases, the symmetries (4-116a), (4-120) and (4-123) are imposed to reduce the arbitrariness in the multiplicity separation and to yield simpler numerical values for the CG coefficients.

If among $[\nu_1]$, $[\nu_2]$ and $[\nu]$, one is totally symmetric, $[n]$, or totally antisymmetric, $[1^n]$, then the CG coefficients are very simple. From (4-90) and (4-121) we have

$$\begin{aligned} C^{[\nu]m}_{[\nu_1]m_1,[n]1} &= \delta_{\nu\nu_1}\delta_{mm_1} . \\ C^{[n],1}_{[\nu_1]m_1,[\nu_2]m_2} &= \frac{1}{\sqrt{h_{\nu_1}}}\delta_{\nu_1\nu_2}\delta_{m_1 m_2} . \\ C^{[\tilde{\nu}]\tilde{m}}_{[\nu_1]m_1,[1^n]1} &= \Lambda^{\nu_1}_{m_1}\delta_{\nu_1\nu}\delta_{m_1 m} . \\ C^{[1^n]1}_{[\nu_1]m_1,[\tilde{\nu}]_2\tilde{m}_2} &= \frac{1}{\sqrt{h_{\nu_1}}}\Lambda^{\nu_1}_{m_1}\delta_{\nu_1\nu_2}\delta_{m_1 m_2} . \end{aligned} \tag{4-124}$$

From this the totally symmetric or antisymmetric state is expressed as

$$\Psi^{[n]} = \frac{1}{\sqrt{h_\nu}} \sum_m \varphi_m^{[\nu]}(x)\psi_m^{[\nu]}(\xi) \,,$$
$$\Psi^{[1^n]} = \frac{1}{\sqrt{h_\nu}} \sum_m \Lambda_m^\nu \varphi_m^{[\nu]}(x)\psi_{\tilde{m}}^{[\tilde{\nu}]}(\xi) \,. \tag{4-125}$$

The symmetry relation given above can be checked by using CG coefficients in Sec. 4.13. For example, (4-117) can be checked by using Table 4.13-3e and 4.13-3f:

$$\sqrt{\frac{1}{6}} C^{[311]\tau,2}_{[311]1,[32]2} = \sqrt{\frac{1}{5}} C^{[32]\tau,2}_{[311]1,[311]2} \,, \quad \sqrt{\frac{1}{6}} \begin{pmatrix} -\sqrt{\frac{1}{10}} \\ -\sqrt{\frac{1}{60}} \end{pmatrix} = \sqrt{\frac{1}{5}} \begin{pmatrix} -\sqrt{\frac{1}{12}} \\ -\sqrt{\frac{1}{72}} \end{pmatrix} \,, \quad \tau = \begin{cases} \alpha \\ \beta \end{cases} .$$

Similarly, (4-123) can be checked by the following example:

$$C^{[221]\tau,5}_{[311]1,[311]6} = C^{[\tilde{32}]\tau,\tilde{1}}_{[311]1,[311]\tilde{1}} = \varepsilon_6 \Lambda_1^{[32]} \Lambda_1^{[311]} C^{[32]\tau,1}_{[311]1,[311]1} = \begin{pmatrix} \sqrt{\frac{4}{12}} \\ -\sqrt{\frac{4}{72}} \end{pmatrix} , \tau = \begin{pmatrix} \alpha \\ \beta \end{pmatrix}$$

4.13. Tables of the CG Coefficients for S_3-S_5

Two sets of CG coefficients for the permutation group $S_2 - S_6$ have been produced. One was computed by Schindler (1978), where only the CG coefficients for the so-called working triplets of irreps are calculated. The other was computed by Chen ([10] 1981). All the CG coefficients of $S_2 - S_6$, except those for $[321]\times[321] \to [321]^5$, along with the program, have been published.

The absolute phase of the permutation group CG coefficients tabulated in Schindler (1968) and Chen ([10] 1981) are all chosen randomly.

Now we adopted the phase convention (4-113) for reasons to be explained in Sec. 7.16. The CG coefficients of $S_3 - S_5$ are listed in Tables 4.13-1 to 4.13-3. The CG coefficients for $[321] \times [321] \to [321]^5$ is given in Gao (1985), the CG coefficients for the remaining Kronecker product of S_6 can be found from the $S_6 \supset S_5$ isoscalar factors (Chen [16] 1984) and the S_5 CG coefficient Table listed below.

The CG coefficients tabulated here and in Gao (1985) differ from the Schindler (1968) results in the following respects. (a) Ours have a consistent absolute phase convention. (b) Our multiplicity separation is based, whenever possible, on the imposition of the symmetries, while theirs is based on an *ad hoc* choice. (c) Our tables give the exact values of the coefficients instead of the approximate values to 16 decimal places.

Some remarks on Table 4.13 are now in order:

1. Each table with a given $[\nu_1]$ and $[\nu_2]$ is divided into two tables according to the values of λ_2, with that $\lambda_2 = 1$ (-1) part at the upper (lower) half of the table.

2. The second column N gives the normalization factors. The values listed in the tables are the squares of the CG coefficients. The asterisk denotes a negative CG coefficient and a blank denotes a zero CG coefficient. For example, from Table 4.13-3b we have

$$C^{[41]2}_{[41]3,[32]2} = \left\langle \begin{matrix} 1235 \\ 4 \end{matrix} \middle| \begin{matrix} 1245 & 124 \\ 3 & 35 \end{matrix} \right\rangle = -\sqrt{\frac{1}{45}} \,.$$

3. Each table can be read either horizontally or vertically, e.g., from Table 4.13-3b one has

$$\Psi_2^{[41]} = \sqrt{\frac{1}{45}} (\sqrt{15}\varphi_1^{[41]}\psi_1^{[32]} + \sqrt{4}\varphi_2^{[41]}\psi_1^{[32]} - \varphi_3^{[41]}\psi_2^{[32]} + \sqrt{12}\varphi_3^{[41]}\psi_4^{[32]}$$
$$- \varphi_4^{[41]}\psi_3^{[32]} + \sqrt{12}\varphi_4^{[41]}\psi_5^{[32]}) \,,$$
$$\varphi_1^{[41]}\psi_4^{[32]} = -\sqrt{\frac{18}{48}}\Psi_4^{[32]} + \sqrt{\frac{10}{16}}\Psi_1^{[221]} \,.$$

4. In the sub-tables 3e and 3f, α and β are the multiplicity labels, e.g., $[311\alpha]$ and $[311\beta]$ represent $[311]\ \tau = \alpha$ and $[311]\tau = \beta$, respectively.

5. Using the symmetries given in Sec. 4.12, we can obtain the other CG coefficients not listed in the tables. The phase factors Λ^{ν}_{m} are given in Table 4.4-1.

6. The applications of the CG coefficients of the permutation group are discussed in Sec. 7.9 and in Chen ([10] 1981).

Table 4.13. Tables of the CG coefficients of the permutation groups S_3 - S_5.

S_3 1 $[21]\times[21]=[3]_s+[21]_s+[1^3]_a$

	N	(11)	(22)
[3]	2	1	1
[21]1	2	1	*1
		(12)	(21)
[21]2	2	*1	*1
[1³]	2	1	*1

S_4 2a $[31]\times[31]=[4]_s+[31]_s+[22]_s+[211]_a$

	N	(11)	(12)	(21)	(22)	(33)
[4]	3	1			1	1
[31]1	6	4			*1	*1
2	6		*1	*1	2	*2
[22]1	6		2	2	1	*1
[211]1	2		1	*1		
		(13)	(31)	(23)	(32)	
[31]3	6	*1	*1	*2	*2	
[22]2	6	2	2	*1	*1	
[211]2	2	1	*1			
3	2			1	*1	

S_4 2b $[31]\times[22]=[31]+[211]$

	N	(11)	(21)	(32)
[31]1	2		1	1
2	4	2	1	*1
[211]1	4	2	*1	1
		(12)	(22)	(31)
[31]3	4	2	*1	*1
[211]2	4	2	1	1
3	2		*1	1

S_4 2c $[22]\times[22]=[4]_s+[22]_s+[1^4]_a$

	N	(11)	(22)
[4]	2	1	1
[22]1	2	1	*1
		(12)	(21)
[22]2	2	*1	*1
[1⁴]	2	1	*1

[1] $[\nu]_s$ denotes the representation belongs to the symmetric product $[[\nu_1]\times[\nu_1]]_s$, $[\nu]_a$ denotes the representation belong to the antisymmetric product $[[\nu_1]\times[\nu_1]]_a$.

[2] N is the normalization coefficient, * denotes that the coefficient is negative. As in Table 3b, $C^{[41]2}_{[41]3,[32]2} = -\sqrt{\frac{1}{45}}$.

cont'd

S_5 3*a* $[41]\times[41]=[5]_s+[41]_s+[32]_s+[311]_a$

	N	(11)	(12)	(13)	(21)	(22)	(23)	(31)	(32)	(33)	(44)
[5]	4	1				1				1	1
[41]1	12	9				*1				*1	*1
2	36		*3		*3	20				*5	*5
3	36			*3			*5	*3	*5	10	*10
[32]1	36		15		15	4				*1	*1
2	36			15			*1	15	*1	2	*2
4	6						2		2	1	*1
[311]1	2		1		*1						
2	2			1				*1			
4	2						1		*1		
		(14)	(24)	(34)	(41)	(42)	(43)				
[41]4	36	*3	*5	*10	*3	*5	*10				
[32]3	36	15	*1	*2	15	*1	*2				
5	6		2	*1		2	*1				
[311]3	2	1			*1						
5	2		1			*1					
6	2			1			*1				

S_5 3*b* $[41]\times[32]=[41]+[32]+[311]+[221]$

	N	(11)	(12)	(14)	(21)	(22)	(24)	(31)	(32)	(34)	(43)	(45)
[41]1	3				1				1		1	
2	45	15			4				*1	12	*1	12
3	45		15			*1	12	*1	2	6	*2	*6
[32]1	72	12			20				*5	*15	*5	*15
2	144		24			*10	*30	*10	20	*15	*20	15
4	48			*18		*10		*10	*5		5	
[311]1	40	20			*12				3	*1	3	*1
2	80		40			6	*2	6	*12	*1	12	1
4	16					*2	6	2		*3		3
[221]1	16			10		*2		*2	*1		1	
3	16					6	2	*6		*1		1
		(13)	(15)	(23)	(25)	(33)	(35)	(41)	(42)	(44)		
[41]4	45	15		*1	12	*2	*6	*1	*2	*6		
[32]3	144	24		*10	*30	*20	15	*10	*20	15		
5	48		*18	*10		5		*10	5			
[311]3	80	40		6	*2	12	1	6	12	1		
5	16			*2	6		3	2		3		
6	8					*1	*3		1	3		
[221]2	16		10	*2		1		*2	1			
4	16			6	2		1	*6		1		
5	8					3	*1		*3	1		

cont'd

S_5 3c [311] × [41] = [41] + [32] + [311] + [221] + [21³]

	N	(11)	(12)	(13)	(21)	(22)	(23)	(34)	(41)	(42)	(43)	(54)	(64)
[41]1	3		1				1	1					
2	3	*1									1	1	
3	3				*1					*1			1
[32]1	48	20	*12				3	3			5	5	
2	48			3	20	3	*6	6		*5			5
4	96			*2		*2	*1	1		30	*15	15	30
[311]1	48	12	20				*5	*5			3	3	
2	48			*5	12	*5	10	*10		*3			3
4	48			3		*3			*12	5	10	*10	5
[221]1	96			30		30	15	*15		2	*1	1	2
3	48			*5		5			20	3	6	*6	3
[21³]1	3			1		*1			1				
		(14)	(24)	(31)	(32)	(33)	(44)	(51)	(52)	(53)	(61)	(62)	(63)
[41]4	3			*1					*1				*1
[32]3	48	3	6	20	3	6			*5				*5
5	96	*2	1		*2	1	15		30	15			*30
[311]3	48	*5	*10	12	*5	*10			*3				*3
5	48	3			*3		*10	*12	5	*10			*5
6	48		3			*3	5			*5	*12	*20	
[221]2	96	30	*15		30	*15	1		2	1			*2
4	48	*5			5		*6	20	3	*6			*3
5	48		*5			5	3			*3	20	*12	
[21³]2	3	1			*1			1					
3	3		1			*1					1		
4	3						1			*1		1	

S_5 3d [32] × [32] = $[5]_s + [41]_s + [32]_s + [311]_a + [221]_s + [21^3]_a$

	N	(11)	(12)	(14)	(21)	(22)	(24)	(33)	(35)	(41)	(42)	(44)	(53)	(55)
[5]	5	1				1		1				1		1
[41]1	30	4				4		4				*9		*9
2	18	4				*1	*3	*1	*3		*3		*3	
3	36		*2	*6	*2	4	*3	*4	3	*6	*3		3	
[32]1	36	16				*4	3	*4	3		3		3	
2	72		*8	6	*8	16	3	*16	*3	6	3		*3	
4	24		2		2	1		*1				9		*9
[311]1	4						1		1		*1		*1	
2	8			2			1		*1	*2	*1		1	
4	40		8	6	*8		*3		3	*6	3		*3	
[221]1	8		2		2	1		*1				*1		1
3	8			2			*1		1	2	*1		1	
[21³]1	20		6	*2	*6		1		*1	2	*1		1	
		(13)	(15)	(23)	(25)	(31)	(32)	(34)	(43)	(45)	(51)	(52)	(54)	
[41]4	36	*2	*6	*4	3	*2	*4	3	3		*6	3		
[32]3	72	*8	6	*16	*3	*8	*16	*3	*3		6	*3		
5	24	2		*1		2	*1			*9			*9	
[311]3	8		2		*1			*1	1		*2	1		
5	40	8	6		3	*8		3	*3		*6	*3		
6	20			4	*3		*4	3	*3			3		
[211]2	8	2		*1		2	*1			1			1	
4	8		2		1			1	1		2	1		
5	4				*1			1	1			*1		
[21³]2	20	6	*2		*1	*6		*1	1		2	1		
3	10			3	1		*3	*1	1			*1		
4	2									1			*1	

cont'd

S_5 3e $[311]\times[32]=[41]+[32]+[311]^2+[221]+[21^3]$

	N	(11)	(12)	(14)	(21)	(22)	(24)	(33)	(35)	(41)	(42)	(44)	(53)	(55)	(63)	(65)
[41]1	3	1				1		1								
2	60	*12				3	*1	3	*1		*5	15	*5	15		
3	120		6	*2	6	*12	*1	12	1	10		*15		15	*10	*30
[32]1	24						5		5		4	3	4	3		
2	48			10			5		*5	*8		*3		3	8	*6
4	48		*10		*10	*5		5		*6	3		*3		*6	
[311α]1	40	16				*4	*3	*4	*3			5		5		
2	80		*8	*6	*8	16	*3	*16	3			*5		5		*10
4	80			10			*5		5	*8	*16	3	16	*3	*8	*6
[311β]1	60	4				*1	12	*1	12		*15		*15			
2	60		*1	12	*1	2	6	*2	*6	15					*15	
4	60		*15		15					1	2	6	*2	*6	1	*12
[221]1	48		6		6	3		*3		*10	5		*5		*10	
3	48		8	*6	*8		3		*3			5		*5		*10
[21³]1	120		10	30	*10		*15		15	6	12	1	*12	*1	6	*2

	N	(13)	(15)	(23)	(25)	(31)	(32)	(34)	(43)	(45)	(51)	(52)	(54)	(61)	(62)	(64)
[41]4	120	6	*2	12	1	6	12	1		15	10		15		10	30
[32]3	48		10		*5			*5		3	*8		3		*8	6
5	48	*10		5		*10	5		*3		*6	*3			6	
[311α]3	80	*8	*6	*16	3	*8	*16	3		5			5			10
5	80		10		5			5	16	*3	*8	16	*3		8	6
6	40				*5			5	*4	*3		4	3	16		
[311β]3	60	*1	12	*2	*6	*1	*2	*6			15				15	
5	60	*15				15			*2	*6	1	*2	*6		*1	12
6	60			*15			15		1	*12		*1	12	*4		
[221]2	48	6		*3		6	*3		*5		*10	*5			10	
4	48	8	*6		*3	*8		*3		*5			*5			10
5	24			4	3		*4	*3		*5			5			
[21³]2	120	10	30		15	*10		15	*12	*1	6	*12	*1		*6	2
3	60			5	*15		*5	15	3	*1		*3	1	*12		
4	3								*1			1		*1		

cont'd

S_5 $3f$ $[311]\times[311]=[5]_s+[41]_s+[32]_s^2+[311]_a+[221\alpha]_s+[221\beta]_a+[21^3]_a+[1^5]_s$

	N	(11)	(12)	(14)	(21)	(22)	(24)	(33)	(35)	(36)	(41)	(42)	(44)	(53)	(55)	(56)	(63)	(65)	(66)
[5]	6	1				1		1					1		1				1
[41]1	6	1				1		1					*1		*1				*1
2	72	20				*5	*3	*5	*3			*3	5	*3	5				*20
3	72		*5	3	*5	10		*10		*3	3		10		*10	*5	*3	*5	
[32α]1	12	4				*1		*1					*1		*1				4
2	12		*1		*1	2		*2					*2		2	1		1	
4	96		*6	10	*6	*3	*5	3	5	10	10	*5	3	5	*3	6	10	6	
[32β]1	72	4				*1	15	*1	15			15	1	15	1				*4
2	72		*1	*15	*1	2		*2		15	*15		2		*2	*1	15	*1	
4	12		2		2	1		*1					1		*1	2		2	
[311]1	4						1		1			*1		*1					
2	4			*1						1	1						*1		
4	4		1		*1											1		*1	
[221α]1	96		10	6	10	5	*3	*5	3	6	6	*3	*5	3	5	*10	6	*10	
3	12			*1			*2		2	*1	*1	*2		2			*1		
[221β]1	12			2			*1		1	2	*2	1		*1			*2		
3	72		*15	1	15		2		*2	1	*1	*2		2		15	*1	*15	
$[21^3]1$	72		3	5	*3		10		*10	5	*5	*10		10		*3	*5	3	

	N	(13)	(15)	(16)	(23)	(25)	(26)	(31)	(32)	(34)	(43)	(45)	(46)	(51)	(52)	(54)	(61)	(62)	(64)
[41]4	72	*5	3		*10		3	*5	*10			*10	5	3		*10		3	5
[32α]3	12	*1			*2			*1	*2			2	*1			2			*1
5	96	*6	10		3	5	*10	*6	3	5	5	*3	*6	10	5	*3		*10	*6
[32β]3	72	*1	*15		*2		*15	*1	*2			*2	1	*15		*2		*15	1
5	12	2			*1			2	*1			*1	*2			*1			*2
[311]3	4		*1				*1							1				1	
5	4	1						*1					*1						1
6	4				1				*1			1				*1			
[221α]2	96	10	6		*5	3	*6	10	*5	3	3	5	10	6	3	5		*6	10
4	12		*1			2	1			2	2			*1	2			1	
5	12			4		1				*1	*1				1		4		
[221β]2	12		2			1	*2			1	*1			*2	*1			2	
4	72	*15	1			*2	*1	15		*2	2		*15	*1	2			1	15
5	72			*4	*15	*1			15	1	*1	15			1	*15	4		
$[21^3]2$	72	3	5			*10	*5	*3		*10	10		3	*5	10			5	*3
3	72			*20	3	*5			*3	5	*5	*3			5	3	20		
4	6			1		*1				1	*1				1		*1		
$[1^5]$	6			1		*1				1	1				*1		1		

4.14. Outer-Product of the Permutation Group and the Littlewood Rule

Suppose that there are two subsystems with identical particles $1, 2, \ldots, n_1$ and $n_1+1, \ldots n_1 + n_2 = n$ which have the permutation symmetry $[\nu_1]$ and $[\nu_2]$, respectively. The question posed in this section is how to construct a wave function of the total system with the permutation symmetry $[\nu]$. Let the wave functions of these two subsystems be denoted as follows:

$$\begin{aligned} \psi_{m_1}^{[\nu_1]}(1,2,\ldots,n_1) &= \psi_{m_1}^{[\nu_1]}(\omega_1^0) = |Y^{[\nu_1]}(\omega_1^0)\rangle, \\ \psi_{m_2}^{[\nu_2]}(n_1+1,\ldots,n) &= \psi_{m_2}^{[\nu_2]}(\omega_2^0) = |Y_{m_2}^{[\nu_2]}(\omega_2^0)\rangle. \\ (\omega_1^0) = (1,2,\ldots,n_1), &\quad (\omega_2^0) = (n_1+1,\ldots,n), \\ m_1 = 1,2,\ldots,h_{\nu_1}, &\quad m_2 = 1,2,\ldots,h_{\nu_2}. \end{aligned} \tag{4-126}$$

They are the Yamanouchi basis of the permutation groups $S_{n_1} = S_{n_1}(1,2,\ldots,n_1)$ and $S_{n_2} = S_{n_2}(n_1+1,\ldots,n)$, respectively. $Y_{m_i}^{(\nu_i)}(\omega_i^0)$ represents a Young tableau associated with the number (ω_i^0). Their products are denoted by

$$\begin{aligned} |m_1 m_2 \omega^0\rangle &= \psi_{m_1}^{[\nu_1]}(\omega_1^0)\psi_{m_2}^{[\nu_2]}(\omega_2^0) = |Y_{m_1}^{[\nu_1]}(\omega_1^0)Y_{m_2}^{[\nu_2]}(\omega_2^0)\rangle, \\ (\omega^0) &= (\omega_1^0, \omega_2^0) = (1,2,\ldots n) . \end{aligned} \tag{4-127}$$

For example

$$|Y_1^{[21]}(123)\rangle = \left| \begin{array}{|c|c|} \hline 1 & 2 \\ \hline 3 \\ \cline{1-1} \end{array} \right\rangle, \quad |Y_2^{[21]}(456)\rangle = \left| \begin{array}{|c|c|} \hline 4 & 6 \\ \hline 5 \\ \cline{1-1} \end{array} \right\rangle . \tag{4-128}$$

Define $h_3 = \binom{n}{n_1}$ *normal order sequences*,

$$\begin{aligned} (\omega) = (\omega_1, \omega_2) \quad (\omega_1) &= (a_1 a_2 \ldots a_{n_1}), \quad (\omega_2) = (a_{n_1+1} \ldots a_n), \\ a_1 < a_2 < \ldots < a_{n_1}, &\quad a_{n_1+1} < a_{n_1+2} < \ldots < a_n . \end{aligned} \tag{4-129}$$

a_i represents any one of the numbers 1,2,..., n. For example $n_1 = 3, n_2 = 2, h_3 = \binom{5}{3} = 10$. The ten normal order sequences are listed in Table 4.14-1.

Table 4.14-1. Normal order sequences (ω) with their ordinals and parity δ_ω.

$\omega(\delta\omega)$	1(+)	2(−)	3(+)	4(−)	5(+)	6(−)	7(+)	8(+)	9(−)	10(+)
(ω)	(123,45)	(124,35)	(134,25)	(234,15)	(125,34)	(135,24)	(235,14)	(145,23)	(245,13)	(345,12)

The ordering of the sequences (ω) is specified in the following way. We regard the part (ω_1) in $(\omega) = (\omega_1, \omega_2)$ as a vector of length n_1. If the last nonzero component of the vector $(\omega_1) - (\overline{\omega}_1)$ is less than zero, then we say that $(\omega) < (\overline{\omega})$. The motivation for this ordering is to ensure that the ordering of the totally symmetric Gel'fand basis vector $\psi^{[n_1]}(\omega_1)$ is consistent with the convention in Sec. 7.4. To each normal order sequence (ω) we associated a parity factor $\delta_\omega, \delta_\omega (= \pm 1)$ being the parity of the permutation operator which transforms the normal order sequence (ω^0) into (ω).

The left coset decomposition of S_n with respect to the subgroup $S_{n_1} \times S_{n_2}$ is denoted by [cf. (2-106b)],

$$S_n = \sum_{\omega=1}^{h_3} Q_\omega (S_{n_1} \times S_{n_2}), \tag{4-130}$$

where the left coset representatives Q_ω are just the so-called order preserving permutation (MacFarlane 1960),

$$Q_\omega = \begin{pmatrix} \omega^0 \\ \omega \end{pmatrix}, \quad Q_\omega(\omega^0) = (\omega). \tag{4-131}$$

Applying the h_3 Q_ω's to (4-127), we get altogether $N = h_{\nu_1} h_{\nu_2} h_3$ basis vectors

$$\begin{aligned} Q_\omega|m_1 m_2 \omega^0\rangle &= |m_1 m_2 \omega\rangle \\ &= \psi^{[\nu_1]}_{m_1}(\omega_1)\psi^{[\nu_2]}_{m_2}(\omega_2) = |Y^{[\nu_1]}_{m_1}(\omega_1)Y^{[\nu_2]}_{m_2}(\omega_2)\rangle, \\ m_i &= 1, 2, \ldots, h_{\nu_i}, \quad \omega = 1, 2, \ldots, h_3, \end{aligned} \tag{4-132}$$

where $Y^{[\nu_i]}_{m_i}(\omega_i)$ denotes a generalized Young tableau formed by filling the Young diagram $[\nu_i]$ with the numbers (ω_i) according to the order specified by the Yamanouchi symbol m_i.

Now let us investigate the condition for these N states to be linearly independent. In Sec. 4.7, it was pointed out that the quantum numbers $(\nu_i m_i)$ are not sufficient to specify the basis functions $\psi^{[\nu_i]}_{m_i}$ uniquely. Besides $(\nu_i m_i)$, we need the Weyl tableaux W_i. Therefore in more exact form, (4-132) should be rewritten as

$$|m_1 m_2 \omega\rangle = \psi^{[\nu_1]}_{m_1}(\omega_1, W_1)\psi^{[\nu_2]}_{m_2}(\omega_2, W_2). \tag{4-133}$$

The vectors of (4-133) are orthogonal in the quantum numbers m_1 and m_2, but not necessrily in ω, unless all the single particle states in the Weyl tableau W_1 are different from those in W_2. Under the above condition we have

$$\langle m_1' m_2' \omega' | m_1 m_2 \omega\rangle = \delta_{m_1 m_1'} \delta_{m_2 m_2'} \delta_{\omega\omega'}, \tag{4-134}$$

i.e., the N basis vectors are orthonormal. If there is any overlap between the single particle states in W_1 and those in W_2, the number of independent basis vectors will be less than N (see Sec. 4.17). In the following we assume that (4-134) is true. According to Sec. 2.11, the N basis vectors of (4-132) carry the induced rep of S_n, $([\nu_1] \times [\nu_2]) \uparrow S_n$, which is also called the outer-product of the irreps $[\nu_1]$ and $[\nu_2]$, denoted as $[\nu_1] \otimes [\nu_2]$. It is a reducible rep of S_n. The decomposition of the induced rep into irreps of S_n is determined by the *Littlewood rule* (or the *outer-product reduction rule*) [cf. (2-106f)]

$$([\nu_1] \times [\nu_2]) \uparrow S_n = \sum_\nu \{\nu_1 \nu_2 \nu\}[\nu]. \tag{4-135}$$

The integers $\{\nu_1 \nu_2 \nu\}$, the times of occurrence of $[\nu]$ in the induced rep, are determined by the *Littlewood rule*:

1. Draw the Young diagrams $[\nu_1]$ and $[\nu_2]$.

2. Among these two, choose the more complicated Young diagram as the base (this is just for convenience, in principle we may choose either one as the base), and fill up the other Young diagram with a's in the first row, b's in the second and so on.

3. Add the boxes labeled a to the base Young diagram, and enlarge it in all possible ways subject to the rule that no two a's appear in the same column and that the resultant diagram be regular.

4. Repeat the process with the boxes labeled b, under the further restriction that at no stage does the total number of b's exceed the total number of a's counting from right to left and from top to bottom. Then continue the process in the same way with c, d, etc.

5. The final number of Young diagrams $[\nu]$ formed in this way gives the coefficient $\{\nu_1 \nu_2 \nu\}$ in the Littlewood rule (4-135).

Example 1: 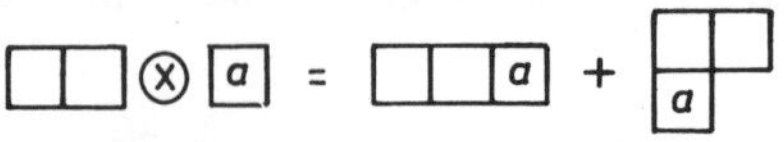

Therefore $[2]\otimes[1] = [3] + [21]$.

Example 2: 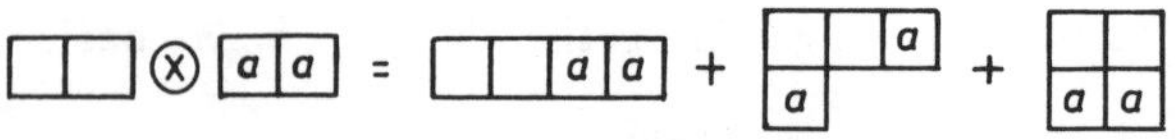

Thus $[2]\otimes[2] = [4]+ [31] + [22]$. Notice that $[2] \times [2] \neq ([2] \otimes [1]) \otimes [1]$.

Example 3: Outer-product reduction of $[21]\times[21]$. The base Young diagram is 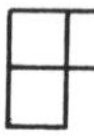, and the second Young diagram is 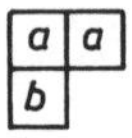. According to step 3, we have the following enlarged Young diagrams.

(1)

(i) (ii) (iii) (iv)

Adding to the box labeled b to (i) we get the following two diagrams,

(2)

Notice that the diagram 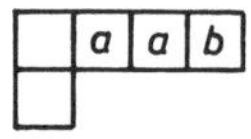 is not permissible, since the letter sequence read from this diagram is (baa), i.e., b occurs ahead of a. On the other hand, from (ii) we have

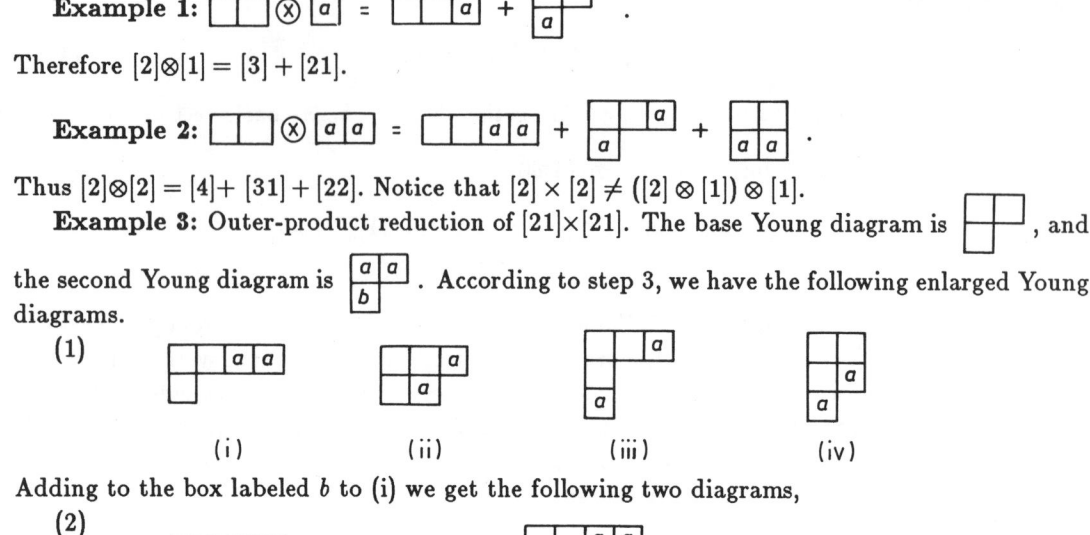

(3)

Here the first diagram is permissible, since the letter sequence is (aba).

From (iii), we have

(4)

Similarly, adding b to (iv), we get the diagrams [222] and [2211]. Finally we have the required result:

$$[21] \otimes [21] =[42] + [411] + [33] + 2[321] + [31^3] + [2^3] + [2^2 1^2] \, . \tag{4-136}$$

Itzykson (1966) has calculated the values $\{\nu_1\nu_2\nu\}$ for S_n with $n \leq 8$. According to the fact that the total dimensions of both sides of (4-135) must be equal, one has

$$h_{\nu_1} h_{\nu_2} h_3 = \sum_{\nu} \{\nu_1\nu_2\nu\} h_\nu . \tag{4-137}$$

This equation provides a useful check for the calculation of the Littlewood rule.

We will prove later that (4-135) still holds when all the Young diagrams are replaced by their conjugate diagrams:

$$[\tilde{\nu}_1] \otimes [\tilde{\nu}_2] = \sum_{\tilde{\nu}} \{\nu_1\nu_2\nu\} [\tilde{\nu}] . \tag{4-138}$$

Table 4.14-2 gives the outer-product reduction rule for $S_3 - S_5$.

Table 4.14-2. Outer-Product Reduction Rule.

S_2 group ①

	[3]	[21]	
↑ $[2]\otimes[1]$	1	1	$[11]\otimes[1]$ ↓
	$[1^3]$	[21]	

① The left(right) heading matches the top(bottom) one.

S_4 group

	[4]	[31]	[22]	[211]	$[1^4]$	
↑ $[3]\otimes[1]$	1	1				$[1^3]\otimes[1]$ ↓
$[21]\otimes[1]$		1	1	1		
$[2]\otimes[2]$	1	1	1			$[1^2]\otimes[1^2]$
$[2]\otimes[11]$		1		1		
	$[1^4]$	[211]	[22]	[31]	[4]	

S_5 group

	[5]	[41]	[32]	[311]	[221]	$[21^3]$	$[1^5]$	
↑ $[4]\otimes[1]$	1	1						$[1^4]\otimes[1]$ ↓
$[31]\otimes[1]$		1	1	1				$[211]\otimes[1]$
$[22]\otimes[1]$			1		1			
$[3]\otimes[2]$	1	1	1					$[1^3]\otimes[11]$
$[21]\otimes[2]$		1	1	1	1			$[21]\otimes[11]$
$[3]\otimes[11]$		1		1				$[1^3]\otimes[2]$
	$[1^5]$	$[21^3]$	[221]	[311]	[32]	[41]	[5]	

Ex. 4.6. Check the outer-product reduction for S_5 by using the Littlewood rule.

4.15. The Calculations of the IDC of S_n

The N basis vectors of (4-132) can be combined linearly into the irreducible basis of S_n

$$\Psi_m^{[\nu]\tau} = \sum_{m_1 m_2 \omega} C^{(\nu)\tau,m}_{\nu_1 m_1 \omega_1,\nu_2 m_2 \omega_2} |m_1 m_2 \omega\rangle, \tag{4-139}$$

where the coefficients

$$C^{(\nu)\tau,m}_{\nu_1 m_1 \omega_1,\nu_2 m_2 \omega_2} \equiv C^{(\nu)\tau,m}_{\nu_1 m_1,\nu_2 m_2,\omega} \tag{4-140}$$

is called the $([\nu_1]\otimes[\nu_2])\uparrow[\nu]$ *induction coefficient* (IDC), or the *outer-product reduction coefficient* (ORC). Expressed in terms of the generalized Young tableaux, (4-140) can be written as

$$|Y_m^{[\nu]\tau}(\omega^0)\rangle = \sum_{m_1\omega_1 m_2\omega_2} C^{(\nu)\tau,m}_{\nu_1 m_1 \omega_1,\nu_2 m_2 \omega_2} |Y_{m_1}^{(\nu_1)}(\omega_1)\rangle |Y_{m_2}^{(\nu_2)}(\omega_2)\rangle, \tag{4-141}$$

where (ω^0) can be omitted. From (4-141), one gets

$$C^{(\nu)\tau,m}_{\nu_1 m_1 \omega_1,\nu_2 m_2 \omega_2} = \langle Y_m^{(\nu)\tau} | Y_{m_1}^{(\nu_1)}(\omega_1) Y_{m_2}^{(\nu_2)}(\omega_2)\rangle. \tag{4-142}$$

From (4-28), (4-134) and (4-139) one obtains the eigenequations satisfied by the IDC,

$$\sum_{m_1' m_2' \omega'} \left[\langle m_1 m_2 \omega | C(f) | m_1' m_2' \omega'\rangle - \lambda_f \delta_{m_1 m_1'} \delta_{m_2 m_2'} \delta_{\omega\omega'}\right] C^{(\nu)\tau,m}_{\nu_1 m_1',\nu_2 m_2',\omega'} = 0,$$
$$f = n, n-1, \ldots 2, \quad \tau = 1, 2, \ldots, \{\nu_1 \nu_2 \nu\}\ . \tag{4-143}$$

From (4-143) we can get both the outer-product reduction rule and the IDC.

According to (2-106h), the matrix element of the permutation R of S_n in the induced rep is

$$\langle m_1 m_2 \omega | R | m_1' m_2' \omega'\rangle = \mathcal{G}_{\omega\omega'}(R) D^{(\nu_1)}_{m_1 m_1'}(p_1) D^{(\nu_2)}_{m_2 m_2'}(p_2)\ , \tag{4-144a}$$

$$\mathcal{G}_{\omega\omega'}(R) = \begin{cases} 1\,, & \text{if } Q_\omega^{-1} R Q_{\omega'} = p_1 p_2 \in S_{n_1} \times S_{n_2} \\ 0\,, & \text{otherwise} \end{cases}\,, \tag{4-144b}$$

where $p_1 \in S_{n_1}$ and $p_2 \in S_{n_2}$. For example,

$$\langle m_1 m_2 (146, 235) | (46) | m_1' m_2' (134, 256)\rangle = D^{(\nu_1)}_{m_1 m_1'}(23) D^{(\nu_2)}_{m_2 m_2'}(45)\ . \tag{4-145}$$

Having found the matrix elements of R, we can use (4-94) to calculate the matrix elements of $C(f)$ recursively. $C(f)$ is a real symmetric matrix. The IDC is obtained by diagonalizing the matrices $\mathbf{C}(n), \ldots \mathbf{C}(2)$ simultaneously.

Phase convention: The relative phase is determined by the Yamanouchi phase convention, while the overall phase is fixed by requiring that the IDC with $m_1 = m_2 = m = 1$ and with the smallest possible index ω be positive

$$C^{[\nu]_\tau, 1}_{[\nu_1]1, [\nu_2]1, \omega}\big|_{\omega=\min} > 0\ . \tag{4-146}$$

Example 1: Find the IDC of S_3 for $[2]\otimes[1]$. The products of the irreducible basis vectors of $S_2(12)$ and $S_3(3)$ are designated by

$$\varphi_1 = \big|\ \boxed{1}\boxed{2}\ \ \boxed{3}\ \big\rangle \tag{4-147}$$

There are three normal order sequences: $(\omega) = (12,\ 3),\ (13,\ 2)$ and $(23,\ 1)$. Thus $h_3 = 3$. The corresponding order-preserving permutations are $Q_\omega = e, (23), (123)$. Under the action of Q_ω, we have $N = h_{\nu_1} h_{\nu_2} h_3 = 1{\bullet}1{\bullet}3 = 3$ orthonormal vectors

$$|1\rangle = \big|\ \boxed{1}\boxed{2}\ \ \boxed{3}\ \big\rangle, \quad |2\rangle = \big|\ \boxed{1}\boxed{3}\ \ \boxed{2}\ \big\rangle, \quad |3\rangle = \big|\ \boxed{2}\boxed{3}\ \ \boxed{1}\ \big\rangle\ . \tag{4-148}$$

They carry the induced rep of S_3. From Table 4.15-1, it is easy to construct the matrix representatives of $C(2) = (12)$ and $C(3) = (12) + (23) + (13)$. A simultaneous diagonalization of $C(2)$ and $C(3)$ gives the IDC listed in Table 4.15-2.

Table 4.15-1

φ_i \ $p\varphi_i$ \ p	(12)	(13)	(23)
1	1	3	2
2	3	2	1
3	2	1	3

Example 2: Find the IDC of S_4 for $[21]\otimes[1]$. The products of the irreducible basis vectors of $S_3(123)$ and $S_1(4)$ are

$$\left|\ \begin{array}{ll}\boxed{1}\boxed{2} \\ \boxed{3} & \end{array}\ \boxed{4}\ \right\rangle \text{ and } \left|\ \begin{array}{ll}\boxed{1}\boxed{3} \\ \boxed{2} & \end{array}\ \boxed{4}\ \right\rangle \tag{4-149a}$$

There are four normal order sequences (ω) which are listed in Table 4.15-2 along with their indices ω and parities δ_ω.

Table 4.15-2

$\omega(\delta_\omega)$	1(+)	2(−)	3(+)	4(−)
(ω)	(123, 4)	(124, 3)	(134, 2)	(234, 1)

Under the operations of the $\binom{n}{n_1} = 4$ order-preserving permutations Q_ω, the basis vectors of (4-149a) give rise to eight orthonormal vectors φ_i listed in Table 4.15-3.

Table 4.15-3. The basis $|m_1 m_2 \omega\rangle$ of the induced rep $([21]\otimes[1])\uparrow S_4$.

φ_i	φ_1	φ_2	φ_3	φ_4	φ_5	φ_6	φ_7	φ_8
$m_1 m_2 \omega$	111	112	113	114	211	212	213	214
$Y^{[21]}_{m1}(\omega_1)Y^{[1]}(\omega_2)$	$\begin{array}{ll}1&2\\3\end{array}\ 4$	$\begin{array}{ll}1&2\\4\end{array}\ 3$	$\begin{array}{ll}1&3\\4\end{array}\ 2$	$\begin{array}{ll}2&3\\4\end{array}\ 1$	$\begin{array}{ll}1&3\\2\end{array}\ 4$	$\begin{array}{ll}1&4\\2\end{array}\ 3$	$\begin{array}{ll}1&4\\3\end{array}\ 2$	$\begin{array}{ll}2&4\\3\end{array}\ 1$

From this table we immediately find the eigenfunctions of $C(2)$, i.e.,

$$\begin{aligned}
&\lambda_2 = 1:\ \psi_1 = \varphi_1\,,\quad \psi_2 = \varphi_2,\quad \psi_3 = \sqrt{\frac{1}{2}}(\varphi_3 + \varphi_4),\ \psi_4 = \sqrt{\frac{1}{2}}(\varphi_7 + \varphi_8)\,,\\
&\lambda_2 = -1:\ \phi_1 = \sqrt{\frac{1}{2}}(\varphi_3 - \varphi_4),\ \phi_2 = \varphi_5,\ \phi_3 = \varphi_6,\ \phi_4 = \sqrt{\frac{1}{2}}(\varphi_7 - \varphi_8)\,.
\end{aligned} \tag{4-149b}$$

Using (4-144), (4-149), and the results of Ex. 2.8 we can also find the representations of $C(f)$ in the basis $\psi_1, \dots, \psi_4$, denoted by $C^{(+)}(f)$.

$$\mathbf{C}^{(+)}(4) = \begin{pmatrix} 0 & 1 & -\sqrt{\frac{1}{2}} & \sqrt{\frac{3}{2}} \\ 1 & 0 & \sqrt{2} & 0 \\ -\sqrt{\frac{1}{2}} & \sqrt{2} & 1 & 0 \\ \sqrt{\frac{3}{2}} & 0 & 0 & 1 \end{pmatrix}, \quad \mathbf{C}^{(+)}(3) = \begin{pmatrix} 0 & 0 & 0 & 0 \\ 0 & 1 & \sqrt{2} & 0 \\ 0 & \sqrt{2} & 2 & 0 \\ 0 & 0 & 0 & 0 \end{pmatrix}. \tag{4-149c}$$

A simultaneous diagonalization of $\mathbf{C}^{(+)}(4)$ and $\mathbf{C}^{(+)}(3)$ gives the IDC with $\lambda_2 = +1$. Similarly, we can obtain the IDC with $\lambda_2 = -1$ (see Table 4.17-3d).

The calculation of IDC by a computer

The EFM for calculating the IDC can be easily implemented on a computer. The procedure of calculation is identical to that in Sec. 4.11. However the vector $\mathbf{U}^\tau$ in (4-112b) should be understood as the IDC, and (4-114) should be replaced by

$$C^{\nu_\tau m'}_{\nu_1 a, \nu_2 b, \omega} = \frac{1}{D^{(\nu)}_{m'm}(T)} \sum_{cd\omega'} [D^{(\nu_1)}_{ac}(p_1) D^{(\nu_2)}_{bd}(p_2) - D^{(\nu)}_{mm}(T)\delta_{ac}\delta_{bd}\delta_{\omega\omega'}] C^{\nu_\tau m}_{\nu_1 c, \nu_2 d, \omega'}\,. \tag{4-150}$$

Programs for calculating the IDC and the tables of IDC for $S_3 - S_6$ are available (Chen [10] 1981, Chen [31] 1985). The IDC of $S_3 - S_5$ are listed in Table 4.17.

A much better algorithm for the IDC will be discussed in Sec. 7.17 which computes the IDC of S_n recursively.

4.16. The Properties of IDC

1. Unitarity

$$\begin{aligned}
&\sum_{m_1 m_2 \omega} C^{\nu_\tau m}_{\nu_1 m_1, \nu_2 m_2, \omega} C^{\nu'_{\tau'} m'}_{\nu_1 m_1, \nu_2 m_2, \omega} = \delta_{\nu\nu'}\delta_{\tau\tau'}\delta_{mm'}\,,\\
&\sum_{\nu\tau m} C^{\nu_\tau m}_{\nu_1 m_1, \nu_2 m_2, \omega} C^{\nu_\tau m}_{\nu_1 m'_1, \nu_2 m'_2, \omega'} = \delta_{m_1 m'_1}\delta_{m_2 m'_2}\delta_{\omega\omega'}\,.
\end{aligned} \tag{4-151}$$

From (4-151b) we get the inverse of (4-139), namely

$$|m_1 m_2 \omega\rangle = \sum_{\nu\tau m} C^{\nu_\tau m}_{\nu_1 m_1, \nu_2 m_2, \omega} \Psi^{[\nu]_\tau}_m \,. \tag{4-152}$$

2. Since $\{\nu_1\nu_2\nu\} = \{\nu_2\nu_1\nu\}$, we have

$$C^{\nu_\tau m}_{\nu_1 m\omega_1, \nu_2 m_2\omega_2} = \varepsilon_1 C^{\nu_\tau m}_{\nu_2 m_2\omega_2, \nu_1 m_1\omega_1} \,. \tag{4-153a}$$

The phase ε_1 is determined by the overall phase convention (4-146). Clearly

$$\varepsilon_1 = \text{sign}(C^{[\nu]_\tau 1}_{[\nu_1]1, [\nu_2]1, \omega}|_{\omega_{\max}}) \,. \tag{4-153b}$$

The values of ε_1 for $S_2 - S_5$ are listed in Table A2 in the Appendix, and in Table 12 of Chen [10] (1981).

For the case $\nu_1 = \nu_2$ if the irrep ν occurs only once, then the IDC fulfill the following relation

$$C^{\nu m}_{\nu_1 m_1\omega_1, \nu_1 m_2\omega_2} = \delta_\nu C^{\nu m}_{\nu_1 m_2\omega_2, \nu_1 m_1\omega_1} \,, \tag{4-154a}$$

where $\delta_\nu = 1$ or -1. As in the case of the CG coefficients, we say that the representation with $\delta_\nu = 1$ (-1) belongs to the symmetric (antisymmetric) product. If ν occurs more than once, then through a proper linear combinations the IDC can be made to satisfy

3. $$C^{\nu_\tau m}_{\nu_1 m_1\omega_1, \nu_1 m_2\omega_2} = \delta_{\nu_\tau} C^{\nu_\tau m}_{\nu_1 m_2\omega_2, \nu_1 m_1\omega_1}, \quad \delta_{\nu_\tau} = \pm 1 \,. \tag{4-154b}$$

We now proceed to derive an important property of the IDC. From (4-144a) and (4-67), we have

$$\langle \tilde{m}_1 \tilde{m}_2 \omega | R_a | \tilde{m}'_1 \tilde{m}'_2 \omega' \rangle = \delta_{p_1} \delta_{p_2} \Lambda^{\nu_1}_{m_1} \Lambda^{\nu_2}_{m_2} \Lambda^{\nu_1}_{m'_1} \Lambda^{\nu_2}_{m'_2} \langle m_1 m_2 \omega | R_a | m'_1 m'_2 \omega' \rangle \,, \tag{4-155}$$

where δ_{p_1} and δ_{p_2} are the parities of the permutations p_1 and p_2 determined from (4-144b). Putting $R = R_a$ in (4-144b) and equating the parities of both sides of (4-144b), we get

$$\delta_{p_1} \delta_{p_2} = \delta_a \delta_\omega \delta_{\omega'} \,.$$

Inserting this into (4-155), we find

$$\langle \tilde{m}_1 \tilde{m}_2 \omega | R_a | \tilde{m}'_1 \tilde{m}'_2 \omega' \rangle = \delta_a \delta_\omega \delta_{\omega'} \Lambda^{\nu_1}_{m_1} \Lambda^{\nu_2}_{m_2} \Lambda^{\nu_1}_{m'_1} \Lambda^{\nu_2}_{m'_2} \langle m_1 m_2 \omega | R_a | m'_1 m'_2 \omega \rangle \,. \tag{4-156a}$$

From (4-143) we have

$$\sum_{m'_1 m'_2 \omega'} \langle \tilde{m}_1 \tilde{m}_2 \omega | C(f) | \tilde{m}'_1 \tilde{m}'_2 \omega' \rangle C^{\tilde{\nu}_\tau \tilde{m}}_{\tilde{\nu}_1 \tilde{m}'_1, \tilde{\nu}_2 \tilde{m}'_2, \omega'} = \lambda_f C^{\tilde{\nu}_\tau, \tilde{m}}_{\tilde{\nu}_1 \tilde{m}_1, \tilde{\nu}_2 \tilde{m}_2, \omega} \,. \tag{4-156b}$$

Substituting (4-156a) into (4-156b), one obtains

$$\sum_{m'_1 m'_2 \omega'} \langle m_1 m_2 \omega | C(f) | m'_1 m'_2 \omega' \rangle \delta_a \delta_\omega \delta_{\omega'} \Lambda^{\nu_1}_{m_1} \Lambda^{\nu_2}_{m_2} \Lambda^{\nu_1}_{m'_1} \Lambda^{\nu_2}_{m'_2} C^{\tilde{\nu}_\tau \tilde{m}}_{\tilde{\nu}_1 \tilde{m}'_1, \tilde{\nu}_2 \tilde{m}'_2, \omega'} = C^{\tilde{\nu}_\tau, \tilde{m}}_{\tilde{\nu}_1 \tilde{m}_1, \tilde{\nu}_2 \tilde{m}_2, \omega} \,. \tag{4-156c}$$

Here $\delta_a = -1$ since $C(f)$ is a 2-cycle class operator. Furthermore, from (4-26b) we know $\tilde{\lambda}_f = -\lambda_f$. Therefore (4-156c) can be rewritten as

$$\sum_{m'_1 m'_2 \omega'} \langle m_1 m_2 \omega | C(f) | m'_1 m'_2 \omega' \rangle \delta_{\omega'} \Lambda^{\nu_1}_{m'_1} \Lambda^{\nu_2}_{m'_2} C^{\tilde{\nu}_\tau \tilde{m}}_{\tilde{\nu}_1 \tilde{m}'_1, \tilde{\nu}_2 \tilde{m}'_2, \omega'} = \tilde{\lambda}_f \delta_\omega \Lambda^{\nu_1}_{m_1} \Lambda^{\nu_2}_{m_2} C^{\tilde{\nu}_\tau, \tilde{m}}_{\tilde{\nu}_1 \tilde{m}_1, \tilde{\nu}_2 \tilde{m}_2, \omega} \tag{4-157}$$

and we have

$$C^{(\tilde{\nu})\tau,\tilde{m}}_{\tilde{\nu}_1\tilde{m}_1,\tilde{\nu}_2\tilde{m}_2,\omega} = \eta^{\nu_\tau}_m \delta_\omega \Lambda^{\nu_1}_{m_1} \Lambda^{\nu_2}_{m_2} \sum_{\tau'} a_{\tau\tau'} C^{(\nu)\tau',m}_{\nu_1 m_1 \nu_2 m_2,\omega} , \tag{4-158}$$

where $a_{\tau\tau'}$ are coefficients, and $\eta^{\nu_\tau}_m$ are phase factors. If ν_1, ν_2 and ν are not self-conjugate simultaneously, we can always choose the solution of the eigenequations so that

$$C^{(\tilde{\nu})\tau,\tilde{m}}_{\tilde{\nu}_1\tilde{m}_1,\tilde{\nu}_2\tilde{m}_2,\omega} = \eta^{\nu_\tau}_m \Lambda^{\nu_1}_{m_1} \Lambda^{\nu_2}_{m_2} C^{(\nu)\tau,m}_{\nu_1 m_1,\nu_2,m_2,\omega} . \tag{4-159}$$

The phase factor $\eta^{\nu_\tau}_m$ can be determined in the following way. From (4-67) we have

$$\langle \Psi^{[\tilde{\nu}]\tau}_{\tilde{m}} | R_a | \Psi^{[\tilde{\nu}]\tau}_{\tilde{m}'} \rangle = \delta_a \Lambda^\nu_m \Lambda^\nu_{m'} \langle \Psi^{[\nu]\tau}_m | R_a | \Psi^{[\nu]\tau}_{m'} \rangle . \tag{4-160}$$

Using (4-156a) and (4-159) it follows that

$$\begin{aligned} \langle \Psi^{[\tilde{\nu}]\tau}_{\tilde{m}} | R_a | \Psi^{[\tilde{\nu}]\tau}_{\tilde{m}'} \rangle &= \eta^{\nu_\tau}_m \eta^{\nu_\tau}_{m'} \delta_a \sum C^{[\nu]\tau,m}_{\nu_1 m_1,\nu_2 m_2,\omega} C^{[\nu]\tau,m'}_{\nu_1 m'_1,\nu_2 m'_2,\omega'} \langle m_1 m_2 \omega | R_a | m'_1 m'_2 \omega' \rangle \\ &= \eta^{\nu_\tau}_m \eta^{\nu_\tau}_{m'} \delta_a \langle \Psi^{[\nu]\tau}_m | R_a | \Psi^{[\nu]\tau}_{m'} \rangle . \end{aligned} \tag{4-161a}$$

Comparing (4-160) with (4-161), we gather that

$$\eta^{\nu_\tau}_m = \varepsilon_2 \Lambda^\nu_m , \tag{4-161b}$$

where $\varepsilon_2 = \pm 1$ is dictated by the convention of the absolute phase of $\Psi^{[\nu]\tau}_m$. Inserting (4-161b) into (4-159), we arrive at an important symmetry of the IDC:

4. $$C^{(\tilde{\nu})\tau,\tilde{m}}_{\tilde{\nu}_1\tilde{m}_1,\tilde{\nu}_2\tilde{m}_2,\omega} = \varepsilon_2 \delta_\omega \Lambda^\nu_m \Lambda^{\nu_1}_{m_1} \Lambda^{\nu_2}_{m_2} C^{(\nu)\tau,m}_{\nu_1 m_1,\nu_2 m_2,\omega} , \tag{4-162a}$$

when ν_1, ν_2 and ν are not self-conjugate simultaneously. The phase factor ε_2 is again determined by the overall phase convention. Obviously

$$\varepsilon_2 = \Lambda^{\nu_1}_{h_{\nu_1}} \Lambda^{\nu_2}_{h_{\nu_2}} \Lambda^\nu_{h_\nu} \text{sign}(\delta_\omega C^{[\nu]\tau,h_\nu}_{[\nu_1]h_{\nu_1},[\nu_2]h_{\nu_2},\omega} |_{\omega=\min}) . \tag{4-162b}$$

If ν_1, ν_2 and ν are self-conjugate simultaneously, then (4-162a) is replaced by

$$C^{\nu_\tau \tilde{m}}_{\nu_1\tilde{m}_1,\nu_2\tilde{m}_2,\omega} = \delta_\omega \Lambda^\nu_m \Lambda^{\nu_1}_{m_1} \Lambda^{\nu_2}_{m_2} \sum_{\tau'} a_{\tau\tau'} C^{\nu_{\tau'} m}_{\nu_1 m_1,\nu_2 m_2,\omega} . \tag{4-162c}$$

Equation (4-162a) shows that if $[\nu_1] \otimes [\nu_2] = \sum_\nu \{\nu_1\nu_2\nu\}[\nu]$, then $[\tilde{\nu}] \otimes [\tilde{\nu}_2] = \sum_\nu \{\nu_1\nu_2\nu\}[\tilde{\nu}]$, i.e., (4-138) holds.

It is worth mentioning here some special cases. When $[\nu]$ is totally symmetric (in such cases $[\nu_1]$ and $[\nu_2]$ are necessarily also symmetric), the IDC is very simple

$$C^{[n],1}_{[n_1]1,[n_2]1,\omega} = \binom{n}{n_1}^{-\frac{1}{2}} . \tag{4-163a}$$

From the symmetry (4-162a), we obtain the IDC for the totally antisymmetric irrep,

$$C^{[1^n],1}_{[1^{n_1}]1,[1^{n_2}]1,\omega} = \delta_\omega \binom{n}{n_1}^{-\frac{1}{2}} . \tag{4-163b}$$

4.17. Tables of the IDC for S_3–S_5

Some remarks about the IDC tables are in order

1. In Tables 4.17-1 to 4.17-3, the generalized Young tableaux $Y^{\nu_1}_{m_1}(\omega_1)Y^{\nu_2}_{m_2}(\omega_2)$ are used as the table headings, while in Tables 4.17-4 to 4.17-8 the indices $(m_1 m_2 \omega)$ are used as the table headings. These two headings are interchangeable (see the example in Table 4.15-3).

2. The second column N is normalization factor. The entries are the squares of the IDC. The asterisk denotes a negative IDC, and the blank denotes a zero IDC. Thus for instance, from Tables 4.17-5a, 5d and (4-142) we have

$$C^{[41]3}_{[21]2,[2]1,\omega=10} = \left\langle \begin{array}{|c|c|c|c|}\hline 1 & 2 & 4 & 5 \\ \hline 3 \\ \cline{1-1}\end{array} \;\middle|\; \begin{array}{|c|c|}\hline 3 & 5 \\ \hline 4 \\ \cline{1-1}\end{array} \;\; \begin{array}{|c|c|}\hline 1 & 2 \\ \hline\end{array} \right\rangle = -\sqrt{\frac{12}{180}} \; .$$

3. In Tables 4.17-3a, 3b, 4b, 5b, 5c, the tables for $[\nu_1]\otimes[\nu_2]$ and $[\tilde{\nu}_1]\otimes[\tilde{\nu}_2]$ are put together, one above and one below.

4. Each table may be read either horizontally or vertically.

5. By utilizing the symmetries (4-153) and (4-162), we can obtain other IDC not listed in the tables. Λ^{ν}_{m} are listed in Table 4.4-1.

6. In the above discussion it is assumed that the single particle states in the Weyl tableau W_1 are entirely different from those in the tableau W_2. If this is not true, then we can prove that the basis

$$\Psi^{[\nu]\tau}_{m} = \sum_{m_1 m_2 \omega} C^{[\nu]\tau m}_{\nu_1 m_1 \omega_1, \nu_2 m_2 \omega_2} \psi^{[\nu_1]}_{m_1}(\omega_1, W_1)\psi^{[\nu_2]}_{m_2}(\omega_2, W_2) \; , \tag{4-164a}$$

is still the Yamanouchi basis $[\nu]m$ of S_n (provided it is nonzero); but it is no longer normalized. We use the projection operator to prove this:

$$\begin{aligned}\Psi^{[\nu]\tau}_{m} &= \text{const } P^{[\nu]\overline{m}}_{m}\left[\psi^{[\nu_1]}_{m_1}(\omega_1, W_1)\psi^{[\nu_2]}_{m_2}(\omega_2, W_2)\right] \\ &= \sum_{m_1 m_2 \omega} C^{[\nu]\tau, m}_{\nu_1 m_1 \omega_1, \nu_2 m_2 \omega_2} \psi^{[\nu_1]}_{m_1}(\omega_1, W_1)\psi^{[\nu_2]}_{m_2}(\omega, W_2) \; .\end{aligned} \tag{4-164b}$$

Since the projection operator $P^{[\nu]\overline{m}}_{m}$ only affects the coordinate indices of $\psi^{[\nu_1]}_{m_1}$ and $\psi^{[\nu_2]}_{m_2}$, $\Psi^{[\nu]\tau}_{m}$ is the Yamanouchi basis $[\nu]m$ no matter whether the Weyl tableaux W_1 and W_2 have common single particle states or not.

As an example, suppose $\psi(12)$ and $\psi(34)$ are the orbital wave functions of two deuterons, with $[\nu_1] = [\nu_2] = [2]$. Using (4-164a) and Table 4.17-3b, and renormalizing the resulting functions, we obtain the Yamanouchi basis of S_4:

$$\begin{aligned}\Psi^{[4]} &= \sqrt{\frac{1}{3}}[\psi(12)\psi(34) + \psi(13)\psi(24) + \psi(14)\psi(23)] \; , \\ \Psi^{[22]}_{1} &= \sqrt{\frac{1}{6}}[2\psi(12)\psi(34) - \psi(13)\psi(24) - \psi(14)\psi(23)] \; , \\ \Psi^{[22]}_{2} &= \sqrt{\frac{1}{2}}[\psi(13)\psi(24) - \psi(14)\psi(23)] \; ,\end{aligned} \tag{4-164c}$$

while $\Psi^{[31]}_{m}$ are identically zero.

Table 4.17 The $([\nu_1] \otimes [\nu_2]) \uparrow [\nu]$ IDC of the permutation groups.

S_2 1 [1]⊗[1] = [2]⊕[11]

		1, 2	2, 1
[2]	2	1	1
[11]	2	1	*1

Table heading

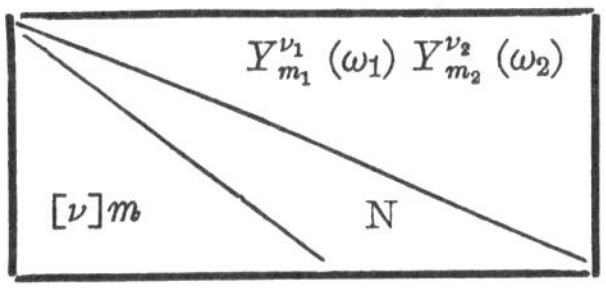

$Y^{\nu_1}_{m_1}(\omega_1)\, Y^{\nu_2}_{m_2}(\omega_2)$
$[\nu]m$
N

S_3 2 [2]⊗[1] = [3]⊕[21]

		12, 3;	13, 2;	23, 1
[3]	3	1	1	1
[21]1	6	4	*1	*1
2	2		1	*1
[21]1	2		1	1
2	6	4	1	*1
[1³]	3	1	*1	1
		1 2, 3	1 3, 2	2 3, 1

[11]⊗[1] = [21]⊕[1³]

S_4 3*a* [3]⊗[1] = [4]⊕[31]

		123, 4;	124, 3;	134, 2;	234, 1
[4]	4	1	1	1	1
[31]1	12	9	*1	*1	*1
2	6		4	*1	*1
3	2			1	*1
[211]1	2			1	1
2	6		4	1	*1
3	12	9	1	*1	1
[1⁴]	4	1	*1	1	*1
		1 2, 4 3	1 2, 3 4	1 3, 2 4	2 3, 1 4

[1³]⊗[1] = [211]⊕[1⁴]

S_4 3*b* [2]⊗[2] = [4]⊕[31]⊕[22]

		12, 34;	34, 12;	23, 14;	14, 23;	13, 24;	24, 13
[4]	6	1	1	1	1	1	1
[31]1	6	1	*1	1	*1	1	*1
2	12	4	*4	*1	1	*1	1
3	4			*1	1	1	*1
[22]1	12	4	4	*1	*1	*1	*1
2	4			*1	*1	1	1
[22]1	4			1	1	1	1
2	12	4	4	*1	*1	1	1
[211]1	4			1	*1	1	*1
2	12	4	*4	*1	1	1	*1
3	6	*1	1	*1	1	1	*1
[1⁴]	6	1	1	1	1	*1	*1
		1 3 2, 4	3 1 4, 2	2 1 3, 4	1 2 4, 3	1 2 3, 4	2 1 4, 3

[11]⊗[11] = [22] + [211] + [1⁴]

cont'd

S_4 3c [2]⊗[11]=[31]⊕[211]

Table heading

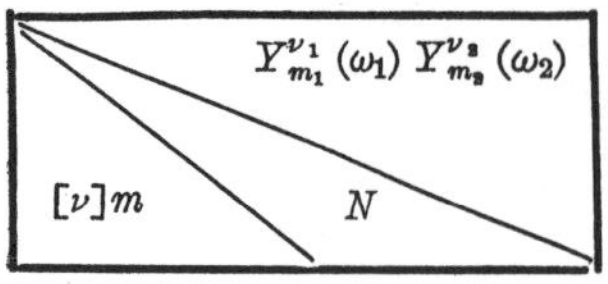

		$12,{}^{3}_{4};$	$34,{}^{1}_{2};$	$23,{}^{1}_{4};$	$14,{}^{2}_{3};$	$13,{}^{2}_{4};$	$24,{}^{1}_{3}$
[31]1	3	1		1		1	
2	24	*4		1	9	1	9
3	8		4	1	*1	*1	1
[211]1	8	4		*1	1	*1	1
2	24		4	*9	*1	9	1
3	3		1		1		*1

S_4 3d [21]⊗[1]=[31]⊕[22]⊕[211]

		$\begin{smallmatrix}12\\3\end{smallmatrix},4;$	$\begin{smallmatrix}12\\4\end{smallmatrix},3;$	$\begin{smallmatrix}13\\4\end{smallmatrix},2;$	$\begin{smallmatrix}23\\4\end{smallmatrix},1$	$\begin{smallmatrix}13\\2\end{smallmatrix},4;$	$\begin{smallmatrix}14\\2\end{smallmatrix},3;$	$\begin{smallmatrix}14\\3\end{smallmatrix},2;$	$\begin{smallmatrix}24\\3\end{smallmatrix},1$
[31]1	3		1	1	1				
2	96	36	4	*1	*1			27	27
3	32			1	*1	12	12	3	*3
[22]1	16	4	4	*1	*1			*3	*3
2	16			3	*3	4	*4	*1	1
[211]1	32	12	*12	3	3			*1	*1
2	96			*27	27	36	*4	*1	1
3	3						1	*1	1

S_5 4a $f_1=4,\ f_2=1$

$\omega(\delta\omega)$	1(+)	2(−)	3(+)	4(−)	5(+)
$(\omega)=(\omega_1,\ \omega_2)$	(1234, 5)	(1235, 4)	(1245, 3)	(1345, 2)	(2345, 1)

S_5 4b [4]⊗[1]=[5]⊕[41]

Table heading

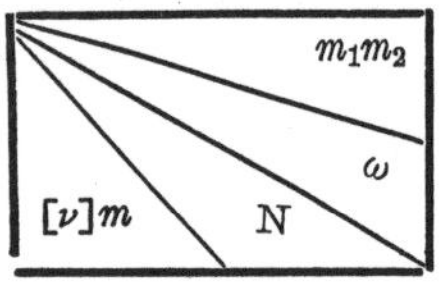

		1,1				
		1	2	3	4	5
[5]	5	1	1	1	1	1
[41]1	20	16	*1	*1	*1	*1
2	12		9	*1	*1	*1
3	6			4	*1	*1
4	2				1	*1
$[21^3]$1	2				1	1
2	6			4	1	*1
3	12		9	1	*1	1
4	20	16	1	*1	1	*1
$[1^5]$	5	1	*1	1	*1	1

$[1^4]$⊗[1]=$[21^3]$⊕$[1^5]$

cont'd

S_5 4c [31]⊗[1]=[41]⊕[32]⊕[311]

		1,1					2,1					3,1				
		1	2	3	4	5	1	2	3	4	5	1	2	3	4	5
[41]1	4		1	1	1	1										
2	540	144	9	*1	*1	*1			128	128	128					
3	270			4	*1	*1	72	72	8	*2	*2				54	54
4	90				1	*1				2	*2	24	24	24	6	*6
[32]1	27	9	9	*1	*1	*1			*2	*2	*2					
2	432			128	*32	*32	144	*36	*4	1	1				*27	*27
3	144				32	*32				*1	1	48	*12	*12	*3	3
4	16							4	4	*1	*1				*3	*3
5	16									3	*3		4	*4	*1	1
[311]1	45	18	*18	2	2	2			*1	*1	*1					
2	1440			*512	128	128	576	*36	*4	1	1				*27	*27
3	480				*128	128				*1	1	192	*12	*12	*3	3
4	32							12	*12	3	3				*1	*1
5	96									*27	27		36	*4	*1	1
6	3													1	*1	1

S_5 4d [211]⊗[1]=[311]⊕[221]⊕[21^3]

		1,1					2,1					3,1				
		1	2	3	4	5	1	2	3	4	5	1	2	3	4	5
[311]1	3			1	1	1										
2	96		36	4	*1	*1				27	27					
3	32				1	*1		12	12	3	*3					
4	480	192	12	*12	3	3				*1	*1				128	128
5	1440				*27	27	576	36	*4	*1	1			512	128	*128
6	45								1	*1	1	18	18	2	*2	2
[221]1	16		4	4	*1	*1				*3	*3					
2	16				3	*3		4	*4	*1	1					
3	144	48	12	*12	3	3				*1	*1				*32	*32
4	432				*27	27	144	36	*4	*1	1			*128	*32	32
5	27								2	*2	2	9	*9	*1	1	*1
[21^3]1	90	24	*24	24	*6	*6				2	2				*1	*1
2	270				54	*54	72	*72	8	2	*2			*4	*1	1
3	540								*128	128	*128	144	*9	*1	1	*1
4	4												1	*1	1	*1

cont'd

S_5 4e [22]⊗[1]=[32]⊕[221]

		1,1					2,1				
		1	2	3	4	5	1	2	3	4	5
[32]1	3			1	1	1					
2	96		36	4	*1	*1				27	27
3	32				1	*1		12	12	3	*3
4	32	16	4	4	*1	*1				*3	*3
5	32				3	*3	16	4	*4	*1	1
[221]1	32	16	*4	*4	1	1				3	3
2	32				*3	3	16	*4	4	1	*1
3	32		12	*12	3	3				*1	*1
4	96				*27	27		36	*4	*1	1
5	3								1	*1	1

S_5 5a $f_1=3,\ f_2=2$

ω $(\omega)=(\omega_1,\ \omega_2)$	1(+) (123, 45)	2(−) (124, 35)	3(+) (134, 25)	4(−) (234, 15)	5(+) (125, 34)
ω $(\omega)=(\omega_1,\ \omega_2)$	6(−) (135, 24)	7(+) (235, 14)	8(+) (145, 23)	9(−) (245, 13)	10(+) (345, 12)

S_5 5b [3]⊗[2]=[5]⊕[41]⊕[32]

		1,1									
		1	2	3	4	5	6	7	8	9	10
[5]	10	1	1	1	1	1	1	1	1	1	1
[41]1	60	9	9	9	9	*4	*4	*4	*4	*4	*4
2	36	9	*1	*1	*1	4	4	4	*4	*4	*4
3	18		4	*1	*1	4	*1	*1	1	1	*4
4	6			1	*1		1	*1	1	*1	
[32]1	18	9	*1	*1	*1	*1	*1	*1	1	1	1
2	36		16	*4	*4	*4	1	1	*1	*1	4
3	12			4	*4		*1	1	*1	1	
4	12					4	*1	*1	*1	*1	4
5	4						1	*1	*1	1	
[221]1	4						1	1	1	1	
2	12					4	1	*1	*1	1	4
3	12			4	4		1	1	*1	*1	
4	36		16	4	*4	4	1	*1	1	*1	*4
5	18	9	1	*1	1	*1	1	*1	1	*1	1
$[21^3]1$	6			1	1		*1	*1	1	1	
2	18		4	1	*1	*4	*1	1	*1	1	4
3	36	9	1	*1	1	4	*4	4	*4	4	*4
4	60	*9	9	*9	9	4	*4	4	4	*4	4
$[1^5]$	10	1	*1	1	*1	1	*1	1	1	*1	1

$[1^3]\otimes[11]=[221]\oplus[21^3]\oplus[1^5]$

cont'd

S_5 5c $[3]\otimes[11]=[41]\oplus[311]$

		1,1									
		1	2	3	4	5	6	7	8	9	10
[41]1	4	1	1	1	1						
2	60	*9	1	1	1	16	16	16			
3	30		*4	1	1	*4	1	1	9	9	
4	10			*1	1		*1	1	*1	1	4
[311]1	15	9	*1	*1	*1	1	1	1			
2	120		64	*16	*16	*4	1	1	9	9	
3	40			16	*16		*1	1	*1	1	4
4	8					4	*1	*1	1	1	
5	24						9	*9	*1	1	4
6	3								1	*1	1
[311]1	3								1	1	1
2	24						9	9	1	1	*4
3	8					4	1	*1	1	*1	
4	40			16	16		1	1	*1	*1	4
5	120		64	16	*16	4	1	*1	*9	9	
6	15	9	1	*1	1	1	*1	1			
[21³]1	10			1	1		*1	*1	1	1	*4
2	30		4	1	*1	*4	*1	1	9	*9	
3	60	9	1	*1	1	*16	16	*16			
4	4	1	*1	1	*1						

$[1^3]\otimes[2]=[311]\oplus[21^3]$

S_5 5d $[21]\otimes[2]=[41]\oplus[32]\oplus[311]\oplus[221]$

		1,1										2,1									
		1	2	3	4	5	6	7	8	9	10	1	2	3	4	5	6	7	8	9	10
[41]1	6					1	1	1	1	1	1										
2	90		16	16	16	1	1	1	*1	*1	*1								12	12	12
3	180	36	4	*1	*1	4	*1	*1	1	1	*4			27	27		27	27	3	3	*12
4	60			1	*1		1	*1	1	*1		12	12	3	*3	12	3	*3	3	*3	
[32]1	18		1	1	1	1	1	1	*1	*1	*1								*3	*3	*3
2	576	36	4	*1	*1	64	*16	*16	16	16	*64			27	27		*108	*108	*12	*12	48
3	192			1	*1		16	*16	16	*16		12	12	3	*3	*48	*12	12	*12	12	
4	192	36	36	*9	*9	16	*4	*4	*4	*4	16			*27	*27						
5	64			9	*9		4	*4	*4	4		12	*12	*3	3						
[311]1	30		3	3	3	*3	*3	*3	3	3	3								*1	*1	*1
2	960	108	12	*3	*3	*192	48	48	*48	*48	192			81	81		*36	*36	*4	*4	16
3	320			3	*3		*48	48	*48	48		36	36	9	*9	*16	*4	4	*4	4	
4	64	12	*12	3	3									*1	*1		4	4	*4	*4	16
5	192			*27	27							36	*4	*1	1	16	4	*4	*36	36	
6	6												1	*1	1	1	*1	1			
[221]1	64	4	4	*1	*1	*16	4	4	4	4	*16			*3	*3						
2	64			3	*3		*12	12	12	*12		4	*4	*1	1						
3	64	12	*12	3	3									*1	*1		*4	*4	4	4	*16
4	192			*27	27							36	*4	*1	1	*16	*4	4	36	*36	
5	6												1	*1	1	*1	1	*1			

cont'd

S_5 5e [21]⊗[11]=[32]⊕[311]⊕[221]⊕[21³]

		1,1										2,1									
		1	2	3	4	5	6	7	8	9	10	1	2	3	4	5	6	7	8	9	10
[32]1	6		1	1	1	1	1	1													
2	192	36	4	*1	*1	*16	4	4	36	36				27	27						
3	64			1	*1		*4	4	*4	4	16	12	12	3	*3						
4	64	*4	*4	1	1									3	3		12	12	12	12	
5	64			*3	3							*4	4	1	*1	16	4	*4	*4	4	16
[311]1	6		1	1	1	*1	*1	*1													
2	192	36	4	*1	*1	16	*4	*4	*36	*36				27	27						
3	64			1	*1		4	*4	4	*4	*16	12	12	3	*3						
4	320	*36	36	*9	*9	16	*4	*4	4	4				3	3		48	48	*48	*48	
5	960			81	*81		36	*36	*4	4	16	*108	12	3	*3	192	48	*48	48	*48	*192
6	30								1	*1	1		*3	3	*3	*3	3	*3	3	*3	3
[221]1	64	12	12	*3	*3									*9	*9		4	4	4	4	
2	192			27	*27							36	*36	*9	9	16	4	*4	*4	4	16
3	192	*12	12	*3	*3	48	*12	*12	12	12				1	1		*16	*16	16	16	
4	576			27	*27		108	*108	*12	12	48	*36	4	1	*1	*64	*16	16	*16	16	64
5	18								3	*3	3		*1	1	*1	1	*1	1	*1	1	*1
[21³]1	60	12	*12	3	3	12	*3	*3	3	3				*1	*1		1	1	*1	*1	
2	180			*27	27		27	*27	*3	3	12	36	*4	*1	1	4	1	*1	1	*1	*4
3	90								12	*12	12		16	*16	16	*1	1	*1	1	*1	1
4	6															1	*1	1	1	*1	1

S_6 6a $f_1=5$, $f_2=1$

ω(δω) (ω)=(ω₁, ω₂)	1(+) (12345, 6)	2(−) (12346, 5)	3(+) (12356, 4)	4(−) (12456, 3)	5(+) (13456, 2)	6(−) (23456, 1)

S_6 6b [5]⊗[1]=[6]⊕[51]

		1,1					
		1	2	3	4	5	6
[6]1	6	1	1	1	1	1	1
[51]1	30	25	*1	*1	*1	*1	*1
2	20		16	*1	*1	*1	*1
3	12			9	*1	*1	*1
4	6				4	*1	*1
5	2					1	*1

cont'd

S_6 7*a* $f_1=4$, $f_2=2$

$\omega(\delta\omega)$ $(\omega)=(\omega_1, \omega_2)$	1(+) (1234, 56)	2(−) (1235, 46)	3(+) (1245, 36)	4(−) (1345, 26)	5(+) (2345, 16)	6(+) (1236, 45)	7(−) (1246, 35)	8(+) (1346, 25)
	9(−) (2346, 15)	10(+) (1256, 34)	11(−) (1356, 24)	12(+) (2356, 14)	13(+) (1456, 23)	14(−) (2456, 13)	15(+) (3456, 12)	

S_6 **7*b*** $[4]\otimes[2] = [6]\oplus[51]\oplus[42]$

		1,1														
		1	2	3	4	5	6	7	8	9	10	11	12	13	14	15
[6]	15	1	1	1	1	1	1	1	1	1	1	1	1	1	1	1
[51]1	30	4	4	4	4	4	*1	*1	*1	*1	*1	*1	*1	*1	*1	*1
2	80	16	*1	*1	*1	*1	9	9	9	9	*4	*4	*4	*4	*4	*4
3	48		9	*1	*1	*1	9	*1	*1	*1	4	4	4	*4	*4	*4
4	24			4	*1	*1		4	*1	*1	4	*1	*1	1	1	*4
5	8				1	*1			1	*1		1	*1	1	*1	
[42]1	240	144	*9	*9	*9	*9	*9	*9	*9	*9	4	4	4	4	4	4
2	144		81	*9	*9	*9	*9	1	1	1	*4	*4	*4	4	4	4
3	72			36	*9	*9		*4	1	1	*4	1	1	*1	*1	4
4	24				9	*9			*1	1		*1	1	*1	1	
5	18						9	*1	*1	*1	*1	*1	*1	1	1	1
6	36							16	*4	*4	*4	1	1	*1	*1	4
7	12								4	*4		*1	1	*1	1	
8	12										4	*1	*1	*1	*1	4
9	4											1	*1	*1	1	

S_6 8*a* $f_1=3, f_2=3$

$\omega(\delta\omega)$ $(\omega)=(\omega_1, \omega_2)$	1(+) (123, 456)	2(−) (124, 356)	3(+) (134, 256)	4(−) (234, 156)	5(+) (125, 346)	6(−) (135, 246)	7(+) (235, 146)	8(+) (145, 236)	9(−) (245, 136)	10(+) (345, 126)
	11(−) (126, 345)	12(+) (136, 245)	13(−) (236, 145)	14(−) (146, 235)	15(+) (246, 135)	16(−) (346, 125)	17(+) (156, 234)	18(−) (256, 134)	19(+) (356, 124)	20(−) (456, 123)

S_6 8*b* $[3]\otimes[3]=[6]\oplus[51]\oplus[42]\oplus[33]$

		1,1																			
		1	2	3	4	5	6	7	8	9	10	11	12	13	14	15	16	17	18	19	20
[6]	20	1	1	1	1	1	1	1	1	1	1	1	1	1	1	1	1	1	1	1	1
[51]1	20	1	1	1	1	1	1	1	1	1	1	*1	*1	*1	*1	*1	*1	*1	*1	*1	*1
2	120	9	9	9	9	*4	*4	*4	*4	*4	*4	4	4	4	4	4	4	*9	*9	*9	*9
3	72	9	*1	*1	*1	4	4	4	*4	*4	*4	4	4	4	*4	*4	*4	1	1	1	*9
4	36		4	*1	*1	4	*1	*1	1	1	*4	4	*1	*1	1	1	*4	1	1	*4	
5	12			1	*1		1	*1	1	*1			1	*1	1	*1		1	*1		
[42]1	120	9	9	9	9	*4	*4	*4	*4	*4	*4	*4	*4	*4	*4	*4	*4	9	9	9	9
2	72	9	*1	*1	*1	4	4	4	*4	*4	*4	*4	*4	*4	4	4	4	*1	*1	*1	9
3	36		4	*1	*1	4	*1	*1	1	1	*4	*4	1	1	*1	*1	4	*1	*1	4	
4	12			1	*1		1	*1	1	*1			*1	1	*1	1		*1	1		
5	36	9	*1	*1	*1	*1	*1	*1	1	1	1	1	1	1	*1	*1	*1	*1	*1	*1	9
6	72		16	*4	*4	*4	1	1	*1	*1	4	4	*1	*1	1	1	*4	*4	*4	16	
7	24			4	*4		*1	1	*1	1			1	*1	1	*1		*4	4		
8	24					4	*1	*1	*1	*1	4	4	*1	*1	*1	*1	4				
9	8						1	*1	*1	1			1	*1	*1	1					
[33]1	36	9	*1	*1	*1	*1	*1	*1	1	1	1	*1	*1	*1	1	1	1	1	1	1	*9
2	72		16	*4	*4	*4	1	1	*1	*1	4	*4	1	1	*1	*1	4	4	4	*16	
3	24			4	*4		*1	1	*1	1			*1	1	*1	1		4	*4		
4	24					4	*1	*1	*1	*1	4	*4	1	1	1	1	*4				
5	8						1	*1	*1	1			*1	1	1	*1					

4.18. $S_{n_1+n_2} \supset S_{n_1} \otimes S_{n_2}$ Irreducible Basis

4.18.1. *The $S_{n_1+n_2} \supset S_{n_1} \otimes S_{n_2}$ subduced basis*

An irrep $[\nu]$ of S_n is reducible with respect to its subgroup $S_{n_1} \otimes S_{n_2}, n_1 + n_2 = n$. The process of the reduction is known as subduction and is denoted by [cf. (2-106a)]

$$[\nu] \downarrow (S_{n_1} \otimes S_{n_2}) = \sum_{\nu} \{\nu_1\nu_2\nu\}([\nu_1],[\nu_2]) \ . \tag{4-165a}$$

According to the Frobenious reciprocity theorem, (2-106g), the multiplicities $\{\nu_1\nu_2\nu\}$ in (4-165a) and (4-135) are identical.

The subduced basis is the $S_n \supset S_{n_1} \otimes S_{n_2}$ basis

$$\left| \begin{matrix} [\nu] \\ \tau\nu_1 m_1 \nu_2 m_2 \end{matrix} \right\rangle \equiv \left| [\nu], \begin{matrix} \tau[\nu_1][\nu_2] \\ m_1 m_2 \end{matrix} \right\rangle \ , \quad \tau = 1, 2, \ldots, \{\nu_1\nu_2\nu\} \ . \tag{4-165b}$$

It belongs to the irrep $[\nu]$ of S_n, and at the same time is the Yamanouchi basis $[\nu_1]m_1$ of $S_{n_1} = S_{n_1}(1, 2, \ldots, n_1)$ and $[\nu_2]m_2$ of $S_{n_2} = S_{n_2}(n_1 + 1, \ldots, n)$.

The basis (4-165b) will be called the non-standard basis of S_n and is used in many instances. The set of quantum numbers $(\tau[\nu_1]m_1[\nu_2]m_2)$ now serves as the component indices of the non-standard basis.

According to (3-265), the non-standard basis obeys the following eigenequations

$$\begin{pmatrix} C(n_1+n_2) \\ C(n_1) \\ C(s_1) \\ C'(n_2) \\ C'(s_2) \end{pmatrix} \left| [\nu], \begin{matrix} \tau[\nu_1] & [\nu_2] \\ m_1 & m_2 \end{matrix} \right\rangle = \begin{pmatrix} \nu \\ \nu_1 \\ m_1 \\ \nu_2 \\ m_2 \end{pmatrix} \left| [\nu], \begin{matrix} \tau[\nu_1] & [\nu_2] \\ m_1 & m_2 \end{matrix} \right\rangle , \tag{4-166}$$

$$\begin{aligned} (C'(n_2); C'(s_2)) &= (C'(n_2);\ C'(n_2-1), \ldots, C'(2)) \ , \\ C'(f) &= \sum_{i>j=n_1+1}^{n_1+f} (ij), \quad f = n_2, n_2 - 1, \ldots 2 \ , \end{aligned} \tag{4-167}$$

where $(C(n_1), C(s_1))$ is the CSCO-II of S_{n_1}, and $C(n_1 + n_2)$ is the CSCO-I (instead of the 2-cycle class operator) of S_n.

Ex. 4.7. From the basis (4-32a), construct the $S_4 \supset S_2(12) \otimes S_2(34)$ basis.

4.18.2. *Transformation between the standard and non-standard basis of S_n*

We usually expand the non-standard basis of S_n in terms of the standard basis of S_n,

$$\left| [\nu], \begin{matrix} \tau[\nu_1][\nu_2] \\ m_1 m_2 \end{matrix} \right\rangle = \sum_m \left| \begin{matrix} [\nu] \\ m \end{matrix} \right\rangle \left\langle \begin{matrix} [\nu] \\ m \end{matrix} \right| \left. [\nu], \begin{matrix} \tau[\nu_1][\nu_2] \\ m_1 m_2 \end{matrix} \right\rangle , \tag{4-168a}$$

where the expansion coefficients is called the $[\nu] \downarrow ([\nu_1] \otimes [\nu_2])$ *subduction coefficients* (SDC), or the transformation coefficients between the standard and non-standard basis of S_n (SNSTC).

Since the standard basis of S_n is also a standard basis of the subgroup S_{n_1}, the SDC is zero unless the tableau formed by the first n_1 numbers in $Y_m^{(\nu)}$ is identical to the Young tableau $Y_{m_1}^{(\nu_1)}$. Therefore the sum over m in (4-168a) is restricted to the Yamanouchi numbers $r_n, r_{n-1}, \ldots, r_{n_1+1}$. Obviously, the SDC are independent of the quantum number m_1, while the partition $[\nu_1]$ is determined by $[\nu]m$ and the particle number n_2. Consequently, the SDI depends only on $[\nu], m, \tau, [\nu_2]$ and m_2 and its notation can be streamlined to

$$\left\langle \begin{matrix}[\nu]\\ m\end{matrix} \middle| [\nu], \begin{matrix}\tau[\nu_1][\nu_2]\\ m_1 m_2\end{matrix} \right\rangle = \left\langle [\nu]m \middle| \tau, [\nu_2]m_2 \right\rangle . \tag{4-169}$$

or expressed as

$$\left\langle \begin{matrix}[\nu]\\ m\end{matrix} \middle| [\nu], \begin{matrix}\tau[\nu_1][\nu_2]\\ m_1 m_2\end{matrix} \right\rangle = \left\langle Y_m^{[\nu]} \middle| [\nu]\tau, Y_{m_1}^{[\nu_1]}(\omega_1^0) Y_{m_2}^{[\nu_2]}(\omega_2^0) \right\rangle = \left\langle Y_m^{[\nu]} \middle| \tau, Y_{m_2}^{[\nu_2]}(\omega_2^0) \right\rangle . \tag{4-170}$$

If the solutions of (4-166) are not unique for a given $\nu, \nu_1, m_1, \nu_2, m_2$, we must use the additional label τ to distinguish them. We may choose appropriate linear combinations so that the solutions with different τ are orthogonal. Consequently the non-standard basis (4-165b) forms an orthonormal complete set and the SDC fulfill the unitarity conditions

$$\sum_{\nu_2 m_2 \tau} \left\langle \begin{matrix}[\nu]\\ m\end{matrix} \middle| [\nu], \begin{matrix}\tau[\nu_1][\nu_2]\\ m_1 m_2\end{matrix} \right\rangle \left\langle \begin{matrix}[\nu]\\ m'\end{matrix} \middle| [\nu], \begin{matrix}\tau[\nu_1][\nu_2]\\ m_1 m_2\end{matrix} \right\rangle = \delta_{mm'} , \tag{4-171a}$$

$$\sum_{m} \left\langle \begin{matrix}[\nu]\\ m\end{matrix} \middle| [\nu], \begin{matrix}\tau[\nu_1][\nu_2]\\ m_1 m_2\end{matrix} \right\rangle \left\langle \begin{matrix}[\nu]\\ m\end{matrix} \middle| [\nu], \begin{matrix}\tau'[\nu_1][\nu_2']\\ m_1 m_2'\end{matrix} \right\rangle = \delta_{\nu_2\nu_2'} \delta_{m_2 m_2'} \delta_{\tau\tau'} . \tag{4-171b}$$

The inverse of (4-168a) is

$$\left| \begin{matrix}[\nu]\\ m\end{matrix} \right\rangle = \sum_{\nu_2 m_2 \tau} \left| [\nu], \begin{matrix}\tau[\nu_1][\nu_2]\\ m_1 m_2\end{matrix} \right\rangle \left\langle \begin{matrix}[\nu]\\ m\end{matrix} \middle| [\nu], \begin{matrix}\tau[\nu_1][\nu_2]\\ m_1 m_2\end{matrix} \right\rangle . \tag{4-168b}$$

Another way of constructing the $S_n \supset S_{n_1} \otimes S_{n_2}$ basis is by means of the CG coefficients of the unitary group (see Sec. 7.13).

4.18.3. The calculation of the SDC

The SDC is an important coefficient. In Secs. 7.14 and 7.15 we will see that the Racah coefficients and 9-ν coefficients of any SU_n group can be expressed in terms of them and in Sections 7.16 and 7.17 will show that the calculation of the $SU_{mn} \supset SU_m \times SU_n$ ISF and $SU_{m+n} \supset SU_m \otimes SU_n$ ISF also involves these coefficients. In the past, the projection operator method was used to calculate the SDC (see Kaplan 1961) which is rather cumbersome. We can now use the following two methods to calculate them.

1. *Eigenfunction method:* Each term on the right-hand side of (4-168a) is already an irreducible basis of S_n and S_{n_1}; therefore we only need to combine them into the standard basis $[\nu_2]m_2$ of S_{n_2}. In other words, to find the SDC we merely need to diagonalize the CSCO-II of S_{n_2} in the Yamanouchi basis $\left|\begin{matrix}[\nu]\\ m\end{matrix}\right\rangle$ with fixed $[\nu_1]m_1$:

$$\sum_{m'} \left[\left\langle \begin{matrix}[\nu]\\ m\end{matrix} \middle| \begin{matrix}C'(n_2)\\ C'(s_2)\end{matrix} \middle| \begin{matrix}[\nu]\\ m'\end{matrix} \right\rangle - \begin{pmatrix}\nu_2\\ m_2\end{pmatrix} \delta_{mm'} \right] \left\langle \begin{matrix}[\nu]\\ m'\end{matrix} \middle| [\nu], \begin{matrix}\tau[\nu_1][\nu_2]\\ m_1 m_2\end{matrix} \right\rangle = 0 . \tag{4-172}$$

In practical calculations, for a given set $[\nu], [\nu_1]$ and m_1(m_1 can be chosen arbitrarily, for example $m_1 = 1$), we construct the matrix representative of the CSCO-II $(C'(n_2), C'(s_2))$, or its variation $M = \sum_{f=2}^{n_2} k_f C'(f)$ (see (4-27a)], of S_{n_2} and diagonalize the matrices (or matrix). The eigenvectors are the required SDC. In order that the basis vectors $\left|[\nu], \begin{smallmatrix}\tau[\nu_1] & [\nu_2] \\ m_1 & m_2\end{smallmatrix}\right\rangle$ with different m_2 have the correct phase, we again use (4-35). In our case it reads

$$\left|[\nu], \begin{matrix}\tau[\nu_1][\nu_2] \\ m_1 m_2'\end{matrix}\right\rangle = \frac{1}{D^{[\nu_2]}_{m_2' m_2}(T_2')}\left[T - D^{[\nu_2]}_{m_2 m_2}(T_2')\right]\left|[\nu], \begin{matrix}\tau[\nu_1][\nu_2] \\ m_1 m_2\end{matrix}\right\rangle ,$$

$$T_2 = (i, i+1) \in S_{n_2}(n_1+1, \ldots, n_1+n_2); \quad T_2' = (i-n_1,\ i-n_1+1) \in S_{n_2}(1, 2\ldots, n_2) . \tag{4-173}$$

From (4-173) we have

$$\left\langle \begin{matrix}[\nu] \\ m\end{matrix}\middle|[\nu], \begin{matrix}\tau[\nu_1][\nu_2] \\ m_1 m_2'\end{matrix}\right\rangle = \frac{1}{D^{[\nu_2]}_{m_2' m_2}(T_2')}\sum_{m'}\left[D^{[\nu]}_{m'm}(T_2) - D^{[\nu_2]}_{m_2 m_2}(T_2')\delta_{mm'}\right]\left\langle \begin{matrix}[\nu] \\ m'\end{matrix}\middle|[\nu], \begin{matrix}\tau[\nu_1][\nu_2] \\ m_1 m_2\end{matrix}\right\rangle . \tag{4-174}$$

It follows that for a given $[\nu], [\nu_1]$ and $[\nu_2]$ and starting with the SDC of the component m_2 we can obtain the SDC of other components m_2' with the correct relative phases by using (4-174). The above algorithm can be easily realized in a computer (Chen [9], 1983).

2. *Using IDC to find the SDC:* In analogy with (3-199), i.e.,

$$P^{[\nu]m'}_{m} = \sqrt{\frac{h_\nu}{n!}}\sum_{a=1}^{n!}\left\langle \begin{matrix}[\nu] \\ m\end{matrix}\middle| R_a \middle| \begin{matrix}[\nu] \\ m'\end{matrix}\right\rangle R_a , \tag{4-175a}$$

we introduce a shift operator from the non-standard basis to the standard basis of S_n:

$$P^{[\nu],\tau[\nu_1]m_1'[\nu_2]m_2'}_{m} = \sqrt{\frac{h_\nu}{n!}}\sum_{Q_\omega p_1 p_2}\left\langle \begin{matrix}[\nu] \\ m\end{matrix}\middle| Q_\omega p_1 p_2 \middle| [\nu], \begin{matrix}\tau[\nu_1][\nu_2] \\ m_1' m_2'\end{matrix}\right\rangle Q_\omega p_1 p_2 . \tag{4-175b}$$

Here we used the decomposition (4-130). Inserting a unit operator

$$\sum_{\nu_1' m_1 \nu_2' m_2 \tau'}\left|[\nu], \begin{matrix}\tau'[\nu_1'][\nu_2'] \\ m_1 m_2\end{matrix}\right\rangle\left\langle[\nu], \begin{matrix}\tau'[\nu_1'][\nu_2'] \\ m_1 m_2\end{matrix}\right| = 1$$

between Q_ω and p_1 inside the brackets in (4-175b), we have

$$P^{[\nu],\tau[\nu_1]m_1'[\nu_2]m_2'}_{m} = \sqrt{\frac{h_\nu}{h_{\nu_1}h_{\nu_2}}}\sqrt{\frac{n_1!n_2!}{n!}}\sum_{m_1 m_2 \omega}\left\langle \begin{matrix}[\nu] \\ m\end{matrix}\middle| Q_\omega \middle| [\nu], \begin{matrix}\tau[\nu_1][\nu_2] \\ m_1 m_2\end{matrix}\right\rangle Q_\omega P^{[\nu_1]m_1'}_{m_1} P^{[\nu_2]m_2'}_{m_2} . \tag{4-176}$$

In deriving this result, we used (4-175a) and the relation

$$\left\langle[\nu], \begin{matrix}\tau'\nu_1'\nu_2' \\ m_1 m_2\end{matrix}\middle| p_1 p_2 \middle| [\nu], \begin{matrix}\tau\nu_1\nu_2 \\ m_1' m_2'\end{matrix}\right\rangle = \delta_{\tau\tau'}\delta_{\nu_1\nu_1'}\delta_{\nu_2\nu_2'}D^{[\nu_1]}_{m_1 m_1'}(p_1)D^{[\nu_2]}_{m_2 m_2'}(p_2) .$$

Applying (4-176) to a normal order state $\Phi_0 = |i_1 i_2, \ldots i_n\rangle$ with n distinct single particle states, and using (3-225), we obtain

$$\begin{aligned}\Psi^{[\nu]\tau'}_{m} &= P^{[\nu],\tau[\nu_1]m_1'[\nu_2]m_2'}_{m}\Phi_0 \\ &= \sqrt{\frac{h_\nu}{h_{\nu_1}h_{\nu_2}}}\sqrt{\frac{n_1!n_2!}{n!}}\sum_{m_1 m_2 \omega}\left\langle \begin{matrix}[\nu] \\ m\end{matrix}\middle| Q_\omega \middle| [\nu], \begin{matrix}\tau[\nu_1][\nu_2] \\ m_1 m_2\end{matrix}\right\rangle \psi^{[\nu_1]}_{m_1}(\omega_1)\psi^{[\nu_2]}_{m_2}(\omega_2) .\end{aligned} \tag{4-177}$$

The label τ' on the left-hand side of (4-177) is used to distinguish the different linearly independent functions $\Psi_m^{[\nu]\tau'}$ resulting from using different superscripts in the projection operator. We can choose $\tau' = \tau$, since τ' and τ have the same range and the choice of the additional label is arbitrary. Comparing (4-139) with (4-177), we obtain another expression for the IDC, i.e.,

$$C^{[\nu]\tau,m}_{\nu_1 m_1,\nu_2 m_2,\omega} = \sqrt{\frac{h_\nu}{h_{\nu_1} h_{\nu_2}}}\sqrt{\frac{n_1!n_2!}{n!}}\left\langle \begin{matrix}[\nu]\\ m\end{matrix}\middle| Q_\omega \middle| [\nu],\ \begin{matrix}\tau[\nu_1][\nu_2]\\ m_1 m_2\end{matrix}\right\rangle . \tag{4-178a}$$

Letting $Q_\omega = e$ (identity), we obtain a relation between the SDC and the IDC,

$$\left\langle \begin{matrix}[\nu]\\ m\end{matrix}\middle| [\nu],\ \begin{matrix}\tau[\nu_1][\nu_2]\\ m_1 m_2\end{matrix}\right\rangle = \sqrt{\frac{h_{\nu_1} h_{\nu_2}}{h_\nu}}\sqrt{\frac{n!}{n_1!n_2!}}\, C^{[\nu]\tau,m}_{\nu_1 m_1,\nu_2 m_2,\omega^0} . \tag{4-179a}$$

Therefore once the IDC is known, the SDI can be found immediately, and vice versa. In Chen ([31] 1986), the IDC is computed from the SDC by means of (4-178a).

We can also define the transformation coefficients from the standard basis of S_n to the irreducible basis classified according to the group chain $S_n \supset S_{n_1}(\omega_1) \otimes S_{n_2}(\omega_2)$:

$$\left\langle \begin{matrix}[\nu]\\ m\end{matrix}\middle| [\nu],\ \begin{matrix}\tau[\nu_1][\nu_2]\\ m_1 m_2\end{matrix}\right\rangle_{(\omega)} = \left\langle Y_m^{[\nu]} \middle| [\nu]\tau,\ Y_{m_1}^{[\nu_1]}(\omega_1) Y_{m_2}^{[\nu_2]}(\omega_2)\right\rangle . \tag{4-178b}$$

From (4-178a) we have a relation between the generalized SDC (Shi 1983) and the IDC:

$$\left\langle \begin{matrix}[\nu]\\ m\end{matrix}\middle| [\nu],\ \begin{matrix}\tau[\nu_1][\nu_2]\\ m_1 m_2\end{matrix}\right\rangle_{(\omega)} = \sqrt{\frac{h_{\nu_1} h_{\nu_2}}{h_\nu}}\sqrt{\frac{n!}{n_1!n_2!}}\, C^{[\nu]\tau,m}_{\nu_1 m_1,\nu_2 m_2,\omega} . \tag{4-179b}$$

Note that according to (4-179a) the phase convention for IDC implies a phase convention for SDC:

$$\left.\left\langle \begin{matrix}[\nu]\\ m\end{matrix}\middle| [\nu],\ \begin{matrix}\tau[\nu_1][\nu_2]\\ m_1\ \ 1\end{matrix}\right\rangle\right|_{m=\min} > 0 . \tag{4-179c}$$

The IDC symmetries (4-162a, c) together with $\delta_{\omega^0} = 1$ and (4-179a) allow us to obtain the following symmetries for SDC:

(i)

$$\left\langle \begin{matrix}[\tilde\nu]\\ \tilde m\end{matrix}\middle| [\tilde\nu],\ \begin{matrix}\tau\tilde\nu_1\tilde\nu_2\\ \tilde m_1 \tilde m_2\end{matrix}\right\rangle = \varepsilon_3 \Lambda^\nu_m \Lambda^{\nu_1}_{m_1} \Lambda^{\nu_2}_{m_2} \left\langle \begin{matrix}[\nu]\\ m\end{matrix}\middle| [\nu],\ \begin{matrix}\tau\nu_1\nu_2\\ m_1 m_2\end{matrix}\right\rangle , \tag{4-180a}$$

when ν_1, ν_2 and ν are not self-conjugate simultaneously;

(ii)

$$\left\langle \begin{matrix}[\nu]\\ \tilde m\end{matrix}\middle| [\nu],\ \begin{matrix}\tau\nu_1\nu_2\\ \tilde m_1 \tilde m_2\end{matrix}\right\rangle = \Lambda^\nu_m \Lambda^{\nu_1}_{m_1} \Lambda^{\nu_2}_{m_2} \sum_{\tau'} a_{\tau\tau'} \left\langle \begin{matrix}[\nu]\\ m\end{matrix}\middle| [\nu],\ \begin{matrix}\tau'\nu_1\nu_2\\ m_1 m_2\end{matrix}\right\rangle , \tag{4-180b}$$

when ν_1, ν_2 and ν are all self-conjugate. According to (4-179c) and (4-180a), we have

$$\varepsilon_3 = \operatorname{sign}\left(\Lambda^\nu_m \Lambda^{\nu_1}_{m_1} \Lambda^{\nu_2}_{h_{\nu_2}} \left.\left\langle \begin{matrix}[\nu]\\ m\end{matrix}\middle| [\nu],\ \begin{matrix}\tau[\nu_1][\nu_2]\\ m_1 h_{\nu_2}\end{matrix}\right\rangle\right|_{m=\min} \right) . \tag{4-180c}$$

Under the Jahn's phase convention (see the footnote to Eq. (4-67)) (4-180a) simplifies to

$$\left\langle \begin{matrix}[\tilde\nu]\\ \tilde m\end{matrix}\middle| [\tilde\nu],\ \begin{matrix}\tau[\tilde\nu][\tilde\nu_2]\\ \tilde m_1 \tilde m_2\end{matrix}\right\rangle = \varepsilon_3' \left\langle \begin{matrix}[\nu]\\ m\end{matrix}\middle| [\nu],\ \begin{matrix}\tau[\nu_1][\nu_2]\\ m_1 m_2\end{matrix}\right\rangle . \tag{4-181}$$

3. *Special cases:* A number of cases now command our attention. (a) $n_2 = 1$.

$$\left\langle \begin{matrix}[\nu] \\ m\end{matrix} \middle| [\nu], \begin{matrix}[\nu_1][1] \\ m_1\end{matrix} \right\rangle = \delta_{(m)_1 m_1} , \tag{4-182a}$$

where $(m)_1$ is the Young tableau resulting from deleting the box with the number n in $Y_m^{[\nu]}$. (b) $n_2 = 2$.

The SDC for $n_2 = 2$ can be denoted simply by $\left\langle [\nu]m \middle| [\nu_2] \right\rangle$. (i) Suppose that n and $n-1$ are either in the same row or in the same column of the Young tableau $Y_m^{[\nu]}$; then

$$\left\langle \begin{matrix}[\nu] \\ m\end{matrix} \middle| [\nu], \begin{matrix}[\nu_1][\nu_2] \\ m_1\end{matrix} \right\rangle = \delta_{(m)_1 m} , \tag{4-182b}$$

where $(m)_1$ is the Young tableau resulting from deleting boxes with the numbers $n-1$ and n in $Y_m^{[\nu]}$. (ii) When $n-1$ and n are neither in the same row nor in the same column, from (4-166) and (4-14b) we know that to obtain the SDC one only needs to diagonalize the transposition $(n-1, n)$ in the 2-dimensional space with the basis vectors $|Y_m^{[\nu]}\rangle$ and $|Y_{m'}^{[\nu]}\rangle = |(n-1,n)Y_m^{[\nu]}\rangle$. Assuming that the Yamanouchi symbol m is larger than m', from (4-17) we obtain the matrix representative of $(n-1, n)$ in the space $\{Y_m^{[\nu]}, Y_{m'}^{[\nu]}\}$, i.e.,

$$D(n, n-1) = \begin{pmatrix} -\frac{1}{\sigma} & \frac{\sqrt{\sigma^2-1}}{\sigma} \\ \frac{\sqrt{\sigma^2-1}}{\sigma} & \frac{1}{\sigma} \end{pmatrix} , \quad \begin{array}{ll} \begin{array}{|c|c|c|c|} \hline \; & \; & \; & n' \\ \hline \; & n \\ \cline{1-2} \end{array} & Y_m^{[\nu]} , \\ \begin{array}{|c|c|c|c|} \hline \; & \; & \; & n \\ \hline \; & n' \\ \cline{1-2} \end{array} & Y_{m'}^{[\nu]} , \end{array} \quad n' = n-1 , \tag{4-183}$$

where σ is the axial distance from $n-1$ to n in $Y_{m'}^{[\nu]}, \sigma > 0$. A diagonalization of (4-183) gives the SDC listed below:

The SDC $\langle [\nu]\ \mathrm{m} | [\nu_2] \rangle$, $[\nu_2]=[2],[11]$.

	[2]	[11]
$[\nu]m$	$\left(\frac{\sigma-a}{2\sigma}\right)^{1/2}$	$\left(\frac{\sigma+1}{2\sigma}\right)^{1/2}$
$[\nu]m'$	$\left(\frac{\sigma+1}{2\sigma}\right)^{1/2}$	$-\left(\frac{\sigma-1}{2\sigma}\right)^{1/2}$

(4-184)

(c) From (4-179a) and (4-163), we get the SDC for the irreps $[n]$ and $[1^n]$;

$$\left\langle [n] \middle| [n];[n_1][n_2] \right\rangle = 1 \quad \left\langle [1^n] \middle| [1^n];[1^{n_1}][1^{n_2}] \right\rangle = 1 .$$

(d) Some SDC are trivial; for example

$$\left\langle \begin{array}{|c|c|c|} \hline 1 & 2 & 3 \\ \hline 4 & 5 \\ \cline{1-2} 6 \\ \cline{1-1} \end{array} \middle| [321]; \begin{array}{|c|c|c|} \hline 1 & 2 & 3 \\ \hline \end{array} \begin{array}{|c|c|} \hline 4 & 5 \\ \hline 6 \\ \cline{1-1} \end{array} \right\rangle = 1, \quad \left\langle \begin{array}{|c|c|c|} \hline 1 & 4 & 6 \\ \hline 2 & 5 \\ \cline{1-2} 3 \\ \cline{1-1} \end{array} \middle| [321]; \begin{array}{|c|} \hline 1 \\ \hline 2 \\ \hline 3 \\ \hline \end{array} \begin{array}{|c|c|} \hline 4 & 6 \\ \hline 5 \\ \cline{1-1} \end{array} \right\rangle = 1 . \tag{4-185}$$

Horie (1964) has given an algebraic expression for the special SDC when $[\nu_1]$ and $[\nu_2]$ are totally symmetric.

Tables 4.18-1 to 4.18-3 list the SDC for $S_3 - S_6$. The trivial SDC such as those shown in (4-185) are omitted. N is the normalization factor. The entries are the squares of the SDC, with * designating a negative SDC. For instance, from Table 2a one has

$$\left\langle \begin{array}{|c|c|c|c|}\hline 1&2&4&5\\ \hline 3 \\ \cline{1-1}\end{array} \right| [41]\ \begin{array}{|c|c|}\hline 1&2\\ \hline\end{array}\ \left. \begin{array}{|c|c|}\hline 3&4\\ \hline 5 \\ \cline{1-1}\end{array} \right\rangle = \left\langle \begin{matrix}[41]\\ 3\end{matrix} \right| [41]; \left. \begin{matrix}[2]&[21]\\ 1&1\end{matrix} \right\rangle = -\sqrt{\frac{2}{18}}$$

Notice that the SDC for $[321]\downarrow([21]\otimes[21])$ are not unique. What is listed in Table 4.18 3e(iv) is different from that in Chen ([9], 1983), and satisfies the symmetry

$$\left\langle \begin{matrix}[321]\\ \tilde{m}\end{matrix} \right| [321], \left. \begin{matrix}\alpha[21][21]\\ \tilde{m}_1\tilde{m}_2\end{matrix} \right\rangle = \Lambda^{[321]}_{m}\Lambda^{[21]}_{m_1}\Lambda^{[21]}_{m_2} \left\langle \begin{matrix}[321]\\ m\end{matrix} \right| [321], \left. \begin{matrix}\beta[21][21]\\ m_1 m_2\end{matrix} \right\rangle .$$

4.18.4. *Tables of the SDC for $S_3 - S_6$*

Table 4.18. The SDC for $S_3 - S_6$.

1a $[\nu]=[31]$

$[\nu];\nu_1,\nu_2 m_2$ \ $[\nu]m$	N	123 4	124 3	134 2
$[31];1,234$	9	1	2	6
$[31];1,{}^{23}_{4}$	36	32	*1	*3
$1,{}^{24}_{3}$	4		3	*1

1b $[\nu]=[22]$

	N	12 34	13 24
$[22];1,{}^{23}_{4}$	4	1	3
$1,{}^{24}_{3}$	4	3	*1

2a $[\nu]=[41]$

	N	1234 5	1235 4	1245 3	1345 2
$[41];1,2345$	48	3	5	10	30
$1,{}^{234}_{5}$	144	135	*1	*2	*6
$1,{}^{235}_{4}$	36		32	*1	*3
$1,{}^{245}_{3}$	4			3	*1
$[41];12,345$	18	3	5	10	
$12,{}^{34}_{5}$	18	15	*1	*2	
$12,{}^{35}_{4}$	3		2	*1	

2b $[\nu]=[32]$

	N	123 45	124 35	134 25	125 34	135 24
$[32];1,{}^{234}_{5}$	9	1	2	6		
$1,{}^{235}_{4}$	144	32	*1	*3	27	81
$1,{}^{245}_{3}$	16		3	*1	9	*3
$1,{}^{23}_{45}$	48	32	*1	*3	*3	*9
$1,{}^{24}_{35}$	16		9	*3	*3	1
$[32];12,345$	9	1	2		6	
$12,{}^{34}_{5}$	9	2	4		*3	
$12,{}^{35}_{4}$	3	2	*1			

2c $[\nu]=[311]$

$[\nu],\nu_1,\nu_2 m_2$ \ m	N	1	2	3	4	5	6
[311]; [1], [31]1	9	1	2	6			
[31]2	576	*32	1	3	135	405	
[31]3	192		*9	3	*15	5	160
[211]1	192	160	*5	*15	3	9	
[211]2	576		*405	*135	*3	1	32
[211]3	9				6	*2	1
[311]; [2], [21]1	3	1	2				
[21]2	18	*2	1		15		
$[1^3]$	18	10	*5		3		
[11], [21]1	18			15		*1	*2
[21]2	3					2	*1
[3]	18			3		5	10

cont'd

3*a* $[\nu]=[51]$

$[\nu], \nu_1, \nu_2 m_2$ \ m	N	1	2	3	4	5
[51]; [1] [5]	50	2	3	5	10	30
[1] [41] 1	1200	1152	*3	*5	*10	*30
[41] 2	144		135	*1	*2	*6
[41] 3	36			32	*1	*3
[41] 4	4				3	*1
[51]; [2] [4]	20	2	3	5	10	
[2] [31] 1	180	162	*3	*5	*10	
[31] 2	18		15	*1	*2	
[31] 3	3			2	*1	
[51]; [3] [3]	10	2	3	5		
[21] 1	40	32	*3	*5		
[21] 2	8		5	*3		

3*b* (i) $[\nu]=[42]$

	N	1	2	3	4	5	6	7	8	9
[42]; [1] [41] 1	48	3	5	10	30					
[41] 2	1296	135	*1	*2	*6	128	256	768		
[41] 3	324		32	*1	*3	64	*2	*6	54	162
[41] 4	36			3	*1		6	*2	18	*6
[42]; [1] [32] 1	162	135	*1	*2	*6	*2	*4	*12		
[32] 2	1296		1024	*32	*96	*32	1	3	*27	*81
[32] 3	144			96	*32		*3	1	*9	3
[32] 4	48					32	*1	*3	*3	*9
[32] 5	16						9	*3	*3	1

3*b* (ii) $[\nu]=[42]$

$[\nu]\nu_1 m_1 \nu_2 m_2$ \ m	N	1	2	3	5	6	8
[42]; [2], [4]	108	3	5	10	10	20	60
[2] [31] 1	108	15	25	50	*2	*4	*12
[31] 2	54	15	*1	*2	8	16	*12
[31] 3	9		2	*1	4	*2	
[2] [22] 1	27	15	*1	*2	*2	*4	3
[22] 2	9		4	*2	*2	1	
[42]; [3] [3]	18	3	5		10		
[21] 1	72	15	25		*32		
[21] 2	8	5	*3				
[21] 1 [3]	9			1		2	6
[21] 1	36			32		*1	*3
[21] 2	4					3	*1

3*c* (i) $[\nu]=[411]$

		1	2	3	4	5	6	7	8	9	10
[411]; [1] [41] 1	48	3	5	10	30						
[41] 2	3600	*135	1	2	6	384	768	2304			
[41] 3	1800		*64	2	6	*96	3	9	405	1215	
[41] 4	200			*6	2		*9	3	*15	5	160
[411]; [1] [311] 1	450	405	*3	*6	*18	2	4	12			
[311] 2	14400		12288	*384	*1152	*32	1	3	135	405	
[311] 3	4800			3456	*1152		*9	3	*15	5	160
[311] 4	192					160	*5	*15	3	9	
[311] 5	576						405	*135	*3	1	32
[311] 6	9								6	*2	1

cont'd

3c(ii) [ν]=[411]

	N	1	2	3	4	5	6	7	8	9	10
[411]; [2][31]1	18	3	5	10							
[31]2	180	*15	1	2		54	108				
[31]3	60		*4	2		*6	3		45		
[211]1	60	45	*3	*6		2	4				
[211]2	180		108	*54		*2	1		15		
[211]3	18					10	*5		3		
[1²][4]	20				2			3		5	10
[31]1	180				162			*3		*5	*10
[31]2	18							15		*1	*2
[31]3	3									2	*1
[411]; [3][21]1	8	3	5								
[21]2	40	*5	3			32					
[1³]	10	5	*3			2					
[21]1, [3]	10			2			3		5		
[21]1[21]1	40			32			*3		*5		
[21]2	8						5		*3		
[21]2, [3]	10				2			3		5	
[21]1	40				32			*3		*5	
[21]2	8							5		*3	

3d [ν]=[33]

	N	1	2	3	4	5
[33]; [1][32]1	9	1	2	6		
[32]2	144	32	*1	*3	27	81
[32]3	16		3	*1	9	*3
[32]4	48	32	*1	*3	*3	*9
[32]5	16		9	*3	*3	1
[2][31]1	9	1	2		6	
[31]2	9	2	4		*3	
[31]3	3	2	*1			
[21]1[21]1	4		1		3	
[21]2	4		3		*1	
[21]2, [21]1	4			1		3
[21]2	4			3		*1

3e(i) [ν]=[321], $n_1=1$, $n_2=5$

	N	1	2	3	4	5	6	7	8	9	10	11	12	13	14	15	16
[321]; [1][32]1	36	1	2	6			3	6	18								
[32]2	2304	128	*4	*12	108	324	*96	3	9	405	1215						
[32]3	256		12	*4	36	*12		*9	3	*15	5	160					
[32]4	768	*32	1	3	3	9							45	135	135	405	
[32]5	256		*9	3	3	*1							45	*15	*15	5	160
[1][311]1	36	3	6	18			*1	*2	*6								
[311]2	2304	384	*12	*36	324	972	32	*1	*3	*135	*405						
[311]3	768		108	*36	324	*108		9	*3	15	*5	*160					

3e(ii) [ν]=[321], $n_1=2$, $n_2=4$

	N	1	2	4	6	7	9	12	14
[321]; [2][31]1	9	1	2	6					
[31]2	576	*2	*4	3	54	108		405	
[31]3	192	*2	1		*6	3	45		135
[22]1	32	2	4	*3	6	12		*5	
[22]2	32	6	*3		*2	1	15		*5
[211]1	192	30	60	*45	*10	*20		27	
[211]2	576	270	*135		10	*5	*75		81
[211]3	18				10	*5	3		

cont'd

3e(iii) $[\nu]=[321]$, $n_1=2$, $n_2=4$

	N	3	5	8	10	11	13	15	16
[321]; [11] [31]1	18	0	0	3	5	10	0	0	0
[31]2	576	81	0	75	*5	*10	0	135	270
[31]3	192	0	27	0	20	*10	45	60	*30
[22]1	32	5	0	15	*1	*2	0	*3	*6
[22]2	32	0	5	0	12	*6	*3	*4	2
[211]1	192	135	0	*45	3	6	0	*1	*2
[211]2	576	0	405	0	*108	54	*3	*4	2
[211]3	9	0	0	0	0	0	6	*2	1

3e(iv) $[\nu]=[321]$, $n_1=3$, $n_2=3$

m	N	2	4	7	9	12	14	
↑ [321]; [21]1[3]	32	1	3	3	5	5	15	[21]2[3]
[321]α; [21]1[21]1	64	9	27	3	5	*5	*15	α; [21]2, [21]1
[21]2	64	3	*1	25	*15	15	*5	[21]2
[321]β; [21]1[21]1	64	5	15	*15	*25	1	3	β; [21]2, [21]1
[21]2	64	*15	5	*5	3	27	*9	[21]2 ↓
[321]; [21]1[1³]	32	15	*5	*5	3	3	*1	[21]2, [1³]
	N	3	5	8	10	13	15	

Table 4.18-4. The phase factor ε_3 in Eq. (4-180a).

ν_1	ν_2	ν	ε_3	ν_1	ν_2	ν	ε_3	ν_1	ν_2	ν	ε_3	ν_1	ν_2	ν	ε_3
[1]	[3]	[31]	1	[1]	[5]	[51]	1	[3]	[3]	[42]	1	[3]	[1³]	[411]	1
[1]	[21]		1	[1]	[41]		1	[3]	[21]		−1	[21]	[3]		1
[1]	[4]	[41]	−1	[2]	[4]		−1	[21]	[3]		1	[21]	[21]		1
[1]	[31]		1	[2]	[31]		1	[21]	[21]		1	[1]	[32]	[33]	1
[2]	[3]		1	[3]	[3]		1	[1]	[41]	[411]	−1	[2]	[31]		−1
[2]	[21]		1	[3]	[21]		1	[1]	[311]		1	[21]	[21]		1
[1]	[31]	[32]	1	[1]	[41]	[42]	−1	[2]	[31]		1				
[1]	[22]		1	[1]	[32]		1	[2]	[221]		1				
[2]	[3]		1	[2]	[4]		1	[11]	[4]		1				
[2]	[21]		−1	[2]	[31]		1	[11]	[31]		−1				
				[2]	[22]		1	[3]	[21]		−1				

Ex. 4.8. Prove that

$$P^{(\nu_1)k_1}_{m_1} P^{(\nu_2)k_2}_{m_2}$$
$$= \sum_{\nu m k \tau} \left(\frac{f_1! f_2! h_\nu}{f! h_{\nu_1} h_{\nu_2}} \right)^{\frac{1}{2}} \left\langle \begin{matrix} [\nu] \\ m \end{matrix} \middle| [\nu], \begin{matrix} \tau \nu_1 \nu_2 \\ m_1 m_2 \end{matrix} \right\rangle \left\langle \begin{matrix} [\nu] \\ k \end{matrix} \middle| [\nu], \begin{matrix} \tau \nu_1 \nu_2 \\ k_1 k_2 \end{matrix} \right\rangle P^{(\nu)k}_m .$$

4.19. $S_n \supset S_{n_1} \otimes S_{n_2}$ Isoscalar Factor*

4.19.1 The $S_n \supset S_{n-1}$ ISF

Let us first introduce notations. The labeling of the irreducible bases of the three groups $S_n(x), S_n(\xi), S_n(q)$ and their subgroups $S_{n-1}(x), S_{n-1}(\xi), S_{n-1}(q)$ is listed in Table 4.19-1.

Table 4.19-1.

$S_n(x)$	$S_{n-1}(x)$	$S_n(\xi)$	$S_{n-1}(\xi)$	$S_n(q)$	$S_{n-1}(q)$
$\left\| \begin{matrix} [\sigma] \\ m_1 \end{matrix} \right\rangle = \left\| \begin{matrix} \sigma \\ [\sigma']m_1' \end{matrix} \right\rangle$	$\left\| \begin{matrix} [\sigma'] \\ m_1' \end{matrix} \right\rangle$	$\left\| \begin{matrix} [\mu] \\ m_2 \end{matrix} \right\rangle = \left\| \begin{matrix} [\mu] \\ [\mu']m_2' \end{matrix} \right\rangle$	$\left\| \begin{matrix} [\mu'] \\ m_2' \end{matrix} \right\rangle$	$\left\| \begin{matrix} [\nu] \\ m \end{matrix} \right\rangle = \left\| \begin{matrix} [\nu] \\ [\nu']m' \end{matrix} \right\rangle$	$\left\| \begin{matrix} [\nu'] \\ m' \end{matrix} \right\rangle$

Here m_1, m_1', m_2, etc. can be understood either as the Young tableaux or the indices of the Young tableaux. From the branching law we have

$$[\sigma]m_1 = [\sigma][\sigma']m_1', \quad [\mu]m_2 = [\mu][\mu']m_2', \quad [\nu]m = [\nu][\nu']m' . \tag{4-186}$$

Let $Y^{\sigma'}_{m_1'}, Y^{\mu'}_{m_2'}$ and $Y^{\nu'}_{m'}$ be the Young tableaux after dropping the last box with the number n in the Young tableaux $Y^{\sigma}_{m_1}, Y^{\mu}_{m_2}$ and Y^{ν}_{m}, respectively. For example

$$\left| \begin{matrix} [42] \\ 9 \end{matrix} \right\rangle = \left| \begin{matrix} 1\,3\,5\,6 \\ 2\,4 \end{matrix} \right\rangle = \left| \begin{matrix} [42] \\ [32]5 \end{matrix} \right\rangle , \quad \left| \begin{matrix} [411] \\ 5 \end{matrix} \right\rangle = \left| \begin{matrix} 1\,2\,3\,6 \\ 4 \\ 5 \end{matrix} \right\rangle = \left| \begin{matrix} [411] \\ [311]1 \end{matrix} \right\rangle .$$

To construct irreducible basis vectors of $S_n(q)$, we first use the CG coefficients $C^{[\nu']\beta',m'}_{\sigma' m_1', \mu' m_2'}$ of S_{n-1} to combine linearly the products of the irreducible basis vectors of $S_n(x)$ and $S_n(\xi)$ into the irreducible basis $[\nu']m'$ of $S_{n-1}(q)$, i.e.,

$$|(\sigma'\mu')\beta'\rangle \equiv \left[\left| \begin{matrix} [\sigma] \\ [\sigma'] \end{matrix} \right\rangle \left| \begin{matrix} [\mu] \\ [\mu'] \end{matrix} \right\rangle \right]^{[\nu']\beta'}_{m'} \equiv \sum_{m_1' m_2'} C^{[\nu']\beta',m'}_{\sigma' m_1', \mu' m_2'} \left| \begin{matrix} [\sigma] \\ [\sigma']m_1' \end{matrix} \right\rangle \left| \begin{matrix} [\mu] \\ [\mu']m_2' \end{matrix} \right\rangle , \tag{4-187}$$

and then employ the $S_n \supset S_{n-1}$ ISF $C^{[\nu]\beta,[\nu']\beta'}_{\sigma\sigma',\mu\mu'}$ to combine (4-187) into the irreducible basis $[\nu]m$ of $S_n(q)$,

$$\left| \begin{matrix} [\nu]\beta \\ m \end{matrix} \right\rangle = \left| \begin{matrix} [\nu]\beta \\ [\nu']m' \end{matrix} \right\rangle = \sum_{\sigma'\mu'\beta'} C^{[\nu]\beta,[\nu']\beta'}_{\sigma\sigma',\mu\mu'} \left[\left| \begin{matrix} [\sigma] \\ [\sigma'] \end{matrix} \right\rangle \left| \begin{matrix} [\mu] \\ [\mu'] \end{matrix} \right\rangle \right]^{[\nu']\beta'}_{m'} . \tag{4-188}$$

Therefore the CG coefficients of S_n are expressed as [cf. (3-303)]

$$C^{[\nu]\beta,m}_{[\sigma]m_1,[\mu]m_2} = \sum_{\beta'} C^{[\nu]\beta,[\nu']\beta'}_{\sigma\sigma,\mu\mu'} C^{[\nu']\beta',m'}_{[\sigma']m_1',[\mu']m_2'} . \tag{4-189a}$$

For example we have

$$C^{[42]\beta,9}_{[411]5,[411]7} = \sum_{\beta'} C^{[42]\beta,[32]\beta'}_{[411][311],[411][311]} C^{[32]\beta',5}_{[311]1,[311]3} \ .$$

From (4-189a) we find

$$C^{[\nu]\beta,[\nu']\beta'}_{\sigma\sigma',\mu\mu} = \sum_{m'_1 m'_2} C^{[\nu]\beta,m}_{[\sigma]m_1,[\mu]m_2} C^{[\nu']\beta',m'}_{[\sigma']m'_1,[\mu']m'_2} \ . \tag{4-189b}$$

If the multiplicity label β' is redundant, we have from (4-189a) a simpler form for the ISF; namely,

$$C^{[\nu]\beta,[\nu']}_{\sigma\sigma',\mu\mu'} = C^{[\nu]\beta,m}_{[\sigma]m_1[\mu]m_2} / C^{[\nu']m'}_{[\sigma']m'_1,[\mu']m'_2} \ . \tag{4-189c}$$

The $S_n \supset S_{n-1}$ ISF satisfy the unitarity condition (3-305) and (3-306). The inverse of (4-188) is

$$\left[\left|\begin{matrix}[\sigma]\\ [\sigma']\end{matrix}\right\rangle \left|\begin{matrix}[\mu]\\ [\mu']\end{matrix}\right\rangle\right]^{[\nu']\beta'}_{m'} = \sum_{\nu\beta} C^{[\nu]\beta[\nu']\beta'}_{\sigma\sigma',\mu\mu'} \left|\begin{matrix}[\nu]\beta\\ [\nu']m'\end{matrix}\right\rangle \ . \tag{4-190}$$

We now proceed to derive the eigenequation to be satisfied by the $S_n \supset S_{n-1}$ ISF. Using the relation $C(n) = C(n-1) + \sum_{i=1}^{n-1}(in)$ and the fact that (4-188) is an eigenfunction of $C(n)$ and $C(n-1)$ with the eigenvalues ν and ν', respectively, we obtain

$$\sum_{i=1}^{n-1}(in)\left|\begin{matrix}[\nu]\beta\\ [\nu']m'\end{matrix}\right\rangle = (\nu - \nu')\left|\begin{matrix}[\nu]\beta\\ [\nu']m'\end{matrix}\right\rangle \ . \tag{4-191}$$

Thus, to find the $S_n \supset S_{n-1}$ ISF, it suffices to diagonalize the operator $\sum_{i=1}^{n-1}(in)$ in the basis $|(\sigma'\mu')\beta'\rangle$ of (4-187), i.e.,

$$\sum_{\sigma'\mu'\beta'}\left[\sum_{i=1}^{n-1}\langle(\overline{\sigma}'\overline{\mu}')\overline{\beta}'|(in)|(\sigma'\mu')\beta'\rangle - (\nu-\nu')\delta_{\overline{\beta}'\beta'}\delta_{\overline{\sigma}'\sigma'}\delta_{\overline{\mu}'\mu'}\right] C^{[\nu]\beta,[\nu']\beta'}_{\sigma\sigma',\mu\mu'} = 0 \ ,$$
$$\beta = 1,2,\ldots(\sigma\mu\nu), \quad \beta' = 1,2,\ldots(\sigma'\mu'\nu') \ . \tag{4-192}$$

By using the fact that the matrix elements of $\Sigma_{i=1}^{n-1}(in)$ are independent of the quantum number m', as well as the identity $(in) = (i,n-1)(n-1,n)(i,n-1)$ and the orthonormality of the irreducible matrix elements, we have

$$\begin{aligned}&\langle(\overline{\sigma}'\overline{\mu}')\overline{\beta}'|\sum_{i=1}^{n-1}(in)|(\sigma'\mu')\beta'\rangle\\ &= \frac{n-1}{h_{\nu'}}\sum_{\nu''} h_{\nu''}\langle(\overline{\sigma}'\overline{\mu}')\overline{\beta}'|(n-1,n)|(\sigma'\mu')\beta'\rangle \ ,\end{aligned} \tag{4-193a}$$

where we used the branching law $m' = [\nu'']m''$. With the help of (4-187), the matrix elements of $(n-1,n)$ can be expressed in terms of the CG coefficients of S_{n-1} and the irreducible matrix elements of S_n:

$$\begin{aligned}&\langle(\overline{\sigma}'\overline{\mu}')\overline{\beta}'|(n-1,n)|(\sigma'\mu')\beta'\rangle\\ &= \sum_{\overline{m}'_1\overline{m}'_2 m'_1 m'_2} C^{[\nu']\overline{\beta}'m'}_{\overline{\sigma}'\overline{m}'_1,\overline{\mu}'\overline{m}'_2} C^{[\nu']\beta',m'}_{\sigma'm'_1,\mu'm'_2} D^{[\sigma]}_{\overline{m}_1 m_1}(n-1,n) D^{[\mu]}_{\overline{m}_2 m_2}(n-1,n) \ .\end{aligned} \tag{4-193b}$$

Since the dimensionality of the irreps of S_n increases very rapidly with n, (4-193b) is not suitable for computing the matrix element for large n. To simplify it we use the factorization (4-189a) for the S_{n-1} CG coefficients and obtain

$$\langle(\overline{\sigma}'\overline{\mu}')\overline{\beta}'|(n-1,n)|(\sigma'\mu')\beta'\rangle = M(\overline{\sigma}'\overline{\mu}'\overline{\beta}'\nu'\nu'',\sigma'\mu'\beta'\nu'\nu'') \ , \tag{4-193c}$$

$$M(\overline{\sigma}'\overline{\mu}'\overline{\beta}'\overline{\nu}'\nu'',\sigma'\mu'\beta'\nu'\nu'')$$

$$= \sum_{\sigma''\mu''\beta''} C^{[\overline{\nu}']\overline{\beta}',[\nu'']\beta''}_{\overline{\sigma}'\sigma'',\overline{\mu}\mu''} C^{[\nu']\beta',[\nu'']\beta''}_{\sigma'\sigma'',\mu'\mu''} D^{[\sigma]}_{\overline{\sigma}'\sigma'',\sigma'\sigma''}(n-1,n) D^{[\mu]}_{\overline{\mu}'\mu'',\mu'\mu''}(n-1,n) \ , \tag{4-193d}$$

where σ'', μ'' and ν'' refer to the group S_{n-2}. Notice that the right-hand side of (4-193c) is a special case ($\overline{\nu}' = \nu'$) of (4-193d). In Eq. (4-193d) only the $S_{n-1} \supset S_{n-2}$ ISF are involved instead of the S_{n-1} CG coefficients, and the summation runs only over $\sigma''\mu''\beta''$ instead of $\overline{m}'_1\overline{m}'_2 m'_1 m'_2$. Therefore the hurdle of the high dimensionality of the permutation group was removed and we now can compute the $S_n \supset S_{n-1}$ ISF for much higher n than it was possible before. A program in Fortran-77 was written based on this algorithm (Novoselsky 1988).

For a given $[\sigma]$, $[\mu]$ and $[\nu']$, we set up an eigenequation (4-192) of the order $l = \sum_{\sigma'\mu'}(\sigma'\mu'\nu')$. The order of the eigenequation (4-93) satisfied by the CG coefficients of S_n is equal to $L = h_\sigma h_\mu$ Notice that l is much smaller than L. For example, for S_6, $l_{\max} = 13$, while $L_{\max} = 256$. Therefore it is much easier to calculate the CG coefficients in terms of the ISF by using (4-189a) than to calculate the CG coefficients directly from (4-93).

The $S_n \supset S_{n-1}$ ISF is precisely the K coefficient defined by Hamermesh (1962) and Harvey (1981). A recursive formula for the K coefficient was given by Hamermesh.

Example 1: Use (4-189c) to find $S_5 \supset S_4$ ISF. From Table 4.13-3 we have

$$C^{[221],[22]}_{[41][4],[32][22]} = \left\langle \begin{array}{|c|c|}\hline 1 & 2 \\ \hline 3 & 4 \\ \hline 5 \\ \cline{1-1}\end{array} \middle| \begin{array}{|c|c|c|c|}\hline 1 & 2 & 3 & 4 \\ \hline 5 \\ \cline{1-1}\end{array} \ \begin{array}{|c|c|c|}\hline 1 & 2 & 5 \\ \hline 3 & 4 \\ \cline{1-2}\end{array} \right\rangle \Big/ \left\langle \begin{array}{|c|c|}\hline 1 & 2 \\ \hline 3 & 4 \\ \hline\end{array} \middle| \begin{array}{|c|c|c|c|}\hline 1 & 2 & 3 & 4 \\ \hline\end{array} \ \begin{array}{|c|c|}\hline 1 & 2 \\ \hline 3 & 4 \\ \hline\end{array} \right\rangle$$

$$= C^{[221]1}_{[41]1,[32]4} / C^{[22]1}_{[4]1,[22]1} = \sqrt{\frac{10}{16}} \Big/ 1 = \sqrt{\frac{5}{8}} \ .$$

Example 2: Find the $S_4 \supset S_3$ ISF $C^{[31],[21]}_{[31][\sigma'],[22][\mu']}$. Acccording to the branching law, we know $[\sigma'] = [3]$, $[21]$ and $[\mu'] = [21]$. In solving (4-191), m' can take any permissible value. We put $m' = 1$ (the first component). Let

$$\varphi_1 = \left[\left| \begin{matrix}[31] \\ [3]\end{matrix} \right\rangle \left| \begin{matrix}[22] \\ [21]\end{matrix} \right\rangle \right]^{[21]}_1 = \left| \begin{matrix}1\,2\,3 \\ 4\end{matrix} \right\rangle \left| \begin{matrix}1\,2 \\ 3\,4\end{matrix} \right\rangle , \tag{4-194a}$$

$$\varphi_2 = \left[\left| \begin{matrix}[31] \\ [21]\end{matrix} \right\rangle \left| \begin{matrix}[22] \\ [21]\end{matrix} \right\rangle \right]^{[21]}_1 = -\sqrt{\frac{1}{2}} \left[\left| \begin{matrix}1\,2\,4 \\ 3\end{matrix} \right\rangle \left| \begin{matrix}1\,2 \\ 3\,4\end{matrix} \right\rangle - \left| \begin{matrix}1\,3\,4 \\ 2\end{matrix} \right\rangle \left| \begin{matrix}1\,3 \\ 2\,4\end{matrix} \right\rangle \right] \ . \tag{4-194b}$$

Here we used (4-187) and the S_3 CG coefficients, Table 4.13-1. With the help of the irreducible matrix elements of S_4 (Table 4.4-2), the representative of the operator $[(14)+(24)+(32)]$ in the basis $\{\varphi_1, \varphi_2\}$ is $\left(\begin{smallmatrix}0 & 2 \\ 2 & 0\end{smallmatrix}\right)$. Its eigenvalues are $\nu = 2, -2$, corresponding to $[31]$ and $[211]$. Its eigenvectors are $\left(\sqrt{\frac{1}{2}}, \sqrt{\frac{1}{2}}\right)$ and $\left(\sqrt{\frac{1}{2}}, -\sqrt{\frac{1}{2}}\right)$. Therefore we have

$$\left| \begin{matrix}[31] \\ [21]1\end{matrix} \right\rangle = \left| \begin{matrix}124 \\ 3\end{matrix} \right\rangle = \sqrt{\frac{1}{2}}(\varphi_1 + \varphi_2) \ , \quad \left| \begin{matrix}[211] \\ [21]1\end{matrix} \right\rangle = \left| \begin{matrix}12 \\ 3 \\ 4\end{matrix} \right\rangle = \sqrt{\frac{1}{2}}(\varphi_1 - \varphi_2) \ . \tag{4-194c}$$

Substituting (4-194) into (4-194c), it is readily seen that the result is consistent with the CG coefficients of S_4 given in Table 4.13-2b.

In Sec. 7.16 it will be shown that the $S_n \supset S_{n-1}$ ISF is precisely the $SU_{mn} \supset SU_m \times SU_n$ ISF (or the single-particle CFP for the group chain $SU_{mn} \supset SU_m \times SU_n$).

4.19.2. Phase convention

1. Overall phase: The partitions are ordered according to their lengths. For example, for S_5, the partitions arranged in ascending order are [5], [41], [32], [311], [221], [21^3], and [1^5] and the partition [5] is said to be smaller than [41]. The pairs of partitions ([σ'], [μ']) are ordered similarly. For the case of S_5, the order is: ([5], [5]), ([5], [41]), ..., ([5], [1^5]), ([41], [5]), ([41], [41]), ..., ([1^5], [1^5]). We demand that for a given [ν] and β, the first non-vanishing component of the vector $\{C^{[\nu]\beta,[\nu']\beta'}_{\sigma\sigma',\mu\mu'}\}$ with the smallest [ν'] be positive, the component of the vector being $\beta'\sigma'\mu'$.

It is easily recognized that this phase convention is consistent with the overall phase convention (4-146) of the CG coefficients.

2. Relative phase: By requiring that the relative phase be the Yamanouchi phase, we have from (4-188) and (4-193c).

$$D^{[\nu]}_{\overline{m}m}(n-1,n) = \left\langle \begin{matrix}[\nu]\beta \\ \overline{m}\end{matrix} \middle| p_{nn-1} \middle| \begin{matrix}[\nu]\beta \\ m\end{matrix} \right\rangle = \sum_{\sigma'\mu'\beta'\overline{\sigma}'\overline{\mu}\overline{\beta}'} C^{(\nu)\beta,(\nu')\beta'}_{\sigma\sigma',\mu\mu'} \times C^{(\nu)\beta,(\overline{\nu}')\overline{\beta}'}_{\sigma\overline{\sigma}',\mu\overline{\mu}'} M(\overline{\sigma}'\overline{\mu}'\overline{\beta}'\overline{\nu}'\nu'', \sigma'\mu'\beta'\nu'\nu'') \ . \tag{4-195a}$$

Therefore

$$C^{(\nu)\beta,(\overline{\nu}')\overline{\beta}'}_{\sigma\overline{\sigma}',\mu\overline{\mu}'} = [D^{[\nu]}_{\overline{m}m}(n-1,n)]^{-1} \sum_{\sigma'\mu'\beta'} M(\overline{\sigma}'\overline{\mu}'\overline{\beta}'\overline{\nu}'\nu'', \sigma'\mu'\beta'\nu'\nu'') C^{(\nu)\beta,(\nu')\beta'}_{\sigma\sigma',\mu\mu'} \ , \tag{4-195b}$$

with $[\nu]\overline{m} = [\nu][\overline{\nu}'][\nu'']m''$ and $[\nu]m = [\nu][\nu'][\nu'']m''$. Eq. (4-195c) can be justified by multiplying both sides by $C^{(\nu)\beta,(\overline{\nu}')\overline{\beta}'}_{\sigma\overline{\sigma}',\mu\overline{\mu}'}$, and summing over $\overline{\sigma}'\overline{\mu}'\overline{\beta}\,'$.

Now the procedure for calculating the ISF can be described as follows: With σ and μ given for each [ν] and β we only need to get the ISF $C^{[\nu]\beta,(\nu')\beta'}_{\sigma\sigma'\mu\mu'}$ with the smallest ν' from the eigenequation (4-192) and the remaining ISF can be calculated from (4-195b). For a practical calculation of (4-195b), $\overline{m}'$ can be chosen as the first component, $\overline{m}' = 1$ and $\overline{m}$ is determined by $\overline{m}'$ and $\overline{\nu}'$ through $[\nu]\overline{m} = [\nu][\overline{\nu}']\overline{m}'$, while $[\nu]m$ is determined by $Y^{(\nu)}_m = (n-1,n)Y^{(\nu)}_{\overline{m}}$.

4.19.3. The properties of the $S_n \supset S_{n-1}$ ISF

1. The unitarity relations (3-305) and (3-306) now read as follows:

$$\sum_{\sigma'\mu'\beta'} C^{(\nu)\beta,(\nu')\beta'}_{\sigma\sigma',\mu\mu'} C^{(\overline{\nu})\overline{\beta},(\nu')\beta'}_{\sigma\sigma',\mu\mu'} = \delta_{\nu\overline{\nu}}\delta_{\beta\overline{\beta}} \ ,$$
$$\sum_{\nu\beta} C^{(\nu)\beta,(\nu')\beta'}_{\sigma\sigma',\mu\mu'} C^{(\nu)\beta,(\nu')\overline{\beta}'}_{\sigma\overline{\sigma}',\mu\overline{\mu}'} = \delta_{\sigma'\overline{\sigma}'}\delta_{\mu'\overline{\mu}'}\delta_{\beta'\overline{\beta}'} \ . \tag{4-196a}$$

Using (4-189) and the properties (4-116)-(4-123) of the CG coefficients, we obtain the following properties of the ISF (Chen [16], 1984, Gao 1985);

2.

$$C^{(\nu)\beta,(\nu')\beta'}_{\sigma\sigma',\mu\mu'} = \varepsilon_1' C^{(\nu)\beta,(\nu')\beta'}_{\mu\mu',\sigma\sigma'} \tag{4-196b}$$

3.

$$\sqrt{\frac{h_{\nu'}}{h_\nu}} C^{(\nu)\beta,(\nu')\beta'}_{\sigma\sigma',\mu\mu'} = \varepsilon_2' \sqrt{\frac{h_{\sigma'}}{h_\sigma}} C^{(\sigma)\beta,(\sigma')\beta'}_{\nu\nu',\mu\mu'} = \varepsilon_3' \sqrt{\frac{h_{\mu'}}{h_\mu}} C^{(\mu)\beta,(\mu')\beta'}_{\sigma\sigma',\nu\nu'} \ , \tag{4-196c}$$

4.

$$\begin{aligned}C^{(\nu)\beta,(\nu')\beta'}_{\sigma\sigma',\mu\mu'} &= \varepsilon'_4\Lambda^{\sigma}_{\sigma'}\Lambda^{\mu}_{\mu'}C^{(\nu)\beta,(\nu')\beta'}_{\tilde{\sigma}\tilde{\sigma}',\tilde{\mu}\tilde{\mu}'} \\ &= \varepsilon'_5\Lambda^{\nu}_{\nu'}\Lambda^{\mu}_{\mu'}C^{(\tilde{\nu})\beta,(\tilde{\nu}')\beta'}_{\sigma\sigma',\tilde{\mu}\tilde{\mu}'} = \varepsilon'_6\Lambda^{\sigma}_{\sigma'}\Lambda^{\nu}_{\nu'}C^{(\tilde{\nu})\beta,(\tilde{\nu}')\beta'}_{\tilde{\sigma}\tilde{\sigma}',\mu\mu'} ,\end{aligned} \tag{4-196d}$$

where we used the factorization property of the phase factor (Butler 1979),

$$\Lambda^{\nu}_{m} = \Lambda^{\nu}_{\nu'}\Lambda^{\nu'}_{m'}, \quad \Lambda^{\nu}_{\nu'} = (-1)^{n_b} , \tag{4-196'}$$

where n_b is the number of boxes below the box labeled with n in the Young tableau Y^{ν}_{m} of S_n.

The phase factors ε'_i in the above equations can be expressed in terms of the phase factors ε_i appearing in the symmetries of the CG coefficients.

$$\varepsilon'_i = \varepsilon_i(\sigma\mu\nu_\beta)\varepsilon_i(\sigma'\mu'\nu'_{\beta'}), \quad i = 1, 2, \ldots, 6 . \tag{4-196e}$$

For the multiplicity-free case, the above symmetries are satisfied automatically. For the cases with multiplicities, we can choose the eigensolutions of (4-192) properly so that ISF fulfill these symmetries. Imposing these symmetries on the ISF leads to a partial, or sometimes even a total removal of the arbitrariness in the choice of the eigensolutions for the non-multiplicity-free case.

4.19.4. Special case

$$C^{[\nu],[\nu']}_{[n][n-1],[\mu][\mu']} = \delta_{\nu\mu}\delta_{\nu'\mu'}, \qquad C^{[\nu],[\nu']}_{[1^n][1^{n-1}],[\mu][\mu']} = \delta_{\nu\tilde{\mu}}\delta_{\nu'\tilde{\mu}'} , \tag{4-196f}$$

$$C^{[n][n-1]}_{[\sigma][\sigma'],[\mu][\mu']} = (h_{\sigma'}/h_{\sigma})^{\frac{1}{2}}\delta_{\sigma\mu}\delta_{\sigma'\mu'}, \quad C^{[1^n][1^{n-1}]}_{[\sigma][\sigma'],[\mu][\mu']} = \Lambda^{\sigma}_{\sigma'}(h_{\sigma'}/h_{\sigma})^{\frac{1}{2}}\delta_{\sigma\tilde{\mu}}\delta_{\sigma'\tilde{\mu}'} . \tag{4-196g}$$

Equation (4-196f) is obvious, while (4-196g) comes from (4-196f) and (4-196c).

4.19.5. Tables of the $S_n \supset S_{n-1}$ ISF

Here we list the $S_n \supset S_{n-1}$ ISF for $n \leq 5$. The $S_n \supset S_{n-1}$ ISF for $n \leq 6$ was given in Chen [16] 1984.

The meaning of the heading is as follows:

$[\sigma][\mu]$	$[\nu']$
$\diagdown$	$[\nu]_\beta$
$[\sigma'][\mu']$	

or

$[\sigma][\mu]$	$[\nu']$
$\diagdown$	$[\nu]_\beta$
$[\sigma'][\mu']$ $[\sigma'][\mu']$	

for $\beta' > 1$

The trivial ISF which are obtainable from (4-196f) and (4-196g) are not included in the table. The tables are arranged in the order of $[\sigma], [\mu], [\nu']$. All the entries represent the square values of the ISF. A minus entry signifies a negative ISF.

Table 4.19. Tables of the $S_n \supset S_{n-1}$ ISF for n=3-5
(i.e. tables of the $SU_{mn} \supset SU_m \times SU_n$ single particle CFP for arbitrary m and n).

S_3: 1*a* [21] × [21]

[21][21]	[2]	
$\sigma'\mu'$ \ ν	[3]	[21]
[2][2]	1/2	1/2
[11][11]	1/2	−1/2

1*b*

[21][21]	[11]	
$\sigma'\mu'$ \ ν	[21]	$[1^3]$
[2][11]	−1/2	1/2
[11][2]	−1/2	−1/2

S_4: 2*a* [31] × [31]

[31][31]	[3]	
	[4]	[31]
[3][3]	1/3	2/3
[21][21]	2/3	−1/3

2*b*

[31][31]	[21]		
	[31]	[22]	[211]
[3][21]	−1/6	1/3	1/2
[21][3]	−1/6	1/3	−1/2
[21][21]	2/3	1/3	

2*c*

[31][31]	$[1^3]$
	[211]
[21][21]	1

3*a* [31] × [22]

[31][22]	[3]
	[31]
[21][21]	1

3*b*

[31][22]	[21]	
	[31]	[211]
[3][21]	1/2	1/2
[21][21]	1/2	−1/2

3*c*

[31][22]	$[1^3]$
	[211]
[21][21]	−1

4*a* [31] × [211]

[31][211]	[3]
	[31]
[21][21]	1

4*b*

[31][211]	[21]		
	[31]	[22]	[211]
[3][21]	−1/2	1/3	1/6
[21][21]		−1/3	2/3
[21]$[1^3]$	1/2	1/3	1/6

4*c*

[31][211]	$[1^3]$	
	[211]	$[1^4]$
[3]$[1^3]$	−2/3	1/3
[21][21]	−1/3	−2/3

5*a* [22] × [22]

[22][22]	[3]
	[4]
[21][21]	1

5*b*

[22][22]	[21]
	[22]
[21][21]	1

5*c*

[22][22]	$[1^3]$
	$[1^4]$
[21][21]	1

6*a* [22] × [211]

[22][211]	[3]
	[31]
[21][21]	1

6*b*

[22][211]	[21]	
	[31]	[211]
[21][21]	−1/2	1/2
[21]$[1^3]$	−1/2	−1/2

6*c*

[22][211]	$[1^3]$
	[211]
[21][21]	1

7*a* [211] × [211]

[211][211]	[3]	
	[4]	[31]
[21][21]	2/3	1/3
$[1^3][1^3]$	1/3	−2/3

cont'd

7b

[211][211]	[21]		
	[31]	[22]	[211]
[21][21]	2/3	1/3	
[21][1^3]	−1/6	1/3	1/2
[1^3][21]	−1/6	1/3	−1/2

7c

[211][211]	[1^3]
	[211]
[21][21]	1

S_5: 8a [41] × [41]

[41][41]	[4]	
	[5]	[41]
[4][4]	1/4	3/4
[31][31]	3/4	−1/4

8b

[41][41]	[31]		
	[41]	[32]	[311]
[4][31]	−1/12	5/12	1/2
[31][4]	−1/12	5/12	−1/2
[31][31]	5/6	1/6	

8c

[41][41]	[22]
	[32]
[31][31]	1

8d

[41][41]	[211]
	[311]
[31][31]	1

9a [41] × [32]

[41][32]	[4]
	[41]
[31][31]	1

9b

[41][32]	[31]		
	[41]	[32]	[311]
[4][31]	1/3	1/6	1/2
[31][31]	2/15	5/12	−9/20
[31][22]	8/15	−5/12	−1/20

9c

[41][32]	[22]	
	[32]	[221]
[4][22]	−3/8	5/8
[31][31]	−5/8	−3/8

9d

[41][32]	[211]	
	[311]	[221]
[31][31]	−1/4	3/4
[31][22]	3/4	1/4

10a [41] × [311]

[41][311]	[4]
	[41]
[31][31]	1

10b

[41][311]	[31]		
	[41]	[32]	[311]
[4][31]	−1/3	5/12	1/4
[31][31]		−3/8	5/8
[31][211]	2/3	5/24	1/8

10c

[41][311]	[22]	
	[32]	[221]
[31][31]	*1/16	15/16
[31][211]	15/16	1/16

10d

[41][311]	[211]		
	[311]	[221]	[21^3]
[4][211]	−1/4	5/12	1/3
[31][31]	−1/8	5/24	−2/3
[31][211]	5/8	3/8	

10e

[41][311]	[1^4]
	[21^3]
[31][211]	1

cont'd

S_5: **11a** [41] × [221]

[41][221]	[31]	
	[32]	[311]
[31][22]	1/4	3/4
[31][211]	3/4	−1/4

11b

[41][221]	[22]	
	[32]	[221]
[4][22]	−5/8	3/8
[31][211]	−3/8	−5/8

11c

[41][221]	[211]		
	[311]	[221]	$[21^3]$
[4][211]	1/2	−1/6	1/3
[31][22]	1/20	−5/12	−8/15
[31][211]	9/20	5/12	−2/15

11d

[41][221]	$[1^4]$
	$[21^3]$
[31][211]	−1

12a $[41] \times [21^3]$

$[41][21^3]$	[31]
	[311]
[31][211]	1

12b

$[41][21^3]$	[22]
	[221]
[31][211]	1

12c

$[41][21^3]$	[211]		
	[311]	[221]	$[21^3]$
[4][211]	1/2	−5/12	1/12
[31][211]	0	1/6	5/6
$[31][1^4]$	1/2	5/12	−1/12

12d

$[41][21^3]$	$[1^4]$	
	$[21^3]$	$[1^5]$
$[4][1^4]$	−3/4	1/4
[31][211]	−1/4	−3/4

13a [32] × [32]

[32][32]	[4]	
	[5]	[41]
[31][31]	3/5	2/5
[22][22]	2/5	−3/5

13b

[32][32]	[31]		
	[41]	[32]	[311]
[31][31]	1/3	2/3	0
[31][22]	−1/3	1/6	1/2
[22][31]	−1/3	1/6	−1/2

13c

[32][32]	[22]	
	[32]	[221]
[31][31]	1/4	3/4
[22][22]	3/4	−1/4

13d

[32][32]	[211]		
	[311]	[221]	$[21^3]$
[31][31]	2/5	0	3/5
[31][22]	3/10	1/2	−1/5
[22][31]	−3/10	1/2	1/5

13e

[32][32]	$[1^4]$
	$[21^3]$
[22][22]	1

14a [32] × [311]

[32][311]	[4]
	[41]
[31][31]	1

14b

[32][311]	[31]			
	[41]	[32]	[311]α	[311]β
[31][31]	−3/10	0	3/5	1/10
[31][211]	−1/6	1/3	0	−1/2
[22][31]	−1/30	5/12	−3/20	2/5
[22][211]	1/2	1/4	1/4	0

cont'd

14c

[32] [311]	[22]	
	[32]	[221]
[31] [31]	−5/8	3/8
[31] [211]	−3/8	−5/8

14d

[32] [311]	[211]			
	[311]α	[311]β	[221]	[21^3]
[31] [31]	0	1/2	−1/3	1/6
[31] [211]	−3/5	1/10	0	−3/10
[22] [31]	1/4	0	−1/4	−1/2
[22] [211]	3/20	2/5	5/12	−1/30

14e

[32] [311]	[1^4]
	[21^3]
[31] [211]	1

15a [32] × [221]

[32] [221]	[4]
	[41]
[22] [22]	1

15b

[32] [221]	[31]		
	[41]	[32]	[311]
[31] [22]	−1/5	1/2	3/10
[31] [211]	3/5	0	2/5
[22] [211]	1/5	1/2	−3/10

15c

[32] [221]	[22]	
	[32]	[221]
[31] [211]	3/4	1/4
[22] [22]	−1/4	3/4

15d

[32] [221]	[211]		
	[311]	[221]	[21^3]
[31] [22]	−1/2	1/6	1/3
[31] [211]	0	2/3	−1/3
[22] [211]	1/2	1/6	1/3

15e

[32] [221]	[1^4]	
	[21^3]	[1^5]
[31] [211]	−2/5	3/5
[22] [22]	3/5	2/5

16a [32] × [21^3]

[32] [21^3]	[31]	
	[32]	[311]
[31] [211]	3/4	1/4
[22] [211]	−1/4	3/4

16b

[32] [21^3]	[22]	
	[32]	[221]
[31] [211]	3/8	5/8
[22] [1^4]	5/8	−3/8

16c

[32] [21^3]	[211]		
	[311]	[221]	[21^3]
[31] [211]	9/20	−5/12	2/15
[31] [1^4]	−1/2	−1/6	1/3
[22] [211]	1/20	5/12	8/15

16d

[32] [21^3]	[1^4]
	[21^3]
[32] [211]	1

17a [311] × [311]

[311] [311]	[4]	
	[5]	[41]
[31] [31]	1/2	1/2
[211] [211]	1/2	−1/2

17b

[311] [311]	[31]			
	[41]	[32]α	[32]β	[311]
[31] [31]	5/12	1/2	1/12	0
[31] [211]	−1/12	0	5/12	1/2
[211] [31]	−1/12	0	5/12	−1/2
[211] [211]	5/12	−1/2	1/12	0

17c

[311] [311]	[22]			
	[32]α	[32]β	[221]α	[221]β
[31] [31]	−3/16	1/2	5/16	0
[31] [211]	5/16	0	3/16	1/2
[211] [31]	5/16	0	3/16	−1/2
[211] [211]	3/16	1/2	−5/16	0

cont'd

17d

[311] [311]	[211]			
	[311]	[221]α	[221]β	[21³]
[31] [31]	1/2	0	−5/12	1/12
[31] [211]	0	−1/2	1/12	5/12
[211] [31]	0	−1/2	−1/12	−5/12
[211] [211]	1/2	0	5/12	−1/12

17e

[311] [311]	[1⁴]	
	[21³]	[1⁵]
[31] [211]	1/2	1/2
[211] [31]	−1/2	1/2

18a [311] × [221]

[311] [221]	[4]
	[41]
[211] [211]	1

18b

[311] [221]	[31]			
	[41]	[32]	[311]α	[311]β
[31] [22]	1/2	1/4	1/4	0
[31] [211]	1/6	−1/3	0	1/2
[211] [22]	1/30	−5/12	3/20	−2/5
[211] [211]	3/10	0	−3/5	−1/10

18c

[311] [221]	[22]	
	[32]	[221]
[31] [211]	−3/8	5/8
[211] [211]	5/8	3/8

18d

[311] [221]	[211]			
	[311]α	[311]β	[221]	[21³]
[31] [22]	3/20	2/5	−5/12	1/30
[31] [211]	−3/5	1/10	0	3/10
[211] [22]	−1/4	0	−1/4	−1/2
[211] [211]	0	−1/2	−1/3	1/6

18e

[311] [221]	[1⁴]
	[21³]
[31] [211]	−1

19a [311] × [21³]

[311] [21³]	[4]
	[41]
[211] [211]	1

19b

[311] [21³]	[31]		
	[41]	[32]	[311]
[31] [211]	−2/3	5/24	1/8
[211] [211]	0	−3/8	5/8
[211] [1⁴]	1/3	5/12	1/4

19c

[311] [21³]	[32]	
	[32]	[221]
[31] [211]	−15/16	1/16
[211] [211]	−1/16	−15/16

19d

[311] [21³]	[211]		
	[311]	[221]	[21³]
[31] [211]	−5/8	3/8	0
[31] [1⁴]	−1/4	−5/12	1/3
[211] [211]	−1/8	−5/24	−2/3

19e

[311] [21³]	[1⁴]
	[21³]
[31] [211]	−1

20a [221] × [221]

[221] [221]	[4]	
	[5]	[41]
[22] [22]	2/5	3/5
[211] [211]	3/5	−2/5

20b

[221] [221]	[31]		
	[41]	[32]	[311]
[22] [211]	−1/3	1/6	1/2
[211] [22]	−1/3	1/6	−1/2
[211] [211]	1/3	2/3	0

cont'd

20c

[221] [221]	[22]	
	[32]	[221]
[22] [22]	3/4	1/4
[211] [211]	1/4	−3/4

20d

[221] [221]	[211]		
	[311]	[221]	[21³]
[22] [211]	3/10	−1/2	1/5
[211] [22]	−3/10	−1/2	−1/5
[211] [211]	2/5	0	−3/5

20e

[221] [221]	[1⁴]
	[21³]
[22] [22]	−1

21a

[221] [21³]	[4]
	[41]
[211] [211]	1

21b

[221] [21³]	[31]		
	[41]	[32]	[311]
[22] [211]	−8/15	5/12	1/20
[211] [211]	−2/15	−5/12	9/20
[211] [1⁴]	−1/3	−1/6	−1/2

21c

[221] [21³]	[22]	
	[32]	[221]
[22] [1⁴]	3/8	5/8
[211] [211]	5/8	−3/8

21d

[221] [21³]	[211]	
	[311]	[221]
[22] [211]	−3/4	1/4
[211] [211]	1/4	3/4

22a [21³] × [21³]

[21³] [21³]	[4]	
	[5]	[41]
[211] [211]	3/4	1/4
[1⁴] [1⁴]	1/4	−3/4

22b

[21³] [21³]	[31]		
	[41]	[32]	[311]
[211] [211]	5/6	1/6	0
[211] [1⁴]	−1/12	5/12	1/2
[1⁴] [211]	−1/12	5/12	−1/2

22c

[21³] [21³]	[22]
	[32]
[211] [211]	1

22d

[21³] [21³]	[211]
	[311]
[211] [211]	1

4.19.6. *The $S_{n_1+n_2} \supset S_{n_1} \otimes S_{n_2}$ ISF**

We introduce the following notation to denote the irreps of the nine groups:

$$\begin{pmatrix} \sigma' & \mu' & \nu' \\ \sigma'' & \mu'' & \nu'' \\ \sigma & \mu & \nu \end{pmatrix}, \quad \begin{pmatrix} S_{n_1}(x) & S_{n_1}(\xi) & S_{n_1}(q) \\ S_{n_2}(x) & S_{n_2}(\xi) & S_{n_2}(q) \\ S_n(x) & S_n(\xi) & S_n(q) \end{pmatrix} . \tag{4-197}$$

For example, $[\mu'']$ and $[\sigma]$ are the irreps of $S_{n_2}(\xi)$ and $S_n(x)$, respectively, with $n = n_1 + n_2$. The $S_n \supset S_{n_1} \otimes S_{n_2}$ bases in the x, ξ and q spaces are denoted by

$$\left| \begin{matrix} [\sigma] \\ \theta[\sigma']m_1'[\sigma'']m_1'' \end{matrix} \right\rangle, \quad \left| \begin{matrix} [\mu] \\ \varphi[\mu']m_2'[\mu'']m_2'' \end{matrix} \right\rangle, \quad \left| \begin{matrix} [\nu] \\ \tau[\nu']m'[\nu'']m'' \end{matrix} \right\rangle, \tag{4-198}$$
$$\theta = 1,2,\ldots\{\sigma'\sigma''\sigma\}, \quad \varphi = 1,2,\ldots\{\mu'\mu''\mu\}, \quad \tau = 1,2,\ldots\{\nu'\nu''\nu\} .$$

The former two in (4-198) can be linearly combined into the third one via the following two steps:

1. Use the CG coefficients of S_{n_1} and S_{n_2} to combine them into the irreducible basis $[\nu']m'$ and $[\nu'']m''$ of $S_{n_1}(q)$ and $S_{n_2}(q)$, respectively,

$$\left|(\sigma'\sigma'')_\theta(\mu'\mu'')_\varphi\beta'\beta''\right\rangle = \left[\left|\begin{matrix}[\sigma]\\ \theta[\sigma'][\sigma'']\end{matrix}\right\rangle\left|\begin{matrix}[\mu]\\ \varphi[\mu'][\mu'']\end{matrix}\right\rangle\right]^{[\nu']_{\beta'}[\nu'']_{\beta''}}_{m'm''}$$

$$= \sum_{m_1'm_2'm_1''m_2''} C^{[\nu']\beta',m'}_{\sigma'm_1',\mu'm_2'} C^{[\nu'']\beta'',m''}_{\sigma''m_1'',\mu''m_2''}\left|\begin{matrix}[\sigma]\\ \theta[\sigma']m_1'[\sigma'']m_1''\end{matrix}\right\rangle\left|\begin{matrix}[\mu]\\ \varphi[\mu']m_2'[\mu'']m_2''\end{matrix}\right\rangle,$$

$$\beta' = 1,2,\ldots(\sigma'\mu'\nu'), \quad \beta'' = 1,2,\ldots(\sigma''\mu''\nu'') . \tag{4-199}$$

2. Use the $S_n \supset S_{n_1}\otimes S_{n_2}$ ISF to combine (4-199) into a basis belonging to the irrep $[\nu]$ of $S_n(q)$.

$$\left|\begin{matrix}[\nu]\beta\\ \tau[\nu']m'[\nu'']m''\end{matrix}\right\rangle = \sum_{\substack{\sigma'\sigma''\theta\beta'\\ \mu'\mu''\varphi\beta''}} C^{[\nu]\beta,\tau[\nu']\beta'[\nu'']\beta''}_{[\sigma]\theta\sigma'\sigma'',[\mu]\varphi\mu'\mu''}\left[\left|\begin{matrix}[\sigma]\\ \theta[\sigma'][\sigma'']\end{matrix}\right\rangle\left|\begin{matrix}[\mu]\\ \varphi[\mu'][\mu'']\end{matrix}\right\rangle\right]^{[\nu']_{\beta'}[\nu'']_{\beta''}}_{m'm''},$$

$$\beta = 1,2,\ldots(\sigma\mu\nu) . \tag{4-200}$$

The inverse of (4-200) is

$$\left[\left|\begin{matrix}[\sigma]\\ \theta[\sigma'][\sigma'']\end{matrix}\right\rangle\left|\begin{matrix}[\mu]\\ \varphi[\mu'][\mu'']\end{matrix}\right\rangle\right]^{[\nu']_{\beta'}[\nu'']_{\beta''}}_{m'm''} = \sum_{\nu\beta\tau} C^{[\nu]\beta,\tau[\nu']\beta'[\nu'']\beta''}_{[\sigma]\theta\sigma'\sigma'',[\mu]\varphi\mu'\mu''}\left|\begin{matrix}[\nu]_\beta\\ \tau[\nu']m'[\nu'']m''\end{matrix}\right\rangle . \tag{4-201}$$

If a basis in the x-space is totally symmetric, then it is necessarily also a $S_n \supset S_{n_1}\otimes S_{n_2}$ irreducible basis, i.e., $|[n]\rangle = \left|\begin{matrix}[n]\\ [n_1][n_2]\end{matrix}\right\rangle$. In such a special case we have:

$$\left|\begin{matrix}[\nu]\\ \tau[\nu']m'[\nu'']m''\end{matrix}\right\rangle^q = \delta_{\nu\mu}\delta_{\nu'\mu'}\delta_{\nu''\mu''}\delta_{\tau\varphi}\delta_{m'm_2'}\delta_{m''m_2''}\left|\begin{matrix}[n]\\ [n_1][n_2]\end{matrix}\right\rangle^x\left|\begin{matrix}[\mu]\\ \varphi[\mu']m_2'[\mu'']m_2''\end{matrix}\right\rangle^\xi . \tag{4-202}$$

Comparing it with (4-200), we have

$$C^{[\nu],\tau[\nu'][\nu'']}_{[n][n_1][n_2],[\mu]\varphi[\mu'][\mu'']} = \delta_{\nu\mu}\delta_{\nu'\mu'}\delta_{\nu''\mu''}\delta_{\tau\varphi} . \tag{4-203}$$

Analogously we have also

$$C^{[\nu],\tau[\nu'][\nu'']}_{[1^n][1^{n_1}][1^{n_2}],[\mu]\varphi\mu'\mu''} = \delta_{\nu\tilde{\mu}}\delta_{\nu'\tilde{\mu}'}\delta_{\nu''\tilde{\mu}''}\delta_{\tau\varphi} . \tag{4-204}$$

In Sec. 7.16 it will be proved that the $S_{n_1+n_2}\supset S_{n_1}\otimes S_{n_2}$ ISF is the n_2-particle CFP in the group chain $SU_{mn}\supset SU_m\times SU_n$.

4.20. Appendix: Derivation of Yamanouchi Matrix Elements by the EFM

In this section we use the ket $|\lambda\rangle = |\lambda_n\lambda_{n-1}\ldots\lambda_2\rangle$ to denote a Yamanouchi basis vector. From the relations

$$[(n-1,n), C(f)] = 0, \quad f = n, n-2, \ldots, 2 , \tag{4-205a}$$

$$[(n-1,n), C(n-1)] \neq 0 \ , \tag{4-205b}$$

we know that the permutation $(n-1,n)$ has non-vanishing matrix elements only between the states $|\lambda_n \lambda'_{n-1} \lambda_{n-2} \dots \lambda_2\rangle$ and $|\lambda_n \lambda_{n-1} \lambda_{n-2} \dots \lambda_2\rangle$, i.e.,

$$\langle \lambda'_n \lambda'_{n-1} \dots \lambda'_2 | (n-1,n) | \lambda_n \lambda_{n-1} \dots \lambda_2 \rangle = \text{const.}\ \delta_{\lambda'_n \lambda_n} \delta_{\lambda'_{n-2} \lambda_{n-2}} \dots \delta_{\lambda'_2 \lambda_2} \ . \tag{4-206}$$

We can easily establish the following identity relations

$$C(n) = C(n-1) + \sum_{i=1}^{n-1} (i,n) \ , \tag{4-207a}$$

$$\sum_{i=1}^{n-2} (i, n-1) = C(n-1) - C(n-2) \ , \tag{4-207b}$$

$$C(n) = C(n-1) + (n-1,n) \sum_{i=1}^{n-2} (i, n-1)(n-1,n) + (n-1,n) \ . \tag{4-207c}$$

Using (4-207b) and (4-205a), (4-207c) becomes

$$C(n) = C(n-1) - C(n-2) + (n-1,n) C(n-1)(n-1,n) + (n-1,n) \ , \tag{4-208a}$$

or written in a more elegant form

$$\left[C(n-1), (n-1,n)\right]_+ = \left(C(n) + C(n-2)\right)(n-1,n) - 1 \ , \tag{4-208b}$$

where $[A,B]_+ = AB + BA$.

Inserting (4-208b) between the two Yamanouchi basis vectors $|\lambda'\rangle$ and $|\lambda\rangle$ and using (4-28b) and (4-206), we obtain

$$\begin{gathered} \langle \lambda'_n \lambda'_{n-1} \dots \lambda'_2 | (n-1,n) | \lambda_n \lambda_{n-1} \dots \lambda_2 \rangle = \delta_{\lambda'_n \lambda_n} \delta_{\lambda'_{n-2} \lambda_{n-2}} \dots \delta_{\lambda'_2 \lambda_2} \mu^{-1} \ , \\ \mu = \lambda_n - \lambda'_{n-1} - \lambda_{n-1} + \lambda_{n-2} \ . \end{gathered} \tag{4-209}$$

From (4-209) we obtain the diagonal matrix element of the permutation $(n-1,n)$

$$\langle \lambda_n \lambda_{n-1} \dots \lambda_2 | (n-1,n) | \lambda_n \lambda_{n-1} \dots \lambda_2 \rangle = \sigma^{-1} \tag{4-210a}$$

$$\sigma = \lambda_n - 2\lambda_{n-1} + \lambda_{n-2} \tag{4-210b}$$

and the off-diagonal matrix elements

$$\langle \lambda_n \lambda'_{n-1} \dots \lambda_2 | (n-1,n) | \lambda_n \lambda_{n-1} \dots \lambda_2 \rangle = \begin{cases} 0, & \text{for} \quad \mu \neq 0 \\ b, & \phantom{\text{for}} \quad \mu = 0 \end{cases} \ , \tag{4-211}$$

where the coefficient b is to be decided upon.

Before going on to consider the constant b, we first examine the conditions under which the permutation $(n-1,n)$ has non-vanishing off-diagonal matrix elements. Let $Y_{\lambda'}$ and Y_λ be the Young tableaux corresponding to the basis vectors $|\lambda_n \lambda'_{n-1} \lambda_{n-2} \dots \lambda_2\rangle$ and $|\lambda_n \lambda_{n-1} \lambda_{n-2} \dots \lambda_2\rangle$ which differ only in the second quantum number. Clearly, $Y_{\lambda'}$ and Y_λ must have the same Young diagrams $[\nu]$ and $[\nu'']$ for the numbers $1, \dots, n$ and $1, \dots n-2$, respectively. If the numbers n and $n-1$ are at the same row or column of the Young tableau Y_λ, the above condition implies that $Y_{\lambda'}$ and Y_λ must be identical, i.e., $(n-1,n)$ does not have non-vanishing matrix elements between two different states $|\lambda\rangle$ and $|\lambda'\rangle$ if one of them is symmetric or antisymmetric in the indices $n-1$ and n. On the other hand, if n and $n-1$ are not at the same row or column of Y_λ, the above condition means that $Y_{\lambda'}$ and Y_λ differ only in the interchange of the positions of the numbers n and $n-1$.

Summarizing, only when n and $n-1$ are not in the same row or column of a Young tableau Y_λ, and only for a unique $Y_{\lambda'} = (n-1,n)Y_\lambda$, does the off-diagonal element $\langle\lambda'|(n-1,n)|\lambda\rangle$ differ from zero.

According to the identity $(n-1,n)^2 = 1$ and the above discussion, we immediately have

$$\langle\lambda|(n-1,n)|\lambda\rangle^2 + \langle\lambda|(n-1,n)|\lambda'\rangle\langle\lambda'|(n-1,n)|\lambda\rangle = 1\,. \tag{4-212}$$

Therefore the constant b is given by the equation

$$|b|^2 = 1 - 1/\sigma^2\,. \tag{4-213a}$$

Under the Yamanouchi phase convention, we have

$$\langle\lambda'|(n-1,n)|\lambda\rangle = \frac{\sqrt{\sigma^2-1}}{|\sigma|}\,. \tag{4-213b}$$

Finally, we are going to prove that the constant σ defined by (4-210b) is exactly the axial distance from $n-1$ to n in the Young tableau Y_λ.

Suppose that the row numbers of the indices n and $n-1$ in Y_λ are l and l', respectively. Then the row numbers of n and $n-1$ in Y'_λ are l' and l, respectively. It follows from (4-210b) and (4-3a) that

$$\begin{aligned}\sigma &= \frac{1}{2}[(f_l + f_{l'}) - 2(f_{l-1} + f_{l'}) + (f_{l-1} + f_{l'-1})] \\ &= \frac{1}{2}[(f_l - f_{l-1}) - (f_{l'} - f_{l'-1})]\,,\end{aligned} \tag{4-214}$$

where

$$f_l = \nu_l(\nu_l - 2l)\,.$$

Therefore

$$\sigma = (\nu_l - l) - (\nu_{l'} - l')\,. \tag{4-215}$$

This is just the Eq. (4-16), since $\nu_l = c_n$, $\nu_{l'} = c_{n-1}, l = r_n$ and $l' = r_{n-1}$.

Chapter 5
Lie Groups

The content of Lie groups and Lie algebras is so rich that here we are able only to cover a very small part of it. Our main concern is how to find the infinitesimal operators of the linear transformation groups and how to generalize the new approach to the rep theory of finite groups to that of Lie groups, so that we can have a unified understanding of the rep theory of both finite groups and Lie groups. Using the concept of "representative of a vector" in quantum mechanics, the theorems on roots and weights are reformulated in a much more transparent way. Many theorems are cited without proof. Interested readers are referred to the books by Racah (1951), Wybourne (1974) and Bacry (1977).

5.1. Tensor

5.1.1. Vector (tensor of rank one)

We use superscripts and subscripts to denote the contravariant and covariant indices, respectively. Equations (2-6) and (2-7) are rewritten as

$$\mathbf{x} = x^{\mu}\, \mathbf{u}_{\mu} \;, \tag{5-1}$$

$$\mathbf{u}'_{\mu} = B^{\nu}_{\mu}\mathbf{u}_{\nu}, \quad x'^{\mu} = A^{\mu}_{\nu}x^{\nu} \;, \tag{5-2a, b}$$

where the dummy suffix summation convention is used. Notice that when A and B are written in matrix form, the superscript (subscript) of $A^{\mu}_{\nu}(B^{\nu}_{\mu})$ is the row index, i.e.,

$$A^{\mu}_{\nu} = A_{\mu\nu}, \quad B^{\nu}_{\mu} = B_{\mu\nu}, \quad B = \tilde{A}^{-1} \,. \tag{5-3}$$

For nonlinear transformations, contravariant and covariant coordinates are defined as those which satisfy the following relations:

$$dx'^{\mu} = A^{\mu}_{\nu}dx^{\nu}, \quad A^{\mu}_{\nu} = \frac{\partial x'^{\mu}}{\partial x^{\nu}} \,. \tag{5-4a, b}$$

$$dx'_{\mu} = B^{\nu}_{\mu}dx_{\nu} \,. \tag{5-5}$$

From (5-5) and (5-3) we have

$$\mathbf{u}_{\mu}dx^{\mu} = \mathbf{u}'_{\mu}dx'^{\mu} = \mathbf{u}_{\nu}B^{\nu}_{\mu}dx'^{\mu} = \mathbf{u}_{\nu}dx^{\nu} \;;$$

therefore $dx^{\nu} = B^{\nu}_{\mu}dx'^{\mu}$ and

$$B^{\nu}_{\mu} = \frac{\partial x'_{\mu}}{\partial x_{\nu}} = \frac{\partial x^{\nu}}{\partial x'^{\mu}} \,. \tag{5-6}$$

Quantities which transform as dx^μ (dx_μ) are called contravariant (covariant) vectors, i.e.,

$$V'^\mu = A^\mu_\nu V^\nu , \quad V'_\mu = B^\nu_\mu V_\nu . \tag{5-7a, b}$$

It is easy to prove that the derivative of a scalar Φ with respect to the contravariant (covariant) variable is a covariant (contravariant) vector. For example,

$$\frac{\partial\Phi}{\partial x'^\mu} = \frac{\partial x^\nu}{\partial x'^\mu}\frac{\partial\Phi}{\partial x^\nu} = B^\nu_\mu \frac{\partial\Phi}{\partial x^\nu} . \tag{5-8}$$

Therefore the differential operators $\partial_\mu = \frac{\partial}{\partial x^\mu}$ and $\partial^\mu = \frac{\partial}{\partial x_\mu}$ can be regarded as covariant and contravariant vectors, respectively.

5.1.2. Tensors with rank higher than one

The tensors $T^{\mu\nu}, T_{\mu\nu}$ and T^μ_ν are called the contravariant, covariant, and mixed tensors of rank 2, respectively, if they satisfy the following transformation rules:

$$T'^{\mu\nu} = A^\mu_\rho A^\nu_\sigma T^{\rho\sigma} , \quad T'_{\mu\nu} = B^\rho_\mu B^\sigma_\nu T_{\rho\sigma} ,$$
$$T'^\mu_\nu = A^\mu_\rho B^\sigma_\nu T^\rho_\sigma . \tag{5-9a}$$

The above definition can be generalized to higher rank tensors. For example, for the mixed tensor of rank three we have

$$T'^\tau_{\rho\sigma} = B^\mu_\rho B^\nu_\sigma A^\tau_\lambda T^\lambda_{\mu\nu} . \tag{5-9b}$$

5.1.3. Metric tensor

The metric tensor g_{ij} defined in (2-16) is a hermitian matrix. The scalar product of two vectors $\mathbf{x}$ and $\mathbf{y}$ in the space L is defined by

$$(\mathbf{x}, \mathbf{y}) = x^{\mu *} g_{\mu\nu} y^\nu . \tag{5-10}$$

It belongs to the so-called *sesquilinear metric* (Gilmore, 1974). Another kind of metric is called the *bilinear metric*, in which the scalar product is defined by

$$(\mathbf{x}, \mathbf{y}) = x^\mu g_{\mu\nu} y^\nu , \tag{5-11a}$$

and the metric tensor

$$g_{\mu\nu} = (\mathbf{u}_\mu, \mathbf{u}_\nu) \tag{5-11b}$$

is a second-rank covariant tensor both for real and complex vector space, while the inverse matrix of $(g_{\mu\nu})$ is a second-rank contravariant tensor

$$g^{\sigma\nu} = (g^{-1})_{\nu\sigma} = M^{\sigma\nu}/\det(g_{\mu\nu}) , \tag{5-12a}$$

$$g_{\mu\nu} g^{\sigma\nu} = \delta^\sigma_\mu , \tag{5-12b}$$

where $M^{\sigma\nu}$ are the co-factors of the elements $g_{\sigma\nu}$ in the matrix g.

Using (5-12a), (2-111) becomes

$$|\overline{\varphi}_\sigma\rangle = |\varphi^\sigma\rangle = g^{\sigma\nu}|\varphi_\nu\rangle . \tag{5-12c}$$

Therefore the dual basis $|\overline{\varphi}_\sigma\rangle$ is a contravariant basis. Notice the difference between (5-10) and (5-11a). For real vectors the difference disappears.

5.1.4. Metric space

If in a space the contravariant and covariant vectors are obtainable from one another by the metric tensor instead of being independent, then the space is called a *metric space*. In a metric space we have

$$V_\mu = g_{\mu\nu} V^\nu , \qquad V^\mu = g^{\mu\nu} V_\nu . \tag{5-13}$$

Therefore the scalar product (5-11a) can be written as

$$(\mathbf{x}, \mathbf{y}) = x^\mu y_\mu = x_\mu y^\mu . \tag{5-11c}$$

From (5-13) it is seen that the metric tensor can be used to raise or lower indices of vectors. This rule also holds for higher rank tensors, as for example

$$T^\nu_\mu = g_{\rho\mu} T^{\rho\nu} .$$

A summation over a pair of superscript and subscript is called a contraction, which reduces the rank of a tensor by two; for example, $U_{\nu\rho} = T^\mu_{\mu\nu\rho} (\equiv \sum_\mu T^\mu_{\mu\nu\rho})$ is a tensor of rank 2.

For the orthonormal bases, $g_{\mu\nu} = \delta_{\mu\nu}$. Then (5-13) shows that $V^\mu = V_\mu$. In other words the covariant and contravariant vectors coincide for orthonormal bases.

From now on we will only deal with metric spaces.

Example: Consider the transformation from the Cartesian coordinates (x, y) to the polar coordinates (r, θ). For the Cartesian coordinates (x, y), the metric tensor is the unit matrix,

$$g = \begin{pmatrix} 1 & 0 \\ 0 & 1 \end{pmatrix} , \tag{5-14}$$

and the covariant and contravariant are identical, i.e., $dx^1 = dx_1 = dx, dx^2 = dx_2 = dy$. In the polar coordinate system, let r, θ be regarded as the contravariant coordinates, $dx'^1 = dr, dx'^2 = d\theta$. From $x = r\cos\theta, y = r\sin\theta$, we have

$$\begin{pmatrix} dx \\ dy \end{pmatrix} = \begin{pmatrix} \cos\theta & -r\sin\theta \\ \sin\theta & r\cos\theta \end{pmatrix} \begin{pmatrix} dr \\ d\theta \end{pmatrix}, \quad A^{-1} = \begin{pmatrix} \cos\theta & -r\sin\theta \\ \sin\theta & r\cos\theta \end{pmatrix} . \tag{5-15}$$

From A^{-1} we find

$$B = \tilde{A}^{-1} = \begin{pmatrix} \cos\theta & \sin\theta \\ -r\sin\theta & r\cos\theta \end{pmatrix} , \quad A = \begin{pmatrix} \cos\theta & \sin\theta \\ -\frac{1}{r}\sin\theta & \frac{1}{r}\cos\theta \end{pmatrix} . \tag{5-16}$$

From (5-9a), (5-14) and (5-16), we obtain the covariant metric tensor in the polar coordinate system:

$$(g'_{\mu\nu}) = g' = Bg\tilde{B} = \begin{pmatrix} 1 & 0 \\ 0 & r^2 \end{pmatrix} . \tag{5-17}$$

Moreover from (5-4a), (5-5) and (5-16) we have

$$\begin{pmatrix} dx'^1 \\ dx'^2 \end{pmatrix} = \begin{pmatrix} dr \\ d\theta \end{pmatrix} = \begin{pmatrix} \cos\theta & \sin\theta \\ -\frac{\sin\theta}{r} & \frac{\cos\theta}{r} \end{pmatrix} \begin{pmatrix} dx \\ dy \end{pmatrix} , \quad \begin{pmatrix} dx'_1 \\ dx'_2 \end{pmatrix} = \begin{pmatrix} \cos\theta & \sin\theta \\ -r\sin\theta & r\cos\theta \end{pmatrix} \begin{pmatrix} dx \\ dy \end{pmatrix} . \tag{5-18}$$

It is easy to verify from (5-17) and (5-18) that

$$\begin{pmatrix} dx'_1 \\ dx'_2 \end{pmatrix} = \begin{pmatrix} 1 & 0 \\ 0 & r^2 \end{pmatrix} \begin{pmatrix} dx'^1 \\ dx'^2 \end{pmatrix} .$$

The invariant ds^2 is

$$ds^2 = dx^\mu g_{\mu\nu} dx^\nu = dx^2 + dy^2 = dx'^\mu g'_{\mu\nu} dx'^\nu = d^2r + r^2 d^2\theta .$$

5.2. Definition of Lie Group and Some Examples

A Lie group is a special kind of continuous group. The group elements $R(a)$ are labeled by r real parameters $a^1, a^2, \ldots, a^r$,

$$R(a) = R(a^1, a^2, \ldots, a^r) \ . \tag{5-19}$$

The parameters a^ρ may vary over a finite or an infinite range. The space of the r parameters is called the *group-parameter space*. A group G is called a Lie group of order r if $R(a)$ obeys the following five postulates:

1. The identity element $R(a_0)$ exists, i.e.,

$$R(a_0)R(a) = R(a)R(a_0) = R(a), \quad \text{for any } R(a) \in G \ . \tag{5-20}$$

The parameters a_0 of the identity element are usually taken as zero, i.e., $R(a_0) = R(0)$.

2. For any a we can find $\overline{a}$ such that

$$R(\overline{a})R(a) = R(a)R(\overline{a}) = R(0) \ ,$$

i.e., for every $R(a)$ an inverse exists:

$$R(\overline{a}) = R^{-1}(a) \ . \tag{5-21}$$

3. For given parameters a and b, we can find c in the set of parameters such that

$$R(c) = R(b)R(a) \ , \tag{5-22}$$

where the parameters c are real functions of the real parameters a and b,

$$c = \varphi(a, b) \ . \tag{5-23}$$

Equation (5-23) is called the *combination law of group parameters* and tells us that the group is closed.

4. Associativity.

$$\begin{aligned} R(a)[R(b)R(c)] &= [R(a)R(b)]R(c) \ , \\ \varphi(\varphi(c, b), a) &= \varphi(c, \varphi(b, a)) \ . \end{aligned} \tag{5-24}$$

5. The parameters c in (5-23) are analytic functions of a and b and $\overline{a}$ in (5-21) are analytic functions of a.

A Lie group is said to be *compact* if its parameters are bounded.

Example 1: The real linear transformation group $GL(2, R)$ in two-dimensional space

$$\begin{pmatrix} x' \\ y' \end{pmatrix} = \begin{pmatrix} a_{11} & a_{12} \\ a_{21} & a_{22} \end{pmatrix} \begin{pmatrix} x \\ y \end{pmatrix}, \quad R(a) = \begin{pmatrix} a_{11} & a_{12} \\ a_{21} & a_{22} \end{pmatrix} \ . \tag{5-25}$$

The totality of all 2×2 nonsingular matrices $R(a)$ forms a real linear transformation group under matrix multiplication. Its elements are labeled by four real parameters $(a_{11}, a_{12}, a_{21}, a_{22})$. The order is four. If we restrict ourselves to the transformations with $\det(R(a)) = 1$, the corresponding group is called the special real linear transformation group of dimension two, and designated by $SL(2, R)$.

Example 2: The complex linear transformation group $GL(2, C)$ in two-dimensional space. If the parameters a in (5-25) are complex, $R(a)$ form a complex linear transformation group of dimension 2. Let $a_{kl} = b_{kl} + ic_{kl}$, where b_{kl} and c_{kl} are real. Therefore its elements are characterized by eight real parameters, $a^1 = b_{11}, \ldots, a^8 = c_{22}$. The order is eight.

Example 3: The group SU_2. If the matrices in (5-25) are unitary, i.e.,

$$U = \begin{pmatrix} a & b \\ c & d \end{pmatrix} = \begin{pmatrix} a_{11} & a_{12} \\ a_{21} & a_{22} \end{pmatrix} , \tag{5-26a}$$

$$U^{\dagger} U = 1 , \tag{5-27}$$

then the totality of matrices (5-26a) forms the unitary group U_2. If we further restrict ourselves to

$$\det (U) = 1 , \tag{5-28}$$

the corresponding group is called the special unitary group SU_2.

From (5-26a) and (5-28) we have

$$U^{-1} = \begin{pmatrix} d & -b \\ -c & a \end{pmatrix} . \tag{5-26b}$$

Moreover from (5-27), (5-26a) and (5-28), we have

$$d = a^*, \quad c = -b^* , \quad |a|^2 + |b|^2 = 1 .$$

Therefore the most general form of the group elements of SU_2 is

$$U = \begin{pmatrix} e^{i\xi} \cos \eta & -e^{i\varsigma} \sin \eta \\ e^{i\varsigma} \sin \eta & e^{-i\xi} \cos \eta \end{pmatrix} . \tag{5-29}$$

It contains three real parameters ξ, η and ς. Thus the order of SU_2 is three.

Example 4: The 2-dimensional rotation group R_2. A point $P(x, y)$ in the $x - y$ plane goes over to another point $P'(x', y')$ through a rotation of angle φ about z axis. From Fig. 5.2, we have

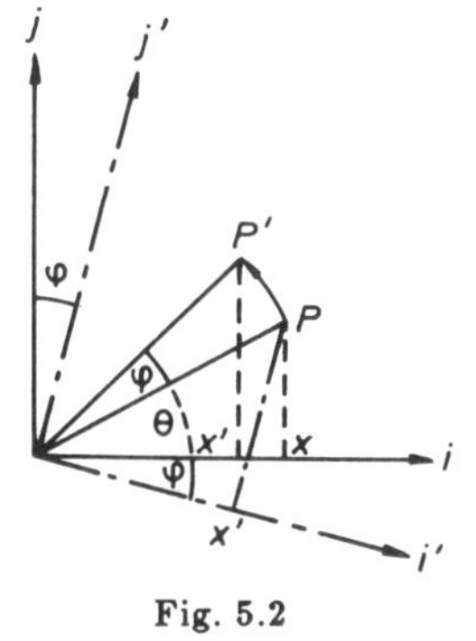

Fig. 5.2

$$\begin{pmatrix} x' \\ y' \end{pmatrix} = \begin{pmatrix} r\cos(\varphi + \theta) \\ r\sin(\varphi + \theta) \end{pmatrix} = R_z(\varphi) \begin{pmatrix} x \\ y \end{pmatrix} , \tag{5-30a}$$

$$R_z(\varphi) = \begin{pmatrix} \cos\varphi & -\sin\varphi \\ \sin\varphi & \cos\varphi \end{pmatrix} . \tag{5-30b}$$

$R_z(\varphi)(\varphi = 0 - 2\pi)$ constitute the 2-dimensional rotation group R_2. $R_z(\varphi)$ in (5-30b) is identical to the rep $D(\varphi)$ of (2-60) carried by the basis $\varphi_1(x) = x$ and $\varphi_2(x) = y$.

There is only one real parameter, thus $r = 1$. $R_z(\varphi)$ of (5-30b) is an orthogonal matrix, i.e.,

$$R_z(\varphi)\tilde{R}_z(\varphi) = I , \quad \tilde{R}_z^{-1} = R_z(\varphi) . \tag{5-31}$$

Consequently, R_2 is also called the special orthogonal group SO_2 of dimension 2.

Comparing (5-27) with (5-31) it is seen that if the unitary transformations are restricted to be real, the unitary group degenerates to the orthogonal group. For example, (5-29) goes over to (5-30b) when $\xi = \varsigma = 0$, and $\eta = \varphi$.

From Fig. 5.2 it is seen that if the point P is kept fixed and the coordinates axes are rotated through the angle $-\varphi$, then the same relation (5-30) holds between the coordinates x', y' of the same point P in the new axes i' and j', and its old coordinates x, y.

It is easy to see that the hierachies of the groups are $GL(2,C) \supset RL(2,R) \supset R_2, GL(2,C) \supset U_2 \supset SU_2 \supset R_2$.

Example 5: The 3-dimensional rotation group R_3. In analogy with (5-31), the transformation matrices for rotations through angles α, β, γ about the x, y, z axes, respectively, are

$$R_x(\alpha) = \begin{pmatrix} 1 & 0 & 0 \\ 0 & \cos\alpha & -\sin\alpha \\ 0 & \sin\alpha & \cos\alpha \end{pmatrix}, \quad R_y(\beta) = \begin{pmatrix} \cos\beta & 0 & \sin\beta \\ 0 & 1 & 0 \\ -\sin\beta & 0 & \cos\beta \end{pmatrix},$$

$$R_z(\gamma) = \begin{pmatrix} \cos\gamma & -\sin\gamma & 0 \\ \sin\gamma & \cos\gamma & 0 \\ 0 & 0 & 1 \end{pmatrix}. \tag{5-32}$$

Alternatively, we can first rotate the point P through angle γ about the z axis, then rotate through angle β about the y axis, and finally rotate through angle α about the z axis. (α, β, γ) are the Euler angles defined by Rose (1957). After these three rotations, the point P goes over to the point P'. The relation between the coordinates of P and P' is

$$\begin{pmatrix} x' \\ y' \\ z' \end{pmatrix} = \mathcal{D}(\alpha\beta\gamma) \begin{pmatrix} x \\ y \\ z \end{pmatrix}, \tag{5-33}$$

$$\begin{aligned} \mathcal{D}(\alpha,\beta,\gamma) &= R(\alpha,\beta,\gamma) = R_z(\alpha)R_y(\beta)R_z(\gamma) \\ &= \begin{pmatrix} \cos\alpha\cos\beta\cos\gamma - \sin\alpha\sin\gamma, & -\cos\alpha\cos\beta\sin\gamma - \sin\alpha\cos\gamma, & \cos\alpha\sin\beta \\ \sin\alpha\cos\beta\cos\gamma + \cos\alpha\sin\gamma, & -\sin\alpha\cos\beta\sin\gamma + \cos\alpha\cos\gamma, & \sin\alpha\sin\beta \\ -\sin\beta\cos\gamma, & \sin\beta\sin\gamma, & \cos\beta \end{pmatrix}. \end{aligned} \tag{5-34}$$

$\mathcal{D}$ is an orthogonal matrix $\mathcal{D}^{-1} = \tilde{\mathcal{D}} \cdot R_3$ is also called the special orthogonal group SO_3 of dimension three.

Let us now consider a new coordinate system $\overline{xyz}$ which is obtained from rotating successively the original coordinate system through angles γ, β and α about the axes z, y, and z, respectively (see the diagram in Bohr 1969, p. 76). From the discussion in Example 4 we know that the relation between the old and new coordinates is given by

$$\begin{pmatrix} \bar{x} \\ \bar{y} \\ \bar{z} \end{pmatrix} = \mathcal{D}^{-1}(\alpha\beta\gamma) \begin{pmatrix} x \\ y \\ z \end{pmatrix} = \tilde{\mathcal{D}}(\alpha\beta\gamma) \begin{pmatrix} x \\ y \\ z \end{pmatrix}. \tag{5-35}$$

5.3. Lie Algebra

5.3.1. Generators of Lie groups

The great contribution of Sophus Lie to the theory of Lie groups was to consider those elements which differ infinitesimally from the identity, and to show that from them one can obtain most of the properties of the Lie group.

We begin with the Taylor expansion of the group elements $R(a)$,

$$R(a) = R(0) + a^\rho X_\rho + \dots , \tag{5-36}$$

where

$$X_\rho = \left(\frac{\partial R(a)}{\partial a^\rho} \right)_{a=0} , \tag{5-37}$$

are called generators of the Lie group. For a Lie group of order r, there are r linearly independent generators. To explore the neighborhood of the identity, we only need retain terms linear in a in (5-36), i.e.,

$$R(a) = 1 + a^\rho X_\rho . \tag{5-38}$$

The inverse element is

$$R^{-1}(a) = 1 - a^\rho X_\rho . \tag{5-39}$$

Suppose that there are two infinitesimal elements and each has only one nonvanishing parameter,

$$R(a) = 1 + \varepsilon X_\rho , \quad R(b) = 1 + \varepsilon X_\sigma . \tag{5-40}$$

According to the definition of the Lie group,

$$R(a)R(b) = R(c) = 1 + C^\tau X_\tau ,$$

$$R(b)R(a) = R(c') = 1 + C'^\tau X_\tau$$

$$\therefore \quad [R(a), \; R(b)] = \varepsilon^2 C^\tau_{\rho\sigma} X_\tau ,$$

$$C^\tau_{\rho\sigma} = ({}^\tau - C'^\tau)/\varepsilon^2 . \tag{5-41}$$

On the other hand from (5-40) we have

$$[R(a), \; R(b)] = \varepsilon^2 [X_\rho, X_\sigma] . \tag{5-42}$$

Comparing (5-41) with (5-42), we get an important relation:

$$[X_\rho, X_\sigma] = C^\tau_{\rho\sigma} X_\tau , \tag{5-43}$$

namely, the commutator of two generators is a linear combination of the r generators. $C^\tau_{\rho\sigma}$ are called the *structure constants* of the Lie group. They have the following two properties.

1. They are antisymmetric with respect to the subscripts.

$$C^\tau_{\rho\sigma} = -C^\tau_{\sigma\rho} . \tag{5-44}$$

2. According to the Jacobi identity

$$[[X_\rho, X_\sigma], X_\tau] + [[X_\sigma, X_\tau], X_\rho] + [[X_\tau, X_\rho], X_\sigma] = 0 , \tag{5-45a}$$

we have

$$C^\mu_{\rho\sigma} C^\nu_{\mu\tau} + C^\mu_{\sigma\tau} C^\nu_{\mu\rho} + C^\mu_{\tau\rho} C^\nu_{\mu\sigma} = 0 . \tag{5-45b}$$

The r generators span a real r-dimensional vector space $\mathcal{L}_r$. Any vector in the space can be expressed as $a^\rho X_\rho$. The product of two basis vectors in the space is defined by their commutator (5-43). The set $\{X_\rho\}$ is thus closed under linear combinations and multiplications defined by (5-43), i.e., $\{X_\rho\}$ constitutes an algebra and is called the *Lie algebra* corresponding to the given Lie group. If a^ρ are real, it is called a real algebra, otherwise it is a complex Lie algebra.

Thanks to the contribution of Lie, the searching of the irreps of the Lie group with infinite number of elements is reduced to that of the Lie algebra with finite number of elements. Having found irreps of the Lie algebra, the irreps of the Lie group are also known accordingly. Therefore the Lie algebra plays a crucial role in the theory of Lie groups. For a given Lie group, we always first find its corresponding Lie algebra. In physical problems, it often occurs that a certain kind of Lie algebra emerges naturally; nevertheless, the corresponding Lie group does not have a simple physical meaning. In such cases, we only deal with the Lie algebra and do not bother about the related Lie group at all.

In the space $\mathcal{L}_r = \{X_\rho : \rho = 1, 2, \ldots r\}$, any vector can be expressed as

$$X = a^\rho X_\rho \, , \tag{5-46}$$

where a^ρ can be thought of as the coordinates of an abstract vector X. According to (5-2), the basis vectors and the coordinates transform in the following ways:

$$X'_\rho = B^\sigma_\rho X_\sigma \, , \quad a'^\rho = A^\rho_\sigma a^\sigma , \quad A = \tilde{B}^{-1} \, . \tag{5-47a, b, c}$$

In the new coordinate system with the basis $\{X'_\rho\}$, the structure constants are $C'^\tau_{\rho\sigma}$,

$$[X'_\rho, X'_\sigma] = C'^\tau_{\rho\sigma} X'_\tau \, . \tag{5-48}$$

From (5-9b), the relation between the new and old structure constants is

$$C'^\tau_{\rho\sigma} = B^\mu_\rho B^\nu_\sigma A^\tau_\lambda C^\lambda_{\mu\nu} \, . \tag{5-49}$$

Equation (5-48) shows that the Lie algebra of the same Lie group may take different forms due to the different choices of the group parameters. This point merits special attention when we are dealing with the classification of Lie algebras.

Example 1: The group $GL(2, R)$. Using (5-25) and (5-37) we get the four generators

$$\begin{aligned} X_1 &= e_{11} = \begin{pmatrix} 1 & 0 \\ 0 & 0 \end{pmatrix} , \quad X_2 = e_{12} = \begin{pmatrix} 0 & 1 \\ 0 & 0 \end{pmatrix} , \\ X_3 &= e_{21} = \begin{pmatrix} 0 & 0 \\ 1 & 0 \end{pmatrix} , \quad X_4 = e_{22} = \begin{pmatrix} 0 & 0 \\ 0 & 1 \end{pmatrix} . \end{aligned} \tag{5-50}$$

It is easy to show that they obey the following commutators:

$$[e_{\alpha\beta}, e_{\gamma\delta}] = \delta_{\beta\gamma} e_{\alpha\delta} - \delta_{\alpha\delta} e_{\gamma\beta} \, . \tag{5-51}$$

Example 2: The group SO_2. From (5-30b) and (5-37) we obtain

$$X_\varphi = \begin{pmatrix} 0 & -1 \\ 1 & 0 \end{pmatrix} . \tag{5-52}$$

Example 3: The group SO_3. From (5-32) and (5-37) we have

$$X_1 = \begin{pmatrix} 0 & 0 & 0 \\ 0 & 0 & -1 \\ 0 & 1 & 0 \end{pmatrix} , \quad X_2 = \begin{pmatrix} 0 & 0 & 1 \\ 0 & 0 & 0 \\ -1 & 0 & 0 \end{pmatrix} , \quad X_3 = \begin{pmatrix} 0 & -1 & 0 \\ 1 & 0 & 0 \\ 0 & 0 & 0 \end{pmatrix} . \tag{5-53a}$$

They obey the commutation relations

$$[X_1, X_2] = X_3 \quad \text{cyclic in 1,2,3} \, . \tag{5-53b}$$

5.4. Finite Transformations

Equation (5-38) is the expression for infinitesimal transformations. Now let us find the expression for finite transformations.

Consider first the single parameter group SO_2. The counterparts of (5-36) and (5-37) are

$$R(\delta\varphi) = 1 + \delta\varphi X_\varphi , \quad X_\varphi = \begin{pmatrix} 0 & -1 \\ 1 & 0 \end{pmatrix} , \tag{5-54}$$

Let the infinitesimal angle $\delta\varphi = \varphi/N$, where N is an arbitrarily large number. Therefore

$$R(\delta\varphi) \cong \left(1 + \frac{\varphi}{N} X_\varphi\right) .$$

Applying $R(\delta\varphi)$ N times, we obtain the finite rotation

$$\begin{aligned} R(\varphi) &\cong \left(1 + \frac{\varphi}{N} X_\varphi\right)^N = \sum_{n=0}^{N} \binom{N}{n} \left(\frac{\varphi}{N} X_\varphi\right)^n \\ &\underset{N\to\infty}{\longrightarrow} 1 + \varphi X_\varphi + \frac{\varphi^2}{2!} X_\varphi^2 + \frac{\varphi^3}{3!} X_\varphi^3 + \ldots \\ &= \cos\varphi \begin{pmatrix} 1 & 0 \\ 0 & 1 \end{pmatrix} + \sin\varphi \begin{pmatrix} 0 & -1 \\ 1 & 1 \end{pmatrix} = \begin{pmatrix} \cos\varphi & -\sin\varphi \\ \sin\varphi & \cos\varphi \end{pmatrix} . \end{aligned} \tag{5-55}$$

It goes back to the familiar form (5-30b). Equation (5-55) can be written formally as

$$R(\varphi) = e^{\varphi X_\varphi} . \tag{5-56}$$

In the above discussion we ignored the unchanged z-component. If the z-component is included, then the generator X_φ in (5-52) goes over to X_3 in (5-53a). Letting $X_3 = -iJ_z$, we get the representative matrix of the operator J_z in the Cartesian coordinate system as shown in (5-58b). The group elements of SO_2 thus take the well-known form

$$R_z(\varphi) = e^{-i\varphi J_z} . \tag{5-57}$$

Analogously, for SO_3 we introduce

$$X_1 = -iJ_x , \quad X_2 = -iJ_y , \quad X_3 = -iJ_z . \tag{5-58a}$$

From (5-53) we have

$$J_x = \begin{pmatrix} 0 & 0 & 0 \\ 0 & 0 & -i \\ 0 & i & 0 \end{pmatrix} , \quad J_y = \begin{pmatrix} 0 & 0 & i \\ 0 & 0 & 0 \\ -i & 0 & 0 \end{pmatrix} , \quad J_z = \begin{pmatrix} 0 & -i & 0 \\ i & 0 & 0 \\ 0 & 0 & 0 \end{pmatrix} . \tag{5-58b}$$

$$[J_x, J_y] = iJ_z , \quad \text{cyclic in } x, y, z . \tag{5-59}$$

$J_{x,y,z}$ are the three components of angular momentum. Equation (5-58b) is their matrix representation in the 3-dimensional Cartesian basis.

The rotation operators corresponding to (5-32) are

$$R_x(\alpha) = e^{-i\alpha J_x} , \quad R_y(\beta) = e^{-i\beta J_y} , \quad R_z(\gamma) = e^{-\gamma\gamma J_z} . \tag{5-60}$$

The operator for a rotation through angle φ about an axis $\mathbf{n}$ with orientation angle (θ', φ') can be expressed as

$$R_{\mathbf{n}}(\varphi) = e^{-i\varphi\mathbf{n}\cdot\mathbf{J}} = \exp\left[-i\varphi(J_x \sin\theta' \cos\varphi' + J_y \sin\theta' \sin\varphi' + J_z \cos\theta')\right] . \tag{5-61}$$

Such a rotation can be written as a product of two rotations

$$R_{\mathbf{n}}(\varphi) = R(\varphi', \theta', 0) R(\varphi, -\theta', -\varphi') , \tag{5-62}$$

namely, first rotate the $\mathbf{n}$ axis onto the z axis, then rotate through angle φ about the z axis, and finally bring the z axis back to the $\mathbf{n}$ axis. Using (5-34) we can get the matrix form of the rotation $R_{\mathbf{n}}(\varphi)$ in the 3-dimensional space x, y and z.

The transition from the infinitesimal transformation (5-54) to the finite transformation can be extended to the more general case

$$R(\delta a) \cong 1 + a^\rho X_\rho, \quad R(a) = \exp(a^\rho X_\rho) . \tag{5-63a, b}$$

It should be mentioned that it is not always possible to write the finite transformation in the form (5-63b). If the transformation can be put in this form, then the group parameters a^ρ are said to be *canonical*. For example in (5-61) $a_x = \varphi \sin\theta' \cos\varphi', a_y = \varphi \sin\theta' \sin\varphi', a_z = \varphi\cos\theta'$ are canonical parameters. If we choose the Euler angles α, β and γ as the group parameters of SO_3, from (5-34) and (5-60), we have

$$R(\alpha, \beta, \gamma) = e^{-i\alpha J_z} e^{-i\beta J_y} e^{-i\gamma J_z} . \tag{5-64}$$

Since J_y and J_z do not commute,

$$R(\alpha, \beta, \gamma) \neq e^{-i(\alpha J_z + \beta J_y + \gamma J_z)} .$$

Therefore the Euler angle α, β and γ are not canonical parameters.

5.5. Correspondence between Lie Groups and Lie Algebras

The classifications of Lie groups and Lie algebras are in one-to-one correspondence. This correspondence is based on the following two relations (5-65) and (5-66). Let R_ρ, R_σ be two infinitesimal elements. Making an expansion of (5-63b) and retaining terms up to ε^2, we obtain

$$R_\rho \cong 1 + \varepsilon X_\rho + \frac{\varepsilon^2}{2!} X_\rho^2, \quad R_\sigma \cong 1 + \varepsilon X_\sigma + \frac{\varepsilon^2}{2!} X_\sigma^2 .$$

Therefore

$$[R_\rho, R_\sigma] = \varepsilon^2 [X_\rho, X_\sigma] = \varepsilon^2 C^\tau_{\rho\sigma} X_\tau . \tag{5-65}$$

$$R_\rho R_\sigma R_\rho^{-1} R_\sigma^{-1} = 1 + \varepsilon^2 [X_\rho, X_\sigma] = 1 + \varepsilon^2 C^\tau_{\rho\sigma} X_\tau . \tag{5-66}$$

According to the above two relations it is easy to establish the following correspondences:

Lie groups	Lie algebras
1a. Abelian Lie groups $[R_\rho, R_\sigma] = 0 ,\quad \rho, \sigma = 1, 2, \ldots, r .$ (5-67a)	1b. Abelian Lie algebras $[X_\rho, X_\sigma] = 0 ,\quad \rho, \sigma = 1, 2, \ldots, r .$ (5-67b)
2a. Subgroups G_s of a Lie group G. Let $X_i, X_j, \ldots, X_k$ be the generators of G_s. Let $R_i = 1 + \varepsilon X_i$, $R_j = 1 + \varepsilon X_j$. Therefore $R_i R_j \in G_s .$ (5-68a)	2b. Subalgebras A_s of a Lie algebra A. By (5-68a), $[R_i, R_j]$ is an element of the group algebra of G_s. Using (5-65) we know that $X_i, X_j, \ldots, X_k$ form a subalgebra A_s of A, i.e., $[X_i, X_j] \in A_s .$ (5-68b)

3a. Invariant subgroup.
If $R_i, R_j, \ldots, R_k$ belong to an invariant subgroup G_s, one has from (1-28)

$$R_\rho R_i R_\rho^{-1} \in G_s \ , R_i \in G_s \ , \quad \rho = 1, 2, \ldots, r$$

Thus

$$R_\rho R_i R_\rho^{-1} R_i^{-1} \in G_s \ . \tag{5-69a}$$

3b. Invariant subalgebras.
From (5-69a) and (5-66) it is known that

$$[X_a, X_\rho] = C_{a\rho}^b X_b \ , \quad a, b = i, j, \ldots, k \ , \quad \rho = 1, 2, \ldots, r \ . \tag{5-69b}$$

The algebra $X_i, \ldots, X_k$ is called the invariant subalgebra of A.

4a. Simple Lie group.
The Lie group which has no invariant subgroups is a simple Lie group.
5a. Semi-simple Lie group.
The Lie group which has no Abelian invariant subgroups is a semi-simple Lie group.

4b. Simple Lie algebra
The Lie algebra which has no invariant subalgebra is a simple Lie algebra.
5b. Semi-simple Lie algebra.
The Lie algebra which has no Abelian invariant subalgebras is a semi-simple Lie algebra.

6a. **Theorem 5.1:** A semi-simple Lie group is a direct product of a set of simple Lie groups,

$$G = G_1 \times G_2 \times \ldots \times G_n \ , \tag{5-70a}$$

where G_i are simple and $[G_i, G_j] = 0$.

6b. **Theorem 5.1′:** A semi-simple Lie algebra is a direct sum of a set of simple Lie algebras,

$$A = A_1 \oplus A_2 \oplus \ldots \oplus A_n \ , \tag{5-70b}$$

where A_i are simple, $[A_i, A_j] = 0$ and the intersections between any A_i and A_j are zeroes.

7. A *Compact Lie algebra* is one corresponding to a compact Lie group.

It is important to distinguish between the semi-simple and non-semi-simple Lie groups, since Abelian invariant subgroups, though apparently the easiest to deal with, can actually be most troublesome from the point of view of representations. Fortunately, in most cases of physical applications we only deal with semi-simple Lie groups. Below we mainly concern ourselves with semi-simple Lie groups. (The criteria for semi-simple Lie group is given in Sec. 5.12.)

In a semi-simple Lie algebra the maximum number of linearly independent generators, denoted as $H_1, \ldots, H_l$, which commute with one another, is called the *rank* of the Lie algebra or the rank of the corresponding Lie group, designated by l. (An equivalent definition of the rank will be given in Sec. 5.17.)

Naturally, any Lie group must be at least of rank 1.

Example 1: For SO_2, there is only one generator J_z. Naturally J_z commutes with itself. Therefore SO_2 is an Abelian group with rank $l = 1$.

Example 2: For SO_3, there are three generators J_x, J_y and J_z. Each of them only commutes with itself. SO_3 is a non-Abelian group of rank 1.

Both SO_2 and SO_3 are simple.

5.6 Linear Transformation Groups

In Secs. 5.2 and 5.3, we gave the general definitions of Lie groups and Lie algebras. In Sec. 5.2 we also gave some simple examples. Below we will extend these examples to the general linear transformation groups. These groups are the most useful ones in physics. Assume $R(a) = R(a^1, a^2, \ldots, a^r)$ is an n-dimensional linear transformation,

$$x \xrightarrow{R(a)} x' = R(a)x \ , \tag{5-71a}$$

or equivalently

$$x'^{\alpha} = R_{\alpha\beta}(a)x^{\beta}, \quad \alpha = 1, 2, \ldots, n\,. \tag{5-71b}$$

Here x may be real or complex. The set of all $n \times n$ matrices $R(a)$ forms a linear transformation group in n-dimensional space. It can be further classified into the following categories:

1. $GL(n, C)$, the *general complex linear transformation group.* The matrix elements $R_{\alpha\beta}(a)$ are complex numbers. The group contains $2n^2$ real parameters; therefore the order is $r = 2n^2$.

2. $GL(n, R)$, the *general real linear transformation group.* The matrix elements are restricted to real numbers. There are n^2 real parameters. The order is $r = n^2$.

3. $SL(n, C), SL(n, R)$, the *special linear transformation groups.* These two groups are obtained from $GL(n, C)$ and $GL(n, R)$ by requiring that the determinants of the transformations be unity. Their orders are equal to $2n^2 - 2$ and $n^2 - 1$, respectively. Obviously we have

$$GL(n, C) \supset SL(n, C) \supset SL(n, R), \quad GL(n, R) \supset SL(n, R).$$

4. U_n and SU_n, the *unitary group* and *unimodular unitary group* in n dimensions. Restricting matrices $R(a)$ to be unitary, i.e.,

$$R(a)R^{\dagger}(a) = R^{\dagger}(a)R(a) = I\,, \tag{5-72a}$$

we get the unitary group U_n of order $r = n^2$. The unitary group is compact, since by (5-72a) the matrix elements $|R_{\alpha\beta}(a)| \leq 1$. The condition (5-72a) also stipulates that

$$\det R(a) = \exp(i\varphi)\,. \tag{5-72b}$$

Demanding that the determinants of $R(a)$ equal unity, we obtain the unimodular unitary group SU_n of order $r = n^2 - 1$. The unitarity (5-72a) ensures that the quantity $\sum_{\alpha=1}^{n} |x^{\alpha}|^2$ is an invariant under the unitary transformation,

$$\sum_{\alpha=1}^{n} |x^{\alpha}|^2 = \sum_{\alpha=1}^{n} |x'^{\alpha}|^2\,. \tag{5-73}$$

The fundamental role of unitary groups in quantum mechanics is easily understood when one realizes that the probability nature of this theory requires a preservation of squares of absolute values of various inner product of wave functions.

5. The group $U(n, m)$. All the linear transformations which keep the quantity

$$\sum_{\alpha=1}^{n} |x^{\alpha}|^2 - \sum_{\beta=n+1}^{n+m} |x^{\beta}|^2 \tag{5-74}$$

invariant form the group $U(n, m)$ with order $r = (n+m)^2$. $U(n, m)$ is a noncompact group. Obviously, $U_n = U(n, 0) = U(0, n)$, $GL(n, m) \supset U(n, m)$. Similarly we can define the group $SU(n, m)$, with order $r = (n+m)^2 - 1$.

6. The complex orthogonal group $O(n, C)$. All the complex linear transformations which leave $\sum_{\alpha=1}^{n} (x^{\alpha})^2$ invariant form the complex orthogonal group. From

$$\sum_{\alpha=1}^{n} (x'^{\alpha})^2 = \sum_{\alpha\beta\beta'} R_{\alpha\beta}R_{\alpha\beta'}x^{\beta}x^{\beta'} = \sum_{\beta=1}^{n} (x^{\beta})^2\,, \tag{5-75}$$

we have

$$\sum_{\alpha} R_{\alpha\beta}R_{\alpha\beta'} = \delta_{\beta\beta'}\,. \tag{5-76a}$$

Thus $R(a)$ are orthogonal matrices,

$$\tilde{R}(a)R(a) = 1 \, . \tag{5-76b}$$

$O(n, C)$ has $n(n-1)/2$ complex parameters (see Sec. 5.8), therefore it is of order $r = n(n-1)$. From (5-76b) we have

$$\det(\tilde{R}(a))\det(R(a)) = 1 \, , \quad \det(R(a)) = \pm 1 \, . \tag{5-76c}$$

The transformation matrices of $O(n, C)$ can be divided into two sets, one is associated with $\det(R(a)) = +1$, and the other with $\det(R(a)) = -1$. The set with determinant $+1$ forms a subgroup–the unimodular complex orthogonal group $SO(n, C)$, representing proper rotations. We can decompose the group $O(n, C)$ into cosets with respect to the subgroup $SO(n, C)$, i.e.,

$$O(n, C) = SO(n, C) \oplus SO(n, C) \times I \, , \tag{5-77}$$

where I is the space inversion operator.

The quotient group $O(n, C)/SO(n, C)$ is a group of order 2. The set with determinant -1 represents rotation-reflection. Any element of $SO(n, C)$ can be reached from the identity via continuous paths in parameter space, while the elements with $\det(R(a)) =$ -1 cannot. In other words, the group $O(n, C)$ consists of two disconnected parts and we cannot go from one part to the other continuously.

7. The real orthogonal group O_n. Restricting the matrices of $O(n, C)$ to be real leads to the real orthogonal group, denoted by O_n or $O(n)$, which is of order $r = \frac{1}{2}n(n-1)$. By further requiring $\det(R(a)) = 1$, we get the unimodular orthogonal group SO_n. It is still of order $\frac{1}{2}n(n-1)$. Similarly, the group O_n also consists of two disconnected parts. Obviously we have

$$O(n, C) \supset SO(n, C) \supset SO_n \, , \quad O_n \supset SO_n \, .$$

8. The group $O(n, m)$. All the real linear transformation which leave the quantity

$$\sum_{\alpha=1}^{n} (x^\alpha)^2 - \sum_{\beta=n+1}^{n+m} (x^\beta)^2 \tag{5-78}$$

invariant form the group $O(n, m)$ with order $r = \frac{1}{2}[n(n-1) + m(m-1)] + nm$. $O(n, m)$ is a noncompact group. The Lorentz group $O(3, 1)$ is a special case of $O(n, m)$.

Obviously, we have $O_n = O(n, 0) = O(0, n)$.

9. The complex sympletic, real sympletic and unitary sympletic groups $SP(2n, C)$, $SP(2n, R)$, and SP_{2n}. Suppose $\mathbf{x} = (x^1, \ldots, x^n; x^{-1}, \ldots, x^{-n})$ and $\mathbf{y} = (y^1, \ldots, y^n; y^{-1}, \ldots y^{-n})$ are two column vectors with dimension $2n$ and $R(a)$ are $2n \times 2n$ matrices, which transform $\mathbf{x}$ and $\mathbf{y}$ into $\mathbf{x}'$ and $\mathbf{y}'$:

$$\mathbf{x}' = R(a)\mathbf{x}, \quad \mathbf{y}' = R(a)\mathbf{y} \, . \tag{5-79a}$$

The sympletic group is the set of all $2n \times 2n$ linear transformations $R(a)$ which leave the skew-symmetric bilinear form

$$\sum_{\alpha=1}^{n} (x^\alpha y^{-\alpha} - x^{-\alpha} y^\alpha) \tag{5-79b}$$

invariant. If the $2n\times 2n$ matrices $R(a)$ are complex (real), it is called the *complex* (real) *sympletic group* of order $2n(2n+1)(n(2n+1))$. If the complex matrices $R(a)$ are unitary, it is called the *unitary sympletic group* SP_{2n}. We have

$$GL(2n, C) \supset SP(2n, C) \supset SP(2n, R), \quad SP(2n, C) \supset SP_{2n}, \quad SU_{2n} \supset SP_{2n} \, .$$

$SP(2n, C)$ and $SP(2n, R)$ are noncompact, while SP_{2n} is compact.

5.7. Infinitesimal Operators for Linear Transformation Groups

If $x^1, x^2, \ldots, x^n$ are subjected to an infinitesimal transformations

$$x' = R(a)x, \quad R(a) = 1 + \mathcal{A}(a) \,, \tag{5-80a}$$

$$\mathcal{A}(a) = \sum_{\alpha\beta} a_{\alpha\beta} e_{\alpha\beta} \,, \tag{5-81}$$

where $a_{\alpha\beta}$ are infinitesimal quantities and $e_{\alpha\beta}$ are the $n \times n$ matrices defined in (2-4). $e_{\alpha\beta}$ obey the commutator (5-51) and the following relation

$$e_{\alpha\beta} e_{\gamma\delta} = \delta_{\beta\gamma} e_{\alpha\delta} \,. \tag{5-82}$$

Equation (5-80a) can also be rewritten as

$$x'^{\alpha} = x^{\alpha} + a_{\alpha\beta} x^{\beta} \,. \tag{5-80b}$$

Under the transformation (5-80b), an arbitrary function $\psi(x)$ goes over to

$$\psi'(x) = \psi(x') = \psi(x^{\alpha} + a_{\alpha\beta} x^{\beta}) = \psi(x) + a_{\alpha\beta} x^{\beta} \frac{\partial}{\partial x^{\alpha}} \psi(x) \,. \tag{5-83}$$

Defining

$$E_{\beta\alpha} = x^{\beta} \frac{\partial}{\partial x^{\alpha}} \,, \tag{5-84}$$

Eq. (5-83) reads

$$\psi'(x) = (1 + a_{\alpha\beta} E_{\beta\alpha}) \psi(x) = (1 + a^{\rho} X_{\rho}) \psi(x) \,, \tag{5-85}$$

and (5-80b) can be expressed as

$$x'^{\alpha} = \left(1 + \sum_{\beta} a_{\alpha\beta} E_{\beta\alpha} \right) x^{\alpha} \,. \tag{5-80c}$$

From this we obtain a simple method for finding the infinitesimal operators of the linear transformation group:

1. First find the infinitesimal matrix $\mathcal{A}$ in the infinitesimal transformations (5-80a), i.e.,

$$\mathcal{A} = \sum_{\alpha\beta} a_{\alpha\beta} e_{\alpha\beta} \,. \tag{5-86a}$$

Notice that not all the parameters $a_{\alpha\beta}$ are independent, except for the group $GL(n, R)$ or $GL(n, C)$.

2. Replacing $e_{\alpha\beta}$ by the differential operator $E_{\beta\alpha}$ of (5-84); Eq. (5-86a) becomes

$$\sum_{\alpha\beta} a_{\alpha\beta} E_{\beta\alpha} \,. \tag{5-86b}$$

3. Expressing the parameters $a_{\alpha\beta}$ in terms of the independent parameters $a^1, a^2, \ldots, a^r$ and setting in turn

$$a^{\rho} = 1, \quad a^{\sigma} = 0, \quad \text{for } \sigma \neq \rho, \quad \rho = 1, 2, \ldots, r,$$

we can obtain the r infinitesimal operators X_{ρ}.

Example. Find the infinitesimal operators of SO_3. According to (5-76), we have

$$\tilde{R}(a)R(a) = [1 + \tilde{\mathcal{A}}(a)[1 + \mathcal{A}(a)] \cong 1 + \tilde{\mathcal{A}}(a) + \mathcal{A}(a) = 1 .$$

Therefore

$$\tilde{\mathcal{A}}(a) + \mathcal{A}(a) = 0 . \tag{5-87}$$

Thus $\mathcal{A}(a)$ is an antisymmetric matrix, and must have the form:

$$\begin{aligned} \mathcal{A}(a) &= \begin{pmatrix} 0 & a^3 & -a^2 \\ -a^3 & 0 & a^1 \\ a^2 & -a^1 & 0 \end{pmatrix} \\ &= a^1(e_{23} - e_{32}) + a^2(e_{31} - e_{13}) + a^3(e_{12} - e_{21}) \\ &\rightarrow a^1(E_{32} - E_{23}) + a^2(E_{13} - E_{31}) + a^3(E_{21} - E_{12}) . \end{aligned} \tag{5-88}$$

Letting $(a^1, a^2, a^3) = (1,0,0), (0,1,0)$ and $(0, 0, 1)$, we obtain the three infinitesimal operators of SO_3.

$$\begin{aligned} X_1 &= E_{32} - E_{23} = z\frac{\partial}{\partial y} - y\frac{\partial}{\partial z} , \\ X_2 &= E_{13} - E_{31} = x\frac{\partial}{\partial z} - z\frac{\partial}{\partial x} , \\ X_3 &= E_{21} - E_{12} = y\frac{\partial}{\partial x} - x\frac{\partial}{\partial y} . \end{aligned} \tag{5-89}$$

If we set $X_k = -iJ_k$, we get the differential form of the angular momentum operators:

$$J_x = i\left(z\frac{\partial}{\partial y} - y\frac{\partial}{\partial z}\right) , \quad \text{cyclic in } x, y, z . \tag{5-90}$$

From the above example it is seen that the infinitesimal operators can be written down immediately once the infinitesimal matrix $\mathcal{A}$ was known.

Comparing (5-58b) with (5-90), the former represents the angular momentum operators in the Cartesian basis, while the latter describe the angular momentum operators acting on wave function $\psi(x, y, z)$. It is easy to show that the differential operators $E_{\alpha\beta}$ in (5-84) and the matrices $e_{\alpha\beta}$ in (2-4) have the same commutators:

$$[E_{\alpha\beta}, E_{\gamma\delta}] = \delta_{\beta\gamma}E_{\alpha\delta} - \delta_{\alpha\delta}E_{\gamma\beta} . \tag{5-91}$$

If the n single particle states $\varphi_1, \varphi_2, \ldots, \varphi_n$ are chosen as the basis in an n-dimensional space, the linear transformation (5-80c) becomes

$$\varphi'_a = \varphi_a + \sum_\beta a_{\alpha\beta}\varphi_\beta . \tag{5-92}$$

Introduce the creation operator $C^\dagger_\alpha$ and annihilation operator C_α by

$$C^\dagger_\alpha|0\rangle = \varphi_\alpha , \quad C_\alpha|0\rangle = 0 . \tag{5-93}$$

They obey the commutators

$$C^\dagger_\alpha C^\dagger_\beta \pm C^\dagger_\beta C^\dagger_\alpha = 0 , \tag{5-94}$$

$$C_\alpha C^\dagger_\beta \pm C^\dagger_\beta C_\alpha = \delta_{\alpha\beta} , \tag{5-95}$$

where the plus sign is for fermions and minus sign for bosons. Equation (5-92) can be rewritten as

$$\varphi'_\alpha = \left(1 + \sum_\beta a_{\alpha\beta} C^\dagger_\beta C_\alpha\right) \varphi_a \ . \tag{5-96}$$

Comparing (5-96) with (5-80c), we see that acting on the single particle states, the infinitesimal operators $E_{\beta\alpha}$ take the form,

$$E_{\beta\alpha} = C^\dagger_\beta C_\alpha \ . \tag{5-97}$$

In the future, we no longer distinguish between the generator $e_{\alpha\beta}$ and the infinitesimal operator $E_{\alpha\beta}$. Depending on the case under study $e_{\alpha\beta}$ may take the following different forms:

$$e_{\alpha\beta}: \quad x^\alpha \frac{\partial}{\partial x^\beta}, \quad C^\dagger_\alpha C_\beta \ , \quad \overset{\beta\text{-th column}}{\begin{pmatrix} & \vdots & \\ \cdots & 1 & \cdots \\ & \vdots & \end{pmatrix}} \ \alpha\text{-th row} \ . \tag{5-98}$$

The form (5-97) is the most convenient one for calculating commutators.

5.8. Metric Tensor in n-Dimensional Space and Infinitesimal Operators

In Sec. 5.7 it was demonstrated that once the infinitesimal matrix $\mathcal{A}(a)$ is known, the infinitesimal operators can be found immediately. We are now going to discuss the problem of how to construct the infinitesimal matrix from the metric tensor $g_{\alpha\beta}$ in n-dimensional space.

Given a set of basis vectors $\{\mathbf{u}_\alpha\}$ and a metric tensor $g_{\alpha\beta}$ in an n-dimensional space, an invariant is specified, either in the form of

$$x^{\alpha *} g_{\alpha\beta} x^\beta = x'^{\alpha *} g_{\alpha\beta} x'^\beta \tag{5-99a}$$

for the sesquilinear metric, or

$$x^\alpha g_{\alpha\beta} x^\beta = x'^\alpha g_{\alpha\beta} x'^\beta \tag{5-99b}$$

for the bilinear metric. From Sec. 5.6 we know that all the transformations $x' = R(a)x$ which leave a quantity invariant form a Lie group. Therefore a metric tensor corresponds to a Lie group. For the sesquilinear metric, from (5-99a) we have

$$x^{\alpha *} g_{\alpha\delta} x^\delta = x'^{\beta *} g_{\beta\gamma} x'^\gamma = R^*_{\beta\alpha}(a) x^{\alpha *} g_{\beta\gamma} R_{\gamma\delta}(a) x^\delta \ . \tag{5-100}$$

Therefore

$$g_{\alpha\delta} = R^*_{\beta\alpha}(a) g_{\beta\gamma} R_{\gamma\delta}(a) \ . \tag{5-101}$$

i.e.,

$$g = R^\dagger(a) g R(a) \ . \tag{5-102}$$

Substituting

$$R(a) = 1 + iB(a) \tag{5-102'}$$

into (5-102), where $B(a)$ is an infinitesimal matrix, and neglecting the second order term, we have

$$B^\dagger(a) g = g\, B(a) \ . \tag{5-103}$$

Analogously, for the bilinear metric, from (5-99b) we have

$$g = \tilde{R}(a)\, g\, R(a) \ . \tag{5-104}$$

Letting $R = 1 + \mathcal{A}(a)$ we have

$$\tilde{\mathcal{A}}(a)g = -g\mathcal{A}(a) \ . \tag{5-105}$$

which may be written as

$$a_{\alpha\beta}g_{\alpha\gamma} = -g_{\beta\alpha}a_{\alpha\gamma} \ . \tag{5-106}$$

Equations (5-103) and (5-105) from our starting points for determining the infinitesimal matrices from the given metric tensors.

5.8.1. Unitary groups

Comparing (5-99a) with (5-73) it is seen that the sesquilinear symmetric metric

$$g_{\alpha\beta} = \delta_{\alpha\beta} \tag{5-107}$$

corresponds to the unitary group. According to (5-103) and (5-107), the infinitesimal matrix $B(a)$ must be hermitian, i.e.,

$$B^\dagger(a) = B(a) \ . \tag{5-108}$$

As an example, for the group $U_3, B(a)$ takes the form

$$B(a) = \begin{pmatrix} c_1 & a_1 - ib_1 & a_2 - ib_2 \\ a_1 + ib_1 & c_2 & a_3 - ib_3 \\ a_2 + ib_2 & a_3 + ib_3 & c_3 \end{pmatrix} \ . \tag{5-109}$$

By setting one of the nine parameters $a_1, \ldots, c_3$ equal to 1 and all the others equal zeros successively, we obtain the following nine infinitesimal operators.

$$X_1 = \begin{pmatrix} 0 & 1 & 0 \\ 1 & 0 & 0 \\ 0 & 0 & 0 \end{pmatrix} , \quad X_2 = \begin{pmatrix} 0 & -i & 0 \\ i & 0 & 0 \\ 0 & 0 & 0 \end{pmatrix} , \quad X_3 = \begin{pmatrix} 0 & 0 & 1 \\ 0 & 0 & 0 \\ 1 & 0 & 0 \end{pmatrix} ,$$
$$X_4 = \begin{pmatrix} 0 & 0 & -i \\ 0 & 0 & 0 \\ i & 0 & 0 \end{pmatrix} , \quad X_5 = \begin{pmatrix} 0 & 0 & 0 \\ 0 & 0 & 1 \\ 0 & 1 & 0 \end{pmatrix} , \quad X_6 = \begin{pmatrix} 0 & 0 & 0 \\ 0 & 0 & -i \\ 0 & i & 0 \end{pmatrix} ,$$
$$X_7' = \begin{pmatrix} 1 & 0 & 0 \\ 0 & 0 & 0 \\ 0 & 0 & 0 \end{pmatrix} , \quad X_8' = \begin{pmatrix} 0 & 0 & 0 \\ 0 & 1 & 0 \\ 0 & 0 & 0 \end{pmatrix} , \quad X_9' = \begin{pmatrix} 0 & 0 & 0 \\ 0 & 0 & 0 \\ 0 & 0 & 1 \end{pmatrix} . \tag{5-110}$$

For unitary groups we often choose real infinitesimal operators $e_{\alpha\beta}$ [see (5-81)],

$$R(a) = 1 - i\sum_{\alpha\beta} a_{\alpha\beta}e_{\alpha\beta} \ . \tag{5-111}$$

Now the parameters $a_{\alpha\beta}$ become complex and obey $a^*_{\alpha\beta} = a_{\beta\alpha}$.

It is readily seen that the unit matrix $I = \sum_{\alpha=1}^n e_{\alpha\alpha}$ is an invariant subalgebra of U_n; therefore U_n is not semi-simple.

Since $x_\alpha x^\alpha = \sum_\alpha (x^\alpha)^* x^\alpha$ = invariant for U_n, the covariant variables x_α are the complex conjugate of the contravariant variables x^α,

$$x_\alpha = x^{\alpha *} \ . \tag{5-112}$$

We can also introduce the mixed-type metric tensor g^β_α through the invariant

$$x^\alpha x_\alpha = x^\alpha g^\beta_\alpha x_\beta \ . \tag{5-113}$$

Therefore

$$g_\alpha^\beta = \delta_{\alpha\beta} \,. \tag{5-114}$$

It is easy to show that g_α^β is an invariant, i.e.,

$$g_\alpha'^\beta = R_{\beta\delta}R^*_{\alpha\gamma}g_\gamma^\delta = R_{\beta\delta}R^*_{\alpha\delta} = \delta_{\alpha\beta} \,. \tag{5-115}$$

A mixed tensor can be constructed out of p contravariant vectors and q covariant vectors:

$$T^{\alpha\beta\ldots\gamma}_{ij\ldots k} = x^\alpha y^\beta \ldots z^\gamma u_i v_j \ldots w_k \,. \tag{5-116}$$

An antisymmetric contravariant tensor is defined by

$$\varepsilon^{i_1 i_2 \ldots i_n} = \begin{array}{ll} 1, & \text{if } p = \begin{pmatrix} 1 & 2 \ldots n \\ i_1 & i_2 \ldots i_n \end{pmatrix} \text{ is } \begin{array}{l} \text{even} \,, \\ \text{odd} \,, \end{array} \\ -1 & \\ 0 & \text{if any two indices are equal} \end{array} \tag{5-117}$$

An antisymmetric covariant tensor $\varepsilon_{i_1,i_2\ldots i_n}$ can be defined in the same way. Under the U_n transformation

$$\begin{aligned} \varepsilon^{i_1 i_2 \ldots i_n} &\to \varepsilon'^{i_1 i_2 \ldots i_n} \\ &= \sum_{j_1 \ldots j_n} R_{i_1 j_1}(a) R_{i_2 j_2}(a) \ldots R_{i_n j_n}(a) \varepsilon^{j_1 j_2 \ldots j_n} \\ &= \det\big(R(a)\big)\varepsilon^{i_1 i_2 \ldots i_n} \,. \end{aligned} \tag{5-118}$$

Thus the antisymmetric contravariant tensor $\varepsilon^{i_1,i_2\ldots i_n}$ is at the same time an invariant of SU_n. The same is true for the antisymmetric covariant tensor $\varepsilon_{i_1 i_2 \ldots i_n}$.

5.8.2. Infinitesimal operators of SU_n

The group SU_n demands that $\det\big(R(a)\big) = 1$. For infinitesimal elements we have

$$\det\big(R(a)\big) = \det(1 + iB) = 1 + i\sum_\alpha B_{\alpha\alpha} + \ldots = 1 \,. \tag{5-119}$$

Therefore

$$\text{Trace}\, B = 0 \,. \tag{5-120}$$

Thus for SU_2, B takes the form

$$B = \begin{pmatrix} c & a - ib \\ a + ib & -c \end{pmatrix} \,. \tag{5-121}$$

Letting $(a, b, c) = (1,0,0), (0,1,0), (0,0,1)$, we get three infinitesimal operators

$$\sigma_x = \begin{pmatrix} 0 & 1 \\ 1 & 0 \end{pmatrix} , \quad \sigma_y = \begin{pmatrix} 0 & -i \\ i & 0 \end{pmatrix} , \quad \sigma_z = \begin{pmatrix} 1 & 0 \\ 0 & -1 \end{pmatrix} \,. \tag{5-122}$$

These are the Pauli matrices.

Of the nine operators $X_1, \ldots, X_9$ in (5-110), the first six are already traceless, while X_7', X_8' and X_9' can be combined into a unit matrix and two traceless matrices, which can be chosen as

$$X_7 = \begin{pmatrix} 1 & 0 & 0 \\ 0 & -1 & 0 \\ 0 & 0 & 0 \end{pmatrix} , \quad X_8 = \sqrt{\frac{1}{3}} \begin{pmatrix} 1 & 0 & 0 \\ 0 & 1 & 0 \\ 0 & 0 & -2 \end{pmatrix} \tag{5-123}$$

$X_1, \ldots, X_6, X_7, X_8$ are the eight infinitesimal operators used by Gell-Mann (1964) in the quark model. The choice of traceless matrices is not unique.

For the group SU_n, the infinitesimal operators can also be chosen to be real. The nondiagonal operators are $e_{\alpha\beta}$, while the diagonal ones can be chosen in many ways. The most usual choices are as follows:

1.

$$h_i = e_{ii} - \frac{1}{n}I = \begin{pmatrix} -1/n & & & & 0 \\ & \ddots & & & \\ & & (n-1)/n & & \\ & & & \ddots & \\ & & & & -1/n \end{pmatrix} \quad i\text{-th row} \tag{5-124}$$

$$\sum_{i=1}^{n} h_i = 0 . \tag{5-125}$$

Clearly only $(n-1)$ h_i's are independent.

2.

$$h_i = \left(\sum_{j=1}^{i} e_{jj} - i e_{i+1,i+1} \right) /(i+1), \quad i = 1, 2, \ldots, n-1. \tag{5-126}$$

For SU_4 for instance, we have

$$h_1 = \frac{1}{2}\begin{pmatrix} 1 & 0 & 0 & 0 \\ 0 & -1 & 0 & 0 \\ 0 & 0 & 0 & 0 \\ 0 & 0 & 0 & 0 \end{pmatrix} . \quad h_2 = \frac{1}{3}\begin{pmatrix} 1 & 0 & 0 & 0 \\ 0 & 1 & 0 & 0 \\ 0 & 0 & -2 & 0 \\ 0 & 0 & 0 & 0 \end{pmatrix} , \quad h_3 = \frac{1}{4}\begin{pmatrix} 1 & 0 & 0 & 0 \\ 0 & 1 & 0 & 0 \\ 0 & 0 & 1 & 0 \\ 0 & 0 & 0 & -3 \end{pmatrix} . \tag{5-127}$$

SU_n is simple and belongs to the classical group A_{n-1} with rank $l = n-1$, since we have $n-1$ commuting infinitesimal operators.

5.8.3. The group $U(n,m)$

From (5-74) and (5-99a) one sees that the matrix tensor is

$$g_{\alpha\beta} = \delta_\alpha \delta_{\alpha\beta}, \quad \delta_\alpha = \begin{cases} 1 , & \alpha = 1, 2, \ldots, n \\ -1 , & \alpha = n+1, n+2, \ldots, \ldots, n+m \end{cases} . \tag{5-128}$$

According to (5-103) and (5-128), the infinitesimal elements of the group $U(n,m)$ must take the form

$$R(a) = 1 + i \left(\begin{array}{c|c} B_n & B_{nm} \\ \hline B_{mn} & B_m \end{array}\right) , \tag{5-129}$$

$$B_n = B_n^\dagger , \quad B_m = B_m^\dagger , \quad B_{nm} = -B_{mn}^\dagger , \tag{5-130}$$

where B_n, B_m and B_{nm} are $n\times n$, $m\times m$ and $n\times m$ matrices, respectively. From (5-129), it is easy to write down the infinitesimal operators of $U(n,m)$. $U(n,m)$ is of order $r = n^2 + m^2 + 2nm = (n+m)^2$. For the group $SU(n,m)$, we have the further restriction

$$\mathrm{Tr}\, B_n + \mathrm{Tr}\, B_m = 0 , \tag{5-131}$$

so that

$$r = (n+m)^2 - 1 . \tag{5-132}$$

5.8.4. Orthogonal group O_n

1. Cartesian basis

Comparing (5-75) with (5-99b), we know that the bilinear symmetric metric

$$g_{\alpha\beta} = \delta_{\alpha\beta} \tag{5-133}$$

corresponds to the orthogonal group. Therefore the covariant and contravariant vectors coincide. From (5-105) and (5-133) we know that $\mathcal{A}$ is an antisymmetric matrix

$$\tilde{\mathcal{A}} = -\mathcal{A} \tag{5-134}$$

with $r = \frac{1}{2}n(n-1)$ real parameters. Consequently

$$\mathcal{A} = \sum_{\alpha>\beta=1}^{n} a_{\beta\alpha}(e_{\beta\alpha} - e_{\alpha\beta}) \to \sum_{\alpha>\beta=1} a_{\beta\alpha}(E_{\alpha\beta} - E_{\beta\alpha}) \ . \tag{5-135}$$

The infinitesimal operators of O_n follow from (5-135) and (5-84):

$$L_{\alpha\beta} = -L_{\beta\alpha} = x_\alpha \frac{\partial}{\partial x_\beta} - x_\beta \frac{\partial}{\partial x_\alpha} = C_\alpha^\dagger C_\beta - C_\beta^\dagger C_\alpha \ . \tag{5-136}$$

Using (5-91) we obtain the commutator relation

$$[L_{\alpha\beta}, L_{\gamma\delta}] = \delta_{\beta\gamma} L_{\alpha\delta} + \delta_{\alpha\delta} L_{\beta\gamma} + \delta_{\beta\delta} L_{\gamma\alpha} + \delta_{\alpha\gamma} L_{\delta\beta} \ . \tag{5-137}$$

Equation (5-137) is the extension of (5-89) of O_3.

2. Spherical basis

In physics we often use the spherical basis. For example in three-dimensional space we take

$$x^1 = -\sqrt{\frac{1}{2}}(x+iy), \quad x^0 = z, \quad x^{-1} = \sqrt{\frac{1}{2}}(x-iy) \tag{5-138}$$

as the contravariant variables. They are the three components of an $l = 1$ wave function.

For the general cases, let ψ_{jm} be the eigenfunctions of $\mathbf{J}^2$ and J_z and

$$x^\alpha = \psi_{j\alpha}, \quad \text{for odd} \quad n = 2j+1 \ , \tag{5-139a}$$

$$x^\alpha = \psi_{j,\alpha\mp1/2}, \quad \text{for even} \quad n = 2j+1, \ \alpha \neq 0 \ . \tag{5-139b}$$

In the last equation the negative (positive) sign is for $\alpha > 0 (\alpha < 0)$. The O_n invariant is

$$\sum_{\alpha=-[\frac{n}{2}]}^{[\frac{n}{2}]} (-1)^\alpha x^\alpha x^{-\alpha} = \text{Invariant} \ , \tag{5-140}$$

where $[\frac{n}{2}]$ is the integer part of $\frac{n}{2}$. Notice that $\alpha \neq 0$ when n is even. Therefore, we have the metric tensor

$$g_{\alpha\beta} = (-1)^\alpha \delta_{\alpha,-\beta} \ . \tag{5-141}$$

From (5-106) and (5-141), we have

$$(-1)^\gamma a_{-\gamma\beta} = -(-1)^\beta a_{-\beta\gamma} \ . \tag{5-142}$$

Changing the indices $-\gamma \to \alpha, \gamma \to -\alpha$, yields

$$(-1)^\alpha a_{\alpha\beta} = -(-1)^\beta a_{-\beta,-\alpha} \ . \tag{5-143a}$$

Therefore $a_{00} = 0$ and

$$\begin{aligned} \mathcal{A}(a) &= \sum_{\alpha\beta} a_{\beta\alpha} e_{\beta\alpha} = \sum_{\beta\geq 0,\alpha} (a_{\beta\alpha} e_{\beta\alpha} + a_{-\alpha-\beta} e_{-\alpha-\beta}) \\ &= \sum_{\beta\geq 0,\alpha} (-1)^\alpha a_{\beta\alpha}[(-1)^\alpha e_{\beta\alpha} - (-1)^\beta e_{-\alpha-\beta}] \\ &\to \sum_{\beta\geq 0,\alpha} (-1)^\alpha a_{\beta\alpha}[(-1)^\alpha E_{\alpha\beta} - (-1)^\beta E_{-\beta-\alpha}] \ . \end{aligned} \tag{5-143b}$$

The infinitesimal operators of O_n under the metric (5-142) are

$$L_{\alpha\beta} = -L_{-\beta-\alpha} = (-1)^\alpha x^\alpha \frac{\partial}{\partial x^\beta} - (-1)^\beta x^{-\beta} \frac{\partial}{\partial x^{-\alpha}} \ . \tag{5-144a}$$

They satisfy the commutators

$$[L_{\alpha\beta}, L_{\gamma\delta}] = (-1)^\gamma \delta_{\beta\gamma} L_{\alpha\delta} + (-1)^\delta \delta_{\alpha\delta} L_{-\beta-\gamma} + (-1)^\gamma \delta_{-\alpha\gamma} L_{-\delta\beta} + (-1)^\delta \delta_{-\beta\delta} L_{\gamma-\alpha} \ . \tag{5-144b}$$

Among these are $[\frac{n}{2}]$ commuting operators:

$$H_\alpha = L_{\alpha\alpha}, \quad \alpha = 1, 2, \dots, [\frac{n}{2}] \ . \tag{5-144c}$$

The group O_{2l+1} belongs to the classical group B_l of rank l and the group O_{2l} belongs to the classical group D_l of rank l (see Sec. 5.19).

Comparing (5-140) with the two-particle wave function of zero orbital angular momentum

$$\Psi_{L=0} = \sum_{m=-l}^{l} (-1)^{l-m} \psi_{lm}(1)\psi_{l-m}(2) \ , \tag{5-145}$$

we know that the $L = 0$ state (5-145) is invariant under the O_{2l+1} transformation. However, for half-integer j, the zero angular momentum state

$$\Psi_{J=0} = \sum_{m=-j}^{j} (-1)^{j-m} \psi_{jm}(1)\psi_{j-m}(2) \tag{5-146}$$

is not invariant under the O_{2j+1}, transformations. The group which keeps (5-146) invariant is the sympletic group $SP(2j, C)$ (see 5.8.6).

Example 1: The group SO_2. The index α only takes two values, +1 and -1. From (5-144a) we obtain a single infinitesimal operator

$$-L_{11} = \begin{pmatrix} 1 & 0 \\ 0 & -1 \end{pmatrix} = \sigma_z \ .$$

σ_z forms an Abelian Lie algebra.

Example 2: The group SO_3. From (5-144a) we obtain three infinitesimal operators L_{10}, L_{-10} and L_{11}, designated as $-J_+ - J_-$ and $-J_0$, respectively,

$$J_+ = \begin{pmatrix} 0 & 1 & 0 \\ 0 & 0 & 1 \\ 0 & 0 & 0 \end{pmatrix} , \quad J_- = \begin{pmatrix} 0 & 0 & 0 \\ 1 & 0 & 0 \\ 0 & 1 & 0 \end{pmatrix} , \quad J_0 = \begin{pmatrix} 1 & 0 & 0 \\ 0 & 0 & 0 \\ 0 & 0 & -1 \end{pmatrix} \ . \tag{5-147}$$

They obey the commutators

$$[J_+, J_-] = J_0, \quad [J_0, J_\pm] = \pm J_\pm \ . \tag{5-148}$$

Comparison with the known commutators of angular momentum, we know that

$$J_\pm = \sqrt{\frac{1}{2}}[J_x \pm iJ_y], \quad J_0 = J_z \ . \tag{5-149}$$

J_0 forms a subalgebra of the Lie algebra (J_+, J_-, J_0).

5.8.5 *The real orthogonal group $O(n,m)$*

The metric tensor is determined by the invariant (5-78),

$$g_{\alpha\beta} = \delta_\alpha \delta_{\alpha\beta}\,, \quad \delta_\alpha = \begin{cases} 1, & \alpha = 1,2,\ldots,n \\ -1, & \alpha = n+1,\ldots,n+m\,. \end{cases} \tag{5-150}$$

From (5-106) and (5-150) we have

$$\delta_\alpha a_{\alpha\beta} = -\delta_\beta a_{\beta\alpha}\,. \tag{5-151}$$

Therefore

$$A(a) = \begin{pmatrix} A_n & A_{nm} \\ \tilde{A}_{nm} & A_m \end{pmatrix}, \quad A_n = -\tilde{A}_n, \quad A_m = -\tilde{A}_m\,. \tag{5-152}$$

The antisymmetric matrices A_n and A_m have $\frac{1}{2}n(n-1)$ and $\frac{1}{2}m(m-1)$ parameters, respectively and A_{nm} has nm parameters. Therefore the order of the group $O(n,m)$ is

$$r = \frac{1}{2}[n(n-1) + m(m-1)] + mn = \frac{1}{2}(m+n)(m+n-1)\,. \tag{5-153}$$

With the help of (5-151), we can immediately write down the infinitesimal operators of $O(n,m)$, i.e.,

$$L_{\alpha\beta} = \delta_\alpha x^\alpha \frac{\partial}{\partial x^\beta} - \delta_\beta x^\beta \frac{\partial}{\partial x^\alpha} = \delta_\alpha C_\alpha^\dagger C_\beta - \delta_\beta C_\beta^\dagger C_\alpha\,. \tag{5-154}$$

Example 3: The Lorentz group $O(3,1)$. Here $n=3, m=1$ and $r=6$. The six infinitesimal operators are

$$L_{ij} = x^i \frac{\partial}{\partial x^j} - x^j \frac{\partial}{\partial x^i}, \quad i,j = 1,2,3\,.$$

$$L_{i4} = x^i \frac{\partial}{\partial x^4} + x^4 \frac{\partial}{\partial x^i}, \quad i = 1,2,3\,. \tag{5-155}$$

The complex orthogonal group $O(n,m,C)$ can be handled in the same way. The infinitesimal matrices $A(a)$ still has the form (5-152). However now $A(a)$ is complex and the number of parameters is doubled.

5.8.6. *Sympletic group*

1. The real sympletic group $SP(2n,R)$

From the invariant (5-79) we obtain the metric tensor

$$\begin{aligned} g_{\alpha\beta} &= \varepsilon_{\alpha\beta} = -\varepsilon_{\beta\alpha} \\ \varepsilon_{\alpha\beta} &= \begin{cases} \delta_\alpha, & \alpha = -\beta\,, \\ 0, & \alpha \neq -\beta\,, \end{cases} \quad \delta_\alpha = -\delta_{-\alpha} = \begin{cases} 1, & \alpha > 0\,, \\ -1, & \alpha < 0\,. \end{cases} \end{aligned} \tag{5-156}$$

Therefore the bilinear antisymmetric metric corresponds to the sympletic group. The infinitesimal matrix satisfies the conditions

$$\delta_\alpha a_{\alpha\beta} = -\delta_\beta a_{-\beta-\alpha}\,. \tag{5-157}$$

Therefore $A(a)$ must be of the form

$$A = \overset{\quad 1\ldots n,\qquad -1\ldots -n}{\begin{pmatrix} A_1 & A_2 \\ A_3 & -\tilde{A}_1 \end{pmatrix}}, \quad A_2 = \tilde{A}_2, \quad A_3 = \tilde{A}_3\,. \tag{5-158}$$

Notice that $\mathcal{A}(a)$ is traceless, therefore the sympletic matrices are unimodular. The infinitesimal operators are

$$L_{\alpha\beta} = \delta_\alpha x^\alpha \frac{\partial}{\partial x^\beta} - \delta_\beta x^{-\beta} \frac{\partial}{\partial x^{-\alpha}} ,$$

$$L_{\alpha\beta} = L_{-\beta-\alpha}, \quad \alpha, \beta = \pm 1, \ldots, \pm n . \tag{5-159}$$

The matrix $\mathcal{A}_1$ has n^2 parameters and the symmetric matrices $\mathcal{A}_2$ and $\mathcal{A}_3$ each has $\frac{1}{2}n(n+1)$ parameters. The group $SP(2n, R)$ is of order

$$r = n^2 + n(n+1) = n(2n+1) . \tag{5-160}$$

The generators obey the commutators

$$[L_{\alpha\beta}, L_{\gamma\delta}] = \delta_{\beta\gamma}\delta_\beta L_{\alpha\delta} - \delta_{\alpha\delta}\delta_\alpha L_{\gamma\beta} + \delta_{\beta\overline{\delta}}\delta_\beta L_{\alpha\overline{\gamma}} - \delta_{\alpha\overline{\gamma}}\delta_\alpha L_{\overline{\beta}\gamma} \tag{5-161}$$

where $\overline{\delta} = -\delta$ etc.

Among the r infinitesimal operators of (5-159), the following n operators H_i commute with one another:

$$H_i = L_{ii}, \quad i = 1, 2, \ldots, n . \tag{5-159$'$}$$

Therefore the rank of $SP(2n, R)$ is n.

Example 4: $SP(4, R)$ has ten parameters. From (5-158) we have

$$\mathcal{A} = \begin{array}{c} \\ 1 \\ 2 \\ -1 \\ -2 \end{array} \begin{array}{c} \begin{array}{cccc} 1 & 2 & -1 & -2 \end{array} \\ \begin{pmatrix} e_1 & \alpha & \gamma & \beta \\ \overline{\alpha} & e_2 & \beta & \delta \\ \overline{\gamma} & \overline{\beta} & -e_1 & \overline{\alpha} \\ \overline{\beta} & \overline{\delta} & -\alpha & -e_2 \end{pmatrix} \end{array} . \tag{5-162}$$

The ten infinitesimal operators of $SP(4, R)$ are

$$\begin{aligned}
H_1 &= x^1 \frac{\partial}{\partial x^1} - x^{-1}\frac{\partial}{\partial x^{-1}} (= L_{11}), & H_2 &= x^2 \frac{\partial}{\partial x^2} - x^{-2}\frac{\partial}{\partial x^{-2}} (= L_{22}), \\
E_\alpha &= x^1 \frac{\partial}{\partial x^2} - x^{-2}\frac{\partial}{\partial x^{-1}} (= L_{12}), & E_\beta &= x^1 \frac{\partial}{\partial x^{-2}} + x^2\frac{\partial}{\partial x^{-1}} (= L_{1\overline{2}}), \\
E_\gamma &= x^1 \frac{\partial}{\partial x^{-1}} (= L_{1\overline{1}}/2), & E_\delta &= x^2 \frac{\partial}{\partial x^{-2}} (= L_{2\overline{2}}/2), \\
E_{\overline{\alpha}} &= x^2 \frac{\partial}{\partial x^1} - x^{-1}\frac{\partial}{\partial x^{-2}} (= L_{21}), & E_{\overline{\beta}} &= x^{-1} \frac{\partial}{\partial x^2} + x^{-2}\frac{\partial}{\partial x^1} (= L_{\overline{1}2}), \\
E_{\overline{\gamma}} &= x^{-1} \frac{\partial}{\partial x^1} (= L_{\overline{1}1}/2), & E_{\overline{\delta}} &= x^{-2} \frac{\partial}{\partial x^2} (= L_{\overline{2}2}/2) .
\end{aligned} \tag{5-163}$$

They are identical to (5-159) except for multiplicative factors.

2. The complex sympletic group $SP(2n, C)$.

The infinitesimal matrix $\mathcal{A}(a)$ of $SP(2n, C)$ still takes the form of (5-158), but now $\mathcal{A}(a)$ is a complex matrix. The infinitesimal operators can still take the form of (5-159), but the parameters $a_{\alpha\beta}$ are now complex. The infinitesimal group elements are expressed as

$$R(a) = 1 + \sum_{\alpha\beta}{}' a_{\alpha\beta} L_{\alpha\beta} . \tag{5-164}$$

The sum runs only over the independent parameters. $SP(2n, C)$ is of order $r = 2n(2n+1)$.

3. Unitary sympletic group SP_{2n}

The infinitesimal group elements of SP_{2n} can be expressed as

$$R(a) = 1 + i \sum_{\alpha\beta}' a_{\alpha\beta} L_{\alpha\beta} , \tag{5-165}$$

where $L_{\alpha\beta}$ are still given by (5-159) and the parameters $a_{\alpha\beta}$ fulfill the hermitian condition

$$a_{\alpha\beta} = a^*_{\beta\alpha} .$$

The infinitesimal matrix takes the form

$$\mathcal{A} = i \begin{pmatrix} \mathcal{A}_1 & \mathcal{A}_2 \\ \mathcal{A}_2^\dagger & -\mathcal{A}_1^* \end{pmatrix} , \quad \mathcal{A}_1 = \mathcal{A}^\dagger, \ \mathcal{A}_2 = \tilde{\mathcal{A}}_2 , \tag{5-166}$$

with order $r = n(2n+1)$ and rank $l = n$. As an example, for the group SP_4, we have

$$\mathcal{A} = i \begin{pmatrix} e_1 & \alpha & \gamma & \beta \\ \alpha^* & e_2 & \beta & \delta \\ \gamma^* & \beta^* & -e_1 & -\alpha^* \\ \beta^* & \delta^* & -\alpha & -e_2 \end{pmatrix} , \tag{5-167}$$

where e_1, e_2 are real and $\alpha, \beta, \gamma, \delta$ are complex. The infinitesimal operators of SP_4 are

$$H_1, \ H_2, \ (E_\rho + E_{\overline{\rho}}) , \ \ i(E_\rho + E_{\overline{\rho}}), \quad \rho = \alpha, \beta, \gamma, \delta , \tag{5-168}$$

where $H_1, H_2, E_\alpha, E_{\overline{\alpha}}, \ldots$ are given by (5-163).

4. The invariant of the sympletic group often takes another form, i.e.

$$\sum_{\alpha=-n}^{n} x^a (-1)^\alpha \varepsilon_{\alpha\beta} y^\beta = \text{Invariant} . \tag{5-169}$$

The corresponding metric tensor is

$$g_{\alpha\beta} = (-1)^\alpha \varepsilon_{\alpha\beta} . \tag{5-170}$$

Equation (5-157) is to be modified to

$$(-1)^\alpha \delta_\alpha a_{\alpha\beta} = -(-1)^\beta \delta_\beta a_{-\beta-\alpha} . \tag{5-171}$$

The infinitesimal operators become

$$L_{\alpha\beta} = (-1)^\alpha \delta_\alpha x^\alpha \frac{\partial}{\partial x^\beta} - (-1)^\beta \delta_\beta x^{-\beta} \frac{\partial}{\partial x^{-\alpha}} . \tag{5-172}$$

If we choose the single-particle wave functions with half-integer j, (5-139b), as the basis for the defining rep of the sympletic group $SP(2j+1, C)$, then the sympletic group $SP(2j+1, C)$ (and therefore its sub-groups $SP(2j+1, R)$ and SP_{2j+1}) leaves the $J = 0$ two-particle state (5-146) invariant. For $j = 3/2$ for instance, from (5-146) and (5-139b), we have

$$\begin{aligned} \Psi_{J=0} &= \frac{1}{2}(\psi_{\frac{3}{2}}\varphi_{-\frac{3}{2}} - \psi_{\frac{1}{2}}\varphi_{-\frac{1}{2}} + \psi_{-\frac{1}{2}}\varphi_{\frac{1}{2}} - \psi_{-\frac{3}{2}}\varphi_{\frac{3}{2}}) \\ &= \frac{1}{2}(x^2 y^{-2} - x^1 y^{-1} + x^{-1} y^1 - x^{-2} y^2) . \end{aligned}$$

The sympletic groups $SP(2n,C), SP(2n,R)$ and SP_{2n} all belong to the classical group C_n.

Ex. 5.1. If the invariant of SO_n is chosen as $\Sigma_{i=-[\frac{n}{2}]}^{[\frac{n}{2}]} x^i x^{-i}$ = invariant, show that the SO_n generators are given by

$$X_{ii}=0, \quad X_{ik}=-X_{ki}=C_i^\dagger C_{-k}-C_k^\dagger C_{-i} \ ,$$
$$i\neq k=\begin{cases}0,\pm1,\pm2,\ldots\pm l\\ \pm1,\pm2,\ldots\pm 2\end{cases}, \quad \text{for } n=\begin{cases}\text{odd}\\ \text{even}\end{cases},$$

and that they satisfy the commutators
$[X_{ik},X_{mn}]=\delta_{k+m}X_{in}-\delta_{k+n}X_{im}+\delta_{i+n}X_{km}-\delta_{i+m}X_{kn}$,
where $\delta_{k+m}=\delta_{k,-m}$.

Ex. 5.2. Show that the generators of SP_{2n} can be chosen as

$$X_{ik}=X_{ki}=\delta_i C_i^\dagger C_{-k}+\delta_k C_k^\dagger C_{-i} \ ,$$

where $\delta_i=+1(-1)$ for $i>0(<0)$ and that they satisfy the commutators

$$[X_{ik},X_{mn}]=\delta_m\delta_{k+m}X_{in}+\delta_n\delta_{k+n}X_{im}+\delta_m\delta_{i+m}X_{kn}+\delta_n\delta_{i+n}X_{km} \ .$$

5.9. Infinitesimal Operators in Group Parameter Space

We use $u(R_a)$ or $u(a)$ to designate functions on the group manifold. The action of Lie group elements on $u(R_a)$ is still defined by (2-87), namely

$$R_b^{-1}u(R_a)=u(R_bR_a) \ . \tag{5-173}$$

Let R_b be an infinitesimal element

$$R_b=R(\delta a)=1+\delta a^\rho X_\rho(a) \ , \tag{5-174a}$$

$$R_b^{-1}=R^{-1}(\delta a)=1-\delta a^\rho X_\rho(a) \ , \tag{5-174b}$$

$$R_bR_a=R(\delta a)R(a)=R(a+da) \ . \tag{5-175}$$

According to the combination law (5-23), we have

$$a^\sigma+da^\sigma=\varphi^\sigma(a,\delta a) \ , \tag{5-176a}$$
$$da^\sigma=\mu_\rho^\sigma(a)\delta a^\rho \ , \tag{5-176b}$$
$$\mu_\rho^\sigma(a)=\frac{\partial\varphi^\sigma(a,b)}{\partial b^\rho}\Big|_{b=0} \ . \tag{5-176c}$$

Rewriting $u(R_a)$ and $u(R_bR_a)$ in (5-173) as $u(a)$ and $u(a+da)$, respectively, and using (5-173), (5-174b) and (5-176), we obtain

$$(1-\delta a^\rho X_\rho(a))u(a)=u(a+da)=u(a)+\frac{\partial u(a)}{\partial a^\sigma}da^\sigma=u(a)+\frac{\partial u(a)}{\partial a^\sigma}\mu_\rho^\sigma(a)\delta a^\rho \ .$$

Since δa^ρ are independent, it follows that

$$X_\rho(a)u(a)=-\mu_\rho^\sigma(a)\frac{\partial}{\partial a^\sigma}u(a) \ . \tag{5-177a}$$

Therefore the infinitesimal operators X_ρ in group parameter space are the differential operators

$$X_\rho(a) = -\mu_\rho^\sigma(a)\frac{\partial}{\partial a^\sigma} \ . \tag{5-177b}$$

Racah (1951) and Eisenhart (1933) called $A_\rho(a) = -X_\rho(a)$ the infinitesimal operators of the first parameter group, while we use the same name for $X_\rho(a)$.

A simple method for obtaining the infinitesimal operators of the first parameter group was given by Xu (1986) based on the generating coordinate method (GCM).

Example: The infinitesimal operator of SO_2. SO_2 is a one-parameter Abelian group. Let $R(\varphi_c) = R(\varphi_b)R(\varphi_a)$; then $\varphi_c = \varphi_a + \varphi_b$. From (5-177b) and (5-176c) we obtain the infinitesimal operators

$$X_\varphi = -\frac{\partial(\varphi_a + \varphi_b)}{\partial\varphi_b}\frac{\partial}{\partial\varphi} = -\frac{\partial}{\partial\varphi} \ .$$

Letting $X = -iJ_z$, we have

$$J_z = \frac{1}{i}\frac{\partial}{\partial\varphi} \ . \tag{5-177c}$$

The example of SO_3 is given in Sec 6.1.

5.10. Isomorphism and Anti-Isomorphism of Lie Groups and Lie Algebras

The Lie algebras $\{X_\rho\}$ and $\{Y_\rho\}$ are said to be isomorphic if they are in one-to-one correspondence and have the same structure constants:

$$[X_\rho, X_\sigma] = C_{\rho\sigma}^\tau X_\tau \ , \tag{5-178a}$$

$$[Y_\rho, Y_\sigma] = C_{\rho\sigma}^\tau Y_\tau \ . \tag{5-178b}$$

Theorem 5.2: Two Lie groups are locally isomorphic in the neighborhood of the identity and are homomorphic globally, if their Lie algebras are isomorphic.

For example, the Lie algebra $\{\sigma_x/2, \sigma_y/2, \sigma_z/2\}$ of SU_2 is isomorphic to the Lie algebra $\{J_x, J_y, J_z\}$ of SO_3; thus the Lie groups SU_2 and SO_3 are locally isomorphic in the neighborhood of the identity.

Below we are going to study the behavior of the groups SO_3 and SU_2 over the whole range of their parameters. With the help of the Pauli matrices (5-122), we easily get the matrix representative of the operator $\boldsymbol{\sigma}\cdot\mathbf{n}$ in the basis $\binom{1}{0}$, $\binom{0}{1}$, which are eigenfunctions of σ_z,

$$\boldsymbol{\sigma}\cdot\mathbf{n} = \begin{pmatrix} \cos\theta & \sin\theta e^{-i\varphi} \\ \sin\theta e^{i\varphi} & -\cos\theta \end{pmatrix} , \tag{5-179a}$$

where θ and φ are the orientation angles of $\mathbf{n}$. The eigenfunctions of $\boldsymbol{\sigma}\cdot\mathbf{n}$ are

$$\chi_{\frac{1}{2}} = \begin{pmatrix} \cos\frac{\theta}{2}e^{-\frac{i}{2}\varphi} \\ \sin\frac{\theta}{2}e^{\frac{i}{2}\varphi} \end{pmatrix} , \quad \chi_{-\frac{1}{2}} = \begin{pmatrix} -\sin\frac{\theta}{2}e^{-\frac{i}{2}\varphi} \\ \cos\frac{\theta}{2}e^{\frac{i}{2}\varphi} \end{pmatrix} . \tag{5-179b}$$

Under the SU_2 transformation [(5-29)], the spin wave functions and the operator $\boldsymbol{\sigma}\cdot\mathbf{n}$ transform as follows

$$\begin{pmatrix} a \\ b \end{pmatrix} \to \begin{pmatrix} a' \\ b' \end{pmatrix} = U\begin{pmatrix} a \\ b \end{pmatrix} , \tag{5-180a}$$

$$(\boldsymbol{\sigma}\cdot\mathbf{n}) \to (\boldsymbol{\sigma}\cdot\mathbf{n})' = U(\boldsymbol{\sigma}\cdot\mathbf{n})U^{-1} \ . \tag{5-180b}$$

The most general form of the transformation matrix of SU_2 is given by (5-29). Putting $\xi = \alpha/2, \eta = \xi = 0$, we have

$$U = A_z(\alpha) = \begin{pmatrix} e^{-i\alpha/2} & 0 \\ 0 & e^{i\alpha/2} \end{pmatrix} . \tag{5-181}$$

According to (5-179) and (5-181), we have

$$(\boldsymbol{\sigma}\bullet\mathbf{n})' = \boldsymbol{\sigma}\bullet\mathbf{n}' = \begin{pmatrix} \cos\theta & \sin\theta e^{-i(\varphi+\alpha)} \\ \sin\theta e^{i(\varphi+\alpha)}, & -\cos\theta \end{pmatrix} . \tag{5-182}$$

$$\chi'_{\frac{1}{2}} = \begin{pmatrix} \cos\frac{\theta}{2}\cdot e^{-\frac{i}{2}(\varphi+\alpha)} \\ \sin\frac{\theta}{2}\cdot e^{\frac{i}{2}(\varphi+\alpha)} \end{pmatrix} , \quad \chi'_{-\frac{1}{2}} = \begin{pmatrix} -\sin\frac{\theta}{2}\cdot e^{-\frac{i}{2}(\varphi+\alpha)} \\ \cos\frac{\theta}{2}\cdot e^{\frac{i}{2}(\varphi+\alpha)} \end{pmatrix} . \tag{5-183}$$

Evidently, $\chi'_{\pm\frac{1}{2}}$ are eigenfunctions of $\boldsymbol{\sigma}\bullet\mathbf{n}'$. Thus the effect of the SU_2 transformation (5-181) on the operator $\boldsymbol{\sigma}\bullet\mathbf{n}$ is equivalent to rotating the unit vector $\mathbf{n}$ through angle α about z axis in three-dimensional space. Therefore the SU_2 matrix $A_z(\alpha)$ of (5-181) corresponds to the SO_3 matrix $R_z(\alpha)$ of (5-30b).

Analogously, in (5-29), by putting $\xi = \varsigma = 0, \eta = \beta/2$, we have

$$U = A_y(\beta) = \begin{pmatrix} \cos\frac{\beta}{2} & -\sin\frac{\beta}{2} \\ \sin\frac{\beta}{2} & \cos\frac{\beta}{2} \end{pmatrix} . \tag{5-184}$$

For simplicity, we let $\varphi = 0$ in (5-179), i.e., let $\mathbf{n}$ be in the xz plane. From (5-179), (5-180) and (5-184) we obtain

$$\boldsymbol{\sigma}\bullet\mathbf{n} = \begin{pmatrix} \cos\theta & \sin\theta \\ \sin\theta & -\cos\theta \end{pmatrix} \rightarrow \boldsymbol{\sigma}\bullet\mathbf{n}' = \begin{pmatrix} \cos(\theta+\beta) & \sin(\theta+\beta) \\ \sin(\theta+\beta) & -\cos(\theta+\beta) \end{pmatrix} , \tag{5-185}$$

$$\chi_{\frac{1}{2}} = \begin{pmatrix} \cos\frac{\theta}{2} \\ \sin\frac{\theta}{2} \end{pmatrix} \rightarrow \chi'_{\frac{1}{2}} = \begin{pmatrix} \cos\left(\frac{\theta+\beta}{2}\right) \\ \sin\left(\frac{\theta+\beta}{2}\right) \end{pmatrix} , \tag{5-186}$$

$$\chi_{-\frac{1}{2}} = \begin{pmatrix} -\sin\frac{\theta}{2} \\ \cos\frac{\theta}{2} \end{pmatrix} \rightarrow \chi'_{-\frac{1}{2}} = \begin{pmatrix} -\sin\left(\frac{\theta+\beta}{2}\right) \\ \cos\left(\frac{\theta+\beta}{2}\right) \end{pmatrix} . \tag{5-187}$$

It follows that the SU_2 transformation $A_y(\beta)$ of (5-184) corresponds to the rotation $R_y(\beta)$ (5-32). In summary, the SU_2 matrix which corresponds to the three-dimensional rotation matrix $\mathcal{D}(\alpha\beta\gamma) = R_z(\alpha)R_y(\beta)R_z(\gamma)$ is equal to

$$D^{\frac{1}{2}}(\alpha\beta\gamma) = A_z(\alpha)A_y(\beta)A_z(\gamma) = \begin{pmatrix} e^{-\frac{i}{2}(\alpha+\gamma)}\cos\frac{\beta}{2} & -e^{-\frac{i}{2}(\alpha-\gamma)}\sin\frac{\beta}{2} \\ e^{\frac{i}{2}(\alpha-\gamma)}\sin\frac{\beta}{2} & e^{\frac{i}{2}(\alpha+\gamma)}\cos\frac{\beta}{2} \end{pmatrix} . \tag{5-188}$$

In fact it is just the $j = 1/2$ representation matrix of SO_3 (Rose 1957).

Moreover, it is easy to prove that if the SU_2 matrices A and B correspond to the rotation matrices R_A and R_B, respectively, then the product AB corresponds to $R_A R_B$. The proof is as follows:

$$A(\boldsymbol{\sigma}\bullet\mathbf{n})A^{-1} = \boldsymbol{\sigma}\bullet R_A\mathbf{n}, \quad B(\boldsymbol{\sigma}\bullet\mathbf{n})B^{-1} = \boldsymbol{\sigma}\bullet R_B\mathbf{n} ,$$

$$AB(\boldsymbol{\sigma}\bullet\mathbf{n})(AB)^{-1} = AB(\boldsymbol{\sigma}\bullet\mathbf{n})B^{-1}A^{-1} = A(\boldsymbol{\sigma}\bullet R_B\mathbf{n})A^{-1} = \boldsymbol{\sigma}\bullet R_A R_B\mathbf{n} .$$

It must be emphasized that the correspondence between the SO_2 and SO_3 group is not one to one. For example, from (5-181) and (5-32) we see that

$$A_z(\alpha+2\pi) = -A_z(\alpha), \quad R_z(\alpha+2\pi) = R_z(\alpha) . \tag{5-189}$$

Rotations through angle α and $\alpha+2\pi$ about any fixed axis are physically identical. Nevertheless they correspond to different elements of the group SU_2. Therefore the isomorphism between the Lie algebras of SU_2 and SO_3 only ensures that SU_2 and SO_3 group are isomorphic in the neighborhood of the identity. They are homomorphic over the whole range of parameters.

The rotation group can be generalized in the following way: we will distinguish the rotation through angle α and $\alpha + 2\pi$ about an axis, but not distinguish the rotations through angle α and $\alpha + 4\pi$. The generalized rotation group is termed as the *double rotation group*, denoted as $SO_3^\dagger$. In this way, we achieved a one-to-one correspondence between the elements of SU_2 and $SO_3^\dagger$. Now each matrix of SU_2 corresponds uniquely to a "rotation" of $SO_3^\dagger$. $SO_3^\dagger$ is isomorphic to SU_2 and SU_2 is called the *universal covering group* of SO_3.

We thus see that the rotation matrices form an unfaithful rep of SU_2, and the homomorphism of SU_2 on SO_3 has Z_2 as its kernel. Conversely, any element of SO_3 can be represented by an SU_2 matrix up to sign. Therefore, we are dealing with a projective rep of SO_3. We can also say that the matrices of SU_2 form a *double-valued rep* of SO_3.

In general, each SO_n group has a universal covering group, called the *spin group*. For SO_3, as we have said it is SU_2; for SO_4, it is the group $SU_2 \times SU_2$; for SO_5 and SO_6, they are SP_4 and SU_4 respectively.

The importance of the universal covering group is that all the reps of a group G can be found from a study of the single-valued reps of its universal covering group.

In summary we have the following isomorphism for compact Lie groups.

$$SO_3 \approx SU_2/Z_2, \quad SO_4 \approx [SU_2 \times SU_2]/Z_2 ,$$
$$SO_5 \approx SP_4/Z_2, \quad SO_6 \approx SU_4/Z_2 .$$

The Lie algebras $\{X_\rho\}$ and $\{Z_\rho\}$ are said to be anti- isomorphic if they have a one-to-one correspondence and corresponding to (5-178a), Z_ρ satisfy the commutator relation

$$[Z_\rho, Z_\sigma] = -C^\tau_{\rho\sigma} Z_\tau . \tag{5-190}$$

Two Lie groups are locally anti-isomorphic in the neighborhood of the identity if their Lie algebras are anti-isomorphic. Thus, if $R(a)$ and $S(a)$ are the elements of the Lie groups generated by $\{X_\rho\}$ and $\{Z_\rho\}$, then in the neighborhood of the identity corresponding to

$$R(a)R(b) = R(c) ,$$

we have

$$S(b)S(a) = S(c) . \tag{5-191}$$

Clearly, if $\{Z_\rho\}$ is anti-isomorphic to $\{X_\rho\}$, then $\{-Z_\rho\}$ is isomorphic to $\{X_\rho\}$.

5.11. Invariant Integration

In the derivation of the theorem for finite groups, we often used the following property

$$\sum_{a=1}^{g} u(R_a) = \sum_{a=1}^{g} u(R_b^{-1} R_a) , \tag{5-192}$$

where R_b is a definite element. (5-192) implies that we attached equal weights to all elements R of a finite group (see (2.73)).

In order to extend the theorems of finite groups to the case of Lie groups, (5-192) must be expressed as an integral over the group parameters a. For this purpose, we have to introduce the *density function* $\rho(a) = \rho(a^1, a^2, \ldots, a^r)$, depending on the r parameters of the Lie group and demand that the integral of the function $u(R_a)$ over the whole domain of the r parameters be invariant under the following parameter transformations

$$\int_G u(R_a)\rho(a)da = \int_G u(R_b^{-1}R_a)\rho(a)da = \int_G u(R_aR_b^{-1})\rho(a)da \ , \qquad da = da^1da^2\ldots da^r \ . \tag{5-193}$$

It can be proved that for compact Lie groups we can find such density functions (Hamermesh 1962). The density function is also called the *weight function.*

The definition of scalar product of functions on the group manifold can be generalized to

$$\langle u_1|u_2\rangle = \sum_a u_1^*(R_a)u_2(R_a) \rightarrow \tag{5-194}$$

$$\langle u_1|u_2\rangle = \int_G u_1^*(a)u_2(a)\rho(a)da \ . \tag{5-195}$$

Weyl proved that the class operator $C(\varphi) = C(\varphi^1, \varphi^2, \ldots, \varphi^l)$ of a compact Lie group with rank l only depends on l class parameters $\varphi^1, \varphi^2, \ldots, \varphi^l$. $C(\varphi)$ is the integration of all the elements belonging to the same class.

$$C(\varphi) = \int_{R(a')\in C(\varphi)} R(a')\rho(a')da' \ . \tag{5-196}$$

The number of elements, g_i, belonging to the class i is replaced by the volume $g(\varphi)$ occupied by the elements of the class φ in the group parameter space,

$$g(\varphi) = \int_{R(a')\in C(\varphi)} \rho(a')da' \ . \tag{5-197}$$

In the class parameter space, we can analogously define a density function

$$\rho(\varphi) = \rho(\varphi^1, \varphi^2, \ldots, \varphi^l) \ , \tag{5-198}$$

so that the scalar product of functions on classes can be extended to

$$\langle q_1|q_2\rangle = \sum_i g_i q_1^*(C_i)q_2(C_i) \rightarrow \tag{5-199}$$

$$\langle q_1|q_2\rangle = \int g(\varphi)q_1^*(\varphi)q_2(\varphi)\rho(\varphi)d\varphi, \qquad d\varphi = d\varphi^1d\varphi^2\ldots d\varphi^l \ . \tag{5-200}$$

The total number of elements of a finite group is replaced by the total *volume of the Lie group*

$$g = \sum_{a=1}^{g} 1 \rightarrow g = \int \rho(a)da = \int \rho(\varphi)d\varphi \int \rho(a')da' \ . \tag{5-201}$$

The method for finding the density function $\rho(a)$ is given by Hamermesh (1962). A simpler method for obtaining the density function of SO_3 will be given in Sec. 6.3.

5.12. Representation of Compact Lie groups

The definition of reps of Lie groups is the same as that of finite groups. A rep is said to be reducible if through a similarity transformation all the rep matrices can be brought to the following form,

$$D(R) = \left(\begin{array}{c|c} D^{(1)}(R) & A(R) \\ \hline 0 & D^{(2)}(R) \end{array}\right) . \tag{5-202}$$

The rep is said to be fully reducible if all the submatrices $A(R) = 0$.

For finite groups, reducible reps are necessarily fully reducible. However reducible reps of Lie groups may not be fully reducible. For example, under the translation operation $x' = x + a$, the basis functions $\varphi_1(x) = 1$ and $\varphi_2(x) = x$ go over to

$$\begin{pmatrix} \varphi_1'(x) \\ \varphi_2'(x) \end{pmatrix} = \begin{pmatrix} 1 \\ x+a \end{pmatrix} = \begin{pmatrix} 1 & 0 \\ a & 1 \end{pmatrix} \begin{pmatrix} 1 \\ x \end{pmatrix} . \tag{5-203a}$$

Thus a rep for the translation group is found to be

$$D(a) = \begin{pmatrix} 1 & a \\ 0 & 1 \end{pmatrix} . \tag{5-203b}$$

The rep $D(a)$ is reducible, but not fully reducible, i.e., we cannot find a similarity transformation which will bring $D(a)$ into diagonal form. Only for fully reducible cases can the study of group reps be reduced to that of all the irreps of the group. For semi-simple Lie groups we have Theorem 5.3.

Theorem 5.3 (Weyl's theorem): Any reducible rep of a semi-simple Lie algebra is fully reducible.

For compact Lie groups, all the theorems which were derived for finite groups in Chapter 3 still hold (Hamermesh 1962). For example, every rep of a compact group is equivalent to a unitary rep. Every rep of a compact group is fully reducible to a sum of irreps, all of which have finite dimensions, and the regular rep contains all irreps.

As we have said, the problem of finding reps of a Lie group can be reduced to that of finding reps of the corresponding Lie algebra, namely, finding r matrices $D(X_\rho)$ for the r infinitesimal operators X_ρ so that they obey the Lie algebra relation

$$[D(X_\rho), D(X_\sigma)] = C^{\tau}_{\rho\sigma} D(X_\tau) . \tag{5-204}$$

5.12.1. Fundamental representation

The linear transformations $R(a)$ in n-dimensional space evidently form a rep by themselves of the linear transformation group. This n-dimensional rep is called the *fundamental* or *defining rep*. Analogously, the r infinitesimal generators (they are $n \times n$ matrices) form the fundamental rep of the Lie algebra. For example, the eight infinitesimal generators $X_1, X_2 \ldots, X_8$ of (5-110) and (5-123) form the fundamental rep of the SU_3 Lie algebra.

5.12.2. Adjoint representation

For finite groups, corresponding to each element R_a of a group G, we can define an operator $\overset{\circ}{R}_a$ by

$$\overset{\circ}{R}_a S = R_a S R_a^{-1} = T, \quad S \in G . \tag{5-205}$$

The set of operators $\overset{\circ}{R}_a$ also forms a rep of the group G, known as the *adjoint rep*. Let S be an infinitesimal operator

$$S = 1 + \varepsilon X_\rho \ . \tag{5-206a}$$

Then the element T in (5-205) is also an infinitesimal operator and can be written as

$$T = 1 + \varepsilon \sum_\sigma \mathcal{D}^{(\nu_0)}_{\sigma\rho}(a) X_\sigma \ . \tag{5-206b}$$

Substituting (5-206) into (5-205) and noting that $\overset{\circ}{R}(a)e = R(a)eR^{-1}(a) = e$, we obtain

$$\overset{\circ}{R}(a)X_\rho = R(a)X_\rho R^{-1}(a) = \sum_\sigma \mathcal{D}^{(\nu_0)}_{\sigma\rho}(a) X_\sigma \ . \tag{5-207}$$

It shows that the r infinitesimal operators $X_1, X_2, \ldots, X_r$ carry an r-dimensional rep of the Lie group G, which is called the *adjoint rep* of the Lie group and designated by (ν_0). The adjoint rep is quite useful in the following.

Suppose that under a Lie group G the r functions $\psi^{(\nu_0)}_\rho$ transform among themselves according to the adjoint rep $\mathcal{D}^{(\nu_0)}$,

$$R(a)\psi^{(\nu_0)}_\rho = \sum_\sigma \mathcal{D}^{(\nu_0)}_{\sigma\rho}(a)\psi^{(\nu_0)}_\sigma \ , \tag{5-208a}$$

$$\mathcal{D}^{(\nu_0)}_{\sigma\rho}(a) = \langle \psi^{(\nu_0)}_\sigma | R(a) | \psi^{(\nu_0)}_\rho \rangle \equiv \langle \nu_0\sigma | R(a) | \nu_0\rho \rangle \ . \tag{5-208b}$$

Letting

$$R(a) = 1 + \delta a^\tau X_\tau, \quad \overset{\circ}{R}(a) = 1 + \delta a^\tau \overset{\circ}{X}_\tau \ , \tag{5-209}$$

Eq. (5-208b) becomes

$$\mathcal{D}^{(\nu_0)}_{\sigma\rho}(a) = \delta_{\sigma\rho} + \delta a^\tau \mathcal{D}^{(\nu_0)}_{\sigma\rho}(X_\tau) \ , \tag{5-210a}$$

$$\mathcal{D}^{(\nu_0)}_{\sigma\rho}(X_\tau) = \langle \nu_0\sigma | X_\tau | \nu_0\rho \rangle \ . \tag{5-210b}$$

Substituting (5-209) and (5-210) into (5-207), we get

$$\overset{\circ}{X}_\tau X_\rho = [X_\tau, X_\rho] = \sum_\sigma \mathcal{D}^{(\nu_0)}_{\sigma\rho}(X_\tau) X_\sigma \ . \tag{5-211}$$

$\overset{\circ}{X}_\tau$ is called the *adjoint operator* of X_τ, which transforms an element X_ρ of the algebra into the element $[X_\tau, X_\rho]$.

Comparing (5-211) with (5-43), we obtain the representatives of the infinitesimal operators X_τ in the adjoint rep:

$$\mathcal{D}^{(\nu_0)}_{\sigma\rho}(X_\tau) = C^\sigma_{\tau\rho} \ . \tag{5-212a}$$

Theorem 5.4: The adjoint rep of a simple algebra is irreducible.

Proof: Since according to Sec. 5.5, a simple Lie algebra $A = \{X_\tau : \tau = 1, 2, \ldots, r\}$ means that among its r infinitesimal operators, we cannot find a subset $A_s = (X_i, X_j, \ldots, X_k), k < r$, such that it is closed under the adjoint operations of the r infinitesimal operators X_τ,

$$\overset{\circ}{X}_\tau X_i \equiv [X_\tau, X_i] \in A_s \ . \tag{5-212b}$$

If the adjoint rep is reducible, then (5-212b) is true, and it contradicts the assumption that the Lie algebra is simple.

Theorem 5.5: The adjoint rep of a semi-simple algebra is faithful.

5.12.3. Metric tensor in the r-dimensional vector space

The r infinitesimal operators $X_\alpha, \alpha = 1, 2, \dots, r$, can be considered as a covariant basis in the r-dimensional space $\mathcal{L}_r$[1)]. The scalar product of any two vectors A and B in $\mathcal{L}_r$ is required to be independent of the original choice of the basis $\{X_\alpha\}$, i.e., it must be defined in terms of a matrix invariant. Therefore, the scalar product of two basis vectors X_α and X_β is defined by

$$g_{\alpha\beta} = (X_\alpha, X_\beta) = \operatorname{tr}(\mathring{X}_\alpha \mathring{X}_\beta) = C^\sigma_{\alpha\rho} C^\rho_{\beta\sigma} . \tag{5-213a}$$

The covariant tensor $g_{\alpha\beta}$ of rank two is the metric tensor in the r-dimensional space $\mathcal{L}_r$. Under the basis transformation (5-47a), it becomes

$$g'_{\alpha\beta} = C'^\sigma_{\alpha\rho} C'^\rho_{\beta\sigma}, \quad g' = Bg\tilde{B} . \tag{5-213b}$$

The scalar product of any two vectors $X_A = a^\alpha X_\alpha$ and $X_B = b^\beta X_\beta$ is given by

$$(X_A, X_B) = a^\alpha b^\beta (X_\alpha, X_\beta) = a^\alpha g_{\alpha\beta} b^\beta . \tag{5-214}$$

Theorem 5.6 (Cartan's Theorem): The necessary and sufficient condition for a Lie algebra to be semi-simple is that the determinant of the metric tensor g does not vanish,

$$\det|g_{\alpha\beta}| \neq 0 . \tag{5-215}$$

From the structure constants $C^\sigma_{\alpha\rho}$ of a Lie algebra, we can construct g, and from (5-215) we can check whether it is semi-simple or not.

We restrict ourselves to semi-simple groups; therefore from $g_{\alpha\beta}$ we can find the contravariant metric tensor $g^{\alpha\beta}$. $g_{\alpha\beta}$ and $g^{\alpha\beta}$ can be used to lower or raise indices:

$$X^\alpha = g^{\alpha\beta} X_\beta , \quad X_\alpha = g_{\alpha\beta} X^\beta . \tag{5-216}$$

Theorem 5.7: The necessary and sufficient condition for a semi-simple algebra to be *compact* is that $\det|g_{\alpha\beta}| < 0$.

5.13. The Invariants and Casimir Operators of Lie Groups

If $I(X_\rho)$ is an operator built out of the r infinitesimal operators and if it is invariant under the group G,

$$I(X'_\rho) = I(R(a) X_\rho R^{-1}(a)) = R(a) I(X_\rho) R^{-1}(a) = I(X_\rho) , \tag{5-217a}$$

namely,

$$[R(a), I(X_\rho)] = 0 , \tag{5-217b}$$

then $I(X_\rho)$ is called an *invariant* of the group G. In other words, $I(X_\rho)$ is a scalar with respect to the group operation.

1) Care must be taken to distinguish the r-dimensional space $\mathcal{L}_r$ from the previous n-dimensional space. For example the eight infinitesimal operators of SU_3, $E_{ij} = x^i \frac{\partial}{\partial x^j} - \frac{1}{3}\delta^i_j x^i \frac{\partial}{\partial x^i}$, $i = 1,2,3$, are the second rank mixed tensor in the ordinary space of dimension n (=3), while in the r (=8)-dimensional space, they are the covariant basis vectors X_α.

It is easy to prove that the invariants must commute with all the infinitesimal operators of G. Assuming that $R(a)$ in (5-217) are infinitesimal transformations,

$$I(X'_\rho) = \exp(\delta a^\sigma X_\sigma) I(X_\rho) \exp(-\delta a^\tau X_\tau) = I(X_\rho) , \tag{5-218}$$

where δa^σ are infinitesimal quantities. Expanding the exponential functions, we obtain

$$[X_\sigma, I(X_\rho)] = 0, \quad \sigma = 1, 2, \ldots, r . \tag{5-219}$$

Conversely, if an operator $I(X_\rho)$ commutes with all the infinitesimal operators, then $I(X_\rho)$ is an invariant of the group G.

With the help of the metric tensor, it is easy to find a quadratic invariant

$$C = X_\alpha g^{\alpha\beta} X_\beta = X_\alpha X^\alpha , \tag{5-220}$$

called the Casimir operator. Using (5-43) and (5-44) it can be shown that

$$[X_\sigma, C] = 0 , \quad \sigma = 1, 2, \ldots, r . \tag{5-221}$$

Utilizing tensor contractions, one can construct higher order invariants. For example

$$C^{(3)} = C^{\beta_2}_{\alpha_1\beta_1} C^{\beta_3}_{\alpha_2\beta_2} C^{\beta_1}_{\alpha_3\beta_3} X^{\alpha_1} X^{\alpha_2} X^{\alpha_3} , \tag{5-222}$$

etc., which are called *generalized Casimir operators.*

Invariant operators are not restricted to the Casimir operators. There are many ways to construct invariants (see Sec. 7.2).

It can be shown that for a semi-simple Lie group of rank l, there exist l invariants $I_1(X_\rho)$, $\ldots, I_l(X_\rho)$ (Racah 1951). The set of operators

$$C = (I_1(X_\rho), \ldots, I_l(X_\rho)) \tag{5-223}$$

forms a CSCO in the class parameter space. We call it the *CSCO of the first kind* (CSCO-I) of the Lie group.

As an example, for the group SO_3 with rank one, there is only one invariant $\mathbf{J}^2 = J_x^2 + J_y^2 + J_z^2$. $\mathbf{J}^2$ is the CSCO-I of SO_3.

5.14. Intrinsic Lie Groups

5.14.1. Definition and intepretation of the intrinsic Lie group

In analogy with finite groups, for any Lie group $G = \{R(a)\}$ we can define an intrinsic Lie group $\overline{G} = \{\overline{R}(a)\}$. The action of the intrinsic group element $\overline{R}(b)$ on any element $R(a)$ in the group space is defined by

$$\overline{R}(b) R(a) = R(a) R(b) . \tag{5-224}$$

The groups $\overline{G}$ and G are commuting and anti-isomorphic. Thus their corresponding Lie algebras obey

$$\begin{gathered} [X_\tau, \overline{X}_\rho] = 0, \quad \tau, \rho = 1, 2, \ldots, r . \\ [X_\tau, X_\rho] = C^\sigma_{\tau\rho} X_\sigma , \quad [\overline{X}_\tau, \overline{X}_\rho] = -C^\sigma_{\tau\rho} \overline{X}_\sigma . \end{gathered} \tag{5-225}$$

From (5-224) we obtain a relation between the intrinsic group element $\overline{R}(b)$ and $R(b)$:

$$\overline{R}(b) = R(a) R(b) R(a)^{-1}, \quad \text{when } R(b) \text{ acts on } R(a) . \tag{5-226a}$$

It should be emphasized that (5-226a) is the defining equation for the operator $\overline{R}(b)$ rather than an identity relation [see the remarks after (3-144)].

We first prove that (5-226a) holds when acting on any function $u(a)$ on the group manifold

$$\overline{R}(b)u(a) = R(a)R(b)R^{-1}(a)u(a) , \tag{5-226b}$$

Proof: According to (3-318), the left-hand side of (5-226b) is

$$\overline{R}(b)u(a) = \overline{R}(b)u\big(R(a)\big) = u\big(R(a)R^{-1}(b)\big) , \tag{5-226c}$$

while according to (2-87b), the right-hand side of (5-226b) is

$$R(a)R(b)R^{-1}(a)u(a) = u\left[\big(R(a)R(b)R^{-1}(a)\big)^{-1}R(a)\right] = u\big(R(a)R^{-1}(b)\big) .$$

Therefore (5-226b) is true. Equation (5-226b) is an identity in the sense that it holds for any group parameters a and b.

Let $R(b)$ be an infinitesimal element,

$$R(b) = 1 + \delta b^\rho X_\rho . \tag{5-226d}$$

Its corresponding intrinsic element is

$$\overline{R}(b) = 1 + \delta b^\rho \overline{X}_\rho . \tag{5-226e}$$

When acting on the function $u(a), X_\rho \to X_\rho(a)$ and $\overline{X}_\rho \to \overline{X}_\rho(a)$. The differential operators of $X_\rho(a)$ and $\overline{X}_\rho(a)$ are given by (5-177b) and (5-234), respectively. Inserting (5-226d) and (5-226e) into (5-226b), we obtain

$$\big(1 + \delta b^\rho \overline{X}_\rho(a)\big)u(a) = R(a)\big(1 + \delta b^\rho X_\rho(a)\big)R^{-1}(a)u(a) .$$

Since $u(a)$ is an arbitrary function on the group manifold, it follows that

$$\overline{X}_\rho(a) = R(a)X_\rho(a)R^{-1}(a) . \tag{5-227}$$

On the other hand, the r infinitesimal operators $\{X_\rho\}$ carry the adjoint rep (ν_0) of the group G. From (5-207) and (5-227) we have

$$\overline{X}_\rho(a) = \sum_\sigma D^{(\nu_0)}_{\sigma\rho}(a)X_\sigma(a) . \tag{5-228a}$$

The inverse of (5-228a) is

$$X_\sigma(a) = \sum_\rho \big(D^{(\nu_0)}(a)\big)^{-1}_{\rho\sigma}\overline{X}_\rho(a) . \tag{5-228b}$$

Equation (5-228) gives the relation between the infinitesimal operators of the Lie group G and the intrinsic Lie group $\overline{G}$. Since the group parameters a in (5-227) and (5-228) are variables, it follows that

$$[X_\rho(a), D^{(\nu_0)}(a)] \neq 0, \quad [\overline{X}_\rho(a), D^{(\nu_0)}(a)] \neq 0 .$$

[For the explicit form see (5-259) and (5-262)]. If we let the parameters a in (5-227) be constants a_0, and $\overline{X}_\rho \to X'_\rho$, we obtain

$$X'_\rho(a) = R(a_0)X_\rho(a)R^{-1}(a_0) = \sum_\sigma D^{(\nu_0)}_{\sigma\rho}(a_0)X_\sigma(a) . \tag{5-229}$$

Now a_0 are constants and we have

$$[X_\rho(a), \mathcal{D}^{(\nu_0)}(a_0)] = [X'_\rho(a), \mathcal{D}^{(\nu_0)}(a_0)] = 0 \ .$$

It must be stressed that although (5-227) and (5-229) are similar in appearance, they are really very different. In order to give a geometrical interpretation of (5-229), we consider the r infinitesimal operators X_ρ as components of an abstract vector X in a fixed coordinate system of an r-dimensional space $\mathcal{L}_r$. Then (5-229) shows that $X'(a)$ are the new components of the vector X in another fixed system rotated through a given angle a_0 with respect to the original one. $X'_\rho(a)$ and $X_\rho(a)$ are members of the same Lie algebra. In other words, $\{X'_\rho\}$ and $\{X_\rho\}$ are isomorphic but not commuting.

In contrast to (5-229), (5-227) can be regarded as a transformation from the fixed coordinate system to the moving (or body-fixed, or intrinsic) coordinate system, where the parameter a are variables instead of constants, since the "orientation angles" a of a moving frame with respect to the fixed one are dynamic variables. It is precisely this fact that makes the Lie algebra $\{X_\rho\}$ and $\{\overline{X}_\rho\}$ commuting and anti-isomorphic. Therefore, the infinitesimal operators $\overline{X}_\rho$ of the intrinsic group $\overline{G}$ can be thought of as the components of the same vector X in the intrinsic frame. The foregoing physical interpretations can be seen more clearly in the example of the group SO_3 to be discussed in Sec. 6.1.

Finally, we want to point out that the relation (5-228a) between the infinitesimal operators of the first and second parameter groups is a generalization of Eisenhart's equation (14.10) (Eisenhart 1933). For details, see Chen([27] 1983).

5.14.2. Infinitesimal operators of intrinsic groups in group parameter space

From (5-226c) we have

$$\overline{R}_b^{-1} u(R_a) = u(R_a R_b) \ . \tag{5-230}$$

In parallel with (5-174)-(5-177) we may write

$$\begin{aligned} R_b = R(\delta a) &= 1 + \delta a^\rho X_\rho(a) \ , \\ \overline{R}_b^{-1} = \overline{R}^{-1}(\delta a) &= 1 - \delta a^\rho \overline{X}_\rho(a) \ , \end{aligned} \tag{5-231}$$

$$\begin{aligned} R_a R_b = R(a)R(\delta a) &= R(a + da) \ , \\ a^\sigma + da^\sigma &= \varphi^\sigma(\delta a, a) \ , \end{aligned} \tag{5-232}$$

$$da^\sigma = \overline{\mu}^\sigma_\rho(a)\delta a^\rho \ , \quad \overline{\mu}^\sigma_\rho(a) = \frac{\partial \varphi^\sigma(b,a)}{\partial b^\rho}\Big|_{b=0} \ . \tag{5-233}$$

Therefore the infinitesimal operators of the intrinsic Lie group in group parameter space are the differential operators

$$\overline{X}_\rho(a) = -\overline{\mu}^\sigma_\rho(a)\frac{\partial}{\partial a^\sigma} \ . \tag{5-234}$$

Racah and Eisenhart called $B_\rho(a) = -\overline{X}_\rho(a)$ the infinitesimal operators of the second parameter group, while we use the same name for $\overline{X}_\rho(a)$.

The definition of the intrinsic state $\Phi_0(X)$ is also identical to the case of finite groups, namely

$$\overline{R}(a)\Phi_0(X) = R(a)\Phi_0(X) \ . \tag{5-235}$$

For further discussion, see Chen ([27] 1983) and Sec. 6.6.

5.15. The CSCO Approach to the Rep Theory of Lie Group

In Chapter 3, the theory of finite group reps was summarized into seven basic theorems. These theorems also hold for compact Lie groups.

We first prove that the CSCO-I of the Lie group G and the intrinsic Lie group $\overline{G}$ are equal. To this end we only need to prove that the invariants of G and $\overline{G}$ are equal, i.e.,

$$I_i(X_\rho) = I_i(\overline{X}_\rho), \quad i = 1, 2, \ldots, l \, . \tag{5-236}$$

$I_i(X_\rho)$ are the polynomials of the r infinitesimal operators. From (5-227) we have

$$\begin{aligned} R(a)X_\rho X_\sigma \ldots X_\tau R^{-1}(a) &= R(a)X_\rho R^{-1}(a)R(a)X_\sigma R^{-1}(a) \ldots R(a)X_\tau R^{-1}(a) \\ &= \overline{X}_\rho \overline{X}_\sigma \ldots \overline{X}_\tau \, . \end{aligned} \tag{5-237}$$

Therefore

$$R(a)I_i(X_\rho)R^{-1}(a) = I_i(\overline{X}_\rho) \, , \tag{5-238}$$

and thus (5-236) follows from (5-217a) and (5-238).

Suppose that $G(s_1)$ is a subgroup of G of rank l_1, and its CSCO-I is designated by

$$C(s_1) = [I_1^{(s_1)}(X_\sigma) \ldots I_{l_1}^{(s_1)}(X_\sigma)] \, , \tag{5-239}$$

where X_σ are the infinitesimal operators of the subgroup $G(s_1)$. Similarly $\overline{G}(s_1)$ is the subgroup of the intrinsic group $\overline{G}$, and its CSCO-I is designated by

$$\overline{C}(s_1) = [I_1^{(s_1)}(\overline{X}_\sigma), \ldots, I_{l_1}^{(s_1)}(\overline{X}_\sigma)] \, . \tag{5-240}$$

Now $[C(s_1), \overline{C}(s_1)] = 0$, but $C(s_1) \neq \overline{C}(s_1)$. Suppose further that the Lie group G has a group chain $G \supset G(s), G(s) = G(s_1) \supset G(s_2) \supset \ldots$ We still use $C(s)$ to designate the set of operators

$$C(s) = (C(s_1), C(s_2), \ldots) \, , \tag{5-241a}$$

where $C(s_i)$ are the CSCO-I of $G(s_i)$. For the intrinsic group chain $\overline{G} \supset \overline{G}(s)$, $\overline{G}(s) = \overline{G}(s_1) \supset \overline{G}(s_2) \supset \ldots$, we have

$$\overline{C}(s) = (\overline{C}(s_1), \overline{C}(s_2), \ldots) \, . \tag{5-241b}$$

If the set of operators $K = (C, C(s), \overline{C}(s))$ is a complete set of commuting operators in the group-parameter space, then K is called the *CSCO-III* of the Lie group. The corresponding group chain is called a canonical subgroup chain, and $(C, C(s))$ is called the *CSCO-II* of the Lie group.

As is known a CSCO in the space of functions of r variables should consist of r operators. Therefore the CSCO-III for a Lie group of order r contains r operators. Among them, the CSCO-I has already provided l operators; therefore the operator set $C(s)$ should contain $(r-l)/2$ operators, while the CSCO-II contains $(r+l)/2$ operators.

According to the above discussion and the generalization of the meaning of the order g, the number g_i of elements in the class i, the class operator C_i and the scalar product, etc., given by (5-195)-(5-201), we can transplant all the formulas for rep theory of finite groups to the case of compact Lie groups. The seven theorems in the summary of Chapter 3 can be reformulated as follows.

Theorem I: In the class space, the eigenoperator $P^{(\nu)}$ of the CSCO-I of a Lie group G is the projection operator onto the irrep (ν) of G,

$$\begin{aligned} CP^{(\nu)} &= \nu P^{(\nu)} \, , \\ P^{(\nu)} &= \frac{h_\nu}{g} \int \chi^{(\nu)}(\varphi)^* C(\varphi)\rho(\varphi) d\varphi \end{aligned} \tag{5-242}$$

where $\chi^{(\nu)}(\varphi) = \chi^{(\nu)}(\varphi^1, \varphi^2, \ldots, \varphi^l)$ is the character depending on l class parameters $\varphi^1, \varphi^2, \ldots, \varphi^l$ (see Sec. 5.28). The projection operators still obey the relation

$$P^{(\nu)} P^{(\mu)} = \delta_{\nu\mu} P^{(\nu)} .$$

Theorem II: In the class-parameter space, the eigenfunctions of the CSCO-I of a Lie group G are the complex conjugates of the characters:

$$C\chi^{(\nu)}(\varphi)^* = \nu\chi^{(\nu)}(\varphi)^* . \tag{5-243}$$

The eigenfunctions of the CSCO-I obey the orthonormality and completeness:

$$\begin{aligned} &\int \frac{g(\varphi)}{g} \chi^{(\nu)}(\varphi)^* \chi^{(\nu')}(\varphi)\rho(\varphi)d\varphi = \delta_{\nu\nu'} , \\ &\sum_\nu \frac{g(\varphi)}{g} \chi^{(\nu)}(\varphi)^* \chi^{(\nu)}(\varphi')\rho(\varphi) = \delta(\varphi - \varphi') , \\ &\delta(\varphi - \varphi') = \delta(\varphi^1 - \varphi'^1) \ldots \delta(\varphi^l - \varphi'^l) . \end{aligned} \tag{5-244}$$

Theorem III, IV and V in Sec. 3.19 do not need any modification, however, the orthogonality of $\psi_m^{(\nu)\kappa}$ on the intrinsic quantum number κ fails (see (6-73)).

Theorem VI: In group space, the eigenoperators of the CSCO-III of a Lie group G are the generalized projection operators $P_{mk}^{(\nu)}$,

$$\begin{aligned} &\begin{pmatrix} C \\ C(s) \\ \overline{C}(s) \end{pmatrix} P_{mk}^{(\nu)} = \begin{pmatrix} \nu \\ m \\ k \end{pmatrix} P_{mk}^{(\nu)} , \\ &P_{mk}^{(\nu)} = \frac{h_\nu}{g} \int D_{mk}^{(\nu)*}(a) R(a)\rho(a)da . \end{aligned} \tag{5-245a}$$

They still satisfy the relations

$$P_{mk}^{(\nu)} P_{lj}^{(\mu)} = \delta_{\nu\mu}\delta_{kl} P_{mj}^{(\nu)} , \tag{5-245b}$$

$$(P_{mk}^{(\nu)})^\dagger = P_{km}^{(\nu)} . \tag{5-245c}$$

Theorem VII: In group-parameter space, the eigenfunctions of the CSCO-III of a Lie group G are the complex conjugate of the irreducible matrix elements $D_{mk}^{(\nu)}(a)$

$$\begin{pmatrix} C \\ C(s) \\ \overline{C}(s) \end{pmatrix} D_{mk}^{(\nu)*}(a) = \begin{pmatrix} \nu \\ m \\ k \end{pmatrix} D_{mk}^{(\nu)*}(a) . \tag{5-246}$$

The eigenfunctions of the CSCO-III obey the orthonormality and completeness:

$$\begin{aligned} &\frac{h_\nu}{g} \int D_{mk}^{(\nu)*}(a) D_{m'k'}^{(\nu')}(a)\rho(a)da = \delta_{\nu\nu'}\delta_{mm'}\delta_{kk'} , \\ &\sum_{\nu mk} \frac{h_\nu}{g} D_{mk}^{(\nu)*}(a) D_{mk}^{(\nu)}(a')\rho(a) = \delta(a - a') , \\ &\delta(a - a') = \delta(a^1 - a'^1) \ldots \delta(a^r - a'^r) . \end{aligned} \tag{5-247}$$

The decomposition theorem for the regular rep of finite groups also holds for compact Lie groups. If an arbitrary function $f(a)$ (i.e., it has no symmetry at all) on the group manifold satisfies the square integrability condition, i.e. if

$$\frac{1}{g} \int |f(a)|^2 \rho(a)da$$

is finite, then the function $f(a)$ forms the basis of the regular rep of the Lie group (Gel'fand 1963). $f(a)$ can be expanded in terms of the orthonormal complete set of functions $\psi_m^{(\nu)k}(a)$

$$\psi_m^{(\nu)k}(a) = \sqrt{\frac{h_\nu}{g}} D_{mk}^{(\nu)*}(a) ,$$

$$f(a) = \sum_{\nu mk} b_{mk}^{(\nu)} \psi_m^{(\nu)k}(a) , \quad b_{mk}^{(\nu)} = \int \psi_m^{(\nu)k}(a)^* f(a)\rho(a) da . \tag{5-248}$$

Therefore the regular rep contains all irreps (ν) of the Lie group G, and the number of occurrence of each irrep equals its dimension h_ν.

5.16. Irreducible Tensors of Lie Groups and Intrinsic Lie Groups

The definition of irreducible tensors of Lie groups is the same as that for finite groups (Sec. 3.17), namely, if a set of h_ν operators transforms under the group G as

$$T_m'^{(\nu)} = R(a) T_m^{(\nu)} R^{-1}(a) = \sum_t D_{tm}^{(\nu)} T_t^{(\nu)} , \tag{5-249}$$

then $T_m^{(\nu)}$ is called the irreducible tensor of G. Letting $R(a)$ be an infinitesimal element, we obtain another equivalent definition of the irreducible tensor from (5-249), i.e.,

$$[X_\rho, T_m^{(\nu)}] = \sum_t D_{tm}^{(\nu)}(X_\rho) T_t^{(\nu)} . \tag{5-250}$$

It is worth noting that here for $\nu = \nu_0$ (the adjoint rep) the irreducible matrix $D^{(\nu_0)}$ is not necessarily identical with $\mathcal{D}^{(\nu_0)}$ in Sec. 5.14. $\mathcal{D}^{(\nu_0)}$ is the irreducible matrix determined by the infinitesimal operators X_ρ. $\mathcal{D}^{(\nu_0)}$ is equivalent to $D^{(\nu_0)}$; however, they are generally not the same. For given $D^{(\nu_0)}$, only through a proper linear combination of X_ρ, can we combine them into the ν_0 irreducible tensors $T_m^{(\nu_0)}$ which transform as (5-250). From now on, it is assumed that X_ρ have been chosen to be identical with $T_\rho^{(\nu_0)}$. Under this convention, all the expressions (5-208)- (5-211), (5-228) and (5-229) remain valid after replacing $\mathcal{D}^{(\nu_0)}$ with $D^{(\nu_0)}$. For example, (5-211) may be rewritten as

$$[X_\rho, X_\tau] = \sum_\sigma D_{\sigma\tau}^{(\nu_0)}(X_\rho) X_\sigma . \tag{5-251}$$

Because of the anti-isomorphism between $\{\overline{X}_\rho\}$ and $\{X_\rho\}$, we have

$$[\overline{X}_\rho, \overline{X}_\tau] = -\sum_\sigma D_{\sigma\tau}^{(\nu_0)}(X_\rho) \overline{X}_\sigma . \tag{5-252}$$

Comparing (5-251) with (5-252), a natural extension is to define $\{\overline{X}_\rho\}$ to be the ν_0-irreducible tensor $\{\overline{T}_\rho^{(\nu_0)}\}$ of the intrinsic group $\overline{G}$. Therefore a general definition of the ν-irreducible tensor $\overline{T}_k^{(\nu)}$ of the intrinsic group $\overline{G}$ is

$$[\overline{X}_\rho, \overline{T}_k^{(\nu)}] = -\sum_i D_{tk}^{(\nu)}(X_\rho) \overline{T}_t^{(\nu)} . \tag{5-253}$$

In analogy with the irreducible tensors of the group G, two irreducible tensors $\overline{T}_{m_1}^{(\nu_1)}$ and $\overline{T}_{m_2}^{(\nu_2)}$ of $\overline{G}$ can be combined into a third irreducible tensor of $\overline{G}$ in terms of the CG coefficients of the group G:

$$\overline{T}_m^{(\nu)\theta} = \sum_{m_1 m_2} C_{\nu_1 m_1, \nu_2 m_2}^{(\nu)\theta m} \overline{T}_{m_1}^{(\nu_1)} \overline{T}_{m_2}^{(\nu_2)} .$$

Let us prove now that the irreducible matrix elements $D^{(\nu)*}_{mk}(a)$ and $D^{(\nu)}_{mk}(a)$ are the irreducible tensors of the groups G and $\overline{G}$, respectively, i.e.,

$$D^{(\nu)*}_{mk}(a) = T^{(\nu)}_{m}\,, \quad D^{(\nu)}_{mk}(a) = \overline{T}^{(\nu)}_{k}\ . \tag{5-254}$$

Regarding $D^{(\nu)}_{mk}(a)$ as a function $u(a)$ on the group manifold and using (5-173) and (5-174), we get

$$\begin{aligned}(1-\delta a^{\rho}X_{\rho})D^{(\nu)}_{mk}(a) &= D^{(\nu)}_{mk}((1+\delta a^{\rho}X_{\rho})R(a))\\ &= D^{(\nu)}_{mk}(R(a)) + \delta a^{\rho}\sum_{t} D^{(\nu)}_{mt}(X_{\rho})D^{(\nu)}_{tk}(R(a))\ .\end{aligned}$$

Thus

$$X_{\rho}D^{(\nu)}_{mk}(a) = -\sum_{t}\langle\nu m|X_{\rho}|\nu t\rangle D^{(\nu)}_{tk}(a)\ , \tag{5-255a}$$

$$\langle\nu m|X_{\rho}|\nu t\rangle = D^{(\nu)}_{mt}(X_{\rho})\ . \tag{5-256}$$

Similarly, from (5-230) and (5-231) we have

$$\overline{X}_{\rho}D^{(\nu)}_{mk}(a) = -\sum_{t}\langle\nu t|X_{\rho}|\nu k\rangle D^{(\nu)}_{mt}(a)\ . \tag{5-255b}$$

If the function $u(R_a)$ in (5-173) and (5-230) is chosen as $\tilde{D}^{-1(\nu)}_{mk}(R_a) = \tilde{D}^{(\nu)}_{mk}(R_a^{-1})$, we have

$$X_{\rho}\tilde{D}^{-1(\nu)}_{mk}(a) = \sum_{t}\langle\nu t|X_{\rho}|\nu m\rangle \tilde{D}^{-1(\nu)}_{tk}(a)\ , \tag{5-257a}$$

$$\overline{X}_{\rho}\tilde{D}^{-1(\nu)}_{mk}(a) = \sum_{t}\langle\nu k|X_{\rho}|\nu t\rangle \tilde{D}^{-1(\nu)}_{mt}(a)\ . \tag{5-257b}$$

If $D^{(\nu)}$ is unitary, the above equations become

$$X_{\rho}D^{(\nu)*}_{mk}(a) = \sum_{t}\langle\nu t|X_{\rho}|\nu m\rangle D^{(\nu)*}_{tk}(a)\ , \tag{5-258a}$$

$$\overline{X}_{\rho}D^{(\nu)*}_{mk}(a) = \sum_{t}\langle\nu k|X_{\rho}|\nu t\rangle D^{(\nu)*}_{mt}(a)\ . \tag{5-258b}$$

If $D^{(\nu)}_{mk}(a)$, etc. are regarded as operators, (5-255) and (5-258) should be rewritten as

$$[X_{\rho}, D^{(\nu)}_{mk}(a)] = -\sum_{t}\langle\nu m|X_{\rho}|\nu t\rangle D^{(\nu)}_{tk}(a)\ , \tag{5-259a}$$

$$[\overline{X}_{\rho}, D^{(\nu)}_{mk}(a)] = -\sum_{t}\langle\nu t|X_{\rho}|\nu k\rangle D^{(\nu)}_{mt}(a)\ . \tag{5-259b}$$

$$[X_{\rho}, D^{(\nu)*}_{mk}(a)] = \sum_{t}\langle\nu t|X_{\rho}|\nu m\rangle D^{(\nu)*}_{tk}(a)\ , \tag{5-260a}$$

$$[\overline{X}_{\rho}, D^{(\nu)*}_{mk}(a)] = \sum_{t}\langle\nu k|X_{\rho}|\nu t\rangle D^{(\nu)*}_{mt}(a)\ . \tag{5-260b}$$

Comparing (5-260a) with (5-250), and (5-259b) with (5-253), we see that (5-254) holds.

By means of the Wigner-Eckart theorem (3-319), we have

$$D^{(\nu)}_{tm}(X_\rho) = \langle \nu t|X_\rho|\nu m\rangle = \sum_\theta \langle \nu\|X\|\nu\rangle^\theta C^{(\nu)\theta,t}_{\nu m,\nu_0\rho} \, , \tag{5-261}$$

where $\theta = 1, 2, \ldots, (\nu_0\nu\nu)$ is the multiplicity label. Using (5-261), Eqs. (5-259) and (5-260) can now be recast into the following form:

$$[X_\rho, D^{(\nu)}_{mk}(a)] = -\sum_{\theta,t} \langle \nu\|X\|\nu\rangle^\theta C^{(\nu)\theta,m}_{\nu t,\nu_0\rho} D^{(\nu)}_{tk}(a) \, , \tag{5-262a}$$

$$[\overline{X}_\rho, D^{(\nu)}_{mk}(a)] = -\sum_{\theta t} \langle \nu\|X\|\nu\rangle^{(\theta)} C^{(\nu)\theta,t}_{\nu k,\nu_0\rho} D^{(\nu)}_{mt}(a) \, , \tag{5-262b}$$

$$[X_\rho, D^{(\nu)*}_{mk}(a)] = \sum_{\theta t} \langle \nu\|X\|\nu\rangle^{(\theta)} C^{(\nu)\theta,t}_{\nu m,\nu_0\rho} D^{(\nu)*}_{tk}(a) \tag{5-263a}$$

$$[\overline{X}_\rho, D^{(\nu)*}_{mk}(a)] = \sum_{\theta t} \langle \nu\|X\|\nu\rangle^{(\theta)} C^{(\nu)\theta,k}_{\nu t,\nu_0\rho} D^{(\nu)*}_{mt}(a) \, . \tag{5-263b}$$

Before ending this section, we give an important relation between the matrix elements X_ρ and $\overline{X}_\rho$. Letting R_a and $\overline{R}_a$ in (3-198) be infinitesimal operators, we obtain the required relation

$$\langle \nu m|X_\rho|\nu k\rangle = \langle \nu k|\overline{X}_\rho|\nu m\rangle \, . \tag{5-264}$$

5.17. The Cartan-Weyl Basis

Let A be an element of the semi-simple Lie algebra,

$$A = a^\mu X_\mu \, . \tag{5-265}$$

We see the eigenvector of the operator

$$\overset{\circ}{A} X = aX \, , \tag{5-266a}$$

i.e.,

$$[A, X] = aX \, . \tag{5-266b}$$

Theorem 5.8: If A has the maximum number of distinct eigenvalues, then only the eigenvalue zero is degenerate.

The degeneracy l of the eigenvalue zero is called the *rank* of the Lie algebra. Denoting the l independent eigenvectors associated with the eigenvalue zero by $H_i, i = 1, 2, \ldots, l$, we have

$$[A, H_i] = 0 \, . \tag{5-267}$$

Since $[A, A] = 0$, A is necessarily of the form $A = \lambda^i H_i$. We can choose A as one of the operators H_i, say, $A = H_1$. Using Jacobi's identity, we can show that $H_1, \ldots H_l$ form a subalgebra, called the *Cartan subalgebra*. It can be further shown that it is an Abelian subalgebra,

$$[H_i, H_j] = 0 \, . \tag{5-268a}$$

The remaining $r - l$ nondegenerate eigenvectors are denoted by E_α,

$$[H_1, E_\alpha] = \alpha_1 E_\alpha \ . \tag{5-268b}$$

Applying the Jacobi identity to H_1, H_i, E_α and using (5-268), we get

$$\begin{aligned}[H_1,[H_i,E_\alpha]] &= [H_i,[H_1,E_\alpha]] + [[H_1,H_i],E_\alpha] \\ &= \alpha_1[H_i,E_\alpha] \ .\end{aligned}$$

Since α_1 is nondegenerate, $[H_i, E_\alpha]$ must be proportional to E_α,

$$[H_i, E_\alpha] = \alpha_i E_\alpha \ . \tag{5-268c}$$

Therefore E_α is a simultaneous eigenvector of $H_1, \ldots H_l$ corresponding to the set of eigenvalues $\alpha = (\alpha_1, \alpha_2, \ldots, \alpha_l)$. α can be regarded as a vector in an l-dimensional space and is called *root vector*, or simply the *root.*

Applying the Jacobi identity once again, we find

$$\begin{aligned}[H_1,[E_\alpha,E_\beta]] &= [E_\alpha,[H_1,E_\beta]] + [[H_1,E_\alpha],E_\beta] \\ &= (\alpha_1+\beta_1)[E_\alpha,E_\beta] \ .\end{aligned} \tag{5-269a}$$

This shows that $[E_\alpha, E_\beta]$ is an eigenvector of H_1 associated with the root $\alpha + \beta$, if $\alpha + \beta$ is a nonvanishing root, i.e.,

$$[E_\alpha, E_\beta] = N_{\alpha\beta} E_{\alpha+\beta} \ . \tag{5-269b}$$

If $\beta = -\alpha$, then $[E_\alpha, E_\beta]$ is a linear combination of H_i, i.e.,

$$[E_\alpha, E_{-\alpha}] = C^i_{\alpha-\alpha} H_i \ . \tag{5-269c}$$

Theorem 5.9: If α is a nonvanishing root of a semi-simple Lie algebra, then $-\alpha$ is also a root.

By choosing the normalization of E_α appropriately, the commutator (5-269c) can be recast into the form $[E_\alpha, E_{-\alpha}] = \alpha^i H_i$, where α^i are the contravariant components of the vector α.

In summary, the commutation relations for a semi-simple Lie algebra can be written in the following *standard form*, often referred to as the *Cartan-Weyl basis*,

$$\begin{aligned} &[H_i, H_k] = 0, &&[\mathbf{H}, E_\alpha] = \boldsymbol{\alpha} E_\alpha \ , \\ &[E_\alpha, E_{-\alpha}] = \boldsymbol{\alpha}\bullet\mathbf{H}, &&[E_\alpha, E_\beta] = \begin{cases} N_{\alpha\beta}E_{\alpha+\beta}, & \text{if } \alpha+\beta \text{ is a root .} \\ 0\ , & \text{if } \alpha+\beta \text{ is not a root .}\end{cases} \end{aligned} \tag{5-270}$$

$$\begin{aligned} &\mathbf{H} = (H_1, \ldots, H_l), \qquad \boldsymbol{\alpha} = (\alpha_1, \ldots, \alpha_l)\ , \\ &\boldsymbol{\alpha}\bullet\mathbf{H} = \alpha^i H_i \ . \end{aligned} \tag{5-271}$$

Example: The group SU_3. After renormalizing the infinitesimal operators of SU_3 given in (5-110) and (5-123), we have

$$\begin{aligned} &H_1 = \frac{1}{2\sqrt{3}}(n_1 - n_2), \quad H_2 = \frac{1}{6}(n_1 + n_2 - 2n_3) \quad , \\ &E_\alpha = \frac{1}{\sqrt{6}} C_1^\dagger C_3, \qquad E_\beta = \frac{1}{\sqrt{6}} C_3^\dagger C_2, \qquad E_\gamma = \frac{1}{\sqrt{6}} C_1^\dagger C_2 \ , \\ &E_{-\alpha} = \frac{1}{\sqrt{6}} C_3^\dagger C_1, \qquad E_{-\beta} = \frac{1}{\sqrt{6}} C_2^\dagger C_3, \qquad E_{-\gamma} = \frac{1}{\sqrt{6}} C_2^\dagger C_1 \ , \end{aligned} \tag{5-272a}$$

where $n_i = C_i^\dagger C_i$. Using the fact that $[n_i, C_i^\dagger] = C_i^\dagger$ and $[n_i, C_i] = -C_i$, we can easily obtain the standard commutators for SU_3 in the form (5-270) with $N_{\alpha\beta} = \frac{1}{\sqrt{6}}$ and the root vectors shown in Fig. 5.17, such a diagram is called a *root diagram.*

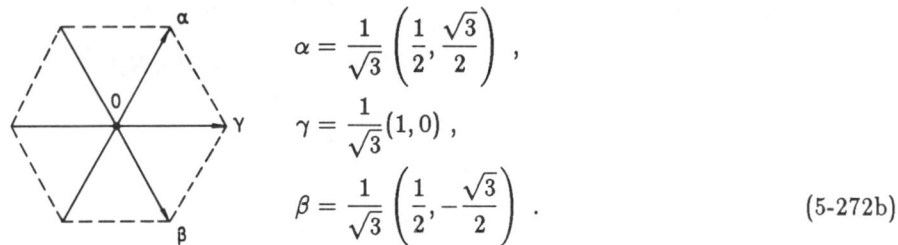

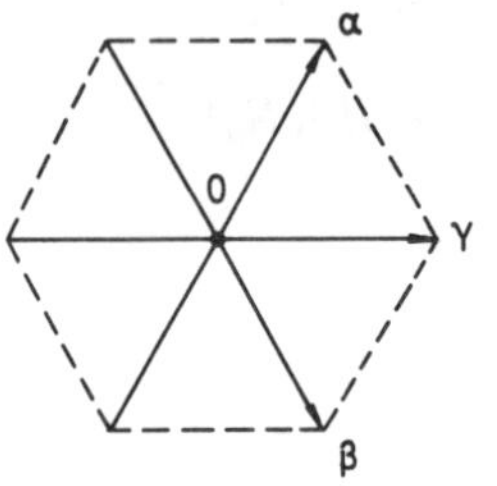

$$\alpha = \frac{1}{\sqrt{3}}\left(\frac{1}{2}, \frac{\sqrt{3}}{2}\right) ,$$

$$\gamma = \frac{1}{\sqrt{3}}(1,0) ,$$

$$\beta = \frac{1}{\sqrt{3}}\left(\frac{1}{2}, -\frac{\sqrt{3}}{2}\right) . \tag{5-272b}$$

Fig. 5.17. The root diagram of SU_3.

5.18. Theorems on Roots

We use (α, β) to denote the scalar product of the root vectors α and β, $(\alpha,\beta) = \alpha^i \beta_i = |\alpha||\beta|\cos\theta_{\alpha\beta}$.

Theorem 5.10: If α and β are roots, then $n = 2(\alpha,\beta)/(\alpha,\alpha)$ is an integer and $\beta - \frac{2\alpha(\alpha,\beta)}{(\alpha,\alpha)}$ is also a root.

If $n > 0$, then according to

$$[E_\alpha, E_{\beta - n\alpha}] = N_{\alpha,\beta-n\alpha} E_{\beta-(n-1)\alpha} ,$$

we know that $\beta - (n-1)\alpha$ is also a root. Thus we have a string of roots,

$$\beta, \beta - \alpha, \ \beta - 2\alpha, \dots, \beta - n\alpha ,$$

called the β-string containing α.

Analogously, if $n < 0$, then we have a string of roots

$$\beta, \beta + \alpha, \ \beta + 2\alpha, \dots, \beta + |n|\alpha .$$

Theorem 5.11: If α is a root, then $m\alpha$ is not a root for integer $|m| > 1$.

Proof: $[E_\alpha, E_\alpha] = 0$, hence 2α is not a root. If 3α were a root, then we would have the string of roots $\alpha, 2\alpha, 3\alpha$, and 2α would be a root. QED.

Theorem 5.12: If α and β are two roots, then the only possible values of $\frac{2(\alpha,\beta)}{(\alpha,\alpha)}$ are

$$m = \frac{2(\alpha,\beta)}{(\alpha,\alpha)} = 0, \pm 1, \pm 2, \pm 3 . \tag{5-273}$$

Therefore, the β-string containing α have at most four roots.

Proof: Suppose that we have a string of five roots $\beta, \beta - \alpha, \dots, \beta - 4\alpha$. By relabeling the roots, they can be written as $\beta - 2\alpha, \beta - \alpha, \beta, \beta + \alpha, \beta + 2\alpha$.

Since $(\beta + 2\alpha) \mp \beta = \begin{cases} 2\alpha \\ 2(\beta+\alpha) \end{cases}$ are not roots, from Theorem 5.10, it follows that

$$(\beta + 2\alpha, \beta) = 0 . \tag{5-274a}$$

Similarly, since $(\beta - 2\alpha) \mp \beta = \begin{cases} -2\alpha \\ 2(\beta - \alpha) \end{cases}$ are not roots, we have

$$(\beta - 2\alpha, \beta) = 0 \ . \tag{5-274b}$$

Adding (5-274a) and (5-274b) leads to $(\beta, \beta) = 0$, which implies that $\beta = 0$.

5.19. Root Diagram

In a space of dimension l, we can "draw" a diagram, called a root diagram. Each simple Lie algebra is associated uniquely with a root diagram. The null vector will correspond to H_i. A root diagram carries the following important messages: If in the diagram we have three weights α, β and γ, satisfying $\alpha + \beta = \gamma$, then $[E_\alpha, E_\beta]$ is equal to E_γ within a multiplicative factor, and if $\alpha + \beta$ is not a root, then E_α and E_β are commuting.

Theorem 5.12, i.e., (5-273), stipulates the angles and lengths of roots. Let $\theta_{\alpha\beta}$ be the angle between the root vectors α and β. We have

$$\cos\theta_{\alpha\beta} = \frac{(\alpha, \beta)}{\sqrt{(\alpha,\alpha)(\beta,\beta)}} = \frac{\sqrt{mn}}{2} \ , \tag{5-275a}$$

where

$$m = \frac{2(\alpha,\beta)}{(\alpha,\alpha)} , \quad n = \frac{2(\alpha,\beta)}{(\beta,\beta)} \ . \tag{5-275b}$$

It suffices to consider the case with $\cos\theta_{\alpha\beta} > 0(\theta_{\alpha\beta} < 90)$, since if $(\alpha, \beta) < 0$, then $(-\alpha, \beta) > 0$. By Eq. (5-273), the only possible values of $\theta_{\alpha\beta}$ are as follows:

m	n	$\cos\theta_{\alpha\beta}$	$\theta_{\alpha\beta}$	$\|\alpha\|^2/\|\beta\|^2$
1	1	1/2	60°	1
1	2	$\sqrt{2}/2$	45°	2
1	3	$\sqrt{3}/2$	30°	3
2	2	1	0°	1
1	0	0	90°	0
0	0	0	90°	indefinite

Example: For $l = 1$, there are only two nonzero roots, $\pm\alpha$, corresponding to the diagram

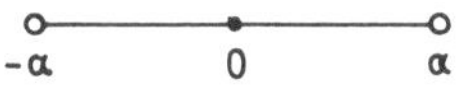

For $l = 2$, we have the following diagram. (i) $\theta_{\alpha\beta} = 60^\circ$. The root diagram is a hexagon, as shown in Fig. 5.19-1. There are eight root vectors (including two null roots), corresponding to the su_3 Lie algebra. (ii) $\theta_{\alpha\beta} = 45^\circ$. Two diagrams arise, namely, Figs. 5.19-2 and 5.19-3. The former corresponds to the Lie algebra B_2. There are 10 root vectors (including two null vectors), which we may associate with the roots of the so_5 Lie algebra (for the proof, see Ex. 5.3).

The latter corresponds to the Lie algebra C_2, which is isomorphic to B_2. Since their root diagrams differ only by a rotation through 45°. C_2 is associated with the roots of the sp_4 Lie algebra (see Ex. 5.4).

(iii) $\theta_{\alpha\beta} = 90^\circ$. The diagram is shown in Fig. 5.19-4 and is the root diagram of D_2, which we may associate with the so_4 Lie algebra (see Ex. 5.3). It can be decomposed into two sets of mutually orthogonal roots. According to Theorem 5.1, the root diagram for a semi-simple non-simple group can be decomposed into two orthogonal subdiagrams, each corresponding to a

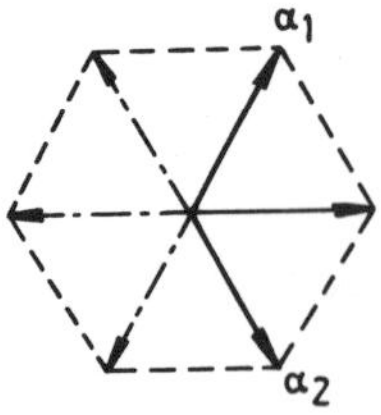

Fig. 5.19-1. Root diagram for A_2.

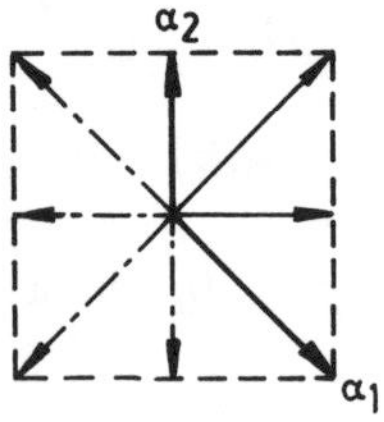

Fig. 5.19-2. Root diagram for B_2.

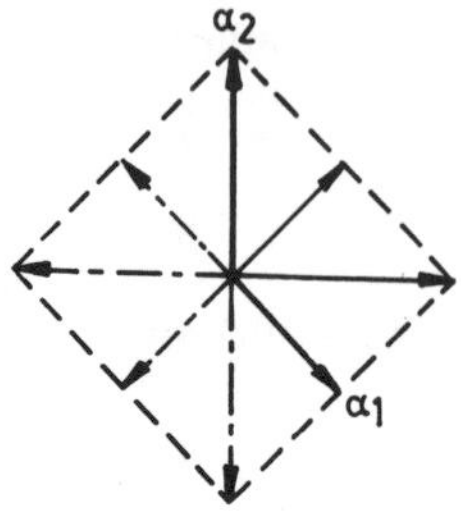

Fig. 5.19-3. Root diagram for C_2.

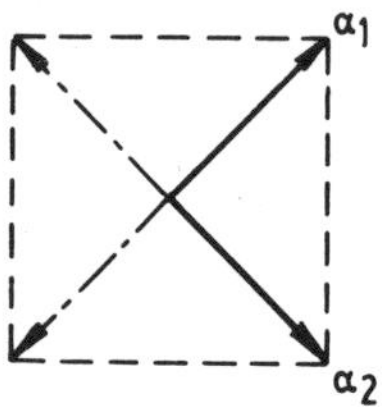

Fig. 5.19-4. Root diagram for D_2.

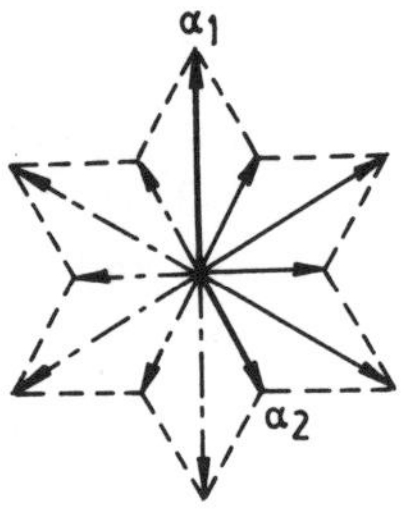

Fig. 5.19-5. Root diagram for G_2.

simple group. Here we have $D_2 = A_1 \oplus A_1$, indicating that the Lie algebra of so_4 is isomorphic to $so_3 \oplus so_3$.

Finally $\theta_{\alpha\beta} = 30°$. The diagram is shown in Fig. 5.19-5 which is the root diagram of the Lie algebra G_2. There are 12 non-zero roots and two null roots.

For Lie algebras A_{l-1}, B_l, C_l and D_l with rank greater than 2, which correspond to the four classical Lie groups SU_l, SO_{2l+1}, SP_{2l} and SO_{2l}, respectively, the non null root vectors and the positive root set Σ^+ (see Sec. 5.20) can be expressed in terms of the mutually orthogonal unit vectors

$$\mathbf{e}_1 = (1,0,\ldots,0), \quad \mathbf{e}_2 = (0,1,0,\ldots,0),\ldots, \quad \mathbf{e}_l = (0,\ldots 0,1)$$

1. A_{l-1} $[r = (l-1)^2 - 1]$

$$\text{Root vectors: } \mathbf{e}_i - \mathbf{e}_j, \quad i \neq j = 1,2,\ldots,l, \qquad \text{(5-276a)}$$
$$\Sigma^+ : \{\mathbf{e}_i - \mathbf{e}_j\}_1^l, \quad i < j .$$

2. B_l $[r = l(2l+1)]$

$$\text{Root vectors: } \pm\mathbf{e}_i, \ \pm\mathbf{e}_i \pm \mathbf{e}_j, \quad i \neq j = 1,2,\ldots,l, \qquad \text{(5-276b)}$$
$$\Sigma^+ : \{\mathbf{e}_i, \mathbf{e}_i \pm \mathbf{e}_j\}_1^l, \quad i < j .$$

3. C_l $[r = l(2l+1)]$

$$\text{Root vectors: } 2\mathbf{e}_i, \pm\mathbf{e}_i + \mathbf{e}_j, \quad i \neq j = 1,2,\ldots,l, \qquad \text{(5-276c)}$$
$$\Sigma^+ : \{2\mathbf{e}_i, \ \mathbf{e}_i \pm \mathbf{e}_j\}_1^l, \quad i < j .$$

4. D_l $[r = l(2l-1)]$

$$\text{Root vectors: } \pm\mathbf{e}_i \pm \mathbf{e}_j \quad i \neq j = 1,2,\ldots,l, \qquad \text{(5-276d)}$$
$$\Sigma^+ : \{\mathbf{e}_i \pm \mathbf{e}_j\}_1^l, \quad i < j .$$

5.20. The Dynkin Diagram and the Simple Root Representation

Dynkin has invented an ingenious scheme to draw the root diagrams for the Lie algebras of any rank. The key point is to focus our attention on the linearly independent root vectors, called the simple roots. The two-dimensional diagram for the simple roots, called the Dynkin diagram, contains all the necessary information about the root vectors.

A root vector α^+ is called a *positive root* if in some given bases the first nonvanishing component of α^+ is positive. The positive root set is designated Σ^+. The solid lines in Fig. 5.19, for instance, represent the positive roots for the ordinary Cartesian basis.

A positive root is said to be *simple* if it cannot be decomposed into the sum of two positive roots. For example, the thick solid lines in Fig. 5.19 represent the simple roots.

For a semi-simple Lie algebra of rank l, there are just l simple roots, denoted by $\alpha_1, \alpha_2, \dots,$ α_l, which form a basis for the l-dimensional root space. The metric tensor is

$$g_{ij} = (\alpha_i, \alpha_j) \ . \tag{5-277}$$

Any root vector $\mathbf{v}$ can be expressed as

$$\mathbf{v} = \sum_{i=1}^{l} v^i_{SRS} \alpha_i \ , \tag{5-278a}$$

and

$$\{\mathbf{v}\}_{SRS} \equiv \{v^1, v^2, \dots, v^l\}_{SRS} \tag{5-278b}$$

is referred to as the representative of the root vector $\mathbf{v}$ in the *simple roots representation (SRS)* (in the sense used in quantum mechanics), or simply the SRS representative of $\mathbf{v}$. The curly brackets are reserved for the SRS representatives.

The basis $\{\alpha_i\}$ is not an orthonormal one. We may introduce the dual basis $\{\overline{\alpha}^i\}$ such that

$$(\overline{\alpha}^j, \alpha_i) = \delta_{ij}, \quad \sum_{i=1}^{l} |\alpha_i)(\overline{\alpha}^i| = 1 \ . \tag{5-279}$$

Therefore

$$v^i_{SRS} = (\overline{\alpha}^i, \mathbf{v}) \ . \tag{5-278c}$$

If β is a positive root, then its SRS coordinates $(\overline{\alpha}^i, \beta)$ are all non-negative. The set of l simple roots is designated Π.

Theorem 5.13: If α and β are two simple roots, then $\alpha - \beta$ is not a root.

Proof: Suppose that $\alpha - \beta = \gamma$ were a positive root, then $\alpha = \beta + \gamma$, violating the hypothesis that α is a simple root. Suppose that $\alpha - \beta = \gamma$ were a negative root, then $\alpha = \beta + (-\gamma)$, again violating the hypothesis that α is a simple root. Hence $\alpha - \beta$ cannot be a root.

Theorem 5.14: If α and β are two simple roots, then $\theta_{\alpha\beta}$ can only equal to 90°, 120°, 135°, or 150°. If $|\alpha| < |\beta|$, then we have the following table

θ	$\cos\theta_{\alpha\beta}$	$\|\beta\|^2/\|\alpha\|^2$		Dynkin diagram	$\frac{2(\alpha,\beta)}{(\alpha,\alpha)}$	$\frac{2(\alpha,\beta)}{(\beta,\beta)}$
120°	$-1/2$	1	A_2		-1	-1
135°	$-\sqrt{2}/2$	2	$B_2(C_2)$		-2	-1
150°	$-\sqrt{3}/2$	3	G_2		-3	-1
90°	0		D_2		0	0

(5-280)

where we associate the simple roots α and β with circles joined by one, two, or three lines for $(|\beta|/|\alpha|)^2 = 1, 2, 3$, respectively, or left unjoined for $\theta_{\alpha\beta} = 90°$. Circles corresponding to the shorter roots are filled, while those to the longer roots are left open (for any semi-simple Lie algebra, there exist simple roots of at most two distinct lengths).

These are the prototype of the *Dynkin diagram* (they are the Dynkin diagrams for $A_2, B_2 \approx C_2, G_2$ and D_2, respectively). The Dynkin diagrams and simple roots for A_l, B_l, C_l, and D_l are shown in Fig. 5.20-1.

Apart from these four infinite series of diagrams, there are only five other possible diagrams, which corresponds to the exceptional Lie algebra G_2, F_4, E_6, E_7, and E_8. The Dynkin diagrams for the last four are shown in Fig. 5.20-2.

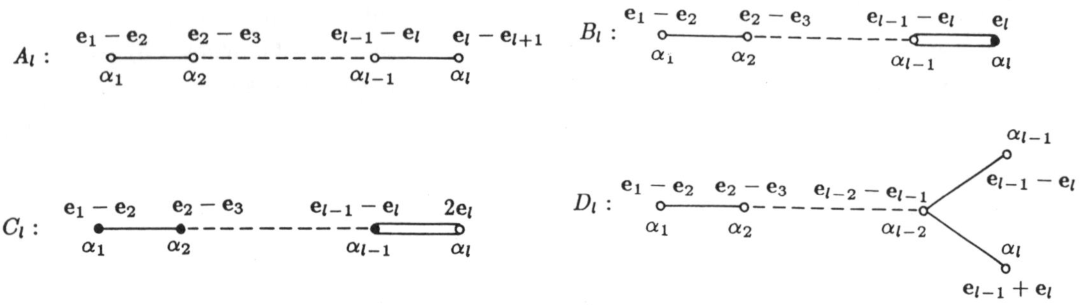

Fig. 5.20-1. The Dynkin diagrams of A_l, B_l, C_l and D_l.

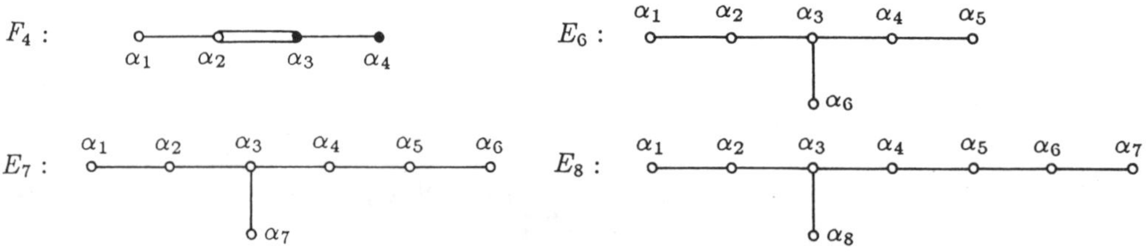

Fig. 5.20.2. The Dynkin diagrams of F_4, E_6, E_7 and E_8.

From Fig. 5.20-1, the following isomorphism is readily seen: $A_1 \approx B_1 \approx C_1, B_2 \approx C_2, D_2 \approx A_1 \oplus A_1, A_3 \approx D_3$.

5.21. The Cartan Matrix

The Cartan matrix for a given Lie algebra with the simple roots $\alpha_1, \alpha_2, \ldots, \alpha_l$ is defined by

$$A_{ij} = \frac{2(\alpha_i, \alpha_j)}{(\alpha_i, \alpha_i)} \, . \tag{5-281}$$

The diagonal matrix elements of A are always equal to 2, while the off diagonal matrix elements are restricted to the values 0, $-1, -2$ and -3. Given any Dynkin diagram, using (5-280), we can readily construct its associated Cartan matrix. For example, for G_2, we have

α_1  α_2

$$A = \begin{array}{c} \begin{array}{cc} \alpha_1 & \alpha_2 \end{array} \\ \begin{pmatrix} 2 & -1 \\ -3 & 2 \end{pmatrix} \end{array} . \tag{5-281'}$$

Similarly for A_l, B_l, C_l and D_l, we have

$$A_l : A = \begin{bmatrix} 2 & -1 & 0 & & & & \\ -1 & 2 & -1 & & & & \\ 0 & -1 & 2 & & & & \\ & & & \ddots & & & \\ & & & & 2 & -1 & 0 \\ & & & & -1 & 2 & -1 \\ & & & & 0 & -1 & 2 \end{bmatrix}, \quad B_l : A = \begin{bmatrix} 2 & -1 & 0 & & & & \\ -1 & 2 & -1 & & & & \\ 0 & -1 & 2 & & & & \\ & & & \ddots & & & \\ & & & & 2 & -1 & 0 \\ & & & & -1 & 2 & -1 \\ & & & & 0 & -2 & 2 \end{bmatrix}$$
(5-282a,b)

$$C_l : A = \begin{bmatrix} 2 & -1 & 0 & & & & \\ -1 & 2 & -1 & & & & \\ 0 & -1 & 2 & & & & \\ & & & \ddots & & & \\ & & & & 2 & -1 & 0 \\ & & & & -1 & 2 & -2 \\ & & & & 0 & -1 & 2 \end{bmatrix}, \quad \begin{matrix} D_l : A \\ (l \geq 4) \end{matrix} = \begin{bmatrix} 2 & -1 & 0 & & & & \\ -1 & 2 & -1 & & & & \\ 0 & -1 & 2 & & & & \\ & & & \ddots & & & \\ & & & & -1 & 0 & 0 \\ & & & & 2 & -1 & -1 \\ & & & & -1 & 2 & 0 \\ & & & & -1 & 0 & 2 \end{bmatrix},$$
(5-283c,d)

Ex. 5.3. For the so_5 and so_4 Lie algebras, we have $H_1 = \frac{1}{\sqrt{6}}(n_1 - n_{-1}), H_2 = \frac{1}{\sqrt{6}}(n_2 - n_{-2})$, $E_\alpha = E_{ik} = \frac{1}{\sqrt{6}}(C_i^\dagger C_{-k} - C_k^\dagger C_{-i})$, for $i \neq k$. Show that (i)

$$[\mathbf{H}, E_\alpha] = \boldsymbol{\alpha} E_\alpha, \quad [E_\alpha, E_{-\alpha}] = -\boldsymbol{\alpha}\bullet\mathbf{H}, \quad \boldsymbol{\alpha} = \frac{1}{\sqrt{6}}(\mathbf{e}_i + \mathbf{e}_k),$$

with $\mathbf{e}_{-i} = -\mathbf{e}_i$ and $\mathbf{e}_0 = 0$, and (ii) the so_5 and so_4 Lie algebras are associated with the Lie algebra B_2 and D_2, respectively.

Ex. 5.4. For the sp_4 Lie algebra, we have $H_1 = n_1 - n_{-1}, H_2 = n_2 - n_{-2}, E_{ik} = \delta_i C_i C_{-k} + \delta_k C_k C_{-i}, (ik) = \pm(1,1), \pm(1,2), \pm(1,-2), \pm(2,2)$. Find the commutators of these operators, draw the corresponding root diagram and show that it is the root diagram of C_2.

Ex. 5.5. Construct the Cartan matrices for the exceptional Lie algebras F_4, E_6, E_7 and E_8 with the Dynkin diagrams shown in Fig. 5.20-2.

Ex. 5.6. Prove that the roots shown in Fig. 5.20-1 are simple.

5.22. Theorems on Weights

As pointed out in Sec. 5.15, the CSCO-II for a Lie group of order r and rank l contains $(r+l)/2$ operators. Among them, l operators are the CSCO-I of $G, I_1, \ldots, I_l$. The other l operators can be chosen as the elements of the Cartan subalgebra, $H_1, \ldots, H_l$. The irreducible basis of G is the eigenvector of the CSCO-II of G, and can be denoted by

$$\psi_m^{(\nu)} = \psi_{\Lambda_1 \ldots \Lambda_l, \xi}^{(I_1, \ldots I_l)}, \tag{5-283a}$$

where ξ designates the quantum numbers of the $(r-3l)/2$ remaining operators. For brevity, the irreducible basis (5-283a) is simple written as ψ_Λ,

$$H_i \psi_\Lambda = \Lambda_i \psi_\Lambda, \quad \Lambda = (\Lambda_1, \Lambda_2, \ldots, \Lambda_l). \tag{5-283b}$$

The l-dimensional vector Λ is called the *weight vector*, or *weight* of the eigenket ψ_Λ. Comparing (5-270) with (5-283b) we know that the roots are the weights of the adjoint rep.

A weight is said to be *positive* if its first nonvanishing component is positive. A weight Λ is said to be higher than Λ', if $\Lambda - \Lambda'$ is positive. The highest weight in an irrep Λ is called the *highest weight.* A weight is said to be *simple* if it has no degeneracy.

Theorem 5.15: Every rep has at least one weight.
Theorem 5.16: Eigenvector with distinct weights are linearly independent.
Theorem 5.17: If a rep is irreducible, then its highest weight is simple.
Theorem 5.18: Two irreps are equivalent if and only if they have the same highest weight.

Therefore, we can use the highest weight Λ to label an irrep of a Lie group.

Theorem 5.19: If ψ_Λ is a vector of weight Λ, then $E_\beta\psi_\Lambda$ is of weight $\Lambda + \beta$.

Proof: $H_i E_\beta \psi_\Lambda = ([H_i, E_\beta] + E_\beta H_i)\psi_\Lambda = (\beta_i + \Lambda_i)E_\beta\psi_\Lambda$.

E_β is called the step operator, and is a generalization of $J_\pm$. In parallel with Theorem 5.10, we have

Theorem 5.20: Let Λ be one of the weights of an irrep and α be a root; then $n = 2(\Lambda, \alpha)/(\alpha, \alpha)$ is an integer and $\Lambda' = \Lambda - n\alpha$ is a weight of this irrep.

The weight $\Lambda - n\alpha$ is the image of Λ with respect to the hyperplane S_α through the origin and perpendicular to the root α (see Fig. 5.22-1). The finite group generated by the reflections $S_{\alpha_i}, i = 1, 2, \ldots, l$, is called the *Weyl reflection group.*

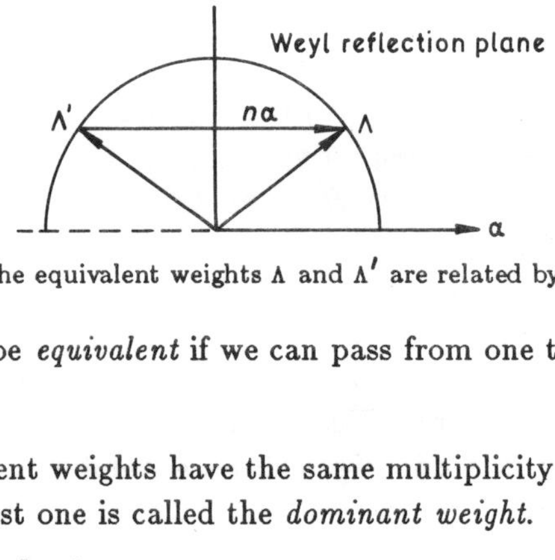

Fig. 5.22-1. The equivalent weights Λ and Λ' are related by a reflection.

Two weights are said to be *equivalent* if we can pass from one to the other by an operation of the Weyl group.

Theorem 5.21: Equivalent weights have the same multiplicity (or degeneracy). In a set of equivalent weights, the highest one is called the *dominant weight.*

Weights of Kronecker product

Let $\Delta(\Lambda)$ be the *complete set of weights* (CSW) for the irrep Λ. Since the eigenvalues of H_i are additive quantum numbers, the CSW of the Kronecker product $(\Lambda \times \Lambda')$ is

$$\Delta(\Lambda \times \Lambda') = \Delta(\Lambda) + \Delta(\Lambda') \ , \tag{5-284a}$$

i.e., an arithemetic sum where every weight of $\Delta(\Lambda')$ is added to every weight of $\Delta(\Lambda)$. Equation (3-284a) is a generalization of $M = m + m'$ for eigenvalues of J_z.

Suppose that $|1\rangle, |2\rangle, \ldots |n\rangle$ carry the defining rep [1] of a linear transformation group, the weights of which are assumed to be $\Lambda_1 > \Lambda_2 > \ldots > \Lambda_n$. The antisymmetric states

$$\left| \begin{array}{|c|} \hline i_1 \\ \hline i_2 \\ \hline \vdots \\ \hline i_k \\ \hline \end{array} \right\rangle \tag{5-284b}$$

carry a $\binom{n}{k}$-dimensional rep $M^{[1^k]}$ of G with the CSW

$$\Lambda_{i_1} + \Lambda_{i_2} + \ldots + \Lambda_{i_k}, \quad i_1 < i_2 < \ldots < i_k \ . \tag{5-284c}$$

The highest weight in $M^{[1^k]}$ is

$$\Lambda = \Lambda_1 + \Lambda_2 + \ldots + \Lambda_k \ , \tag{5-284d}$$

which corresponds to an irrep designated $(M^{[1^k]})_{\text{h.w.}}$.

Analogously, the symmetric states

$$\left| \begin{array}{|c|c|c|c|}\hline i_1 & i_2 & \ldots & i_k \\ \hline \end{array} \right\rangle \tag{5-284e}$$

carry a $\binom{n+k-1}{k}$-dimensional rep $M^{[k]}$ of G with the CSW

$$\Lambda_{i_1} + \Lambda_{i_2} + \ldots + \Lambda_{i_k}, \quad i_1 \le i_2 \le \ldots \le i_k \ . \tag{5-284f}$$

The highest weight in $M^{[k]}$ is $k\Lambda_1$, corresponding to an irrep designated by $(M^{[k]})_{\text{h.w.}}$.

Example: The SU_3 group.

1. Irrep [1]. The three flavor quarks u, d, and s carry the irrep [1] of SU_3. The Cartan subalgebra consists of

$$H_1 = \frac{1}{2\sqrt{3}}(n_u - n_d), \quad H_2 = \frac{1}{6}(n_u + n_d - 2n_s) \ . \tag{5-285a}$$

The three weights are

$$\begin{aligned} \Lambda_u &= \frac{1}{\sqrt{3}}\Big(\frac{1}{2}, \frac{1}{2\sqrt{3}}\Big) = \frac{1}{3}(2\alpha, \beta) \ , \\ \Lambda_d &= \frac{1}{\sqrt{3}}\Big(-\frac{1}{2}, \frac{1}{2\sqrt{3}}\Big) = \frac{1}{3}(-\alpha - 2\beta) \\ \Lambda_s &= \frac{1}{\sqrt{3}}\Big(0, -\frac{1}{\sqrt{3}}\Big) = \frac{1}{3}(-\alpha + \beta) \ , \end{aligned} \tag{5-285b}$$

where $\Lambda_u > \Lambda_s > \Lambda_d$, while α and β are given by (5-272b).

2. Irrep $[1^2]$. The basis vectors are $\tilde{u} = \begin{array}{|c|}\hline d \\ \hline s \\ \hline \end{array}$, $\tilde{d} = \begin{array}{|c|}\hline u \\ \hline s \\ \hline \end{array}$, $\tilde{s} = \begin{array}{|c|}\hline u \\ \hline d \\ \hline \end{array}$ with the weights

$$\Lambda_{\tilde{u}} = -\Lambda_u, \quad \Lambda_{\tilde{d}} = -\Lambda_d, \quad \Lambda_{\tilde{s}} = -\Lambda_s \ . \tag{5-286}$$

3. Irrep [21]. The basis vectors for the irrep [21] are $\begin{smallmatrix} uu \\ d \end{smallmatrix}$ etc., (see Sec. 7.5) as shown on the outer perimeter and the origin of Fig. 5.22-2. The weights are just the roots of SU_3. The weights for the SU_3 irreps [1], [11] and [21] are shown in Fig. 5.22-2.

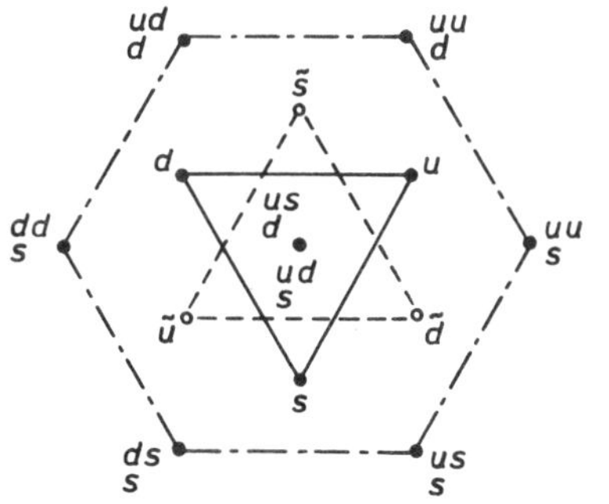

Fig. 5.22-2. The weights of SU_3 for the irreps [1], [11] and [21]. Notice that Λ_u, Λ_d, and Λ_s are equivalent weights; and the six weights on the outer perimeter are equivalent weights.

The SU_3 weights can also be expressed in the basis $\mathbf{e}_1$, $\mathbf{e}_2$ and $\mathbf{e}_3$. From Fig. 5.20-1, we have $\alpha_1 = (\mathbf{e}_1 - \mathbf{e}_2)$, $\alpha_2 = (\mathbf{e}_2 - \mathbf{e}_3)$. Using (5-285b) we get

$$\Lambda_u = \left(\frac{2}{3}, -\frac{1}{3}, -\frac{1}{3}\right), \quad \Lambda_d = \left(-\frac{1}{3}, \frac{2}{3}, -\frac{1}{3}\right), \quad \Lambda_s = \left(-\frac{1}{3}, -\frac{1}{3}, \frac{2}{3}\right) \ , \tag{5-287a}$$

or

$$\Lambda_i = \mathbf{e}_i - \frac{1}{3}(\mathbf{e}_1 + \mathbf{e}_2 + \mathbf{e}_3) , \tag{5-287b}$$

where $i = 1, 2$, and 3 refer to u, d and s, respectively. Λ_i are eigenvalues of the elements of the Cartan subalgebra,

$$H_i = n_i - \frac{1}{3}(n_1 + n_2 + n_3) . \tag{5-287c}$$

Note that the ordering of the weights in (5-287c) is $\Lambda_u > \Lambda_d > \Lambda_s$, differing from that in (5-285b).

5.23. The Dynkin Representation

For each root vector or weight vector $\mathbf{v}$ we associate a set of numbers

$$(\mathbf{v})^{\mathrm{DYN}} \equiv (v_1, \ldots , v_l)^{\mathrm{DYN}} , \tag{5-288a}$$

$$v_i^{\mathrm{DYN}} = 2\frac{(\alpha_i, \mathbf{v})}{(\alpha_i, \alpha_i)} , \tag{5-288b}$$

and call it the representative of $\mathbf{v}$ in the *Dynkin* (DYN) *representation*. Henceforth, the round brackets will be used to denote the DYN representatives. By Theorems 5.10 and 5.20, the coordinates v_i^{DYN} of any root or weight vector $\mathbf{v}$ are integers.

Theorem 5.22: If Λ is the highest weight of a representation (either reducible or irreducible), then its DYN coordinates a_i in $\Lambda = (a_1, a_2, \ldots , a_l)$ are all non-negative.

The DYN representative $(a_1, \ldots a_l)$ of a highest weight is called the *Dynkin label* of the irrep.

Theorem 5.23: If $\mathbf{R}$ is half of the sum of the positive roots,

$$\mathbf{R} = \frac{1}{2}\sum_{\alpha>0} \alpha , \tag{5-288c}$$

then

$$(\mathbf{R})^{\mathrm{DYN}} = (1, 1, \ldots , 1) . \tag{5-288d}$$

For a Lie group of rank l, there are l *basic irreps* with the highest weights denoted by $\mathbf{M}^j, j = 1, 2, \ldots , l$ whose DYN representatives are $(1, 0, \ldots , 0), (0, 1, 0, \ldots , 0), \ldots$ and $(0, \ldots , 0, 1)$, i.e.

$$(\mathbf{M}^j)^{\mathrm{DYN}} = (m_1^j, m_2^j, \ldots , m_l^j) , \tag{5-289a}$$

$$m_i^j = 2\frac{(\alpha_i, \mathbf{M}^j)}{(\alpha_i, \alpha_i)} = \delta_{ij} , \quad i, j = 1, 2, \ldots , l . \tag{5-289b}$$

Obviously, $\mathbf{M}^1, \mathbf{M}^2, \ldots , \mathbf{M}^l$ are the basis vectors of the DYN representation, i.e.,

$$\mathbf{v} = \sum_{i=1}^{l} v_i^{\mathrm{DYN}} \mathbf{M}^i . \tag{5-290a}$$

The weight $\mathbf{M}^i$ will be referred to as the i-th *basic weight*.

As with Eq. (5-279), we can introduce the dual basis $\overline{\mathbf{M}}_i$,

$$(\overline{\mathbf{M}}_i, \mathbf{M}^j) = \delta_{ij}, \quad \sum_{i=1}^{l} |\mathbf{M}^i)(\overline{\mathbf{M}}_i| = 1 . \tag{5-291}$$

Thus

$$v_i^{\rm DYN} = (\overline{\mathbf{M}}_i, \mathbf{v}) \ . \tag{5-290b}$$

It is easy to see that the two basic weights of SU_3 are $\mathbf{M}^1 = \Lambda_u$ and $\mathbf{M}^2 = \Lambda_{\tilde{d}}$ as given in (5-285) and (5-286).

Theorem 5.24: The simple roots α_j and basic weight $\mathbf{M}^j, j = 1, 2, \ldots, l$, form a bi-orthogonal basis.

Proof: The proof is trivial, since from (5-289b) and (5-281) we have

$$(\mathbf{M}^j, \alpha_i) = N_i \delta_{ij}, \quad N_i = (\alpha_i, \alpha_i)/2, \quad i, j = 1, 2, \ldots, l \ . \tag{5-292}$$

Therefore, $\mathbf{M}^j$ is orthogonal to α_i for $i \neq j$.

Figure 5.23 illustrates this fact for the Lie algebra of rank two (for the weight diagrams of B_2 and C_2, see Ex. 5.13, and for that of G_2, see Fig. 5.24-6).

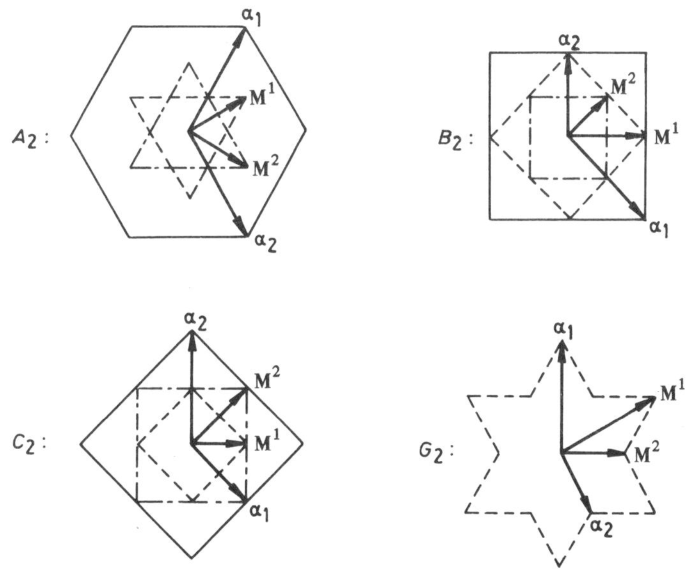

Fig. 5.23. Bi-orthogonal basis formed by simple roots and basic weights.

From (5-292) we infer that

1. $$\overline{\alpha}^i = (N_i)^{-1}\mathbf{M}^i \ , \quad \overline{\mathbf{M}}_i = (N_i)^{-1}\alpha_i \tag{5-293a,b}$$

2. $$A_{ij} = (\overline{\mathbf{M}}_i, \alpha_j) \ , \tag{5-294a}$$

which in turn means that

$$\alpha_j = \sum_i A_{ij}\mathbf{M}^i \ . \tag{5-294b}$$

Comparing it with (5-290a), we have

$$(\alpha_j)^{\rm DYN} = (A_{1j}, \ldots, A_{lj}) \ , \tag{5-294c}$$

which shows that the *j-th column of the Cartan matrix A gives the DYN representative of the simple root* α_j.

3. The inverse of (5-294b) is

$$\mathbf{M}^j = \sum_i (A^{-1})_{ij}\alpha_i \ . \tag{5-295a}$$

Therefore we have

$$(A^{-1})_{ij} = (\overline{\alpha}^i, \mathbf{M}^j) \ , \tag{5-295b}$$

$$\{\mathbf{M}^j\}_{\mathrm{SRS}} = \{(A^{-1})_{1j}, \dots, (A^{-1})_{lj}\} \ , \tag{5-295c}$$

which shows that the *j-th column of the inverse of the Cartan matrix, A^{-1}, gives the SRS representative of the basic weight* $\mathbf{M}^j$.

Equations (5-294c) and (5-295c) can be put into a more explicit form, namely,

$$A = \begin{array}{c} \begin{array}{cccc} \alpha_1 & \alpha_2 & & \alpha_l \ (\mathrm{DYN}) \end{array} \\ \begin{bmatrix} A_{12} & A_{12} & \dots & A_{1l} \\ A_{21} & A_{22} & \dots & A_{2l} \\ \vdots & \vdots & \vdots & \vdots \\ A_{l1} & A_{l2} & \dots & A_{ll} \end{bmatrix} \end{array}, \qquad A^{-1} = \begin{array}{c} \begin{array}{cccc} \mathbf{M}^1 & \mathbf{M}^2 & & \mathbf{M}^l \ \{\mathrm{SRS}\} \end{array} \\ \begin{bmatrix} A^{-1}_{11} & A^{-1}_{12} & \dots & A^{-1}_{1l} \\ A^{-1}_{21} & A^{-1}_{22} & \dots & A^{-1}_{2l} \\ \vdots & \vdots & \vdots & \vdots \\ A^{-1}_{l1} & A^{-1}_{l2} & \dots & A^{-1}_{ll} \end{bmatrix} \end{array}, \tag{5-296}$$

where $A^{-1}_{ij} \equiv (A^{-1})_{ij}$. The interpretation of the Cartan matrix and its inverse A^{-1} as shown in (5-296) is very useful and should be kept well in mind.

From (5-278c) and (5-293a) we obtain the counterpart of (5-288b), i.e.,

$$v^i_{\mathrm{SRS}} = 2\frac{(\mathbf{M}^i, \mathbf{v})}{(\alpha_i, \alpha_i)} \ . \tag{5-297}$$

Since the coordinates and the basis vectors transform contragrediently, from (5-295a) we have

$$v_i^{\mathrm{DYN}} = \sum_j A_{ij} v^j_{\mathrm{SRS}} \ . \tag{5-298}$$

Equation (5-298) and its inverse can be written as

$$(\mathbf{v})^{\mathrm{DYN}} = A\{\mathbf{v}\}_{\mathrm{SRS}} \ , \quad \{\mathbf{v}\}_{\mathrm{SRS}} = A^{-1}(\mathbf{v})^{\mathrm{DYN}} \ , \tag{5-299a, b}$$

where $(\mathbf{v})^{\mathrm{DYN}}$ and $\{\mathbf{v}\}_{\mathrm{SRS}}$ are to be understood as column vectors.

The inverse Cartan matrices associated with the Dynkin diagrams for the Lie algebras A_l, B_l, C_l and D_l given in Fig. 5.20-1 are as follows:

$$A_l: \quad \begin{matrix} i \rightarrow, \\ k \\ \downarrow \end{matrix} \quad A^{-1} = \frac{1}{l+1}\left[\begin{array}{cccc|cc} \ddots & & (l+1-i)k & & 2 & 1 \\ & \ddots & & & 4 & 2 \\ & & \ddots & & 6 & 3 \\ (l+1-k)i & & & \ddots & \vdots & \vdots \\ \hline 2 & 4 & 6 & \dots & 2l-2, & l-1 \\ 1 & 2 & 3 & \dots & l-1, & l \end{array}\right], \quad B_l: \quad A^{-1} = \frac{1}{2}\left[\begin{array}{cc|cc} \ddots & 2k & 2 & 1 \\ & \ddots & 4 & 2 \\ & & 6 & 3 \\ & & \vdots & \vdots \\ \hline & 2i & 2l-2, & l-1 \\ & & 2l-2, & l \end{array}\right], \tag{5-300a,b}$$

$$D_l (l \geq 4) \qquad A^{-1}_{C_l} = \tilde{A}^{-1}_{B_l}, \quad A^{-1} = \frac{1}{4}\left[\begin{array}{cccc|cc} \ddots & & 4k & & 2 & 2 \\ & \ddots & & & 4 & 4 \\ & & \ddots & & 6 & 6 \\ & 4i & & \ddots & \vdots & \vdots \\ \hline 2 & 4 & 6 & \dots & l, & l-2 \\ 2 & 4 & 6 & \dots & l-2, & l \end{array}\right]. \tag{5-300c,d}$$

For example, for $A_3, A^{-1} = \frac{1}{4}\begin{pmatrix} 3 & 2 & 1 \\ 2 & 4 & 2 \\ 1 & 2 & 3 \end{pmatrix}$, $\mathbf{M}^1 = \frac{1}{4}(3\alpha_1 + 2\alpha_2 + \alpha_3), \mathbf{M}^2 = \frac{1}{4}(2\alpha_1 + 4\alpha_2 + 2\alpha_3), \mathbf{M}^3 = \frac{1}{4}(\alpha_1 + 2\alpha_2 + 3\alpha_3)$.

For computing the dimensions of irreps, or the eigenvalues of the Casimir operators, etc., we often come across scalar products of the form $(\Lambda, \Lambda), (\Lambda, \beta)$ and (α, β). The evaluation of the scalar product can be carried out either in the DYN or SRS representation, or preferably, for one vector in the DYN representation and the other in the SRS representation. It is easy to show that

$$(\mathbf{u}, \mathbf{v}) = \sum_{ij} u^i_{\mathrm{SRS}} g_{ij} v^j_{\mathrm{SRS}} \equiv \{\mathbf{u}\}_{\mathrm{SRS}} \bullet \{\mathbf{v}\}_{\mathrm{SRS}} , \tag{5-301a}$$

$$= \sum_i N_i u^{\mathrm{DYN}}_i \, v^i_{\mathrm{SRS}} \equiv (\mathbf{u})^{\mathrm{DYN}} \bullet \{\mathbf{v}\}_{\mathrm{SRS}} , \tag{5-301b}$$

$$= \sum_i N_i u^i_{\mathrm{SRS}} v^{\mathrm{DYN}}_i \equiv \{\mathbf{u}\}_{\mathrm{SRS}} \bullet (\mathbf{v})^{\mathrm{DYN}} . \tag{5-301c}$$

The dimension of an irrep is given by the Weyl formula

$$\dim(\Lambda) = \prod_{\alpha^+ \in \Sigma^+} \frac{(K, \alpha^+)}{(R, \alpha^+)} , \tag{5-302a}$$

$$K = \Lambda + R . \tag{5-302b}$$

The *Casimir operator* is given by (Wybourne 1974)

$$C = g^{\rho\sigma} X_\rho X_\sigma = g^{ik} H_i H_k + \sum_\alpha g^{\alpha-\alpha} E_\alpha E_{-\alpha} . \tag{5-303a}$$

Suppose that $|\Lambda\rangle$ is the highest weight state, $E_{\alpha^+}|\Lambda\rangle = 0$. Under proper normalization, $g^{\alpha-\alpha}=1$ and $[E_\alpha, E_{-\alpha}] = \alpha^i H_i$. Therefore

$$C|\Lambda\rangle = \left(g^{ik} \Lambda_i \Lambda_k + \sum_{\alpha > 0} [E_\alpha, E_{-\alpha}] \right) |\Lambda\rangle$$
$$= \left[(\Lambda, \Lambda) + \sum_{\alpha > 0} (\alpha, \Lambda) \right] |\Lambda\rangle .$$

Hence, the eigenvalue of the Casimir operator is given by

$$C = (\Lambda, \Lambda) + 2(R, \Lambda) = K^2 - R^2 , \tag{5-303b}$$

$$K^2 = (K, K), \quad R^2 = (R, R) . \tag{5-303c}$$

Using (5-301c) and (5-288c), we have

$$C = \sum_{k=1}^{l} \Lambda^k_{\mathrm{SRS}} (\Lambda^{\mathrm{DYN}}_k + 2)(\alpha_k, \alpha_k)/2 . \tag{5-304}$$

For example, let us consider the SU_3 group. Let the DYN representative of a highest weight be $\Lambda = (a_1, a_2)$. The inverse of the Cartan matrix is $A^{-1} = \frac{1}{3}\begin{pmatrix} 2 & 1 \\ 1 & 2 \end{pmatrix}$. According to (5-299b), the SRS representative of Λ is

$$\begin{pmatrix} w^1 \\ w^2 \end{pmatrix} = \frac{1}{3}\begin{pmatrix} 2 & 1 \\ 1 & 2 \end{pmatrix} \begin{pmatrix} a_1 \\ a_2 \end{pmatrix} = \frac{1}{3}\begin{pmatrix} 2a_1 + a_2 \\ a_1 + 2a_2 \end{pmatrix} .$$

With the normalization $(\alpha_i, \alpha_i)/2 = 1$, the eigenvalue of C is

$$C = \frac{1}{3}[(a_1 + 2)(2a_1 + a_2) + (a_2 + 2)(a_1 + 2a_2)]$$
$$= 2\left[\frac{1}{3}(a_1 - a_2)^2 + a_1 + a_2 + a_1 a_2\right] . \tag{5-305}$$

We can show that the DYN representative of a weight Λ or a root β is the eigenvalue set of the commuting operators

$$H_{\alpha_i} = 2\frac{\boldsymbol{\alpha}_i \bullet \mathbf{H}}{(\alpha_i, \alpha_i)} = 2\sum_j \frac{\alpha_i^{(j)} H_j}{(\alpha_i, \alpha_i)} , \tag{5-306}$$

$$H_{\alpha_i}|\psi_\Lambda\rangle = a_i|\psi_\Lambda\rangle, \quad a_i = 2\frac{(\alpha_i, \Lambda)}{(\alpha_i, \alpha_i)} , \tag{5-307a}$$

$$[H_{\alpha_i}, E_\beta] = b_i E_\beta, \quad b_i = 2\frac{(\alpha_i, \beta)}{(\alpha_i, \alpha_i)} . \tag{5-307b}$$

Equations (5-307) follow from (5-306), (5-270) and (5-283). From (5-307b) we have

$$[H_{\alpha_i}, E_{\alpha_j}] = A_{ij} E_{\alpha_j} , \tag{5-307c}$$

which is consistent with (5-294c). Comparing (5-306) with (5-270), we see that through proper normalization of the step operators $E_{\pm\alpha_i}$ one has the commutator

$$[E_{\alpha_i}, E_{-\alpha_i}] = H_{\alpha_i} . \tag{5-308}$$

The generators satisfying the following commutators are called the *Chevalley basis* of a Lie algebra.

$$[H_{\alpha_i}, H_{\alpha_j}] = 0 , \qquad [H_{\alpha_i}, E_{\alpha_j}] = A_{ij} E_{\alpha_j} , \tag{5-309a,b}$$

$$[E_{\alpha_i}, E_{-\alpha_i}] = H_{\alpha_i} , \quad [E_\beta, E_\gamma] = \pm(p+1) E_{\beta+\gamma} , \tag{5-309c,d}$$

where $p = m$ if $(\beta + \gamma)$ is a root, and $p = -1$ if $(\beta + \gamma)$ is not a root, and $m > 0$ is the greatest integer for which $\gamma - m\beta$ is a root (Weybourne 1974).

The step operators associated with the simple roots, $E_{\alpha_1}, \dots, E_{\alpha_l}$ are called the *generators of the Lie algebra.* Given these generators, the commuting operators H_{α_i} are obtained from (5-308), while the step operators associated with the non-simple roots can be obtained from recursive use of the commutator (5-309d).

There are two methods for finding the Chevalley basis:

1. Use (5-306).

2. (a) Identify the step operators E'_{α_i} associated with the simple roots α_i, which may differ from E_{α_i} of the Chevalley basis by multiplicative factors. (b) Compute the commutators $[E'_{\alpha_i}, E'_{-\alpha_i}] = H'_{\alpha_i}$, and $[H'_{\alpha_i}, E'_{\alpha_j}] = \mathcal{N}_i A_{ij} E'_{\alpha_j}$. (c) Then the Chevalley basis is: $H_{\alpha_i} = H'_{\alpha_i}/\mathcal{N}_i, E_{\pm\alpha_i} = E'_{\pm\alpha_i}/\sqrt{\mathcal{N}_i}$.

Example: The Chevalley basis of su_3

Method 1. From H_1, H_2, with α and β given by (5-272b), we obtain

$$H_\alpha = \sqrt{3}H_1 + 3H_2 = n_1 - n_3 ,$$
$$H_\beta = \sqrt{3}H_1 - 3H_2 = n_3 - n_2 ,$$
$$E_\alpha = C_1^\dagger C_3, \quad E_\beta = C_3^\dagger C_2 . \tag{5-310}$$

Method 2. From (5-272a), we have

$$E'_\alpha = \sqrt{\frac{1}{6}}C_1^\dagger C_3, \quad E'_\beta = \sqrt{\frac{1}{6}}C_3^\dagger C_2 ,$$

$$H'_\alpha = [E'_\alpha, E'_{-\alpha}] = \frac{1}{6}(n_1 - n_3), \quad H'_\beta = [E'_\beta, E'_{-\beta}] = \frac{1}{6}(n_3 - n_2) ,$$

$$\left[\begin{pmatrix} H'_\alpha \\ H'_\beta \end{pmatrix}, E'_\alpha\right] = \frac{1}{6}\begin{pmatrix} 2 \\ -1 \end{pmatrix} E'_\alpha, \quad \left[\begin{pmatrix} H'_\alpha \\ H'_\beta \end{pmatrix}, E'_\beta\right] = \frac{1}{6}\begin{pmatrix} -1 \\ 2 \end{pmatrix} E'_\beta . \tag{5-311}$$

The Cartan matrix for A_2 is $\begin{pmatrix} 2 & -1 \\ -1 & 2 \end{pmatrix}$. Therefore $\mathcal{N}_\alpha = \mathcal{N}_\beta = \frac{1}{6}$, and the Chevalley for su_3 is: $H_\alpha = 6H'_\alpha, H_\beta = 6H'_\beta, E_\alpha = \sqrt{6}E'_\alpha, E_\beta = \sqrt{6}E'_b$ etc., in agreement with (5-310).

Ex. 5.7. Construct the Chevalley basis for the Lie algebras B_2 and C_2 given in Exs. 5.3 and 5.4.

5.24. Alogrithms for Computing the Roots and Weights

Using the definition on the Dynkin representative of a root or weight vector, Theorems 5.10 and 5.20 can be reformulated as follows:

Theorem 5.25: If there is a root β (or a weight Λ) whose DYN representative is $(m_1, m_2, \ldots, m_l)$, then there exist l strings of roots (weights),

$$\beta, \beta - \theta_i\alpha_i, \ldots, \beta - m_i\alpha_i , \quad i = 1, 2, \ldots, l$$
$$(\Lambda, \Lambda - \theta_i\alpha_i, \ldots, \Lambda - m_i\alpha_i), \quad \theta_i = \text{sign } (m_i) . \tag{5-312}$$

This theorem provides us with the following two algorithms for roots and weights.

Algorithm 1: The steps for obtaining all positive roots from the Dynkin diagram are:

1. Construct the Cartan matrix A and write down the DYN representatives of the l simple roots,

$$\alpha_j = (A_{1j}, \ldots, A_{lj}) , \quad j = 1, 2, \ldots, l . \tag{5-313}$$

2. Starting from $\alpha_1 = (A_{11}, A_{21}, \ldots, A_{l1})$ for each *negative coordinates* A_{i1}, we have a string of positive roots,

$$\alpha_1, \alpha_1 + \alpha_i, \alpha_1 + 2\alpha_i, \ldots, \alpha_1 - A_{i1}\alpha_i . \tag{5-314}$$

3. Using (5-313), compute the DYN representative of the i-th terminal root,

$$\beta \equiv \alpha_1 - A_{i1}\alpha_i = (b_1, b_2, \ldots, b_l) . \tag{5-315}$$

Continue this process until all the coordinates of all the terminal roots are non-negative.

4. Repeat the above steps for the simple roots $\alpha_2, \ldots, \alpha_l$, ignoring in each step the roots which have already appeared in the foregoing steps.

In this way, we can easily obtain all the positive roots and the result can be checked by Eq. (5-288c).

Example 1: Computing the positive roots of the Lie algebra G_2. From the Cartan matrix (5-281′) we have the simple roots

$$\alpha_1 = (2, -3), \quad \alpha_2 = (-1, 2) .$$

From α_1 we find: $\alpha_1, \alpha_1 + \alpha_2, \alpha_1 + 2\alpha_2, \alpha_1 + 3\alpha_2 = (-1, 3)$; $\alpha_1 + 3\alpha_2 + \alpha_1 = 2\alpha_1 + 3\alpha_2 = (1, 0)$. From α_2 we have: $\alpha_2, \alpha_2 + \alpha_1$ (duplicated).

As a check note that $R = \frac{1}{2}[\alpha_1+\alpha_2+(\alpha_1+\alpha_2)+(\alpha_1+2\alpha_2)+(\alpha_1+3\alpha_2)+(2\alpha_1+3\alpha_2)] = (1,1)$.

Example 2: Computing the positive roots of the Lie algebra B_3. From the Cartan matrix (5-282b) we have the simple roots

$$\alpha_1 = (2,-1,0), \quad \alpha_2 = (-1,2,-2), \quad \alpha_3 = (0,-1,2) \ .$$

From α_1 we find: $\alpha_1, \alpha_1+\alpha_2 = (1,1,-2)$; $\alpha_1+\alpha_2+\alpha_3, \alpha_1+\alpha_2+2\alpha_3 = (1,-1,2)$; $\alpha_1+2\alpha_2+2\alpha_3 = (0,1,0)$.
Similarly from $\alpha_2 : \alpha_2, \alpha_2+\alpha_3, \alpha_2+2\alpha_3 = (-1,0,2)$; $\alpha_1+\alpha_2+2\alpha_3$ (duplicated).
From $\alpha_3 : \alpha_3, \alpha_3+\alpha_2$ (duplicated).
Check: $R = \frac{1}{2}[\alpha_1+\alpha_2+\alpha_3+(\alpha_1+\alpha_2)+(\alpha_2+\alpha_3)+(\alpha_1+\alpha_2+\alpha_3)+(\alpha_2+2\alpha_3)+(\alpha_1+\alpha_2+2\alpha_3)+(\alpha_1+2\alpha_2+2\alpha_3)] = \frac{1}{2}(5\alpha_1+8\alpha_2+9\alpha_3) = (1,1,1)$.

Algorithm 2: The steps for computing the complete set of weights are as follows:

1. Same as step 1 for Algorithm 1.

2. Starting from the highest weight $\Lambda = (a_1, a_2, \dots, a_l)$, for each $a_i > 0$, we have a string of weights

$$\lambda^1(=\Lambda), \lambda^2, \dots, \lambda^{a_i+1}, \quad \lambda^n = \Lambda - (n-1)\alpha_i \ . \tag{5-316a}$$

3. Compute the DYN representative of each weight in the string. For each weight $\lambda^m = (m_1, \dots, m_l)$, if $m_k > 0$, then we have a string of weights

$$\lambda^m, \lambda^m - \alpha_k, \dots, \lambda^m - m_k\alpha_k \ . \tag{5-316b}$$

4. Continue the string-searching process for each weight which has not already appeared in the previous steps, until the lowest weight is reached.

5. If the weight system contains the null weight, then by the symmetry between the positive and negative weights, it suffices to compute only the positive weights.

Example 3: Compute the weight system for the irrep $\Lambda = (1,0)$ of G_2.

α_1 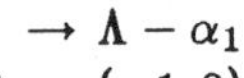 α_2,
$$A = \begin{pmatrix} 2 & -1 \\ -3 & 2 \end{pmatrix}, \quad A^{-1} = \begin{pmatrix} 2 & 1 \\ 3 & 2 \end{pmatrix}, \quad \alpha_1 = (2,-3), \quad \alpha_2 = (-1,2) \ .$$

From the first column of A^{-1}, we have $\Lambda = 2\alpha_1 + 3\alpha_2$.

Following the steps of Algorithm 2, we can easily obtain

$$\begin{array}{ccccccccc}
 & & & & & \nearrow & \Lambda-\alpha_1-3\alpha_2 \ (2,-3) & \rightarrow & \Lambda-2\alpha_1-3\alpha_2 \ (0,0) \ \rightarrow \\
\Lambda & \rightarrow & \Lambda-\alpha_1 & \rightarrow & \Lambda-\alpha_1-\alpha_2 & \rightarrow & \Lambda-\alpha_1-2\alpha_2 & & \\
(1,0) & & (-1,3) & & (0,1) & & (1,-1) & & \\
 & & & & & \searrow & \Lambda-2\alpha_1-2\alpha_2 \ (-1,2) & \rightarrow & \Lambda-2\alpha_1-3\alpha_2 \ (0,0) \ \rightarrow
\end{array}$$

$$\begin{array}{l}
\rightarrow \Lambda-3\alpha_1-3\alpha_2 \ (-2,3) \ \searrow \\
\qquad \Lambda-3\alpha_1-4\alpha_2 \ (-1,1) \rightarrow \Lambda-3\alpha_1-5\alpha_2 \ (0,-1) \rightarrow \Lambda-3\alpha_1-6\alpha_2 \ (1,-3) \rightarrow \Lambda-4\alpha_1-6\alpha_2 \ (-1,0) \ . \\
\rightarrow \Lambda-2\alpha_1-4\alpha_2 \ (1,-2) \ \nearrow
\end{array} \tag{5-317}$$

The weights $\mathbf{M}$ can be grouped into *layers*, as shown in (5-317), according to their layer index $L(M)$,

$$L(M) = \frac{1}{2}[\delta(\Lambda) - \delta(M)] \ , \tag{5-318a}$$

$$\delta(M) = 2\sum_{i=1}^{l} w^i = \sum_{i=1}^{l} a_i r_i \ , \tag{5-318b}$$

where $\{w^1, w^2, \ldots, w^l\}$ and $(a_1, a_2, \ldots, a_l)$ are the SRS and DYN representatives of $\mathbf{M}$, respectively, and r_i is the *height* of the weight system of the i-th basic rep. According to (5-295c) and (5-318),

$$r_i = 2\sum_{k=1}^{l} (A^{-1})_{ki} \ . \tag{5-318'}$$

The values of r_i are listed in (5-319). $L(M)$ is always an integer and corresponds to the number of simple roots that have to be subtracted from Λ in order to get M.

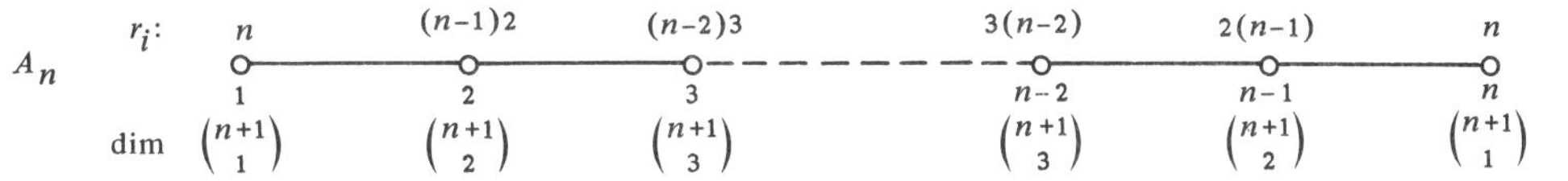

(5-319a)

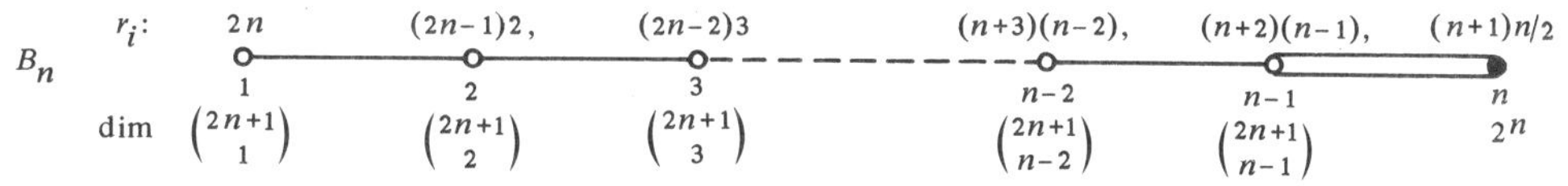

(5-319b)

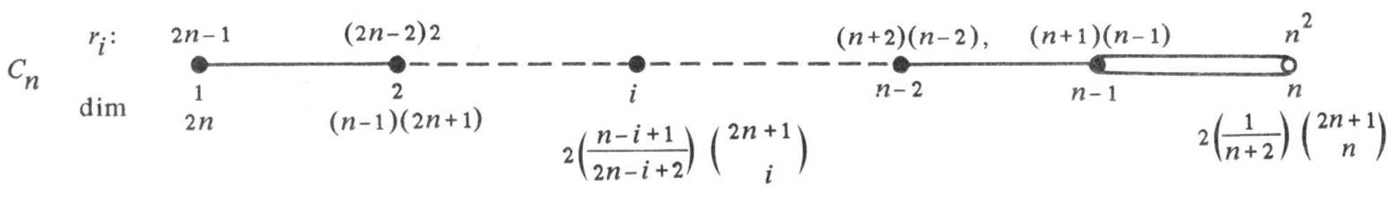

(5-319c)

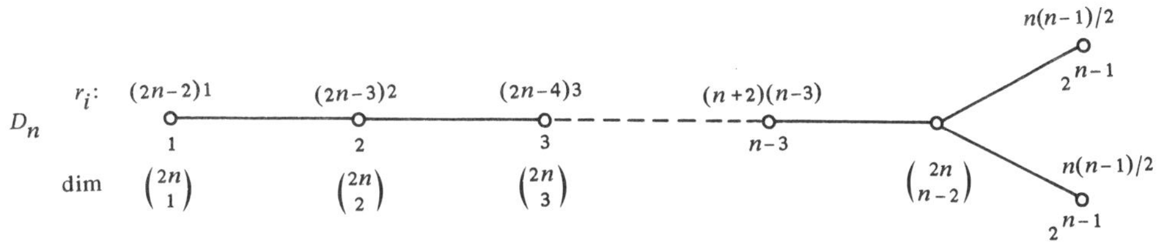

(5-319d)

If Λ is the highest weight, then $\delta(\Lambda)$ is called the *height of the irrep*, which equals to the number of layers minus one.

The *power* or *level* $\delta(\Lambda)$ of a weight Λ is half of the layer $L(\Lambda)$, i.e., $\delta(\Lambda) = \frac{1}{2}L(\Lambda)$.

In (5-317), the highest and lowest weights belong to the zeroth and tenth layers, respectively. The height of the irrp (1,0) of G_2 is $T = 2[(A^{-1})_{11} + (A^{-1})_{21}] = 2(2+3) = 10$. The irrep (1,0) is the adjoint rep of G_2 with dimension 14. Thus we can infer that the null weight has the multiplicity two. In general case, Algorithm 2 alone is not enough for determining the multiplicity of weights. For this we need the help of (5-321b).

Similarly, we can find the positive weights for the irrep (0,1) of G_2. Collecting the positive weights for the irreps (1,0) and (0,1), and arranging them as columns of a matrix, we obtain

$$W^{(1)} = \begin{pmatrix} 1 & -1 & 0 & 1 & 2 & -1 \\ 0 & 3 & 1 & -1 & -3 & 2 \end{pmatrix} , \quad W^{(2)} = \begin{pmatrix} 0 & 1 & -1 \\ 1 & -1 & 2 \end{pmatrix} . \tag{5-319'}$$

$W^{(i)}$ will be referred to as the *weight system* of the basic rep M^i.

Algorithms 1 and 2 are more effective than the methods in Wybourne (1974) and Laskar (1977).

The weights for an irrep (a_1, a_2) of any Lie algebra of rank 2 form "concentric" polygons, which are invariant under the corresponding Weyl group. According to Fig. 5.19 and a Theorem proved in Eq. (8-17), we know that the Weyl groups for $A_2, B_2(C_2)$ and G_2 are the point groups C_{3v}, C_{4v} and C_{6v} respectively. The perimeter of the weights is a polygon consisting of sides with lengths equal to $a_1|\boldsymbol{\alpha}_1|$ and $a_2|\boldsymbol{\alpha}_2|$ alternatively. If both a_1 and a_2 are non-zero, then it is a hexagon, octagon and dodecahedron for A_2, B_2 (or C_2) and G_2, respectively (see Fig. 5.24), and if one of a_1 and a_2 is zero, then it is a triangle, quadrangle and hexagon for A_2, B_2 (or C_2) and G_2, respectively.

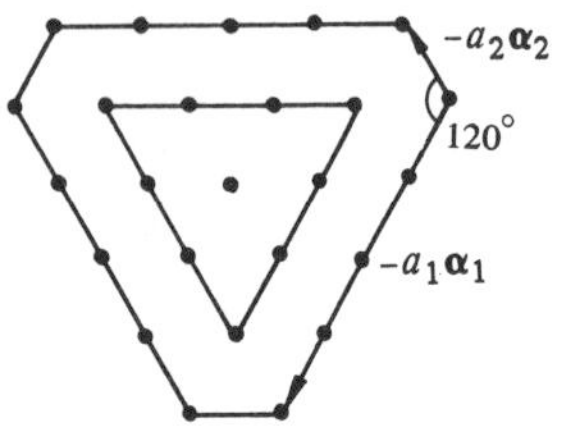

Fig. 5.24-1. Weight diagram for the A_2 irrep (41)

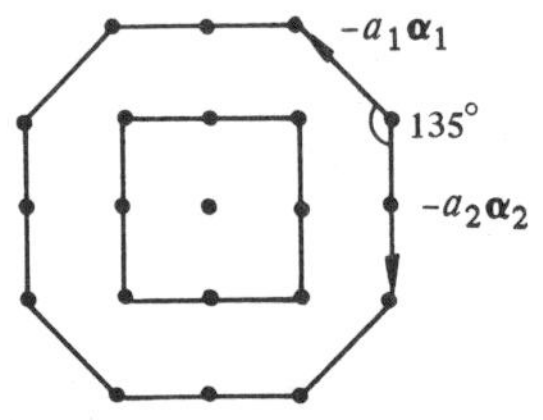

Fig. 5.24-2. Weight diagram for the B_2 irrep (12)

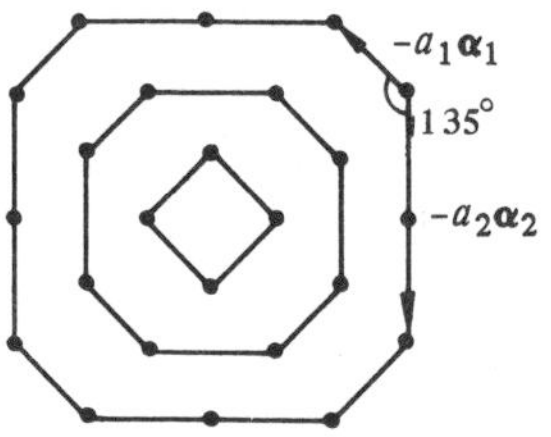

Fig. 5.24-3 Weight diagram for the C_2 irrep (12)

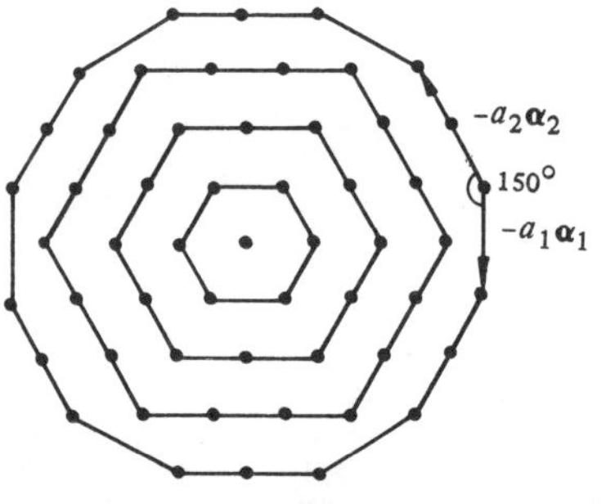

Fig. 5.24-4. Weight diagram for the G_2 irrep (12)

The foregoing procedure exemplified in (5-317) suggests the following diagrammatic method for constructing the weight diagram of the irrep (a_1a_2) for any Lie algebra of rank 2.

1. In the root diagram draw the simple roots $\boldsymbol{\alpha}_1$ and $\boldsymbol{\alpha}_2$, as well as the basic weights $\mathbf{M}^1$ and $\mathbf{M}^2$.

2. Mark the point $\Lambda = a_1\mathbf{M}^1 + a_2\mathbf{M}^2$. Starting from the two sides $-a_1\boldsymbol{\alpha}_1$ and $-a_2\boldsymbol{\alpha}_2$ originating from the point Λ, we can easily draw the perimeter polygon, as shown in Figs. 5.24.

3. The weight diagram results from a special "chess game." The rule for this "chess" is that (a) the chessman can only take two kinds of steps, i.e., moving ahead through either $-\boldsymbol{\alpha}_1$ or $-\boldsymbol{\alpha}_2$; (b) the chessman is not allowed to go outside the perimeter. Then starting from the point $\Lambda = a_1\mathbf{M}^1 + a_2\mathbf{M}^2$, all the possible stopover points of the chessman yield the weight diagram of the irrep (a_1a_2).

As an example, in Fig. 5.24-6, we show the vectors $\boldsymbol{\alpha}_1, \boldsymbol{\alpha}_2, \mathbf{M}^1$ and $\mathbf{M}^2$ of G_2 with solid lines. Figure 5.24-6 is the weight diagram for the irrep (1,0) of G_2 which results from playing the "G_2 chess." As is seen, there is a one-to-one correspondence between Fig. 5.24-6 and Eq. (5-317).

We know that A_2 is a subalgebra of G_2. The simple roots and basic weights of A_2 are also shown in Fig. 5.24-5.

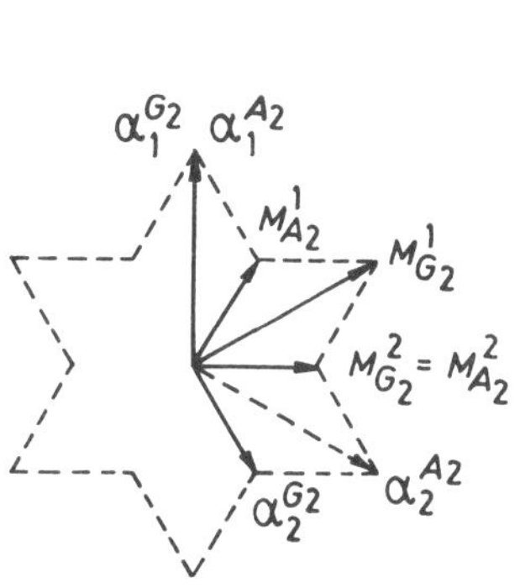

Fig. 5.24-5. The simple roots and basic weights of A_2 and G_2

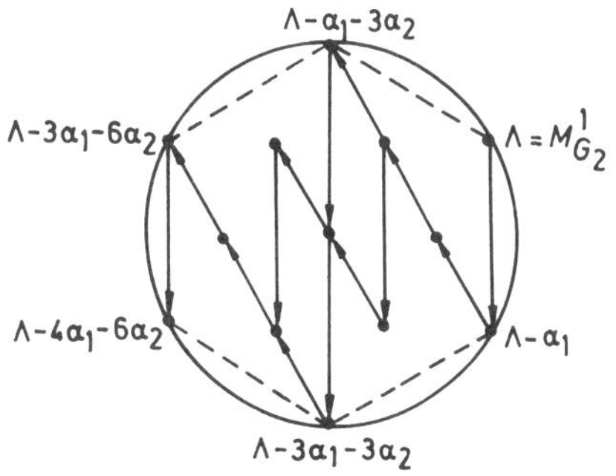

Fig. 5.24-6. The weight diagram of the irrep (1,0) of G_2 resulting from playing "G_2 chess."

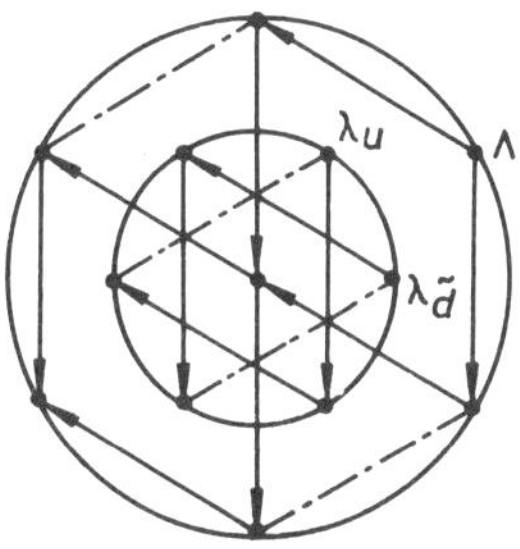

Fig. 5.24-7. The weight diagrams of Fig. 5.24-6 decomposed into three weight diagrams (11), (10) and (01) of A_2 under the "A_2 chess rule."

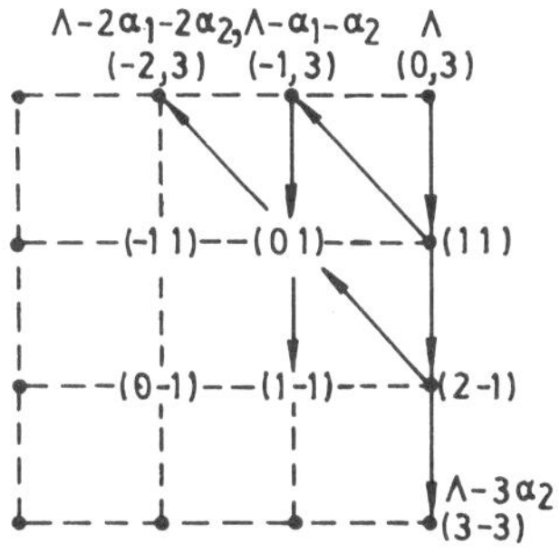

Fig. 5.24-8. The weight diagram of the irrep (03) of B_2.

Now let us take a look at what happens if starting from the same point $\Lambda = \mathbf{M}^1_{G_2} = \mathbf{M}^1_{A_2} + \mathbf{M}^2_{A_2}$, we play "$A_2$ chess" instead of "G_2 chess." We immediately see that the chessman can only reach eight weights (the null weights being counted twice), which constitute the weight diagram of the irrep (11) of A_2, as shown in Fig. 5.24-7. Starting from a weight other than these weights, say λ_u in Fig. 5.24-7, by playing "A_2 chess" again, we obtain a triangle weight diagram, corresponding to the irrep (10) of A_2. Similarly, we obtain another triangle belonging to the irrep (01) of A_2. It is thus seen that the fourteen weights of the irrep (10) of G_2, under the "A_2 chess" rule, group into three disconnected subweight diagrams, corresponding to the subduction

$$\begin{array}{lccc} & G_2 & & A_2 \\ & (10) & \longrightarrow & (11)+(10)+(01) \; . \\ \dim & 14 & = & 8\; + \;3\; + \;3 \end{array} \tag{5-320}$$

Freudenthal formula

The multiplicity n_M of a weight M in the irrep Λ can be calculated from the Freudenthal recursive formula (Freudenthal 1969),

$$[(\Lambda + R, \Lambda + R) - (M + R, M + R)]n_M = 2\sum_{\beta>0} \sum_{k=1,2,\ldots} (M + k\beta, \beta)n_{M+k\beta} \; . \tag{5-321a}$$

It can be recast into a more suitable form,

$$(C_\Lambda - C_M)n_M = \sum_{\beta>0} \sum_{k=1,2,\ldots} 2(M+k\beta,\beta)n_{M+k\beta} , \qquad (5\text{-}321\text{b})$$

where the sum over k terminates when $M + k\beta$ exceeds the highest weight, $C_\Lambda(C_M)$ is the eigenvalue of the Casimir operator for the highest weight Λ (the weight M). If $\{w^1,\ldots,w^l\}$ and $(a_1\ldots a_l)$ are its SRS and DYN are representatives of Λ, respectively, one has by using (5-301) and (5-288c),

$$C_\Lambda = (\Lambda,\Lambda) + 2(R,\Lambda) = \sum_k N_k w^k(a_k+2) . \qquad (5\text{-}322\text{a})$$

Suppose $M' = M + k\beta = (m'_1,\ldots,m'_l)$ and $\beta = \{\beta^1,\ldots,\beta^l\}$, then again using (5-301), one has

$$2(M+k\beta,\beta) = 2(m'_1,\ldots,m'_l) \bullet \{\beta^1,\ldots,\beta^l\}$$

$$= \sum_{i=1}^l (\alpha_i,\alpha_i)m'_i\beta^i . \qquad (5\text{-}322\text{b})$$

According to Theorems 5.21 and 5.17, it suffices to calculate the multiplicities of the weights which are not equivalent to the highest weight, and among the equivalent weights, only the multiplicity of the weight which belongs to the lowest layer. The multiplicities of weights are calculated recursively from low layers to high layers. To calculate the multiplicity of a weight M, we first noted down all the weights M' whose layer indices are lower than that of M. The weights M' are of the form

$$M' = \sum_i b_i\alpha_i + M ,$$

with $b_i \geq 0$. Secondly, we check whether $\sum_i b_i\alpha_i = k\beta$, where β is a positive root. If not, then M' is ignored, and if yes, its contribution to the sum is calculated by (5-322b).

Example 4: Determine the weights and their multiplicities of the irrep (03) of B_2:

$$\alpha_1 \!=\!=\!=\! \alpha_2 , \quad \alpha_1 = (2,-2), \quad \alpha_2 = (-1,2) .$$

Using Algorithm 1, it is trivial to get the positive roots:

$$\Sigma^+ : \begin{array}{cccc} \alpha_1 & \alpha_2 & \alpha_1+\alpha_2 & \alpha_1+2\alpha_2 \\ \{10\}, & \{01\}, & \{11\}, & \{12\} \end{array} . \qquad (5\text{-}323)$$

Using Algorithm 2, we can find the positive weights listed below, where both the DYN and SRS representatives of each weight are given, which are to be used for computing (5-322).

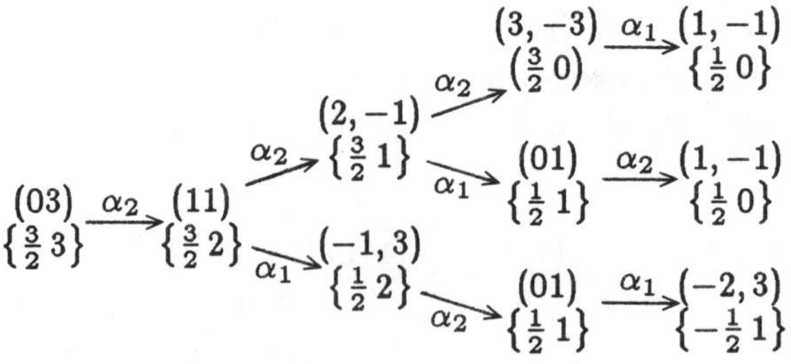

$$\begin{array}{c}
(03) \xrightarrow{\alpha_2} (11) \xrightarrow{\alpha_2} \begin{array}{c}(2,-1)\\ \{\frac{3}{2}1\}\end{array} \xrightarrow{\alpha_2} \begin{array}{c}(3,-3)\\ (\frac{3}{2}0)\end{array} \xrightarrow{\alpha_1} \begin{array}{c}(1,-1)\\ \{\frac{1}{2}0\}\end{array} \\
\{\tfrac{3}{2}3\} \quad \{\tfrac{3}{2}2\} \xrightarrow{\alpha_1} \begin{array}{c}(-1,3)\\ \{\frac{1}{2}2\}\end{array} \xrightarrow{\alpha_1} \begin{array}{c}(01)\\ \{\frac{1}{2}1\}\end{array} \xrightarrow{\alpha_2} \begin{array}{c}(1,-1)\\ \{\frac{1}{2}0\}\end{array} \\
\xrightarrow{\alpha_2} \begin{array}{c}(01)\\ \{\frac{1}{2}1\}\end{array} \xrightarrow{\alpha_1} \begin{array}{c}(-2,3)\\ \{-\frac{1}{2}1\}\end{array}
\end{array} . \qquad (5\text{-}324)$$

The weight diagram is shown in Fig. 5.24-8. From the diagram it is clear that we only need to determine the multiplicity of the weight $M = (01)$, which is in the third layer. Using (5-322), Eq. (5-321b) reads

$$[C_{(03)} - C_{(01)}]n_{(01)} = 2[(03) \bullet \{12\} + (11) \bullet \{11\} + (-1,3) \bullet \{01\} + (2,-1) \bullet \{10\}],$$
$$8n_{(01)} = 6 + (2+1) + 3 + 4 ,$$

which gives $n_{(01)} = 2$. Hence each of the four inner weights in Fig. 5.24-8 has the multiplicity 2, and the dimension of the irrep (03) of B_2 is equal to $12 + 2 \times 4 = 20$.

The dominant weight multiplicities for reps of simple Lie algebras have been tabulated by Bremner (1985).

Ex. 5.8. Show that $2\mathbf{R} = \sum_i r_i \boldsymbol{\alpha}_i$ for A_l and D_l; $2\mathbf{R}(B_l) = \sum_i r_i(C_l)\boldsymbol{\alpha}_i$ and $2\mathbf{R}(C_l) = \sum_i r_i(B_l)\boldsymbol{\alpha}_i$.

Ex. 5.9. Find the positive roots as well as eigenvalues of the Casimir operators of A_3 and C_3.

Ex. 5.10. Find the weight systems for the irreps (100) and (001) of B_3.

Ex. 5.11. Find the weight systems for the irreps $M^i, i = 1, 2, 3, 4$ of D_4.

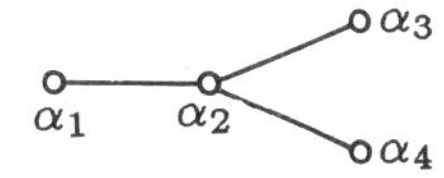

Ex. 5.12. Find the multiplicities for the weights of the irrep (11) of A_2 and irrep (200) (the adjoint rep) of C_3.

Ex. 5.13. Draw the weight diagrams for the irreps (10), (01), (11) and (21) of B_2 and C_2.

Ex. 5.14. Draw the weight diagrams for the irreps (30) and (31) of A_2.

Ex. 5.15. Using the diagramatic method, show that the weight diagram of the irrep (02) of G_2 decomposes in the following way,

$$\begin{array}{ccc} G_2 & & A_2 \\ (02) & \longrightarrow & (20) + (02) + (11) + (10) + (01) + (00)\ . \end{array}$$

Ex. 5.16. Derive the dimension formulas for the irreps of SP_4, SO_5, SO_6 and SO_8.

Ex. 5.17. Find the DYN representaive of the highest weight of the adjoint rep of B_3.

5.25. The Fundamental Weight System

In this section, we deal only with the Lie algebras A_l, B_l, C_l and D_l.

Using Algorithm 2, we can easily obtain the weight systems W, which will be referred to as the *weight matrices*, for the fundamental (defining) reps $\mathbf{M}^1$ of A_l, B_l, C_l and D_l.

$$A_l : W = \begin{array}{c} \begin{array}{ccccc} \lambda^1 & \lambda^2 & & \lambda^l & \lambda^{l+1}\ \text{(DYN)} \end{array} \\ \left[\begin{array}{ccccc} 1 & -1 & \cdots & 0 & 0 \\ 0 & 1 & \cdots & 0 & 0 \\ 0 & 0 & \cdots & \cdot & \cdot \\ & & & \cdot & \cdot \\ \cdot & \cdot & & \cdot & \cdot \\ \cdot & \cdot & & 0 & 0 \\ \cdot & \cdot & & -1 & 0 \\ 0 & 0 & \cdot & 1 & -1 \end{array}\right] \end{array}, \quad B_l : W = \begin{array}{c} \begin{array}{ccccc} \lambda^1 & \lambda^2 & & \lambda^{l-1} & \lambda^l\ \text{(DYN)} \end{array} \\ \left[\begin{array}{ccccc} 1 & -1 & \cdots & 0 & 0 \\ 0 & 1 & \cdots & 0 & 0 \\ 0 & 0 & \cdots & \cdot & \cdot \\ & & & \cdot & \cdot \\ \cdot & \cdot & & \cdot & \cdot \\ \cdot & \cdot & & -1 & 0 \\ \cdot & \cdot & & 1 & -1 \\ 0 & 0 & \cdots & 0 & 2 \end{array}\right] \end{array}, \tag{5-325a, b}$$

$$C_l : W = \begin{array}{c} \begin{array}{ccccc} \lambda^1 & \lambda^2 & & \lambda^{l-1} & \lambda^l\ \text{(DYN)} \end{array} \\ \left[\begin{array}{ccccc} 1 & -1 & \cdots & 0 & 0 \\ 0 & 1 & \cdots & 0 & 0 \\ 0 & 0 & \cdots & \cdot & \cdot \\ \cdot & \cdot & & \cdot & \cdot \\ \cdot & \cdot & & -1 & 0 \\ \cdot & \cdot & & 1 & -1 \\ 0 & 0 & \cdots & 0 & 1 \end{array}\right] \end{array}, \quad \begin{array}{l} D_l : W = \\ l \geq 4 \end{array} \begin{array}{c} \begin{array}{cccccc} \lambda^1 & \lambda^2 & & \lambda^{l-2} & \lambda^{l-1} & \lambda^l\ \text{(DYN)} \end{array} \\ \left[\begin{array}{cccccc} 1 & -1 & \cdots & 0 & 0 & 0 \\ 0 & 1 & \cdots & 0 & 0 & 0 \\ 0 & 0 & \cdots & \cdot & \cdot & \cdot \\ & & & \cdot & \cdot & \cdot \\ \cdot & \cdot & & \cdot & \cdot & \cdot \\ \cdot & \cdot & & -1 & 0 & 0 \\ \cdot & \cdot & & 1 & -1 & 0 \\ 0 & 0 & \cdots & 0 & 1 & -1 \\ 0 & 0 & \cdots & 0 & 1 & 1 \end{array}\right] \end{array} \tag{5-325c,d}$$

where the j-th column of W gives the DYN representative of the j-th weight of the irrep $\mathbf{M}^1$,

$$(\lambda^j)^{\mathrm{DYN}} = (W_{1j}, \dots, W_{nj}) , \tag{5-326a}$$

where

$$n = \begin{cases} l+1, \\ l, \end{cases} \quad \text{for} \begin{cases} A_l \\ B_l, C_l, D_l . \end{cases} \tag{5-326b}$$

That is,

$$\lambda^j = \sum_i W_{ij} \mathbf{M}^i , \tag{5-327a}$$

$$W_{ij} = (\overline{\mathbf{M}}_i, \lambda^j) = 2\frac{(\alpha_i, \lambda^j)}{(\alpha_i, \alpha_i)} . \tag{5-327b}$$

In the process of constructing the weight matrices (5-325), we obtain incidentally,

$$\begin{matrix} A_l : \\ B_l \quad (C_l) : \\ D_l : \end{matrix} \ \alpha_i = \lambda_i - \lambda_{i+1},\ i = \begin{cases} 1, 2, \dots, l; \\ 1, 2, \dots, l-1; \quad \alpha_l = \lambda_l (\alpha_l = 2\lambda_l) ; \\ 1, 2, \dots, l-2; \quad \alpha_{l-1} = \lambda_{l-1} - \lambda_l, \\ \qquad \alpha_l = \lambda_{l-1} + \lambda_l \end{cases} . \tag{5-327'}$$

5.26. Fundamental Weight System Representation and Cartesian Representation

The representation with $(\lambda^1, \dots, \lambda^n)$ as the basis vectors is called the Fundamental weight system (FWS) representation,

$$\mathbf{v} = \sum_{i=1}^{n} v_i^{\mathrm{FWS}} \lambda^i , \tag{5-328a}$$

with the convention that square brackets are used for denoting a FWS representative,

$$[\mathbf{v}]^{\mathrm{FWS}} \equiv [v_1, \dots, v_n]^{\mathrm{FWS}} . \tag{5-328b}$$

From (5-325a) we see that of the $l+1$ weights λ^i for A_l, only l weights are linearly independent,

$$\lambda^1 + \lambda^2 + \ldots + \lambda^{l+1} = 0 . \tag{5-328'}$$

The roots and weights can also be represented in the Cartesian (CAR) representation, the basis vectors of which are just the n mutually orthogonal unit vectors $\mathbf{e}_1, \mathbf{e}_2, \dots, \mathbf{e}_n$. We use $|i\rangle$ to denote the i-th single particle state. The basis vectors of the fundamental (or defining) reps and the Cartan subalgebras of A_l, B_l, C_l and D_l are

$$\begin{aligned} &A_l : \quad \{|i\rangle : i = 1, 2, \dots, l+1\} , \quad H_i = n_i - \tfrac{1}{l+1} \sum_{j=1}^{l+1} n_j , \\ &B_l : \quad \{|i\rangle : i = 0, \pm 1, \pm 2, \dots, \pm l\} , \ H_i = n_i - n_{-i} , \\ &C_l, D_l : \{|i\rangle : i = \pm 1, \pm 2, \dots \pm l\}, \quad H_i = n_i - n_{-i} \end{aligned} \tag{5-329}$$

It is easily seen that

$$\mathbf{H}|i\rangle = \lambda^i |i\rangle , \tag{5-330}$$

$$\lambda^i = \begin{cases} \mathbf{e}_i - \frac{1}{l+1}\mathbf{p}, & i = 1, 2, \dots, l+1 \qquad A_l \\ \mathbf{e}_i, & i = 1, 2, \dots, l \qquad B_l, C_l, D_l , \end{cases} \tag{5-331a,b}$$

where $\mathbf{p}$ is a sum of the $l+1$ unit vectors. The vector $\mathbf{p}$ is perpendicular to the l simple roots $\mathbf{e}_i - \mathbf{e}_{i+1}$, $i = 1, 2, \ldots l$, of A_l, and thus perpendicular to any root or weight vectors $\mathbf{v}$ of A_l,

$$\mathbf{p} = \mathbf{e}_1 + \mathbf{e}_2 + \ldots + \mathbf{e}_{l+1}, \quad (\mathbf{p}, \mathbf{v}) = 0 . \tag{5-331c}$$

Notice that Eqs. (5-327′), (5-331) and Figs. 5.20-1 are self consistent.

The metric tensor for the FWS representation is

$$A_l : \; g^{ij}_{\text{FWS}} = (\lambda^i, \lambda^j) = \begin{cases} l/(l+1) , & i = j \\ & \text{for} \\ -1/(l+1) , & i \neq j , \end{cases} \tag{5-332a}$$

$$B_l, C_l, D_l : \; (\lambda^i, \lambda^j) = \delta_{ij} . \tag{5-332b}$$

Hence we see that the CAR representation is identical with the FWS representation for B_l, C_l and D_l. For A_l, the metric tensor of (5-332a) is a singular matrix, therefore we cannot define the dual basis $\{\overline{\lambda}_i\}$ as in the usual way. Put differently, we notice that for A_l, the expansion (5-328a) is not unique. To ensure uniqueness, we impose the condition that

$$\sum_{i=1}^{l+1} v_i^{\text{FWS}} = 0 . \tag{5-333}$$

Substituting (5-331a) into (5-328a), it is seen that under the constraint (5-333), for A_l we also have

$$\mathbf{v} = \sum_{i=1}^{l+1} v_i^{\text{FWS}} \mathbf{e}_i, \quad v_i^{\text{FWS}} = (\mathbf{e}_i, \mathbf{v}) . \tag{5-334a}$$

Due to (5-331a),

$$v_i^{\text{FWS}} = (\mathbf{e}_i, \mathbf{v}) = (\lambda^i, \mathbf{v}) . \tag{5-334b}$$

Using (5-334a), the scalar product of any root(s) and/or weight(s) of A_l can be easily calculated. Therefore we reached the important conclusion that for all the cases of A_l, B_l, C_l and D_l, the representatives of any root or weight in the FWS and CAR representations are identical,

$$[\mathbf{v}]^{\text{FWS}} = [\mathbf{v}]^{\text{CAR}} , \tag{5-335a}$$

and that the scalar product of the root and/or weight vectors can always be calculated from

$$(\mathbf{u}, \mathbf{v}) = \sum_{i=1}^{n} u_i^{\text{FWS}} v_i^{\text{FWS}} , \tag{5-335b}$$

in spite of the fact that the basis (5-331a) for the FWS representation of A_l is not orthonormal.

By (5-331), (5-327b) can be rewritten as

$$W_{ij} = 2\frac{(\alpha_i, \mathbf{e}_j)}{(\alpha_i, \alpha_i)} = (\overline{\mathbf{M}}_i, \mathbf{e}_j) . \tag{5-336}$$

The significance of the FWS representation is that the FWS representative $[l_1, l_2, \ldots, l_n]$ of a highest weight Λ is just the *Cartan-Weyl*, or *natural label* for the irreps.

In the DYN representation, because of (5-326a), Eq. (5-328a) reads

$$v_i^{\text{DYN}} = \sum_{j=1}^{n} W_{ij} v_j^{\text{FWS}} . \tag{5-337a}$$

Combining (5-299a) and (5-337a), we get

$$(\mathbf{v})^{\mathrm{DYN}} = A\{\mathbf{v}\}_{\mathrm{SRS}} = W[\mathbf{v}]^{\mathrm{FWS}} . \tag{5-337b}$$

Equation (5-337b) shows that the weight matrix W is the transformation matrix from the Cartan-Weyl label $[l_1, l_2, \ldots, l_n]$ to the Dynkin label $(a_1, a_2, \ldots, a_l)$.

Now we are in a position to examine the meaning of the basic irreps $\mathbf{M}^i$ of A_l, B_l, C_l and D_l. According to (5-284d) and (5-325), we have

$$(\mathbf{M}^1)^{[1^k]}_{\text{h.w.}} = \mathbf{M}^k , \quad k = \begin{cases} 1,2,\ldots,l, & A_l , \\ 1,2,\ldots,l-1, & B_l , \\ 1,2,\ldots,l-2, & D_l , \end{cases} \tag{5-338a}$$

$$(\mathbf{M}^1)^{[1^k]}_{\text{h.w.}} = \mathbf{M}^k , \quad k = 1,2,\ldots,l, \qquad C_l . \tag{5-338b}$$

According to the dimension formula for the basic reps $\mathbf{M}^i$ shown in (5-319), for the cases of (5-338a), we have

$$\dim (\mathbf{M}^1)^{[1^k]} = \dim (\mathbf{M}^k) . \tag{5-339}$$

Therefore the antisymmetric reps in (5-338a) are irreducible. Stated differently, the basic rep $\mathbf{M}^k$ for the cases (5-338a) is just the k-th antisymmetric power of the defining rep $\mathbf{M}^1$,

$$(\mathbf{M}^1)^{[1^k]} = \mathbf{M}^k, \quad k = \begin{cases} 1,2,\ldots,l, & A_l , \\ 1,2,\ldots,l-1 , & B_l , \\ 1,2,\ldots,l-2 , & D_l . \end{cases} \tag{5-340}$$

The reps which can be constructed from the powers of the defining rep $\mathbf{M}^1$ are called the *vector* or *true reps*. The basic rep M^l of B_l, and the basic reps $\mathbf{M}^{l-1}$ and $\mathbf{M}^l$ of D_l cannot be constructed in this way, and they are referred to as the *spinor reps*.

Remark: For D_4, the basic reps $\mathbf{M}^1, \mathbf{M}^3$ and $\mathbf{M}^4$ are carried into one another by automorphisms of D_4, and we must identify one of them as the defining rep and the other two with the spinor reps.

The Kronecker product of two spinor reps gives vector reps, while that of a spinor rep and a vector rep gives spinor reps. For example, we have

$$B_l : (\mathbf{M}^l)^{[2]}_{\text{h.w.}} = (\mathbf{M}^1)^{[1^l]}_{\text{h.w.}}, \quad (\mathbf{M}^l)^{[1^2]}_{\text{h.w.}} = \mathbf{M}^{l-1} , \tag{5-341a}$$

$$D_l : (\mathbf{M}^l)^{[2]}_{\text{h.w.}} = (\mathbf{M}^1)^{[1^l]}_{\text{h.w.}} ,$$

$$(\mathbf{M}^{l-1} \times \mathbf{M}^l)_{\text{h.w.}} = (\mathbf{M}^1)^{[1^{l-1}]}_{\text{h.w.}} ,$$

$$(\mathbf{M}^{l-1})^{[1^2]}_{\text{h.w.}} = (\mathbf{M}^l)^{[1^2]}_{\text{h.w.}} = \mathbf{M}^{l-2} . \tag{5-341b}$$

In the Dynkin diagrams shown in Fig. 5.20-1, the simple roots are expressed in terms of the Cartesian basis $\{\mathbf{e}_i\}$. Let us define the *root matrix* R whose j-th column gives the CAR representaive of α_j, according to (5-335a), which is just the FWS representative of α_j. Therefore,

$$\boldsymbol{\alpha}_j = \sum_{i=1}^{n} R_{ij}\mathbf{e}_i = \sum_{i=1}^{n} R_{ij}\lambda^i , \tag{5-342a}$$

$$[\boldsymbol{\alpha}_j]^{\mathrm{FWS}} = [R_{1j}, \ldots, R_{nj}] \ , \tag{5-342b}$$

$$R_{ij} = (\mathbf{e}_i, \boldsymbol{\alpha}_j) = (\lambda^i, \boldsymbol{\alpha}_j) \ . \tag{5-342c}$$

From Eq. (5-327′) we have

$$A_l : R = \begin{array}{c} \\ \lambda^1 \\ \lambda^2 \\ \cdot \\ \cdot \\ \cdot \\ \lambda^l \\ \lambda^{l+1} \\ \text{(DYN)} \end{array} \begin{array}{c} \begin{array}{ccccc} \alpha_1 & \alpha_2 & & \alpha_{l-1} & \alpha_l \end{array} {}^{[\mathrm{FWS}]} \\ \left[\begin{array}{ccccc} 1 & 0 & \cdots & 0 & 0 \\ -1 & 1 & \cdots & 0 & 0 \\ 0 & -1 & \cdots & \cdot & \cdot \\ \cdot & \cdot & & \cdot & \cdot \\ \cdot & \cdot & & 1 & 0 \\ 0 & 0 & \cdots & -1 & 1 \\ 0 & 0 & \cdots & 0 & -1 \end{array}\right] \\ \\ \end{array} \ , \quad B_l : R = \begin{array}{c} \\ \lambda^1 \\ \lambda^2 \\ \cdot \\ \cdot \\ \cdot \\ \lambda^{l-1} \\ \lambda^l_{C_l} \\ \text{(DYN)} \end{array} \begin{array}{c} \begin{array}{ccccc} \alpha_1 & \alpha_2 & & \alpha_{l-1} & \alpha_1 \end{array} {}^{[\mathrm{FWS}]} \\ \left[\begin{array}{ccccc} 1 & 0 & \cdots & 0 & 0 \\ -1 & 1 & \cdots & 0 & 0 \\ 0 & -1 & \cdots & \cdot & \cdot \\ \cdot & \cdot & & \cdot & \cdot \\ \cdot & \cdot & & 0 & 0 \\ \cdot & \cdot & \cdots & 1 & 0 \\ 0 & 0 & \cdots & -1 & 1 \end{array}\right] \\ \\ \end{array}$$

(5-343a,b)

$$C_l : R = \begin{array}{c} \\ \lambda^1 \\ \lambda^2 \\ \cdot \\ \cdot \\ \cdot \\ \lambda^{l-1} \\ \lambda^l_{B_l} \\ \text{(DYN)} \end{array} \begin{array}{c} \begin{array}{ccccc} \alpha_1 & \alpha & & \alpha_{l-1} & \alpha_l \end{array} {}^{[\mathrm{FWS}]} \\ \left[\begin{array}{ccccc} 1 & 0 & \cdots & 0 & 0 \\ -1 & 1 & \cdots & 0 & 0 \\ 0 & -1 & \cdots & \cdot & \cdot \\ \cdot & \cdot & & \cdot & \cdot \\ \cdot & \cdot & & 0 & 0 \\ 0 & 0 & \cdots & 1 & 0 \\ 0 & 0 & \cdots & -1 & 2 \end{array}\right] \\ \\ \end{array} \ , \quad \begin{array}{l} D_l : R = \\ l \geq 4 \end{array} \begin{array}{c} \\ \lambda^1 \\ \lambda^2 \\ \cdot \\ \cdot \\ \cdot \\ \lambda^{l-1} \\ \lambda^l \\ \text{(DYN)} \end{array} \begin{array}{c} \begin{array}{ccccc} \alpha_1 & \alpha_2 & & \alpha_{l-1} & \alpha_1 \end{array} {}^{[\mathrm{FWS}]} \\ \left[\begin{array}{ccccc} 1 & 0 & \cdots & 0 & 0 \\ -1 & 1 & \cdots & 0 & 0 \\ 0 & -1 & \cdots & \cdot & \cdot \\ \cdot & \cdot & & \cdot & \cdot \\ \cdot & \cdot & & 0 & 0 \\ \cdot & \cdot & \cdots & 1 & 1 \\ 0 & 0 & \cdots & -1 & 1 \end{array}\right] \\ \\ \end{array} \ ,$$

(5-343c,d)

Comparing (5-325) with (5-343), one sees that the root matrices are simply related to the weight matrices by

$$R = \tilde{W} \ , \quad \text{for } A_l \text{ and } D_l \ , \tag{5-344a}$$

$$R^{B_l} = \tilde{W}^{C_l}, \quad R^{C_l} = \tilde{W}^{B_l} \ . \tag{5-344b}$$

From (5-326a) and (5-344) one knows that for A_l and D_l, the j-th row of the root matrix R is just the DYN representative of the weight λ^j, while the j-th row of the root matrix of $B_l(C_l)$ is just the DYN representative of the weight λ^j of $C_l(B_l)$. These facts are explicitly shown in (5-343) by the column headings of the matrices (noting that $\lambda^i_{B_l} = \lambda^i_{C_l}$, for $i = 1, 2, \ldots, l-1$). For example, from (5-343b,c) we have

$$\begin{aligned} B_l :\ & (\lambda^l)^{\mathrm{DYN}} = (0, \ldots, 0, -1, 2) \ , \\ & [\alpha_l]^{\mathrm{FWS}} = [0, 0, \ldots, 0, 1] \qquad (\text{i.e., } \alpha_l = \lambda^l) \ ; \\ C_l :\ & (\lambda^l)^{\mathrm{DYN}} = (0, \ldots, 0, -1, 1) \ , \\ & [\alpha_l]^{\mathrm{FWS}} = [0, 0, \ldots, 0, 2] \qquad (\text{i.e., } \alpha_l = 2\lambda^l) \ . \end{aligned} \tag{5-345}$$

Equation (5-344) can also be proved directly from (5-336) and (5-342c).

From (5-294c) and (5-326), Eq. (5-342a) reads, in the DYN representation,

$$A_{ij} = \sum_{k=1}^{n} W_{ik} R_{kj} \ , \quad \text{i.e.,}\ A = WR \ . \tag{5-346a}$$

Equation (5-346a) can also be derived from the representation transformation,

$$(\overline{\mathbf{M}}_i, \alpha_j) = \sum_{k=1}^{n} (\overline{\mathbf{M}}_i, \mathbf{e}_k)(\mathbf{e}_k, \alpha_j) \ . \tag{5-346b}$$

The inverse of (5-327a) is

$$\mathbf{M}^j = \sum_{i=1}^{n} (W^{-1})_{ij} \lambda^i = \sum_{i=1}^{n} (W^{-1})_{ij} \mathbf{e}_i \ . \tag{5-347a}$$

Therefore

$$[\mathbf{M}^j]^{\mathrm{FWS}} = [(W^{-1})_{1j}, \dots , (W^{-1})_{nj}] \ , \tag{5-347b}$$

$$(W^{-1})_{ij} = (\mathbf{e}_i, \mathbf{M}^j) = (\lambda^i , \mathbf{M}^j) \ . \tag{5-347c}$$

The inverse of (5-342a) is

$$\lambda^j = \sum_{i=1}^{l} (R^{-1})_{ij} \alpha_i \ , \quad j = 1, 2, \dots , n \ , \tag{5-348a}$$

which implies that

$$\{\lambda^j\}^{\mathrm{SRS}} = \{(R^{-1})_{1j}, \dots , (R^{-1})_{lj}\} \tag{5-348b}$$

$$(R^{-1})_{ij} = (\overline{\alpha}^i, \lambda^j) = (\overline{\alpha}^i, \mathbf{e}_j) \ . \tag{5-348c}$$

The weight matrix W of A_l in (5-325a) is an $l \times (l+1)$ rectangular matrix. To find its inverse, we employ the following trick. Uniting (5-337a) and (5-333) into one matrix equation, one gets

$$\begin{bmatrix} v_1^{\mathrm{DYN}} \\ \cdot \\ \cdot \\ v_1^{\mathrm{DYN}} \\ 0 \end{bmatrix} = \mathcal{W} \begin{bmatrix} v_1^{\mathrm{FWS}} \\ \cdot \\ \cdot \\ v_l^{\mathrm{FWS}} \\ v_{l+1}^{\mathrm{FWS}} \end{bmatrix} \ , \quad \mathcal{W} = \left[\begin{array}{c} W \\ \hline 1 1 \dots 1 \end{array} \right] \ , \tag{5-349}$$

where $\mathcal{W}$ is a $(l+1) \times (l+1)$ matrix. Finding $\mathcal{W}^{-1}$ and deleting its last column, $(\frac{1}{l+1}, \dots , \frac{1}{l+1})$, we obtain W^{-1}. The inverse W^{-1} for A_l, B_l, C_l and D_l are given in (5-350), where the relations given by (5-347b), (5-348b) and (5-344) are shown explicitly by the row and column headings of the matrices.

$$A_l : W^{-1} = \begin{array}{c} \\ \lambda^1 \\ \lambda^2 \\ \cdot \\ \cdot \\ \cdot \\ \\ \lambda^l \\ \lambda^{l+1} \\ \{\mathrm{SRS}\} \end{array} \begin{array}{c} \begin{array}{cccccc} \mathbf{M}^1 & \mathbf{M}^2 & \mathbf{M}^3 & & \mathbf{M}^{l-1} & \mathbf{M}^l \ [\mathrm{FWS}] \end{array} \\ \begin{bmatrix} l & l-1 & l-2 & \cdots & 2 & 1 \\ -1 & l-1 & l-2 & \cdots & 2 & 1 \\ -1 & -2 & l-2 & \cdots & \cdot & \cdot \\ \cdot & \cdot & -3 & & \cdot & \cdot \\ \cdot & \cdot & \cdot & & \cdot & \cdot \\ \cdot & \cdot & \cdot & & 2 & 1 \\ -1 & -2 & \cdot & & -1+1 & 1 \\ -1 & -2 & -3 & \cdots & -l+1 & -l \end{bmatrix} \\ \\ \end{array} \times \frac{1}{l+1} \ , \tag{5-350a}$$

$$
B_l : W^{-1} = \begin{array}{c} \\ \lambda^1 \\ \lambda^2 \\ \cdot \\ \cdot \\ \cdot \\ \\ \lambda^{l-1} \\ \lambda^l_{C_l} \\ \{\mathrm{SRS}\} \end{array}
\begin{array}{c}
\begin{array}{cccccc} \mathbf{M}^1 & \mathbf{M}^2 & & \mathbf{M}^{l-1} & \mathbf{M}^l & [\mathrm{FWS}] \end{array} \\
\left[\begin{array}{ccccc}
1 & 1 & \cdots & 1 & 1/2 \\
0 & 1 & \cdots & 1 & 1/2 \\
0 & 0 & \cdots & 1 & 1/2 \\
\cdot & \cdot & & \cdot & \cdot \\
\cdot & \cdot & & \cdot & \cdot \\
\cdot & \cdot & & \cdot & \cdot \\
0 & 0 & \cdots & 1 & 1/2 \\
0 & 0 & \cdots & 0 & 1/2
\end{array}\right] \\ \\
\end{array} , \tag{5-350b}
$$

$$
C_l : W^{-1} = \begin{array}{c} \\ \lambda^1 \\ \lambda^2 \\ \cdot \\ \cdot \\ \cdot \\ \\ \lambda^{l-1} \\ \lambda^l_{B_l} \\ \{\mathrm{SRS}\} \end{array}
\begin{array}{c}
\begin{array}{cccccc} \mathbf{M}^1 & \mathbf{M}^2 & & \mathbf{M}^{l-1} & \mathbf{M}^l & [\mathrm{FWS}] \end{array} \\
\left[\begin{array}{ccccc}
1 & 1 & \cdots & 1 & 1 \\
0 & 1 & \cdots & 1 & 1 \\
0 & 0 & \cdots & 1 & 1 \\
\cdot & \cdot & & \cdot & \cdot \\
\cdot & \cdot & & \cdot & \cdot \\
\cdot & \cdot & & \cdot & \cdot \\
0 & 0 & \cdots & 1 & 1 \\
0 & 0 & \cdots & 0 & 1
\end{array}\right] \\ \\
\end{array} , \tag{5-350c}
$$

$$
\begin{array}{c} D_l : W^{-1} \\ l \geq 4 \end{array} = \begin{array}{c} \\ \lambda^1 \\ \lambda^2 \\ \cdot \\ \cdot \\ \cdot \\ \\ \\ \\ \lambda^{l-1} \\ \lambda^l \\ \{\mathrm{SRS}\} \end{array}
\begin{array}{c}
\begin{array}{ccccccc} \mathbf{M}^1 & \mathbf{M}^2 & & \mathbf{M}^{l-2} & \mathbf{M}^{l-1} & \mathbf{M}^l & [\mathrm{FWS}] \end{array} \\
\left[\begin{array}{cccccc}
1 & 1 & \cdots & 1 & 1/2 & 1/2 \\
0 & 1 & \cdots & 1 & 1/2 & 1/2 \\
0 & 0 & \cdots & 1 & 1/2 & 1/2 \\
\cdot & \cdot & & \cdot & \cdot & \cdot \\
\cdot & \cdot & & \cdot & \cdot & \cdot \\
\cdot & \cdot & & \cdot & \cdot & \cdot \\
\cdot & \cdot & & \cdot & \cdot & \cdot \\
\cdot & \cdot & & 1 & 1/2 & 1/2 \\
0 & 0 & \cdots & 0 & 1/2 & 1/2 \\
0 & 0 & \cdots & 0 & -1/2 & 1/2
\end{array}\right] \\ \\
\end{array} . \tag{5-350d}
$$

Notice that the j-th column of W^{-1} gives the Cartan-Weyl label of the j-th basic weight $\mathbf{M}^j$ and that the Cartan Weyl labels for spinor reps $\mathbf{M}^l$ of B_l and $\mathbf{M}^{l-1}$ and $\mathbf{M}^l$ of D_l involve half integers.

It is interesting to compare (5-296) [or its explicit forms (5-283) and (5-300)], (5-343) and (5-350), which give the various representatives of the three sets of basic vectors α_j, λ^j and $\mathbf{M}^j$, i.e., $(\alpha_j)^{\mathrm{DYN}}$, $\{\mathbf{M}^j\}^{\mathrm{SRS}}$; $[\alpha_j]^{\mathrm{FWS}}$, $(\lambda^j)^{\mathrm{DYN}}$ and $[\mathbf{M}^j]^{\mathrm{FWS}}$, $\{\lambda^j\}^{\mathrm{SRS}}$.

From (5-337b) and (5-346a) we obtain the following transformation relations,

$$
\{\mathbf{v}\}_{\mathrm{SRS}} = A^{-1}(\mathbf{v})^{\mathrm{DYN}} = R^{-1}[\mathbf{v}]^{\mathrm{FWS}} , \tag{5-351a}
$$

$$
[\mathbf{v}]^{\mathrm{FWS}} = W^{-1}(\mathbf{v})^{\mathrm{DYN}} = R\{\mathbf{v}\}_{\mathrm{SRS}} , \tag{5-351b}
$$

where

$$
R^{-1} = \tilde{W}^{-1}, \quad \text{for } A_l \text{ and } D_l , \tag{5-351c}
$$

$$
(R^{B_l})^{-1} = (\tilde{W}^{C_l})^{-1} , \quad (R^{C_l})^{-1} = (\tilde{W}^{B_l})^{-1} , \tag{5-351d}
$$

i.e., for A_l and D_l, R^{-1} is equal to the transpose of W^{-1}, while R^{-1} of B_l is equal to the transpose of W^{-1} of C_l and vice versa.

Equation (5-351b) shows that W^{-1} is the transformation matrix from the Dynkin label $(a_1, a_2, \ldots a_l)$ to the Cartan-Weyl label $[l_1 l_2 \ldots l_n]$.

The dimension formula (5-302) can also be evaluated in the FWS representation. By using (5-276) and (5-335b), we have

$$A_l: \quad \dim(\Lambda) = \prod_{j>i=1}^{l+1} \left(\frac{p_i - p_j}{g_i - g_j}\right) , \tag{5-352a}$$

$$B_l, C_l: \quad \dim(\Lambda) = \prod_{i=1}^{l} \frac{p_i}{g_i} \prod_{j>i=1}^{l} \left(\frac{p_i - p_j}{g_i - g_j}\right) \left(\frac{p_i + p_j}{g_i + g_j}\right) , \tag{5-352b}$$

$$D_l: \quad \dim(\Lambda) = \prod_{j>i=1}^{l} \left(\frac{p_i - p_j}{g_i - g_j}\right) \left(\frac{p_i + p_j}{g_i + g_j}\right) , \tag{5-352c}$$

where

$$\begin{aligned} & p_i = l_i + g_i , \\ & [\Lambda]^{\mathrm{FWS}} = [l_1, \ldots, l_n] , \quad [\mathbf{R}]^{\mathrm{FWS}} = [g_1, \ldots, g_n] . \end{aligned} \tag{5-352d}$$

By using (5-351b), (5-350) and (5-288d) we can find

$$\begin{aligned} & A_l: g_i = \frac{l}{2} - i + 1 , \quad B_l: g_i = l - i + \frac{1}{2} , \\ & C_l: g_i = l - i + 1 , \quad D_l: g_i = l - i . \end{aligned} \tag{5-352e}$$

Similarly, from (5-303b) we can calculate the eigenvalues of the Casimir operators:

$$C_{\mathrm{SU}_n} = \sum_{i=1}^{n} l_i(l_i + n + 1 - 2i) , \qquad C_{\mathrm{SO}_{2n+1}} = \sum_{i=1}^{n} l_i(l_i + 2n + 1 - 2i) , \tag{5-352f}$$

$$C_{\mathrm{SP}_{2n}} = \sum_{i=1}^{n} l_i(l_i + 2n + 2 - 2i) , \quad C_{\mathrm{SO}_{2n}} = \sum_{i=1}^{n} l_i(l_i + 2n - 2i) . \tag{5-352g}$$

Putting (5-328) and (5-334) together, we have

$$\mathbf{v} = \sum_{i=1}^{n} \mathbf{e}_i(\mathbf{e}_i, \mathbf{v}) = \sum_{i=1}^{n} \lambda^i(\lambda^i, \mathbf{v}) = \sum_{i=1}^{n} \lambda^i(\mathbf{e}_i, \mathbf{v}) = \sum_{i=1}^{n} \mathbf{e}_i(\lambda^i, \mathbf{v}) . \tag{5-353a}$$

Consequently, in root or weight space, we have the equalities

$$\begin{aligned} 1 &= \sum_{i=1}^{n} |\mathbf{e}_i)(\mathbf{e}_i| = \sum_{i=1}^{n} |\lambda^i)(\lambda^i| \\ &= \sum_{i=1}^{n} |\lambda^i)(\mathbf{e}_i| = \sum_{i=1}^{n} |\mathbf{e}_i)(\lambda^i| . \end{aligned} \tag{5-353b}$$

Using (5-331a), we obtain from (5-353b)

$$|\mathbf{p})(\mathbf{p}| = 0 \ , \tag{5-354}$$

i.e., the projection operator $|\mathbf{p})(\mathbf{p}|$ is a null operator in root or weight space of A_l.

From (5-350a) we see that the matrix W^{-1} of A_l involves fractions, and thus the Cartan-Weyl label $[l_1 l_2 \ldots l_n]$ for irreps of A_l also involves fractions. To avoid this awkward situation, we introduce a modified FWS representation, denoted as FWS$'$, whose basis vectors are $\lambda^1, \lambda^2, \ldots, \lambda^l$, instead of $\lambda^1, \lambda^2, \ldots, \lambda^{l+1}$. Using (5-328$'$), we can easily obtain the transformation from the FWS to the FWS$'$ representation,

$$\mathbf{v} = \sum_{i=1}^{l} v_i^{\mathrm{FWS}'} \lambda^i \ , \tag{5-355a}$$

where

$$v_i^{\mathrm{FWS}'} = v_i^{\mathrm{FWS}} - v_{l+1}^{\mathrm{FWS}} \ , \quad i = 1, 2, \ldots, l \ . \tag{5-355b}$$

Let

$$[\Lambda]^{\mathrm{FWS}'} = [k_1 k_2 \ldots k_l] \ . \tag{5-355c}$$

One has from Eq. (3-555b) that

$$k_i = l_i - l_n, \quad i = 1, 2, \ldots l; \quad n = l + 1 \ . \tag{5-355d}$$

Letting ν_i be the number of particles in the state $|i\rangle$, from (5-329) we get the relation between the Cartan-Weyl label $[l_1 l_2 \ldots l_l]$ of the SU$_n$ group and the partition label $[\nu] = [\nu_1 \nu_2 \ldots \nu_n]$ of the U_n group,

$$l_i = \nu_i - \frac{N}{l+1}, \quad N = \sum_{i=1}^{l+1} \nu_i \ . \tag{5-355e}$$

From (5-355d) and (5-355e) we also have

$$k_i = \nu_i - \nu_n, \quad i = 1, 2, \ldots, l \ . \tag{5-355f}$$

Therefore $[k_1 k_2 \ldots k_l]$ represents a Young diagram resulting from deleting the ν_n columns of length n in the Young diagram $[\nu_1 \nu_2 \ldots \nu_n]$ (cf. (7-71)).

As with (5-337a), Eq. (5-355a) becomes, in the DYN representation,

$$v_i^{\mathrm{DYN}} = \sum_{j=1}^{l} W_{ij} v_j^{\mathrm{FWS}'} \ , \tag{5-356a}$$

$$(\mathbf{v})^{\mathrm{DYN}} = W'[\mathbf{v}]^{\mathrm{FWS}'} \ , \tag{5-356b}$$

where W' is a submatrix of W given in (5-325a), and is precisely the weight matrix of C_l,

$$W' = W^{C_l} \ . \tag{5-356c}$$

Now Eq. (5-337b) is replaced by

$$(\mathbf{v})^{\mathrm{DYN}} = A\{\mathbf{v}\}_{\mathrm{SRS}} = W'[\mathbf{v}]^{\mathrm{FWS}'} \ , \tag{5-357a}$$

which in turn implies that

$$\{\mathbf{v}\}_{\mathrm{SRS}} = A^{-1}(\mathbf{v})^{\mathrm{DYN}} = R'^{-1}[\mathbf{v}]^{\mathrm{FWS}'} , \tag{5-357b}$$

where

$$R'^{-1} = A^{-1}W' . \tag{5-358a}$$

Using (5-295b) and (5-336), from (5-358a) we have

$$\begin{aligned}(R'^{-1})_{ik} &= \sum_{j=1}^{l}(\overline{\alpha}^i, \mathbf{M}^j)(\overline{\mathbf{M}}_j, \mathbf{e}_k) = (\overline{\alpha}^i, \mathbf{e}_k)\\ &= (R^{-1})_{ik}, \quad i,k = 1,2,\ldots,l .\end{aligned} \tag{5-358b}$$

Therefore, R'^{-1} results from deleting the last column of R^{-1}.

From (5-357a,b) we obtain

$$[\mathbf{v}]^{\mathrm{FWS}'} = (W^{C_l})^{-1}(\mathbf{v})^{\mathrm{DYN}} = R'\{\mathbf{v}\}_{\mathrm{SRS}} . \tag{5-357c}$$

It should be emphasized that although R'^{-1} is a submatrix of R^{-1}, there is no such simple relation between the matrix R' and R. In fact, R' has to be found by inverting the matrix R'^{-1}. Equations (5-356) and (5-357) show that the transformation between the FWS′ and DYN representations for A_l is identical with that between the FWS and DYN representations for C_l.

Writing out explicitly, the relation (5-357c) between the FWS′ representative $[k_1, k_2 \ldots k_l]$ and the DYN representative $(a_1 a_2 \ldots a_l)$ of a highest weight Λ is

$$\begin{aligned} k_1 &= a_1 + a_2 + \ldots + a_l ,\\ k_2 &= a_2 + \ldots + a_l ,\\ &\ldots\\ k_{l-1} &= a_{l-1} + a_l ,\\ k_l &= a_l . \end{aligned} \tag{5-357d}$$

The relation (5-357d) between the Young diagram $[k_1 k_2 \ldots k_l]$ and the label $(a_1 a_2 \ldots a_l)$ is vividly illustrated in the following Weyl tableaux of the highest weights.

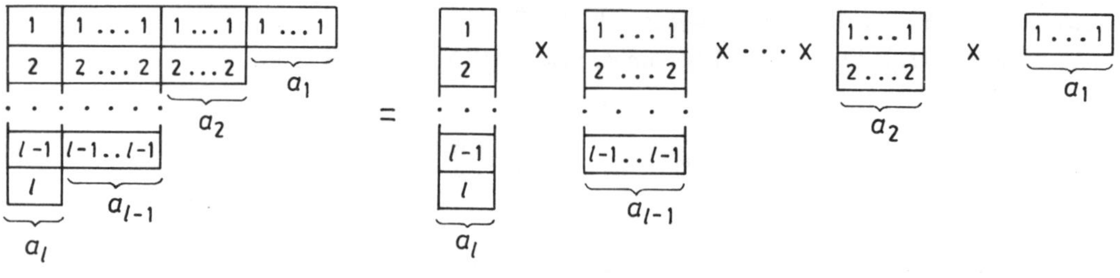

As an example, in the following we list the matrices W, R, W', R' and their inverses for A_2,

$$W = \tilde{R} = \begin{pmatrix} 1 & -1 & 0 \\ 0 & 1 & -1 \end{pmatrix} , \qquad \tilde{W}^{-1} = \frac{1}{3}\begin{pmatrix} 2 & -1 & -1 \\ 1 & 1 & -2 \end{pmatrix} , \tag{5-358a}$$

$$W' = \begin{pmatrix} 1 & -1 \\ 0 & 1 \end{pmatrix} , \quad W'^{-1} = \begin{pmatrix} 1 & 1 \\ 0 & 1 \end{pmatrix} , \quad R' = \begin{pmatrix} 1 & 1 \\ -1 & 2 \end{pmatrix} , \quad R'^{-1} = \frac{1}{3}\begin{pmatrix} 2 & -1 \\ 1 & 1 \end{pmatrix} , \tag{5-358b}$$

$$WW^{-1} = W'W'^{-1} = W'^{-1}W' = \begin{pmatrix} 1 & 0 \\ 0 & 1 \end{pmatrix}, W^{-1}W = \begin{pmatrix} 2/3 & -1/3 & -1/3 \\ -1/3 & 2/3 & -1/3 \\ -1/3 & -1/3 & 2/3 \end{pmatrix}. \quad (5\text{-}358\text{c})$$

Notice that $W^{-1}W$ is not a unit matrix, a peculiarity to be discussed in the next section.

Ex. 5.18. Prove Eqs. (5-338) and (5-341).
Ex. 5.19. Derive Eq. (5-352).
Ex. 5.20. Show that the Cartan-Weyl irrep labels of SU_4 and SO_6 are related as shown below.

$$SU_4 \qquad SO_6 \qquad SU_4 \qquad SO_6$$

$$[l_1 + l_2, l_1 - l_3, l_2 - l_3] \leftrightarrow [l_1 l_2 l_3] \quad \underset{\alpha_1}{\circ}\!\!-\!\!\underset{\alpha_2}{\circ}\!\!-\!\!\underset{\alpha_3}{\circ} \qquad \underset{\alpha_1}{\circ}\!\!\!<\!\!\begin{smallmatrix}\circ\,\alpha_2\\[2pt]\circ\,\alpha_3\end{smallmatrix}$$

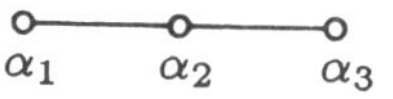

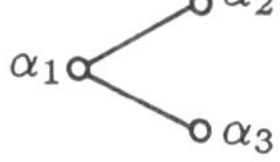

Ex. 5.21. Find the SRS and CAR representatives of the weights for the irrep $\mathbf{M}^1$ of D_4.
Ex. 5.22. The dimensions of the basic reps of C_3 and C_4 are listed below:

$$C_3: \underset{\substack{\alpha_1 \\ 6}}{\bullet}\!\!-\!\!\underset{\substack{\alpha_2 \\ 14}}{\bullet}\!\!\Longleftarrow\!\!\underset{\substack{\alpha_3 \\ 14}}{\circ} \qquad C_4: \underset{\substack{\alpha_1 \\ 8}}{\bullet}\!\!-\!\!\underset{\substack{\alpha_2 \\ 27}}{\bullet}\!\!-\!\!\underset{\substack{\alpha_3 \\ 48}}{\bullet}\!\!\Longleftarrow\!\!\underset{\substack{\alpha_4 \\ 42}}{\circ}$$

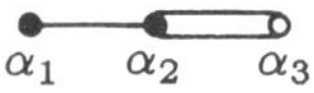

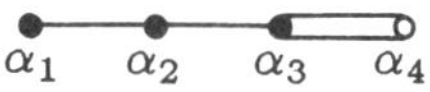

Show that

$$C_3: (\mathbf{M}^1)^{[1^2]} = \mathbf{M}^2 + \mathbf{M}^0, \quad (\mathbf{M}^1)^{[1^3]} = \mathbf{M}^3 + \mathbf{M}^1,$$
$$C_4: (\mathbf{M}^1)^{[1^2]} = \mathbf{M}^2 + \mathbf{M}^0, \quad (\mathbf{M}^1)^{[1^3]} = \mathbf{M}^3 + \mathbf{M}^1,$$
$$(\mathbf{M}^1)^{[1^4]} = \mathbf{M}^4 + \mathbf{M}^2 + \mathbf{M}^0,$$

where $\mathbf{M}^0$ is the identity rep.

5.27. Comparison of the Different Representations

We have introduced various representations for the classical groups, i.e., the SRS and DYN representations for all classical Lie groups, and the extra two representations, the FWS and CAR representations for A_l, B_l, C_l and D_l. The SRS representation has the advantage of being intuitive and the SRS representative of a weight M is closely related to the power $\delta(M)$ of the weight, $\delta(M) = 2\sum_{i=1}^{l} w^i$. The DYN representation is especially important due to the fact that the two basic theorems on roots and weights take very simple forms in the DYN representation and thus it is most suitable for computing the root and weight systems. The SRS and DYN representations are complementary or dual to each other, since their bases are dual bases up to a normalization factor. For B_l, C_l and D_l, the FWS representation coincides with the CAR representation, with the obvious merit that its basis vectors are orthonormal. For A_l, the FWS representation is similar but not identical with the CAR representation, as shown in (5-332a), (5-353a) and (5-335). The FWS representation has the virtue that its "effective metric tensor" is a unit matrix, as seen from (5-335b), but has the drawback that the representative of a weight is not a set of integers or half integers. Then comes the fifth representation for A_l, the modified FWS, or FWS' representation. The FWS' representation has the virtue that a weight is represented by a set of integers, which is just the partition if the weight is a highest weight, but now its "effective metric tensor" is no longer a unit matrix, i.e.,

$$(\mathbf{u}, \mathbf{v}) \neq \sum_{i=1}^{l} u_i^{\text{FWS}'} v_i^{\text{FWS}'}. \quad (5\text{-}359)$$

Despite the striking similarities between the FWS and CAR representations, there are subtle and profound differences between the bases of the two representations, which can be misleading.

For example, from (5-336), it seems that we might have

$$\mathbf{e}_j = \sum_{i=1}^{l} W_{ij}\mathbf{M}^i = \sum_{i=1}^{l}(\overline{\mathbf{M}}_i, \mathbf{e}_j)\mathbf{M}^i \ . \tag{5-360}$$

Comparing this with (5-327a), it appears that for A_l we also have $\lambda^i = \mathbf{e}_i$ which contradicts (5-331a). In fact (5-360) is false, since if it were true, then in the $(l+1)$-dimensional space $(\mathbf{e}_1, \dots, \mathbf{e}_{l+1})$ we would have

$$\sum_{i=1}^{l} |\mathbf{M}^i)(\overline{\mathbf{M}}_i| = 1 \ , \tag{5-361}$$

which means that the l basic vectors $\mathbf{M}^1, \dots, \mathbf{M}^l$ form a complete set of vectors in the $(l+1)$-dimensional space. But this is impossible. In fact, (5-361) holds only in the l-dimensional root or weight space spanned by $(\lambda^1, \dots, \lambda^{l+1})$, or $(\alpha_1, \dots \alpha_l)$, or $(\mathbf{M}^1 \dots, \mathbf{M}^l)$.

Using (5-336) and (5-347c), we have

$$(WW^{-1})_{ik} = \sum_{j=1}^{l+1}(\overline{\mathbf{M}}_i, \mathbf{e}_j)(\mathbf{e}_j, \mathbf{M}^k) = (\overline{\mathbf{M}}_i, \mathbf{M}^k) = \delta_{ik}, \quad i,k = 1,2,\dots,l \ . \tag{5-362a}$$

On the other hand,

$$(W^{-1}W)_{ik} = \sum_{j=1}^{l}(\mathbf{e}_i, \mathbf{M}^j)(\overline{\mathbf{M}}_j, \mathbf{e}_k) \neq (\mathbf{e}_i, \mathbf{e}_k) = \delta_{ik}, \quad i,k = 1,2,\dots,l+1 \ , \tag{5-362b}$$

since

$$|\mathbf{M}^j)(\mathbf{M}_j| \begin{cases} \neq 1, & \text{in the space}(\mathbf{e}_1, \dots, \mathbf{e}_{l+1}) \ , \\ = 1, & \text{in the space}(\lambda^1, \dots, \lambda^{l+1}) \ . \end{cases}$$

However, by (5-347c) and (5-327b), we have

$$(W^{-1}W)_{ik} = \sum_{j=1}^{l}(\lambda^i, \mathbf{M}^j)(\overline{\mathbf{M}}_j, \lambda^k) = (\lambda^i, \lambda^k) = g^{ik}_{\mathrm{FWS}} \ . \tag{5-363}$$

An example of (5-363) is given in (5-358c). Therefore, although WW^{-1} is an $l \times l$ unit matrix, $W^{-1}W$ is not a unit matrix. Equation (5-363) can be written as

$$W^{-1}W = I - \frac{1}{n}N, \quad N = \begin{bmatrix} 1 \, 1 \cdots 1 \\ 1 \, 1 \cdots 1 \\ \vdots \ \vdots \quad \ \vdots \\ 1 \, 1 \cdots 1 \end{bmatrix} \ , \quad n = l+1 \ , \tag{5-364}$$

where I is an $n \times n$ unit matrix. According to (5-328′), $[\mathbf{v}]^{\mathrm{FWS}} = [11, \dots 1]$ is a null vector, therefore N is equivalent to a null matrix. Hence we see that although $W^{-1}W$ is not a unit matrix in appearance, it is equivalent to an $n \times n$ unit matrix.

A more rigorous way to treat the representation transformation for A_l is to artificially expand the l-dimensional root space into a "pseudo" n-dimensional space, with the convention that $\mathbf{M}^n = \boldsymbol{\alpha}_n = 0, v^n_{\mathrm{SRS}} = v^{\mathrm{DYN}}_n = 0$. Define the $n \times n$ matrices,

$$\mathcal{W} = \tilde{\mathcal{R}} = \begin{bmatrix} W \\ \\ 11\dots1 \end{bmatrix} \ , \quad \mathcal{R}^{-1} = \tilde{\mathcal{W}}^{-1} = \begin{bmatrix} R^{-1} \\ \\ \frac{1}{n}\frac{1}{n}\dots\frac{1}{n} \end{bmatrix} \ ,$$

$$\mathcal{A} = \begin{bmatrix} A & \\ & n \end{bmatrix} \ , \quad \mathcal{A}^{-1} = \begin{bmatrix} A^{-1} & \\ & n^{-1} \end{bmatrix} \ . \tag{5-365}$$

We have

$$A = \mathcal{W}\mathcal{R}\ , \quad \mathcal{W}\mathcal{W}^{-1} = \mathcal{W}^{-1}\mathcal{W} = \mathfrak{I}\ , \tag{5-366}$$

where $\mathfrak{I}$ is an $n \times n$ unit matrix. Then in place of (5-337b), in the "pseudo" n-dimensional space we have

$$(\mathbf{v})^{\mathrm{DYN}} = \mathcal{A}\{\mathbf{v}\}_{\mathrm{SRS}} = \mathcal{W}[\mathbf{v}]^{\mathrm{FWS}}\ . \tag{5-367}$$

From (5-366) and (5-367), we can easily obtain the other two relations corresponding to (5-351a, b). By (5-333) and the special structure of the matrices in (5-365), as well as the convention that $v^n_{\mathrm{SRS}} = v^{\mathrm{DYN}}_n = 0$, we see that the following equivalences hold,

$$\mathcal{R} \doteq R, \quad \mathcal{R}^{-1} \doteq R^{-1}, \quad \mathcal{W} \doteq W, \quad \mathcal{W}^{-1} \doteq W^{-1}, \quad \mathcal{A} \doteq A, \quad \mathcal{A}^{-1} \doteq A^{-1}\ , \tag{5-368}$$

where "$\doteq$" means equivalence, i.e., $\mathcal{R}$ and R have the same effect, etc. Thus we are led back to Eqs. (5-337b) and (5-351a,b).

Introducing the column vectors

$$\mathbf{M} = \begin{bmatrix} \mathbf{M}^1 \\ \vdots \\ \mathbf{M}^l \end{bmatrix}\ , \quad \boldsymbol{\alpha} = \begin{bmatrix} \alpha_1 \\ \vdots \\ \alpha_l \end{bmatrix}\ , \quad \boldsymbol{\lambda} = \begin{bmatrix} \lambda^1 \\ \vdots \\ \lambda^n \end{bmatrix}\ , \tag{5-369}$$

the transformations between the coordinate vectors and that between the basis vectors for the three representations are summarized in the following:

$$\begin{aligned} &1.\ (\mathbf{v})^{\mathrm{DYN}} = A\{\mathbf{v}\}_{\mathrm{SRS}} &&= W[\mathbf{v}]^{\mathrm{FWS}}\ , \\ &\qquad \mathbf{M} = \tilde{A}^{-1}\boldsymbol{\alpha} &&= \tilde{W}^{-1}\boldsymbol{\lambda} \end{aligned} \tag{5-370}$$

$$\begin{aligned} &2.\ \{\mathbf{v}\}_{\mathrm{SRS}} = A^{-1}(\mathbf{v})^{\mathrm{DYN}} &&= R^{-1}[\mathbf{v}]^{\mathrm{FWS}}\ , \\ &\qquad \boldsymbol{\alpha} = \tilde{A}\mathbf{M} &&= \tilde{R}\,\boldsymbol{\lambda}\ , \end{aligned} \tag{5-371}$$

$$\begin{aligned} &3.\ [\mathbf{v}]^{\mathrm{FWS}} = W^{-1}(\mathbf{v})^{\mathrm{DYN}} &&= R\{\mathbf{v}\}_{\mathrm{SRS}}\ , \\ &\qquad \boldsymbol{\lambda} = \tilde{W}\mathbf{M} &&= \tilde{R}^{-1}\boldsymbol{\alpha}\ . \end{aligned} \tag{5-372}$$

For A_l and D_l, the transformations take the more symmetric form,

$$(\mathbf{v})^{\mathrm{DYN}} = W[\mathbf{v}]^{\mathrm{FWS}}, \quad [\mathbf{v}]^{\mathrm{FWS}} = \tilde{W}\{\mathbf{v}\}_{\mathrm{SRS}}\ , \tag{5-373a}$$

$$\boldsymbol{\alpha} = W\boldsymbol{\lambda}, \qquad \boldsymbol{\lambda} = \tilde{W}\mathbf{M}\ . \tag{5-373b}$$

The three representatives for $\alpha_j, \mathbf{M}^j$ and λ^j are listed below.

	SRS	DYN	FWS
$\alpha_j:$	$\{0\ldots010\ldots0\}$	$(A_{1j}\ldots A_{lj})$	$[R_{1j}\ldots R_{lj}]\ .$
$\mathbf{M}^j:$	$\{(A^{-1})_{1j}\ldots(A^{-1})_{lj}\},$	$(0\ldots010\ldots0),$	$[(W^{-1})_{1j}\ldots(W^{-1})_{lj}].$
$\lambda^j:$	$\{(R^{-1})_{1j}\ldots(R^{-1})_{lj}\},$	$(W_{1j}\ldots W_{lj}),$	$[0\ldots010\ldots0]\ .$

Comparing (5-372) with (5-357d), noting (5-350) and remembering that the two-particle states (5-145) and (5-146) are invariants of orthogonal group and sympletic group, respectively, we conclude that the Cartan-Weyl label $[\Lambda]^{\rm FWS} = [l_1 l_2 \ldots]$ for the vector representation of B_l and D_l (for any irreps of C_l) is the permutational symmetry with the smallest number of particles in which the orthogonal (sympletic) symmetry is first encountered.

The labeling schemes for simple roots and irreps of the classical Lie algebras have been studied by Wybourne (1974), Laskar (1977) and King (1981). Equations (5-371)-(5-373) and (5-357) here cover Eqs. (12.22)- (12.24), and (12.28) of Wybourne (1974), and Table 9 of King (1981). Extending the CAR representation to the case of the exceptional Lie algebras is possible. However, the choice of a particular Cartesian representation for exceptional Lie algebra is not unique (King 1981).

5.28. The Characters and CG Series of Lie Algebras

5.28.1. The characters of Lie groups

The character $\chi(\Lambda, \phi)$ of the irrep Λ for a semi-simple Lie group is a function of the l class parameters $\phi^1, \phi^2, \ldots \phi^l$ (Weyl, 1946),

$$\chi(\Lambda, \phi) = \sum_{\rm m} \gamma^{\Lambda}({\rm m}) \exp\left(i \sum_j m_j \phi^j \right) , \tag{5-374}$$

where the sum is extended over all distinct weights $[{\rm m}] = [m_1 m_2 \ldots]$ (in the FWS representation) of the rep Λ and $\gamma^{\Lambda}({\rm m})$ is the multiplicity of each weight. Weyl has shown that the character (5-374) can be calculated by

$$\chi(\Lambda, \phi) = \frac{\xi(K, \phi)}{\xi(R, \phi)} \tag{5-375a}$$

$$\xi(K, \phi) = \sum_S \delta_S \exp\left[i \sum_j (SK)_j \phi^j \right] , \tag{5-375b}$$

where K and R are defined in (5-302b) and (5-288c), the sum runs over all the elements S of the Weyl reflection group, and δ_S is $+1$ or -1 according as the number of reflection is even or odd.

For the Lie algebra A_l, B_l, C_l and D_l, the characters have simple explicity expressions (Weyl, 1946),

$$\chi(\Lambda, \phi) = \frac{\det |f(p_i \phi^k)|}{\det |f(g_j \phi^k)|} \tag{5-376a}$$

where both the numerator and denominator are determinants of $n \times n$ matrices ($n = l + 1$ for A_l, and $n = l$ for B_l, C_l and D_l) with matrix elements $f(p_j \phi^k)$ and $f(g_j, \phi^k)$ in the j-th row and k-th column. p_j and g_j are given in (5-352), and the function $f(p_j \phi^k)$ has the following form

	A_l	B_l and C_l	D_l (for $l_n = 0$)
$f(p_j \phi^k)$	$\exp(i p_j \phi^k)$	$\sin(p_j \phi^k)$	$\cos(p_j \phi^k)$

The character for the spinor representation ($l_n \neq 0$) of D_l is given by (Boerner, p. 248)

$$\chi(\Lambda, \phi) = \frac{\det|\cos(p_j \phi^k)| + \det|i \sin(p_j \phi^k)|}{\det|\cos(g_j \phi^k)|} \tag{5-376c}$$

Using $p_i = l_i + g_i$ and (5-355e), the character for the SU_n group can be written in terms of the partition $[\nu]$,

$$\chi([\nu], \phi) = \exp\left[-i\sum_{j=1}^{n}\nu_j\right]\frac{\det|\exp\left[i(\nu_j + g_j)\phi^k\right]|}{\det|\exp(ig_j\phi^k)|} . \qquad (5\text{-}376d)$$

5.28.2. The CG series of Lie groups

The CG series of a Lie algebra can be determined from the weight theorems. The CG series is denoted by

$$\Lambda \times \Lambda' = \sum_{\Lambda''}(\Lambda\Lambda'\Lambda'')(\Lambda'') . \qquad (5\text{-}377)$$

The coefficients $(\Lambda\Lambda'\Lambda'')$ can be decided upon in the following way:

From (5-284a) and (5-377) we know that the CSW of the Kronecker product $(\Lambda \times \Lambda')$, $\Delta(\Lambda \times \Lambda')$, is decomposed as follows:

$$\Delta(\Lambda \times \Lambda') = \sum_{\Lambda''}\oplus(\Lambda\Lambda'\Lambda'')\Delta(\Lambda'') . \qquad (5\text{-}378)$$

The highest weight of $\Delta(\Lambda \times \Lambda')$ is

$$\Lambda_1'' = \Lambda + \Lambda' , \qquad (5\text{-}379)$$

and is necessarily simple. This implies that $(\Lambda\Lambda'\Lambda_1'') = 1$. Using Algorithm 2, we can find the CSW $\Delta(\Lambda_1'')$ for the irrep Λ_1''. Subtracting $\Delta(\Lambda_1'')$ from $\Delta(\Lambda \times \Lambda')$, we get the set of weights $\Delta_1 = \Delta(\Lambda \times \Lambda') - \Delta(\Lambda_1'')$. Using Theorem 5.22 we can find the highest weight Λ_2'' in the set Δ_1. Suppose that the number of the weight Λ_2'' in Δ_1 is m_2, then $(\Lambda\Lambda'\Lambda_2'') = m_2$. Find the CSW $\Delta(\Lambda_2'')$ and subtract $m_2\Delta(\Lambda_2'')$ from Δ_1 to get Δ_2. If the highest weight Λ_3'' occurs m_3 times in Δ_2, then $(\Lambda\Lambda'\Lambda_3'') = m_3$. Continue this process, until all the weights in $\Delta(\Lambda \times \Lambda')$ are exhausted.

Example: Find the CG series for (10)× (01) of A_2. It is convenient to work in the DYN representation. The CSW for the irreps (10) and (01) are

$$\Delta(10) = \{(10), (-1, 1), (0, -1)\}, \quad \Delta(01) = \{(01), (1, -1), (-1, 0)\} .$$

Thus we have

$$((10) \times (01)) = \{(11), (2, -1), (00)^3, (-1, 2), (-2, 1), (1, -2), (-1, -1)\} .$$

The highest weight here is (11), with the CSW

$$(11) = \{(11), (2, -1), (00)^2, (-1, 2), (-2, 1), (1, -2), (-1, -1)\} ,$$

while

$$\Delta((10) \times (01)) - \Delta(11) = (00) .$$

Therefore the CG series is

$$(10) \times (01) = (11) + (00) .$$

The foregoing method is straightforward but tedious and becomes untractable for irreps of high dimensions. A simpler method is provided by Weyl (1946) which is based on the decomposition, (3-273b), of the character $\chi(\Lambda, \phi)\chi(\Lambda', \phi)$ of the Kronecker product rep. Using the expression (5-375a) for $\chi(\Lambda, \phi)$ and (5-374) for $\chi(\Lambda'; \phi)$, it can be shown (Weyl 1946, Judd 1963 and Racah 1964) that

$$\chi(\Lambda, \phi)\chi(\Lambda', \phi) = \sum_{\mathbf{m}'} \gamma^{\Lambda'}(\mathbf{m}')\chi(\Lambda + \mathbf{m}', \phi) , \tag{5-380a}$$

$$\chi(\Lambda + \mathbf{m}', \phi) = \chi(l_1 + m_1', l_2 + m_2', \dots, \phi) , \tag{5-380b}$$

where the sum is extended over all distinct weights $[\mathbf{m}'] = [m_1' m_2' \dots]$ of the irrep Λ'. Notice that the symbol $[l_1 + m_1', l_2 + m_2', \dots]$ may not be a permissible irrep label. In such cases, we need to use Eq. (5-376) to express $\chi(\Lambda + \mathbf{m}', \phi)$ in an acceptable form. For example, for the group SO_5 from (5-352f) we have $g_1 = \frac{3}{2}, g_2 = \frac{1}{2}$, and from (5-376b) we have

$$\chi(l_1, l_2, \phi) = \begin{vmatrix} \sin\left(l_1 + \frac{3}{2}\right)\phi^1 & \sin\left(l_1 + \frac{3}{2}\right)\phi^2 \\ \sin\left(l_2 + \frac{1}{2}\right)\phi^1 & \sin\left(l_2 + \frac{1}{2}\right)\phi^2 \end{vmatrix} \Bigg/ \begin{vmatrix} \sin\left(\frac{3}{2}\right)\phi^1 & \sin\left(\frac{3}{2}\right)\phi^2 \\ \sin\left(\frac{1}{2}\right)\phi^1 & \sin\left(\frac{1}{2}\right)\phi^2 \end{vmatrix} . \tag{5-381}$$

From (5-378) we immediately know that

$$\chi(l_1, l_2) = -\chi(l_2 - 1, l_1 + 1) = -\chi(l_1, -l_2 - 1) = -\chi(-l_1 - 3, l_2) , \tag{5-382a}$$

$$\chi(l_1, l_2 + 1) = \chi\left(l_1, -\frac{1}{2}\right) = 0 . \tag{5-382b}$$

Example: Find the CG series for $[l_1 \frac{1}{2}] \times [11]$ of the group SO_5.

The irrep [11] of SO_5 has 10 weights, $\mathbf{m}' = \pm[11], \pm[10], \pm[01], \pm[1,-1], [00]^2$. Using (5-380) and (5-382) we get

$$\left[l_1 \frac{1}{2}\right] \times [11] = \left[l_1 + 1, \frac{3}{2}\right] + \left[l_1 \frac{3}{2}\right] + \left[l_1 - 1, \frac{3}{2}\right] + \left[l_1 + 1, \frac{1}{2}\right] + 2\left[l_1 \frac{1}{2}\right] + \left[l_1 - 1, \frac{1}{2}\right] . \tag{5-382c}$$

Ex. 5.23. Find the CG series (10) × (10) of A_2.
Ex. 5.24. Find the CG series (01) × (01) and (10) × (01) of G_2.
Ex. 5.25. Compute the CG series [21] × [21] of SU_3
Ex. 5.26. Compute the CG series for [210]×[100] of SP_6 and SO_7.
Ex. 5.27. Compute the CG series for $[1000] \times [\frac{1}{2}\frac{1}{2}\frac{1}{2}\frac{1}{2}]$ of SO_8.

Chapter 6
The Rotation Group

The rotation group SO_3 in three-dimensional space is the simplest non-Abelian Lie group and is the one most familiar to us. The rep theory of SO_3 can be most easily understood from the CSCO approach. In fact the new approach to the rep theory of finite groups stems from the SO_3 group. The purpose of this chapter is to use the group SO_3 as example to illustrate how the rep theory for finite groups and Lie groups can be unified by the CSCO approach. Regarding the other topics in the rep theory of the rotation group, such as the angular momentum theory, the D functions, the CG coefficients and irreducible tensors of SO_3, etc., readers are referred to Rose (1957), Edmonds (1957), Fano (1959), and Biendenharn (1984).

In Sec. 5.10 we introduced the double rotation groups $SO_2^\dagger$ or $SO_3^\dagger$. Written out explicitly, we have

$$SO_2 = \{e^{-i\varphi j_z} : 0 \le \varphi \le 2\pi\} \ , \tag{6-1a}$$

$$SO_2^\dagger = \{e^{-i\varphi J_z} : 0 \le \varphi \le 4\pi\} \ , \tag{6-1b}$$

$$\begin{aligned} SO_3 &= \{R_{\mathbf{n}(\theta'\varphi')}(\varphi) : 0 \le \theta' \le \pi, \quad 0 \le \varphi' \le 2\pi, \quad 0 \le \varphi \le \pi\} \\ &= \{R(\alpha\beta\gamma) : 0 \le \alpha, \gamma \le 2\pi, \quad 0 \le \beta \le \pi\} \ , \end{aligned} \tag{6-1c}$$

$$\begin{aligned} SO_3^\dagger &= \{R_{\mathbf{n}(\theta'\varphi')}(\varphi) : 0 \le \theta' \le \pi, \quad 0 \le \varphi' \le 2\pi, \quad 0 \le \varphi \le 2\pi\} \\ &= \{R(\alpha\beta\gamma) : 0 \le \alpha, \gamma \le 2\pi, \ 2n\pi \le \beta \le (2n+1)\pi, n = 0, 1\} \ . \end{aligned} \tag{6-1d}$$

In the following we shall discuss the reps for both the rotation group and the double rotation group.

6.1. The Differential Operators of $J_{x,y,z}$ and $\overline{J}_{x,y,z}$ in Group Parameter Space

Using (5-177b) and the combination law (5-23) of the group parameters, we can find the differential form for the infinitesimal operators J_x, J_y and J_z in group parameter space. We choose the Euler angle α, β and γ as parameters, and consider the product of two successive rotations

$$R(\alpha_3\beta_3\gamma_3) = R(\alpha_2\beta_2\gamma_2)R(\alpha_1\beta_1\gamma_1) \ . \tag{6-1a$'$}$$

However the combination law for the Euler angles α, β and γ,

$$(\alpha_3\beta_3\gamma_3) = \varphi(\alpha_1\beta_1\gamma_1, \alpha_2\beta_2\gamma_2) , \tag{6-1b'}$$

is not easy to work out explicitly. By the homomorphism of SU_2 to SO_3, one can obtain implicitly the combination law for the SO_3 group parameters in terms of those for the SU_2 group parameters. Therefore let us consider the product of two successive SU_2 transformations

$$\begin{pmatrix} c_0 - ic_3, & -c_1 - ic_2 \\ c_1 - ic_2, & c_0 + ic_3 \end{pmatrix} = \begin{pmatrix} b_0 - ib_3, & -b_1 - ib_2 \\ b_1 - ib_2, & b_0 + ib_3 \end{pmatrix} \begin{pmatrix} a_0 - ia_3, & -a_1 - ia_2 \\ a_1 - ia_2, & a_0 + ia_3 \end{pmatrix} . \tag{6-2a}$$

From (5-188) one gets the relations between the group parameters of SO_3 and SU_2:

$$\begin{aligned} a_0 &= \cos\frac{\beta_1}{2}\cos\frac{\alpha_1+\gamma_1}{2} , & a_1 &= \sin\frac{\beta_1}{2}\cos\frac{\gamma_1-\alpha_1}{2} , \\ a_2 &= \sin\frac{\beta_1}{2}\sin\frac{\gamma_1-\alpha_1}{2} , & a_3 &= \cos\frac{\beta_1}{2}\sin\frac{\alpha_1+\gamma_1}{2} , \\ b_0 &= \cos\frac{\beta_2}{2}\cos\frac{\alpha_2+\gamma_2}{2} , & b_1 &= \sin\frac{\beta_2}{2}\cos\frac{\gamma_2-\alpha_2}{2} , \\ b_2 &= \sin\frac{\beta_2}{2}\sin\frac{\gamma_2-\alpha_2}{2} , & b_3 &= \cos\frac{\beta_2}{2}\sin\frac{\alpha_2+\gamma_2}{2} , \end{aligned} \tag{6-3}$$

$$\begin{aligned} \alpha_3 &= \tan^{-1}\frac{c_1c_3 - c_2c_0}{c_1c_0 + c_2c_3} , \quad \beta_3 = 2\sin^{-1}\sqrt{c_1^2 + c_2^2} , \\ \gamma_3 &= \tan^{-1}\frac{c_1c_3 + c_2c_0}{c_1c_0 + c_2c_3} . \end{aligned} \tag{6-4}$$

From (6-2a) one immediately obtains the combination law for the SU_2 group parameters

$$\begin{aligned} c_0 &= a_0b_0 - a_1b_1 - a_2b_2 - a_3b_3 , & c_1 &= a_0b_1 + a_1b_0 + a_2b_3 - a_3b_2 , \\ c_2 &= a_0b_2 - a_1b_3 + a_2b_0 + a_3b_1 , & c_3 &= a_0b_3 + a_1b_2 - a_2b_1 + a_3b_0 . \end{aligned} \tag{6-2b}$$

According to (5-177b) and (5-176c)

$$-X_\alpha = \left(\frac{\partial\alpha_3}{\partial\alpha_2}\right)_0 \frac{\partial}{\partial\alpha} + \left(\frac{\partial\beta_3}{\partial\alpha_2}\right)_0 \frac{\partial}{\partial\beta} + \left(\frac{\partial\gamma_3}{\partial\alpha_2}\right)_0 \frac{\partial}{\partial\gamma} , \tag{6-5a}$$

$$-X_\beta = \left(\frac{\partial\alpha_3}{\partial\beta_2}\right)_0 \frac{\partial}{\partial\alpha} + \left(\frac{\partial\beta_3}{\partial\beta_2}\right)_0 \frac{\partial}{\partial\beta} + \left(\frac{\partial\gamma_3}{\partial\beta_2}\right)_0 \frac{\partial}{\partial\gamma} , \tag{6-5b}$$

$$-X_\gamma = \left(\frac{\partial\alpha_3}{\partial\gamma_2}\right)_0 \frac{\partial}{\partial\alpha} + \left(\frac{\partial\beta_3}{\partial\gamma_2}\right)_0 \frac{\partial}{\partial\beta} + \left(\frac{\partial\gamma_3}{\partial\gamma_2}\right)_0 \frac{\partial}{\partial\gamma} , \tag{6-5c}$$

where the subscript "0" indicates that the values are to be evaluated at $\alpha_2 = \beta_2 = \gamma_2 = 0$. Using

$$\frac{\partial\alpha_3}{\partial\alpha_2} = \sum_{ij} \frac{\partial\alpha_3}{\partial c_i}\frac{\partial c_i}{\partial b_j}\frac{\partial b_j}{\partial\alpha_2}$$

etc., as well as (6-2) and (6-5), we can find the differential operators for X_α, X_β and X_γ. With the aid of

$$\alpha X_\alpha = -i\alpha J_z, \quad \beta X_\beta = -i\beta J_y, \quad \gamma X_\gamma = -i\gamma J_z ,$$

we obtain

$$J_z = \frac{1}{i}\frac{\partial}{\partial\alpha} , \tag{6-6a}$$

$$J_y = \frac{1}{i}\left(-\sin\alpha\cot\beta\frac{\partial}{\partial\alpha} + \cos\alpha\frac{\partial}{\partial\beta} + \frac{\sin\alpha}{\sin\beta}\frac{\partial}{\partial\gamma}\right) . \tag{6-6b}$$

We can only get two differential operators J_y and J_z from (6-5a), since both X_α and X_γ correspond to the same J_z. The third operator can be found by using the commutator $[J_y, J_z] = iJ_x$,

$$J_x = \frac{1}{i}\left(-\cos\alpha\cot\beta\frac{\partial}{\partial\alpha} - \sin\alpha\frac{\partial}{\partial\beta} + \frac{\cos\alpha}{\sin\beta}\frac{\partial}{\partial\gamma}\right) . \tag{6-6c}$$

Equations (6-6) give the infinitesimal operators of the first parameter group, and are identical to Eq. (39) of Eisenberg (1970, p. 98). Therefore the infinitesimal operators of the first parameter group of SO_3 are precisely the well known differential forms of the angular momentum operators J_x, J_y and J_z in the group parameter space.

Similarly, the differential operators of the generators of the intrinsic group $\overline{SO}_3$ can be found by using (5-234). Because of the anti-isomorphism between the groups $\overline{SO}_3$ and SO_3, it is known that, corresponding to the SO_3 group element (5-64), the group element of $\overline{SO}_3$ is

$$\overline{R}(\alpha\beta\gamma) = e^{-i\gamma\overline{J}_z}e^{-i\beta\overline{J}_y}e^{-i\alpha\overline{J}_z} . \tag{6-7}$$

The counterpart of (6-5) is

$$\begin{aligned} -\overline{X}_\alpha &= \left(\frac{\partial\alpha_3}{\partial\alpha_1}\right)_0\frac{\partial}{\partial\alpha} + \left(\frac{\partial\beta_3}{\partial\alpha_1}\right)_0\frac{\partial}{\partial\beta} + \left(\frac{\partial\gamma_3}{\partial\alpha_1}\right)_0\frac{\partial}{\partial\gamma} , \\ \left(\frac{\partial\alpha_3}{\partial\alpha_1}\right) &= \sum_{i,j}\frac{\partial\alpha_3}{\partial c_i}\frac{\partial c_i}{\partial a_j}\frac{\partial a_j}{\partial\alpha_1} , \\ &\cdots \end{aligned} \tag{6-8}$$

where the subscript "0" means that the values are to be evaluated at $\alpha_1 = \beta_1 = \gamma_1 = 0$. We finally obtain

$$\begin{aligned} \overline{J}_x &= \frac{1}{i}\left(\cos\gamma\cot\beta\frac{\partial}{\partial\gamma} + \sin\gamma\frac{\partial}{\partial\beta} - \frac{\cos\gamma}{\sin\beta}\frac{\partial}{\partial\alpha}\right) = -J_x(\alpha\leftrightarrow\gamma) , \\ \overline{J}_y &= \frac{1}{i}\left(-\sin\gamma\cot\beta\frac{\partial}{\partial\gamma} + \cos\gamma\frac{\partial}{\partial\beta} + \frac{\sin\gamma}{\sin\beta}\frac{\partial}{\partial\alpha}\right) = J_y(\alpha\leftrightarrow\gamma) , \\ \overline{J}_z &= \frac{1}{i}\frac{\partial}{\partial\gamma} = J_z(\alpha\leftrightarrow\gamma) . \end{aligned} \tag{6-9}$$

This is identical with Eq. (29) in Eisenberg (1970, p. 96).

It is easy to verify that the operators in (6-6) and (6-9) satisfy (5-228a), namely,

$$\begin{pmatrix} \overline{J}_x \\ \overline{J}_y \\ \overline{J}_z \end{pmatrix} = \tilde{\mathcal{D}}(\alpha,\beta,\gamma)\begin{pmatrix} J_x \\ J_y \\ J_z \end{pmatrix} , \tag{6-10}$$

where $\tilde{\mathcal{D}}$ is the transpose of the matrix $\mathcal{D}(\alpha\beta\gamma)$ of the adjoint representation ($J = 1$) of SO_3 (see (5-34)).

By comparing (6-10) with (5-35), we find that the infinitesimal operators $\overline{J}_{x,y,z}$ of the intrinsic group $\overline{SO}_3$ are the components $J_{1,2,3}$ of angular momentum in the intrinsic frame. We thus conclude that the projections $J_{1,2,3}$ of angular momentum in the intrinisic frame are the generators of the intrinsic rotation group $\overline{SO}_3$, just as the projections $J_{x,y,z}$ of angular momentum in the fixed frame constitute the generators of SO_3.

Our naming $\overline{G}$ the intrinsic group and the interpretation given in Sec. 5.14 for the generators $\overline{X}_\rho$ of $\overline{G}$ all originated from the SO_3 case.

Using (6-6) and (6-9), we can easily check (5-236), that is, the CSCO-I of SO_3 and $\overline{SO}_3$ are equal:

$$J_x^2 + J_y^2 + J_z^2 = \overline{J}_x^2 + \overline{J}_y^2 + \overline{J}_z^2 \, .$$

Louck (1976, 1970, 1965) studied the transformation between the laboratory (fixed) frame and the intrinsic frame. Louck (1970) and Biedenharn (1968) carried this kind of study to the case of U_n group.

6.2. Irreps of the SO_2 Group

The group SO_2 is Abelian with the elements $R(\varphi) = e^{-i\varphi J_z}$. Each element is a class by itself. Therefore the class operator $C(\varphi)$ is still $e^{-i\varphi J_z}$. The generator of SO_2 is J_z. Since SO_2 is a Lie group of rank one, its CSCO-I consists of a single invariant. Obviously, J_z is the CSCO-I of SO_2.

The CSCO-I of a finite group consists of few class operators, while the CSCO-I of a Lie group of rank l consists of l invariants $\{I_i(X_\rho) : i = 1, 2, \ldots, l\}$. The following question may now be raised: what is the relation between the two? For finite groups we proved that all the class operators of a group G must be a CSCO-I of G. An extension of this conclusion to the Lie group implies that all the class operators (the number of which is infinite) of a Lie group must form a CSCO-I of the Lie group. For instance, the infinite number of class operators

$$C(\varphi) = e^{-iJ_z\varphi}, \quad 0 \leq \varphi \leq 2\pi \, , \tag{6-11}$$

of the group SO_2 form a CSCO-I of SO_2. On the other hand we know that the property of a Lie group is decided by the behavior of those elements which lie in the neighborhood of the identity. Therefore to form a CSCO-I of a Lie group, one merely needs to take the class operators in the neighborhood of the identity.

Expanding the class operator $C(\varphi)$ in (6-11) around $\varphi = 0$, one gets

$$C(\varphi) = 1 - i\varphi J_z + \ldots . \tag{6-12}$$

A simple relation is thus revealed between the CSCO-I $C(\varphi)$ of SO_2 borrowed from the finite group rep theory and the CSCO-I J_z of SO_2 obtained from the Lie group rep theory: By making the Taylor expansion of the former in the neighborhood of the identity, one can get the latter. For the Abelian group SO_2, the CSCO-I J_z is also the CSCO-III of SO_2. According to Theorem VII in Sec. 3.20, the eigenfunctions $\chi^{(m)*}(\varphi)$ of J_z in the group parameter space are irreducible matrix elements (or characters). From (5-177c), we know that $J_z = -i\frac{\partial}{\partial\varphi}$. $\chi^{(m)*}(\varphi)$ satisfies the eigen equation

$$-i\frac{\partial}{\partial\varphi}\chi^{(m)*}(\varphi) = m\chi^{(m)*}(\varphi) \, . \tag{6-13}$$

The following two cases are to be considered separately.

1. SO_2. Since $\varphi = 0$ and 2π correspond to the same point in the group parameter space, $\chi^{(m)}(\varphi)$ has to obey the periodic condition

$$\chi^{(m)}(\varphi = 0) = \chi^{(m)}(\varphi = 2\pi) \, . \tag{6-14}$$

With the condition (6-14), the eigensolution to (6-13) is

$$\chi^{(m)}(\varphi) = e^{-im\varphi}, \quad m = 0, \pm 1, \pm 2, \ldots \tag{6-15}$$

Clearly, the eigenfunctions $\chi^{(m)}(\varphi)$ form a complete set of orthogonal functions in the interval $(0,2\pi)$ with the weight function

$$\rho(\varphi) = 1 . \tag{6-16}$$

The group volume of SO_2 is

$$g = \int_o^{2\pi} d\varphi = 2\pi . \tag{6-17}$$

Consequently the orthonormality and completeness (5-247) becomes

$$\frac{1}{2\pi} \int_0^{2\pi} \chi^{(m)*}(\varphi)\chi^{(m')}(\varphi) d\varphi = \delta_{mm'} , \tag{6-18}$$

$$\frac{1}{2\pi} \sum_{m=0,\pm 1,\ldots} \chi^{(m)*}(\varphi)\chi^{(m)}(\varphi') = \delta(\varphi - \varphi') , \tag{6-19}$$

and the projection operator (5-245a) reads

$$P^{(m)} = \frac{1}{2\pi} \int_0^{2\pi} e^{-i(J_z - m)\varphi} d\varphi . \tag{6-20}$$

2. $SO_2^{\dagger}$. According to (6-1b), $\varphi = 0$ and $\varphi = 4\pi$ now correspond to the same group element instead of $\varphi = 0$ and $\varphi = 2\pi$. The boundary condition (6-14) is replaced by

$$\chi^{(m)}(\varphi = 0) = \chi^{(m)}(\varphi = 4\pi) . \tag{6-21}$$

The eigensolution to (6-21) is still $\chi^{(m)}(\varphi) = e^{-im\varphi}$, however m now takes integers as well as half integers

$$\chi^{(m)}(\varphi) = e^{-im\varphi} , \quad m = 0, \pm\frac{1}{2}, \pm 1, \pm\frac{3}{2} , \ldots \tag{6-22}$$

The half integer rep is called the spinor rep or double-valued rep of SO_2. The weight function is the same as before, $\rho(\varphi) = 1$, but the group volume becomes $g = 4\pi$. The orthonormality and completeness become

$$\begin{aligned} &\frac{1}{4\pi} \int_0^{4\pi} \chi^{(m)*}(\varphi)\chi^{(m')}(\varphi) d\varphi = \delta_{mm'} , \\ &\frac{1}{4\pi} \sum_{m=0,\pm\frac{1}{2},\pm 1,\ldots} \chi^{(m)*}(\varphi)\chi^{(m)}(\varphi') = \delta(\varphi - \varphi') . \end{aligned} \tag{6-23}$$

The projection operator is

$$P^{(m)} = \frac{1}{4\pi} \int_0^{4\pi} e^{-i(J_z - m)\varphi} d\varphi . \tag{6-24}$$

6.3. The CSCO-I and Characters of SO_3

In Chapter 5 it was shown that the CSCO-I of SO_3 is $\mathbf{J}^2$. We now proceed to investigate the relationship between $\mathbf{J}^2$ and the class operators of SO_3.

According to Sec. 1.5, all the rotations about any axes $\mathbf{n}(\theta', \varphi')$ through the same angle φ belong to the same class. The rotation operators can be written as

$$R_{\mathbf{n}(\theta'\varphi')}(\varphi) = e^{-i(\mathbf{n}\cdot\mathbf{J})\varphi} , \quad 0 \le \theta' \le \pi , 0 \le \varphi' \le 2\pi , 0 \le \varphi \le \pi . \tag{6-25}$$

Using (5-196), we obtained the class operators of SO_3,

$$C(\varphi) = \int_0^{2\pi} d\varphi' \int_0^{\pi} \sin\theta' d\theta' \exp\left[-i(J_x \sin\theta' \cos\varphi' + J_y \sin\theta' \sin\varphi' + J_z \cos\theta')\varphi\right] . \tag{6-26}$$

From small φ, we have the following expansion

$$\begin{aligned} C(\varphi) = \int_0^{2\pi} d\varphi' \int_0^{\pi} & \{1 - i(J_x \sin\theta' \cos\varphi' + J_y \sin\theta' \sin\varphi' + J_z \cos\theta')\varphi \\ & - \frac{1}{2}(J_x^2 \sin^2\theta' \cos^2\varphi' + J_y^2 \sin^2\theta' \sin^2\varphi' + J_z^2 \cos^2\theta')\varphi^2 \\ & - \frac{1}{2}[(J_xJ_y + J_yJ_x)\sin^2\theta' \cos\varphi' \sin\varphi' + (J_zJ_x + J_xJ_z)\sin\theta' \cos\theta' \cos\varphi' \\ & + (J_yJ_z + J_zJ_y)\sin\theta' \cos\theta' \sin\varphi']\varphi^2 + \dots\} \sin\theta' d\theta' d\varphi' \\ = 4\pi & \left[1 - \frac{1}{3!}(J_x^2 + J_y^2 + J_z^2)\varphi^2 + \dots\right] . \end{aligned} \tag{6-27}$$

Here we once again have a simple relation between the CSCO-I $C(\varphi)$ of SO_3 extrapolated from the finite group rep theory and the CSCO-I $\mathbf{J}^2 = J_x^2 + J_y^2 + J_z^2$ of SO_3 obtained from the Lie group rep theory.

According to (5-197), the volume of the elements belonging to the class φ is

$$g(\varphi) = \int_0^{2\pi} d\varphi' \int_0^{\pi} \sin\theta' d\theta' = 4\pi . \tag{6-28}$$

One sees that $g(\varphi)$ is independent of φ.

To find the characters and the density function in the class parameter space, we have to know the differential operator of $\mathbf{J}^2$ in the class parameter space. Let us first find the infinitesimal operators of the first parameter group of SO_3 with $(\theta', \varphi', \varphi)$ as the group parameters. In analogy with (6-1), we consider the product of two successive rotations

$$R(\theta_3'\varphi_3'\varphi_3) = R(\theta_2'\varphi_2'\varphi_2)R(\theta_1'\varphi_1'\varphi_1) . \tag{6-29}$$

The combination law for the parameters $(\theta', \varphi', \varphi)$ is still not easy to write out explicitly and is again obtained implicitly through the parameters (6-2a) of the SU_2 group. The relations between these two sets of parameters are given by Smirnov (1951).

$$\begin{aligned} a_0 &= \cos(\varphi_1/2) , & a_1 &= \sin\theta_1' \cos\varphi_1' \sin(\varphi_1/2) , \\ a_2 &= \sin\theta_1' \sin\varphi_1' \sin(\varphi_1/2) , & a_2 &= \cos\theta_1' \sin(\varphi_1/2) . \end{aligned} \tag{6-30}$$

As in (6-5a), we have

$$-X_\varphi = \left(\frac{\partial\theta_3'}{\partial\varphi_2}\right)_0 \frac{\partial}{\partial\theta'} + \left(\frac{\partial\varphi_3'}{\partial\varphi_2}\right)_0 \frac{\partial}{\partial\varphi'} + \left(\frac{\partial\varphi_3}{\partial\varphi_2}\right)_0 \frac{\partial}{\partial\varphi} , \tag{6-31}$$

where the subscript "0" indicates $\varphi_2 = 0$. It can be shown that the infinitesimal operators $X_{\theta'}$ and $X_{\varphi'}$ corresponding to the parameters θ' and φ' are zero. Using

$$\left(\frac{\partial \theta'_3}{\partial \varphi_2}\right)_0 = \left(\frac{\partial \theta'_3}{\partial c_i}\right)_0 \frac{\partial c_i}{\partial b_j} \left(\frac{\partial b_j}{\partial \varphi_2}\right)_0 ,$$
$$\cdots\cdots\cdots \tag{6-32}$$

as well as (6-2b) and (6-30), we can find the operator X_φ. From $\exp(\varphi X_\varphi) = \exp[-i(\mathbf{J}\cdot\mathbf{n})\varphi]$ we get

$$X_\varphi = -i(\sin\theta' \cos\varphi' J_x + \sin\theta' \sin\varphi' J_y + \cos\theta' J_z) . \tag{6-33}$$

Comparing (6-31) with (6-33) we finally have

$$\begin{aligned}
J_x &= \frac{1}{i}\left[\sin\theta' \cos\varphi' \frac{\partial}{\partial\varphi} + \frac{1}{2}\left(\sin\varphi' + \cos\theta' \cos\varphi' \cot\frac{\varphi}{2}\right)\frac{\partial}{\partial\theta'}\right.\\
&\quad \left. + \frac{\cos\theta' \cos\varphi' \sin(\varphi/2) - \sin\varphi' \cos(\varphi/2)}{2\sin\theta' \sin(\varphi/2)} \frac{\partial}{\partial\varphi'}\right] ,\\
J_y &= \frac{1}{i}\left[\sin\theta' \sin\varphi' \frac{\partial}{\partial\varphi} + \frac{1}{2}\left(-\cos\varphi' + \cos\theta' \sin\varphi' \cot\frac{\varphi}{2}\right)\frac{\partial}{\partial\theta'}\right.\\
&\quad \left. + \frac{\cos\theta' \sin\varphi' \sin(\varphi/2) + \cos\varphi' \cos(\varphi/2)}{2\sin\theta' \sin(\varphi/2)} \frac{\partial}{\partial\varphi'}\right] ,\\
J_z &= \frac{1}{i}\left[\cos\theta' \frac{\partial}{\partial\varphi} - \frac{1}{2}\sin\theta' \cot(\varphi/2)\frac{\partial}{\partial\theta'} - \frac{1}{2}\frac{\partial}{\partial\varphi'}\right] ,\\
J^2 &= -\frac{1}{\sin^2(\varphi/2)}\frac{\partial}{\partial\varphi}\left(\sin^2\frac{\varphi}{2}\frac{\partial}{\partial\varphi}\right) - \frac{1}{4\sin^2(\varphi/2)\sin\theta'}\frac{\partial}{\partial\theta'}\left(\sin\theta'\frac{\partial}{\partial\theta'}\right)\\
&\quad - \frac{1}{4\sin^2(\varphi/2)\sin^2\theta'}\frac{\partial^2}{\partial\varphi'^2} .
\end{aligned} \tag{6-34}$$

According to Theorem II in Sec. 3.20, the character $\chi^{(\nu)}(\varphi)$ of SO_3 satisfies the eigenfunction

$$J^2\chi^{(\nu)}(\varphi) = \nu\chi^{(\nu)}(\varphi) , \tag{6-35a}$$

i.e.,

$$-\frac{1}{\sin^2(\varphi/2)}\frac{\partial}{\partial\varphi}\left(\sin^2\frac{\varphi}{2}\frac{\partial}{\partial\varphi}\right)\chi^{(\nu)}(\varphi) = \nu\chi^{(\nu)}(\varphi) , \tag{6-35b}$$

or expressed differently

$$-\frac{1}{\rho(\varphi)}\frac{d}{d\varphi}\left(\rho(\varphi)\frac{d}{d\varphi}\right)\chi^{(\nu)}(\varphi) = \nu\chi^{(\nu)}(\varphi) , \tag{6-35c}$$

$$\rho(\varphi) = 4\sin^2(\varphi/2) = 2(1-\cos\varphi) . \tag{6-36}$$

We will see that $\rho(\varphi)$ is the density function in class parameter space. The factor 4 in front of $\sin^2(\varphi/2)$ is introduced so that the density function is the same as that used by Hamermesh (1962).[1)]

[1)] The density function can be enlarged or reduced by any constant factor without affecting the ratios $\rho(\varphi)/g$ and $\rho(a)/g$, and therefore without affecting the final results.

From (6-36) and (5-201), the group volume of SO_3 is

$$g = \int_0^{\pi} \rho(\varphi) d\varphi \int_0^{2\pi} d\varphi' \int_0^{\pi} \sin\theta' d\theta' = 8\pi^2 . \tag{6-37}$$

Let

$$u^{(\nu)}(\varphi) = \sin\frac{\varphi}{2}\chi^{(\nu)}(\varphi) . \tag{6-38}$$

Equation (6-35b) becomes

$$\frac{d^2u^{(\nu)}(\varphi)}{d^2\varphi} = -\left(j+\frac{1}{2}\right)^2 u^{(\nu)}(\varphi) ,$$
$$\left(j+\frac{1}{2}\right)^2 = \nu + \frac{1}{4}, \quad \nu = j(j+1) . \tag{6-39}$$

Its solution is $u^{(\nu)}(\varphi) \equiv u^{(j)}(\varphi) = \sin(j+1/2)\varphi$ [another solution $\cos(j+1/2)\varphi$ is discarded, since $\chi^{(\nu)}(\varphi) = \cos(j+\frac{1}{2})\varphi/\sin\frac{\varphi}{2}$ diverges at $\varphi = 0$]. The eigenvalues j are to be decided by the boundary conditions.

From (6-35c) and making use of the Green theorem,

$$\rho(\varphi)\left(\chi^{(j')}\frac{d\chi^{(j)}}{d\varphi} - \chi^{(j)}\frac{d\chi^{(j')}}{d\varphi}\right)\Bigg|_a^b = [j(j+1) - j'(j'+1)]\int_a^b \chi^{(j')}\chi^{(j)}\rho(\varphi)d\varphi , \tag{6-40}$$

$$\chi^{(j)}(\varphi) = \frac{\sin(j+1/2)\varphi}{\sin(\varphi/2)} . \tag{6-41}$$

If $\chi^{(j)}$ or $\frac{d\chi^{(j)}}{d\varphi}$ vanishes at the boundaries $\varphi = a$ and $\varphi = b$, it follows from (6-40) that the functions $\chi^{(j)}$ form an orthogonal complete set in the interval $[a, b]$ with the weight $\rho(\varphi)$,

$$\int_a^b \chi^{(j)}(\varphi)\chi^{(j')}(\varphi)\rho(\varphi)d\varphi = 0, \quad \text{for } j \neq j' . \tag{6-42}$$

Equation (6-41) does not vanish at $\varphi = 0, \pi$, or 2π; therefore the boundary condition has to be chosen as

$$\frac{d\chi^{(j)}(\varphi)}{d\varphi} = 0, \quad \text{for } \varphi = a, b . \tag{6-43a}$$

From (6-41) we obtain

$$\frac{d\chi^{(j)}(\varphi)}{d\varphi} = \frac{\left(j+\frac{1}{2}\right)\cos\left(j+\frac{1}{2}\right)\varphi\sin\frac{\varphi}{2} - \frac{1}{2}\sin\left(j+\frac{1}{2}\right)\varphi\cos\frac{\varphi}{2}}{\sin^2\frac{\varphi}{2}} . \tag{6-43'}$$

Using the L'Hopital rule we find its extrema at $\varphi = 0$ or 2π, i.e.,

$$\frac{d\chi^{(j)}(\varphi)}{d\varphi}\Big|_{\varphi=0 \text{ or } 2\pi} = -j(j+1)\frac{\sin\left(j+\frac{1}{2}\right)\varphi}{\cos\frac{\varphi}{2}}\Big|_{\varphi=0 \text{ or } 2\pi} . \tag{6-43b}$$

Using the L'Hopital rule again, we get its extremum at $\varphi = \pi$

$$\frac{d\chi^{(j)}(\varphi)}{d\varphi}\Big|_{\varphi=\pi} = (2j+1)j(j+1)\frac{\cos\left(j+\frac{1}{2}\right)\varphi}{\sin\frac{\varphi}{2}}\Big|_{\varphi=\pi} . \tag{6-43c}$$

The boundary a for φ is always chosen to be zero, while the boundary b may be chosen as π or 2π depending on whether the group being considered is SO_3 or $SO_3^\dagger$.

1. The case of SO_3. The range of φ is specified by (6-25), i.e., $a = 0$ and $b = \pi$. It follows from (6-43b) and (6-43c) that the quantum number j are necessarily integers, $j = 0, 1, 2, \ldots$. Therefore $\chi^{(j)}(\varphi)$ form an orthogonal complete set of functions in the interval $[0, \pi]$. Putting in the norm we have

$$\frac{1}{2\pi}\int_0^\pi \chi^{(j)}(\varphi)\chi^{(j')}(\varphi)\rho(\varphi)d\varphi = \delta_{jj'}, \quad j, j' = 0, 1, 2, \ldots . \tag{6-44a}$$

$$\frac{1}{2\pi}\sum_{j=0,1,2,\ldots} \chi^{(j)}(\varphi)\chi^{(j)}(\varphi')\rho(\varphi) = \delta(\varphi - \varphi') . \tag{6-44b}$$

This is the realization of (5-244) for the group SO_3. Now $g = 8\pi^2$ and $g(\varphi) = 4\pi$.

It is seen that the weight function (6-36) in the differential equation (6-35b) is just the density function and (6-41) is the irreducible character of SO_3.

The purpose of using this awkward way to find the character of SO_3 is to show the applicability of the theorems for the finite groups to the Lie groups. Later Eq. (6-64) will give a very simple way for determining the character of SO_3.

2. The case of $SO_3^\dagger$. For the double group $SO_3^\dagger$, (6-1d), the range of φ is from zero to 2π. Therefore the boundaries in (6-43b) become $a = 0$ and $b = 2\pi$. From (6-43b) we obtain $j = 0, 1/2, 1, 3/2, \ldots$. The functions $\chi^{(j)}(\varphi)$ of (6-41) form an orthogonal complete set in the interval $[0, 2\pi]$:

$$\frac{1}{4\pi}\int_0^{2\pi} \chi^{(j)}(\varphi)\chi^{(j')}(\varphi)\rho(\varphi)d\varphi = \delta_{jj'}, \quad j, j' = 0, \frac{1}{2}, 1, \frac{3}{2}, \ldots \tag{6-45a}$$

$$\frac{1}{4\pi}\sum_{j=0,\frac{1}{2},1,\frac{3}{2},\ldots} \chi^{(j)}(\varphi)\chi^{(j)}(\varphi')\rho(\varphi) = \delta(\varphi - \varphi') . \tag{6-45b}$$

For integer j, (6-45) goes back to (6-44).

Remark: The orthonormality (6-44a) remains true when both j and j' are half integers, but is not true when j is an integer and j' is a half integer. In order to restore the orthogonal property for the latter case, one has to extend the interval of φ from $[0, \pi]$ to $[0, 2\pi]$ and (6-44a) is replaced by (6-45a).

Setting $\varphi = 0$ in (6-41), we obtain the character of the identity, i.e., the dimension of the irrep j of SO_3,

$$h_\nu = h_j = 2j + 1 .$$

From (5-242) we obtain the projection operator for the group SO_3 and $SO_3^\dagger$:

$$SO_3 : P^{(j)} = \frac{2j+1}{2\pi^2}\int_0^\pi \sin\left(j + \frac{1}{2}\right)\varphi \sin\frac{\varphi}{2} C(\varphi)d\varphi, \quad j = 0, 1, 2, \ldots \tag{6-46a}$$

$$SO_3^\dagger : P^{(j)} = \frac{2j+1}{4\pi^2}\int_0^{2\pi} \sin\left(j + \frac{1}{2}\right)\varphi \sin\frac{\varphi}{2} C(\varphi)d\varphi, \quad j = 0, 1/2, 1, 3/2, \ldots \tag{6-46b}$$

6.4. The CSCO-III and Irreducible Matrix Element of SO_3

We now return back to the group-parameter space. SO_3 is a Lie group with three parameters. The CSCO-III should consist of three operators. Besides $\mathbf{J}^2$, we need to find another two operators.

We choose the group chain $SO_3 \supset SO_2$ to classify the irreducible basis of SO_3. The corresponding intrinsic group chain is $\overline{SO}_3 \supset \overline{SO}_2$. J_z and $\overline{J}_z$ are the CSCO-I of SO_2 and $\overline{SO}_2$, respectively. Let

$$K = (C, C(s), \overline{C}(s)) = (\mathbf{J}^2, J_z, \overline{J}_z)\ .$$

K is the CSCO-III of SO_3.

Choosing the Euler angles as the group parameters, we obtain from (6-6) and (6-9) the following differential operators,

$$J^2 = -\frac{1}{\sin\beta}\frac{\partial}{\partial\beta}\left(\sin\beta\frac{\partial}{\partial\beta}\right) - \frac{1}{\sin^2\beta}\left(\frac{\partial^2}{\partial\alpha^2} + \frac{\partial^2}{\partial\gamma^2}\right) - \frac{2\cos\beta}{\sin^2\beta}\frac{\partial^2}{\partial\alpha\partial\gamma}\ , \tag{6-47a}$$

$$J_z = \frac{1}{i}\frac{\partial}{\partial\alpha},\quad \overline{J}_z = \frac{1}{i}\frac{\partial}{\partial\gamma}\ . \tag{6-47b}$$

The simultaneous eigenfunctions of $\mathbf{J}^2, J_z$ and $\overline{J}_z$ are the complex conjugate of the well known D-functions

$$\begin{pmatrix} \mathbf{J}^2 \\ J_z \\ \overline{J}_z \end{pmatrix} D^{(j)*}_{mk}(\alpha\beta\gamma) = \begin{pmatrix} j(j+1) \\ m \\ k \end{pmatrix} D^{(j)*}_{mk}(\alpha\beta\gamma)\ . \tag{6-48}$$

For a given set of quantum numbers (j, m, k), there is only one solution $D^{(j)*}_{mk}(\alpha\beta\gamma)$. Equation (6-48) can be treated by the separation of variables method:

$$D^{(j)*}_{mk}(\alpha\beta\gamma) = e^{im\alpha}d^{(j)}_{mk}(\beta)e^{ik\gamma}\ , \tag{6-49}$$

$$\left[-\frac{d}{d\beta}\sin\beta\frac{d}{d\beta} + \frac{1}{\sin\beta}(m^2 + k^2 - 2mk\cos\beta)\right] d^{(j)}_{mk}(\beta) = j(j+1)\sin\beta\, d^{(j)}_{mk}(\beta)\ . \tag{6-50}$$

The density function (or the weight function) $\rho(\alpha\beta\gamma)$ can also be factorized, $\rho(\alpha\beta\gamma) = \rho(\alpha)\rho(\beta)\rho(\gamma)$. Comparing (6-47b) with (6-13), we know that $\rho(\alpha) = \rho(\gamma) = 1$. Comparing (6-50) with (6-35c), we have $\rho(\beta) = \sin\beta$. Therefore

$$\rho(\alpha\beta\gamma) = \sin\beta\ . \tag{6-51}$$

1. The case of SO_3. From (6-1c), the group volume is

$$g = \int_0^{2\pi} d\alpha \int_0^{2\pi} d\gamma \int_0^{\pi} \sin\beta d\beta = 8\pi^2\ ,$$

in agreement with (6-37). The eigenfunctions

$$\psi^{(j)k}_m(\alpha\beta\gamma) = \sqrt{\frac{2j+1}{8\pi^2}} D^{(j)*}_{mk}(\alpha\beta\gamma),\quad j = 0, 1, 2, 3, \ldots \tag{6-52}$$

of the CSCO-III form an orthonormal set of functions in the intervals $[0, 2\pi]$ for α and γ, and $[0, \pi]$ for β, with the weight $\sin\beta$,

$$\int_0^{2\pi} d\alpha \int_0^{2\pi} d\gamma \int_0^{\pi} \sin\beta d\beta \psi^{(j)k}_m(\alpha\beta\gamma)^* \psi^{(j')k'}_{m'}(\alpha\beta\gamma) = \delta_{jj'}\delta_{mm'}\delta_{kk'}\ , \tag{6-53a}$$

$$\sum_{j=0,1,2,\ldots}\ \sum_{m=-j}^{j}\ \sum_{k=-j}^{j} \psi^{(j)k}_m(\alpha\beta\gamma)^* \psi^{(j)k}_m(\alpha\beta\gamma)\sin\beta = \delta(\alpha - \alpha')\delta(\beta - \beta')\delta(\gamma - \gamma')\ . \tag{6-53b}$$

The generalized projection operator $P_{mk}^{(j)}$ of (5-245a) now becomes

$$P_{mk}^{(j)} = \frac{2j+1}{8\pi^2} \int_0^{2\pi} d\alpha \int_0^{2\pi} d\gamma \int_0^{\pi} \sin\beta d\beta D_{mk}^{(j)*}(\alpha\beta\gamma) R(\alpha\beta\gamma) \ . \tag{6-54}$$

2. The case of $SO_3^\dagger$. From (6-1d), the group volume of $SO_3^\dagger$ is

$$g = 2\int_0^{2\pi} d\alpha \int_0^{2\pi} d\gamma \int_0^{\pi} \sin\beta d\beta = 16\pi^2 \ , \tag{6-55}$$

$$\psi_m^{(j)k}(\alpha\beta\gamma) = \frac{\sqrt{2j+1}}{4\pi} D_{mk}^{(j)*}(\alpha\beta\gamma) \quad j = 0, \frac{1}{2}, 1, \frac{3}{2}, \ldots \tag{6-56}$$

They form an orthonormal set in the intervals $[0, 2\pi]$ for α and γ, $[0, \pi]$ and $[2\pi, 3\pi]$ for β with the weight $\sin\beta$,

$$\int_0^{2\pi} d\alpha \int_0^{2\pi} d\gamma \left[\int_0^{\pi} + \int_{2\pi}^{3\pi}\right] \sin\beta d\beta \psi_m^{(j)k}(\alpha\beta\gamma)^* \psi_{m'}^{(j')k'}(\alpha\beta\gamma) = \delta_{jj'}\delta_{mm'}\delta_{kk'} \ , \tag{6-57a}$$

$$\sum_{j=0,\frac{1}{2},1,\ldots} \sum_m \sum_k \psi_m^{(j)k}(\alpha\beta\gamma)^* \psi_m^{(j)k}(\alpha'\beta'\gamma') \sin\beta = \delta(\alpha-\alpha')\delta(\beta-\beta')\delta(\gamma-\gamma') \ . \tag{6-57b}$$

The generalized projection operator is

$$P_{mk}^{(j)} = \frac{2j+1}{16\pi^2} \int_0^{2\pi} d\alpha \int_0^{2\pi} d\gamma \left[\int_0^{\pi} + \int_{2\pi}^{3\pi}\right] \sin\beta d\beta D_{mk}^{(j)*}(\alpha\beta\gamma) R(\alpha\beta\gamma) \ . \tag{6-58}$$

The projection operators of (6-54) or (6-58) satisfy the following eigenequation

$$\begin{pmatrix} J^2 \\ J_z \\ \overline{J}_z \end{pmatrix} P_{mk}^{(j)} = \begin{pmatrix} j(j+1) \\ m \\ k \end{pmatrix} P_{mk}^{(j)} \ . \tag{6-59}$$

6.5. The CSCO-II and Irreducible Bases of SO_3

$(\mathbf{J}^2, J_z)$ is the CSCO-II of SO_3, and is the complete set of commuting operators in the (θ, φ) space. The differential operators of $\mathbf{J}^2$ and J_z in the (θ, φ) space are well known in quantum mechanics, i.e.,

$$\begin{aligned} J^2 &= -\frac{1}{\sin\theta}\left(\frac{\partial}{\partial\theta}\sin\theta\frac{\partial}{\partial\theta} + \frac{1}{\sin\theta}\frac{\partial}{\partial\varphi^2}\right) \ , \\ J_z &= \frac{1}{i}\frac{\partial}{\partial\varphi} \ . \end{aligned} \tag{6-60}$$

The simultaneous eigenfunctions of the CSCO-II are the spherical harmonic functions $Y_{lm}(\theta, \varphi)$,

$$\begin{pmatrix} J^2 \\ J_z \end{pmatrix} Y_{lm}(\theta, \varphi) = \begin{pmatrix} l(l+1) \\ m \end{pmatrix} Y_{lm}(\theta, \varphi) \ . \tag{6-61}$$

$Y_{lm}(\theta, \varphi)$ form an orthonormal set in the interval $[0, \pi]$ for θ and $[0, 2\pi]$ for φ,

$$\int_0^{2\pi} d\varphi \int_0^{\pi} \sin\theta d\theta Y_{lm}^*(\theta, \varphi) Y_{l'm'}(\theta, \varphi) = \delta_{ll'}\delta_{mm'} \ , \tag{6-62a}$$

$$\sum_{l=0,1,2,\ldots} \sum_m Y^*_{lm}(\theta,\varphi) Y_{lm}(\theta',\varphi') \sin\theta = \delta(\theta-\theta')\delta(\varphi-\varphi') \,. \tag{6-62b}$$

Y_{lm} is the $SO_3 \supset SO_2$ irreducible basis.

Analogously, the $SO_3^\dagger \supset SO_2^\dagger$ (or $SU_2 \supset SO_2$) irreducible basis satisfies the following eigenequations

$$\begin{pmatrix} J^2 \\ J_z \end{pmatrix} \psi_m^{(j)} = \begin{pmatrix} j(j+1) \\ m \end{pmatrix} \psi_m^{(j)} \,. \tag{6-63}$$

Incidentally, we can use the fact that $\psi_m^{(j)}$ is the $SO_3 \supset SO_2$ irreducible basis to calculate the simple characters of SO_3. Since the characters only depend on classes, it suffices to calculate the characters of the rotation $R_z(\varphi)$

$$\begin{aligned} \chi^{(j)}(\varphi) &= \sum_{m=-j}^{j} \langle \psi_m^{(j)} | e^{-iJ_z\varphi} | \psi_m^{(j)} \rangle = \sum_{m=-j}^{j} e^{-im\varphi} \\ &= [e^{ij\varphi} + e^{-ij\varphi} + e^{i(j-1)\varphi} + \ldots](e^{i\varphi/2} - e^{-i\varphi/2})(e^{i\varphi/2} - e^{-i\varphi/2})^{-1} \\ &= \frac{\sin(j+\frac{1}{2})\varphi}{\sin(\varphi/2)} \,, \end{aligned} \tag{6-64}$$

in agreement with (6-41). In fact, it is the standard method to calculate the characters of SO_3 or SU_2.

6.6. The Intrinsic State of SO_3

In the case when a given irrep j occurs more than once, we need to introduce the intrinsic quantum numbers to distinguish these equivalent irreps. To this purpose, we first introduce the intrinsic state of SO_3. The definition of the intrinsic state $\Phi_0(X)$ is given by (5-235), i.e.,

$$\overline{R}(\alpha\beta\gamma)\Phi_0(X) = R(\alpha\beta\gamma)\Phi_0(X) \,. \tag{6-65}$$

To elucidate the physical meaning of the intrinsic state, we consider the collective rotation of a deformed nucleus. A Hartree-Fock (HF) state of a nucleus with an open shell is in general not spherical; in most cases it is prolate. The HF states may have different orientations in the space, such as Φ_1, Φ_{a_0} etc. (see Fig. 6.6-1), where $\Phi_{a_0}(X) = R(a_0)\Phi_1(X)$, or more generally, $\Phi_a(X) = R(a)\Phi_1(X)$ and $R(a)$ is a rotation operator. All these HF states are energy degenerate.

We can define three kinds of coordinate axes: (i) the fixed (external) axes x, y, z, (ii) the symmetry axes x_0, y_0, z_0 of the oblate nucleus, (iii) the intrinsic (body-fixed) axes 1,2,3. From (6-65) it is known that the HF state whose intrinsic axes coincide with the external ones is the intrinsic state. Therefore the choice of the intrinsic state depends on the choice of the intrinsic axes. There are two possible choices.

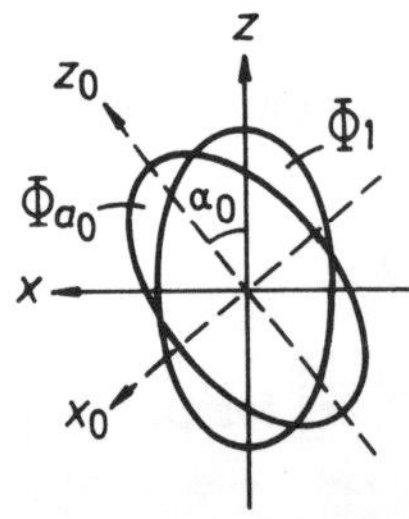

Fig. 6.6-1

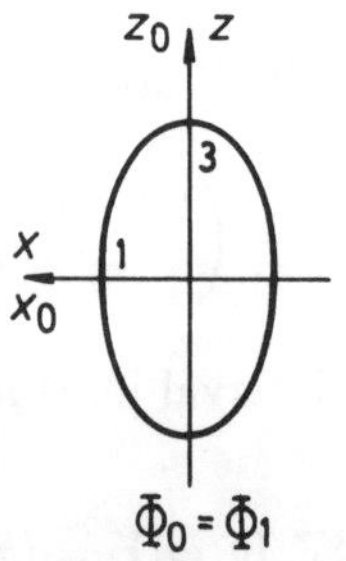

Fig. 6.6-2

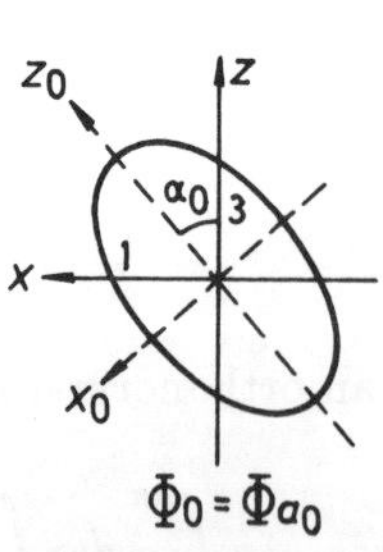

Fig. 6.6-3

1. Choose the symmetric axes of the oblate nucleus as the intrinsic axes. In such a case, the axes (x_0, y_0, z_0) coincide with the axes (1,2,3), and the state Φ_1 (see Fig. 6.6-2) is our intrinsic state.

2. The intrinsic axes (1,2,3) do not coincide with the symmetry axes (x_0, y_0, z_0). Let a_0 be the orientation angle of the axes (x_0, y_0, z_0) with respect to the axes (1,2,3). In this case, the HF state $\Phi_{a_0}(X)$ (see Fig. 6.6-3) is our intrinsic state.

It is thus seen that the arbitrariness in the choice of the intrinsic state stems from the freedom in the choice of the intrinsic axes. In principle any state $\Phi_a(X)$ can be chosen as the intrinsic state. Obviously the first choice is the most convenient one and afterwards we always assume that the symmetry axes of the prolate nucleus are the intrinsic axes and Φ_1 is the intrinsic state. All other HF state $\Phi_a(X)$ can be generated from Φ_0 through the rotation $R(a)$:

$$\Phi_a(X) = R(a)\Phi_0(X) . \tag{6-66}$$

Equations (6-65) and (6-66) specify the action of the intrinsic group element on any state $\Phi_a(X)$.

$$\overline{R}(b)\Phi_a(X) = \overline{R}(b)R(a)\Phi_0(X) = R(a)R(b)\Phi_0(X) = R(a)\Phi_b(X) .$$

In Elliott's SU_3 model for nuclear rotation (Elliott 1958, see Sec. 9.8), the so called leading state $\phi((\lambda\mu)\overline{\varepsilon}\overline{\Lambda}\overline{\nu})$ is defined as the intrinsic state with $\overline{\varepsilon} = 2\lambda + \mu, \overline{\Lambda} = \mu/2$ and $\overline{\nu} = \mu$. The physical meaning of Elliott's intrinsic state is identical to that discussed here.

Now the meaning of (6-65) becomes very clear: $\overline{R}(\alpha\beta\gamma)$ and $R(\alpha\beta\gamma)$ are rotation operators about the intrinsic and external axes, respectively; they have the same effect on the intrinsic state Φ_0, since Φ_0 is the HF state whose symmetry axes, which have been chosen as the intrinsic axes, coincide with the external ones.

6.7. The Projection State of SO_3

Suppose that a nonspherical HF state has axial symmetry. Although it does not have definite total angular momentum, it has definite $J_z = K$. From (6-65) one has

$$\overline{J}_z\Phi_0^{(K)}(X) = J_z\Phi_0^{(K)}(X) = K\Phi_0^{(K)}(X) , \tag{6-67}$$

namely, the z-component of angular momentum of the HF state $\Phi_0(X)$, whose symmetry axes coincide with the external ones, is just the third component of the angular momentum in the intrinsic frame.

Applying the projection operator $P_{MK'}^{(J)}$ on the intrinsic state $\Phi_0^{(K)}(X)$ yields a state with definite angular momentum J and z-component M, if it does not vanish. Now let us study the problem of the intrinsic component of the total angular momentum $\mathbf{J}$ of the projected state. It was pointed out in Sec. 3.13 that not every element of the intrinsic group has a definite action on the state in the configuration space, but only the class operators of certain intrinsic subgroups have definite actions. The method for finding these class operators is as follows: First find all the operators which leave the intrinsic state $\Phi_0^{(K)}(X)$ unchanged. For instance, for the intrinsic state $\Phi_0^{(K)}(X)$ in (6-67), these operators are seen to be

$$R_z'(\varphi)\Phi_0^{(K)}(X) = \Phi_0^{(K)}(X) ,$$

$$R_z'(\varphi) = e^{-i\varphi(J_z-K)} . \tag{6-68}$$

This means that $\Phi_0^{(K)}(X)$ is an axially symmetric state.

According to Sec. 3.13, only the intrinsic group elements $\overline{R}_z(\varphi) = e^{-i\overline{J}_z\varphi}$, and thus the operator $\overline{J}_z$ have definite meaning.

From (6-59) one has

$$\overline{J}_z P^{(J)}_{MK'}\Phi^{(K)}_0(X) = (\overline{J}_z P^{(J)}_{MK'})\Phi^{(K)}_0(X) = K' P^{(J)}_{MK'}\Phi^{(K)}_0(X) \,.$$

On the other hand, from (6-67) one has

$$\overline{J}_z P^{(J)}_{MK'}\Phi^{(K)}_0(X) = P^{(J)}_{MK'}\overline{J}_z\Phi^{(K)}_0(X) = K P^{(J)}_{MK'}\Phi^{(K)}_0(X) \,.$$

Comparing the above two equations, one gets

$$P^{(J)}_{MK'}\Phi^{(K)}_0(X) = \delta_{KK'}\Psi^{(J)K}_M(X) \,, \tag{6-69}$$

$$\Psi^{(J)K}_M(X) = \frac{2J+1}{8\pi^2}\int D^{(J)*}_{MK}(\alpha\beta\gamma)R(\alpha\beta\gamma)\Phi^{(K)}_0(X)\sin\beta d\alpha d\beta d\gamma \,. \tag{6-70}$$

Equation (6-69) shows that for an axially symmetric state, the third component of its angular momentum in the intrinsic frame is determined entirely by the z-component of the angular momentum of the intrinsic state. This is precisely the mathematical expression of the well known fact in nuclear physics that the angular momentum of a nucleus along its symmetry axis comes solely from the intrinsic motion of nucleons, in other words a nucleus has no collective rotation about its symmetry axis (Bohr 1975).

According to the above discussion, the projection state $\Psi^{(J)K}_M$ satisfies the simultaneous eigenequations

$$\begin{pmatrix} J^2 \\ J_z \\ \overline{J}_z \end{pmatrix} \Psi^{(J)K}_M(X) = \begin{pmatrix} J(J+1) \\ M \\ K \end{pmatrix} \Psi^{(J)K}_M(X) \,. \tag{6-71}$$

The intrinsic state

$$\Phi\left((\lambda\mu),\ \overline{\varepsilon} = 2\lambda+\mu,\ \overline{\Lambda} = \mu/2,\ \overline{\nu} = \mu\right) \,,$$

in Elliott's SU_3 model does not have definite J_z values. The range of J_z is $K = \mu, \mu-2, \ldots, 1$ or 0. Using the projection operator one can likewise pick out the component with given J, M, K from the intrinsic state

$$\Psi^{(J)}_{MK} = P^{(J)}_{MK}\phi((\lambda\mu)\overline{\varepsilon}\overline{\Lambda}\overline{\nu}), \quad K = \mu,\ \mu-2, \ldots, 1 \quad \text{or}\, 0 \,. \tag{6-72}$$

The intrinsic quantum number K is used to distinguish the different states with the same J and M, which differ in their intrinsic excitation states. The energy levels with the same quantum number K form a rotation band. Therefore the intrinsic quantum numbers K may be used to characterize the rotation bands (Elliott 1958, Arima 1978).

Another point worth mentioning is that the projected states (6-70) are orthogonal with respect to J and M, but not orthogonal with respect to the intrinsic quantum number K. From (5-245b) and (5-245c) one has

$$\left\langle\Psi^{(J')K'}_{M'}|\Psi^{(J)K}_M\right\rangle = \left\langle\Phi_0|P^{(J')}_{K'M'}P^{(J)}_{MK}\middle|\Phi_0\right\rangle = \delta_{JJ'}\delta_{MM'}\left\langle\Phi_0|P^{(J)}_{K'K}|\Phi_0\right\rangle \,. \tag{6-73}$$

6.8. Irreducible Tensors of SO_3 and $\overline{SO}_3$

6.8.1. *The irreducible tensor of the adjoint rep of SO_3 and $\overline{SO}_3$*

The infinitesimal operators J_x, J_y and J_z of SO_3 form a basis for the adjoint rep of SO_3. As pointed out in Sec. 5.16, when the irreducible matrix elements are given by the solutions of (6-48), in other words, when the $SO_3 \supset SO_2$ irreducible basis is used, only through a suitable linear combination of J_x, J_y and J_z can we obtain the irreducible tensor $T_\rho^{(1)} = J_\rho$, $\rho = 1, 0, -1$, of the adjoint rep of SO_3,

$$J_1 = -\sqrt{\frac{1}{2}}(J_x + iJ_y),\ J_0 = J_z,\ J_{-1} = \sqrt{\frac{1}{2}}(J_x - iJ_y)\ . \tag{6-74}$$

Similarly, the irreducible tensor of the adjoint rep of $\overline{SO}_3$ is $\overline{T}^{(1)} = \overline{J}_\rho, \rho = 1, 0, -1$,

$$\overline{J}_1 = -\frac{1}{\sqrt{2}}(\overline{J}_x + i\overline{J}_y),\ \overline{J}_0 = \overline{J}_z,\ \overline{J}_{-1} = \frac{1}{\sqrt{2}}(\overline{J}_x - i\overline{J}_y)\ . \tag{6-75}$$

From (5-228) one has

$$\overline{J}_\rho = \sum_\sigma D_{\sigma\rho}^{(1)}(\alpha\beta\gamma)J_\sigma, \quad J_\sigma = \sum_\rho D_{\sigma\rho}^{(1)*}(\alpha\beta\gamma)\overline{J}_\rho\ , \tag{6-76}$$

where $D^{(1)}$ is the irreducible matrix for $j = 1$ (for explicit expression see Rose 1957). From (5-264) we obtain the relation between the matrix elements of J_ρ and $\overline{J}_\rho$

$$\langle jm'|J_\rho|jm\rangle = \langle jm|\overline{J}_\rho|jm'\rangle\ . \tag{6-77}$$

According to angular momentum theory (Rose 1957) and (6-77),

$$\langle jm|J_z|jm\rangle = \langle jm|\overline{J}_z|jm\rangle = m\ , \tag{6-78a}$$

$$\langle jm \pm 1|J_z \pm iJ_y|jm\rangle = \langle jm|\overline{J}_x \pm i\overline{J}_y|jm \pm 1\rangle = [(j \mp m)(j \pm m + 1)]^{1/2}\ . \tag{6-78b}$$

We stress that (6-78a) does not imply $\overline{J}_z = J_z$ and (6-78b) does not mean $(\overline{J}_x \pm i\overline{J}_y) = (J_x \pm iJ_y)^\dagger$. This point will become more clear by rewriting (6-78) in the following form:

$$\langle\psi_m^{(j)}|J_z|\psi_m^{(j)}\rangle = \langle\psi^{(j)m}|\overline{J}_z|\psi^{(j)m}\rangle\ , \tag{6-79a}$$

$$\langle\psi_{m\pm1}^{(j)}|J_z \pm iJ_y|\psi_m^{(j)}\rangle = \langle\psi^{(j)m}|\overline{J}_x \pm i\overline{J}_y|\psi^{(j)m\pm1}\rangle\ , \tag{6-79b}$$

where $\psi_m^{(j)}$ and $\psi^{(j)m}$ are the $SO_3 \supset SO_2$ and $\overline{SO}_3 \supset \overline{SO}_2$ irreducible basis, respectively.

6.8.2. *Irreducible tensor of SO_3 and $\overline{SO}_3$ for general cases*

SO_3 is a simply reducible group, therefore the multiplicity label θ in (5-261) is redundant. The reduced matrix element of the angular momentum J is (see Rose 1957)

$$\langle j\|J\|j\rangle = \sqrt{j(j+1)}\ . \tag{6-80}$$

From (5-250), (5-253), (5-261) and (6-80) we obtain the definition for the irreducible tensor of SO_3 and $\overline{SO}_3$,

$$[J_\rho, T_m^{(\nu)}] = \sum_t D_{tm}^{(\nu)}(J_\rho)T_t^{(\nu)} = \sqrt{\nu(\nu+1)}C_{\nu m,1\rho}^{\nu,m+\rho}T_{m+\rho}^{(\nu)}\ , \tag{6-81a}$$

$$[\overline{J}_\rho, \overline{T}_M^{(\nu)}] = -\sum_t D_{tm}^{(\nu)}(J_\rho)\overline{T}_t^{(\nu)} = -\sqrt{\nu(\nu+1)}C_{\nu m,1\rho}^{\nu,m+\rho}\overline{T}_{m+\rho}^{(\nu)}\ , \tag{6-81b}$$

where $\nu = j$. From (6-81) one has

$$[J_\rho, J_\sigma] = -\sqrt{2} C^{1,\rho+\sigma}_{1\rho,1\sigma} J_{\rho+\sigma} , \tag{6-82a}$$

$$[\overline{J}_\rho, \overline{J}_\sigma] = \sqrt{2} C^{1,\rho+\sigma}_{1\rho,1\sigma} \overline{J}_{\rho+\sigma} . \tag{6-82b}$$

Using the expression for the CG coefficients of SO_3 (see Rose 1957), one obtains from (6-81)

$$\begin{gathered}[J_z, T^{(\nu)}_m] = m T^{(\nu)}_m, \quad [\overline{J}_z, \overline{T}^{(\nu)}_m] = -m\overline{T}^{(\nu)}_m , \\ [J_x \pm iJ_y, T^{(\nu)}_m] = [(\nu \mp m)(\nu \pm m + 1)]^{1/2} T^{(\nu)}_{m\pm1} , \\ [\overline{J}_x \pm i\overline{J}_y, \overline{T}^{(\nu)}_m] = -[(\nu \mp m)(\nu \pm m + 1)]^{1/2} \overline{T}^{(\nu)}_{m\pm1} .\end{gathered} \tag{6-83}$$

From (5-262) and (5-263) it follows that

$$[J_\rho, D^{(j)}_{mk}(\alpha\beta\gamma)] = -\sqrt{j(j+1)} C^{jm}_{jm-\rho,1\rho} D^{(j)}_{m-\rho,k}(\alpha\beta\gamma) , \tag{6-84a}$$

$$[\overline{J}_\rho, D^{(j)}_{mk}(\alpha\beta\gamma)] = -\sqrt{j(j+1)} C^{jk+\rho}_{jk,1\rho} D^{(j)}_{m,k+\rho}(\alpha\beta\gamma) , \tag{6-84b}$$

$$[J_\rho, D^{(j)*}_{mk}(\alpha\beta\gamma)] = \sqrt{j(j+1)} C^{jm+\rho}_{jm,1\rho} D^{(j)*}_{m+\rho,k}(\alpha\beta\gamma) , \tag{6-85a}$$

$$[\overline{J}_\rho, D^{(j)*}_{mk}(\alpha\beta\gamma) = \sqrt{j(j+1)} C^{jk}_{jk-\rho,1\rho} D^{(j)*}_{m,k-\rho}(\alpha\beta\gamma) . \tag{6-85b}$$

One sees, by comparing (6-81) with (6-84) and (6-85), that

$$D^{(j)*}_{mk}(\alpha\beta\gamma) = T^{(j)}_m, \quad D^{(j)}_{mk}(\alpha\beta\gamma) = \overline{T}^{(j)}_k . \tag{6-86a,b}$$

From (6-86a) it is seen that there are $2j+1$ independent tensor operators $T^{(j)}$ with the same irrep label j, which are enumerated by assigning the intrinsic quantum number k to each of them. This is consistent with a theorem proved by Louck (1970) which asserts that the number of the linearly independent irreducible tensor operators belonging to the same irrep (ν) equals the dimension of the irrep.

Now let us look at the physical meaning of the intrinsic irreducible tensor defined by (6-81b). Using (6-87), (6-84b) and (6-81), it is easy to show that if the irreducible tensor $T^{(\nu)}_m$ of SO_3 is a scalar under rotation $\overline{R}(\alpha\beta\gamma)$ about the intrinsic axes,

$$[\overline{J}_\rho, T^{(\nu)}_m] = 0 , \tag{6-87}$$

then the tensor

$$T'^{(\nu)}_m = \sum_{m'} D^{(\nu)}_{m'm}(\alpha\beta\gamma) T^{(\nu)}_{m'} \tag{6-88}$$

is the intrinsic irreducible tensor $\overline{T}^{(\nu)}_m$. $T'^{(\nu)}_m$ is nothing else but the operator $T^{(\nu)}_m$ expressed in the intrinsic frame.

Some important operators in physics satisfy the requirement (6-87); these include the angular momentum operator $T^{(1)}_\rho = J_\rho$ and the multipole operator

$$T^{(\lambda)}_\mu = \sum_{i=1}^{A} r_i^\lambda Y_{\lambda\mu}(\theta_i\varphi_i) . \tag{6-89}$$

The corresponding intrinsic irreducible tensors are

$$J'_\rho = \overline{J}_\rho \ ,$$

$$T'^{(\lambda)}_\mu = \overline{T}^{(\lambda)}_\mu = \sum_{i=1}^{A} r'^{\lambda}_i Y_{\lambda\mu}(\theta'_i \varphi'_i) \ , \tag{6-90}$$

where $r_i, \theta_i, \varphi_i (r'_i, \theta'_i, \varphi'_i)$ are the coordinates of the i-th particle in the fixed (intrinsic) frame.

Note that $T'^{(\nu)}_m$ as defined by (6-88) is in general not necessarily an intrinsic irreducible tensor $\overline{T}^{(\nu)}_m$, since the transformation property $[\overline{J}_\rho, T'^{(\nu)}_m]$ of $T'^{(\nu)}_m$ under the intrinsic rotation also depends on the transformation property $[\overline{J}_\rho, T^{(\nu)}_m]$ of the tensor $T^{(\nu)}_m$ under the intrinsic rotation and there is no direct connection between $[\overline{J}_\rho, T'^{(\nu)}_m]$ and $[\overline{J}_\rho, T^{(\nu)}_m]$. For example, suppose that $T^{(\nu)}_m = D^{(\nu)*}_{mk_0}(\alpha\beta\gamma)$, with k_0 fixed. From (6-88) it is seen that

$$T'^{(\nu)}_m = \delta_{m,k_0} \ .$$

Therefore $T'^{(\nu)}_m$ is a scalar with respect to either the external or intrinsic rotation, rather than the intrinsic irreducible tensor $\overline{T}^{(\nu)}_m$.

We abstract the concept of the intrinsic group from the concrete physical problem and then turn back to use it in treating the problem of the collective rotation of nuclei about the intrinsic axes. Some puzzling aspects of the problem now become transparent. All the relations given by Bohr (1975) Appendix IA-b have been easily obtained in this chapter as a special case of the results in the previous chapter. So it may be said that the intrinsic group $\overline{SO_3}$ provides an appropriate mathematical formalism for the description of nuclear rotation about the intrinsic axes.

Chapter 7

The Unitary Groups

Unitary groups have extensive applications in physics. These applications fall mainly into the following two categories.

1. The unitary group is the symmetry group or approximate symmmetry group of the Hamiltonian of a system. In such a case, an eigenfunction of the Hamiltonian belongs to a certain irrep of the unitary group and thus the states of the system can be labeled by the irrep labels of the unitary group and its subgroups. For example, the Hamiltonian of a nucleus has the SU_2 symmetry due to the charge independence of the nuclear force if the Coulomb forces between protons are neglected and the nuclear energy levels are characterized by the isospin T and its third component T_z, which are the irrep labels of SU_2 and SO_2, respectively. The Hamiltonian of a nucleus exhibits SU_4 symmetry if the two-body interaction does not depend on the spin and isospin of the two nucleons, as is for light nuclei. An isotropic harmonic oscillator in three-dimensional space has SU_3 symmetry. Elementary particles obey approximate SU_3 or SU_4, or SU_5 symmetry in flavor space, and SU_3 symmetry in color space.

2. The unitary group is not a symmetry group of the Hamiltonian. In such cases the purpose of introducing the unitary group is to define a complete set of basis functions classified according to the irreps of the unitary group and its subgroups, so that the Hamiltonian can be conveniently diagonalized in this basis. If the symmetry group G of the Hamiltonian is a subgroup of SU_n, we will naturally choose the $SU_n \supset G$ irreducible basis to facilitate the diagonalization. For example we use the group chains $SU_{2l+1} \supset SO_3$ or $SU_{2j+1} \supset SO_3$ and $SU_4 \supset SU_2 \times SU_2$ in the nuclear shell model.

In physics we generally use the special unitary group SU_n. However in the presentation of theorems, it is more convenient to use the group U_n. The theory of the semi-simple group applies to the group U_n (Moshinsky 1963), although U_n is not a semi-simple group.

The rep theory for the unitary group to be introduced below can be easily extended to the so-called graded (or super) unitary group $SU(m/n)$ (Chen [4] 1983, [5] 1983, [6] 1984, [7] 1984, [8] 1984, [13] 1984 and [16] 1984).

7.1. Unitary Groups in Coordinate Space and State Space

Suppose that there are f particles with the coordinate indices $a_j = 1, 2, \ldots, f$ and n single particle states $m_1, m_2, \ldots, m_n$ (or $\alpha, \beta, \gamma, \ldots \delta$). We use $i_j (j = 1, 2, \ldots, f)$ to denote any one of the states $m_1, \ldots, m_n$. The n single particle states $\varphi_{m_1}, \varphi_{m_2}, \ldots, \varphi_{m_n}$ span the space V_n, the fundamental rep space of the unitary group in state space $\mathcal{U}_n$ or $S\mathcal{U}_n$. Under the operation of the group element $\mathcal{R}(b)$ of $\mathcal{U}_n$ they transform as follows

$$\mathcal{R}(b)\varphi_{m_i} = \sum_{j=1}^{n} \mathcal{D}_{ji}(b)\varphi_{m_j} = \sum_{j=1}^{n} R_{ij}\varphi_{m_j} , \tag{7-1}$$

where b are the group parameters, and $\mathcal{D}(b)$ is an $n \times n$ unitary matrix. The matrix $\mathcal{D}(b)$ is related to R in (5-80a) by $\mathcal{D}_{ji} = R_{ij}$. φ_m is a tensor of rank one. According to Sec. 5.4, $\mathcal{R}(b)$ can be written as

$$\mathcal{R}(b) = \exp[-i \sum_{\alpha\beta} b_{\alpha\beta} e_{\alpha\beta}] , \tag{7-2a}$$

$$e_{\alpha\beta} = c_\alpha^\dagger c_\beta . \tag{7-2b}$$

$e_{\alpha\beta}$ are the infinitesimal operators acting on the single particle states,

$$e_{\alpha\beta}\varphi_\gamma = \delta_{\beta\gamma}\varphi_\alpha . \tag{7-3a}$$

They satisfy the relation

$$e_{\alpha\beta}e_{\gamma\delta} = \delta_{\beta\gamma}e_{\alpha\delta} . \tag{7-3b}$$

The n^f f-particle product states are designated by

$$|i_1 i_2 \ldots i_f\rangle = \varphi_{i_1}^{a_1}\varphi_{i_2}^{a_2} \cdots \varphi_{i_f}^{a_f}, \quad i_1, i_2, \ldots, i_f = m_1, m_2, \ldots m_n . \tag{7-4}$$

The action of the group element $\mathcal{R}(b)$ on the f-particle product states is defined as inducing a unitary transformation on all the single particle states:

$$\begin{aligned}\mathcal{R}(b)|i_1 i_2 \ldots i_f\rangle &= (\mathcal{R}^{(a_1)}(b)\varphi_{i_1}^{a_1})(\mathcal{R}^{(a_2)}(b)\varphi_{i_2}^{a_2}) \ldots (\mathcal{R}^{(a_f)}(b)\varphi_{i_f}^{a_f}) \\ &= \sum_{i'_1 i'_2 \ldots i'_f} \mathcal{D}_{i'_1 i_1}(b) \ldots \mathcal{D}_{i'_f i_f}(b)|i'_1 i'_2 \ldots i'_f\rangle .\end{aligned} \tag{7-5}$$

It follows that the n^f product states (7-4) form an f-th rank tensor in n-dimensional space. The space spanned by (7-4) is called the *tensor space*, hereafter denoted by V_n^f, and it generates a reducible rep for both the groups $\mathcal{U}_n$ and S_f. The group element $\mathcal{R}^{(a_i)}(b)$ is

$$\mathcal{R}^{(a_i)}(b) = \exp\left[-i \sum_{\alpha\beta} b_{\alpha\beta} e_{\alpha\beta}^{(a_i)}\right] , \tag{7-6a}$$

$$e_{\alpha\beta}^{(a_i)} = c_\alpha^\dagger(a_i) c_\beta(a_i) , \tag{7-6b}$$

where $c_\alpha^\dagger(a_i)$ and $c_\beta(a_i)$ are the creation and annihilation operators of the particle a_i,

$$c_\alpha^\dagger(a_i)|0\rangle = \varphi_\alpha^{a_i} \tag{7-7a}$$

$$e_{\alpha\beta}^{(a_i)}\varphi_\gamma^{(a_j)} = \delta_{ij}\delta_{\beta\gamma}\varphi_\alpha^{(a_i)} . \tag{7-7b}$$

Since the operators $e_{\alpha\beta}^{(a_i)}$ with different a_i are commuting, we have the commutator

$$[e_{\alpha\beta}^{(a_i)}, e_{\gamma\delta}^{(a_j)}] = \delta_{ij}(\delta_{\beta\gamma}e_{\alpha\delta}^{(a_i)} - \delta_{\alpha\delta}e_{\gamma\beta}^{(a_i)}) . \tag{7-7c}$$

From (7-5) and (7-6) we obtain the expression for the elements of the unitary group acting on many-particle states

$$\mathcal{R}(b) = \prod_{i=1}^{f} \mathcal{R}^{(a_i)}(b) = \exp\Big(-i\sum_{\alpha\beta} b_{\alpha\beta}\mathcal{E}_{\alpha\beta}\Big) , \tag{7-8a}$$

$$\mathcal{E}_{\alpha\beta} = \sum_{i=1}^{f} e^{(a_i)}_{\alpha\beta} . \tag{7-8b}$$

$\mathcal{E}_{\alpha\beta}$ are the generators of the unitary group acting on many-particle states. Using (7-7c) it is easy to show that

$$[\mathcal{E}_{\alpha\beta}, \mathcal{E}_{\gamma\delta}] = \delta_{\beta\gamma}\mathcal{E}_{\alpha\delta} - \delta_{\alpha\delta}\mathcal{E}_{\gamma\beta} . \tag{7-9}$$

In other words, the operators $\mathcal{E}_{\alpha\beta}$ obey the same commutator as the operators $e_{\alpha\beta}$, however, (7-3b) is no longer true. Now $\mathcal{E}_{\alpha\beta}\mathcal{E}_{\gamma\delta} \neq \delta_{\beta\gamma}\mathcal{E}_{\alpha\delta}$, due to the occurrence of cross terms like $e^{(a_i)}_{\alpha\beta}e^{(a_j)}_{\gamma\delta}$.

Letting the group parameter $b_{\alpha\beta} = \varepsilon$ be infinitesimal, and setting all other parameters $b_{\alpha'\beta'} = 0$ for $\alpha' \neq \alpha$ and $\beta' \neq \beta$, from (7-8b) and (7-7b) we obtain

$$\begin{aligned}\mathcal{E}_{\alpha\beta}(\varphi^{a_1}_{i_1}\varphi^{a_2}_{i_2}\dots\varphi^{a_f}_{i_f}) = {} & \delta_{i_1\beta}\varphi^{a_1}_{\alpha}\varphi^{a_2}_{i_2}\dots\varphi^{a_f}_{i_f} + \delta_{i_2\beta}\varphi^{a_1}_{i_1}\varphi^{a_2}_{\alpha}\dots\varphi^{a_f}_{i_f} + \dots \\ & + \delta_{i_f\beta}\varphi^{a_1}_{i_1}\varphi^{a_2}_{i_2}\dots\varphi^{a_f}_{\alpha} .\end{aligned} \tag{7-10}$$

We may note that (7-10) still holds when some a_i's are equal. $\mathcal{E}_{\alpha\alpha}$ is the number operator for the state α,

$$\hat{n}_\alpha = \mathcal{E}_{\alpha\alpha} = \sum_{i=1}^{f} c^\dagger_\alpha(a_i)c_\alpha(a_i) . \tag{7-11}$$

Analogously, we can define the unitary group U_n or SU_n in coordinate space. The single particle states $\varphi^{a_i}_m (i = 1, 2, \dots, n)$ carry the fundamental rep of U_n. Under the action of the group element $R(b)$ of U_n, $\varphi^{a_i}_m$ transform as

$$R(b)\varphi^{a_i}_m = \sum_{j=1}^{n} D_{ji}(b)\varphi^{a_j}_m . \tag{7-12}$$

The infinitesimal operators of U_n are

$$E_{a_1a_2} = \sum_{\alpha=1}^{n} e^{(\alpha)}_{a_1a_2} , \tag{7-13}$$

$$e^{(\alpha)}_{a_1a_2} = c^\dagger_\alpha(a_1)c_\alpha(a_2) . \tag{7-14}$$

The operator $e^{(\alpha)}_{a_1a_2}$ changes the coordinate index a_2 into a_1 of a particle in the state α,

$$e^{(\alpha)}_{a_1a_2}\varphi^{a_3}_{\beta} = \delta_{\alpha\beta}\delta_{a_2a_3}\varphi^{a_1}_{\alpha} . \tag{7-15}$$

In parallel to (7-7c), (7-9) and (7-10), we have

$$[e^{(\alpha)}_{ab}, e^{(\beta)}_{cd}] = \delta_{\alpha\beta}(\delta_{bc}e^{(\alpha)}_{ad} - \delta_{ad}e^{(\alpha)}_{cb}) , \tag{7-16}$$

$$[E_{ab}, E_{cd}] = \delta_{bc}E_{ad} - \delta_{ad}E_{cb} . \tag{7-17}$$

$$E_{ab}(\varphi_{i_1}^{a_1}\varphi_{i_2}^{a_2}\dots\varphi_{i_n}^{a_n})$$
$$= \delta_{a_1 b}\varphi_{i_1}^{a}\varphi_{i_2}^{a_2}\dots\varphi_{i_n}^{a_n} + \delta_{a_2 b}\varphi_{i_1}^{a_1}\varphi_{i_2}^{a}\dots\varphi_{i_n}^{a_n} + \dots + \delta_{a_n b}\varphi_{i_1}^{a_1}\varphi_{i_1}^{a_2}\dots\varphi_{i_n}^{a} . \tag{7-18}$$

As in Chapter 4, we may define the coordinate permutation group S_f and the state permutation group $\mathcal{S}_f$, which permute the subscripts of a and those of i, respectively. Obviously, the groups defined in different spaces commute, i.e.,

$$[S_f, \mathcal{S}_f] = 0, \quad [U_n, \mathcal{U}_n] = 0 ,$$
$$[S_f, \mathcal{U}_n] = 0, \quad [\mathcal{S}_f, U_n] = 0 , \tag{7-19}$$

and the groups defined in the same space do not commute,

$$[S_f, U_n] \neq 0, \quad [\mathcal{S}_f, \mathcal{U}_n] \neq 0 . \tag{7-20}$$

We use $\mathcal{S}_f$ and $\mathcal{U}_n$ as examples to illustrate (7-20). As we know, the state permutation operators $\wp$ of $\mathcal{S}_f$ have definite meaning when acting on the product state (7-4) with all the single particle state $i_1, i_2, \dots, i_f$ different. However, after the action of the group elements of $\mathcal{U}_n$ on the product states, identical single particle states will occur in the product states and the permutation operator $\wp$ becomes meaningless (only the class operators of $\mathcal{S}_f$ remain meaningful). Therefore $\mathcal{S}_f$ does not commute with $\mathcal{U}_n$.

7.2. Relations between the CSCO-I and Generators of Unitary Groups and Permutation Groups

7.2.1. The Gel'fand invariants

The following n operators were introduced by Gel'fand (1950)

$$\Im_k^{(n)} = \sum_{i_1 i_2, \dots i_k = 1}^{n} \mathcal{E}_{i_1 i_2}\mathcal{E}_{i_2 i_3}\dots\mathcal{E}_{i_k i_1} , \quad k = n, n-1, \dots, 1 . \tag{7-21}$$

It can be proved that $\Im_k^{(n)}$ satisfy the condition (5-219), i.e.,

$$[\mathcal{E}_{ij}, \Im_k^{(n)}] = 0 , \quad i, j = 1, 2, \dots, n . \tag{7-22}$$

Consequently, $\{\Im_k^{(n)}\}$ is a set of invariants of $\mathcal{U}_n$. $\Im_1^{(n)}$ is the number operator

$$\Im_1^{(n)} = \sum_{i=1}^{n} \mathcal{E}_{ii} = f . \tag{7-23}$$

Gel'fand showed that the eigenvalues of the n invariants $\{\Im_k^{(n)}\}$ uniquely label an irrep of $\mathcal{U}_n$; therefore, $\{\Im_k^{(n)}\}$ is the CSCO-I of $\mathcal{U}_n$.

From (5-124), the infinitesimal operators of $S\mathcal{U}_n$ are known to be

$$\mathcal{E}'_{ij} = \mathcal{E}_{ij} - \delta_{ij}\frac{1}{n}\sum_{k=1}^{n} \mathcal{E}_{kk} . \tag{7-24}$$

The CSCO-I of $S\mathcal{U}_n$ consists of the $n-1$ invariants

$$\Im_k'^{(n)} = \sum_{i_1 i_2 \dots i_k = 1}^{n} \mathcal{E}'_{i_1 i_2}\mathcal{E}'_{i_2 i_3}\dots\mathcal{E}'_{i_k i_1} , \quad k = n, n-1, \dots, 2 . \tag{7-25a}$$

$\Im_k^{\prime(n)}$ can be expressed in terms of $\Im_k^{(n)}$. For example, from (7-24) we have

$$\Im_1^{\prime(n)} = 0 \,, \quad \Im_2^{\prime(n)} = \Im_2^{(n)} - \frac{1}{n}\left(\sum_{i=1}^{n} \mathcal{E}_{ii}\right)^2 = \Im_2^{(n)} - \frac{f^2}{n} \,. \tag{7-25b}$$

The above relations hold as well for the unitary group U_n or SU_n in coordinate space.

In view of the functional relationship between the CSCO-I of $S\mathcal{U}_n$ and $\mathcal{U}_n$, any eigenfunction $\psi^{(\nu)}$ of the CSCO-I of $\mathcal{U}_n$ is necessarily an eigenfunction of the CSCO-I of $S\mathcal{U}_n$, and thus if $\psi^{(\nu)}$ belongs to the irrep (ν) of $\mathcal{U}_n$, it also belongs to an irrep of $S\mathcal{U}_n$, and vice versa. In the following we will prove that irreps of $\mathcal{U}_n$ remain irreducible when we go to the subgroup $S\mathcal{U}_n$. The proof is as follows.

From (7-5) it is known that the irreducible matrix elements $D^{(\nu)}(\mathcal{R})$ of an irrep (ν) of $\mathcal{U}_n$ are homogeneous polynomials of degree f in the transformation matrix elements R_{ij},

$$D^{(\nu)}(\mathcal{R}) = F^{(\nu)}(R_{ij}) \,. \tag{7-26a}$$

The unitary transformation $|R_{ij}|$ can be made unimodular by the scale transformation

$$R_{ij} = (\det R)^{1/n} R'_{ij} \,. \tag{7-26b}$$

From (7-26a) and (7-26b) we know that the representation matrices of $\mathcal{U}_n$ and those of $S\mathcal{U}_n$ are related by the following

$$D^{(\nu)}(\mathcal{U}_n) = (\det R)^{f/n} D^{(\nu)}(S\mathcal{U}_n) \,. \tag{7-26c}$$

Therefore an irrep (ν) of $\mathcal{U}_n$ remains irreducible for $S\mathcal{U}_n$.

7.2.2. The relation between the CSCO-I of permutation groups and unitary groups

Partensky (1972) defined the operators

$$P_k^{a_1 a_2 \ldots a_k} = \frac{1}{k!} \sum_{i_1 i_2 \ldots i_k = 1}^{n} \left(e_{i_1 i_2}^{(a_1)} e_{i_2 i_3}^{(a_2)} \ldots e_{i_k i_1}^{(a_k)}\right)_s \,, \tag{7-27}$$

where the subscript s means symmetrization in the indices $a_1, a_2, \ldots a_k$,

$$\left(e_{i_1 i_2}^{(a_1)} e_{i_2 i_3}^{(a_2)} \ldots e_{i_k i_1}^{(a_k)}\right)_s = \sum_{p \in S_k} p\left(e_{i_1 i_2}^{(a_1)} e_{i_2 i_3}^{(a_2)} \ldots e_{i_k i_1}^{(a_k)}\right) \,.$$

We now prove that acting on the basis $\varphi_{m_1}^{a_1} \varphi_{m_2}^{a_2} \ldots \varphi_{m_k}^{a_k}$, $P_k^{a_1 a_2 \ldots a_k}$ is equivalent to the average k-cycle class operator $C^{(k)}(k)$ of S_k,

$$P_k^{a_1 a_2 \ldots a_k} = C^{(k)}(k) = \frac{1}{(k-1)!} C_{(k)}(k) \,. \tag{7-28}$$

From (7-7b) we have

$$\sum_{i_1 i_2 = 1}^{n} e_{i_1 i_2}^{(a_1)} e_{i_2 i_1}^{(a_2)} \varphi_{m_1}^{(a_1)} \varphi_{m_2}^{(a_2)} = \sum_{i_1 i_2 = 1}^{n} \delta_{i_1 m_2} \delta_{i_2 m_1} \varphi_{i_1}^{(a_1)} \varphi_{i_2}^{(a_2)} = \varphi_{m_2}^{(a_1)} \varphi_{m_1}^{(a_2)} \,.$$

Therefore

$$\sum_{i_1 i_2 = 1}^{n} e_{i_1 i_2}^{(a_1)} e_{i_2 i_1}^{(a_2)} = (a_1 a_2) \,. \tag{7-29a}$$

It is easy to show that in general we have

$$\sum_{i_1 i_2 \ldots i_k=1}^{n} e_{i_1 i_2}^{(a_1)} e_{i_2 i_3}^{(a_2)} \ldots e_{i_k i_1}^{(a_k)} = (a_1 a_2 \ldots a_k) \ . \tag{7-29b}$$

A symmetrization of (7-29b) gives rise to (7-28). For example,

$$P_3^{a_1 a_2 a_3} = \frac{1}{2} C_{(3)}(3) = \frac{1}{2}[(123) + (132)] \ .$$

Partensky defined another operator

$$P_k^f = \binom{f}{k}^{-1} \sum_{a_1 > \ldots > a_k = 1}^{f} P_k^{a_1 a_2 \ldots a_k} = \left[\binom{f}{k}(k-1)!\right]^{-1} C_{(k)}(f) = \frac{1}{g_{(k)}} C_{(k)}(f) \ ,$$

$$g_{(k)} = \binom{f}{k}(k-1)! \ , \tag{7-30}$$

where $C_{(k)}(f)$ is the k-cycle class operator of S_f, and $g_{(k)}$ is the number of elements in the class. For example

$$\begin{aligned} P_3^4 &= \frac{1}{4}[P_3^{a_1 a_2 a_3} + P_3^{a_1 a_2 a_4} + P_3^{a_1 a_3 a_4} + P_3^{a_2 a_3 a_4}] \\ &= \frac{1}{8}[(123) + (132) + (124) + (142) + (134) + (143) + (234) + (243)] \ . \end{aligned}$$

Now we turn to derive a relation between P_k^f and the invariants of $\mathcal{U}_n$. From (7-30), (7-27) and (7-21) we have

$$\begin{aligned} P_2^f &= \frac{2}{f(f-1)} \sum_{i_1 i_2 = 1}^{n} \sum_{a_1 > a_2 = 1}^{f} e_{i_1 i_2}^{(a_1)} e_{i_2 i_1}^{(a_2)} = \frac{1}{f(f-1)} \sum_{i_1 i_2 = 1}^{n} \left[\sum_{a_1 a_2 = 1}^{f} e_{i_1 i_2}^{(a_1)} e_{i_2 i_1}^{(a_2)} - \sum_{a_1 = 1}^{f} e_{i_1 i_2}^{(a_1)} e_{i_2 i_1}^{(a_1)}\right] \\ &= \frac{1}{f(f-1)}\left[\Im_2^{(n)} - n \sum_{i_1 = 1}^{n} \sum_{a_1 = 1}^{f} e_{i_1 i_1}^{(a_1)}\right] = \frac{1}{f(f-1)}[\Im_2^{(n)} - nf] \ . \end{aligned} \tag{7-31}$$

Let us introduce the function

$$F^{(n)} = \prod_{j > i = 1}^{n} (1 - \delta_{a_i a_j}) \ . \tag{7-32a}$$

For example

$$F^{(3)} = 1 - (\delta_{a_1 a_2} + \delta_{a_1 a_3} + \delta_{a_2 a_3}) + 2\delta_{a_1 a_2}\delta_{a_2 a_3} \ . \tag{7-32b}$$

Combining (7-30), (7-27), (7-32), (7-3b) and (7-21),

$$\begin{aligned} P_3^f &= \frac{1}{f(f-1)(f-2)} \sum_{i_1 i_2 i_3 = 1}^{n} \sum_{a_1 > a_2 > a_3 = 1}^{f} (e_{i_1 i_2}^{a_1} e_{i_2 i_3}^{a_2} e_{i_3 i_1}^{a_3})_s \\ &= \frac{1}{f(f-1)(f-2)} \sum_{i_1 i_2 i_3 = 1}^{n} \sum_{a_1 a_2 a_3 = 1}^{f} e_{i_1 i_2}^{a_1} e_{i_2 i_3}^{a_2} e_{i_3 i_1}^{a_3} F^{(3)} \\ &= [\Im_3^{(n)} - 2n\Im_2^{(n)} + (n^2+1)f - f^2]/[f(f-1)(f-2)] \ , \end{aligned} \tag{7-33}$$

where we used $\sum_{i_1=1}^{n}\sum_{a_1=1}^{f} e_{i_1 i_1}^{(a_1)} = \sum_{i=1}^{n} n_i = f, n_i$ being the number of particles in the state i [see (7-11)].

According to (7-30), (7-31) and (7-33) we obtain

$$\Im_2^{(n)} = 2C_{(2)}(f) + nf \ ,$$
$$\Im_3^{(n)} = 3C_{(3)}(f) + 4nC_{(2)}(f) + (n^2-1)f + f^2 \ . \tag{7-34}$$

In the same way we can prove that in the tensor space V_n^f, $\Im_k^{(n)}$ is a function of the class operators $C_{(k)}(f), \ldots, C_{(2)}(f)$ of S_f, i.e.,

$$\Im_k^{(n)} = F_k^{(n)}(C_{(k)}(f)\ , C_{(k-1)}(f), \ldots, C_{(2)}(f)) \ . \tag{7-35}$$

It was noted in Sec. 3.2 that all the class operators of S_f are functions of the CSCO-I of S_f. Furthermore, the CSCO-I of S_f and $\mathcal{S}_f$ are equal, $C(f)$[1] $= \mathcal{C}(f)$, therefore the invariants of $\mathcal{U}_n$ are functions of $C(f)$ or $\mathcal{C}(f)$,

$$\Im_k^{(n)} = F_k^{(n)}(C(f)) = F_k^{(n)}(\mathcal{C}(f)) \ . \tag{7-36}$$

Analogously, it can be shown that the invariants $I_k^{(n)}$ of U_n are functions of the CSCO-I of S_f, i.e., $I_k^{(n)} = F_k^{(n)}(C(f))$. Combining it with (7-36) leads to an important relation

$$I_k^{(n)} = \Im_k^{(n)} = F_k^{(n)}(C(f)) = F_k^{(n)}(\mathcal{C}(f)) \ . \tag{7-37}$$

In other words, in tensor space, among the CSCO's $\{I_k^{(n)}\}, \{\Im_k^{(n)}\}, C(f)$ and $\mathcal{C}(f)$ of the four groups $U_n, \mathcal{U}_n, S_f$ and $\mathcal{S}_f$, only one is functionally independent. Therefore in the space V_n^f, an eigenfunction of any one of the four CSCO's must be eigenfunction of the other three CSCO's. Thus we can use the same irrep label (ν) for the four groups $U_n, \mathcal{U}_n, S_f$ and $\mathcal{S}_f$ and we have the following theorem:

Theorem 7.1: In the tensor space V_n^f, if $\psi^{(\nu)}$ belongs to the irrep (ν) of one of the four groups $U_n, \mathcal{U}_n, S_f$ and $\mathcal{S}_f$, then $\psi^{(\nu)}$ also belongs to the irrep (ν) of any one of the other three groups.

There are a variety of ways to label the irreps of the four groups. The most extensively used is the partition

$$[\nu] = [\nu_1\nu_2\ldots\nu_n] = [m_{in}] \equiv [m_{1n}, m_{2n}, \ldots, m_{nn}] \ .$$

We may use the eigenvalues of $C(f)$ or $I_k^{(n)}$ as the irrep label as well. Partensky (1972) proved that the eigenvalues of $I_k^{(n)}$ are related to the partition through the following

$$I_k^{(n)} = \sum_{i=1}^{n} m_{in} f_k^{(n)}(i) \ , \tag{7-38a}$$

where $f_k^{(n)}(i)$ is obtained from the recursive formula

$$f_{k+1}^{(n)}(i) = q_{in} f_k^{(n)}(i) - \sum_{j=1}^{i-1} f_k^{(n)}(j) \ ,$$
$$q_{in} = m_{in} + n - i, \quad f_1^{(n)}(i) \equiv 1 \ . \tag{7-38b}$$

[1]In Secs. 7.2–7.5, $C(f)$ represents the CSCO-I of S_f, while $C_{(k)}(f)$ represents the k-cycle class operators of S_f.

For example

$$I_1^{(n)} = f, \quad I_2^{(n)} = \sum_{i=1}^{n} m_{in}(m_{in} - 2i + n + 1) \ . \tag{7-38c}$$

From (7-38c), (7-34) and (7-37) we obtain the eigenvalues of the two- and three-cycle class operators as functions of the partition

$$\lambda_{(2)}(f) = \frac{f}{2} + \frac{1}{2}\sum_i \nu_i(\nu_i - 2i) \ ,$$

$$\lambda_3(f) = \frac{1}{3}\left\{2f - \frac{3}{2}f^2 + \sum_i \nu_i \left[\nu_i^2 - \left(3i - \frac{3}{2}\right)\nu_i + 3i(i-1)\right]\right\} \ . \tag{7-38d}$$

7.2.3 Relations between the generators of unitary groups and permutation groups

Suppose the coordinates of the particles are all different. From (7-29a) we obtain the coordinate permutation operator

$$p_{ab} = \sum_{\alpha,\beta=1}^{n} e_{\alpha\beta}^{(a)} e_{\beta\alpha}^{(b)} \ . \tag{7-39a}$$

Setting $n = 2$ in (7-39a) and using the Pauli matrices

$$\sigma_x = e_{12} + e_{21}, \quad \sigma_y = i(e_{21} - e_{12}) \ ,$$
$$\sigma_z = (e_{11} - e_{22}) \ , \quad 1 = e_{11} + e_{22} \ , \tag{7-39b}$$

we obtain the well known Dirac equality

$$p_{ab} = \frac{1}{2}(1 + \boldsymbol{\sigma}_a \cdot \boldsymbol{\sigma}_b) \ . \tag{7-39c}$$

Expressed in terms of the generators of $S\mathcal{U}_n$, (7-39a) reads

$$p_{ab} = \sum_{\alpha\beta=1}^{n} e_{\alpha\beta}^{\prime(a)} e_{\beta\alpha}^{\prime(b)} + \frac{1}{n} \ ,$$

$$e'_{\alpha\beta} = e_{\alpha\beta} - \delta_{\alpha\beta}\frac{1}{n}\sum_{\gamma} e_{\gamma\gamma} \ . \tag{7-39d}$$

Equation (7-39d) is the extension of Dirac's equality (7-39c).

The relation between the generators p_{ab} of S_f and E_{ab} of U_n is

$$E_{ab}E_{ba} = 1 + p_{ab} \ . \tag{7-40}$$

This may be verified directly by applying it to the state $\varphi_\alpha^a \varphi_\beta^b$ and noting (7-18).

When the single particle states are all different, the state permutation operator $\wp_{\alpha\beta}$ is expressed as

$$\wp_{\alpha\beta} = \sum_{a,b=1}^{f} e_{ab}^{(\alpha)} e_{ba}^{(\beta)} \ , \tag{7-41}$$

and the counterpart of (7-40) is

$$\mathcal{E}_{\alpha\beta}\mathcal{E}_{\beta\alpha} = 1 + \wp_{\alpha\beta} \ . \tag{7-42}$$

From (7-40), (7-42), one also sees that S_f does not commute with U_n, and $\mathcal{S}_f$ does not commute with $\mathcal{U}_n$.

7.3. The CSCO-II and CSCO-III of U_n and SU_n

The following discussion is equally valid for the unitary groups in coordinate space and state space. We assume here that U_n stands for the unitary group in state space for convenience.

The group U_n has $r = n^2$ parameters, therefore we need to find n^2 commuting operators to form the CSCO-III of U_n. We are going to show that the CSCO of the group chain

$$U_n \supset U_{n-1} \supset \ldots \supset U_2 \supset U_1 \tag{7-43}$$

and the CSCO of the corresponding intrinsic group chain

$$\overline{U}_n \supset \overline{U}_{n-1} \supset \ldots \supset \overline{U}_2 \supset \overline{U}_1 \tag{7-44}$$

constitute the CSCO-III of U_n.

According to (5-236), U_n and its intrinsic group $\overline{U}_n$ have the same CSCO, which provides n commuting operators

$$I_k^{(n)} = \overline{I}_k^{(n)} \ , k = n, n-1, \ldots, 2, 1 \ . \tag{7-45a}$$

$$I_k^{(n)} = \sum_{i_1 \ldots i_k = 1}^{n} E_{i_1 i_2} E_{i_2 i_3} \ldots E_{i_k i_1} \ . \tag{7-45b}$$

We still use C to represent the CSCO-I,

$$C = \{I_k^{(n)}\} \equiv (I_n^{(n)}, I_{n-1}^{(n)}, \ldots, I_2^{(n)}, \ I_1^{(n)}) \ , \tag{7-46}$$

and use $C(s)$ and $\overline{C}(s)$ to represent the CSCO of the subgroup chains (7-43) and (7-44),

$$\begin{aligned} C(s) &= (\{I_k^{(n-1)}\}, \ldots, \{I_k^{(2)}\} \ , I_1^{(1)}) \ , \\ \overline{C}(s) &= (\{\overline{I}_k^{(n-1)}\} \ , \ldots, \{\overline{I}_k^{(2)}\} \ \overline{I}_1^{(1)}) \ , \end{aligned} \tag{7-47a}$$

where

$$\left.\begin{aligned} I_k^{(m)} &= \sum_{i_1 \ldots i_k = 1}^{m} E_{i_1 i_2} E_{i_2 i_3} \ldots E_{i_k i_1} \ , \\ \overline{I}_k^{(m)} &= \sum_{i_1 \ldots i_k = 1}^{m} \overline{E}_{i_1 i_2} \overline{E}_{i_2 i_3} \ldots \overline{E}_{i_k i_1} \ . \end{aligned}\right. \quad m = n-1, \ldots, 2, 1 \ . \tag{7-47b}$$

$C(s)$ contains $\sum_{m=1}^{n-1} m = \frac{1}{2} n(n-1)$ operators. Therefore the set of operators

$$K = (C, C(s), \overline{C}(s)) \tag{7-48}$$

contains n^2 operators. It is easy to show that the n^2 operators commute with one another. Gel'fand proved that they are independent. Therefore K is the CSCO-III of U_n, and the group chain (7-43) is a canonical group chain.

Similarly, it can be shown that a canonical group chain of SU_n is

$$SU_n \supset SU_{n-1} \times U_1 \supset SU_{n-2} \times U_1 \supset \ldots \supset SU_2 \times U_1 \supset U_1 \ . \tag{7-49}$$

SU_n and $\overline{SU}_n$ have the same CSCO which contains $n-1$ operators:

$$C = \{I_k'^{(n)}\} = (I_n'^{(n)}, I_{n-1}'^{(n)}, \ldots, I_2'^{(n)}) \ . \tag{7-50}$$

Each link $SU_l \times U_1$ in the group chain (7-49) provides l operators, i.e., the $l-1$ operators of the CSCO of SU_l and the CSCO $E_{l+1,l+1}$ of the group U_1 which acts on the single particle state φ_{l+1}. However $E_{l+1,l+1}$ does not meet the traceless condition (5-120) and thus is not an infinitesimal operator of SU_n. To remedy this shortcoming, we take a suitable linear combination of the operators $E_{11}, \ldots, E_{ll}$ and $E_{l+1,l+1}$ so that it is traceless. We usually take the form of (5-126), i.e.,

$$H_l = \frac{1}{l+1}\left(\sum_{i=1}^{l} E_{ii} - lE_{l+1,l+1}\right) = \frac{1}{l+1}\left(\sum_{i=1}^{l} \hat{n}_i - l\hat{n}_{l+1}\right), \quad l = n-1, \ldots, 1 , \qquad (7\text{-}51)$$

where (7-11) has been used.

H_l obviously commute with one another and with the invariants $I_k^{\prime(m)}$ of $SU_m (m = n, n-1, \ldots, 2)$, since from (7-25) it is known that acting on any f-particle product state the operator $I_k^{\prime(m)}$ will not change the particle number in any single particle state.

The CSCO of the subgroup chain (7-49) and its corresponding intrinsic subgroup chain are

$$C(s) = (\{I_k^{\prime(n-1)}\}, \ldots, I_2^{\prime(2)}; H_{n-1}, \ldots, H_2, H_1) . \qquad (7\text{-}52a)$$

$$\overline{C}(s) = (\{\overline{I}_k^{\prime(n-1)}\}, \ldots, \overline{I}_2^{\prime(2)}; \overline{H}_{n-1}, \ldots, \overline{H}_2, \overline{H}_1) , \qquad (7\text{-}52b)$$

each consisting of $\frac{1}{2}n(n-1)$ operators. Equation (7-50), together with (7-52) provide $r = n^2 - 1$ operators which form the CSCO-III, $K = (C, C(s), \overline{C}(s))$, of the group SU_n, while $(C, C(s))$ forms the CSCO-II of SU_n.

Example 1: The SU_2 group. The spin-up and spin-down states $\varphi_{1/2}$ and $\varphi_{-1/2}$ carry the fundamental rep of SU_2. The generators of SU_2 are the Pauli matrices σ_x, σ_y and σ_z [Eq. (7-39b)]. The CSCO-I of SU_2 consists of a single operator $I'^{(2)}$. From (7-24), (7-25) and (7-39b) we find

$$I'^{(2)} = 2\mathbf{J}^2 = 2(J_x^2 + J_y^2 + J_z^2) , \qquad (7\text{-}53)$$

where $\mathbf{J} = \frac{1}{2}\boldsymbol{\sigma}$.

The irreps of SU_2 can be labeled either by the partitions $[\nu] = [m_{12}, m_{22}]$ or the eigenvalues of $J^2 = I'^{(2)}/2$. From (7-25b) and (7-38c) we have

$$\mathbf{J}^2 = \frac{1}{2}(m_{12} - m_{22})\left[\frac{1}{2}(m_{12} - m_{22}) + 1\right] .$$

Since $\mathbf{J}^2 = J(J+1)$, we get the relation between J and the partition $[\nu] = [m_{12}, m_{22}]$, i.e.,

$$J = (m_{12} - m_{22})/2 . \qquad (7\text{-}54a)$$

We generally use J instead of $J(J+1)$ to label the SU_2 irreps. The partition $[\nu] = [m_{12}, m_{22}]$ can be expressed in terms of J and the particle number $f = m_{12} + m_{22}$ as

$$[\nu] = \left[\frac{f}{2} + J, \frac{f}{2} - J\right] . \qquad (7\text{-}54b)$$

The meaning of (7-54) will be seen more clearly in Eq. (7-72b).

A canonical group chain of SU_2 is $SU_2 \supset U_1$. The CSCO-II and CSCO-III of SU_2 are (J^2, J_z) and $(J^2, J_z, \overline{J}_z)$, respectively (see Chapter 5).

Example 2: SU_3 group. The three flavor quarks u, d, and s carry the fundamental rep of SU_3, while $(u,d), (d,s)$ and (s,u) carry the fundamental reps of the three kinds of SU_2

subgroups, corresponding to the $I-, U-$ and $V-$spin reps, respectively. A canonical group chain of SU_3 is $SU_3 \supset SU_2 \times U_1 \supset SO_2$. The eight generators of SU_3 are given in (5-110) and (5-123). The CSCO-I of SU_3 is $C = (I_3'^{(3)}, I_2'^{(2)})$. The CSCO-I of SU_2 and SO_2 are taken as the isospin I^2 and its projection I_z, that for U_1 is chosen as the hypercharge

$$Y = \frac{1}{3}(E_{11} + E_{22} - 2E_{33}) ,$$

which is identical to H_2 in (7-51). The physical meaning of the hypercharge will be seen in (7-72c). For SU_3 its CSCO-II is

$$(C, C(s)) = (I_3'^{(3)},\ I_2'^{(3)};\ I^2,\ I_z\ , Y) . \tag{7-55a}$$

The irreducible bases of SU_3 are eigenfunctions of the CSCO-II and are labeled by the five quantum numbers $(pq)II_zY$,

$$\begin{pmatrix} I_3'^{(3)} \\ I_2'^{(3)} \\ I^2 \\ I_z \\ Y \end{pmatrix} \psi^{(pq)}_{II_zY} = \begin{pmatrix} p \\ q \\ I(I+1) \\ I_z \\ Y \end{pmatrix} \psi^{(pq)}_{II_zY} . \tag{7-55b}$$

We will show in (7-71) that the irrep of SU_3 can be labeled by the partition $[\nu] = [m_{13}, m_{23}]$. On the other hand it was shown in Sec. 5.23 that to label irreps of SU_3 we only need to use two integers (a_1, a_2), or (λ, μ), which are related to the partition $[\nu]$ by $\lambda = m_{13} - m_{23}$ and $\mu = m_{23}$ [see (5-357d)]. The eigenvalues p and q of the CSCO-I of SU_3 are functions of λ and μ. For example, from (7-38c) and (7-25b), and noting that the particle number $f = m_{13}+m_{23} = \lambda+2\mu$, we obtain [cf. Eq. (5-305)],

$$q = 2\left[\frac{1}{3}(\lambda - \mu)^2 + \lambda + \mu + \lambda\mu\right] . \tag{7-55c}$$

Later we will use $\psi^{(\lambda\mu)}_{II_zY}$ to denote an irreducible basis of SU_3.

The CSCO-III of SU_3 consists of the eight operators

$$K = (C, C(s), \overline{C}(s)) = (I_3'^{(3)}, I_2'^{(3)}, I^2, I_z, Y, \overline{I}^2, \overline{I}_z, \overline{Y}) . \tag{7-55d}$$

The complex conjugates of the common eigenfunctions of K in the group-parameter space yield the irreducible matrix elements of SU_3,

$$D^{(pq)}_{II_zY, \overline{I}\overline{I}_z\overline{Y}}(a^1 a^2 \ldots a^8) , \tag{7-55e}$$

where $a^1, \ldots, a^8$ are group parameters. For further discussion of the irreducible matrice of SU_3, readers are referred to Akyemapong (1972).

7.4. The Gel'fand Basis and Gel'fand Matrix Elements

The $U_n \supset U_{n-1} \supset \ldots \supset U_2 \supset U_1$ (or $SU_n \supset SU_{n-1} \times U_1 \supset \ldots \supset SU_2 \times U_1$) irreducible bases are referred to as the *Gel'fand bases* of $U_n(SU_n)$ (Gel'fand 1950, Baird 1963), which are usually labeled by the Gel'fand tableaux,

$$\left| \begin{pmatrix} [\nu] \\ (m) \end{pmatrix} \right\rangle = \left| \begin{pmatrix} m_{1n} & m_{2n} & \ldots & m_{nn} \\ & m_{1n-1} \quad m_{n-1n-1} & & \\ & \cdots & & \\ & m_{12} m_{22} & & \\ & m_{11} & & \end{pmatrix} \right\rangle , \tag{7-56}$$

with $[\nu] = [m_{11}m_{12}\dots m_{nn}]$. For $SU_n, m_{nn} = 0$, [see (7-71)]. Equation (7-56) belongs to the irrep $[m_{1l}m_{2l}\dots m_{ll}]$ of $U_l, l = n, n-1, \dots 1$. m_{ij} are positive integers and obey the "betweeness condition"

$$\begin{matrix} m_{i,j+1} & \geq & m_{i+1,j+1} \\ & \rotatebox{-45}{$\geq$} \quad m_{ij} \quad \rotatebox{45}{$\geq$} & \end{matrix} .$$

The Gel'fand bases are eigenfunctions of the CSCO-II $(C, C(s))$ of U_n [see Eqs. (7-46) and (7-47)].

We can define a so-called lexical ordering for the Gel'fand basis vectors by considering a Gel'fand tableau as a vector,

$$\mathbf{m} = (m_{1n} \dots m_{nn}, m_{1n-1} \dots m_{n-1n-1}, \dots m_{12}, m_{22}, m_{11}) .$$

We shall say that the basic vector $\left|\binom{[\nu]}{(m)}\right\rangle$ precedes the vector $\left|\binom{[\nu]}{(m')}\right\rangle$, if the first nonvanishing component of the vector $\mathbf{m} - \mathbf{m}'$ is positive. For example, see Table 7.5-2.

The matrix elements of the infinitesimal operators $E_{n,n-1}$ of the group U_n in the Gel'fand basis were first given by Gel'fand (1950), and rederived by Baird (1963). For example

$$E_{21}\begin{pmatrix} m_{12} & & m_{22} \\ & m_{11} & \end{pmatrix} = [(m_{11} - m_{22})(m_{12} - m_{11} + 1)]^{1/2} \begin{pmatrix} m_{12} & & m_{22} \\ & m_{11} - 1 & \end{pmatrix} , \tag{7-57a}$$

$$\begin{aligned} &E_{32}\begin{pmatrix} m_{13} & & m_{23} & & m_{33} \\ & m_{12} & & m_{22} & \\ & & m_{11} & & \end{pmatrix} \\ &= \left[\frac{m_{11} - m_{12}}{m_{22} - m_{12} - 1}\right]^{\frac{1}{2}} \left[(-)\frac{(m_{13} - m_{12} + 1)(m_{23} - m_{12})(m_{33} - m_{12} - 1)}{(m_{22} - m_{12})}\right]^{\frac{1}{2}} \\ &\times \begin{pmatrix} m_{13} & & m_{23} & & m_{33} \\ & m_{12} - 1, & & m_{22} & \\ & & m_{11} & & \end{pmatrix} \\ &+ \left[\frac{m_{11} - m_{22} + 1}{m_{12} - m_{22} + 1}\right]^{\frac{1}{2}} \left[(-)\frac{(m_{13} - m_{22} + 2)(m_{23} - m_{22} + 1)(m_{33} - m_{22})}{(m_{12} - m_{22} + 2)}\right]^{\frac{1}{2}} \\ &\times \begin{pmatrix} m_{13} & & m_{23} & & m_{33} \\ & m_{12}, & & m_{22} - 1 & \\ & & m_{11} & & \end{pmatrix} . \end{aligned} \tag{7-57b}$$

Using (7-17), from the matrices of the adjacent infinitesimal operators $E_{i,i-1}$, we can obtain those of the non-adjacent infinitesimal operators.

The formula for the Gel'fand matrix elements of the operator $E_{n,n-1}$ become very cumbersome for large n. Paldus (1974) gave a simplified formula for the two-columned irreps encountered in the many-electron problem. Since any entry in the Gel'fand tableau now satisfies the conditions $2 \geq m_{ij} \geq 0$, we can specify each row i of such a Gel'fand tableau by three integers

a_i, b_i and c_i, which designate the numbers of 2's, 1's and 0's in the partition $[m_{1i}m_{2i}\dots m_{ii}]$, and replace the Gel'fand tableau by the *Paldus tableau*

$$\left|\begin{matrix} a_n & b_n & c_n \\ a_{n-1} & b_{n-1} & c_{n-1} \\ \dots & \dots & \dots \\ a_1 & b_1 & c_1 \end{matrix}\right\rangle = \left|\begin{pmatrix} m_{1n}m_{2n}\dots m_{nn} \\ m_{1n-1}\dots m_{n-1n-1} \\ \dots \\ m_{11} \end{pmatrix}\right\rangle . \tag{7-58a}$$

For example, for the group U_5

$$\left|\begin{matrix} 1 & 1 & 3 \\ 1 & 1 & 2 \\ 1 & 1 & 1 \\ 0 & 2 & 0 \\ 0 & 1 & 0 \end{matrix}\right\rangle = \left|\begin{pmatrix} 2\ 1\ 0\ 0\ 0 \\ 2\ 1\ 0\ 0 \\ 2\ 1\ 0 \\ 1\ 1 \\ 1 \end{pmatrix}\right\rangle .$$

For the bases (7-58a), Paldus gave the following simple formula

$$E_{i+1,i}\left|\begin{matrix} \dots & \dots & \dots \\ a_{i+1} & b_{i+1} & c_{i+1} \\ a_i & b_i & c_i \\ a_{i-1} & b_{i-1} & c_{i-1} \\ \dots & \dots & \dots \end{matrix}\right\rangle = \left[\frac{b_i(b_i+1)}{(b_{i-1}+1)(b_{i+1}+1)}\right]^{1/2}\left|\begin{matrix} \dots & \dots & \dots \\ a_{i+1} & b_{i+1} & c_{i+1} \\ a_i & b_i-1 & c_i+1 \\ a_{i-1} & b_{i-1} & c_{i-1} \\ \dots & \dots & \dots \\ \dots & \dots & \dots \end{matrix}\right\rangle$$

$$+\left[\frac{(b_i+1)(b_i+2)}{(b_{i-1}+1)(b_{i+1}+1)}\right]^{1/2}\left|\begin{matrix} \dots & \dots & \dots \\ a_{i+1} & b_{i+1} & c_{i+1} \\ a_i-1 & b_i+1 & c_i \\ a_{i-1} & b_{i-1} & c_{i-1} \\ \dots & \dots & \dots \end{matrix}\right\rangle . \tag{7-58b}$$

The first term at the right-hand side vanishes unless $b_i \neq 0$ and $c_{i+1}-1 = c_i = c_{i-1}$, while the second term vanishes unless $a_i \neq 0$ and $a_{i+1} = a_i = a_{i-1}+1$. Also the convention $a_0 = b_0 = c_0 = 0$ is used.

The Gel'fand basis is widely used due to the following attractive features:

1. It is an orthonormal basis

$$\left\langle \begin{pmatrix} [\nu'] \\ (m') \end{pmatrix} \middle| \begin{pmatrix} [\nu] \\ (m) \end{pmatrix} \right\rangle = \delta_{\nu\nu'}\delta_{(m)(m')} \ .$$

2. It is multiplicity free, since $U_n \supset U_{n-1} \supset \dots \supset U_2 \supset U_1$ is a canonical group chain. In other words, a basis vector is uniquely labeled by a Gel'fand tableau.

3. In this basis, the matrix elements of the infinitesimal operator of U_n have algebraic formulas.

4. It is closely related to the Yamanouchi basis of the permutation group. As will be seen, the Gel'fand bases for unitary groups are the quasi-standard basis of the permutation group.

Nevertheless, it also suffers from a serious drawback, namely, more often then not, it is not the basis we need in physics, except in some rare cases (for example in the quark model of elementary particles; for details see Sec. 7.5).

Ex. 7.1. Using (7-57) show that the representation matrix of $E_{32}E_{23}$ in the basis $\left\{\left|\begin{pmatrix} 2 & & 1 & & 0 \\ & 2 & & 0 & \\ & & 1 & & \end{pmatrix}\right\rangle, \left|\begin{pmatrix} 2 & & 1 & & 0 \\ & 1 & & 1 & \\ & & 1 & & \end{pmatrix}\right\rangle\right\}$ is equal to $I + D^{[21]}(23)$, where I is a unit matrix and $D^{[21]}(23)$ is the irreducible matrix for the permutation p_{23}.

Ex. 7.2. Using $E_{13} = [E_{12}, E_{23}]$ calculate $E_{13} \left| \begin{pmatrix} 2 & & 1 & & 0 \\ & 1 & & 1 & \\ & & 1 & & \end{pmatrix} \right\rangle$.

7.5. The Gel'fand Basis of Unitary Groups and Quasi-Standard Basis of Permutation Groups

7.5.1. The CSCO-II of unitary groups and CSCO of the broken chains of permutation groups

We now proceed to extend (7-37) to the more general expression involving the invariants of subgroups of $\mathcal{U}_n$ and the class operators of subgroups of $\mathcal{S}_f$. We still consider the configuration $(m_1)^{f_1}(m_2)^{f_2}\ldots(m_n)^{f_n}$ with $\sum_{i=1}^{n} f_i = f$.

We define the following operator in analogy to (7-31)

$$P_k^{(f-f_n)} = \binom{f-f_n}{k}^{-1} \frac{1}{k!} \sum_{a_1>a_2>\ldots>a_k=1}^{f} \sum_{i_1 i_2 \ldots i_k=1}^{n-1} (e_{i_1 i_2}^{(a_1)} e_{i_2 i_3}^{(a_2)} \ldots e_{i_k i_1}^{(a_k)})_s \,. \tag{7-59}$$

Applying $P_k^{(f-f_n)}$ to the normal order state

$$|\omega\rangle = |\overbrace{m_1 \ldots m_1}^{f_1} \overbrace{m_2 \ldots m_2}^{f_2} \ldots \overbrace{m_n \ldots m_n}^{f_n}\rangle \tag{7-60}$$

and using (7-7b), we get

$$\begin{aligned} P_k^{(f-f_n)}|\omega\rangle &= \binom{f-f_n}{k}^{-1} \frac{1}{k!} \sum_{a_1>a_2>\ldots>a_k=1}^{f-f_n} (a_1 a_2 \ldots a_k)_s |\omega\rangle \\ &= \frac{1}{g_k'} \sum_{a_1>a_2>\ldots>a_k=1}^{f-f_n} (a_1(a_2\ldots a_k)_s)|\omega\rangle \\ &= \frac{1}{g_k'} C_{(k)}(f-f_n)|\omega\rangle = \frac{1}{g_k'} \mathcal{C}_{(k)}(f-f_n)|\omega\rangle \\ &= \mathcal{C}^{(k)}(f-f_n)|\omega\rangle \end{aligned} \tag{7-61}$$

where (3-110) has been used in the last step, and $g_k' = \binom{f-f_n}{k}(k-1)!$ [see (1-23)], and $C_{(k)}(f-f_n)$ is the k-cycle class operator of the group $S_{f-f_n}(1,2,\ldots,f-f_n)$, while $\mathcal{C}^{(k)}(f-f_n)$ is the average k-cycle class operator of the state permutation group $\mathcal{S}_{f-f_n}(i_1 i_2 \ldots i_{f-f_n})$.

Similarly we can prove that acting on a non-normal ordered state

$$|\tilde{\omega}\rangle = p|\omega\rangle = \begin{pmatrix} 1 & 2 & \ldots & f \\ p_1 & p_2 & \ldots & p_f \end{pmatrix} |\omega\rangle \tag{7-62}$$

$P_k^{(f-f_n)}$ gives

$$\begin{aligned} P_k^{f-f_n}|\tilde{\omega}\rangle &= \binom{f-f_n}{k}^{-1} \frac{1}{k!} \sum_{a_1>a_2>\ldots>a_k}' (a_1 a_2 \ldots a_k)_s |\tilde{\omega}\rangle \\ &= \frac{1}{g_k'} C_{(k)}(f-f_n)'|\tilde{\omega}\rangle = \mathcal{C}^{(k)}(f-f_n)|\tilde{\omega}\rangle \end{aligned} \tag{7-63}$$

where the prime in the summation symbol means that the indices $a_1, a_2, \ldots, a_k$ are to be taken from among the indices $p_1, p_2, \ldots, p_{f-f_n}$, and $C_{(k)}(f-f_n)'$ is the k-cycle class operator of the permutation group $S_{f-f_n}(p_1, p_2 \ldots, p_{f-f_n})$. For example, for $n=3, f=6, f_n=2$,

$$|\omega\rangle = |m_1 m_1 m_2 m_2 m_3 m_3\rangle \ , \ p = \begin{pmatrix} 1\,2\,3\,4\,5\,6 \\ 6\,2\,1\,5\,3\,4 \end{pmatrix} \ ,$$
$$|\tilde{\omega}\rangle = p|\omega\rangle = |m_2 m_1 m_3 m_3 m_2 m_1\rangle \ .$$

From (7-59) and (7-10) we have

$$\begin{aligned} P_2^4|\tilde{\omega}\rangle &= \left(\frac{1}{6}\sum_{a_1>a_2=1}^{6}\ \sum_{i_1 i_2=m_1}^{m_2} e^{(a_1)}_{i_1 i_2} e^{(a_2)}_{i_2 i_1}\right)(\varphi^{(1)}_{m_2}\varphi^{(2)}_{m_1}\varphi^{(3)}_{m_3}\varphi^{(4)}_{m_3}\varphi^{(5)}_{m_2}\varphi^{(6)}_{m_1}) \\ &= \frac{1}{6}[(12)+(15)+(16)+(25)+(26)+(56)]|\tilde{\omega}\rangle \\ &= \frac{1}{6}C_{(2)}(4)'|\tilde{\omega}\rangle = \frac{1}{6}\mathcal{C}_{(2)}(4)|\tilde{\omega}\rangle \end{aligned}$$

where $C_{(2)}(4)'$ is the two-cycle class operator of $S_4(1256)$. Therefore, acting on any state $p|\omega\rangle$, the operator $P_k^{f-f_n}$ is always equivalent to the average k-cycle class operator $\mathcal{C}^{(k)}(f-f_n)$ of the state permutation group $\mathcal{S}_{f-f_n}$. On the other hand, in analogy with (7-31) and (7-33), $P_k^{f-f_n}$ can be expressed in terms of the invariants $\mathfrak{S}_k^{(n-1)}, \ldots, \mathfrak{S}_2^{(n-1)}$ and $\mathfrak{S}_1^{(n-1)}$ of the group $\mathcal{U}_{n-1}$, and thus $\mathfrak{S}_k^{(n-1)}$, turning things round, can be expressed in terms of the class operators $\mathcal{C}_{(k)}(f-f_n), \ldots, \mathcal{C}_{(2)}(f-f_n)$ of $\mathcal{S}_{f-f_n}$, i.e., in terms of the CSCO-I of $\mathcal{S}_{f-f_n}$,

$$\mathfrak{S}_k^{(n-1)} = F_k^{(n-1)}(\mathcal{C}(f-f_n)) \ . \tag{7-64a}$$

In the same way, we can prove that

$$\begin{aligned} &\mathfrak{S}_k^{(n-2)} = F_k^{(n-2)}(\mathcal{C}(f-f_n-f_{n-1})) \ , \\ &\cdots \\ &\mathfrak{S}_k^{(2)} = F_k^{(2)}(\mathcal{C}(f_1+f_2)) \ , \\ &\mathfrak{S}^{(1)} = f \ . \end{aligned} \tag{7-64b}$$

According to Sec. 4.8, the quasi-standard basis vectors $\psi^{(\nu)}$ are eigenfunctions of the operator set $(\mathcal{C}(f), \mathcal{C}(f-f_n), \ldots, \mathcal{C}(f_1+f_2), \mathcal{C}(f_1))$, the CSCO of the broken chain $\mathcal{S}_f \supset \mathcal{S}_{f-f_n} \supset \ldots \supset \mathcal{S}_{f_1+f_2} \supset \mathcal{S}_{f_1}$. From (7-37) and (7-64) it is seen that the quasi-standard basis vectors $\psi^{(\nu)\kappa}$ are eigenfunctions of the CSCO-II of $\mathcal{U}_n$. Therefore we have the following theorem.

Theorem 7.2: The quasi-standard basis of the permutation group is the Gel'fand basis of the unitary group.

Since the Yamanouchi basis is a special case of the quasi-standard basis, the Yamanouchi basis is also a Gel'fand basis of the unitary group. For example

$$|Y_m^{[\nu]}\rangle = \left|\ \begin{array}{|c|c|c|}\hline \alpha & \beta & \delta \\ \hline \gamma \\ \cline{1-1}\end{array}\ \right\rangle = \left|\begin{pmatrix} 3\ 1\ 0\ 0 \\ 2\ 1\ 0 \\ 2\ 0 \\ 1 \end{pmatrix}\right\rangle = \left|\begin{pmatrix} [\nu] \\ (m) \end{pmatrix}\right\rangle \ . \tag{7-65}$$

Based on the above facts, we immediately obtain from (7-40) the relation between the Yamanouchi matrix elements $D^{[\nu]}_{m'm}(ij)$ of the permutation group and Gel'fand matrix elements of the unitary group:

$$D^{[\nu]}_{m'm}(ij) = \sum_{(m'')} \left\langle \begin{pmatrix} [\nu] \\ (m') \end{pmatrix} \middle| E_{ij} \middle| \begin{pmatrix} [\nu] \\ (m'') \end{pmatrix} \right\rangle \left\langle \begin{pmatrix} [\nu] \\ (m) \end{pmatrix} \middle| E_{ij} \middle| \begin{pmatrix} [\nu] \\ (m'') \end{pmatrix} \right\rangle - \delta_{m'm} \, . \tag{7-66}$$

From (7-66) we also know that the Yamanouchi phase convention that the nondiagonal matrix elements $D^{[\nu]}_{m'm}(i, i-1)$ be positive is consistent with the *Gel'fand-Biedenharn phase convention* of the unitary group that the matrix elements of $E_{i,i-1}$ be positive.

7.5.2. The labeling and finding of the Gel'fand basis

The Weyl tableaux used to label the quasi-standard basis of the state-permutation group in Chapter 4 can now be used to label the Gel'fand basis of the unitary group. In view of the one-to-one correspondence between the Weyl tableau and Gel'fand symbol (see Sec. 4.8), the dimension $h_\nu(\mathcal{U}_n)$ of an irrep $[\nu]$ of $\mathcal{U}_n$ can be decided upon by counting the total allowed Weyl tableaux associated with the given partition $[\nu]$, i.e., $h_\nu(\mathcal{U}_n)$ is equal to the total number of Weyl tableaux filled with the n single particle states $m_1, m_2, \ldots, m_n$ in all possible ways. For example, for $n = 3, m_1 = u, m_2 = d, m_3 = s$, according to the rule for writing Weyl tableaux, we obtain the eight Weyl tableaux associated with the partition $[21]$ as follows

$$\begin{array}{|c|c|}\hline u & u \\ \hline d \\ \cline{1-1}\end{array} \quad \begin{array}{|c|c|}\hline u & d \\ \hline d \\ \cline{1-1}\end{array} \quad \begin{array}{|c|c|}\hline u & u \\ \hline s \\ \cline{1-1}\end{array} \quad \begin{array}{|c|c|}\hline u & d \\ \hline s \\ \cline{1-1}\end{array} \quad \begin{array}{|c|c|}\hline d & d \\ \hline s \\ \cline{1-1}\end{array} \quad \begin{array}{|c|c|}\hline u & s \\ \hline d \\ \cline{1-1}\end{array} \quad \begin{array}{|c|c|}\hline u & s \\ \hline s \\ \cline{1-1}\end{array} \quad \begin{array}{|c|c|}\hline d & s \\ \hline s \\ \cline{1-1}\end{array} \tag{7-67}$$

Therefore the dimension of the irrep $[21]$ of $\mathcal{SU}_3$ is $h_{[21]}(\mathcal{SU}_3) = 8$.

From Sec. 4.8 we know that when two identical state indices appear in the same column of a Weyl tableau $W_k^{[\nu]}$, the Gel'fand basis vector $|W_k^{[\nu]}\rangle = 0$. Consequently, the Young diagram of $\mathcal{U}_n$ or $\mathcal{SU}_n$ can have at most only n rows.

A formula for the dimensionality of $\mathcal{SU}_n$ irreps is as follows (Hamermesh 1962)

$$h_{[\nu]}(\mathcal{U}_n) = \prod_{1 \le i < j \le n} \frac{(\nu_i - i - \nu_j + j)}{j - i} \, . \tag{7-68}$$

Thus for instance, $h_{(\lambda\mu)}(\mathcal{SU}_3) = \frac{1}{2}(\lambda+1)(\mu+1)(\lambda+\mu+2)$.

A simple formula was given by Robinson (1961). As illustration we have

$$\begin{aligned} h_{[422]}(\mathcal{U}_n) &= \begin{array}{|c|c|c|c|}\hline n & n+1 & n+2 & n+3 \\ \hline n-1 & n \\ \cline{1-2} n-2 & n-1 \\ \cline{1-2}\end{array} \Bigg/ \begin{array}{|c|c|c|c|}\hline 6 & 5 & 2 & 1 \\ \hline 3 & 2 \\ \cline{1-2} 2 & 1 \\ \cline{1-2}\end{array} \\ &= \frac{n^2(n+1)(n+2)(n+3)(n-1)(n-2)(n-1)}{6{\cdot}5{\cdot}2{\cdot}1{\cdot}3{\cdot}2{\cdot}2{\cdot}1} \, , \end{aligned} \tag{7-69}$$

where the denominator is the product of the hook lengths defined in (4-4b), while the numerator is a product of integers. The rule for filling the integers in a Young diagram is as follows: Fill all the boxes along the diagonal of the Young diagram with integer n, and fill consecutively the subsequent boxes in the same rows with $n+1, n+2$, etc., and in the same columns with $n-1, n-2$, etc. The dimension of the adjoint rep $[21^{n-2}]$ of $\mathcal{SU}_n$ is easily found to be $n^2 - 1$ by Robinson's formula,

$$h_{[21^{n-2}]}(\mathcal{SU}_n) = \begin{array}{|c|c|}\hline n & n+1 \\ \hline n-1 \\ \cline{1-1} \vdots \\ \cline{1-1} 3 \\ \cline{1-1} 2 \\ \cline{1-1}\end{array} \Bigg/ \begin{array}{|c|c|}\hline n & 1 \\ \hline n-2 \\ \cline{1-1} \vdots \\ \cline{1-1} 2 \\ \cline{1-1} 1 \\ \cline{1-1}\end{array} = \frac{(n+1)!}{n{\cdot}(n-2)!} = n^2 - 1 \, .$$

SU_n is a simple group and has n^2-1 infinitesimal operators. According to Sec. 5.5, the n^2-1 infinitesimal operators carry an irrep which is just the adjoint rep $[21^{n-2}]$.

Now we turn to study the transformation property of the totally anti- symmetric state $[1^n]$ under group elements $\mathcal{R}$ of $\mathcal{U}_n$.

$$\mathcal{R}\left|\begin{array}{|c|}\hline m_1\\\hline m_2\\\hline \vdots\\\hline m_n\\\hline\end{array}\right\rangle = \mathcal{R}\left[\frac{1}{\sqrt{n!}}\sum_{m_1\ldots m_n}\varepsilon_{m_1\ldots m_n}\varphi^{a_1}_{m_1}\cdots\varphi^{a_n}_{m_n}\right]$$

$$= \frac{1}{\sqrt{n!}}\sum_{\substack{m_1\ldots m_n\\ m'_1\ldots m'_n}}\varepsilon_{m_1\ldots m_n}\mathcal{D}_{m'_1 m_1}\cdots\mathcal{D}_{m'_n m_n}\varphi^{a_1}_{m'_1}\cdots\varphi^{a_n}_{m'_n} \tag{7-70a}$$

$$= \det(\mathcal{D})|[1^n]\rangle = e^{i\varphi}|[1^n]\rangle$$

where (5-72b) and the following expression for the determinant $\det(\mathcal{D})$ have been used,

$$\det(\mathcal{D}) = \sum_{i_1\ldots i_n}\varepsilon^{i_1\ldots i_n}\varepsilon^{i'_1\ldots i'_n}\mathcal{D}_{i'_1 i_1}\cdots\mathcal{D}_{i'_n i_n}\ . \tag{7-70b}$$

Analogously we can show that

$$\mathcal{R}|m^n\rangle = e^{im\varphi}|m^n\rangle\ . \tag{7-70c}$$

Suppose that the group $\mathcal{U}_n$ has an irrep $[\nu]=[\nu_1\nu_2\ldots\nu_n]$. Annexing a single column of n boxes to the Young diagram $[\nu]$ gives a new representation $[\nu']=[\nu_1+1,\nu_2+1,\ldots,\nu_n+1]$. Obviously, the irreps $[\nu]$ and $[\nu']$ of $\mathcal{U}_n$ have the same number of Weyl tableaux and thus have the same dimension. From (7-70a) it is known that the only change in the representation matrices due to the annexing is that they are multiplied by a common factor $e^{i\varphi}$. Similarly, the irreps $[\nu']=[\nu_1+m,\nu_2+m,\ldots,\nu_n+m]$ and $[\nu]$ of $\mathcal{U}_n$ have the same dimension and their representation matrices differ only by a common factor $e^{im\varphi}$, i.e.,

$$D^{[\nu_1\nu_2\ldots]}(\mathcal{U}_n) = e^{-im\varphi}D^{[\nu_1+m,\nu_2+m,\ldots]}(\mathcal{U}_n)\ . \tag{7-70d}$$

If we are dealing with the special unitary group $S\mathcal{U}_n$, $\det(\mathcal{D})=e^{i\varphi}=1$, and we have

$$D^{[\nu_1\nu_2\ldots]}(S\mathcal{U}_n) = D^{[\nu_1+m,\nu_2+m\ldots]}(S\mathcal{U}_n)\ . \tag{7-70e}$$

Therefore for the SU_n group, the irreps $[\nu_1\ldots\nu_n]$ and $[\nu_1+m,\ldots,\nu_n+m]$ are equivalent, and we only need consider those Young diagrams which have at most $n-1$ rows:

$$[\nu_1\nu_2\ldots\nu_n] \equiv [\nu_1-\nu_n,\nu_2-\nu_n,\ldots,\nu_{n-1}-\nu_n,0] \tag{7-71}$$

and the irrep $[1^n]$ or $[m^n]$ becomes the identity rep designated as $[0]$.

In atomic or nuclear physics, the state belonging to the irrep $[1^n]$ corresponds to a closed shell state. Under the group SU_n, it behaves, like a vacuum state and thus can be ignored.

We can use the the three labeling schemes for the quasi-standard basis of the permutation group to characterize the Gel'fand basis, since these two bases are identical. Besides, we often use other quantum numbers to label the Gel'fand basis. For example, for SU_2 group, we often use the quantum numbers J and J_z, which are related to the Gel'fand symbol by

$$J=\frac{1}{2}(m_{12}-m_{22}),\quad J_z=m_{11}-\frac{1}{2}(m_{12}+m_{22})\ . \tag{7-72a}$$

The first equation is just the equation (7-54). We can easily use the Weyl tableaux to derive (7-72a). Let us fill the boxes in the first row of a Young diagram $[m_{12}m_{22}]$ with the spin up states α, and the second row with the spin down states β. The state corresponding to this Weyl tableau has the maximum possible J_z value for the irrep $[m_{12}, m_{22}]$, i.e., $J_z = \frac{1}{2}(m_{12} - m_{22})$. Thus the total spin must be $J = \frac{1}{2}(m_{12} - m_{22})$.

A general Weyl tableau corresponding to the Gel'fand symbol $\begin{pmatrix} m_{12} & m_{22} \\ m_{11} & \end{pmatrix}$ is as follows.

$$\begin{pmatrix} m_{12} \; m_{22} \\ m_{11} \end{pmatrix} = \begin{array}{|c|c|} \hline m_{11}\,\alpha\text{'s} & (m_{12}-m_{11})\,\beta\text{'s} \\ \hline m_{22}\,\beta\text{'s} & \multicolumn{1}{c}{} \\ \cline{1-1} \end{array} \quad . \tag{7-72b}$$

The corresponding state has $J = \frac{1}{2}(m_{12} - m_{22})$ and

$$J_z = \frac{1}{2}[m_{11} - (m_{12} - m_{11} + m_{22})] = m_{11} - \frac{1}{2}(m_{12} + m_{22}) \ .$$

In the SU_4 quark model of elementary particles, it is assumed that there exist four kinds of flavor quarks, u, d, s and c, carrying the fundamental rep of SU_4. The quantum numbers of these flavor quarks are listed in Table 7.5-1. Q is the charge, I and I_z are the isospin and its z component, B, S, C and Y are the baryon number, strangeness, charm and hypercharge, respectively, while Z is a quantity introduced by Haacke (1976), I_z, Y and Z can be expressed in terms of the quark numbers n_u, n_d, n_s and n_c:

$$\begin{aligned} I_z &= \frac{1}{2}(n_u - n_d) \ , \\ Y &= B + S = \frac{1}{3}(n_u + n_d - 2n_s), \quad (\text{only for}\, u, d, s) \\ Z &= \frac{1}{4}(n_u + n_d + n_s - 3n_c) \ . \end{aligned} \tag{7-72c}$$

Table 7.5-1. The quantum numbers of the flavor quarks u, d, s and c.

	Q	I	I_z	B	S	C	Y	Z
u	$\frac{3}{2}$	$\frac{1}{2}$	$\frac{1}{2}$	$\frac{1}{3}$	0	0	$\frac{1}{3}$	$\frac{1}{4}$
d	$-\frac{1}{3}$	$\frac{1}{2}$	$-\frac{1}{2}$	$\frac{1}{3}$	0	0	$\frac{1}{3}$	$\frac{1}{4}$
s	$-\frac{1}{3}$	0	0	$\frac{1}{3}$	-1	0	$-\frac{2}{3}$	$\frac{1}{4}$
c	$\frac{2}{3}$	0	0	$\frac{1}{3}$	0	1	0	$-\frac{3}{4}$

We usually order the single quark states according to the sequence u, d, s and c, then I_z, Y and Z are the eigenvalues of the commuting operators H_1, H_2 and H_3 in (7-51).

The Gel'fand basis of SU_4 can be labeled in three ways: using a Gel'fand symbol, a Weyl tableau, and the quantum numbers $[\nu](\lambda\mu)IYZI_z$. Their mutual relations are given in (7-73).

$$\begin{pmatrix} m_{14} & & m_{24} & & m_{34} & & 0 \\ & m_{13} & & m_{23} & & m_{33} & \\ & & m_{12} & & m_{22} & & \\ & & & m_{11} & & & \end{pmatrix} \leftrightarrow \begin{array}{|l|} \hline m_{11}\ 1\text{'s},\ (m_{12}-m_{11})\ 2\text{'s},\ (m_{13}-m_{12})\ 3\text{'s},\ (m_{14}-m_{13})\ 4\text{'s} \\ \hline m_{22}\ 2\text{'s},\ (m_{23}-m_{22})\ 3\text{'s},\ (m_{24}-m_{23})\ 4\text{'s} \\ \hline m_{33}\ 3\text{'s},\ (m_{34}-m_{33})\ 4\text{'s} \\ \hline \end{array} \tag{7-73a}$$

Gel'fand basis **Weyl tableau(1,2,3,4 denote the single quark states)**

$$[\nu] = [m_{14}m_{24}m_{34}] \, , \qquad \lambda = (m_{13} - m_{23}), \quad u = (m_{23} - m_{33})$$

$$I = \frac{1}{2}(m_{12} - m_{22}) \, , \qquad I_z = m_{11} - \frac{1}{2}(m_{12} - m_{22})$$

$$Y = (m_{12} + m_{22}) - \frac{2}{3}(m_{13} + m_{23} + m_{33}) \, , \quad Z = m_{13} + m_{23} + m_{33} - \frac{3}{4}(m_{14} + m_{24} + m_{34}) \, . \tag{7-73b}$$

In elementary particle physics one often uses the dimensions to label irreps of SU_n. For example, we use (20) to label the 20-dimensional irrep [21] of SU_4, and (8) the 8-dimensional irrep [21] of SU_3, etc. Table A.1 in the Appendix gives the correspondence between dimensions and partitions of groups from $SU_3 - SU_6$.

As an example, in Table 7.5-2 we list the three labeling schemes for the eight-dimensional irrep (the adjoint rep of SU_3).

Table 7.5-2. Three labeling schemes for the eight-dimensional irrep of SU_3.

particles	P	N	Σ^+	Σ^0	Σ^-	Λ	Ξ^0	Ξ^-
(I, I_z, Y)	$(\frac{1}{2}\ \frac{1}{2}, 1)$	$(\frac{1}{2}\ -\frac{1}{2}, 1)$	$(1\ 1,\ 0)$	$(1\ 0,\ 0)$	$(1\ -1,\ 0)$	$(0\ 0,\ 0)$	$(\frac{1}{2}\ \frac{1}{2},\ -1)$	$(\frac{1}{2}\ -\frac{1}{2},\ -1)$
Weyl tableau	u u / d	u d / d	u u / s	u d / s	d d / s	u s / d	u s / s	d s / s
Gel'fand symbol	$\begin{pmatrix} 2\ 1\ 0 \\ 2\ 1 \\ 2 \end{pmatrix}$	$\begin{pmatrix} 2\ 1\ 0 \\ 2\ 1 \\ 1 \end{pmatrix}$	$\begin{pmatrix} 2\ 1\ 0 \\ 2\ 0 \\ 2 \end{pmatrix}$	$\begin{pmatrix} 2\ 1\ 0 \\ 2\ 0 \\ 1 \end{pmatrix}$	$\begin{pmatrix} 2\ 1\ 0 \\ 2\ 0 \\ 0 \end{pmatrix}$	$\begin{pmatrix} 2\ 1\ 0 \\ 1\ 1 \\ 1 \end{pmatrix}$	$\begin{pmatrix} 2\ 1\ 0 \\ 1\ 0 \\ 1 \end{pmatrix}$	$\begin{pmatrix} 2\ 1\ 0 \\ 1\ 0 \\ 0 \end{pmatrix}$

In elementary particle physics, the SU_3 irreducible basis vectors are usually plotted in the weight diagrams with the hypercharge Y as the y-axis and the third component I_z of the isospin as the x-axis (de Swart 1963). Figure 7.5-1 gives the weight diagrams for the irreps [21], [3] and $[1^3]$, with the dimensions 8, 10 and 1, respectively.

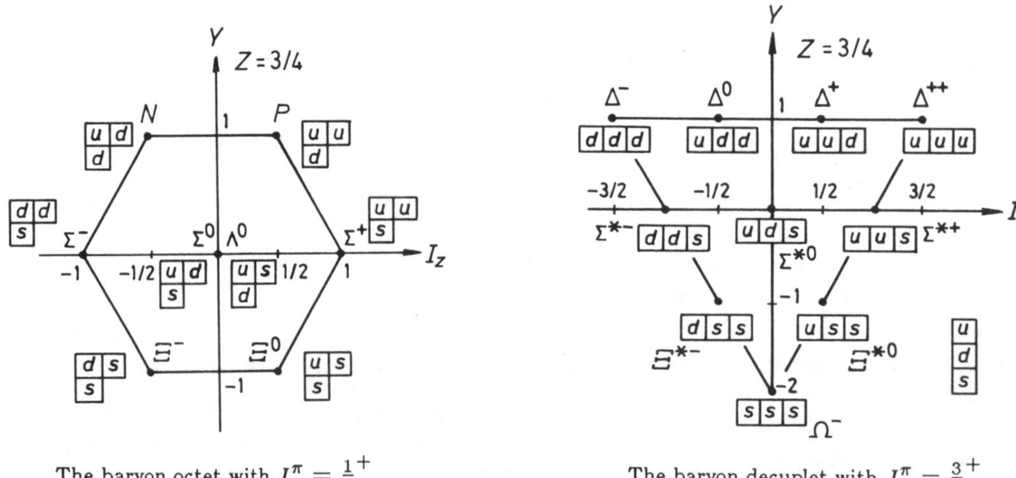

The baryon octet with $J^\pi = \frac{1}{2}^+$. The baryon decuplet with $J^\pi = \frac{3}{2}^+$.

Fig.7.5-1. The weight diagrams of SU_3 for the irreps [21], [3] and $[1^3]$.

The particles belonging to the same irrep of SU_3 form a multiplet and they should have the same baryon number, spin and parity. The baryon octet ($[\nu] = [21]$) and decuplet ($[\nu] = [3]$)

have $J^\pi = \frac{1}{2}^+$ and $\frac{3}{2}^+$, respectively. Table 7.5-3 gives the masses of the members of the octet and decuplet. In the SU_3 model, the strongest interactions are assumed to be invariant under SU_3. In the absence of any other interactions, the members of the same multiplet should have the same mass. The SU_3 symmetry is broken by some unknown weaker interactions and the mass degeneracy of the particles belonging to the same unitary multiplet is removed, as is seen from Table 7.5-3. Okubo (1962) assumed that this symmetry-breaking interaction is the irreducible tensor $T^{[21]}_{000}$, i.e., the $I = I_z = Y = 0$ component of the irrep [21] of SU_3, and derived the following mass formulas, known as the *Gell-Mann-Okubo* mass formula (for the derivation see Lichtenberg 1978, p. 179),

$$\frac{1}{2}(M_N + M_\Xi) = \frac{3}{4}M_\Lambda + \frac{1}{4}M_\Sigma \ ,$$
$$M_{\Omega^-} - M_{\Xi^*} = M_{\Xi^*} - M_{\Sigma^*} = M_{\Sigma^*} - M_\Delta \ .$$

The Gell-Mann-Okubo formula works rather well for baryons.

Table 7.5-3. The masses of the members of the octet and decuplet (in MeV).

octet	P	N	Σ^+	Σ^0	Σ^-	Λ^0	Ξ^0	Ξ^-		
	938	940	1189	1192	1197	1116	1315	1321		
decuplet	Δ^{++}	Δ^+	Δ^0	Δ^-	Σ^{*+}	Σ^{*0}	Σ^{*-}	Ξ^{*0}	Ξ^{*-}	Ω^-
	1231	1231	1232	1239	1382	1381	1386	1532	1535	1672

If one further takes into account the electromagnetic interaction, which breaks the SU_2 symmetry, then the masses of the different members of the same isomultiplet are split up. The SU_2 symmetry-breaking interaction is relatively minor, as is seen from Table 7.5-3 that the different members of an isomultiplet have masses which differ by less than 10 MeV.

The irrep [21] of SU_4 is 20-dimensional. The weight diagrams of SU_4 are three-dimensional with I_z, Y and Z as the coordinate axes. They can be resolved into plane diagrams according to the quantum number Z. For the irrep [21], the diagram with $Z = 3/4$ is just the first diagram in Fig. 7.5-1, while those for $Z = -1/4$ and $-5/4$ are shown in Fig. 7.5-2.

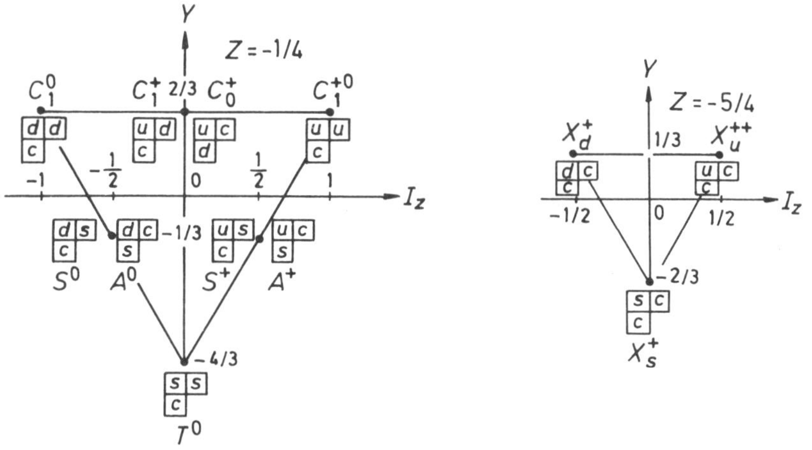

Fig. 7.5-2. The weight diagrams of SU_4 for the irrep [21].

Now let us summarize the method of decomposing the reducible tensor (7-4) of U_n into the irreducible basis of $\mathcal{U}_n$ and S_f. With the notation used in Chapter 4, the reducible tensor (7-4) is expressed as

$$\Phi_a = p_a|\omega\rangle = p_a|i_1 i_2 \dots i_f\rangle \ . \tag{7-74}$$

In the first place, according to the different configurations $(m_1)^{f_1}(m_2)^{f_2}\dots(m_n)^{f_n}$, the n^f basis vectors (7-74) are divided into several groups, and each group contains $\mathcal{N}_{f_1 f_2 \dots f_n}$ vectors where

$$\mathcal{N}_{f_1 f_2 \dots f_n} = f!/(f_1! f_2! \dots f_n!) \ . \tag{7-75a}$$

$$n^f = \sum_{f_1+f_2+\dots+f_n=f} \mathcal{N}_{f_1 f_2 \dots f_n} \ . \tag{7-75b}$$

Then for each configuration $(m_1)^{f_1}(m_2)^{f_2}\cdots(m_n)^{f_n}$, we proceed to reduce the $\mathcal{N}_{f_1 f_2 \dots f_n}$ basis vectors. Since the Gel'fand basis of the unitary group is the quasi-standard basis of the permutation group, the reduction problem for the unitary group is thus converted into that of the permutation groups. Each irreducible basis vector resulting from the reduction of (7-74) is labeled uniquely by a Young tableau Y_m and a Weyl tableau W_κ:

$$\left|\begin{matrix}[\nu] \\ Y_m, W_\kappa\end{matrix}\right\rangle = \sum_{a=1}^{\mathcal{N}_{f_1 \dots f_n}} u_{\nu m \kappa, a}\Phi_a, \qquad \begin{matrix} m = 1, 2, \dots, h_\nu(S_f) \\ \kappa = 1, 2, \dots, \tau^\nu_{f_1 f_2 \dots f_n} \end{matrix} \ , \tag{7-76}$$

where $h_\nu(S_f)$ is the dimension of the irrep $[\nu]$ of S_f and $\tau^\nu_{f_1 \dots f_n}$ is the number of times the irrep $[\nu]$ of S_f occurs in the configuration $(m_1)^{f_1}(m_2)^{f_2}\dots(m_n)^{f_n}$. From the eigenequation (4-76) or (4-77), we can find the coefficients $u_{\nu m \kappa, a}$.

Under the action of the group S_f, (7-76) only changes its Young tableau Y_m,

$$p\left|\begin{matrix}[\nu] \\ Y_m, W_\kappa\end{matrix}\right\rangle = \sum_{m'=1}^{h_\nu(S_f)} D^{[\nu]}_{m'm}(p)\left|\begin{matrix}[\nu] \\ Y_{m'}, W_\kappa\end{matrix}\right\rangle \ , \tag{7-77a}$$

and under the action of the group $\mathcal{U}_n$, (7-76) only changes its Weyl tableau W_κ,

$$\mathcal{E}_{\alpha\beta}\left|\begin{matrix}[\nu] \\ Y_m, W_\kappa\end{matrix}\right\rangle = \sum_{\kappa'=1}^{h_\nu(\mathcal{U}_n)} \left\langle\begin{matrix}[\nu] \\ W_{\kappa'}\end{matrix}\right|\mathcal{E}_{\alpha\beta}\left|\begin{matrix}[\nu] \\ W_\kappa\end{matrix}\right\rangle\left|\begin{matrix}[\nu] \\ Y_m, W_{\kappa'}\end{matrix}\right\rangle \ , \tag{7-77b}$$

where $h_\nu(\mathcal{U}_n)$ is the dimension of the irrep $[\nu]$ of $\mathcal{U}_n$. The first factor in the right-hand side of (7-77b) is the Gel'fand matrix element. Obviously, the dimension $h_\nu(\mathcal{U}_n)$ is related to $\tau^\nu_{f_1 \dots f_n}$ by

$$h_\nu(\mathcal{U}_n) = \sum_{f_1+f_2+\dots+f_n=f} \tau^\nu_{f_1 f_2 \dots f_n} \ . \tag{7-78}$$

The total number of the irreducible basis vectors is still n^f:

$$n^f = \sum_\nu h_\nu(S_f) h_\nu(\mathcal{U}_n) \ , \tag{7-79}$$

where the summation over ν includes all possible partitions $[\nu] = [\nu_1 \nu_2 \dots \nu_f]$ of the integer f.

The n^f irreducible basis vectors $\left|\begin{matrix}[\nu] \\ Y_m W_\kappa\end{matrix}\right\rangle$ can be placed in a rectangular array in which the rows are labeled by Y_m and the columns by W_κ. The array is in block-diagonal form and consists of N blocks, N being the class number of S_f and each belongs to the irrep $[\nu]$ of S_f and $\mathcal{U}_n$. The basis vectors in any particular row of a block span the irrep $[\nu]$ of $\mathcal{U}_n$. There are $h_\nu(S_f)$ rows which indicates that the irrep $[\nu]$ of $\mathcal{U}_n$ occurs $h_\nu(S_f)$ times. Similarly, the basis vectors in each column of the block span the irrep $[\nu]$ of S_f. There are $h_\nu(\mathcal{U}_n)$ columns which implies that the irrep $[\nu]$ of S_f occurs $h_\nu(\mathcal{U}_n)$ times.

For example, suppose $f = 3$. The $3^3 = 27$ irreducible basis vectors of S_3 and $\mathcal{U}_3$ are placed in the rectangular array shown in Fig. 7.5-3.

The explicit form of the 27 irreducible basis vectors of Fig. 7.5-3 was already given in Chapter 3. By setting $\alpha = u, \beta = d$ and $\gamma = s$ in Table 3.9 and also by referring to Fig. 7.5-1 we obtain the wave functions of Σ^{*0}, Σ^0 and Λ^0 listed in Table 7.5-4.

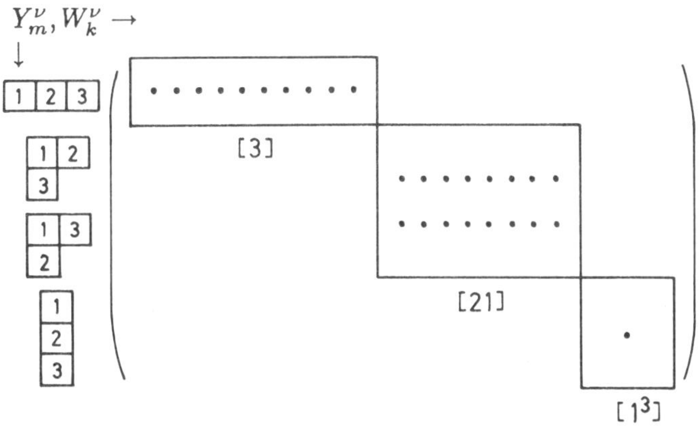

$3^3 = 1 \times 10 + 2 \times 8 + 1 \times 1$

For the tableux W^ν_k, see Fig. 7.5-1

Fig. 7.5-3. The irreducible basis of S_3 and $S\mathcal{U}_3$.

Table 7.5-4. The baryon $S\mathcal{U}_3$ wave functions.

ν, m, k		$\|uds\rangle$	$\|dus\rangle$	$\|sdu\rangle$	$\|usd\rangle$	$\|sud\rangle$	$\|dsu\rangle$
3, 1, 1	$\left\| \boxed{1\,2\,3}, \Sigma^{*0} \right\rangle$	$\frac{1}{\sqrt{6}}$	$\frac{1}{\sqrt{6}}$	$\frac{1}{\sqrt{6}}$	$\frac{1}{\sqrt{6}}$	$\frac{1}{\sqrt{6}}$	$\frac{1}{\sqrt{6}}$
0, 1, 1	$\left\| \begin{smallmatrix}1&2\\3\end{smallmatrix}, \Sigma^0 \right\rangle$	$\frac{1}{\sqrt{3}}$	$\frac{1}{\sqrt{3}}$	$-\frac{1}{\sqrt{12}}$	$-\frac{1}{\sqrt{12}}$	$-\frac{1}{\sqrt{12}}$	$-\frac{1}{\sqrt{12}}$
0, −1, 1	$\left\| \begin{smallmatrix}1&3\\2\end{smallmatrix}, \Sigma^0 \right\rangle$	0	0	$-\frac{1}{2}$	$\frac{1}{2}$	$-\frac{1}{2}$	$\frac{1}{2}$
0, 1, −1	$\left\| \begin{smallmatrix}1&2\\3\end{smallmatrix}, \Lambda^0 \right\rangle$	0	0	$-\frac{1}{2}$	$\frac{1}{2}$	$\frac{1}{2}$	$-\frac{1}{2}$
0, −1, −1	$\left\| \begin{smallmatrix}1&3\\2\end{smallmatrix}, \Lambda^0 \right\rangle$	$\frac{1}{\sqrt{3}}$	$-\frac{1}{\sqrt{3}}$	$\frac{1}{\sqrt{12}}$	$\frac{1}{\sqrt{12}}$	$-\frac{1}{\sqrt{12}}$	$-\frac{1}{\sqrt{12}}$
−3, −1, −1	$\left\| \begin{smallmatrix}1\\2\\3\end{smallmatrix}, [0] \right\rangle$	$\frac{1}{\sqrt{6}}$	$-\frac{1}{\sqrt{6}}$	$-\frac{1}{\sqrt{6}}$	$-\frac{1}{\sqrt{6}}$	$\frac{1}{\sqrt{6}}$	$\frac{1}{\sqrt{6}}$

By setting (α, β) in Table 3.4-1 equal to $(u, d), (u, s)$ and (d, s) respectively, from (3-90) we obtain the wave functions of the elementary particles P, Σ^+ and Σ^- listed in Table 7.5-5. Similarly, from Ex. 6 in Sec. 3.4, we obtain the wave functions of N, Ξ^0 and Ξ^- listed in Table 7.5-6.

The irreducible basis for the totally symmetric rep [3] is

$$\left| \boxed{1\,2\,3} \quad \boxed{\alpha\,\alpha\,\beta} \right\rangle = \frac{1}{\sqrt{3}}(|\alpha\alpha\beta\rangle + |\alpha\beta\alpha\rangle + |\beta\alpha\alpha\rangle) \,. \tag{7-80}$$

The SU_4 baryon wave functions can be obtained from Table 7.5-4 by replacing the state labels (u, d, s) with (u, s, c) and (d, s, c), and from Table 7.5-5 and 7.5-6 by replacing (u, d) with (u, c), (d, c) and (s, c).

Table 7.5-5. The baryon SU_3 wave functions.

	$\|uud\rangle$	$\|udu\rangle$	$\|duu\rangle$
	$\|uus\rangle$	$\|usu\rangle$	$\|suu\rangle$
	$\|dds\rangle$	$\|dsd\rangle$	$\|sdd\rangle$
[12/3] ; P, Σ^+, Σ^-	$\frac{2}{\sqrt{6}}$	$-\frac{1}{\sqrt{6}}$	$-\frac{1}{\sqrt{6}}$
[13/2] ; P, Σ^+, Σ^-	0	$\frac{1}{\sqrt{2}}$	$-\frac{1}{\sqrt{2}}$

Table 7.5-6

	$\|udd\rangle$	$\|dud\rangle$	$\|ddu\rangle$
	$\|uss\rangle$	$\|sus\rangle$	$\|ssu\rangle$
	$\|dss\rangle$	$\|sds\rangle$	$\|ssd\rangle$
[12/3] ; N, Ξ^0, Ξ^-	$\frac{1}{\sqrt{6}}$	$\frac{1}{\sqrt{6}}$	$-\frac{2}{\sqrt{6}}$
[13/2] ; N, Ξ^0, Ξ^-	$\frac{1}{\sqrt{2}}$	$-\frac{1}{\sqrt{2}}$	0

In the above discussions, the coordinates of the f particles are assumed to be all different. If the coordinates of different particles are allowed to be identical, then the standard basis $\left|\begin{matrix}[\nu]\\ Y_m W_\kappa\end{matrix}\right\rangle$ of S_f goes over to the quasi-standard basis $\left|\begin{matrix}[\nu]\\ W_m W_\kappa\end{matrix}\right\rangle$ of S_f (or the Gel'fand basis of U_n), where W_m is the Weyl tableau in coordinate space. Under the action of the infinitesimal operators E_{ij} and $\mathcal{E}_{\alpha\beta}$, the $[h_\nu(U_n)]^2$ irreducible basis vectors of U_n and $\mathcal{U}_n$ transform as follows:

$$E_{ij}\left|\begin{matrix}[\nu]\\ W_m, W_\kappa\end{matrix}\right\rangle = \sum_{m'=1}^{h_\nu(U_n)} \left\langle\begin{matrix}[\nu]\\ W_{m'}\end{matrix}\right| E_{ij} \left|\begin{matrix}[\nu]\\ W_m\end{matrix}\right\rangle \left|\begin{matrix}[\nu]\\ W_{m'}, W_\kappa\end{matrix}\right\rangle , \tag{7-81a}$$

$$\mathcal{E}_{\alpha\beta}\left|\begin{matrix}[\nu]\\ W_m, W_\kappa\end{matrix}\right\rangle = \sum_{\kappa'=1}^{h_\nu(U_n)} \left\langle\begin{matrix}[\nu]\\ W_{\kappa'}\end{matrix}\right| \mathcal{E}_{\alpha\beta} \left|\begin{matrix}[\nu]\\ W_\kappa\end{matrix}\right\rangle \left|\begin{matrix}[\nu]\\ W_m, W_{\kappa'}\end{matrix}\right\rangle . \tag{7-81b}$$

From (7-81) it is seen that as soon as any one of the $[h_\nu(U_n)]^2$ irreducible basis vectors is known, all the others can be found with the help of the Gel'fand matrix elements. However this method is not as simple as the EFM.

Starting with the next section, we will be mainly concerned with the unitary group in state space and we will change our notation from $\mathcal{U}_n(S\mathcal{U}_n)$ to $U_n(SU_n)$, in order to be consistent with the conventional notation.

Ex. 7.3. Construct the analogy of Table 7.5-2 for the irrep [3] of SU_3.

Ex. 7.4. Consider six quarks in the configuration $(u)^2(d)^2(s)^2$ and write out the possible Gel'fand basis vectors and their corresponding Gel'fand symbols.

7.6. The Contragredient Representation

According to Sec. 2.4 the rep $\tilde{D}^{-1}$ is called the contragredient rep of D. If D is unitary, then $\tilde{D}^{-1} = D^*$. It can be shown that the contragredient rep of the irrep $[\nu] = [\nu_1\nu_2 \ldots \nu_n]$ is $[\nu'] = [\nu_1', \nu_2', \ldots, \nu_n']$, where

$$\nu_1' = \nu_1 - \nu_n,\ \nu_2' = \nu_1 - \nu_{n-1}, \ldots\ \nu_p' = \nu_1 - \nu_{n-p+1}, \ldots, \nu_n' = 0 . \tag{7-82}$$

The relation (7-82) is easily committed to memory by noting that putting the Young diagram $[\nu]$ and the upside down diagram of $[\nu']$ together, we get the Young diagram $[\nu_1^n]$, i.e., the identity rep $[0]$ of SU_n. For instance, corresponding to the irrep $[\nu] = [43221]$ of SU_5, the contragredient rep $[\nu'] = [3221]$, as shown in the Fig. 7.6-1.

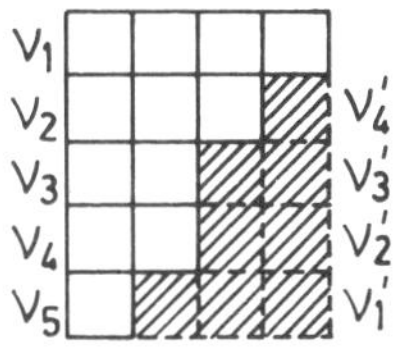

Fig. 7.6-1. The irrep $[\nu] = [43221]$ and its contragredient rep $[\nu'] = [3221]$ of SU_5.

A general proof of (7-82) is given by Elliott (1979). We will only discuss a special case of (7-82), i.e., $[\nu] = [1]$, and $[\nu'] = [1^{n-1}]$.

Suppose $\varphi_{i_1} \dots \varphi_{i_n}$ and $\psi^{i_1} \dots \psi^{i_n}$ span the irreps $[1]$ and $[1^{n-1}]$ of SU_n, respectively. The basis of the anti-symmetric rep $[1^{n-1}]$ can be expressed in terms of the determinant

$$\psi^{i_1} = \sum_{i_2 \dots i_n} \varepsilon^{i_1 i_2 \dots i_n} \varphi_{i_2} \dots \varphi_{i_n} \ . \tag{7-83}$$

Under the SU_n transformation

$$\begin{aligned} \psi^{i_1} \to \psi'^{i_1} &= \sum_{i_2 \dots i_n} \sum_{i'_2 \dots i'_n} \varepsilon^{i_1 i_2 \dots i_n} D_{i'_2 i_2} \dots D_{i'_n i_n} \varphi_{i'_2} \dots \varphi_{i'_n} \\ &= \sum_{i'_2 \dots i'_n} M_{i'_1 i_1} \varepsilon^{i'_1 i'_2 \dots i'_n} \varphi_{i'_2} \dots \varphi_{i'_n} \\ &= \sum_{i'_1} (D^{-1})_{i_1 i'_1} \det(D) \psi^{i'_1} = \sum_{i'_1} (\tilde{D}^{-1})_{i'_1 i_1} \psi^{i'_1} \ , \end{aligned} \tag{7-84}$$

where $M_{i'_1 i_1}$ is the cofactor of the element $D_{i'_1 i_1}$. In deriving (7-84) the following relations were used:

$$(D^{-1})_{i_1 i'_1} = M_{i'_1 i_1} / \det(D) \ , \tag{7-85a}$$

$$M_{i'_1 i_1} = \sum_{i_2 \dots i_n} \varepsilon^{i_1 i_2 \dots i_n} \varepsilon^{i'_1 i'_2 \dots i'_n} D_{i'_2 i_2} \dots D_{i'_n i_n} \ . \tag{7-85b}$$

Equation (7-84) shows that the basis $\psi^{i_1} \dots \psi^{i_n}$ of the irrep $[1^{n-1}]$ of SU_n indeed transform according to the contragredient rep $\tilde{D}^{-1}$.

In atomic or nuclear physics, $[\nu] = [1]$ corresponds to a single particle state, while $[1^{n-1}]$ corresponds to a single hole state, and $[1^n]$ is the closed shell state. The relation between $[\nu_1 \nu_2 \dots \nu_n]$ and its contragredient rep $[\nu'_1 \nu'_2 \dots \nu'_n]$ is that of particle-hole conjugation.

If a partition $[\nu] = [\nu_1, \nu_2 \dots]$ satisfies

$$\nu_p + \nu_{n-p+1} = \nu_1 \ , \quad p = 1, 2, \dots, n \ , \tag{7-86}$$

then $[\nu'] = [\nu]$, i.e., the irrep $[\nu]$ and its contragredient rep are equivalent. Thus for the SU_5 group, we have $[\nu] = [\nu'] = [4321], [\nu] = [\nu'] = [4222], [\nu] = [\nu'] = [21^3]$, etc.

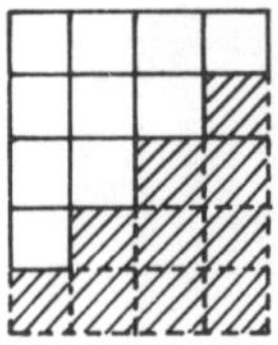

$[\nu] = [\nu'] = [4321]$,

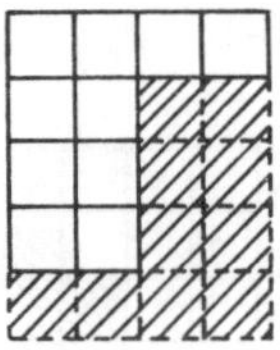

$[\nu] = [\nu'] = [4222]$

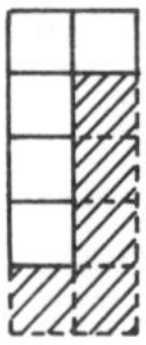

$[\nu] = [\nu'] = [2111]$

It is easily seen that the adjoint rep $[\nu_0] = [21^{n-1}]$ of SU_n obeys the condition (7-86). Therefore $D^{[\nu_0]} = D^{[\nu_0']} = (\tilde{D}^{[\nu_0]})^{-1} = (D^{[\nu_0]})^*$, i.e., the adjoint rep is equivalent to a real rep. The adjoint rep [2] of SU_2 (corresponding to $J = 1$), for instance, is equivalent to the real rep $\mathcal{D}(\alpha\beta\gamma)$ of (5-34).

It follows from (7-82) that for the SU_3 group, the contragredient rep of the irrep $(\lambda\mu)$ is $(\mu\lambda)$.

7.7. The CG Coefficients of SU$_n$ Group

Suppose $[\nu_1]$ and $[\nu_2]$ are two irreps of SU_n. Their CG series is denoted by

$$[\nu_1] \times [\nu_2] = \sum_{\nu} \{\nu_1\nu_2\nu\}[\nu] \ . \tag{7-87}$$

The two irreducible basis $\left|\begin{matrix}[\nu_1]\\ W_1\end{matrix}\right\rangle$ and $\left|\begin{matrix}[\nu_2]\\ W_2\end{matrix}\right\rangle$ can be coupled to another irreducible basis $\left|\begin{matrix}[\nu]_\tau\\ W\end{matrix}\right\rangle$ of SU_n by means of the CG coefficients

$$\left|\begin{matrix}[\nu]_\tau\\ W\end{matrix}\right\rangle = \sum_{W_1W_2} C^{\nu_\tau W}_{\nu_1W_1,\nu_2W_2}\left|\begin{matrix}[\nu_1]\\ W_1\end{matrix}\right\rangle\left|\begin{matrix}[\nu_2]\\ W_2\end{matrix}\right\rangle ,$$
$$\tau = 1, 2, \ldots, \{\nu_1\nu_2\nu\} \ , \tag{7-88a}$$

where W_1, W_2 and W are component indices.

The CG coefficients of SU_n satisfy the unitarity condition

$$\sum_{W_1W_2} C^{\nu_\tau W}_{\nu_1W_1,\nu_2W_2} C^{\nu'_{\tau'}W'}_{\nu_1W_1,\nu_2W_2} = \delta_{\nu\nu'}\delta_{\tau\tau'}\delta_{WW'} \ , \tag{7-89a}$$

$$\sum_{\nu\tau W} C^{\nu_\tau W}_{\nu_1W_1,\nu_2W_2} C^{\nu_\tau W}_{\nu_1W_1',\nu_2W_2'} = \delta_{W_1W_1'}\delta_{W_2W_2'} \ . \tag{7-89b}$$

The inverse of (7-88a) is

$$\left|\begin{matrix}[\nu_1]\\ W_1\end{matrix}\right\rangle\left|\begin{matrix}[\nu_2]\\ W_2\end{matrix}\right\rangle = \sum_{\nu\tau W} C^{\nu_\tau W}_{\nu_1W_1,\nu_2W_2}\left|\begin{matrix}[\nu]_\tau\\ W\end{matrix}\right\rangle \ . \tag{7-88b}$$

7.7.1. *The CG coefficients of* U$_n$ *and the IDC of the permutation group*

The CG coefficients of a group depend on the particular basis we used. In this section we restrict ourselves to the CG coefficients for the Gel'fand basis of SU_n. The CG coefficients of SU_n for other kinds of basis will be discussed in Sec. 7.13.

The problem of the CG coefficients of SU_2 was solved a long time ago by Wigner (see Rose 1957), which are the well known coupling coefficients of angular momentum. The CG coefficients of SU_n with $n \geq 3$ are usually calculated by Racah's infinitesimal operator method (Racah 1951). Systematic tables are available only for the CG coefficients of SU_3 (de Swart 1963) and SU_4 (Rabl 1975, Haacke 1976). Biedenharn (1967) discussed a canonical definition of the CG coefficients for U_n.

As n gets larger, the calculation of the SU_n CG coefficients by Racah's method become increasingly unwieldly. In the following, we are going to introduce a new method for evaluating the CG coefficients of the SU_n Gel'fand basis from the IDC of the permutation group (Chen [25], 1978). The distinguished advantage of the new method is that the calculation is rank independent. We first discuss the U_n CG coefficients.

1. The case without identical single particle states in the Weyl tableaux. The outer-product in Sec. 4.14 refers to the coordinate- permutation group S_f. We now generalize it to the state-permutation group $\mathcal{S}_f$. First we assume that the f single particle states are all distinct. In such a case, the groups $\mathcal{S}_f$ and S_f are on the same footing. Interpreting the indices of the particle coordinates as indices of the single particle states, the standard basis $|Y_m^{(\nu)\tau}(\omega^0)\rangle$ and $|Y_{m_i}^{(\nu_i)}(\omega_i)\rangle$ of the permutation groups become the Gel'fand bases $\left|\begin{smallmatrix}[\nu]\tau\\ W\end{smallmatrix}\right\rangle$ and $\left|\begin{smallmatrix}[\nu_i]\\ W_i\end{smallmatrix}\right\rangle$; therefore (4-141) can be rewritten as

$$\left|\begin{matrix}[\nu]_\tau\\ W\end{matrix}\right\rangle = \sum_{m_1\omega_1 m_2\omega_2} C^{\nu_\tau m}_{\nu_1 m_1\omega_1,\nu_2 m_2\omega_2}\left|\begin{matrix}[\nu_1]\\ W_1\end{matrix}\right\rangle\left|\begin{matrix}[\nu_2]\\ W_2\end{matrix}\right\rangle . \tag{7-90}$$

Comparing (7-88a) with (7-90), we have

$$C^{\nu_\tau m}_{\nu_1 m_1\omega_1,\nu_2 m_2\omega_2} = C^{\nu_\tau W}_{\nu_1 W_1,\nu_2 W_2} , \tag{7-91}$$

i.e., the IDC of S_f are just the CG coefficients of the U_f Gel'fand bases without identical single particle states in their Weyl tableaux. Thus the IDC tables are also the tables of the CG coefficients of U_f. By means of index substitutions, we can further obtain the CG coefficients of U_n with arbitrary $n \geq f$. Suppose that we have n single particle states $\alpha, \beta, \gamma, \ldots, \delta$ with the convention that the state are ordered as $\alpha < \beta < \gamma < \ldots < \delta$. A series of tables for the CG coefficients of U_n can be obtained by making the following index substitution for the Young tableaux $Y_m^{(\nu)}, Y_{m_1}^{(\nu_1)}$ and $Y_{m_2}^{(\nu_2)}$ in the IDC table of the permutation group $S_f (f \leq n)$:

$$1 \to i_1 , \quad 2 \to i_2 , \ldots , f \to i_f ,$$

where $i_1, i_2, \ldots, i_f$ are the f single particle states chosen arbitrarily from among the n single particle states $\alpha, \beta, \gamma, \ldots, \delta$ under the restriction that $i_1 < i_2 < i_3 < \ldots < i_f$. As an example, take $n = 6$. We have six single particle states $\alpha, \beta, \gamma, \delta, \varepsilon, \varphi$. $\binom{6}{4} = 15$ tables of the U_6 CG coefficients are obtained from the following index substitution for each IDC table in Table 4.17:

$$\begin{aligned}(1234) \to &(\alpha\beta\gamma\delta), (\alpha\beta\gamma\varepsilon), (\alpha\beta\gamma\varphi), (\alpha\beta\delta\varepsilon), (\alpha\beta\delta\varphi) ,\\ &(\alpha\beta\varepsilon\varphi), (\alpha\gamma\delta\varepsilon), (\alpha\gamma\delta\varphi), (\alpha\gamma\varepsilon\varphi), (\alpha\delta\varepsilon\varphi),\\ &(\beta\gamma\delta\varepsilon), (\beta\gamma\delta\varphi), (\beta\gamma\varepsilon\varphi), (\beta\delta\varepsilon\varphi), (\gamma\delta\varepsilon\varphi).\end{aligned}$$

Thus it can be seen that simply by index substitutions, each IDC table gives rise to an infinite number of CG coefficient tables of unitary groups.

2. The case with identical single particle states in the Weyl tableaux. If we let the coordinate indices in (4-142a) go over to the state indices according to the following:

$$\begin{aligned}(12\ldots f) &\to (i_1 i_2 \ldots i_f), \quad i_1 \leq i_2 \leq \ldots \leq i_f ,\\ (\omega^0) \to (\overline{\omega}^0) &= (i_1 i_2 \ldots i_f), \quad (\omega_1) = (a_1 \ldots a_{f_1}) \to (\overline{\omega}_1) = (i_{a_1} \ldots i_{a_{f_1}}) ,\\ (\omega_2) &= (a_{f_1+1} \ldots a_f) \to (\overline{\omega}_2) = (i_{a_{f_1+1}} \ldots i_{a_f}) ,\end{aligned} \tag{7-92}$$

then the normal order sequences (ω^0) and (ω_i) go over to the normal order states $(\overline{\omega}^0)$ and $(\overline{\omega}_i)$, respectively.

It can be shown [see (5-12) in Chen [31] 1986] that (4-141) still holds when there are identical single particle states in the normal order state $(\overline{\omega}^0) = (i_1 i_2 \ldots i_f)$, namely

$$P^{[\nu]_\tau m}|\overline{\omega}^0\rangle = \sum_{\substack{m_1 m_2 \\ \omega_1 \omega_2}} C^{\nu_\tau m}_{\nu_1 m_1 \omega_1, \nu_2 m_2 \omega_2} Q_\omega P^{[\nu_1]m_1}|i_1 \ldots i_{f_1}\rangle P^{[\nu_2]m_2}|i_{f_1+1} \ldots i_f\rangle \,. \tag{7-93}$$

With identical single particle states, $|Y_m^{[\nu]\tau}(\overline{\omega}^0)\rangle$ and $|Y_{m_i}^{[\nu_i]}(\overline{\omega}_i)\rangle$ are the unnormalized Gel'fand bases of U_n. From (4-81a) and (4-86) we have

$$|Y_m^{[\nu]}(\overline{\omega}^0)\rangle = R^{[\nu]m}(\overline{\omega}^0)\left|\begin{matrix}[\nu] \\ W\end{matrix}\right\rangle \tag{7-94a}$$

$$|Y_{m_i}^{[\nu_i]}(\overline{\omega}_i)\rangle = R^{[\nu_i]m_i}(\overline{\omega}_i)\left|\begin{matrix}[\nu_i] \\ W_i\end{matrix}\right\rangle \,. \tag{7-94b}$$

It must be stressed that in (7-94a) there may be several m which give rise to the same Weyl tableau W, and in (7-94b) there may be several m_i which give rise to the same Weyl tableaux W_i. A Gel'fand basis vector vanishes whenever there are identical single particle states appearing in the same column of the Weyl tableau.

Substituting (7-94) into (7-93) and making use of (7-88), we get the general relation between the CG coefficients of U_n and the IDC of the permutation groups,

$$C^{\nu_\tau W}_{\nu_1 W_1, \nu_2 W_2} = \frac{1}{R^{[\nu]m}(\overline{\omega}^0)} \sum_{\substack{m_1 m_2 \\ \omega_1 \omega_2}}{}' R^{[\nu_1]m_1}(\overline{\omega}_1) R^{[\nu_2]m_2}(\overline{\omega}_2) C^{\nu_\tau m}_{\nu_1 m_1 \omega_1, \nu_2 m_2 \omega_2} \,, \tag{7-95}$$

where the prime in the summation symbol indicates that the sum runs over only those m_i which will give rise to the same Weyl tableaux W_i, and the relations between $(\overline{\omega}^0)$, $(\overline{\omega}_i)$ and (ω^0), (ω_i) are specified by (7-92).

When there is no identical single particle states in $(\overline{\omega}^0)$, the correspondence between $m_i \overline{\omega}_i$ and W_i is one-to-one, and only one term at the right-hand sides of (7-95) survives, therefore (7-95) reduces to the special case (7-91).

An important conclusion reached from the above discussion is that the *value of the U_n CG coefficients do not depend explicitly on n*, and once the IDC are known, with the help of some universal (and positive) normalization constants $R^{[\nu]m}(\omega)$, which can be evaluated easily by the method of Chen ([31] 1987), the CG coefficients of any U_n group can be found from (7-95) with one stroke, without the need of calculating them for each n at a time.

Note that the left-hand side of (7-95) is independent of the quantum number m. Therefore in practical calculations, m can take any possible value as long as (7-94a) is satisfied.

The relation (7-95) between the CG coefficients of U_n and the IDC of the permutation group also tells us about the CG series of SU_n. Namely, what irreps $[\nu]$ are contained in the product of two irreps $[\nu_1]$ and $[\nu_2]$ of SU_n is determined by the Littlewood rule under the restriction that one has to disregard the Young diagrams $[\nu]$ of more than n rows and delete the columns of length n in any Young diagram $[\nu]$.

For example, from (4-136) we get the CG series for the groups U_3 and SU_3 as follows:

$$\begin{aligned} U_3 &: [21] \times [21] = [42] + [411] + [33] + 2[321] + [222] \,, \\ SU_3 &: [21] \times [21] = [42] + [3] + [33] + 2[21] + [0] \,. \end{aligned} \tag{7-96}$$

As a dimension check[2], $8 \times 8 = 27 + 10 + 1 + 2 \times 8 + 1$. Similarly, for SU_4 we have

$$SU_4 : [21] \times [21] = [42] + [411] + [33] + 2[321] + [222] + [2] + [11] \,. \tag{7-97}$$

We note again that $20 \times 20 = 126 + 70 + 50 + 2 \times 64 + 10 + 10 + 6$.

7.7.2. The procedure for evaluating the SU_n CG coefficients

1. Pick out the typical normal order states. Suppose the single particle states $\alpha, \beta, \gamma, \ldots, \varepsilon$ span the fundamental rep of U_n. There are altogether $\binom{n+f-1}{f}$ normal order states $|\bar{\omega}^0\rangle = |i_1, i_2 \ldots i_f\rangle, i = \alpha, \beta, \gamma, \ldots, \varepsilon$. We only need to calculate the CG coefficients for some typical normal order states. Thus for $f = 4$ and $n = 4$, there are 35 normal order states, while the typical ones are only the following eight states:

$$|\alpha\beta\gamma\delta\rangle\,, \quad |\alpha\alpha\beta\gamma\rangle\,, \quad |\alpha\beta\beta\gamma\rangle\,, \quad |\alpha\beta\gamma\gamma\rangle\,, \quad |\alpha\alpha\beta\beta\rangle\,, \quad |\alpha\alpha\alpha\beta\rangle\,, \quad |\alpha\beta\beta\beta\rangle\,, \quad |\alpha\alpha\alpha\alpha\rangle\,.$$

The CG coefficients for the first case are simply the IDC; the CG coefficients for the last case are trivial, while the CG coefficients for cases 5, 6 and 7, are just the SU_2 CG coefficients which have been tabulated (Rotenberg 1959). Therefore only the CG coefficients for the three normal order states $|\bar{\omega}^0\rangle = |\alpha\alpha\beta\gamma\rangle, |\alpha\beta\beta\gamma\rangle$ and $|\alpha\beta\gamma\gamma\rangle$ need to be calculated.

Once the CG coefficients for the above typical cases are known, all the CG coefficients of any U_n group for four-particle system are obtained merely through index substitutions. For example, the CG coefficients of U_5 can be obtained from those of U_4 by setting

$$\begin{aligned} |\bar{\omega}^0\rangle &= |\alpha\beta\gamma\delta\rangle = |1234\rangle,\ |1235\rangle,\ |1245\rangle,\ |1345\rangle,\ |2345\rangle\,, \\ |\bar{\omega}^0\rangle &= \ |\alpha\alpha\beta\gamma\rangle = |1123\rangle,\ |1124\rangle,\ |2234\rangle,\ \ldots\ \ |3345\rangle\,. \end{aligned} \tag{7-98}$$

2. Change the headings in the IDC tables. By making the following substitutions $1 \to i_1, 2 \to i_2, \ldots, f \to i_f$, the Young tableaux in the headings of the IDC tables become the Weyl tableaux:

$$Y_m^{[\nu]}(\bar{\omega}^0) \to W\,, \quad Y_{m_i}^{[\nu_i]}(\bar{\omega}_i) \to W_i\,. \tag{7-99}$$

3. Divide the entries in the row $[\nu]m$ of the IDC table with $R^{[\nu]m}(\bar{\omega}^0)$, and multiply the entries in the column labeled $m_1\omega_1 m_2\omega_2$ with $R^{[\nu_1]m_1}(\bar{\omega}_1)R^{[\nu_2]m_2}(\bar{\omega}_2)$. The norms $R^{[\nu]m}(\omega)$ for $S_f, f \leq 5$, are listed in Table 4.8.

4. Finally the CG coefficients are obtained by summing up the entries under the same heading (W_1, W_2).

To illustrate, Table 7.7-1a gives the IDC of S_3 (the table is taken from Table 4.17-2). By letting $(\bar{\omega}^0) = (\alpha\alpha\beta)$, the Young tableaux in Table 7.7-1a go over to the Weyl tableaux in Table 7.7-1b, where in the third row there are two $\alpha's$ in the same column so that $\left|{}^{\alpha\beta}_{\alpha}\right\rangle = 0$.

Dividing the first and second row in Table 7.7-1a with $R^{[3]}(\alpha\alpha\beta) = R^{[21]}(\alpha\alpha\beta) = \sqrt{2}$, and multiplying the first, second and third columns by $R^{[2]}(\alpha\alpha) = \sqrt{2}, R^{[2]}(\alpha\beta) = 1$ and $R^{[2]}(\alpha\beta) = 1$, respectively, we get Table 7.7-1b. The CG coefficients of U_n result from adding thé second and third columns in Table 7.7-1b and are shown in Table 7.7-1c.

[2]The dimension tables are given in Table A.1 in the Appendix.

Table 7.7-1a. [2]× [1] IDC of S_3.

	[1 2] [3]	[1 3] [2]	[2 3] [1]
[1 2 3]	$\frac{1}{\sqrt{3}}$	$\frac{1}{\sqrt{3}}$	$\frac{1}{\sqrt{3}}$
[1 2 / 3]	$\sqrt{\frac{2}{3}}$	$-\frac{1}{\sqrt{6}}$	$-\frac{1}{\sqrt{6}}$
[1 3 / 2]	0	$\frac{1}{\sqrt{2}}$	$-\frac{1}{\sqrt{2}}$

Table 7.7-1b. The intermediate step.

	[α α] [β]	[α β] [α]	[α β] [α]
[α α β]	$\sqrt{\frac{1}{3}}$	$\sqrt{\frac{1}{6}}$	$\sqrt{\frac{1}{6}}$
[α α / β]	$\sqrt{\frac{2}{3}}$	$-\sqrt{\frac{1}{12}}$	$-\sqrt{\frac{1}{12}}$
[α β / α]			

Table 7.7-1c. The CG coefficients of U_n.

	[α α] [β]	[α β] [α]
[α α β]	$\sqrt{\frac{1}{3}}$	$\sqrt{\frac{2}{3}}$
[α α / β]	$\sqrt{\frac{2}{3}}$	$-\sqrt{\frac{1}{3}}$

As can be seen, the coefficients in Table 7.7-1c are precisely the SU_2 CG coefficients (including the phase).

Analogously, from the [21]×[1] IDC in Table 4.17-3d, we find the CG coefficients of U_n given in Table 7.7-2.

Table 7.7-2. [21]× [1] CG coefficients of U_n.

[21]⊗[1]

	$\alpha\alpha/\beta$, γ	$\alpha\alpha/\gamma$, β	$\alpha\beta/\gamma$, α	$\alpha\gamma/\beta$, α
$\alpha\alpha\beta/\gamma$	0	$\sqrt{\frac{1}{3}}$	$\sqrt{\frac{2}{3}}$	0
$\alpha\alpha\gamma/\beta$	$\sqrt{\frac{3}{8}}$	$\sqrt{\frac{1}{24}}$	$-\sqrt{\frac{1}{48}}$	$\frac{3}{4}$
$\alpha\alpha/\beta/\gamma$	$\sqrt{\frac{3}{8}}$	$-\sqrt{\frac{3}{8}}$	$\frac{\sqrt{3}}{4}$	$-\frac{1}{4}$
$\alpha\alpha/\beta\gamma$	$\frac{1}{2}$	$\frac{1}{2}$	$-\sqrt{\frac{1}{8}}$	$-\sqrt{\frac{3}{8}}$

[21]⊗[1]

	$\alpha\beta/\beta$, γ	$\alpha\beta/\gamma$, β	$\beta\beta/\gamma$, α	$\alpha\gamma/\beta$, β
$\alpha\beta\beta/\gamma$	0	$\sqrt{\frac{2}{3}}$	$\sqrt{\frac{1}{3}}$	0
$\alpha\beta\gamma/\beta$	$\sqrt{\frac{3}{8}}$	$\sqrt{\frac{1}{48}}$	$-\sqrt{\frac{1}{24}}$	$\frac{3}{4}$
$\alpha\beta/\beta/\gamma$	$\sqrt{\frac{3}{8}}$	$-\frac{\sqrt{3}}{4}$	$\sqrt{\frac{3}{8}}$	$-\frac{1}{4}$
$\alpha\beta/\beta\gamma$	$\frac{1}{2}$	$\sqrt{\frac{1}{8}}$	$-\frac{1}{2}$	$-\sqrt{\frac{3}{8}}$

[21]⊗[1]

	$\alpha\beta/\gamma$, γ	$\alpha\gamma/\beta$, γ	$\alpha\gamma/\gamma$, β	$\beta\gamma/\gamma$, α
$\alpha\beta\gamma/\gamma$	$\sqrt{\frac{1}{2}}$	0	$\frac{1}{2}$	$\frac{1}{2}$
$\alpha\gamma\gamma/\beta$	0	$\frac{\sqrt{3}}{2}$	$\sqrt{\frac{1}{8}}$	$-\sqrt{\frac{1}{8}}$
$\alpha\gamma/\beta/\gamma$	0	$\frac{1}{2}$	$-\sqrt{\frac{3}{8}}$	$\sqrt{\frac{3}{8}}$
$\alpha\beta/\gamma\gamma$	$\sqrt{\frac{1}{2}}$	0	$-\frac{1}{2}$	$-\frac{1}{2}$

7.7.3 Phase convention

1. The relative phases of the IDC of S_f and CG coefficients of U_n. The Young-Yamanouchi phase convention for the IDC ensures that the CG coefficients of U_n calculated from (7-95) fulfil the Gel'fand-Biedenharn phase convention (see Sec. 7.5).

2. The overall phase. By (7-95), the overall phases of the CG coefficients of SU_n and the IDC of the permutation group are closely related. The overall phase convention (4-148) for the IDC ensures that the CG coefficients of SU_n derived from (7-95) satisfy the Baird-Biedenharn (1965) phase convention which demands that the CG coefficient associated with the highest weight (HW) be positive, i.e.,

$$C^{\nu_\tau HW}_{\nu_1 HW_1,\nu_2 W_2} > 0 \ . \tag{7-100}$$

The highest weight state $|[\nu]HW\rangle$ is defined as the state formed by filling the Young diagram $[\nu]$ in the following way: The boxes in the first row are all filled with 1 (or α), and the boxes in the second row with 2 (or β), etc.

The convention (7-100) is a generalization of the Condon-Shortley phase convention $C^{JJ}_{j_1j_1,j_2J-j_1} > 0$, for the CG coefficients of SU_2. The phase convention (7-100) ensures that *our U_n CG coefficients are also the SU_n CG coefficients* (Chen [31] 1987).

From (7-95) and (4-153a) it is known that the CG coefficients of SU_n has the symmetry

$$C^{\nu_\tau W}_{\nu_1 W_1,\nu_2 W_2} = \varepsilon_1 C^{\nu_\tau,W}_{\nu_2 W_2,\nu_1 W_1} \ . \tag{7-101}$$

The phase factor ε_1 is given by (4-153b) and tabulated in Table A.2 in the Appendix.

The CG coefficients for the SU_n Gel'fand basis are tabulated in a rank independent way in Chen ([31], 1987) for partitions of integers up to six.

Ex. 7.5. Construct the CG series of SU_3 for $[31] \times [22]$.

Ex. 7.6. Construct the four-quark state $\left|\begin{matrix} uus \\ d \end{matrix}\right\rangle$ in terms of P, $\Sigma^+, \Sigma^0, \Lambda^0$ and a single quark state.

Ex. 7.7. Change the component indices and the Weyl tableaux in Table 7.7-2 to the quantum number $\mathrm{II_zY}$.

7.8 The CG Coefficients of SU_n and the $S_f \supset S_{f_1} \otimes S_{f_2}$ Irreducible Basis

Let

$$\left|\begin{matrix} [\nu_i] \\ m_i\omega_i^0, \ W_i \end{matrix}\right\rangle = |Y^{[\nu_i]}_{m_i}(\omega_i^0), W_i\rangle \ ,$$

$$(\omega_1^0) = (1, 2, \ldots, f_1) \ , \quad (\omega_2^0) = (f_1+1, \ldots, f_1+f_2) \ , \tag{7-102a}$$

be the irreducible basis $[\nu_i]m_i$ of the permutation group $S_{f_i} = S_{f_i}(\omega_i^0)$ and the irreducible basis $[\nu_i]W_i$ of the unitary group SU_n.

Theorem 7.3: The CG coefficients of SU_n are the indirect coupling coefficients for $S_f \supset S_{f_1} \otimes S_{f_2}$ irreducible basis ($f = f_1 + f_2$), namely,

$$\left|\begin{matrix} [\nu] \\ \theta[\nu_1]m_1[\nu_2]m_2, \ W \end{matrix}\right\rangle = \sum_{W_1W_2} C^{[\nu]_\theta,W}_{\nu_1 W_1,\nu_2 W_2} \left|\begin{matrix} [\nu_1] \\ m_1\omega_1^0, \ W_1 \end{matrix}\right\rangle \left|\begin{matrix} [\nu_2] \\ m_2\omega_2^0, \ W_2 \end{matrix}\right\rangle \ . \tag{7-102b}$$

Proof: According to the definition of the CG coefficients of SU_n, the right-hand sides of (7-102b) belongs to the irrep $[\nu]$ of SU_n, and in view of Theorem 7.1 it must also belong to the irrep $[\nu]$ of S_f. Besides, the quantum numbers $[\nu_i]m_i$ are fixed in (7-102b), therefore (7-102b) remains the irreducible basis $[\nu_1]m_1$ of S_{f_1} and $[\nu_2]m_2$ of S_{f_2}. In other words, (7-102b) is the irreducible basis $|\nu W\rangle$ of SU_n and the $S_f \supset S_{f_1} \otimes S_{f_2}$ irreducible basis.

Using (4-168b), we can transfer the nonstandard basis of S_f to the standard basis of S_f:

$$\left| \begin{matrix} [\nu] \\ m, W \end{matrix} \right\rangle = \sum_{\nu_2 m_2 \theta W_1 W_2} \left\langle \begin{matrix} [\nu] \\ m \end{matrix} \right| [\nu], \begin{matrix} \theta[\nu_1][\nu_2] \\ m_1 \quad m_2 \end{matrix} \bigg\rangle C^{[\nu]_\theta, W}_{\nu_1 W_1, \nu_2 W_2} \left| \begin{matrix} [\nu_1] \\ m_1 \omega_1^0, W_1 \end{matrix} \right\rangle \left| \begin{matrix} [\nu_2] \\ m_2 \omega_2^0, W_2 \end{matrix} \right\rangle . \qquad (7\text{-}102c)$$

7.9. The $\mathbf{SU}_{mn} \supset \mathbf{SU}_m \times \mathbf{SU}_n$ Irreducible Basis

7.9.1. *The CG coefficients of S_f and the $SU_{mn} \supset SU_m \times SU_n$ irreducible basis*

A particle usually has several kinds of degrees of freedom. For example, a nucleon has the degrees of freedom in the orbital space V^x, spin space V^σ and isospin space V^τ. A nucleon with orbital angular momentum l has altogether $n = 4(2l+1)$ states.

In many cases, the permutation symmetry $[\nu]$ in orbital space, the total spin S and the total isospin T are good, or approximately good quantum numbers (for example, when the nuclear force is the Serber force, $V_{ij} = v(r_{ij})(W + M^x p_{ij})$ where W and M are constants, and ${}^x p_{ij}$ is the permutation operator in orbital space). Therefore it is most convenient to use basis functions with definite space symmetry $[\nu]$, definite spin S and isospin T, namely, we need $SU_{4(2l+1)} \supset SU_{2l+1} \times SU_2 \times SU_2$ irreducible basis, rather than the Gel'fand basis classified according to the group chain $SU_n \supset U_{n-1} \supset \ldots \supset U_2 \supset U_1$. In this section, we will give a general method for constructing such a type of basis. We begin with two types of degrees of freedom, i.e., consider the $SU_{mn} \supset SU_m \times SU_n$ irreducible basis. For example the $SU_4 \supset SU_2 \times SU_2$ in nuclear physics, the $SU_6 \supset SU_3^{(f)} \times SU_2^{(\sigma)}$ or $SU_8 \supset SU_4^{(f)} \times SU_2^{(\sigma)}$ in particle physics, where f denotes flavor space.

Consider the following three kinds of f-particle product states:

$$x-\text{space:} \quad \varphi = \varphi_{i_1}(x_1)\ldots\varphi_{i_f}(x_f), \quad i_1 \ldots i_f = 1, 2, \ldots m , \qquad (7\text{-}103a)$$

$$\xi-\text{space:} \quad x = \chi_{\alpha_1}(\xi_1)\ldots\chi_{\alpha_f}(\xi_f), \quad \alpha_1 \ldots \alpha_f = 1, 2, \ldots n , \qquad (7\text{-}103b)$$

$$q-\text{space:} \quad \psi = \psi_{s_1}(q_1)\ldots\psi_{s_f}(q_f), \quad s = (i,\alpha) = (1,1), \ldots (1,n), (2,1), \ldots (m,n) , \qquad (7\text{-}103c)$$

where

$$q = (x, \xi), \quad \psi_s(q) = \varphi_i(x)\chi_\alpha(\xi) ,$$

$$\begin{pmatrix} \varphi_1 \ldots \varphi_m \\ \chi_1 \ldots \chi_n \\ \psi_{1,1} \ldots \psi_{m,n} \end{pmatrix} \text{ carry the fundamental rep of } \begin{pmatrix} SU_m \\ SU_n \\ SU_{mn} \end{pmatrix} .$$

To avoid repetitive statements, here we have used a straightforward shorthand notation.

For example, in the SU_6 model of elementary particles, x represents flavor space with φ_1, φ_2 and φ_3 representing the quark states u, d and s, while ξ represents spin space, $\chi_1(\chi_2)$ being the spin-up (spin-down) state $\chi_\alpha(\chi_\beta)$, and in q space, $s = (u\alpha), (u\beta), (d\alpha), (d\beta), (s\alpha), (s\beta)$ carry the fundamental rep of SU_6.

The group elements of SU_m, SU_n and SU_{mn} consist of the following matrices, respectively,

$$U_{ij}(a^1 \ldots a^{r_m}), \quad U_{\alpha\beta}(b^1 \ldots b^{r_n}), \quad U_{i\alpha,j\beta}(c^1 \ldots c^r) ,$$
$$r_m = m^2 - 1, \quad r_n = n^2 - 1, \quad r = (mn)^2 - 1 , \qquad (7\text{-}104a)$$

where r_m, r_n and r are orders of the three groups. $U_{ij}, U_{\alpha\beta}$ and $U_{i\alpha,j\beta}$ are $m\times m, n\times n$ and $mn\times mn$ matrices, respectively. Notice that the group $SU_m\times SU_n$ consists of the unitary matrices

$$U_{i\alpha,j\beta}(a^1\ldots a^{r_m}, b^1\ldots b^{r_n}) = U_{ij}(a^1\ldots a^{r_m})U_{\alpha\beta}(b^1\ldots b^{r_n}) \tag{7-104b}$$

with the order $r_m + r_n = m^2+n^2-2$. Obviously, $SU_m\times SU_n$ is a subgroup of SU_{mn}.

The f-particle product states $\begin{pmatrix}\varphi\\ \chi\\ \psi\end{pmatrix}$ carry a reducible rep $\begin{pmatrix}S_f(x)\\ S_f(\xi)\\ S_f(q)\end{pmatrix}$ as well as of $\begin{pmatrix}SU_m\\ SU_n\\ SU_{mn}\end{pmatrix}$.

Assume that

$$\varphi^{[\nu_1]}_{m_1}(x, W_1) = \left|\begin{matrix}[\nu_1]\\ m_1, W_1\end{matrix}\right\rangle \tag{7-105a}$$

is the standard basis $[\nu_1]m_1$ of $S_f(x)$ and the irreducible basis $[\nu_1]W_1$ of SU_m, and let

$$\chi^{[\nu_2]}_{m_2}(\xi, W_2) = \left|\begin{matrix}[\nu_2]\\ m_2\ , W_2\end{matrix}\right\rangle \tag{7-105b}$$

be the standard basis $[\nu_2]m_2$ of $S_f(\xi)$ and the irreducible basis $[\nu_2]W_2$ of SU_n, where W_1 and W_2 are component indices of the irreps of SU_m and SU_n, respectively. For clarity, we assume that they are labels for the Gel'fand bases of SU_m and SU_n, although the following discussion is independent of classification of the irreducible bases of SU_m and SU_n. Therefore W_1 and W_2 may represent two Weyl tableaux filled with the state indices $i_1, i_2, \ldots, i_f$ and $\alpha_1, \alpha_2\ldots, \alpha_f$, respectively. To illustrate, we have

$$\left|\begin{matrix}[\nu_1]\\ m_1=1, W_1\end{matrix}\right\rangle = \left|\begin{array}{|c|c|}\hline u & d\\ \hline s & \\ \cline{1-1}\end{array}\right\rangle_1 = \left|\begin{array}{|c|c|}\hline 1 & 2\\ \hline 3 & \\ \cline{1-1}\end{array}\ \begin{array}{|c|c|}\hline u & d\\ \hline s & \\ \cline{1-1}\end{array}\right\rangle,$$

$$\left|\begin{matrix}[\nu_2]\\ m_2=2, W_2\end{matrix}\right\rangle = \left|\begin{array}{|c|c|}\hline \alpha & \alpha\\ \hline \beta & \\ \cline{1-1}\end{array}\right\rangle_2 = \left|\begin{array}{|c|c|}\hline 1 & 3\\ \hline 2 & \\ \cline{1-1}\end{array}\ \begin{array}{|c|c|}\hline \alpha & \alpha\\ \hline \beta & \\ \cline{1-1}\end{array}\right\rangle.$$

where the state $\left|\begin{matrix}[\nu_1]\\ m_1\ W_1\end{matrix}\right\rangle$ and $\left|\begin{matrix}[\nu_2]\\ m_2\ W_2\end{matrix}\right\rangle$ are linear combinations of the product states φ and χ, respectively.

In terms of the CG coefficients of permutation groups, (7-105) can be coupled into the Yamanouchi basis $[\nu]m$ of the group $S_f(q)$

$$\left|\begin{matrix}[\nu]\\ m, \beta[\nu_1]W_1[\nu_2]W_2\end{matrix}\right\rangle = \left|\begin{matrix}[\nu]\beta[\nu_1][\nu_2]\\ m,\quad W_1W_2\end{matrix}\right\rangle = \sum_{m_1m_2} C^{\nu_\beta m}_{\nu_1m_1,\nu_2m_2}\left|\begin{matrix}[\nu_1]\\ m_1W_1\end{matrix}\right\rangle\left|\begin{matrix}[\nu_2]\\ m_2W_2\end{matrix}\right\rangle,$$

$$\beta = 1, 2, \ldots, (\nu_1\nu_2\nu)\ . \tag{7-106a}$$

We now are going to prove that (7-106a) is the $[\nu]$ irreducible basis of SU_{mn}. According to the definition of the CG coefficients, (7-106a) is an eigenfunction of the CSCO-I $C(f)$ of $S_f(q)$. On account of (7-37), the invariants $I^{(mn)}_k$ of the group SU_{mn} for the basis (7-103c) are functions of the operator set $C(f)$ of $S_f(q)$:

$$I^{(mn)}_k = F^{(mn)}_k(C(f)), \quad k = mn, mn-1, \ldots, 3, 2\ . \tag{7-107}$$

Equation (7-107) remains true for the basis (7-106a), since (7-106a) are linear combinations of (7-103c). Therefore (7-106a) are also eigenfunctions of the CSCO $\{I^{(mn)}_k\}$ of SU_{mn}, which in turn means that (7-106a) are the bases of the irrep $[\nu]$ of SU_{mn}. Furthermore, since ν_1W_1 and ν_2W_2 in the right-hand side of (7-106a) are fixed, the left-hand side remains the irreducible basis of SU_m and SU_n. Therefore

$$\left|\begin{matrix}[\nu], \beta[\nu_1][\nu_2]\\ m,\quad W_1W_2\end{matrix}\right\rangle \text{ belongs to } \begin{matrix}SU_{mn} & \supset & SU_m & \times & SU_n, & S_f(q)\\ [\nu] & & [\nu_1]W_1 & & [\nu_2]W_2 & [\nu]m\ .\end{matrix}$$

Thus we have

Theorem 7.4: The CG coefficients of the permutation group are the indirect coupling coefficients for the $SU_{mn} \supset SU_m \times SU_n$ irreducible basis.

Consequently, by utilizing the CG coefficients of the permutation group, we can easily construct the $SU_{mn} \supset SU_m \times SU_n$ irreducible bases for arbitrary m and n. Thus from (4-124) and (7-106a), we obtain the $SU_{mn} \supset SU_m \times SU_n$ basis for the totally anti-symmetric rep of $S_f(q)$ and SU_{mn},

$$\left|[1^f],\, \begin{matrix}[\nu_1][\tilde{\nu}_1] \\ W_1 W_2\end{matrix}\right\rangle = \sum_{m_1} \frac{\Lambda^{\nu_1}_{m_1}}{\sqrt{h_{\nu_1}}} \left|\begin{matrix}[\nu_1] \\ m_1 W_1\end{matrix}\right\rangle \left|\begin{matrix}[\tilde{\nu}_1] \\ \tilde{m}_1 W_2\end{matrix}\right\rangle . \tag{7-108}$$

Using the unitarity of the CG coefficients, (7-106a) can be inverted,

$$\left|\begin{matrix}[\nu_1] \\ m_1, W_1\end{matrix}\right\rangle \left|\begin{matrix}[\nu_2] \\ m_2, W_2\end{matrix}\right\rangle = \sum_{\nu\beta m} C^{\nu_\beta m}_{\nu_1 m_1, \nu_2 m_2} \left|\begin{matrix}[\nu], \beta[\nu_1][\nu_2] \\ m \quad W_1 W_2\end{matrix}\right\rangle . \tag{7-106b}$$

Utilizing the CG coefficients given in Table 4.13-1, the $SU_{mn} \supset SU_m \times SU_n$ (such as $SU_6 \supset SU_3 \times SU_2, SU_8 \supset SU_4 \times SU_2$, etc.) baryon wave functions in the quark model can be expressed in terms of the following general formulas:

$$\begin{aligned}
&\left|[3],\, \begin{matrix}[21]\,[21] \\ W_1\ W_2\end{matrix}\right\rangle = \sqrt{1/2}\left[\left|\begin{matrix}[21] \\ W_1\end{matrix}\right\rangle_1 \left|\begin{matrix}[21] \\ W_2\end{matrix}\right\rangle_1 + \left|\begin{matrix}[21] \\ W_1\end{matrix}\right\rangle_2 \left|\begin{matrix}[21] \\ W_2\end{matrix}\right\rangle_2\right], \\
&\left|[3],\, \begin{matrix}[3]\,[3] \\ W_1\ W_2\end{matrix}\right\rangle = \left|\begin{matrix}[3] \\ W_1\end{matrix}\right\rangle \left|\begin{matrix}[3] \\ W_2\end{matrix}\right\rangle, \\
&\left|[21],\, \begin{matrix}[3]\,[21] \\ W_1\ W_2\end{matrix}\right\rangle_m = \left|\begin{matrix}[3] \\ W_1\end{matrix}\right\rangle \left|\begin{matrix}[21] \\ W_2\end{matrix}\right\rangle_m, \quad \left|[21],\, \begin{matrix}[21]\,[3] \\ W_1\ W_2\end{matrix}\right\rangle_m = \left|\begin{matrix}[21] \\ W_1\end{matrix}\right\rangle_m \left|\begin{matrix}[3] \\ W_2\end{matrix}\right\rangle, \\
&\left|[21],\, \begin{matrix}[21]\,[21] \\ W_1\ W_2\end{matrix}\right\rangle_1 = \sqrt{1/2}\left[\left|\begin{matrix}[21] \\ W_1\end{matrix}\right\rangle_1 \left|\begin{matrix}[21] \\ W_2\end{matrix}\right\rangle_1 - \left|\begin{matrix}[21] \\ W_1\end{matrix}\right\rangle_2 \left|\begin{matrix}[21] \\ W_2\end{matrix}\right\rangle_2\right], \\
&\left|[21],\, \begin{matrix}[21]\,[21] \\ W_1\ W_2\end{matrix}\right\rangle_2 = -\sqrt{1/2}\left[\left|\begin{matrix}[21] \\ W_1\end{matrix}\right\rangle_1 \left|\begin{matrix}[21] \\ W_2\end{matrix}\right\rangle_2 + \left|\begin{matrix}[21] \\ W_1\end{matrix}\right\rangle_2 \left|\begin{matrix}[21] \\ W_2\end{matrix}\right\rangle_1\right], \\
&\left|[21],\, \begin{matrix}[1^3]\,[21] \\ W_1\ W_2\end{matrix}\right\rangle_m = \Lambda^{[21]}_m \left|\begin{matrix}[1^3] \\ W_1\end{matrix}\right\rangle \left|\begin{matrix}[21] \\ W_2\end{matrix}\right\rangle_{\tilde{m}} \\
&\left|[1^3],\, \begin{matrix}[1^3]\,[3] \\ W_1\ W_2\end{matrix}\right\rangle = \left|\begin{matrix}[1^3] \\ W_1\end{matrix}\right\rangle \left|\begin{matrix}[3] \\ W_2\end{matrix}\right\rangle, \\
&\left|[1^3],\, \begin{matrix}[21]\,[21] \\ W_1\ W_2\end{matrix}\right\rangle = \sqrt{1/2}\left[\left|\begin{matrix}[21] \\ W_1\end{matrix}\right\rangle_1 \left|\begin{matrix}[21] \\ W_2\end{matrix}\right\rangle_2 - \left|\begin{matrix}[21] \\ W_1\end{matrix}\right\rangle_2 \left|\begin{matrix}[21] \\ W_2\end{matrix}\right\rangle_1\right].
\end{aligned} \tag{7-109}$$

Here $m = 1, 2$ denotes the Young tableaux $\begin{array}{|c|c|}\hline 1 & 2 \\ \hline 3 \\ \cline{1-1}\end{array}$ and $\begin{array}{|c|c|}\hline 1 & 3 \\ \hline 2 \\ \cline{1-1}\end{array}$ respectively, and $\tilde{m}$ represents the conjugate Young tableau.

Example: Find the $SU_6 \supset SU_3 \times SU_2$ baryon wave functions in the flavor-spin space. We choose the isospin representation, $i = u, d, s$. Suppose we need to find the proton wave function with spin 1/2 and projection $-1/2$, belonging to the 56-dimensional irrep (i.e., $[\nu] = [3]$) of SU_6.

According to Table 7.5-2, the Weyl tableau of the proton is $\begin{array}{|c|c|}\hline u&u\\\hline d\\\cline{1-1}\end{array}$, and the Weyl tableau corresponding to spin 1/2 and projection $-1/2$ is $\begin{array}{|c|c|}\hline \alpha&\beta\\\hline \beta\\\cline{1-1}\end{array}$. Setting $W_1 = \begin{array}{|c|c|}\hline u&u\\\hline d\\\cline{1-1}\end{array}$ and $W_2 = \begin{array}{|c|c|}\hline \alpha&\beta\\\hline \beta\\\cline{1-1}\end{array}$ in the first equation of (7-109), we obtain the required wave function

$$
\begin{aligned}
|[3], P_{1/2-1/2}\rangle &= \left|[3], \begin{array}{|c|c|}\hline u&u\\\hline d\\\cline{1-1}\end{array}\ \begin{array}{|c|c|}\hline \alpha&\beta\\\hline \beta\\\cline{1-1}\end{array}\right\rangle \\
&= \sqrt{\frac{1}{2}}\left(\left|\begin{array}{|c|c|}\hline u&u\\\hline d\\\cline{1-1}\end{array}\right\rangle_1\left|\begin{array}{|c|c|}\hline \alpha&\beta\\\hline \beta\\\cline{1-1}\end{array}\right\rangle_1 + \left|\begin{array}{|c|c|}\hline u&u\\\hline d\\\cline{1-1}\end{array}\right\rangle_2\left|\begin{array}{|c|c|}\hline \alpha&\beta\\\hline \beta\\\cline{1-1}\end{array}\right\rangle_2\right).
\end{aligned}
\tag{7-110a}
$$

Using Tables 7.5-5 and 7.5-6, the above wave function can be expressed in terms of the product states

$$
\begin{aligned}
|[3], P_{1/2-1/2}\rangle = \sqrt{\frac{1}{2}}\Big[&\frac{1}{6}(2|uud\rangle - |udu\rangle - |duu\rangle)(|\alpha\beta\beta\rangle + |\beta\alpha\beta\rangle - 2|\beta\beta\alpha\rangle) \\
&+\frac{1}{2}(|udu\rangle - |duu\rangle)(|\alpha\beta\beta\rangle - |\beta\alpha\beta\rangle)\Big] \,.
\end{aligned}
\tag{7-110b}
$$

We also have

$$
|[3], \Delta^{+}_{3/2,1/2}\rangle = |[3]\,, \begin{array}{|c|c|c|}\hline u&u&d\\\hline\end{array}\ \begin{array}{|c|c|c|}\hline \alpha&\alpha&\beta\\\hline\end{array}\rangle = |\begin{array}{|c|c|c|}\hline u&u&d\\\hline\end{array}\rangle|\begin{array}{|c|c|c|}\hline \alpha&\alpha&\beta\\\hline\end{array}\rangle. \tag{7-110c}
$$

Analogously, from the fourth equation of (7-109), we get the wave function for the excited states of the proton with spin 3/2 and projection $-1/2$, belonging to the 70-dimensional irrep (i.e., $[\nu] = [21]$) of SU_6

$$
\left|\begin{matrix}[21]\\ m, P_{3/2-1/2}\end{matrix}\right\rangle = \left|\begin{array}{|c|c|}\hline u&u\\\hline d\\\cline{1-1}\end{array}\right\rangle_m\left|\begin{array}{|c|c|c|}\hline \alpha&\beta&\beta\\\hline\end{array}\right\rangle. \tag{7-111a}
$$

Letting the tableau W_1 in (7-109) be $\begin{array}{|c|c|}\hline u&u\\\hline c\\\cline{1-1}\end{array}$, $\begin{array}{|c|c|}\hline u&s\\\hline c\\\cline{1-1}\end{array}$, etc. we immediately get the $SU_8 \supset SU_4 \times SU_2$ wave function. Thus we have

$$
|[3], S^{+}_{1/2,1/2}\rangle = \sqrt{\frac{1}{2}}\left(\left|\begin{array}{|c|c|}\hline u&s\\\hline c\\\cline{1-1}\end{array}\right\rangle_1\left|\begin{array}{|c|c|}\hline \alpha&\alpha\\\hline \beta\\\cline{1-1}\end{array}\right\rangle_1 + \left|\begin{array}{|c|c|}\hline u&s\\\hline c\\\cline{1-1}\end{array}\right\rangle_2\left|\begin{array}{|c|c|}\hline \alpha&\alpha\\\hline \beta\\\cline{1-1}\end{array}\right\rangle_2\right) \,. \tag{7-111b}
$$

In the above we considered only the wave function of the proton in flavor-spin space. In orbital space, the three constituent quarks of the proton, which is the ground state of the three-quark system, must be in the totally symmetric state $\psi^{[3]}(x)$. If there were no other degrees of freedom, the total wave function of the proton would be $\psi^{[3]}(x) \times |[3], P_{\frac{1}{2}S_z}\rangle$, which is totally symmetric. However, the quarks are assumed to be fermions, and the total wave function has to be anti-symmetric. To eliminate this contradiction, one assumes that the quark has a new degree for freedom, called the color degree of freedom. Besides flavor, a quark also has one of the three *colors*, red, green or blue, denoted by r, g, b. The quarks r, g and b carry the fundamental rep of the SU_3 group in color space. One also assumes that observable baryons and mesons are all colorless, i.e., in the color singlet $|[0]\rangle^{(c)}$ of the color SU_3. For a three-quark system, the color singlet is

$$
|[0]\rangle^{(c)} = |[1^3]\rangle^{(c)} = \left|\begin{array}{|c|}\hline r\\\hline g\\\hline b\\\hline\end{array}\right\rangle \,. \tag{7-112a}
$$

Therefore, the totally anti-symmetric state of proton is

$$
|P_{\frac{1}{2}S_z}\rangle = \psi^{[3]}(x)|[1^3]\rangle^{(c)}|[3], P_{\frac{1}{2}S_z}\rangle \,, \tag{7-112b}
$$

while the totally anti-symmetric state of the particle Δ (the excited state of the nucleon) is

$$|\Delta_{\frac{3}{2}S_z}) = \psi^{[3]}(x)|[1^3]\rangle^{(c)}|[3], \Delta_{\frac{3}{2}S_z}\rangle. \tag{7-112c}$$

7.9.2. The irreps $([\nu_1],[\nu_2])$ of the groups SU_m and SU_2 contained in the irrep $[\nu]$ of SU_{mn}.

From (7-106) we know that the induction rule

$$([\nu_1]\times[\nu_2]) \uparrow SU_{mn} = \sum_{\nu}(\nu_1\nu_2\nu)D^{[\nu]}(SU_{mn}) , \tag{7-113a}$$

as well as the subduction rule

$$[\nu]\downarrow(SU_m \times SU_n) = \sum_{\nu}(\nu_1\nu_2\nu)(D^{[\nu_1]}(SU_m), D^{[\nu_2]}(SU_n)) , \tag{7-113b}$$

are governed by the CG series of the permutation group, so long as we ignore the entire columns of length mn in the Young diagram $[\nu]$ of (7-113a), and entire columns of length $m(n)$ in the Young diagram $[\nu_1]([\nu_2])$ of (7-113b).

For example, from the CG series (Table 4.10) and (4-90), we find the irreps $[\nu_1]$ and $[\nu_2]$ of SU_3 and SU_2 contained in the irrep [3] and [21] of SU_6 as follows:

$$\begin{aligned}[3] \rightarrow & ([3],[3]) + ([21],[21]) = ([3],[3]) + ([21],[1]) , \\ & \text{dimension: } 56 = (10,4) + (8,2) .\end{aligned} \tag{7-114a}$$

$$\begin{aligned}[21] \rightarrow & ([3],[21]) + ([21],[3]) + ([21],[21]) + ([1^3],[21]) \\ & = ([3],[1]) + ([21],[3]) + ([21],[1]) + ([1^3],[1]) , \\ & \text{dimension: } 70 = (10,2) + (8,4) + (8,2) + (1,2) .\end{aligned} \tag{7-114b}$$

The equality of the dimensionalities offers a useful check for the correctness of the decomposition, as shown in Eq. (7-114).

7.9.3. Representation transformation between the $SU_{mn} \supset SU_m \times SU_n$ irreducible basis and the SU_{mn} Gel'fand basis

The $SU_{mn} \supset SU_m \times SU_n$ irreducible basis in (7-106a) is expressed in terms of the products of wave functions in x and ξ spaces. It can also be expressed in terms of the Gel'fand bases of SU_{mn} and the Yamanouchi bases of $S_f(q)$, $\left|\begin{matrix}[\nu] \\ m, W\end{matrix}\right\rangle$,

$$\left|\begin{matrix}[\nu] & \beta[\nu_1] & [\nu_2] \\ m & W_1 & W_2\end{matrix}\right\rangle = \sum_{W}\left|\begin{matrix}[\nu] \\ m, W\end{matrix}\right\rangle\left\langle\begin{matrix}[\nu] \\ W\end{matrix}\right|[\nu], \begin{matrix}\beta[\nu_1][\nu_2] \\ W_1W_2\end{matrix}\right\rangle . \tag{7-115}$$

The transformation coefficients, or the subduction coefficients (SDC) of SU_{mn}, are obviously independent of m, therefore the label m in (7-115) can be ignored and (7-115) takes the following form,

$$\left|[\nu], \begin{matrix}\beta[\nu_1][\nu_2] \\ W_1W_2\end{matrix}\right\rangle = \sum_{W}\left|\begin{matrix}[\nu] \\ W\end{matrix}\right\rangle\left\langle\begin{matrix}[\nu] \\ W\end{matrix}\right|[\nu], \begin{matrix}\beta[\nu_1][\nu_2] \\ W_1W_2\end{matrix}\right\rangle . \tag{7-116a}$$

The SU_{mn} SDC obey the unitarity relation

$$\sum_{W}\left\langle\begin{matrix}[\nu] \\ W\end{matrix}\right|[\nu], \begin{matrix}\beta[\nu_1] & [\nu_2] \\ W_1 & W_2\end{matrix}\right\rangle\left\langle\begin{matrix}[\nu] \\ W\end{matrix}\right|[\nu], \begin{matrix}\beta'[\nu_1'] & [\nu_2'] \\ W_1' & W_2'\end{matrix}\right\rangle = \delta_{\beta\beta'}\delta_{\nu_1\nu_1'}\delta_{W_1W_1'}\delta_{W_2W_2'} ,$$

$$\sum_{\beta\nu_1\nu_2,W_1W_2} \left\langle \begin{matrix}[\nu]\\ W\end{matrix} \middle| [\nu], \begin{matrix}\beta[\nu_1][\nu_2]\\ W_1W_2\end{matrix} \right\rangle \left\langle \begin{matrix}[\nu]\\ W'\end{matrix} \middle| [\nu], \begin{matrix}\beta[\nu_1][\nu_2]\\ W_1W_2\end{matrix} \right\rangle = \delta_{WW'} \ . \tag{7-116b}$$

The inverse of (7-116a) is

$$\left| \begin{matrix}[\nu]\\ W\end{matrix} \right\rangle = \sum_{\nu_1\nu_2\beta W_1W_2} \left| [\nu], \begin{matrix}\beta[\nu_1][\nu_2]\\ W_1W_2\end{matrix} \right\rangle \left\langle \begin{matrix}[\nu]\\ W\end{matrix} \middle| [\nu], \begin{matrix}\beta[\nu_1][\nu_2]\\ W_1W_2\end{matrix} \right\rangle \ . \tag{7-116c}$$

The SU_{mn} SDC in (7-115) can be found by using Racah's step operator method, and has been given for the $SU_6 \supset SU_3 \times SU_2$ and $SU_8 \supset SU_4 \times SU_2$ baryon wave functions for various kinds of symmetries (Chen, [26] 1978).

Ex. 7.8. Find the $SU_6 \supset SU_3 \times SU_2$ wave function $|[3], \Sigma^0_{\frac{1}{2}\frac{1}{2}}\rangle$.

Ex. 7.9. Find the $SU_3 \times SU_2$ content in the irrep [31] of SU_6.

7.10. The $\mathrm{SU}_{n_1n_2n_3} \supset \mathrm{SU}_{n_1} \times \mathrm{SU}_{n_2} \times \mathrm{SU}_{n_3}$ Irreducible Bases and the Racah Coefficients of Permutation Groups

Suppose a particle has three kinds of degrees of freedom $x_i (i = 1,2,3)$. In the space V_{x_i} there are n_i single particle states. We have two ways to construct the $SU_n \supset SU_{n_1} \times SU_{n_2} \times SU_{n_3} (n = n_1n_2n_3)$ irreducible basis.

1. $SU_n \supset (SU_{n_1n_2} \supset SU_{n_1} \times SU_{n_2}) \times SU_{n_3}$ basis. From (7-106a), this kind of basis can be expressed as

$$\left| \begin{matrix}[\nu]\\ m,\end{matrix} \left(\begin{matrix}[\nu_{12}] & [\nu_3]\\ [\nu_1]W_1[\nu_2]W_2, & W_3\end{matrix} \right) \right\rangle^{\beta_{12}\beta}$$
$$= \sum_{m_1m_2m_3m_{12}} C^{[\nu_{12}]\beta_{12}m_{12}}_{\nu_1m_1,\nu_2m_2} C^{[\nu]\beta,m}_{[\nu_{12}]m_{12},[\nu_3]m_3} \left| \begin{matrix}[\nu_1]\\ m_1W_1\end{matrix} \right\rangle^{x_1} \left| \begin{matrix}[\nu_2]\\ m_2W_2\end{matrix} \right\rangle^{x_2} \left| \begin{matrix}[\nu_3]\\ m_3W_3\end{matrix} \right\rangle^{x_3} , \tag{7-117}$$

which is the Yamanouchi basis $[\nu]m$ of $S_f(q)$ with $q = (x_1, x_2, x_3)$, the irreducible basis $[\nu_i]W_i$ of SU_{n_i}, and belongs to the irreps $[\nu]$ and $[\nu_{12}]$ of SU_n and $SU_{n_1n_2}$, respectively.

2. $SU_n \supset SU_{n_1} \times (SU_{n_2n_3} \supset SU_{n_2} \times SU_{n_3})$ irreducible basis. Similarly we have

$$\left| \begin{matrix}[\nu]\\ m,\end{matrix} \left(\begin{matrix}[\nu_1] & [\nu_{23}]\\ W_1, & [\nu_2]W_2[\nu_3]W_3\end{matrix} \right) \right\rangle^{\beta_{23}\beta'}$$
$$= \sum_{m_1m_2m_3m_{23}} C^{[\nu]\beta',m}_{\nu_1m_1,\nu_{23}m_{23}} C^{[\nu_{23}]\beta_{23},m_{23}}_{\nu_2m_2,\nu_3m_3} \left| \begin{matrix}[\nu_1]\\ m_1W_1\end{matrix} \right\rangle^{x_1} \left| \begin{matrix}[\nu_2]\\ m_2W_2\end{matrix} \right\rangle^{x_2} \left| \begin{matrix}[\nu_3]\\ m_3W_3\end{matrix} \right\rangle^{x_3} , \tag{7-118}$$

which is the Yamanouchi basis $[\nu]m$ of $S_f(q)$, the irreducible basis $[\nu_i]W_i$ of SU_{n_i}, and belongs to the irreps $[\nu]$ and $[\nu_{23}]$ of SU_n and $SU_{n_2n_3}$, respectively.

For instance, suppose x_1 is the orbital space, $n_1 = 2l+1$; x_2 and x_3 are spin and isospin spaces, $n_2 = n_3 = 2$. Using (7-118) and (4-124), a totally anti-symmetric state is constructed,

$$\left| [1^f] \left(\begin{matrix}[\nu_1] & [\nu_{23}]\\ W_1, & \beta[\nu_2]W_2[\nu_3]W_3\end{matrix} \right) \right\rangle = \left| [1^f], \begin{matrix}[\nu_1] & [\tilde{\nu}_1]\\ \alpha LM, & \beta SM_STM_T\end{matrix} \right\rangle$$
$$= \sum_{m_1m_2m_3} \frac{\Lambda^{\nu_1}_{m_1}}{\sqrt{h_{\nu_1}(S_f)}} C^{[\tilde{\nu}_1]\beta,\tilde{m}_1}_{\nu_2m_2,\nu_3m_3} \left| \begin{matrix}[\nu_1]\\ m_1, \alpha LM\end{matrix} \right\rangle^{x_1} \left| \begin{matrix}[\nu_2]\\ m_2, \ SM_S\end{matrix} \right\rangle^{x_2} \left| \begin{matrix}[\nu_3]\\ m_3, \ TM_T\end{matrix} \right\rangle^{x_3} \tag{7-119}$$

where $\beta = \beta_{23}, [\nu_2] = [\frac{f}{2}+S, \frac{f}{2}-S], [\nu_3] = [\frac{f}{2}+T, \frac{f}{2}-T]$. Equation (7-119) is the $SU_{4(2l+1)} \supset SU_{2l+1} \times (SU_4 \supset SU_2 \times SU_2)$ basis.

DeGrand (1976) used the $SU_{12} \supset SU_3^f \times SU_2^s \times SU_2^p$ irreducible basis in the study of the bag model of elementary particles, where f, s and p stand for flavor, spin and pseudo-spin. Since the particles state must be a singlet $|[1^3]\rangle$ in color space, the irrep of SU_{12} is $[\nu]=[3]$, while $[\nu_1]=[\nu_2]=[\nu_3]=[21]$. Now the quantum numbers β, ν_{12} and β_{12} are redundant. From (7-117) and Table 4.13-1 we obtain the required wave function

$$
\begin{aligned}
\left|[\nu];\,\begin{matrix}[\nu_1][\nu_2][\nu_3]\\ W_1,W_2,W_3\end{matrix}\right\rangle &= \left|[3];\ \begin{array}{|c|c|}\hline u&d\\\hline s\\\cline{1-1}\end{array}\ \begin{array}{|c|c|}\hline \alpha&\alpha\\\hline \beta\\\cline{1-1}\end{array}\ \begin{array}{|c|c|}\hline a&a\\\hline b\\\cline{1-1}\end{array}\right\rangle \\
&= \sum_{m_1m_2m} C^{[21]m}_{[21]m_1,[21]m_2}\left|\begin{array}{|c|c|}\hline u&d\\\hline s\\\cline{1-1}\end{array}\right\rangle_{m_1}\left|\begin{array}{|c|c|}\hline \alpha&\alpha\\\hline \beta\\\cline{1-1}\end{array}\right\rangle_{m_2}\left|\begin{array}{|c|c|}\hline a&a\\\hline b\\\cline{1-1}\end{array}\right\rangle_{m} \\
&= \frac{1}{2}\left[\left(\left|\begin{array}{|c|c|}\hline u&d\\\hline s\\\cline{1-1}\end{array}\right\rangle_1\left|\begin{array}{|c|c|}\hline \alpha&\alpha\\\hline \beta\\\cline{1-1}\end{array}\right\rangle_1-\left|\begin{array}{|c|c|}\hline u&d\\\hline s\\\cline{1-1}\end{array}\right\rangle_2\left|\begin{array}{|c|c|}\hline \alpha&\alpha\\\hline \beta\\\cline{1-1}\end{array}\right\rangle_2\right)\left|\begin{array}{|c|c|}\hline a&a\\\hline b\\\cline{1-1}\end{array}\right\rangle_1\right] \\
&\quad -\frac{1}{2}\left[\left(\left|\begin{array}{|c|c|}\hline u&d\\\hline s\\\cline{1-1}\end{array}\right\rangle_1\left|\begin{array}{|c|c|}\hline \alpha&\alpha\\\hline \beta\\\cline{1-1}\end{array}\right\rangle_2+\left|\begin{array}{|c|c|}\hline u&d\\\hline s\\\cline{1-1}\end{array}\right\rangle_2\left|\begin{array}{|c|c|}\hline \alpha&\alpha\\\hline \beta\\\cline{1-1}\end{array}\right\rangle_1\right)\left|\begin{array}{|c|c|}\hline a&a\\\hline b\\\cline{1-1}\end{array}\right\rangle_2\right],
\end{aligned}
\tag{7-120}
$$

where a and b denote the pseudo-spin up and down states. Equation (7-120) is identical with DeGrand's result.

The bases of (7-117) and (7-118) differ by a unitary transformation

$$
\begin{aligned}
&\left|\begin{matrix}[\nu]\\ m,\end{matrix}\left(\begin{matrix}[\nu_{12}] & [\nu_3]\\ [\nu_1]W_1[\nu_2]W_2, & W_3\end{matrix}\right)\right\rangle^{\beta_{12}\beta} \\
&= \sum_{\nu_{23}\beta_{23}\beta'} U(\nu_1\nu_2\nu\nu_3,\nu_{12}\nu_{23})^{\beta_{12}\beta}_{\beta_{23}\beta'}\left|\begin{matrix}[\nu]\\ m,\end{matrix}\left(\begin{matrix}[\nu_1] & [\nu_{23}]\\ W_1,[\nu_2] & W_2[\nu_3]W_3\end{matrix}\right)\right\rangle^{\beta_{23}\beta'},
\end{aligned}
\tag{7-121}
$$

where U is the Racah coefficient of the permutation group. From (7-117) and (7-118) we have

$$
U(\nu_1\nu_2\nu\nu_3;\nu_{12}\nu_{23})^{\beta_{12}\beta}_{\beta_{23}\beta'} = \sum_{\text{fix } m} C^{[\nu_{12}]\beta_{12},m_{12}}_{\nu_1m_1,\nu_2m_2}C^{[\nu]\beta,m}_{\nu_{12}m_{12},\nu_3m_3}C^{[\nu_{23}]\beta_{23},m_{23}}_{\nu_2m_2,\nu_3m_3}C^{[\nu]\beta'm}_{\nu_1m_1,\nu_{23}m_{23}}\,. \tag{7-122}
$$

The values of U are independent of m, W_1, W_2 and W_3. If we ignore the unitary group and pay attention only to the permutation group, (7-121) can be written in the familiar form of angular momentum theory,

$$
\begin{aligned}
&|(([\nu_1][\nu_2])[\nu_{12}],[\nu_3]:[\nu]m\rangle^{\beta_{12}\beta} \\
&\sum_{\nu_{23}\beta_{23}\beta'} U(\nu_1\nu_2\nu\nu_3;\nu_{12}\nu_{23})^{\beta_{12}\beta}_{\beta_{23}\beta'}|[\nu_1]([\nu_2][\nu_3])[\nu_{23}]:[\nu]m\rangle^{\beta_{23}\beta'}\,.
\end{aligned}
\tag{7-123}
$$

Compared with the Racah coefficients of the group SU_2, the only difference here is the occurrence of the multiplicity labels $\beta_{12}, \beta_{23}, \beta$ and β' due to the non-simple reducibility of the permutation group. The Racah coefficients satisfy the unitarity relations

$$
\sum_{\nu_{23}\beta_{23}\beta'} U(\nu_1\nu_2\nu\nu_3;\nu_{12}\nu_{23})^{\beta_{12}\beta}_{\beta_{23}\beta'}U(\nu_1\nu_2\nu\nu_3;\bar\nu_{12}\nu_{23})^{\bar\beta_{12}\bar\beta}_{\beta_{23}\beta'} = \delta_{\beta_{12}\bar\beta_{12}}\delta_{\beta\bar\beta}\delta_{\nu_{12}\bar\nu_{12}}\,, \tag{7-124a}
$$

$$
\sum_{\nu_{12}\beta_{12}\beta} U(\nu_1\nu_2\nu\nu_3;\nu_{12}\nu_{23})^{\beta_{12}\beta}_{\beta_{23}\beta'}U(\nu_1\nu_2\nu\nu_3;\nu_{12}\bar\nu_{23})^{\beta_{12}\beta}_{\bar\beta_{23}\bar\beta'} = \delta_{\beta_{23}\bar\beta_{23}}\delta_{\beta'\bar\beta'}\delta_{\nu_{23}\bar\nu_{23}}\,. \tag{7-124b}
$$

The formulas in the SU_2 Racah algebra (Rose, 1957) can all be extended to the Racah coefficients of the permutation group. For example,

$$
C^{[\nu_{12}]\beta_{12}m_{12}}_{\nu_1m_1,\nu_2m_3}C^{[\nu]\beta,m}_{\nu_{12}m_{12},\nu_3m_3} = \sum_{\beta_{23}\beta'} U(\nu_1\nu_2\nu\nu_3;\nu_{12}\nu_{23})^{\beta_{12}\beta}_{\beta_{23}\beta'}C^{[\nu_{23}]\beta_{23},m_{23}}_{\nu_2m_2,\nu_3m_3}C^{[\nu]\beta',m}_{\nu_1m_1,\nu_{23}m_{23}}\,, \tag{7-125}
$$

$$\sum_{\beta'} U(\nu_1\nu_2\nu\nu_3;\nu_{12}\nu_{23})^{\beta_{12}\beta}_{\beta_{23}\beta'} C^{[\nu]\beta',m}_{\nu_1 m_1,\nu_{23}m_{23}} = \sum_{m_2m_3m_{12}} C^{[\nu_{12}]\beta_{12},m_{12}}_{\nu_1 m_1,\nu_2 m_2} C^{[\nu]\beta,m}_{\nu_{12}m_{12},\nu_3 m_3} C^{[\nu_{23}]\beta_{23},m_{23}}_{\nu_2 m_2,\nu_3 m_3} . \quad (7\text{-}126)$$

The generalized formulas of the SU_2 Racah algebra (Chen [1] 1965) can also be further extended to the permutation group. Letting

$$\{CCCC\} \equiv C^{[\nu_{12}]\beta_{12},m_{12}}_{\nu_1 m_1,\nu_2 m_2} C^{[\nu]\beta,m}_{\nu_{12}m_{12},\nu_3 m_3} C^{[\nu]\beta' m}_{\nu_1 m_1,\nu_{23}m_{23}} C^{[\nu_{23}]\beta_{23},m_{23}}_{\nu_2 m_2,\nu_3 m_3} . \quad (7\text{-}127)$$

Equations (13) and (14) in Chen (1965) are generalized to

$$\sum_{\text{fix } m_\kappa} \{CCCC\} = \frac{h_\nu}{h_{\nu_\kappa}} U(\nu_1\nu_2\nu\nu_3;\nu_{12}\nu_{23})^{\beta_{12}\beta}_{\beta_{23}\beta'} \quad (7\text{-}128a)$$

$$\sum_{\text{fix } m_a m_b m_\kappa} \{CCCC\}\left\{C^{[\nu_\kappa]\beta_\kappa,m_\kappa}_{\nu_a m_a,\nu_b m_b}\right\}^{-1}$$
$$= \frac{h_\nu}{h_{\nu_\kappa}} \sum_{\beta_\kappa} C^{[\nu_\kappa]\beta_\kappa,m_\kappa}_{\nu_a m_a,\nu_b m_b} U(\nu_1\nu_2\nu\nu_3;\nu_{12}\nu_{23})^{\beta_{12}\beta}_{\beta_{23}\beta'} , \quad (7\text{-}128b)$$

where the summation in (7-128a) is over all the m's except m_κ, while in (7-128b) it is over all the m's except m_a, m_b and m_κ, and $\nu_\kappa m_\kappa$ may be any one of $\nu_1 m_1$, $\nu_2 m_2$, $\nu_3 m_3$, $\nu_{12}m_{12}$, $\nu_{23}m_{23}$ and νm, while h_ν and h_{ν_κ} are the dimensions of the irreps $[\nu]$ and $[\nu_\kappa]$ of the permutation group, respectively. The coefficient $C^{[\nu_\kappa]\beta_\kappa,m_\kappa}_{\nu_a m_a,\nu_b m_b}$ in (7-128b) may be any one of the four C's in (7-127).

Using (4-189a), the Racah coefficients of S_f can be expressed in terms of those of S_{f-1} and the $S_f \supset S_{f-1}$ ISF,

$$U(\nu_1\nu_2\nu\nu_3;\nu_{12}\nu_{23})^{\beta_{12}\beta}_{\beta_{23}\beta'} = \sum_{\substack{\nu_1'\nu_2'\nu_3'\nu_{12}'\nu_{23}' \\ \theta_{12}\theta_{23}\theta\theta'}} U(\nu_1'\nu_2'\nu'\nu_3';\nu_{12}'\nu_{23}')^{\theta_{12}\theta}_{\theta_{23}\theta'}$$
$$\times C^{\nu_{12}\beta_{12},\nu_{12}'\theta_{12}}_{\nu_1\nu_1',\nu_2\nu_2'} C^{\nu\beta,\nu'\theta}_{\nu_{12}\nu_{12}',\nu_3\nu_3'} C^{\nu_{23}\beta_{23},\nu_{23}'\theta_{23}}_{\nu_2\nu_2',\nu_3\nu_3'} C^{\nu\beta',\nu'\theta'}_{\nu_1\nu_1',\nu_{23}\nu_{23}'} , \quad (7\text{-}129)$$

where ν' and ν_i' refer to the group S_{f-1} and the summation is carried out under fixed ν'. From (7-126) we have

$$\sum_{\beta'} U(\nu_1\nu_2\nu\nu_3;\nu_{12}\nu_{23})^{\beta_{12}\beta}_{\beta_{23}\beta'} C^{\nu\beta',\nu'\theta'}_{\nu_1\nu_1',\nu_{23}\nu_{23}'} = \sum_{\substack{\nu_2'\nu_3'\nu_{12}' \\ \theta_{12}\theta_{23}\theta}} U(\nu_1'\nu_2'\nu'\nu_3';\nu_{12}'\nu_{23}')^{\theta_{12}\theta}_{\theta_{23}\theta'}$$
$$\times C^{\nu_{12}\beta_{12},\nu_{12}'\theta_{12}}_{\nu_1\nu_1',\nu_2\nu_2'} C^{\nu\beta,\nu'\theta}_{\nu_{12}\nu_{12}',\nu_3\nu_3'} C^{\nu_{23}\beta_{23},\nu_{23}'\theta_{23}}_{\nu_2\nu_2',\nu_3\nu_3'} . \quad (7\text{-}130)$$

Algebraic expressions have been given for some special Racah coefficients of the permutation group (Vanagas 1972).

7.11. $\mathbf{SU}_{n_1n_2n_3n_4} \supset \mathbf{SU}_{n_1} \times \mathbf{SU}_{n_2} \times \mathbf{SU}_{n_3} \times \mathbf{SU}_{n_4}$ Irreducible Basis and the 9ν Coefficients of the Permutation Group*

Let $n = n_1n_2n_3n_4$. There are two possible ways of constructing the SU_n irreducible basis.

1. The $SU_n \supset (SU_{n_1n_2} \supset SU_{n_1} \times SU_{n_2}) \times (SU_{n_3n_4} \supset SU_{n_3} \times SU_{n_4})$ irreducible basis.

$$\left|\begin{matrix}[\nu] \\ m\end{matrix}, \left(\begin{matrix}[\nu_{12}] & & [\nu_{34}] \\ [\nu_1]w_1[\nu_2]w_2 & , & [\nu_3]w_3[\nu_4]w_4\end{matrix}\right)\right\rangle^{\tau_{12}\tau_{34}\tau} = \sum_{\substack{m_1m_2m_3m_4 \\ m_{12}m_{34}}} C^{[\nu_{12}]\tau_{12},m_{12}}_{\nu_1 m_1,\nu_2 m_2} C^{[\nu_{34}]\tau_{34},m_{34}}_{\nu_3 m_3,\nu_4 m_4}$$
$$\times C^{[\nu]\tau,m}_{\nu_{12}m_{12},\nu_{34}m_{34}} \left|\begin{matrix}[\nu_1] \\ m_1w_1\end{matrix}\right\rangle^{x_1} \left|\begin{matrix}[\nu_2] \\ m_2w_2\end{matrix}\right\rangle^{x_2} \left|\begin{matrix}[\nu_3] \\ m_3w_3\end{matrix}\right\rangle^{x_3} \left|\begin{matrix}[\nu_4] \\ m_4w_4\end{matrix}\right\rangle^{x_4} . \quad (7\text{-}131)$$

2. The $SU_n \supset (SU_{n_1n_3} \supset SU_{n_1} \times SU_{n_3}) \times (SU_{n_2n_4} \supset SU_{n_2} \times SU_{n_4})$ irreducible basis

$$\left|\begin{matrix}[\nu] \\ m,\end{matrix}\left(\begin{matrix}[\nu_{13}] & [\nu_{24}] \\ [\nu_1]w_1[\nu_3]w_3, & [\nu_2]w_2[\nu_4]w_4\end{matrix}\right)\right\rangle^{\tau_{13}\tau_{24}\tau'} = \sum_{\substack{m_1m_2m_3m_4 \\ m_{13}m_{24}}} C^{[\nu_{13}]\tau_{13}m_{13}}_{\nu_1m_1,\nu_3m_3} C^{[\nu_{24}]\tau_{24}m_{24}}_{\nu_2m_2,\nu_4m_4}$$
$$\times C^{[\nu]\tau' m}_{\nu_{13}m_{13},\nu_{24}m_{24}} \left|\begin{matrix}[\nu_1] \\ m_1w_1\end{matrix}\right\rangle^{x_1} \left|\begin{matrix}[\nu_2] \\ m_2w_2\end{matrix}\right\rangle^{x_2} \left|\begin{matrix}[\nu_3] \\ m_3w_3\end{matrix}\right\rangle^{x_3} \left|\begin{matrix}[\nu_4] \\ m_4w_4\end{matrix}\right\rangle^{x_4} . \tag{7-132}$$

The bases (7-131) and (7-132) are related by a unitary transformation

$$\left|\begin{matrix}[\nu] \\ m,\end{matrix}\left(\begin{matrix}[\nu_{12}] & [\nu_{34}] \\ [\nu_1]w_1[\nu_2]w_2, & [\nu_3]w_3[\nu_4]w_4\end{matrix}\right)^{\tau_{12}\tau_{34}\tau}\right.$$
$$= \sum_{\substack{\nu_{13}\nu_{24}, \\ \tau_{13}\tau_{24}\tau'}} \begin{pmatrix}\nu_1 & \nu_2 & \nu_{12} \\ \nu_3 & \nu_4 & \nu_{34} \\ \nu_{13} & \nu_{24} & \nu\end{pmatrix}^{\tau_{12}\tau_{34}\tau}_{\tau_{13}\tau_{24}\tau'} \left|\begin{matrix}[\nu] \\ m\end{matrix}, \left(\begin{matrix}[\nu_{13}] & [\nu_{24}] \\ [\nu_1]w_1[\nu_3]w_3, & [\nu_2]w_2[\nu_4]w_4\end{matrix}\right)\right\rangle^{\tau_{13}\tau_{24}\tau'} . \tag{7-133}$$

The coefficients in (7-133) are the 9ν coefficients of the permutation group, and can be expressed as

$$\begin{pmatrix}\nu_1 & \nu_2 & \nu_{12} \\ \nu_3 & \nu_4 & \nu_{34} \\ \nu_{13} & \nu_{24} & \nu\end{pmatrix}^{\tau_{12}\tau_{34}\tau}_{\tau_{13}\tau_{24}\tau'} = \sum_{\text{fix } m} \{CCCCCC\} . \tag{7-134a}$$

$$\{CCCCCC\} = C^{[\nu_{12}]\tau_{12},m_{12}}_{\nu_1m_1,\nu_2m_2} C^{[\nu_{34}]\tau_{34},m_{34}}_{\nu_3m_3,\nu_4m_4} C^{[\nu]\tau,m}_{\nu_{12}m_{12},\nu_{34}m_{34}} C^{[\nu_{13}]\tau_{13},m_{13}}_{\nu_1m_1,\nu_3m_3} C^{[\nu_{24}]\tau_{24},m_{24}}_{\nu_2m_2,\nu_4m_4}$$
$$\times C^{[\nu]\tau',m}_{\nu_{13}m_{13},\nu_{24}m_{24}} , \tag{7-134b}$$

where the summation is over all m's except m. Equations (9) and (10) in Chen (1965) can be extended to

$$\sum_{\text{fix } m_k} \{CCCCCC\} = \frac{h_\nu}{h_{\nu_k}} \begin{pmatrix}\nu_1 & \nu_2 & \nu_{12} \\ \nu_3 & \nu_4 & \nu_{34} \\ \nu_{13} & \nu_{24} & \nu\end{pmatrix}^{\tau_{12}\tau_{34}\tau}_{\tau_{13}\tau_{24}\tau'} . \tag{7-135a}$$

$$\sum_{\text{fix } m_am_bm_k} \{CCCCCC\}\{C^{[\nu_k]\tau_k,m_k}_{\nu_am_a,\nu_bm_b}\}^{-1} = \frac{h_\nu}{h_{\nu_k}} \sum_{\tau_k} C^{[\nu_k]\tau_k,m_k}_{\nu_am_a,\nu_bm_b} \begin{pmatrix}\nu_1 & \nu_2 & \nu_{12} \\ \nu_3 & \nu_4 & \nu_{34} \\ \nu_{13} & \nu_{24} & \nu\end{pmatrix}^{\tau_{12}\tau_{34}\tau}_{\tau_{13}\tau_{24}\tau'} , \tag{7-135b}$$

where the summation in (7-135a) is over all m's except m_k, while in (7-135b) it extends over all m's except m_a, m_b and m_k.

In analogy with (7-130), the 9ν coefficients of S_f are expressible in terms of those of S_{f-1} and the $S_f \supset S_{f-1}$ ISF,

$$\begin{pmatrix}\nu_1 & \nu_2 & \nu_{12} \\ \nu_3 & \nu_4 & \nu_{34} \\ \nu_{13} & \nu_{24} & \nu\end{pmatrix}^{\tau_{12}\tau_{34}\tau}_{\tau_{13}\tau_{24}\tau'} = \sum_{\text{fix } \nu'} \begin{pmatrix}\nu_1' & \nu_2' & \nu_{12}' \\ \nu_3' & \nu_4' & \nu_{34}' \\ \nu_{13}' & \nu_{24}' & \nu'\end{pmatrix}^{\theta_{12}\theta_{34}\theta}_{\theta_{13}\theta_{24}\theta'} C^{\nu_{12}\tau_{12},\nu_{12}'\theta_{12}}_{\nu_1\nu_1',\nu_2\nu_2'}$$
$$\times C^{\nu_{34}\tau_{34},\nu_{34}'\theta_{34}}_{\nu_3\nu_3',\nu_4\nu_4'} C^{\nu\tau,\nu'\theta}_{\nu_{12}\nu_{12}',\nu_{34}\nu_{34}'} C^{\nu_{13}\tau_{13},\nu_{13}'\theta_{13}}_{\nu_1\nu_1',\nu_3\nu_3'} C^{\nu_{24}\tau_{24},\nu_{24}'\theta_{24}}_{\nu_2\nu_2',\nu_4\nu_4'} C^{\nu\tau',\nu'\theta'}_{\nu_{13}\nu_{13}',\nu_{24}\nu_{24}'} , \tag{7-136}$$

where the sum runs over $\nu_1'\nu_2'\nu_3'\nu_4'\nu_{12}'\nu_{34}'\nu_{13}'\nu_{24}'\theta_{12}\theta_{34}\theta_{13}\theta_{24}\theta\theta'$, but with ν' fixed. It is readily seen that

$$\begin{pmatrix}\nu_1 & \nu_2 & \nu_{12} \\ 0 & \nu_3 & \nu_3 \\ \nu_1 & \nu_{23} & \nu\end{pmatrix}^{\tau_{12}\tau}_{\tau_{23}\tau'} = U(\nu_1\nu_2\nu\nu_3; \nu_{12}\nu_{23})^{\tau_{12}\tau}_{\tau_{23}\tau'} . \tag{7-137}$$

The values of the 9ν coefficients are independent of $w_1w_2w_3w_4m$. If we are only interested in the permutation group, (7-133) can be written as

$$|(\nu_1\nu_2)\nu_{12},(\nu_3\nu_4)\nu_{34:}\nu m)\rangle^{\tau_{12}\tau_{34}\tau} = \sum_{\tau_{13}\tau_{24}\tau'} \begin{pmatrix} \nu_1 & \nu_2 & \nu_{12} \\ \nu_3 & \nu_4 & \nu_{34} \\ \nu_{13} & \nu_{24} & \nu \end{pmatrix}^{\tau_{12}\tau_{34},\tau}_{\tau_{13}\tau_{24}\tau'} |(\nu_1\nu_3)\nu_{13},(\nu_2\nu_4)\nu_{24:}\nu m\rangle^{\tau_{13}\tau_{24}\tau'} . \qquad (7\text{-}138)$$

7.12. $SU_{m+n} \supset SU_m \otimes SU_n$ Irreducible Basis

Suppose that there are two sets of single particle states with m and n states, and there are f_1 and f_2 particles occupying the m states of the first set and the n states of the second set, respectively. For example, in the mixed configuration $(l_1)^{f_1}(l_2)^{f_2}$, there are f_1 particles in the orbital l_1 and f_2 particles in the orbital l_2; thus $m = 2l_1 + 1$ and $n = 2l_2 + 1$. Now we need to construct the $SU_{m+n} \supset SU_m \otimes SU_n$ irreducible bases for the $f_1 + f_2$ particles.

7.12.1. The IDC of permutation groups and $SU_{m+n} \supset SU_m \otimes SU_n$ irreducible bases

We have three kinds of product states $\varphi^{(1)}, \varphi^{(2)}$ and φ with particle numbers f_1, f_2 and $f = f_1 + f_2$, respectively:

$$\varphi^{(1)} = \varphi_{i_1}(1)\dots\varphi_{i_{f_1}}(f_1), i = 1,2,\dots,m . \qquad (7\text{-}139a)$$

$$\varphi^{(2)} = \varphi_{j_1}(f_1+1)\dots\varphi_{j_{f_2}}(f),\ \ j = m+1,\dots,m+n, f = f_1+f_2 . \qquad (7\text{-}139b)$$

$$\varphi = \varphi_{k_1}(1)\dots\varphi_{k_f}(f), \quad k = 1,2,\dots,m+n . \qquad (\text{-}139c)$$

The $\begin{pmatrix}\varphi^{(1)}\\ \varphi^{(2)}\\ \varphi\end{pmatrix}$ carry a reducible rep of $\begin{pmatrix}S_{f_1}\\ S_{f_2}\\ S_f\end{pmatrix}$ and $\begin{pmatrix}SU_m\\ SU_n\\ SU_{m+n}\end{pmatrix}$. In the following we will use the symbol S_{f_2} to denote the permutation group $S_{f_2}(f_1+1,\dots,f_1+f_2)$.

The group elements of SU_m, SU_n, SU_{m+n} and $SU_m \otimes SU_n$ are of the following forms

$$\left(\begin{array}{c|c} U_1 & 0 \\ \hline 0 & I \end{array}\right), \quad \left(\begin{array}{c|c} I & 0 \\ \hline 0 & U_2 \end{array}\right), \quad \begin{pmatrix} u_{11} & \dots & u_{1,m+n} \\ \vdots & & \vdots \\ u_{m+n,1} & \dots & u_{m+n,m+n} \end{pmatrix}, \quad \left(\begin{array}{c|c} U_1 & 0 \\ \hline 0 & U_2 \end{array}\right) . \qquad (7\text{-}140)$$

Their orders are $r_m = m^2 - 1, r_n = n^2 - 1, r = (m+n)^2 - 1$ and $r' = r_m + r_n$, respectively. Evidently, $SU_m \otimes SU_n$ is a subgroup of SU_{m+n}. $\binom{\varphi^{(1)}}{\varphi^{(2)}}$ can be linearly combined into the Yamanouchi basis $\binom{(\nu_1 m_1)}{(\nu_2 m_2)}$ of $\binom{S_{f_1}}{S_{f_2}}$ and the Gel'fand basis $\binom{(\nu_1 w_1)}{(\nu_2 w_2)}$ of $\binom{SU_m}{SU_n}$, the Weyl tableaux w_1 and w_2 being filled with the state indices $i_1\dots i_f$ and $j_1\dots j_f$, respectively. The aforementioned bases are designated by

$$\left|\begin{matrix}[\nu_1]\\ m_1\omega_1^0, w_1\end{matrix}\right\rangle = \varphi^{[\nu_1]}_{m_1}(\omega_1^0, w_1), \quad \left|\begin{matrix}[\nu_2]\\ m_2\omega_2^0, w_2\end{matrix}\right\rangle = \varphi^{[\nu_2]}_{m_2}(\omega_2^0, w_2) ,$$

respectively. Using the IDC of permutation groups, they can be coupled into the Yamanouchi basis (νm) of S_f,

$$\begin{aligned}\left|\begin{matrix}[\nu]\\ m,\beta[\nu_1]w_1[\nu_2]w_2\end{matrix}\right\rangle &\equiv \left|\begin{matrix}[\nu],\ \beta[\nu_1][\nu_2]\\ m \quad w_1w_2\end{matrix}\right\rangle \\ &= \sum_{\substack{m_1\omega_1\\ m_2\omega_2}} C^{[\nu]\beta m}_{\nu_1 m_1\omega_1,\nu_2 m_2\omega_2}\begin{pmatrix}\omega_0\\ \omega\end{pmatrix}\left[\left|\begin{matrix}[\nu_1]\\ m_1\omega_1^0, w_1\end{matrix}\right\rangle\left|\begin{matrix}[\nu_2]\\ m_2\omega_2^0, w_2\end{matrix}\right\rangle\right] \\ &= \sum_{\substack{m_1\omega_1\\ m_2\omega_2}} C^{[\nu]\beta,m}_{\nu_1 m_1\omega_1,\nu_2 m_2\omega_2}\left|\begin{matrix}[\nu_1]\\ m_1\omega_1, w_1\end{matrix}\right\rangle\left|\begin{matrix}[\nu_2]\\ m_2\omega_2, w_2\end{matrix}\right\rangle .\end{aligned} \qquad (7\text{-}141)$$

In analogous to the discussion following (7-106), it can be proved that (7-141) belongs to the irrep $[\nu]$ of SU_{m+n}, and at the same time it is the irreducible basis $\nu_1 w_1 (\nu_2 w_2)$ of the group $SU_m(SU_n)$. Hence we have

Theorem 7.5: The IDC of permutation groups are the indirect coupling coefficients for the $SU_{m+n} \supset SU_m \otimes SU_n$ irreducible basis.

Therefore it is easy to construct the $SU_{m+n} \supset SU_m \otimes SU_n$ irreducible bases for arbitrary m and n by means of the IDC.

The inverse of (7-141) is

$$\left|\begin{matrix}[\nu_1] \\ m_1 w_1, w_1\end{matrix}\right\rangle \left|\begin{matrix}[\nu_2] \\ m_2 w_2, w_2\end{matrix}\right\rangle = \sum_{\nu\beta m} C^{[\nu]\beta m}_{\nu_1 m_1 w_1, \nu_2 m_2 w_2} \left|\begin{matrix}[\nu], & \beta[\nu_1][\nu_2] \\ m & w_1 w_2\end{matrix}\right\rangle . \tag{7-142}$$

7.12.2. The content of irreps $([\nu_1],[\nu_2])$ of $SU_m \otimes SU_n$ in the irrep of SU_{m+n}

Equations (7-141) and (7-142) show that the SU_{m+n} subduction rule

$$[\nu] \downarrow (SU_m \otimes SU_n) = \sum_{\nu_1\nu_2} \{\nu_1\nu_2\nu\}(D^{[\nu_1]}(SU_m), D^{[\nu_2]}(SU_n)) , \tag{7-141'}$$

and the $SU_m \times SU_n$ induction rule

$$([\nu_1] \times [\nu_2]) \uparrow SU_{m+n} = \sum_{\nu} \{\nu_1\nu_2\nu\} D^{[\nu]}(SU_{m+n}) , \tag{7-142'}$$

are determined by the Littlewood rule, so long as we disregard the Young diagrams $[\nu_1]([\nu_2])$ of more than $m(n)$ rows and columns of length $m(n)$ in (7-141'), and the Young diagram $[\nu]$ of more than $m+n$ rows and columns of length $m+n$ in (7-142').

For example, from Table 4.14-2 we obtain the irreps $([\nu_1],[\nu_2])$ of (SU_3, SU_2) contained in the irrep [32] of SU_5 as follows:

$$\begin{aligned} [32] \to\ & ([32],[0]) + ([0],[1]) + ([31],[1]) + ([1],[2]) \\ & + ([22],[1]) + ([1],[0]) + ([3],[2]) + ([2],[3]) \\ & + ([21],[2]) + ([2],[1]) + ([21],[0]) + ([11],[1]), \\ \text{dimension } 175 \to\ & (15,1) + (1,2) + (15,2) + (3,3) + (6,2) + (3,1) \\ & + (10,3) + (6,4) + (8,3) + (6,2) + (8,1) + (3,2) . \end{aligned} \tag{7-143}$$

7.12.3. The representation transformation between the $SU_{m+n} \supset SU_m \otimes SU_n$ irreducible basis and the Gel'fand basis of SU_{m+n}

The $SU_{m+n} \supset SU_m \otimes SU_n$ irreducible bases (7-141) can be expressed in terms of the SU_{m+n} Gel'fand bases:

$$\left|\begin{matrix}[\nu] \\ \beta[\nu_1]W_1[\nu_2]W_2\end{matrix}\right\rangle = \left|[\nu], \begin{matrix}\beta[\nu_1][\nu_2] \\ W_1\ W_2\end{matrix}\right\rangle = \sum_{W} \left|\begin{matrix}[\nu] \\ W\end{matrix}\right\rangle \left\langle \begin{matrix}[\nu] \\ W\end{matrix}\right|[\nu], \begin{matrix}\beta[\nu_1][\nu_2] \\ W_1\ W_2\end{matrix}\right\rangle . \tag{7-144}$$

The transformation coefficients $\left\langle \begin{matrix}[\nu] \\ W\end{matrix}\middle|[\nu], \begin{matrix}\beta[\nu_1][\nu_2] \\ W_1\ W_2\end{matrix}\right\rangle$, or the subduction coefficients (SDC) of SU_{m+n}, also satisfy the unitarity relation (7-116), and are closely related to the SDC $\left\langle \begin{matrix}[\nu] \\ m\end{matrix}\middle|[\nu], \begin{matrix}\beta[\nu_1][\nu_2] \\ m_1\ m_2\end{matrix}\right\rangle$

of the permutation group. Following the line of reasoning in Sec. 7.7 which led to (7-95), from (7-94) and (4-168a) we obtain

$$\left|[\nu],\, \begin{matrix}\beta[\nu_1][\nu_2] \\ W_1\, W_2\end{matrix}\right\rangle R^{[\nu_1]m_1}(\overline{\omega}\,{}^0_1)R^{[\nu_2]m_2}(\overline{\omega}\,{}^0_2)$$

$$= \sum_m \left|\begin{matrix}[\nu] \\ W\end{matrix}\right\rangle R^{[\nu]m}(\overline{\omega}\,{}^0)\left\langle \begin{matrix}[\nu] \\ m\end{matrix}\right|\left.[\nu],\, \begin{matrix}\beta[\nu_1][\nu_2] \\ m_1\, m_2\end{matrix}\right\rangle . \tag{7-145}$$

Comparing (7-144) with (7-145), we obtain the relation between the two coefficients as follows:

$$\left\langle \begin{matrix}[\nu] \\ W\end{matrix}\right|\left.[\nu],\, \begin{matrix}\beta[\nu_1][\nu_2] \\ W_1\, W_2\end{matrix}\right\rangle$$

$$= [R^{[\nu_1]m_1}(\overline{\omega}\,{}^0_1)R^{[\nu_2]m_2}(\overline{\omega}\,{}^0_2)]^{-1}\sum_m{}' \; R^{[\nu]m}(\overline{\omega}\,{}^0)\left\langle \begin{matrix}[\nu] \\ m\end{matrix}\right|\left.[\nu],\, \begin{matrix}\beta[\nu_1][\nu_2] \\ m_1\, m_2\end{matrix}\right\rangle , \tag{7-146a}$$

where $(\overline{\omega}\,{}^0)$, $(\overline{\omega}\,{}^0_1)$ and $(\overline{\omega}\,{}^0_2)$ are the normal order states in the Weyl tableaux W, W_1 and W_2, respectively, i.e., the Young tableaux $Y^{[\nu]}_m(\overline{\omega}\,{}^0)$, $Y^{[\nu_1]}_{m_1}(\overline{\omega}\,{}^0_1)$ and $Y^{[\nu_2]}_{m_2}(\overline{\omega}\,{}^0_2)$ in state space. The prime in the summation symbol indicates that the sum is restricted to those m which give rise to the same Weyl tableau W.

Equation (7-146a) shows that once the SDC of the permutation group are known, the SDC of SU_{m+n} can be found for arbitrary m and n. Again, the values of the SU_{m+n} SDC do not depend on m and n explicitly.

In the case when the single particle states in the Weyl tableau W are all different, only one term survives in the summation of (7-146a) and all the normalization constants R become unity; hence

$$\left\langle \begin{matrix}[\nu] \\ W\end{matrix}\right|\left.[\nu],\, \begin{matrix}\beta[\nu_1][\nu_2] \\ W_1\, W_2\end{matrix}\right\rangle = \left\langle \begin{matrix}[\nu] \\ m\end{matrix}\right|\left.[\nu],\, \begin{matrix}\beta[\nu_1][\nu_2] \\ m_1\, m_2\end{matrix}\right\rangle . \tag{7-146b}$$

For example, by letting the ordinals 1,2,3,4 in the Young tableaux of Table 4.18 be the state indices $\alpha, \beta, \gamma, \delta$, respectively, the Young tableaux become the Weyl tableaux and the S_f SDC $\left\langle \begin{matrix}[\nu] \\ m\end{matrix}\right|\left.[\nu]\,,\, \begin{matrix}\beta[\nu_1][\nu_2] \\ m_1\, m_2\end{matrix}\right\rangle$ become the SU_{m+n} SDC $\left\langle \begin{matrix}[\nu] \\ W\end{matrix}\right|\left.[\nu],\, \begin{matrix}\beta[\nu_1][\nu_2] \\ W_1\, W_2\end{matrix}\right\rangle$ such as

$$\left|[31],\, \begin{array}{|c|}\hline \alpha \\ \hline\end{array}\; \begin{array}{|c|c|}\hline \beta & \gamma \\ \hline \delta \\ \cline{1-1}\end{array}\right\rangle = \frac{\sqrt{8}}{3}\left|\begin{array}{|c|c|c|}\hline \alpha & \beta & \gamma \\ \hline \delta \\ \cline{1-1}\end{array}\right\rangle - \frac{1}{6}\left|\begin{array}{|c|c|c|}\hline \alpha & \beta & \delta \\ \hline \gamma \\ \cline{1-1}\end{array}\right\rangle - \frac{1}{\sqrt{12}}\left|\begin{array}{|c|c|c|}\hline \alpha & \gamma & \delta \\ \hline \beta \\ \cline{1-1}\end{array}\right\rangle .$$

Following the steps similar to those for calculating the SU_n CG coefficients from the permutation group IDC, we can obtain the SU_{m+n} SDC from the permutation group SDC by using (7-146a) and the normalization coefficients given in Table 4.8.

Thus by letting $(\omega^0) = (\alpha\beta\beta\delta)$ in Table 4.18-1, the Young tableaux in Table 4.18-1 become the Weyl tableaux in Table 7.12-1. It is seen that there are two β's in the tableau W_2 at the third row, thus

$$\left|[31]\,,\, \begin{array}{|c|}\hline \alpha \\ \hline\end{array}\; \begin{array}{|c|c|}\hline \beta & \gamma \\ \hline \delta \\ \cline{1-1}\end{array}\right\rangle = 0\,.$$

In Table 4.18-1, dividing the first and second rows by $R^{[3]}(\beta\beta\delta) = \sqrt{2}$ and $R^{[21]1}(\beta\beta\delta) = \sqrt{2}$, respectively, and multiplying the first, second and third columns by $R^{[31]1}(\alpha\beta\beta\delta) = \sqrt{2}, R^{[31]2}$ $(\alpha\beta\beta\delta) = \sqrt{\frac{1}{2}}$, and $R^{[31]3}(\alpha\beta\beta\delta) = \sqrt{\frac{3}{2}}$, respectively, we obtain Table 7.12-1; on adding the second and third columns which have the same heading, we obtain the SU_{m+n} SDC shown in Table 7.12-2.

Table 7.12-1. (Intermediate step)

	$\begin{array}{lll}\alpha&\beta&\beta\\\delta\end{array}$	$\begin{array}{lll}\alpha&\beta&\delta\\\beta\end{array}$	$\begin{array}{lll}\alpha&\beta&\delta\\\beta\end{array}$
[31], α $\begin{array}{lll}\beta&\beta&\delta\end{array}$	$-\frac{1}{3}$	$-\frac{\sqrt{2}}{6}$	$-\frac{\sqrt{2}}{2}$
[31], α $\begin{array}{ll}\beta&\beta\\\delta\end{array}$	$\frac{\sqrt{8}}{3}$	$-\frac{1}{12}$	$-\frac{1}{4}$
[31], α $\begin{array}{ll}\beta&\delta\\\beta\end{array}$	0	0	0

Table 7.12-2. $\left\langle {[\nu] \atop W} \middle| [\nu], {[\nu_1] \atop W_1} {[\nu_2] \atop W_1} \right\rangle$

	$\begin{array}{lll}\alpha&\beta&\beta\\\delta\end{array}$	$\begin{array}{lll}\alpha&\beta&\delta\\\beta\end{array}$
[31], α $\begin{array}{lll}\beta&\beta&\delta\end{array}$	$-\frac{1}{3}$	$-\frac{\sqrt{8}}{3}$
[31], α $\begin{array}{ll}\beta&\beta\\\delta\end{array}$	$\frac{\sqrt{8}}{3}$	$-\frac{1}{3}$

Furthermore, by letting the normal order state $(\overline{\omega}^0) = (\alpha\beta\beta\delta)$ in Table 7.11-2 be equal to other possible states, we obtain the SDC for arbitrary SU_{m+n}. This is illustrated by setting

$$(\overline{\omega}_0) = (\alpha\beta\beta\delta) = (1223), (1224), (1225), (1334), (1335), \\ (1445), (2334), (2335), (2445), (3445)$$

we obtain the SDC of SU_5. Consequently, each SDC table gives an infinite number of coefficients with the same structure. The SU_{m+n} SDC are tabulated in Chen ([31] 1987).

7.13. The Isoscalar Factors and the Fractional Parentage Coefficients

7.13.1. Isoscalar factors

1. The Gel'fand basis. The factorization formula (3-303) of the CG coefficients also applies to the Lie group. Since the Gel'fand basis is the $SU_n \supset SU_{n-1} \otimes U_1 \supset \ldots \supset SU_2 \otimes U_1 \supset SO_2$ irreducible basis, its CG coefficients can be decomposed into a product of a set of $SU_i \supset SU_{i-1} \times U_1$ ISF (summing over the intermediate multiplicity labels, if there are any). The $SU_i \supset SU_{i-1} \times U_1$ ISF is also called the $SU_{i-1} singlet factor$, abbreviated as SU_{i-1} SF. Thus the CG coefficients of SU_n for the Gel'fand basis can be expressed schematically as

$$(SU_n\text{CGC}) = (SU_{n-1}\text{SF})(SU_{n-2}\text{SF})\ldots(SU_2\text{SF})(SU_2\text{CGC})\,. \tag{7-147}$$

The SU_2 CG coefficients are just the SU_1SF. Therefore the evaluation of the CG coefficients of SU_n is reduced to that of SU_iSF. From (3-303) we get:

Example 1: The CG coefficients of SU_3:

$$\begin{array}{ccc} C^{[\mu]_\theta,IYI_z}_{[\mu_1]I_1Y_1I_{1z},[\mu_2]I_2Y_2I_{2z}} = & C^{[\mu]_\theta,IY}_{[\mu_1]I_1Y_1,[\mu_2]I_2Y_2} & C^{I,I_z}_{I_1I_{1z},I_2I_{2z}} \\ SU_3\text{CGC} & SU_2\text{SF} & SU_2\text{CGC} \end{array} \tag{7-148}$$

Example 2: The CG coefficients of SU_4:

$$\begin{array}{cccc} C^{[\nu]_\tau,[\mu]ZIYI_z}_{[\nu_1][\mu_1]Z_1I_1Y_1I_{1z},[\nu_2][\mu_2]Z_2I_2Y_2I_{2z}} = \sum\limits_\theta & C^{[\nu]_\tau,[\mu]_\theta Z}_{[\nu_1][\mu_1]Z_1,[\nu_2][\mu_2]Z_2} & C^{[\mu]_\theta,IY}_{[\mu_1]I_1Y_1,[\mu_2]I_2Y_2} & C^{I,I_z}_{I_1I_{1z},I_2I_{2z}} \\ & SU_3\text{SF} & SU_2\text{SF} & SU_2\text{CGC} \end{array} \tag{7-149}$$

The tables of the SU_2SF and SU_3SF are given by de Swart (1963) and Haacke (1976), respectively. A new method for calculating the SU_nSF will be given in Sec. 7.17.

2. The $SU_{mn} \supset SU_m \times SU_n$ and $SU_{m+n} \supset SU_m \otimes SU_n$ irreducible bases. Both irreducible bases will be designated by

$$\left| \begin{array}{c} [\nu] \\ \beta[\sigma]W_1[\mu]W_2 \end{array} \right\rangle, \quad \begin{array}{l} \beta = 1, 2, \ldots, (\nu_1\nu_2\nu)\ \text{for}\ SU_{mn}\,, \\ \beta = 1, 2, \ldots, \{\nu_1\nu_2\nu\}\ \text{for}\ SU_{m+n}\,, \end{array} \tag{7-150}$$

where ν, σ and μ are the irrep labels for SU_{mn} (or SU_{m+n}), SU_m and SU_n, respectively.
Setting $\Lambda = [\sigma][\mu], m = W_1W_2$ in (3-300) we get

$$\left| \begin{matrix} [\nu]\tau \\ \beta[\sigma]W_1[\mu]W_2 \end{matrix} \right\rangle = \sum C^{[\nu]_\tau,\beta[\sigma][\mu]W_1W_2}_{[\nu']\beta'\sigma'\mu'W_1'W_2',[\nu'']\beta''\sigma''\mu''W_1''W_2''} \times \left| \begin{matrix} [\nu'] \\ \beta'[\sigma']W_1'[\mu']W_2' \end{matrix} \right\rangle \left| \begin{matrix} [\nu''] \\ \beta''[\sigma'']W_1''[\mu'']W_2'' \end{matrix} \right\rangle , \tag{7-151}$$

where the sum runs over $\beta'\sigma'\mu'W_1'W_2'\beta''\sigma''\mu''W_1''W_2''$. Analogously, by index substitution, from (3-303) we obtain the factorization formula for the $SU_{mn}(SU_{m+n})$ CG coefficients

$$C^{[\nu]_\tau,\beta[\sigma][\mu]W_1W_2}_{[\nu']\beta'\sigma'\mu'W_1'W_2',[\nu'']\beta''\sigma''\mu''W_1''W_2''} = \sum_{\theta\varphi} C^{[\nu]_\tau,\beta[\sigma]_\theta[\mu]_\varphi}_{[\mu']\beta'\sigma'\mu',[\nu'']\beta''\sigma''\mu''} C^{[\sigma]_\theta W_1}_{\sigma'W_1'\sigma''W_1''} C^{[\mu]_\varphi W_2}_{\mu'W_2',\mu''W_2''} ,$$
$$\theta = 1, 2, \ldots, \{\sigma'\sigma''\sigma\}, \quad \varphi = 1, 2, \ldots, \{\mu'\mu''\mu\} , \tag{7-152}$$

where the first factor at the right-hand side is the $SU_{mn} \supset SU_m \times SU_n$ or $SU_{m+n} \supset SU_m \otimes SU_n$ ISF, and the second and third factors are the SU_m and SU_n CG coefficients, respectively. Equation (7-152) can be expressed schematically as

$$(SU_{mn} \supset SU_m \times SU_n\text{CGC}) = (SU_{mn} \supset SU_m \times SU_n\text{ISF})(SU_m\text{CGC})(SU_n\text{CGC}),$$
$$(SU_{m+n} \supset SU_m \otimes SU_n\text{CGC}) = (SU_{m+n} \supset SU_m \otimes SU_n\text{ISF})(SU_m\text{CGC})(SU_n\text{CGC}) . \tag{7-153}$$

By index substitutions in (3-305) and (3-306), we obtain the unitarity relation of the ISF. For fixed $[\sigma]$ and $[\mu]$ we have

$$\sum_{\substack{\beta'\sigma'\mu'\theta \\ \beta''\sigma''\mu''\varphi}} C^{[\nu]_\tau,\beta[\sigma]_\theta[\mu]_\varphi}_{[\nu']\beta'[\sigma'][\mu'],[\nu'']\beta''[\sigma''][\mu'']} C^{[\overline{\nu}]\overline{\tau},\overline{\beta}[\sigma]_\theta[\mu]_\varphi}_{[\nu']\beta'[\sigma'][\mu'],[\nu'']\beta''[\sigma''][\mu'']} = \delta_{\nu\overline{\nu}}\delta_{\beta\overline{\beta}}\delta_{\tau\overline{\tau}} , \tag{7-154a}$$

$$\sum_{\nu\beta\tau} C^{[\nu]\tau,\beta[\sigma]_\theta[\mu]_\varphi}_{[\nu']\beta'[\sigma'][\mu'],[\nu'']\beta''[\sigma''][\mu'']} C^{[\nu]\tau,\beta[\sigma]\overline{\theta}[\mu]\overline{\varphi}}_{[\nu']\overline{\beta}'[\overline{\sigma}'][\overline{\mu}'],[\nu'']\overline{\beta}''[\overline{\sigma}''][\overline{\mu}'']}$$
$$= \delta_{\beta'\overline{\beta}'}\delta_{\beta''\overline{\beta}''}\delta_{\sigma'\overline{\sigma}'}\delta_{\sigma''\overline{\sigma}''}\delta_{\mu'\overline{\mu}'}\delta_{\mu''\overline{\mu}''}\delta_{\theta\overline{\theta}}\delta_{\varphi\overline{\varphi}} . \tag{7-154b}$$

The calculation methods for the $SU_{mn} \supset SU_m \times SU_n$ and $SU_{m+n} \supset SU_m \otimes SU_n$ ISF are given in Sec. 7.16 and Sec. 7.17.

Example 3: The $SU_4 \supset SU_2 \times SU_2$ ISF. Now the labels θ and φ are redundant, and the irreps of SU_2 are labeled by the spin S or isospin T. Equation (7-152) takes the following form:

$$C^{[\nu]_\tau,\beta STM_SM_T}_{[\nu_1]\beta_1S_1T_1M_{S_1}M_{T_1},[\nu_2]\beta_2S_2T_2M_{S2}M_{T2}} = C^{[\nu]\tau,\beta ST}_{[\nu_1]\beta_1S_1T_1,[\nu_2]\beta_2S_2T_2} C^{SM_S}_{S_1M_{S_1},S_2M_{S_2}} C^{TM_T}_{T_1M_{T_1},T_2M_{T_2}} \tag{7-155}$$

Example 4: The $SU_6 \supset SU_3 \times SU_2$ ISF. $[\nu], [\mu]$ and S are used to label the irreps of SU_6, SU_3 and SU_2, respectively. The $SU_6 \supset SU_3 \times SU_2$ CG coefficients are expressed as

$$C^{[\nu]_\tau,\beta[\mu]IYSI_zS_z}_{[\nu_1]\beta_1[\mu_1]I_1Y_1S_1I_{1z}S_{1z},[\nu_2]\beta_2[\mu_2]I_2Y_2S_2I_{2z}S_{2z}}$$
$$= \sum_\theta C^{[\nu]_\tau,\beta[\mu]_\theta S}_{[\nu_1]\beta_1[\mu_1]S_1,[\nu_2]\beta_2[\mu_2]S_2} C^{[\mu]_\theta,IYI_z}_{[\mu_1]I_1Y_1I_{1z},[\mu_2]I_2Y_2I_{2z}} C^{S,S_z}_{S_1S_{1z}S_2S_{2z}} \tag{7-156}$$

Example 5: The $SU_{2l+1} \supset SO_3$ ISF. The $SU_{2l+1} \supset SO_3$ irreducible basis is denoted by $\left|\begin{matrix}[\nu]\\ \alpha LM\end{matrix}\right\rangle$, where α takes care of the multiple occurrence of the irrep L of SO_3 in the irrep $[\nu]$ of SU_{2l+1}. We have

$$\begin{array}{ll} C^{[\nu]_\tau,\alpha LM}_{[\nu_1]\alpha_1L_1M_1,[\nu_2]\alpha_2L_2M_2} & = C^{[\nu]_\tau,\alpha L}_{[\nu_1]\alpha_1L_1,[\nu_2]\alpha_2L_2} C^{LM}_{L_1M_1,L_2M_2} \, . \\ SU_{2l+1} \supset SO_3\text{CGC}, & SU_{2l+1} \supset SO_3\text{ISF}, \ SO_3\text{CGC} \end{array} \tag{7-157}$$

Equations (3-302) and (3-307) now take the following form:

$$\left|\begin{matrix}[\nu]_\tau\\ \alpha LM\end{matrix}\right\rangle = \sum_{\alpha_1L_1\alpha_2L_2} C^{[\nu]_\tau,\alpha L}_{[\nu_1]\alpha_1L_1,[\nu_2]\alpha_2L_2} \left[\left|\begin{matrix}[\nu_1]\\ \alpha_1L_1\end{matrix}\right\rangle \left|\begin{matrix}[\nu_2]\\ \alpha_2L_2\end{matrix}\right\rangle\right]^L_M , \tag{7-158a}$$

$$\left[\left|\begin{matrix}[\nu_1]\\ \alpha_1L_1\end{matrix}\right\rangle \left|\begin{matrix}[\nu_2]\\ \alpha_2L_2\end{matrix}\right\rangle\right]^L_M = \sum_{\nu\tau\alpha} C^{[\nu]_\tau,\alpha L}_{[\nu_1]\alpha_1L_1,[\nu_2]\alpha_2L_2} \left|\begin{matrix}[\nu]_\tau\\ \alpha LM\end{matrix}\right\rangle , \tag{7-158b}$$

where the square brackets denote angular momentum coupling.

7.13.2. The orbital fractional parentage coefficients (CFP)

1. The single particle CFP. Suppose that there are $n-1$ particles in the orbital l and they are in a state $\left|\begin{matrix}l^{n-1}[\nu_1]\\ m_1,\alpha_1L_1M_1\end{matrix}\right\rangle$, which is the Yamanouchi basis $[\nu_1]m_1$ of S_{n-1} and the $SU_{2l+1} \supset SO_3$ irreducible basis $\left|\begin{matrix}[\nu_1]\\ \alpha_1L_1M_1\end{matrix}\right\rangle$. Our aim is to construct a similar state $\left|\begin{matrix}l^n[\nu]\\ m,\alpha LM\end{matrix}\right\rangle$ for the n-particle system. Since $\left|\begin{matrix}l^{n-1}[\nu_1]\\ m_1,\alpha_1L_1M_1\end{matrix}\right\rangle \psi_{lm_l}(n)$ form a complete set, we have

$$\left|\begin{matrix}l^n[\nu]\\ m,\alpha LM\end{matrix}\right\rangle = \sum_{\alpha_1L_1} (l^{n-1}[\nu_1]\alpha_1L_1, l|\}l^n[\nu]\alpha L) \left[\left|\begin{matrix}l^{n-1}[\nu_1]\\ m_1,\alpha_1L_1\end{matrix}\right\rangle \psi_l(n)\right]^L_M . \tag{7-159a}$$

The coefficients $(l^{n-1}[\nu_1]\alpha_1L_1, l|\}l^n[\nu]\alpha L)$ are called the *orbital single-particle CFP* and $\left|\begin{matrix}l^{n-1}[\nu_1]\\ m_1,\alpha_1L_1\end{matrix}\right\rangle$ are called the parent states. The quantum number ν_1m_1 and νm are related by $[\nu]m = [\nu][\nu_1]m_1$, in other words, by deleting the last box in $Y^{[\nu]}_m$, the resulting tableau is $Y^{[\nu_1]}_{m_1}$.

On the other hand, according to (7-158a), we can use the $SU_{2l+1} \supset SO_3$ ISF to couple $\left|\begin{matrix}l^{n-1}[\nu_1]\\ m_1\alpha_1L_1\end{matrix}\right\rangle$ and $\psi^{[1]}_{lm_l}(n)$ into another $SU_{2l+1} \supset SO_3$ irreducible basis,

$$\left|\begin{matrix}l^n[\nu]\\ m,\alpha LM\end{matrix}\right\rangle = \sum_{\alpha_1L_1} C^{[\nu]\alpha L}_{[\nu_1]\alpha_1L_1,[1]l} \left[\left|\begin{matrix}l^{n-1}[\nu_1]\\ m_1,\alpha_1L_1\end{matrix}\right\rangle \psi^{[1]}_l(n)\right]^L_M . \tag{7-159b}$$

Since the ISF are independent of the quantum numbers m_1 and M, (7-159b) can be simply written as

$$|[\nu]\alpha L\rangle = \sum_{\alpha_1L_1} C^{[\nu]\alpha L}_{[\nu_1]\alpha_1L_1,[1]l} |[\nu_1]\alpha_1L_1\rangle\psi(n) , \tag{7-159c}$$

where the symbol, $[\ \ ]^L_M$, for angular momentum coupling is omitted. Examples of (7-159c) are given in (7-191) and (7-197).

By the definition of ISF, the left-hand side of (7-159b) must belong to the irrep $[\nu]$ of SU_{2l+1}, which implies by Theorem 7.1, that it also belongs to the irrep $[\nu]$ of S_n. In addition, once the quantum numbers $[\nu], [\nu_1]$ and $[m_1]$ are given, the quantum number m is also fixed. Therefore the left-hand side of (7-159b) is the Yamanouchi basis $[\nu]m$. Comparing (7-159a) with (7-159b) we know that the $SU_{2l+1} \supset SO_3$ ISF are just the orbital CFP,

$$(l^{n-1}[\nu_1]\alpha_1 L_1, l|\}l^n[\nu]\alpha L) = C^{[\nu],\alpha L}_{[\nu_1]\alpha_1 L_1,[1]l} \,. \tag{7-160}$$

They satisfy the unitarity relation

$$\sum_{\alpha_1 L_1} C^{[\nu],\alpha L}_{[\nu_1]\alpha_1 L_1[1]l} C^{[\nu']\alpha' L}_{[\nu_1]\alpha_1 L_1,[1]l} = \delta_{\nu\nu'}\delta_{\alpha\alpha'} \,, \tag{7-161a}$$

$$\sum_{\nu\alpha} C^{[\nu],\alpha L}_{[\nu_1]\alpha_1 L_1,[1]l} C^{[\nu],\alpha L}_{[\nu_1]\alpha_1' L_1',[1]l} = \delta_{\alpha_1\alpha_1'}\delta_{L_1 L_1'} \,. \tag{7-161b}$$

The inverse of (7-159b) is

$$\left[\left|\begin{matrix} l^{n-1}[\nu_1] \\ m_1, \alpha_1 L_1 \end{matrix}\right\rangle \psi_l(n)\right]^L_M = \sum_{\nu\alpha} C^{[\nu],\alpha L}_{[\nu_1]\alpha_1 L_1,[1]l} \left|\begin{matrix} l^n[\nu] \\ m, \alpha L M \end{matrix}\right\rangle \,. \tag{7-162}$$

Between the groups SU_{2l+1} and SU_3, we can insert another subgroup or a subgroup chain of SU_{2l+1} which contains SO_3 as its subgroup and whose irrep labels may be served as the additional label α. For example, for the d shell, one usually inserts the group SO_5 between SU_5 and SO_3. The $SU_5 \supset SO_5 \supset SO_3$ irreducible basis of an n-particle system is

$$\left|\begin{matrix} d^n[\nu] \\ m, (\omega) L M \end{matrix}\right\rangle = \sum_{\omega_1 L_1} C^{[\nu],(\omega)L}_{[\nu_1](\omega_1)L_1,[1](\omega_2)d} \left[\left|\begin{matrix} d^{n-1}[\nu_1] \\ m_1, (\omega_1) L_1 \end{matrix}\right\rangle \psi_d(n)\right]^L_M \,, \tag{7-163}$$

where (ω) is the irrep label of SO_5. The first factor at the right-hand side of (7-163) is the $SU_5 \supset SO_5 \supset SO_3$ ISF which can be factorized as

$$C^{[\nu],(\omega)L}_{[\nu_1](\omega_1)L_1,[1](\omega_2)d} = C^{[\nu],(\omega)}_{[\nu_1](\omega_1),[1](\omega_2)} C^{(\omega),L}_{(\omega_1)L_1,(\omega_2)d} \,. \tag{7-164}$$

Further discussion can be found in Jahn (1950, 1951).

The definition of ISF can be extended to the mixed configuration case, such as the $2s, 1d$ shell. In analogy with (7-163), Elliott's $SU_6 \supset SU_3 \supset SO_3$ irreducible basis (Elliott 1958) is now expressed as

$$\left|\begin{matrix} [\nu] \\ m, \beta(\lambda\mu) K L M \end{matrix}\right\rangle$$

$$= \sum_{\beta_1\lambda_1\mu_1 K_1 L_1 L_2} C^{[\nu],\beta(\lambda\mu)}_{[\nu_1]\beta_1(\lambda_1\mu_1),[1](20)} C^{(\lambda\mu),KL}_{(\lambda_1\mu_1)K_1 L_1,(20)L_2} \left[\left|\begin{matrix} [\nu_1] \\ m_1, \beta_1(\lambda_1\mu_1) K_1 L_1 \end{matrix}\right\rangle \psi_{L_2}(n)\right]^L_M \,, \tag{7-165}$$

where $[\nu]$, $(\lambda\mu)$ are the irrep labels of SU_6 and SU_3, respectively, $L_2 = 0, 2$. The first and second factors in (7-165) are the $SU_6 \supset SU_3$ and $SU_3 \supset SO_3$ ISF, respectively. Notice that a single particle in the $2s, 1d$ shell belongs to the irrep (20) of SU_3.

2. Many-particle CFP. We now consider the CFP for separating n_2 particles from an n-particle system. We still use the symbols $(\omega_1^0) = (1, 2, \ldots, n_1)$ and $(\omega_2^0) = (n_1 + 1, \ldots, n), n =$

n_1+n_2, and use the $SU_{2l+1}\supset SO_3$ classification scheme as an example. The generalization to other group chains is straightforward.

In the foregoing discussion, we generally used the Yamanouchi basis. However, in discussing the many-particle CFP, it is more convenient to use the $S_n\supset S_{n_1}\otimes S_{n_2}$ irreducible basis. The $S_n\supset S_{n_1}\otimes S_{n_2}$ and $SU_{2l+1}\supset SO_3$ irreducible basis can be expanded in terms of the $SU_{2l+1}\supset SO_3$ ISF,

$$\left|\begin{matrix}[\nu]\\ \tau\nu_1 m_1\nu_2 m_2,\alpha LM\end{matrix}\right\rangle=\sum_{\alpha_1L_1\alpha_2L_2} C^{[\nu]\tau,\alpha L}_{\nu_1\alpha_1L_1,\nu_2\alpha_2L_2}\left[\left|\begin{matrix}l^{n_1}[\nu_1]\\ m_1\omega_1^0,\alpha_1L_1\end{matrix}\right\rangle\left|\begin{matrix}l^{n_2}[\nu_2]\\ m_2\omega_2^0,\alpha_2L_2\end{matrix}\right\rangle\right]^L_M, \tag{7-166}$$

where the first factor in the right-hand side of (7-166) is the $SU_{2l+1}\supset SO_3$ ISF, and is also called the *many-particle CFP*, and often written as

$$(l^{n_1}[\nu_1]\alpha_1L_1,l^{n_2}[\nu_2]\alpha_2L_2|\}l^n[\nu]\tau\alpha L)=C^{[\nu]_\tau,\alpha L}_{\nu_1\alpha_1L_1,\nu_2\alpha_2L_2}\,. \tag{7-167}$$

If the Yamanouchi basis is used, we have

$$\left|\begin{matrix}l^n[\nu]\\ m,\alpha LM\end{matrix}\right\rangle=\sum_{\substack{\nu_2m_2\tau\\ \alpha_1L_1\alpha_2L_2}}\left\langle\begin{matrix}[\nu]\\ m\end{matrix}\right|[\nu],\left.\begin{matrix}\tau\nu_1\nu_2\\ m_1m_2\end{matrix}\right\rangle C^{[\nu]_\tau,\alpha L}_{\nu_1\alpha_1L_1,\nu_2\alpha_2L_2}\left[\left|\begin{matrix}l^{n_1}[\nu_1]\\ m_1\omega_1^0,\alpha_1L_1\end{matrix}\right\rangle\left|\begin{matrix}l^{n_2}[\nu_2]\\ m_2\omega_2^0,\alpha_2L_2\end{matrix}\right\rangle\right]^L_M. \tag{7-168}$$

For a general $G\supset G(s)$ irreducible basis, (7-166) and (7-168) are replaced by (7-102b) and (7-102c).

In summary, the so-called CFP is either an ISF or a certain kind of CG coefficient.

For instance, the two-particle orbital CFP given by Harvey (1981, Table 4) are just the well known SU_2 coefficients.

3. The two-particle CFP. The one-particle and two-particle CFP are the most useful ones, and the latter can be expressed in terms of the former. Let $Y^{[\nu']}_{m'}(Y^{[\nu_1]}_{m_1})$ be the Young tableau resulting from deleting the box $n(n-1)$ in the Young tableau $Y^{[\nu]}_m(Y^{[\nu']}_{m'})$, respectively. From (7-159b) we have

$$\begin{aligned}\left|\begin{matrix}l^n[\nu]\\ m,\alpha LM\end{matrix}\right\rangle&=\sum_{\alpha'L'}C^{[\nu],\alpha l}_{[\nu']\alpha'L',[1]l}\left[\left|\begin{matrix}l^{n-1}[\nu']\\ m',\alpha'L'\end{matrix}\right\rangle\psi_l(n)\right]^L_M\\ &=\sum_{\alpha'L'\alpha_1L_1}C^{[\nu],\alpha L}_{[\nu']\alpha'L',[1]l}C^{[\nu'],\alpha'L'}_{[\nu_1]\alpha_1L_1,[1]l}\left[\left[\left|\begin{matrix}l^{n-2}[\nu_1]\\ m_1,\alpha_1L_1\end{matrix}\right\rangle\psi_l(n-1)\right]^{L'}\psi_l(n)\right]^L_M\\ &=\sum_{\alpha'L'\alpha_1L_1L_2}C^{[\nu],\alpha L}_{[\nu']\alpha'L',[1]l}C^{[\nu'],\alpha'L'}_{[\nu_1]\alpha_1L_1,[1]l}U(L_1lLl;L'L_2)\\ &\quad\times\left[\left|\begin{matrix}l^{n-2}[\nu_1]\\ m_1\omega_1^0,\alpha_1L_1\end{matrix}\right\rangle\left|\begin{matrix}l^2[\nu_2]\\ m_2\omega_2^0,\alpha_2L_2\end{matrix}\right\rangle\right]^L_M C^{[\nu_2]L_2}_{[1]l,[1]l}\,,\end{aligned} \tag{7-169}$$

where $[\nu_2]=[2]$ or $[11]$ and is one-dimensional; therefore $m_2=1$ and will be ignored henceforth. Setting $n_2=2$ in (7-167) and comparing it with (7-169), we get an expression for the two-particle CFP:

$$C^{[\nu]\alpha L}_{[\nu_1]\alpha_1L_1,[\nu_2]\alpha_2L_2}=\left\langle\begin{matrix}[\nu]\\ m\end{matrix}\right|[\nu],\left.\begin{matrix}[\nu_1][\nu_2]\\ m_1\end{matrix}\right\rangle^{-1}\sum_{\alpha'L'}C^{[\nu]\alpha L}_{[\nu']\alpha'L',[1]l}C^{[\nu']\alpha'L'}_{[\nu_1]\alpha_1L_1,[1]l}\times U(L_1lLl;L'L_2)C^{[\nu_2]L_2}_{[1]l,[1]l}\,. \tag{7-170}$$

Notice that the two-particle CFP in (7-170) are independent of the choice of the quantum number m_1.

Example 6: For the $1p$ shell, $[\nu]=[42], [\nu_1]=[31]$, there are two possible choices:

(i) $\left|\begin{matrix}[\nu]\\ m\end{matrix}\right\rangle = \Big|$ 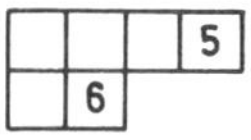 $\Big\rangle$, $[\nu']=[41]$

(ii) $\left|\begin{matrix}[\nu]\\ m\end{matrix}\right\rangle = \Big|$ 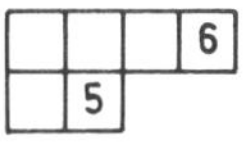 $\Big\rangle$, $[\nu']=[32]$ (7-171a)

Correspondingly, we have

$$\text{(i)}\ C^{[42],\alpha L}_{[31]L_1,[2]L_2} = \left(\sqrt{\frac{\sigma-1}{2\sigma}}\right)^{-1} \sum_{L'} C^{[42],\alpha L}_{[41]L',[1]1} C^{[41],L'}_{[31]L_1,[1]1} U(L_1 1 L 1; L' L_2) .$$

$$\text{(ii)}\ C^{[42],\alpha L}_{[31]L_1,[2]L_2} = \left(\sqrt{\frac{\sigma+1}{2\sigma}}\right)^{-1} \sum_{L'} C^{[42],\alpha L}_{[32]L',[1]1} C^{[32]L'}_{[31]L_1,[1]1} U(L_1 1 L 1; L' L_2) , \tag{7-171b}$$

where (4-184) has been used. σ is the axial distance (here $\sigma=3$). Using the one-particle CFP given by Jahn (1951), it is easily verified that (7-171a) and (7-171b) give identical results.

For the $j-j$ coupling, we usually take $SU_{2j+1} \supset SP_{2j+1} \supset SO_3$ irreducible basis. In analogy with (7-166), we have

$$\left|\begin{matrix}[\nu]\\ \nu_1 m_1 \nu_2 m_2, \alpha J M\end{matrix}\right\rangle = \sum_{\alpha_1 J_1 \alpha_2 J_2} C^{[\nu]\alpha J}_{\nu_1\alpha_1 J_1,\nu_2\alpha_2 J_2} \left[\left|\begin{matrix} j^{n_1}[\nu_1]\\ m_1\omega_0^1, \alpha_1 J_1\end{matrix}\right\rangle \left|\begin{matrix} j^{n_2}[\nu_2]\\ m_2\omega_0^2, \alpha_2 J_2\end{matrix}\right\rangle\right]^J_M , \tag{7-172}$$

where $\alpha=(v\alpha')$, v is the irrep label of SP_{2j+1}, called the seniority (see Sec. 9.5) and α' is an additional quantum number. Here we dropped the redundant index τ, since the Young diagrams $[\nu_1]$ and $[\nu_2]$ are at most of two columns, and according to the Littlewood rule, their product is multiplicity free. The $SU_{2j+1} \supset SP_{2j+1} \supset SO_3$ ISF in (7-172) is the CFP for the $j-j$ coupling,

$$(j^{n_1}[\nu_1]\alpha_1 J_1, j^{n_2}[\nu_2]\alpha_2 J_2 | \} j^n[\nu]\alpha J) = C^{[\nu]\alpha J}_{\nu_1\alpha_1 J_1,\nu_2\alpha_2 J_2} . \tag{7-173}$$

For further discussion, see de-Shalit (1963).

7.13.3. The spin-isospin CFP

1. One-particle CFP. As was the case with (7-159b), one has

$$\left|\begin{matrix}\gamma^n[\tilde{\nu}]\\ \tilde{m}, \beta S T M_S M_T\end{matrix}\right\rangle = \sum_{\beta_1 S_1 T_1} C^{[\tilde{\nu}]\beta ST}_{[\tilde{\nu}_1]\beta_1 S_1 T_1,[1]\frac{1}{2}\frac{1}{2}} \left[\left|\begin{matrix}\gamma^{n-1}[\tilde{\nu}_1]\\ \tilde{m}_1, \beta_1 S_1 T_1\end{matrix}\right\rangle \psi_{\frac{1}{2}\frac{1}{2}}(n)\right]^{ST}_{M_S M_T} . \tag{7-174}$$

Equation (7-174) is the Yamanouchi basis $[\tilde{\nu}]\tilde{m}$ of S_n and the $SU_4 \supset SU_2 \times SU_2$ irreducible basis $[\tilde{\nu}]ST$. The $SU_4 \supset SU_2 \times SU_2$ ISF are just the spin-isospin CFP, and are usually designated by

$$(\gamma^{n-1}[\tilde{\nu}_1]\beta_1 S_1 T_1, \gamma | \} \gamma^n[\tilde{\nu}]\beta S T) = C^{[\tilde{\nu}]\beta ST}_{[\tilde{\nu}_1]\beta_1 S_1 T_1,[1]\frac{1}{2}\frac{1}{2}} . \tag{7-175}$$

2. Many-particle CFP. Analogously to (7-166), one has

$$\left|\begin{matrix}[\tilde{\nu}]\\ \tau\tilde{\nu}_1\tilde{m}_1\tilde{\nu}_2\tilde{m}_2, \beta S T M_S M_T\end{matrix}\right\rangle$$

$$= \sum_{\beta_1 S_1 T_1 \beta_2 S_2 T_2} C^{[\tilde{\nu}]\tau,\beta ST}_{\tilde{\nu}_1\beta_1 S_1 T_1,\tilde{\nu}_2\beta_2 S_2 T_2} \left[\left|\begin{matrix}\gamma^{n_1}[\tilde{\nu}_1]\\ \tilde{m}_1\omega_1^0, \beta_1 S_1 T_1\end{matrix}\right\rangle \left|\begin{matrix}\gamma^{n_2}[\tilde{\nu}_2]\\ \tilde{m}_2\omega_2^0, \beta_2 S_2 T_2\end{matrix}\right\rangle\right]^{ST}_{M_S M_T} . \tag{7-176}$$

The many-particle CFP are usually denoted by

$$(\gamma^{n-n_2}[\tilde{\nu}_1]\beta_1 S_1 T_1,\ \gamma^{n_2}[\tilde{\nu}_2]\beta_2 S_2 T_2|\}\gamma^n[\tilde{\nu}]\beta ST) = C^{[\tilde{\nu}],\beta ST}_{[\tilde{\nu}_1]\beta_1 S_1 T_1,[\tilde{\nu}_2]\beta_2 S_2 T_2} \ . \tag{7-177}$$

3. Two-particle CFP. In parallel to (7-170) we have

$$C^{[\tilde{\nu}]\beta ST}_{[\tilde{\nu}_1]\beta_1 S_1 T_1,[\tilde{\nu}_2]S_2 T_2} = \left\langle \begin{matrix}[\tilde{\nu}]\\ \tilde{m}\end{matrix} \middle| [\tilde{\nu}],\ \begin{matrix}[\tilde{\nu}_1][\tilde{\nu}_2]\\ \tilde{m}_1\end{matrix} \right\rangle^{-1} \sum_{\beta' S' T'} C^{[\tilde{\nu}]\beta ST}_{[\tilde{\nu}]\beta' S' T',[1]\frac{1}{2}\frac{1}{2}}$$

$$\times C^{[\tilde{\nu}']\beta' S' T'}_{[\tilde{\nu}_1]\beta_1 S_1 T_1,[1]\frac{1}{2}\frac{1}{2}} U\left(S_1\frac{1}{2}S\frac{1}{2};S'S_2\right) U(T_1\frac{1}{2}T\frac{1}{2};T'T_2) \ . \tag{7-178}$$

The $SU_4 \supset SU_2 \times SU_2$ CFP for the spin-isospin case can be easily extended to the $SU_6 \supset SU_3 \times SU_2$ case. Using (3-302), we have

$$\left| \begin{matrix}\gamma^n[\tilde{\nu}]\\ \tau\tilde{\nu}_1\tilde{m}_1\tilde{\nu}_2\tilde{m}_2,\beta[\mu]IYI_Z,SS_Z\end{matrix} \right\rangle = \sum C^{[\tilde{\nu}]_\tau,\beta[\mu]_\theta S}_{[\tilde{\nu}_1]\beta_1[\mu_1]S_1,[\tilde{\nu}_2]\beta_2[\mu_2]S_2}$$

$$\times C^{[\mu]_\theta IY}_{[\mu_1]I_1Y_1,[\mu_2]I_2Y_2} \left[\left| \begin{matrix}\gamma^{n_1}[\tilde{\nu}_1]\\ \tilde{m}_1\omega_1^0,\beta_1[\mu_1]I_1Y_1S_1\end{matrix} \right\rangle \left| \begin{matrix}\gamma^{n_2}[\tilde{\nu}_2]\\ \tilde{m}_2\omega_2^0,\beta_2[\mu_2]I_2Y_2S_2\end{matrix} \right\rangle \right]^{IYS}_{I_Z S_Z} , \tag{7-179}$$

where the sum runs over to $\beta_1[\mu_1]S_1I_1Y_1\beta_2[\mu_2]S_2I_2Y_2$ and θ, and $Y = Y_1 + Y_2$. The $SU_6 \supset (SU_2^I \otimes U_1) \times SU_2^S$ ISF is

$$C^{[\tilde{\nu}]_\tau,\beta[\mu],IYS}_{[\tilde{\nu}_1]\beta_1[\mu_1]I_1Y_1S_1;[\tilde{\nu}_2]\beta_2[\mu_2]I_2Y_2S_2} = \sum_\theta C^{[\tilde{\nu}]_\tau,\beta[\mu]_\theta S}_{[\tilde{\nu}_1]\beta_1[\mu_1]S_1,[\tilde{\nu}_2]\beta_2[\mu_2]S_2} C^{[\mu]_\theta IY}_{[\mu_1]I_1Y_1,[\mu_2]I_2Y_2} \ . \tag{7-180}$$

For two particle CFP in parallel to (7-178), we have

$$C^{[\tilde{\nu}],\beta[\mu]S}_{[\tilde{\nu}_1]\beta_1[\mu_1]S_1,[\tilde{\nu}_2]\beta_2[\mu_2]S_2} = \left\langle \begin{matrix}[\tilde{\nu}]\\ \tilde{m}\end{matrix} \middle| [\tilde{\nu}]\ \begin{matrix}[\tilde{\nu}_1][\tilde{\nu}_2]\\ \tilde{m}_1\end{matrix} \right\rangle^{-1} \sum_{\beta'\mu'S'} C^{[\tilde{\nu}],\beta[\mu]S}_{[\tilde{\nu}']\beta'[\mu']S',[1]}$$

$$\times C^{[\tilde{\nu}_1'],\beta'[\mu']S'}_{[\tilde{\nu}_1]\beta_1[\mu_1]S_1,[1]} U\left(S_1\frac{1}{2}S\frac{1}{2},S'S_2\right) U([\mu_1][1][\mu][1];[\mu'][\mu_2]) , \tag{7-181}$$

where the last factor is the SU_3 Racah coefficients (see Sec. 7.14).

7.13.4. The total CFP

Starting from (7-108a), and making use of (4-168b), (4-180a) and (4-171b), the totally anti-symmetric wave function for n particles can be expressed as

$$\left| \begin{matrix}l^n[\nu]\\ \alpha\beta LSTM_S M_T\end{matrix} \right\rangle^a = \sum_{\tau\nu_1 m_1\nu_2 m_2} \frac{\Lambda^{\nu_1}_{m_1}\Lambda^{\nu_2}_{m_2}}{\sqrt{h_\nu}} \left| \begin{matrix}l^n[\nu]\\ \tau\nu_1 m_1\nu_2 m_2,\alpha LM\end{matrix} \right\rangle \left| \begin{matrix}\gamma^n[\tilde{\nu}]\\ \tau\tilde{\nu}_1\tilde{m}_1\tilde{\nu}_2\tilde{m}_2,\beta STM_S M_T\end{matrix} \right\rangle . \tag{7-182}$$

Substituting (7-166) and (7-176) into (7-182), we get

$$\left| \begin{matrix}l^n[\nu]\\ \alpha\beta LSTMM_S M_T\end{matrix} \right\rangle^a = \sum (l^{n-2}[\nu_1]\alpha_1\beta_1 L_1S_1T_1, l^{n_2}[\nu_2]\alpha_2\beta_2 L_2S_2T_2|\}l^n[\nu]\alpha\beta LST)$$

$$\times \left[\left| \begin{matrix}l^{n-n_2}[\nu_1]\\ \alpha_1\beta_1L_1S_1T_1\end{matrix} \right\rangle^a_{\omega_1^0} \left| \begin{matrix}l^{n_2}[\nu_2]\\ \alpha_2\beta_2L_2S_2T_2\end{matrix} \right\rangle^a_{\omega_2^0} \right]^{LST}_{MM_S M_T} , \tag{7-183}$$

where the sum runs over to $[\nu_1]\alpha_1\beta_1L_1S_1T_1[\nu_2]\alpha_2\beta_2L_2S_2$ and T_2. The *total CFP* for separating

n_2 particles out of n particles is as follows

$$(l^{n-n_2}[\nu_1]\alpha_1\beta_1 L_1 S_1 T_1, l^{n_2}[\nu_2]\alpha_2\beta_2 L_2 S_2 T_2|\}l^n[\nu]\alpha\beta LST)$$
$$= \sum_{\tau} \sqrt{\frac{h_{\nu_1}h_{\nu_2}}{h_\nu}} C^{[\nu]_\tau,\alpha L}_{[\nu_1]\alpha_1 L_1,[\nu_2]\alpha_2 L_2} C^{[\tilde{\nu}]_\tau,\beta ST}_{[\tilde{\nu}_1]\beta_1 S_1 T_1,[\tilde{\nu}_2]\beta_2 S_2 T_2}, \tag{7-184}$$

where $(h_{\nu_1}h_{\nu_2}/h_\nu)^{\frac{1}{2}}$ is called the weight factor; as will be seen later, in fact it is the $SU_{4(2l+1)} \supset SU_{2l+1} \times SU_4$ ISF. Therefore, the so-called total CFP are precisely the $SU_{4(2l+1)} \supset (SU_{2l+1} \supset SO_3) \times (SU_4 \supset SU_2 \times SU_2)$ ISF. Using the Racah's factorization lemma (3-303) and also (7-244), they can be factorized as

$$C^{[1^n],[\nu]\alpha L[\tilde{\nu}]\beta ST}_{[1^{n_1}][\nu_1]\alpha_1 L_1[\tilde{\nu}_1]\beta_1 S_1 T_1,[1^{n_2}][\nu_2]\alpha_2 L_2[\tilde{\nu}_2]\beta_2 S_2 T_2}$$
$$= \sum_{\tau} \sqrt{\frac{h_{\nu_1}h_{\nu_2}}{h_\nu}} C^{[\nu]_\tau,\alpha L}_{[\nu_1]\alpha_1 L_1,[\nu_2]\alpha_2 L_2} C^{[\tilde{\nu}]_\tau,\beta ST}_{[\tilde{\nu}_1]\beta_1 S_1 T_1,[\tilde{\nu}_2]\beta_2 S_2 T_2}. \tag{7-185}$$

From the example we see the power of the Racah's factorization lemma which actually enables us to write down the expression (7-185) for the total CFP directly, without the foregoing step-by-step derivation. Further discussion of the total CFP will be given in Sec. 7.16.

7.13.5. Eigenfunction method for evaluating the CFP

Several methods are available for evaluating the CFP or ISF. Here we only introduce the EFM. We first change notation and rewrite (7-159b) as

$$\psi\begin{pmatrix} [\nu] \\ m,\ \alpha L \end{pmatrix} = \sum_{\alpha_1 L_1} C^{[\nu],\alpha L}_{[\nu_1]\alpha_1 L_1,[1]} \left[\psi\begin{pmatrix} [\nu_1] \\ m_1,\ \alpha_1 L_1 \end{pmatrix} \psi_l(n)\right]^L_M, \tag{7-186}$$

which is already the Yamanouchi basis $[\nu_1]m_1$ of S_{n-1}. Therefore the requirement for it to be the Yamanouchi basis $[\nu]m$ of S_n is equivalent to the requirement that it be an eigenfunction of the 2-cycle class operator $C(n)$ of S_n. In analogous to (4-191), we obtain the eigenequation satisfied by $\psi\begin{pmatrix} [\nu] \\ m, \alpha L \end{pmatrix}$ as follows

$$C'(n)\psi\begin{pmatrix} [\nu] \\ m,\ \alpha L \end{pmatrix} = (\nu - \nu_1)\psi\begin{pmatrix} [\nu] \\ m,\ \alpha L \end{pmatrix},$$
$$C'(n) = \sum_{i=1}^{n-1} p_{in}. \tag{7-187}$$

Consequently, to obtain the CFP $C^{[\nu]\alpha L}_{[\nu_1]\alpha_1 L_1,[1]}$ one only needs to diagonalize the operator $C'(n)$ in the basis $\left[\psi\begin{pmatrix} [\nu_1] \\ m_1,\alpha_1 L_1 \end{pmatrix} \psi_l(n)\right]^L_M$ with fixed ν_1, m_1 and L. To obtain the matrix elements of $C'(n)$, we use (7-159) to expand the wave functions of $n-1$ particles. The matrix elements of the transposition $p_{n,n-1}$ is expressed as

$$\left\langle \left[\psi\begin{pmatrix} [\nu_1] \\ m_1,\alpha_1 L_1 \end{pmatrix} \psi_l(n)\right]^L_M \middle| P_{n,n-1} \middle| \left[\psi\begin{pmatrix} [\nu_1] \\ m_1',\alpha_2 L_2 \end{pmatrix} \psi_l(n)\right]^L_M \right\rangle$$
$$= \delta_{m_1 m_1'} \sum_{\alpha' L'} C^{[\nu_1],\alpha_1 L_1}_{[\nu']\alpha' L',[1]} C^{[\nu_1],\alpha_2 L_2}_{[\nu']\alpha' L',[1]} \langle L' L_1 L | p_{n,n-1} | L' L_2 L\rangle$$

$$= \delta_{m_1 m_1'} \langle \alpha_1 L_1 | p_{n,n-1} | \alpha_2 L_2 \rangle^{(L)}_{\nu_1 m_1}{}^{3)} \tag{7-188a}$$

where $|L'L_1L\rangle$ is a short hand for

$$|L'L_1L\rangle = \left[\left[\psi\begin{pmatrix}[\nu'] \\ m', \alpha' L'\end{pmatrix} \psi_l(n-1)\right]^{L_1} \psi_l(n)\right]^L .$$

It can be shown that

$$\langle L'L_1L|p_{n,n-1}|L'L_2L\rangle = (-1)^{L'+L+L_1+L_2} U(lL'Ll; L_1L_2) . \tag{7-188b}$$

$$\begin{aligned}\langle \alpha_1 L_1 | p_{n,n-1} | \alpha_2 L_2 \rangle^{(L)}_{\nu_1 m_1} &= \langle \alpha_2 L_2 | p_{n,n-1} | \alpha_1 L_1 \rangle^{(L)}_{\nu_1 m_1} \\ &= \sum_{\alpha' L'} C^{[\nu_1],\alpha_1 L_1}_{[\nu']\alpha' L',[1]} C^{[\nu_1],\alpha_2 L_2}_{[\nu']\alpha' L',[1]} \langle L'L_1L|p_{n,n-1}|L'L_2L\rangle .\end{aligned} \tag{7-188c}$$

(Numerical tables of $\langle L'L_1L|p_{n,n-1}|L'L_2L\rangle$ for $l = 1$ are given by Jahn 1951.) Using $p_{in} = p_{i,n-1}p_{n-1,n}p_{i,n-1}$ and (7-188a) one gets

$$\begin{aligned}&\langle \alpha_1 L_1 | p_{in} | \alpha_2 L_2 \rangle^{(L)}_{\nu_1 m_1} = \langle \alpha_2 L_2 | p_{in} | \alpha_1 L_1 \rangle^{(L)}_{\nu_1 m_1} \\ &= \left\langle \left[\psi\begin{pmatrix}[\nu_1] \\ m_1, \ \alpha_1 L_1\end{pmatrix} \psi_l(n)\right]^L_M \middle| p_{in-1}p_{nn-1}p_{in-1} \middle| \left[\psi\begin{pmatrix}[\nu_1] \\ m_1, \alpha_2 L_2\end{pmatrix} \psi_l(n)\right]^L_M \right\rangle \\ &= \sum_{m_1'} [D^{[\nu_1]}_{m_1' m_1}(i, n-1)]^2 \langle \alpha_1 L_1 | p_{nn-1} | \alpha_2 L_2 \rangle^{(L)}_{\nu_1 m_1'} .\end{aligned} \tag{7-188d}$$

From (7-186), (7-187) and (7-188) we obtain the eigenequation satisfied by the CFP.

$$\sum_{\alpha_2 L_2} [\langle \alpha_1 L_1 | C'(n) | \alpha_2 L_2 \rangle^L_{\nu_1} - \delta_{\alpha_1\alpha_2}\delta_{L_1 L_2}(\nu - \nu_1)] C^{[\nu]\alpha L}_{[\nu_1]\alpha_2 L_2,[1]} = 0 , \tag{7-189a}$$

where

$$\langle \alpha_1 L_1 | C'(n) | \alpha_2 L_2 \rangle^L_{\nu_1} = \sum_{i=1}^{n-1} \langle \alpha_1 L_1 | p_{in} | \alpha_2 L_2 \rangle^{(L)}_{\nu_1 m_1} . \tag{7-189b}$$

Notice that although each term in the right-hand side of (7-189b) depends on m_1 their sum is independent of m_1. Dividing (7-189b) by h_{ν_1}, summing over m_1 and using (7-188d), we can show that

$$\langle \alpha_1 L_1 | C'(n) | \alpha_2 L_2 \rangle^L_{\nu_1} = \frac{n-1}{h_{\nu_1}} \sum_{m_1} \langle \alpha_1 L_1 | p_{nn-1} | \alpha_2 L_2 \rangle^{(L)}_{\nu_1 m_1} . \tag{7-189c}$$

Furthermore, using the fact that the matrix elements of p_{nn-1} in the above are independent of the component index m' of the irrep ν' of S_{n-2}, Eq. (7-189b) can be further simplified as

$$\langle \alpha_1 L_1 | C'(n) | \alpha_2 L_2 \rangle^L_{\nu_1} = \frac{n-1}{h_{\nu_1}} \sum_{\nu_1'} h_{\nu_1'} \langle \alpha_1 L_1 | p_{nn-1} | \alpha_2 L_2 \rangle^{(L)}_{\nu_1 \nu_1' m_1'} . \tag{7-189'}$$

Equation (7-189′) provides a most effective way for computing the matrix of $C'(n)$.

---

[3] When $[\nu_1] \neq [\nu_1']$, the matrix elements

$$\left\langle \left[\psi\begin{pmatrix}[\nu_1] \\ m_1, \alpha_1 L_1\end{pmatrix} \psi_l(n)\right]^L_M \middle| p_{n,n-1} \middle| \left[\psi\begin{pmatrix}[\nu_1'] \\ m_1', \alpha_2 L_2\end{pmatrix} \psi_l(n)\right]^L_M \right\rangle$$

$$= \sum_{\alpha' L'} C^{[\nu_1]\alpha_1 L_1}_{[\nu']\alpha' L',[1]} C^{[\nu_1']\alpha_2 L_2}_{[\nu']\alpha' L',[1]} \left\langle L'L_1L|p_{n,n-1}|L'L_2L\right\rangle$$

are not vanishing for $m_1 \neq m_1'$, see Jahn (1951, Sec. 5).

If ν is a single root of the secular equation of (7-189a), the additional label α is redundant, and if ν is a τ- fold root, then there are τ sets of eigensolutions $C^{[\nu]\alpha L}_{[\nu_1]\alpha_1 L_1,[1]}, \alpha = 1,2,\ldots,\tau$, which can be made orthogonal with respect to the index α.

From $[\nu]m = [\nu][\nu_1]m_1$ and (7-186) it is seen that different $[\nu_1]$ correspond to different components m of the irrep $[\nu]$. In the procedure described above, the eigenequation (7-189a) is solved individually for each $[\nu_1]$. As a consequence, the relative phases of the CFP $C^{[\nu]\alpha L}_{[\nu_1]\alpha_1 L_1,[1]}$ with respect to $C^{[\nu]\alpha L}_{[\overline{\nu}_1]\overline{\alpha}_1\overline{L}_1,[1]}$ are chosen randomly. What is more serious is that for the multiplicity-not-free cases, the multiplicity label α is also chosen randomly for each $[\nu_1]$. Therefore, the basis vectors $\psi\begin{pmatrix}[\nu]\\ m,\alpha L\end{pmatrix}$ and $\psi\begin{pmatrix}[\nu]\\ \overline{m},\alpha L\end{pmatrix}$ constructed from (7-186) in terms of the CFP $C^{[\nu]\alpha L}_{[\nu_1]\alpha_1 L_1,[1]}$ and $C^{[\nu]\alpha L}_{[\overline{\nu}_1]\overline{\alpha}_1\overline{L}_1,[1]}$, respectively, are in general not the two partners of the same irrep $[\nu]$. These two shortcomings can be remedied by the following technique.

From (7-186), we have

$$D^{[\nu]}_{\overline{m}m}(n-1,n) = \left\langle \psi\begin{pmatrix}[\nu]\\ \overline{m},\alpha L\end{pmatrix}\middle| p_{n,n-1} \middle| \psi\begin{pmatrix}[\nu]\\ m,\alpha L\end{pmatrix}\right\rangle = \sum_{\alpha_1 L_1 \overline{\alpha}_1 \overline{L}_1 \alpha' L'} C^{[\nu]\alpha L}_{[\overline{\nu}_1]\overline{\alpha}_1\overline{L}_1,[1]}$$
$$\times C^{[\overline{\nu}_1]\overline{\alpha}_1\overline{L}_1}_{[\nu']\alpha' L',[1]} C^{[\nu]\alpha L}_{[\nu_1]\alpha_1 L_1,[1]} C^{[\nu_1]\alpha_1 L_1}_{[\nu']\alpha' L',[1]} \langle L'\overline{L}_1 L|p_{n-1,n}|L' L_1 L\rangle \ . \qquad (7\text{-}189\text{d})$$

As in our derivation of (4-195c) from (4-195a), we have from (7-189d)

$$C^{[\nu]\alpha L}_{[\overline{\nu}_1]\overline{\alpha}_1\overline{L}_1,[1]} = [D^{[\nu]}_{\overline{m}m}(n-1,n)]^{-1}$$
$$\times \sum_{\substack{\alpha_1 L_1\\ \alpha' L'}} \left[C^{[\overline{\nu}_1]\overline{\alpha}_1\overline{L}_1}_{[\nu']\alpha' L',[1]} C^{[\nu_1]\alpha_1 L_1}_{[\nu']\alpha' L',[1]} \langle L'\overline{L}_1 L|p_{n-1,n}|L' L_1 L\rangle\right] C^{[\nu]\alpha L}_{[\nu_1]\alpha_1 L_1,[1]} \ . \qquad (7\text{-}189\text{e})$$

Thus the correct procedure for computing the CFP is that for each possible $[\nu]$, we take the CFP $C^{[\nu]\alpha L}_{[\nu_1]\alpha_1 L_1,[1]}$ from the eigenvectors of (7-189a) only for a given $[\nu_1]$, while all the other CFP $C^{[\nu]\alpha L}_{[\overline{\nu}_1]\overline{\alpha}_1\overline{L}_1,[1]}$, for $\overline{\nu}_1 \neq \nu_1$, are to be calculated from (7-189e).

Equations (7-189) and (7-188c) provide a recursive way for calculating the CFP. This method is simpler than Jahn's (1951) method.

Example 7: Find the CFP for the 1p shell, $n = 3, [\nu_1] = [2]$ and $L = 1$. Since the irrep $[\nu_1] = [2]$ constains $L_1 = 0,2$, we now need to diagonalize $C'(3)$ in the basis $\varphi_1 = \left[|[2]S\rangle\psi(3)\right]^P$ and $\varphi_2 = \left[|[2]D\rangle\psi(3)\right]^P$. The CFP for $n = 2$ is equal to one, therefore from (7-188c) we have

$$\langle\alpha_1 L_1|p_{23}|\alpha_2 L_2\rangle^{(L)}_{[2]} = \langle PL_1P|p_{23}|PL_2P\rangle \ .$$

Using the Racah coefficients table, we find the matrix $\langle PL_1P|p_{23}|PL_2P\rangle$ as follows:

PL_1 \\ PL_2	PS	PD
PS	$\frac{1}{3}$	$\frac{\sqrt{5}}{3}$
PD	$\frac{\sqrt{5}}{3}$	$\frac{1}{6}$

The eigenvalue corresponding to $[\nu_1] = [2]$ is $\nu_1 = 1$. Using (7-189′) Eq. (7-189a) becomes

$$\begin{pmatrix} \frac{5}{3}-\nu & \frac{2\sqrt{5}}{3} \\ \frac{2\sqrt{5}}{3} & \frac{4}{3}-\nu \end{pmatrix} \begin{pmatrix} C^{[\nu]P}_{[2]S,[1]} \\ C^{[\nu]P}_{[2]D,[1]} \end{pmatrix} = 0 \ . \qquad (7\text{-}190)$$

From (7-190) we find two single roots $\nu = 0$ and 3, corresponding to $[\nu] = [21]$ and $[3]$, respectively, and the CFP listed below

$[\nu]L$ \ $[\nu_1]L_1$	$[2]S$	$[2]D$
$[3]P$	$\frac{\sqrt{5}}{3}$	$\frac{2}{3}$
$[21]P$	$\frac{2}{3}$	$-\frac{\sqrt{5}}{3}$

In our example, (7-159b) takes the form

$$\left|\,\boxed{1}\boxed{2}\boxed{3}\,,PM\right\rangle = \left[\left(\frac{\sqrt{5}}{3}\left|\,\boxed{1}\boxed{2}\,,S\right\rangle + \frac{2}{3}\left|\,\boxed{1}\boxed{2}\,,D\right\rangle\right)\psi(3)\right]^P_M , \tag{7-191a}$$

$$\left|\,\begin{array}{l}\boxed{1}\boxed{2}\\ \boxed{3}\end{array}\,,PM\right\rangle = \left[\left(\frac{2}{3}\left|\,\boxed{1}\boxed{2}\,,S\right\rangle - \frac{\sqrt{5}}{3}\left|\,\boxed{1}\boxed{2}\,,D\right\rangle\right)\psi(3)\right]^P_M . \tag{7-191b}$$

or more succinctly

$$|[3]P\rangle = \left(\frac{\sqrt{5}}{3}|[2]S\rangle + \frac{2}{3}|[2]D\rangle\right)\psi(3) , \tag{7-191c}$$

$$|[21]P\rangle = \left(\frac{2}{3}|[2]S\rangle - \frac{\sqrt{5}}{3}|[2]D\rangle\right)\psi(3) . \tag{7-191d}$$

The rest of the CFP for $n = 3$ are trivial. They are either equal to one or zero, as determined by the angular momentum coupling rule. For example we have

$$|[21]D\rangle = C_1\Big[|[2]D\rangle\psi_p(3)\Big]^D + C_2\Big[|[2]S\rangle\psi_p(3)\Big]^D = \Big[|[2]D\rangle\psi_p(3)\Big]^D . \tag{7-192a}$$

$$|[21]D\rangle = \Big[|[11]P\rangle\psi(3)\Big]^D, \quad |[21]P\rangle = \Big[|[11]P\rangle\psi(3)\Big]^P . \tag{7-192b}$$

Example 8: Find the CFP for the $1p$ shell, $n = 4, [\nu_1] = [21], L = D$. From Table IC-4 in Bohr (1969), it is known that the irrep $[21]$ contains $L_1 = P, D$. Thus we need to diagonalize $C'(4)$ in the basis (φ_1, φ_2), or (χ_1, χ_2):

$$\varphi_1 = \left[\left|\,\begin{array}{l}\boxed{1}\boxed{2}\\ \boxed{3}\end{array}\,,P\right\rangle\psi(4)\right]^D , \quad \varphi_2 = \left[\left|\,\begin{array}{l}\boxed{1}\boxed{2}\\ \boxed{3}\end{array}\,,D\right\rangle\psi(4)\right]^D ,$$

$$\chi_1 = \left[\left|\,\begin{array}{l}\boxed{1}\boxed{3}\\ \boxed{2}\end{array}\,,P\right\rangle\psi(4)\right]^D , \quad \chi_2 = \left[\left|\,\begin{array}{l}\boxed{1}\boxed{3}\\ \boxed{2}\end{array}\,,D\right\rangle\psi(4)\right]^D , \tag{7-193}$$

and using the CFP (7-191b), as well as (7-192), we expand the three-particle states in (7-193) as follow:

$$\varphi_1 = \frac{2}{3}|SPD\rangle - \frac{\sqrt{5}}{3}|DPD\rangle, \quad \varphi_2 = |DDD\rangle ,$$
$$\chi_1 = |PPD\rangle , \qquad \chi_2 = |PDD\rangle . \tag{7-194}$$

From (7-188b), or Table 1 given by Jahn (1951), we find the matrix representatives of p_{34} in the bases (φ_1, φ_2) and (χ_1, χ_2):

$$((\alpha_1 L_1 | p_{34} | \alpha_2 L_2\rangle_{[21]m_1=1}) = \mathcal{D}_1(34) = \begin{pmatrix} 1/2 & -\sqrt{3}/6 \\ -\sqrt{3}/6 & 5/6 \end{pmatrix} .$$

$$((\alpha_1 L_1 | p_{34} | \alpha_2 L_2\rangle_{[21]m_1=2}) = \mathcal{D}_2(34) = \begin{pmatrix} 1/2 & \sqrt{3}/2 \\ \sqrt{3}/2 & -1/2 \end{pmatrix} . \tag{7-195}$$

From (7-189′) and (7-195), we obtain the matrix representative of $C'(4)$ in the basis (φ_1, φ_2) or (χ_1, χ_2),

$$\mathcal{D}_1(C'(4)) = \mathcal{D}_2(C'(4)) = \begin{pmatrix} 3/2 & \sqrt{3}/2 \\ \sqrt{3}/2 & 1/2 \end{pmatrix} . \tag{7-196}$$

By diagonalizing (7-196), the CFP $(p^3, p|\}p^4) = C^{[\nu]L}_{[\nu_1]L_1,[1]}$ are found to be

$[\nu]L$ \ $[\nu_1]L_1$	$[21]P$	$[21]D$
$[31]D$	$\sqrt{3}/2$	$1/2$
$[22]D$	$-1/2$	$\sqrt{3}/2$

i.e.,

$$\left| \begin{smallmatrix} 1 & 2 & 4 \\ 3 & & \end{smallmatrix} , DM \right\rangle = \left(\frac{\sqrt{3}}{2} \left| \begin{smallmatrix} 1 & 2 \\ 3 & \end{smallmatrix} , P \right\rangle + \frac{1}{2} \left| \begin{smallmatrix} 1 & 2 \\ 3 & \end{smallmatrix} , D \right\rangle \right) \psi(4) ,$$

$$\left| \begin{smallmatrix} 1 & 2 \\ 3 & 4 \end{smallmatrix} , DM \right\rangle = \left(-\frac{1}{2} \left| \begin{smallmatrix} 1 & 2 \\ 3 & \end{smallmatrix} , P \right\rangle + \frac{\sqrt{3}}{2} \left| \begin{smallmatrix} 1 & 2 \\ 3 & \end{smallmatrix} , D \right\rangle \right) \psi(4) . \tag{7-197a}$$

or written succinctly

$$\|[31]D\rangle = \left(\frac{\sqrt{3}}{2} \|[21]P\rangle + \frac{1}{2} \|[21]D\rangle \right) \psi(4) ,$$

$$\|[22]D\rangle = \left(-\frac{1}{2} \|[21]P\rangle + \frac{\sqrt{3}}{2} \|[21]D\rangle \right) \psi(4) . \tag{7-197b}$$

The above technique can be easily extended to the calculation of any $SU_n \supset G$ CFP. Such as:

1. The $SU_3 \supset SU_2$ CFP. Corresponding to (7-186), we have

$$\psi \begin{pmatrix} & [\nu] \\ m, & IYI_Z \end{pmatrix} = \sum_{I_1 Y_1} C^{[\nu]IY}_{[\nu_1]I_1Y_1,[1]} \left[\psi \begin{pmatrix} & [\nu_1] \\ m_1, & I_1 Y_1 \end{pmatrix} \psi(n) \right]^I_{I_z} , \tag{7-198}$$

where $\nu_1 m_1 I Y$ are fixed. The matrix elements in (7-188c) are replaced by the following

$$\begin{aligned} \langle \alpha_1 L_1 | p_{n-1,n} | \alpha_2 L_2 \rangle^L &\to \langle I_1 Y_1 | p_{n-1,n} | I_2 Y_2 \rangle^I \\ &= \sum_{I'Y'} C^{[\nu_1]I_1Y_1}_{[\nu']I'Y',[1]} C^{[\nu_1]I_2Y_2}_{[\nu']I'Y',[1]} \langle I' I_1 I | p_{n-1,n} | I' I_2 I \rangle . \end{aligned} \tag{7-199}$$

2. $SU_4 \supset SU_2 \times SU_2$ CFP

$$\psi\begin{pmatrix} [\nu] \\ m, \beta STM_S M_T \end{pmatrix} = \sum_{\beta_1 S_1 T_1} C^{[\nu]\beta ST}_{[\nu_1]\beta_1 S_1 T_1,[1]} \left[\psi\begin{pmatrix} [\nu_1] \\ m_1, \beta_1 S_1 T_1 \end{pmatrix} \psi(n)\right]^{ST}_{M_S M_T}, \tag{7-200}$$

$$\begin{aligned} &\langle \alpha_1 L_1 | p_{n-1,n} | \alpha_2 L_2 \rangle \to \langle \beta_1 S_1 T_1 | p_{n-1n} | \beta_2 S_2 T_2 \rangle \\ &= \sum_{\beta' S' T'} C^{[\nu_1]\beta_1 S_1 T_1}_{[\nu']\beta' S' T',[1]} C^{[\nu_1]\beta_2 S_2 T_2}_{[\nu']\beta' S' T',[1]} \langle S' S_1 S | p_{n-1n} | S' S_2 S \rangle \langle T' T_1 T | p_{n-1n} | T' T_2 T \rangle \, . \end{aligned} \tag{7-201}$$

3. $SU_6 \supset SU_3 \times SU_2$ CFP.

$$\psi\begin{pmatrix} [\nu] \\ m, \beta(\lambda\mu)S \end{pmatrix} = \sum_{\beta_1(\lambda_1\mu_1)S_1\theta} C^{[\nu]\beta(\lambda\mu)_\theta S}_{[\nu_1]\beta_1(\lambda_1\mu_1)S_1,[1]} \left[\psi\begin{pmatrix} [\nu_1] \\ m_1, \ \beta_1(\lambda_1\mu_1)S_1 \end{pmatrix} \psi(n)\right]^{(\lambda\mu)_\theta S} . \tag{7-202}$$

Compared with (7-179), here we ignored the component indices of the irreps $(\lambda\mu)$ and S, since the $SU_6 \supset SU_3 \times SU_2$ CFP are independent of these quantum numbers.

$$\begin{aligned} &\langle \alpha_1 L_1 | p_{n-1,n} | \alpha_2 L_2 \rangle^L \to \langle \beta_1(\lambda_1\mu_1) S_1 | p_{n-1,n} | \beta_2(\lambda_2\mu_2) S_2 \rangle^{(\lambda\mu)S} \\ &= \sum_{\beta'(\lambda'\mu')S'} C^{[\nu_1]\beta_1(\lambda_1\mu_1)S_1}_{[\nu']\beta'(\lambda'\mu')S',[1]} C^{[\nu_1]\beta_2(\lambda_2\mu_2)S_2}_{[\nu']\beta'(\lambda'\mu')S',[1]} \langle S' S_1 S | p_{n-1,n} | S' S_2 S \rangle \\ &\qquad \times \langle (\lambda'\mu')(\lambda_1\mu_1)(\lambda\mu) | p_{n-1,n} | (\lambda'\mu')(\lambda_2\mu_2)(\lambda\mu) \rangle \, , \end{aligned} \tag{7-203}$$

where

$$\begin{aligned} &\langle (\lambda'\mu')(\lambda_1\mu_1)(\lambda\mu) | p_{n-1,n} | (\lambda'\mu')(\lambda_2\mu_2)(\lambda\mu) \rangle \\ &= (-)^{\lambda'-\mu'+\lambda-\mu+\lambda_1-\mu_1+\lambda_2-\mu_2} U((10)(\lambda'\mu')(\lambda\mu)(10);(\lambda_1\mu_1)(\lambda_2\mu_2)) \, . \end{aligned} \tag{7-204}$$

Below we list some references where either numerical tables, or algebraic expression, or computer codes of various kinds of CFP or ISF can be found.

1. $SU_3 \supset SO_3$. Tables of one-particle CFP are found in Jahn (1951). For two-particle CFP, see Elliott (1953). Tables of $SU_3 \supset SO_3$ ISF are given by Sun (1965). For algebraic expression of the $SU_3 \supset SO_3$ ISF, consult Vergados (1968) and Horie (1964). Akiyama (1973) has given a computer code for $SU_3 \supset SO_3$ ISF.

2. $SU_3 \supset SU_2 \otimes U_1$ ISF. For numerical tables, see de Swart (1963). Algebraic expressions are found in Hecht (1965).

3. $SU_4 \supset SU_3 \otimes U_1$ ISF. Numerical tables are given by Rabl (1975), Haacke (1976).

4. $SU_4 \supset SU_2 \times SU_2$. For one-particle CFP, see Jahn (1951). Two-particle CFP are given by Elliott (1953) and Harvey (1981). Algebraic expression may be found in Hecht (1969).

5. $SU_6 \supset SU_3 \times SU_2$ ISF. Numerical tables are collected in So (1979), Strottman (1979), Zhang (1977), Machacek (1976), Cook (1965).

6. $SU_6 \supset SU_3$ ISF. Numerical tables are given by Akiyama (1966). Algebraic expression are compiled in Hecht (1965). Braunschweig (1978) gave a computer code.

7. $SU_{15} \supset SU_3$ ISF. A computer code is given by Wu (1983).

8. $SU_{mn} \supset SU_m \times SU_n$ CFP for arbitrary m and n. Numerical tables are found in Chen ([16], 1984) and Wu (1987). A computer code for one-particle CFP is given by Novoselsky [1] (1988).

9. $SU_{m+n} \supset SU_m \otimes SU_n$ single particle CFP for arbitrary m and n. For numerical tables, see Chen ([7], 1984). A computer code for one-particle CFP is given by Novoselsky [1] (1988).

10. $SU_{2j+1} \supset SP_{2j+1} \supset SO_3$ and $SU_{2l+1} \supset SO_{2l+1} \supset SO_3$ CFP Tables: a) for totally anti-symmetric and totally symmetric states respectively. See Bayman (1966). b) for arbitrary symmetries, see Novoselsky (1988).

11. The elementary reduced Wigner coefficients (analytic expression): Le Blanc (1987).

12. The $SO_5 \supset U_2$ ISF analytic expression: Hecht (1988).

7.14. $\mathbf{S}_f \supset \mathbf{S}_{f_1} \otimes \mathbf{S}_{f_2} \otimes \mathbf{S}_{f_3}$ Irreducible Basis and $\mathbf{SU}_n$ Racah Coefficients*

The contents of this section are very similar to those of Sec. 7.10. The only difference is that the unitary and permutation groups interchange their roles.

We begin with the construction of the $S_f \supset S_{f_1} \otimes S_{f_2} \otimes S_{f_3}$ irreducible basis, with $f = f_1 + f_2 + f_3, S_{f_i} \equiv S_{f_i}(\omega_i^0)$, the definition of (ω_1^0) and (ω_2^0) being the same as in the previous section, $(\omega_3^0) = (f_1 + f_2 + 1, \ldots, f_1 + f_2 + f_3)$. There are two ways of constructing such a basis.

1. The $S_f \supset (S_{f_1+f_2} \supset S_{f_1} \otimes S_{f_2}) \otimes S_{f_3}$ irreducible basis. According to (7-102b), it can be expressed as

$$\left| \begin{matrix} [\nu] \\ W, \end{matrix} \begin{pmatrix} [\nu_{12}] & [\nu_3] \\ \nu_1 m_1 \nu_2 m_2 & , m_3 \end{pmatrix} \right\rangle^{\tau_{12}\tau} = \sum_{W_1 W_2 W_3 W_{12}} C^{[\nu_{12}]\tau_{12}, W_{12}}_{\nu_1 W_1, \nu_2 W_2} C^{[\nu]\tau, W}_{\nu_{12} W_{12}, \nu_3 W_3} \left| \begin{matrix} [\nu_1] \\ m_1 \omega_1^0, \ W_1 \end{matrix} \right\rangle \left| \begin{matrix} [\nu_2] \\ m_2 \omega_2^0, \ W_2 \end{matrix} \right\rangle \left| \begin{matrix} [\nu_3] \\ m_3 \omega_3^0, \ W_3 \end{matrix} \right\rangle . \tag{7-205a}$$

It is an irreducible basis $[\nu]W$ of SU_n and Yamanouchi basis $[\nu_i]m_i$ of S_{f_i}, and it belongs to the irrep $[\nu]$ and $[\nu_{12}]$ of S_f and $S_{f_1+f_2}$, respectively.

2. The $S_f \supset S_{f_1} \otimes (S_{f_2+f_3} \supset S_{f_2} \otimes S_{f_3})$ irreducible basis. In this case we have

$$\left| \begin{matrix} [\nu] \\ W, \end{matrix} \begin{pmatrix} [\nu_1] & [\nu_{23}] \\ m_1, & \nu_2 m_2 \nu_3 m_3 \end{pmatrix} \right\rangle^{\tau_{23}\tau'} = \sum_{W_1 W_2 W_3 W_{23}} C^{[\nu_{23}]\tau_{23}, W_{23}}_{\nu_2 W_2, \nu_3 W_3} C^{[\nu]\tau', W}_{\nu_1 W_1, \nu_{23} W_{23}} \left| \begin{matrix} [\nu_1] \\ m_1 \omega_1^0, \ W_1 \end{matrix} \right\rangle \left| \begin{matrix} [\nu_2] \\ m_2 \omega_2^0, \ W_2 \end{matrix} \right\rangle \left| \begin{matrix} [\nu_3] \\ m_3 \omega_3^0, \ W_3 \end{matrix} \right\rangle . \tag{7-205b}$$

The bases in (7-205a) and (7-205b) are related by a unitarity transformation:

$$\left| \begin{matrix} [\nu] \\ W, \end{matrix} \begin{pmatrix} [\nu_{12}] & [\nu_3] \\ \nu_1 m_1 \nu_2 m_2, & m_3 \end{pmatrix} \right\rangle^{\tau_{12}\tau} = \sum_{\nu_{23}\tau_{23}\tau'} U(\nu_1 \nu_2 \nu \nu_3, \nu_{12} \nu_{23})^{\tau_{12}\tau}_{\tau_{23}\tau'} \left| \begin{matrix} [\nu] \\ W, \end{matrix} \begin{pmatrix} [\nu_1] & [\nu_{23}] \\ m_1, & \nu_2 m_2 \nu_3 m_3 \end{pmatrix} \right\rangle^{\tau_{23}\tau'} , \tag{7-206a}$$

where U is the Racah coefficients of SU_n, the value of which is independent of W, m_1, m_2 and m_3. Therefore if we are only interested in the unitary group, (7-206a) can be cast into a form familiar in angular momentum theory

$$|(([\nu_1][\nu_2])[\nu_{12}], [\nu_3] : [\nu] W\rangle^{\tau_{12}, \tau} = \sum_{\nu_{23}\tau_{23}\tau'} U(\nu_1 \nu_2 \nu \nu_3, \nu_{12} \nu_{23})^{\tau_{12}\tau}_{\tau_{23}\tau'} |[\nu_1]([\nu_2][\nu_3])[\nu_{23}]; \ [\nu] W\rangle^{\tau_{23}\tau'} . \tag{7-206b}$$

From (7-204) and (7-205) we obtain

$$U(\nu_1 \nu_2 \nu \nu_3; \nu_{12} \nu_{23})^{\tau_{12}\tau}_{\tau_{23}\tau'} = \sum_{\text{fix } W} C^{[\nu_{12}]\tau_{12}, W_{12}}_{\nu_1 W_1, \nu_2 W_2} C^{[\nu]\tau, W}_{\nu_{12} W_{12}, \nu_3 W_3} C^{[\nu_{23}]\tau_{23}, W_{23}}_{\nu_2 W_2, \nu_3 W_3} C^{[\nu]\tau', W}_{\nu_1 W_1, \nu_{23} W_{23}} . \tag{7-207}$$

The formula (7-124) to (7-130) for the Racah coefficients of permutation groups are also valid for the unitary group SU_n, if we made the following index substitutions: $m \to W, m_i \to W_i, \tau \to \beta$, and the dimensions h_ν, h_{ν_κ}, etc. are understood to be the dimensions of irreps of SU_n.

The CG coefficients of SU_n depend on the choice of the classification scheme for the irreducible basis, while the Racah coefficients do not. The latter only depend on the irreps.

Example 1: The Racah coefficients of SU_3.

a. Under the $SU_3 \supset SU_2 \times U_1$ irreducible basis, by substituting (7-148) into (7-207) we have

$$U(\mu_1\mu_2\mu\mu_3, \mu_{12}\mu_{23})^{\tau_{12}\tau}_{\tau_{23}\tau'} = \sum_{\overline{W}_1\overline{W}_2\overline{W}_3\overline{W}_{12}\overline{W}_{23}} U(I_1I_2II_3; I_{12}I_{23})\{CCCC\} \,, \tag{7-208a}$$

where

$$\{CCCC\} = C^{[\mu_{12}]\tau_{12}\overline{W}_{12}}_{\mu_1\overline{W}_1,\mu_2\overline{W}_2} C^{[\mu]_\tau\overline{W}}_{\mu_{12}\overline{W}_{12},\mu_3\overline{W}_3} C^{[\mu]\tau'\overline{W}}_{\mu_1\overline{W}_1,\mu_{23}\overline{W}_{23}} C^{[\mu_{23}]\tau_{23},\overline{W}_{23}}_{\mu_2\overline{W}_2,\mu_3\overline{W}_3} \,, \tag{7-208b}$$

with $\overline{W} = IY, \overline{W}_i = I_iY_i$.

b. In the $SU_3 \supset SO_3 \supset SO_2$ basis, the CG coefficients can be factorized as

$$C^{[\mu]_\tau\kappa LM}_{[\mu_1]\kappa_1L_1M_1,[\mu_2]\kappa_2L_2M_2} = C^{[\mu]_\tau\kappa L}_{[\mu_1]\kappa_1L_1,[\mu_2]\kappa_2L_2} C^{LM}_{L_1M_1,L_2M_2} \,, \tag{7-209}$$

where the index κ is the multiplicity label for reducing the irrep $[\mu]$ of SU_3 into the irreps (L) of SO_3. The Racah coefficient of SU_3 is now expressed as

$$U(\mu_1\mu_2\mu\mu_3; \mu_{12}\mu_{23})^{\tau_{12}\tau}_{\tau_{23}\tau'} = \sum_{\overline{W}_1\overline{W}_2\overline{W}_3\overline{W}_{12}\overline{W}_{23}} U(L_1L_2LL_3; L_{12}L_{23})\{CCCC\} \,, \tag{7-210}$$

where $\{CCCC\}$ is still given by (7-208b) with the understanding that now $\overline{W} = \kappa L$ and $\overline{W}_i = \kappa_i L_i$.

If all the multiplicity labels $\tau_{12}, \tau_{23}, \tau$ and τ' are redundant, then the Racah coefficients evaluated from (7-208a) and (7-210) must be equal to within a phase factor. For the case when any of the multiplicity labels have more than one possible value, the Racah coefficients of (7-208a) and (7-210) may differ by a linear transformation.

Example 2: The Racah coefficients of SU_4. We choose the $SU_4 \supset SU_2 \times SU_2$ irreducible basis. Substituting (7-155) into (7-207), we have

$$\begin{aligned} &U(\nu_1\nu_2\nu\nu_3; \nu_{12}\nu_{23})^{\tau_{12}\tau}_{\tau_{23}\tau'} \\ &\quad = \sum_{\overline{W}_1\overline{W}_2\overline{W}_3\overline{W}_4} U(S_1S_2SS_3; S_{12}S_{23})U(T_1T_2TT_3; T_{12}T_{23})\{CCCC\} \,, \end{aligned} \tag{7-211a}$$

where $\{CCCC\}$ is still given by (7-208b) with $\overline{W} = \beta ST, \overline{W}_i = \beta_i S_i T_i$. As in (7-130) we have

$$\begin{aligned} &\sum_{\tau'} U(\nu_1\nu_2\nu\nu_3; \nu_{12}\nu_{23})^{\tau_{12}\tau}_{\tau_{23}\tau'} C^{[\nu]_{\tau'}\overline{W}}_{\nu_1\overline{W}_1,\nu_{23}\overline{W}_{23}} \\ &\quad = \sum_{\overline{W}_2\overline{W}_3\overline{W}_{12}} U(S_1S_2SS_3; S_{12}S_{23})U(T_1T_2TT_3; T_{12}T_{23}) \\ &\quad \times C^{[\nu_{12}]\tau_{12}\overline{W}_{12}}_{\nu_1\overline{W}_1,\nu_2\overline{W}_2} C^{[\nu]_\tau\overline{W}}_{\nu_{12}\overline{W}_{12},\nu_3\overline{W}_3} C^{[\nu_{23}]\tau_{23}\overline{W}_{23}}_{\nu_2\overline{W}_2,\nu_3\overline{W}_3} \,. \end{aligned} \tag{7-211b}$$

Draayer (1973), and Sun (1980) discussed the Racah coefficients of SU_3; Hecht (1969), the Racah coefficients of SU_4; Kaplan (1961) and Le Blanc (1987), the Racah coefficients of SU_n; Derome (1965, 1966), the Racah algebra for an arbitrary group.

7.15. $S_f \supset S_{f_1} \otimes S_{f_2} \otimes S_{f_3} \otimes S_{f_4}$ Irreducible Basis and the 9ν Coefficients of SU_n*

7.15.1. The 9ν coefficients of SU_n

Let us introduce the following symbols for particle numbers:

$$f_{12} = f_1 + f_2, \quad f_{34} = f_3 + f_4, \quad f_{13} = f_1 + f_3, \quad f_{24} = f_2 + f_4\,,$$
$$f_{123} = f_{12} + f_3, \quad f = f_{12} + f_{34} = f_{13} + f_{24}\,. \tag{7-212a}$$

Let the four normal order sequences be

$$(\omega_1^0) = (1, 2, \ldots, f_1), \qquad (\omega_2^0) = (f_1 + 1, \ldots, f_{12})\,,$$
$$(\omega_3^0) = (f_{12} + 1, \ldots, f_{123}), \quad (\omega_4^0) = (f_{123} + 1, \ldots, f)\,. \tag{7-212b}$$

The permutation groups $S_{f_i}(\omega_i^0)$ are designated S_{f_i}.

We can use the following two ways to obtain the $S_f \supset S_{f_1} \otimes S_{f_2} \otimes S_{f_3} \otimes S_{f_4}$ irreducible basis.

1. The $S_f \supset (S_{f_{12}} \supset S_{f_1} \otimes S_{f_2}) \otimes (S_{f_{34}} \supset S_{f_3} \otimes S_{f_4})$ irreducible basis:

$$\left| \begin{matrix} [\nu] \\ W_1 \end{matrix} \left(\begin{matrix} [\nu_{12}] & [\nu_{34}] \\ \nu_1 m_1 \nu_2 m_2, & \nu_3 m_3 \nu_4 m_4 \end{matrix} \right) \right\rangle^{\tau_{12}\tau_{34}\tau} = \sum_{\substack{W_1 W_2 W_3 W_4 \\ W_{12} W_{34}}} C^{[\nu_{12}]\tau_{12}, W_{12}}_{\nu_1 W_1, \nu_2 W_2} C^{[\nu_{34}]\tau_{34}, W_{34}}_{\nu_3 W_3, \nu_4 W_4}$$
$$\times C^{[\nu]_\tau, W}_{\nu_{12} W_{12}, \nu_{34} W_{34}} \left| \begin{matrix} [\nu_1] \\ m_1 \omega_1^0, W_1 \end{matrix} \right\rangle \left| \begin{matrix} [\nu_2] \\ m_2 \omega_2^0, W_2 \end{matrix} \right\rangle \left| \begin{matrix} [\nu_3] \\ m_3 \omega_3^0, W_3 \end{matrix} \right\rangle \left| \begin{matrix} [\nu_4] \\ m_4 \omega_4^0, W_4 \end{matrix} \right\rangle . \tag{7-213a}$$

The left-hand side of (7-213a) is the irreducible basis $[\nu]W$ and $[\nu_i]m_i$ of SU_n and S_{f_i}, respectively, and it also belongs to the irrep $[\nu], [\nu_{12}]$ and $[\nu_{34}]$ of $S_f, S_{f_{12}}$ and $S_{f_{34}}$, respectively.

2. The $S_f \supset (S_{f_{13}} \supset S_{f_1} \otimes S_{f_3}) \otimes \otimes (S_{f_{24}} \supset S_{f_2} \otimes S_{f_4})$ irreducible basis

$$\left| \begin{matrix} [\nu] \\ W, \end{matrix} \left(\begin{matrix} [\nu_{13}] & [\nu_{24}] \\ \nu_1 m_1 \nu_3 m_3, & \nu_2 m_2 \nu_4 m_4 \end{matrix} \right) \right\rangle\!\!\right\rangle^{\tau_{13}\tau_{24}\tau'} = \sum_{\substack{W_1 W_2 W_3 W_4 \\ W_{13} W_{24}}} C^{[\nu_{13}]\tau_{13} W_{13}}_{\nu_1 W_1, \nu_3 W_3} C^{[\nu_{24}], \tau_{24} W_{24}}_{\nu_2 W_2, \nu_4 W_4}$$
$$\times C^{[\nu]\tau' W}_{\nu_{13} W_{13}, \nu_{24} W_{24}} \left| \begin{matrix} [\nu_1] \\ m_1 \omega_1^0, W_1 \end{matrix} \right\rangle \left| \begin{matrix} [\nu_2] \\ m_2 \omega_2^0, W_2 \end{matrix} \right\rangle \left| \begin{matrix} [\nu_3] \\ m_3 \omega_3^0, W_3 \end{matrix} \right\rangle \left| \begin{matrix} [\nu_4] \\ m_4 \omega_4^0, W_4 \end{matrix} \right\rangle , \tag{7-213b}$$

where we use the symbol $|\,\rangle\rangle$ to distinguish (7-213b) from (7-218) to be given below. The bases (7-213a) and (7-213b) differ by a unitary transformation

$$\left| \begin{matrix} [\nu] \\ W, \end{matrix} \left(\begin{matrix} [\nu_{12}] & [\nu_{34}] \\ \nu_1 m_1 \nu_2 m_2, & \nu_3 m_3 \nu_4 m_4 \end{matrix} \right) \right\rangle^{\tau_{12}\tau_{34}\tau}$$
$$= \sum_{\substack{\nu_{13}\nu_{24}, \\ \tau_{13}\tau_{24}\tau'}} \begin{pmatrix} \nu_1 & \nu_2 & \nu_{12} \\ \nu_3 & \nu_4 & \nu_{34} \\ \nu_{13} & \nu_{24} & \nu \end{pmatrix}^{\tau_{12}\tau_{34}\tau}_{\tau_{13}\tau_{24}\tau'} \left| \begin{matrix} [\nu] \\ W, \end{matrix} \left(\begin{matrix} [\nu_{13}] & [\nu_{24}] \\ \nu_1 m_1 \nu_3 m_3, & \nu_2 m_2 \nu_4 m_4 \end{matrix} \right) \right\rangle\!\!\right\rangle^{\tau_{13}\tau_{24}\tau'} . \tag{7-214a}$$

The first factor in the left-hand side of (7-214a) is a 9ν coefficient of SU_n, analogous of the $9j$ coefficients of SU_2, which is independent of the indices $W, m_1, \ldots, m_4$. Therefore if we focus our attention solely on the unitary group, (7-214a) can be rewritten as

$$|(\nu_1\nu_2)\nu_{12}, (\nu_3\nu_4)\nu_{34} : [\nu]W\rangle^{\tau_{12}\tau_{34}\tau}$$
$$= \sum_{\substack{\nu_{13}\nu_{24}\tau' \\ \tau_{13}\tau_{24}}} \begin{pmatrix} \nu_1 & \nu_2 & \nu_{12} \\ \nu_3 & \nu_4 & \nu_{34} \\ \nu_{13} & \nu_{24} & \nu \end{pmatrix}^{\tau_{12}\tau_{34}\tau}_{\tau_{13}\tau_{24}\tau'} |(\nu_1\nu_3)\nu_{13}, (\nu_2\nu_4)\nu_{24} : [\nu]W\rangle^{\tau_{13}\tau_{24}\tau'} . \tag{7-214b}$$

As with (7-136), the 9ν coefficients of SU_n can be expressed in terms of the 9μ coefficients of a subgroup G_s of SU_n and the $SU_n \supset G_s$ ISF. For example, in the $SU_4 \supset SU_2 \times SU_2$ classification, the 9ν coefficientfs of SU_4 can be expressed as

$$\begin{pmatrix} \nu_1 & \nu_2 & \nu_{12} \\ \nu_3 & \nu_4 & \nu_{34} \\ \nu_{13} & \nu_{24} & \nu \end{pmatrix}^{\tau_{12}\tau_{34}\tau}_{\tau_{13}\tau_{24}\tau'} = \sum_{\text{fix}\,\overline{W}} \begin{pmatrix} S_1 & S_2 & S_{12} \\ S_3 & S_4 & S_{34} \\ S_{13} & S_{24} & S \end{pmatrix} \begin{pmatrix} T_1 & T_2 & T_{12} \\ T_3 & T_4 & T_{34} \\ T_{13} & T_{24} & T \end{pmatrix}$$

$$\times C^{[\nu_{12}]\tau_{12},\overline{W}_{12}}_{\nu_1\overline{W}_1,\nu_2\overline{W}_2} C^{[\nu_{34}]\tau_{34},\overline{W}_{34}}_{\nu_3\overline{W}_3,\nu_4\overline{W}_4} C^{[\nu]\tau,\overline{W}}_{\nu_{12}\overline{W}_{12},\nu_{34}\overline{W}_{34}} C^{[\nu_{13}]\tau_{13}\overline{W}_{13}}_{\nu_1\overline{W}_1,\nu_3\overline{W}_3} C^{[\nu_{24}]\tau_{24}\overline{W}_{24}}_{\nu_2\overline{W}_2,\nu_4\overline{W}_4} C^{[\nu]\tau'\overline{W}}_{\nu_{13}\overline{W}_{13},\nu_{24}\overline{W}_{24}}, \tag{7-215}$$

where $\overline{W} = \beta ST$, and $\overline{W}_i = \beta_i S_i T_i$.

Equations (7-134)-(7-138) are also applicable to the group SU_n after the index substitutions $m \to W, m_i \to W_i$, and the re-interpretation of the h's as the dimensions of the irreps of SU_n.

7.15.2. Evaluation of the Racah coefficients and 9ν coefficients of SU_n

For high order unitary groups, due to the difficulty in the calculation of the CG coefficients, it is unfeasible to evaluate the Racah or 9ν coefficients by means of (7-207) or (7-134). This section and Sec. 9.4 will give two other methods for calculating these coefficients.

Let us first consider the 9ν coefficients. Define the permutation operator

$$P = \begin{pmatrix} \omega_2^0 & \omega_3^0 \\ \omega_3^0 & \omega_2^0 \end{pmatrix} = \begin{pmatrix} f_1+1, & f_1+2, & \ldots, & f_{12}, & f_{12}+1, & \ldots, & f_{123} \\ f_{12}+1, & f_{12}+2, & \ldots, & f_{123}, & f_1+1, & \ldots, & f_{12} \end{pmatrix} . \tag{7-216}$$

As an example, for $f_1 = 2, f_2 = 3, f_3 = 4, f_{12} = 5, f_{13} = 6, f_{123} = 9$, we have

$$P = \begin{pmatrix} 3 & 4 & 5 & 6 & 7 & 8 & 9 \\ 6 & 7 & 8 & 9 & 3 & 4 & 5 \end{pmatrix} = (3695847) .$$

Let $S'_{f_3} = S_{f_3}(\omega'_3), S'_{f_2} = S_{f_2}(\omega'_2)$, where

$$(\omega'_3) = (f_1+1, f_1+2, \ldots, f_{13}), \quad (\omega'_2) = (f_{13}+1, \ldots, f_{123}) . \tag{7-217}$$

The $S_f \supset (S'_{f_{13}} \supset (S_{f_1} \otimes S'_{f_3})) \otimes (S'_{f_{24}} \supset (S'_{f_2} \otimes S_{f_4}))$ irreducible basis can be expressed as

$$\left| \begin{matrix} [\nu] \\ W, \end{matrix} \begin{pmatrix} [\nu_{13}] & [\nu_{24}] \\ \nu_1 m_1 \nu_3 m_3, & \nu_2 m_2 \nu_4 m_4 \end{pmatrix} \right\rangle^{\tau_{13}\tau_{24}\tau'} = \sum_{\substack{W_1W_2W_3W_4 \\ W_{13}W_{24}}} C^{[\nu_{13}]\tau_{13},W_{13}}_{\nu_1W_1,\nu_3W_3} C^{[\nu_{24}]\tau_{24},W_{24}}_{\nu_2W_2,\nu_4W_4} C^{[\nu]\tau'W}_{\nu_{13}W_{13},\nu_{24}W_{24}}$$

$$\times \left| \begin{matrix} [\nu_1] \\ m_1\omega_1^0, W_1 \end{matrix} \right\rangle \left| \begin{matrix} [\nu_2] \\ m_2\omega'_2, W_2 \end{matrix} \right\rangle \left| \begin{matrix} [\nu_3] \\ m_3\omega'_3, W_3 \end{matrix} \right\rangle \left| \begin{matrix} [\nu_4] \\ m_4\omega_4^0, W_4 \end{matrix} \right\rangle . \tag{7-218}$$

Notice that (7-218) and (7-213) have exactly the same quantum numbers and they differ only in the indices of the particles belonging to the irrep $[\nu_2]$ and $[\nu_3]$. Using (7-214a) and (7-218), the 9ν coefficients can be expressed as (Kramer 1967),

$$\begin{pmatrix} \nu_1 & \nu_2 & \nu_{12} \\ \nu_3 & \nu_4 & \nu_{34} \\ \nu_{13} & \nu_{24} & \nu \end{pmatrix}^{\tau_{12}\tau_{34}\tau}_{\tau_{13}\tau_{24}\tau'} = \left\langle [\nu] \begin{pmatrix} [\nu_{12}] & [\nu_{34}] \\ \nu_1 m_1 \nu_2 m_2, & \nu_3 m_3 \nu_4 m_4 \end{pmatrix}^{\tau_{12}\tau_{34}\tau} \right.$$

$$\times \left| \begin{pmatrix} \omega_2^0 & \omega_3^0 \\ \omega_3^0 & \omega_2^0 \end{pmatrix} \right| [\nu] \left. \begin{pmatrix} [\nu_{13}] & [\nu_{24}] \\ \nu_1 m_1 \nu_3 m_3, & \nu_2 m_2 \nu_4 m_4 \end{pmatrix}^{\tau_{13}\tau_{24}\tau'} \right\rangle , \tag{7-219}$$

where we supressed the index W. Using the SDC in (4-168a), the non-standard bases of the permutation groups in (7-219) can be expanded in terms of the standard bases, as for instance

$$\left|[\nu]\left(\begin{matrix}[\nu_{12}] & [\nu_{34}] \\ \nu_1 m_1 \nu_2 m_2, & \nu_3 m_3 \nu_4 m_4\end{matrix}\right)\right\rangle^{\tau_{12}\tau_{34}\tau} = \sum_{m m_{12} m_{34}} \left|\begin{matrix}[\nu] \\ m\end{matrix}\right\rangle \left\langle \begin{matrix}[\nu] \\ m\end{matrix} \middle| [\nu] \begin{matrix}\tau[\nu_{12}][\nu_{34}] \\ m_{12}\ m_{34}\end{matrix}\right\rangle$$
$$\times \left\langle \begin{matrix}[\nu_{12}] \\ m_{12}\end{matrix} \middle| [\nu_{12}], \begin{matrix}\tau_{12}[\nu_1][\nu_2] \\ m_1\ m_2\end{matrix}\right\rangle \left\langle \begin{matrix}[\nu_{34}] \\ m_{34}\end{matrix} \middle| [\nu_{34}], \begin{matrix}\tau_{34}[\nu_3][\nu_4] \\ m_3\ m_4\end{matrix}\right\rangle . \tag{7-220}$$

Interchanging $2 \leftrightarrow 3$, we find a similar expression for (7-218). Combining (7-219) and (7-220) we obtain

$$\left(\begin{matrix}\nu_1 & \nu_2 & \nu_{12} \\ \nu_3 & \nu_4 & \nu_{34} \\ \nu_{13} & \nu_{24} & \nu\end{matrix}\right)^{\tau_{12}\tau_{34}\tau}_{\tau_{13}\tau_{24}\tau'} = \sum_{\substack{m_{12}m_{34}m \\ m_{13}m_{24}m'}} D^{[\nu]}_{mm'}(P) \left\langle \begin{matrix}[\nu] \\ m\end{matrix} \middle| [\nu], \begin{matrix}\tau[\nu_{12}][\nu_{34}] \\ m_{12}\ m_{34}\end{matrix}\right\rangle$$
$$\times \left\langle \begin{matrix}[\nu_{12}] \\ m_{12}\end{matrix} \middle| [\nu_{12}], \begin{matrix}\tau_{12}[\nu_1][\nu_2] \\ m_1\ m_2\end{matrix}\right\rangle \left\langle \begin{matrix}[\nu_{34}] \\ m_{34}\end{matrix} \middle| [\nu_{34}], \begin{matrix}\tau_{34}[\nu_3][\nu_4] \\ m_3\ m_4\end{matrix}\right\rangle \left\langle \begin{matrix}[\nu] \\ m'\end{matrix} \middle| [\nu], \begin{matrix}\tau'[\nu_{13}][\nu_{24}] \\ m_{13}\ m_{24}\end{matrix}\right\rangle$$
$$\times \left\langle \begin{matrix}[\nu_{13}] \\ m_{13}\end{matrix} \middle| [\nu_{13}], \begin{matrix}\tau_{13}[\nu_1][\nu_3] \\ m_1\ m_3\end{matrix}\right\rangle \left\langle \begin{matrix}[\nu_{24}] \\ m_{24}\end{matrix} \middle| [\nu_{24}], \begin{matrix}\tau_{24}[\nu_2][\nu_4] \\ m_2\ m_4\end{matrix}\right\rangle , \tag{7-221a}$$

where the sum is carried out under fixed m_1, m_2, m_3 and m_4.

By letting $f_3 = 0, [\nu_3] = [0]$, and $P = e$ (identity), from (7-137) and (7-221a) we obtain an expression for the SU_n Racah coefficients (Kramer 1968):

$$U(\nu_1\nu_2\nu\nu_3;\nu_{12}\nu_{23})^{\tau_{12}\tau}_{\tau_{23}\tau'} = \sum_{m_{12}m_{23}m} \left\langle \begin{matrix}[\nu] \\ m\end{matrix} \middle| [\nu], \begin{matrix}\tau[\nu_{12}][\nu_3] \\ m_{12}\ m_3\end{matrix}\right\rangle \left\langle \begin{matrix}[\nu_{12}] \\ m_{12}\end{matrix} \middle| [\nu_{12}], \begin{matrix}\tau_{12}[\nu_1][\nu_2] \\ m_1\ m_2\end{matrix}\right\rangle$$
$$\times \left\langle \begin{matrix}[\nu] \\ m\end{matrix} \middle| [\nu], \begin{matrix}\tau'[\nu_1][\nu_{23}] \\ m_1\ m_{23}\end{matrix}\right\rangle \left\langle \begin{matrix}[\nu_{23}] \\ m_{23}\end{matrix} \middle| [\nu_{23}], \begin{matrix}\tau_{23}[\nu_2][\nu_3] \\ m_2 m_3\end{matrix}\right\rangle , \tag{7-221b}$$

where the sum is carried out with m_1, m_2 and m_3 fixed.

Since the SDC of the permutation groups do not depend on n, we reach a significant conclusion that the *Racah coefficients as well as the 9ν coefficients of the group SU_n do not depend on n explicitly*. In other words, they only depend on the partition labels. Therefore we can tabulate the Racah coefficients and 9ν coefficients of the group SU_n for arbitrary n instead of each table referring to only one particular n.

The transformation coefficients of permutation groups can be easily calculated by the EFM. Consequently we can use (7-221a) and (7-221b) to calculate the Racah coefficients and 9ν coefficients of the group SU_n for arbitrary n. The SU_n Racah coefficients have been calculated in this way and tabulated in Chen ([31], 1987).

Utilizing the symmetries (4-180) and (4-67) of the SDC, and the irreducible matrix elements of S_f, we get from (7-221) the following two symmetries:

$$\left(\begin{matrix}\tilde\nu_1 & \tilde\nu_2 & \tilde\nu_{12} \\ \tilde\nu_3 & \tilde\nu_4 & \tilde\nu_{34} \\ \tilde\nu_{13} & \tilde\nu_{24} & \tilde\nu\end{matrix}\right)^{\tau_{12}\tau_{34}\tau}_{\tau_{13}\tau_{24}\tau'} = \varepsilon \cdot \left(\begin{matrix}\nu_1 & \nu_2 & \nu_{12} \\ \nu_3 & \nu_4 & \nu_{34} \\ \nu_{13} & \nu_{24} & \nu\end{matrix}\right)^{\tau_{12}\tau_{34}\tau}_{\tau_{13}\tau_{24}\tau'} , \tag{7-222a}$$

$$U(\tilde\nu_1\tilde\nu_2\tilde\nu\tilde\nu_3,\ \tilde\nu_{12}\tilde\nu_{23})^{\tau_{12}\tau}_{\tau_{23}\tau'} = \varepsilon U(\nu_1\nu_2\nu\nu_3;\nu_{12}\nu_{23})^{\tau_{12}\tau}_{\tau_{23}\tau'} \tag{7-222b}$$

where $\varepsilon = \pm 1$ is a phase depending on our phase choices.

Hecht (1975) studied the 9ν coefficients of SU_3 and SU_4; Kukulin (1967), the 9ν coefficients of SU_4; and Kramer (1969), the 9ν coefficients of SU_n.

7.16. $\mathbf{SU}_{mn} \supset \mathbf{SU}_m \times \mathbf{SU}_n$ CFP

7.16.1. $SU_{mn} \supset SU_m \times SU_n$ CFP and $S_{f_1+f_2} \supset S_{f_1} \otimes S_{f_2}$ ISF

In Sec. 7.12 we introduced a method for calculating the $SU_{mn} \supset SU_m \times SU_n$ CFP, where the Racah coefficients of SU_m and SU_n must be known beforehand. However, for $m > 3$, no tables of SU_m Racah coefficients are available. Therefore we have to seek after new methods.

As with (4-197), we introduce the following symbols to denote the irreps of the three unitary groups and nine permutation groups:

$$\begin{pmatrix} \sigma' & \mu' & \nu'_{\beta'} \\ \sigma'' & \mu'' & \nu''_{\beta''} \\ \sigma_\theta & \mu_\varphi & \nu_{\tau,\beta} \end{pmatrix}, \quad \begin{pmatrix} SU_m & SU_n & SU_{mn} \\ SU_m & SU_n & SU_{mn} \\ SU_m & SU_n & SU_{mn} \end{pmatrix}, \quad \begin{pmatrix} S_{f_1}(x) & S_{f_1}(\xi) & S_{f_1}(q) \\ S_{f_2}(x) & S_{f_2}(\xi) & S_{f_2}(q) \\ S_f(x) & S_f(\xi) & S_f(q) \end{pmatrix} \tag{7-223}$$

By making the following index substitutions,

$$\Lambda \to \sigma\mu, \quad m \to W_1 W_2, \quad \theta \to \theta\varphi, \quad 1 \to ', 2 \to '' \ .$$

We get the following equations from (3-300), (3-301) and (3-307),

$$\left| \begin{matrix} [\nu]\tau \\ \beta\sigma W_1 \mu W_2 \end{matrix} \right\rangle = \sum_{\substack{\beta'\sigma'\mu'\theta \\ \beta''\sigma''\mu''\varphi}} C^{[\nu]_\tau,\beta[\sigma]_\theta[\mu]_\varphi}_{\nu'\beta'\sigma'\mu',\nu''\beta''\sigma''\mu''} \left[\left| \begin{matrix} [\nu'] \\ \beta'\sigma'\mu' \end{matrix} \right\rangle \left| \begin{matrix} [\nu''] \\ \beta''\sigma''\mu'' \end{matrix} \right\rangle \right]^{[\sigma]_\theta[\mu]_\varphi}_{W_1 W_2} . \tag{7-224a}$$

$$\left[\left| \begin{matrix} [\nu'] \\ \beta'\sigma'\mu' \end{matrix} \right\rangle \left| \begin{matrix} [\nu''] \\ \beta''\sigma''\mu'' \end{matrix} \right\rangle \right]^{[\sigma]_\theta[\mu]_\varphi}_{W_1 W_2}$$
$$= \sum_{W'_1 W'_2 W''_1 W''_2} C^{[\sigma]_\theta, W_1}_{\sigma' W'_1, \sigma'' W''_1} C^{[\mu]_\varphi, W_2}_{\mu' W'_2, \mu'' W''_2} \left| \begin{matrix} [\nu'] \\ \beta'\sigma' W'_1 \mu' W'_2 \end{matrix} \right\rangle \left| \begin{matrix} [\nu''] \\ \beta''\sigma'' W''_1 \mu'' W''_2 \end{matrix} \right\rangle . \tag{7-225}$$

$$\left[\left| \begin{matrix} [\nu'] \\ \beta'\sigma'\mu' \end{matrix} \right\rangle \left| \begin{matrix} [\nu''] \\ \beta''\sigma''\mu'' \end{matrix} \right\rangle \right]^{[\sigma]_\theta[\mu]_\varphi}_{W_1 W_2} = \sum_{\nu\tau\beta} C^{[\nu]_\tau,\beta[\sigma]_\theta[\mu]_\varphi}_{\nu'\beta'\sigma'\mu',\nu''\beta''\sigma''\mu''} \left| \begin{matrix} [\nu]\tau \\ \beta\sigma W_1 \mu W_2 \end{matrix} \right\rangle . \tag{7-226}$$

Equation (7-226) is the inverse of (7-224a). The first factor at the right-hand side of (7-224a) is the $SU_{mn} \supset SU_m \times SU_n$ ISF and is independent of W_1 and W_2. Therefore (7-224a) can be written succinctly as

$$\left| \begin{matrix} [\nu]\tau \\ \beta\sigma\mu \end{matrix} \right\rangle = \sum_{\substack{\beta'\sigma'\mu'\theta \\ \beta''\sigma''\mu''\varphi}} C^{[\nu]\tau,\beta[\sigma]\theta[\mu]\varphi}_{\nu'\beta'\sigma'\mu',\nu''\beta''\sigma''\mu''} \left[\left| \begin{matrix} [\nu'] \\ \beta'\sigma'\mu' \end{matrix} \right\rangle \left| \begin{matrix} [\nu''] \\ \beta''\sigma''\mu'' \end{matrix} \right\rangle \right]^{[\sigma]\theta[\mu]\varphi} . \tag{7-224b}$$

Attaching the Young tableaux $Y^{(\nu')}_{m'}(\omega^0_1)$ and $Y^{(\nu'')}_{m''}(\omega^0_2)$ with $(\omega^0_1) = (1, 2, \ldots, f_1)$ and $(\omega^0_2) = (f_1 + 1, \ldots, f)$ to the two irreducible basis vectors at the right-hand side of (7-224a), it reads:

$$\left| \begin{matrix} [\nu] \\ \tau\nu' m' \nu'' m'', \beta\sigma W_1 \mu W_2 \end{matrix} \right\rangle = \sum_{\substack{\beta'\sigma'\mu'\theta \\ \beta''\sigma''\mu''\varphi}} C^{[\nu]_\tau,\beta[\sigma]\theta[\mu]\varphi}_{\nu'\beta'\sigma'\mu',\nu''\beta''\sigma''\mu''}$$
$$\times \left[\left| \begin{matrix} [\nu'] \\ m'\omega^0_1, \beta'\sigma'\mu' \end{matrix} \right\rangle \left| \begin{matrix} [\nu''] \\ m''\omega^0_2, \beta''\sigma''\mu'' \end{matrix} \right\rangle \right]^{[\sigma]_\theta[\mu]_\varphi}_{W_1 W_2} . \tag{7-227}$$

The left-hand side of (7-227) is still the $SU_{mn} \supset SU_m \times SU_n$ basis. It belongs to the irrep $[\nu]$ of SU_{mn}. According to Theorem 7.1, it must also belong to the irrep $[\nu]$ of the permutation group $S_f(q)$. On the other hand, $[\nu']m'$ and $[\nu'']m''$ at the right-hand side of (7-227) are fixed. Therefore (7-227) is also an $S_f(q) \supset S_{f_1}(q) \otimes S_{f_2}(q)$ basis.

With the help of (7-225), the last factor in (7-227) can be put into the form

$$(I) \equiv \left[\left|\begin{matrix}[\nu'] \\ m'\omega_1^0, \beta'\sigma'\mu'\end{matrix}\right\rangle\left|\begin{matrix}[\nu''] \\ m''\omega_2^0, \beta''\sigma''\mu''\end{matrix}\right\rangle\right]^{[\sigma]_\theta[\mu]_\varphi}_{W_1W_2}$$

$$= \sum_{W_1'W_2'W_1''W_2''} C^{[\sigma]_\theta, W_1}_{\sigma'W_1', \sigma''W_1''} C^{[\mu]\varphi, W_2}_{\mu'W_2', \mu''W_2''} \left|\begin{matrix}[\nu'] \\ m'\omega_1^0, \beta'\sigma'W_1'\mu'W_2'\end{matrix}\right\rangle\left|\begin{matrix}[\nu''] \\ m''\omega_2^0, \beta''\sigma''W_1''\mu''W_2''\end{matrix}\right\rangle .$$

Using (7-106a), the $SU_{mn} \supset SU_m \times SU_n$ bases can be further expanded; thus

$$(I) = \sum_{\substack{W_1'W_2'W_1''W_2'' \\ m_1'm_2'm_1''m_2''}} C^{[\sigma]\theta W_1}_{[\sigma']W_1'[\sigma'']W_1''} C^{[\mu]\varphi W_2}_{[\mu']W_2', [\mu'']W_2''} C^{[\nu']\beta' m'}_{[\sigma']m_1' [\mu']m_2'} C^{[\nu'']\beta'', m''}_{[\sigma'']m_1'', [\mu'']m_2''}$$

$$\times \left|\begin{matrix}[\sigma'] \\ m_1'\omega_1^0, W_1'\end{matrix}\right\rangle\left|\begin{matrix}[\mu'] \\ m_2'\omega_1^0, W_2'\end{matrix}\right\rangle\left|\begin{matrix}[\sigma''] \\ m_1''\omega_2^0, W_1''\end{matrix}\right\rangle\left|\begin{matrix}[\mu''] \\ m_2''\omega_2^0, W_2''\end{matrix}\right\rangle$$

$$= \sum_{m_1'm_2'm_1''m_2''} C^{[\nu']\beta', m'}_{[\sigma']m_1', [\mu']m_2'} C^{[\nu'']\beta'' m''}_{[\sigma'']m_1'', [\mu'']m_2''} \left|\begin{matrix}[\sigma] \\ \theta[\sigma']m_1'[\sigma'']m_1'', W_1\end{matrix}\right\rangle\left|\begin{matrix}[\mu] \\ \varphi[\mu']m_2'[\mu'']m_2'', W_2\end{matrix}\right\rangle , \tag{7-228}$$

where (7-102b) has been used in the last step. Comparing (4-200) and (4-199) with (7-227) and (7-228), one derives an important relation,

$$C^{[\nu]\tau, \beta[\sigma]\theta[\mu]\varphi}_{[\nu']\beta'\sigma'\mu', [\nu'']\beta''\sigma''\mu''} = C^{[\nu]\beta, \tau[\nu']\beta'[\nu'']\beta''}_{[\sigma]\theta\sigma'\sigma'', [\mu]\varphi\mu'\mu''} . \tag{7-229}$$

Or expressed in the form of overlap integrals,

$$\left\langle\begin{matrix}[\nu] \\ \tau\nu'm'\nu''m'', \beta\sigma W_1\mu W_2\end{matrix}\middle|\left[\left|\begin{matrix}[\sigma] \\ \theta\sigma'\sigma'', W_1\end{matrix}\right\rangle\left|\begin{matrix}[\mu] \\ \varphi\mu'\mu'', W_2\end{matrix}\right\rangle\right]^{[\nu']_{\beta'}[\nu'']_{\beta''}}_{m'm''}\right\rangle$$

$$= \left\langle\begin{matrix}[\nu] \\ \tau\nu'm'\nu''m'', \beta\sigma W_1\mu W_2\end{matrix}\middle|\left[\left|\begin{matrix}[\nu'] \\ m', \beta'\sigma'\mu'\end{matrix}\right\rangle\left|\begin{matrix}[\nu''] \\ m'', \beta''\sigma''\mu''\end{matrix}\right\rangle\right]^{[\sigma]_\theta[\mu]_\varphi}_{W_1W_2}\right\rangle . \tag{7-230}$$

Thus we proved that the $SU_{mn} \supset SU_m \times SU_n$ ISF (or the f_2-particle CFP) are just the $S_{f_1+f_2} \supset S_{f_1} \otimes S_{f_2}$ ISF and that the values of the $SU_{mn} \supset SU_m \times SU_n$ ISF do not depend on m and n explicitly, since the values of $S_{f_1+f_2} \supset S_{f_1} \otimes S_{f_2}$ ISF are independent of m and n.

Remark: More precisely, from (7-229) we can only identify the $U_{mn} \supset U_m \times U_n$ CFP with the $S_f \supset S_{f_1} \otimes S_{f_2}$ ISF. However the overall phase convention in Sec. 4.19 ensures that CFP calculated from (7-229) is also the $SU_{mn} \supset SU_m \times SU_n$ CFP (Chen, [16] 1984).

Setting $[\nu''] = [\sigma''] = [\mu''] = [1]$ in (7-229), ignoring the redundant labels $\tau, \theta, \varphi, \beta'', \sigma''$ and μ'', and using (4-189b), we obtain an expression for the $SU_{mn} \supset SU_m \times SU_n$ one-particle CFP in terms of the CG coefficients of the permutation groups S_f and S_{f-1},

$$C^{[\nu], \beta\sigma\mu}_{[\nu']\beta'\sigma'\mu', [1]} = C^{[\nu]\beta, [\nu']\beta'}_{[\sigma]\sigma', [\mu]\mu'} = \sum_{m_1'm_2'} C^{[\nu]\beta, m}_{\sigma m_1, \mu m_2} C^{[\nu']\beta', m'}_{\sigma' m_1', \mu' m_2'} . \tag{7-231a}$$

When β' is redundant, it reduces to (4-189c), i.e.,

$$C^{[\nu],\beta\sigma\mu}_{[\nu']\sigma'\mu',[1]} = C^{[\nu]\beta,m}_{\sigma m_1,\mu m_2} / C^{[\nu'],m'}_{\sigma' m'_1,\mu' m'_2} \tag{7-231b}$$

From (7-231a) and (4-196g) we have

$$C^{[1^n][\sigma][\tilde{\sigma}]}_{[1^{n-1}][\sigma'],[\tilde{\sigma}'][1]} = \Lambda^{\sigma}_{\sigma'} \sqrt{h_{\sigma'}/h_{\sigma}} \tag{7-232}$$

7.16.2. The evaluation of the $SU_{mn} \supset SU_m \times SU_n$ many-particle CFP

Making use of (4-168a) and (4-92a), we have

$$\left| \begin{matrix} [\nu]\beta \\ \tau\nu' m' \nu'' m'' \end{matrix} \right\rangle = \sum_{m m_1 m_2}^{\text{fix}\, m'm''} \left\langle \begin{matrix} [\nu] \\ m \end{matrix} \middle| [\nu], \begin{matrix} \tau\nu'\nu'' \\ m'm'' \end{matrix} \right\rangle C^{[\nu]\beta,m}_{\sigma m_1,\mu m_2} \varphi^{\sigma}_{m_1} \psi^{\mu}_{m_2} . \tag{7-233}$$

By using the factorization formula (4-189a) of the CG coefficients f_2 times, we get

$$C^{[\nu]\beta,m}_{\sigma m_1,\mu m_2} = \sum_{\beta'} I^{\nu m\beta,\nu'\beta'}_{\sigma m_1 \sigma',\mu m_2 \mu'} C^{[\nu']\beta',m'}_{\sigma' m'_1,\mu' m'_2} , \tag{7-234a}$$

where I is a sum of products of the $S_f \supset S_{f-1}$ ISF, $S_{f-1} \supset S_{f-2}$ ISF,... and $S_{f_1+1} \supset S_{f_1}$ ISF,

$$I^{\nu m\beta,\nu'\beta'}_{\sigma m_1 \sigma',\mu m_2 \mu'} = \sum_{\overline{\beta}\bar{\bar{\beta}}\cdots\widehat{\beta}} C^{\nu\beta,\overline{\nu}\overline{\beta}}_{\sigma\overline{\sigma},\mu\overline{\mu}} C^{\overline{\nu}\overline{\beta},\bar{\bar{\nu}}\bar{\bar{\beta}}}_{\overline{\sigma}\bar{\bar{\sigma}},\overline{\mu}\bar{\bar{\mu}}} \cdots C^{\hat{\nu}\hat{\beta},\nu'\beta'}_{\widehat{\sigma}\sigma',\hat{\mu}\mu'} . \tag{7-234b}$$

with $[\sigma]m_1 = [\sigma][\overline{\sigma}][\bar{\bar{\sigma}}]\ldots[\widehat{\sigma}][\sigma']m'_1$, $[\mu]m_2 = [\mu][\overline{\mu}][\bar{\bar{\mu}}]\ldots[\widehat{\mu}][\mu']m'_2$.

Notice that I not only depends on $\nu, \beta, \nu'\beta', \sigma, \sigma', \mu$ and μ', but also depends on m_1, m_2 and m, i.e., on the positions of the numbers $f, f-1, \ldots, f_1$ in the Young tableaux $Y^{\sigma}_{m_1}, Y^{\mu}_{m_2}$ and Y^{ν}_{m}.

From (7-234a) we have

$$I^{\nu m\beta,\nu'\beta'}_{\sigma m_1 \sigma',\mu m_2 \mu'} = \sum_{m'_1 m'_2}^{\text{fix}\, m'} C^{[\nu]\beta,m}_{\sigma m_1,\mu m_2} C^{[\nu']\beta',m'}_{\sigma' m'_1,\mu' m'_2} . \tag{7-234c}$$

Using (4-168b) and (7-234), (7-233) reads

$$\begin{aligned} \left| \begin{matrix} [\nu]\beta \\ \tau\nu' m' \nu'' m'' \end{matrix} \right\rangle &= \sum^{\text{fix}\, m'_1 m'_2} \left\langle \begin{matrix} [\nu] \\ m \end{matrix} \middle| [\nu], \begin{matrix} \tau\nu'\nu'' \\ m'm'' \end{matrix} \right\rangle \left\langle \begin{matrix} [\sigma] \\ m_1 \end{matrix} \middle| [\sigma], \begin{matrix} \theta\sigma'\sigma'' \\ m'_1 m''_1 \end{matrix} \right\rangle \left\langle \begin{matrix} [\mu] \\ m_2 \end{matrix} \middle| [\mu], \begin{matrix} \varphi\mu'\mu'' \\ m'_2 m''_2 \end{matrix} \right\rangle \\ &\times I^{\nu m\beta,\nu'\beta'}_{\sigma m_1 \sigma',\mu m_2 \mu'} \left[\sum_{m'_1 m'_2} C^{[\nu']\beta',m'}_{\sigma' m'_1,\mu' m'_2} \left| \begin{matrix} [\sigma] \\ \theta\sigma' m'_1 \sigma'' m''_1 \end{matrix} \right\rangle \left| \begin{matrix} [\mu] \\ \varphi\mu' m'_2 \mu'' m''_2 \end{matrix} \right\rangle \right] , \end{aligned} \tag{7-235a}$$

where the first sum runs over $m, m_1, m_2, m''_1, m''_2, \sigma'', \mu'', \beta', \theta$ and φ under fixed m'_1 and m'_2. In deriving (7-235a), we used the fact that the SDC and $I^{\nu m\beta,\nu'\beta'}_{\sigma m_1 \sigma',\mu m_2 \mu'}$ do not depend on m'_1 and m'_2. The square bracket term in (7-235a) can be expressed as

$$[\ ] = \sum_{\nu''\beta'' m''} C^{[\nu'']\beta'',m''}_{\sigma'' m''_1,\mu'' m''_2} \left[\left| \begin{matrix} [\sigma] \\ \theta\sigma'\sigma'' \end{matrix} \right\rangle \left| \begin{matrix} [\mu] \\ \varphi\mu'\mu'' \end{matrix} \right\rangle \right]^{[\nu']\beta'[\nu'']\beta''}_{m'm''} \tag{7-235b}$$

Comparing (7-235) with (4-200), and using (7-229) we finally get

$$C^{[\nu]\tau,\beta[\sigma]\theta[\mu]\varphi}_{[\nu']\beta'\sigma'\mu',[\nu'']\beta''\sigma''\mu''} = \sum_{mm_1m_2m_1''m_2''}^{\text{fix}\, m'm''m_1'm_2''} I^{\nu m\beta,\nu'\beta'}_{\sigma m_1\sigma',\mu m_2\mu'} C^{[\nu'']\beta'',m''}_{\sigma''m_1'',\mu''m_2''}$$

$$\times \left\langle \begin{matrix}[\nu]\\ m\end{matrix} \middle| [\nu], \begin{matrix}\tau\nu'\nu''\\ m'm''\end{matrix} \right\rangle \left\langle \begin{matrix}[\sigma]\\ m_1\end{matrix} \middle| [\sigma], \begin{matrix}\theta\sigma'\sigma''\\ m_1'm_1''\end{matrix} \right\rangle \left\langle \begin{matrix}[\mu]\\ m_2\end{matrix} \middle| [\mu] \begin{matrix}\varphi\mu'\mu''\\ m_2'm_2''\end{matrix} \right\rangle . \quad \text{(7-236a)}$$

Using (7-234c) this can be rewritten as

$$C^{[\nu]\tau,\beta[\sigma]\theta[\mu]\varphi}_{[\nu']\beta'\sigma'\mu',[\nu'']\beta''\sigma''\mu''} = \sum_{mm_1m_2m_1''m_2''}^{\text{fix}\, m'm''} C^{[\nu]\beta,m}_{\sigma m_1,\mu m_2} C^{[\nu']\beta',m'}_{\sigma'm_1',\mu'm_2'} C^{[\nu'']\beta'',m''}_{\sigma''m_1'',\mu''m_2''}$$

$$\times \left\langle \begin{matrix}[\nu]\\ m\end{matrix} \middle| [\nu], \begin{matrix}\tau\nu'\nu''\\ m'm''\end{matrix} \right\rangle \left\langle \begin{matrix}[\sigma]\\ m_1\end{matrix} \middle| [\sigma], \begin{matrix}\theta\sigma'\sigma''\\ m_1'm_1''\end{matrix} \right\rangle \left\langle \begin{matrix}[\mu]\\ m_2\end{matrix} \middle| [\mu] \begin{matrix}\varphi\mu'\mu''\\ m_2'm_2''\end{matrix} \right\rangle . \quad \text{(7-236b)}$$

We can infer from (7-236a) that

$$\sum_{\tau} \left\langle \begin{matrix}[\nu]\\ m\end{matrix} \middle| [\nu] \begin{matrix}\tau\nu'\nu''\\ m'm''\end{matrix} \right\rangle C^{[\nu]\tau,\beta[\sigma]\theta[\mu]\varphi}_{[\nu']\beta'\sigma'\mu',[\nu'']\beta''\sigma''\mu''} = \sum_{m_1m_2m_1''m_2''}^{\text{fix}\, m_1'm_2'} I^{\nu m\beta,\nu'\beta'}_{\sigma m_1\sigma',\mu m_2\mu'} C^{[\nu'']\beta'',m''}_{\sigma''m_1'',\mu''m_2''}$$

$$\times \left\langle \begin{matrix}[\sigma]\\ m_1\end{matrix} \middle| [\sigma], \begin{matrix}\theta\sigma'\sigma''\\ m_1'm_1''\end{matrix} \right\rangle \left\langle \begin{matrix}[\mu]\\ m_2\end{matrix} \middle| [\mu], \begin{matrix}\varphi[\mu'][\mu'']\\ m_2'm_2''\end{matrix} \right\rangle . \quad \text{(7-237)}$$

When τ is redundant, (7-237) gives a simpler expression for the $SU_{mn} \supset SU_m \times SU_n$ ISF, i.e.,

$$C^{[\nu],\beta[\sigma]\theta[\mu]\varphi}_{[\nu']\beta'\sigma'\mu',[\nu'']\beta''\sigma''\mu''} = \left\langle \begin{matrix}[\nu]\\ m\end{matrix} \middle| [\nu], \begin{matrix}\nu'\nu''\\ m'm''\end{matrix} \right\rangle^{-1} \sum_{m_1m_2m_1''m_2''}^{\text{fix}\, m_1'm_2'} I^{\nu m\beta,\nu'\beta'}_{\sigma m_1\sigma',\mu m_2\mu'}$$

$$\times C^{[\nu'']\beta'',m''}_{\sigma''m_1'',\mu''m_2''} \left\langle \begin{matrix}[\sigma]\\ m_1\end{matrix} \middle| [\sigma] \begin{matrix}\theta\sigma'\sigma''\\ m_1'm_1''\end{matrix} \right\rangle \left\langle \begin{matrix}[\mu]\\ m_2\end{matrix} \middle| [\mu] \begin{matrix}\varphi\mu'\mu''\\ m_2'm_2''\end{matrix} \right\rangle . \quad \text{(7-236c)}$$

Equations (7-236a) and (7-236c) offer a convenient method for calculating the many-particle CFP from the one-particle CFP and the SDC, while (7-236b) and the following simplified formulas are used to calculate the CFP from the CG coefficients and the SDC of permutation groups.

1. When the multiplicity labels τ and β' are redundant, we have

$$C^{[\nu],\beta[\sigma]\theta[\mu]\varphi}_{\nu'\sigma'\mu',\nu''\beta''\sigma''\mu''} = \left[\left\langle \begin{matrix}[\nu]\\ m\end{matrix} \middle| [\nu], \begin{matrix}\nu'\nu''\\ m'm''\end{matrix} \right\rangle C^{[\nu']m'}_{\sigma'm_1',\mu'm_2'} \right]^{-1}$$

$$\times \sum_{m_1m_2m_1''m_2''}^{\text{fix}\, m_1'm_2'} C^{[\nu]\beta,m}_{\sigma m_1,\mu m_2} C^{[\nu'']\beta'',m''}_{\sigma''m_1'',\mu''m_2''} \left\langle \begin{matrix}[\sigma]\\ m_1\end{matrix} \middle| [\sigma] \begin{matrix}\theta\sigma'\sigma''\\ m_1'm_1''\end{matrix} \right\rangle \left\langle \begin{matrix}[\mu]\\ m_2\end{matrix} \middle| [\mu] \begin{matrix}\varphi\mu'\mu''\\ m_2'm_2''\end{matrix} \right\rangle . \quad \text{(7-238a)}$$

By the symmetry (7-242b), the case when φ and β' (or θ and β') are redundant can be converted to the above case.

2. When the multiplicity label β' is redundant, we obtain

$$C^{[\nu]\tau,\beta[\sigma]\theta[\mu]\varphi}_{\nu'\sigma'\mu',\nu''\beta''\sigma''\mu''} = (C^{[\nu'],m'}_{\sigma'm_1',\mu'm_2'})^{-1} \sum_{mm_1m_2}^{\text{fix}\, m'm''m_1'm_2'} C^{[\nu]\beta,m}_{\sigma m_1,\mu m_2} C^{[\nu'']\beta'',m''}_{\sigma''m_1'',\mu''m_2''}$$

$$\times \left\langle \begin{matrix}[\nu]\\ m\end{matrix} \middle| [\nu], \begin{matrix}\tau\nu'\nu''\\ m'm''\end{matrix} \right\rangle \left\langle \begin{matrix}[\sigma]\\ m\end{matrix} \middle| [\sigma], \begin{matrix}\theta\sigma'\sigma''\\ m_1'm_1''\end{matrix} \right\rangle \left\langle \begin{matrix}[\mu]\\ m_2\end{matrix} \middle| [\mu], \begin{matrix}\varphi\mu'\mu''\\ m_2'm_2''\end{matrix} \right\rangle . \quad \text{(7-238b)}$$

The case when β'' is redundant can be converted to the above case by using (7-239a).

For two-particle CFP, (7-236a) and (7-236c) reduce to

$$C^{[\nu],\beta\sigma\mu}_{[\nu']\beta'\sigma'\mu',[\nu'']\sigma'\mu''} = \langle[\nu]m|[\nu'']\rangle^{-1} \sum_{m_1 m_2 \overline{\beta}}^{\text{fix } m_1' m_2'} C^{\nu\beta,\overline{\nu}\overline{\beta}}_{\sigma\overline{\sigma},\mu\overline{\mu}} C^{\overline{\nu}\overline{\beta},\nu'\beta'}_{\overline{\sigma}\sigma',\overline{\mu}\mu'} \langle[\sigma]m_1|[\sigma'']\rangle\langle[\mu]m_2|[\mu'']\rangle \,, \quad (7\text{-}238c)$$

where $[\nu''] = [\sigma''] \times [\mu'']$, and the simpler notation $\langle[\nu]m|[\nu'']\rangle$ is used for the SDC. Equation (7-238c) is a simplified version of the formula given by Harvey (1981 p. 235) (what he defined as the K coefficients are just the $SU_{mn} \supset SU_m \times SU_n$ ISF).

7.16.3. The symmetries of the $SU_{mn} \supset SU_m \times SU_n$ ISF

From (3-328)-(330a) we obtain the following three symmetries:

1. $$C^{[\nu]\tau,\beta[\sigma]\theta[\mu]\varphi}_{[\nu']\beta'\sigma'\mu',[\nu'']\beta''\sigma''\mu''} = \varepsilon_1 C^{[\nu]\tau,\beta[\sigma]\theta[\mu]\varphi}_{[\nu'']\beta''\sigma''\mu'',[\nu']\beta'\sigma'\mu'} \,, \quad (7\text{-}239a)$$

where ε_1 (or ε_i in the following) is a phase factor, depending on all partitions, $\varepsilon_1 = \pm 1$,

2. $$C^{[\nu]\tau\beta\sigma\mu}_{[\nu']\beta'\sigma'\mu',[\nu'']\beta''\sigma''\mu''} = \varepsilon_2 C^{[\overline{\nu}]\tau,\beta\overline{\sigma}\overline{\mu}}_{[\overline{\nu}']\beta'\overline{\sigma}'\overline{\mu}',[\overline{\nu}'']\beta''\overline{\sigma}''\overline{\mu}''} \,. \quad (7\text{-}239b)$$

3. $$\sqrt{\frac{h_\sigma(SU_m)h_\mu(SU_n)}{h_\nu(SU_{mn})}} C^{[\nu],\beta\sigma\mu}_{[\nu']\beta'\sigma'\mu',[\nu'']\beta''\sigma''\mu''}$$
$$= \varepsilon_3 \sqrt{\frac{h_{\sigma'}(SU_m)h_{\mu'}(SU_n)}{h_{\nu'}(SU_{mn})}} C^{[\overline{\nu}'],\beta'\overline{\sigma}'\overline{\mu}'}_{[\overline{\nu}]\beta\overline{\sigma}\overline{\mu},[\nu'']\beta''\sigma''\mu''} \,, \quad (7\text{-}240)$$

where $h_\sigma(SU_m)$ is the dimension of the irrep $[\sigma]$ of SU_m, $[\overline{\nu}]$, $[\overline{\sigma}]$ and $[\overline{\mu}]$ are the contragredient reps of $[\nu]$, $[\sigma]$ and $[\mu]$, respectively.

4. From (4-121), (4-180a) and (7-236b) we have

$$C^{[\nu]\tau,\beta[\sigma]\theta[\mu]\varphi}_{[\nu']\beta'\sigma'\mu',[\nu'']\beta''\sigma''\mu''} = \varepsilon_4 C^{[\nu]\tau,\beta[\tilde{\sigma}]\theta[\tilde{\mu}]\varphi}_{[\nu']\beta'\tilde{\sigma}'\tilde{\mu}',[\nu'']\beta''\tilde{\sigma}''\tilde{\mu}''} = \varepsilon_5 C^{[\tilde{\nu}]\tau,\beta[\tilde{\sigma}]\theta[\mu]\varphi}_{[\tilde{\nu}']\beta'\tilde{\sigma}'\mu',[\tilde{\nu}'']\beta''\tilde{\sigma}''\mu''}$$
$$= \varepsilon_6 C^{[\tilde{\nu}]\tau,\beta[\sigma]\theta[\tilde{\mu}]\varphi}_{[\tilde{\nu}']\beta'\sigma'\tilde{\mu}',[\tilde{\nu}'']\beta''\sigma''\tilde{\mu}''} \,. \quad (7\text{-}241)$$

where $[\tilde{\sigma}]$ is the conjugate of the Young diagram $[\sigma]$, etc.

5. We first rewrite (7-236b) in the following form,

$$C^{[\nu]\tau,\beta[\sigma]\theta[\mu]\varphi}_{[\nu']\beta'\sigma'\mu',[\nu'']\beta''\sigma''\mu''} = \frac{1}{h_{\nu'}h_{\nu''}} \sum_{m'm''} \sum_{m m_1 m_2 m_1'' m_2''} C^{[\nu]\beta,m}_{\sigma m_1,\mu m_2} C^{[\nu']\beta',m'}_{\sigma' m_1',\mu m_2'} C^{[\nu'']\beta'',m''}_{\sigma'' m_1'',\mu'' m_2''}$$
$$\times \left\langle \begin{matrix}[\nu]\\ m\end{matrix}\middle|[\nu],\begin{matrix}\tau\nu'\nu''\\ m'm''\end{matrix}\right\rangle \left\langle \begin{matrix}[\sigma]\\ m_1\end{matrix}\middle|[\sigma]\begin{matrix}\theta\sigma'\sigma''\\ m_1'm_1''\end{matrix}\right\rangle \left\langle \begin{matrix}[\mu]\\ m_2\end{matrix}\middle|[\mu]\begin{matrix}\varphi\mu'\mu''\\ m_2'm_2''\end{matrix}\right\rangle \,. \quad (7\text{-}242a)$$

From (4-177) and (7-242a) it can be shown that

$$\sqrt{\frac{h_{\nu'}h_{\nu''}}{h_\nu}} C^{[\nu]\tau,\beta[\sigma]\theta[\mu]\varphi}_{[\nu']\beta'\sigma'\mu',[\nu'']\beta''\sigma''\mu''} = \varepsilon_7 \sqrt{\frac{h_{\sigma'}h_{\sigma''}}{h_\sigma}} C^{[\sigma]\theta,\beta[\nu]\tau[\mu]\varphi}_{[\sigma']\beta'\sigma'\mu',[\sigma'']\beta''\sigma''\nu''}$$
$$= \varepsilon_8 \sqrt{\frac{h_{\mu'}h_{\mu''}}{h_\mu}} C^{[\mu]\varphi,\beta[\sigma]\theta[\nu]\tau}_{[\mu']\beta'\sigma'\nu',[\mu'']\beta''\sigma''\nu''} \,, \quad (7\text{-}242b)$$

where h_ν is the dimension of the irrep $[\nu]$ of S_f, etc.

6. $$C^{[\nu]\beta,[\sigma][\mu]}_{[\nu']\beta'[\sigma'][\mu'],[0][0][0]} = \delta_{\nu\nu'}\delta_{\beta\beta'}\delta_{\sigma\sigma'}\delta_{\mu\mu'} \,, \quad (7\text{-}243a)$$

$$C^{[0],[0][0]}_{[\nu']\beta'\sigma'\mu',[\nu'']\beta''\sigma''\mu''} = \sqrt{\frac{h_{\sigma'}(SU_m)h_{\mu'}(SU_n)}{h_{\nu'}(SU_{mn})}}\delta_{\nu'\overline{\nu}''}\delta_{\beta'\beta''}\delta_{\sigma'\overline{\sigma}''}\delta_{\mu'\overline{\mu}''} \,. \tag{7-243b}$$

7. From (4-203) and (4-204) we have

$$C^{[\nu]\tau,[f][\mu]\varphi}_{[\nu'][f_1][\mu'],[\nu''][f_2][\mu'']} = \delta_{\nu\mu}\delta_{\nu'\mu'}\delta_{\nu''\mu''}\delta_{\tau\varphi} \,, \tag{7-243c}$$

$$C^{[\nu]\tau,[1^f][\mu]\varphi}_{[\nu'][1^{f_1}][\mu'],[\nu''][1^{f_2}][\mu'']} = \delta_{\nu\tilde{\mu}}\delta_{\nu'\tilde{\mu}'}\delta_{\nu''\tilde{\mu}''}\delta_{\tau\varphi} \,. \tag{7-243d}$$

8. From (7-236c), (4-180a) and (4-171b), we obtain

$$C^{[f],[\sigma]\theta[\mu]\varphi}_{[f_1]\sigma'\mu',[f_2]\sigma''\mu''} = \sqrt{\frac{h_{\sigma'}h_{\sigma''}}{h_\sigma}}\delta_{\sigma\mu}\delta_{\sigma'\mu'}\delta_{\sigma''\mu''}\delta_{\theta\varphi} \,. \tag{7-244a}$$

$$C^{[1^f],[\sigma]\theta[\mu]\varphi}_{[1^{f_1}]\sigma'\mu',[1^{f_2}]\sigma''\mu''} = \sqrt{\frac{h_{\sigma'}h_{\sigma''}}{h_\sigma}}\delta_{\sigma\tilde{\mu}}\delta_{\sigma'\tilde{\mu}'}\delta_{\sigma''\tilde{\mu}''}\delta_{\theta\varphi} \,. \tag{7-244b}$$

The symmetries (7-239), (7-240) and (7-243a, b) come from the unitary group, while the symmetries (7-241), (7-242), (7-243c,d) and (7-244) come from the permutation group. It is seen that the interplay of these two groups greatly deepens our understanding of the symmetries of the ISF of the two groups. For example, for the $SU_4 \supset SU_2 \times SU_2$ ISF, we obtain from (7-239b) and (7-240)

$$\begin{aligned} C^{[221][41][32]}_{[21][3][21],[2][2][2]} &= \varepsilon_2 C^{[4322][74][65]}_{[221][41][32],[222][42][42]} \\ &= \varepsilon_3' \left(\sqrt{\frac{3\times 3}{h_{[222]}(SU_4)}}\Big/\sqrt{\frac{4\times 2}{h_{[221]}(SU_4)}}\right) C^{[222][42][42]}_{[21][3][21],[21][3][21]} \,. \end{aligned}$$

Here the equivalence (7-71) has been used. It is seen that the two-particle CFP for a five-particle system, and the six-particle CFP for an eleven-particle system, and the three-particle CFP for a six-particle system are related to one another; this is hardly understandable from the point of view of the permutation group. $C^{[221][41][32]}_{[21][3][21],[2][2][2]}$ is also the $SU_6 \supset SU_3 \times SU_2$ ISF, therefore it also has the following symmetries:

$$\begin{aligned} C^{[221][41][32]}_{[21][3][21],[2][2][2]} &= \varepsilon_4 C^{[4^3 32^2][874][10,9]}_{[2^4 1][4^2 1][54],[2^5][4^2 2][64]} \\ &= \varepsilon_3 \left(\sqrt{\frac{2h_{[33]}(SU_3)}{h_{[2^4 1]}(SU_6)}}\Big/\sqrt{\frac{2h_{[41]}(SU_3)}{h_{[221]}(SU_6)}}\right) C^{[2^4 1][4^2 1][54]}_{[2^3 1][43][43],[2][2][2]} \,. \end{aligned}$$

Example 1: The application of (7-244b). The $SU_{12} \supset SU_4 \times SU_3$ ISF evaluated with much labour by Matveev (1978) can be easily found from (7-244b):

$$C^{[1^6][33][\widetilde{33}]}_{[1^3][3][\tilde{3}],[1^3][3][\tilde{3}]} = \sqrt{h_{[3]}h_{[3]}/h_{[33]}} = \sqrt{1/5} \,,$$

$$C^{[1^6][33][\widetilde{33}]}_{[1^3][21][21],[1^3][21][21]} = \sqrt{h_{[21]}h_{[21]}/h_{[33]}} = \sqrt{4/5} \,.$$

7.16.4. *More examples*

In the following we will use examples to check the validity of the conclusion that the $SU_{mn} \supset SU_m \times SU_n$ ISF do not depend on m and n explicitly.

Example 2: Single-particle CFP. The $SU_{mn} \supset SU_m \times SU_n$ single-particle CFP are listed in Table 4.19 for particle number $f \leq 5$, and in Chen ([16] 1984) for $f \leq 6$.

Table 4.19-13d is reproduced in Table 7.16, supplemented with the headings for the $SU_4 \supset SU_2 \times SU_2$ and $SU_6 \supset SU_3 \times SU_2$ CFP. It is seen that except for the phase, the $SU_{mn} \supset SU_m \times SU_n$ CFP listed in Table 7.16 are identical not only with the $SU_4 \supset SU_2 \times SU_2$ CFP (Jahn 1951) but also with the $SU_6 \supset SU_3 \times SU_2$ CFP (So 1979). The phase in Jahn's $SU_4 \supset SU_2 \times SU_2$ CFP has not been systematically determined. In Sec. 4.19 we have give a systematic way to determine the phase of the $SU_{mn} \supset SU_m \times SU_n$ ISF.

Table 7.16. $SU_4 \supset SU_2 \times SU_2$ *ISF* $C^{[\nu],ST}_{[\nu']S'T',[1]}$
$SU_6 \supset SU_3 \times SU_2$ *ISF* $C^{[\nu],(\lambda\mu)S}_{[\nu'](\lambda'\mu')S',[1]}$
$SU_{mn} \supset SU_m \times SU_n$ *ISF* $C^{[\nu],\sigma\mu}_{[\nu']\sigma'\mu',[1]}$

$SU_4 \supset SU_2 \times SU_2$	$SU_6 \supset SU_3 \times SU_2$	$SU_{mn} \supset SU_m \times SU_n$		$[\nu']=[211]$		
${}^{2S+1,2T+1}\Gamma$	$(\lambda\mu)S$	$[\sigma]$	$[\mu]$		$[\nu]$	
${}^{22}\Gamma$	$(12)1/2$	$[32]$	$[32]$	$[311]$	$[221]$	$[2111]$
${}^{2S'+1,2T'+1}\Gamma$	$(\lambda'\mu')S'$	$[\sigma']$	$[\mu']$			
${}^{33}\Gamma$	$(21)1$	$[31]$	$[31]$	$\sqrt{2/5}$	0	$\sqrt{3/5}$
${}^{31}\Gamma$	$(21)0$	$[31]$	$[22]$	$\sqrt{3/10}$	$\sqrt{1/2}$.	$-\sqrt{1/5}$
${}^{13}\Gamma$	$(02)1$	$[22]$	$[31]$	$-\sqrt{3/10}$	$\sqrt{1/2}$	$\sqrt{1/5}$

Example 3: Two-particle CFP. Let us use (7-238c) to find the two-particle CFP $C^{[221],[41][32]}_{[21][3][21],[2][2][2]}$. The value m in (7-238c) can be chosen arbitrarily. We take $m = 1$, i.e.,

$$[221]m = \begin{array}{|c|c|}\hline 1 & 2 \\ \hline 3 & 4 \\ \hline 5 & \\ \cline{1-1}\end{array} \quad . \text{ Therefore}$$

$$C^{[221],[41][32]}_{[21][3][21],[2][2][2]} = \left\langle \begin{matrix} 1\ 2 \\ 3\ 4 \\ 5 \end{matrix} \middle| [2] \right\rangle^{-1}$$
$$\times \left\{ \left\langle \begin{matrix} 1\ 2\ 3\ 4 \\ 5 \end{matrix} \middle| [2] \right\rangle \left\langle \begin{matrix} 1\ 2\ 4 \\ 3\ 5 \end{matrix} \middle| [2] \right\rangle C^{[221][22]}_{[41][4],[32][31]} C^{[22][21]}_{[4][3],[31][21]} \right.$$
$$+ \left\langle \begin{matrix} 1\ 2\ 3\ 5 \\ 4 \end{matrix} \middle| [2] \right\rangle \left\langle \begin{matrix} 1\ 2\ 4 \\ 3\ 5 \end{matrix} \middle| [2] \right\rangle C^{[221][22]}_{[41][31],[32][31]} C^{[22][21]}_{[31][3],[31][21]}$$
$$+ \left\langle \begin{matrix} 1\ 2\ 3\ 4 \\ 5 \end{matrix} \middle| [2] \right\rangle \left\langle \begin{matrix} 1\ 2\ 5 \\ 3\ 4 \end{matrix} \middle| [2] \right\rangle C^{[221][22]}_{[41][4],[32][22]} C^{[22],[21]}_{[4][3],[22][21]}$$
$$\left. + \left\langle \begin{matrix} 1\ 2\ 3\ 5 \\ 4 \end{matrix} \middle| [2] \right\rangle \left\langle \begin{matrix} 1\ 2\ 5 \\ 3\ 4 \end{matrix} \middle| [21] \right\rangle C^{[221][22]}_{[41][31],[32][22]} C^{[22][21]}_{[31][3],[22][21]} \right\} .$$

Using the formula (4-184) and one-body ISF Table 4.19, we obtain

$$C^{[221][41][32]}_{[21][3][21],[2][2][2]} = 2\left\{ 0 + \sqrt{\frac{5}{8}}\sqrt{\frac{1}{4}}\left(-\sqrt{\frac{3}{8}}\right)\sqrt{\frac{1}{3}} + \sqrt{\frac{3}{8}}\sqrt{\frac{3}{4}}\sqrt{\frac{5}{8}}(1) + 0 \right\} = \sqrt{\frac{5}{16}} .$$

This is seen to be identical, to within a phase factor, to the $SU_4 \supset SU_2 \times SU_2$ CFP tabulated by Elliott (1953),

$$C^{[221],[41][32]}_{[21][3][21],[2][2][2]} = C^{[\tilde{32}]^{42}\Gamma}_{[21]^{42}\Gamma,[2]^{33}\Gamma} = \sqrt{\frac{5}{16}} .$$

In summary, the values of the $SU_{mn} \supset SU_m \times SU_n$ CFP depend only on the partitions rather than on m and n. The reason we failed to recognize this obvious fact is because we usually use specific quantum numbers for a given m and n instead of partitions to label the irreps of SU_m and SU_n. Now we know that each $SU_{mn} \supset SU_m \times SU_n$ CFP with a particular m and n gives an infinite number of $SU_{m'n'} \supset SU_{m'} \times SU_{n'}$ CFP with $m' = m, m+1, \ldots$ and $n' = n, n+1, \ldots$ However, not every $SU_{mn} \supset SU_m \times SU_n$ CFP can be deduced from the $SU_{(m-1)n} \supset SU_{m-1} \times SU_n$ CFP or $SU_{m(n-1)} \supset SU_m \times SU_{n-1}$ CFP. The reason is that the group SU_m and SU_n have more possible partitions than the groups SU_{m-1} and SU_{n-1}. The partitions of SU_m can have at most m rows, while those of SU_{m-1}, only $m-1$ rows. Therefore the $SU_{mn} \supset SU_m \times SU_n$ CFP with $[\sigma'], [\sigma'']$ of $[\sigma]$ of m rows cannot be deduced from the $SU_{(m-1)n} \supset SU_{m-1} \times SU_n$ CFP.

At long last we give an example to illustrate the use of (7-238a).

Example 4: Find the three-particle CFP $A = C^{[21^4][222][41]}_{[1^3][21][21],[1^3][21][21]}$. In applying (7-238a), the indices m_1, m_2, m', m'' and m can be chosen arbitrarily. A suitable choice will reduce the labour involved in the calculation. We take $m'_1 = 1, m'_2 = 2$ and $m = 5$. From (7-238a) we have

$$A = \left(\left\langle \begin{matrix}[21^4]\\5\end{matrix} \middle| [1^3][1^3] \right\rangle \sqrt{1/2}\right)^{-1} \sum_{m_1 m_2 m''_1 m''_2} C^{[21^4]5}_{[222]m_1,[42]m_2}$$
$$\times \left\langle \begin{matrix}[222]\\m_1\end{matrix} \middle| \begin{matrix}[21]\,[21]\\1 \quad m''_1\end{matrix} \right\rangle \left\langle \begin{matrix}[42]\\m_2\end{matrix} \middle| \begin{matrix}[21]\,[21]\\2 \quad m''_2\end{matrix} \right\rangle C^{[1^3]}_{[21]m''_1,[21]m''_2}$$

Using the symmetries (4-122) and (4-180a), as well as the CG coefficients of S_3, we get

$$A = \left\langle \begin{matrix}[51]\\1\end{matrix} \middle| [3][3] \right\rangle^{-1} \sum_{m_1 m_2 m''_2} C^{[51]1}_{[33]m_1,[42]m_2} \left\langle \begin{matrix}[33]\\m_1\end{matrix} \middle| \begin{matrix}[21]\,[21]\\2 \quad m''_2\end{matrix} \right\rangle \left\langle \begin{matrix}[42]\\m_2\end{matrix} \middle| \begin{matrix}[21]\,[21]\\2 \quad m''_2\end{matrix} \right\rangle .$$

From Table 4.18 we know that m_1 and m_2 range over $m_1 = 3, 5, m_2 = 4, 7, 9$. From the table of CG coefficients (Chen, [10] 1981), $C^{[51]1}_{[33]3,[42]7} = C^{[51]1}_{[33]5,[42]9} = \sqrt{1/5}$. we finally obtain

$$C^{[\widetilde{51}],[222][41]}_{[1^3][21][21],[1^3][21][21]} = \frac{2}{3} .$$

All the $SU_{mn} \supset SU_m \times SU_n$ CFP have been tabulated in Wu (1988) for systems with up to 6 particles.

7.16.5. $SU_{4(2l+1)} \supset (SU_{2l+1} \supset SO_3) \times (SU_4 \supset SU_2 \times SU_2)$ *ISF and total CFP*

Equation (7-185) already gives an expression between the total CFP and the ISF. Now let us further consider the problem of constructing wave functions for a system of f particles with a definite symmetry $[\nu]$ under the $SU_{4(2l+1)}$ transformations, out of the totally anti-symmetric wave functions of the subsystems with f_1 and f_2 particles. To simplify notations, we ignore the additional labels α and β (or equivalently, we absorb α and β into the labels L and ST, respectively). Suppose

$$\left| \begin{matrix}[\nu]\\ \varepsilon[\sigma]L[\mu]ST\end{matrix} \right\rangle \text{ belong to } \begin{matrix} SU_{4(2l+1)} \supset (SU_{2l+1} \supset SO_3) \times (SU_4 \supset SU_2 \times SU_2) \\ [\nu] \qquad\quad [\sigma] \qquad\quad L \qquad\quad [\mu] \qquad S \qquad T \end{matrix}, \tag{7-245}$$

where $\varepsilon = 1, 2, \ldots, (\sigma\mu\nu)$. By means of the ISF, the basis in (7-245) can be expressed in terms of the totally anti-symmetric states of the f_1 and f_2 particles

$$\left| \begin{matrix}[\nu]_\tau\\ \varepsilon[\sigma]L,[\mu]ST\end{matrix} \right\rangle = \sum_{\substack{\sigma' L' S' T'\\ \sigma'' L'' S'' T''}} C^{[\nu]\tau,\varepsilon[\sigma]L[\mu]ST}_{[1^{f_1}]\sigma' L' \tilde{\sigma}' S' T',[1^{f_2}]\sigma'' L'' \tilde{\sigma}'' S'' T''}$$
$$\times \left[\left| \begin{matrix}[1^{f_1}]\\ [\sigma']L'[\tilde{\sigma}']S'T'\end{matrix} \right\rangle \left| \begin{matrix}[1^{f_2}]\\ [\sigma'']L''[\tilde{\sigma}'']S''T''\end{matrix} \right\rangle \right]^{LST} , \tag{7-246a}$$

where the index τ is redundant, since from the outer product reduction rule $\{[1^{f_1}][1^{f_2}][\nu]\} \le 1$. The coefficients in (7-246a) are the $SU_{4(2l+1)} \supset (SU_{2l+1} \supset SO_3) \times (SU_4 \supset SU_2 \times SU_2)$ ISF and they can be factorized as

$$C^{[\nu],\varepsilon[\sigma]L,[\mu]ST}_{[1^{f_1}]\sigma' L'\tilde{\sigma}' S'T',[1^{f_2}]\sigma'' L''\tilde{\sigma}'' S''T''} = \sum_{\theta\varphi} C^{[\nu],\varepsilon[\sigma]_\theta[\mu]_\varphi}_{[1^{f_1}]\sigma'\tilde{\sigma}',[1^{f_2}]\sigma''\tilde{\sigma}''} \, C^{[\sigma]_\theta,L}_{[\sigma']L',[\sigma'']L''} \, C^{[\mu]_\varphi,ST}_{[\tilde{\sigma}']S'T',[\tilde{\sigma}'']S''T''} \; . \tag{7-247}$$

The first factor in the right-hand side is the $SU_{4(2l+1)} \supset SU_{2l+1} \times SU_4$ ISF.

Setting $\Lambda \to$ LST and $\beta \to \varepsilon\sigma\mu$ in (3-307), we obtain the inverse of (7-246a),

$$\left[\left|\begin{matrix}[1^{f_1}]\\ [\sigma']L',[\tilde{\sigma}']S'T'\end{matrix}\right\rangle\left|\begin{matrix}[1^{f_2}]\\ [\sigma'']L'',[\tilde{\sigma}'']S''T''\end{matrix}\right\rangle\right]^{LST}$$

$$= \sum_{\nu\varepsilon\sigma\mu} C^{[\nu],\varepsilon[\sigma]L[\mu]ST}_{[1^{f_1}]\sigma' L'\tilde{\sigma}' S'T',[1^{f_2}]\sigma'' L''\tilde{\sigma}'' S''T''} \left|\begin{matrix}[\nu]\\ \varepsilon[\sigma]L,[\mu]ST\end{matrix}\right\rangle , \tag{7-246b}$$

where the product of the totally anti-symmetric states of f_1 and f_2 particles is expressed in terms of the states of the $f_1 + f_2$ particles. Notice that the sum in (7-246b) runs over all possible $[\nu]$, instead of only the anti-symmetric term $[1^f]$. One generally uses

$$(l^{f_1}[\sigma']L'S'T', l^{f_2}[\sigma'']L''S''T''|\}l^f[\sigma]LST)$$

to designate the total CFP, with the symbol $(\,|\})$ stressing the nonunitarity of the total CFP. Now it is seen that the total CFP is a special case of the ISF of (7-247) with $[\nu] = [1^f]$ and $[\mu] = [\tilde{\sigma}]$,

$$(l^{f_1}[\sigma']L'S'T', l^{f_2}[\sigma'']L''S''T''|\}l^f[\sigma]LST) = C^{[1^f],[\sigma]L[\tilde{\sigma}]ST}_{[1^{f_1}]\sigma' L'\tilde{\sigma}' S'T';[1^{f_2}]\sigma'' L''\tilde{\sigma}'' S''T''} , \tag{7-248}$$

while the $SU_{4(2l+1)} \supset (SU_{2l+1} \supset SO_3) \times (SU_4 \supset SU_2 \times SU_2)$ ISF of (7-247) satisfy unitarity.

7.17 THE $SU_{m+n} \supset SU_m \otimes SU_n$ CFP*

The remark in the beginning of Sec. 7.16 also applies to the $SU_{m+n} \supset SU_m \otimes SU_n$ CFP. We now prove that: (a) the $S_{f_1+f_2} \supset S_{f_1} \otimes S_{f_2}$ outer product ISF are the $SU_{m+n} \supset SU_m \otimes SU_n$ CFP for separating f_2 particles; (b) the values of the $SU_{m+n} \supset SU_m \otimes SU_n$ CFP are thus independent of m and n; (c) the $SU_{m+n} \supset SU_m \otimes SU_n$ CFP can be expressed in terms of the IDC and SDC of the permutation group.

A more straightforward exposition of this problem is given in (Chen [7] 1984).

7.17.1 *The $S_f \supset S_{f-1}$ outer product ISF (The $SU_f \supset SU_{f-1} \otimes U_1$ ISF)*

In parallel to subsection 4.19.1, the question we are facing now is how to calculate the IDC of S_f if those of S_{f-1} are known. In analogous with Table 4.19-1, we introduce the following notations to designate the irreducible bases of the unitary groups and permutation groups.

Table 7.17-1. Notation for the irreducible bases of S_f and SU_f.

SU_f, $S_{f_1}(\omega_1)$, $(\omega_1) = (\omega_1' f)$	SU_{f-1}, $S_{f_1-1}(\omega_1')$	SU_f, $S_{f_2}(\omega_2)$, $(\omega_2) = (\omega_2' f)$	SU_{f-1}, $S_{f_2-1}(\omega_2')$	SU_f, S_f	SU_{f-1}, S_{f-1}
$\left\lvert\begin{matrix}[\sigma]\\ m_1\omega_1\end{matrix}\right\rangle = \left\lvert\begin{matrix}[\sigma]\\ [\sigma']m_1'\omega_1'\end{matrix}\right\rangle$	$\left\lvert\begin{matrix}[\sigma']\\ m_1'\omega_1'\end{matrix}\right\rangle$	$\left\lvert\begin{matrix}[\mu]\\ m_2\omega_2\end{matrix}\right\rangle = \left\lvert\begin{matrix}[\mu]\\ [\mu']m_2'\omega_2'\end{matrix}\right\rangle$	$\left\lvert\begin{matrix}[\mu']\\ m_2'\omega_2'\end{matrix}\right\rangle$	$\left\lvert\begin{matrix}[\nu]\\ m\end{matrix}\right\rangle = \left\lvert\begin{matrix}[\nu]\\ [\nu']m'\end{matrix}\right\rangle$	$\left\lvert\begin{matrix}[\nu']\\ m'\end{matrix}\right\rangle$

As with (4-186) we have

$$[\sigma]m_1\omega_1 = [\sigma][\sigma']m_1'\omega_1', \quad [\mu]m_2\omega_2 = [\mu][\mu']m_2'\omega_2', \quad [\nu]m = [\nu][\nu']m' \; . \tag{7-249}$$

However, we now have to distinguish between two cases: the particle index f is either in (ω_1), or in (ω_2). Suppose that f is in (ω_1), then after deleting the box f, the generalized Young tableau $\left|\begin{matrix}\sigma\\ m_1\omega_1\end{matrix}\right\rangle$ goes over to $\left|\begin{matrix}\sigma'\\ m_1'\omega_1'\end{matrix}\right\rangle$, otherwise, it remains $\left|\begin{matrix}\sigma\\ m_1\omega_1\end{matrix}\right\rangle$. For example

$$\left|\begin{matrix}[\sigma]\\ m_1\omega_1\end{matrix}\right\rangle = \left|\begin{matrix}[21]\\ m_1=1,\omega=(135)\end{matrix}\right\rangle = \left|\begin{array}{|c|c|}\hline 1&3\\\hline 5\\\cline{1-1}\end{array}\right\rangle \xrightarrow{\text{delete } 6} \left|\begin{matrix}[\sigma']\\ m_1'\omega_1'\end{matrix}\right\rangle = \left|\begin{matrix}[\sigma]\\ m_1\omega_1\end{matrix}\right\rangle \tag{7-250a}$$

$$\left|\begin{matrix}[\sigma]\\ m_1\omega_1\end{matrix}\right\rangle = \left|\begin{matrix}[21]\\ m_1=2,\ (\omega)=(256)\end{matrix}\right\rangle = \left|\begin{array}{|c|c|}\hline 2&6\\\hline 5\\\cline{1-1}\end{array}\right\rangle \xrightarrow{\text{delete } 6} \left|\begin{array}{|c|}\hline 2\\\hline 5\\\hline\end{array}\right\rangle$$

$$= \left|\begin{matrix}[\sigma']\\ m_1'\omega_1'\end{matrix}\right\rangle = \left|\begin{matrix}[11]\\ m_1'=1,\ \omega_1'=(25)\end{matrix}\right\rangle . \tag{7-250b}$$

We first use the IDC of S_{f-1} (i.e., the CG coefficients of SU_{f-1}), $C^{[\nu']\beta',m'}_{\sigma'm_1'\omega_1',\mu'm_2'\omega_2'}$ to combine the irreducible bases of $S_{f_1}(\omega_1)$ and $S_{f_2}(\omega_2)$ (i.e., the SU_f Gel'fand bases) into the standard basis $[\nu']m'$ of S_{f-1} (i.e., the SU_{f-1} Gel'fand basis),

$$|(\sigma'\mu')\beta'\rangle \equiv \left[\left|\begin{matrix}[\sigma]\\ [\sigma']\end{matrix}\right\rangle\left|\begin{matrix}[\mu]\\ [\mu']\end{matrix}\right\rangle\right]^{[\nu']\beta}_{m'} = \sum_{m_1'\omega_1'm_2'\omega_2'} C^{[\nu']\beta',m'}_{\sigma'm_1'\omega_1',\mu'm_2'\omega_2'}\left|\begin{matrix}[\sigma]\\ [\sigma']m_1'\omega_1'\end{matrix}\right\rangle\left|\begin{matrix}[\mu]\\ [\mu']m_2'\omega_2'\end{matrix}\right\rangle , \tag{7-251}$$

and next use the $S_f \supset S_{f-1}$ outer product ISF $C^{[\nu]\beta,[\nu']\beta'}_{\sigma\sigma',\mu\mu'}$ to combine (7-251) linearly into the irreducible basis of S_f (i.e., the special Gel'fand basis of SU_f),

$$\left|\begin{matrix}[\nu]\beta\\ m\end{matrix}\right\rangle = \left|\begin{matrix}[\nu]\beta\\ [\nu']m'\end{matrix}\right\rangle = \sum_{\sigma'\mu'\beta'} C^{[\nu]\beta,[\nu']\beta'}_{\sigma\sigma',\mu\mu'}\left[\left|\begin{matrix}[\sigma]\\ [\sigma']\end{matrix}\right\rangle\left|\begin{matrix}[\mu]\\ [\mu']\end{matrix}\right\rangle\right]^{[\nu']\beta}_{m'} . \tag{7-252}$$

Equation (7-252) has exactly the same form as (4-188), and its inverse expansion is identical to (4-190). Comparing (7-251) and (7-252) with (4-140), and noting that the sum over $\sigma'm_1'\omega_1'$ and $\mu'm_2'\omega_2'$ are equivalent to the sum over $m_1\omega_1$ and $m_2\omega_2$, respectively, we get the factorization formula for the IDC of S_f

$$C^{[\nu]\beta,m}_{\sigma m_1,\mu m_2,\omega} = \sum_{\beta'} C^{[\nu]\beta,[\nu']\beta'}_{\sigma\sigma',\mu\mu'} C^{[\nu']\beta'm'}_{\sigma'm_1',\mu'm_2',\omega'} , \tag{7-253}$$

where $(\omega) = (\omega_1,\omega_2)$ and $(\omega') = (\omega_1',\omega_2')$, and the quantum numbers have to satisfy (7-249).

Since the left-hand side contains the SU_f CG coefficients, and the last factor in (7-253) is the SU_{f-1} CG coefficient, the $S_f \supset S_{f-1}$ outer product ISF are just the $SU_f \supset SU_{f-1}$ ISF. From (7-253) we obtain

$$C^{[\nu]\beta,[\nu']\beta'}_{\sigma\sigma',\mu\mu'} = \sum_{m_1'm_2'\omega'} C^{[\nu]\beta,m}_{\sigma m_1,\mu m_2,\omega} C^{[\nu']\beta',m'}_{\sigma'm_1',\mu'm_2',\omega'} . \tag{7-254a}$$

When the multiplicity label β' is redundant, from (7-253) we get

$$C^{[\nu]\beta,[\nu']}_{\sigma\sigma',\mu\mu'} = C^{[\nu]\beta.m}_{\sigma m_1,\mu m_2,\omega} / C^{[\nu'],m'}_{\sigma'm_1',\mu'm_2',\omega'} . \tag{7-254b}$$

For example from (7-254b) and Table 4.17-5d and Table 8f(1) of IDC (Chen [10] 1981) we have

$$C^{[321]\alpha,[311]}_{[21][2],[21][21]} = \left\langle {}^{(\alpha)}\begin{array}{|c|c|c|}\hline 1&3&5\\\hline 2&6\\\cline{1-2} 4\\\cline{1-1}\end{array} \middle| \begin{array}{|c|c|}\hline 1&3\\\hline 6\\\cline{1-1}\end{array}\,,\ \begin{array}{|c|c|}\hline 2&4\\\hline 5\\\cline{1-1}\end{array} \right\rangle \Bigg/ \left\langle \begin{array}{|c|c|c|}\hline 1&3&5\\\hline 2\\\cline{1-1} 4\\\cline{1-1}\end{array} \middle| \begin{array}{|c|c|}\hline 1&3\\\hline\end{array}\,,\ \begin{array}{|c|c|}\hline 2&4\\\hline 5\\\cline{1-1}\end{array} \right\rangle$$

$$= \sqrt{\frac{48}{3480}} \Bigg/ \sqrt{\frac{48}{320}} = \sqrt{\frac{1}{12}}\,,$$

$$C^{[321]\beta,[32]}_{[21][21],[21][2]} = \left\langle {}^{(\beta)}\begin{array}{|c|c|c|}\hline 1&2&4\\\hline 3&5\\\cline{1-2} 6\\\cline{1-1}\end{array} \middle| \begin{array}{|c|c|}\hline 1&2\\\hline 3\\\cline{1-1}\end{array}\,,\ \begin{array}{|c|c|}\hline 4&5\\\hline 6\\\cline{1-1}\end{array} \right\rangle \Bigg/ \left\langle \begin{array}{|c|c|c|}\hline 1&2&4\\\hline 3&5\\\cline{1-2}\end{array} \middle| \begin{array}{|c|c|}\hline 1&2\\\hline 3\\\cline{1-1}\end{array}\,,\ \begin{array}{|c|c|}\hline 4&5\\\hline\end{array} \right\rangle$$

$$= -\sqrt{\frac{48}{768}} \Bigg/ \sqrt{\frac{48}{320}} = -\sqrt{\frac{5}{12}}\,. \tag{7-255}$$

The $S_f \supset S_{f-1}$ outer product ISF still satisfies the eigenequation (4-192). Equations (4-193) and (4-195) remain true under the following replacements: $n \to f$ and

$$\begin{aligned}&\langle(\overline{\sigma}'\overline{\mu}')\overline{\beta}'|(f-1,f)|(\sigma'\mu')\beta'\rangle\\ &= \sum_{\overline{m}'_1\overline{m}'_2\overline{\omega}'m'_1m'_2\omega'} C^{[\nu']\overline{\beta}'\overline{m}'}_{\overline{\sigma}'\overline{m}'_1,\overline{\mu}'\overline{m}'_2,\overline{\omega}'} C^{[\nu']\beta',m'}_{\sigma'm'_1,\mu'm'_2,\omega'} \langle \overline{m}_1\overline{m}_2\overline{\omega}|(f-1,f)|m_1m_2\omega\rangle\,,\end{aligned} \tag{7-256a}$$

where

$$\langle \overline{m}_1\overline{m}_2\overline{\omega}|(f-1,f)|m_1m_2\omega\rangle = \langle \psi^{[\sigma]}_{\overline{m}_1}(\overline{\omega}_1)\psi^{[\mu]}_{\overline{m}_2}(\overline{\omega}_2)|(f-1,f)|\psi^{[\sigma]}_{m_1}(\omega_1)\psi^{[\mu]}_{m_2}(\omega_2)\rangle\,, \tag{7-256b}$$

and

$$M(\overline{\sigma}'\overline{\mu}'\overline{\beta}'\overline{\nu}'\nu'',\sigma'\nu'\beta'\nu'\nu'') = \sum_{\sigma''\mu''\beta''} C^{[\overline{\nu}']\beta',[\nu'']\beta''}_{\overline{\sigma}'\sigma'',\overline{\mu}'\mu''} C^{[\nu']\beta',[\nu'']\beta''}_{\sigma'\sigma'',\mu'\mu''}$$

$$\times \begin{cases} \delta_{\overline{\mu}'\mu''} D^{[\sigma]}_{\overline{\sigma}'\sigma'',\sigma'\sigma''}(f_1-1,f_1)\,, & \text{if } (f-1,f) \text{ is in } (\omega_1)\,,\\ \delta_{\overline{\sigma}'\sigma''} D^{[\mu]}_{\overline{\mu}'\mu'',\mu'\mu''}(f_2-1,f_2)\,, & \text{if } (f-1,f) \text{ is in } (\omega_2)\,,\\ 1, \text{ if } (f-1,f) \text{ is not in the same } (\omega_i) \text{ and} & \\ \quad \text{if } |\Psi^{\sigma}_{\overline{\sigma}'\sigma''}\Psi^{\mu}_{\overline{\mu}'\mu''}\rangle = (f-1,f)|\Psi^{\sigma}_{\sigma'\sigma''}\Psi^{\mu}_{\mu'\mu''}\rangle\,, & \\ 0\,, \text{ otherwise}\,, & \end{cases} \tag{7-256c}$$

where σ'',μ'' and ν'' are the Young diagrams after deleting the index $f-1$ from the Young tableaux $\sigma'm'_1(\overline{\sigma}'\overline{m}'_1), \mu'm'_2(\overline{\mu}'\overline{m}'_2)$ and $\nu'm'$, respectively.

Equations (4-192) and (7-254) also provide a recursive approach to the calculation of the IDC. The degree of the eigenequation (4-192) satisfied by the $S_f \supset S_{f-1}$ outer product ISF is much lower than the degree of (4-143) satisfied by the IDC of S_f. Therefore the recursive approach to the IDC is more convenient than the direct approach. An efficient program was written for computing the outer product ISF based on this algorithm (Novoselsky, [1] 1988).

Example 1: Find the $S_4 \supset S_3$ outer product ISF $C^{[\nu][\nu']}_{\sigma\sigma',\mu\mu'}$ for $[\nu'] = [21], [\mu] = [1]$ and $[\sigma] = [21]$.

Step 1. According to the branching law, it is known that the possible values of $([\sigma'], [\mu'])$ are ([21], [0]), ([2], [1]) and ([11], [1]). Since the $S_f \supset S_{f-1}$ outer product ISF is independent of the component index m' in (7-252), we can set $m' = 1$. From (7-251) and Table 4.17-2 for the IDC of S_3, we can construct the following three basis vectors:

$$\psi = \left[\left| \begin{matrix}[21]\\ [21]\end{matrix} \right\rangle \left| \begin{matrix}[1]\\ [0]\end{matrix} \right\rangle \right]^{[21]}_1 = \left| \begin{array}{|c|c|}\hline 1 & 2 \\ \hline 3 \\ \cline{1-1}\end{array}, \boxed{4} \right\rangle = \psi_1 \,,$$

$$\psi_2 = \left[\left| \begin{matrix}[21]\\ [2]\end{matrix} \right\rangle \left| \begin{matrix}[1]\\ [1]\end{matrix} \right\rangle \right]^{[21]}_1 = \sqrt{\frac{1}{6}} \left(\left| \begin{array}{|c|c|}\hline 1 & 2 \\ \hline 4 \\ \cline{1-1}\end{array}, \boxed{3} \right\rangle - \left| \begin{array}{|c|c|}\hline 1 & 3 \\ \hline 4 \\ \cline{1-1}\end{array}, \boxed{2} \right\rangle - \left| \begin{array}{|c|c|}\hline 2 & 3 \\ \hline 4 \\ \cline{1-1}\end{array}, \boxed{1} \right\rangle \right)$$

$$= \sqrt{\frac{2}{3}}\psi_2 - \sqrt{\frac{1}{3}}\,\psi_3 \,,$$

$$\psi_3 = \left[\left| \begin{matrix}[21]\\ [11]\end{matrix} \right\rangle \left| \begin{matrix}[1]\\ [1]\end{matrix} \right\rangle \right]^{[21]}_1 = \sqrt{\frac{1}{2}} \left(\left| \begin{array}{|c|c|}\hline 1 & 4 \\ \hline 3 \\ \cline{1-1}\end{array}, \boxed{2} \right\rangle + \left| \begin{array}{|c|c|}\hline 2 & 4 \\ \hline 3 \\ \cline{1-1}\end{array}, \boxed{1} \right\rangle \right) = \psi_4 \,, \tag{7-257}$$

where $(\psi_1, \ldots, \psi_4)$ is the basis defined in (4-149b). From (4-149c) we obtain the matrix representative of the operator $C'(4) = C(4) - C(3)$ in the basis $(\psi_1, \ldots, \psi_4)$,

$$C'(4) = C^{(+)}(4) - C^{(+)}(3) = \begin{pmatrix} 0 & 1 & -\sqrt{\frac{1}{2}} & \sqrt{\frac{3}{2}} \\ 1 & -1 & 0 & 0 \\ -\sqrt{\frac{1}{2}} & 0 & -1 & 0 \\ \sqrt{\frac{3}{2}} & 0 & 0 & 1 \end{pmatrix} . \tag{7-258}$$

From (7-257) and (7-258) we find the matrix representative of $C'(4)$ in the basis $(\varphi_1, \varphi_2, \varphi_3)$,

$$M = \begin{pmatrix} 0 & \sqrt{\frac{3}{2}} & \sqrt{\frac{3}{2}} \\ \sqrt{\frac{3}{2}} & -1 & 0 \\ \sqrt{\frac{3}{2}} & 0 & 1 \end{pmatrix} . \tag{7-259}$$

A diagonalization of M yields the three eigenvalues 2, 0, -2, corresponding to the irreps $[\nu] = [31], [22], [211]$, and three eigenvectors, giving the $S_4 \supset S_3$ outer product ISF listed in Table 7.17-2. This table is the same as Haacke's Table II for the $SU_4 \supset SU_3$ ISF up to phase factors.

Table 7.17-2. $S_4 \supset S_3$ outer product ISF and $SU_n \supset SU_{n-1}$ ISF $C^{[\nu][\nu']}_{\sigma\sigma',\mu\mu'}$.

$[\nu']$ $[\nu]$ / $\sigma'\sigma', \mu\mu'$	[31]	[21] [22]	[211]
[21][21], [1][0]	$\sqrt{3/8}$	1/2	$\sqrt{3/8}$
[21][11], [1][1]	3/4	$\sqrt{3/8}$	$-1/4$
[21][2], [1][1]	1/4	$-\sqrt{3/8}$	$-3/4$

7.17.2. *The $S_f \supset S_{f_{12}} \otimes S_{f_{34}}$ outer product ISF ($SU_f \supset SU_{f_{12}} \otimes SU_{f_{34}}$ ISF)*

This subsection is in parallel to Sec. 4.19.6. The notation for the particle numbers are the same as that given in (7-212a). Divide the number $1, 2, \ldots, f$ into four sets (ω_1'), (ω_2'), (ω_3') and

(ω_4'), each of them being a normal order sequence:

$$(\omega_1') = (a_1, \dots, a_{f_1}), \quad (\omega_2') = (a_{f_1+1}, \dots, a_{f_{12}}) \,, \tag{7-260}$$

$$(\omega_1'') = (a_{f_{12}+1}, \dots, a_{f_{123}}), \quad (\omega_2'') = (a_{f_{123}+1}, \dots, a_f) \,,$$

$$(\omega') = (\omega_1', \omega_2'), (\omega'') = (\omega_1'', \omega_2''), \quad (\omega_1) = (\omega_1', \omega_1''), (\omega_2) = (\omega_2', \omega_2'') \,, \tag{7-261}$$

where (ω_1') and (ω_2') are the normal order sequences taken from $(1, 2, \dots, f_{12})$ and (ω_1'') and (ω_2'') are those taken from $(f_{12}+1, \dots, f)$ [see (4-129)]. The single particle states with the particle numbers $1, 2, \dots, f_{12}$ and $f_{12}+1, \dots f$ span the defining rep of $SU_{f_{12}}$ and $SU_{f_{34}}$, respectively. We shall use the following notation to denote the nine irreps of the three unitary groups along with the corresponding particle numbers and normal order sequences:

$$\begin{pmatrix} \sigma' & \mu' & \nu'_{\beta'} \\ \sigma'' & \mu'' & \nu''_{\beta''} \\ \sigma_\theta & \mu_\varphi & \nu_{\tau,\beta} \end{pmatrix} \begin{pmatrix} SU_{f_{12}} & SU_{f_{12}} & SU_{f_{12}} \\ SU_{f_{34}} & SU_{f_{34}} & SU_{f_{34}} \\ SU_f & SU_f & SU_f \end{pmatrix} \begin{pmatrix} f_1 & f_2 & f_{12} \\ f_3 & f_4 & f_{34} \\ f_{13} & f_{24} & f \end{pmatrix} \begin{pmatrix} \omega_1' & \omega_2' & \omega' \\ \omega_1'' & \omega_2'' & \omega'' \\ \omega_1 & \omega_2 & \omega \end{pmatrix} \tag{7-262}$$

$$\begin{array}{lll} \beta' = 1, 2, \dots \{\sigma'\mu'\nu'\}, & \beta'' = 1, 2, \dots, \{\sigma''\mu''\nu''\}, & \beta = 1, 2, \dots \{\sigma\mu\nu\}, \\ \theta = 1, 2, \dots \{\sigma'\sigma''\sigma\}, & \varphi = 1, 2, \dots \{\mu'\mu''\mu\}, & \tau = 1, 2, \dots \{\nu'\nu''\nu\} \end{array} .$$

For example $[\nu'']$ labels the irrep of $SU_{f_{34}}$ with f_{34} particles, and the irreducible basis of $SU_{f_{34}}$ and $S_{f_{34}}$ is $|Y_{m''}^{(\nu'')}(\omega'')\rangle = \left| \begin{matrix} [\nu''] \\ m''\omega'' \end{matrix} \right\rangle$.

We introduce the following three nonstandard bases of the permutation groups $S_{f_{13}}, S_{f_{24}}$ and S_f:

$$\left| \begin{matrix} [\sigma] \\ \theta[\sigma']m_1'[\sigma'']m_1'' \end{matrix} \right\rangle, \quad \left| \begin{matrix} [\mu] \\ \varphi[\mu']m_2'[\mu'']m_2'' \end{matrix} \right\rangle, \quad \left| \begin{matrix} [\nu] \\ \tau[\nu']m'[\nu'']m'' \end{matrix} \right\rangle \tag{7-263}$$

$$S_{f_{13}} \supset S_{f_1} \otimes S_{f_3} \qquad S_{f_{24}} \supset S_{f_2} \otimes S_{f_4} \quad S_f \supset S_{f_{12}} \otimes S_{f_{34}} .$$

They are also the $SU_f \supset SU_{f_{12}} \otimes SU_{f_{34}}$ irreducible bases. For example, $\left| [42], \begin{array}{|c|c|c|} \hline 1 & 2 & 4 \\ \hline 3 \\ \cline{1-1} \end{array} \quad \begin{array}{|c|c|} \hline 5 & 6 \\ \hline \end{array} \right\rangle$ is the $SU_6 \supset SU_4 \otimes SU_2$ basis, belonging to the irreps [42], [31] and [2] of SU_6, SU_4 and SU_2, respectively.

The former two in (7-263) can be linearly combined into the third one through the following two steps:

1. Use the IDC of $S_{f_{12}}$ (i.e., the CGC of $SU_{f_{12}}$) and $S_{f_{34}}$ (i.e., the CGC of SU_{34}) to combine them into the standard basis $[\nu']m'$ of $S_{f_{12}}$ (i.e., the Gel'fand basis of $SU_{f_{12}}$) and $[\nu'']m''$ of $S_{f_{34}}$ (i.e., the Gel'fand basis of $SU_{f_{34}}$),

$$\begin{aligned} |(\sigma'\sigma'')_\theta(\mu'\mu'')_\varphi\beta\beta'\rangle &\equiv \left[\left| \begin{matrix} [\sigma] \\ \theta[\sigma'][\sigma''] \end{matrix} \right\rangle \left| \begin{matrix} [\mu] \\ \varphi[\mu'][\mu''] \end{matrix} \right\rangle \right]^{[\nu']_{\beta'}[\nu'']_{\beta''}}_{m'm''} \\ &= \sum_{m_1'm_2'\omega'm_1''m_2'',\omega''} C^{[\nu']\beta',m'}_{\sigma'm_1',\mu'm_2',\omega'} C^{[\nu'']\beta'',m''}_{\sigma''m_1'',\mu''m_2''\omega''} \left| \begin{matrix} [\sigma] \\ \theta\sigma'm_1'\omega_1', \sigma''m_1''\omega_1'' \end{matrix} \right\rangle \left| \begin{matrix} [\mu] \\ \varphi\mu'm_2'\omega_2', \mu''m_2''\omega_2'' \end{matrix} \right\rangle . \end{aligned} \tag{7-264}$$

2. Use the $S_f \supset S_{f_{12}} \otimes S_{f_{34}}$ outer product ISF to combine (7-264) into the $S_f \supset S_{f_{12}} \otimes S_{f_{34}}$

irreducible basis (i.e., the $SU_f \supset SU_{f_{12}} \otimes SU_{f_{34}}$ basis)

$$\left|\begin{matrix}[\nu]\beta \\ \tau[\nu']m'[\nu'']m''\end{matrix}\right\rangle = \sum_{\substack{\theta\sigma'\sigma''\beta' \\ \varphi\mu'\mu''\beta''}} C^{[\nu]\beta,\tau[\nu']\beta'[\nu'']\beta''}_{[\sigma]\theta\sigma'\sigma'',[\mu]\varphi\mu'\mu''} \left[\left|\begin{matrix}[\sigma] \\ \theta[\sigma'][\sigma'']\end{matrix}\right\rangle \left|\begin{matrix}[\mu] \\ \varphi[\mu'][\mu'']\end{matrix}\right\rangle\right]^{[\nu']_{\beta'}[\nu'']_{\beta''}}_{m'm''} . \tag{7-265}$$

which has the same form as (4-200) and its inverse expansion is identical to (4-201).

7.17.3. The $\mathcal{SU}_{m+n} \supset \mathcal{SU}_n$ ISF and $S_f \supset S_{f_{12}} \otimes S_{f_{34}}$ outer product ISF

In the above discussion, we only considered the permutation group and the unitary group in coordinate space. Now we want to take into account the unitary groups in state space which are designated by the script letter $\mathcal{U}$. We introduce the following symbols to denote the irreps of $\mathcal{SU}_m, \mathcal{SU}_n, \mathcal{SU}_{m+n}$ and the permutation groups:

$$\begin{pmatrix}\sigma' & \mu' & \nu'_{\beta'} \\ \sigma'' & \mu'' & \nu''_{\beta''} \\ \sigma_\theta & \mu_\varphi & \nu_{\tau,\beta}\end{pmatrix}, \begin{pmatrix}\mathcal{SU}_m & \mathcal{SU}_n & \mathcal{SU}_{m+n} \\ \mathcal{SU}_m & \mathcal{SU}_n & \mathcal{SU}_{m+n} \\ \mathcal{SU}_m & \mathcal{SU}_n & \mathcal{SU}_{m+n}\end{pmatrix}, \begin{pmatrix}f_1 & f_2 & f_{12} \\ f_3 & f_4 & f_{34} \\ f_{13} & f_{24} & f\end{pmatrix} . \tag{7-266}$$

Let

$$\left|\begin{matrix}[\nu'] \\ \beta'\sigma'W_1'\mu'W_2'\end{matrix}\right\rangle, \left|\begin{matrix}[\nu''] \\ \beta''\sigma''W_1''\mu''W_2''\end{matrix}\right\rangle, \left|\begin{matrix}[\nu] \\ \beta\sigma W_1\mu W_2\end{matrix}\right\rangle \tag{7-267}$$

be the $\mathcal{SU}_{m+n} \supset \mathcal{SU}_m \otimes \mathcal{SU}_n$ bases for particles $(1, 2, \ldots, f_{12})$, $(f_{12}+1, \ldots, f)$ and $(1, 2, \ldots, f)$, respectively. The $\mathcal{SU}_{m+n} \supset \mathcal{SU}_m \otimes \mathcal{SU}_n$ ISF are defined by an expansion identical to (7-224). Equations (7-225)-(7-227) are still valid. Analogously, we can prove (Chen [7] 1984) that the $\mathcal{SU}_{m+n} \supset \mathcal{SU}_m \otimes \mathcal{SU}_n$ ISF are precisely the $S_f \supset S_{f_{12}} \otimes S_{f_{34}}$ outer product ISF, i.e.,

$$\underset{\mathcal{SU}_{m+n} \supset \mathcal{SU}_m \times \mathcal{SU}_n \text{ISF}}{C^{[\nu]_\tau,\beta[\sigma]_\theta[\mu]_\varphi}_{\nu'\beta'\sigma'\mu',\nu''\beta''\sigma''\mu''}} = \underset{S_f \supset S_{f_{12}} \otimes S_{f_{34}} \text{outer product ISF}}{C^{[\nu]\beta,\tau[\nu']\beta'[\nu'']\beta''}_{\sigma\theta\sigma'\sigma'',\mu\varphi\mu'\mu''}} . \tag{7-268}$$

Therefore the values of the $\mathcal{SU}_{m+n} \supset \mathcal{SU}_m \otimes \mathcal{SU}_n$ ISF do not depend explicitly on m and n, and all the remarks about the $\mathcal{SU}_{mn} \supset \mathcal{SU}_m \times \mathcal{SU}_n$ ISF apply to the $\mathcal{SU}_{m+n} \supset \mathcal{SU}_m \otimes SU_n$ ISF as well.

A special case of the $\mathcal{SU}_{m+n} \supset \mathcal{SU}_m \otimes \mathcal{SU}_n$ ISF is the $\mathcal{SU}_n \supset \mathcal{SU}_{n-1} \otimes \mathcal{U}_1$ ISF, or abbreviated as $\mathcal{SU}_n \supset \mathcal{SU}_{n-1}$ ISF, or the $\mathcal{SU}_{n-1}$ *singlet factor*. For $\mathcal{SU}_n \supset \mathcal{SU}_{n-1} \times \mathcal{U}_1$ ISF, the irreps $[\mu'], [\mu'']$ and $[\mu]$ of the group $\mathcal{U}_1$ must be totally symmetric, i.e., $[\mu'] = [f_2], [\mu''] = [f_4]$ and $[\mu] = [f_{24}]$, and (7-268) becomes

$$\underset{\mathcal{SU}_{n-1} SF}{C^{[\nu]_\tau,[\sigma]_\theta[f_{24}]}_{[\nu'][\sigma'][f_2],[\nu''][\sigma''][f_4]}} = \underset{S_f \supset S_{f_{12}} \otimes S_{f_{34}} \text{outer product } ISF}{C^{[\nu],\tau[\nu'][\nu'']}_{[\sigma]\theta[\sigma'][\sigma''],[f_{24}][f_2][f_4]}} . \tag{7-269}$$

where we dropped the redundant indices φ, β, β' and β''.

7.17.4. The evaluation of $\mathcal{SU}_{m+n} \supset \mathcal{SU}_m \otimes \mathcal{SU}_n$ ISF

In analogy with the derivation of (7-236b), the $\mathcal{SU}_{m+n} \supset \mathcal{SU}_m \otimes \mathcal{SU}_n$ ISF can be expressed in terms of the IDC of $S_{f_{12}}, S_{f_{34}}$ and S_f, and the SDC of $S_{f_{13}}, S_{f_{24}}$ and S_f,

$$C^{[\nu]_\tau,\beta[\sigma]_\theta[\mu]_\varphi}_{[\nu']\beta'\sigma'\mu',\nu''\beta''\sigma''\mu''} = \sum_{mm_1m_2\omega m_1''m_2''}^{\text{fix } m'm''} C^{[\nu]\beta,m}_{\sigma m_1,\mu m_2,\omega} C^{[\nu']\beta',m'}_{\sigma'm_1',\mu'm_2',\omega'} C^{[\nu'']\beta'',m''}_{\sigma''m_1'',\mu''m_2'',\omega''}$$

$$\times \left\langle\begin{matrix}[\nu] \\ m\end{matrix}\middle|[\nu], \begin{matrix}\tau\nu'\nu'' \\ m'm''\end{matrix}\right\rangle \left\langle\begin{matrix}[\sigma] \\ m_1\end{matrix}\middle|[\sigma] \begin{matrix}\theta\sigma'\sigma'' \\ m_1'm_1''\end{matrix}\right\rangle \left\langle\begin{matrix}[\mu] \\ m_2\end{matrix}\middle|[\mu], \begin{matrix}\varphi\mu'\mu'' \\ m_2'm_2''\end{matrix}\right\rangle . \tag{7-270}$$

Noting that $(\omega_0) = (\omega_1, \omega_2)$, we know from (7-261) that the sum over ω implies a sum over ω' and ω''.

For the two-particle CFP, (7-270) can be further simplified

(a) $[\mu''] = [0]$,

$$C^{[\nu],\beta\sigma\mu}_{[\nu']\sigma'\mu',[\nu''][\sigma''][0]} = \delta_{\nu''\sigma''}\left(\langle[\nu]m|[\nu'']\rangle C^{[\nu']m'}_{\sigma'm'_1,\mu'm'_2,\omega'}\right)^{-1} \sum_{m_1}^{\text{fix } m'_1} C^{[\nu]\beta,m}_{\sigma m_1,\mu m_2,\omega}\langle[\sigma]m_1\|[\sigma'']\rangle \,. \tag{7-271a}$$

For $[\sigma''] = [0]$ we have a similar result.

(b) $[\sigma''] = [\mu''] = [1]$

$$C^{[\nu]\beta[\sigma][\mu]}_{[\nu']\sigma'\mu',[\nu''][1][1]} = (\langle[\nu]m|[\nu'']\rangle C^{[\nu']m'}_{\sigma'm'_1,\mu'm'_2,\omega'})^{-1} \sum_{m_1 m_2 \omega''}^{\text{fix } m'_1 m'_2 \omega'} C^{[\nu]\beta,m}_{\sigma m_1,\mu m_2,\omega} C^{[\nu'']}_{[1],[1],\omega''} \,. \tag{7-271b}$$

Example 2: Single-particle CFP. For the single-particle CFP, the expression (7-224b) can be written concisely as

$$\left|\begin{matrix}[\nu]\\ \beta\sigma\mu\end{matrix}\right\rangle = \sum_{\beta'\sigma'} C^{[\nu],\beta\sigma\mu}_{\nu'\beta'\sigma'\mu,[1][1][0]}\left|\begin{matrix}[\nu']\\ \beta'\sigma'\mu\end{matrix}\right\rangle|\psi(f)\rangle + \sum_{\beta'\mu'} C^{[\nu],\beta\sigma\mu}_{\nu'\beta'\sigma\mu',[1][0][1]}\left|\begin{matrix}[\nu']\\ \beta'\sigma\mu'\end{matrix}\right\rangle|\varphi(f)\rangle \,, \tag{7-272}$$

where $\psi(f)$ belongs to the defining rep [1] of SU_m and $\varphi(f)$ belongs to that of SU_n.

From the IDC tables, Table 4.17-5d, and Table 8f(1) (Chen [10] 1981), and using (7-254b) we obtain

$$\begin{aligned}\left|\begin{matrix}[321]\\ \alpha[21][21]\end{matrix}\right\rangle &= \sqrt{\frac{9}{20}}\left(\left|\begin{matrix}[32]\\ [2][21]\end{matrix}\right\rangle|\psi(6)\rangle + \left|\begin{matrix}[32]\\ [21][2]\end{matrix}\right\rangle|\varphi(6)\rangle\right)\\ &+ \sqrt{\frac{1}{20}}\left(\left|\begin{matrix}[32]\\ [11][21]\end{matrix}\right\rangle|\psi(6)\rangle + \left|\begin{matrix}[32]\\ [21][11]\end{matrix}\right\rangle|\varphi(6)\rangle\right) \;;\end{aligned}$$

$$\begin{aligned}\left|\begin{matrix}[321]\\ \beta[21][21]\end{matrix}\right\rangle &= \frac{1}{2}\left(-\left|\begin{matrix}[32]\\ [2][21]\end{matrix}\right\rangle|\psi(6)\rangle + \left|\begin{matrix}[32]\\ [21][2]\end{matrix}\right\rangle|\varphi(6)\rangle\right.\\ &+ \left.\left|\begin{matrix}[32]\\ [11][21]\end{matrix}\right\rangle|\psi(6)\rangle - \left|\begin{matrix}[32]\\ [21][11]\end{matrix}\right\rangle|\psi(6)\rangle\right) \,.\end{aligned}$$

7.17.5. Symmetries of the $SU_{m+n} \supset SU_m \otimes SU_n$ ISF

The symmetries (7-239) and (7-240) still hold for the $SU_{m+n} \supset SU_m \otimes SU_n$ ISF, with the exception that $h_\nu(SU_{mn})$ and $h_{\nu'}(SU_{mn})$ should be replaced by $h_\nu(SU_{m+n})$ and $h_{\nu'}(SU_{m+n})$.

For example, for $SU_4 \supset SU_2 \otimes SU_2$ ISF, the symmetries (7-239b) and (7-240) tell us that

$$\begin{aligned}C^{[321][21][21]}_{[31][11][2],[11][1][1]} &= \varepsilon_2 C^{[4321][32][32]}_{[332][22][31],[11][1][1]}\\ &= \varepsilon_3\left(\sqrt{\frac{1\times 3}{h_{[332]}(SU_4)}}\Big/\sqrt{\frac{2\times 2}{h_{[321]}(SU_4)}}\right) C^{[332][22][31]}_{[321][21][21],[11][1][1]} \,.\end{aligned}$$

If the first coefficient is regarded as the $SU_5 \supset SU_3 \otimes SU_2$ ISF, we have

$$\begin{aligned}C^{[321][21][21]}_{[31][11][2],[11][1][1]} &= \varepsilon_2 C^{[4^2321][432][32]}_{[3^32][32^2][31],[1^3][11][1]}\\ &= \varepsilon_3\left(\sqrt{\frac{3\times 3}{h_{[3^32]}(SU_5)}}\Big/\sqrt{\frac{2\times 8}{h_{[321]}(SU_5)}}\right) C^{[3^32][322][31]}_{[3321][321][21],[11][1][1]} \,.\end{aligned}$$

In addition, the $\mathcal{SU}_{m+n} \supset \mathcal{SU}_m \otimes \mathcal{SU}_n$ ISF also has the following symmetries:

1. From (7-269a) we know that the $\mathcal{SU}_{m+n} \supset \mathcal{SU}_m \otimes \mathcal{SU}_n$ ISF is invariant under the interchange of indices,

$$\mu' \leftrightarrow \sigma'', \ \nu' \leftrightarrow \sigma, \ \nu'' \leftrightarrow \mu, \ \beta \leftrightarrow \tau, \ \beta' \leftrightarrow \theta, \ \beta'' \leftrightarrow \varphi , \tag{7-273a}$$

which can be written in a more symmetric form as

$$\begin{pmatrix} \sigma' & \mu' & \nu'_{\beta'} \\ \sigma'' & \mu'' & \nu''_{\beta''} \\ \sigma_\theta & \mu_\varphi & \nu_{\tau,\beta} \end{pmatrix} \leftrightarrow \begin{pmatrix} \sigma' & \sigma'' & \sigma_\theta \\ \mu' & \mu'' & \mu_\varphi \\ \nu'_{\beta'} & \nu''_{\beta''} & \nu_{\beta,\tau} \end{pmatrix} . \tag{7-273b}$$

2. By using the symmetry of the IDC, (4-162a), and the symmetry of the SDC, Eq. (4-180a), one gets from (7-270)

$$C^{[\nu]_\tau,\beta[\sigma]_\theta[\mu]_\varphi}_{\nu'\beta'\sigma'\mu',\nu''\beta''\sigma''\mu''} = \varepsilon C^{[\tilde{\nu}]_\tau,\beta[\tilde{\sigma}]_\theta[\tilde{\mu}]\varphi}_{\tilde{\nu}'\beta'\tilde{\sigma}'\tilde{\mu},\tilde{\nu}''\beta''\tilde{\sigma}''\tilde{\mu}''} , \tag{7-273c}$$

where ε is a phase factor depending on the phase convention.

The relation between the $\mathcal{SU}_{m+n} \supset \mathcal{SU}_m \otimes \mathcal{SU}_n$ ISF and the 9ν coefficients of $\mathcal{SU}_{m+n}$ is given in (9-26).

The single-particle CFP for $\mathcal{SU}_{m+n} \supset \mathcal{SU}_m \otimes \mathcal{SU}_n$ have been tabulated for systems with up to six particles (Chen [7] 1984), and to twelve particles (Novoselsky, [1] 1988).

7.18. The SU$_n$ Singlet Factor

Since the $SU_{m+n} \supset SU_m \otimes SU_n$ ISF do not depend on m and n explicitly, the SU_n SF (singlet factor) also do not depend on n explicitly, a conclusion which has already been obtained by Chen ([30] 1979) through a different route. Therefore any given SU_n SF gives an infinite number of SU_m SF,

$$(SU_m\text{SF}) = (SU_n\text{SF}), \quad m = n+1, n+2, \ldots \tag{7-274a}$$

Of course, not every SU_n SF can be deduced from the SU_{n-1} SF. We call an SU_n SF derivable, if it can be deduced from the SU_{n-1} SF through the equation SU_n SF $= SU_{n-1}$ SF, otherwise we call it underivable. It is very simple to see whether an SU_n SF $C^{[\nu]\tau,[\sigma]\theta[f_{24}]}_{[\nu'][\sigma'][f_2],[\nu''][\sigma''][f_4]}$ is derivable or not. If the Young diagram $[\sigma]$ has r rows, then the SU_r SF is underivable, and the SU_n SF with $n > r$ is derivable, i.e.,

$$(SU_n\text{SF}) = (SU_r\text{SF}), \quad \text{for}\, n = r+1, r+2, \ldots \tag{7-274b}$$

Equations (7-147) and (7-274) show that the calculation of the SU_n CG coefficients in the Gel'fand basis is reduced to that of few underivable SU_r SF, which can be calculated from (7-270).

The obvious fact that the SU_n SF are in fact independent of n has escaped our attention for many years. The reason for this oversight is that we generally use specific quantum numbers for specific n, rather than the partitions, to label the irreps of SU_n. Table 7.18-1 gives the labeling schemes which are in common use.

Table 7.18-1. The usual labeling schemes for irreps $[\nu]$, $[\sigma]$ and $[\mu]$.

	$[\nu]$	$[\sigma]$	$[\mu]$
SU_2 CGC	isospin I		$I_z = \frac{1}{2}(n_1 - n_2)$ 1)
$SU_3 \supset SU_2 \otimes U_1$ ISF	$(\lambda\mu)$ or dim. of SU_3	$2I+1$	$Y = \frac{1}{3}\,(n_1 + n_2 - 2n_3)$
$SU_4 \supset SU_3 \otimes U_1$ ISF	dim. of SU_4	dim. of SU_3	$Z = \frac{1}{4}\,(n_1 + n_2 + n_3 - 3n_4)$

1) Here $[\sigma] = [n_1]$, $[\mu] = [n_2]$.

In the past, the calculation and tabulation of the SU_n SF were done for one n at a time. With the discovery of the relation (7-274), by using the partitions as labels, we can revolutionalize the tabulation of the SU_n SF, so that from a given table we can read out an infinite number of SU_n SF of the same type. Tables 7.18-2 and 7.18-3 are two examples of such tabulations (more examples are given in Chen ([30] 1979). From them one can read out all the SU_n SF of the same type with $n = r, r+1, r+2 \dots$, where r is the number of rows of the blank Young diagram $[\sigma]$ (without being filled with the letter n) in the table headings. For example, Table 7.18-2 covers the SU_2 CGC table (Rotenberg 1959), the SU_2 SF table (Swart 1963), and the SU_3 SF table (Haacke 1976), and Table 7.18-3 covers the SU_2 SF and SU_3 SF table.

7.19. Second Quantized Expression for the CFP

The total CFP can be written in second quantized form which is quite useful. We use the round brackets, $(\,|, \text{or}\, |)$, to denote a second quantized state, while the ordinary bracket, $\langle \,|, \text{or}\, | \,\rangle$, an ordinary quantized state. The following discussions are valid both for fermions and bosons.

7.19.1. One-particle CFP

Suppose that we have n identical fermions in a state $\left| \begin{matrix} [1^n] \\ \alpha JM \end{matrix} \right)$ which is an $SU_n \supset SO_3$ irreducible basis, α being additional quantum numbers which may be the irrep labels of another group (or groups) inserted between SU_N and SO_3. Let $b^\dagger_{jm}$ and b_{jm} be creation and annihilation operators. Applying the number operator

$$\hat{n} = \sum_{jm} b^\dagger_{jm} b_{jm} \tag{7-275}$$

to the state $\left| \begin{matrix} [1^n] \\ \alpha JM \end{matrix} \right)$, we have

$$n \left| \begin{matrix} [1^n] \\ \alpha JM \end{matrix} \right) = \sum_{jm} b^\dagger_{jm} b_{jm} \left| \begin{matrix} [1^n] \\ \alpha JM \end{matrix} \right) .$$

Inserting a unit operator between $b^\dagger_{jm}$ and b_{jm},

$$\begin{aligned} n \left| \begin{matrix} [1^n] \\ \alpha JM \end{matrix} \right) &= \sum_{\alpha_1 J_1 M_1 jm} b^\dagger_{jm} \left| \begin{matrix} [1^{n-1}] \\ \alpha_1 J_1 M_1 \end{matrix} \right) \left(\begin{matrix} [1^{n-1}] \\ \alpha_1 J_1 M_1 \end{matrix} \right| b_{jm} \left| \begin{matrix} [1^n] \\ \alpha JM \end{matrix} \right) \\ &= \sum_{\alpha_1 J_1 M_1 jm} b^\dagger_{jm} \left| \begin{matrix} [1^{n-1}] \\ \alpha_1 J_1 M_1 \end{matrix} \right) \left(\begin{matrix} [1^n] \\ \alpha JM \end{matrix} \right| b^\dagger_{jm} \left| \begin{matrix} [1^{n-1}] \\ \alpha_1 J_1 M_1 \end{matrix} \right) \\ &= \sum_{\alpha_1 J_1 j} (-1)^{J_1 + j - J} \left[b^\dagger_j \left| \begin{matrix} [1^{n-1}] \\ \alpha_1 J_1 \end{matrix} \right) \right]^J_M \left(\begin{matrix} [1^n] \\ \alpha J \end{matrix} \right\| b^\dagger_j \left\| \begin{matrix} [1^{n-1}] \\ \alpha_1 J_1 \end{matrix} \right) . \end{aligned} \tag{7-276}$$

Multiplying (7-276) from the left with $\left(\begin{matrix} [1^n] \\ \alpha JM \end{matrix} \right|$, we obtain

$$n = \sum_{\alpha_1 J_1 j} \left(\begin{matrix} [1^n] \\ \alpha J \end{matrix} \right\| b^\dagger_j \left\| \begin{matrix} [1^{n-1}] \\ \alpha_1 J_1 \end{matrix} \right)^2 . \tag{7-277}$$

The total anti-symmetric state can be expanded by using the one-particle CFP,

$$\left| \begin{matrix} [1^n] \\ \alpha JM \end{matrix} \right\rangle = \sum_{\alpha_1 J_1 j} \langle [1^n] \alpha J | \{ [1^{n-1}] \alpha_1 J_1, j \rangle \left[\left| \begin{matrix} [1^{n-1}] \\ \alpha_1 J_1 \end{matrix} \right\rangle \psi_j(n) \right]^J_M , \tag{7-278}$$

Table 7.18-3 $SU_n \supset SU_{n-1} \otimes U_1$ ISF $C^{[\nu]\tau,\,[31][2]}_{[21][\sigma'][\mu'],\,[21][\sigma''][\mu'']}$, $\tau = D, F$

SU_2SF λ, Y 3, 0	SU_3SF $(h_\sigma), Z$ $(15), -\frac{1}{2}$		$[\sigma'][\mu'], [\sigma''][\mu'']$	[321]D 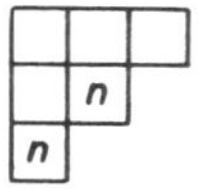 (8D) $\underline{64D}$	[321]F 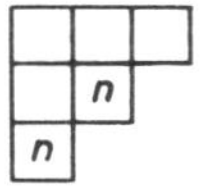 (8F) $\underline{64F}$	[411] 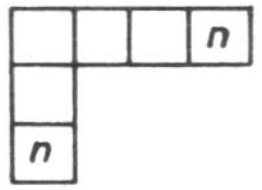 (10) $\underline{70}$	[33]  (10*) $\underline{50}$	[42] 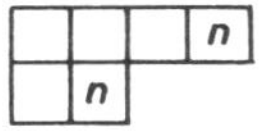 (27) $\underline{126}$
$\lambda'Y, \lambda''Y''$ 1, 0; 3, 0	$(h_{\sigma'})Z', (h_{\sigma''})Z''$ $(3^*) -\frac{1}{4}, (6) -\frac{1}{4}$	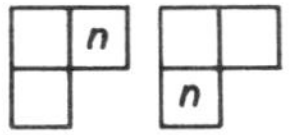	[11][1], [2][1]	$\frac{1}{\sqrt{5}}$	0	$-\frac{1}{2}$	$-\frac{1}{2}$	$\sqrt{\frac{3}{10}}$
2, −1; 2, 1	$(3) -\frac{5}{4}, (8)\frac{3}{4}$	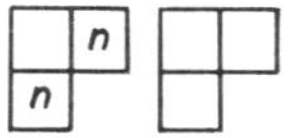	[1][2], [21][0]	$-\sqrt{\frac{3}{10}}$	$\frac{1}{\sqrt{6}}$	$\frac{1}{\sqrt{6}}$	$-\frac{1}{\sqrt{6}}$	$\frac{1}{\sqrt{5}}$
2, 1; 2, −1	$(8)\frac{3}{4}, (3) -\frac{5}{4}$	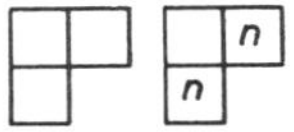	[21][0], [1][2]	$-\sqrt{\frac{3}{10}}$	$-\frac{1}{\sqrt{6}}$	$-\frac{1}{\sqrt{6}}$	$\frac{1}{\sqrt{6}}$	$\frac{1}{\sqrt{5}}$
3, 0; 1, 0	$(6) -\frac{1}{4}, (3^*) -\frac{1}{4}$		[2][1], [11][1]	$\frac{1}{\sqrt{5}}$	0	$\frac{1}{2}$	$\frac{1}{2}$	$\sqrt{\frac{3}{10}}$
3, 0; 3, 0	$(6) -\frac{1}{4}, (6) -\frac{1}{4}$		[2][1], [2][1]	0	$\sqrt{\frac{2}{3}}$	$-\frac{1}{\sqrt{6}}$	$\frac{1}{\sqrt{6}}$	0

Table 7.18-2 $SU_n \supset SU_{n-1} \otimes U_1$ ISF $C^{[\nu];[3][3]}_{[3][\sigma'][\mu'],[3][\sigma''][\mu'']}$

(SU_2CGC)	(SU_2SF)	(SU_3SF)		[33] 0	[42] 1	[6] 3	[51] 2
				□□□ / nnn (10*) $\underline{50}$	□□□n / nn (27) $\underline{126}$	□□□nnn (28) $\underline{84''}$	□□□nn / n (35) $\underline{140}$
0	4, −1	(10), $-\frac{3}{2}$	$[\sigma'][\mu']$ $[\sigma''][\mu'']$				
$\frac{3}{2}\ -\frac{3}{2},\ \frac{3}{2}\ \frac{3}{2}$	1, −2; 4, 1	$(1)-\frac{9}{4}$, $(10)\frac{3}{4}$	[0] [3] [3] [0] nnn □□□	$-\frac{1}{2}$	$\frac{3}{2\sqrt{5}}$	$\frac{1}{2\sqrt{5}}$	$-\frac{1}{2}$
$\frac{3}{2}\ -\frac{1}{2},\ \frac{3}{2}\ \frac{1}{2}$	2, −1; 3, 0	$(3)-\frac{5}{4}$, $(6)-\frac{1}{4}$	[1] [2] [2] [1] □nn □□n	$\frac{1}{2}$	$-\frac{1}{2\sqrt{5}}$	$\frac{3}{2\sqrt{5}}$	$-\frac{1}{2}$
$\frac{3}{2}\ \frac{1}{2},\ \frac{3}{2}\ -\frac{1}{2}$	3, 0; 2, −1	$(6)-\frac{1}{4}$, $(3)-\frac{5}{4}$	[2] [1] [1] [2] □□n □nn	$-\frac{1}{2}$	$-\frac{1}{2\sqrt{5}}$	$\frac{3}{2\sqrt{5}}$	$\frac{1}{2}$
$\frac{3}{2}\ \frac{3}{2},\ \frac{3}{2}\ -\frac{3}{2}$	4, 1; 1, −2	$(10)\frac{3}{4}$, $(1)-\frac{9}{4}$	[3] [0] [0] [3] □□□ nnn	$\frac{1}{2}$	$-\frac{3}{2\sqrt{5}}$	$\frac{1}{2\sqrt{5}}$	$\frac{1}{2}$

1) The meaning of the table heading is

SU_1SF (SU_2CGC) I_z	(SU_2SF) λ, Y	(SU_3SF) $(h_\sigma)Z$	$\begin{pmatrix}[\nu'] \\ [\sigma'][\mu']\end{pmatrix}\begin{pmatrix}[\nu''] \\ [\sigma''][\mu'']\end{pmatrix}$ \\ $\begin{pmatrix}[\nu] \\ [\sigma][\mu]\end{pmatrix}$	diagram I (h_ν) $\underline{H_\nu}$	SU_1SF SU_2SF SU_3SF
$I'I'_zI''I''_z$	$\lambda'Y'$, $\lambda''Y''$	$(h_{\sigma'})Z'$, $(h_{\sigma''})Z''$	diagram	(entries)	

1) Where the irreps of SU_2, SU_3 and SU_4 are labeled by $2I+1$, $\dim(h)$, and $\dim(H)$, respectively (for dimension tables, see Table A1 in the Appendix).
2) The total Young diagram and the one formed by the blank boxes are the irrep labels of SU_n and SU_{n-1} respectively. respectively.

Comparing (7-276) with (7-278), we see that the reduced matrix element is proportional to the CFP,

$$\left(\begin{matrix}[1^n]\\ \alpha J\end{matrix}\middle\| b_j^\dagger \middle\| \begin{matrix}[1^{n-1}]\\ \alpha_1 J_1\end{matrix}\right) = \text{const}\,\langle [1^n]\alpha J\{|[1^{n-1}]\alpha_1 J_1, j\rangle\ . \tag{7-279}$$

From (7-277), (7-279) and the normalization of the CFP, we obtain

$$\text{const} = \sqrt{n}\ .$$

Thus

$$\left(\begin{matrix}[1^n]\\ \alpha J\end{matrix}\middle\| b_j^\dagger \middle\| \begin{matrix}[1^{n-1}]\\ \alpha_1 J_1\end{matrix}\right\rangle = \binom{n}{1}^{1/2} \langle [1^n]\alpha J\{|[1^{n-1}]\alpha_1 J_1, j\rangle\ . \tag{7-280}$$

The physical implication of (7-280) is clear; it says that the probabilities for separating *any one fermion* from an n-fermion antisymmetric state is equal to n times of the probability for separating *the n-th fermion* from the same state.

7.19.2. Two-particle CFP

It is easy to verify that for both fermions and bosons, we have the identity

$$\sum_{j_1, m_1 j_2 m_2} b_{j_1 m_1}^\dagger b_{j_2 m_2}^\dagger b_{j_2 m_2} b_{j_1 m_1} = \hat{n}^2 - \hat{n}\ . \tag{7-281}$$

Applying (7-281) to the state $\left|\begin{matrix}[1^n]\\ \alpha J M\end{matrix}\right\rangle$, we get

$$n(n-1)\left|\begin{matrix}[1^n]\\ \alpha J M\end{matrix}\right) = \sum_{j_1 m_1 j_2 m_2 \alpha_1 J_1 M_1} b_{j_1 m_1}^\dagger b_{j_2 m_2}^\dagger \left|\begin{matrix}[1^{n-2}]\\ \alpha_1 J_1 M_1\end{matrix}\right)\left(\begin{matrix}[1^{n-2}]\\ \alpha_1 J_1 M_1\end{matrix}\middle| b_{j_2 m_2} b_{j_1 m_1} \middle| \begin{matrix}[1^n]\\ \alpha J M\end{matrix}\right)\ . \tag{7-282}$$

Using the relation

$$\sum_{j_1 j_2} = \frac{2}{1+\delta_{j_1 j_2}} \sum_{j_1 \le J_2}\ , \tag{7-283}$$

and the normalized pair creation operator

$$A^\dagger(j_1 j_2 J_2 M_2) = \sqrt{\frac{1}{1+\delta_{j_1 j_2}}}\,[b_{j_1}^\dagger b_{j_2}^\dagger]_{M_2}^{J_2}\ , \tag{7-284}$$

(7-282) can be written as

$$\begin{aligned} n(n-1)\left|\begin{matrix}[1^n]\\ \alpha J M\end{matrix}\right) &= 2\sum_{j_1\le j_2}\sum_{\alpha J_1 J_2} (-1)^{J_1+J_2-J}\left[A^\dagger(j_1 j_2 J_2)\left|\begin{matrix}[1^{n-2}]\\ \alpha_1 J_1\end{matrix}\right)\right]_M^J \\ &\times \left(\begin{matrix}[1^n]\\ \alpha J\end{matrix}\middle\| A^\dagger(j_1 j_2 J_2) \middle\| \begin{matrix}[1^{n-2}]\\ \alpha_1 J_1\end{matrix}\right)\ . \end{aligned} \tag{7-285}$$

Therefore

$$\frac{n(n-1)}{2} = \sum_{j_1\le j_2}\sum_{\alpha J_1 J_2} \left(\begin{matrix}[1^n]\\ \alpha J\end{matrix}\middle\| A^\dagger(j_1 j_2 J_2) \middle\| \begin{matrix}[1^{n-2}]\\ \alpha_1 J_1\end{matrix}\right)^2\ . \tag{7-286}$$

On the other hand, from the expansion in terms of the CFP we have

$$\left|\begin{matrix}[1^n]\\ \alpha JM\end{matrix}\right\rangle = \sum_{j_1\le j_2,\alpha J_1 J_2} \langle [1^n]\alpha J_2\{|[1^{n-2}]\alpha_1 J_1\,,[1^2](j_1j_2)J_2\rangle \left[\left|\begin{matrix}[1^{n-2}]\\ \alpha_1 J_1\end{matrix}\right\rangle\left|\begin{matrix}[1^2]\\ (j_1j_2)J_2\end{matrix}\right\rangle\right]^J_M . \tag{7-287}$$

where the sum is restricted to $j_1 \le j_2$, otherwise the states in the right-hand side of (7-287) would be overcomplete. Comparing (7-285) with (7-287), we know that

$$\left(\begin{matrix}[1^n]\\ \alpha J\end{matrix}\left\|A^\dagger(j_1j_2J_2)\right\|\begin{matrix}[1^{n-2}]\\ \alpha_1 J_1\end{matrix}\right) = \text{const}\,\langle [1^n]\alpha J\{|[1^{n-2}]\alpha_1 J_1, [1^2](j_1j_2)J_2\rangle\ . \tag{7-288}$$

From (7-286) and (7-288) and the normalization of CFP, it follows that

$$\sum_{j_1\le j_2,\alpha J_1 J_2} \langle [1^n]\alpha J\{|[1^{n-2}]\alpha_1 J_1, [1^2](j_1j_2)J_2\rangle^2 = 1\ , \tag{7-289}$$

we finally get the required relation

$$\left(\begin{matrix}[1^n]\\ \alpha J\end{matrix}\left\|A^\dagger(j_1j_2J_2)\right\|\begin{matrix}[1^{n-2}]\\ \alpha_1 J_1\end{matrix}\right) = \binom{n}{2}^{1/2} \langle [1^n]\alpha J\{|[1^{n-2}]\alpha_1 J_1, [1^2](j_1j_2)J_2\rangle \tag{7-290}$$

The physical implication of (7-290) is again very clear.

7.19.3 The CFP in the interacting boson model

In the interacting boson model (IBM) of nuclei (Arima 1978 and Sec. 9.9), there are two kinds of bosons b_l with $l = 0, 2$, which carry the defining rep of SU_6 and belong to the irrep (20) of SU_3. The two-boson state in the $SU_6 \supset SU_3 \supset SO_3$ group chain can be expressed as

$$\left|\begin{matrix}[2]\\ (\lambda_2\mu_2)L_2M_2\end{matrix}\right) = \sum_{l_1\le l_2} C^{(\lambda_2\mu_2)L_2}_{(20)l_1,(20)l_2} A^\dagger(l_1l_2L_2M_2)\ . \tag{7-291a}$$

The first factor at the right-hand side is the $SU_3 \supset SO_3$ ISF. According to Elliott (1958), the permissible values of $(\lambda_2, \mu_2)L_2$ are $(40)S, D, G$ and $(02)S, D$. The inverse of (7-291a) is

$$A^\dagger(l_1l_2L_2M_2) = \sum_{\lambda_2\mu_2} C^{(\lambda_2\mu_2)L_2}_{(20)l_1,(20)l_2}\left|\begin{matrix}[2]\\ (\lambda_2\mu_2)L_2M_2\end{matrix}\right)\ . \tag{7-291b}$$

For one-particle CFP we have

$$\left(\begin{matrix}[n]\\ (\lambda\mu)L\end{matrix}\left\|b_l^\dagger\right\|\begin{matrix}[n-1]\\ (\lambda_1\mu_1)L_1\end{matrix}\right) = \binom{n}{1}^{1/2} C^{[n],(\lambda\mu)L}_{[n-1](\lambda_1\mu_1)L_1,[1](20)l}\ . \tag{7-292}$$

For two-particle CFP, we need a relation similar to (7-285), i.e.,

$$n(n-1)\left|\begin{matrix}[\nu]\\ (\lambda\mu)LM\end{matrix}\right) = 2\sum_{l_1\le l_2}\sum_{(\lambda_1\mu_1)L_1L_2}(-1)^{L_1+L_2-L}\left[A^\dagger(l_1l_2L_2)\left|\begin{matrix}[n-2]\\ (\lambda_1\mu_1)L_1\end{matrix}\right)\right]^L_M$$
$$\times\left(\begin{matrix}[n]\\ (\lambda\mu)L\end{matrix}\left\|A^\dagger(l_1l_2L_2)\right\|\begin{matrix}[n-2]\\ (\lambda_1\mu_1)L_1\end{matrix}\right)\ . \tag{7-293}$$

By using (7-291), (7-293) becomes

$$n(n-1)\left|\begin{matrix}[n]\\(\lambda\mu)LM\end{matrix}\right) = 2\sum_{(\lambda_1\mu_1)L_1(\lambda_2\mu_2)L_2}(-1)^{L_1+L_2-L}\left[\left|\begin{matrix}[2]\\(\lambda_2\mu_2)L_2\end{matrix}\right)\left|\begin{matrix}[n-2]\\(\lambda_1\mu_1)L_1\end{matrix}\right)\right]^L_M$$
$$\times\left(\begin{matrix}[n]\\(\lambda\mu)L\end{matrix}\middle\|\begin{matrix}[2]\\(\lambda_2\mu_2)L_2\end{matrix}\middle\|\begin{matrix}[n-2]\\(\lambda_1\mu_1)L_1\end{matrix}\right)\ . \tag{7-294}$$

Therefore

$$\frac{n(n-1)}{2} = \sum_{(\lambda_1\mu_1)L_1(\lambda_2\mu_2)L_2}\left(\begin{matrix}[n]\\(\lambda\mu)L\end{matrix}\middle\|\begin{matrix}[2]\\(\lambda_2\mu_2)L_2\end{matrix}\middle\|\begin{matrix}[n-2]\\(\lambda_1\mu_1)L_1\end{matrix}\right)^2\ . \tag{7-295}$$

Thus the second quantized form of the two-particle CFP in the IBM is

$$\left(\begin{matrix}[n]\\(\lambda\mu)L\end{matrix}\middle\|\begin{matrix}[2]\\(\lambda_2\mu_2)L_2\end{matrix}\middle\|\begin{matrix}[n-2]\\(\lambda_1\mu_1)L_1\end{matrix}\right) = \binom{n}{2}^{1/2} C^{[n](\lambda\mu)L}_{[n-2](\lambda_1\mu_1)L_1,[2](\lambda_2\mu_2)L_2}\ . \tag{7-296}$$

The extension of (7-292) and (7-296) to other group chains is trivial.

Chapter 8
The Point Groups

This chapter deals with the elementary concepts and definitions relating to point groups, with emphasis on the application of the EFM to them, and the introduction of the so-called representation group, which is quite useful for treating the double-valued reps of the point groups as well as the single-valued and double-valued reps of space groups. Readers are referred to McWeeny (1963) and Cotton (1971) for more detailed presentations of the point groups. We shall use the Schönflies notation.

8.1. Basic Operations of Point Groups

Many objects (or systems) either microscopic or macroscopic, exhibit some form of symmetry. For example, nuclei, atoms, molecules, snow flakes, equilateral triangles, hexagonal nuts, cubes, regular pyramids or prisms, etc. The operation which brings a system into a position indistinguishable from that which it originally occupied is called a *symmetry operation* of the system. The operation does not change the distance between any two points inside the system. All such operations form a group, called the *symmetry group* of the system.

The symmetry group of a finite system is called a *point group*, since all the symmetry operations must leave at least one point of the system unmoved.

The Cartesian coordinates x, y, z carry the fundamental rep of the three-dimensional orthogonal group O_3. Since the point group is a subgroup of O_3, x, y, z also carry a rep D for the point group, which is a *faithful* but in general reducible rep of the point group G.

The atoms in a molecule are said to be *equivalent* if they interchange with one another under the symmetry operations. Equivalent atoms are necessarily identical chemically.

8.1.1. Basic operations and their faithful reps

The basic operations of point groups are as follows:

1. *Proper (or pure) rotation*

A rotation through angle $2\pi/l$ about an axis $\mathbf{n}$ is designated as $C_l^{\mathbf{n}}$. The maximum l is called the *order* (or *fold*) of the rotation axis. A point group may contain several rotation axes. The one with the highest order is called the principle axis. One usually chooses the principle axis as the z-axis. From (2-60b) we get the faithful rep of C_l^z,

$$D(C_l^z) = \begin{pmatrix} \cos\frac{2\pi}{l} & -\sin\frac{2\pi}{l} & 0 \\ \sin\frac{2\pi}{l} & \cos\frac{2\pi}{l} & 0 \\ 0 & 0 & 1 \end{pmatrix} . \tag{8-1}$$

To simplify notation, the superscript z in C_l^z will be ignored when no confusion will arise. We have

$$C_l = e^{-i(\frac{2\pi}{l})J_z}, \quad C_l^k = e^{-i(\frac{2k\pi}{l})J_z} = C_{l/k} = (C_l)^k ,$$
$$C_l^{l-1} = C_l^{-1}, \quad C_l^l = e . \tag{8-2}$$

The set $\{C_l^k : k = 1, 2, \ldots , l\}$ forms the cyclic group. Evidently, the determinant of the natural rep of any proper rotation is equal to $+1$.

Suppose that there are two-fold axes with an angle φ between them, both are perpendicular to an l-fold principle axis. Choose one of the two-fold axes as the x-axis and the other in the direction $\mathbf{n}$ (see Fig. 8.1-1). Clearly, we have

$$C_2^{(\varphi)} \equiv C_2^{\mathbf{n}_\varphi} = R_z(\varphi) C_2^x R_z^{-1}(\varphi) , \tag{8-3}$$

$$D(C_2^x) = \begin{pmatrix} 1 & & 0 \\ & -1 & \\ 0 & & -1 \end{pmatrix} , \tag{8-4}$$

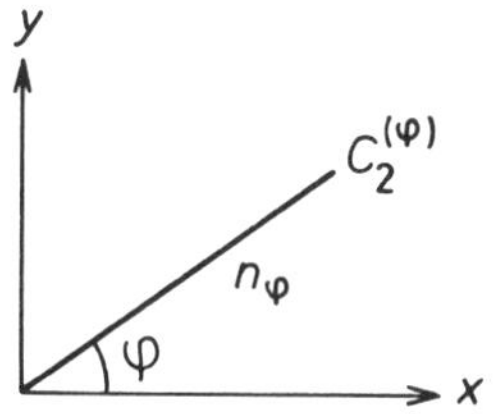

Fig. 8.1-1

where $R_z(\varphi)$ is a rotation through an angle φ about the z-axis.

Two operations A and B of a group G are said to be *equivalent* if there exists an operation C in G that

$$B = CAC^{-1} . \tag{8-5}$$

If an axis A is sent into another axis B under an operation C of G, then the axes A and B are said to be *equivalent.*

For example, in Fig. 8.1-1, $C_2^{\mathbf{n}_\varphi}$ and C_2^x are equivalent operations, while $\mathbf{n}_\varphi$ and x are equivalent axes, if $R_z(\varphi)$ is an element of the point group G. From (5-32), we obtain the faithful rep of $R_z(\varphi)$,

$$D(R_z(\varphi)) = \begin{pmatrix} \cos\varphi & -\sin\varphi & 0 \\ \sin\varphi & \cos\varphi & 0 \\ 0 & 0 & 1 \end{pmatrix} . \tag{8-6}$$

From (8-3)–(8-6) we immediately have

$$D(C_2^{(\varphi)}) = \begin{pmatrix} \cos 2\varphi & \sin 2\varphi & 0 \\ \sin 2\varphi & -\cos 2\varphi & 0 \\ 0 & 0 & -1 \end{pmatrix} , \tag{8-7}$$

and from (8-4) and (8-7) we see that

$$D(C_2^{(\varphi)}) D(C_2^x) = D(R_z(2\varphi)) . \tag{8-8a}$$

Since D is a faithful rep of G, we have the important relation:

$$C_2^{(\varphi)} C_2^x = R_z(2\varphi) , \tag{8-8b}$$

namely, successive rotations through angle π about two axes making an angle φ with one another is equivalent to a rotation through angle 2φ about an axis perpendicular to the first two axes. Therefore if a group G has two-fold axes making an angle φ with one another, then there must exist an n-fold axis with $n = 2\pi/2\varphi = \pi/\varphi$. Conversely, if a group G has an n-fold axis and one two-fold axis, then there must exist n two-fold axes, with angle $\varphi = \pi/n$ between any two adjacent axes.

Letting $\varphi = \pi/2$ in Eq. (8-7), we get

$$D(C_2^y) = \begin{pmatrix} -1 & & 0 \\ & 1 & \\ 0 & & -1 \end{pmatrix} . \tag{8-9}$$

From (8-4) and (8-9) it is seen that $[D(C_2^x), D(C_2^y)] = 0$. Thus

$$[C_2^x, C_2^y] = 0 , \tag{8-10}$$

namely, two rotations about perpendicular axes through angle π commute with one another.

If there is a rotation in the group G which carries a rotation axis $\mathbf{n}$ into -$\mathbf{n}$, we say that $\mathbf{n}$ is a two-sided axis.

2. *Reflection across a horizontal plane*

The operation σ_h is a reflection across a plane (xy plane) perpendicular to the principle axis and passing through the origin. There is only one operation σ_h. Below we often use $\sigma^{\mathbf{n}}$ to denote a reflection plane with $\mathbf{n}$ as the normal direction. Thus $\sigma_h = \sigma^z$. We have

$$D(\sigma^z) = D(\sigma_h) = \begin{pmatrix} 1 & & 0 \\ & 1 & \\ 0 & & -1 \end{pmatrix} , \tag{8-11a}$$

$$\det(D(\sigma_h)) = -1 . \tag{8-11b}$$

According to (8-6) and (8-11), $[D(\sigma^Z), D(R_z(\varphi))] = 0$. Therefore

$$[\sigma^Z, R_z(\varphi)] = 0 . \tag{8-12}$$

That is, a rotation commutes with the reflection across a plane which is perpendicular to the rotation axis.

Any plane molecule has σ_h symmetry with the molecular plane as the reflection plane.

3. *Reflection across a vertical plane*

σ_v denotes a reflection across a plane containing the principle axis. Suppose that a group G has two vertical reflection planes, $\sigma^{\mathbf{n}} = \sigma^{(\varphi)}$ and $\sigma^y = \sigma^{(0)}, \varphi$ being the angle between the normal $\mathbf{n}$ and the y axis (or between the intersection of $\sigma^{\mathbf{n}}$ with the xy plane and the x axis, see Fig. 8.1-2). From Fig. 8.1-2 one sees that

$$\sigma^{(\varphi)} = \sigma^{\mathbf{n}} = R_z(\varphi)\sigma^y R_z(-\varphi) , \tag{8-13}$$

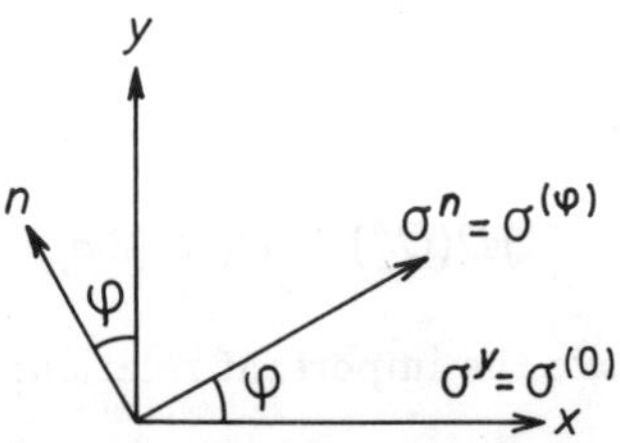

Fig. 8.1-2

$$D(\sigma^{(0)}) = D(\sigma^y) = \begin{pmatrix} 1 & & 0 \\ & -1 & \\ 0 & & 1 \end{pmatrix} . \tag{8-14}$$

Combining (8-13), (8-14) and (8-6) we get

$$D(\sigma^{(\varphi)}) = \begin{pmatrix} \cos 2\varphi & \sin 2\varphi & 0 \\ \sin 2\varphi & -\cos 2\varphi & 0 \\ 0 & 0 & 1 \end{pmatrix} . \tag{8-15}$$

From (8-14) and (8-15) we have

$$D(\sigma^{(\varphi)})D(\sigma^{(0)}) = D(R_z(2\varphi)) , \tag{8-16}$$

therefore

$$\sigma^{(\varphi)}\sigma^{(0)} = R_z(2\varphi) . \tag{8-17}$$

That is, the product of two vertical reflections across planes making an angle φ with each other is equivalent to a rotation through an angle 2φ about the z-axis.

Consequently, if a group G has two vertical reflection planes making an angle φ with each other, then there must exist an n-fold axis with $n = \pi/\varphi$. Conversely, if a group G has an n-fold axis and a vertical reflection plane, then there must exist n vertical reflection planes with angle $\varphi = \pi/n$ between any two adjacent planes.

Letting $\varphi = \pi/2$ in (8-15), we get

$$D(\sigma^x) = D(\sigma^{(\pi/2)}) = \begin{pmatrix} -1 & & 0 \\ & 1 & \\ 0 & & 1 \end{pmatrix} . \tag{8-18}$$

From (8-14) and (8-18), we have $[D(\sigma^{(\pi/2)}), D(\sigma^{(0)})] = 0$. Therefore

$$[\sigma^{(\pi/2)}, \sigma^{(0)}] = 0 . \tag{8-19}$$

This shows that two reflection operations commute if their reflection planes are perpendicular.

If $R_z(\varphi)$ in (8-13) is an element of the group G, then σ^{n} and σ^x are said to be *equivalent planes.*

From the products of the basic operations (1) and (2), we obtain another two operations:

4. *The inversion I*

This operation is characterized by

$$I = C_2^z\sigma_h = \sigma_h C_2^z, \quad \sigma^z = \sigma_h = C_2^z I . \tag{8-20}$$

From (8-6), (8-11) and (8-20) we have

$$D(I) = \begin{pmatrix} -1 & & 0 \\ & -1 & \\ 0 & & -1 \end{pmatrix} . \tag{8-21}$$

Since $D(I)$ equals a unit matrix times (-1), the inversion I commutes with any elements of the point group.

5. *Improper rotation*

The improper rotation S_n^z is defined as a product of a proper rotation and a horizontal reflection,

$$S_n^z = C_n^z \sigma_h = \sigma_h C_n^z = C_n^z \sigma^z \ . \tag{8-22}$$

From (8-1) and (8-11) we find the faithful rep of an improper rotation,

$$D(S_n^z) = \begin{pmatrix} \cos\frac{2\pi}{n} & -\sin\frac{2\pi}{n} & 0 \\ \sin\frac{2\pi}{n} & \cos\frac{2\pi}{n} & 0 \\ 0 & 0 & -1 \end{pmatrix} . \tag{8-23}$$

According to (8-20) and (8-22), S_n^z can be expressed as

$$S_n^z = C_n^z C_2^z I \ . \tag{8-24a}$$

Using (8-2) we get

$$C_n^z C_2^z = \exp\left[-2\pi i\left(\frac{2+n}{2n}\right)J_z\right] = \exp\left[i\frac{2\pi}{\kappa}J_z\right] = \overline{C}_\kappa^z \ , \tag{8-24b}$$

where

$$\kappa = \frac{2n}{n-2} \ , \tag{8-24c}$$

$$\overline{C}_\kappa^z = (C_\kappa^z)^{-1} \ . \tag{8-24d}$$

It follows that

$$S_3 = \overline{C}_6 \times I, \quad S_4 = \overline{C}_4 \times I, \quad S_6 = \overline{C}_3 \times I \ . \tag{8-25}$$

From (8-22) we know that when n=even, $\{S_n^k : k = 1, 2, \ldots, n\}$ forms an Abelian group, denoted by S_n,

$$S_n^k = (S_n)^k = \begin{cases} C_n^k \\ C_n^k \sigma_h \end{cases} , \quad \text{for } k = \begin{cases} \text{even} \\ \text{odd} \end{cases} . \tag{8-26a}$$

From (8-24a) we get

$$S_2 = S_n^{n/2} = I, \quad \text{for}\, n/2 = \text{odd} \ . \tag{8-26b}$$

Therefore, the groups S_n for $n = 4k+2, k = 0, 1, \ldots$ must contain the inversion.

A rotation axis or a reflection plane is called a *symmetry element* of the point group.

Equation (8-5) tells us that a set of equivalent operations constitutes a class of a point group. Concerning the classes of point groups we have the following obvious results.

1. The identity operation, inversion I, and horizontal reflection σ_h each forms a class by itself.

2. Proper (or improper) rotations through the same angle about equivalent axes belong to the same class.

3. Equivalent reflections belong to the same class.

4. For a two-sided axis, a rotation (proper or improper) and its inverse belong to the same class.

For a point group, we often use the symbol "nX" to designate a class, or a class operator, where n is the number of elements in the class and X is an operator belonging to the class. For example, in Table 8.3-4, $3C_2 = C_2^{(0^\circ)} + C_2^{(120^\circ)} + C_2^{(240^\circ)}$.

8.2. Some Commonly Used Point Groups

The classification and generators of some of the most commonly used point groups are listed in Table 8.2-1. McWeeny (1963) listed the group elements and classes for them. Figure 8.2-1 gives the four types of molecules with the symmetries C_{3v}, D_3, D_{3h} and D_{3d}. Figure 8.2-2 gives their projections in the xy plane, along with the symmetry operations (not including the operations C_3, C_3^2).

Table 8.2-1. Some commonly used point groups.[1]

group	order	class numbers	inter-relations	generators	symmetry objects
C_1	2	2		e	
C_i	2	2		I	
C_s	2	2	$C_2 = C_{1h}$	σ_h	
C_n	n	n		C_n	
C_{nv}	$2n$	$\begin{cases} 3+n/2 & \text{e.}n \\ (3+n)/2 & \text{o.}n \end{cases}$	$C_{nv} \sim D_n$	$C_n, \sigma_v^{(1)}$	a regular n-sided pyramid
C_{nh}	$2n$	$2n$	$C_{nh} = \begin{cases} C_n \times C_s, & n = \text{e. or o.} \\ C_n \times C_i, & n = \text{e.} \end{cases}$	C_n, σ_h	
S_n	n	n		S_n	
D_n	$2n$	$\begin{cases} (n+6)/2 & \text{e.}n \\ (n+3)/2 & \text{o.}n \end{cases}$	$D_n \sim C_{nv}$	$C_n, C_2^{(1)}$	
D_{nh}	$4n$	$\begin{cases} n+6 & \text{e.}n \\ n+3 & \text{o.}n \end{cases}$	$D_{nh} = \begin{cases} D_n \times C_s, & n = \text{e. or o.} \\ D_n \times C_i, & n = \text{e.} \end{cases}$	$C_n, C_2^{(1)}, \sigma_h$	a regular n-sided prism
D_{nd}	$4n$	$n+3$	$D_{nd} = \begin{cases} C_{nv} \times C_i \\ D_n \times C_i, \end{cases} \; n = \text{o.}$	$C_n, C_2^{(1)}, \sigma_v^{(1)}$	
T	12	4		$C_2^z, C_3^{(1)}$	cube (proper rotations only)
T_d	24	5		$S_4^z, C_3^{(1)}$	tetrahedron
T_h	24	8	$T_h = T \times C_i$	$C_2^z, C_3^{(1)}, I$	
O	24	5		$C_4^z, C_3^{(1)}$	cube, (octahedron, proper rot. only)
O_h	48	10	$O_h = O \times C_i$	$C_4^z, C_3^{(1)}, I$	cube, octahedron

[1]e. = even, o. = odd.

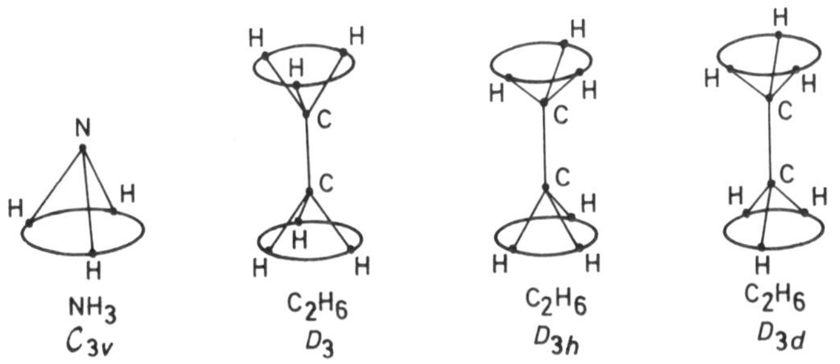

Fig. 8.2-1. Molecules having the C_{3v}, D_3, D_{3h} and D_{3d} symmetries.

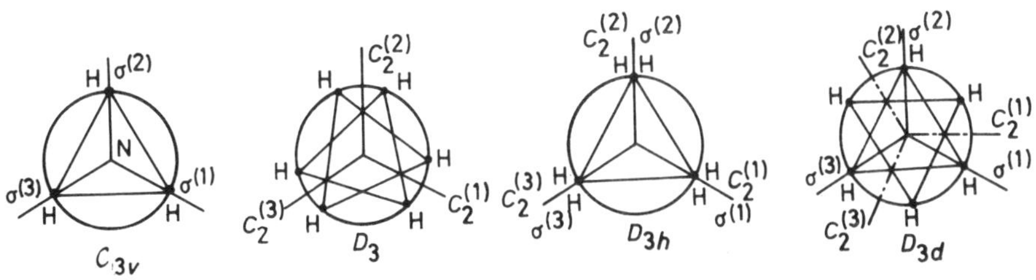

Fig. 8.2-2. The projections of Fig. 8.2-1 onto the xy plane.

The groups C_{nv}, D_n, D_{nh} and D_{nd} are called *axial groups* or *dihedral groups* and are similar to the above case with $n = 3$.

Figure 8.2-3 shows a tetrahedron inscribed in a cube. The pure rotation group of a regular tetrahedron is denoted by T. It is seen that the group T has three 2-fold axes passing the midpoints of two non-adjacent sides, C_2^x, C_2^y and C_2^z and four 3-fold axes $C_3^{(i)}, i = 1, 2, 3, 4$ (starting from the origin and pointing to the four corners 1,2,3,4 of the tetrahedron).

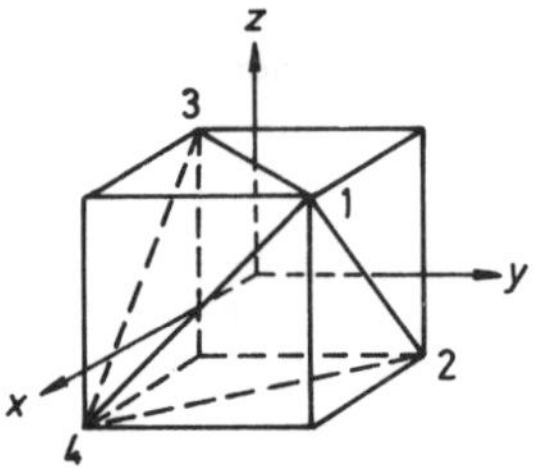

Fig. 8.2-3. The tetrahedron group T_d.

According to the transformations of the vertices under the symmetry operations, it is easy to find the correspondence between the point group and permutation group, as shown in Table 8.2-2. The notation given in the second row of Table 8.2-2 was used by McWeeny (1963). For example, $C_3^{\bar{x}y\bar{z}}$ is a three-fold axis passing the origin and having equal angles with respect to the $\bar{x}(= -x), y$ and $\bar{z}(= -z)$ axes. From the first half of Table 8.2-2, one sees that T is isomorphic to the alternative group $\mathcal{A}_4$ and has four classes partitioned by the solid and dotted vertical lines. We can obtain products of the symmetry operations by means of the multiplication rule of the permutation group. For example, from (234)(143)=(123), we know that $C_3^{(1)} C_3^{(2)} = \overline{C}_3^{(4)} \equiv (C_3^{(4)})^{-1}$.

Table 8.2-2 also shows that e, C_2^x, C_2^y, and C_2^z form a subgroup D_2 of T and D_2 is isomorphic to the four group.

Table 8.2-2. The isomorphism of T to $\mathcal{A}_4$, and T_d to S_4.

e	C_2^x	C_2^y	C_2^z	$C_3^{(1)}$ C_3^{xyz}	$C_3^{(2)}$ $C_3^{\bar{x}y\bar{z}}$	$C_3^{(3)}$ $C_3^{\bar{x}\bar{y}z}$	$C_3^{(4)}$ $C_3^{x\bar{y}\bar{z}}$	$\overline{C}_3^{(1)}$	$\overline{C}_3^{(2)}$	$\overline{C}_3^{(3)}$	$\overline{C}_3^{(4)}$
e	(14) (23)	(12) (34)	(13) (24)	(234)	(143)	(124)	(132)	(243)	(134)	(142)	(123)

S_4^z	$\overline{S}_4^z$	S_4^x	$\overline{S}_4^x$	S_4^y	$\overline{S}_4^y$	$\sigma^{(12)}$ σ^{xz}	$\sigma^{(23)}$ $\sigma^{y\bar{z}}$	$\sigma^{(34)}$ $\sigma^{x\bar{z}}$	$\sigma^{(13)}$ σ^{xy}	$\sigma^{(14)}$ σ^{yz}	$\sigma^{(24)}$ $\sigma^{x\bar{y}}$
(1234)	(1432)	(1342)	(1243)	(1423)	(1324)	(12)	(23)	(34)	(13)	(14)	(24)

The group T_d is the symmetry group of a regular tetrahedron. In addition to the twelve elements of the group T, T_d contains another twelve elements which can be expressed as $S_4^z T$. From Table 8.2-2 one sees that T_d is isomorphic to the permutation group S_4 and has five classes partitioned by the solid vertical lines. $\sigma^{(12)}$ is a reflection plane containing the side (34) and the midpoint of its opposite side (12). Under the reflection $\sigma^{(ij)}$, the vertices i and j interchange with one another. The notation $\sigma^{\alpha\beta}(\alpha, \beta = x, y, z, \bar{x}, \bar{y}, \bar{z})$ denotes a reflection plane whose normal lies in the $\alpha\beta$ plane and has equal angle with respect to the axes α and β.

Octahedron group: The group O is the pure rotation group of a regular octahedron, or a cube. The former can be inscribed in the latter (see Fig. 8.2-4).

The group O has the following symmetry operations:

1. Three 4-fold axes, C_4^x, C_4^y and C_4^z.

2. Four 3-fold axes passing the center of two opposite faces of the octahedron, or equivalently, joining the nearest and fartherest cube corners. The four corners of the cube above the origin

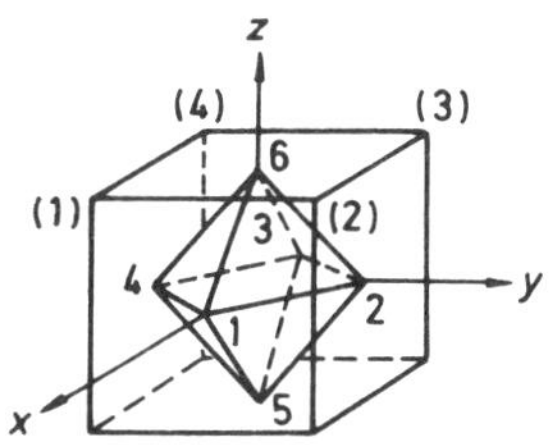

Fig. 8.2-4. The octahedron group O.

are labeled by (1), (2), (3) and (4). The four 3-fold axes are denoted by $C_3^{(i)}, i = 1,2,3,4$, or written as

$$C_3^{(1)} = C_3^{x\bar{y}z}, C_3^{(2)} = C_3^{xyz}, C_3^{(3)} = C_3^{\bar{x}yz}, C_3^{(4)} = C_3^{\bar{x}\bar{y}z} \, . \tag{8-27c}$$

3. Six 2-fold axes passing the midpoints of two opposite sides of the octahedron, such as (see Table 8.2-3) $C_2^{xy}, C_2^{x\bar{y}}, \ldots$ Here C_2^{xy} is a two-fold axis lying in the xy plane and having the same angle with respect to the axes x and y.

The group O has 24 elements falling into five classes. The group T is a subgroup of O. The isomorphism between the group O and a subgroup of S_6 is shown in Table 8.2-3.

By annexing the inversion I to the group O, we get the group O_h with 48 elements: $O_h = (e \oplus I)O$.

From (8-20) we have

$$\sigma^i = C_2^i \cdot I, \quad i = x, y, z, xy, xz, \ldots \, . \tag{8-28}$$

In Table 8.2-3, the first row forms the group T; the first and second rows form the group O, while the first and third row form the group T_h.

Crystal point groups: In crystals, there are only 32 possible point groups (see Sec. 10.2), namely $C_n, C_{nh}, C_{nv}, D_n, D_{nh}$, for $n = 2,3,4,6$; $D_{2d}, D_{3d}, C_1, C_i, C_s, S_4, S_6, T, T_h, T_d, O$ and O_h.

The group tables for the crystal point group O can be found from Table 10.24-1.

Molecular point groups: Apart from the 32 crystal point groups, the molecular point groups also have the following groups: $C_5, C_7, C_8, C_{5h}, C_{5v}, D_5, D_{5h}, D_{5d}, D_{4d}, D_{6d}, D_{8h}, D_{\infty h}, C_{\infty v}$, and the isocahedron groups I and I_h. The group $C_{\infty v}$ is the symmetry group of linear molecules without symmetry centers such as CO (see Fig. 8.2-5), and $D_{\infty h}$ is the symmetry group of linear molecules with symmetry centres, such as H_2, O_2, CO_2, etc., (see Fig. 8.2-6). $C_{\infty v}$ and $D_{\infty h}$ contain the following symmetry operations:

$$\begin{aligned} C_{\infty v} &= \{R^{(z)}(\varphi), \ \sigma^{(\varphi)}\}, \ \varphi = 0 - 2\pi \, , \\ D_{\infty h} &= \{R_z(\varphi), \ \sigma^{(\varphi)}, \ C_2^{(\varphi)}, \ S_z(\varphi), I\}, \quad \varphi = 0 - 2\pi \, . \end{aligned} \tag{8-29}$$

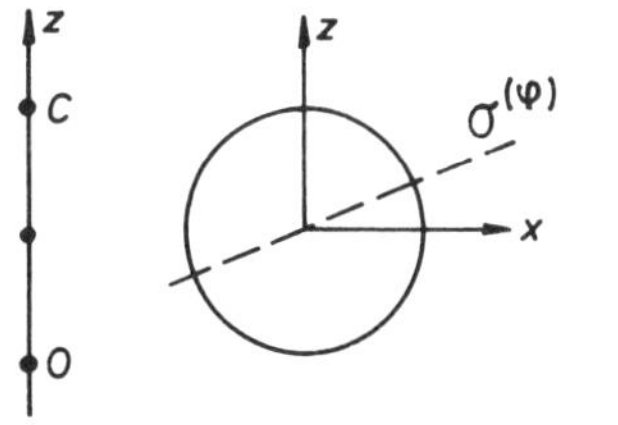

Fig. 8.2-5. The molecule CO with $C_{\infty v}$ symmetry.

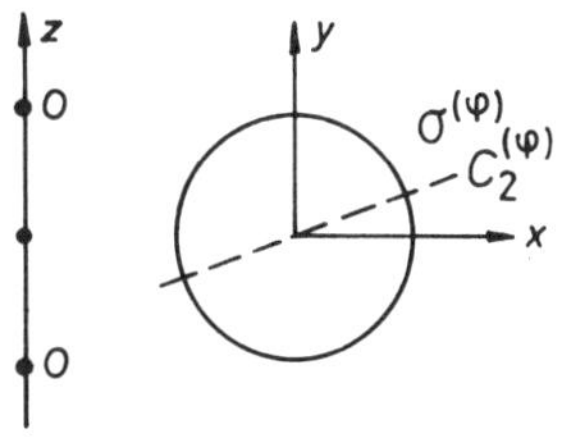

Fig. 8.2-6. The molecule O_2 with $D_{\infty h}$ symmetry.

Table 8.2-3. The elements of O_h and the isomorphism of O_h to a subgroup of S_6.

$O\{T$	e	C_2^x	C_2^y	C_2^z	$C_3^{(1)}$	$C_3^{(2)}$	$C_3^{(3)}$	$C_3^{(4)}$	$\overline{C}_3^{(1)}$	$\overline{C}_3^{(2)}$	$\overline{C}_3^{(3)}$	$\overline{C}_3^{(4)}$
	e	(24) (56)	(13) (56)	(13) (24)	(164) (235)	(126) (345)	(154) (236)	(125) (346)	(146) (253)	(162) (354)	(145) (263)	(152) (364)
	C_2^{xy}	$C_2^{x\bar{y}}$	C_2^{xz}	$C_2^{x\bar{z}}$	C_2^{yz}	$C_2^{y\bar{z}}$	C_4^x	C_4^y	C_4^z	$\overline{C}_4^x$	$\overline{C}_4^y$	$\overline{C}_4^z$
	(12) (34) (56)	(14)(23) (56)	(16)(35)(24)	(15)(36)(24)	(13) (26) (45)	(13)(25)(46)	(2645)	(1536)	(1234)	(2546)	(1635)	(1432)
	I	σ^x	σ^y	σ^z	$S_6^{(1)}$	$S_6^{(2)}$	$S_6^{(3)}$	$S_6^{(4)}$	$\overline{S}_6^{(1)}$	$\overline{S}_6^{(2)}$	$\overline{S}_6^{(3)}$	$\overline{S}_6^{(4)}$
	(13) (24) (56)	(13)	(24)	(56)	(126345)	(152364)	(125346)	(162354)	(154362)	(146325)	(164352)	(145326)
	σ^{xy}	$\sigma^{x\bar{y}}$	σ^{xz}	$\sigma^{x\bar{z}}$	σ^{yz}	$\sigma^{y\bar{z}}$	S_4^x	S_4^y	S_4^z	$\overline{S}_4^x$	$\overline{S}_4^y$	$\overline{S}_4^z$
	(14) (23)	(12) (34)	(15) (36)	(16) (35)	(25) (46)	(26) (45)	(2645) (13)	(1536) (24)	(1234) (56)	(2546) (13)	(1635) (24)	(1432) (56)

The subgroup chains for the 32 crystal point groups are shown in Fig. 8.2-7 (taken from Koster 1963).

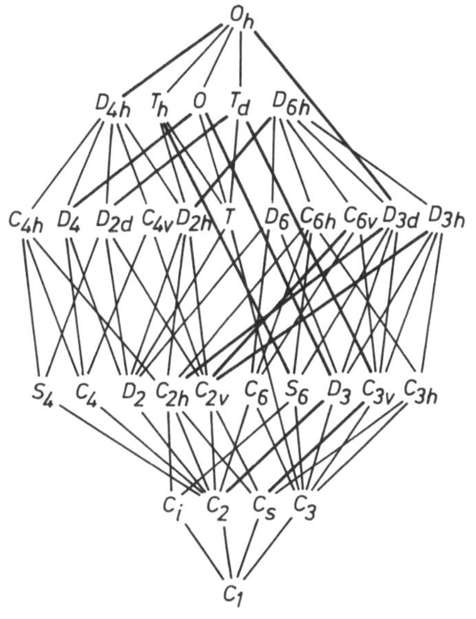

Fig. 8.2-7. Subgroup chains of the 32 point groups.

8.3. The CSCO-I and CSCO-II of Point Groups

8.3.1. The conventional labeling for point group irreps (Mullikan Notation)

1. Use A or B to denote one-dimensional irreps; E, two-dimensional irreps; T or F, 3-dimensional irreps.
2. $A(B)$ denotes irreps with character $\chi(C_n) = 1(-1)$.
3. A_1 or B_1 denotes irreps irreps with $\chi(C_2) = 1$ or $\chi(\sigma_v) = 1$, where C_2 is a 2-fold axis perpendicular to the principal axis. A_2 or B_2 denotes irreps with $\chi(C_2) = -1$ or $\chi(\sigma_v) = -1$.
4. A', B', E' and T' are associated with $\chi(\sigma_h) = 1$, while A'', B'', E'' and T'' are associated with $\chi(\sigma_h) = 1$.
5. A_g, B_g, E_g and T_g are associated with $\chi(I) = 1$, while A_u, B_u, E_u and T_u are associated with $\chi(I) = -1$.

Since the character tables of point groups are all known (Hamermesh 1962, Koster 1963, Cotton 1971), we can use the method of Sec. 3.12 to find the CSCO of the point groups quite easily. The CSCO of direct product groups can be found by using (3-37).

The CSCO of the cyclic group $\mathcal{C}_n$, $\mathcal{S}_n$ and $\mathcal{C}_{nh}$ are as follows:

$$\begin{aligned}
&\mathcal{C}_n : C = C_n^z, && \lambda^{(\nu)} = \exp(-2\pi\nu i/n), && \nu = 0,1,2,\dots,n-1 \\
&\mathcal{S}_{2n} : C = S_{2n}^z, && \lambda^{(\nu)} = \pm\exp(-\pi\nu i/n) && \nu = 0,1,2,\dots,2n-1 \\
&\mathcal{C}_{nh} : C = (C_n^z, I), && \lambda^{(\nu)} = (e^{-2\pi\nu i/n}, \pm 1), && \text{for } n = \text{even}, \\
&\qquad C = (C_n^z, \sigma_h), && \lambda^{(\nu)} = (e^{-2\pi\nu i/n}, \pm 1), && \text{for } n = \text{odd}
\end{aligned} \tag{8-30}$$

The CSCO-I and CSCO-II of the point groups are listed in Tables 8.3-1 to 8.3-17. The first column gives the conventional irrep labels. The second column gives the eigenvalues of the CSCO-I and the third column (for non-abelian groups) gives the eigenvalues of the CSCO, $C(s)$, of the subgroup chain $G(s)$. The third column (for abelian groups) or fourth column (for non-abelian groups) lists the $G \supset G(s)$ irreducible bases which are simultaneous eigenfunctions of the CSCO-II $(C, C(s))$ of G. (For the method of finding these bases see Sec. 8.4).

The tables for G and $G\times \mathcal{C}_s$, or G and $G\times \mathcal{C}_i$, are listed together. For example, for D_n and $D_{nh}=D_n\times \mathcal{C}_s$; $\mathcal{C}_{nv}$ and $D_{nd}=\mathcal{C}_{nv}\times \mathcal{C}_i(n=\text{odd})$; T and $T_h=T\times \mathcal{C}_i$; O and $O_h=O\times \mathcal{C}_i$ etc. In the table for $G\times \mathcal{C}_s(G\times \mathcal{C}_i)$, by simply ignoring the CSCO $\sigma_h(I)$ for $\mathcal{C}_s(\mathcal{C}_i)$, we are left with the CSCO of the group G. By dropping the primes or the subscripts g or u in the first column, we are left with the conventional label for the group G. For example, from Table 8.3-5, we know that the CSCO of $\mathcal{C}_{3v}$ is $3\sigma_v$ with the eigenvalues 3, −3, 0, corresponding to the irreps A_1, A_2 and E, respectively. When there are several kinds of σ_v (or C_2), we use diagrams to specify explicitly which operations are included in C or $C(s)$.

Table 8.3-1. D_2, $D_{2h}=D_2\times \mathcal{C}_i$.[1,2,3]

	C_2^x	C_2^y	I	eigenfunctions
A_g	1	1	1	x^2, y^2, z^2
B_{1g}	−1	−1	1	xy, R_z
B_{2g}	−1	1	1	xz, R_y
B_{3g}	1	−1	1	yz, R_x
A_u	1	1	−1	xyz
B_{1u}	−1	−1	−1	z
B_{2u}	−1	1	−1	y
B_{3u}	1	−1	−1	x

[1] R_x, R_y, R_z are components of an axial vector.
[2] $D_2=(e, C_2^x, C_2^y, C_2^z)$.
[3] $D_{2h}=(e, C_2^x, C_2^y, C_2^z, I, \sigma^x, \sigma^y, \sigma^z)$.

Table. 8.3-2. $\mathcal{C}_{2v}$.

	C_2^z	σ^y	eigenfunctions
A_1	1	1	z, x^2, y^2, z^2
A_2	1	−1	R_z, xy
B_1	−1	1	x, R_y, xz
B_2	−1	−1	y, R_x, yz

$\mathcal{C}_{2v}=(e, \sigma^x, \sigma^y, C_2^z)$

Table 8.3-3. D_{2d}.

	$2C_2'$	2σ	σ^y	eigenfunctions
A_1	2	2		x^2+y^2, z^2
A_2	−2	−2		R_z
B_1	2	−2		x^2-y^2
B_2	−2	2		z, xy
E	0	0	(1, −1)	(x, y), (xz, yz), $(R_y, -R_x)$

Table 8.3-4. D_3, $D_{3h}=D_3\times \mathcal{C}_i$.

	$3C_2$	σ_h	$C_2^x=C_2^{(1)}$	eigenfunctions
A_1'	3	1		x^2+y^2, z^2
A_2'	−3	1		R_z
E'	0	1	(1, −1)	(x, y), $(x^2-y^2, -2xy)$
A_1''	3	−1		zR_z,
A_2''	−3	−1		z
E''	0	−1	(1, −1)	(R_x, R_y), $(yz, -xz)$

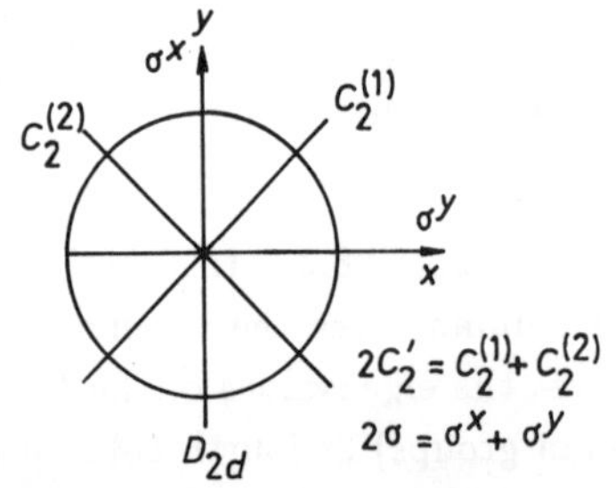

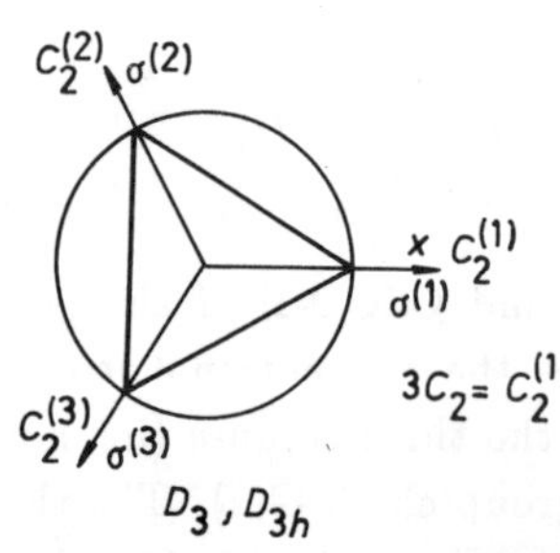

Table 8.3-5. C_{3v}, $D_{3d} = C_{3v} \times C_i$.[1,2]

	$3\sigma_\nu$	I	σ^y	eigenfunctions
A_{1g}	3	1		x^2+y^2, z^2
A_{2g}	-3	1		R_z
E_g	0	1	$(1, -1)$	$(R_y, -R_x)$, (xz, yz) $(x^2-y^2, -2xy)$
A_{2u}	3	-1		z,
A_{1u}	-3	-1		y^3-3x^2y
E_u	0	-1	$(1, -1)$	(x, y)

[1] C_{3v} (solid lines)
[2] D_{3d} (both solid and dotted lines)

Table 8.3-6. C_{4v}.

	$2C^z$	2σ	σ^y	eigenfunctions
A_1	2	2		x^2+y^2, z, z^2
A_2	2	-2		R_z
B_1	-2	2		x^2-y^2
B_2	-2	-2		xy
E	0	0	$(1, -1)$	(x, y), $(R_y, -R_x)$ (xz, yz)

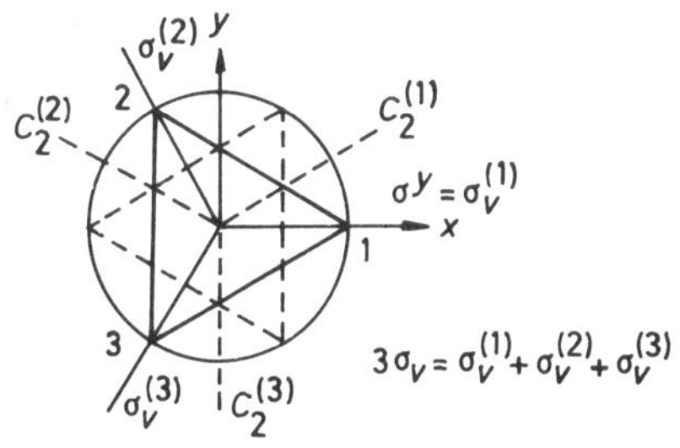

C_{3v} (solid lines)
D_{3d} (both solid and dotted lines)

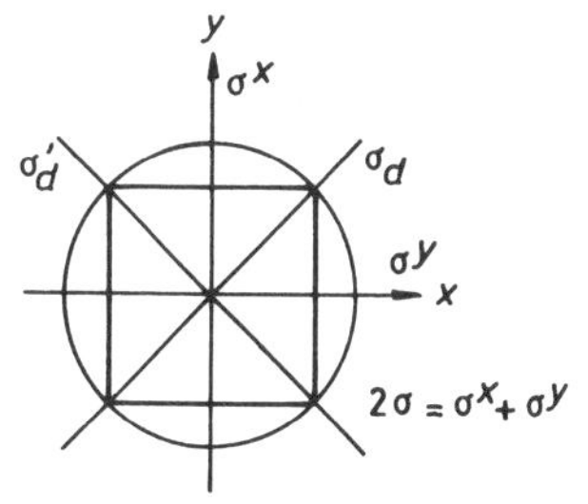

C_{4v}

Table 8.3-7. D_{4d}.

	$2S_8^z$	4σ	σ^y	eigenfunctions
A_1	2	4		x^2+y^2, z^2
A_2	2	-4		R_z
B_1	-2	-4		
B_2	-2	4		z
E_1	$\sqrt{2}$	0	$(1, -1)$	(x, y)
E_2	0	0	$(1, -1)$	$(x^2-y^2, -2xy)$
E_3	$-\sqrt{2}$	0	$(1, -1)$	$(R_y, -R_x)$ (xz, yz)

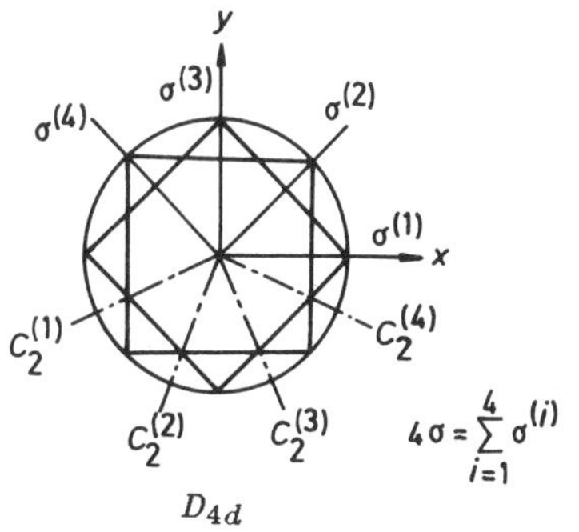

D_{4d}

Table 8.3-8. D_4, $D_{4h} = D_4 \times C_i$.

	$2C_2$	$2C_2'$	I	C_2^x	eigenfunctions
A_{1g}	2	2	1		x^2+y^2, z^2
A_{2g}	-2	-2	1		R_z
B_{1g}	2	-2	1		x^2-y^2
B_{2g}	-2	2	1		xy
E_g	0	0	1	$(1, -1)$	(R_x, R_y), $(yz, -xz)$
A_{1u}	2	2	-1		
A_{2u}	-2	-2	-1		z
B_{1u}	2	-2	-1		
B_{2u}	-2	2	-1		
E_u	0	0	-1	$(1, -1)$	(x, y)

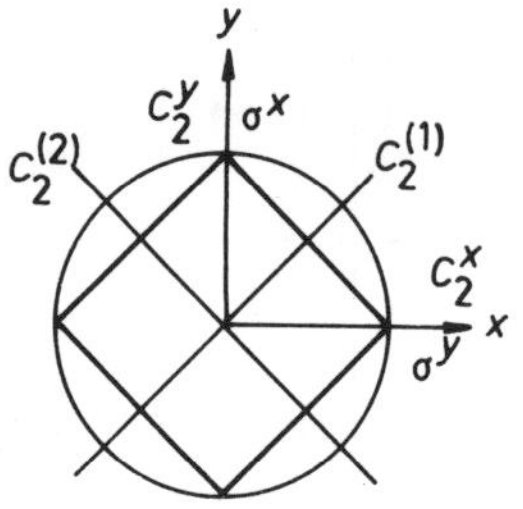

D_4 and D_{4h}

$2C_2 = C_2^x + C_2^y$

$2C_2' = C_2^{(1)} + C_2^{(2)}$

Table 8.3-9. D_5, $D_{5h} = D_5 \times C_s$.

	$2C_5^z$	$2C_z$	σ_h	C_2^x	eigenfunctions
A_1'	2	5	1		x^2+y^2, z^2
A_2'	2	-5	1		R_z
E_1'	$2\cos 72^\circ$	0	1	$(1\ -1)$	(x, y)
E_2'	$2\cos 144^\circ$	0	1	$(1, -1)$	$(x^2-y^2, -2xy)$
A_1''	2	5	-1		
A_2''	2	-5	-1		z
E_1''	$2\cos 72^\circ$	0	-1	$(1, -1)$	(R_x, R_y), $(yz, -xz)$
E_2''	$2\cos 144^\circ$	0	-1	$(1, -1)$	

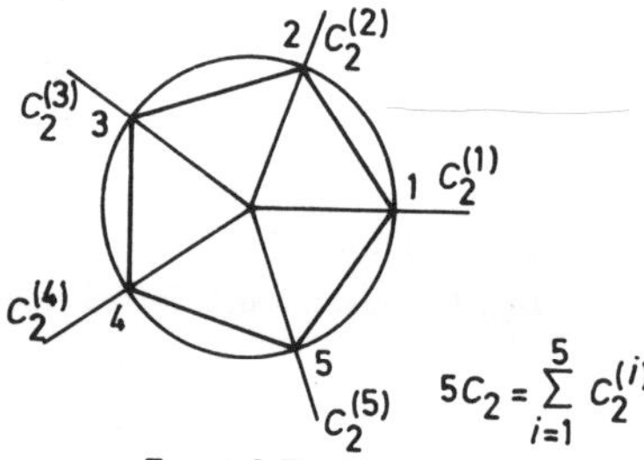

D_5 and D_{5h}

$5C_2 = \sum_{i=1}^{5} C_2^{(i)}$

Table 8.3-10. C_{5v}, $D_{5d} = C_{5v} \times C_i$.

	$2C_5^z$	$5\sigma_v$	I	σ_y	eigenfunctions
A_{1g}	2	5	1		x^2+y^2, z^2
A_{2g}	2	-5	1		R_z
E_{1g}	$2\cos 72^\circ$	0	1	$(1\ -1)$	$(R_y, -R_x)$, (xz, yz)
E_{2g}	$2\cos 144^\circ$	0	1	$(1, -1)$	$(x^2-y^2, -2xy)$
A_{2u}	2	5	-1		z
A_{1u}	2	-5	-1		
E_{1u}	$2\cos 72^\circ$	0	-1	$(1, -1)$	(x,y)
E_{2u}	$2\cos 144^\circ$	0	-1	$(1, -1)$	

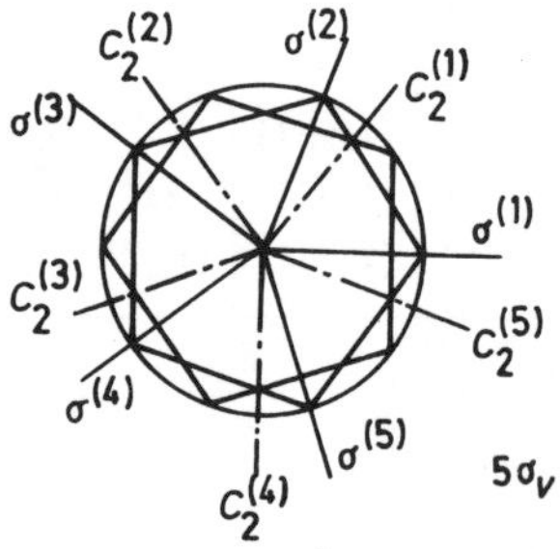

$5\sigma_v = \sum_{i=1}^{5} \sigma^{(i)}$

C_{5v} and D_{5d}

Table 8.3-11. $C_{6\nu}$.

	$2C_6^z$	$3\sigma_\nu$	σ^y	eigenfunctions
A_1	2	3		z, x^2+y^2, z^2
A_2	2	-3		R_z
B_1	-2	3		
B_2	-2	-3		
E_1	1	0	$(1, -1)$	$(x, y)(R_y, -R_x)$ (xz, yz)
E_2	-1	0	$(1, -1)$	$(x^2-y^2, -2xy)$

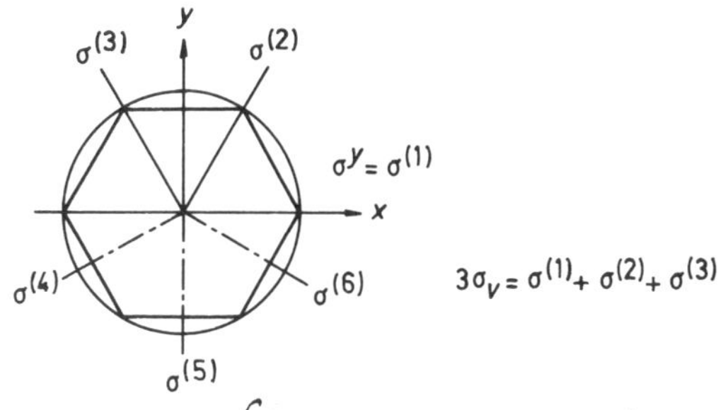

C_{6v}

Table 8.3-12. D_{6d}.

	$2S_{12}^z$	$6\sigma_d$	σ^y	eigenfunctions
A_1	2	6		x^2+y^2, z^2
A_2	2	-6		R_z
B_1	-2	-6		
B_2	-2	6		z
E_1	$\sqrt{3}$	0	$(1, -1)$	(x, y)
E_2	1	0	$(1, -1)$	$(x^2-y^2, -2xy)$
E_3	0	0	$(1, -1)$	
E_4	-1	0	$(1, -1)$	
E_5	$-\sqrt{3}$	0	$(1, -1)$	$(R_y, -R_x)$, (xz, yz)

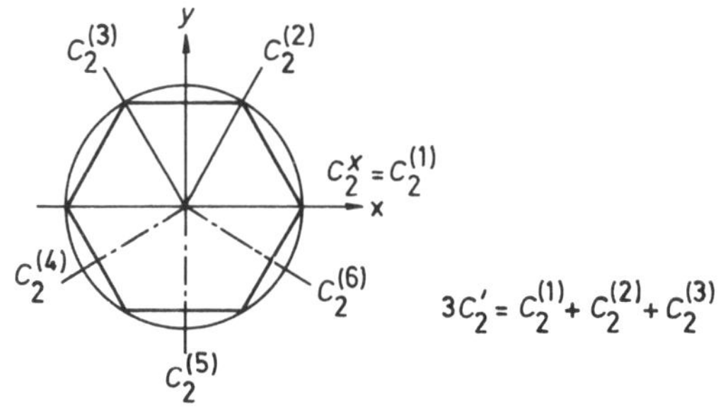

D_{6h}

Table 8.3-13. D_6, $D_{6h} = D_6 \times C_s$.[1,2,3]

	$2C_6^z$	$3C_2'$	C_2^x	eigenfunctions
A_{1g}	2	3		x^2+y^2, z^2
A_{2g}	2	-3		R_z
B_{1g}	-2	3		
B_{2g}	-2	-3		
E_{1g}	1	0	$(1, -1)$	(R_x, R_y), $(yz, -xz)$
E_{2g}	-1	0	$(1, -1)$	$(x^2-y^2, -2xy)$
A_{1u}	2	3		
A_{2u}	2	-3		z
B_{1u}	-2	3		
B_{2u}	-2	-3		
E_{1u}	1	0	$(1, -1)$	(x, y)
E_{2u}	-1	0	$(1, -1)$	

Table 8.3-14. $C_{\infty v}$, $D_{\infty v} = C_{\infty v} \times C_i$.

	$2C^z(\varphi)$	$\sigma^{(0)}$	I	eigenfunctions
Σ_g^+	1	1	1	z^2, x^2+y^2
Σ_g^-	1	-1	1	R_z
Π_g	$2\cos\varphi$	—	1	(R_x, R_y), (xz, yz)
Δ_g	$2\cos 2\varphi$	—	1	
⋮				
Σ_u^+	1	1	-1	z
Σ_u^-	1	-1	-1	
Π_u	$2\cos\varphi$	—	-1	(x, y) or $(e^{i\varphi}, e^{-i\varphi})$
Δ_u	$2\cos 2\varphi$	—	-1	

[1] Now there an infinite number of reflection planes $\sigma^{(\varphi)} = (\varphi = 0 - 2\pi)$. $\sigma^{(0)} = \sigma^y$.

[2] The bases of the two dimensional reps $\Pi, \Delta \ldots$, are linear combinations of those of the two one-dimensional reps complex conjugate to one another. $\Sigma^{\pm}$ denotes $\sigma^{(0)} = \pm 1$.

[3] The symbol "—" means that the basis is not an eigenvector of $\sigma^{(0)}$.

Table 8.3-15. T, $T_h = T \times C_i$.

	$4C_3$	I	(C_2^x, C_2^y)	eigenfunctions
A_g	4	1		$x^2+y^2+z^2$
E_g	$\begin{cases} e^{2\pi i/3} \\ e^{-2\pi i/3} \end{cases}$	1 1		$(2z^2-x^2-y^2,\ \sqrt{3}\,(x^2-y^2))$
T_g	0	1	$(1\ -1,\ -1\ 1,\ -1\ -1)$	(R_x, R_y, R_z), (yz, zx, xy)
A_u	4	-1		
E_u	$\begin{cases} e^{2\pi i/3} \\ e^{-2\pi i/3} \end{cases}$	-1 -1		
T_u	0	-1	$(1\ -1,\ -1\ 1,\ -1\ -1)$	(x, y, z)

Table 8.3-16. T_d.[1,2]

	C	$C(s)$	eigenfunctions	$C(s)$	eigenfunctions
	$6\sigma_d$	(C_2^x, C_2^y)	$T_d \supset D_2$ basis	S_4^z	$T_d \supset S_4$ basis
A_1	6	$(1, 1)$	F^{A_1}	1	F^{A_1}
A_2	-6	$(1, 1)$	F^{A_2}	-1	F^{A_2}
E	0	——	——	$(1, -1)$	$(2z^2-x^2-y^2,\ \sqrt{3}\,(x^2-y^2))$
T_1	-2	$((1, -1), (-1, 1), (-1, -1))$	(R_x, R_y, R_z)	$(-i, i, 1)$	$\left(-\sqrt{\frac{1}{2}}(R_x+iR_y), \sqrt{\frac{1}{2}}(R_x-iR_y), R_z\right)$
T_2	2	$((1, -1), (-1, 1), (-1, -1))$	(x, y, z), (yz, xz, xy)	$(-i, i, -1)$	$\left(-\sqrt{\frac{1}{2}}(x+iy), \sqrt{\frac{1}{2}}(x-iy), z\right)$

[1] $F^{A_1} = x^2+y^2+z^2$, $F^{A_2} = x^4(y^2-z^2) + y^4(z^2-x^2) + z^4(x^2-y^2)$, $R_x = x(y^2-z^2)$, cyclic in x, y, z.
[2] Here "—" means that the $T_d \supset D_2$ basis is not an eigenvector of (C_2^x, C_2^y).

Table 8.3-17. O, $O_h = O \times C_i$.[1]

	C		$C(s)$	eigenfunctions	$C(s)$	eigenfunctions
	$6C_2$	I	(C_2^x, C_2^y)	$O_h \supset D_2$ basis	C_4^z	$O_h \supset C_4$ basis
A_{1g}	6	1	$(1, 1)$	F^{A_1}	1	F^{A_1}
A_{2g}	-6	1	$(1, 1)$	F^{A_2}	-1	F^{A_2}
E_g	0	1	——	——	$(1, -1)$	$(2z^2-x^2-y^2,\ \sqrt{3}\,(x^2-y^2))$
T_{1g}	-2	1	$((1\ -1), (-1\ 1), (-1\ -1))$	(R_x, R_y, R_z)	$(-i, i, 1)$	$\left(-\sqrt{\frac{1}{2}}(R_x+iR_y), \sqrt{\frac{1}{2}}(R_x-iR_y), R_z\right)$
T_{2g}	2	1	$((1\ -1), (-1\ 1), (-1\ -1))$	(yz, xz, xy)	$(-i, i, -1)$	$\left(-\sqrt{\frac{1}{2}}(yz+ixz), \sqrt{\frac{1}{2}}(yz-ixz), xy\right)$
A_{1u}	6	-1	$(1\ 1)$	$xyz \cdot F^{A_2}$	1	$xyz \cdot F^{A_2}$
A_{2u}	-6	-1	$(1\ 1)$	xyz	-1	xyz
E_u	0	-1	——	——	$(1, -1)$	$(xyz(2z^2-x^2-y^2),\ \sqrt{3}\,xyz(x^2-y^2))$
T_{1u}	-2	-1	$((1\ -1), (-1\ 1), (-1\ -1))$	(x, y, z)	$(-i, i, 1)$	$\left(-\sqrt{\frac{1}{2}}(x+iy), \sqrt{\frac{1}{2}}(x-iy), z\right)$
T_{2u}	2	-1	$((1\ -1), (-1\ 1), (-1\ -1))$	$(x(y^2-z^2), y(z^2-x^2), z(x^2-y^2))$	$(-i, i, -1)$	

[1] See footnote of Table 8.3-16.

8.3.2. The CSCO-II for the commonly used point groups

The set of operators $C(s)$ in the CSCO-II of point groups can be chosen according to the practical problem concerned. For finite groups, usually the explicit form of irreducible bases in terms of the lowest possible polynomials of x, y and z are given. By looking at which operators' eigenfunctions they are, we can find the corresponding operator set $C(s)$.

For example, the 2-dimensional irreducible basis function of axial groups are usually chosen as (x, y). They are eigenfunctions of C_2^x or σ^y, with the eigenvalues $(1, -1)$,

$$C_2^x \begin{pmatrix} x \\ y \end{pmatrix} = \begin{pmatrix} x \\ -y \end{pmatrix}, \quad \sigma^y \begin{pmatrix} x \\ y \end{pmatrix} = \begin{pmatrix} x \\ -y \end{pmatrix} . \tag{8-31}$$

Therefore C_2^x or σ^y is the set of operator $C(s)$ for axial groups. Thus we have

$$G = D_n, \ D_{nh}, \ G(s) = C_2, \ C(s) = C_2^x \ .$$
$$G = C_{nv}, \ D_{nd}, \ G(s) = C_s, \ C(s) = \sigma^y \ .$$

For the group T_d, the three basis functions of the irrep T_2 are often chosen as (x, y, z). They are eigenfunctions of (C_2^x, C_2^y) associated with the eigenvalues $(1, -1)$, $(-1, 1)$ and $(-1, -1)$, respectively. Consequently, (C_2^x, C_2^y) is the $C(s)$ for the irrep T_2. From Table 8.3-1 we know that (C_2^x, C_2^y) is the CSCO of the group D_2. Therefore, (x, y, z) is the $T_d \supset D_2$ irreducible basis. Under certain circumstances, it is more suitable to use the complex functions $-\frac{1}{\sqrt{2}}(x + iy)$, $\frac{1}{\sqrt{2}}(x - iy)$ and z (i.e., rY_{11}, rY_{1-1} and rY_{10}) as the irreducible basis of the irrep T_2. According to (8-23), they are eigenfunctions of S_4^z with the eigenvalues $-i, i$, and -1, respectively. Thus S_4^z is our $C(s)$. Since S_4^z is the class operator of the cyclic group S_4, Y_{11}, Y_{1-1}, and Y_{10} are $T_d \supset S_4$ irreducible basis.

8.4. Irreducible Matrix Elements and Irreducible Basis of Point Groups G

We could use the method of Sec. 3.13 to get the irreducible matrices of point groups by decomposing the regular reps. However, with simple polynomials of x, y and z as basis, the irreducible bases of the point groups are known, as listed in Table 8.3. Using this fact, we can find some general formulas for the irreducible matrix elements of the point group without decomposing the regular rep.

8.4.1. Irreducible matrices

With the polar vector (x, y, z) or axial vector (R_x, R_y, R_z) as basis, the matrix representatives of the operators $C_2^x, C_2^y, C_2^z, \sigma^x, \sigma^y$, and σ^z are given by the following relations

$$C_2^x \begin{pmatrix} A_x \\ A_y \\ A_z \end{pmatrix} = \begin{pmatrix} A_x \\ -A_y \\ -A_z \end{pmatrix}, C_2^y \begin{pmatrix} A_x \\ A_y \\ A_z \end{pmatrix} = \begin{pmatrix} -A_x \\ A_y \\ -A_z \end{pmatrix}, C_2^z \begin{pmatrix} A_x \\ A_y \\ A_z \end{pmatrix} = \begin{pmatrix} -A_x \\ -A_y \\ A_z \end{pmatrix}, \tag{8-32a}$$

$$\sigma^x \begin{pmatrix} A_x \\ A_y \\ A_z \end{pmatrix} = \pm \begin{pmatrix} -A_x \\ A_y \\ A_z \end{pmatrix}, \sigma^y \begin{pmatrix} A_x \\ A_y \\ A_z \end{pmatrix} = \pm \begin{pmatrix} A_x \\ -A_y \\ A_z \end{pmatrix}, \sigma^z \begin{pmatrix} A_x \\ A_y \\ A_z \end{pmatrix} = \pm \begin{pmatrix} A_x \\ A_y \\ -A_z \end{pmatrix}, \tag{8-32b}$$

where in (8-32a), **A** is a polar or axial vector, while in (8-32b), the plus (minus) sign is for polar (axial) vector. (Note that $C_2^x = C_2^{(0)}, C_2^y = C_2^{(90°)}, \sigma^x = \sigma^{(90°)}, \sigma^y = \sigma^{(0)}$, and $\sigma^z = \sigma_h$).

The irreps of axial groups are at most two-dimensional. The possible irreducible bases are as follows:

1. The rep $D^{(1)}$ with (x, y) as basis. From (8-6), (8-7), (8-15) and (8-23) we obtain the following irreducible matrices,

$$D^{(1)}(R_z(\varphi)) = D^{(1)}(S_z(\varphi)) = \begin{pmatrix} \cos\varphi & -\sin\varphi \\ \sin\varphi & \cos\varphi \end{pmatrix} , \tag{8-33a}$$

$$D^{(1)}(C_2^{(\varphi)}) = D^{(1)}(\sigma^{(\varphi)}) = \begin{pmatrix} \cos 2\varphi & \sin 2\varphi \\ \sin 2\varphi & -\cos 2\varphi \end{pmatrix} . \tag{8-33b}$$

2. The rep $D^{(2)}$ with (R_x, R_y) as basis. Because the axial vectors (R_x, R_y) and the polar vectors (x, y) transform in the same way under rotations but differently under reflections [see (8-32)], we obtain from (8-33)

$$\begin{aligned} D^{(2)}(R_z(\varphi)) &= -D^{(2)}(S_z(\varphi)) = D^{(1)}(R_z(\varphi)) , \\ D^{(2)}(C_2^{(\varphi)}) &= -D^{(2)}(\sigma^{(\varphi)}) = D^{(1)}(C_2^{(\varphi)}) . \end{aligned} \tag{8-34}$$

3. The rep $D^{(3)}$ with $(R_y, -R_x)$ as basis. With σ^y as $C(s)$ and the axial vectors as basis, the first component of the two-dimensional irreducible basis should be taken as R_y, which is associated with the eigenvalue $+1$. The second component can be taken either as R_x or $-R_x$. For some groups, such as D_{2d}, C_{3v}, C_{4v}, etc., (x, y) and (R_y, R_x) belong to the same irrep. To ensure a consistent phase, the sign of R_x cannot be chosen arbitrarily. Instead, we have to choose $-R_y$. The reason is as follows: Using $D_{11}^{(3)} = D_{22}^{(2)}, D_{22}^{(3)} = D_{11}^{(2)}, D_{ij}^{(3)} = -D_{ji}^{(2)}$, for $(ij) = (12), (21)$, we obtain from (8-34) and (8-33)

$$\begin{aligned} D^{(3)}(R_z(\varphi)) &= -D^{(3)}(S_z(\varphi)) = D^{(1)}(R_z(\varphi)) , \\ -D^{(3)}(C_2^{(\varphi)}) &= D^{(3)}(\sigma^{(\varphi)}) = D^{(1)}(\sigma^{(\varphi)}) . \end{aligned} \tag{8-35}$$

We know that the groups D_{2d}, C_{3v}, C_{4v}, etc. do not contain the operations $C_2^{(\varphi)}$ and $S_z(\varphi)$. From (8-35) it is seen that for these groups, the reps carried by (x, y) and $(R_y, -R_x)$ are exactly the same, and this is the reason why we choose $(R_y, -R_x)$ as basis rather than (R_y, R_x).

4. The rep $D^{(4)}$ with $(x^2 - y^2, -2xy)$ as basis:

$$x^2 - y^2 = \frac{1}{\sqrt{2}}(Y_{22} + Y_{2-2}), \quad -2xy = \frac{i}{\sqrt{2}}(Y_{22} - Y_{2-2}) .$$

Using (8-33), we can derive the irreducible matrices with $(x^2 - y^2, -2xy)$ as basis. [The reason for choosing $(x^2 - y^2, -2xy)$ instead of $(x^2 - y, 2xy)$ as basis is the same as given in (8-35)]. For example, from (8-33a) we have

$$x' = x\cos\varphi + y\sin\varphi, \quad y' = -x\sin\varphi + y\cos\varphi .$$

Therefore

$$\begin{pmatrix} x'^2 - y'^2 \\ -2x'y' \end{pmatrix} = \begin{pmatrix} \cos 2\varphi & -\sin 2\varphi \\ \sin 2\varphi & \cos 2\varphi \end{pmatrix} \begin{pmatrix} x^2 - y^2 \\ -2xy \end{pmatrix} .$$

It follows that

$$D^{(4)}(R_z(\varphi)) = D^{(4)}(S_z(\varphi)) = \begin{pmatrix} \cos 2\varphi & \sin 2\varphi \\ -\sin 2\varphi & \cos 2\varphi \end{pmatrix} , \tag{8-36a}$$

$$D^{(4)}(C_2^{(\varphi)}) = D^{(4)}(\sigma^{(\varphi)}) = \begin{pmatrix} \cos 4\varphi & -\sin 4\varphi \\ -\sin 4\varphi & -\cos 4\varphi \end{pmatrix} . \tag{8-36b}$$

For certain groups, there might be two or three among the above four bases belonging to the same irrep. Examples can be found in Table 8.3-3, -4, -5, -6 and -11. Letting the angle φ in (8-33) to (8-36) be equal to some specific value determined by the group G, we immediately get the 2-dimensional irreps of G. For example, according to Table 8.3-5 and the associated diagram, we obtain the irrep E of the group $\mathcal{C}_{3v}$ with (x, y) or $(x^2 - y^2, -2xy)$ as basis

$$
\begin{gathered}
D^E(e) = \begin{pmatrix} 1 & 0 \\ 0 & 1 \end{pmatrix}, \quad D^{(E)}(C^z(120°)) = \begin{pmatrix} -1/2 & -\sqrt{3}/2 \\ \sqrt{3}/2 & -1/2 \end{pmatrix}, \\
D^E(C^z(240°)) = \begin{pmatrix} -1/2 & \sqrt{3}/2 \\ -\sqrt{3}/2 & -1/2 \end{pmatrix}, \\
D^E(\sigma^y) = \begin{pmatrix} 1 & 0 \\ 0 & -1 \end{pmatrix}, \quad D^{(E)}(\sigma^{(120°)}) = \begin{pmatrix} -1/2 & -\sqrt{3}/2 \\ -\sqrt{3}/2 & 1/2 \end{pmatrix}, \\
D^E(\sigma^{(240°)}) = \begin{pmatrix} -1/2 & \sqrt{3}/2 \\ \sqrt{3}/2 & 1/2 \end{pmatrix}.
\end{gathered} \tag{8-37}
$$

According to the transformation of the coordinates $(x, y, z)[(yz, xz, xy)]$ under the operation of the group O, we can obtain the irrep $T_1[T_2]$ of the group O. From Fig. 8.2-4, we get Table 8.4-1.

Table 8.4-1. The transformation of (x,y,z) under some operations of the group O.

R	C_2^{xy}	$C_2^{x\bar{y}}$	C_2^{xz}	$C_2^{x\bar{z}}$	C_2^{yz}	$C_2^{y\bar{z}}$	C_4^z
$R(xyz)$	$yx\bar{z}$	$\bar{y}\,\bar{x}\,\bar{z}$	$z\bar{y}x$	$\bar{z}\,\bar{y}\,\bar{x}$	$\bar{x}zy$	$\bar{x}\,\bar{z}\,\bar{y}$	$\bar{y}xz$

From the above table, we have

$$
C_4^z \begin{pmatrix} x \\ y \\ z \end{pmatrix} = \begin{pmatrix} -y \\ x \\ z \end{pmatrix} = \begin{pmatrix} 0 & -1 & 0 \\ 1 & 0 & 0 \\ 0 & 0 & 1 \end{pmatrix} \begin{pmatrix} x \\ y \\ z \end{pmatrix}.
$$

Therefore the representation matrix for C_4^z in the irrep T_1 is [cf. (2-59)]

$$
D^{T_1}(C_4^z) = \begin{pmatrix} 0 & -1 & 0 \\ 1 & 0 & 0 \\ 0 & 0 & 1 \end{pmatrix}.
$$

From Table 8.4-1 we can also get

$$
C_4^z \begin{pmatrix} yz \\ xz \\ xy \end{pmatrix} = \begin{pmatrix} xz \\ -yz \\ -yx \end{pmatrix} = \begin{pmatrix} 0 & 1 & 0 \\ -1 & 0 & 0 \\ 0 & 0 & -1 \end{pmatrix} \begin{pmatrix} yz \\ xz \\ xy \end{pmatrix}.
$$

Thus

$$
D^{T_2}(C_4^z) = \begin{pmatrix} 0 & 1 & 0 \\ -1 & 0 & 0 \\ 0 & 0 & -1 \end{pmatrix}.
$$

Analogously, we can obtain

$$D^{T_i}(C_2^{xy}) = (-1)^i \begin{pmatrix} 0 & -1 & 0 \\ -1 & 0 & 0 \\ 0 & 0 & 1 \end{pmatrix} , \quad D^{T_i}(C_2^{x\bar{y}}) = (-1)^i \begin{pmatrix} 0 & 1 & 0 \\ 1 & 0 & 0 \\ 0 & 0 & 1 \end{pmatrix} ,$$

$$D^{T_i}(C_2^{xz}) = (-1)^i \begin{pmatrix} 0 & 0 & -1 \\ 0 & 1 & 0 \\ -1 & 0 & 0 \end{pmatrix} , \quad D^{T_i}(C_2^{x\bar{z}}) = (-1)^i \begin{pmatrix} 0 & 0 & 1 \\ 0 & 1 & 0 \\ 1 & 0 & 0 \end{pmatrix} ,$$

$$D^{T_i}(C_2^{yz}) = (-1)^i \begin{pmatrix} 1 & 0 & 0 \\ 0 & 0 & -1 \\ 0 & -1 & 0 \end{pmatrix} , \quad D^{T_i}(C_2^{y\bar{z}}) = (-1)^i \begin{pmatrix} 1 & 0 & 0 \\ 0 & 0 & 1 \\ 0 & 1 & 0 \end{pmatrix} ,$$

$$D^{T_i}(C_4^{z}) = (-1)^i \begin{pmatrix} 0 & 1 & 0 \\ -1 & 0 & 0 \\ 0 & 0 & -1 \end{pmatrix} , \quad i = 1, 2, \tag{8-38a}$$

where the basis for the irrep $T_1[T_2]$ of O is $(x, y, z)[(yz, xz, xy)]$.

By using the irreducible basis listed in Table 8.3-16, as well as Eqs. (8-15) and (8-23), we can obtain the irreducible matrices of the six reflection planes σ^{ij} and S_4^z of the group T_d, which are again expressed by Eq. (8-38a) with the following substitutions:

$$C_2^{ij} \to \sigma^{ij}, \quad ij = xy, x\bar{y}, xz, x\bar{z}, yz, y\bar{z} \quad C_4^z \to \overline{S}_4^z .$$

However one should notice that $(x, y, z)[(yz, xz, xy)]$ carries the irrep $T_2[T_1]$ of the group T_d.

For later applications, we still need the following matrices for the irrep E of the groups T_d and O,

$$T_d: \quad D^E(\sigma^{x\bar{z}}) = \begin{pmatrix} -\frac{1}{2} & \frac{\sqrt{3}}{2} \\ \frac{\sqrt{3}}{2} & \frac{1}{2} \end{pmatrix} , \quad O: \quad D^E(C_2^{x\bar{z}}) = \begin{pmatrix} -\frac{1}{2} & \frac{\sqrt{3}}{2} \\ \frac{\sqrt{3}}{2} & \frac{1}{2} \end{pmatrix} . \tag{8-38b}$$

The irreducible matrices of the group O_h can be written down immediately from those of the group O, by using the following formulas,

$$D^{T_{ig}} = D^{T_i}, \quad D^{T_{iu}} = -D^{T_i}, i = 1, 2, \quad D^{E_g} = D^E, \quad D^{E_u} = -D^E , \tag{8-39}$$

8.4.2. Irreducible basis

The $G \supset G(s)$ irreducible basis $\psi_\kappa^{(\mu)}$ of a point group G satisfies the eigenequations

$$\begin{pmatrix} C \\ C(s) \end{pmatrix} \psi_\kappa^{(\mu)} = \begin{pmatrix} \mu \\ \kappa \end{pmatrix} \psi_\kappa^{(\mu)} . \tag{8-40}$$

According to Sec. 3.13, for irreps with dimension greater than one, it suffices to find the solution to (8-40) for one component of the irrep μ. Choosing appropriate operators T and using the irreducible matrices $D(T)$ given above, we can obtain the other components $\psi_\kappa^{(\mu)}$ of the same irrep μ from the following equation

$$\psi_{\kappa'}^{(\mu)} = \frac{1}{D_{\kappa'\kappa}^{(\mu)}(T)} [T - D_{\kappa\kappa}^{(\mu)}(T)] \psi_\kappa^{(\mu)} . \tag{8-41}$$

For the axial group, the operator T can be chosen as $C_2^{(\varphi)}$ or $\sigma^{(\varphi)}$, φ can take any permissible value so long as $D_{\kappa'\kappa}^{(\mu)}(C_2^{(\varphi)})$ or $D_{\kappa'\kappa}^{(\mu)}(\sigma^{(\varphi)})$ is nonzero.

From Table 8.3-16 (Table 8.3-17) it is seen that the third component of the irrep T_i in the $O \supset D_2(T_d \supset D_2)$ classification is also an eigenfunction of $C_4^z(S_4^z)$, which is the CSCO of $C_4(S_4)$. Therefore, for the group $O(T_d)$, we can choose $(C, C(s)) = (6C_2, C_4^z)[6\sigma, S_4^z)]$. By solving Eq. (8-40), we can obtain the first component of the irrep E, and the third component of the irrep T_i. By further using Eq. (8-42) given below, we can find the second component of the irrep E in the $O \supset C_4(T_d \supset S_4)$ classification, and the first and second components of the irrep T_i in the $O \supset D_2(T_d \supset D_2)$ classification (for the ordering of the components, see Tables 8.3-16 and 8.3-17).

Using Eq. (8-38), (8-41) can be put into the following explicit form tailored for the groups O and T_d.

The group T_d The group O

$$T_d \supset S_4^z\colon\ \psi_2^E = \sqrt{\tfrac{1}{3}}[2\sigma^{x\bar{z}} + 1]\psi_1^E\ , \qquad O \supset C_4^z\colon\ \psi_2^E = \sqrt{\tfrac{1}{3}}[2C_2^{x\bar{z}} + 1]\psi_1^E\ , \tag{8-42a}$$

$$T_d \supset D_2\colon\ \psi_1^{T_i} = (-1)^i \sigma^{x\bar{z}} \psi_3^{T_i}\ , \qquad O \supset D_2\colon\ \psi_1^{T_i} = (-1)^i C_2^{x\bar{z}} \psi_3^{T_i}, i = 1, 2, \tag{8-42b}$$

$$\psi_2^{T_i} = (-1)^i \sigma^{y\bar{z}} \psi_3^{T_i}\ . \qquad \psi_2^{T_i} = (-1)^i C_2^{y\bar{z}} \psi_3^{T_i}\ .$$

8.4.3. The splitting of atomic levels and $O_3 \supset G \supset G(s)$ basis

The symmetry of an atom placed in a crystal field with the point group symmetry G is lowered from O_3 to G, and its energy levels, originally characterized by l, will be splitted. According to (3-337d), the number of the sublevels is given by the subduction of the irrep $D^{(l)}$ of O_3 with respect to G,

$$D^{(l)} \downarrow G = \sum_{\mu} \oplus \tau_\mu D^{(\mu)}(G)\ , \tag{8-43a}$$

where τ_μ is determined by (3-250), i.e.,

$$\tau_\mu = \sum_i (g_i/g)\chi_i^{l\downarrow G}(\chi_i^\mu)^*\ . \tag{8-43b}$$

According to (6-64) and the fact that under the inversion $I, Y_{lm} \to (-)^l Y_{lm}$, we have

$$\chi^l(\varphi) = \frac{\sin(l+1/2)\varphi}{\sin(\varphi/2)} \times \begin{cases} 1 & \text{proper rotations}\ , \\ (-1)^l & \text{improper rotations}\ . \end{cases} \tag{8-44a}$$

Letting $\varphi = \pi/2$ in (8-44a) and noting that $\sigma^x = IC_2^x$, we have

$$\chi^l(C_2) = (-1)^l\ , \qquad \chi^l(\sigma) = 1\ . \tag{8-44b}$$

Setting φ to be some specific values compatible with the group elements of G, we can find from (8-44) the characters $\chi_i^{l\downarrow G}$ for the subduced rep of G.

Example 1: The splitting of atomic levels in the C_{4v} crystal field. The character $\chi^{l\downarrow C_{4v}}$ for the subduced rep of C_{4v} is listed in Table 8.4-2. From Table 8.4-2, and the simple character of C_{4v} found in Exercise 3.17, we can calculate the multiplicities τ_μ as listed in Table 8.4-3, which is called compatibility table.

Table 8.4-2. $\chi^{l\downarrow C_{4v}}$

l	e	C_2^z	$2C_4^z$	2σ	$2\sigma'$
0	1	1	1	1	1
1	3	−1	1	1	1
2	5	1	−1	1	1
3	7	−1	−1	1	1
4	9	1	1	1	1

Table 8.4-3. Compatibility table for reps of O_3 and the point group C_{4v}.

0	A_1
1	$A_1 + E$
2	$A_1 + B_1 + B_2 + E$
3	$A_1 + B_1 + B_2 + 2E$
4	$2A_1 + A_2 + B_1 + B_2 + 2E$

As mentioned in Sec. 3.19, the $O_3 \supset G \supset G(s)$ irreducible basis $\phi^l_{\mu\kappa}$ is the approximate (without considering the mixing of different orbitals l) perturbed wave function. The point group symmetry-adapted basis $\phi^l_{\mu\kappa}$ can be expanded in terms of $r^l Y_{lm}$, or simply in terms of Y_{lm}, since r is invariant,

$$\phi^{(l)}_{\mu\kappa} = \sum_{m=-l}^{l} a^{\mu}_{\kappa,m} Y_{lm} , \tag{8-45a}$$

where $a^{\mu}_{\kappa,m}$ are the $O_3 \downarrow G$ subduction coefficients, and are determined by

$$\sum_{m'} \left[\left\langle Y_{lm} \left| \begin{matrix} C \\ C(s) \end{matrix} \right| Y_{lm'} \right\rangle - \begin{pmatrix} \mu \\ \kappa \end{pmatrix} \delta_{mm'} \right] a^{\mu}_{\kappa,m'} = 0 . \tag{8-45b}$$

The coefficients $a^{\mu}_{\kappa,m}$ can also be calculated by the projection operator method and have been tabulated in Bradley (1972, Table 2.2-6, pp. 64-70).

The $O_3 \supset G \supset G(s)$ irreducible basis is also expressible in terms of homogeneous polynomials of x, y and z.

$$\phi^{(l)}_{\mu\kappa} = \sum_{\alpha,\beta=0,1}^{l} a^{\mu}_{\kappa,\alpha\beta} x^{\alpha} y^{\beta} z^{l-\alpha-\beta} , \tag{8-45c}$$

where the summation is restricted to $\alpha + \beta \le l$. Suppose that C is a class operator, $C = \Sigma_i R_i$. Then

$$C(x^{\alpha} y^{\beta} z^{l-\alpha-\beta}) = \sum_i (R_i x)^{\alpha} (R_i y)^{\beta} (R_i z)^{l-\alpha-\beta} . \tag{8-46}$$

Using (8-46) and the irreducible matrix elements (8-6), (8-7), (8-15), (8-23) and (8-38), we can find the matrix elements of the operators C and $C(s)$ in the basis $|\alpha\beta\rangle = x^{\alpha} y^{\beta} z^{l-\alpha-\beta}$. From the eigenequations

$$\sum_{\gamma\delta} \left[\left\langle \alpha\beta \left| \begin{matrix} C \\ C(s) \end{matrix} \right| \gamma\delta \right\rangle - \delta_{\alpha\gamma}\delta_{\beta\delta} \begin{pmatrix} \mu \\ \kappa \end{pmatrix} \right] a^{\mu}_{\kappa,\gamma\delta} = 0 , \tag{8-47}$$

we can find the coefficients $a^{\mu}_{\kappa,\alpha\beta}$.

Another way of constructing the basis (8-45) is to use the CG coefficients of the point group (see Sec. 8.5).

Example 2: The group C_n. The CSCO of C_n is

$$C = C_n = R_z(2\pi/n) = \exp(-2\pi i J_z/n) . \tag{8-48a}$$

It is easy to find the eigenfunctions of the operator C in the polar coordinate system,

$$\begin{gathered} C\phi^{(\mu)} = \lambda^{\mu}\phi^{(\mu)}, \ C = \exp\left(-\frac{2\pi}{n}\frac{\partial}{\partial\varphi}\right) , \\ \phi^{(\mu)} = \sqrt{\frac{1}{2\pi}} e^{i\mu\varphi}, \ \lambda^{\mu} = e^{-2\pi\mu i/n}, \ \mu = 0, \ \pm 1, \ \pm 2, \dots, \pm \left[\frac{n}{2}\right] , \end{gathered} \tag{8-48b}$$

where $[n/2]$ is the integer part of $n/2$. The one-dimensional irreducible basis (8-48b) is complex. For convenience, one usually combines the bases of two irreps complex conjugate to one another into two real basis vectors which carry a 2-dimensional reducible rep, denoted by the symbol E,

$$\phi_1^E = \cos\mu\varphi, \quad \phi_2^E = \sin\mu\varphi \ . \tag{8-48c}$$

For example, for $\mu = 1, (\phi_1^E, \phi_2^E) = (x, y)$; for $\mu = 2, (\phi_1^E, \phi_2^E) = (x^2 - y^2, -2xy), \ldots$

Example 3: The group D_n and D_{nh}. The CSCO-II of D_n or D_{nh} contains the following operators,

$$\begin{aligned} C = e^{-2\pi i J_z/n} + e^{2\pi i J_z/n} &= 2\cos(2\pi J_z/n) \ , \\ C(s) &= C_2^x \ . \end{aligned} \tag{8-49a}$$

In polar coordinates, under the operation $C_2^x, \varphi \to -\varphi$. Therefore

$$\left.\begin{aligned} \begin{pmatrix} C \\ C(s) \end{pmatrix} \cos\mu\varphi &= \begin{pmatrix} 2\cos(2\pi\mu/n) \\ 1 \end{pmatrix} \cos\mu\varphi \ , \\ \begin{pmatrix} C \\ C(s) \end{pmatrix} \sin\mu\varphi &= \begin{pmatrix} 2\cos(2\pi\mu/n) \\ -1 \end{pmatrix} \sin\mu\varphi \ , \end{aligned}\right. \qquad \mu = 1, 2, 3, \ldots, \left[\frac{n}{2}\right] \ . \tag{8-49b}$$

This shows that $(\cos\mu\varphi, \pm\sin\mu\varphi)$ carries a two-dimensional irrep E of D_n or D_{nh}. The sign in front of $\sin\mu\varphi$ is determined by the relative phase convention, as disccused before.

It is thus seen that a 2-dimensional irrep of D_n or D_{nh} decomposes into two irreps of the group $\mathcal{C}_n$, which are complex conjugate to one another. To turn things round, we can first find the irreducible bases of $\mathcal{C}_n$, and then pick its real and imaginary parts which give the 2-dimensional irreducible basis of D_n or D_{nh}.

Example 4: The group O_h. The CSCO-II of the group O_h is $(6C_2, I, C_4^z)$,

$$6C_2 = C_2^{xy} + C_2^{x\bar{y}} + C_2^{xz} + C_2^{x\bar{z}} + C_2^{yz} + C_2^{y\bar{z}} \ . \tag{8-50}$$

Let us consider the quadratic form of x, y and z. From (8-38a) and (8-46) one finds

$$(6C_2)xy = 2xy, \quad C_4^z xy = -xy \ .$$

Note that $(6C_2)xy \neq [(6C_2)x][(6C_2)y]$. Consulting Table 8.3-17, we know that xy is the third component of the irrep T_{2g}. Interchanging x, y and z cyclically, we get another two components. Finally we have

$$\begin{aligned} \phi_1^{T_{2g}} = yz = \frac{i}{\sqrt{2}}(Y_{21} + Y_{2-1}), \quad \phi_2^{T_{2g}} &= zx = -\frac{1}{\sqrt{2}}(Y_{21} - Y_{2-1}) \ , \\ \phi_3^{T_{2g}} = xy = \frac{-i}{\sqrt{2}}(Y_{22} - Y_{2-2}) \ , & \end{aligned} \tag{8-51}$$

where we ignored the factor r^2.

According to

$$\begin{aligned} (6C_2)x^2 = (6C_2)y^2 = (6C_2)z^2 &= 2(x^2 + y^2 + z^2) \ , \\ C_4^z x^2 = y^2, \quad C_4^z y^2 = x^2, \quad C_4^z z^2 &= z^2, \end{aligned}$$

we find

$$\phi^{A_{1g}} = (x^2 + y^2 + z^2) = Y_{00}, \quad \phi_1^{E_g} = (2z^2 - x^2 - y^2) = Y_{20} \ . \tag{8-52a}$$

From (8-42) and (8-38a) we get

$$\phi_2^{E_g} = \sqrt{\frac{1}{3}}(2C_2^{xz}+1)(2z^2-x^2-y^2) = \sqrt{3}(x^2-y^2) = \sqrt{\frac{1}{2}}(Y_{22}+Y_{2-2}) \,. \tag{8-52b}$$

Analogously, we can find eigenfunctions consisting of higher order polynomials of x, y and z, as listed in Table 8.3-17.

Tables 8.3-1 to 8.3-17 give the CSCO-II of the point groups along with their eigenfunctions. The CSCO-II of the groups G and $G \times C_i$ or $G \times C_s$ are listed in the same table. The remarks about the relation of the CSCO-I of G and $G \times C_i$ or $G \times C_s$ also applies to the CSCO-II. For example, from Table 8.3-5 we obtain the following table for the group C_{3v}.

Table 8.4-4. The CSCO-II of C_{3v} and its eigenfunctions.

	$3\sigma_v$	σ^y	
A_1	3		x^2+y^2, z^2
A_2	-3		z, R_z
E	0	$(1, -1)$	(x, y), $(R_y - R_x)$, (xz, yz), $(x^2-y^2, -2xy)$

Ex. 8.1. Use (8-33) to evaluate the rep matrices for the irrep E of C_{4v} and D_{4h}, and compare with the results of Exercise 3.14.

Ex. 8.2. From the third component xy for the irrep T_1 of the group T_d, use (8-24b) to find the first and second components.

Ex. 8.3. Find the compatibility table for the group C_{6v}, $l = 0, 1, 2, 3, 4, 5$ (the characters of C_{6v} are given in Table 3.11-2).

Ex. 8.4. Show that for C_{4v} and C_{6v} we have

$$D^{4n+k} = n \text{ reg } \oplus D^k, \quad (k = 0, 1, 2, 3) \quad \text{for } C_{4v}$$
$$D^{6n+k} = n \text{ reg } \oplus D^k, \quad (k = 0, 1, 2, 3, 4, 5) \quad \text{for } C_{6v}$$

where "reg" stands for the regular representation of the group.

8.5. The CG Coefficients of Point Groups

8.5.1. The CG series of point groups

The CG series of the 32 point groups,

$$D^{(\mu_1)} \times D^{(\mu_2)} = \sum_{\mu} \oplus(\mu_1\mu_2\mu) D^{(\mu)} \,, \tag{8-53a}$$

were given by Koster (1963). As an example, Table 8.5-1 gives the CG series for the groups T_d and O. Due to the relation $(\mu_1\mu_2\mu) = (\mu_2\mu_1\mu)$, only the right upper triangle of the table is shown. For groups with real characters, the coefficients $(\mu_1\mu_2\mu)$ satisfy the relation (3-276). If A is the identity rep, then

$$D^{(\mu)} \times D^A = D^{(\mu)} \,. \tag{8-53b}$$

Table 8.5-1. The CG series of the groups T_d and O.

	A_1	A_2	E	T_1	T_2
A_1	A_1	A_2	E	T_1	T_2
A_2		A_1	E	T_2	T_1
E			A_1+A_2+E	T_1+T_2	T_1+T_2
T_1				$A_1+E+T_1+T_2$	$A_2+E+T_1+T_2$
T_2					$A_1+E+T_1+T_2$

8.5.2. *The CG coefficients of point groups*

The CG coefficients of point groups are defined through the equation

$$\psi_{\kappa}^{(\mu)\tau}(x_1,x_2)=\sum_{\kappa_1\kappa_2} C^{(\mu)\tau\kappa}_{\mu_1\kappa_1,\mu_2\kappa_2}\varphi_{\kappa_1}^{(\mu_1)}(x_1)\varphi_{\kappa_2}^{(\mu_2)}(x_2)\ . \tag{8-54}$$

For the single-valued reps of the 32 crystal point groups, the multiplicity label τ is redundant. However, for the double-valued reps, the multiplicity can be larger than one (Koster 1963).

The CG coefficients of point groups can be easily calculated by the EFM [see (3-293)]. As an example, let us apply Eq. (3-293) to the Kronecker product $T_2\times T_2$ of the group T_d.

According to (3-294), the matrix elements of the CSCO-II, (C, S_4^z), of T_d in the uncoupled rep can be expressed as

$$\langle i'j'|C|ij\rangle=\sum_{\alpha} D^{(T_2)}_{i'i}(\sigma^{\alpha})D^{(T_2)}_{j'j}(\sigma^{\alpha})\ ,$$

$$\langle ii|C|jj\rangle=\langle ij|C|ji\rangle=\sum_{\alpha}[D^{(T_2)}_{ij}(\sigma^{\alpha})]^2\ , \tag{8-55a}$$

where the sum runs over $\alpha=(12),(13),(14),(23),(24)$, and (34);

$$\langle i'j'|S_4|ij\rangle=D^{(T_2)}_{i'i}(S_4^z)D^{(T_2)}_{j'j}(S_4^z)\ . \tag{8-55b}$$

From (8-38a) and (8-55) we obtain the representatives of C and S_4^z in the uncoupled rep, which are block diagonalized with the submatrices as

$$\begin{matrix} & |11\rangle|22\rangle|33\rangle \\ \mathbf{C}= & \begin{pmatrix} 2 & 2 & 2\\ 2 & 2 & 2\\ 2 & 2 & 2\end{pmatrix}\end{matrix}\ ,\quad \begin{matrix} & |11\rangle|22\rangle|33\rangle \\ \mathbf{S}_4^z= & \begin{pmatrix} 0 & 1 & 0\\ 1 & 0 & 0\\ 0 & 0 & 1\end{pmatrix}\end{matrix}\ , \tag{8-55c}$$

$$\begin{matrix} & |12\rangle|21\rangle \\ \mathbf{C}= & \begin{pmatrix} 0 & 2\\ 2 & 0\end{pmatrix}\end{matrix}\ ,\qquad \begin{matrix} & |12\rangle|21\rangle \\ \mathbf{S}_4^z= & \begin{pmatrix} 0 & -1\\ -1 & 0\end{pmatrix}\end{matrix}\ . \tag{8-55d}$$

By diagonalizing the two matrices in (8-55c) simultaneously and consulting Table 8.3-16, we obtain the following three eigenvectors,

$$\Psi^{A_1}=\sqrt{\frac{1}{3}}(|11\rangle+|22\rangle+|33\rangle)\ ,$$

$$\Psi_1^E=\sqrt{\frac{1}{6}}(-|11\rangle-|22\rangle+2|33\rangle)\ ,\quad \Psi_2^E=\sqrt{\frac{1}{2}}(|11\rangle-|22\rangle)\ . \tag{8-56a}$$

The relative phase of the basis vectors of the irrep E cannot be fixed from the eigenequations. To ensure a consistent relative phase, we take a solution, say Ψ_1^E, from (8-55c), and then use (8-42) and (8-38a) to find out Ψ_2^E, i.e.,

$$\Psi_2^E = \sqrt{\frac{1}{3}}(2\sigma^{x\bar{z}}+1)\sqrt{\frac{1}{6}}(-|11\rangle - |22\rangle + 2|33\rangle) = \sqrt{\frac{1}{2}}(|11\rangle - |22\rangle)\ .$$

Similarly, a simultaneous diagonalization of the two matrices in (8-55d) gives

$$\Psi_3^{T_1} = \sqrt{\frac{1}{2}}(|12\rangle - |21\rangle)\ , \quad \Psi_3^{T_2} = \sqrt{\frac{1}{2}}(|12\rangle + |21\rangle)\ . \tag{8-56b}$$

By changing 1, 2, and 3 cyclically, we obtain

$$\Psi_1^{T_1} = \sqrt{\frac{1}{2}}(|23\rangle - |32\rangle)\ , \quad \Psi_1^{T_2} = \sqrt{\frac{1}{2}}(|23\rangle + |32\rangle)\ ,$$
$$\Psi_2^{T_1} = \sqrt{\frac{1}{2}}(|31\rangle - |13\rangle)\ , \quad \Psi_2^{T_2} = \sqrt{\frac{1}{2}}(|31\rangle + |13\rangle)\ . \tag{8-56c}$$

The CG coefficients of the group O and T_d are exactly the same and are listed in Table 8.5-2.

The CG coefficients of point groups are generally used to couple irreducible bases of two particles or two systems. However, they can also be used to couple two irreducible bases of the same particle. Letting $x_1 = x_2 = x$, in (8-54), if the resulting expression is not zero, then it is an unnormalized irreducible basis of the group G,

$$\psi_\kappa^{(\mu)}(x) = \sum_{\kappa_1\kappa_2} C_{\mu_1\kappa_1,\mu_2\kappa_2}^{\mu,\kappa}\varphi_{\kappa_1}^{(\mu_1)}(x)\varphi_{\kappa_2}^{(\mu_2)}(x)\ . \tag{8-57a}$$

For example, noting $|1\rangle = x, |2\rangle = y, |3\rangle = z$, we have from (8-56)

$$\psi^{A_1} = \sqrt{\frac{1}{3}}(x^2+y^2+z^2), \quad \psi_1^E = \sqrt{\frac{1}{6}}(2z^2-x^2-y^2), \quad \psi_2^E = \sqrt{\frac{1}{2}}(x^2-y^2)\ ,$$
$$\psi_3^{T_1} = \sqrt{\frac{1}{2}}(xy-yx) = 0, \quad \psi_3^{T_2} = \sqrt{2}xy \quad \text{etc.} \tag{8-57b}$$

They are identical to the bases listed in Table 8.3-16, except for the normalization factors.

Consequently, with the known CG coefficients of a point group G, we can use (8-57a) to construct the $O_3 \supset G \supset G(s)$ irreducible bases $\psi_{\mu\kappa}^l$ of higher l from those of lower l.

The CG coefficients of the point groups have been tabulated by Koster (1963).

Ex. 8.5. Calculate the CG coefficients for the products $E \times E$ and $T_1 \times T_2$ of the group O.

Ex. 8.6. Using the CG coefficients for the product $E \times T_2$ of T_d, show that $\psi_{(T_2)3}^{l=3} = Y_{30}$. Then using (8-42) show that

$$\psi_{(T_2)1}^{l=3} = \frac{1}{4}[\sqrt{5}(Y_{33}+Y_{3-3}) - \sqrt{3}(Y_{31}+Y_{3-1})]\ ,$$
$$\psi_{(T_2)2}^{l=3} = -\frac{1}{4i}[\sqrt{5}(Y_{33}-Y_{3-3}) + \sqrt{3}(Y_{31}-Y_{3-1})]\ .$$

Ex. 8.7. Using Table 8.5-2, show that

$$\psi_{(T_1)3}^l = Y_{2l+1,0}\ , \quad \text{for the group } O\ ,$$
$$\psi_{(T_2)3}^l = Y_{2l+1,0}\ , \quad \text{for the group } T_d\ .$$

Table 8.5-2. The CG coefficients of the groups T_d and O.*

$E\times E$	(11)	(12)	(21)	(22)
A_1	1/2			1/2
A_2		1/2	−1/2	
$(E)1$	−1/2			1/2
$(E)2$		1/2	1/2	

$E\times T_1$	(11)	(21)
$(T_1)1$	−1/4	3/4
$(T_2)1$	−3/4	−1/4

$E\times T_1$	(12)	(22)
$(T_1)2$	−1/4	−3/4
$(T_2)2$	3/4	−1/4

$E\times T_1$	(13)	(23)
$(T_1)3$	1	
$(T_2)3$		1

$E\times T_2$	(11)	(21)
$(T_1)1$	−3/4	−1/4
$(T_2)1$	−1/4	3/4

$E\times T_2$	(12)	(22)
$(T_1)2$	3/4	−1/4
$(T_2)2$	−1/4	−3/4

$E\times T_2$	(13)	(23)
$(T_1)3$		1
$(T_2)3$	1	

$T_i\times T_i$	(11)	(22)	(33)
A_1	1/3	1/3	1/3
$(E)1$	−1/6	−1/6	2/3
$(E)2$	1/2	−1/2	0

$i = 1,2,$

$T_i\times T_i$	(23)	(32)
$(T_1)1$	1/2	−1/2
$(T_2)1$	1/2	1/2

$T_i\times T_i$	(13)	(31)
$(T_1)2$	−1/2	1/2
$(T_2)2$	1/2	1/2

$T_i\times T_i$	(12)	(21)
$(T_1)3$	1/2	−1/2
$(T_2)3$	1/2	1/2

*Here $-\frac{1}{2}$ should read $-\sqrt{1/2}$, etc.

8.6. Molecular Orbital Theory

1. A molecule is a many-electron system. One of the successful theories in describing molecules is the so called *molecular orbital theory*. The essence of the theory is as follows: Suppose that there is a molecule consisting of n atoms. Let us first strip several, say n, electrons off the molecule and then consider a single electron moving around the molecular skleton — an n-valent ion. One of the possible states of a single electron in the skeleton is termed a *molecular orbital.*

Let $\mathbf{R}_i, i = 1, 2, \ldots, n$ be the position vectors of the n nuclei and $\mathbf{r}$ be the position vector of the electron, as shown in Fig. 8.6. Suppose that $\phi^l_{\mu_2\kappa_2}$ is the atomic orbital wave function (single electron wave function) in the $O_3 \supset G \supset G(s)$ classification. The approximation obtained by choosing linear combinations of the n atomic orbitals as the molecular orbitals,

$$\psi = \sum_{i=1}^{n}\sum_{\kappa_2} b_i\phi^l_{\mu_2\kappa_2}(i), \quad \phi^l_{\mu_2\kappa_2}(i) \equiv \phi^l_{\mu_2\kappa_2}(\mathbf{r}-\mathbf{R}_i) , \tag{8-58}$$

is called the atomic orbital linear combination approximation for molecular orbitals. Equation (8-58) shows that the electron is not localized to a particular nucleus. Obviously ψ must be an irreducible basis $(\mu)\kappa$ of the point group G of the molecule,

$$\psi^{(\mu)\tau}_{\kappa} = \sum_{\kappa_2 i} b^{(\mu)\tau}_{\kappa,\kappa_2 i}\phi^l_{\mu_2\kappa_2}(\mathbf{r}-\mathbf{R}_i) , \tag{8-59}$$

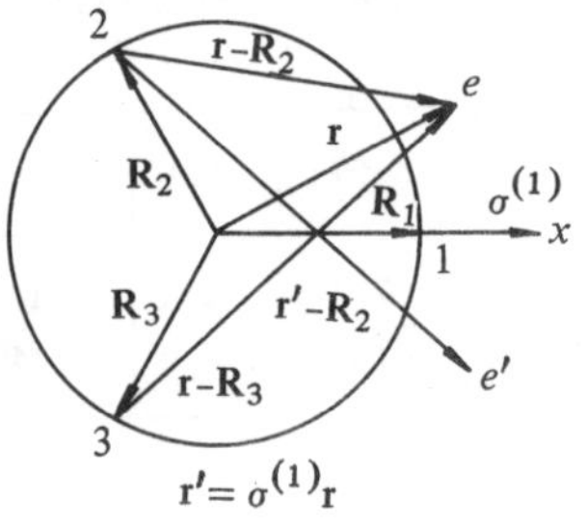

Fig. 8.6.

where τ is a multiplicity label. The last result is called the *symmetry adapted wave function*, usually designated as *SALC (symmetry adapted linear combination)*, or the *single electron SALC*.

2. The SALC (8-59) is not yet a molecular orbital wave function. We have to consider interactions between various atomic orbitals, i.e., to diagonalize the single electron Hamiltonian H in the basis (8-59). The molecular orbital wave function $\psi_\kappa^{(\mu)}(E)$ with energy E is a linear combination of $\psi_\kappa^{(\mu)\tau}$ in (8-59),

$$\psi_\kappa^{(\mu)}(E) = \sum_\tau a_{E,\tau} \psi_\kappa^{(\mu)\tau} . \tag{8-60a}$$

Due to (3-320), it suffices to diagonalize the Hamiltonian within the same irrep μ. The expansion coefficients $a_{E,\tau}$ satisfy the eigenequation

$$\sum_{\tau'} [\langle \psi^{(\mu)\tau} | H | \psi^{(\mu)\tau'} \rangle - \delta_{\tau\tau'} E] a_{E,\tau'} = 0 , \tag{8-60b}$$

where the matrix elements

$$\langle \psi^{(\mu)\tau} | H | \psi^{(\mu)\tau'} \rangle = \langle \psi_\kappa^{(\mu)\tau} | H | \psi_\kappa^{(\mu)\tau'} \rangle , \tag{8-60c}$$

are independent of the quantum number κ.

The order of the eigenequation (8-60b) is equal to the number of times that the irrep μ occurs. Therefore, in the case that the irrep μ occurs only once, the SALC (8-59) obtained merely from symmetry consideration is already an eigenfunction of the Hamiltonian H.

3. Many compounds, such as the complexes, have a metal atom in the center and several equivalent atoms, called *ligands*, surrounding the center. In such cases, the first step to construct the SALC is to combine linearly the orbitals of the surrounding atoms into the $G \supset G(s)$ irreducible bases (i.e., the SALC), and the second step is to find the $O_3 \supset G \supset G(s)$ irreducible basis for the central atom and, finally, consider the interaction among the various atomic orbitals, namely, combine linearly the SALC (possible more than one) of the surrounding atoms and the wave function of the central atom, all belonging to the same irrep μ. By solving the secular equation, we can obtain the molecular orbitals. An electron in any molecular orbital belongs to the surrounding atoms as well as to the central atom.

4. Having found the molecular orbitals, we now fill, in turn, the n electrons which were originally stripped off, to the molecular orbitals according to the ordering of the energy levels. Each energy level can accommodate at most $2h_\mu$ electrons, where h_μ is the dimension of the irrep μ, and the factor 2 comes from the two possible spin states of the electron. One of the possible ways of filling the molecular orbitals is called a configuration. The configuration with the lowest orbitals all occupied is called the ground state configuration. The others are called excited configurations. One configuration may correspond to several n-electron product wave functions.

5. Combine linearly these n-electron product wave functions into the $G \supset G(s)$ irreducible basis $\Phi_{\kappa}^{(\mu)\beta}$, called the *many-electron SALC*, β being the additional quantum numbers. $\Psi_{\kappa}^{(\mu)\beta}$ is an approximate molecule wave function, corresponding to the extreme single particle model in nuclear physics.

6. Further consider the interaction between the electrons. By diagonalizing the Hamiltonian in the basis $\Psi_{\kappa}^{(\mu)\beta}$ with the same μ and κ but different β (notice that β includes the configuration label), we can obtain a better approximation to the molecular wave functions. This corresponds to the configuration interaction approximation in atomic or nuclear physics.

From the above discussions, we see that the molecular orbital theory is very close to the atomic or nuclear shell model and can be termed as the *molecular shell model.* A more detailed discuslsion of point (5) will be given in Sections 9.11.

8.7. Single Electron SALC

Let us first consider the case when all the atomic orbitals are s orbitals (i.e., $l = 0$). Suppose that S is an element of the group G. From (2-58) we have

$$\begin{aligned} S\varphi(i) &= S\varphi(\mathbf{r}-\mathbf{R}_i) = \varphi(S^{-1}\mathbf{r}-\mathbf{R}_i) = \varphi(S^{-1}(\mathbf{r}-S\mathbf{R}_i)) \\ &= \varphi(\mathbf{r}-S\mathbf{R}_i) = \varphi(\mathbf{r}-\mathbf{R}_j) = \varphi(j) \ , \end{aligned} \tag{8-61a}$$

where we used the fact that the s orbital wave function only depends on the length of $\mathbf{r}-\mathbf{R}_i$, $\varphi(\mathbf{r}-\mathbf{R}_i) = \varphi(|\mathbf{r}-\mathbf{R}_i|)$. $\mathbf{R}_j = S\mathbf{R}_i$ is the position vector resulting from applying the operator S to the position vector $\mathbf{R}_i$ and is the position vector of the vertex j. From (8-61a) it is seen that under the group operations, the s-state atomic orbitals $\varphi(i)$ transform just as the vertices. For example, in Fig. 8.6-1, under the reflection $\sigma^{(1)}$, $\varphi(2)$ goes over to $\varphi(3)$,

$$\begin{aligned} \sigma^{(1)}\varphi(2) &= \sigma^{(1)}\varphi(\mathbf{r}-\mathbf{R}_2) = \varphi(\mathbf{r}'-\mathbf{R}_2) \\ &= \varphi(\sigma^{(1)}(\mathbf{r}-\sigma^{(1)}\mathbf{R}_2)) = \varphi(\mathbf{r}-\mathbf{R}_3) = \varphi(3) \ . \end{aligned} \tag{8-61b}$$

Assume that under the operation S, the vertex i goes over to j and let

$$D_{kj}(S) = \delta_{kj} \ . \tag{8-62a}$$

Thus

$$S\varphi(i) = \sum_j D_{ji}(S)\varphi(j) \ . \tag{8-62b}$$

Due to the overlap between atomic orbitals, $\varphi(i)$ are not orthogonal,

$$g_{ij} = \langle\varphi(i)|\varphi(j)\rangle \neq \delta_{ij} \ . \tag{8-63}$$

In quantum chemistry, the overlap integrals (i.e., the metric tensor) g_{ij} are usually designated S_{ij}. g_{ij} becomes δ_{ij} only when the atoms i and j are infinitely far apart from each other. According to (2-15), we can introduce the dual basis $\{\overline{\varphi}(i)\}$. The quantity $D_{ji}(S)$ in (8-62b) can be written as

$$D_{ji}(S) = \langle\overline{\varphi}(j)|S|\varphi(i)\rangle \ . \tag{8-64}$$

By using the isomorphism between the point group and the subgroup of a permutation group (such as Eq. (3-21) for the group C_{6v}, Table 8.2-2 and 8.2-3 for T_d and O, respectively), it is very easy to find $D_{ji}(S)$. Examples will be given below.

Since $\{\varphi(i)\}$ is not orthonormal, according to (2-121a), in general the rep carried by it is not unitary. However, the rep given by (8-62a) or (8-64) is still unitary,

$$D_{ij}(S^{-1}) = D_{ji}(S) = D^{\dagger}_{ij}(S) \ .$$

A diagonalization of the CSCO-II of G in the basis $\{\varphi(i)\}$ gives the irreducible basis of G,

$$\psi^{(\mu)\tau}_{\kappa} = \sum_{i} u^{(\mu)\tau}_{\kappa,i}\varphi(i) \ , \tag{8-65}$$

while the coefficients $u^{(\mu)\tau}_{\kappa,i}$ are determined by the eigenequations

$$\sum_{i}\left[\left\langle\overline{\varphi}(j)\left|\begin{matrix}C\\ C(s)\end{matrix}\right|\varphi(i)\right\rangle - \begin{pmatrix}\mu\\ \kappa\end{pmatrix}\delta_{ij}\right]u^{(\mu)\tau}_{\kappa,i} = 0 \ . \tag{8-66}$$

From (2-133a) and (2-136b) we have the orthonormal and completeness conditions for $u^{(\mu)\tau}_{\kappa,i}$,

$$\sum_{ij} u^{(\mu)\tau *}_{\kappa,i} g_{ij} u^{(\mu')\tau'}_{\kappa',j} = \delta_{\mu\mu'}\delta_{\tau\tau'}\delta_{\kappa\kappa'} \ , \tag{8-67a}$$

$$\sum_{\mu\tau\kappa i} u^{(\mu)\tau^*}_{\kappa,i} g_{ij} u^{(\mu)\tau}_{\kappa,l} = \delta_{jl} \ . \tag{8-67b}$$

Suppose that the overlaps between the atomic orbitals $\{\varphi(i)\}$ were negligible, i.e.,

$$\mathring{g}_{ij} = \langle\mathring{\varphi}(i)|\mathring{\varphi}(j)\rangle = \delta_{ij} \ .$$

In such a case, the SALC would be

$$\mathring{\psi}^{(\mu)\tau}_{\kappa} = \sum_{i} a^{(\mu)\tau}_{\kappa,i}\mathring{\varphi}(i) \ , \tag{8-68}$$

where $a^{(\mu)\tau}_{\kappa,i}$ is a unitary matrix. Because $a^{(\mu)\tau}_{\kappa,i}$ still satisfy (8-66), $u^{(\mu)\tau}_{\kappa,i}$ and $a^{(\mu)\tau}_{\kappa,i}$ differ only by a multiplicative factor,

$$u^{(\mu)\tau}_{\kappa,i} = N^{(\mu)\tau}_{\kappa} a^{(\mu)\tau}_{\kappa,i} \ . \tag{8-69}$$

Moreover, from the normalization we have

$$N^{(\mu)\tau}_{\kappa} = \left(\sum_{ij} a^{(\mu)\tau^*}_{\kappa,j} g_{ij} a^{(\mu)\tau}_{\kappa,j}\right)^{-1/2} \ . \tag{8-70}$$

Therefore

$$\psi^{(\mu)\tau}_{\kappa} = N^{(\mu)\tau}_{\kappa}\sum_{i} a^{(\mu)\tau}_{\kappa,i}\varphi(i) \ . \tag{8-71a}$$

Using the unitarity of $a^{(\mu)\tau}_{\kappa,i}$, we get

$$\varphi(i) = \sum_{(\mu)\tau\kappa}(a^{*(\mu)\tau}_{\kappa,i}/N^{(\mu)\tau}_{\kappa})\psi^{(\mu)\tau}_{\kappa} \ . \tag{8-72}$$

Comparing this with (2-137a), we find

$$v^{(\mu)\tau}_{\kappa,i} = (a^{(\mu)\tau}_{\kappa,i})^*/N^{(\mu)\tau}_{\kappa} \ . \tag{8-73a}$$

According to (2-134),

$$v_{\kappa,i}^{(\mu)\tau} = \sum_j g_{ij} u_{\kappa j}^{(\mu)\tau} = \sum_j g_{ij} N_\kappa^{(\mu)\tau} a_{\kappa j}^{(\mu)\tau} . \qquad \text{(8-73b)}$$

In the following, for simplicity, we assume that $g_{ij} = \delta_{ij}$, i.e., let

$$\psi_\kappa^{(\mu)\tau} = \sum_i a_{\kappa,i}^{(\mu)\tau} \varphi(i) . \qquad \text{(8-71b)}$$

In order to return to the nonorthogonal basis $\{\varphi(i)\}$, all we need to do is to multiply the right-hand side of (8-71b) by the coefficients $N_\kappa^{(\mu)\tau}$.

Since the group formed by the permutations of the vertices under the group G is isomorphic to a subgroup of a permutation group, the method for finding the irreducible basis of the permutation group can be used for finding the SALC (8-71a).

Example 1: The group C_{3v}. From the isomorphism (1-15) between the group C_{3v} and the permutation group S_3, according to (3-82) and (3-90), we can write down the eigenfunctions of $3\sigma = \sigma^{(1)} + \sigma^{(2)} + \sigma^{(3)}$, and $\sigma^y = \sigma^{(1)}$, i.e., the $C_{3v} \supset C_s$ irreducible basis,

$$\psi^{A_1} = \sqrt{\frac{1}{3}}[\varphi(1) + \varphi(2) + \varphi(3)] N^{A_1} , \qquad \text{(8-74)}$$

$$\psi_1^E = \sqrt{\frac{1}{6}}[2\varphi(1) - \varphi(2) - \varphi(3)] N_1^E ,$$

$$\psi_2^E = \sqrt{\frac{1}{2}}[\varphi(2) - \varphi(3)] N_2^E . \qquad \text{(8-75)}$$

Example 2: The group T_d. From Table 8.2-2 we can find the operation of the CSCO-II, $(6\sigma, S_4^z)$, of T_d on the function $\varphi(i)$,

$$C\varphi(i) = 3\varphi(i) + \varphi(j) + \varphi(k) + \varphi(l), \quad \text{cyclic in } i, j, k, l = 1, 2, 3, 4,$$

$$S_4^z \varphi(i) = \varphi(i+1), \quad i = 1, 2, 3, 4, \qquad \text{(8-76)}$$

with the convention that $\varphi(5) \equiv \varphi(1)$. Solving the eigenequation of (C, S_4^z), we get the irreducible bases,

$$\psi^{A_1} = \psi^{(6)} = \frac{1}{2}[\varphi(1) + \varphi(2) + \varphi(3) + \varphi(4)] ,$$

$$\psi_3^{T_2} = \psi_{-1}^{(2)} = \frac{1}{2}[\varphi(1) - \varphi(2) + \varphi(3) - \varphi(4)] . \qquad \text{(8-77a)}$$

From (8-42) and Table 8.2-2 we obtain the two other components of the irrep T_2,

$$\psi_1^{T_2} = (34)\psi_3^{T_2} = \frac{1}{2}[\varphi(1) - \varphi(2) - \varphi(3) + \varphi(4)] ,$$

$$\psi_2^{T_2} = (23)\psi_3^{T_2} = \frac{1}{2}[\varphi(1) + \varphi(2) - \varphi(3) - \varphi(4)] . \qquad \text{(8-77b)}$$

If an irrep μ occurs more than once (i.e., the eigenvalue (μ, κ) in (8-66) is degenerate), we can use the intrinsic operator set $\overline{C}(s)$ to distinguish the equivalent irreps (Chen [19] 1979).

We now introduce another method given by Fieck (1977) for finding the coefficients $a_{\kappa,i}^{(\mu)}$ in (8-71b). Fieck proved that apart from normalization factors, there is a simple relation between the values of the $O_3 \supset G \supset G(s)$ irreducible basis at the vertex i, $\phi^l_{\mu\kappa}(\mathbf{R}_i)$, and the coefficient $a^{(\mu)}_{\kappa,i}$,

$$a^{(\mu)l}_{\kappa,i} = \phi^{(l)}_{\mu\kappa}(\mathbf{R}_i) , \tag{8-78}$$

where $\mathbf{R}_i$ is the position vector of the vertex i. The multiplicity τ is larger than one if different choices of l in the right-hand side of (8-78) leads to different values of $a^{(\mu)l}_{\kappa,i}$, and in such cases, l can serve as the multiplicity label. Needless to say, to find $a^{(\mu)l}_{\kappa,i}$ from (8-78), it is most convenient to choose l as small as possible. Let us now prove (8-78).

From our assumption, $\phi^l_{\mu\kappa}$ transform according to the irrep $D^{(\mu)}$ under the point group G,

$$S\phi^{(l)}_{\mu\kappa}(\mathbf{r}) = \sum_{\kappa'} D^{(\mu)}_{\kappa'\kappa}(S)\phi^{(l)}_{\mu\kappa'}(\mathbf{r}) . \tag{8-79}$$

Letting $\mathbf{r} = \mathbf{R}_j$, we get

$$S\phi^{(l)}_{\mu\kappa}(\mathbf{R}_j) = \sum_{\kappa'} D^{(\mu)}_{\kappa'\kappa}(S)\phi^{(l)}_{\mu\kappa'}(\mathbf{R}_i) . \tag{8-80}$$

On the other hand, using (8-62b) and noting that D in (8-62b) is a real and symmetric matrix, the left-hand side of (8-80) can be put into the following form,

$$S\phi^{(l)}_{\mu\kappa}(\mathbf{R}_j) = \phi^{(l)}_{\mu\kappa}\left(\sum_i D_{ij}(S^{-1})\mathbf{R}_i\right) = \sum_i D_{ji}(S)\phi^{(l)}_{\mu\kappa}(\mathbf{R}_i) . \tag{8-81}$$

Comparing (8-80) with (8-81), we have

$$\sum_{\kappa'} D^{(\mu)}_{\kappa'\kappa}(S)\phi^{(l)}_{\mu\kappa'}(\mathbf{R}_j) = \sum_i D_{ji}(S)\phi^{(l)}_{\mu\kappa}(\mathbf{R}_i) . \tag{8-82}$$

It is this relation that we will use below for proving (8-78).

To prove (8-78), it suffices to show that

$$\psi^{(\mu)l}_{\kappa}(\mathbf{r}) = N\sum_i \phi^{(l)}_{\mu\kappa}(\mathbf{R}_i)\varphi(i) \tag{8-83}$$

is the $G \supset G(s)$ irreducible basis $(\mu)\kappa$, where N is a normalization constant. Noting that $\phi^l_{\mu\kappa}(\mathbf{R}_i)$ in (8-83) are coefficients and the group operation S only effects the function $\varphi(i)$, and using (8-82) we get

$$\begin{aligned} S\psi^{(\mu)l}_{\kappa}(\mathbf{r}) &= N\sum_{ij} \phi^{(l)}_{\mu\kappa}(\mathbf{R}_i)D_{ji}(S)\varphi(j) \\ &= N\sum_{\kappa' j} D^{(\mu)}_{\kappa'\kappa}(S)\varphi^{(l)}_{\mu\kappa'}(\mathbf{R}_j)\varphi(j) = \sum_{\kappa'} D^{(\mu)}_{\kappa'\kappa}(S)\psi^{(\mu)l}_{\kappa'}(\mathbf{r}) . \end{aligned} \tag{8-84}$$

It should be stressed that for the multiplicity-not-free cases, Eq. (8-83) with different l are not orthogonal.

Example 3: The group C_{3v} and T_d. According to the coordinates of the vertices in the diagram associated with Tables 8.3-5, and in Fig. 8.2-3, we obtain the values of x, y, and z listed in Tables 8.7-1 and 8.7-2, respectively. From Table 8.3-5 and Table 8.3-16 we get the $O_3 \supset G \supset G(s)$ irreducible basis $\phi^l_{\mu\kappa}(\mathbf{r})$. Then letting $\mathbf{r} = \mathbf{R}_1, \mathbf{R}_2, \ldots$, we obtain the values of $\phi^l_{\mu\kappa}(\mathbf{R}_i)$ listed in Tables 8.7-1 and 8.7-2.

Table 8.7-1. The values of $\phi^l_{\mu\kappa}$ $(\mathbf{R}_i)$ for the group C_{3v}.

$\phi^{(l)}_{\mu\varkappa}(\boldsymbol{r})$	$\phi^{(l)}_{\mu,\varkappa}(\boldsymbol{R}_1)$	$\phi^{(l)}_{\mu,\varkappa}(\boldsymbol{R}_2)$	$\phi^{(l)}_{\mu,\varkappa}(\boldsymbol{R}_3)$
$\phi^A=x^2+y^2$	1	1	1
$\phi^{l=1}_{E,1}=x$	1	$-1/2$	$-1/2$
$\phi^{l=1}_{E,2}=y$	0	$\sqrt{3}/2$	$-\sqrt{3}/2$
$\phi^{l=2}_{E,1}=x^2-y^2$	1	$-1/2$	$-1/2$
$\phi^{l=2}_{E,2}=-2xy$	0	$\sqrt{3}/2$	$-\sqrt{3}/2$

Table 8.7-2. The values of $\phi^l_{\mu\kappa}$ $(\mathbf{R}_i)$ for the group T_d.

$\phi^{(l)}_{\mu,\varkappa}(\boldsymbol{r})$	$\phi^{(l)}_{\mu\varkappa}(\boldsymbol{R}_1)$,	$\phi^{(l)}_{\mu\varkappa}(\boldsymbol{R}_2)$,	$\phi^{(l)}_{\mu\varkappa}(\boldsymbol{R}_3)$,	$\phi^{(l)}_{\mu\varkappa}(\boldsymbol{R}_4)$
$\phi_{A_1}=x^2+y^2+z^2$	3	3	3	3
$\phi^{l=1}_{T_2,1}=x$	1	-1	-1	1
$\phi^{l=1}_{T_2,2}=y$	1	1	-1	-1
$\phi^{l=1}_{T_2,3}=z$	1	-1	1	-1

Comparing Table 8.7-1 with Eq. (8-75), and Table 8.7-2 with Eq. (8-77), one sees that the EFM and Eq. (8-78) give identical results (including the phase). In Table 8.7-1, we deliberately listed the values of $\phi^{l=1}_{\mu\kappa}$ as well as $\phi^{l=2}_{\mu\kappa}$. The fact that they lead to the same values of $\phi^l_{\mu\kappa}(\mathbf{R}_i)$ shows that here the irrep E occurs only once.

Now we consider the cases with $l \neq 0$.

Suppose that the atomic orbital wave function $\phi^{l_2}_{\mu_2\kappa_2}(i) = \phi^{l_2}_{\mu_2\kappa_2}(\mathbf{r}-\mathbf{R}_i)$ belongs to the irreps l_2, μ_2, κ_2 of the group chain $O_3 \supset G \supset G(s)$. Let us first consider the action of the group elements S on the function

$$\begin{aligned} S\phi^{l_2}_{\mu_2\kappa_2}(i) &= \phi^{l_2}_{\mu_2\kappa_2}(S^{-1}\mathbf{r}-\mathbf{R}_i) = \phi^{l_2}_{\mu_2\kappa_2}(S^{-1}(\mathbf{r}-S\mathbf{R}_i)) \\ &= \sum_{\kappa_2'} D^{(\mu_2)}_{\kappa_2'\kappa_2}(S)\phi^{l_2}_{\mu_2\kappa_2'}(\mathbf{r}-S\mathbf{R}_i) \\ &= \sum_{\kappa_2'} D^{(\mu_2)}_{\kappa_2'\kappa_2}(S)\phi^{l_2}_{\mu_2\kappa_2'}(\mathbf{r}-\sum_j D_{ji}(S)\mathbf{R}_j) \ . \end{aligned} \tag{8-85}$$

Since in (8-85) there is only one nonvanishing term in the summation over j, the summation symbol can be shifted to the left,

$$S\phi^{l_2}_{\mu_2\kappa_2}(i) = \sum_{j\kappa_2'} D^{(\mu_2)}_{\kappa_2'\kappa_2}(S)D_{ji}(S)\phi^{l_2}_{\mu_2\kappa_2'}(j) \ . \tag{8-86}$$

This shows that $\phi^{l_2}_{\mu_2\kappa_2}(i)$ transform according to $D^{(\mu_2)} \otimes D$ under the group G.

Introducing

$$\varphi_{\mu_1\kappa_1,\mu_2\kappa_2} = \sum_i \phi^{l_1}_{\mu_1\kappa_1}(\mathbf{R}_i)\phi^{l_2}_{\mu_2\kappa_2}(i) \ , \tag{8-87}$$

and using (8-86) and (8-82), we obtain

$$\begin{aligned} S\varphi_{\mu_1\kappa_1,\mu_2\kappa_2} &= \sum_{ij\kappa_2'} \phi^{l_1}_{\mu_1\kappa_1}(\mathbf{R}_i)D^{(\mu_2)}_{\kappa_2'\kappa_2}(S)D_{ji}(S)\phi^{l_2}_{\mu_2\kappa_2'}(j) \\ &= \sum_{\kappa_2'\kappa_1'j} D^{(\mu_1)}_{\kappa_1'\kappa_1}(S)D^{(\mu_2)}_{\kappa_2'\kappa_2}(S)\phi^{l_1}_{\mu_1\kappa_1'}(\mathbf{R}_j)\phi^{l_2}_{\mu_2\kappa_2'}(j) \\ &= \sum_{\kappa_2'\kappa_1'} D^{(\mu_1)}_{\kappa_1'\kappa_1}(S)D^{(\mu_2)}_{\kappa_2'\kappa_2}(S)\varphi_{\mu_1\kappa_1',\mu_2\kappa_2'} \ . \end{aligned} \tag{8-88}$$

Therefore, under the group $G, \varphi_{\mu_1\kappa_1,\mu_2\kappa_2}$ transform according to $D^{(\mu_1)} \otimes D^{(\mu_2)}$. Using the CG coefficients of the point group, we immediately find the SALC

$$\psi^{(\mu)\tau}_{\kappa} = \sum_{\kappa_1\kappa_2} C^{(\mu)\theta,\kappa}_{\mu_1\kappa_1,\mu_2\kappa_2}\varphi_{\mu_1\kappa_1,\mu_2\kappa_2} \ , \tag{8-89a}$$

i.e.,

$$\psi_{\kappa}^{(\mu)\tau}(\mathbf{r}) = N \sum_{\kappa_1\kappa_2 i} C^{(\mu)\theta,\kappa}_{\mu_1\kappa_1,\mu_2\kappa_2} \phi^{l_1}_{\mu_1\kappa_1}(\mathbf{R}_i)\phi^{l_2}_{\mu_2\kappa_2}(\mathbf{r}-\mathbf{R}_i) , \tag{8-89b}$$

where N is a normalization factor and τ is the multiplicity label,

$$\tau = \mu_1, \mu_2, \theta, l_1 . \tag{8-90}$$

The choice of l_1 has been discussed after Eq. (8-78). Comparing (8-59) and (8-89), we know that the coefficients in (8-59) are

$$b^{(\mu)\tau}_{\kappa,\kappa_2 i} = N \sum_{\kappa_1} C^{(\mu)\theta,\kappa}_{\mu_1\kappa_1,\mu_2\kappa_2} \phi^{l_1}_{\mu_1\kappa_1}(\mathbf{R}_i) . \tag{8-91}$$

When μ_2 is one-dimensional, (8-89b) reduces to

$$\psi_{\kappa}^{(\mu)\tau}(\mathbf{r}) = N \sum_{i} \phi^{l_1}_{\mu_1\kappa_1}(\mathbf{R}_i)\phi^{l_2}_{\mu_2}(\mathbf{r}-\mathbf{R}_i) ,$$

$$(\mu) = (\mu_1) \times (\mu_2) . \tag{8-92}$$

When (μ_2) is an identity rep, (8-89) or (8-92) reduces to (8-83).

Example 4: The group T_d for the p atomic orbitals. Suppose that the four equivalent atomic orbitals are p orbitals. The p_x, p_y and p_z orbital wave functions are proportional to x, y and z, respectively. According to Table 8.3-16, they belong to the irrep T_2 of the group T_d. Therefore, in (8-89b) we should let $\{\phi^{l_2}_{\mu_2\kappa_2}(i)\} = \{\phi^{(p)}_{T_2\kappa_2}(i)\} = \{x(i), y(i), z(i)\}$. From (8-89b), Table 8.7-2 and Table 8.5-2, we obtain the SALC for the p orbital of the group T_d, as listed in Table 8.7-3.

Table 8.7-3. The SALC for the p orbitals of the group T_d, $|(\mu_1\mu_2)\mu,\kappa\rangle = N\sum_{\kappa_1\kappa_2 i} C^{\mu,\kappa}_{\mu_1\kappa_1,\mu_2\kappa_2}\phi^{l_1=1}_{\mu_1\kappa_1}(\mathbf{R}_i)\,\phi^{l_2=1}_{\mu_2\kappa_2}(i)$.

$(\mu_1, \mu_2)\mu, \kappa$	$x(1)$	$x(2)$	$x(3)$	$x(4)$	$y(1)$	$y(2)$	$y(3)$	$y(4)$	$z(1)$	$z(2)$	$z(3)$	$z(4)$
$(T_2, T_2)A_1$	$\sqrt{\frac{1}{12}}$	$-\sqrt{\frac{1}{12}}$	$-\sqrt{\frac{1}{12}}$	$\sqrt{\frac{1}{12}}$	$\sqrt{\frac{1}{12}}$	$\sqrt{\frac{1}{12}}$	$-\sqrt{\frac{1}{12}}$	$-\sqrt{\frac{1}{12}}$	$\sqrt{\frac{1}{12}}$	$-\sqrt{\frac{1}{12}}$	$\sqrt{\frac{1}{12}}$	$-\sqrt{\frac{1}{12}}$
$(T_2, T_2)E$, 1	$-\sqrt{\frac{1}{24}}$	$\sqrt{\frac{1}{24}}$	$\sqrt{\frac{1}{24}}$	$-\sqrt{\frac{1}{24}}$	$-\sqrt{\frac{1}{24}}$	$-\sqrt{\frac{1}{24}}$	$\sqrt{\frac{1}{24}}$	$\sqrt{\frac{1}{24}}$	$\sqrt{\frac{1}{6}}$	$-\sqrt{\frac{1}{6}}$	$\sqrt{\frac{1}{6}}$	$-\sqrt{\frac{1}{6}}$
E, 2	$\sqrt{\frac{1}{8}}$	$-\sqrt{\frac{1}{8}}$	$-\sqrt{\frac{1}{8}}$	$\sqrt{\frac{1}{8}}$	$-\sqrt{\frac{1}{8}}$	$-\sqrt{\frac{1}{8}}$	$\sqrt{\frac{1}{8}}$	$\sqrt{\frac{1}{8}}$				
$(T_2, T_2)T_1$, 1					$-\sqrt{\frac{1}{8}}$	$\sqrt{\frac{1}{8}}$	$-\sqrt{\frac{1}{8}}$	$\sqrt{\frac{1}{8}}$	$\sqrt{\frac{1}{8}}$	$\sqrt{\frac{1}{8}}$	$-\sqrt{\frac{1}{8}}$	$-\sqrt{\frac{1}{8}}$
T_1, 2	$\sqrt{\frac{1}{8}}$	$-\sqrt{\frac{1}{8}}$	$\sqrt{\frac{1}{8}}$	$-\sqrt{\frac{1}{8}}$					$-\sqrt{\frac{1}{8}}$	$\sqrt{\frac{1}{8}}$	$\sqrt{\frac{1}{8}}$	$-\sqrt{\frac{1}{8}}$
T_1, 3	$-\sqrt{\frac{1}{8}}$	$-\sqrt{\frac{1}{8}}$	$\sqrt{\frac{1}{8}}$	$\sqrt{\frac{1}{8}}$	$\sqrt{\frac{1}{8}}$	$-\sqrt{\frac{1}{8}}$	$-\sqrt{\frac{1}{8}}$	$\sqrt{\frac{1}{8}}$				
$(T_2, T_2)T_2$, 1					$\sqrt{\frac{1}{8}}$	$-\sqrt{\frac{1}{8}}$	$\sqrt{\frac{1}{8}}$	$-\sqrt{\frac{1}{8}}$	$\sqrt{\frac{1}{8}}$	$\sqrt{\frac{1}{8}}$	$-\sqrt{\frac{1}{8}}$	$-\sqrt{\frac{1}{8}}$
T_2, 2	$\sqrt{\frac{1}{8}}$	$-\sqrt{\frac{1}{8}}$	$\sqrt{\frac{1}{8}}$	$-\sqrt{\frac{1}{8}}$					$\sqrt{\frac{1}{8}}$	$-\sqrt{\frac{1}{8}}$	$-\sqrt{\frac{1}{8}}$	$\sqrt{\frac{1}{8}}$
T_2, 3	$\sqrt{\frac{1}{8}}$	$\sqrt{\frac{1}{8}}$	$-\sqrt{\frac{1}{8}}$	$-\sqrt{\frac{1}{8}}$	$\sqrt{\frac{1}{8}}$	$-\sqrt{\frac{1}{8}}$	$-\sqrt{\frac{1}{8}}$	$\sqrt{\frac{1}{8}}$				
$(A_1, T_2)T_2$, 1	$\frac{1}{2}$	$\frac{1}{2}$	$\frac{1}{2}$	$\frac{1}{2}$								
T_2, 2					$\frac{1}{2}$	$\frac{1}{2}$	$\frac{1}{2}$	$\frac{1}{2}$				
T_2, 3									$\frac{1}{2}$	$\frac{1}{2}$	$\frac{1}{2}$	$\frac{1}{2}$

Equation (8-89) gives a general method for finding the single electron SALC. It is most suitable for the cases with high l atomic orbitals. However, for some simple cases, it is more convenient to use the projection operator method, or the EFM. Below, we shall give some examples for the axial groups. Assume that the z axis is the principle axis and the atomic orbital is p_z.

1) The group $C_n(C_{nh})$. Suppose that n equivalent atoms sit on the vertices of a regular polygonal with n sides, such as the n Carbon atoms in the molecule C_nH_n with each carbon atom contributing one p_z orbital $\varphi(i)$.

From (8-48) it is known that the eigenvalues of the group operator, or the class operator, C_n^j, of the group C_n are

$$\lambda_j^{(\mu)} = \exp(-2\pi\mu j i/n) \ . \tag{8-93a}$$

According to (3-68), the character is $\chi_j^\mu = \lambda_j^\mu$. From (3-235a) and (8-93a), we obtain the projection operator of the group C_n,

$$P^{(\mu)} = \sum_{j=1}^{n} \exp(2\pi\mu j i/n) C_n^j \ , \tag{8-93b}$$

where we ignored a constant factor. Applying it to any atomic orbital, say $\varphi(1)$, and using

$$C_n\varphi(1) = \varphi(2), \quad C_n^j\varphi(1) = \varphi(j+1) \ , \tag{8-94}$$

with the convention $\varphi(n+1) \equiv \varphi(1)$, we find the irreducible basis of C_n,

$$\psi^{(\mu)} = \sqrt{\frac{1}{n}} \sum_{j=1}^{n} e^{i(j-1)\mu\beta}\varphi(j) \ , \tag{8-95}$$

$$\beta = 2\pi/n, \mu = 0, \pm 1, \pm 2, \dots, \pm \left[\frac{n}{2}\right] \ .$$

2) The group $D_n(D_{nh})$. In analogy with Sec. 8.4, by combining the two conjugate irreducible bases of (8-95) into two real functions, we obtain the two-dimensional irreducible basis of the group $D_n(D_{nh})$,

$$\psi_1^{E_\mu} = \sqrt{\frac{2}{n}} \sum_{j=1}^{n} \sin((j-1)\mu\beta)\varphi(j) \ , \tag{8-96a}$$

$$\mu = 1, 2, \dots, \left[\frac{n}{2}\right] \ .$$

$$\psi_2^{E_\mu} = \sqrt{\frac{2}{n}} \sum_{j=1}^{n} \cos((j-1)\mu\beta)\varphi(j) \ , \tag{8-96b}$$

Letting $\mu = 0$ (or $\mu = 0, n/2$, when n is even) we get one (or two, for even n) one-dimensional irreps) with the bases,

$$\psi^A = \sqrt{\frac{1}{n}} \sum_{j=1}^{n} \varphi(j) \ , \qquad \text{when } n \text{ is even or odd} \ , \tag{8-97a}$$

$$\psi^B = \sqrt{\frac{1}{n}} \sum_{j=1}^{n} (-)^{j+1}\varphi(j) \ , \quad \text{when } n \text{ is even.} \tag{8-97b}$$

We choose the two-fold axis C_2^x passing through the vertex 1 as the operator $C(s)$ (see the diagrams D_n and D_{nh} associated with Table 8.3). Since $\varphi(i)$ are p_z orbitals, we have

$$C_2^x\varphi(1) = -\varphi(1), \ C_2^x\varphi(i) = -\varphi(n-i+2), \ \sigma_h\varphi(i) = -\varphi(i) \ . \tag{8-98}$$

Consequently, $C_2^x\psi_1^{E_\mu} = \psi_1^{E_\mu}, C_2^x\psi_2^{E_\mu} = -\psi_2^{E_\mu}$, i.e., $\psi_1^{E_\mu}$ and $\psi_2^{E_\mu}$ are the first and second components of the irrep E. (If $\varphi(i)$ are the s orbitals, the order is reversed, i.e., $\psi_1^{E_\mu}$ and $\psi_2^{E_\mu}$ in (8-96) are the second and first components, respectively.)

For example, letting $n = 5, \beta = 72°$, we obtain the SALC for the group D_{5h},

$$\begin{aligned}
\psi_1^{E_1''} &= \sqrt{\frac{2}{5}}[\sin\beta(\varphi(2)-\varphi(5)) + \sin 2\beta(\varphi(3)-\varphi(4))] \,, \\
\psi_2^{E_1''} &= \sqrt{\frac{2}{5}}[\varphi(1) + \cos\beta(\varphi(2)+\varphi(5)) + \cos 2\beta(\varphi(3)+\varphi(4))] \,, \\
\psi_1^{E_2''} &= \sqrt{\frac{2}{5}}[\sin 2\beta(\varphi(2)-\varphi(5)) - \sin\beta(\varphi(3)-\varphi(4))] \,, \\
\psi_2^{E_2''} &= \sqrt{\frac{2}{5}}[\varphi(1) + \cos 2\beta(\varphi(2)+\varphi(5)) + \cos\beta(\varphi(3)+\varphi(4))] \,, \\
\psi^{A_2''} &= \sqrt{\frac{1}{5}}\sum_{i=1}^{5}\varphi(i) \,.
\end{aligned} \tag{8-99}$$

For $n = 6, \beta = 60°$, we have the SALC for the group D_{6h},

$$\begin{aligned}
\psi^{A_{2u}} &= \sqrt{\frac{1}{6}}\sum_{i=1}^{6}\varphi(i) \,, \\
\psi^{B_{1g}} &= \sqrt{\frac{1}{6}}[\varphi(1)-\varphi(2)+\varphi(3)-\varphi(4)+\varphi(5)-\varphi(6)] \,, \\
\psi_1^{E_{1g}} &= \frac{1}{2}[\varphi(2)+\varphi(3)-\varphi(5)-\varphi(6)] \,, \\
\psi_2^{E_{1g}} &= \sqrt{\frac{1}{12}}[2\varphi(1)+\varphi(2)-\varphi(3)-2\varphi(4)-\varphi(5)+\varphi(6)] \,, \\
\psi_1^{E_{2u}} &= \frac{1}{2}[\varphi(2)-\varphi(3)+\varphi(5)-\varphi(6)] \,, \\
\psi_2^{E_{2u}} &= \sqrt{\frac{1}{12}}[2\varphi(1)-\varphi(2)-\varphi(3)+2\varphi(4)-\varphi(5)-\varphi(6)] \,.
\end{aligned} \tag{8-100}$$

It is still not easy to use (8-60b) to find the molecular orbitals. Further approximations are needed. The crudiest one is the *Hückel approximation*, which contains the following three approximations:

1. The atomic orbital are orthogonal,

$$g_{ij} = \langle\varphi(i)|\varphi(j)\rangle = \delta_{ij} \tag{8-101}$$

2. The diagonal elements of H are equal to a constant,

$$\langle\varphi(i)|H|\varphi(i)\rangle = E_0 \,. \tag{8-102}$$

3. Only the nondiagonal elements of H between the neighboring atomic orbitals are not zero, and are equal to $-F$,

$$\langle\varphi(i)|H|\varphi(j)\rangle = \begin{cases} -F & \text{when } i \text{ and } j \text{ are neighbouring orbitals,} \\ 0 & \text{otherwise ,} \end{cases} \tag{8-103}$$

where the constant $F > 0$. The Hückel approximation is a very crude one. Nevertheless, it has met with success. Though it is not reliable quantitatively, it often gives correct qualitative results.

Example 1: Find the energy levels of the molecule C_nH_n. From (8-96) we have

$$\begin{aligned}E_\mu &= \frac{2}{n}\left\langle \sum_{j=1}^{n} \cos((j-1)\mu\beta)\varphi(j)|H|\sum_{j'=1}^{n}\cos((j'-1)\mu\beta)\varphi(j')\right\rangle \\ &= E_0 - \frac{4}{n}F\sum_{j=1}^{n}\cos((j-1)\mu\beta)\cos(j\mu\beta) \\ &= E_0 - \frac{4}{n}F[\cos(\mu\beta)\sum_{j=1}^{n}\cos^2(j\mu\beta) + \sin\beta\sum_{j=1}^{n}\sin(j\mu\beta)\cos(j\mu\beta)] \\ &= E_0 - 2F\cos(\mu\beta)\ . \end{aligned} \tag{8-104}$$

The energy levels of (8-104) for D_{nh} molecular orbitals are shown in Fig. 8.7-1, while those for D_{6h} are shown in Fig. 8.7-2. In these diagrams, we use small letters to denote the irrep labels.

There is a simple method for drawing such energy levels: Draw a regular polygon with n sides, as shown in Fig. 8.7-2. Then the lines connecting the two vertices on the same level give the energy levels.

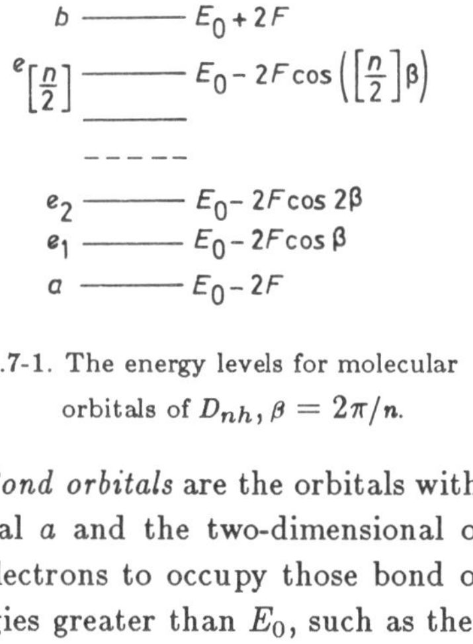

Fig. 8.7-1. The energy levels for molecular orbitals of D_{nh}, $\beta = 2\pi/n$.

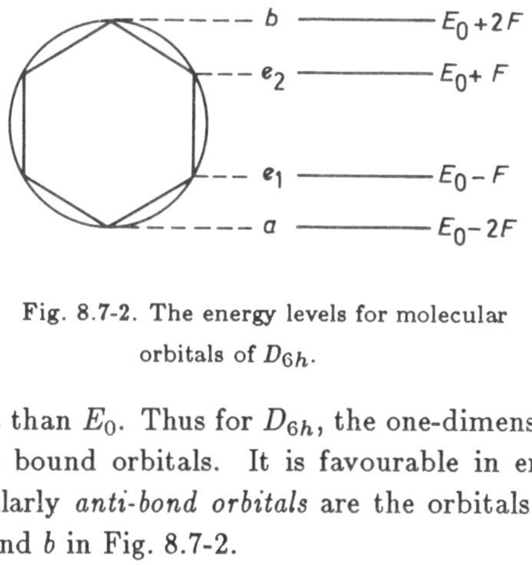

Fig. 8.7-2. The energy levels for molecular orbitals of D_{6h}.

Bond orbitals are the orbitals with energies less than E_0. Thus for D_{6h}, the one-dimensional orbital a and the two-dimensional orbital e_1 are bound orbitals. It is favourable in energy for electrons to occupy those bond orbitals. Similarly *anti-bond orbitals* are the orbitals with energies greater than E_0, such as the orbitals e_2 and b in Fig. 8.7-2.

Example 2: The ground state energy of the benzene molecule. Figure 8.7-2 gives the energy levels of a single electron moving around a benzene molecule with six electrons stripped off. Let us now put the six electrons on the lowest orbitals in Fig. 8.7-2 and obtain the ground state configuration $(a)^2(e_1)^4$. The energy of the ground state is

$$E_g = 2(E_0 - 2F) + 4(E_0 - F) = 6E_0 - 8F\ .$$

We see that the energy is lowered by the amount $8F$ due to the interaction between the atomic orbitals. This is the reason why the molecule benzene is very stable.

Example 3: The naphthalene molecule. The naphthalene molecule has the symmetry D_{2h}, as shown in Fig. 8.7-3. Each atom in the vertices contributes one p_z orbital. It is readily seen that under the group D_{2h}, the atomic orbitals $\varphi(1), \varphi(2), \ldots, \varphi(10)$ fall into three groups, $(\varphi(1), \varphi(4), \varphi(5), \varphi(8))$, $(\varphi(2), \varphi(3), \varphi(6), \varphi(7))$ and $(\varphi(9), \varphi(10))$, each transforming among themselves. According to Table 8.3-1, the CSCO of D_{2h} is (C_2^x, C_2^y, I). It is easy to find their simultaneous eigenfunctions, and the result is shown in Table 8.7-4.

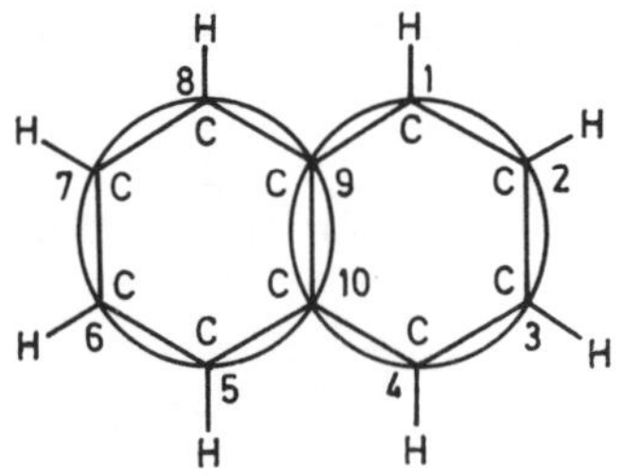

Fig. 8.7-3. The naphthalene molecule with the symmetry D_{2h}.

Table 8.7-4. The SALC for naphthalene $C_{10}H_8$.

C_2^x	C_2^y	I		$\varphi(2)$ $\varphi(1)$	$\varphi(3)$ $\varphi(4)$	$\varphi(6)$ $\varphi(5)$	$\varphi(7)$ $\varphi(8)$	$\varphi(9)$	$\varphi(10)$
-1	-1	-1	$\psi^{B_{1u}}$	$\frac{1}{2}$	$\frac{1}{2}$	$\frac{1}{2}$	$\frac{1}{2}$		
1	1	-1	ψ^{A_u}	$\frac{1}{2}$	$-\frac{1}{2}$	$\frac{1}{2}$	$-\frac{1}{2}$		
-1	1	1	$\psi^{B_{2g}}$	$\frac{1}{2}$	$\frac{1}{2}$	$-\frac{1}{2}$	$-\frac{1}{2}$		
1	-1	1	$\psi^{B_{3g}}$	$\frac{1}{2}$	$-\frac{1}{2}$	$-\frac{1}{2}$	$\frac{1}{2}$		
1	-1	1	$\psi^{B_{3g}}$					$\sqrt{\frac{1}{2}}$	$-\sqrt{\frac{1}{2}}$
-1	-1	-1	$\psi^{B_{1u}}$					$\sqrt{\frac{1}{2}}$	$\sqrt{\frac{1}{2}}$

8.8. Double Point Group

In the previous sections, we considered only the single-valued reps of point groups. However, it is often necessary to take into account the spin degree of freedom of electrons if the spin-orbit coupling is not negligible. We then need to consider the spinor reps, or double-valued reps of the point group, instead of the vector, or single-valued reps. In Chapter 6 we introduced the double rotation group $SO_3^\dagger$ for obtaining the spinor reps of the rotation group. Similarly, we introduce the double point group $\mathbf{G}^\dagger$. In the following we use boldfaced $\mathbf{G}$ to denote an abstract group, while the ordinary G will denote the matrix group formed by the representation matrices of the elements of $\mathbf{G}$.

Any point group is isomorphic either to a pure rotation point group, or isomorphic to the direct product of a pure rotation point group and the inversion group (e, I). Consequently, it suffices to consider the double point group of the pure rotation group.

Corresponding to the point group

$$\mathbf{G} = \{\gamma_i : i = 1, 2, \ldots, |G|\}, \quad \gamma_i = R_{\mathbf{n}_i}(\varphi_i) , \tag{8-105a}$$

we have the double point group

$$\mathbf{G}^\dagger = \{\gamma_i, \tilde{\gamma}_i : i = 1, 2, \ldots, |G|\}, \quad \tilde{\gamma}_i = R_{\mathbf{n}_i}(\varphi_i + 2\pi) . \tag{8-105b}$$

The double point group is a subgroup of the double rotation group $\mathbf{SO}_3^\dagger$ [(6-1d)]. The latter can also be expressed as

$$\mathbf{SO}_3^\dagger = \{\pm D^{1/2}(\alpha\beta\gamma) : R(\alpha\beta\gamma) \in \mathbf{SO}_3\} , \tag{8-106a}$$

while $\mathbf{G}^\dagger$ as

$$\mathbf{G}^\dagger = \{\pm D^{1/2}(\alpha\beta\gamma) : R(\alpha\beta\gamma) \in \mathbf{G}\} . \tag{8-106b}$$

The group table of $G^\dagger$ can be constructed in the following way: first determine the Euler angles α, β and γ for each element of the point group G; then find their matrix representatives $D^{1/2}(\alpha\beta\gamma)$ (see Table 6.1, in Bradley 1972, pp. 422). The multiplication table of these matrices gives the group table of $G^\dagger$. The multiplication table for D_2^+ is given below where we use $-\gamma_i$ to denote $\tilde{\gamma}_i$.

Example: $D_2^\dagger$.

Table 8.8-1. Group multiplication table of $D_2^\dagger$.

	e	C_{2x}	C_{2y}	C_{2z}
	γ_1	γ_2	γ_3	γ_4
$(\alpha\beta\gamma)$:	(000)	$(0\pi\pi)$	$(0\pi 0)$	(00π)
$D^{1/2}(\alpha\beta\gamma)$:	$\begin{pmatrix}1&0\\0&1\end{pmatrix}$	$\begin{pmatrix}0&-i\\-i&0\end{pmatrix}$	$\begin{pmatrix}0&-1\\1&0\end{pmatrix}$	$\begin{pmatrix}-1&0\\0&i\end{pmatrix}$
	1	2	3	4
	2	-1	4	-3
	3	-4	-1	2
	4	3	-2	-1

Remarks: Note that $(\gamma_1, \gamma_2, \gamma_3, \gamma_4)$ does not form a group. One quarter of the group table of $G^\dagger$ is enough. From this, it is easy to form the other three quarters of the group table by inspection.

The group tables of the point group $O^\dagger$ and $D_6^\dagger$ are given in Bradley (1972, Table 6.2, pp. 423), and the former is reproduced in Table 10.24-1. The correspondence between the McWeeny notation and Bradley notation for the point group operators is also shown in Table 10.24-1. In the following we switch to Bradley's notation.

There is a set of rules derived by Opechowski governing the separation of the group $G^\dagger$ into classes, and the evaluation of the character table of $G^\dagger$ (see Bradley 1972, pp. 420). However, these rules are of little use for constructing the irreducible matrices.

Usually, the double point group $G^\dagger$ is regarded as an abstract group of order $2|G|$, and the irreps of $G^\dagger$ can be found in the usual way. Among the irreps of $G^\dagger$, some are identical to those of the point group G, which are just the single-valued reps of G, while others are not included in the irreps of G, which are the double-valued reps of G.

For example, the group $D_2^\dagger$ is of order 8, and has five classes with the class operators,

$$C_1 = e, \quad C_2 = \tilde{e}, \quad C_3 = C_{2x} + \tilde{C}_{2x}, \quad C_4 = C_{2y} + \tilde{C}_{2y}, \quad C_5 = C_{2z} + \tilde{C}_{2z} \,. \tag{8-107}$$

Using the EFM of Chapter 3, we can find its characters and irreps listed in Tables 8.8-2 and 8.8-3. It is seen that $D_2^\dagger$ has five irreps: the four one-dimensional reps are the single-valued reps of D_2, for which $D(\tilde{\gamma}_i) = D(\gamma_i)$, while the two-dimensional rep is the double-valued rep of D_2, for which $D(\tilde{\gamma}_i) = -D(\gamma_i)$.

The above method, though straightforward, is not economic. The single-valued irreps of point groups are all known. All we need are the double-valued irreps of G. Can we have a simple method for getting only double-valued irreps of G?

In nature there are only two kinds of reps, the vector and spinor reps. In the vector, or single-valued, rep space L_v rotations through the angles φ and $\varphi + 2\pi$ are identical, thus

$$\tilde{\gamma}_i = \gamma_i, \quad G^\dagger \to G = \{\gamma_i : i = 1, 2, \ldots, |G|\} \,. \tag{8-108a}$$

In the spinor, or double-valued rep space L_s, according to (8-105b) and (5-181), we have

$$\tilde{\gamma}_i = -\gamma_i, \quad G^\dagger \to G^\dagger = \{\gamma_i, -\gamma_i : i = 1, 2, \ldots, |G|\} \,. \tag{8-108b}$$

Table 8.8-2. Characters of $D_2^\dagger$.

	e	$\tilde{e}$	$2C_{2x}$	$2C_{2y}$	$2C_{2z}$
A	1	1	1	1	1
B_1	1	1	-1	-1	1
B_2	1	1	-1	1	-1
B_3	1	1	1	-1	-1
$\overline{E}$	2	-2	0	0	0

Table 8.8-3. Double valued rep $\overline{E}$ of $\mathbf{D}_2$.[1]

e	C_{2x}	C_{2y}	C_{2z}
$\begin{pmatrix}1&0\\0&1\end{pmatrix}$	$\begin{pmatrix}i&0\\0&-i\end{pmatrix}$	$\begin{pmatrix}0&1\\-1&0\end{pmatrix}$	$\begin{pmatrix}0&i\\i&0\end{pmatrix}$

[1]Here $D^{\overline{E}}(\tilde{\gamma}_i) = -D^{\overline{E}}(\gamma_i)$, $i = 1,2,3,4$.

Equation (8-108) shows that both in the vector space L_v and the spinor space L_s, the number of the linearly independent (L.I.) elements of the matrix groups G, which are the representations of the double point group $\mathbf{G}^\dagger$, are equal to $|G|$. In L_v, the L.I. elements form a group which is just the point group. In L_s, the L.I. elements do not form a group. Their multiplication relations are

$$\gamma_i\gamma_j = \theta(i,j)\gamma_{ij}, \quad i,j = 1,2,\ldots,|G|, \quad \theta(i,j) = \pm 1 . \tag{8-108c}$$

This shows that $\{\gamma_i : i = 1,2,\ldots,|G|\}$ forms a projective rep [cf. (2-45′)] of the point group $\mathbf{G}$. We thus see that obtaining the double-valued irreps of $\mathbf{G}$ is equivalent to finding the projective rep of $\mathbf{G}$.

As we see, the elements of the matrix group $G^\dagger$ are not linearly independent. How do we handle a group with *linearly dependent group elements*? This leads to the definition of the so-called representation (rep) group. In the following section, we shall show that in order to get the double-valued reps of $\mathbf{G}$, we need only treat the rep group with the "effective order" of $|G|$ which is the number of L.I. elements of the matrix group $G^\dagger$, rather than $2|G|$.

8.9. The Rep Group

For establishing a unified approach to both vector and projective reps, or the single-valued and double-valued reps, we introduce the *representation group* or *rep group*. The key point of the rep group approach is that we focus our attention to the L.I. elements of a (matrix) group instead of the whole elements.

Suppose that there is an abstract group $\mathbf{G}$,

$$\mathbf{G} = \{\hat{R}_s : s = 1,2,\ldots,|G|\} . \tag{8-109}$$

The representation of $\mathbf{G}$ in a rep space L is

$$G = \{R_s : s = 1,2,\ldots,|G|\} , \tag{8-110}$$

where R_s can be understood either as operators or matrices and they are in general not linearly independent. Suppose that there are only $|g|$ L.I. matrices which form the set (called the *fundamental set*),

$$F = \{R_1, R_2, \ldots, R_{|g|}\} , \tag{8-111}$$

and the others are related to them as

$$R_i^{(l)} = \exp(2\pi li/m)R_i, \quad l = 0,1,\ldots,m-1, \quad |G| = m|g| , \tag{8-112}$$

then

$$G = \{R_i^{(l)} : i = 1,2,\ldots,|g|, \quad l = 0,1,\ldots,m-1\} \equiv \{R_i^{(l)} : i = 1,2,\ldots,|g|\}_m \tag{8-113}$$

is said to be a *rep group*.

The essential difference between an abstract group **G** and a rep group G is that for the group algebra of **G** all the elements are linearly independent, while for G only $|G|/m$ elements are linearly independent. The former is a special case of the latter with $m = 1$.

The space spanned by $R_1, R_2, \ldots, R_{|g|}$ is called the *group space* L_g *of the rep group* G. The multiplication relation of G is totally specified by the $|g| \times |g|$ group table.

$$R_i R_j = \eta(i,j) R_{ij} \quad |\eta(i,j)| = 1 \ . \tag{8-114}$$

Suppose that the rep group G has N classes, N being also the number of classes of the abstract group **G**. As for an abstract group, we can construct N class operators by forming the algebraic sums of all operators belonging to the same classes. However, due to the linear dependence of the group elements, the N class operators of the rep group G are not linearly independent. Some of the class operators may be null operators, and some may differ by only phase factors. Let $C_1(= e), C_2, \ldots, C_n$ be the L.I. class operators of G, $n \le N$ (the equality holds only when $m = 1$, i.e., for an abstract group).

The vector space spanned by the n L.I. class operators $C_1, C_2, \ldots, C_n$ is referred to as the *class space* L_n *of the rep group* G. They form the *class algebra* of the rep group G with the multiplication rule

$$C_i C_j = \sum_{k=1}^{n} C_{ij}^k C_k, \quad i,j = 1,2,\ldots,n \ . \tag{8-115}$$

Notice that now the structure constants C_{ij}^k may be imaginary.

Suppose that there is a group $\mathbf{G}_0$ of order $|G_0| = |g|$ and with elements γ_i,

$$G_0 = \{\gamma_i : i = 1,2,\ldots,|g|\} \ , \tag{8-116a}$$

if $\mathbf{G}_0$ has the multiplication relation

$$\gamma_i \gamma_j = \gamma_{ij}, \quad i,j = 1,2,\ldots,|g| \ , \tag{8-116b}$$

then the rep group G is an m-fold covering group of $\mathbf{G}_0$, and under the mapping $R_i \to \gamma_i$, the fundamental set F of G is a projective rep of the group $\mathbf{G}_0$.

Thanks to (8-112), an irrep ν of the rep group G may be specified by giving explicitly only the irreducible matrices for the $|g|$ L.I. elements. Hence for simplicity, we shall just say that

$$D^{(\nu)}(G) = \{D^{(\nu)}(R_i) : i = 1,2,\ldots,|g|\} \ , \tag{8-117}$$

is an irrep of the rep group G. $D^{(\nu)}(G)$ is clearly a projective irrep of the group $\mathbf{G}_0$ under the mapping $R_i \leftrightarrow \gamma_i$, namely, each irrep of the rep group G gives a projective irrep of $\mathbf{G}_0$. Therefore, the construction of projective irreps of the group $\mathbf{G}_0$ for the factor system $\eta(i,j)$ can be replaced by the construction of the vector irreps of the rep group G, which is, as will be seen later, as easy as that for a finite group of order $|g| = |G|/m$.

On the other hand, the irreps and irreducible bases of the abstract group **G** in the rep space L is obviously identical to those of the rep group G, and thus the former task can be replaced by the latter one. It is much easier to work with the rep group G than with the abstract group **G**. Before proceeding with the problem of constructing irreps of a rep group, we first extend the definitions on the regular rep, intrinsic group, etc. to the rep group case.

The rep generated by the group space L_g is called the *regular rep* of G, namely

$$\begin{aligned} R_j R_k &= \sum_{i=1}^{|g|} D_{ik}(R_j) R_i \quad j,k = 1,2,\ldots,|g| \ , \\ D_{ik}(R_j) &= \eta(j,k)\delta_{i,jk} \ . \end{aligned} \tag{8-118}$$

Now consider the intrinsic group $\overline{G}$ of a rep group G. We first define $|g|$ operator $\overline{R}_j$ in the group space L_g by

$$\overline{R}_j R_k = R_k R_j = \eta(k,j) R_{kj}, \quad j,k = 1,2,\ldots,|g| , \tag{8-119a}$$

and define

$$\overline{R}_j^{(l)} = \exp(2\pi l i/m)\overline{R}_j, \quad j = 1,2,\ldots,|g|, \quad l = 0,1,\ldots,m-1 . \tag{8-119b}$$

Then

$$\overline{G} = \{\overline{R}_j^{(l)} : j = 1,2,\ldots |g|, \quad l = 0,1,\ldots,m-1\} , \tag{8-119c}$$

forms the *intrinsic group* $\overline{G}$ of the rep group G. The regular rep $D(\overline{G})$ of the intrinsic group $\overline{G}$ is defined by

$$\overline{R}_j R_k = \sum_{i=1}^{|g|} D_{ik}(\overline{R}_j) R_i, \quad j,k = 1,2,\ldots,|g| ,$$
$$D_{ik}(\overline{R}_j) = \eta(k,j)\delta_{i,kj} . \tag{8-120}$$

The new approach to the rep theory of (abstract) groups can be easily extended to the rep group (see the extensive review by Chen [15] 1985). All the definitions of the CSCO-I, -II and -III of **G**, all the formulas and conclusions remain valid for the rep group under the substitutions

$$g(= |G|) \to |g| = |G|/m, \quad N \to n . \tag{8-121}$$

For instance the Burnside Theorem may be extended to

Theorem 8.1: A rep group G with n linearly independent class operators contains n and only n, instead of N, inequivalent irreps.

Theorem 8.2: The regular rep of a rep group G contains n inequivalent irreps, and the number of times each irrep occurs is equal to its dimension,

$$|g| = \sum_{\nu=1}^{n} h_\nu^2 . \tag{8-122}$$

Therefore, the irreps and irreducible bases of the rep group can be found just as those of abstract groups. The only differences are:

1. Now the elements of the regular matrices of the group G or its intrinsic group $\overline{G}$ may be imaginary, while for an abstract group, they can only be 1 or zero.

2. For a nontrivial rep group (i.e., $m > 1$ instead of $m = 1$), there is no identity rep, while for an abstract group, there must exist the identity rep.

3. For an abstract group **G**, the identity is never chosen as a member in the CSCO-I, while for a rep group G, if there is only one L.I. class operator, it must be the identity and is the CSCO of the rep group.

Example: Find the double-valued irrep of $\mathbf{D}_2$. In the spinor rep space, $\tilde{\gamma}_i = -\gamma_i$. From (8-107) we have

$$C_1 = e, \quad C_2 = -e, \quad C_3 = C_4 = C_5 = 0 .$$

This shows that the rep group $D_2^\dagger = \{\pm e, \pm C_{2x}, \pm C_{2y}, \pm C_{2z}\}$ has only one L.I. class operator. We thus immediately know that there is only one double-valued irrep for $\mathbf{D}_2$. Now $|g| = 4$, and from (8-122) we have $4 = 2^2$. Therefore this double-valued irrep is two-dimensional.

The CSCO-I of the rep group $D_2^\dagger$ is $C = e$ and the diagonalization of C is, of course, unnecessary. The operator set $C(s)$ can be chosen as γ_2. $(\gamma_2, \overline{\gamma}_2)$ is the CSCO-III of the rep group. By reading the group Table 8.8-1 horizontally and vertically, we can write down the regular reps of γ_2 and the intrinsic element $\overline{\gamma}_2$, respectively;

$$D(\gamma_2) = (2\bar{1}4\bar{3}) = \begin{pmatrix} 0 & -1 & 0 & 0 \\ 1 & 0 & 0 & 0 \\ 0 & 0 & 0 & -1 \\ 0 & 0 & 1 & 0 \end{pmatrix}, \quad D(\overline{\gamma}_2) = (2\bar{1}\bar{4}3) = \begin{pmatrix} 0 & -1 & 0 & 0 \\ 1 & 0 & 0 & 0 \\ 0 & 0 & 0 & 1 \\ 0 & 0 & -1 & 0 \end{pmatrix} . \tag{8-123}$$

We can find the common eigenvectors $\{u_{\nu mk,i} : i = 1, \ldots, 4\}$ of $D(\gamma_2)$ and $D(\overline{\gamma}_2)$, and adjust the phases of the eigenvectors by using (3-201). Then from (3-198) we can obtain the irrep of the rep group D_2, which is just that shown in Table 8.8-3.

The advantage of the rep group approach to this problem is now very clear. Originally, we need to treat a group of order $m|g| = 8$, and now we only need to treat a rep group with the "effective order" $|g| = 4$. For cases with large m and large $|g|$ (e.g., for the single-valued reps of space groups, $m = 2, 3, 4, 6$, while for the double-valued reps of space groups, $m = 4, 6, 8, 12$, and $|g|$ can be as high as 48), a tremendous simplification results from using the rep group approach.

8.10. Symmetry Adapted Basis for Double Point Groups

The $\mathbf{G} \supset \mathbf{G}(s)$ symmetry-adapted basis $\Psi_\kappa^{(\mu)}$ in the spinor rep space still obey the eigenequation (8-40), where $(C, C(s))$ is to be understood as the CSCO-II of the rep group G. Let $u_+(u_-)$ be the spin wave function for spin up (down). The basis $\Psi_\kappa^{(\mu)}$ can be found in the following way. Start with an arbitrary function $\Psi_1 = Y_{lm}u_-$. Apply the $|g|$ group operators of the rep group G to Ψ_1,

$$R\Psi_1 = (RY_{lm})(Ru_-) , \tag{8-124}$$

and pick out the linearly independent ones, say $\Psi_1, \ldots, \Psi_n$. The transformation of $u_{\pm 1}$ under the point group operations is given in Bradley (1972, Table 6.7, pp. 446). By diagonalizing the CSCO-II of the rep group $G, (C, C(s))$, in the basis $(\Psi_1, \ldots, \Psi_n)$ we can get the symmetry-adapted basis $\Psi_\kappa^{(\mu)} = \Sigma_{i=1}^n a_i \Psi_i$.

Example: Find the $\mathbf{D}_3 \supset (\mathbf{e}, \mathbf{C}'_{21})$ spinor irreducible bases. The three two-fold axes of $\mathbf{D}_3$ are shown in Fig. 8.10. The corresponding Euler angles are (see Bradley 1972, Table 2.1, pp. 54):

$$C'_{21} : (\pi, \pi, 0); \quad C'_{22} : \left(\frac{5}{3}\pi, \pi, 0\right); \quad C'_{23} : \left(\frac{1}{3}\pi, \pi, 0\right) .$$

The transformation of $u_\pm$ under $C'_{2i}, i = 1, 2, 3$, is shown in Table 8.10-1.

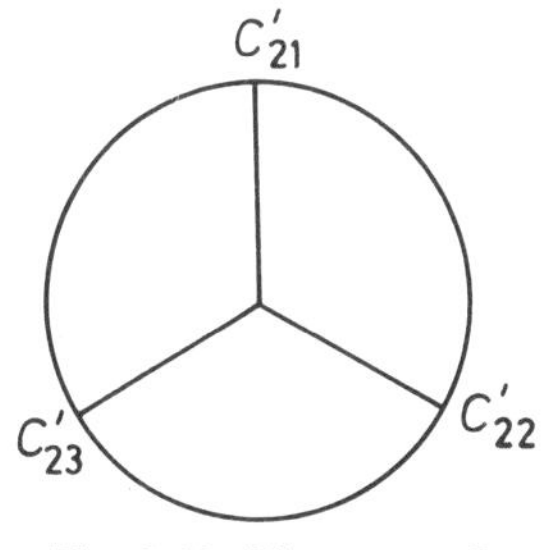

Fig. 8.10. The group D_3.

Table 8.10-1

	u_-	u_+
C'_{21}	$-iu_+$	$-iu_-$
C'_{22}	$-\varepsilon^* u_+$	εu_-
C'_{23}	εu_+	$-\varepsilon^* u_-$

$\varepsilon^* = \exp(\pi i/6)$.

Applying the group operators of D_3 to $\Psi_1 = Y_{lm}u_-$, we have

$$\begin{aligned} &C_3\Psi_1 = \exp[i(-120°m + 60°)](Y_{lm}u_-), \quad C_3^{-1}\Psi_1 = \exp[i(120°m - 60°)](Y_{lm}u_-) , \\ &C'_{21}\Psi_1 = -i(-)^{l+m}(Y_{l-m}u_+) , \\ &C'_{22}\Psi_1 = -\exp[-i(120°m + 30°)](-)^{l+m}(Y_{l-m}u_+) , \\ &C'_{23}\Psi_1 = \exp[i(120°m + 30°)](-)^{l+m}(Y_{l-m}u_+) . \end{aligned} \tag{8-125}$$

We see that there are only two linearly independent functions,

$$\Psi_m = Y_{lm} u_- \ , \qquad \Phi_m = (-)^{l+m} Y_{l-m} u_+ \ . \tag{8-126}$$

The CSCO-I of the rep group D_3 is $C = C'_{21} + C'_{22} + C'_{23}$ (see Exercise 8.8), and $C(s) = C'_{21}$. Using (8-125), one easily found that

$$\begin{aligned} &C\Psi_m = -ib_m\Phi_m, && C\Phi_m = -ib_m\Psi_m \ , \\ &b_m = 1 - 2\sin(120^\circ m + 30^\circ) \ , \\ &C(s)\Psi_m = -i\Phi_m, && C(s)\Psi_m = -i\Psi_m \ . \end{aligned}$$

Therefore we have: for $m = 2 \bmod(+3)$, $b_m = 3$,

$$\Gamma_5: \quad \Psi_i^{(3i)} = \Psi_m - \Phi_m, \quad \Gamma_6: \quad \Psi_i^{(-3i)} = \Psi_m + \Phi_m \ ; \tag{8-127}$$

for $m = 0, 1 \bmod(+3)$, $b_m = 0$,

$$\Gamma_4: \quad \Psi_i^{(0)} = \Psi_m - \Phi_m \ , \tag{8-128}$$

$$\Psi_{-i}^{(0)} = \theta_m(\Psi_m + \Phi_m) \ , \tag{8-129}$$

where Γ_4, Γ_5 and Γ_6 are the irrep labels used by Koster (1963). The phase θ_m is determined by using (8-41),

$$\theta_m = \begin{cases} i \ , \\ -i \ , \end{cases} \quad \text{for } m = \begin{cases} 0 \ , \\ 1 \ . \end{cases} \tag{8-130}$$

Ex. 8.8. Find the CSCO-I and characters of the rep group $D_3^\dagger$. The group table is given below.

Table 8.10-2. Group table of $D_3^\dagger$.

e	C_3	C_3^2	C'_{21}	C'_{22}	C'_{23}
1	2	3	4	5	6
2	−3	1	−6	−4	−5
3	1	−2	5	−6	−4
4	−5	−6	−1	2	3
5	−6	−4	3	−1	2
6	−4	−5	2	3	−1

Ex. 8.9. Find the double-valued irreps of $\mathbf{D}_3$ in the $\mathbf{D}_3 \supset (\mathbf{e},\ C'_{21})$ basis.

Chapter 9
Applications of Group Theory to Many-Body Systems

Group theory plays a significant role in treating a many-body system. In this chapter we shall restrict ourselves to its applications in the nuclear shell model (both for pure and mixed configurations), the quark model, the dynamic symmetry model of nuclei and the molecular shell model. We only sketch the bare bones and leave the details to the references.

In Secs. 9.1-9.4, we use the round brackets $|\ )$ to denote anti-symmetrized states.

9.1. Pure Configuration Shell Model

Consider f particles in the orbital l. The question is how to evaluate the matrix elements of one-body and two-body operators in the anti-symmetric states (7-182).

9.1.1. One-body operator

By using the single-particle CFP (7-184), the matrix elements of a one-body operator can be reduced to

$$\begin{aligned}&\left(\begin{matrix} l^f[\nu] \\ \alpha\beta LSTM \end{matrix}\left|\sum_{i=1}^{f} Q(i)\right|\begin{matrix} l^f[\nu'] \\ \alpha'\beta'L'S'T'M' \end{matrix}\right) = f\left(\begin{matrix} l^f[\nu] \\ \alpha\beta LSTM \end{matrix}\left|Q(f)\right|\begin{matrix} l^f[\nu'] \\ \alpha'\beta'L'S'T'M' \end{matrix}\right)\\ &= f\sum_{\nu_1\alpha_1\beta_1L_1S_1T_1}\frac{h_{\nu_1}}{\sqrt{h_\nu h_{\nu'}}}C^{[\nu]\alpha L}_{[\nu_1]\alpha_1L_1,[1]}C^{[\tilde\nu]\beta ST}_{[\tilde\nu_1]\beta_1S_1T_1,[1]}C^{[\nu']\alpha' L'}_{[\nu_1]\alpha_1L_1,[1]}C^{[\tilde\nu']\beta' S'T'}_{[\tilde\nu_1]\beta_1S_1T_1,[1]}\\ &\times\left\langle\left[\psi\left(\begin{matrix} l^{f-1}[\nu_1] \\ \alpha_1\beta_1L_1S_1T_1\end{matrix}\right)\psi(f)\right]^{LST}_{M}\left|Q(f)\right|\left[\psi\left(\begin{matrix} l^{f-1}[\nu_1] \\ \alpha_1\beta_1L_1S_1T_1\end{matrix}\right)\psi(f)\right]^{L'S'T'}_{M'}\right\rangle,\end{aligned}\tag{9-1}$$

where M stands collectively for M_L, M_S and M_T, and $\psi(f)$ is the total wave function of the f-th particle. In the process of deriving (9-1), we first used the anti-symmetry property of the wave functions, and then recalled the fact that the operator $Q(f)$ of the f-th particle does not change the quantum $\alpha_1\beta_1L_1S_1T_1$ of the first $f-1$ particles. Integrating over the coordinates of the $f-1$ particles, the last factor in (9-1) becomes

$$\begin{aligned}&\sum_{\substack{M_1M_{S_1}M_{T_1}\\ m_lm_sm_tm_l'm_s'm_t'}} C^{LM_L}_{L_1M_1,lm_l}C^{SM_S}_{S_1M_{s_1},1/2m_s}C^{TM_T}_{T_1M_{T_1},1/2m_t}C^{L'M_L'}_{L_1M_1,lm_l'}C^{S'M_s'}_{S_1M_{s_1},1/2m_s'}\\ &\times C^{T'M_T'}_{T_1M_{T_1},1/2m_t'}\langle\psi_{m_lm_sm_t}|Q|\psi_{m_l'm_s'm_t'}\rangle\ .\end{aligned}\tag{9-2}$$

It is seen that in terms of the single-particle CFP, the evaluation of matrix elements of a one-body operator for a many-body system can be reduced to that for a one-body system without using the explicit form of the many-body wave function.

A generalization of the above assertion is that in terms of the f_2-particle CFP, the calculation of matrix elements of f_2-body operators for a many-body system can be simplified to that for a f_2-body system.

In the case when Q acts only on the orbital space and is an irreducible tensor $Q_\mu^{(\lambda)}$ of the rotation group, using the unitarity of the CFP's we obtain from (9-1), (9-2) and (3-319)

$$\left(\begin{matrix} l^f[\nu] \\ \alpha\beta LST \end{matrix}\left\| \sum_{i=1}^{f} Q^{(\lambda)}(i) \right\| \begin{matrix} l^f[\nu'] \\ \alpha'\beta'L'S'T' \end{matrix}\right)$$

$$= \delta_{\nu\nu'}\delta_{\beta\beta'}\delta_{SS'}\delta_{TT'} f \sum_{\nu_1\alpha_1 L_1} \frac{h_{\nu_1}}{h_\nu} C^{[\nu]\alpha L}_{[\nu_1]\alpha_1 L_1,[1]} C^{[\nu]\alpha' L'}_{[\nu_1]\alpha_1 L_1,[1]} U(L_1 l L\lambda; L'l)\langle l\|Q^{(\lambda)}\|l\rangle \,. \tag{9-3}$$

For instance, for the kinetic energy operator T by setting $\lambda = 0$ in (9-3) and using (7-161) we get

$$\left(\begin{matrix} l^f[\nu] \\ \alpha\beta LST \end{matrix}\left\| \sum_{i=1}^{f} T(i) \right\| \begin{matrix} l^f[\nu'] \\ \alpha'\beta'L'S'T' \end{matrix}\right) = \delta_{\nu\nu'}\delta_{\alpha\alpha'}\delta_{\beta\beta'}\delta_{LL'}\delta_{SS'}\delta_{TT'} f\langle l\|T\|l\rangle \,. \tag{9-4}$$

9.1.2. Two-body operator

Assuming that the interaction potential V_{ij} is central and using the two-body CFP [let $n_2 = 2$ in (7-183)], we have

$$\left(\begin{matrix} l^f[\nu] \\ \alpha\beta LST \end{matrix}\left\| \sum_{i>j=1}^{f} V_{ij} \right\| \begin{matrix} l^f[\nu'] \\ \alpha'\beta'LST \end{matrix}\right) = \binom{f}{2} \sum \frac{h_{\nu_1}}{\sqrt{h_\nu h_{\nu'}}} C^{[\nu]\alpha L}_{[\nu_1]\alpha_1 L_1,[\nu_2]L_2} C^{[\nu']\alpha' L}_{[\nu_1]\alpha_1 L_1[\nu_2]L_2}$$

$$\times C^{[\tilde\nu]\beta ST}_{[\tilde\nu_1]\beta_1 S_1 T_1,[\tilde\nu_2]S_2T_2} C^{[\tilde\nu']\beta' ST}_{[\tilde\nu_1]\beta_1 S_1 T_1,[\tilde\nu_2]S_2T_2} \left(\begin{matrix} l^2[\nu_2] \\ L_2S_2T_2 \end{matrix}\left\| V_{f-1,f} \right\| \begin{matrix} l^2[\nu_2] \\ L_2S_2T_2 \end{matrix}\right) \,, \tag{9-5}$$

where the sum runs over $[\nu_1][\nu_2]\alpha_1\beta_1 L_1 S_1 T_1 L_2 S_2$ and T_2. For further discussions, see Elliott (1957).

Equation (9-5) can be easily extended to the $SU_6 \supset SU_3 \times SU_2$ bases used in the hyper-nuclear physics. The anti-symmetrized wave function (7-182) now reads

$$\left|\begin{matrix} l^f[\nu] \\ \alpha LM, \beta[\mu]IYI_zSS_z \end{matrix}\right) = \sum_m \frac{\Lambda_m^\nu}{\sqrt{h_\nu}} \left|\begin{matrix} l^f[\nu] \\ m, \alpha LM \end{matrix}\right\rangle \left|\begin{matrix} \gamma^f[\tilde\nu] \\ \tilde m, \beta[\mu]IYI_zSS_z \end{matrix}\right\rangle \,. \tag{9-6}$$

From (7-167) and (7-180) we obtain

$$\left(\begin{matrix} l^f[\nu] \\ \alpha L, \beta[\mu]IYS \end{matrix}\left\| \sum_{i>j=1}^{f} V_{ij} \right\| \begin{matrix} l^f[\nu] \\ \alpha L, \beta[\mu]IYS \end{matrix}\right) = \binom{f}{2} \sum \frac{h_{\nu_1}}{h_\nu}$$

$$\times \left(C^{[\nu]\alpha L}_{[\nu_1]\alpha_1 L_1[\nu_2]L_2} C^{[\tilde\nu],\beta[\mu]_\theta S}_{[\tilde\nu_1]\beta_1[\mu_1]S_1,[\tilde\nu_2][\mu_2]S_2} C^{[\mu]_\theta IY}_{[\mu_1]I_1Y_1,[\mu_2]I_2Y_2} \right)^2$$

$$\times \left(\begin{matrix} l^2[\nu_2] \\ L_2[\mu_2]I_2Y_2S_2 \end{matrix}\left\| V_{f-1,f} \right\| \begin{matrix} l^2[\nu_2] \\ L_2[\mu_2]I_2Y_2S_2 \end{matrix}\right) \,, \tag{9-7}$$

where the sum runs over $\nu_1\mu_1\nu_2\mu_2\alpha_1\beta_1 L_1 I_1 Y_1 S_1 L_2 I_2 Y_2 S_2$ and θ. For off-diagonal matrix elements we have similar expressions.

9.2. Anti-symmetric Wave Functions for a A + B System

Let A and B be two nuclei, or two shells in the same nucleus with f_1 and f_2 particles, respectively and $f = f_1 + f_2$. The relevant quantum numbers for the system A, B and A + B are shown in Table 9.2,

Table 9.2.

	A	B	A + B
symmetry in orbital space	$[\nu_1]$	$[\nu_2]$	$[\nu]$
orbital angular momentum	L_1	L_2	L
spin	S_1	S_2	S
isospin	T_1	T_2	T
symmetry in spin space	$[\sigma_1]$	$[\sigma_2]$	$[\sigma]$
symmetry in isospin space	$[\mu_1]$	$[\mu_2]$	$[\mu]$

where

$$[\sigma_i] = \left[\frac{f_i}{2} + S_i, \frac{f_i}{2} - S_i\right], \ [\sigma] = \left[\frac{f}{2} + S, \frac{f}{2} - S\right],$$
$$[\mu_i] = \left[\frac{f_i}{2} + T_i, \frac{f_i}{2} - T_i\right], \ [\mu] = \left[\frac{f}{2} + T, \frac{f}{2} - T\right]. \tag{9-8}$$

The anti-symmetric wave functions of A and B are denoted by $\left|{}^{[\nu_i]}_{L_i\beta_i S_i T_i}\right)$ (see (7-182)). For brevity here and in what follows we often ignore the additional quantum number α_i and the projections M_{L_i}, M_{S_i} and M_{T_i}.

The anti-symmetric wave function for the total system A + B is then

$$\Psi_{LST} = \mathcal{A}\left[\left|\begin{matrix}[\nu_1]\\ L_1\beta_1S_1T_1\end{matrix}\right)_{\omega_1^0}\left|\begin{matrix}[\nu_2]\\ L_2\beta_2S_2T_2\end{matrix}\right)_{\omega_2^0} F(A-B)\right]^{LST}, \tag{9-9}$$

where $(\omega_1^0) = (1, 2, \ldots, f_1)$ and $(\omega_2^0) = (f_1 + 1, \ldots, f)$. The function $F(\text{A}-\text{B})$ is equal to one if A and B represent two shells in a nucleus, and is equal to the relative motion wave function if A and B represent two nuclei in a nuclear reaction, or two clusters in the cluster model,

$$F(\text{A} - \text{B}) = F(\mathbf{R}_\text{A} - \mathbf{R}_\text{B}), \tag{9-10}$$

$\mathbf{R}_\text{A}$ and $\mathbf{R}_\text{B}$ being coordinates of the mass centers of A and B. $\mathcal{A}$ is the anti-symmetrization operator, and can be expressed as

$$\mathcal{A} = \binom{f}{f_1}^{-1/2}\sum_{\omega}\delta_\omega\binom{\omega^0}{\omega}, \tag{9-11}$$

where $\binom{\omega^0}{\omega}$ is the order-preserving operator [(4-131)], and δ_ω is its permutation parity.

The form (9-9) for the anti-symmetric wave function is not convenient for calculating matrix elements of the interactions between the two clusters A and B. Instead, we will construct the anti-symmetric wave function in the following way. First, in orbital space we construct the $SU_{m+n} \supset SU_m \otimes SU_n$ bases (if A and B represent two shells l_1 and l_2, then $m = 2l_1 + 1$ and $n = 2l_2 + 1$),

$$\left|\begin{matrix}[\nu] & \tau[\nu_1][\nu_2]\\ m, LM_L; & L_1\ \ L_2\end{matrix}\right\rangle = \sum_{m_1m_2\omega} C^{[\nu]\tau,m}_{\nu_1m_1,\nu_2m_2,\omega}\binom{\omega^0}{\omega}\left[\left|\begin{matrix}[\nu_1]\\ m_1\omega_1^0, L_1\end{matrix}\right\rangle\left|\begin{matrix}[\nu_2]\\ m_2\omega_2^0,\ L_2\end{matrix}\right\rangle F(A-B)\right]^L_{M_L}. \tag{9-12}$$

Next, in spin-isospin space we construct the $SU_4 \supset SU_2 \times SU_2$ bases,

$$\left|\begin{matrix}[\tilde{\nu}]\\ \tilde{m}, \beta STM_sM_T\end{matrix}\right\rangle = \sum_{m_\sigma m_\mu} C^{[\tilde{\nu}]\beta,\tilde{m}}_{[\sigma]m_\sigma,[\mu]m_\mu} \left|\begin{matrix}[\sigma]\\ m_\sigma, SM_s\end{matrix}\right\rangle \left|\begin{matrix}[\mu]\\ m_\mu, TM_T\end{matrix}\right\rangle . \tag{9-13}$$

Finally we use (4-125) to form the totally anti-symmetric wave function

$$\left|\begin{matrix}[\nu] & \tau[\nu_1][\nu_2]\\ L\beta ST, & L_1\ L_2\end{matrix}\right) = \sum_m \frac{\Lambda^\nu_m}{\sqrt{h_\nu}} \left|\begin{matrix}[\nu] & \tau[\nu_1][\nu_2]\\ m, L; & L_1\ L_2\end{matrix}\right\rangle \left|\begin{matrix}[\tilde{\nu}]\\ \tilde{m}\beta ST\end{matrix}\right\rangle . \tag{9-14}$$

The two anti-symmetric bases (9-9) and (9-14) are related by a unitary transformation

$$\left|\begin{matrix}[\nu] & \tau[\nu_1][\nu_2]\\ L\beta ST, & L_1\ L_2\end{matrix}\right)$$

$$= \sum_{\substack{\beta_1 S_1 T_1\\ \beta_2 S_2 T_2}} C^{[\tilde{\nu}]\tau,\beta ST}_{[\tilde{\nu}_1]\beta_1 S_1 T_1,[\tilde{\nu}_2]\beta_2 S_2 T_2} \mathcal{A} \left[\left|\begin{matrix}[\nu_1]\\ L_1\beta_1 S_1 T_1\end{matrix}\right)_{\omega_1^0} \left|\begin{matrix}[\nu_2]\\ L_2\beta_2 S_2 T_2\end{matrix}\right)_{\omega_2^0} F(A-B)\right]^{LST} . \tag{9-15}$$

Equation (9-15) can be justified by noting that the $SU_4 \supset SU_2 \times SU_2$ ISF ensure that the spin-isospin wave functions at the right-hand side belong to the irrep $[\tilde{\nu}]$ of SU_4, while the anti-symmetrization operator ensures that the right-hand side of (9-15) is totally anti-symmetric. Therefore the orbital wave function must have the symmetry $[\nu]$. By (7-154), the inverse of (9-15) is

$$\mathcal{A} \left[\left|\begin{matrix}[\nu_1]\\ L_1\beta_1 S_1 T_1\end{matrix}\right)_{\omega_1^0} \left|\begin{matrix}[\nu_2]\\ L_2\beta_2 S_2 T_2\end{matrix}\right)_{\omega_2^0} F(A-B)\right]^{LST}$$

$$= \sum_{\nu\beta\tau} C^{[\tilde{\nu}]\tau,\beta ST}_{[\tilde{\nu}_1]\beta_1 S_1 T_1,[\tilde{\nu}_2]\beta_2 S_2 T_2} \left|\begin{matrix}[\nu] & \tau[\nu_1][\nu_2]\\ L\beta ST, & L_1\ L_2\end{matrix}\right) . \tag{9-16}$$

The extension to the $SU_6 \supset SU_3 \times SU_2$ cases for hyper-nuclei is straightforward. All we have to do is to interprete $[\mu]$ in the above equations as the irrep label of SU_3, and T as $[\mu]IY$, replace the spin-isospin wave function (9-13) by the $SU_6 \supset SU_3 \times SU_2$ wave function

$$\left|\begin{matrix}[\tilde{\nu}]\\ \tilde{m}, \beta[\mu]IYI_z, SS_z\end{matrix}\right\rangle = \sum_{m_\sigma m_\mu} C^{[\tilde{\nu}]\beta,\tilde{m}}_{[\sigma]m_\sigma,[\mu]m_\mu} \left|\begin{matrix}[\sigma]\\ m_\sigma,\ SS_z\end{matrix}\right\rangle \left|\begin{matrix}[\mu]\\ m_\mu,\ IYI_z\end{matrix}\right\rangle , \tag{9-17a}$$

and substitute for (9-14) the expression

$$\left|\begin{matrix}[\nu] & \tau[\nu_1][\nu_2]\\ L\beta[\mu]IYS; & L_1\ L_2\end{matrix}\right) = \sum_m \frac{\Lambda^\nu_m}{\sqrt{h_\nu}} \left|\begin{matrix}[\nu] & \tau[\nu_1][\nu_2]\\ m, L; & L_1\ L_2\end{matrix}\right\rangle \left|\begin{matrix}[\tilde{\nu}]\\ \tilde{m}, \beta[\mu]IYS\end{matrix}\right\rangle . \tag{9-17b}$$

Using (7-179), the generalization of (9-15) and (9-16) are as follows:

$$\left|\begin{matrix}[\nu] & \tau[\nu_1][\nu_2]\\ L\beta[\mu]IYS\ , & L_1\ L_2\end{matrix}\right) = \sum_{\beta_1\mu_1 S_1\beta_2\mu_2 S_2} C^{[\tilde{\nu}]\tau,\beta\mu S}_{[\tilde{\nu}_1]\beta_1\mu_1 S_1,[\tilde{\nu}_2]\beta_2\mu_2 S_2}$$

$$\times \mathcal{A} \left[\left|\begin{matrix}[\nu_1]\\ L_1\beta_1[\mu_1]S_1\end{matrix}\right\rangle_{\omega_1^0} \left|\begin{matrix}[\nu_2]\\ L_2\beta_2[\mu_2]S_2\end{matrix}\right\rangle_{\omega_2^0} F(A-B)\right]^{L[\mu]S}_{IY} . \tag{9-17c}$$

$$\mathcal{A} \left[\left|\begin{matrix}[\nu_1]\\ L_1\beta_1[\mu_1]S_1\end{matrix}\right)_{\omega_1^0} \left|\begin{matrix}[\nu_2]\\ L_2\beta_2[\mu_2]S_2\end{matrix}\right)_{\omega_2^0} F(A-B)\right]^{L[\mu]S}_{IY}$$

$$= \sum_{\nu\tau\beta} C^{[\tilde{\nu}]\tau,\beta[\mu]S}_{[\tilde{\nu}_1]\beta_1[\mu_1]S_1,[\tilde{\nu}_2]\beta_2[\mu_2]S_2} \left|\begin{matrix}[\nu] & \tau[\nu_1][\nu_2]\\ L\beta[\mu]IYS, & L_1\ L_2\end{matrix}\right) , \tag{9-17d}$$

where we omitted the quantum numbers M, I_z and S_z.

9.3. Transformation between Symmetry Bases and Physical Bases in the Quark Model

Suppose A and B are the two clusters consisting of f_1 and f_2 quarks, respectively. For clarity in our notation, we assume that there are only two kinds of flavor quarks, u and d. The extension to more kinds of flavor quarks is straightforward. A quark is said to be in the orbital state a (b) if it belongs to the cluster A(B). The wave functions for describing the clusters A and B are taken to be the irreducible bases classified according to the following group chain

	$SU_{24} \supset$	SU_2 $\times$	SU_{12} $(\supset$	SU_3 $\times$	$SU_4)$
		orbital		color	spin-isospin
irrep label:	$[1^{f_i}]$	$[\nu_i]$	$[\tilde{\nu}_i]$	$[\sigma_i]$	$[\mu_i]$

In terms of the CG coefficients of the permutation group, the aforesaid bases are expressed as

$$\left| \begin{matrix} [\nu_i] \\ [\sigma_i] W_i [\mu_i] S_i T_i \end{matrix} \right) = \sum_{m_1 m_2 m} \frac{\Lambda^{\nu_i}_m}{\sqrt{h_{\nu_i}}} C^{[\tilde{\nu}_i],\tilde{m}}_{\sigma_i m_1, \mu_i m_2} \psi^{[\nu_i]}_m(\mathbf{r}) \left| \begin{matrix} [\sigma_i] \\ m_1, W_i \end{matrix} \right\rangle \left| \begin{matrix} [\mu_i] \\ m_2, S_i T_i \end{matrix} \right\rangle , \tag{9-18a}$$

where $\psi(\mathbf{r})$ is the orbtial wave function and W_i are the component indices for irreps of SU_3 in color space. For instance, a nucleon N and its resonance state Δ are described by [cf. (7-112b) and (7-112c)]

$$|N) = \left| \begin{matrix} [3] \\ [1^3][3]\frac{1}{2}, \frac{1}{2} \end{matrix} \right) , \quad |\Delta) = \left| \begin{matrix} [3] \\ [1^3][3]\frac{3}{2}, \frac{3}{2} \end{matrix} \right) . \tag{9-18b}$$

The totally anti-symmetric wave functions of these two clusters can be expressed by

$$\mathcal{A} \left[\left| \begin{matrix} [\nu_1] a^{f_1} \\ \sigma_1 \mu_1 S_1 T_1 \end{matrix} \right)_{\omega_0^1} \left| \begin{matrix} [\nu_2] b^{f_2} \\ \sigma_2 \mu_2 S_2 T_2 \end{matrix} \right)_{\omega_0^2} \right]^{[\sigma] ST}_{W M_S M_T} , \tag{9-19a}$$

where the square bracket indicates the couplings in terms of the SU_3 and SU_2 CG coefficients so that they have definite SU_3 symmetry $[\sigma]W$, definite channel spin S and isospin T. Equation (9-19a) has been referred to as the *physical basis* [Harvey (1981)], since it has a clear physical meaning. Nevertheless, in order to exploit the CFP technique in the evaluation of matrix elements, it is preferable to use the following $SU_{24} \supset SU_2 \times SU_{12} (\supset SU_3 \times SU_4)$ basis for the $n = f_1 + f_2$ quarks,

$$\left| \begin{matrix} [\nu] a^{f_1} b^{f_2} \\ \alpha [\sigma] W [\mu] \beta S T M_S M_T \end{matrix} \right) . \tag{9-19b}$$

Equation (9-19b) is referred to as the *symmetry basis*. In analogy with (9-16), the physical and symmetry basis differ by a unitary transformation

$$\mathcal{A} \left[\left| \begin{matrix} [\nu_1] a^{f_1} \\ \sigma_1 \mu_1 S_1 T_1 \end{matrix} \right)_{\omega_0^1} \left| \begin{matrix} [\nu_2] b^{f_2} \\ \sigma_2 \mu_2 S_2 T_2 \end{matrix} \right)_{\omega_0^2} \right]^{[\sigma] ST}$$
$$= \sum_{\tilde{\nu} \alpha \mu \varphi \beta} C^{[\tilde{\nu}], \alpha [\sigma][\mu] \varphi}_{\tilde{\nu} \sigma_1 \mu_1, \tilde{\nu}_2 \sigma_2 \mu_2} C^{[\mu] \varphi, \beta S T}_{\mu_1 S_1 T_1, \mu_2 S_2 T_2} \left| \begin{matrix} [\nu] a^{f_1} b^{f_2} \\ \alpha [\sigma][\mu] \beta S T \end{matrix} \right) , \tag{9-20a}$$

where the first factor in the right-hand side is the $SU_{12} \supset SU_3 \times SU_4$ ISF, and the second is the $SU_4 \supset SU_2 \times SU_2$ ISF.

The symmetry basis can be expanded in terms of the $SU_{24} \supset SU_2 \times SU_{12}$ ($\supset SU_3 \times SU_4(\supset SU_2 \times SU_2)$) CFP. Using the Racah factorization lemma and (7-244b), we can write down immediately

$$\left|\begin{matrix}[\nu]w\\ \alpha[\sigma][\mu]\beta ST\end{matrix}\right) = \sum \sqrt{\frac{h_{\nu'}h_{\nu''}}{h_\nu}}\, C^{[\tilde{\nu}],\alpha[\sigma][\mu]\varphi}_{\tilde{\nu}'\sigma'\mu',\tilde{\nu}''\sigma''\mu''}\, C^{[\sigma]w}_{\nu'w',\nu''w''}\, C^{[\mu]\varphi,\beta ST}_{\mu'S'T',\mu''S''T''}$$
$$\times \left[\left|\begin{matrix}[\nu']w'\\ \sigma'\mu'S'T'\end{matrix}\right)_{1\ldots n_1} \left|\begin{matrix}[\nu'']w''\\ \sigma''\mu''S''T''\end{matrix}\right)_{n_1+1\ldots n}\right]^{[\sigma]ST}, \qquad \text{(9-20b)}$$

where the sum runs over $\sigma'\mu'w'S'T'\sigma''\mu''w''S''T''$ and φ. w', w'' and w are the Weyl tableaux resulting from filling the Young diagrams $[\nu']$, $[\nu'']$ and $[\nu]$, respectively, with the state indices a and b. $C^{[\nu]w}_{\nu'w',\nu''w''}$ is the SU_2 CG coefficient (i.e., the orbital CFP). It is clearly seen from the previous discussion that in such a calculation, the $SU_{mn} \supset SU_m \times SU_n$ ISF plays a crucial role.

Example: We calculate the transformation matrix between the physical and symmetry bases for the channel $S = T = 2$.

From (9-18b), (9-20a) and (7-238a), and the permutation group CG coefficients tabulated by Chen ([10] 1981), we can calculate the transformation matrix as shown in Table 9.3 (one of the coefficients has already been obtained in Sec. 7.16).

Table 9.3. The transformation matrix between the physical and symmetry bases.

$S=2,\ T=2$	$\left\lvert\begin{matrix}[51]\\ [2^3][42]\end{matrix}\right)$	$\left\lvert\begin{matrix}[33]\\ [2^3][42]\end{matrix}\right)$	$\left\lvert\begin{matrix}[33]\\ [2^3][6]\end{matrix}\right)$	$\left\lvert\begin{matrix}[42]\\ [2^3][51]\end{matrix}\right)$	$\left\lvert\begin{matrix}[42]\\ [2^3][411]\end{matrix}\right)$
$(\Delta\Delta)$	$\sqrt{\frac{20}{45}}$	$-\sqrt{\frac{16}{45}}$	$-\sqrt{\frac{9}{45}}$	0	0
$(N\Delta)_1$	$\sqrt{\frac{5}{45}}$	$-\sqrt{\frac{4}{45}}$	$\sqrt{\frac{36}{45}}$	0	0
$(N\Delta)_2$	0	0	0	1	0
$(CC')_1$	$\sqrt{\frac{4}{9}}$	$\sqrt{\frac{5}{9}}$	0	0	0
$(CC')_2$	0	0	0	0	1

where

$$(N\Delta)_i = \sqrt{\frac{1}{2}}\left[\mathcal{A}(|N)|\Delta)) - (-1)^i \mathcal{A}(|\Delta)|N))\right],$$
$$(CC')_i = \sqrt{\frac{1}{2}}\left[\mathcal{A}(|C)|C')) - (-1)^i \mathcal{A}(|C')|C))\right],$$
$$|C) = \left|\begin{matrix}[3]\\ [21][21]1/2,3/2\end{matrix}\right), \quad |C') = \left|\begin{matrix}[3]\\ [21][21]3/2,1/2\end{matrix}\right). \qquad \text{(9-20c)}$$

(CC′) is the so-called hidden color channel, since the three-quark clusters C as well as C′ are not color singlets. Instead, they have the symmetry [21] in color space. However, the six-quark cluster is in the color singlet $[2^3]$. Table 9.3 agrees with Harvey's (1981) Table 11. For further discussion see Chen ([20] 1982).

9.4. The CFP for a Mixed Configuration

We now consider the mixed configuration $(l_1)^{f_{13}}(l_2)^{f_{24}}$, where f_{13} and f_{24} are nucleon numbers in the orbitals l_1 and l_2 [see (7-212a)]. The spin-isospin CFP are identical to those for pure configurations. For the orbital wave function, we choose the $SU_{m+n} \supset (SU_m \supset SO_3) \otimes (SU_n \supset SO_3)$ basis with $m = 2l_1 + 1$ and $n = 2l_2 + 1$,

$$\left| \begin{matrix} [\nu] & \beta[\sigma][\mu] \\ m, LM, & L_1 \;\; L_2 \end{matrix} \right\rangle \equiv \left| \begin{matrix} [\nu] \\ m, \beta[\sigma][\mu] L_1 L_2 LM \end{matrix} \right\rangle . \tag{9-21}$$

As with (9-12), it can be expressed in terms of the SU_m and SU_n irreducible bases.

The *mixed configuration* CFP for separating f_3 and f_4 particles from the orbitals l_1 and l_2, respectively, is defined as the expansion coefficient in the following equation

$$(I) \equiv \left| \begin{matrix} [\nu] & \beta l_1^{f_{13}}[\sigma] l_2^{f_{24}}[\mu] \\ (\tau[\nu']m'[\nu'']m'')LM, & L_1 \;\; L_2 \end{matrix} \right\rangle = \sum C^{[\nu]\tau,\beta[\sigma][\mu]L_1L_2L}_{[\nu']\beta'[\sigma'][\mu']L_1'L_2'L';[\nu'']\beta''[\sigma''][\mu'']L_1''L_2''L''}$$
$$\times \left[\left| \begin{matrix} [\nu'] \\ m'\omega', \beta' l_1^{f_1}[\sigma'] l_2^{f_{12}-f_1}[\mu'] L_1' L_2' L' \end{matrix} \right\rangle \left| \begin{matrix} [\nu''] \\ m''\omega'', \beta'' l_1^{f_3}[\sigma''] l^{f_{34}-f_3}[\mu''] L_1'' L_2'' L'' \end{matrix} \right\rangle \right]^L_M , \tag{9-22}$$

where the sum runs over $\beta'\sigma'\mu' L_1' L_2' L' \beta''\sigma''\mu'' L_1'' L_2'' L'' f_1$ and f_2; for the meaning of the quantum numbers in (9-22), see (7-261) and (7-266). The left-hand side of (9-22) is the $S_f \supset S_{f_{12}} \otimes S_{f_{34}}$ and $SU_{m+n} \supset SU_m \otimes SU_n$ irreducible basis.

We now proceed to derive an expression for the mixed configuration CFP. We first expand the left-hand side of (9-22) in terms of the $SU_{m+n} \supset SU_m \otimes SU_n$ ISF

$$(I) = \sum_{\substack{\beta'\sigma'\mu'\theta \\ \beta''\sigma''\mu''\varphi}} C^{[\nu]\tau,\beta[\sigma]\theta[\mu]\varphi}_{[\nu']\beta'[\sigma'][\mu'],[\nu'']\beta''[\sigma''][\mu'']}$$
$$\times \left[\left| \begin{matrix} [\nu'] \\ m'\omega', \beta'[\sigma'][\mu'] \end{matrix} \right\rangle \left| \begin{matrix} [\nu''] \\ m''\omega'', \beta''[\sigma''][\mu''] \end{matrix} \right\rangle \right]^{[\sigma]\theta[\mu]\varphi}_{L_1L_2LM} \tag{9-23a}$$

and then the square bracket term in (9-23a) is, in turn, expanded in terms of the $SU_m \supset SO_3 \supset SO_2$ and $SU_n \supset SO_3 \supset SO_2$ CG coefficients; thus we have

$$(I) = \sum C^{[\nu]\tau,\beta[\sigma]\theta[\mu]\varphi}_{[\nu']\beta'[\sigma'][\mu'],[\nu'']\beta''[\sigma''][\mu'']} C^{[\sigma]\theta L_1}_{[\sigma']L_1',[\sigma'']L_1''} C^{[\mu]\varphi L_2}_{[\mu']L_2',[\mu'']L_2''}$$
$$\times C^{L_1M_1}_{L_1'M_1',L_1''M_1''} C^{L_2M_2}_{L_2'M_2',L_2''M_2''} C^{LM}_{L_1M_1,L_2M_2}$$
$$\times \left| \begin{matrix} [\nu'] \\ m'\omega', \beta'[\sigma'][\mu'] L_1' M_1' L_2' M_2' \end{matrix} \right\rangle \left| \begin{matrix} [\nu''] \\ m''\omega'', \beta''[\sigma''][\mu''] L_1'' M_1'' L_2'' M_2'' \end{matrix} \right\rangle . \tag{9-23b}$$

Comparing (9-22) with (9-23b) we get

$$C^{[\nu]\tau,\beta[\sigma][\mu]L_1L_2L}_{[\nu']\beta'[\sigma'][\mu']L_1'L_2'L';[\nu'']\beta''[\sigma''][\mu'']L_1''L_2''L''}$$
$$= \sum_{\theta\varphi} C^{[\nu]\tau,\beta[\sigma]\theta[\mu]\varphi}_{[\nu']\beta'[\sigma'][\mu'],[\nu'']\beta''[\sigma''][\mu'']} C^{[\sigma]\theta L_1}_{[\sigma']L_1',[\sigma'']L_1''} C^{[\mu]\varphi L_2}_{[\mu']L_2',[\mu'']L_2''} \begin{pmatrix} L_1' & L_2' & L' \\ L_1'' & L_2'' & L'' \\ L_1 & L_2 & L \end{pmatrix} . \tag{9-24}$$

This shows that the mixed configuration CFP can be expressed in terms of the $SU_{m+n} \supset SU_m \otimes SU_n$ ISF, the pure-configuration CFP $C^{[\sigma]\theta L_1}_{\sigma' L_1', \sigma'' L_1''}$ and $C^{[\mu]\varphi L_2}_{\mu' L_2', \mu'' L_2''}$ and the $9j$ coefficients of SU_2.

Kukulin (1967) gave another expression for the mixed configuration CFP in terms of the 9ν coefficients of SU_{m+n}, which, after using the property (7-222) of the 9ν coefficients, reads

$$\begin{aligned} &C^{[\nu]\tau,\beta[\sigma][\mu]L_1L_2L}_{[\nu']\beta'[\sigma'][\mu']L_1'L_2'L';[\nu'']\beta''[\sigma''][\mu'']L_1''L_2''L''} \\ &= \varepsilon \binom{f_{13}}{f_3}^{1/2} \binom{f_{24}}{f_2}^{1/2} \binom{f}{f_{34}}^{-1/2} \left(\frac{h_{\sigma'}h_{\sigma''}h_{\mu'}h_{\mu''}h_{\nu}}{h_\sigma h_\mu h_{\nu'} h_{\nu''}}\right)^{1/2} \\ &\times \sum_{\theta\varphi} C^{[\sigma]\theta L_1}_{[\sigma']L_1',[\sigma'']L_1''} C^{[\mu]\varphi L_2}_{[\mu']L_2',[\mu'']L_2''} \begin{pmatrix} L_1' & L_2' & L' \\ L_1'' & L_2'' & L'' \\ L_1 & L_2 & L \end{pmatrix} \begin{pmatrix} [\sigma'] & [\mu'] & [\nu'] \\ [\sigma''] & [\mu''] & [\nu''] \\ [\sigma] & [\mu] & [\nu] \end{pmatrix}^{\beta'\beta''\tau}_{\theta\varphi\beta}, \end{aligned} \tag{9-25}$$

where $\varepsilon = \pm 1$ is a phase factor. Comparing (9-24) with (9-25) we obtain a relation between the $SU_{m+n} \supset SU_m \otimes SU_n$ ISF and SU_{m+n} 9ν coefficients

$$\begin{aligned} C^{[\nu]\tau,\beta[\sigma]\theta[\mu]\varphi}_{[\nu']\beta'[\sigma'][\mu'];[\nu'']\beta''[\sigma''][\mu'']} &= \varepsilon \binom{f_{13}}{f_3}^{1/2} \binom{f_{24}}{f_4}^{1/2} \binom{f}{f_{34}}^{-1/2} \\ &\times \left(\frac{h_{\sigma'}h_{\sigma''}h_{\mu'}h_{\mu''}h_{\nu}}{h_\sigma h_\mu h_{\nu'} h_{\nu''}}\right)^{1/2} \begin{pmatrix} [\sigma'] & [\mu'] & [\nu'] \\ [\sigma''] & [\mu''] & [\nu''] \\ [\sigma] & [\mu] & [\nu] \end{pmatrix}^{\beta'\beta''\tau}_{\theta\varphi\beta}. \end{aligned} \tag{9-26}$$

From (7-229) and (9-26) we obtain a symmetry property of the SU_{m+n} 9ν coefficients

$$\begin{pmatrix} [\sigma'] & [\mu'] & [\nu'] \\ [\sigma''] & [\mu''] & [\nu''] \\ [\sigma] & [\mu] & [\nu] \end{pmatrix}^{\beta'\beta''\tau}_{\theta\varphi\beta} = \begin{pmatrix} [\sigma'] & [\sigma''] & [\sigma] \\ [\mu'] & [\mu''] & [\mu] \\ [\nu'] & [\nu''] & [\nu] \end{pmatrix}^{\theta\varphi\beta}_{\beta'\beta''\tau}. \tag{9-27}$$

As to the evaluation of matrix elements of one-body and two-body operators in terms of the mixed configuration CFP, the reader is referred to Kukulin (1967) and Chen (1988).

Equation (9-26) provides another method for computing the 9ν coefficients of the unitary group, i.e., through the $SU_{m+n} \supset SU_m \otimes SU_n$ ISF.

9.5. The Dynamic Symmetry Models of Nuclei

Nuclear many-body problem is notorious for its difficulty, since the nuclei consists of many strongly interacting particles whose number is neither so small that exact calculation is possible nor large enough for statistical techniques to be applicable. As a consequence, model theories play a decisive role for understanding the nuclei. Among the various nuclear models, the shell model is the underpinning of any microscopic model of nuclei. Its basic idea is that as first approximation, nucleons can be described as moving independently in an average field provided by all other nucleons. The average field produces a series (in fact an infinite number of) single particle levels $nl\,jm$ exhibiting a shell structure, where n is the principle quantum number and $\mathbf{j} = \mathbf{l} + \frac{1}{2}$. The next step is to take into account the residual interaction between nucleons, which has two important ingredients, i.e., the short-range force and long-range force. The former favors spheric shape for nuclei while the latter deformed one. The main feature of the low-lying nuclear spectroscopy can be accounted qualitatively by the interplay of these two kinds of residual interactions. However detailed quantitative calculation is extremely difficult due to the huge dimension of the Hilbert space one meets as soon as many valence nucleons are present.

Therefore the shell model calculation is only restricted to small mass nuclei (the $1p$ shell and $2s, 1d$ shell nuclei) and a very few nuclei in the neighborhood of double magic nuclei.

On the other hand, the relatively simple structure of the low-lying spectra of nuclei prompts the developments of phenomenological collective models. The simplicity in the spectra was interpreted as arising from collective vibration and rotation of the nuclei. Among these the most prominent and successful ones are the Bohr-Mottelson (1953) geometric model and the Arima-Iachello (1976-79) interacting boson model (IBM).

The microscopic interpretation of such collective motion is a challenging problem. However great simplification occurs when the system has a dynamic symmetry, which means, as we have already mentioned in Sec. 3.19.3, that its Hamiltonian can be written in terms of the generators of a Lie algebra and is only a linear function of the Casimir operators of a complete chain of groups. For such cases the eigenvalue and eigenfunction of H can have analytic expressions. The first dynamic symmetry nuclear model is the Elliott model, however it is the IBM which pushes the nuclear dynamic symmetry to the foreground and revives the interest of nuclear physicists in group theory. An important lesson one learns from these pioneering works is that each dynamic symmetry is associated with a particular collective mode.

In the following sections we shall introduce both the boson dynamic symmetry model, i.e., the IBM, and several fermion dynamic symmetry models (FDSM), which include the pairing and neutron-proton pairing models, the Elliott model (Elliott, 1958) and the SO_8 and SP_6 FDSM (Ginocchio 1980, Wu, 1987). From these exactly solvable models one can see clearly the power and elegance of group theory in its application to the many-body problem.

As we will see that all these FDSM's have a common group structure that the dynamic G involves particle number non-conserving operators, which consist of the raising operators and lowering operators, and contains a subgroup $G_1 \times G_2$, where G_1 is a particle number conserving one and is called the core group, while $G_2 = U_1 \sim SO_2$ has the particle number operator as its generator. In most cases discussed here the group chain $G \supset G_1 \times G_2$ has the form of $SO_n \supset SO_{n-2} \times SO_2$, with $n = 3, 5, 7$ and 8. The vector coherent state theory (Rowe 1985, Hecht 1987) provided an elegant way for dealing with such algebras.

In the following we use $|\ \ \rangle$ and $|\ \ )$ to represent a boson and a fermion state, respectively.

9.6. The Quasispin Model

Let us consider n nucleons in a single j-shell and suppose that the short-range residual interaction is the predominant one, which is approximated by the pairing force

$$H_{SU2} = -G_0 S^\dagger S \ , \tag{9-28}$$

$$S^\dagger = S_+ = \sqrt{\frac{\Omega}{2}}[a_j^\dagger a_j^\dagger]_0^0 = \sum_{m>0} s_+^{(m)} \ , \tag{9-29}$$

where $\Omega = j + 1/2$, $S^\dagger$ is the creation operator for a pair of nucleons with $J = 0$, and

$$s_+^{(m)} = (-)^{j-m} a_m^\dagger a_{-m}^\dagger \ , \quad a_m^\dagger \equiv a_{jm}^\dagger \ . \tag{9-30a}$$

Introducing

$$s_0^{(m)} = \frac{1}{2}(\widehat{n}_m + \widehat{n}_{-m} - 1), \quad s_-^{(m)} = (s_+^{(m)})^\dagger \ . \tag{9-30b,c}$$

we have the commutator

$$[s_+^{(m)}, s_-^{(m)}] = 2s_0^{(m)}, \quad [s_0^{(m)}, s_\pm^{(m)}] = \pm s_\pm^{(m)} \ . \tag{9-30d}$$

Thus $(s_+^{(m)} s_-^{(m)}, s_0^{(m)})$ form an SU_2 algebra. From (9-30b) we see that

$$s_0^{(m)} = \begin{cases} 1/2 & (m,-m) \text{ is full} \\ -1/2 \quad \text{if} & (m,-m) \text{ is empty} \\ 0 & (m,-m) \text{ is singly occupied} . \end{cases} \tag{9-31a}$$

Therefore one can associate each pair level $(m, -m)$ with a "quasispin" $\mathbf{s}^{(m)}$ (Kerman, 1961, Eisenberg 1976),

$$\mathbf{s}^{(m)} = \begin{cases} 1/2 & \text{if} \quad (m,-m) \text{ is full or empty} \\ 0 & \phantom{\text{if}} \quad (m,-m) \text{ is singly occupied} . \end{cases} \tag{9-31b}$$

The total quasispin operators are denoted by S_+, S_-, S_0, where $S_- = (S_+)^\dagger$, and

$$S_0 = \sum_{m>0} s_0^{(m)} = \frac{1}{2}(\hat{n} - \Omega), \quad \hat{n} = \sum_m a_m^\dagger a_m . \tag{9-32}$$

The commutators between S_+, S_-, and S_0 are identical with (9-30d). S_+, S_- and S_0 form the quasispin SU_2 algebra. We can take over all the results from the angular momentum theory; for example, its irreps can be labeled by the quasispin S. The Chevalley basis for the quasispin SU_2 is,

$$E_\alpha = S_+, \quad E_{-\alpha} = S_-, \quad H_\alpha = 2S_0 = \hat{n} - \Omega, \quad [E_\alpha, E_{-\alpha}] = H_\alpha . \tag{9-33}$$

The pairing Hamiltonian (9-28) can be expressed in terms of the Casimir operator $\mathbf{S}^2$ of the quasispin SU_2 group,

$$H_{SU2} = -G_0[\mathbf{S}^2 - S_0(S_0 - 1)] . \tag{9-34}$$

Therefore the pairing Hamiltonian has the quasispin SU_2 as its dynamic group.

The particle number non-conserving Lie group SU_2 has the structure of $SO_3 \supset SO_1 \times SO_2$, the core group SO_1 is the trivial identity transformation group and the generator of SO_2 is $S_0 = (\hat{n} - \Omega)/2$. The irreducible basis of the core group SO_1 and the lowest weight (l.w.) state of SU_2 is denoted by $|v, \alpha JM\rangle$ which satisfies,

$$S_-|v, \alpha JM\rangle = 0 , \tag{9-35}$$

where α represents extra quantum numbers other than JM for labeling the v-particle state with angular momentum JM. The quantum number v is called the *seniority* which is the number of nucleons entirely free of S pairs. In the language of coherent state theory, the state $|v\alpha JM\rangle$ is called the intrinsic state. To avoid confusion with the intrinsic state defined by (3-151), we call it coherent intrinsic state. According to (9-32) the weight for the l.w. state is $(S_0)_{\min} = \frac{1}{2}(v - \Omega)$. Hence the highest weight is

$$S = -(S_0)_{\min} = \frac{1}{2}(\Omega - v) . \tag{9-36}$$

Relation (9-36) can be easily understood by noting that when all Ω pair levels are full or empty, S reaches its maximal value $\frac{\Omega}{2}$, a sum of Ω spin-$\frac{1}{2}$ entities; for a state with seniority v, v pair levels are singly occupied, and consequently only $(\Omega - v)$ pair levels contribute to the total quasispin and S is thus reduced to $\frac{1}{2}(\Omega - v)$.

It is more convenient to use the seniority v, which has direct physical meaning, instead of the quasispin S to label irreps of the quasispin SU_2. Similarly we use $n = v, v+2, v+4, \ldots, 2\Omega - v$, instead of $S_0 = -S, -S+1, \ldots, S$, as the component index of the irrep $v = \Omega - 2S$. By applying a power of the raising operator $S^\dagger$ to the coherent intrinsic state, we can obtain other states of the same irrep v, since $S^\dagger$ is the generator of SU_2 and thus does not change the irrep v. Therefore the eigenfunction of H_{SU_2} can be written as

$$|v, n\alpha JM\rangle = \eta_{nv}(S^\dagger)^{(n-v)/2}|v\alpha JM\rangle \; . \tag{9-37}$$

The following equation is very useful for evaluating the norm η_{nv}. Suppose that

$$[A, B] = X, \quad [[A, B], B] = Y, \quad [Y, B] = 0 \; , \tag{9-38}$$

then we have the identity

$$\begin{aligned}
[A, B^n] &= \sum_{i=0}^{n-1} B^i X B^{n-i-1} \\
&= \sum_{i=0}^{n-1} \left[B^i \sum_{j=0}^{n-i-2} B^j Y B^{n-i-j-2} + B^{n-1} X \right] \\
&= \sum_{i=0}^{n-1} (n-i-1) B^{n-2} Y + n B^{n-1} X \\
&= \binom{n}{2} B^{n-2} Y + n B^{n-1} X \; .
\end{aligned} \tag{9-39}$$

Using (9-30d), i.e.

$$[S, S^\dagger] = -2S_0 = \Omega - \hat{n} \; , \tag{9-40}$$

as well as (9-39) and $\hat{n}|v\alpha JM\rangle = v|v\alpha JM\rangle$, we have

$$\begin{aligned}
\eta_{nv}^{-2} &= \langle v\alpha JM|S^{(n-v)/2}(S^\dagger)^{(n-v)/2}|v\alpha JM\rangle \\
&= \left(\frac{n-v}{4}\right)(2\Omega - n - v + 2)\eta_{n-2,v}^{-2} \; .
\end{aligned} \tag{9-41a}$$

By induction we obtain the norm

$$\eta_{nv} = \left[\frac{2^{n-\Omega}(2\Omega - n - v)!!}{(n-v)!!(\Omega - v)!} \right]^{1/2} \; . \tag{9-41b}$$

The eigenvalue of H_{SU2} is obtained from (9-34), (9-32) and (9-36),

$$E_{nv} = -(G_0/4)(n-v)(2\Omega - n - v + 2) \; . \tag{9-42}$$

Since $G_0 > 0$ for attractive pairing force, we see that the state with lowest seniority, i.e., the state with as many as possible S pairs, lies lowest in energy. The highest weight state is

$$|v, n = 2\Omega - v, \alpha, JM\rangle = \frac{1}{(\Omega - v)!}(S^\dagger)^{\Omega - v}|v\alpha JM\rangle \; . \tag{9-43}$$

Notice that the vacuum state $|0\rangle$ belongs to the irrep with the highest quasispin $\Omega/2$ rather than the identity rep of SU_2.

9.7. The Proton-Neutron Quasispin Model

In the previous section we consider only one kind of nucleons. Now we introduce the proton-neutron quasispin model (Parikh 1965, Hecht 1985). The single particle creation operator is denoted by $a^\dagger_{mt_z} \equiv a^\dagger_{jm,\frac{1}{2}t_z}$ with $t_z = 1/2$ for proton and $t_z = -1/2$ for neutron. The shell degeneracy is enlarged to $\Omega = 2j + 1$ and the pairing Hamiltonian is extended to

$$H_{SO5} = -G_0 \sum_{M_T} S^\dagger(M_T) S(M_T) = -G_0 \mathbf{S}^\dagger \cdot \mathbf{S} \ , \tag{9-44}$$

$$S^\dagger(M_T) = \sqrt{\frac{\Omega}{2}} [a^\dagger_{j\frac{1}{2}} a^\dagger_{j\frac{1}{2}}]^{0\,1}_{0\,M_T} \ . \tag{9-45}$$

The operator $S^\dagger(M_T)$ creates a pair of nucleons with $J = 0$ and $T = 1, T_z = M_T$. The total isospin operators are

$$T_\mu = \sqrt{\frac{\Omega}{2}} [a^\dagger_{j\frac{1}{2}} \tilde{a}_{j\frac{1}{2}}]^{0\,1}_{0\,\mu}, \quad T_{\pm 1} = \mp \sqrt{\frac{1}{2}} T_\pm, \quad T_0 = T_z \ , \tag{9-46a}$$

$$\tilde{a}_{jm,\frac{1}{2}\mu} = (-)^{j+m+\frac{1}{2}+\mu} a_{j-m,\frac{1}{2}-\mu} \ . \tag{9-46b}$$

The following operators form the Cartan-Weyl basis for $SO_5 \sim SP_4$,

$$\begin{aligned} &H_1 = \frac{1}{2}(\hat{n} - \Omega), \quad H_2 = T_z = \frac{1}{2}(\hat{n}_1 - \hat{n}_2), \quad E_{01} = \sqrt{\frac{1}{2}} T_+ \\ &E_{11} = \sqrt{\frac{1}{2}} S^\dagger(1), \quad E_{10} = \sqrt{\frac{1}{2}} S^\dagger(0), \qquad E_{1-1} = \sqrt{\frac{1}{2}} S^\dagger(-1) \end{aligned} \tag{9-47}$$

$$\begin{aligned} &E_{-i-k} = (E_{ik})^\dagger, \quad ik = 11, 10, 1-1, 01. \\ &\hat{n} = \hat{n}_1 + \hat{n}_2, \qquad \hat{n}_\mu = \sum_m a^\dagger_{m\mu} a_{m\mu} \ , \end{aligned} \tag{9-48}$$

where $\mu = 1/2$ and $-1/2$ are indexed as 1 and 2 respectively. The Lie algebra has the structure of $so_5 \supset so_3 \oplus so_2$ (or $sp_4 \supset su_2 \oplus u_1$). The core algebra is $su_2 \sim so_3 = (T_1, T_{-1}, T_z)$, while $so_2 = H_1$. The raising operators are $S^\dagger(M_T)$. The commutator relations are

$$[H_1, H_2] = 0, \quad [H_1, E_{ik}] = iE_{ik}, \quad [H_2, E_{ik}] = kE_{ik} \ ,$$

$$[E_{ik}, E_{i'k'}] = \begin{cases} NE_{i+i',k+k'}, \text{if } (i+i', k+k') \text{ is a root, } N = \pm 1, \\ 0, \text{otherwise} \end{cases} \tag{9-49}$$

$$[E_{ik}, E_{-i-k}] = iH_1 + kH_2 \ .$$

The simple roots are $(1, -1)$ and $(0, 1)$ while the Chevalley basis is

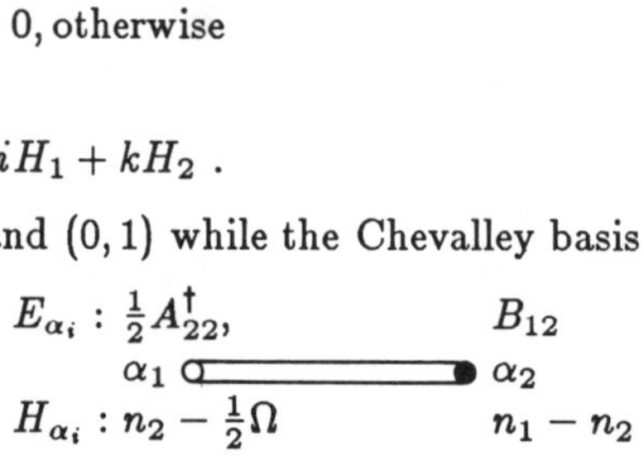

Fig. 9.7. The Chevalley basis of SO_5.

where

$$A^\dagger_{\alpha\beta} = \sum_m (-)^{j-m} a^\dagger_{m\alpha} a^\dagger_{-m\beta} \ , \tag{9-50a}$$

$$B_{\alpha\beta} = \sum_m a^\dagger_{m\alpha} a_{m\beta} - \frac{\Omega}{2} \delta_{\alpha\beta} \ . \tag{9-50b}$$

We see that $E_{\alpha_1} = -E_{1-1}$ and $E_{\alpha_2} = \sqrt{2}E_{01}$. The eigenvalue of (H_1, H_2), denoted as $[L_1, L_2]$, is the Cartan-Weyl label of SO_5, while the eigenvalue (a_1, a_2) of $(H_{\alpha_1}, H_{\alpha_2})$ is the Dynkin label of SO_5. The relation between the two is $L_1 = a_1 + a_2/2, L_2 = a_2/2$. The Casimir operator of SO_5 is

$$\begin{aligned} C_{SO5} &= H_1^2 + H_2^2 + \sum_{\alpha>0}(E_\alpha E_{-\alpha} + E_{-\alpha}E_\alpha) \\ &= \mathbf{S}^\dagger \bullet \mathbf{S} + \mathbf{T}^2 - H_1(-H_1+3) \, . \end{aligned} \tag{9-51}$$

Using (9-51) we have

$$H_{SO5} = -G_0[C_{SO5} + H_1(-H_1+3) - \mathbf{T}^2] \, . \tag{9-52}$$

Therefore H_{SO5} is a function of the Casimir operators of the group chain $SO_5 \supset SO_3$ and thus has a dynamic symmetry. The eigenvalue of C_{SO5} can be found from (5-352f),

$$C_{SO5} = L_1(L_1+3) + L_2(L_2+1) \, . \tag{9-53}$$

The irreducible basis function $|v; tt_z, \alpha JM\rangle$ of the core group SU_2 which are totally free of any S pairs,

$$S(M_T)|v; tt_z, \alpha JM\rangle = 0, \quad \text{for } t_z = t, \dots, -t, \quad \text{and} \quad M_T = 0, \pm 1, \tag{9-54}$$

are called vector coherent states, which are states of v nucleons with isospin t and the z-components t_z, v is the seniority and t is called the reduced isospin.

Applying $(H_{\alpha_1}, H_{\alpha_2})$ to the state $|v; t, -t, \alpha JM\rangle$ we find the lowest weight $[\frac{v-\Omega}{2}, -t]$ for the SO_5 irrep generated by the coherent intrinsic states. The highest weight of SO_5 is

$$[L_1 L_2] = \left[\frac{\Omega - v}{2}, t\right] \, . \tag{9-55}$$

Therefore we can use (v, t) as the irrep label for SO_5. A general irreducible basis can be constructed (Hecht 1985) by acting $p = (n-v)/2$ pair creation operators with the resultant isospin T_p on the vector coherent intrinsic states and by coupling t and T_p to the total isospin T in terms of the SU_2 CG coefficients

$$|n, (v,t)TM_T, \alpha JM\rangle = [P_{pT_p}(\mathbf{S}^\dagger)|v; t, \alpha JM]^T_{M_T} \, , \tag{9-56a}$$

where $P_{pT_pM_p}(\mathbf{S}^\dagger)$, called the raising polynomial, is a product of p pair creation operators $S^\dagger(M_p)$ coupled to $T_p = p, p-2, \dots, 0$ or 1,

$$P_{pT_pM_p}(\mathbf{S}^\dagger) = \mathcal{N}_{pT}(\mathbf{S}^\dagger \bullet \mathbf{S})^{(p-T_p)/2}\mathbf{Y}_{T_pM_p}(\mathbf{S}^\dagger) \, , \tag{9-56b}$$

$$\mathcal{N}_{pT} = \left[\frac{4\pi(p+T_p)!!}{(p-T_p)!!(p+T_p+1)!}\right]^{1/2} \, , \tag{9-56c}$$

where $\mathbf{Y}_{T_pM_p}(\mathbf{S}^\dagger)$ is the solid spheric harmonics $[\mathbf{Y}_{LM}(\mathbf{r}) = r^L Y_{LM}(\theta\varphi)]$. Notice that the states (9-56a) are not normalized, and if several T_p are possible for fixed n, t and T, then the states with different T_p have a nonzero overlap. The problem of constructing orthonormalized wave functions is solved satisfactorily by the vector coherent state theory (Rowe 1985, Hecht 1987).

The eigenvalue of H_{SO5} in the state (9-56a) is obtained from (9-52), (9-47) and (9-55),

$$E_{nvt} = -G_0\left[\frac{1}{4}(n-v)(2\Omega - n - v + 6) + t(t+1) - T(T+1)\right] \, . \tag{9-57}$$

The expression (9-57) is very similar to (9-42). We see that the state with low seniority and low isospin T lies low in energy.

9.8. The Elliott Model

Now we consider the other extreme case where the long-range force predominates. The long-range force can be simulated by the following separable force, called quadrupole-quadrupole $(Q-Q)$ force

$$H_{SU3} = -B_2 \sum_{\mu} (-)^{\mu} P^2_{\mu} P^2_{-\mu} = -B_2 P^2 \bullet P^2 \ , \tag{9-58}$$

$$P^2_{\mu} = \sum_{j=1}^{n} q^{(j)}_{\mu} \ , \tag{9-59a}$$

$$q_{\mu} = \sqrt{\frac{\pi}{10}} i^2 [r^2 Y_{2\mu}(\mathbf{r}) + p^2 Y_{2\mu}(\mathbf{p})] = -\sqrt{\frac{1}{8}} q_{\mu}(\text{Elliott}) \ . \tag{9-59b}$$

Within a single major shell, the operator q_{μ} is equivalent to

$$q_{\mu} = \sqrt{\frac{2\pi}{5}} r^2 i^2 Y_{2\mu}(\theta, \varphi) \ . \tag{9-59c}$$

Notice that we use $i^l Y_{l\mu}$ instead of $Y_{l\mu}$ as the basis function, and in (9-58) we included the one-body term $\sum_j q^{(j)} \bullet q^{(j)}$. Define the dipole operator P^1_{μ} by

$$P^1_{\mu} = \sqrt{3/8} L_{\mu} = \sqrt{3/8} \sum_{j=1}^{n} (\mathbf{r}_j \times \mathbf{p}_j) \ . \tag{9-60}$$

It is convenient to shift to the second quantized form. In the single particle basis $b^{\dagger}_{LM}|0)$, using the Wigner-Eckart theorem the one-body operator T^r_{μ} can be expressed as

$$T^r_{\mu} = \sum_{LL'} \sqrt{\frac{2L+1}{2r+1}} \langle L \| T^r \| L' \rangle (b^{\dagger}_L \tilde{b}_{L'})^r_{\mu} \ . \tag{9-61}$$

Now suppose that the nuclear average field is the harmonic oscillator potential. Using the oscillator wave functions, the reduced matrix elements of P^r_{μ} can be calculated and (9-61) can be written explicitly (Elliott 1958),

$$P^1_{\mu} = \sum_{L} \sqrt{\frac{L(L+1)(2L+1)}{8}} (b^{\dagger}_L \tilde{b}_L)^1_{\mu} \ , \tag{9-62a}$$

$$P^2_{\mu} = \frac{1}{\sqrt{8}} \sum_{L} \left[(2N+3) \sqrt{\frac{L(L+1)(2L+1)}{5(2L-1)(2L+3)}} (b^{\dagger}_L \tilde{b}_L)^2_{\mu} + \sqrt{\frac{6(L+1)(L+2)(N-L)(N+L+3)}{5(2L+3)}} (b^{\dagger}_L \tilde{b}_{L+2} + b^{\dagger}_{L+2} \tilde{b}_L)^2_{\mu} \right] \ . \tag{9-62b}$$

It should be mentioned that since we use $i^l Y_{l\mu}$ as basis, the sign of the first term is opposite to that given by Elliott.

For the $1p$ shell, $N = 1, L = 1$ and the quanta $b^\dagger_{1m}$ is denoted by $c^\dagger_m$. The harmonic oscillator potential has SU_3 as a symmetry group, whose defining basis is $c^\dagger_1|0), c^\dagger_{-1}|0)$ and $c^\dagger_0|0)$. $m = 1, -1$ and 0 is indexed as 1, 2 and 3, respectively. The states $m = 1, -1$ form the defining basis for an SU_2, whose irreps are labeled by the "spin" Λ. Letting $N = 1, L = 1$ in (9-62) we can obtain P^r_μ in terms of $c^\dagger_a c_b$, which can be put in the following unified form,

$$P^r_\mu = \frac{\sqrt{3}}{2} \sum_{ab} (-)^{1+b} C^{r\mu}_{1a,1b} B'_{a,-b}, \quad r = 1, 2 \ . \tag{9-63}$$

$$B'_{ab} = c^\dagger_a c_b \equiv B_{ab} - \frac{1}{2}\delta_{ab} \ . \tag{9-64}$$

The nine operators $B'_{ab}, a, b = 1, -1, 0$ are the generators of U_3. The commutators for P^r_μ are

$$[P^r, P^s]^t_\sigma = \frac{\sqrt{3}}{2}(-)^t[1 - (-)^{r+s+t}]\hat{r}\hat{s} \begin{Bmatrix} r & s & t \\ 1 & 1 & 1 \end{Bmatrix} P^t_\sigma, \quad t = 1, 2, \tag{9-65a}$$

where $\hat{r} = \sqrt{2r+1}$, and

$$[P^r, P^s]^t_\sigma \equiv \sum_{\mu\nu} C^{t\sigma}_{r\mu,s\nu}[P^r_\mu, P^s_\nu] \ . \tag{9-65b}$$

Equation (9-65) is just the commutators for the SU_3 Lie algebra. The weight operators are

$$H_1 = -\sqrt{8}P^2_0 = 2n_0 - n_1 - n_{-1}, \quad H_2 = L_Z = n_1 - n_{-1} \ . \tag{9-66}$$

Clearly, H_1 characterizes the size of deformation, and $H_2/2$ is the third component of the quasispin spin Λ.

In the sd shell, $N = 2, L = 0$ and 2, from (9-62) we have

$$\underline{\underline{P}}^2_\mu = (d^\dagger \tilde{s} + s^\dagger \tilde{d})^2_\mu + \frac{\sqrt{7}}{2}(d^\dagger \tilde{d})^2_\mu \ , \tag{9-67a}$$

$$\underline{\underline{P}}^1_\mu = \frac{\sqrt{15}}{2}(d^\dagger \tilde{d})^1_\mu \ , \tag{9-67b}$$

where we write $d^\dagger_m = b^\dagger_{2m}, s^\dagger = b^\dagger_{00}$. The commutators of $\underline{\underline{P}}^r_\mu$ can be evaluated with the result

$$[\underline{\underline{P}}^1, \underline{\underline{P}}^1]^t_\sigma = \frac{\sqrt{3}}{2}\delta_{t1}\underline{\underline{P}}^1_\sigma, \quad [\underline{\underline{P}}^2, \underline{\underline{P}}^2]^t_\sigma = \frac{\sqrt{15}}{2}\delta_{t1}\underline{\underline{P}}^1_\sigma \tag{9-68a}$$

$$[\underline{\underline{P}}^1, \underline{\underline{P}}^2]^t_\sigma = \sqrt{6}\delta_{t2}\underline{\underline{P}}^2_\sigma \ . \tag{9-68b}$$

Notice that the commutators (9-65) and (9-68) are exactly the same in spite of their different appearances.

The two-quanta creation operator $b^\dagger_{LM}, L = 0, 2$ can be expressed as

$$b^\dagger_{LM} = \sum_{ab} C^{LM}_{1a,1b} A^\dagger_{ab}, \quad A^\dagger_{ab} = c^\dagger_a c^\dagger_b \ . \tag{9-69}$$

Therefore $d^\dagger$ and $s^\dagger$ can be thought of as " compound particles", and $c_a^\dagger$ can be regarded as the "Elliott quark". $s^\dagger$ and $d^\dagger$ are the $SU_3 \supset SO_3$ basis belonging to the SU_3 irrep (20). On the other hand, $A_{ab}^\dagger$ are the $SU_3 \supset SU_2$ basis of the same irrep (20) of SU_3. From (9-69) we have,

$$d_2^\dagger = A_{11}^\dagger, \quad d_1^\dagger = \sqrt{2}A_{13}^\dagger\,, \quad d_0^\dagger = \sqrt{\frac{2}{3}}(A_{12}^\dagger + A_{33}^\dagger)\,, \tag{9-70a}$$

$$d_{-2}^\dagger = A_{22}^\dagger, \quad d_{-1}^\dagger = \sqrt{2}A_{23}^\dagger, \quad s^\dagger = \sqrt{\frac{1}{3}}(2A_{12}^\dagger - A_{33}^\dagger)\,, \tag{9-70b}$$

$$\sqrt{2}A_{12}^\dagger = \frac{1}{\sqrt{3}}(\sqrt{2}s^\dagger + d_0^\dagger), \quad A_{33}^\dagger = \frac{1}{\sqrt{3}}(\sqrt{2}d_0^\dagger - s^\dagger)\,. \tag{9-70c}$$

The Casimir operator of SU_3 is

$$C_{SU3} = \sum_{r=1}^{2} P^r \bullet P^r \tag{9-71a}$$

with the eigenvalue

$$C_{SU3}(\lambda\mu) = (\lambda^2 + \mu^2 + \lambda\mu + 3\lambda + 3\mu)/2\,. \tag{9-71b}$$

H_{SU3} can be written in terms of the Casimir operators of the group chain $SU_3 \supset SO_3$,

$$H_{SU3} = -B_2\left[C_{SU3} - \frac{3}{8}L(L+1)\right]\,. \tag{9-72a}$$

Elliott model exploits the symmetries and degeneracies of the harmonic-oscillator Hamiltonian and has found useful applications in the nuclei at the beginning of the $s-d$ shell, $16 < A < 24$. In the $s-d$ shell, there are six orbital states and four spin-isospin states. The totally anti-symmetric state is adapted to the following group chain,

$$\begin{array}{ccccc} SU_{24} \supset & (SU_6 \supset & SU_3 \supset & SO_3) & \times\, SU_4 \\ [1^n] & [f] & (\lambda\mu) & KLM & [\tilde{f}] \end{array} \tag{9-72b}$$

where $[f]$ is a partition of the integer n, the number of nucleons in the s-d shell, $\lambda = 2h_1 - 2h_2, \mu = 2h_2 - 2h_3, [2h_1, 2h_2, 2h_3]$ is the partition of the total quanta number $2n$, and K is the third component of $\mathbf{L}$ in the intrinsic coordinate system (Sec. 6.7). The $SU_6 \downarrow SU_3$ and $SU_3 \downarrow SO_3$ subduction rules are given in Elliott (1958). The orbital wave function is denoted by $|[f](\lambda\mu)KLM)$. Using (9-72a) we have,

$$H_{SU3}|[f](\lambda\mu)KLM) = E_L^{(\lambda\mu)}|[f](\lambda\mu)KLM) \tag{9-72c}$$

$$E_L^{(\lambda\mu)} = -\frac{B_2}{2}\left[C_{SU3}(\lambda\mu) - \frac{3}{8}L(L+1)\right]\,. \tag{9-72d}$$

The significance of the Elliott model is that for the first time it gave a microscopic description of nuclear collective rotations, and revealed the important fact that the SU_3 dynamic symmetry is associated with the rotational mode.

Similar to Moshinsky (1968), we can find the $SU_3 \supset SU_2 \times U_1$ basis for a given irrep $(\lambda\mu)$ from its highest weight state |h.w.),

$$|(\lambda\mu)\varepsilon\Lambda K) = \left|\begin{pmatrix} 2h_1\,2h_2\,2h_3 \\ h_{12}\,h_{22} \\ h_{11} \end{pmatrix}\right\rangle$$

$$= N_{h_{11}}^{h_{12}h_{22}} N_{h_{12}h_{22}}^{h_1h_2h_3} (B_{21})^{h_{12}-h_{11}} (B_{13})^{h_{12}-2h_2} (L_{23})^{h_{22}-2h_3} |\text{h.w.})\,, \tag{9-73a}$$

where the relation between the quantum numbers $(\lambda\mu), \varepsilon, \Lambda$ and K on the one hand, and the Gel'fand symbol on the other hand, is

$$\lambda = 2h_1 - 2h_2, \quad \mu = 2h_2 - 2h_3, \quad \varepsilon = 8n - 3(h_{12} + h_{22}) \ , \tag{9-73b}$$

$$\Lambda = (h_{12} - h_{22})/2, \qquad K = 2h_{11} - (h_{12} + h_{22}) \ , \tag{9-73c}$$

while

$$L_{23} = B_{23}(B_{11} - B_{22} + 1) - B_{21}B_{13} \ , \tag{9-73d}$$

$$N^{h_{12}h_{22}}_{h_{11}} / N^{h_{12}h_{22}}_{h_{11}-1} = [(h_{12} - h_{11} + 1)(h_{11} - h_{22})]^{1/2} \ , \tag{9-74a}$$

$$N^{h_1h_2h_3}_{h_{12}h_{22}} / N^{h_1h_2h_3}_{h_{12}-1h_{22}} = \left[\frac{(h_{12} - h_{22} + 1)}{(h_{12} - 2h_2)(h_{12} - 2h_3 + 1)(2h_1 - h_{12} + 1)}\right]^{1/2} \ , \tag{9-74b}$$

$$N^{h_1h_2h_3}_{h_{12}h_{22}-1} / N^{h_1h_2h_3}_{h_{12}h_{22}} = [(h_{12} - h_{22} + 2)(2h_1 - h_{22} + 2)(2h_2 - h_{22} + 1)(h_{22} - 2h_3)]^{1/2} \ . \tag{9-74c}$$

The highest weight state $|$ h.w.) can be constructed with the same technique as used in Castanos (1984) for the Sp(6, R) $\supset U_3$ wave functions and the result is as following,

$$\begin{aligned} |\text{h.w.}) = |(\lambda\mu)\varepsilon_h\Lambda_h K_h) &= \left| \begin{pmatrix} 2h_1\, 2h_2\, 2h_3 \\ 2h_2\, 2h_3 \\ 2h_2 \end{pmatrix} \right\rangle \\ &= N_{h_1h_2h_3}(A^\dagger_{33})^{h_1-h_2}(\Delta^\dagger_{22})^{h_2-h_3}(\Delta^\dagger)^{h_3}|0) \ , \end{aligned} \tag{9-75a}$$

where $\varepsilon_h = 2\lambda + \mu$, $\Lambda_h = \mu/2$, $K_h = \mu$, and

$$\Delta^\dagger = \begin{vmatrix} A^\dagger_{11} & A^\dagger_{12} & A^\dagger_{13} \\ A^\dagger_{12} & A^\dagger_{22} & A^\dagger_{23} \\ A^\dagger_{13} & A^\dagger_{23} & A^\dagger_{33} \end{vmatrix} \ , \tag{9-75b}$$

$$N_{h_1h_2h_3} = \left[\frac{h_2!2^{h_2-h_1-h_3}(h_1 - h_3 + 1)!(2h_1 - 2h_2 + 1)!!(2h_2 - 2h_3 + 1)!!}{(h_1 + 1)!(h_1 - h_2)!(h_2 - h_3)!h_3!(2h_2 + 1)!(2h_1 - 2h_3 + 1)!!}\right]^{1/2} \ , \tag{9-75c}$$

and $\Delta^\dagger_{ij}$ is the cofactor of the determinant $\Delta^\dagger = (= \Sigma_{ij} A^\dagger_{ij}\Delta^\dagger_{ij})$.

The $SU_3 \supset SU_2 \times U_1$ state is referred to as the intrinsic state, which is precisely the intrinsic state $\Phi^{(K)}_0(X)$ discussed in Sec. 6.6 and 6.7. By using (6-70) we can project out physical basis with good angular momentum from the intrinsic state.

Since the $Q - Q$ force is attractive, (9-72d) and (9-71b) show that the larger the deformation $\varepsilon_h = 2\lambda + \mu$, the larger the eigenvalue $C_{SU3}(\lambda\mu)$, and the lower the energy of the state.

9.9. The Interacting Boson Model

The Elliott model though elegant, its applicability to heavier nuclei is spoiled by the strong spin-orbit force, which pushes down the level $j = l + 1/2$ of the unique parity in the N-th shell down to the $(N - 1)$-th shell. However most of the rotational nuclei are medium-heavy and heavy nuclei. Arima and Iachello introduced the interacting boson model (IBM) and greatly extended the application region of the Elliott model. By re-interpreting the quanta creation operators $s^\dagger$ and $d^\dagger$ in the Elliott model as the creation operators of s and d bosons which are the approximations of the coherent fermion pairs with $J = 0$ and 2, respectively, we are led to the IBM. In the IBM, an even-even nuclei with n valence nucleons is approximated as a N-boson system, $N = n/2$, interacting through one and two-body forces. Each boson has two possible states, either $L = 0$ or 2, i.e., the s or d boson, and they span the defining rep of the SU_6 group. The IBM Hamiltonian can be conveniently expressed in terms of the boson multipole operators P^r_μ,

$$H_B = \varepsilon_s N_s + \varepsilon_d N_d + B_0 C_{U5} + \sum_{r=1}^{3} B_r P^r(\chi_r)\cdot P^r(\chi_r) \ , \tag{9-76a}$$

$$P^r_\mu(\chi_r) = \delta_{r2}(d^\dagger \tilde{s} + s^\dagger \tilde{d})^2_\mu + (-)^{[(r-1)/2]}\chi_r(d^\dagger \tilde{d})^r_\mu, \quad r = 1,2,3 \ , \tag{9-76b}$$

where N_s and N_d are numbers of s and d bosons, $N = N_s + N_d$ is the total number of bosons, and C_{U5} is the Casimir operator of U_5,

$$C_{U5} = N_d(N_d + 4) \ . \tag{9-77}$$

The group U_6 has three and only three subgroup chains that contain the rotation group SO_3,

$$\begin{array}{ll} & \nearrow U_5 \rightarrow SO_5 \rightarrow SO_3 \\ U_6 \rightarrow & SO_6 \rightarrow SO_5 \rightarrow SO_3 \\ & \searrow SU_3 \rightarrow SO_3 \end{array} \tag{9-78a}$$

Using the method in Sec. 5.23, we can find (Chen 1986) the Chevalley basis for the various subgroups in (9-78a) as well as bases for the basic and identity reps of these subgroups and the results are listed in Tables 9.9 and Figs. 9.9, where

$$\mathcal{B}_{ij} = b^\dagger_i b_j, \quad N_i = b^\dagger_i b_i \tag{9-78b}$$

and the indexing of the boson operators for each case are given in Table 9.9-1b, -2b and -3b.

Table 9.9-1a. The Chevalley basis of SO_5.

	$\mathcal{B}_{12} + \mathcal{B}_{45}$	$\sqrt{2}(\mathcal{B}_{23}+\mathcal{B}_{34})$	$\mathcal{B}_{14} + \mathcal{B}_{25}$	$\sqrt{2}(\mathcal{B}_{13} - \mathcal{B}_{45})$
$H_{\alpha_1} = N_1 - N_2 + N_4 - N_5$	2	−1	0	1
$H_{\alpha_2} = 2(N_2 - N_4)$	−2	2	2	0

Table 9.9-1b. Bases of the basic and identity reps of SO_5.

	$d^\dagger_2$ $b^\dagger_1$	$d^\dagger_1$ $b^\dagger_2$	$d^\dagger_0$ $b^\dagger_3$	$d^\dagger_{-1}$ $b^\dagger_4$	$d^\dagger_{\bar{2}}$ $b^\dagger_5$	$s^\dagger$ $b^\dagger_6$	$d^\dagger \cdot d^\dagger$
$H_{\alpha_1} = N_1 - N_2 + N_4 - N_5$	1	−1	0	1	−1	0	0
$H_{\alpha_2} = 2(N_2 - N_4)$	0	2	0	−2	0	0	0

Table 9.9-2a. The Chevalley basis and roots of SO_6.*

	$\mathcal{B}_{26} + \mathcal{B}_{34}$	$\mathcal{B}_{12} + \mathcal{B}_{45}$	$\mathcal{B}_{23} + \mathcal{B}_{64}$	$\mathcal{B}_{16} - \mathcal{B}_{35}$	$\mathcal{B}_{14} + \mathcal{B}_{25}$	$\mathcal{B}_{13} - \mathcal{B}_{65}$
$H_{\alpha_1} = q + l$	2	−1	0	1	1	−1
$H_{\alpha_2} = p - q$	−1	2	−1	1	0	1
$H_{\alpha_3} = q - l$	0	−1	2	−1	1	1

*$p = N_1 - N_5$, $q = N_2 - N_4$, $l = N_3 - N_6$.

Table 9.9-2b. Bases of the basic and identity reps of SU_3.

	$d_2^\dagger$	$d_1^\dagger$	$\sqrt{\frac{1}{2}}(d_0^\dagger + s^\dagger)$	$d_{-1}^\dagger$	$d_{-2}^\dagger$	$\sqrt{\frac{1}{2}}(d_0^\dagger - s^\dagger)$	$d^\dagger \cdot d^\dagger - s^\dagger \cdot s^\dagger$
	$b_1^\dagger$	$b_2^\dagger$	$b_3^\dagger$	$b_4^\dagger$	$b_5^\dagger$	$b_6^\dagger$	
$H_{\alpha_1} = N_2 + N_3 - N_4 - N_6$	0	1	1	−1	0	−1	0
$H_{\alpha_1} = N_1 - N_2 + N_4 - N_5$	1	−1	0	1	−1	0	0
$H_{\alpha_3} = N_2 - N_3 - N_4 + N_6$	0	1	−1	−1	0	1	0

Table 9.9-3a. The Chevalley basis and roots of SU_3.

	$\sqrt{2}(B_{12} + B_{23}) + B_{64}$	$\sqrt{2}(B_{34} + B_{45}) + B_{26}$	$\sqrt{2}(B_{16} + B_{65}) + B_{24}$
$H_{\alpha_1} = 2(N_1 - N_3) + N_6 - N_4$	2	−1	1
$H_{\alpha_2} = 2(N_3 - N_5) + N_2 - N_6$	−1	2	1

Table 9.9-3b. Basis of the irrep (20) of SU_3.*

$(\varepsilon \Lambda K)$	$(-2\,1\,2)$	$(1\,\frac{1}{2}\,1)$	(400)	$(1\,\frac{1}{2}\,-1)$	$(-2\,1\,-2)$	(-210)
	$A_{11}^\dagger$ $d_2^\dagger$ $b_1^\dagger$	$\sqrt{2}A_{13}^\dagger$ $d_1^\dagger$ $b_2^\dagger$	$A_{33}^\dagger$ $\sqrt{\frac{1}{3}}(\sqrt{2}\,d_0^\dagger - s^\dagger)$ $b_3^\dagger$	$\sqrt{2}A_{23}^\dagger$ $d_{-1}^\dagger$ $b_4^\dagger$	$A_{22}^\dagger$ $d_{-2}^\dagger$ $b_5^\dagger$	$\sqrt{2}A_{12}^\dagger$ $\sqrt{\frac{1}{3}}(d_0^\dagger + \sqrt{2}s^\dagger)$ $b_6^\dagger$
$H_{\alpha_1} = 2(N_1 - N_3) + N_6 - N_4$	2	0	−2	−1	0	1
$H_{\alpha_2} = 2(N_3 - N_5) + N_2 - N_6$	0	1	2	0	−2	−1

* $A_{ij}^\dagger$ are defined by (9-70).

The elements associated with the simple roots in the Chevalley basis can be vividly shown in the following Dynkin diagrams:

E_{α_i} : $B_{12} + B_{45}$ $\qquad$ $\sqrt{2}(B_{23} + B_{34})$

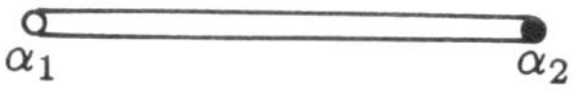

H_{α_i} : $N_1 - N_2 + N_4 - N_5$, $\quad 2(N_2 - N_4)$

Fig. 9.9-1. The Chevalley basis of SO_5.

E_{α_i} : $B_{26} + B_{34}$, $B_{12} + B_{45}$, $B_{23} + B_{64}$

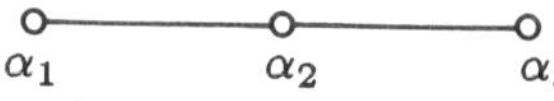

H_{α_i} : $\quad q + l \qquad p - q \qquad q - l$

Fig. 9.9-2. The Chevalley basis of SO_6.
$p = N_1 - N_5$, $q = N_2 - N_4$, $l = N_3 - N_6$

E_{α_i} : $\sqrt{2}(B_{12} + B_{23}) + B_{64}$, $\quad \sqrt{2}(B_{34} + B_{45}) + B_{26}$

H_{α_i} : $2(N_1 - N_3) + N_6 - N_4$, $2(N_3 - N_5) + N_2 - N_6$

Fig. 9.9-3. The Chevalley basis of SU_3.

The irreps of SO_5, SO_6, and SU_3 are labeled by $(\tau_1\tau_2), (\sigma_1\sigma_2\sigma_3)$ and $(\lambda\mu)$, respectively, which are the eigenvalues of $\{H_{\alpha_i}\}$ of the corresponding groups. From Table 9.9 we see that in the (s, d) boson space, $\tau_2 = \sigma_1 = \sigma_3 = 0$, and both λ and μ are even. Therefore the irreps of SO_5 and SO_6 in the IBM are simply labeled by $\tau = \tau_1$ and $\sigma = \sigma_2$.

When the strength parameters take some special values, the Hamiltonian (9-76a) has a dynamic symmetry.

A. U_5 limit:

When $B_2 = 0$, the Hamiltonian (9-76a) becomes

$$H_{U5} = \varepsilon_s N_s + \varepsilon_d N_d + B_0 C_{U5} + \sum_{r=1,3} B_r P^r(\chi_r)\bullet P^r(\chi_r) . \tag{9-79}$$

By keeping $B_r\chi_r$ invariant, χ_r can be chosen freely without affecting the result. Let $\chi_1 = \chi_3 = \sqrt{2}$

$$\wp^r_\mu \equiv P^r_\mu(\sqrt{2}) = (-)^{[(r-1)/2]}\sqrt{2}(d^\dagger\tilde{d})^r_\mu, \quad r = 1,3 . \tag{9-80a}$$

$$\wp^1_\mu = \sqrt{1/5}L_\mu . \tag{9-80b}$$

The ten operators $\wp^r_\mu, r = 1,3$, form the SO_5 Lie algebra with the commutators

$$[\wp^r, \wp^s]^t_\sigma = -\sqrt{2}[1-(-)^{r+s+t}]\ \hat{r}\,\hat{s}(-)^{[t/2]} \begin{Bmatrix} r\,s\,t \\ 2\,2\,2 \end{Bmatrix} \wp^t_\sigma . \tag{9-81}$$

For computing boson commutators the following formulas are useful

$$[b_i, f(b_i^\dagger, b_i)] = \frac{\partial}{\partial b_i^\dagger} f(b_i^\dagger, b_i) , \tag{9-82a}$$

$$[b_i^\dagger, f(b_i^\dagger, b_i)] = -\frac{\partial}{\partial b_i} f(b_i^\dagger, b_i) , \tag{9-82b}$$

where $f(b_i^\dagger, b_i)$ is a function of the boson operators $b_i^\dagger$ and b_i. The Casimir operator of SO_5 is

$$C_{SO5} = \sum_{r=1,3} \wp^r\bullet\wp^r . \tag{9-83}$$

Therefore H_{U5} can be re-written as

$$H_{U5} = \varepsilon_s N_s + \varepsilon_d N_d + B_0 C_{U5} + B_3 C_{SO5} + \frac{1}{5}(B_1 - B_3)\mathbf{L}^2 . \tag{9-84a}$$

It has the $U_5 \supset SO_5 \supset SO_3$ dynamic symmetry and its eigenvalue is

$$E_{U5} = \varepsilon_s N_s + \varepsilon_d N_d + B_0 N_d(N_d+5) + B_3\tau(\tau+3) + \frac{1}{5}(B_1 - B_3)L(L+1) , \tag{9-84b}$$

which resembles the spectra of an anharmonic vibrator.

According to Table 9.9-1b, $(d_2^\dagger)^\tau|0)$ is the highest weight state of the SO_5 irrep $(\tau, 0)$. Therefore

$$|\tau; JM) = |\tau; 2\tau, 2\tau) = \frac{1}{\sqrt{\tau!}}(d_2^\dagger)^\tau|0) . \tag{9-85}$$

The basis function for the irrep τ of SO_5 with JM other than $(2\tau, 2\tau)$ can be obtained by operating a definite function of $\wp^1$ and $\wp^3$ on (9-85),

$$\begin{aligned} |\tau n_\Delta JM) &= f_{n_\Delta JM}(\wp^1, \wp^3)|\tau, 2\tau, 2\tau) \\ &\equiv (d_\mu^\dagger)^\tau_{n_\Delta JM}|0) = \sum_{\sum t_\mu = \tau} C^{n_\Delta JM}_{t_2 \ldots t_{-2}} (d_2^\dagger)^{t_2} \ldots (d_{-2}^\dagger)^{t_{-2}}|0) \ , \end{aligned} \tag{9-86}$$

where n_Δ is an additional quantum number which counts the number of triplets of d bosons coupled to $J = 0$. Obviously we have

$$(d \bullet d)(d_2^\dagger)^\tau|0) = 0 \ . \tag{9-87}$$

Since $(d \bullet d)$ is an invariant of SO_5, i.e., $[\wp^r_\mu, (d \bullet d)] = 0$, from (9-86) and (9-87) we have

$$(d \bullet d)|\tau n_\Delta JM) = 0 \ . \tag{9-88}$$

Therefore the quantum number τ has the physical meaning that it counts the number of d bosons not coupled to $J = 0$, and is termed as the SO_5 *seniority*.

The general eigenfunctions for H_{U5} is

$$|NN_d\tau n_\Delta JM) = \eta_{NN_d\tau}(s^\dagger)^{(N-N_d)}(d^\dagger \bullet d^\dagger)^{(N_d-\tau)/2}|\tau n_\Delta JM) \ , \tag{9-89a}$$

$$\eta_{NN_d\tau} = \left[\frac{(2\tau+3)!!}{(N-N_d)!(N_d+\tau+3)!!(N_d-\tau)!!}\right]^{1/2} \ . \tag{9-89b}$$

The allowed values of N_d, τ and J are

$$\begin{aligned} &N_d = 0, 1, \ldots, N, \ \tau = N_d, N_d - 2, \ldots 0 \quad \text{or} \quad 1 \ , \\ &\tau = 3n_\Delta + \lambda, \qquad n_\Delta = 0, 1, 2, \ldots \\ &J = \lambda, \lambda + 1, \ldots, 2\lambda - 2, 2\lambda \ . \end{aligned} \tag{9-90}$$

B. The O_6 limit:

The condition for the O_6 limit to occur is

$$\varepsilon_s = \varepsilon_d, \quad \chi_2 = B_0 = 0 \ .$$

Besides (9-80) and (9-81), the SO_6 Lie algebra has the following extra generators and commutators

$$\wp^2_\mu = P^2_\mu(0) = (d^\dagger \tilde{s} + s^\dagger \tilde{d})^2_\mu \ , \tag{9-91}$$

$$[\wp^2, \wp^2]^t_\sigma = \sqrt{2}(-)^{[t/2]}\wp^t_\sigma, \ t = 1, 3 \ , \tag{9-92a}$$

$$[\wp^3, \wp^2]^t_\sigma = \sqrt{\frac{14}{5}}\delta_{t2}\wp^2_\sigma \ . \tag{9-92b}$$

$$H_{SO6} = \varepsilon_s N + B_2 C_{SO6} + (B_3 - B_2)C_{SO5} + \frac{1}{5}(B_1 - B_3)\mathbf{L}^2 \tag{9-93a}$$

$$C_{SO6} = \sum_{r=1}^{3} \wp^r \bullet \wp^r \ , \tag{9-94}$$

$$E_{SO6} = \varepsilon_s N + B_2\sigma(\sigma+4) + (B_3 - B_2)\tau(\tau+3) + \frac{1}{5}(B_1 - B_3)L(L+1) \ . \tag{9-93b}$$

The spectra given by (9-93b) corresponds to the γ unstable rotor.

From Table 9.9-2b we see that the state $(d_2^\dagger)^\tau|0)$ also belongs to the irrep $(0\tau 0)$ of SO_6, or the irrep $\sigma = \tau$. The invariant

$$\Im^\dagger = (d^\dagger \bullet d^\dagger - s^\dagger \bullet s^\dagger) \tag{9-95a}$$

of SO_6 is called a generalized pair. Clearly

$$(d\bullet d - s\bullet s)(d_2^\dagger)^\tau|0) = 0 \ . \tag{9-95b}$$

Thus $\sigma(=\tau)$ is the number of bosons totally free of the generalized pairs, and is termed as the SO_6 *seniority*. The state for $\sigma > \tau$ is constructed as follows

$$|N = \sigma, \sigma\tau, n_\Delta JM) = f_{\sigma\tau}(s^\dagger, \Im^\dagger)|\tau n_\Delta JM) \ , \tag{9-96a}$$

where $f_{\sigma\tau}(s^\dagger, \Im^\dagger)$ is a polynomial of $s^\dagger$ and $d_\mu^\dagger$ of order $(\sigma - \tau)$,

$$f_{\sigma\tau}(s^\dagger, \Im^\dagger) = \sum_{p=0} D_p(\sigma\tau)(s^\dagger)^{\sigma-\tau-2p}(\Im^\dagger)^p, \quad f_{\sigma\sigma}(s^\dagger, \Im^\dagger) = 1 \ , \tag{9-96b}$$

and the coefficients $D_p(\sigma\tau)$ are determined by the requirement that there are no generalized pairs in $|\sigma\sigma\tau n_\Delta JM)$,

$$\Im|\sigma\sigma\tau n_\Delta JM) = 0 \ . \tag{9-97}$$

Using the commutators

$$[\Im, \Im^\dagger] = 4(3 + \hat{N}), \quad [[\Im, \Im^\dagger], \Im^\dagger] = 8\Im^\dagger, \quad [\Im, s^\dagger] = 2s, \tag{9-98}$$

from (9-97) one obtains (Ginocchio 1980) a recursive formula for $D_p(\sigma\tau)$,

$$D_{p+1}(\sigma\tau) = \frac{(\sigma-\tau-2p)(\sigma-\tau-2p-1)}{4(p+1)(\sigma-p+1)} D_p(\sigma\tau) \ . \tag{9-99a}$$

By induction and inserting the value of $D_0(\sigma\tau)$, which is obtained from the normalization condition, we find the final expression for $D_p(\sigma\tau)$

$$D_p(\sigma\tau) = \left[\frac{2^{\sigma+1}(\sigma-\tau)!(2\tau+3)!!}{(\sigma+1)!(\sigma+\tau+3)!}\right]^{1/2} \frac{(\sigma+1-p)!}{4^p(\sigma-\tau-2p)!p!} \ . \tag{9-99b}$$

Finally the state with $n > \sigma$ is given by

$$|N\sigma\tau n_\Delta JM) = \xi_{N\sigma}(\Im^\dagger)^{(N-\sigma)/2}|\sigma\sigma\tau n_\Delta JM) \ , \tag{9-100}$$

where

$$\xi_{N\sigma} = \left[\frac{(2\sigma+4)!!}{(N+\sigma+4)!!(N-\sigma)!!}\right]^{1/2} \ . \tag{9-101}$$

The ranges of the quantum numbers are well known,

$$\sigma = N, N-2, \ldots, 0 \text{ or } 1 ,$$
$$\tau = \sigma, \sigma - 1, \ldots, 0 \tag{9-102}$$

C. The U_3 limit

The condition $\varepsilon_s = \varepsilon_d$, $B_3 = 0$ and $\chi_2 = -\sqrt{7}/2$ leads to the following boson Hamiltonian,

$$H_{SU3} = \varepsilon_s N + \sum_{r=1,2} B_r \underline{\underline{P}}^r \bullet \underline{\underline{P}}^r , \tag{9-103a}$$

$$\underline{\underline{P}}^1_\mu = \sqrt{\frac{3}{8}} L_\mu = P^1_\mu \left(\frac{\sqrt{15}}{2}\right), \quad \underline{\underline{P}}^2_\mu = P^2_\mu \left(\frac{-\sqrt{7}}{2}\right) , \tag{9-103b}$$

where $\underline{\underline{P}}^1_\mu$ and $\underline{\underline{P}}^2_\mu$ are precisely the operators given by (9-67) in the Elliott model. Using (9-71a), (9-103a) can be re-written as

$$H_{SU3} = \varepsilon_s N + B_2 C_{SU3} + \frac{3}{8}(B_1 - B_2)\mathbf{L}^2 , \tag{9-103c}$$

$$E_{SU3} = \varepsilon_s N + B_2 C_{SU3}(\lambda\mu) + \frac{3}{8}(B_1 - B_2) L(L+1) . \tag{9-103d}$$

The weight operators are $\hat{\varepsilon} = -\sqrt{8}\underline{\underline{P}}^2$ and $K = L_z$. The boson operators $s^\dagger$ and $d_0^\dagger$ are not the eigenoperators of $\hat{\varepsilon}$. They can be recombined into $b_3^\dagger$ and $b_6^\dagger$ which are eigenoperators of $\hat{\varepsilon}$,

$$b_3^\dagger = \sqrt{\frac{1}{3}}(\sqrt{2} d_0^\dagger - s^\dagger), \quad b_6^\dagger = \sqrt{\frac{1}{3}}(\sqrt{2} s^\dagger + d_0^\dagger) . \tag{9-104a}$$

The weight operators $\hat{\varepsilon}$ and K can now be expressed as

$$\hat{\varepsilon} = -\sum_\mu (-)^\mu |\mu| \hat{N}_\mu + 4\hat{N}_3 - 2\hat{N}_6, \quad K = \sum_\mu \mu \hat{N}_\mu , \tag{9-104b}$$

where $\hat{N}_\mu = d_\mu^\dagger d_\mu$, for $\mu = \pm 1, \pm 2$, and $\hat{N}_i = b_i^\dagger b_i$, for $i = 3, 6$.

The IBM SU_3 wave function can be obtained from the SU_3 wave function in the Elliott model, (9-73)-(9-75), simply by interpreting the s, d quanta as the s, d bosons and replacing the SU_3 lowering operators $B'_{ij} = c_i^\dagger c_j$ by (cf. Table 9.9-3a).

$$B_{21} = \sqrt{2}(b_6^\dagger d_2 + d_{-2}^\dagger b_6) + d_{-1}^\dagger d_1 ,$$
$$B_{23} = \sqrt{2}(d_{-2}^\dagger d_{-1} + d_{-1}^\dagger b_3) + b_6^\dagger d_1 ,$$
$$B_{13} = \sqrt{2}(d_2^\dagger d_1 + d_1^\dagger b_3) + b_6^\dagger d_{-1} . \tag{9-105a}$$

According to (9-70) and (9-104a), the determinant $\Delta^\dagger$ in (9-75b) can be written in terms of the boson operators as

$$\Delta^\dagger = \begin{vmatrix} d_2^\dagger & \sqrt{\frac{1}{2}} b_6^\dagger & \sqrt{\frac{1}{2}} d_1^\dagger \\ \sqrt{\frac{1}{2}} b_6^\dagger & d_{-2}^\dagger & \sqrt{\frac{1}{2}} d_{-1}^\dagger \\ \sqrt{\frac{1}{2}} d_1^\dagger & \sqrt{\frac{1}{2}} d_{-1}^\dagger & b_3^\dagger \end{vmatrix} . \tag{9-105b}$$

9.10. The SO_8 and SP_6 Fermion Dynamic Symmetry Model

The attractive feature of the IBM is that it is simple and has only a few parameters and yet is still powerful enough to correlate a vast amount of empirical data of the low-lying states of even-even nuclei. However a nuclei consists of fermions instead of bosons, one of the shortcomings of the IBM is that it totally neglected the Pauli effect. The SO_8 and SP_6 FDSM (Wu 1986, 1987), which is referred to as the FDSM in this section for brevity, remedies this shortcoming by starting from the fermion dynamic symmetry. This model is motivated by the Ginocchio (1980) model and has been applied to even-even and odd A nuclei with considerable success (Guidry 1986, 1987, Casten 1986, Han 1987, H. Wu 1987). In the FDSM, the normal and abnormal (or unique) parity single particle levels of a physical shell are re-classified in terms of $(ki)jm$, called the $k-i$ basis. The creation operator for the uncoupled $k-i$ basis is denoted by $b^\dagger_{km_k,im_i}$, which is related to the shell model single-particle creation operator $\xi^\dagger_{jm}$ by an SO_3 CG matrix

$$b^\dagger_{km_k,im_i} = \sum_j C^{jm}_{km_k,im_i} \xi^\dagger_{jm} \ , \tag{9-106}$$

where k and i are the pseudo orbital angular momentum and pseudo spin, respectively. The advantage of the $k-i$ basis is that we can easily truncate the huge shell model space to the (S, D) space consisting of only S and D fermion pairs by coupling either k or i to zero. When $k(i)$ are coupled to zero we say that the nucleons are $i(k)$-active. According to the $k-i$ classification, the normal parity levels in a major shell has either $i = 3/2$ or $k = 1$, while the abnormal level has $k = 0$ (and thus $i = j$).

In the simplest version of the FDSM, it is assumed that the nucleons in the abnormal orbit are always paired to $J = 0$, and the number of particles in the normal and abnormal orbits are kept fixed, and thus the S pairs in the abnormal level can be ignored for low-lying states, although the modification of the Ginocchio model by the inclusion of the abnormal level is a crucial step forward. In the following we first study the group structure of the FDSM, and for simplicity in notation, we assume that only a single $k(i)$ value matches with $i = 3/2$ ($k = 1$).

9.10.1. *The generators of SO_8 and SP_6*

A. The i-active case (i=3/2)

(i) Generators in the uncoupled representation

We indexed $m_i = 3/2, 1/2, -1/2, -3/2$ as 1,2,3,4. The following 28 ($= 6+6+16$) operators, $A^\dagger_{m_1m_2}$ (6), $A_{m_1m_2}$ (6) and $B_{m_1m_2}$ (16) are closed under commutation,

$$A^\dagger_{m_1m_2} = \sum_m (-)^{k-m} b^\dagger_{km,\frac{3}{2}m_1} b^\dagger_{k\bar{m},\frac{3}{2}m_2} = -A^\dagger_{m_2m_1} \ , \tag{9-107a}$$

$$B_{m_1m_2} = B'_{m_1m_2} - \frac{\Omega}{4}\delta_{m_1m_2}, \quad \Omega = 2(2k+1) \ , \tag{9-107b}$$

$$B'_{m_1m_2} = \sum_m b^\dagger_{km,\frac{3}{2}m_1} b_{km,\frac{3}{2}m_2}, \quad B_{mm} = n_m \ , \tag{9-107c}$$

where $\overline{m} = -m$, and n_m is the fermion number operator. The commutation relations are

$$[A_{ab}, A^\dagger_{cd}] = \delta_{ad}B_{cd} + \delta_{bc}B_{da} - \delta_{ac}B_{db} - \delta_{bd}B_{ca} \ , \tag{9-108a}$$

$$[B_{ab}, B_{cd}] = \delta_{bc}B_{ad} - \delta_{ad}B_{cb} \ , \tag{9-108b}$$

$$[B_{ab}, A^\dagger_{cd}] = \delta_{bc}A^\dagger_{ad} + \delta_{bd}A^\dagger_{ca} \ , \tag{9-108c}$$

$$[B_{ab}, A_{cd}] = -\delta_{ac}A_{bd} - \delta_{ad}A_{cb} \ . \tag{9-108d}$$

Therefore the 28 operators form a Lie algebra. Using the method given in Sec. 5.23 we can find the Chevalley basis of the algebra as shown in Fig. 9.10-1, which is easily recognized as the SO_8 Lie algebra.

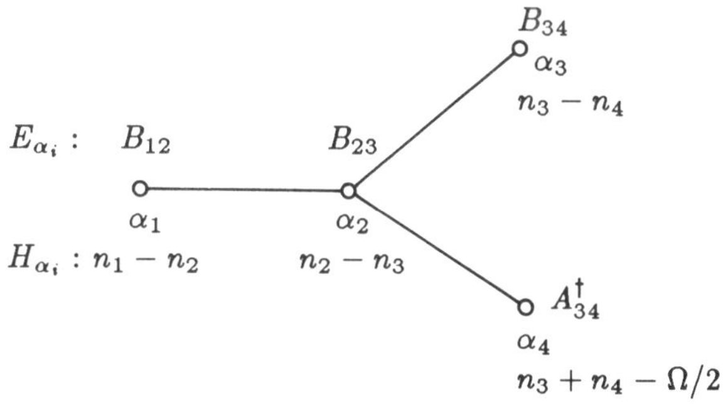

Fig. 9.10-1. The Chevalley basis of SO_8.

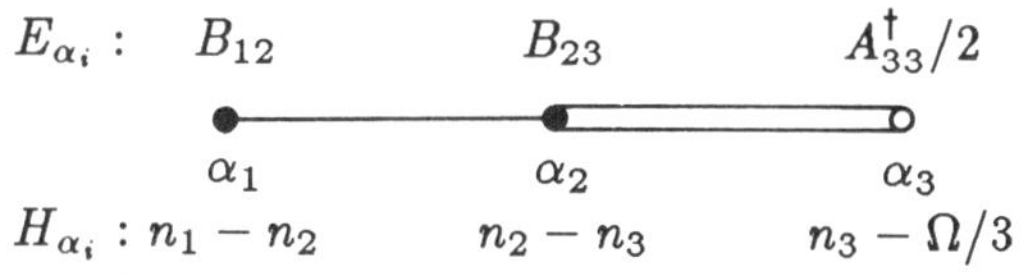

Fig. 9.10-2. The Chevalley basis of SP_6.

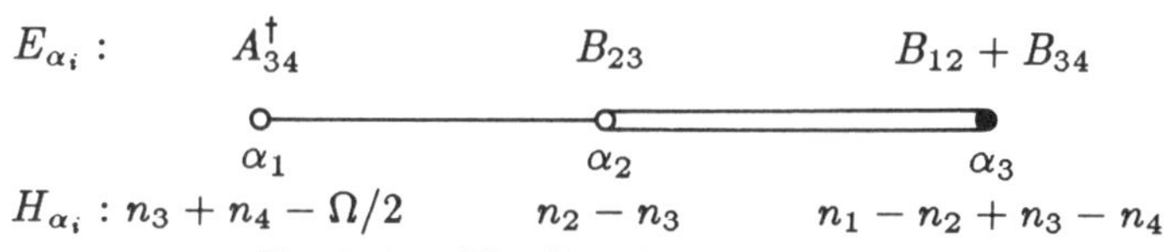

Fig. 9.10-3. The Chevalley basis of SO_7.

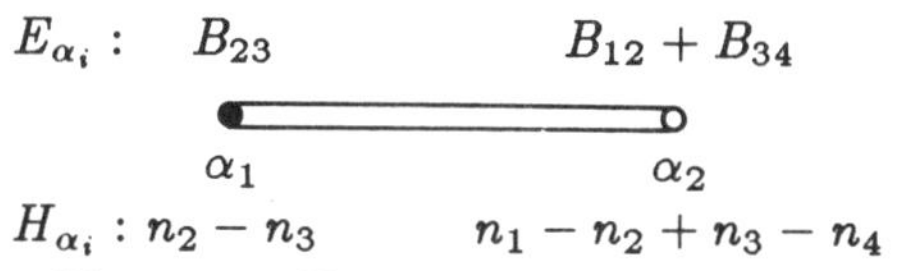

Fig. 9.10-4. The Chevalley basis of SO_5.

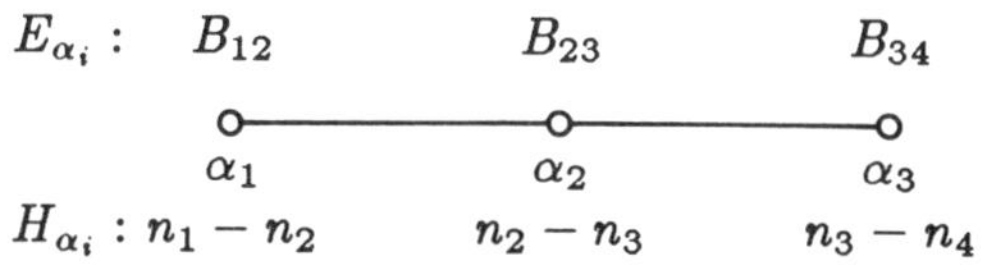

Fig. 9.10-5. The Chevalley basis of $SO_6 \sim SU_4$.

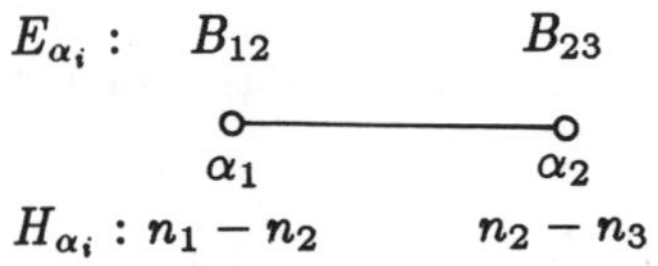

Fig. 9.10-6. The Chevalley basis of SU_3.

(ii) Generators in the coupled representation

$$A^{r\dagger}_{\mu} = \sqrt{\frac{\Omega}{2}}[b^{\dagger}_{k\frac{3}{2}} b^{\dagger}_{k\frac{3}{2}}]^{0r}_{0\mu} = \sum_{ab} C^{r\mu}_{\frac{3}{2}a,\frac{3}{2}b} A^{\dagger}_{ab}, \quad r = 0, 2, \tag{9-109a}$$

$$P^{r}_{\mu} = \sqrt{\frac{\Omega}{2}}[b^{\dagger}_{k\frac{3}{2}} \tilde{b}_{k\frac{3}{2}}]^{0r}_{0\mu} = \sum_{ab} C^{r\mu}_{\frac{3}{2}a,\frac{3}{2}b} (-)^{\frac{3}{2}+b} B'_{a\bar{b}}, \quad r = 0, 1, 2, 3 . \tag{9-109b}$$

The commutation relations are

$$[A^{r}_{\mu}, A^{s\dagger}_{\nu}] = \Omega \delta_{rs}\delta_{\mu\nu} - 2\sum_{t} K^{t\sigma}_{r-\mu, s\nu}(-)^{\mu} P^{t}_{\sigma} , \tag{9-110a}$$

$$[P^{r}_{\mu}, A^{s\dagger}_{\nu}] = \sum_{t} K^{t\sigma}_{r\mu, s\nu} A^{t\dagger}_{\sigma} , \tag{9-110b}$$

$$[P^{r}_{\mu}, P^{s}_{\nu}] = \frac{1}{2}\sum_{t}[(-)^{t} - (-)^{r+s}] K^{t\sigma}_{r\mu, s\nu} P^{t}_{\sigma} , \tag{9-110c}$$

$$K^{t\sigma}_{r\mu, s\nu} = -2\, \hat{r}\, \hat{s}\, C^{t\sigma}_{r\mu, s\nu} \begin{Bmatrix} r & s & t \\ \frac{3}{2} & \frac{3}{2} & \frac{3}{2} \end{Bmatrix} . \tag{9-110d}$$

The building blocks are

$$S^{\dagger} = A^{0\dagger} = \sqrt{\frac{1}{2}}(\overline{A}^{\dagger}_{14} - \overline{A}^{\dagger}_{23}), \quad D^{\dagger}_{\mu} = \overline{A}^{2\dagger}_{\mu} , \tag{9-111a}$$

i.e.

$$D^{\dagger}_{2} = \overline{A}^{\dagger}_{12}, \quad D^{\dagger}_{1} = \overline{A}^{\dagger}_{13}, \quad D^{\dagger}_{0} = \sqrt{\frac{1}{2}}(\overline{A}^{\dagger}_{14} + \overline{A}^{\dagger}_{23}) , \tag{9-111b}$$

$$D^{\dagger}_{-2} = \overline{A}^{\dagger}_{34}, \quad D^{\dagger}_{-1} = \overline{A}^{\dagger}_{24} , \tag{9-111c}$$

where $\overline{A}^{\dagger}_{ab} = \sqrt{2}A^{\dagger}_{ab}$.

B. The k-active case (k=1)

(i) Generators in the uncoupled representation

We index $m_k = 1, -1, -0$ as 1, 2, 3. The following 21 operators, $A^{\dagger}_{m_1 m_2}$ (6), $A_{m_1 m_2}$(6) and

$B_{m_1m_2}(9)$ form a Lie algebra,

$$A^{\dagger}_{m_1m_2} = \sum_{m_i}(-1)^{i-m_i} b^{\dagger}_{1m_1,im_i} b^{\dagger}_{1m_2,i\overline{m}_i} = A^{\dagger}_{m_2m_1} , \tag{9-112a}$$

$$B_{m_1m_2} = B'_{m_1m_2} - \frac{\Omega}{3}\delta_{m_1m_2}, \quad \Omega = \frac{3}{2}(2i+1) , \tag{9-112b}$$

$$B'_{m_1m_2} = \sum_{m_i} b^{\dagger}_{1m_1,im_i} b_{1m_2,im_i}, \quad B_{mm} = n_m , \tag{9-112c}$$

with the commutators

$$[A_{ab}, A^{\dagger}_{cd}] = -\delta_{ad}B_{cb} - \delta_{bc}B_{da} - \delta_{ac}B_{db} - \delta_{bd}B_{ca} . \tag{9-113}$$

The commutation relations $[B_{ab}, B_{cd}], [B_{ab}, A^{\dagger}_{cd}]$ and $[B_{ab}, A_{cd}]$ are the same as (9-108b) - (9-108d).

The Chevalley basis for the algebra and its Dynkin diagram is shown in Fig. 9.10-2, which shows that it is the SP_6 Lie algebra.

(ii) Generators in the coupled representation

$$A^{r\dagger}_{\mu} = \sqrt{\frac{\Omega}{2}}[b^{\dagger}_{1i} b^{\dagger}_{1i}]^{r0}_{\mu 0} = \frac{\sqrt{3}}{2}\sum_{ab} C^{r\mu}_{1a,1b} A^{\dagger}_{ab}, \quad r = 0, 2 , \tag{9-114a}$$

$$P^{r}_{\mu} = \sqrt{\frac{\Omega}{2}}[b^{\dagger}_{1i} \tilde{b}_{1i}]^{r0}_{\mu 0} = \frac{\sqrt{3}}{2}\sum_{ab}(-)^{1+b} C^{r\mu}_{1a,1b} B'_{a\bar{b}}, \quad r = 0, 1, 2 . \tag{9-114b}$$

The commutators of the SP_6 Lie algebra is again given by Eq. (9-110) with $K^{t\sigma}_{r\mu,s\nu}$ being understood as

$$K^{t\sigma}_{r\mu,s\nu} = \sqrt{3}\ \hat{r}\ \hat{s}\ C^{t\sigma}_{r\mu,s\nu} \begin{Bmatrix} r & s & t \\ 1 & 1 & 1 \end{Bmatrix} . \tag{9-114c}$$

The building blocks are

$$S^{\dagger} = A^{0\dagger} = \sqrt{\frac{1}{3}}(2\overline{A}^{\dagger}_{12} - \overline{A}^{\dagger}_{33}), \quad D^{\dagger}_{\mu} = \overline{A}^{2\dagger}_{\mu} , \tag{9-115a}$$

i.e.,

$$D^{\dagger}_{2} = \overline{A}^{\dagger}_{11}, \quad D^{\dagger}_{1} = \sqrt{2}\overline{A}^{\dagger}_{13}, \quad D^{\dagger}_{0} = \sqrt{\frac{2}{3}}(\overline{A}^{\dagger}_{12} + \overline{A}^{\dagger}_{33}) , \tag{9-115b}$$

$$D^{\dagger}_{-2} = \overline{A}^{\dagger}_{22} , \quad D^{\dagger}_{-1} = \sqrt{2}\overline{A}^{\dagger}_{23} , \tag{9-115c}$$

where

$$\overline{A}^{\dagger}_{ab} = \frac{\sqrt{3}}{2} A^{\dagger}_{ab} . \tag{9-115d}$$

Notice that (9-115) is identical with (9-70a,b) in the Elliott model. From (9-115) we have

$$\sqrt{2}\,\overline{A}^{\dagger}_{12} = \sqrt{\frac{1}{3}}(D^{\dagger}_{0} + \sqrt{2}S^{\dagger}), \quad \overline{A}^{\dagger}_{33} = \sqrt{\frac{1}{3}}(\sqrt{2}D^{\dagger}_{0} - S^{\dagger}) . \tag{9-116}$$

It is quite interesting to compare Eqs. (9-114a, b) and Fig. 9.10-2 for SP_6 with (9-45, 46) and Fig. 9.7 for $SP_4(\sim SO_5)$. Mathematically, when the pseudo orbital angular momentum $k=1$ and pseudo spin i in SP_6 go over to the isospin $t=1/2$ and angular momentum j, respectively, the SP_6 degenerates to SP_4.

9.10.2. The SO_8 and SP_6 Hamiltonian

The FDSM Hamiltonian for the normal parity levels has the following general form,

$$H_F = G_0 S^\dagger \bullet S + G_2 D^\dagger \bullet D + \sum_r B_r P^r \bullet P^r \,, \tag{9-117}$$

where the sum over r is from 0 to 3 for SO_8 and from 0 to 2 for SP_6. A nice feature of the Hamiltonian (9-117) is that it decouples the (S, D) subspace from the remaining fermion space (a gigantic one) and analytic solutions of the shell model Schrödinger equation are possible for some limiting cases even for nuclei *in the middle of a shell*. Notice the great resemblance between the FDSM Hamiltonian (9-117) and the IBM Hamiltonian (9-76a).

The group SO_8 and SP_6 have, respectively, three and two subgroup chains containing the group SO_3,

$$\begin{array}{cccc} SO_8 \supset (SO_5 \supset SO_3) \times & (SU_2 \supset U_1) & & \\ (P^3_\mu, P^1_\mu),\ P^1_\mu, & (S^\dagger, S, S_0), & P^0 & \\ (\tau_1\tau_2)\ \ JM & v & N & \end{array} \tag{9-118a}$$

$$\begin{array}{ccc} SO_8 \supset & SO_7 \supset & (SO_5 \supset SO_3) \times U_1 \\ & (D^\dagger_\mu, D_\mu, P^r_\mu, r=0,1,3), & \\ & \{w, \theta_1\theta_2\} & (\tau_1\tau_2)\quad JM \quad N \end{array} \tag{9-118b}$$

$$\begin{array}{cccc} SO_8 \supset & (SO_6 & \supset SO_5 \supset SO_3) \times U_1 & \\ & (P^r_\mu, r=1,2,3) & & \\ \{u, \rho_1\rho_2\rho_3\} & (\sigma_1\sigma_2\sigma_3), & (\tau_1\tau_2), \quad JM & N \end{array} \tag{9-118c}$$

$$\begin{array}{ccc} SP_6 \supset (SU_2 \supset U_1) & \times SO_3 & \\ (S^\dagger, S, S_0)\ P^0 & P^1_\mu & \\ v \quad N & JM & \end{array} \tag{9-119a}$$

$$\begin{array}{ccc} SP_6 & \supset (SU_3 \supset SO_3) \times U_1 & \\ & (P^2_\mu, P^1_\mu) & P^0 \\ \{u, \lambda_0\mu_0\} & (\lambda\mu) \quad JM & N \end{array} \tag{9-119b}$$

where below each group are its generators and the Dynkin irrep labels except the labels for the SO_8 and SP_6, which will be explained later. (Noting that the labels used by Hecht 1987 are the Cartan-Weyl labels.) For both the SO_8 and SP_6 cases,

$$S_0 = (n - \Omega)/2, \quad P^0 = n/2 \,. \tag{9-120}$$

The Casimir operators for the above various groups are [cf. (9-71a), (9-83) and (9-94)],

$$C_{SO8} = S^{\dagger}\bullet S + D^{\dagger}\bullet D + \sum_{r=1}^{3} P^r\bullet P^r + S_0(S_0 - 6) \qquad (9\text{-}121a)$$

$$C_{SO7} = D^{\dagger}\bullet D + \sum_{r=1,3} P^r\bullet P^r + S_0(S_0 - 5) \qquad (9\text{-}121b)$$

$$C_{SO6} = \sum_{r=1}^{3} P^r\bullet P^r \ , \quad C_{SO5} = \sum_{r=1,3} P^r\bullet P^r \ , \qquad (9\text{-}121c)$$

$$C_{SP6} = S^{\dagger}\bullet S + D^{\dagger}\bullet D + \sum_{r=1,2} P^r\bullet P^r + S_0(S_0 - 6) \ , \qquad (9\text{-}121d)$$

$$C_{SU3} = \sum_{r=1}^{2} P^r\bullet P^r, \quad C_{SU2} = S^{\dagger}S + S_0(S_0 - 1) \ . \qquad (9\text{-}121e)$$

Notice that

$$C_{SO8} = C_{SO7} + C_{SO6} - C_{SO5} + S^{\dagger}S - S_0 \ . \qquad (9\text{-}121f)$$

The FDSM Hamiltonian (9-117) can be expressed in terms of the Casimir operators,

$$H_{SO8} = H_0 + (G_0 - G_2)S^{\dagger}S + (B_2 - G_2)C_{SO6} + (B_3 - B_2)C_{SO5} + \frac{1}{5}(B_1 - B_3)\mathbf{J}^2 \qquad (9\text{-}122a)$$

$$= H_0' + (G_0 - B_2)S^{\dagger}S + (G_2 - B_2)(C_{SO7} + \frac{\Omega - n}{2})$$
$$+ (B_3 - B_2)C_{SO5} + \frac{1}{5}(B_1 - B_3)\mathbf{J}^2 \ , \qquad (9\text{-}122b)$$

$$H_{SP6} = H_0'' + (G_0 - G_2)S^{\dagger}S + (B_2 - G_2)C_{SU3} + \frac{3}{8}(B_1 - B_2)\mathbf{J}^2 \ . \qquad (9\text{-}122c)$$

From (9-122a,b) we see that H_{SO8} has the following dynamic symmetries

$$\begin{array}{lcl} SU_2 \times SO_5 & & B_2 = G_2 \ , \\ SO_7 \supset SO_5 & \text{when} & G_0 = B_2 \ , \\ SO_6 \supset SO_5 & & G_0 = G_2 \ , \end{array} \qquad (9\text{-}123)$$

while from (9-122c) we see that H_{SP6} has two dynamic symmetries,

$$\begin{array}{lcl} SU_2 \times SO_3 & \text{when} & B_2 = G_2 \ , \\ SU_3 \supset SO_3 & & G_0 = G_2 \ . \end{array} \qquad (9\text{-}124)$$

It is interesting to compare (9-122a) with (9-84a) and (9-93a), and (9-122c) with (9-103c). The spectra given by the $SO_8 \supset SO_5 \times SU_2$ and $SP_6 \supset SU_2 \times SO_3$ limits are identical or similar to the IBM U_5 limit, while those given by $SO_8 \supset SO_6$ and $SP_6 \supset SU_3$ are identical with the IBM SO_6 and SU_3 limits respectively. The spectra for $SO_7 \supset SO_5$ is in between the IBM U_5 limit and SO_6 limit, and correspond to the transitional nuclei from vibration to γ-unstable rotor. The new dynamic symmetry SO_7 which has no counterpart in the IBM has been verified in experiment (Casten 1986).

9.10.3. The FDSM wave functions

Before going to construct the wave functions, we first list the Chevalley basis and bases for the basic reps of SO_5, SO_6 and SU_3 in Tables 9.10-1 to 9.10-3.

Table 9.10-1a. The Chevalley basis of SO_5.

	B_{23}	$B_{12}+B_{34}$	B_{14}	$B_{13}-B_{24}$
$H_{\alpha_1}=n_2-n_3$	2	−1	0	1
$H_{\alpha_2}=n_1-n_2+n_3-n_4$	−2	2	2	0

Table 9.10-1b. Bases of the basic and identity reps of SO_5.[3]

	$\xi_1^\dagger$	$\xi_2^\dagger$	$\xi_3^\dagger$	$\xi_4^\dagger$	$D_2^\dagger$ $\bar{A}_{12}^\dagger$	$D_1^\dagger$ $\bar{A}_{13}^\dagger$	$D_0^\dagger$ $\sqrt{\frac{1}{2}}(\bar{A}_{14}^\dagger+\bar{A}_{23}^\dagger)$	$D_{-1}^\dagger$ $\bar{A}_{24}^\dagger$	$D_{-2}^\dagger$ $\bar{A}_{34}^\dagger$	$\sqrt{\frac{1}{2}}(\bar{A}_{14}^\dagger-\bar{A}_{23}^\dagger)$	$D^\dagger\cdot D^\dagger$
$H_{\alpha_1}=n_2-n_3$	0	1	−1	0	1	−1	0	1	−1	0	0
$H_{\alpha_2}=n_1-n_2+n_3-n_4$	1	−1	1	−1	0	2	0	−2	0	0	0

[3] $\xi_1^\dagger=b_{m,3/2}^\dagger$, $\xi_2^\dagger=b_{m,1/2}^\dagger$, $\xi_3^\dagger=b_{m,-1/2}^\dagger$, $\xi_4^\dagger=b_{m,-3/2}^\dagger$; $\bar{A}_{ab}^\dagger=\sqrt{2}A_{ab}^\dagger$.

Table 9.10-2a. The Chevalley basis and roots of SO_6.

	B_{12}	B_{23}	B_{34}	B_{13}	B_{14}	B_{24}
$H_{\alpha_1}=n_1-n_2$	2	−1	0	1	1	−1
$H_{\alpha_2}=n_2-n_3$	−1	2	−1	1	0	1
$H_{\alpha_3}=n_3-n_4$	0	−1	2	−1	1	1

Table 9.10-2b. Bases of the basic and identity reps of SO_6.[4]

	$\xi_1^\dagger$	$\xi_2^\dagger$	$\xi_3^\dagger$	$\xi_4^\dagger$	$D_2^\dagger$ $\bar{A}_{12}^\dagger$	$D_1^\dagger$ $\bar{A}_{13}^\dagger$	$\sqrt{\frac{1}{2}}(D_0^\dagger+S^\dagger)$ $\bar{A}_{14}^\dagger$	$D_{-1}^\dagger$ $\bar{A}_{24}^\dagger$	$D_{-2}^\dagger$ $\bar{A}_{34}^\dagger$	$\sqrt{\frac{1}{2}}(D_0^\dagger-S^\dagger)$ $\bar{A}_{23}^\dagger$	$D^\dagger\cdot D^\dagger-S^\dagger\cdot S^\dagger$
$H_{\alpha_1}=n_1-n_2$	1	−1	0	0	0	1	1	−1	0	−1	0
$H_{\alpha_2}=n_2-n_3$	0	1	−1	0	1	−1	0	1	−1	0	0
$H_{\alpha_3}=n_3-n_4$	0	0	1	−1	0	1	−1	−1	0	1	0

[4] See footnote of Table 9.10-1b.

Table 9.10-3a. The Chevalley basis and roots of SU_3.

	B_{12}	B_{23}	B_{13}
$H_{\alpha_1}=n_1-n_2$	2	−1	1
$H_{\alpha_2}=n_2-n_3$	−1	2	1

Table 9.10-3b. Bases of the irreps (10) and (20) of SU_3.[5]

$(\varepsilon\Lambda K)$	$(-1\frac{1}{2}\,1)$	$(-1\frac{1}{2}\,-1)$	(200)	(-212)	$(1\frac{1}{2}\,1)$	(400)	$(1\frac{1}{2}\,-1)$	$(-21\,-2)$	(-210)
	$\xi_1^\dagger$ $b_{1,m}^\dagger$	$\xi_2^\dagger$ $b_{-1,m}^\dagger$	$\xi_3^\dagger$ $b_{0,m}^\dagger$	$D_2^\dagger$ $\bar{A}_{11}^\dagger$	$D_1^\dagger$ $\sqrt{2}\bar{A}_{13}^\dagger$	$\sqrt{\frac{1}{3}}(\sqrt{2}D_0^\dagger-S^\dagger)$ $\bar{A}_{33}^\dagger$	$D_{-1}^\dagger$ $\sqrt{2}\bar{A}_{23}^\dagger$	$D_{-2}^\dagger$ $\bar{A}_{22}^\dagger$	$\sqrt{\frac{1}{3}}(\sqrt{2}D_0^\dagger+S^\dagger)$ $\sqrt{2}\bar{A}_{12}^\dagger$
$H_{\alpha_1}=n_1-n_2$	1	−1	0	2	0	−2	−1	0	1
$H_{\alpha_2}=n_2-n_3$	0	1	−1	0	1	2	0	−2	−1

[5] $\bar{A}_{ab}^\dagger=\frac{\sqrt{3}}{2}A_{ab}^\dagger$

By comparing Table 9.10 with Table 9.9, we see that there is a $1-1$ correspondence between the generators (both in the uncoupled and coupled representations), building blocks, and invariants of the IBM subgroups SO_5, SO_6 and SU_3 and those of the FDSM subgroups SO_5, SO_6 and SU_3, respectively. All the commutators remain true under the simultaneous change of all quantities in one model to their counterparts in the other model. Therefore the irreducible basis of SO_5, SO_6 and SU_3 in the (S, D) subspace of the FDSM can be obtained from the corresponding ones in the IBM simply by changing $s^\dagger \to S^\dagger$, $d^\dagger_\mu \to D^\dagger_\mu$ and by renormalizing the fermion wave functions.

The analytic wave functions for the five limits of the FDSM have been found for some physically important cases (Ginocchio 1980, Lü 1988, Hecht 1987). The construction of the wave function for the $SO_8 \supset SO_5 \times SU_2$ and $SP_6 \supset SO_3 \times SU_2$ limits in the (S, D) subspace is straightforward but tedious (Ginocchio 1980, Lü 1988). In the following we will sketch the scheme for constructing the FDSM wave function in general cases (not necessarily in the (S, D) subspace) for the SO_7, SO_6 and SU_3 limits by using the powerful coherent state technique (Rowe 1985, Hecht 1987).

In Sec. 9.7 we use the seniority v and reduced isospin t of the coherent intrinsic state as the irrep label for SO_5. Now let us generalize this scheme to SO_7, SO_8 and SP_6.

A. The $SO_7 \supset SO_5 \times SO_2$ *(or* $SO_7 \supset SP_4 \times U_1$*) chain*

Let $|w(\theta_1\theta_2)\ \text{l.w.}\rangle$ be a coherent intrinsic state, which is a wave function of w fermions belonging to the irrep $(\theta_1\theta_2)$ of the group SO_5, and is a lowest weight state of SO_5 and SO_7. Using Fig. 9.10-4, in the state $|w(\theta_1\theta_2)\ \text{l.w.}\rangle$ we have,

$$n_2 - n_3 = -\theta_1, \quad n_1 - n_2 + n_3 - n_4 = -\theta_2, \quad w = n_1 + n_2 + n_3 + n_4 \ , \tag{9-125a}$$

$$D_\mu |w(\theta_1\theta_2)\text{l.w.}\rangle = 0 \ . \tag{9-125b}$$

From (9-125a) we have

$$n_3 + n_4 = \frac{1}{2}(w + 2\theta_1 + \theta_2) \ . \tag{9-125c}$$

Equation (9-125b) shows that w is the number of nucleons totally free of D pairs, and thus w is called the *D-pair seniority*. According to Fig. 9.10-3 and (9-125a,c) the weight of SO_7 in $|w(\theta_1\theta_2)\ \text{l.w.}\rangle$ is

$$\left(\frac{w + 2\theta_1 + \theta_2 - \Omega}{2}, -\theta_1, -\theta_2\right) \ . \tag{9-126a}$$

By applying the raising polynomial to $|w(\theta_1\theta_2)\text{l.w.}\rangle$ we can raise the weight and the highest weight is just that of (9-126a) but with opposite sign. Hence the Dynkin label for irreps of SO_7 is

$$\left(\frac{\Omega - w}{2} - \theta_1 - \frac{\theta_2}{2}, \theta_1, \theta_2\right) \leftrightarrow \{w; \theta_1, \theta_2\} \ , \tag{9-126b}$$

where the right-hand side is a short hand notation for the left-hand side.

B. The $SO_8 \supset SO_6 \times SO_2$ *(or* $SO_8 \supset SU_4 \times U_1$*) chain*

By the isomorphism between the Lie algebra SO_6 and SU_4, the irrep of SO_6 or U_4 can be labeled either by the usual partition $[f_1 \ldots f_4]$ or by the Dynkin label $(\rho_1\rho_2\rho_3)$ and the total fermion number u,

$$\begin{aligned} &\rho_1 = f_1 - f_2, \quad \rho_2 = f_2 - f_3, \quad \rho_3 = f_3 - f_4 \ , \\ &u = f_1 + f_2 + f_3 + f_4 \ . \end{aligned} \tag{9-127}$$

The coherent intrinsic state, denoted by $|u(\rho_1\rho_2\rho_3)\ \text{l.w.}\rangle$, is a lowest weight state of SO_6 and SO_8,

$$A^r_\mu|u(\rho_1\rho_2\rho_3)\text{l.w.}\rangle = 0, \quad r = 0, 2\ , \tag{9-128}$$

which shows that u is the number of fermions totally free of S and D pairs, and is called *S, D-pair seniority.* The lowest weight of $SO_6 \sim SU_4$ in the irrep $(\rho_1\rho_2\rho_3)$ is $(-\rho_3 - \rho_2 - \rho_1)$. Using Fig. 9.10-5 we know that in the coherent intrinsic state we have

$$\begin{aligned} &n_1 - n_2 = -\rho_3, \quad n_2 - n_3 = -\rho_2, \quad n_3 - n_4 = -\rho_1, \\ &n_3 + n_4 = \frac{1}{2}(u + \rho_1 + 2\rho_2 + \rho_3)\ . \end{aligned} \tag{9-129}$$

Using Fig. 9.10-1 the weight of SO_8 in the state $|u(\rho_1\rho_2\rho_3)\text{l.w.}\rangle$ is

$$\left(-\rho_3, -\rho_2, -\rho_1, \frac{u-\Omega}{2} + \frac{\rho_1 + 2\rho_2 + \rho_3}{2}\right)\ . \tag{9-130}$$

Therefore the Dynkin label for irreps of SO_8 is

$$\left(\rho_3, \rho_2, \rho_1, \frac{\Omega - u}{2} - \frac{\rho_1 + 2\rho_2 + \rho_3}{2}\right) \leftrightarrow \{u; \rho_1, \rho_2, \rho_3\}\ , \tag{9-131}$$

where the right-hand side is again a short hand notation for the left.

C. The $SP_6 \supset SU_3 \times U_1$ chain

The lowest weight state of SU_3 and SP_6 is denoted by $|u(\lambda_0\mu_0)\text{l.w.}\rangle$ which obeys

$$A^r_\mu|u(\lambda_0\mu_0)\text{l.w.}\rangle = 0, \quad r = 0, 2, \tag{9-132}$$

u is again the S, D-pair seniority. Exactly as in the previous case, we can find the relation between the SP_6 irrep label and the U_3 irrep label $(\lambda_0\mu_0)$ of the intrinsic state $|u(\lambda_0\mu_0)\text{l.w.}\rangle$ as follows,

$$\left(\mu_0, \lambda_0, \frac{\Omega - u}{3} - \frac{2\lambda_0 + \mu_0}{3}\right) \leftrightarrow \{u; \lambda_0\mu_0\}\ . \tag{9-133}$$

Now we proceed to construct the wave functions.

A. The $SO_7 \supset SO_5$ wave function

The basis function belonging to the SO_7 irrep $\{w; \theta_1\theta_2\}$ can be obtained by operating the raising polynomial to the vector intrinsic states $|w(\theta_1\theta_2)\alpha\rangle$, where α denotes the SO_5 subgroup labels

$$|n, \{w, \theta_1\theta_2\}(\tau_1\tau_2), \tau n_\Delta JM\rangle = \wp \left[P^{(\overline{\kappa},\tau)}(D^\dagger_\mu)|w(\theta_1\theta_2)\rangle\right]^{(\tau_1\tau_2)}_{n_\Delta JM}\ , \tag{9-134}$$

where $\wp$ is a normalization constant, $P^{(\overline{\kappa},\tau)}(D^\dagger_\mu)$ is a polynomial of $D^\dagger_\mu$ of degree $\overline{\kappa} = (n - w)/2$ and belongs to the SO_5 irrep $(\tau 0)$, while the square bracket denotes the SO_5 coupling $(\tau 0) \times (\theta_1\theta_2) \rightarrow (\tau_1\tau_2)$.

$$P^{(\overline{\kappa},\tau)}_{n'_\Delta LM}(D^\dagger_\mu) = |\overline{\kappa}\overline{\kappa}\tau n'_\Delta LM)_{d^\dagger_\mu \rightarrow D^\dagger_\mu} \tag{9-135}$$

$$= \left[\frac{(2\tau+3)!!}{(\overline{\kappa}+\tau+3)!!(\overline{\kappa}-\tau)!!}\right]^{1/2} (D^\dagger \bullet D^\dagger)^{(\overline{\kappa}-\tau)/2}(D^\dagger_\mu)^\tau_{n'_\Delta LM}\ . \tag{9-136}$$

Equation (9-135) denotes that the polynomial can be obtained from the U_5 IBM wave function (9-89) by replacing $s^\dagger$ and $d^\dagger_\mu$ with $S^\dagger$ and $D^\dagger_\mu$, respectively, and deleting the boson vaccum $|0)$. Notice that in (9-134) τ is an additional quantum number. Since τ cannot be associated directly with the eigenvalue of a hermitian operator, the state in (9-134) is not orthogonal with respect to τ [see the discussion after (9-56)].

The norms in (9-134) for states of low D-pair seniority w have been calculated by Hecht (1987). It is interesting to note that all the Pauli factors are contained in the norm $\wp$.

B. The $SO_8 \supset SO_6 \supset SO_5$ wave function

Similarly, we have the wave functions for the $SO_8 \supset SO_6 \supset SO_5$ dynamic symmetry,

$$\begin{aligned}&|n = 2p + u, \{u, \rho_1\rho_2\rho_3\}, (\sigma_1\sigma_2\sigma_3), (\tau_1\tau_2), \sigma n_\Delta JM\rangle \\ &= \wp\left[P^{(p,\sigma)}(S^\dagger, D^\dagger_\mu)|u(\rho_1\rho_2\rho_3)\rangle\right]^{(\sigma_1\sigma_2\sigma_3)}_{(\tau_1\tau_2)n_\Delta JM}, \end{aligned} \tag{9-137a}$$

where the raising polynomial is of degree $p = (n-u)/2$ and belongs to the SO_6 irrep $(0\,\sigma\,0)$, and the square bracket denotes the SO_6 coupling $(0\,\sigma\,0) \times (\rho_1\rho_2\rho_3) \to (\sigma_1\sigma_2\sigma_3)$. The polynomial can be obtained from the IBM SO_6 wave function (9-100) by replacing $s^\dagger, d^\dagger_\mu$ with $S^\dagger$, $D^\dagger_\mu$, and deleting the boson vacuum $|0)$,

$$P^{(p,\sigma)}_{\tau n'_\Delta LM}(S^\dagger, D^\dagger_\mu) = |p, \sigma, \tau, n'_\Delta LM)_{s^\dagger \to S^\dagger, d^\dagger_\mu \to D^\dagger_\mu}\,. \tag{9-137b}$$

The norm $\wp$ for states of low S, D-pair seniority u have been found by Hecht (1987). Analogously, the wave function (9-137a) is not orthogonal with respect to the additional index σ.

C. The $SP_6 \supset SU_3 \supset SU_2$ wave function

We choose the weight operators as in the Elliott model Eq. (9-66)

$$\begin{pmatrix}\hat{\varepsilon}\\ K\end{pmatrix} = \begin{pmatrix}-\sqrt{8}P_0^2\\ J_z\end{pmatrix} = \begin{pmatrix}2n_0 - n_1 - n_{-1}\\ n_1 - n_{-1}\end{pmatrix}. \tag{9-138}$$

The wave function for the $SP_6 \supset SU_3 \supset SU_2$ group chain is

$$\begin{aligned}&|n = 2p + u, \{u; \lambda_0\mu_0\}(\lambda\mu), [h]\varepsilon\Lambda K\rangle \\ &= \wp_{h_1h_2h_3}\left[P^{[h]}(A^{r\dagger}_\mu)|u(\lambda_0\mu_0)\rangle\right]^{(\lambda\mu)}_{\varepsilon\Lambda K}, \end{aligned} \tag{9-139}$$

where $\wp_{h_1h_2h_3}$ is a normalization constant, $[h] = [2h_1 2h_2 2h_3]$, $P^{[h]}(A^{r\dagger}_\mu)$ is a polynomial of $A^{r\dagger}_\mu$ of degree $P = 2h_1 + 2h_2 + 2h_3 = (n-u)/2$, and belongs to the irrep $[h]$ of U_3. The square bracket denotes the SU_3 coupling $(2h_1 - 2h_2, 2h_2 - 2h_3) \times (\lambda_0\mu_0) \to (\lambda\mu)$. $P^{[h]}(A^{r\dagger}_\mu)$ can be obtained from (9-73)–(9-75) by the following replacement,

$$A^\dagger_{ab} \to \overline{A}^\dagger_{ab} = \frac{\sqrt{3}}{2}A^\dagger_{ab} \tag{9-140}$$

and interpreting $A^\dagger_{ab}$ as two-fermion creation (9-112). The wave function (9-139) is not orthogonal with respect to $[h]$.

The physical basis, i.e., the $SP_6 \supset SU_3 \supset SO_3$ wave function can be projected out from the intrinsic state $|n, \{u; \lambda_0\mu_0\}(\lambda\mu), [h]\varepsilon\Lambda K\rangle$.

It can be shown (Lü 1988) that for the case $u = 0$, the norm in (9-139) is

$$\wp_{h_1h_2h_3} = \left[\frac{1}{3^N}\prod_{i=1}^{3}\frac{\{\overline{\Omega} - h_i + (i-1)/2\}!}{\{\overline{\Omega} + (i-1)/2\}!}\right]^{1/2} , \tag{9-141}$$

where $\overline{\Omega} = \Omega/3$.

The several most important intrinsic states for even-even nuclei are given below along with the corresponding Weyl tableaux.

i. The ground state band

$$|(2N,0), h.w.\rangle = \left| \begin{array}{|c|c|c|c|}\hline 0 & 0 & \cdots & 0 \\ \hline \end{array} \right\rangle = \wp_{(N,0,0)}\sqrt{\frac{1}{N!}}(\overline{A}^{\dagger}_{33})^{N}|0\rangle$$

$$= \wp_{(N,0,0)}\sqrt{\frac{1}{N!}}\left(\sqrt{\frac{2}{3}}D^{\dagger}_{0} - \sqrt{\frac{1}{3}}S^{\dagger}\right)^{N}|0\rangle , \tag{9-142a}$$

where $N = n/2$ and

$$\wp_{(N,0,0)} = \left[\frac{(\overline{\Omega} - N)!}{3^N\overline{\Omega}!}\right]^{1/2} . \tag{9-142b}$$

ii. The gamma band (K = 2)

$$|(2N-4,2), \text{h.w.}\rangle = \left| \begin{array}{|c|c|c|c|c|}\hline 1 & 1 & 0 & \cdots & 0 \\ \hline 0 & 0 & \multicolumn{3}{c}{} \\ \cline{1-2} \end{array} \right\rangle$$

$$= \wp_{(N-1,1,0)}\left[\frac{2}{(2N-1)(N-2)!}\right]^{1/2}\Delta^{\dagger}_{22}(\overline{A}^{\dagger}_{33})^{N-2}|0\rangle , \tag{9-143a}$$

$$\wp_{(N-1,1,0)} = \left[\frac{(\overline{\Omega} - N + 1)!}{3^N\overline{\Omega}!(\overline{\Omega} + 1/2)}\right]^{1/2} . \tag{9-143b}$$

iii. The beta band (K = 0)

$$|(2N-4,2), \varepsilon = 4N-6, \Lambda = 1, K = 0\rangle = \left| \begin{array}{|c|c|c|c|c|}\hline 1 & -1 & 0 & \cdots & 0 \\ \hline 0 & 0 & \multicolumn{3}{c}{} \\ \cline{1-2} \end{array} \right\rangle$$

$$= -\wp_{(N-1,1,0)}\left[\frac{4}{(2N-1)(N-2)!}\right]^{1/2}(\Delta^{\dagger}_{12})(\overline{A}^{\dagger}_{33})^{N-2}|0\rangle . \tag{9-144}$$

It can be shown (Lü 1988) that in case of $u = 0$ or $w = 0$, when the shell degeneracy $\Omega \to$ infinite, we have $\wp \to \sqrt{\frac{1}{\Omega^N}}$. By letting $\frac{S^{\dagger}}{\sqrt{\Omega}} \to s^{\dagger}$, $\frac{D^{\dagger}_{\mu}}{\sqrt{\Omega}} \to d^{\dagger}_{\mu}$, the FDSM wave functions for the limits of $SO_8 \supset SO_5 \times SU_2, SO_8 \supset SO_7$ and $SP_6 \supset SO_3 \times SU_2$ all go over to the IBM U_5 wave function, while that for $SO_8 \supset SO_6$ and $SP_6 \supset SU_3$ become the IBM SO_6 and SU_3 wave functions, respectively. For example, the intrinsic state for the ground band, β and γ band in the IBM can be simply obtained from (9-142)–(9-144) by ignoring the Pauli $\wp_{h_1h_2h_3}$ and replacing $s^{\dagger}, d^{\dagger}_{\mu}$ by $S^{\dagger}, D^{\dagger}_{\mu}$.

Now we are in a position to discuss the relationship between the several dynamic symmetry models of nuclei introduced in Secs. 5–9.10. The SO_8 or SP_6 FDSM is not only a generalization of the SU_2 pairing model and the Elliott model where only the short range (for the former) or the long range (for the latter) residual interaction has been taken into account, but also a generalization of the IBM, where fermion pairs are approximated as bosons. The SO_8 or SP_6

FDSM in its simplest version has similar dynamic symmetries as the IBM, giving rise to same or similar spectra and yielding identical or up to a Pauli factor transition rates as the IBM (Ginocchio 1980, Chen 1986). However the former has the distinguishing feature that it is a fermion model, no boson approximation is made and the Pauli principle is taken into account fully. When the shell degeneracy Ω become infinite, the Pauli restriction becomes negligible and the FDSM degenerates to the IBM.

Finally, it is interesting to investigate the relation between the three SU_3 models, the Elliott model, the IBM and FDSM. By comparing (9-114b) with (9-63), we see that the realization of the Elliott SU_3 in the $1p$ shell together with interpreting the quanta $c_m^\dagger$ (or the "Elliott quarks", which are bosons) as the fermions $b_{km_k im_i}^\dagger$ (or the "Ginocchio quarks" which are the constituents of the compound particles, the S, D fermion pairs) yields, the FDSM SU_3; while by comparing (9-103b) with (9-67), we see that the realization of the Elliott SU_3 in the s, d shell together with interpreting the quanta s, d as the s, d bosons, gives rise the IBM SU_3. In the Elliott SU_3, there are both SU_3 and SU_6, while in the FDSM SU_3 there is only SU_3 but no SU_6, and in the IBM SU_3, there is only SU_6 but no basic SU_3 (i.e., it does not contain the basic, or defining irrep of SU_3). Schematically, we have

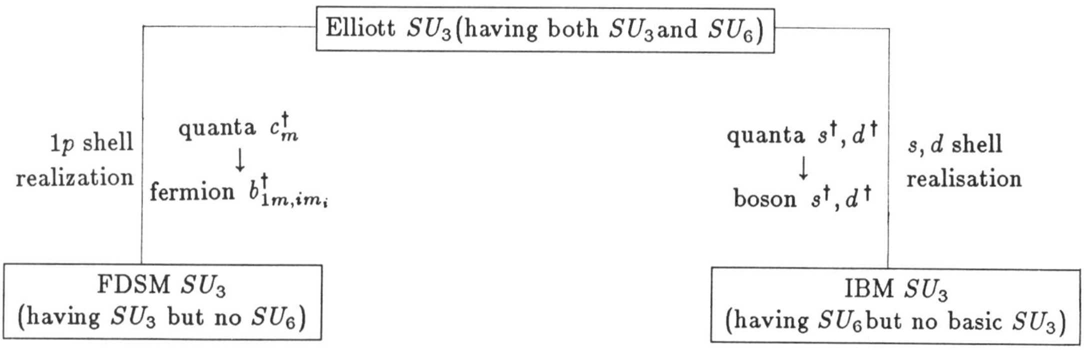

9.11 The Molecular Shell Model

In Secs. 9.1 and 9.2, the nuclear shell model is treated by using Racah's CFP technique. The same technique can also be applid to the molecular shell model (Tang 1966, 1975). A new method was proposed by Paldus (1974, 1976) for the molecular shell model in which the many-electron wave functions are expressed in terms of the Gel'fand bases and the Hamiltonian is diagonalized directly without using the CFP. The most attractive feature of this new approach is that it can easily be translated into a computer program and therefore makes a large-scale configuration-mixing calculation feasible.

9.11.1 The Hamiltonian as a function of infinitesimal operators of the unitary group

Moshinky (1962) pointed out that any one-body or two-body operator can be expressed in terms of the infinitesimal operators of the unitary group. In a second quantized representation, a one-body operator reads

$$F = \sum_{\alpha\beta} \langle\alpha|F|\beta\rangle a_\alpha^\dagger a_\beta \ , \qquad \text{(9-145a)}$$

where $a_\alpha^\dagger(a_\beta)$ are creation (annihilation) operators, while α and β are labels of the single particle states. From (5-97) $E_{\alpha\beta} = a_\alpha^\dagger a_\beta$, (9-145a) can be written as

$$F = \sum_{\alpha\beta} \langle\alpha|F|\beta\rangle E_{\alpha\beta} \ . \qquad \text{(9-145b)}$$

A two-body operator V in second quantized representation may be written as

$$V = \frac{1}{2}\sum_{\alpha\beta\gamma\delta} \langle\alpha\beta|V|\gamma\delta\rangle a^\dagger_\alpha a^\dagger_\beta a_\delta a_\gamma \;;$$
$$\langle\alpha\beta|V|\gamma\delta\rangle = \int \varphi^*_\alpha(1)\varphi^*_\beta(2)V_{12}\varphi_\gamma(1)\varphi_\delta(2)d\tau_1 d\tau_2 \;. \tag{9-146}$$

For fermions we have

$$a^\dagger_\alpha a^\dagger_\beta a_\delta a_\gamma = -a^\dagger_\alpha a^\dagger_\beta a_\delta a_\delta = E_{\alpha\gamma}E_{\beta\delta} - \delta_{\gamma\beta}E_{\alpha\delta}$$

while for bosons we have

$$a^\dagger_\alpha a^\dagger_\beta a_\delta a_\gamma = a^\dagger_\alpha a^\dagger_\beta a_\gamma a_\delta = E_{\alpha\gamma}E_{\beta\delta} - \delta_{\gamma\beta}E_{\alpha\delta} \;.$$

Therefore, regardless of fermions or bosons we always have

$$V = \frac{1}{2}\sum_{\alpha\beta\gamma\delta} \langle\alpha\beta|V|\gamma\delta\rangle E_{\alpha\gamma}E_{\beta\delta} - \frac{1}{2}\sum_{\alpha\beta\delta}\langle\alpha\beta|V|\beta\delta\rangle E_{\alpha\delta} \;. \tag{9-147}$$

9.11.2 Spin-free approximation

Since the "spin-free' is a good approximation in quantum chemistry (Matsen 1974), it is convenient to separate the many-electron wave functions into two parts, one related to spin space, and the other to orbital space. An f-electron[1)] antisymmetric wave function with definite spin (SM_S) and point group symmetry $(\mu)\kappa$ [hereafter referred to as a *many-electron SALC* wave function] can be expressed as

$$\Psi(\beta\mu\kappa, SM_S) = \sum_m \frac{\Lambda^\nu_m}{\sqrt{h_\nu}} \left| \begin{matrix}[\nu]\\ m,\ \beta\mu\kappa\end{matrix} \right\rangle \left| \begin{matrix}[\tilde\nu]\\ \tilde m,\ SM_s\end{matrix}\right\rangle , \tag{9-148}$$

$$[\nu] = \left[2^{\frac{f}{2}-S}, 1^{2S}\right] \;, \quad [\tilde\nu] = \left[\frac{f}{2}+S,\ \frac{f}{2}-S\right] \;. \tag{9-149}$$

where $[\nu]$ labels the irrep of the group SU_n and the permutation group S_f, m is the Yamanouchi number and β the additional quantum numbers. Under the spin-free approximation we have

$$\langle\psi(\beta\mu\kappa, SM_S)|H|\Psi(\beta'\mu\kappa', S'M'_s)\rangle = \delta_{\mu\mu'}\delta_{\kappa\kappa'}\delta_{SS'}\delta_{M_S M'_S}\left\langle \begin{matrix}[\nu]\\ \beta\mu\end{matrix}\right| H \left| \begin{matrix}[\nu]\\ \beta'\mu\end{matrix}\right\rangle , \tag{9-150}$$

$$\left\langle \begin{matrix}[\nu]\\ \beta\mu\end{matrix}\right| H \left| \begin{matrix}[\nu]\\ \beta'\mu\end{matrix}\right\rangle = \left\langle \begin{matrix}[\nu]\\ m, \beta\mu\kappa\end{matrix}\right| H \left| \begin{matrix}[\nu]\\ m,\ \beta'\mu\kappa\end{matrix}\right\rangle \,. \tag{9-151}$$

On the right-hand side of (9-150), only the matrix element in orbital space is left and its value is independent of m and κ. The spin S manifests itself only through the partition $[\nu]$ [see (9-149)].

Suppose that under a given circumstance there are n one-particle levels needed to be accounted for and which span the basic rep of the group U_n. Since the Gel'fand bases of the group U_n, $\left|\begin{matrix}[\nu]\\ W\end{matrix}\right\rangle$, form a complete set, the f-electron wave function $\left|\begin{matrix}[\nu]\\ \beta\mu\kappa\end{matrix}\right\rangle$ can be expanded as follows:

$$\left|\begin{matrix}[\nu]\\ \beta\mu\kappa\end{matrix}\right\rangle = \sum_W \left\langle \begin{matrix}[\nu]\\ W\end{matrix}\middle| \begin{matrix}[\nu]\\ \beta\mu\kappa\end{matrix}\right\rangle \left|\begin{matrix}[\nu]\\ W\end{matrix}\right\rangle . \tag{9-152}$$

---

1) As a first approximation, f is taken as the number of valence electrons.

Substituting this into (9-151), we get

$$\left\langle \begin{matrix} [\nu] \\ \beta\mu \end{matrix} \middle| H \middle| \begin{matrix} [\nu] \\ \beta'\mu \end{matrix} \right\rangle = \sum_{WW'} \left\langle \begin{matrix} [\nu] \\ W' \end{matrix} \middle| \begin{matrix} [\nu] \\ \beta'\mu\kappa \end{matrix} \right\rangle \left\langle \begin{matrix} [\nu] \\ \beta\mu'\kappa \end{matrix} \middle| \begin{matrix} [\nu] \\ W \end{matrix} \right\rangle \left\langle \begin{matrix} [\nu] \\ W \end{matrix} \middle| H \middle| \begin{matrix} [\nu] \\ W' \end{matrix} \right\rangle . \tag{9-153}$$

Here κ can take any permissible value. Using (9-145b), (9-147), and (7-58), we can easily calculate the matrix elements $\left\langle \begin{smallmatrix} [\nu] \\ W \end{smallmatrix} \middle| H \middle| \begin{smallmatrix} [\nu] \\ W' \end{smallmatrix} \right\rangle$. Therefore, the remaining tasks are: (i) to find an appropriate subgroup chain of U_n which can provide additional quantum numbers β so that the irreducible bases $\left| \begin{smallmatrix} [\nu] \\ \beta\mu\kappa \end{smallmatrix} \right\rangle$ can be uniquely specified; (ii) to find the transformation coefficients $\left\langle \begin{smallmatrix} [\nu] \\ W \end{smallmatrix} \middle| \begin{smallmatrix} [\nu] \\ \beta\mu\kappa \end{smallmatrix} \right\rangle$ from the point group symmetry adapted irreducible basis $\left| \begin{smallmatrix} [\nu] \\ \beta\mu\kappa \end{smallmatrix} \right\rangle$ to the Gel'fand basis. These two problems have been solved satisfactorily (Chen 1980, Gao 1987), but will not be presented here due to space limitation.

Chapter 10
The Space Groups

In virtue of the existence of crystalline structure in solids, solid state physics is hardly separable from the theory of space groups. The Hamiltonian of a perfect solid is invariant under the operations of a space group. Therefore the labels of irreps of a space group can be used to characterize the energy levels of a particle or quasi-particle in a crystal, as well as the electronic energy band structure and the phonon dispersion curves in crystalline solids. The knowledge of the irreps and the bases of space groups not only helps us in the understanding of some properties of the solutions of the Hamiltonian, but also provides considerable simplification in the actual calculation of the eigenstates of a Hamiltonian.

The conventional rep theory of space groups is much more involved than that of point groups. The work on the determination of the irreps of a space group was started by Seitz in 1936. Thanks to the efforts of many scientists, tables of the irreps of the 230 space groups are now available [Kovalev 1961, Miller 1967 (reprinted in Cracknell 1979), Bradley 1972, etc.]. Programs for computing the irreps of space groups have been prepared (e.g., Worlton 1973, Neto 1975, Flodmark 1984, Ping 1988).

The conventional approach to the space group rep is far from satisfactory. First, the theory itself is rather complicated and elusive for one who only has a general knowledge of finite group representations. Second, the practical methods for constructing irreps of the little group are mainly Herring's little group method and the projective rep method (Bradley 1972, Birman 1974), or a variation of the projective-rep method (Kovalev 1958). The former two methods require the construction of reps for groups with high orders. For instance, for the projective-rep method, the order can be as high as 192 and 384 for the single-valued and double-valued reps, respectively. Kovalev's approach is rather tedious. Third, for many purposes, e.g., for investigating symmetry change or symmetry breaking in continuous phase transitions (Deonarine 1983), we need to use the irreducible basis adapted to a given group chain. Unfortunately, this requirement is not usually met by the existing tables or programs for the irreps of space groups.

In this chapter, we shall apply the theory for rep groups introduced in Chapter 8 to space groups, and show that the problem of constructing the subgroup symmetry-adapted irreps for space groups is as easy as that for point groups.

The notation for space group operators follows that used by Bradley (1972). The correspondence between McWeeny notation used in Chapter 8 and the Bradley notation for the point group operators is shown in Table 10.21-1.

10.1. The Euclidean Group

10.1.1. Definition of the Euclidean group

Suppose V_3 is a real Euclidean space in three dimensions. Let us find the transformations E which keep the distance between any two points in V_3 unchanged, i.e.,

$$|E\mathbf{x} - E\mathbf{y}| = |\mathbf{x} - \mathbf{y}| . \tag{10-1}$$

E is called a length preserving transformation. All these transformation form a group $\mathbf{E}(3)$ known as the Euclidean group in three dimensions.

The simplest elements in $\mathbf{E}(3)$ are translations $T_\mathbf{a}$:

$$T_\mathbf{a}\mathbf{x} = \mathbf{x} + \mathbf{a} , \tag{10-2}$$

i.e., under the operation of $T_\mathbf{a}$, each point $\mathbf{x}$ in V_3 is shifted to $\mathbf{x}+\mathbf{a}$. All the translations $T_\mathbf{a}$ form a subgroup $\mathbf{T}(3)$ of the group $\mathbf{E}(3)$ known as the translation group. The translation group is Abelian, since

$$T_a T_b = T_b T_a = T_{a+b} . \tag{10-3}$$

Obviously, $T_\mathbf{a}^{-1} = T_{-\mathbf{a}}$.

We can easily find the translation operator $T_\mathbf{a}$ acting on a function $\psi(\mathbf{x})$. Consider an infinitesimal translation δx in the x direction. Under the operator $T_{\delta x}$, $\psi(x)$ goes over to

$$\begin{aligned} \psi'(x) &= T_{\delta x}\psi(x) = \psi(T_{\delta x}^{-1}x) = \psi(x - \delta x) \\ &\cong \psi(x) - \frac{\partial \psi}{\partial x}\delta x \cong e^{-\delta x \frac{\partial}{\partial x}}\psi . \end{aligned} \tag{10-4}$$

From (10-4) the operator for a finite translation $\mathbf{a}$ can be deduced:

$$T_a = e^{-a\cdot\nabla} = e^{-i\hat{k}\cdot a} , \tag{10-5}$$

where $\hat{\mathbf{k}} = -\frac{1}{i}\nabla$ is the momentum operator (choosing the Planck constant $\hbar = 1$) [c.f. the rotation operator $R_\mathbf{n}(\varphi) = \exp(-i\varphi\mathbf{J}\cdot\mathbf{n})$].

10.1.2. Properties of the Euclidean group operators

Suppose that E is an arbitary element of $\mathbf{E}(3)$ and $E\phi = \mathbf{a}$, where $\phi = (0,0,0)$ is the origin. Thus $T_{-\mathbf{a}}E\phi = \phi$, i.e., $T_{-\mathbf{a}}E$ is the operation in $\mathbf{E}(3)$ which leaves the origin unchanged. We denote it by $\alpha = T_{-\mathbf{a}}E$. All the length preserving transformations which keep the origin unchanged form a group $\mathbf{O}(3)$, which is just the orthogonal group in three dimensions. Since $\alpha = T_{-\mathbf{a}}E$, the group element E is uniquely expressed as

$$E = T_\mathbf{a}\alpha . \tag{10-6a}$$

We use the *Seitz notation* $\{\alpha|\mathbf{a}\}$ to represent a general element E of the group $\mathbf{E}(3)$, $\{\varepsilon|0\}$ represents the identity, and $\{\varepsilon|\mathbf{a}\}$ the translation $T_\mathbf{a}$. With this notation, (10-6a) becomes

$$\{\alpha|\mathbf{a}\} = \{\varepsilon|\mathbf{a}\}\{\alpha|0\}, \quad \{\varepsilon|\mathbf{a}\} \in \mathbf{T}(3), \{\alpha|0\} \in O(3) . \tag{10-6b}$$

The operation of $\{\alpha|\mathbf{a}\}$ on any vector $\mathbf{x}$ is defined by

$$\{\alpha|\mathbf{a}\}\mathbf{x} = \alpha\mathbf{x} + \mathbf{a} , \tag{10-7}$$

i.e., performing the rotation α on the vector $\mathbf{x}$ followed by the translation $\mathbf{a}$. Using

$$\{\alpha|\mathbf{a}\}\{\beta|\mathbf{b}\}\mathbf{x} = \{\alpha|\mathbf{a}\}(\beta\mathbf{x} + \mathbf{b}) = \alpha\beta\mathbf{x} + \alpha\mathbf{b} + \mathbf{a} \, , \tag{10-8}$$

we obtain the multiplication law

$$\{\alpha|\mathbf{a}\}\{\beta|\mathbf{b}\} = \{\alpha\beta|\alpha\mathbf{b} + \mathbf{a}\} \, . \tag{10-9}$$

Setting $\{\alpha\beta|\alpha\mathbf{b} + \mathbf{a}\} = \{\varepsilon|0\}$, we find the inverse element of $\{\alpha|\mathbf{a}\}$,

$$\{\alpha|\mathbf{a}\}^{-1} = \{\alpha^{-1}| - \alpha^{-1}\mathbf{a}\} \, . \tag{10-10}$$

From (10-9) we have

$$\{\alpha|\mathbf{a}\} = \{\varepsilon|\mathbf{a}\}\{\alpha|0\} = \{\alpha|0\}\{\varepsilon|\alpha^{-1}\mathbf{a}\} \, . \tag{10-11}$$

Thus translations and rotations (or rotation-reflections) do not commute, unless the translation vector $\mathbf{a}$ is parallel to the rotation axis or the reflection plane, i.e.,

$$[\{\varepsilon|\mathbf{a}\}, \{\alpha|0\}] = 0, \text{ when } \mathbf{a}// \text{ the rotation axis of } \alpha \, , \tag{10-12a}$$

$$[\{\varepsilon|\mathbf{a}\}, \{\sigma|0\}] = 0, \text{ when } \mathbf{a}// \text{ the reflection plane } \sigma \, . \tag{10-12b}$$

From (10-9) we also have

$$\{\alpha|\mathbf{a} + \mathbf{b}\} = \{\varepsilon|\mathbf{a}\}\{\alpha|\mathbf{b}\} = \{\alpha|\mathbf{b}\}\{\varepsilon|\alpha^{-1}\mathbf{a}\} \, . \tag{10-13}$$

With the Seitz notation, (10-5) is rewritten as

$$\{\varepsilon|\mathbf{a}\} = e^{-a\cdot\nabla} = e^{i\hat{k}\cdot a} \, . \tag{10-14}$$

10.2. The Lattice Group

A crystal is formed by arranging atoms or ions in a space lattice

$$L = \{\mathbf{R}_n\} \, , \tag{10-15a}$$

called an *empty lattice*, defined by the set of points (the lattice points)

$$\mathbf{R}_n = n_1\mathbf{t}_1 + n_2\mathbf{t}_2 + n_3\mathbf{t}_3 \, , \tag{10-15b}$$

with integers n_i. $\mathbf{t}_i$ are called *primitive translations* and must not be coplanar. $\mathbf{R}_n$ is called a *lattice vector*. The parallelepiped defined by $\mathbf{t}_1, \mathbf{t}_2$ and $\mathbf{t}_3$ is called the *primitive cell*. Notice that atoms or ions are not necessarily at the lattice points, and generally an array of atoms or ions with a specific relative orientation is associated with each lattice point.

The basis vectors $\mathbf{t}_1, \mathbf{t}_2$ and $\mathbf{t}_3$ are in general not orthogonal to each other. Their scalar product, i.e., the metric tensor is designated by

$$g_{ij} = \mathbf{t}_i \bullet \mathbf{t}_j \, . \tag{10-16}$$

All the rotations α which keep $\mathbf{R}_n$ in the lattice $\{\mathbf{R}_n\}$ form a point group $\mathbf{P}$ which is referred to as the *point group of the empty lattice*, or the *holosymmetric point group* of the crystal system,

$$\mathbf{P} = \{\alpha : \alpha\mathbf{R}_n \in L\} \, . \tag{10-17}$$

The point group $\mathbf{P}$ always contains the space inversion operator I, since if $\mathbf{R}_n$ is a lattice vector, so is $-\mathbf{R}_n$.

The translation group with the lattice vectors $\mathbf{R}_n$ as translation vectors is called the *lattice group* and is denoted by

$$\mathbf{T} = \{\{\varepsilon|\mathbf{R}_n\}\} . \tag{10-18}$$

Obviously, the lattice group is a subgroup of the symmetry group of an infinite crystal, since any point $\mathbf{x}$ in the crystal is equivalent to the points $\mathbf{x} + \mathbf{R}_n$. In reality a crystal is finite but contains a very large number of atoms, $\sim 10^{20}$ per cm^3. Therefore it can be treated as an infinite crystal if what we are concerned with are the bulk properties of the material, such as conductivity and specific heat.

10.3. The Space Group

Together with translation symmetry, a crystal also possesses certain kinds of rotation and reflection symmetries. The complete symmetry group of a crystal is called the *space group*, designated $\mathbf{G}$. The space group is a subgroup of the Euclidean group, whereas the lattice group is a subgroup of the space group, i.e., we have $\mathbf{E}(3) \supset \mathbf{G} \supset \mathbf{T}$. An element of a space group $\mathbf{G}$ is designated by

$$\{\alpha|\mathbf{a}\} = \{\alpha|\mathbf{V}(\alpha) + \mathbf{R}_n\} , \tag{10-19a}$$

where the vector $\mathbf{V}(\alpha)$ associated with the rotation α is called a *nonprimitive*, or *fractional translation*. $\mathbf{V}(\alpha)$ is either zero or a translation which is less than a lattice vector. We always associate $\mathbf{V}(\varepsilon) = 0$ with the identity rotation ε. α is a rotation operator (proper or improper) belonging to the so-called *isogonal point group* $\mathbf{G}_0$ of the space group. As we have said, a crystal is formed by arranging a collection of atoms or ions in an empty lattice. Since both the arrangement and the ions have some symmetries of their own, the crystal point group $\mathbf{G}_0$ has a lower symmetry than the point group $\mathbf{P}$ of the empty lattice, unless the arrangement as well as the ions have a symmetry no lower than that of $\mathbf{P}$. In other words, in general $\mathbf{G}_0$ is a subgroup of $\mathbf{P}$.

A space group $\mathbf{G}$ is designated by

$$\mathbf{G} = \{\{|\alpha|\mathbf{a}\}\} = \{\{\alpha|\mathbf{V}(\alpha) + \mathbf{R}_n\} : \alpha \in \mathbf{G}_0, \mathbf{R}_n \in L\} . \tag{10-19b}$$

Space groups can be divided into two types. The first ones are those for which $\mathbf{V}(\alpha) = 0$ for every α. These are called *simple* or *symmorphic space groups*. There are altogether 73 symmorphic space groups. Clearly, the point group $\mathbf{G}_0 = \{\alpha\}$ is a subgroup of the symmorphic space group.

The others are those for which not all $\mathbf{V}(\alpha)$ are zeroes, called *nonsymmorphic space groups*. There are altogether 157 nonsymmorphic space groups. Notice that the crystal point group $\mathbf{G}_0$ is not a subgroup of the nonsymmorphic space group. $\{\alpha|\mathbf{V}(\alpha)\}$ with nonzero $\mathbf{V}(\alpha)$ is called a *screw rotation* if α is a rotation, or a *glide reflection* if α is a reflecton.

Let us study the restrictions on the nonprimitive translations.

Suppose α is an n-fold axis C_n with a nonprimitive translation $\mathbf{V}$ in parallel with the axis. $\{C_n|0\}$ and $\{\varepsilon|\mathbf{V}\}$ commute on account of (10-12). Therefore

$$\{C_n|\mathbf{V}\}^n = \{C_n^n|n\mathbf{V}\} = \{\varepsilon|n\mathbf{V}\} . \tag{10-20}$$

Thus $n\mathbf{V} = l\mathbf{R}_m$, where $\mathbf{R}_m$ is the shortest lattice vector along the direction of the C_n axis, and $\mathbf{V}$ takes the following form

$$\mathbf{V} = \frac{l}{n}\mathbf{R}_m , \quad l = 0, 1, 2, \ldots , n-1 . \tag{10-21}$$

Similarly, if σ is a reflection plane with a non-primitive translation $\mathbf{V}$ parallel with the plane, then

$$\mathbf{V} = \frac{1}{2}\mathbf{R}_m , \tag{10-22}$$

$\mathbf{R}_m$ being the shortest lattice vector in the direction of $\mathbf{V}$.

The non-primitive translations of the 157 space groups have the following general form

$$\mathbf{V} = \frac{1}{m}(m_1\mathbf{t}_1 + m_2\mathbf{t}_2 + m_3\mathbf{t}_3) , \tag{10-23}$$

$m = 2, 3, 4, 5, 6; m_i = 0, 1, \ldots, m-1$. The space group is a discrete infinite group. However, to define a space group, it suffices to give the primitive translation $\mathbf{t}_i$ and a finite number of elements $\{\alpha|\mathbf{V}(\alpha)\}$.

From the multiplication rule

$$\{\alpha|\mathbf{V}(\alpha) + \mathbf{R}_n\}\{\beta|\mathbf{V}(\beta) + \mathbf{R}_m\} = \{\alpha\beta|\alpha\mathbf{V}(\beta) + \mathbf{V}(\alpha) + \alpha\mathbf{R}_m + \mathbf{R}_n\} , \tag{10-24}$$

one infers that the lattice vectors and non-primitive translations must satisfy the following conditions,

$$\alpha\mathbf{R}_m = \mathbf{R}_l, \quad \mathbf{R}_l \in L , \tag{10-25a}$$

$$\mathbf{V}(\alpha) + \alpha\mathbf{V}(\beta) = \mathbf{V}(\alpha\beta) + \mathbf{R}_{\alpha\beta} . \tag{10-25b}$$

According to

$$\{\alpha|\mathbf{a}\}\{\varepsilon|\mathbf{R}_n\}\{\alpha^{-1}| - \alpha^{-1}\mathbf{a}\} = \{\varepsilon|\alpha\mathbf{R}_n\} , \tag{10-26}$$

and (10-25a), we know that the lattice group $\mathbf{T}$ is an invariant subgroup of the space group.

A symmorphic space group $\mathbf{G}$ is a semi-direct product of the lattice group and the crystal point group, $\mathbf{G} = \mathbf{T} \wedge \mathbf{G}_0$.

10.4. The Point Group P and The Crystal System

Equation (10-25a) shows that the primitive translations $\mathbf{t}_1$, $\mathbf{t}_2$ and $\mathbf{t}_3$ span a three-dimensional rep of the point group $\mathbf{G}_0$,

$$\alpha\mathbf{t}_i = \sum_{j=1}^{3} D_{ji}(\alpha)\mathbf{t}_j . \tag{10-27}$$

$\alpha\mathbf{t}_i = n_1\mathbf{t}_1 + n_2\mathbf{t}_2 + n_3\mathbf{t}_3$ on account of $\alpha\mathbf{t}_i$ belonging to the lattice L. Therefore the matrix elements[1)] $D_{ji}(\alpha)$ in (10-27), as well as its characters $\chi(\alpha)$ are integers. The effect of the point-group operations α on the primitive translations $\mathbf{t}_i$ is given by Bradley (1972, Table 3.2, pp. 84). On the other hand, (5-32) tells us that in the Cartesian bases $\mathbf{i}, \mathbf{j}, \mathbf{k}$ the character of rotation operator is $\chi(\varphi) = \pm 1 + 2\cos\varphi$, φ being the rotation angle, and $+1$ (-1) correspond to proper (improper) rotations. Since the bases $(\mathbf{t}_1, \mathbf{t}_2, \mathbf{t}_3)$ and $(\mathbf{i}, \mathbf{j}, \mathbf{k})$ differ only by a similarity transformation, and the character is invariant under a similarity transformation, we immediately arrive at the condition for the allowable rotation angles,

$$\pm 1 + 2\cos\varphi = \text{integer} . \tag{10-28}$$

Thus φ can only take on the values $2\pi/n, n = 1, 2, 3, 4, 6$. In other words, the group $\mathbf{P}$ can only contain 1-, 2-, 3-, 4- and 6-fold axes. Among the point groups, only the 32 crystal point groups fulfill this requirement. They include the cyclic group $C_n, n = 1, 2, 3, 4, 6$; the

[1)] Since $\mathbf{t}_i$ are not orthogonal, the rep matrix $D(\alpha)$ of the unitary operator α is not a unitary matrix.

group $D_n, n = 2,3,4,6$; the tetrahedral group T and the octahedral group O. The inclusion of inversion and reflection results in the groups C_i, S_4, S_6; $C_{nh}, n = 1,2,3,4,6$; C_{nv} and $D_{nh}, n = 2,3,4,6$; D_{2d}, D_{3d}, T_d, T_h and O_h.

Apart from the restriction (10-28), the following theorem imposes another constraint on the permissible point groups $\mathbf{P}$ of the empty lattice.

Theorem 10.1: If the point group $\mathbf{P}$ of an empty lattice contains an n-fold axis $C_n, n > 2$, then it necessarily contains the group C_{nv}. (For proof, see Tao 1986).

As a result of this restriction, only seven among the 32 crystal point groups survive. They are $C_i, C_{2h} = C_2 \times C_i, D_{2h} = D_2 \times C_i, D_{3d} = D_3 \times C_i, D_{4h} = D_4 \times C_i, D_{6h} = D_6 \times C_i$ and $O_h = O \times C_i$. These are the only allowable point groups $\mathbf{P}$ of the empty lattice, and as expected, they all contain the inversion I. Their interrelation is as follows

$$\begin{array}{ccccccccc} O_h & \supset & D_{4h} & \supset & D_{2h} & \supset & C_{2h} & \supset & C_i \\ & & & & \cap & & \cap & & \\ & & & & D_{6h} & & D_{3d} & & . \end{array} \tag{10-29}$$

Definition 10.1: Two empty lattices are said to belong to the same crystal system if they have the same symmetry point group $\mathbf{P}$.

Thus there are only seven crystal systems. Their names and the conventional forms of the related metric tensors are

$$\begin{array}{cccc} \text{Triclinic} & \text{Monoclinic} & \text{Orthorhomic} & \text{Tetragonal} \\ \begin{pmatrix} g_{11} & g_{12} & g_{13} \\ g_{21} & g_{22} & g_{23} \\ g_{31} & g_{32} & g_{33} \end{pmatrix}, & \begin{pmatrix} g_{11} & g_{12} & 0 \\ g_{21} & g_{22} & 0 \\ 0 & 0 & g_{33} \end{pmatrix}, & \begin{pmatrix} g_{11} & 0 & 0 \\ 0 & g_{22} & 0 \\ 0 & 0 & g_{33} \end{pmatrix}, & \begin{pmatrix} a & b & b \\ b & a & b \\ b & b & a \end{pmatrix}, \\ \text{Trigonal} & \text{Hexagonal} & \text{Cubic} & \\ \begin{pmatrix} a & 0 & 0 \\ 0 & a & 0 \\ 0 & 0 & c \end{pmatrix}, & \begin{pmatrix} a & -a/2 & 0 \\ -a/2 & 0 & 0 \\ 0 & 0 & c \end{pmatrix}, & \begin{pmatrix} a & 0 & 0 \\ 0 & a & 0 \\ 0 & 0 & a \end{pmatrix} & \end{array} . \tag{10-30}$$

10.5. The Bravais Lattices

In the previous section, we discussed the restriction on the allowable point groups $\mathbf{P}$ imposed by the symmetry of the empty lattice. Conversely, the point group $\mathbf{P}$ will also impose restrictions on the possible types of the empty lattices.

Definition 10.2: Two lattices belonging to the same crystal system are said to be of the same type if one of them can be obtained from the other by a continuous deformation such that the Bravais lattice does not undergo a transformation passing through a crystal system of lower symmetry.

For example, the face-oriented cubic Bravais lattice Γ_c^f cannot be deformed into a body-centered cubic Bravais lattice Γ_c^v without passing through the trigonal Bravais lattice. Therefore Γ_c^f and Γ_c^v are of different types.

The seven crystal systems contain 14 types of lattices, called the 14 *Bravais lattices*. Their names, geometric forms along with the primitive translations are given in Fig. 10.5. With these

Fig. 10.5. The 14 Bravais lattices.

In the following a, b and c represent the lengths of the three sides, while α, β and γ represent the angles between the sides **b** and **c**, **c** and **a**, and **a** and **b**, respectively.

1. Triclinic – $P(\Gamma_t), \mathbf{P} = C_i$.
 $a \neq b \neq c$
 $\alpha \neq \beta \neq \gamma$

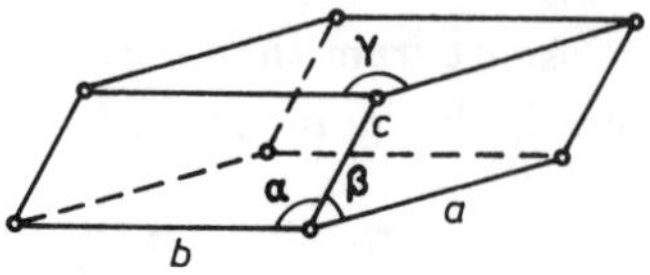

2. Monoclinic $a \neq b \neq c, \alpha = \beta = \pi/2 \neq \gamma, \mathbf{P} = C_{2h}$

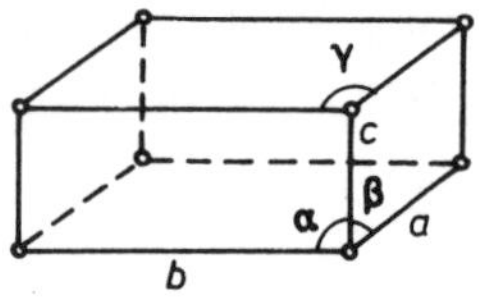

Simple Γ_m
Monoclinic-P
$\mathbf{t}_i$: $(0, -b, 0)$; $(a \sin\gamma, -a\cos\gamma, 0)$; $(0, 0, c)$

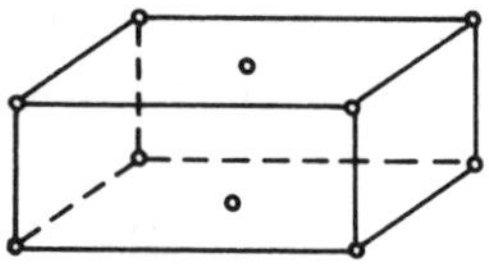

Base-centered Γ_m^b
Monoclinic-B
$\mathbf{t}_i$: $(0 - b, 0)$; $\frac{1}{2}(a\sin\gamma, -a\cos\gamma, -c)$; $\frac{1}{2}(a\sin\gamma, -a\cos\gamma, c)$

3. Orthorhombic $a \neq b \neq c, \alpha = \beta = \gamma = \pi/2, \mathbf{P} = D_{2h}$.

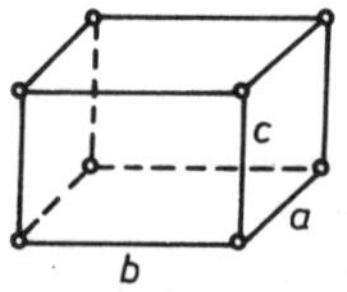

Simple Γ_0
Orthorhombic-P
$\mathbf{t}_i$: $(0, -b, 0)$; $(a, 0, 0)$; $(0, 0, c)$

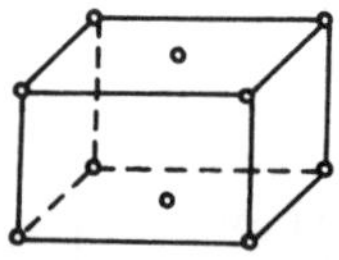

Base-centered Γ_0^b
Orthorombic-C
$\mathbf{t}_i$: $\frac{1}{2}(a, -b, 0)$; $\frac{1}{2}(a, b, 0)$; $(0, 0, c)$

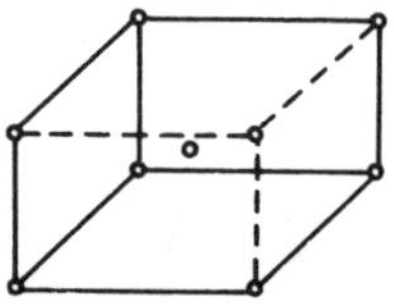

Body-centered Γ_0^v
Orthorombic-I
$\mathbf{t}_i$: $\frac{1}{2}(a, b, c)$; $\frac{1}{2}(-a, -b, c)$; $\frac{1}{2}(a, -b, -c)$;

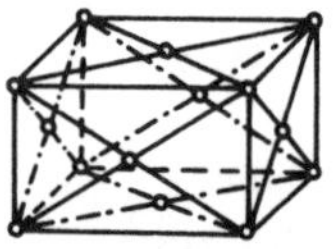

Face-centered Γ_0^f
Orthorombic-F
$\mathbf{t}_i$: $\frac{1}{2}(a, 0, c)$; $\frac{1}{2}(0, -b, c)$; $\frac{1}{2}(a, -b, 0)$

4. Trigonal – $R(\Gamma_{rh}), \mathbf{P} = D_{3d}$.
 $a = b \neq c$,
 $\alpha = \beta = \gamma < 2\pi/3$,
 $\alpha \neq \pi/2$.
 $\mathbf{t}_i : (0, -a, c); \frac{1}{2}(\sqrt{3}\,a, a, 2c); \frac{1}{2}(-\sqrt{3}\,a, a, 2c)$.

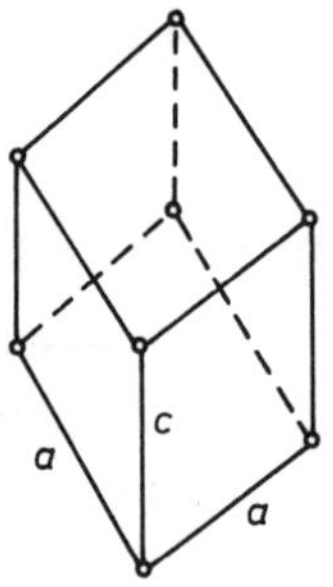

5. Tetragonal, $a = b \neq c, \alpha = \beta = \gamma = \pi/2, \mathbf{P} = D_{4h}$.

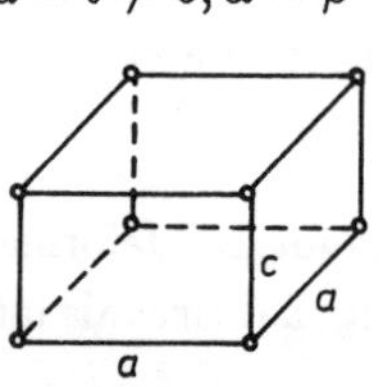

Simple Γ_q
Tetragonal-P
$\mathbf{t}_i$: $(a, 0, 0)$; $(0, a, 0)$; $(0, 0, c)$

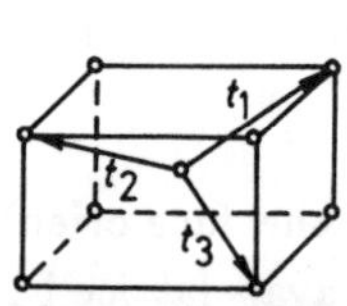

Body-centered Γ_q^v
Tetragonal-I
$\mathbf{t}_i$: $\frac{1}{2}(-a, a, c)$; $\frac{1}{2}(a, -a, c)$; $\frac{1}{2}(a, a, -c)$

6. Hexagonal $-\mathrm{P}$, $\Gamma_h, \mathbf{P} = D_{6h}$
   $a = b \neq c, \alpha = \beta = \pi/2, \gamma = 2\pi/3.$
   $\mathbf{t}_i:$ $(0, -a, 0);$
   $\frac{1}{2}(\sqrt{3}\,a, a, 0);$
   $(0, 0, c).$

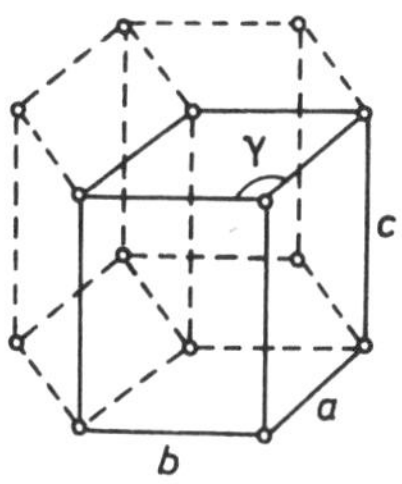

7. Cubic $\mathbf{P} = O_h, a = b = c, \alpha = \beta = \gamma = \pi/2.$

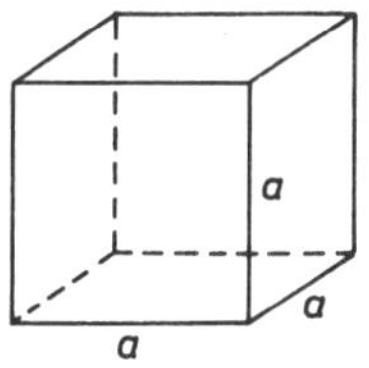

Simple Γ_c
Cubic-P
$\mathbf{t}_i$: $(a, 0, 0)$; $(0, a, 0)$; $(0, 0, a)$

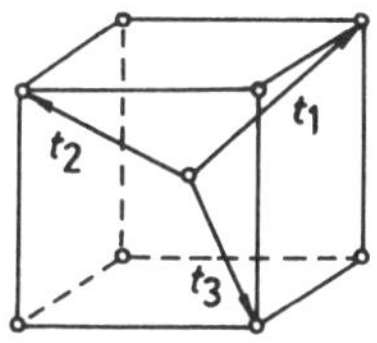

Body-centered Γ_a^v
Cubic-I
$\mathbf{t}_i$: $\frac{1}{2}(-a, a, a)$; $\frac{1}{2}(a, -a, a)$; $\frac{1}{2}(a, a, -a)$.

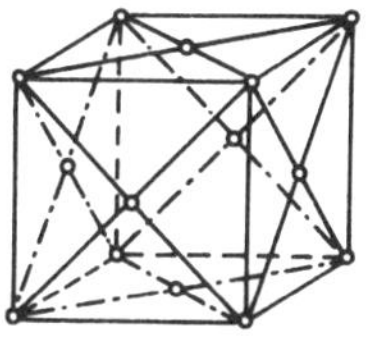

Face-centered Γ_c^f
Cubic-F
$\mathbf{t}_i$: $\frac{1}{2}(0, a, a)$; $\frac{1}{2}(a, 0, a)$, $\frac{1}{2}(a, a, 0)$

primitive translations one can calculate the metric tensor g_{ij} associated with a given Bravais lattice. For instance, for the cubic system we have

$$a^2\begin{pmatrix}1&0&0\\0&1&0\\0&0&1\end{pmatrix}, \quad \frac{a^2}{4}\begin{pmatrix}3&-1&-1\\-1&3&-1\\-1&-1&3\end{pmatrix}, \quad \frac{a^2}{4}\begin{pmatrix}2&1&1\\1&2&1\\1&1&2\end{pmatrix} \tag{10-31}$$

It can be shown that from the 14 Bravais lattices $\{\mathbf{R}_n\}$, the 32 crystal point groups $\mathbf{G}_0 = \{\alpha\}$, and the various kinds of possible nonprimitive translations $\mathbf{V}(\alpha)$, one can construct altogether 230 space groups $\mathbf{G} = \{\alpha|\mathbf{V}(\alpha) + \mathbf{R}_n\}$. These space groups are divided into 32 classes on the basis of the crystal point groups $\mathbf{G}_0$. We use the Schönflies symbol (standing for point group) with superscript to denote a space group. For example, there are 10 space groups associated with the point group O_h and they are designated as $O_h^1, \ldots, O_h^{10}$.

The generators of the 230 space groups are given by Bradley (1972, Table 3.7, pp. 127).

The distribution of the 32 crystal groups among the seven crystal systems is as follows.

1. Triclinic: $\mathbf{P} = C_i$; $\mathbf{G}_0 = C_i, C_1$
2. Monoclinic: $\mathbf{P} = C_{2h}$; $\mathbf{G}_0 = C_{2h}, C_2, C_s$
3. Orthorhombic: $\mathbf{P} = D_{2h}$; $\mathbf{G}_0 = D_{2h}, D_2, C_{2v}$
4. Trigonal: $\mathbf{P} = D_{3d}$; $\mathbf{G}_0 = D_{3d}, D_3, C_{3v}, C_3, S_6$
5. Tetragonal: $\mathbf{P} = D_{4h}$; $\mathbf{G}_0 = D_{4h}, D_{4d}, D_4, C_{4v}, S_4, C_4, D_{2d}$
6. Hexagonal: $\mathbf{P} = D_{6h}$; $\mathbf{G}_0 = D_{6h}, D_6, C_{6h}, C_{6v}, C_6, D_{3h}, C_{3h}$
7. Cubic: $\mathbf{P} = O_h$; $\mathbf{G}_0 = O_h, O, T_d, T, T_h$.

10.6. Operators of the Space Group

10.6.1. The properties of group operators

It must be emphasized that the operators of the Euclidean group or the space group are not linear if they are regarded as coordinate transformation operators. To show this it suffices to consider the translation operators $T_{\mathbf{a}}$. Suppose $\mathbf{z} = c_1\mathbf{x} + c_2\mathbf{y}$, where c_1 and c_2 are arbitrary constants. From (10-2),

$$T_{\mathbf{a}}\mathbf{z} = c_1\mathbf{x} + c_2\mathbf{y} + \mathbf{a} \ . \tag{10-32a}$$

If the operator $T_{\mathbf{a}}$ were a linear operator, it should have been

$$T_{\mathbf{a}}\mathbf{z} = c_1 T_{\mathbf{a}}\mathbf{x} + c_2 T_{\mathbf{a}}\mathbf{y} = c_1\mathbf{x} + c_2\mathbf{y} + (c_1 + c_2)\mathbf{a} \ . \tag{10-32b}$$

However $(c_1 + c_2)\mathbf{a} \neq \mathbf{a}$. Therefore, the coordinate transformation operators $T_{\mathbf{a}}$ are not linear.

Nevertheless, the operators of a space group are linear if they are defined as the *substitutional operators* in a function space, i.e.,

$$\{\alpha|\mathbf{a}\}\psi(x) = \psi(\{\alpha|\mathbf{a}\}^{-1}\mathbf{x}) = \psi(\alpha^{-1}(\mathbf{x} - \mathbf{a})) \ . \tag{10-33}$$

To show this, it again suffices to consider the translational operators. Suppose

$$T_{\mathbf{a}}\psi(\mathbf{x}) = \psi(\mathbf{x} - \mathbf{a}) \ ,$$

then we have

$$T_{\mathbf{a}}[c_1\psi(\mathbf{x}) + c_2\varphi(\mathbf{x})] = c_1\psi(\mathbf{x} - \mathbf{a}) + c_2\varphi(\mathbf{x} - \mathbf{a}) = c_1 T_{\mathbf{a}}\psi(\mathbf{x}) + c_2 T_{\mathbf{a}}\varphi(\mathbf{x}) \ .$$

Therefore the operator $T_{\mathbf{a}} = \{\varepsilon|\mathbf{a}\}$ is linear under the definition of (10-33). The same is true for the operator $\{\alpha|\mathbf{a}\}$.

With this definition, the multiplication rule for the group elements is identical to (10-9) which was used in coordinate space; that is

$$\begin{aligned}\{\alpha|\mathbf{a}\}\{\beta|\mathbf{b}\}\psi(\mathbf{x}) &= \{\alpha|\mathbf{a}\}\psi(\beta^{-1}\mathbf{x} - \beta^{-1}\mathbf{b}) = \psi(\beta^{-1}\{\alpha|\mathbf{a}\}^{-1}\mathbf{x} - \beta^{-1}\mathbf{b}) \\ &= \psi(\{\alpha\beta|\alpha\mathbf{b} + \mathbf{a}\}^{-1}\mathbf{x}) = \{\alpha\beta|\alpha\mathbf{b} + \mathbf{a}\}\psi(\mathbf{x}) \ . \end{aligned} \tag{10-34a}$$

It is worth mentioning that in performing the operations in (10-34a) and the like, one must be very careful to notice the following points:

$$\{\varepsilon|\mathbf{a}\}\psi(\alpha^{-1}\mathbf{x}) = \psi(\alpha^{-1}(\mathbf{x} - \mathbf{a})) \neq \psi(\alpha^{-1}\mathbf{x} - \mathbf{a}) \ , \tag{10-34b}$$

$$\{\alpha|0\}\psi(\mathbf{x} - \mathbf{b}) = \psi(\alpha^{-1}\mathbf{x} - \mathbf{b}) \neq \psi(\alpha^{-1}(\mathbf{x} - \mathbf{b})) \ , \tag{10-34c}$$

$$\{\alpha|0\}\psi(\beta^{-1}\mathbf{x}) = \psi(\beta^{-1}\alpha^{-1}\mathbf{x}) \neq \psi(\alpha^{-1}\beta^{-1}\mathbf{x}) \ . \tag{10-34d}$$

Equation (10-33) can be rewritten as

$$\{\alpha|\mathbf{a}\}\psi(\mathbf{x}) = \psi(\mathbf{x}' - \mathbf{a}') \ , \tag{10-35a}$$

$$\mathbf{x}' - \mathbf{a}' = \alpha^{-1}(\mathbf{x} - \mathbf{a}) \ , \tag{10-35b}$$

or in component form [cf. (2-5a)]

$$x_i' - a_i' = \sum_{j=1}^{3} D_{ij}(\alpha^{-1})(x_j - a_j) = \sum_{j=1}^{3} D_{ji}(\alpha)(x_j - a_j) \ , \tag{10-35c}$$

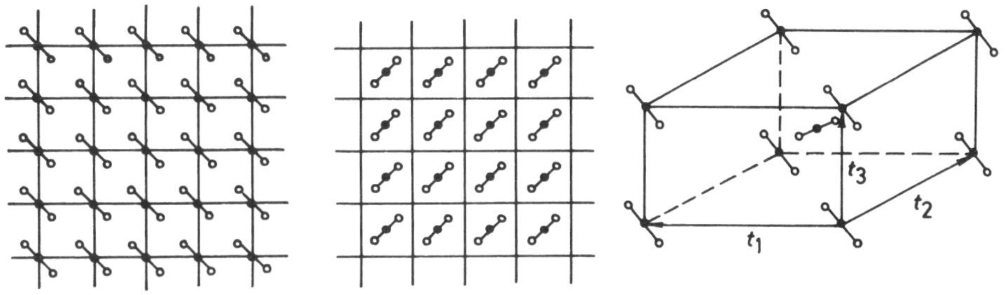

Fig. 10.6-1. The arrangement of atoms in TiO_2 and the primitive cell of D_{4h}^{14}. • denotes the molecule Ti, and ∘, the molecule O. (a) The atoms in the layers 1,3,... (b) The atoms in the layers 0,2,4,... (c) The primitive cell of D_{4h}^{14}.

where $D_{ij}(\alpha)$ are the matrix elements of the point-group operator α in the rep carried by the basis functions x, y and z.

Henceforth we will use only the definition (10-35).

10.6.2. Example: group D_{4h}^{14}

As an example we will give the operators of the space group D_{4h}^{14}. It is the symmetry group of the crystal TiO_2. TiO_2 belongs to the tetragonal system and the arrangement of its atoms is shown in Fig. 10.6-1. Although there is a TiO_2 molecule in the center of each primitive cell, it belongs to the simple tetragonal instead of the body-centered tetragonal, since the molecule in the center has a different orientation from that of the molecules in the corners.

The group operators of D_{4h}^{14} are as follows:

1. The pure translations

$$\mathbf{R}_n = n_1\mathbf{t}_1 + n_2\mathbf{t}_2 + n_3\mathbf{t}_3 \ ,$$

where $\mathbf{t}_1, \mathbf{t}_2$ and $\mathbf{t}_3$ are mutually orthogonal and $t_1 = t_2 = a, t_3 = c$.

2. The operations belonging to the point group D_{2h} (see Fig. 10.6-2)

$$D_{2h}: \{\varepsilon|0\}, \{C_{2z}|0\}, \{C_{2a}|0\}, \{C_{2b}|0\} \ ,$$
$$\{I|0\}, \{\sigma_z|0\}, \{\sigma_a|0\}, \{\sigma_b|0\} \ .$$

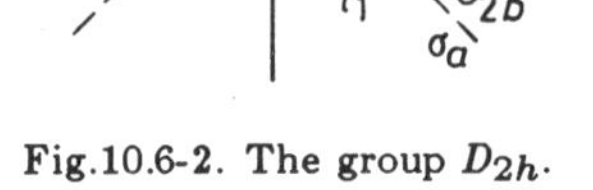

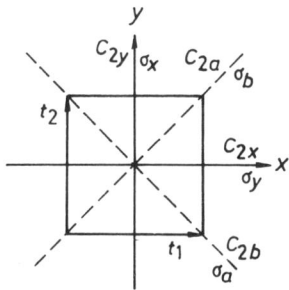

Fig.10.6-2. The group D_{2h}.

3. The operators involving non-primitive translations

$$\{C_{4z}^+|\mathbf{v}\}, \{C_{4z}^-|\mathbf{v}\}, \{C_{2x}|\mathbf{v}\}, \{C_{2y}|\mathbf{v}\}$$
$$\{S_{4z}^-|\mathbf{v}\}, \{S_{4z}^+|\mathbf{v}\}, \{\sigma_x|\mathbf{v}\}, \{\sigma_y|\mathbf{v}\}, \mathbf{v} = \frac{1}{2}(\mathbf{t}_1 + \mathbf{t}_2 + \mathbf{t}_3) \ .$$

According to (10-35) and the matrix elements of point-group operators (such as (8-1), (8-23) etc.), we obtain the following table for the action of the group elements of D_{4h}^{14} on a function $\psi(x, y, z)$.

Table 10.6. The action of the group elements of the space group D_{4h}^{14} on a function $\psi(x,y,z)$.*

$\{\alpha\vert\mathbf{a}\}$	$\{\varepsilon\vert 0\}$	$\{C_{2a}\vert 0\}$	$\{C_{2b}\vert 0\}$	$\{C_{2z}\vert 0\}$	$\{C_{4z}^{+}\vert\mathbf{v}\}$	$\{C_{4z}^{-}\vert\mathbf{v}\}$
$\{\alpha\vert\mathbf{a}\}\psi$	$\psi(x, y, z)$	$\psi(y, x, -z)$	$\psi(-y, -x, -z)$	$\psi(-x, -y, z)$	$\psi(\eta, -\xi, \zeta)$	$\psi(-\eta, \xi, \zeta)$
$\{\alpha\vert\mathbf{a}\}$	$\{I\vert 0\}$	$\{\sigma_a\vert 0\}$	$\{\sigma_b\vert 0\}$	$\{\sigma_z\vert 0\}$	$\{S_{4z}^{-}\vert\mathbf{v}\}$	$\{S_{4z}^{+}\vert\mathbf{v}\}$
$\{\alpha\vert\mathbf{a}\}\psi$	$\psi(-x, -y, -z)$	$\psi(-y, -x, z)$	$\psi(y, x, z)$	$\psi(x, y, -z)$	$\psi(-\eta, \xi, -\zeta)$	$\psi(\eta, -\xi, -\zeta)$
$\{\alpha\vert\mathbf{a}\}$	$\{C_{2x}\vert\mathbf{v}\}$	$\{C_{2y}\vert\mathbf{v}\}$	$\{\sigma_x\vert\mathbf{v}\}$	$\{\sigma_y\vert\mathbf{v}\}$	$\{\varepsilon\vert\mathbf{R}_n\}$	
$\{\alpha\vert\mathbf{a}\}\psi$	$\psi(\xi, -\eta, -\zeta)$	$\psi(-\xi, \eta, -\zeta)$	$\psi(-\xi, \eta, \zeta)$	$\psi(\xi, -\eta, \zeta)$	$\psi(x - n_1a, y - n_2a, z - n_3c)$	

* $\xi = x - \frac{1}{2}$, $\eta = y - \frac{1}{2}$, $\zeta = z - \frac{1}{2}$.

10.7. The Reciprocal Lattice Vectors

Let $(\mathbf{t}_1, \mathbf{t}_2, \mathbf{t}_3)$ be a set of unorthonormal basis vectors. According to (2-15a), we can introduce another set of dual basis vectors $(\mathbf{b}_1, \mathbf{b}_2, \mathbf{b}_3)$, so that

$$\mathbf{b}_i \cdot \mathbf{t}_j = 2\pi\delta_{ij} , \tag{10-36}$$

where the factor 2π is introduced for convenience. It is easily seen that the relation between $\mathbf{b}_i$ and $\mathbf{t}_i$ is

$$\mathbf{b}_i = 2\pi(\mathbf{t}_j \times \mathbf{t}_k)/[\mathbf{t}_i \cdot (\mathbf{t}_j \times \mathbf{t}_k)], \quad i, j, k \quad \text{cyclic.} \tag{10-37}$$

From (5-12c) we know that $\mathbf{t}_i$ is the covariant basis, while $\mathbf{b}_i$ is the contravariant basis, and $\mathbf{b}_i \cdot \mathbf{t}_j = 2\pi\delta_{ij}$ is an invariant. The metric tensor for the contravariant basis is

$$g^{ij} = \mathbf{b}_i \cdot \mathbf{b}_j . \tag{10-38}$$

According to (5-12b), we have

$$|g^{ij}| = 4\pi^2 |g_{ij}|^{-1} . \tag{10-39}$$

Let us define

$$\mathbf{K}_m = m_1\mathbf{b}_1 + m_2\mathbf{b}_2 + m_{33} , \tag{10-40a}$$

with integers m_i. $\mathbf{K}_m$ are referred to as the reciprocal lattice vectors, and the lattice formed by the vectors $\mathbf{K}_m$ is called the *reciprocal lattice*, denoted as

$$L^{-1} = \{\mathbf{K}_m\} . \tag{10-40b}$$

We first prove that the reciprocal lattice and the space lattice have the same symmetry point group $\mathbf{P}$. From (10-15b) and (10-40a) it follows that

$$\mathbf{K}_m \cdot \mathbf{R}_n = 2\pi \sum_{i=1}^{3} n_i m_i = 2\pi \times \text{integer} . \tag{10-41a}$$

Since $\alpha^{-1}\mathbf{R}_n$ is a lattice vector, $\mathbf{K}_m \cdot \alpha^{-1}\mathbf{R}_n = 2\pi\times$ integer. Using the unitarity of the operator α, one has

$$\mathbf{K}_m \cdot \alpha^{-1}\mathbf{R}_n = \alpha\mathbf{K}_m \cdot \mathbf{R}_n = 2\pi \times \text{integer} .$$

The above relation holds for any $\mathbf{R}_n$, thus $\alpha\mathbf{K}_m$ is also a reciprocal lattice vector,

$$\alpha\mathbf{K}_m = \mathbf{K}_l . \tag{10-41b}$$

Consequently, the space lattice and its reciprocal lattice must belong to the same crystal system. However they may not necessarily of the same type. In fact, they are of the same type except for the cubic and orthorhombic crystal systems, for which the reciprocal lattice of the body-centered space lattice is face-centered and vice versa. For the cubic, this can be seen from the following equation,

$$g_I = \frac{a^4}{4} g_F^{-1} , \tag{10-42}$$

where (10-31) has been used. Equation (10-42) agrees with the condition (10-39) apart from a trivial constant factor. Table 10.7 lists the cases for which the reciprocal lattices and their space lattices belong to different crystal types.

Table 10.7. The correspondence between the reciprocal lattices and space lattices when they are of different types.

crystal system	space lattice	reciprocal lattice
orthorhombic	body-centered	face-centered
	face-centered	body-centered
cubic	body-centered	face-centered
	face-centered	body-centered

10.8. Irreps of the Lattice Group

With $\mathbf{t}_i$ as basis

$$\mathbf{r} = \xi_1 \mathbf{t}_1 + \xi_2 \mathbf{t}_2 + \xi_3 \mathbf{t}_3 . \tag{10-43}$$

In analogy with (10-4) we have

$$\begin{aligned} \{\varepsilon|\delta\xi_1 \mathbf{t}_1\}\psi(\xi_1) &= \psi(\xi_1 - \delta\xi_1) \cong \exp\left[-\delta\xi_1 \frac{\partial}{\partial\xi_1}\right]\psi(\xi_1) \\ &= \exp\left[-(\delta\xi_1 \mathbf{t}_1)\cdot\frac{1}{2\pi}\left(\frac{\partial}{\partial\xi_1}\mathbf{b}_1\right)\right]\psi(\xi_1) \end{aligned} \tag{10-44}$$

Thus we know the translation operator

$$\{\varepsilon|\mathbf{R}_n\} = e^{-i\hat{k}\cdot\mathbf{R}_n} = \exp\left[-\left(n_1\frac{\partial}{\partial\xi_1} + n_2\frac{\partial}{\partial\xi_2} + n_3\frac{\partial}{\partial\xi_3}\right)\right] , \tag{10-45a}$$

$$\hat{\mathbf{k}} = -i\nabla = -i\left(\frac{\partial}{\partial\xi_1}\mathbf{b}_1 + \frac{\partial}{\partial\xi_2}\mathbf{b}_2 + \frac{\partial}{\partial\xi_3}\mathbf{b}_3\right)\Big/ 2\pi . \tag{10-45b}$$

The lattice group is Abelian, hence the group operator is just the class operator. Its CSCO can be chosen as the group operators $\{\varepsilon|\mathbf{R}_n\} = \exp(-i\hat{k}\cdot\mathbf{R}_n)$. The irreducible bases of the lattice group are eigenfunctions of the CSCO,

$$\{\varepsilon|\mathbf{R}_n\}\psi_{\mathbf{k}} = \exp(-i\hat{\mathbf{k}}\cdot\mathbf{R}_n)\psi_{\mathbf{k}} = \exp(-i\mathbf{k}\cdot\mathbf{R}_n)\psi_{\mathbf{k}} , \tag{10-46}$$

where $\mathbf{k}$ are eigenvalues of the operator $\hat{\mathbf{k}}$. $\mathbf{k}$ is called the *wave vector* and is used to label irreps of the lattice group. $\mathbf{k}$ can be expressed as

$$\mathbf{k} = p_1\mathbf{b}_1 + p_2\mathbf{b}_2 + p_3\mathbf{b}_3 . \tag{10-47}$$

One of the solutions of (10-46) is easily found to be

$$\psi_{\mathbf{k}}(r) = \exp(i\mathbf{k}\cdot\mathbf{r}) = \exp[2\pi i(p_1\xi_1 + p_2\xi_2 + p_3\xi_3)] . \tag{10-48a}$$

Since $\exp(i\mathbf{K}_m \cdot \mathbf{R}_n) = 1$ from (10-41a), $\psi_{\mathbf{k}+\mathbf{K}_m} = \exp(i(\mathbf{k}+\mathbf{K}_m)\cdot\mathbf{r})$ is also a solution of (10-46). Therefore the irreps $\mathbf{k}$ and $\mathbf{k}+\mathbf{K}_m$ of the lattice group are equivalent and $\mathbf{k}$ and $\mathbf{k}+\mathbf{K}_m$ are referred to as equivalent wave vectors, denoted by

$$\mathbf{k} \doteq \mathbf{k} + \mathbf{K}_m \ . \tag{10-48b}$$

The most general form of the basis belonging to the irrep $\mathbf{k}$ of the lattice group is

$$\varphi^{(\mathbf{k})}(\mathbf{r}) = \sum_{\mathbf{K}_m} V(\mathbf{k}+\mathbf{K}_m)\exp[i(\mathbf{k}+\mathbf{K}_m)\cdot\mathbf{r}] \ , \tag{10-48c}$$

where $V(\mathbf{k}+\mathbf{K}_m)$ are coefficients.

According to (3-348), the projection operator for the lattice group is

$$P^{(\mathbf{k})} = \text{const.} \sum_{\mathbf{R}_n} \exp[i(\mathbf{k}-\hat{\mathbf{k}})\cdot\mathbf{R}_n] \ . \tag{10-49}$$

The CSCO of the lattice group can be conveniently chosen as the momentum operator

$$\hat{\mathbf{k}} = -i\nabla \tag{10-45b}$$

with the eigenvalue $\mathbf{k}$ modulo $\mathbf{K}_m$.

10.9. The Brillouin Zone

On the basis of the above discussion, it follows that we can obtain all the irreps of the lattice group by allowing the components p_1, p_2 and p_3 of the wave vector $\mathbf{k}$ to range over a unit interval.

In order to visualize the region of space in which $\mathbf{k}$ must lie, let us introduce the primitive cell for the reciprocal lattice, which is defined, as in the case for the space lattice, as the parallelepiped formed by the basis vectors $\mathbf{b}_1, \mathbf{b}_2$ and $\mathbf{b}_3$. The primitive cell defined in this way, though simple, has the disadvantage that it does not exhibit the point-group symmetry of the reciprocal lattice. It is more preferable to use the symmetrized primitive cell, known as the *Wigner-Seitz cell*, which is defined as the region bounded by the planes bisecting normally the vectors joining reciprocal lattice points. This region surrounding the origin is called the *Brillouin zone* or the first Brillouin zone. We thus conclude that to obtain all the inequivalent irreps of the lattice group, one only needs to let the wave vector $\mathbf{k}$ runs over all the points in the Brillouin zone.

The vector $\mathbf{k}$ in the Brillouin zone is called the reduced wave vector. Therefore, an irrep of the lattice group is labeled by a point in the Brillouin zone, or by a reduced wave vector $\mathbf{k}$. Some examples of Brillouin zones for two-dimensional lattices are shown in Fig. 10.9. The Brillouin zones or the Wigner-Seitz cells for a number of three-dimensional lattices can be found in Bradley (1972, Figs. 3.2—3.15), or Slater (1962).

The points in a Brillouin zone are divided into two groups.

1. General points: A wave vector $\mathbf{k}$ is called a general point if it does not have any symmetry, i.e., for any operator $\alpha \in \mathbf{G}_0, \alpha\mathbf{k}$ and $\mathbf{k}$ are not equivalent.

2. Special points: A point with a certain kind of symmetry is called a special point. For example, the point that lies on a rotation axis or a reflection plane and therefore is invariant under the rotation or reflection, or the point $\mathbf{k}$ in a surface of the Brillouin zone which may be invariant or goes to its equivalent $\mathbf{k}' = \mathbf{k} + \mathbf{K}_m$ under a rotation or reflection.

Special points are subdivided into (Bradley 1972) two groups.

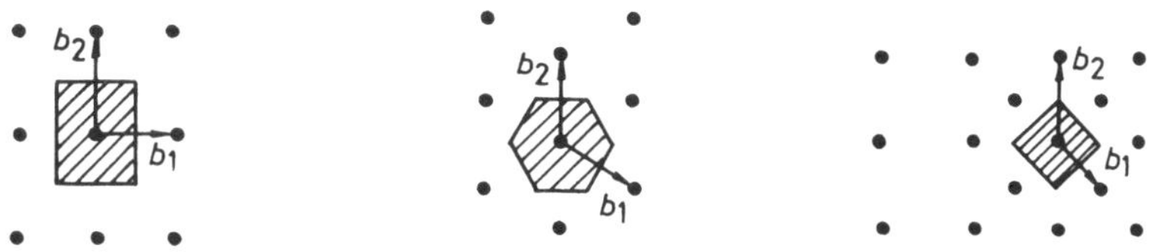

Simple rectangular   Hexagonal   Face-centered square

Fig. 10.9. The Brillouin zones for two-dimensional lattices [taken from Elliott (1979)].

1. Points of symmetry: $\mathbf{k}$ is called a point of symmetry if there exists a neighborhood of $\mathbf{k}$ in which $\mathbf{k}$ is the point with the highest symmetry.

2. Lines or planes of symmetry: $\mathbf{k}$ is called a line (plane) of symmetry if in a sufficiently small neighborhood of $\mathbf{k}$ there is always a line (plane) through $\mathbf{k}$, all points of which have the same symmetry as that of $\mathbf{k}$.

The special point in a Brillouin zone are labeled by the standard solid state physics symbols, e.g., Γ, X, M, etc. (see Bradley 1972, pp. 96-118).

Of all the rotations in the point group $\mathbf{P}$, those which leave the wave vector $\mathbf{k}$ invariant modulo a reciprocal lattice vector form a subgroup of $\mathbf{P}$ designated by $\mathbf{P}(k)$:

$$\mathbf{P}(\mathbf{k}) = \{\alpha \in \mathbf{P} : \alpha\mathbf{k} \doteq \mathbf{k}\} . \tag{10-50}$$

$\mathbf{P}(\mathbf{k})$ is referred to as the symmetric group of the wave vector $\mathbf{k}$. The symmetry group $\mathbf{P}(\mathbf{k})$ are listed in Bradley (1972, Table 3.6). Examples are given in Sec. 10.16.

10.10. The Electron State in a Periodic Potential

Consider an electron moving in a crystal. As a first approximation, we can use the independent particle model, i.e., assume that each electron moves independently in a fixed averaged potential $V(\mathbf{r})$, which has the translation symmetry $V(\mathbf{r}+\mathbf{R}_n) = V(\mathbf{r})$. If the crystal does not have symmetries other than the translational symmetry, then the lattice group is the symmetry group of the electron Hamiltonian, and the electron eigenstates must belong to an irrep $\mathbf{k}$ of the lattice group. We know that $\exp(i\mathbf{k}\cdot\mathbf{r})$ is a basis vector of the irrep $\mathbf{k}$. Suppose $w_\mathbf{k}(\mathbf{r})$ is a basis vector for the identity rep of the lattice group

$$\{\varepsilon|R_n\}w_\mathbf{k}(\mathbf{r}) = w_\mathbf{k}(\mathbf{r}-\mathbf{R}_n) = w_\mathbf{k}(\mathbf{r}) , \tag{10-51a}$$

then the general expression for the basis vector of the irrep $\mathbf{k}$ is

$$\varphi^{(k)}(\mathbf{r}) = w_\mathbf{k}(\mathbf{r})\exp(i\mathbf{k}\cdot\mathbf{r}) . \tag{10-51b}$$

The wave function $\varphi^{(\mathbf{k})}(\mathbf{r})$ for an electron in a crystal is called the *Bloch function*. The function $w_k(\mathbf{r})$ is found from the Schrödinger equation

$$\left[-\frac{\hbar^2}{2M}\nabla^2 + V(\mathbf{r})\right]\varphi^{(\mathbf{k})}(\mathbf{r}) = \varepsilon(\mathbf{k})\varphi^{(k)}(\mathbf{r}) , \tag{10-52a}$$

or

$$\left[\frac{\hbar^2}{2M}(\nabla + i\mathbf{k})^2 + \varepsilon(\mathbf{k}) - V(\mathbf{r})\right] w_\mathbf{k}(\mathbf{r}) = 0 . \tag{10-52b}$$

Since $w_k(\mathbf{r})$ is a periodic function of $\mathbf{r}$, this differential equation needs to be solved only within a single cell by using the periodic boundary condition on $w_\mathbf{k}(\mathbf{r})$ at the edge of the cell.

Group theory has thus enabled us to replace a problem for the whole crystal by one for a single cell.

The periodic function $w_{\mathbf{k}}(\mathbf{r})$ can be expanded into the Fourier series

$$w_k(\mathbf{r}) = \sum_{\mathbf{K}_m} v(\mathbf{k} + \mathbf{K}_m) \exp[i\mathbf{K}_m \cdot \mathbf{r}] . \tag{10-53}$$

We see, by inserting (10-53) into (10-51b), that (10-48c) is another expression for the Bloch function.

10.11. Representation Space of the Space Group

Since the translation group $\mathbf{T}$ is a subgroup of the space group $\mathbf{G}$, naturally we shall choose the group chain $\mathbf{G} \supset \mathbf{T}$ to classify the IRB of $\mathbf{G}$. From (10-48b) we know that

$$\psi_{\mathbf{k}} = \exp[i(\mathbf{k} + \mathbf{K}_m) \cdot \mathbf{r}] \tag{10-54a}$$

carries the irrep $\mathbf{k}$ of $\mathbf{T}$, with $\mathbf{k}$ restricted to the Brillouin zone. The functions of (10-54a) with the same $\mathbf{k}$ but all possible $\mathbf{K}_m$ form an eigenspace $\mathcal{L}_{\mathbf{k}}$ of the translation operator $\{\varepsilon|\mathbf{R}_n\}$,

$$\mathcal{L}_{\mathbf{k}} = \{\exp[i(\mathbf{k} + \mathbf{K}_m) \cdot \mathbf{r}_m] : \mathbf{K}_m \in L^{-1}\} . \tag{10-54b}$$

Our primary task is to find linear combinations of (10-54a) with $\mathbf{k}$ and $\mathbf{K}_m$ such that the combinations form the irreducible bases of the space group $\mathbf{G}$. As a routine procedure, we apply all the group elements of $\mathbf{G}$ to the function $\psi_{\mathbf{k}}$ and pick out the linearly independent functions that will carry a rep of $\mathbf{G}$, and then reduce this rep into irreps of $\mathbf{G}$. From (10-33) and (10-54a),

$$\{\alpha|\mathbf{a}\}\psi_{\mathbf{k}} = \exp[-i\alpha(\mathbf{k} + \mathbf{K}_m) \cdot \mathbf{a}] \exp[i\alpha(\mathbf{k} + \mathbf{K}_m) \cdot \mathbf{r}] . \tag{10-55a}$$

It is seen that the translation part $\{\varepsilon|\mathbf{a}\}$ of the group element $\{\alpha|\mathbf{a}\}$ only affects the phase factor, whereas the rotation part α changes the wave vector $\mathbf{k}+\mathbf{K}_m$ into $\alpha(\mathbf{k}+\mathbf{K}_m)$. Therefore, $\{\alpha|\mathbf{a}\}\psi_{\mathbf{k}}$ belongs to the eigenspace $\mathcal{L}_{\alpha\mathbf{k}}$ of $\{\varepsilon|\mathbf{R}_n\}$. Equation (10-55a) shows that the functions $\{\alpha|\mathbf{V}(\alpha) + \mathbf{R}_n\}\psi_{\mathbf{k}}$ with the same α but different $\mathbf{R}_n$ are linearly dependent. Consequently, although the space group $\mathbf{G}$ has an infinite number of elements, it generates from ψ_k only $|\mathbf{G}_0|$ linearly independent functions, which can be chosen as

$$\begin{aligned}\psi_{\alpha_j \mathbf{k}} &= \{\alpha_j|\mathbf{V}(\alpha_j)\}\psi_{\mathbf{k}} \\ &= \exp[i\alpha_j(\mathbf{k} + \mathbf{K}_m) \cdot (\mathbf{r} - \mathbf{V}(\alpha_j))] , \quad j = 1, 2, \ldots , |\mathbf{G}_0| .\end{aligned} \tag{10-55b}$$

They form a rep space

$$\mathcal{L}(*\mathbf{k}) = \{\psi_{\alpha_i k} : \quad i = 1, 2, \ldots , |\mathbf{G}_0|\} \tag{10-55c}$$

for the space group $\mathbf{G}$. In general $\mathcal{L}(*k)$ is a reducible space of $\mathbf{G}$.

Stated differently, in a rep space with Bloch functions as basis vectors, the space group $\mathbf{G}$ has only $|\mathbf{G}_0|$ linearly independent operators $\{\alpha_i|\mathbf{V}(\alpha_i)\}$, which form a rep space

$$L(*\mathbf{k}) = \{\{\alpha_i|\mathbf{V}(\alpha_i)\} : \quad i = 1, 2, \ldots , |\mathbf{G}_0|\} \tag{10-55d}$$

for the space group $\mathbf{G}$. The space $\mathcal{L}(*\mathbf{k})$ and $L(*\mathbf{k})$ are isomorphic.

10.12. The Little Group G(k)

The problem we face now is how to reduce this $|\mathbf{G}_0|$-dimensional rep of $\mathbf{G}$. According to the procedure introduced in Chapter 3 we need first to find the CSCO of the space group, and then to find the eigenvectors of the CSCO in the space $\mathcal{L}(*\mathbf{k})$. However, this procedure proves to be unsuitable, due to the fact that the class operator of the space group has a rather complicated structure. We have to resort to another strategy. We first sandwich a group $\mathbf{H}$ between the space group $\mathbf{G}$ and the translation group $\mathbf{T}$; then we determine the $\mathbf{H} \supset \mathbf{T}$ IRB, and finally we get the $\mathbf{G} \supset \mathbf{H} \supset \mathbf{T}$ IRB.

Of the $|\mathbf{G}_0|$ operations in the point group $\mathbf{G}_0$, all the rotations γ which leave the wave vector $\mathbf{k}$ invariant modulo a reciprocal lattice vector, i.e.,

$$\gamma \mathbf{k} = \mathbf{k} + \mathbf{K}_m \doteq \mathbf{k} \ , \tag{10-56}$$

form a subgroup of $\mathbf{G}_0$ which is designated as

$$\mathbf{G}_0(\mathbf{k}) = \{\gamma \in \mathbf{G}_0 : \gamma\mathbf{k} \doteq \mathbf{k}\} \tag{10-57a}$$

and is called the little co-group.

It is easily recognized that the little co-group $\mathbf{G}_0(\mathbf{k})$ is the intersection of the symmetry group $\mathbf{P}(\mathbf{k})$ of $\mathbf{k}$ and the isogonal point group $\mathbf{G}_0$, i.e.,

$$\mathbf{G}_0(\mathbf{k}) = \mathbf{P}(\mathbf{k}) \cap \mathbf{G}_0 \ . \tag{10-57b}$$

All the elements $\{\gamma|\mathbf{V}(\gamma) + \mathbf{R}_n\}$, for $\gamma \in \mathbf{G}_0(\mathbf{k})$ and $\mathbf{R}_n \in L$, form another space group designated as

$$\mathbf{G}(\mathbf{k}) = \{\{\gamma_i|\mathbf{V}(\gamma_i) + \mathbf{R}_n\} : \quad i = 1, 2, \ldots, |\mathbf{G}_0(\mathbf{k})|, \mathbf{R}_n \in L\} \ . \tag{10-58}$$

$\mathbf{G}(\mathbf{k})$ is referred to as the little group, or the group of the wave vector $\mathbf{k}$. $\mathbf{G}(\mathbf{k})$ is a subgroup of $\mathbf{G}$ and contains $\mathbf{T}$ as its subgroup. Therefore, the subgroup $\mathbf{G}(\mathbf{k})$ is a candidate for the group $\mathbf{H}$ to be sandwiched between $\mathbf{G}$ and $\mathbf{T}$. Another way of saying this is that for any subgroup $\mathbf{H}$ of the space group $\mathbf{G}$, the point group of $\mathbf{H}$ must be the symmetry group of a certain wave vector $\mathbf{k}$. Thus we can use the wave vector $\mathbf{k}$ to label this subgroup, that is, use $\mathbf{G}(\mathbf{k})$ to denote $\mathbf{H}$.

The order of $\mathbf{G}_0(\mathbf{k})$ is a divisor of the order of $\mathbf{G}_0$,

$$q = |\mathbf{G}_0|/|\mathbf{G}_0(\mathbf{k})| \ , \tag{10-59}$$

where q is an integer.

The $|\mathbf{G}_0(\mathbf{k})|$ linearly independent functions

$$\psi_{\gamma_i\mathbf{k}} = \{\gamma_i|\mathbf{V}(\gamma_i)\}\psi_{\mathbf{k}} \tag{10-60a}$$

form a space

$$\mathcal{L}(\mathbf{k}) = \{\psi_{\gamma_i\mathbf{k}} : \quad i = 1, 2, \ldots, |\mathbf{G}_0(\mathbf{k})|\} \ . \tag{10-60b}$$

$\mathcal{L}(\mathbf{k})$ is a subspace of $\mathcal{L}_{\mathbf{k}}$ as well as a subspace of $\mathcal{L}(*\mathbf{k})$, and is a $|\mathbf{G}_0(\mathbf{k})|$-dimensional rep space of the little group $\mathbf{G}(\mathbf{k})$. $\mathcal{L}(\mathbf{k})$ is in general reducible. By decomposing $\mathcal{L}(\mathbf{k})$, we can get the $\mathbf{G}(\mathbf{k}) \supset \mathbf{T}$ IRB.

Since $\mathcal{L}(\mathbf{k})$ is an eigenspace of $\{\varepsilon|\mathbf{R}_n\}$, in $\mathcal{L}(\mathbf{k})$ we have

$$\{\varepsilon|\mathbf{R}_n\} = \exp(-i\mathbf{k} \cdot \mathbf{R}_n) \cdot \mathbf{I} \ , \tag{10-61}$$

where $\mathbf{I}$ is a unit matrix. Therefore, in the space $\mathcal{L}(\mathbf{k})$, the translation operator $\{\varepsilon|\mathbf{R}_n\}$ commutes with any operator of $\mathbf{G}(\mathbf{k})$.

10.13. The Representation Groups $\mathbf{G_k}$ and $\mathbf{G'_k}$

10.13.1. The rep group $\mathbf{G_k}$

From

$$\{\gamma_j|\mathbf{V}(\gamma_j)+\mathbf{R}_n\} = \{\varepsilon|\mathbf{R}_n\}\{\gamma|\mathbf{V}(\gamma_j)\} \tag{10-62a}$$

and Eq. (10-61), we can see that the group operators of $\mathbf{G(k)}$ in the space $\mathcal{L}(\mathbf{k})$ are related by

$$\{\gamma_j|\mathbf{V}(\gamma_j)+\mathbf{R}_n\} = \varepsilon(j,n)\{\gamma_j|\mathbf{V}(\gamma_j)\}\ , \tag{10-62b}$$

$$\varepsilon(j,n) = \varepsilon(n) = \exp(-i\mathbf{k}\cdot\mathbf{R}_n)\ . \tag{10-62c}$$

To avoid notational clumsiness, we use the same symbol $\{\gamma|\mathbf{V}(\gamma)\}$ to denote both the group element and the corresponding operator or representative matrix in $\mathcal{L}(\mathbf{k})$.

Equation (10-62b) shows that in the rep space $\mathcal{L}(\mathbf{k})$ the little group $\mathbf{G(k)}$ has only $|\mathbf{G}_0(\mathbf{k})|$ linearly independent operators $\{\gamma_i|\mathbf{V}(\gamma_i)\}, i=1,2,\ldots,|\mathbf{G}_0(\mathbf{k})|$.

Using (10-24), (10-25b) and (10-62b) we obtain the multiplication relation for these independent operators,

$$\{\gamma_i|\mathbf{V}(\gamma_i)\}\{\gamma_j|\mathbf{V}(\gamma_j)\} = \mu(i,j)\{\gamma_{ij}|\mathbf{V}(\gamma_{ij})\}\ , \tag{10-63a}$$

$$\begin{aligned}\mu(i,j) &= \exp(-i\mathbf{k}\cdot\mathbf{R}_{ij}) \\ &= \exp\{-i\mathbf{k}\cdot[\mathbf{V}(\gamma_i)+\gamma_i\mathbf{V}(\gamma_j)-\mathbf{V}(\gamma_{ij})]\}\ .\end{aligned} \tag{10-63b}$$

By identifying $\mathcal{L}(\mathbf{k}), |\mathbf{G}_0(\mathbf{k})|, \{\gamma_i|\mathbf{V}(\gamma_i)\}$, and $\mu(i,j)$ with L, $|\mathrm{g}|$, R_i, and $\eta(i,j)$, respectively, in Sec. 8.9 we see that all the distinct operators $\varepsilon(i,n)\{\gamma_i|\mathbf{V}(\gamma_i)\}$ form a rep group [cf. Eq. (8-113)],

$$\mathbf{G_k} = \{\{\gamma_i|V(\gamma_i)\}: \quad i=1,2,\ldots,|\mathbf{G}_0(\mathbf{k})|\}_m\ , \tag{10-63c}$$

where m is an integer depending on $\mathbf{k}$ (see discussion below).

The rep group $\mathbf{G}_k$ can be regarded as a faithful rep of an abstract group $\widehat{\mathbf{G}}_\mathbf{k}$. The abstract group could be, for example, the so-called central extension $\overline{\mathbf{G}}^{\mathbf{k}*}$ defined by Schur (see Bradley 1972). Notice that what is called the representation group by Birman (1974) is just another name for the central extension, and thus differs from our definition for the rep group.

From (10-62b) it is clear that the irreducible basis of the little group $\mathbf{G(k)}$ in the space $\mathcal{L}(\mathbf{k})$ is identical to that of the rep group $\mathbf{G_k}$; their representation matrices are related by

$$D^{(\mathbf{k})(\nu)}(\{\gamma|\mathbf{V}(\gamma)+\mathbf{R}_n\}) = e^{-i\mathbf{k}\cdot\mathbf{R}_n}D^{(\mathbf{k})(\nu)}(\{\gamma|\mathbf{V}(\gamma)\})\ , \tag{10-64}$$

where $(\mathbf{k})(\nu)$ is the label for the irrep of $\mathbf{G(k)}$ or $\mathbf{G_k}$. It is to be noted that, in this paper, the symbol $D^{(j)}(X)$ is always regarded as the representative matrix of an operator X with respect to a certain basis labeled by the index j, and $D^{(j)}(Y)$ is that for another operator Y, while X and Y may belong to *different* groups. This notation is consistent with the convention used in quantum mechanics and is very convenient.

The irrep $D^{(\mathbf{k}(\nu)}(\{\gamma|\mathbf{c}\})$ is called the small rep of the little group $\mathbf{G(k)}$.

If the wave vector $\mathbf{k}$ is a point of symmetry, then $\mathbf{k}$ is of the form

$$\mathbf{k} = \frac{1}{m}(m_1\mathbf{b}_1+m_2\mathbf{b}_2+m_3\mathbf{b}_3)\ , \tag{10-65a}$$

where m and m_i are integers. In such a case, the phase factor $\mu(i,j)$ in (10-63b) is of the form

$$\mu(i,j) = \exp(-i\mathbf{k}\cdot\mathbf{R}_{ij}) = \exp(2\pi li/m), \quad l=0,1,\ldots m-1\ . \tag{10-65b}$$

Hence the rep group $\mathbf{G_k}$ is an m-fold covering group of $\mathbf{G}_0(\mathbf{k})$ with the elements

$$e^{2\pi li/m}\{\gamma_j|\mathbf{V}(\gamma_j)\}, \quad j=1,2,\ldots,|\mathbf{G}_0(\mathbf{k})|, \quad l=0,1,\ldots,m-1\,. \tag{10-65c}$$

Nevertheless, if $\mathbf{k}$ is a line (or plane) of symmetry, for instance if $\mathbf{k}$ is of the form

$$\mathbf{k}=\frac{1}{m'}(m_1\mathbf{b}_1+m_2\mathbf{b}_2)+p_3\mathbf{b}_3\,, \tag{10-65d}$$

where p_3 is an arbitrary number, say an irrational number, then the phase factor $\mu(i,j)$ will not have the simple form of (10-65b). It is thus seen that the factor system μ has the unpleasant feature that the integer $m=|\mathbf{G_k}|/|\mathbf{G}_0(\mathbf{k})|$ depends on the wave vector $\mathbf{k}$ and may become very large for $\mathbf{k}$ in a line (or plane) of symmetry. To avoid this trouble, we proceed to the next section.

10.13.2. The rep group $\mathbf{G}'_\mathbf{k}$

Let us make the following gauge transformation for the group elements of $\mathbf{G}(\mathbf{k})$:

$$\begin{aligned} R_i &\equiv \{\gamma_i|\mathbf{V}(\gamma_i)\}' \\ &= \exp[i\mathbf{k}\cdot\mathbf{V}(\gamma_i)]\{\gamma_i|\mathbf{V}(\gamma_i)\}\,. \end{aligned} \tag{10-66}$$

It follows from (10-63a) that in the space $\mathcal{L}(\mathbf{k})$ we have

$$R_iR_j=\eta(i,j)R_{ij}\,, \tag{10-67a}$$

$$\eta(i,j)=\exp\{-i\mathbf{k}\cdot[\gamma_i\mathbf{V}(\gamma_j)-\mathbf{V}(\gamma_j)]\}\,. \tag{10-67b}$$

According to Eq. (10-56a)

$$\gamma_i^{-1}\mathbf{k}=\mathbf{k}+\mathbf{K}_{\gamma_i}\,, \tag{10-67c}$$

where $\mathbf{K}_{\gamma_i}$ is a reciprocal lattice vector. Thus

$$\eta(i,j)=\exp[-i\mathbf{K}_{\gamma_i}\cdot\mathbf{V}(\gamma_j)]\,. \tag{10-67d}$$

With Eqs. (10-23) and (10-67d), the phase factor $\eta(j,k)$ is of the form

$$\eta(j,k)=\exp[2\pi i a_{jk}/m], \quad m=2,3,4,6\,, \tag{10-68}$$

where a_{jk} is an integer depending on j and k. Therefore, in the space $\mathcal{L}(\mathbf{k})$, the $m|\mathbf{G}_0(\mathbf{k})|$ operators

$$R_j^{(l)}=\exp(2\pi li/m)R_j\,, \quad j=1,2,\ldots,|\mathbf{G}_0(\mathbf{k})|\,, \quad l=0,1,\ldots,m-1\,, \tag{10-69}$$

form a rep group designated as

$$\mathbf{G}'_k=\{R_i: i=1,2,\ldots,|\mathbf{G}_0(\mathbf{k})|\}_m\,. \tag{10-70}$$

The rep group $\mathbf{G}'_\mathbf{k}$ is an m-fold covering group of the point group $\mathbf{G}_0(\mathbf{k})$ where the integer m depends only on what kind of fractional translation the space group $\mathbf{G}$ has and takes only four possible values, 2,3,4 and 6, for all 230 space groups.

Since R_i differs from $\{\gamma_i|\mathbf{V}(\gamma_i)\}$ only by the phase factor $\exp[i\mathbf{k}\cdot\mathbf{V}(\gamma_i)]$, the groups $\mathbf{G}'_\mathbf{k}, \mathbf{G}_\mathbf{k}$, and $\mathbf{G}(\mathbf{k})$ have identical irreducible bases, and their matrices, upon using (10-64) and (10-66), are related to one another by

$$D^{(\mathbf{k})(\nu)}(\{\gamma_i|\mathbf{V}(\gamma_i)\}) = \bar{e}^{i\mathbf{k}\cdot\mathbf{V}(\gamma_i)} D^{(\mathbf{k})(\nu)}(R_i) , \tag{10-71a}$$

$$D^{(\mathbf{k})(\nu)}(\{\gamma_i|\mathbf{c}_i\}) = e^{-i\mathbf{k}\cdot\mathbf{c}_i} D^{(\mathbf{k})(\nu)}(R_i) , \tag{10-71b}$$

where $D^{(\mathbf{k})(\nu)}(R_i)$ is the irreducible matrix for the element R_i of the rep group $\mathbf{G}'_k$.

For notational convenience, we often use $\Delta(\gamma_i)$ to denote the matrix $D^{(\mathbf{k})(\nu)}(R_i)$, i.e.,

$$\Delta(\gamma_i) \equiv D^{(\mathbf{k})(\nu)}(R_i) = D^{(\mathbf{k})(\nu)}(\{\gamma_i|\mathbf{V}(\gamma_i)\}') . \tag{10-72a}$$

Equation (10-71b) then reads

$$D^{(\mathbf{k})(\nu)}(\{\gamma_i|\mathbf{c}_i\}) = e^{-i\mathbf{k}\cdot\mathbf{c}_i}\Delta(\gamma_i) . \tag{10-71c}$$

It follows from (10-67a) and (10-72a), that

$$\Delta(\gamma_i)\Delta(\gamma_j) \equiv \eta(i,j)\Delta(\gamma_i\gamma_j) . \tag{10-72b}$$

Δ is the projective irrep of the point group $\mathbf{G}_0(\mathbf{k})$.

In summary, the problem of finding the irreps of an infinite group $\mathbf{G}(\mathbf{k})$ is converted into finding that of the rep group $\mathbf{G}_k$ or $\mathbf{G}'_\mathbf{k}$. For those wave vectors $\mathbf{k}$ which are lines or planes of symmetry, we must work with the rep group $\mathbf{G}'_\mathbf{k}$, while for those $\mathbf{k}$ which are points of symmetry, we can work either with the group $\mathbf{G}_\mathbf{k}$ or with $\mathbf{G}'_\mathbf{k}$. However, the multiplication relation (10-67a) for $\mathbf{G}'_\mathbf{k}$ is much simpler than that for $\mathbf{G}_\mathbf{k}$; in the following we shall work only with the rep group $\mathbf{G}'_\mathbf{k}$, irrespective of points of symmetry or line (planes) of symmetry.

The $|\mathbf{G}_0(\mathbf{k})|$ functions

$$\psi_j \equiv R_j \exp[i(\mathbf{k}+\mathbf{K}_m)\cdot\mathbf{r}], \quad j = 1, 2, \ldots, |\mathbf{G}_0(\mathbf{k})| , \tag{10-73a}$$

carry the rep space $\mathcal{L}(\mathbf{k})$ for $\mathbf{G}'_\mathbf{k}$, which coincides with the space $\mathcal{L}(\mathbf{k})$ of (10-60b). The group space of the rep group $\mathbf{G}'_\mathbf{k}$ is denoted by

$$L(\mathbf{k}) = \{R_i : \quad i = 1, 2, \ldots, |\mathbf{G}_0(\mathbf{k})|\} . \tag{10-73b}$$

The spaces $\mathcal{L}(\mathbf{k})$ and $L(\mathbf{k})$ are isomorphic, and it is more convenient to work with the latter. In the following we work mainly with $L(\mathbf{k})$.

10.13.3. Special cases of the rep group $\mathbf{G}'_\mathbf{k}$

The following four cases need to be considered separately.

1. For a general $\mathbf{k}$ point. This is a trivial case, since now $\mathbf{G}'_\mathbf{k} = \mathbf{G}_\mathbf{k} = \{\varepsilon\}$.

2. For the symmorphic space group, or the nonsymmorphic space group whose little group $\mathbf{G}(\mathbf{k})$ is symmorphic for the wave vector $\mathbf{k}$ under consideration. For these cases, $\mathbf{V}(\gamma) \equiv 0$, and according to (10-63b) and (10-67d), the phase factor $\mu(i,j) = \eta(i,j) = 1$, hence

$$\mathbf{G}_\mathbf{k} = \mathbf{G}'_\mathbf{k} = \mathbf{G}_0(\mathbf{k}) , \tag{10-74}$$

i.e., the rep group $\mathbf{G}'_\mathbf{k}$ (or $\mathbf{G}_\mathbf{k}$) is identical to the point group $\mathbf{G}_0(\mathbf{k})$, whose irreps are already known. In passing we point out that in such cases the space $L(\mathbf{k}) = \{R_i\} = \{\gamma_i\}$ is the regular rep space of the point group $\mathbf{G}_0(\mathbf{k})$.

3. For an interior point of a Brillouin zone. When the wave vector $\mathbf{k}$ is not on the surface of the Brillouin zone, the only possibility for $\gamma\mathbf{k} = \mathbf{k} + \mathbf{K}_m$ is that $\mathbf{K}_m = 0$, i.e.

$$\gamma\mathbf{k} = \mathbf{k} \ . \tag{10-75}$$

Comparing this with (10-67c), we know that now $\mathbf{K}_\gamma \equiv 0$, and the phase factor in (10-67a) is again equal to one, $\eta(i,j) \equiv 1$. Therefore, the rep group $\mathbf{G}'_\mathbf{k}$ is isomorphic to the point group $\mathbf{G}_0(\mathbf{k}) = \{\gamma_i\}$. Suppose $D^{(\nu)}$ is the irrep of the point group $\mathbf{G}_0(\mathbf{k})$. Then the irrep of the rep group $\mathbf{G}'_\mathbf{k}$ is

$$D^{(\mathbf{k})(\nu)}(R_i) = D^{(\nu)}(\gamma_i) \ , \tag{10-76a}$$

while the irrep of the little group $\mathbf{G}(\mathbf{k})$ is [Eq. (10-71b)],

$$D^{(\mathbf{k})(\nu)}(\{\gamma|\mathbf{c}\}) = e^{-i\mathbf{k}\cdot\mathbf{c}} D^{(\nu)}(\gamma) \ . \tag{10-76b}$$

Observe that, for the origin point $\mathbf{k} = 0$, we again have

$$\mathbf{G}_\mathbf{k} = \mathbf{G}'_\mathbf{k} \ , \tag{10-77}$$

i.e., the distinction between $\mathbf{G}_\mathbf{k}$ and $\mathbf{G}'_\mathbf{k}$ disappears.

It should also be noted that for the case of interior points, although the rep group $\mathbf{G}'_\mathbf{k}$ and the point group $\mathbf{G}_0(\mathbf{k})$ have identical irreps (10-76a), their irreducible bases do not coincide. For example, if

$$\varphi^{(\nu)}_{\mathbf{k},a} = \sum_i u^{(\nu)}_{a,i} \gamma_i \psi_\mathbf{k} \tag{10-78a}$$

is the irreducible basis of $\mathbf{G}_0(\mathbf{k})$, where $u^{(\nu)}_{a,i}$ are coefficients and $\psi_\mathbf{k} = \exp[i(\mathbf{k}+\mathbf{K}_m)\cdot\mathbf{r}]$, then the corresponding irreducible basis of $\mathbf{G}'_\mathbf{k}$ is

$$\psi^{(\nu)}_{\mathbf{k},a} = \sum_i u^{(\nu)}_{a,i} \{\gamma_i|\mathbf{V}(\gamma_i)\}' \psi_\mathbf{k} \ . \tag{10-78b}$$

4. For a point on the surface of a Brillouin zone and for nonsymmorphic little groups. For cases (1)-(3) above, the irreps of the rep group $\mathbf{G}'_\mathbf{k}$ can be directly taken over from those of the point group $\mathbf{G}_0(\mathbf{k})$. The only case for which the irreps of $\mathbf{G}'_\mathbf{k}$ cannot be obtained in this way and have to be worked out anew is when the wave vector $\mathbf{k}$ is on the surface of a Brillouin zone and its little group $\mathbf{G}(\mathbf{k})$ is nonsymmorphic.

10.14. The Irreducible Basis and Matrices of $\mathbf{G}'_\mathbf{k}$

10.14.1. The group table of $\mathbf{G}'_\mathbf{k}$

The group table of $\mathbf{G}'_\mathbf{k}$ can easily be constructed from the group table $\gamma_\rho\gamma_\sigma = \gamma_{\rho\sigma}$ of the little co-group $\mathbf{G}_0(\mathbf{k})$ by replacing γ's with R's and multiplying the $(\rho\sigma)$ entry with the phase factor,

$$\eta(\rho,\sigma) = \exp[-i(\gamma_\rho^{-1}\mathbf{k} - \mathbf{k}) \cdot \mathbf{V}(\gamma_\sigma)] \ . \tag{10-79a}$$

Since the phase factors $\eta(\rho,\sigma)$ are crucial for the whole process of analysis, we give a simple formula for calculating them. To this end, we introduce the following notations. Let

$$\begin{aligned} \mathbf{b} &= (\mathbf{b}_1, \mathbf{b}_2, \mathbf{b}_3), \quad \mathbf{t} = (\mathbf{t}_1, \mathbf{t}_2, \mathbf{t}_3) \ , \\ \mathbf{p} &= (\mathbf{p}_1, \mathbf{p}_2, \mathbf{p}_3), \quad \tau_\sigma = (\tau_{\sigma 1}, \tau_{\sigma 2}, \tau_{\sigma 3}) \ . \end{aligned} \tag{10-79b}$$

The wave vector $\mathbf{k}$ and the non-primitive translation $\mathbf{V}(\gamma_\sigma)$ can be expressed as

$$\begin{aligned}\mathbf{k} &= \mathbf{p} \cdot \mathbf{b} = p_1\mathbf{b}_1 + p_2\mathbf{b}_2 + p_3\mathbf{b}_3 \ ,\\ \mathbf{V}(\gamma_\sigma) &= \tau_\sigma \cdot \mathbf{t} = \tau_{\sigma 1}\mathbf{t}_1 + \tau_{\sigma 2}\mathbf{t}_2 + \tau_{\sigma 3}\mathbf{t}_3 \ .\end{aligned} \tag{10-79c}$$

Obviously, under rotations, (p_1, p_2, p_3) transforms as $(\mathbf{t}_1, \mathbf{t}_2, \mathbf{t}_3)$, while $(\tau_{\sigma 1}, \tau_{\sigma 2}, \sigma_{\sigma 3})$ transforms as $(\mathbf{b}_1, \mathbf{b}_2, \mathbf{b}_3)$. Furthermore

$$\begin{aligned}\gamma_\rho^{-1}\mathbf{k} &= \gamma_\rho^{-1}(\mathbf{p} \cdot \mathbf{b}) = \mathbf{p} \cdot (\gamma_\rho^{-1}\mathbf{b}) = \mathbf{p}_\rho \cdot \mathbf{b} \ ,\\ \mathbf{p}_\rho &= \gamma_\rho \mathbf{p} = (p_{\rho 1}, p_{\rho 2}, p_{\rho 3}) \ .\end{aligned} \tag{10-79d}$$

The transformed vector $\mathbf{p}_\rho = \gamma_\rho \mathbf{p}$ can be found from Table 3.2 in Bradley by replacing the $\mathbf{t}_i$'s with $\mathbf{p}_i$'s. Using (10-79a), (10-79b) and (10-79d) we finally obtain

$$\eta(\rho, \sigma) = \exp[-2\pi i(\mathbf{p}_\rho - \mathbf{p}) \cdot \tau_\sigma] \ , \tag{10-80a}$$

where

$$(\mathbf{p}_\rho - \mathbf{p}) \cdot \tau_\sigma = \sum_{i=1}^{3}(p_{\rho i} - p_i)\tau_{\sigma i} \ . \tag{10-80b}$$

On the basis of the group table of $G'_\mathbf{k}$, it is easy to determine the classes of $G'_\mathbf{k}$. The number n of linearly independent class operators of $G'_\mathbf{k}$ is less than the number N of classes, $n < N$. Out of the n linearly independent class operators, we can choose a subset which is a CSCO in the n-dimensional class space of $G'_\mathbf{k}$ and this subset is the CSCO-I of the rep group $G'_\mathbf{k}$. The procedure for finding the CSCO-I and the simple characters of $G'_\mathbf{k}$ is exactly the same as that for an abstract group. An example will be given in Sec. 10.15.

Remark: In Sec. XXI of Chen ([31] 1985), it was pointed out that "for computer calculations it is expedient to choose self-adjoint CSCO's, since the diagonalization of hermitian matrices is much easier than that of non-hermitian matrices." From experience (Ping 1988) this observation has been found to be untrue. In fact the difficulty in finding a self-adjoint CSCO far surpasses that in diagonalizing a non-hermitian CSCO. Therefore it is preferable to choose a linear combination of the class operators, regardless of their being self-adjoint or not, as the CSCO of $G'_\mathbf{k}$.

10.14.2. The CSCO-II and CSCO-III of $G'_\mathbf{k}$

To decompose the group space $L(\mathbf{k})$ of $G'_\mathbf{k}$, we need to introduce a suitable group chain $G'_\mathbf{k} \supset G'_s$, which can be assumed to be a canonical one without loss of generality. For simplicity in exposition, here G'_s is assumed to be simply a subgroup of $G'_\mathbf{k}$ instead of a subgroup chain, i.e.,

$$G'_s = \{R_s\}, \quad R_s = \{\gamma_s|\mathbf{V}(\gamma_s)\}' \ . \tag{10-81}$$

We also need the intrinsic group $\overline{G}'_\mathbf{k}$ of the rep group $G'_\mathbf{k}$ defined by

$$\overline{R}_i R_j = R_j R_i \quad \text{for any} \quad R_j \in L(\mathbf{k}) \ . \tag{10-82}$$

$\overline{G}'_\mathbf{k}$ and $G'_\mathbf{k}$ are commutative and anti-isomorphic. Corresponding to the subgroup G'_s of $G'_\mathbf{k}$ is the intrinsic subgroup

$$\overline{G}'_s = \{\overline{R}_s\} \tag{10-83}$$

of $\overline{G}'_\mathbf{k}$. Let C and $C(s)$ be the CSCO of $G'_\mathbf{k}$ and G'_s respectively; then $\overline{C}$ and $\overline{C}(s)$ are the CSCO of $\overline{G}'_\mathbf{k}$ and $\overline{G}'_s$, respectively, where $\overline{C}$ and $\overline{C}(s)$ are obtained from C and $C(s)$ by replacing the elements of $\mathbf{G}'_{\mathbf{k}}$ with the corresponding intrinsic group elements. Furthermore, we still have $\overline{C} = C$.

If $K = (C, C(s), \overline{C}(s))$ is a CSCO in the $|\mathbf{G}_0(\mathbf{k})|$-dimensional space $L(\mathbf{k})$, then K is the CSCO-III of $\mathbf{G}'_{\mathbf{k}}$, while $M = (C, C(s))$ is the CSCO-II of $\mathbf{G}'_{\mathbf{k}}$.

The intrinsic group $\overline{\mathbf{G}}(\mathbf{k})$ of the little group $\mathbf{G}(\mathbf{k})$ is defined by

$$\overline{\mathbf{G}}(\mathbf{k}) = \{\overline{\{\gamma_i|\mathbf{V}(\gamma_i) + \mathbf{R}_n\}}: \quad i = 1, 2, \ldots, |\mathbf{G}_0(\mathbf{k})|, \mathbf{R}_n \in L\} \tag{10-84a}$$

with

$$\overline{\{\gamma_i|\mathbf{c}\}} = e^{-i\mathbf{k}\cdot\mathbf{c}}\overline{R}_i, \quad \mathbf{c} = \mathbf{V}(\gamma_i) + \mathbf{R}_n . \tag{10-84b}$$

The eigenvector of the CSCO-III of $\mathbf{G}'_{\mathbf{k}}$ in the space $L(\mathbf{k})$ is denoted by $P_a^{(\mathbf{k})(\nu)b}$. Eq. (3-169a) now reads

$$\begin{bmatrix} C \\ C(s) \\ \overline{C}(s) \end{bmatrix} P_a^{(\mathbf{k})(\nu)b} = \begin{bmatrix} \nu \\ a \\ b \end{bmatrix} P_a^{(\mathbf{k})(\nu)b}, \quad a, b = a_1, a_2, \ldots, a_{h_\nu} . \tag{10-85}$$

The eigenvector $P_a^{(\mathbf{k})(\nu)b}$ is a linear combination of R_i,

$$P_a^{(\mathbf{k})(\nu)b} = \sum_{i=1}^{|G_0(\mathbf{k})|} u_{ab,i}^{(\mathbf{k})(\nu)} R_i . \tag{10-86}$$

Equations (3-173), (3-176) and (3-180) still hold under the substitutions

$$\mathbf{g} \to |\mathbf{G}_0(\mathbf{k})| , \quad u_{\nu mk,a} \to u_{ab,i}^{(\mathbf{k})(\nu)} . \tag{10-87}$$

Then the coefficients $u_{ab,i}^{(\mathbf{k})(\nu)}$ are the solutions to (3-176).

The eigenvector $P_a^{(\mathbf{k})(\nu)b}$, for $a, b = a_1, \ldots, a_{h_\nu}$ constitutes the $\mathbf{G}'_{\mathbf{k}} \supset \mathbf{G}'_s$ and $\overline{\mathbf{G}}'_{\mathbf{k}} \supset \overline{\mathbf{G}}'_s$ IRB. Eq. (3-173) shows that the number of times that the irrep $(\mathbf{k})(\nu)$ of $\mathbf{G}'_{\mathbf{k}}$ appears in the space $L(\mathbf{k})$ is equal to its dimension h_ν.

One of the advantages of the EFM for constructing the IRB or irreps of $\mathbf{G}(\mathbf{k})$ is that the subgroup chain used to classify the IRB or irrep can be chosen at will without the restriction that the subgroup has to be an invariant subgroup of $\mathbf{G}(\mathbf{k})$.

However, if for some circumstances we are only interested in obtaining irreps of $\mathbf{G}(\mathbf{k})$ without the requirement that they be in a certain classification scheme, then we pay attention only to the operator set $C(s)$, without bothering about its related subgroup chain. In such cases, the eigenvalue of $C(s)$ is used merely to distinguish between the components of an irreducible basis, and $C(s)$ can be chosen differently for different irreps. The choice of $C(s)$ can be arbitrary so long as its eigenvalues can provide enough labels for the basis vectors of the same irrep. It is always desirable that $C(s)$ contains as few operators as possible. For example, for two-dimensional irreps the possible choice of $C(s)$ is a (plane) reflection operator σ, or a two-fold rotation C_2; for irreps with $h_\nu = 3(4)$ it is a three-fold (four-fold) rotation C_3 (C_4); and for irreps with $h_\nu = 6$, it is (C_{2x}, C_{2y}, I) or (C_{31}^+, I) where I is the inversion.

10.14.3. The irreps of $\mathbf{G}'_{\mathbf{k}}$ and the projective irreps of $\mathbf{G}_0(\mathbf{k})$

With the standard phase choice for the eigenvectors $P_a^{(\mathbf{k})(\nu)b}$, the irreducible matrix elements of $\mathbf{G}'_{\mathbf{k}}$ are simply related to the coefficients $u_{ab,i}^{(\mathbf{k})(\nu)}$ by [cf. Eq. (3-198)],

$$\begin{aligned} D_{ab}^{(\mathbf{k})(\nu)}(R_i) &= \Delta_{ab}^{(\nu)}(\gamma_i) \\ &= \left[\frac{|\mathbf{G}_0(\mathbf{k})|}{h_\nu}\right]^{1/2} (u_{ab,i}^{(\mathbf{k})(\nu)})^* . \end{aligned} \tag{10-88}$$

By solving the eigenequation (3-176) we can obtain all the irreps $D^{(\mathbf{k})(\nu)}$ of $\mathbf{G}'_k$, i.e., the projective irreps $\Delta^{(\nu)}$ of the little co-group $\mathbf{G}_0(\mathbf{k})$.

The rule for determining the phases of $P_a^{(\mathbf{k})(\nu)b}$ so that (10-88) holds is identical to that given in Sec. 3.9 under the substitution (10-87).

From (3-200) and (10-88) we get the two orthogonal theorems for the projective irreps of $\mathbf{G}_0(\mathbf{k})$,

$$\frac{h_\nu}{|\mathbf{G}_0(\mathbf{k})|}\sum_{i=1}^{|\mathbf{G}_0(\mathbf{k})|}\Delta_{ab}^{(\nu)}(\gamma_i)^*\Delta_{a'b'}^{(\nu')}(\gamma_i)=\delta_{\nu\nu'}\delta_{aa'}\delta_{bb'}\ , \tag{10-89a}$$

$$\sum_{\nu=1}^{n}\sum_{a,b=1}^{h_\nu}\frac{h_\nu}{|\mathbf{G}_0(\mathbf{k})|}\Delta_{ab}^{(\nu)}(\gamma_i)^*\Delta_{ab}^{(\nu)}(\gamma_j)=\delta_{ij}\ . \tag{10-89b}$$

Inserting (10-88) into (10-86), we get the normalized generalized projection operator,

$$P_a^{(\mathbf{k})(\nu)b}=\left[\frac{h_\nu}{|\mathbf{G}_0(\mathbf{k})|}\right]^{1/2}\sum_{i=1}^{|\mathbf{G}_0(\mathbf{k})|}D_{ab}^{(k)(\nu)}(R_i)^*R_i \tag{10-90}$$

for the rep group $\mathbf{G}'_{\mathbf{k}}$. The generalized projection operator for the rep group $\mathbf{G}'_{\mathbf{k}}$ or for the little group $\mathbf{G}(\mathbf{k})$ is

$$P_{ab}^{(\mathbf{k})(\nu)}=\frac{h_\nu}{|\mathbf{G}_0(\mathbf{k})|}\sum_{i=1}^{|\mathbf{G}_0(\mathbf{k})|}D_{ab}^{(\mathbf{k})(\nu)}(R_i)^*R_i\ . \tag{10-91a}$$

Using Eqs. (10-66) and (10-71a) we can rewrite Eq. (10-91a)

$$P_{ab}^{(\mathbf{k})(\nu)}=\frac{h_\nu}{|\mathbf{G}_0(\mathbf{k})|}\sum_{i=1}^{|\mathbf{G}_0(\mathbf{k})|}D_{ab}^{(\mathbf{k})(\nu)}(\{\gamma_i|\mathbf{V}(\gamma_i)\})^*\{\gamma_i|\mathbf{V}(\gamma_i)\}\ . \tag{10-91b}$$

10.14.4. The irreducible basis of G(k)

The $\mathbf{G}'_{\mathbf{k}}\supset\mathbf{G}'_s$ and $\overline{\mathbf{G}}'_{\mathbf{k}}\supset\overline{\mathbf{G}}'_s$ IRB in the rep space $\mathcal{L}(\mathbf{k})$ is simply given by

$$\begin{aligned}\psi_{\mathbf{k},a}^{(\nu)b}&=P_a^{(\mathbf{k})(\nu)b}\psi_{\mathbf{k}}(\mathbf{x})\\&=\sum_{i=1}^{|\mathbf{G}_0(\mathbf{k})|}u_{ab,i}^{(\mathbf{k})(\nu)}\psi_i\ .\end{aligned} \tag{10-92}$$

Notice that $\psi_{\mathbf{k},a}^{(\nu)b}$ is also the $\mathbf{G}(\mathbf{k})\supset\mathbf{G}(s)$ and $\overline{\mathbf{G}}(\mathbf{k})\supset\overline{\mathbf{G}}(s)$ IRB, where $\mathbf{G}(s)$ is the subgroup of the little group $\mathbf{G}(\mathbf{k})$, which has $\mathbf{G}'_s$ of (10-81) as its rep group.

Under the action of the element $\{\gamma|\mathbf{c}\}$ of $\mathbf{G}(\mathbf{k})[\overline{\{\gamma|\mathbf{c}\}}$ of $\overline{\mathbf{G}}(\mathbf{k})]$, $\psi_{\mathbf{k}a}^{(\nu)b}$ only changes its "external" (intrinsic) quantum number $a(b)$

$$\{\gamma|\mathbf{c}\}\psi_{\mathbf{k},a}^{(\nu)b}=\sum_{a'}D_{a'a}^{(\mathbf{k})(\nu)}(\{\gamma|\mathbf{c}\})\psi_{\mathbf{k},a'}^{(\nu)b}\ , \tag{10-93a}$$

$$\overline{\{\gamma|\mathbf{c}\}}\psi_{\mathbf{k},a}^{(\nu)b}=\sum_{b'}D_{bb'}^{(k)(\nu)}(\{\gamma|\mathbf{c}\})\psi_{\mathbf{k},a}^{(\nu)b'}\ . \tag{10-93b}$$

The h_ν sets of IRB of $\mathbf{G}(\mathbf{k})$, $\{\psi_{\mathbf{k},a}^{(\nu)b}: a=1,2,\ldots,h_\nu\}$, $b=1,2,\ldots,h_\nu$, which carry h_ν equivalent (or identical under the standard phase choice) irreps of $\mathbf{G}(\mathbf{k})$, are distinguished by the intrinsic quantum number b.

10.15. Examples: The Point W OF O_h^7

In this section, we give several examples of the application of the EFM for obtaining the characters and irreps of the rep group $\mathbf{G}'_{\mathbf{k}}$. From these it is then trivial to obtain the characters and small reps of the little group $\mathbf{G}(\mathbf{k})$ by using (10-71b).

Since the cases (1)-(3) in Sec. 10.13 are trivial, we treat here only case (4), i.e., when $\mathbf{k}$ is a surface point and the little group $\mathbf{G}(\mathbf{k})$ is non-symmorphic.

10.15.1. Seeking the CSCO and the characters of the point W of the space group O_h^7

The vector $\mathbf{p}$ for the point W is

$$\mathbf{p} = \left(\frac{1}{2}, \frac{1}{4}, \frac{3}{4}\right) . \tag{10-94}$$

O_h^7 belongs to the face-centered cubic Γ_c^f with the generators

$$\begin{aligned} &\{C_{2z}|0\},\ \{C_{2x}|0\},\ \{C_{31}^+|0\}\ , \\ &\{C_{2a}|\boldsymbol{\tau}\},\ \{I|\boldsymbol{\tau}\},\ \boldsymbol{\tau} = \left(\frac{1}{4}, \frac{1}{4}, \frac{1}{4}\right) . \end{aligned} \tag{10-95}$$

The little co-group is

$$\mathbf{G}_0(W) = D_{2d} = (\varepsilon, C_{2x}, C_{2d}, C_{2f}, \sigma_y, \sigma_z, S_{4x}^-, S_{4x}^+), \quad |\mathbf{G}_0(W)| = 8 . \tag{10-96}$$

Using (10-80a) and Table 10.21-2 under the heading "cubic Γ_c^f" we can calculate the phase factor,

$$\eta_\rho = \exp[-2\pi i(\gamma_\rho \mathbf{p} - \mathbf{p}) \cdot \boldsymbol{\tau}] , \tag{10-97}$$

as shown in the first column of Table 10.15-1. $\eta_\rho = \pm 1, \pm i$. Hence the representation group $\mathbf{G}'_W$ is a four-fold covering group of D_{2d}.

Table 10.15-1. The group table of the rep group $\mathbf{G}'_W$ for the space group O_h^7.

η_ρ	$\{\varepsilon\|0\}'$	$\{C_{2x}\|0\}'$	$\{C_{2d}\|\tau\}'$	$\{C_{2f}\|\tau\}'$	$\{\sigma_y\|\tau\}'$	$\{\sigma_z\|\tau\}'$	$\{S_{4z}^-\|0\}$	$\{S_{4x}^+\|0\}$
1	1	2	3	4	5	6	7	8
−1	2	1	−4	−3	−6	−5	8	7
i	3	4	$i1$	$i2$	$i7$	$i8$	5	6
$-i$	4	3	$-i2$	$-i1$	$-i8$	$-i7$	6	5
$-i$	5	6	−8	−7	−1	−2	4	3
1	6	5	7	8	2	1	3	4
i	7	8	$i6$	$i5$	$i3$	$i4$	2	1
$-i$	8	7	$-i5$	$-i6$	$-i4$	$-i3$	1	2

If we take up the group table of D_{2d}, and multiply the $(\rho\sigma)$ entries by the phase factor η_ρ for those columns σ which have the non-primitive translation τ, then we get the group table of the representation group $\mathbf{G}'_W$, as shown in Table 10.15-1.

By multiplying the j-th column of Table 10.15-1 from the left with the element R_j^{-1} we obtain the class structure of $\mathbf{G}'_W$ shown in Table 10.15-2.

Table 10-15.2. The class structure of G'_W.

R_i \ R_j ($R_j^{-1} R_i R_j$)	1	2	3	4	5	6	7	8
1	1	1	1	1	1	1	1	1
2	2	2	−2	−2	−2	−2	2	2
3	3	−3	3	−3	−*i*4	*i*4	−*i*4	*i*4
4	4	−4	−4	4	*i*3	−*i*3	−*i*3	*i*3
5	5	−5	*i*6	−*i*6	5	−5	−*i*6	*i*6
6	6	−6	−*i*5	*i*5	−6	6	−*i*5	*i*5
7	7	7	*i*8	*i*8	*i*8	*i*8	7	7
8	8	8	−*i*7	−*i*7	−*i*7	−*i*7	8	8

From Table 10.15-2 it is easily seen that G'_W has 14 classes with the class operators

$$\begin{aligned} &C_1 = R_1, \;\; C'_1 = iR_1, \;\; C''_1 = -iR_1, \;\; C'''_1 = -R_1, \\ &C_2 = R_2 - R_2, \;\; C'_2 = iC_2, \;\; C_3 = R_3 - R_3 + iR_4 - iR_4, \\ &C'_3 = iC_3, \;\; C_4 = R_5 - R_5 + iR_6 - iR_6, \;\; C'_4 = iC_4, \\ &C_5 = R_7 + iR_8, \;\; C'_5 = iC_5, \;\; C''_5 = -iC_5, \;\; C'''_5 = -C_5 \, . \end{aligned} \tag{10-98}$$

However, only the class operators C_1 and C_5 are linearly independent. Hence $n = 2$. Using Table 10.15-1 we can easily establish the multiplication relation for the class operators C_1 and C_5,

$$C_5 \begin{bmatrix} C_1 \\ C_5 \end{bmatrix} = \begin{bmatrix} 0 & 1 \\ 2i & 0 \end{bmatrix} \begin{bmatrix} C_1 \\ C_5 \end{bmatrix} \, . \tag{10-99a}$$

Then the representation matrix of C_5 in the class space is

$$D(C_5) = \begin{bmatrix} 0 & 2i \\ 1 & 0 \end{bmatrix} \tag{10-99b}$$

By diagonalizing $D(C_5)$, we know that C_5 has two distinct eigenvalues $\nu = \pm(1+i)$; therefore C_5 is a CSCO-I of G'_W. We could use the eigenvalues $\pm(1+i)$ to label the two irreps of G'_W, but we prefer to use the conventional symbols W_1 and W_2 to label them.

The eigenvectors of $D(C_5)$ with the normalization (3-43) (remembering that $g_1 = 1$ and $g_5 = 2$) are

$$\begin{aligned} &\nu = 1 + i, \qquad \mathbf{q}^{(W_1)} = \sqrt{1/8}(2, 1 - i) \, , \\ &\nu = -(1 + i), \;\; \mathbf{q}^{(W_2)} = \sqrt{1/8}(2, -1 + i) \, . \end{aligned} \tag{10-100}$$

From (3-245) and (10-100) we get the characters of the classes C_1 and C_5. Using (10-98) we in turn get the characters of the remaining classes. The complete character table is shown in Table 10.15-3.

Table 10.15-3. The character table of the representation group G'_W for the space group O_h^7.

ν	(ν)	Class 1	1′	1″	1‴	2	2′	3	3′	4	4′	5	5′	5″	5‴
$1 + i$	W_1	2	$2i$	$-2i$	−2	0	0	0	0	0	0	$1 + i$	$-1 + i$	$1 - i$	$-1 - i$
$-(1 + i)$	W_2	2	$2i$	$-2i$	−2	0	0	0	0	0	0	$-1 - i$	$1 - i$	$-1 + i$	$1 + i$

10.15.2. Seeking the CSCO-I from the existing character table

For a group about which we know nothing except its group table, we can use the foregoing method to obtain the CSCO-I and characters of the group simultaneously. However, if the character table of a group is known, as is the case for all point groups and the 230 space groups, it is trivial to find the CSCO-I of the group by the method given in Sec. 3.12.

Now let us try to find the CSCO-I for the rep group $\mathbf{G}'_X$ of the space group O_h^7. The vector $\mathbf{p}$ for the point X is

$$\mathbf{p} = \left(\frac{1}{2}, \frac{1}{2}, 0\right) .$$

From Table T159 of Kovalev (1961), with slight changes in notation, we can write down the characters of the rep group $\mathbf{G}'_X$. We enter the characters of those elements whose characters are not identically zero in Table 10.15-4.

Table 10.15-4. Character table of the rep group $\mathbf{G}'_X$ for the space group O_h^7.*

(C_1, C_2)	(ν)	$\{\varepsilon\|0\}'$	$\{C_{2x}\|0\}'$	$\{C_{2a}\|\tau\}'$	$\{C_{2b}\|\tau\}'$	$\{\sigma_{da}\|0\}'$	$\{\sigma_{db}\|0\}'$
(0,2)	X_1	2	2	0	0	2	2
(0, −2)	X_2	2	2	0	0	−2	−2
(2*i*, 0)	X_3	2	−2	2*i*	−2*i*	0	0
(−2*i*,0)	X_4	2	−2	−2*i*	2*i*	0	0

*$\mathbf{X}_i$ are related to Kovalev's $\hat{\tau}^i$ by $\mathbf{X}_1 \to \hat{\tau}^3$, $\mathbf{X}_2 \to \hat{\tau}^4$, $\mathbf{X}_3 \to \hat{\tau}^2$, $\mathbf{X}_4 \to \hat{\tau}^1$.

It is seen that $\{C_{2a}|\tau\}'$ and $-\{C_{2b}|\tau\}'$ belong to the same class, and $\{\sigma_{da}|0\}$ and $\{\sigma_{db}|0\}$ belong to the same class. Furthermore, according to (3-67) Table 10.15-4, after deleting the sixth and eighth columns, is (accidently) the eigenvalue table of the class operators. Hence we see that the class operators

$$\begin{aligned} C_1 &= \{C_{2a}|\tau\}' - \{C_{2b}|\tau\}' , \\ C_2 &= \{\sigma_{da}|0\} + \{\sigma_{db}|0\} , \end{aligned} \tag{10-101}$$

have $n = 4$ distinct sets of eigenvalues as shown in the first column of Table 10.15-4. Therefore, (C_1, C_2) is a CSCO-I of $\mathbf{G}'_X$.

10.15.3. Constructing irreps of the rep group $\mathbf{G}'_\mathbf{k}$. The Point W of O_h^7

With $n = 2$, and $|\mathbf{G}_0(W)| = 8$, (8-122) reads

$$8 = 2^2 + 2^2 . \tag{10-102}$$

Therefore, the eight-dimensional rep produced by the space $L(W)$ can be decomposed into two inequivalent irreps with dimension 2, each occurring twice. We choose

$$(C(s), \overline{C}(s)) = (R_7, \overline{R}_7) . \tag{10-103}$$

The matrices of the operators R_7 and $\overline{R}_7$ in the space $L(W)$ can be read out from the seventh row and seventh column, respectively, of Table 10.15-1.

$$\begin{aligned} D(R_7) &= (7, 8, i6, i5, i3, i4, 2, 1) , \\ D(\overline{R}_7) &= (78564321) , \end{aligned} \tag{10-104a}$$

where the shorthand notations stand for

$$D(R_7)=\left(\begin{array}{cc|cc|cc|cc} & & & & & & & 1\\ & & & & & & 1 & \\ \hline & & & & i & & & \\ & & & & & i & & \\ \hline & & & i & & & & \\ & & i & & & & & \\ \hline 1 & & & & & & & \\ & 1 & & & & & & \end{array}\right), \quad D(\overline{R}_7)=\left(\begin{array}{cc|cc|cc|cc} & & & & & & & 1\\ & & & & & & 1 & \\ \hline & & & & & 1 & & \\ & & & & 1 & & & \\ \hline & & 1 & & & & & \\ & & & 1 & & & & \\ \hline 1 & & & & & & & \\ & 1 & & & & & & \end{array}\right) \tag{10-104b}$$

These matrices are very much like the ordinary regular rep matrices, in that in each row and each column there is only one nonvanishing element.

The element $R_7 = \{S_{4x}^{-}|0\}$ obeys the algebraic equation $(R_7)^4 = 1$. Hence the eigenvalues of R_7 are easily found to be $\pm 1, \pm i$ without the necessity of solving the secular equation. The same applies to the eigenvalues of $\overline{R}_7$.

Substituting the eigenvalues $a, b = \pm 1, \pm i$ into the eigenequations of $D(R_7)$ and $D(\overline{R}_7)$, we obtain eight simultaneous eigenvectors of $(R_7, \overline{R}_7)$ corresponding to the eight distinct sets of eigenvalues, $(a, b) = (1,1)(i,1),(1,i)(i,i),(-1,-1),(-i,-1)(-1,-i),(-i,-i)$. Therefore, $(R_7, \overline{R}_7)$ is already a CSCO-III of the group $\mathbf{G}'_W$, and the CSCO-I of $\mathbf{G}'_W$ is redundant in decomposing the rep space $L(\mathbf{k})$. Of course it happens only by accident. If we had chosen $(C(s), \overline{C}(s)) = (R_2, \overline{R}_2)$, the CSCO-III of $\mathbf{G}'_W$ would have had to include the CSCO-I of $\mathbf{G}'_W$. Hence a suitable choice of $C(s)$ can save a lot of labor.

The eigenvectors are listed in Table 10.15-5.

From Eq. (10-85) we know that the first four vectors in Table 10.15-5 belong to an irrep labeled by W_1, and the other four vectors belong to another irrep labeled by W_2.

The phases of the eigenvectors in Table 10.15-5 are determined by the three steps given in Sec. 3.9.3. The phases for the first, fourth, fifth, and eighth rows are determined by step (1). The phases of the second and sixth rows can be chosen arbitrarily according to step (2). The phases of the third and seventh rows are determined by step (3). For example, from the first and second rows of Table 10.15-5 as well as from Table 10.15-1 we can evaluate

$$\begin{aligned} D_{12}^{(W_1)}(R_3) &= \langle \psi_1^{(W_1)1} | R_3 | \psi_i^{(W_1)1} \rangle \\ &= \frac{1}{4} \langle \mathbf{1}+\mathbf{2}+\mathbf{7}+\mathbf{8} \,|\, \mathbf{3} \,|\, \mathbf{3}+\mathbf{4}+\mathbf{5}+\mathbf{6} \rangle = i \ , \end{aligned}$$

where we used obvious abbreviations. Therefore, the coefficient in front of R_3 in the third row $\psi_1^{(W_1)b=2} \equiv \psi_1^{(W_1)i}$ must be proportional to $D_{12}^{(W_1)}(R_3)^* = -i$.

10.16. Irreducible Basis and Representations of the Space Group

10.16.1. The k *star*

We first factorize the space group $\mathbf{G}$ into left cosets with respect to the little group $\mathbf{G}(\mathbf{k})$,

$$\mathbf{G} = \mathbf{G}(\mathbf{k}) + \{\beta_2|\mathbf{V}(\beta_2)\}\mathbf{G}(\mathbf{k}) + \ldots + \{\beta_q|\mathbf{V}(\beta_q)\}\mathbf{G}(\mathbf{k}) \ . \tag{10-105}$$

Suppose that $\psi_{\mathbf{k},a}^{(\nu)}$ is an IRB of the little group $\mathbf{G}(\mathbf{k})$ and let us define

$$\psi_{\mathbf{k}_\sigma a}^{(\nu)} = \{\beta_\sigma|\mathbf{V}(\beta_\sigma)\}\psi_{\mathbf{k},a}^{(\nu)} \ , \quad \sigma = 1, 2, \ldots, q; a = 1, 2, \ldots, h_\nu \ , \tag{10-106}$$

Table 10.15-5. The irreducible bases and irreps of the rep group $\mathbf{G}'_W$ for the space group O_h^7.

$\psi_a^{(v)b}$	N		$\{\varepsilon\|0\}'$ R_1	$\{C_{2x}\|0\}'$ R_2	$\{C_{2d}\|0\}'$ R_3	$\{C_{2f}\|\tau\}'$ R_4	$\{\sigma_y\|\tau\}'$ R_5	$\{\sigma_z\|\tau\}'$ R_6	$\{S_{4z}^-\|0\}'$ R_7	$\{S_{4z}^+\|0\}'$ R_8
$\psi_1^{(W_1)1}$	$\frac{1}{2}$	D_{11}^*	1	1					1	1
$\psi_i^{(W_1)1}$	$\frac{1}{2}$	D_{21}^*			1	1	1	1		
$\psi_1^{(W_1)i}$	$\frac{1}{2}$	D_{12}^*			$-i$	i	-1	1		
$\psi_i^{(W_1)i}$	$\frac{1}{2}$	D_{22}^*	1	-1					$-i$	i
$\psi_{-1}^{(W_2)-1}$	$\frac{1}{2}$	D_{11}^*	1	1					-1	-1
$\psi_{-i}^{(W_2)-1}$	$\frac{1}{2}$	D_{21}^*			1	1	-1	-1		
$\psi_{-1}^{(W_2)-i}$	$\frac{1}{2}$	D_{12}^*			$-i$	i	1	-1		
$\psi_{-i}^{(W_2)-i}$	$\frac{1}{2}$	D_{22}^*	1	-1					i	$-i$
$\Delta^{(W_1)}(\gamma_i)= D^{(W_1)}(R_i)$			$\begin{bmatrix}1&0\\0&1\end{bmatrix}$	$\begin{bmatrix}1&0\\0&-1\end{bmatrix}$	$\begin{bmatrix}0&i\\1&0\end{bmatrix}$	$\begin{bmatrix}0&-i\\1&0\end{bmatrix}$	$\begin{bmatrix}0&-1\\1&0\end{bmatrix}$	$\begin{bmatrix}0&1\\1&0\end{bmatrix}$	$\begin{bmatrix}1&0\\0&i\end{bmatrix}$	$\begin{bmatrix}1&0\\0&-i\end{bmatrix}$
$\Delta^{(W_2)}(\gamma_i)= D^{(W_2)}(R_i)$			$\begin{bmatrix}1&0\\0&1\end{bmatrix}$	$\begin{bmatrix}1&0\\0&-1\end{bmatrix}$	$\begin{bmatrix}0&i\\1&0\end{bmatrix}$	$\begin{bmatrix}0&-i\\1&0\end{bmatrix}$	$\begin{bmatrix}0&1\\-1&0\end{bmatrix}$	$\begin{bmatrix}0&-1\\-1&0\end{bmatrix}$	$\begin{bmatrix}-1&0\\0&-i\end{bmatrix}$	$\begin{bmatrix}-1&0\\0&i\end{bmatrix}$

with the convention that $\{\beta_1|\mathbf{V}(\beta_1)\} = \{\varepsilon|0\}$. The q wave vectors

$$\mathbf{k}_\sigma = \beta_\sigma \mathbf{k}, \quad \sigma = 1, 2, \ldots, q \tag{10-107a}$$

form what is called a star, or set of mutually inequivalent $\mathbf{k}$ vectors

$$*\mathbf{k} = (\mathbf{k}_1, \mathbf{k}_2, \ldots, \mathbf{k}_q) , \tag{10-107b}$$

with $\mathbf{k}_1 \equiv \mathbf{k}$. The wave vector $\mathbf{k}$ is called the canonical wave vector (Birman 1974). Any one of the q wave vectors can serve as the canonical wave vector. If $\mathbf{k}$ is a general point in the Brillouin zone, the $\mathbf{k}$ is called a general star, otherwise it is called a special star.

As an illustration, we consider the simple cubic lattice. The primitive translations are $\mathbf{t}_1 = a\mathbf{i}, \mathbf{t}_2 = a\mathbf{j}$, and $\mathbf{t}_3 = a\mathbf{k}$, while the basis vectors of the reciprocal lattice are $\mathbf{b}_1 = (2\pi/a)\mathbf{i}, \mathbf{b}_2 = (2\pi/a)\mathbf{j}, \mathbf{b}_3 = (2\pi/a)\mathbf{k}$. The Brillouin zone is shown in Fig. 10.16-1, which is the cubic bounded by the planes bisecting normally the vectors $\mathbf{b}_1, \mathbf{b}_2$ and $\mathbf{b}_3$. The special points are labelled by Γ, X, M, R (points of symmetry), and $\Delta, \Sigma, \Lambda, S, Z, T$ (lines or planes of symmetry). Let us find the symmetry group $\mathbf{G}_0(\mathbf{k})$ for each special point.

1. The star Γ. This is the point $\mathbf{k} = 0$. Its symmetry group is $\mathbf{G}_0(\mathbf{k}) = O_h$, with $|\mathbf{G}_0(\mathbf{k})| = |\mathbf{G}_0|$. The star consists of only one point. The little group is the space group itself, $\mathbf{G}(\mathbf{k}) = \mathbf{G}$.

2. Z. This is a general point on the intersection line of a face of the cube and the plane $k_z = 0$. It is sent into itself or an equivalent point under the group $C_{2v} = (e, \sigma_y, \sigma_z, C_{2x})$. Therefore, $\mathbf{G}_0(\mathbf{k}) = \mathcal{C}_{2v}$ and is of order 4. The star consists of $q = 12$ points as shown in

Fig. 10.16-2, where the four solid arms represent four points in the xy plane (the other eight points are in the xz and yz planes), while the dashed arms represent the points equivalent to those represented by the solid arms.

3. Σ. The point is along the bisector of the angle between the axes $\mathbf{k}_x$ and $\mathbf{k}_y$ (see Fig. 10.16-3). The group $\mathbf{G}_0(\mathbf{k}) = C_{2v}, C_{2v} = (e, \sigma_b, \sigma_z, C_{2a})$. $|\mathbf{G}_0(\mathbf{k})| = 4$ and $q = 12$.

4. Δ. This point is on the k_y axis. $\mathbf{G}_0(\mathbf{k}) = C_{4v}$ with k_y as the four-fold axis. $|\mathbf{G}_0(\mathbf{k})| = 8$ and $q = 6$. The star contains six points, four points in the xy plane (see Fig. 10.16-4) and the other two in the directions of $\mathbf{k}_z$ and $-\mathbf{k}_z$ respectively.

5. X. X is the intersection of the $\mathbf{k}_y$ axis with a face of the cube. $\mathbf{G}_0(\mathbf{k}) = D_{4h}$ (with $\mathbf{k}_y$ as the four-fold axis). $|\mathbf{G}_0(\mathbf{k})| = 16$ and $q = 3$. The star contains three points X, X' and X'', as shown in Fig. 10.16-1 and Fig. 10.16-5.

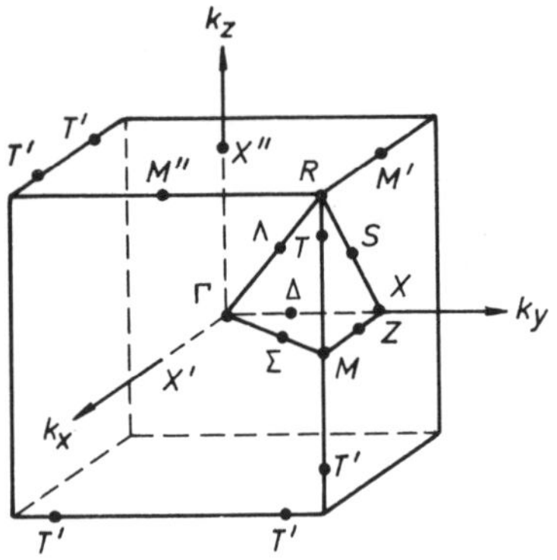

Fig. 10.16-1. The Brillouin zone for simple cubic.

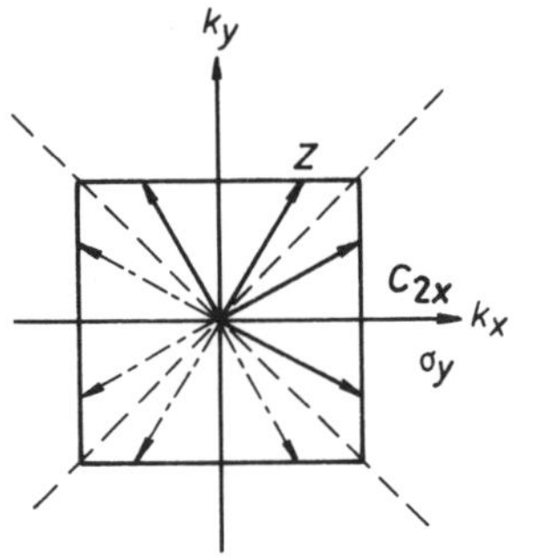

Fig. 10.16-2. The star Z; $q = 12$, $\mathbf{G}_0(\mathbf{k}) = C_{2v}$ $(e, \sigma_y, \sigma_z, C_{2x})$.

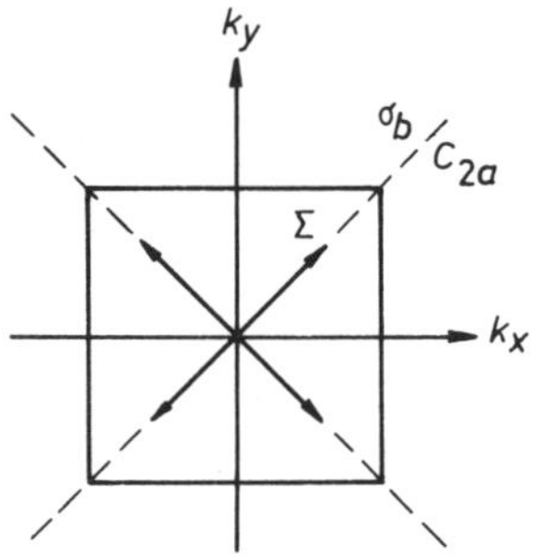

Fig. 10.16-3. The star Σ: $q = 12$, $\mathbf{G}_0(\mathbf{k}) = (e, \sigma_b, \sigma_z, C_{2a})$.

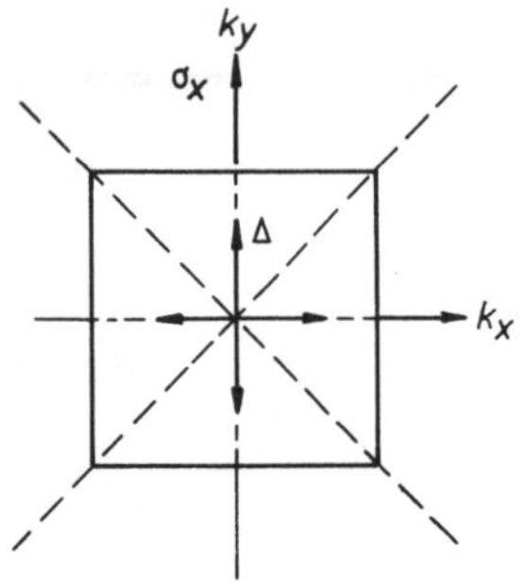

Fig. 10.16-4. The star Δ: $q = 6$, $\mathbf{G}_0(\mathbf{k}) = C_{4v}$ (with $\mathbf{k}_y$ as the four-fold axis).

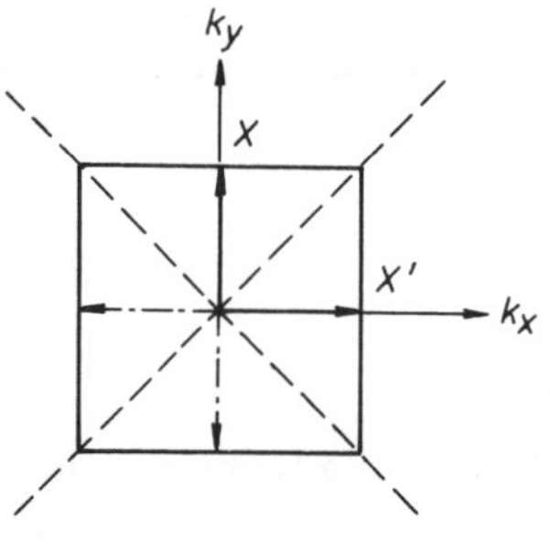

Fig. 10.16-5. The star X: $q = 3$, $\mathbf{G}_0$ $(\mathbf{k}) = D_{4h}$ (with $\mathbf{k}_y$ as the four-fold axis).

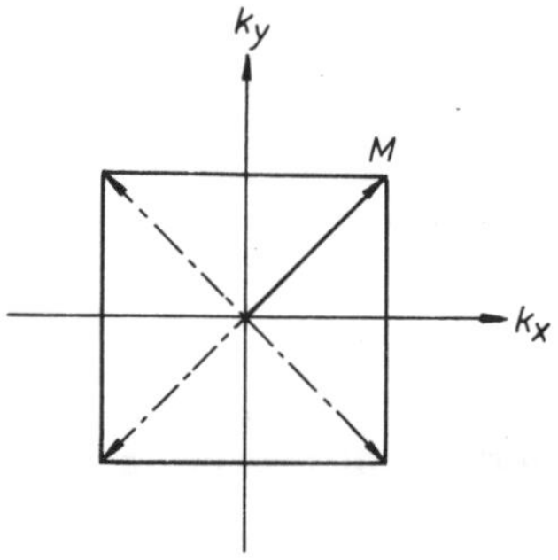

Fig. 10.16-6. The star M: $q = 3$, $\mathbf{G}_0$ $(\mathbf{k}) = D_{4h}$ (with $\mathbf{k}_z$ as the four-fold axis).

6. M. $\mathbf{G}_0(\mathbf{k}) = D_{4h}$ (with $\mathbf{k}_z$ as the four-fold axis (Fig. 10.16-6). $|\mathbf{G}_0(\mathbf{k})| = 16$ and $q = 3$. The star contains three points M, M', and M'' (Fig. 10.16-1).

7. T. $\mathbf{G}_0(\mathbf{k}) = C_{4v}$ (with $\mathbf{k}_z$ as the four-fold axis). $|\mathbf{G}_0(\mathbf{k})| = 8$ and $q = 6$. The star consists of six points (Fig. 10.16-1).

8. Λ. $\mathbf{G}_0(\mathbf{k}) = C_{3v}$ (with the line $\Gamma\Delta$ as the three-fold axis). $|\mathbf{G}_0(\mathbf{k})| = 6$ and $q = 8$.

9. R. $\mathbf{G}_0(\mathbf{k}) = \mathbf{G}_0 = O_h$. The star consists of only one point.

10.16.2. The induced rep

The space

$$\mathcal{L}(\mathbf{k}_\sigma) = \{\psi^{(\nu)}_{\mathbf{k}_\sigma a} : \quad a = 1, 2, \dots, h_\nu\} \tag{10-108a}$$

is isomorphic to the space $\mathcal{L}(\mathbf{k})$, or to the group space $L(\mathbf{k})$ of the rep group $\mathbf{G}'_{\mathbf{k}}$. Clearly, the space $\mathcal{L}(*k)$ defined by (10-55c) is decomposed into a direct sum of the q spaces $\mathcal{L}(\mathbf{k}_\sigma)$

isomorphic to one another,

$$\mathcal{L}(*\mathbf{k}) = \sum_{\sigma=1}^{q} \oplus \mathcal{L}(\mathbf{k}_\sigma) \ . \tag{10-108b}$$

Let us apply an element $\{\alpha|\mathbf{a}\}$ of $\mathbf{G}$ to (10-106),

$$\{\alpha|\mathbf{a}\}\psi^{(\nu)}_{\mathbf{k}_\sigma a} = \{\alpha|\mathbf{a}\}\{\beta_\sigma|\mathbf{v}_\sigma\}\psi^{(\nu)}_{\mathbf{k},a} \ . \tag{10-109}$$

$\{\alpha|\mathbf{a}\}\{\beta_\sigma|V_\sigma\}$ must be an element of $\mathbf{G}$ and must belong to one and only one coset, say the τ-th coset, i.e.,

$$\{\alpha|\mathbf{a}\}\{\beta_\sigma|\mathbf{V}_\sigma\} = \{\beta_\tau|\mathbf{V}_\tau\}\{\gamma|\mathbf{c}\} \ . \tag{10-110}$$

Assembling (10-106), (10-110) and (10-93a), we obtain

$$\begin{aligned}\{\alpha|\mathbf{a}\}\psi^{(\nu)}_{\mathbf{k}_\sigma a} &= \{\beta_\tau|\mathbf{V}_\tau\}\sum_b D^{(\mathbf{k})(\nu)}_{ba}(\{\gamma|\mathbf{c}\})\psi^{(\nu)}_{\mathbf{k},b} \\ &= \sum_b D^{(\mathbf{k})(\nu)}_{ba}(\{\gamma|\mathbf{c}\})\psi^{(\nu)}_{\mathbf{k}_\tau,b} \ ,\end{aligned} \tag{10-111}$$

where $D^{(\mathbf{k})(\nu)}(\{\gamma|\mathbf{c}\})$ is the irreducible matrix of the little group $\mathbf{G}(\mathbf{k})$. Therefore, the qh_ν functions $\psi^{(\nu)}_{\mathbf{k}_\sigma a}$ carry a rep for the space group $\mathbf{G}$, which is called the induced rep and denoted by $D^{(*\mathbf{k})(\nu)}$ (Birman 1974) or $D^{(\mathbf{k})(\nu)} \uparrow \mathbf{G}$ (Bradley 1972). It can be proved that the induced rep $D^{(*\mathbf{k})(\nu)}$ is an irrep of the space group $\mathbf{G}$ (Chen [31] 1985).

The induced rep is of qh_ν dimension, and its matrix elements can be expressed as

$$D^{(*\mathbf{k}),(\nu)}_{\tau b,\sigma a}(\{\alpha|\mathbf{a}\}) = \langle\psi^{(\nu)}_{\mathbf{k}_\tau b}|\{\alpha|\mathbf{a}\}|\psi^{(\nu)}_{\mathbf{k}_\sigma a}\rangle \ . \tag{10-112a}$$

Define

$$R_{\bar{\tau}\alpha\sigma} \equiv \{\beta_\tau|\mathbf{V}_\tau\}^{-1}\{\alpha|\mathbf{a}\}\{\beta_\sigma|\mathbf{V}_\sigma\} \ . \tag{10-112b}$$

It follows from (10-106) and (10-112a) that

$$D^{(*\mathbf{k})(\nu)}_{\tau b,\sigma a}(\{\alpha|\mathbf{a}\}) = \langle\psi^{(\nu)}_{\mathbf{k},b}|R_{\bar{\tau}\alpha\sigma}|\psi^{(\nu)}_{\mathbf{k},a}\rangle \ . \tag{10-112c}$$

According to the left coset decomposition (10-105), we know that $R_{\bar{\tau}\alpha\sigma}$ is either an element of the little group $\mathbf{G}(\mathbf{k})$ or a coset representative $\{\beta|\mathbf{V}(\beta)\}$ times an element of $\mathbf{G}(\mathbf{k})$. For the former, the right-hand side of (10-112c) is just the irreducible matrix element of $\mathbf{G}(\mathbf{k})$, and for the latter, (10-112c) must vanish, since $R_{\bar{\tau}\alpha\sigma} = \{\beta|\mathbf{V}(\beta)\}\{\gamma|\mathbf{c}\}$ will change the wave vector $\mathbf{k}$ into $\mathbf{k}_\beta$, while $\mathbf{k}$ is the label for the irreps of the translational group $\mathbf{T}$, and the bases belonging to different irreps of $\mathbf{T}$ are orthogonal. Hence (10-112c) can be expressed as [cf. (2-106h)]

$$\begin{aligned}D^{(*\mathbf{k})(\nu)}_{\tau b,\sigma a}(\{\alpha|\mathbf{a}\}) &= D^{(\mathbf{k})(\nu)}_{ba}(R_{\bar{\tau}\alpha\sigma}) \\ &= D^{(\mathbf{k})(\nu)}_{ba}(\{\beta_\tau|\mathbf{V}_\tau\}^{-1}\{\alpha|\mathbf{a}\}\{\beta_\sigma|\mathbf{V}_\sigma\}) \ .\end{aligned} \tag{10-112d}$$

It is convenient to write the $qh_\nu \times qh_\nu$ matrix $D^{(*\mathbf{k})(\nu)}$ in block decomposition form (Birman 1974),

$$D^{(*\mathbf{k})(\nu)}(\{\alpha|\mathbf{a}\}) = \begin{bmatrix} D^{(*\mathbf{k})(\nu)}_{(11)} & \cdots & D^{(*\mathbf{k})(\nu)}_{(1q)} \\ \vdots & & \vdots \\ D^{(*\mathbf{k})(\nu)}_{(q1)} & \cdots & D^{(*\mathbf{k})(\nu)}_{(qq)} \end{bmatrix} , \tag{10-113a}$$

where

$$D^{(*\mathbf{k})(\nu)}_{(\tau\sigma)}(\{\alpha|\mathbf{a}\}) = D^{(\mathbf{k})(\nu)}(\{\beta_\tau|\mathbf{V}_\tau\}^{-1}\{\alpha|\mathbf{a}\}\{\beta_\sigma|\mathbf{V}_\sigma\}) \tag{10-113b}$$

is a $(h_\nu \times h_\nu)$ matrix. According to (10-110), for given $\{\alpha|\mathbf{a}\}$ and σ there is only a unique τ that enables $R_{\bar{\tau}\alpha\sigma}$ to belong to the little group $\mathbf{G}(\mathbf{k})$. As a consequence, in each row and each column of the block form (10-113a), only one matrix block differs from zero. Another way of saying this is that for given σ and τ, only those elements of $\mathbf{G}$ which satisfy

$$\{\alpha|\mathbf{V}(\alpha)+\mathbf{R}_n\} = \{\beta_\tau|\mathbf{V}_\tau\}\{\gamma|\mathbf{V}(\gamma)+\mathbf{R}_m\}\{\beta_\sigma|\mathbf{V}_\sigma\}^{-1}, \quad \gamma \in \mathrm{G}_0(\mathbf{k}) \tag{10-113c}$$

have the nonzero submatrices

$$D^{(*\mathbf{k})(\nu)}_{(\tau\sigma)}(\{\alpha|\mathbf{V}(\alpha)+\mathbf{R}_n\}) = D^{(\mathbf{k})(\nu)}(\{\gamma|\mathbf{V}(\gamma)+\mathbf{R}_m\}) \tag{10-113d}$$

Notice that there is a one-to-one correspondence between α and γ on the one hand, and $\mathbf{R}_n$ and $\mathbf{R}_m$ on the other hand.

10.16.3. A simple algorithm for full rep matrices

To go further, we note that (10-113d) though simple in appearance, is not the best form for practical construction of the full matrices of a space group. In the following, we rewrite it in a more appropriate form.

From (10-113b) we have

$$\begin{aligned} &D^{(*\mathbf{k})(\nu)}_{(\tau\sigma)}(\{\alpha|\mathbf{a}\}) \\ &= D^{(\mathbf{k})(\nu)}(\{\beta_\tau^{-1}\alpha\beta_\sigma|\beta_\tau^{-1}(\alpha\mathbf{V}_\sigma - \mathbf{V}_\tau + \mathbf{a})\}) \ . \end{aligned} \tag{10-113e}$$

With the help of (10-71c), this becomes

$$D^{(*\mathbf{k})(\nu)}_{(\tau\sigma)}(\{\alpha|\mathbf{a}\}) = \exp[-i\mathbf{k}_\tau \cdot (\alpha\mathbf{V}_\sigma - \mathbf{V}_\tau + \mathbf{a})]\mathcal{D}_{(\tau\sigma)}(\{\alpha|\mathbf{a}\}) \tag{10-114a}$$

where

$$\mathcal{D}_{(\tau\sigma)}(\{\alpha|\mathbf{a}\}) \equiv \Delta(\beta_\tau^{-1}\alpha\beta_\sigma) \ , \tag{10-114b}$$

with the convention

$$\Delta(\beta_\tau^{-1}\alpha\beta_\sigma) = 0 \quad \text{if} \quad \beta_\tau^{-1}\alpha\beta_\sigma \notin \mathrm{G}_0(\mathbf{k}) \ . \tag{10-114c}$$

In other words,

$$\Delta(\beta_\tau^{-1}\alpha\beta_\sigma) = 0 \tag{10-115a}$$

unless

$$\beta_\tau^{-1}\alpha\beta_\sigma = \gamma \ , \tag{10-115b}$$

or

$$\alpha\beta_\sigma = \beta_\tau\gamma \ . \tag{10-115c}$$

Multiplying (10-115c) from the right by $\mathbf{k}$, we obtain another form of condition (10-115b)

$$\alpha\mathbf{k}_\sigma \doteq \mathbf{k}_\tau \ . \tag{10-115d}$$

Setting $\{\alpha|\mathbf{a}\} = \{\varepsilon|\mathbf{R}_n\}$, from (10-115d) we must have $\sigma = \tau$, while from (10-114) we have

$$\begin{aligned} D^{(*\mathbf{k})(\nu)}_{(\tau\sigma)}(\{\varepsilon|\mathbf{R}_n\}) &= \delta_{\tau\sigma}e^{-i\mathbf{k}_\tau\cdot\mathbf{R}_n}\Delta(\varepsilon) \\ &= \delta_{\tau\sigma}e^{-i\mathbf{k}_\tau\cdot\mathbf{R}_n}\mathbf{I}_\nu \ , \end{aligned} \tag{10-116}$$

where $\mathbf{I}_\nu$ is the $h_\nu \times h_\nu$ unit matrix.

Hence we see that the translation $\{\varepsilon|\mathbf{R}_n\}$ is represented by the diagonal matrix

$$D^{(*\mathbf{k})(\nu)}(\{\varepsilon|\mathbf{R}_n\}) = \begin{bmatrix} e^{-i\mathbf{k}_1\cdot\mathbf{R}_n}\mathbf{I}_\nu & & \\ & \ddots & \\ & & e^{-i\mathbf{k}_q\cdot\mathbf{R}_n}\mathbf{I}_\nu \end{bmatrix} . \tag{10-117}$$

Equation (10-114) gives a very convenient formula for constructing irreps of the space group $\mathbf{G}$ from the irreps $\Delta(\gamma) = D^{(\mathbf{k})(\nu)}(\{\gamma|\mathbf{V}(\gamma)\}')$ of the rep group $\mathbf{G}'_\mathbf{k}$. The procedure for obtaining irreps of $\mathbf{G}$ can be summarized as follows.

1. Following our procedure for (10-113a), we first introduce a matrix $\mathcal{D}(\{\alpha|\mathbf{a}\})$, whose $(\sigma\tau)$ block is the matrix $\mathcal{D}_{(\tau\sigma)}(\{\alpha|\mathbf{a}\})$ defined by (10-114b). To obtain $\mathcal{D}(\{\alpha|\mathbf{a}\})$, let us build up an array with q rows labeled by $\varepsilon, \beta_2^{-1}, \ldots, \beta_q^{-1}$ and q columns labeled by $\varepsilon, \beta_2, \ldots, \beta_q$,

$$\mathcal{D}(\{\alpha|\mathbf{a}\}) = \begin{array}{c} \\ \varepsilon \\ \beta_2^{-1} \\ \vdots \\ \beta_\tau^{-1} \\ \vdots \\ \beta_q^{-1} \end{array} \begin{array}{c} \varepsilon \;\; \beta_2 \;\; \cdots \;\; \beta_\sigma \;\; \cdots \;\; \beta_q \\ \begin{pmatrix} & & & 0 & & \\ & & & 0 & & \\ & & & \vdots & & \\ 0 & 0 & \cdots & \Delta(\gamma) & \cdots & 0 \\ & & & \vdots & & \\ & & & 0 & & \end{pmatrix} \end{array} \quad \text{for} \quad \beta_\tau^{-1}\alpha\beta_\sigma = \gamma . \tag{10-118}$$

Utilizing the point-group multiplication table, we form the products $\beta_\tau^{-1}\alpha\beta_\sigma$ for given τ with varying $\sigma = 1, 2, \ldots$. In each step we check whether $\beta_\tau^{-1}\alpha\beta_\sigma$ is an element of the point group $\mathbf{G}_0(\mathbf{k})$. If not, we put a zero (an $h_\nu \times h_\nu$ null matrix) in the $(\tau\sigma)$ block; if yes, e.g., $\beta_\tau^{-1}\alpha\beta_\sigma = \gamma$, then we put $\Delta(\gamma)$ into the $(\tau\sigma)$ block and zero for all the remaining entries in the τ-th row and σ-th column. We repeat this process for each $\tau = 1, 2, \ldots, q$.

2. Multiplying the nonzero matrices $\Delta(\gamma)$ in Eq. (10-118) by the appropriate phase factors $\exp[-i\mathbf{k}_\tau \cdot (\alpha\mathbf{V}_\sigma - \mathbf{V}_\tau + \mathbf{a})]$, we immediately get the sought-for matrix $D^{(*\mathbf{k})(\nu)}(\{\alpha|\mathbf{a}\})$.

3. The following symmetries (10-119) of the matrix $\mathcal{D}(\{\alpha|\mathbf{a}\})$ can be used either to save work or to check the calculation.

From (10-114b) it is seen that if

$$\mathcal{D}_{(\tau\sigma)}(\{\alpha|\mathbf{a}\}) = \Delta(\gamma), \quad \gamma = \beta_\tau^{-1}\alpha\beta_\sigma , \tag{10-119a}$$

then

$$\mathcal{D}_{(\sigma\tau)}(\{\alpha|\mathbf{a}\}^{-1}) = \Delta(\gamma^{-1}) . \tag{10-119b}$$

Furthermore, if $\alpha = \alpha^{-1}$, we have

$$\mathcal{D}_{(\sigma\tau)}(\{\alpha|\mathbf{a}\}) = \Delta(\gamma^{-1}) , \tag{10-119c}$$

since

$$\mathcal{D}_{(\sigma\tau)}(\{\alpha|\mathbf{a}\}) = \Delta(\beta_\sigma^{-1}\alpha\beta_\tau) = \Delta((\beta_\tau^{-1}\alpha\beta_\sigma)^{-1}) .$$

For examples of these symmetries, see (10-175) and (10-186).

From the foregoing discussion we see clearly that in the process of constructing irreps of the space group we are able to avoid tedious space-group multiplication, and only the much simpler point-group multiplication is required.

Another point that deserves pointing out is that the choice of the coset representative $\{\beta_\sigma|\mathbf{V}_\sigma\}$ is arbitrary, i.e., any element in a coset can be chosen as the representative of that coset. Different choices of the representatives correspond to different conventions for relative phases between the basis vectors $\psi^{(\nu)}_{\mathbf{k}_\sigma a}$ with different σ. For practical purposes, it is always desirable to choose elements that do not associate with non-primitive translations as the coset representatives, since under such a choice (10-114) is simplified as follows:

$$D^{(*\mathbf{k})(\nu)}_{(\tau\sigma)}(\{\alpha|\mathbf{a}\}) = e^{-i\mathbf{k}_\tau\cdot\mathbf{a}}\Delta(\beta_\tau^{-1}\alpha\beta_\sigma)\ , \tag{10-120}$$

which is suitable only when $\mathbf{V}(\beta_\sigma) = \mathbf{V}(\beta_\tau) = 0$.

10.16.4. The $\mathbf{G}\supset\mathbf{G}(\mathbf{k}_\sigma)\supset\mathbf{G}(s_\sigma)\supset\mathbf{T}$ *irreducible basis*

Finally let us take a look at the meaning of the irreducible basis vectors $\psi^{(\nu)}_{\mathbf{k}_\sigma a}$ of the space group $\mathbf{G}$. We have already seen that $\psi^{(\nu)}_{\mathbf{k},a}$ is the $\mathbf{G}\supset\mathbf{G}(\mathbf{k})\supset\mathbf{G}(s)\supset\mathbf{T}$ irreducible basis and obeys the eigenequations

$$\begin{bmatrix} C \\ C(s) \\ \hat{\mathbf{k}} \end{bmatrix} \psi^{(\nu)}_{\mathbf{k},a} = \begin{bmatrix} \nu \\ a \\ \mathbf{k} \end{bmatrix} \psi^{(\nu)}_{\mathbf{k},a}\ , \tag{10-121}$$

where C and $C(s)$ are the CSCO-I of the rep groups $\mathbf{G}'_{\mathbf{k}}$ and $\mathbf{G}'_s$ respectively, and $\hat{\mathbf{k}} = -i\nabla$ is the CSCO-I of $\mathbf{T}$. Since the little group $\mathbf{G}(\mathbf{k})$ and the rep group $\mathbf{G}'_{\mathbf{k}}$ have common irreducible bases and common irrep labels, for convenience in exposition, we shall refer to the CSCO-I C of $\mathbf{G}'_{\mathbf{k}}$ as the CSCO-I of the little group $\mathbf{G}(\mathbf{k})$; similarly, the CSCO-I $C(s)$ of $\mathbf{G}'_s$ will be referred to as the CSCO-I of the subgroup $\mathbf{G}(s)$ of $\mathbf{G}(\mathbf{k})$.

As in (10-56) we may define the little group $\mathbf{G}(\mathbf{k}_\sigma)$ for the wave vector $\mathbf{k}_\sigma = \beta_\sigma\mathbf{k}$ such that under the operations of the rotational part of $\mathbf{G}(\mathbf{k}_\sigma)$, the wave vector $\mathbf{k}_\sigma$ is invariant modulo a reciprocal lattice vector. Clearly the relation between the groups $\mathbf{G}(\mathbf{k}_\sigma)$ and $\mathbf{G}(\mathbf{k})$ is

$$\mathbf{G}(\mathbf{k}_\sigma) = \{\beta_\sigma|\mathbf{V}_\sigma\}\mathbf{G}(\mathbf{k})\{\beta_\sigma|\mathbf{V}_\sigma\}^{-1}\ . \tag{10-122a}$$

Suppose $\{\alpha|\mathbf{a}\}$ is an element of $\mathbf{G}(\mathbf{k}_\sigma)$; then $\{\alpha|\mathbf{a}\}$ is necessarily of the form

$$\{\alpha|\mathbf{a}\} = \{\beta_\sigma|\mathbf{V}_\sigma\}\{\gamma|\mathbf{V}(\gamma)+\mathbf{R}_n\}\{\beta_\sigma|\mathbf{V}_\sigma\}^{-1}\ . \tag{10-123a}$$

From (10-123a) and (10-111) we have

$$\{\alpha|\mathbf{a}\}\psi^{(\nu)}_{\mathbf{k}_\sigma a} = \sum_b D^{(\mathbf{k})(\nu)}_{ba}(\{\gamma|\mathbf{V}(\gamma)+\mathbf{R}_n\})\psi^{(\nu)}_{\mathbf{k}_\sigma b}\ , \tag{10-123b}$$

that is, $\psi^{(\nu)}_{\mathbf{k}_\sigma a}$ is the irreducible basis of the group $\mathbf{G}(\mathbf{k}_\sigma)$. Furthermore, as in (10-122a), we define

$$\mathbf{G}(s_\sigma) \equiv \{\beta_\sigma|\mathbf{V}_\sigma\}\mathbf{G}(s)\{\beta_\sigma|\mathbf{V}_\sigma\}^{-1}\ . \tag{10-122b}$$

Clearly $\mathbf{G}(s_\sigma)$ is a subgroup of $\mathbf{G}(\mathbf{k}_\sigma)$. According to Eq. (10-122), the CSCO-I of $\mathbf{G}(\mathbf{k}_\sigma)$ and $\mathbf{G}(s_\sigma)$ are

$$C(\mathbf{k}_\sigma) = \{\beta_\sigma|\mathbf{V}_\sigma\}C\{\beta_\sigma|\mathbf{V}_\sigma\}^{-1}\ ,$$

$$C(s_\sigma) = \{\beta_\sigma|\mathbf{V}_\sigma\}C(s)\{\beta_\sigma|\mathbf{V}_\sigma\}^{-1}\ . \tag{10-124}$$

With (10-106), (10-121) and (10-124) we have

$$\begin{bmatrix} C(\mathbf{k}_\sigma) \\ C(s_\sigma) \\ \hat{\mathbf{k}} \end{bmatrix} \psi^{(\nu)}_{\mathbf{k}_\sigma a} = \begin{bmatrix} \nu \\ a \\ \mathbf{k}_\sigma \end{bmatrix} \psi^{(\nu)}_{\mathbf{k}_\sigma a}\ . \tag{10-125}$$

Hence we see that the partner (or component) $\psi^{(\nu)}_{\mathbf{k}_\sigma a}$ of the irreducible basis of a space group $\mathbf{G}$ is the $\mathbf{G} \supset \mathbf{G}(\mathbf{k}_\sigma) \supset \mathbf{G}(s_\sigma) \supset \mathbf{T}$ irreducible basis. In other words, the group chains used to classify the irreducible basis vectors of a space group $\mathbf{G}$ vary with the components. This is quite different from the usual case (e.g., the permutation group, rotation group, or unitary group), where the same group chain is always used to classify all the components in a given irrep of the group.

10.17. The Irreducible Basis and Matrices of C^4_{2v}

C^4_{2v} is a non-symmorphic space group, $\mathbf{G}_0 = \mathcal{C}_{2v}$. It belongs to the simple orthorhombic lattice. The three perpendicular primitive translations are

$$\mathbf{t}_1 = a\mathbf{i}, \quad \mathbf{t}_2 = b\mathbf{j}, \quad t_3 = c\mathbf{k} .$$

The group elements are specified by

$$\mathcal{R}_1 = \{\varepsilon|0\}, \quad \mathcal{R}_2 = \{\sigma_x|\mathbf{v}\}, \quad \mathcal{R}_3 = \{\sigma_y|\mathbf{v}\}, \quad \mathcal{R}_4 = \{C_{2z}|0\}, \quad \mathbf{v} = \mathbf{t}_1/2 . \tag{10-126}$$

The arrangement of the atoms in the xy plane is shown in Fig. 10.17-1. The basis vectors for the reciprocal lattice are

$$\mathbf{b}_1 = \frac{2\pi}{a}\mathbf{i}, \quad \mathbf{b}_2 = \frac{2\pi}{b}\mathbf{j}, \quad \mathbf{b}_3 = \frac{2\pi}{c}\mathbf{k} .$$

The Brillouin zone is shown in Fig. 10.17-2.

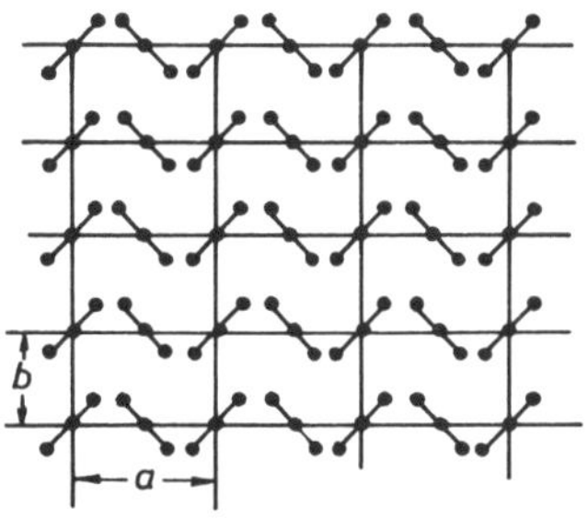

Fig. 10.17-1. The arrangement of the atoms in the xy plane for a crystal with symmetry C^4_{2v}.

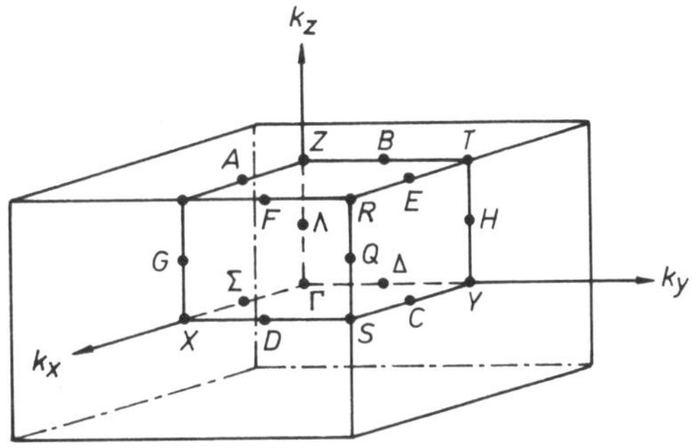

Fig. 10.17-2. The Brillouin zone for the simple orthorhombic lattice

In the following we shall consider the IRB and irreps for four stars.

10.17.1. *General star:* $\mathbf{p} = (p_1, p_2, p_3)$

$\mathbf{G}_0(\mathbf{k}) = \varepsilon, \mathbf{G}(k) = \mathbf{T}$. The irrep of the rep group is $\Delta(\varepsilon) = 1$. The irreducible basis of $\mathbf{G}(\mathbf{k})$ is

$$\psi_\mathbf{k} = \exp[i(\mathbf{k} + \mathbf{K}_m) \cdot \mathbf{r}] = \exp[2\pi i \sum_j \kappa_j \xi_j] , \quad \kappa_j = p_j + m_j . \tag{10-127}$$

The group elements in (10-126) are just the coset representatives $\{\beta_i|\mathbf{V}(\beta_i)\}$. The induced basis, i.e., the IRB of the space group, is

$$\begin{aligned}
\Psi_1 &= \mathcal{R}_1\psi_\mathbf{k} = \exp[2\pi i \sum_j \kappa_j \xi_j] , \\
\Psi_2 &= \mathcal{R}_2\psi_\mathbf{k} = \exp[i(\mathbf{k} + \mathbf{K}_m) \cdot \mathbf{v}] \exp[2\pi i(-\kappa_1\xi_1 + \kappa_2\xi_2 + \kappa_3\xi_3)] , \\
\Psi_3 &= \mathcal{R}_3\psi_\mathbf{k} = \exp[-i(\mathbf{k} + \mathbf{K}_m) \cdot \mathbf{v}] \exp[2\pi i(\kappa_1\xi_1 - \kappa_2\xi_2 + \kappa_3\xi_3)] , \\
\Psi_4 &= \mathcal{R}_4\psi_\mathbf{k} = \exp[2\pi i(-\kappa_1\xi_1 - \kappa_2\xi_2 + \kappa_3\xi_3)] .
\end{aligned} \tag{10-128}$$

The **k** star is

$$(*\mathbf{k}) = (p_1, p_2, p_3), \quad (-p_1, p_2, p_3), \quad (p_1, -p_2, p_3), \quad (-p_1, -p_2, p_3)\ .$$

Using (10-118) we can find the matrices $\mathcal{D}(\{\alpha|\mathbf{a}\})$, e.g.,

$$\mathcal{D}(\{\sigma_x|\mathbf{v}\}) = \left[\begin{array}{cc|cc} & \Delta(\varepsilon) & & \\ \Delta(\varepsilon) & & & \\ \hline & & & \Delta(\varepsilon) \\ & & \Delta(\varepsilon) & \end{array}\right], \quad \mathcal{D}(\{\sigma_y|\mathbf{v}\}) = \left[\begin{array}{cc|cc} & & \Delta(\varepsilon) & \\ & & & \Delta(\varepsilon) \\ \hline \Delta(\varepsilon) & & & \\ & \Delta(\varepsilon) & & \end{array}\right]. \tag{10-129}$$

From (10-114) and (10-129) we get the irreps of the space group C_{2v}^4,

$$D^{*(\mathbf{k})}(\{\varepsilon|\mathbf{R}_n\}) = \left[\begin{array}{cc|cc} e_1 & & & \\ & e_2 & & \\ \hline & & e_3 & \\ & & & e_4 \end{array}\right], \quad D^{(*\mathbf{k})}(\{\sigma_x|\mathbf{v}\}) = \left[\begin{array}{cc|cc} 0 & 1 & & \\ 1 & 0 & & \\ \hline & & 0 & 1 \\ & & 1 & 0 \end{array}\right],$$

$$D^{(*\mathbf{k})}(\{\sigma_y|\mathbf{v}\}) = \left[\begin{array}{cc|cc} & & e_1^2 & 0 \\ & & 0 & 1 \\ \hline 1 & 0 & & \\ 0 & (e_1^*)^2 & & \end{array}\right], \quad D^{(*\mathbf{k})}(\{C_{2z}|0\}) = \left[\begin{array}{cc|cc} & & 0 & 1 \\ & & e_1^2 & 0 \\ \hline 0 & (e_1^*)^2 & & \\ 1 & 0 & & \end{array}\right], \tag{10-130}$$

where $e_j = \exp[-i\mathbf{k}_j \cdot \mathbf{R}_n]$.

10.17.2. The star $\Gamma : \mathbf{p} = (0,0,0)$

$\mathbf{G}_0(\mathbf{k}) = \mathcal{C}_{2v}$ and the little group is the space group itself, $\mathbf{G}(\mathbf{k}) = \mathbf{G}$. The rep group $\mathbf{G}'_\Gamma$ is isomorphic to the point group $\mathcal{C}_{2v} = (\varepsilon, \sigma_x, \sigma_y, C_{2z})$. The irreps of $\mathcal{C}_{2v}$ are known and listed in Table 10.17-1.

Table 10.17-1. Irreps of $\mathcal{C}_{2v}$.

ν	e	σ_x	σ_y	C_{2z}
A_1	1	1	1	1
A_2	1	-1	-1	1
B_2	1	-1	1	-1
B_2	1	1	-1	-1

From (10-92) and Table 10.17-1 we get the irreducible basis of $\mathbf{G}(\mathbf{k}) = \mathbf{G}$,

$$\psi_{\mathbf{k}}^{(A_1)} = \frac{1}{2}(\psi_1 + \psi_2 + \psi_3 + \psi_4) = 2\begin{cases} \cos\mu_1\xi_1 \cos\mu_2\xi_2 \exp(i\mu_3\xi_3)\ , \\ \sin\mu_1\xi_1 \sin\mu_2\xi_2 \exp(i\mu_3\xi_3)\ , \end{cases} \quad \text{for} \quad m_1 = \begin{matrix} \text{even} \\ \text{odd,} \end{matrix}$$

$$\psi_{\mathbf{k}}^{(A_2)} = \frac{1}{2}(\psi_1 - \psi_2 - \psi_3 + \psi_4) = 2\begin{cases} -\sin\mu_1\xi_1 \sin\mu_2\xi_2 \exp(i\mu_3\xi_3)\ , \\ \cos\mu_1\xi_1 \cos\mu_2\xi_2 \exp(i\mu_3\xi_3)\ , \end{cases} \quad \text{for} \quad m_1 = \begin{matrix} \text{even}\ , \\ \text{odd}\ , \end{matrix}$$

$$\psi_{\mathbf{k}}^{(B_1)} = \frac{1}{2}(\psi_1 - \psi_2 + \psi_3 - \psi_4) = 2i\begin{cases} \sin\mu_1\xi_1 \cos\mu_2\xi_2 \exp(i\mu_3\xi_3)\ , \\ \cos\mu_1\xi_1 \sin\mu_2\xi_2 \exp(i\mu_3\xi_3)\ , \end{cases} \quad \text{for} \quad m_1 = \begin{matrix} \text{even}\ , \\ \text{odd}\ , \end{matrix}$$

$$\psi_{\mathbf{k}}^{(B_2)} = \frac{1}{2}(\psi_1 + \psi_2 - \psi_3 - \psi_4) = 2i\begin{cases} \cos\mu_1\xi_1 \sin\mu_2\xi_2 \exp(i\mu_3\xi_3)\ , \\ \sin\mu_1\xi_1 \cos\mu_2\xi_2 \exp(i\mu_3\xi_3)\ , \end{cases} \quad \text{for} \quad m_1 = \begin{matrix} \text{even}\ , \\ \text{odd}\ . \end{matrix} \tag{10-131}$$

where $\mu_i = 2\pi m_i$ and ψ_i are obtainable from Ψ_i in (10-128) by setting $\mathbf{p} = (0,0,0)$.

10.17.3 The star $\Sigma : \mathbf{p} = (p_1, 0, 0)$.

The star Σ consists of two interior points. $\mathbf{G}_0(\mathbf{k}) = \mathcal{C}_s = (\varepsilon, \sigma_y)$. The rep group $\mathbf{G}'_\Sigma$ is isomorphic to the point group C_s. Thus the irreps of the rep group $\mathbf{G}'_\Sigma$ are $\Delta^{(A)}(\varepsilon) = 1$, $\Delta^{(A)}(\sigma_y) = 1; \Delta^{(B)}(\varepsilon) = 1, \Delta^{(B)}(\sigma_y) = -1$.

The generalized projection operators $P_a^{(\mathbf{k})(\nu)b}$ are

$$P^{(\Sigma)(A)} = \frac{1}{2}(R_1 + R_2), \quad P^{(\Sigma)(B)} = \frac{1}{2}(R_1 - R_2) , \tag{10-132}$$

where $R_1 = \{\varepsilon|0\}$, and $R_2 = \exp(i\mathbf{k}\cdot\mathbf{v})\{\sigma_y|\mathbf{v}\}$.

The irreducible basis $\psi_{\mathbf{k},a}^{(\nu)b}$ of the little group $\mathbf{G}(\Sigma)$ is

$$\psi_{\mathbf{k}}^{(A)} = \frac{1}{2}(R_1 + R_2)\psi_{\mathbf{k}} = \exp\{2\pi i[(p_1 + m_1)\xi_1 + m_3\xi_3]\}\begin{cases} \cos(2\pi m_2 \xi_2) \\ i\sin(2\pi m_2 \xi_2) \end{cases} \text{for} \quad m_1 = \begin{matrix} \text{even} , \\ \text{odd} , \end{matrix}$$

$$\psi_{\mathbf{k}}^{(B)} = \frac{1}{2}(R_1 - R_2)\psi_{\mathbf{k}} = \exp\{2\pi i[(p_1 + m_1)\xi_1 + m_3\xi_3]\}\begin{cases} i\sin(2\pi m_2 \xi_2) \\ \cos(2\pi m_2 \xi_2) \end{cases} \text{for} \quad m_1 = \begin{matrix} \text{even} , \\ \text{odd} . \end{matrix} \tag{10-133}$$

The irreps of the space group can be found as follows. Choose $\{\varepsilon|0\}$ and $\{C_{2z}|0\}$ as the coset representatives $\{\beta_i|\mathbf{V}(\beta_i)\}$. The $\mathbf{k}$ star is $(*\mathbf{k}) = (\mathbf{k}, -\mathbf{k})$.

Using (10-118) we can find the matrices $\mathcal{D}(\{\alpha|\mathbf{a}\})$,

$$\mathcal{D}(\{\varepsilon|0\}) = \begin{pmatrix} \Delta(\varepsilon) & \\ & \Delta(\varepsilon) \end{pmatrix} , \qquad \mathcal{D}(\{\sigma_x|\mathbf{v}\}) = \begin{pmatrix} & \Delta(\sigma_y) \\ \Delta(\sigma_y) & \end{pmatrix} ,$$

$$\mathcal{D}(\{\sigma_y|\mathbf{v}\}) = \begin{pmatrix} \Delta(\sigma_y) & \\ & \Delta(\sigma_y) \end{pmatrix} , \quad \mathcal{D}(\{C_{2z}|\mathbf{v}\}) = \begin{pmatrix} & \Delta(\varepsilon) \\ \Delta(\varepsilon) & \end{pmatrix} . \tag{10-134}$$

From (10-134) and (10-120) it is trivial to obtain the irreps of C_{2v}^4. The irreducible bases $\psi_{\mathbf{k}_\sigma a}^{(\nu)b}$ of C_{2v}^4 are

$$\psi_{\mathbf{k}_1}^{(\nu)} = \psi_{\mathbf{k}}^{(\nu)}, \quad \psi_{\mathbf{k}_2}^{(\nu)} = \{C_{2z}|0\}\psi_{\mathbf{k}}^{(\nu)}, \ \nu = A, B . \tag{10-135}$$

10.17.4. The star $X : \mathbf{p} = (\frac{1}{2}, 0, 0)$.

The star X is a point of symmetry on the surface of the Brillouin zone: $\mathbf{G}_0(\mathbf{k}) = \mathbf{G}_o = \mathcal{C}_{2v}, q = 1$ and $\mathbf{G}(\mathbf{k}) = \mathbf{G}$. With the aid of (10-79a) we can easily find the factors $\eta(\rho,\sigma)$ and construct the group table of $\mathbf{G}'_X$, as shown in Table 10.17-2. It resembles very much Table 8.8-1, the group table for the rep group $D_2^\dagger$. The rep group $\mathbf{G}'_X$ has $N = 5$ classes and $n = 1$ linearly independent class operator: $C_1 = R_1, C_2 = -R_1, C_3 = R_2 - R_2, C_4 = R_3 - R_3, C_5 = R_4 - R_4$. Therefore, there is only one inequivalent irrep of dimension 2, denoted by X. Choosing $C(s) = \{C_{2z}|0\}$ and diagonalizing the regular rep matrices of $C(s)$ and $\overline{C}(s)$, we obtain the eigenvectors $P_a^{(\mathbf{k})(\nu)b}$ of the CSCO-III of $\mathbf{G}'_X$ and the irreps of $\mathbf{G}'_X$, as shown in Table 10.17-3.

Table 10.17-2. Group table of $\mathbf{G}'_X$.

R_1 $\{\varepsilon\|0\}$	R_2 $\{\sigma_x\|\mathbf{v}\}'$	R_3 $\{\sigma_y\|\mathbf{v}\}'$	R_4 $\{C_{2z}\|0\}$
1	2	3	4
2	−1	−4	3
3	4	1	2
4	−3	−2	1

Table 10.17-3. Irreps of $\mathbf{G}'_X$.

	R_1	R_2	R_3	R_4
$P_1^{(X)1}$	1			1
$P_{-1}^{(X)1}$		1	1	
$P_{-1}^{(X)-1}$		−1	1	
$P_{-1}^{(X)-1}$	1			−1
$\Delta^{(X)}$	$\begin{pmatrix}1&0\\0&1\end{pmatrix}$	$\begin{pmatrix}0&-1\\1&0\end{pmatrix}$	$\begin{pmatrix}0&1\\1&0\end{pmatrix}$	$\begin{pmatrix}1&0\\0&-1\end{pmatrix}$

The irreducible bases of $\mathbf{G}(\mathbf{k})(=\mathbf{G})$ are (ignoring the normalizations),

$$\begin{aligned}\psi_1^{(X)1} &= \psi_1+\psi_4, & \psi_{-1}^{(X)1} &= \psi_2+\psi_3\ ,\\ \psi_1^{(X)-1} &= -\psi_2+\psi_3, & \psi_{-1}^{(X)-1} &= \psi_1-\psi_4\ ,\end{aligned} \tag{10-136}$$

where $\psi_j = R_j\psi_{\mathbf{k}}$, and are related to Ψ_j in (10-128) by

$$\psi_j = \exp[i\mathbf{k}\cdot\mathbf{V}(\gamma_j)]\Psi_j\Big|_{\mathbf{p}=(\frac{1}{2}00)}\ .$$

10.18. The Clebsch-Gordan Coefficients of Space Groups*

10.18.1. The CG series

Suppose that $D^{(*\mathbf{k})(\nu)}$ and $D^{(*\mathbf{k}')(\nu')}$ are two irreps of a space group $\mathbf{G}$. The Kronecker product of these two irreps can be reduced to a direct sum of the irreps of $\mathbf{G}$,

$$D^{(*\mathbf{k})(\nu)}\otimes D^{(*\mathbf{k}')(\nu')} = \sum_{*\mathbf{k}''}\sum_{\nu''}\oplus(*\mathbf{k}\nu * \mathbf{k}'\nu'|*\mathbf{k}''\nu'')D^{(*\mathbf{k}'')(\nu'')}\ , \tag{10-137}$$

where $(*\mathbf{k}\nu * \mathbf{k}'\nu'|*\mathbf{k}''\nu'')$ is an integer and is the number of times that the irrep $D^{(*\mathbf{k}'')(\nu'')}$ occurs in the Kronecker product rep. Equation (10-137) is referred to as the Clebsch-Gordon series of the space group, and the integer $(*\mathbf{k}\nu * \mathbf{k}'\nu'|*\mathbf{k}''\nu'')$ is called the reduction coefficient (Birman, 1974) or the multiplicity.

Several methods are available for calculating the space-group reduction coefficient (Birman, 1962, 1974; Hsieh 1964; Bradley 1972). A complete and explicit set of tables for the reduction coefficients has been published (Davies, 1979, 1980; Cracknell, 1979).

Let $\psi^{(\nu)}_{\mathbf{k}_\sigma a}$ and $\psi^{(\nu')}_{\mathbf{k}'_{\sigma'}a'}$ be the basis vectors carrying the irreps $D^{(*\mathbf{k})(\nu)}$ and $D^{(*\mathbf{k}')(\nu')}$, respectively. The CG coefficients of the space group are defined as the expansion coefficients in the following equation:

$$\psi^{(\nu'')\theta}_{\mathbf{k}_{\sigma''}a''} = \sum_{\sigma a\sigma' a'}\left[\begin{array}{c|c}\nu\mathbf{k}\nu'\mathbf{k}' & \nu''\mathbf{k}''\theta\\ \sigma a\sigma' a' & \sigma'' a''\end{array}\right]\psi^{(\nu)}_{\mathbf{k}_\sigma a}\psi^{(\nu')}_{\mathbf{k}'_{\sigma'}a'}\ ,\quad \theta=1,2,\dots,(*\mathbf{k}\nu * \mathbf{k}'\nu'|*\mathbf{k}''\nu'')\ , \tag{10-138}$$

where θ is the multiplicity label. The inverse of (10-138) is

$$\psi^{(\nu)}_{\mathbf{k}_\sigma a}\psi^{(\nu')}_{\mathbf{k}'_{\sigma'}a'} = \sum_{\nu''\mathbf{k}''\sigma''a''\theta}\left[\begin{array}{c|c}\nu\mathbf{k}\nu'\mathbf{k}' & \nu''\mathbf{k}''\theta\\ \sigma a\sigma' a' & \sigma'' a''\end{array}\right]\psi^{(\nu'')\theta}_{\mathbf{k}''_{\sigma''}a''}\ . \tag{10-139a}$$

The CG coefficients can be expressed as

$$\left[\begin{array}{c|c}\nu\mathbf{k}\,\nu'\mathbf{k}' & \nu''\mathbf{k}''\theta\\ \sigma a\,\sigma' a' & \sigma'' a''\end{array}\right] = \langle\psi^{(\nu)}_{\mathbf{k}_\sigma a}\psi^{(\nu')}_{\mathbf{k}'_{\sigma'}a'}|\psi^{(\nu'')\theta}_{\mathbf{k}''_{\sigma''}a''}\rangle\ . \tag{10-139b}$$

With the aid of

$$\langle \psi^{(\nu)}_{\mathbf{k}_\sigma a}\psi^{(\nu')}_{\mathbf{k}'_{\sigma'}a'}|\{\varepsilon|\mathbf{R}_n\}|\psi^{(\nu'')\theta}_{\mathbf{k}''_{\sigma''}a''}\rangle = e^{-i\mathbf{k}''_{\sigma''}\cdot\mathbf{R}_n}C$$
$$= \langle\{\varepsilon|-\mathbf{R}_n\}(\psi^{(\nu)}_{\mathbf{k}_\sigma a}\psi^{(\nu')}_{\mathbf{k}'_{\sigma'}a'})|\psi^{(\nu'')\theta}_{\mathbf{k}''_{\sigma''}a''}\rangle = e^{-(\mathbf{k}_\sigma+\mathbf{k}'_{\sigma'})\cdot\mathbf{R}_n}C\ ,$$

where C is the abbreviation for the CG coefficient (10-139b), we know that the CG coefficient vanishes unless

$$\mathbf{k}_\sigma + \mathbf{k}'_{\sigma'} - \mathbf{k}''_{\sigma''} = \mathbf{K}_m\ . \tag{10-140}$$

According to (10-140) we can introduce the wave-vector selection rule,

$$*\mathbf{k}\otimes *\mathbf{k}' = \sum_{*\mathbf{k}''}(*\mathbf{k} * \mathbf{k}'| * \mathbf{k}'') * \mathbf{k}''\ . \tag{10-141}$$

The integers $(*\mathbf{k} * \mathbf{k}'| * \mathbf{k}'')$ are referred to as the wave-vector reduction coefficients. Quite simply we have

$$(*\mathbf{k} * \mathbf{k}'| * \mathbf{k}'') = (*\mathbf{k}' * \mathbf{k}| * \mathbf{k}'')\ . \tag{10-142}$$

Assuming that there are q, q', and q'' points in the stars $*\mathbf{k}, *\mathbf{k}'$ and $*\mathbf{k}''$, respectively, it follows from (10-141) that

$$qq' = \sum_{*\mathbf{k}''}(*\mathbf{k} * \mathbf{k}'| * \mathbf{k}'')q''\ . \tag{10-143a}$$

A simple prescription for determining the wave-vector reduction coefficients are given by Birman (1962) and is recapitulated here.

The $q \bullet q'$ wave vectors $\mathbf{k}_\sigma + \mathbf{k}'_\tau$ can be expressed as

$$\beta_\sigma(k + \mathbf{k}'_{\sigma'}) = \beta_\sigma(\mathbf{k} + \beta_{\sigma'}\mathbf{k}')\ ,\quad \sigma = 1, 2, \ldots, q;\quad \sigma' = 1, 2, \ldots, q'\ . \tag{10-143b}$$

Let us introduce a table, called the wave-vector table, the entry of which in the σ-th row and σ'-th column is $\beta_\sigma(\mathbf{k}+\beta_{\sigma'}\mathbf{k}')$. Suppose that $\mathbf{k}+\beta_{\sigma'}\mathbf{k}'$ belongs to the star $*\mathbf{k}''$, then $\beta_\sigma(\mathbf{k}+\beta_{\sigma'}\mathbf{k}')$ still belongs to the star $*\mathbf{k}''$. In other words, all the wave vectors in the wave-vector table belong to the same star. Therefore, to determine the wave-vector selection rule, it suffices to construct a single row, say the first row, of the wave-vector table. From this row and using (10-143a) we can obtain the wave-vector selection rule. It is convenient to choose $q \geq q'$, i.e., let the number of rows be larger than or equal to the number of columns in the wave-vector table. As examples, we consider the selection rules for the face-centered cubic lattice Γ^f_c.

Example 1: $W(\frac{1}{2}, \frac{1}{4}, \frac{3}{4})\otimes X(\frac{1}{2}, 0, \frac{1}{2})$. According to Table 3.11 in Bradley (1972, pp. 162), $q = q_W = 6, q' = q_X = 3$. The star X consists of three points: $\mathbf{k}'_{\sigma'} = (\frac{1}{2}0\frac{1}{2}), (\frac{1}{2}\frac{1}{2}0)$, and $(0\frac{1}{2}\frac{1}{2})$. Using (10-140) we can construct Table 10.18-1. Consulting Bradley's Table 3.11, we know that $\mathbf{k}_\Delta = (0\frac{1}{4}\frac{1}{4})$ and $q'' = q_\Delta = 6$. Therefore, we have the selection rule

$$*W\otimes *X = 2 * \Delta + *W\ . \tag{10-143c}$$

Example 2: $\Delta(\alpha 0\alpha)\otimes X(\frac{1}{2}0\frac{1}{2})\quad (\alpha \leq \frac{1}{2})$. $q = q_\Delta = 6, q' = q_X = 3$. Similarly, from Table 10.18-2, Eq. (10-143a) and the fact that $q'' = q_Z = 12$, we get

$$*\Delta\otimes *X = *\Delta' + *Z\ . \tag{10-143d}$$

Table 10.18-1. Wave-vector table for $^*W \otimes {}^*X$.

$q \backslash q'$	1	2	3
1	Δ $(0, \frac{1}{4}, \frac{1}{4})$	Δ $(0, \frac{3}{4}, \frac{3}{4})$	W $(\frac{1}{2}, \frac{3}{4}, \frac{1}{4})$

Table 10.18-2. Wave vector table for $*\Delta \otimes {}^*X$.

$q \backslash q'$	1	2	3
1	Δ' $(\frac{1}{2}+\alpha, 0, \frac{1}{2}+\alpha)$	Z $(\frac{1}{2}+\alpha, \frac{1}{2}, \alpha)$	Z $(\alpha, \frac{1}{2}, \frac{1}{2}+\alpha)$

10.18.2. The calculation of the CG coefficients

Setting $\sigma'' = 1$ in (10-138), we have

$$\psi^{(\nu'')\theta}_{\mathbf{k}''a''} = \sum_{\sigma a \sigma' a'} \begin{bmatrix} \nu \mathbf{k} \nu' \mathbf{k}' & \nu'' \mathbf{k}'' \theta \\ \sigma a \sigma' a' & 1a'' \end{bmatrix} |\sigma a \sigma' a'\rangle \ , \tag{10-144a}$$

where

$$|\sigma a \sigma' a'\rangle \equiv \psi^{(\nu)}_{\mathbf{k}_\sigma a} \psi^{(\nu')}_{\mathbf{k}'_{\sigma'} a'} \ . \tag{10-144b}$$

Since $\psi^{(\nu'')\theta}_{\mathbf{k}''a''}$ is the irreducible basis of the rep group $\mathbf{G}'_{\mathbf{k}''}$, it is necessarily an eigenvector of the CSCO-II $(C, C(s))$ of the group $\mathbf{G}'_{\mathbf{k}''}$:

$$\begin{bmatrix} C \\ C(s) \end{bmatrix} \psi^{(\nu'')\theta}_{\mathbf{k}''a''} = \begin{bmatrix} \nu'' \\ a'' \end{bmatrix} \psi^{(\nu'')\theta}_{\mathbf{k}''a''} \ , \tag{10-145a}$$

or, written in matrix form

$$\begin{bmatrix} M(C) \\ M(C(s)) \end{bmatrix} U^{(\nu'')\theta}_{\mathbf{k}''a''} = \begin{bmatrix} \nu'' \\ a'' \end{bmatrix} U^{(\nu'')\theta}_{\mathbf{k}''a''} \ , \tag{10-145b}$$

where $M(C)$ and $M(C(s))$ are the representatives of the operators C and $C(s)$ in the uncoupled representation with the basis $|\sigma a \sigma' a'\rangle$, and the vector

$$U^{(\nu'')\theta}_{\mathbf{k}''a''} = \left\{ \begin{bmatrix} \nu \mathbf{k} \nu' \mathbf{k}' & \nu'' \mathbf{k}'' \theta \\ \sigma a \sigma' a' & 1a'' \end{bmatrix} \right\} \tag{10-146}$$

is the representative of $\psi^{(\nu'')\theta}_{\mathbf{k}''a''}$; (ν'', a'') is the eigenvalue of $(C, C(s))$. In other words, the CG coefficients (10-146) result from a diagonalization of the matrices $M(C)$ and $M(C(s))$ simultaneously. To calculate $M(C)$ and $M(C(s))$, we must first calculate the matrices $M(\{\gamma''|\mathbf{c}''\})$ for the group elements $\{\gamma''|\mathbf{c}''\}'$ contained in the CSCO-II of $\mathbf{G}'_{\mathbf{k}''}$. In this section, to avoid notational clumsiness, we use the abbreviation

$$\{\gamma''|\mathbf{c}''\} \equiv \{\gamma''|\mathbf{V}(\gamma'')\} \ . \tag{10-147}$$

From $\{\gamma''|\mathbf{c}''\}' = e^{i\mathbf{k}''\cdot\mathbf{c}''}\{\gamma''|\mathbf{c}''\}$, we have

$$M(\{\gamma''|\mathbf{c}''\}') = e^{i\mathbf{k}''\cdot\mathbf{c}''} M(\{\gamma''|\mathbf{c}''\}) \ . \tag{10-148}$$

The matrix elements of $M(\{\gamma''|\mathbf{c}''\})$ can be expressed as

$$\begin{aligned} M_{\tau b\tau' b',\sigma a\sigma' a'}(\{\gamma''|\mathbf{c}''\}) &= \langle \tau b\tau' b'|\{\gamma''|\mathbf{c}''\}|\sigma a\sigma' a'\rangle \\ &= D^{(*\mathbf{k})(\nu)}_{\tau b,\sigma a}(\{\gamma''|\mathbf{c}''\}) D^{(*\mathbf{k}')(\nu')}_{\tau' b',\sigma' a'}(\{\gamma''|\mathbf{c}''\}) \,, \end{aligned} \tag{10-149a}$$

or in the form of a direct product of the matrices

$$M(\{\gamma''|\mathbf{c}''\}) = D^{(*\mathbf{k})(\nu)}(\{\gamma''|\mathbf{c}''\}) \otimes D^{(*\mathbf{k}')(\nu')}(\{\gamma''|\mathbf{c}''\}) \,. \tag{10-149b}$$

The matrices $D^{(*\mathbf{k})(\nu)}(\{\gamma''|\mathbf{c}''\})$ and $D^{(*\mathbf{k}')(\nu')}(\{\gamma''|\mathbf{c}''\})$ can be evaluated from the irreps of the rep groups $\mathbf{G}'_{\mathbf{k}}$ and $\mathbf{G}'_{\mathbf{k}'}$ by using (10-114) or (10-120), while the matrices $M(C)$ and $M(C(s))$ can be evaluated by using Eq. (10-149). From the secular equations of $M(C)$ and $M(C(s))$, we can get the eigenvalues $(\nu'' a'')$ and their degeneracies. If the degeneracy of the eigenvalue $(\nu'' a'')$ is d, then it implies that the reduction coefficient is

$$(*\mathbf{k}\nu * \mathbf{k}'\nu' | * \mathbf{k}''\nu'') = d \,. \tag{10-150}$$

Substituting the eigenvalue (ν'', a'') into (10-145b), we can get d orthogonal eigenvectors,

$$U^{(\nu'')\theta}_{\mathbf{k}'' a''} = \left\{ \begin{bmatrix} \nu\mathbf{k}\nu'\mathbf{k}' & \nu''\mathbf{k}''\theta \\ \sigma a\sigma' a' & 1a'' \end{bmatrix} \right\} , \quad \theta = 1, 2, \ldots, d \,, \tag{10-151}$$

where the component index for the vector U is $(\sigma a\sigma' a')$.

10.18.3. Relative phase of the CG coefficients

To ensure that the CG coefficients (10-146) with the same $\nu'', \mathbf{k}''$, and θ, but different a'' have the correct relative phase, we can use the following technique.

Suppose that we have found d orthogonal eigenvector $U^{(\nu'')\theta}_{\mathbf{k}'' a''}, \theta = 1, 2, \ldots d$, for a particular a''. To find the CG vector for the component b'', we search for an element $R_i = \{\gamma_i|\mathbf{v}_i\}'$ of $G'_{\mathbf{k}''}$ which has only one non-vanishing element in the a''-th column of the ν''-th irrep,

$$R_i \psi^{(\nu'')}_{\mathbf{k}'' a''} = \Delta_{a'' a''}(\gamma_i) \psi^{(\nu'')}_{\mathbf{k}'' a''} + \Delta_{b'' a''}(\gamma_i) \psi^{(\nu'')}_{\mathbf{k}'' b''} \,. \tag{10-152}$$

Then we have

$$U^{(\nu'')\theta}_{\mathbf{k}'' b''} = [\Delta_{b'' a''}(\gamma_i)]^{-1} [M(R_i) - \Delta_{a'' a''}(\gamma_i)] U^{(\nu'')\theta}_{\mathbf{k}'' a''} \,. \tag{10-153}$$

By choosing in each step appropriate R_i's which obey the condition (10-152), all the other eigenvectors $U^{(\nu'')}_{\mathbf{k}'' b''}$ can be found in turn from (10-153). In few cases for given a'' and b'' we cannot find an element satisfying (10-153). For such cases, we search for two elements R_i and R_j which only have two nonvanishing off diagonal elements in the a''-th column of ν''-th irrep, i.e.,

$$\begin{aligned} R_i \psi_{a''} &= \Delta_{a'' a''}(\gamma_i) \psi_{a''} + \Delta_{b'' a''}(\gamma_i) \psi_{b''} + \Delta_{c'' a''}(\gamma_i) \psi_{c''} \,, \\ R_j \psi_{a''} &= \Delta_{a'' a''}(\gamma_j) \psi_{a''} + \Delta_{b'' a''}(\gamma_j) \psi_{b''} + \Delta_{c'' a''}(\gamma_j) \psi_{c''} \,, \end{aligned} \tag{10-154}$$

where obvious short-hand notation is used. Solving the linear equations (10-154), we can find both $\psi_{b''}$ and $\psi_{c''}$ from $\psi_{a''}$. This process can be extended further, however in practice Eqs. (10-153) and (10-154) are sufficient.

10.18.4. The full CG coefficients of space groups

Up to now we have found only the CG coefficients $U^{(\nu'')\theta}_{\mathbf{k}''a''}$ corresponding to $\sigma''=1$. It follows from (10-106) that the CG coefficients

$$U^{(\nu'')\theta}_{\mathbf{k}''_{\sigma''}a''} = \left\{\begin{bmatrix} \nu\mathbf{k}\nu'\mathbf{k}' & \nu''\mathbf{k}''\theta \\ \sigma a\sigma' a' & \sigma'' a'' \end{bmatrix}\right\} \tag{10-155}$$

for $\sigma'' \neq 1$ can be found from $U^{(\nu'')\theta}_{\mathbf{k}''a''}$ by the formula

$$\begin{aligned} U^{(\nu'')\theta}_{\mathbf{k}''_{\sigma''}a''} &= M(\{\beta_{\sigma''}|\mathbf{V}_{\sigma''}\})U^{(\nu'')\theta}_{\mathbf{k}''a''} \\ &= D^{(*\mathbf{k})(\nu)}(\{\beta_{\sigma''}|\mathbf{V}_{\sigma''}\}) \otimes D^{(*\mathbf{k}')(\nu')}(\{\beta_{\sigma''}|\mathbf{V}_{\sigma''}\})U^{(\nu'')\theta}_{\mathbf{k}''a''} \,, \end{aligned} \tag{10-156}$$

where $\{\beta_{\sigma''}|\mathbf{V}_{\sigma''}\}$ is the coset representative of the space group $\mathbf{G}$ with regard to the little group $\mathbf{G}(\mathbf{k}'')$:

$$\mathbf{G} = \sum_{\sigma''=1}^{q''} \{\beta_{\sigma''}|\mathbf{V}_{\sigma''}\}\mathbf{G}(\mathbf{k}'') \,. \tag{10-157}$$

Equation (10-156) can be rewritten as

$$\begin{bmatrix} \nu\mathbf{k}\nu'\mathbf{k}' & \nu''\mathbf{k}''\theta \\ \tau b\tau' b' & \sigma'' a'' \end{bmatrix} = \sum_{aa'} D^{(*\mathbf{k})(\nu)}_{\tau b,\sigma a}(\{\beta_{\sigma''}|\mathbf{V}_{\sigma''}\}) D^{(*\mathbf{k}')(\nu')}_{\tau' b',\sigma' a'}(\{\beta_{\sigma''}|\mathbf{V}_{\sigma''}\}) \begin{bmatrix} \nu\mathbf{k}\nu'\mathbf{k}' & \nu''\mathbf{k}''\theta \\ \sigma a\sigma' a' & 1a'' \end{bmatrix} . \tag{10-158}$$

Observe that there is no summation over σ and σ' on the right-hand side of (10-158), since the index $\sigma(\sigma')$ is uniquely specified by $\tau(\tau')$ and $\{\beta_{\sigma''}|\mathbf{V}_{\sigma''}\}$, for which the submatrix

$$D^{(*\mathbf{k})(\nu)}_{(\tau\sigma)}(\{\beta_{\sigma''}|\mathbf{V}_{\sigma''}\})[D^{(*\mathbf{k}')(\nu')}_{(\tau'\sigma')}(\{\beta_{\sigma''}|\mathbf{V}_{\sigma''}\})]$$

does not vanish.

Some remarks:

The following remarks should be added in regard to the eigenfunction method for evaluating CG coefficients of space groups.

1. Here we only need to know *a priori* the wave-vector selection rule (which is easy to work out), but not the CG series.

2. For computer calculations, the CSCO-II of $\mathbf{G}'_{\mathbf{k}''}$ can be appropriately chosen so that it consists of only a single operator.

3. It seems at first sight that the order of the eigenequation (10-145b) equal to $(qh_\nu)\times(q'h_{\nu'})$. But actually, due to the wave-vector selection rule (10-140) the order of (10-145b) is much smaller.

10.18.5. Summary of the eigenfunction method for space group CG coefficients

The scheme for obtaining the CG coefficients of space groups can be summarized as follows.

1. Determine the wave-vector selection rule.

2. Pick out one star $*\mathbf{k}''$ among those for which the wave-vector reduction coefficients $(*\mathbf{k} * \mathbf{k}'| * \mathbf{k}'') \geq 1$ and the CG coefficients are to be calculated.

3. Choose the canonical wave vectors $\mathbf{k}, \mathbf{k}'$, and $\mathbf{k}''$.

4. For each of the canonical wave vectors, choose appropriate coset representatives of $\mathbf{G}$,

$$\begin{aligned} &\mathbf{G} = \sum_{\sigma} \oplus\{\beta_\sigma|\mathbf{V}_\sigma\}\mathbf{G}(\mathbf{k}) \,, \quad \mathbf{G} = \sum_{\sigma'} \oplus\{\beta_{\sigma'}|\mathbf{V}_{\sigma'}\}\mathbf{G}(\mathbf{k}') \,, \\ &\mathbf{G} = \sum_{\sigma''} \oplus\{\beta''|\mathbf{V}_{\sigma''}\}\mathbf{G}(\mathbf{k}'') \,, \end{aligned} \tag{10-159}$$

and find all the $\mathbf{k}$ points in the stars $*\mathbf{k}$ and $*\mathbf{k}'$ according to

$$\begin{aligned} \mathbf{k}_\sigma &= \beta_\sigma \mathbf{k} , \quad \sigma = 1,2,\dots,q , \\ \mathbf{k}'_{\sigma'} &= \beta_{\sigma'} \mathbf{k}', \quad \sigma' = 1,2,\dots,q' . \end{aligned} \tag{10-160}$$

To simplify the calculation, in the following we always choose as the coset representatives elements whose non-primitive translations are zero, i.e., we assume

$$\mathbf{V}_\sigma = \mathbf{V}_{\sigma'} = \mathbf{V}_{\sigma''} = 0 . \tag{10-161}$$

5. Determine all the index pairs $(\sigma\sigma')$ which satisfy

$$\mathbf{k}_\sigma + \mathbf{k}'_{\sigma'} = \mathbf{k}'' + \mathbf{K}_m . \tag{10-162}$$

For convenience in exposition, in the following we assume that there exist only two such pairs $(\sigma\sigma')$ and $(\tau\tau')$.

6. Using the eigenfunction method or consulting an existing table, e.g., the Kovalev (1961) table, find the irreps $D^{(\mathbf{k})(\nu)}$, $D^{(\mathbf{k}')(\nu')}$ and $D^{(\mathbf{k}'')(\nu'')}$ for the rep groups $\mathbf{G}'_{\mathbf{k}}, \mathbf{G}'_{\mathbf{k}'}$, and $\mathbf{G}'_{\mathbf{k}''}$, respectively.

7. Construct the irreducible matrices $D^{(*\mathbf{k})(\nu)}(\{\gamma''|\mathbf{V}(\gamma'')\}$ and $D^{(*\mathbf{k}')(\nu')}(\{\gamma''|\mathbf{V}(\gamma'')\})$ for the elements $\{\gamma''|\mathbf{V}(\gamma'')\}$ of the group $\mathbf{G}_{\mathbf{k}''}$, from which the CSCO-II of $\mathbf{G}_{\mathbf{k}''}$ is composed. However, only the submatrices related to the indices $\sigma, \tau, (\sigma', \tau')$, instead of the full matrices $D^{(*\mathbf{k})(\nu)}(D^{(*\mathbf{k}')(\nu')})$, are required to determine the CG coefficients, e.g.,

$$\left[D^{(*\mathbf{k})(\nu)}(\{\gamma''|\mathbf{V}(\gamma'')\})\right] = \begin{array}{c} \\ \tau \\ \sigma \end{array} \begin{array}{c} \begin{array}{cc} \tau & \sigma \end{array} \\ \begin{pmatrix} D_{(\tau\tau)} & D_{(\tau\sigma)} \\ D_{(\sigma\tau)} & D_{(\sigma\sigma)} \end{pmatrix} \end{array}$$

$$\left[D^{(*\mathbf{k}')(\nu')}(\{\gamma''|\mathbf{V}(\gamma'')\})\right] = \begin{array}{c} \\ \tau' \\ \sigma' \end{array} \begin{array}{c} \begin{array}{cc} \tau' & \sigma' \end{array} \\ \begin{pmatrix} D'_{(\tau'\tau')} & D'_{(\tau'\sigma')} \\ D'_{(\sigma'\tau')} & D'_{(\sigma'\sigma')} \end{pmatrix} \end{array} , \tag{10-163a}$$

where the large square brackets denote a submatrix, and

$$\begin{aligned} D_{(\tau\sigma)} &= \exp[-i\mathbf{k}_\tau \cdot \mathbf{V}(\gamma'')]\Delta(\beta_\tau^{-1}\gamma''\beta_\sigma) , \\ D'_{(\tau'\sigma')} &= \exp[-i\mathbf{k}'_{\tau'} \cdot \mathbf{V}(\gamma'')]\Delta(\beta_{\tau'}^{-1}\gamma''\beta_{\sigma'}) , \end{aligned} \tag{10-163b}$$

from Eq. (10-120).

8. With the help of Eq. (10-149b), construct the representation matrix of $\{\gamma''|\mathbf{V}(\gamma'')\}$ in the uncoupled rep,

$$M(\{\gamma''|\mathbf{V}(\gamma'')\}') = \begin{array}{c} \\ \tau\tau' \\ \sigma\sigma' \end{array} \begin{array}{c} \begin{array}{cc} \tau\tau' & \sigma\sigma' \end{array} \\ \begin{pmatrix} D_{(\tau\tau)} \otimes D'_{(\tau'\tau')} & D_{(\tau\sigma)} \otimes D'_{(\tau'\sigma')} \\ D_{(\sigma\tau)} \otimes D'_{(\sigma'\tau')} & D_{(\sigma\sigma)} \otimes D'_{(\sigma'\sigma')} \end{pmatrix} \end{array} . \tag{10-164a}$$

Then form the matrix

$$M(\{\gamma''|\mathbf{V}(\gamma'')\}') = \exp[i\mathbf{k}'' \cdot \mathbf{V}(\gamma'')] M(\{\gamma''|\mathbf{V}(\gamma'')\}) . \tag{10-164b}$$

The ordering for rows or columns in the matrix M is as follows:

$$|\tau 1\tau' 1\rangle \ldots |\tau 1\tau' h_{\nu'}\rangle, |\tau 2\tau' 1\rangle \ldots |\tau 2\tau' h_{\nu'}\rangle \ldots |\tau h_\nu \tau' h_{\nu'}\rangle \ ,$$
$$|\sigma 1\sigma' 1\rangle \ldots |\sigma 1\sigma' h_{\nu'}\rangle, |\sigma 2\sigma' 1\rangle \ldots |\sigma 2\sigma' h_{\nu'}\rangle \ldots |\sigma h_\nu \sigma' h_{\nu'}\rangle \ . \tag{10-165}$$

Suppose the class operator C_i is contained in the CSCO-II of $\mathbf{G}'_{\mathbf{k}''}$. By adding up the matrices $M(\{\gamma''|\mathbf{V}(\gamma'')\}')$ for all $\{\gamma''|\mathbf{V}(\gamma'')\}'$ belonging to the class i, we obtain the matrix $M(C_i)$. In this way we can obtain the matrices $M(C)$ and $M(C(s))$ of the CSCO-II of $\mathbf{G}'_{\mathbf{k}''}$.

9. By diagonalizing the matrices $M(C)$ and $M(C(s))$ simultaneously, we get the eigenvalues $(\nu''a'')$ and their degeneracies $d = (*\mathbf{k}\nu * \mathbf{k}'\nu' | * \mathbf{k}''\nu'')$. Or equivalently, we first diagonalize the matrix $M(C)$ and get the eigenvalues ν'' and the corresponding degeneracies $m_{\nu''}$. Then the reduction coefficient $(*\mathbf{k}\nu * \mathbf{k}'\nu' | * \mathbf{k}''\nu'') = m_{\nu''}/h_{\nu''}$; $h_{\nu''}$ is the dimension of the irrep $D^{(\mathbf{k}'')(\nu'')}$ of $\mathbf{G}'_{\mathbf{k}''}$.

For each ν'' and a specific but arbitrarily chosen a'', find the eigenvectors $U^{(\nu'')\theta}_{\mathbf{k}''a''}, \theta = 1, 2, \ldots, d$.

10. Using (10-154), obtain all the CG coefficients belonging to $\sigma'' = 1$.

11. Using (10-158), obtain the CG coefficients for $\sigma'' \neq 1$.

12. Pick out another star $*\mathbf{k}''$, choose its canonical wave vector $\mathbf{k}''$ and coset representatives $\{\beta_{\sigma''}|\mathbf{V}_{\sigma''}\}$, then return to step 5 and go through to the end.

10.19. Examples: Obtaining Space Group Clebsch-Gordan Coefficients*

In this section, following the prescription in Sec. 10.18 step by step, we shall give two examples of calculations for the CG coefficients of the non-symmorphic space group O_h^7.

10.19.1. The CG coefficients of O_h^7 for $*X(1) \otimes *X(2) \to *X(\nu'')$

1. The wave-vector selection rule is easily found:

$$*X \otimes *X = 3\Gamma + 2 * X \ . \tag{10-166}$$

2. Pick out $*\mathbf{k}'' = *X$.

3. The canonical wave vectors are chosen as

$$\mathbf{k} = \mathbf{k}' = \mathbf{k}'' = \mathbf{k}_z = \left(\frac{1}{2}, \frac{1}{2}, 0\right) \ , \tag{10-167}$$

with $\mathbf{b}_1, \mathbf{b}_2, \mathbf{b}_3$ as basis.

Notice that the choice in (10-167) is different from that given by Berenson, (1975). They chose

$$\mathbf{k} = \mathbf{k}_x, \quad \mathbf{k}' = \mathbf{k}_y, \quad \mathbf{k}'' = \mathbf{k}_z \ . \tag{10-168}$$

4. The coset representatives of O_h^7 with respect to the little group $\mathbf{G}(X)$ are chosen to be

$$\{\beta_2|\mathbf{V}_2\} = \{C_{31}^+|0\} \ , \quad \{\beta_3|\mathbf{V}_3\} = \{C_{31}^-|0\} \ . \tag{10-169}$$

Under the rotation C_{31}^+, the wave vectors $\mathbf{k}_x, \mathbf{k}_y$ and $\mathbf{k}_z$ transform among themselves cyclically.

The star X contains three $\mathbf{k}$ points, $\mathbf{k}_\sigma = \beta_\sigma \mathbf{k}$,

$$\mathbf{k}_1 = \mathbf{k} = \mathbf{k}_z = \left(\frac{1}{2}, \frac{1}{2}, 0\right) \ , \quad \mathbf{k}_2 = \mathbf{k}_x = \left(0, \frac{1}{2}, \frac{1}{2}\right) \ , \quad \mathbf{k}_3 = \mathbf{k}_y = \left(\frac{1}{2}, 0, \frac{1}{2}\right) \ . \tag{10-170}$$

5. From (10-170) it is readily seen that

$$\mathbf{k}_x + \mathbf{k}_y = \mathbf{k}_z + (001) \ . \tag{10-171}$$

Hence the index pairs are $(\tau\tau') = (23)$ and $(\sigma\sigma') = (32)$.

6. The ray or projective irreps $\Delta^{X_i}(\gamma)$ for the little co-group $\mathbf{G}_0(\mathbf{X})$ are given in Kovalev (1961), Table T159. We list them here in Table 10.19-1. In Table 10.19-1 and in the following we often use the symbols

$$\varepsilon = \begin{bmatrix} 1 & 0 \\ 0 & 1 \end{bmatrix}, \quad \lambda = \begin{bmatrix} 1 & 0 \\ 0 & -1 \end{bmatrix}, \quad \varphi = \begin{bmatrix} 0 & 1 \\ 1 & 0 \end{bmatrix}, \quad \kappa = \begin{bmatrix} 0 & 1 \\ -1 & 0 \end{bmatrix}, \mu = \begin{bmatrix} 1 & 0 \\ 0 & i \end{bmatrix}. \tag{10-172}$$

7. From Eq.(10-101) we know that the CSCO-I of the group $\mathbf{G}'_X$ is

$$C = (C_1, C_2), \quad C_1 = \{C_{2a}|\tau\}' - \{C_{2b}|\tau\}', C_2 = \{\sigma_{da}|0\} + \{\sigma_{ab}|0\}. \tag{10-173}$$

The eigenvalues of (C_1, C_2) are listed in Table 10.19-1. Furthermore, from this table it is seen that apart from the identity the only elements whose matrices are always diagonalized for the irreps $X_1 - X_4$ are $\{C_{4z}^+|\boldsymbol{\tau}\}'$ and $\{C_{4z}^-|\tau\}'$. Hence either $\{C_{4z}^+|\tau\}'$ or $\{C_{4z}^-|\tau\}'$ can be taken as the operator $C(s)$, whose eigenvalues are used to distinguish the two basis vectors of the irreps $X_1 - X_4$. We choose

$$C(s) = \{C_{4z}^+|\tau\}'. \tag{10-174}$$

The eigenvalues of $C(s)$ listed in Table 10.19-1 are inferred from the diagonal elements of the matrices $\Delta^{X_j}(C_{4z}^+)$.

Using (10-114) or (10-163) and Table 10.19-1 we can easily construct the space-group rep $D^{(*X)(\nu)}$. For example,

$$\mathcal{D}(\{C_{4z}^+|\boldsymbol{\tau}\}) = \begin{matrix} & \varepsilon & C_{31}^+ & C_{31}^- \\ \varepsilon \\ C_{31}^- \\ C_{31}^+ \end{matrix} \begin{pmatrix} \Delta(C_{4z}^+) & 0 & 0 \\ 0 & 0 & \Delta(C_{2a}) \\ 0 & \Delta(C_{4z}^-) & 0 \end{pmatrix}, \tag{10-175a}$$

$$\begin{aligned} D^{(*X)(j)}(\{C_{4z}^+|\boldsymbol{\tau}\}) &= -i\mathcal{D}(\{C_{4z}^+|\tau\}) \\ &= \begin{bmatrix} -\lambda i & 0 & 0 \\ 0 & 0 & -\kappa i \\ 0 & -\lambda i & 0 \end{bmatrix}, \quad j = 1, 2. \end{aligned} \tag{10-175b}$$

Similarly we have

$$D^{(*X)(j)}(\{C_{2a}|\tau\}) = \begin{bmatrix} -\kappa i & 0 & 0 \\ 0 & 0 & -\lambda i \\ 0 & -\lambda i & 0 \end{bmatrix},$$

$$D^{(*X)(j)}(\{C_{2b}|\tau\}) = \begin{bmatrix} -\kappa i & 0 & 0 \\ 0 & 0 & -\kappa i \\ 0 & -\kappa i & 0 \end{bmatrix}, \tag{10-175c}$$

$$D^{(*X)(j)}(\{\sigma_{da}|0\}) = (-1)^{j+1} \begin{bmatrix} \varepsilon & 0 & 0 \\ 0 & 0 & \varphi \\ 0 & \varphi & 0 \end{bmatrix},$$

$$D^{(*X)(j)}(\{\sigma_{db}|0\}) = (-1)^{j+1} \begin{bmatrix} \varepsilon & 0 & 0 \\ 0 & 0 & \varepsilon \\ 0 & \varepsilon & 0 \end{bmatrix}, \quad j = 1, 2.$$

8. From (10-164) and (10-175) we have

$$M(C_1) = M(\{C_{2a}|\tau\}') - M(\{C_{2b}|\tau\}') = \begin{matrix} & 23 & 32 \\ 23 & & \begin{matrix} -\lambda & \kappa \\ -\kappa & \lambda \end{matrix} \\ 32 & \begin{matrix} -\lambda & \kappa \\ -\kappa & \lambda \end{matrix} & \end{matrix} \times (i) , \tag{10-176a}$$

$$M(C_2) = -\begin{bmatrix} & & \varepsilon & \varphi \\ & & \varphi & \varepsilon \\ \varepsilon & \varphi & & \\ \varphi & \varepsilon & & \end{bmatrix} , \quad M(C(s)) = i\begin{bmatrix} & & 0 & \lambda \\ & & -\lambda & 0 \\ \kappa & 0 & & \\ 0 & -\kappa & & \end{bmatrix} , \tag{10-176b}$$

$$M(\{C_{2a}|\tau\}') = i\begin{bmatrix} & & -\lambda & 0 \\ & & 0 & \lambda \\ -\lambda & 0 & & \\ 0 & \lambda & & \end{bmatrix} , \quad M(\{\sigma_a|0\}) = \begin{bmatrix} & & 0 & \varphi \\ & & \varphi & 0 \\ 0 & \varphi & & \\ \varphi & 0 & & \end{bmatrix} . \tag{10-176c}$$

By setting $(\tau\tau') = (23), (\sigma\sigma') = (32)$, and $h_\nu = h_{\nu'} = 2$ in (10-165), we obtain the ordering of the product basis vectors

$$\begin{matrix} |x1y1\rangle , & |x1y2\rangle , & |x2y1\rangle , & |x2y2\rangle , \\ |y1x1\rangle , & |y1x2\rangle , & |y2x1\rangle , & |y2x2\rangle , \end{matrix} \tag{10-177}$$

where $|xayb\rangle$ represents the vector $|\psi^{(X_1)}_{\mathbf{k}_x a}\psi^{(X_2)}_{\mathbf{k}_y b}\rangle$.

9. The eigenvalues of $M(C_1)$ and $M(C_2)$ are found to be

$$(C_1, C_2) = (0,2), \quad (0,-2), \quad (2i,0), \quad (-2i,0) ,$$

and each has the degeneracy 2. By comparing with Table 10.19-1, we get the CG series

$$*X(1) \otimes *X(2) = *X(1) \oplus *X(2) \oplus *X(3) \oplus X(4) . \tag{10-178}$$

The eigenvalues of $M(C(s))$ are found to be $\pm 1, \pm i$. Substituting the four sets of eigenvalues of $(C_1, C_2, C(s))$ — $(0,2,1), (0,-2,1), (2i,0,-i)$, and $(-2i,0,-i)$ – into the eigenequation (10-145), we obtain the four eigenvectors listed in the odd rows of Table 10.19-2. They are the CG coefficients for the first component $a'' = 1$.

10. On the basis of (10-154) and Table 10.19-1 the second component of the CG coefficients can be expressed as

$$\begin{aligned} U^{(X_j)}_{\mathbf{k}_z 2} &= \frac{M(\{C_{2a}|\tau\}')}{\Delta^{(X_j)}_{21}(C_{2a})} U^{(X_j)}_{\mathbf{k}_z 1} \\ &= -M(\{C_{2a}|\tau\}') U^{(X_j)}_{\mathbf{k}_z 1} , \quad j = 1,2, \\ U^{(X_j)}_{\mathbf{k}_z 2} &= \frac{M(\{\sigma_{da}|0\})}{\Delta^{(X_j)}_{21}(\sigma_{da})} U^{(X_j)}_{\mathbf{k}_z 1} \\ &= i(-1)^j M(\{\sigma_{da}|0\}) U^{(X_j)}_{\mathbf{k}_z 1} , \quad j = 3,4 . \end{aligned} \tag{10-179}$$

With (10-179) and (10-176c), as well as the odd rows in Table 10.19-2 we can find $U^{(X_j)}_{\mathbf{k}_z 2}$, e.g.,

$$U^{(X_3)}_{\mathbf{k}_z 2} = \frac{i}{\sqrt{8}} \begin{bmatrix} & & & & & & & 1 \\ & & & & & & 1 & \\ & & & & & 1 & & \\ & & & & 1 & & & \\ & & & 1 & & & & \\ & & 1 & & & & & \\ & 1 & & & & & & \\ 1 & & & & & & & \end{bmatrix} \begin{bmatrix} 1 \\ -1 \\ 1 \\ -1 \\ -1 \\ -1 \\ 1 \\ 1 \end{bmatrix} = \frac{1}{\sqrt{8}} \begin{bmatrix} i \\ i \\ -i \\ -i \\ -i \\ i \\ -i \\ i \end{bmatrix} \tag{10-180}$$

The CG coefficients $U^{(X_j)}_{\mathbf{k}_z 2}$ are listed in the even rows of Table 10.19-2.

11. In order to obtain the CG coefficients with $\sigma'' \neq 1$, we in general need to use (10-158). However, for the case where

$$\begin{aligned} &* \mathbf{k} = *\mathbf{k}' = *\mathbf{k}'' , \\ &\{\beta_\sigma|\mathbf{V}_\sigma\} = \{\beta_{\sigma'}|\mathbf{V}_{\sigma'}\} = \{\beta_{\sigma''}|\mathbf{V}_{\sigma''}\} , \end{aligned} \tag{10-181}$$

it is a very simple matter to obtain the CG coefficients with $\sigma'' \neq 1$ from those with $\sigma'' = 1$. For example, in our case here for $\sigma'' = 2, \{\beta_{\sigma''}|\mathbf{V}_{\sigma''}\} = C^+_{31}$,

$$\begin{aligned} \psi^{X(1)}_{\mathbf{k}_x} &= C^+_{31}\psi^{X(1)}_{\mathbf{k}_z} \\ &= \left(\frac{1}{8}\right)^{1/2} (|y1z1\rangle - i|y1z2\rangle - i|y2z1\rangle + |y2z2\rangle - |z1y1\rangle + i|z1y2\rangle + i|z2y1\rangle + |z2y2\rangle) . \end{aligned}$$

Hence the CG coefficients for $\sigma'' = 2$ and 3 result from a cyclic permutation of x, y, and z in Table 10-19-2.

Similarly, by letting $*\mathbf{k} = \Gamma$, we can get the CG coefficients for $*X(1) \otimes *X(2) \to \Gamma(\nu'')$.

10.19.2. The CG coefficients of O_h^7 for $*X(1) \otimes *W(1) \to *\Delta(\nu'')$

We now turn to a more general example for which the three wave-vector stars $*\mathbf{k}, *\mathbf{k}'$, and $*\mathbf{k}''$ are all different and the multiplicity may be larger than one.

1. The wave-vector selection rule is given in (10-143c) i.e.,

$$*X \otimes *W = 2 * \Delta + *W . \tag{10-182a}$$

2. Pick out $*\mathbf{k}'' = *\Delta$.
3. The canonical wave vectors are chosen to be

$$\begin{array}{ccc} \text{Star } X & \text{Star } W & \text{Star } \Delta \\ \mathbf{k} = (\frac{1}{2}, \frac{1}{2}, 0) , & \mathbf{k}' = (\frac{1}{2}, \frac{1}{4}, \frac{3}{4}) , & \mathbf{k}'' = (\frac{1}{4}, \frac{1}{4}, 0) . \end{array} \tag{10-182b}$$

4. The coset representatives and $\mathbf{k}_\sigma$ for star X are identical with (10-169) and (10-170), while the coset representatives $\{\beta_{\sigma'}|\mathbf{V}_{\sigma'}\}$ and $\mathbf{k}'_{\sigma'}$ for star W are listed in Table 10.19-3.

It should be noted that although the wave vector, e.g., $\mathbf{k}' = (\frac{1}{2}, -\frac{1}{4}, \frac{1}{4})$, is equivalent to $(\frac{1}{2}, \frac{3}{4}, \frac{1}{4})$, the former is *not replaceable* by the latter; otherwise errors will be incurred.

5. From

$$\mathbf{k}_2 + \mathbf{k}'_5 = \mathbf{k}'' + (011) , \quad \mathbf{k}_3 + \mathbf{k}'_6 = \mathbf{k}'' + (001) , \tag{10-183}$$

it is seen that the index pairs are $(\tau\tau') = (25)$ and $(\sigma\sigma') = (36)$.

6. The irreps $D^{(X)(\nu)}$ and $D^{(W)(\nu')}$ are given in Tables 10.19-1 and 10.15-5 respectively.

Table 10.19-1. The ray irreps $\Delta^{X_i}(\gamma)$ for the little co-group $\mathbf{G}_0(\mathbf{X})$, i.e., the irreps of the rep group $\mathbf{G}'_X$. $\Delta^{X_2} = \delta^{(2)}\Delta^{X_1}$, $\Delta^{X_4} = \delta^{(4)}\Delta^{X_3}$; $X_1 \to \hat{\tau}^3$, $X_2 \to \hat{\tau}^4$, $X_3 \to \hat{\tau}^2$, $X_4 \to \hat{\tau}^1$, where $\hat{\tau}^i$ are the labels used by Kovalev. From Kovalev (1961).

	(C_1C_2)	$C(s)$		ε	C_{2y}	C_{2z}	C_{2x}	C_{4z}^-	C_{4z}^+	C_{2a}	C_{2b}	I	σ_y	σ_z	σ_x	S_{4z}^+	S_{4z}^-	σ_{da}	σ_{db}
X_1	(0, 2)	$\begin{pmatrix}1\\-1\end{pmatrix}$		ε	φ	ε	φ	λ	λ	κ	κ	κ	λ	κ	λ	φ	φ	ε	ε
X_2	(0, −2)	$\begin{pmatrix}1\\-1\end{pmatrix}$	$\delta^{(2)}$	+	+	+	+	+	+	+	+	−	−	−	−	−	−	−	−
X_3	(2i, 0)	$\begin{pmatrix}-i\\i\end{pmatrix}$		ε	λ	$-\varepsilon$	$-\lambda$	λi	$-\lambda i$	εi	$-\varepsilon i$	κ	$-\varphi$	$-\kappa$	φ	φi	$-\varphi i$	$-\kappa i$	κi
X_4	(−2i, 0)	$\begin{pmatrix}-i\\i\end{pmatrix}$	$\delta^{(4)}$	+	−	+	−	+	+	−	−	+	−	+	−	+	+	−	−

Table 10.19-2. The CG coefficients $\left(\begin{smallmatrix}\nu\mathbf{k}\,\nu'\mathbf{k}'\\ \sigma a\sigma' a'\end{smallmatrix}\middle|\begin{smallmatrix}X(j)\\ k_z a''\end{smallmatrix}\right)$ of O_h^7 for $*X(1) \otimes *X(2)$ and $\sigma'' = 1$. $xayb$ signifies $\mathbf{k}_x a \mathbf{k}_y b$, etc.

	(C_1C_2)	$C(s)$	N	$x1y1$	$x1y2$	$x2y1$	$x2y2$	$y1x1$	$y1x2$	$y2x1$	$y2x2$
$X(1)$	(0, 2)	1	$(\frac{1}{8})^{2/2}$	1	$-i$	$-i$	1	−1	i	i	−1
		−1	$(\frac{1}{8})^{1/2}$	$-i$	1	1	$-i$	i	−1	−1	i
$X(2)$	(0, −2)	1	$(\frac{1}{8})^{1/2}$	1	i	i	1	1	i	i	1
		−1	$(\frac{1}{8})^{1/2}$	i	1	1	i	i	1	1	i
$X(3)$	(2i, 0)	$-i$	$(\frac{1}{8})^{1/2}$	1	−1	1	−1	−1	−1	1	1
		i	$(\frac{1}{8})^{1/2}$	i	i	$-i$	$-i$	$-i$	i	$-i$	i
$X(4)$	(−2i, 0)	$-i$	$(\frac{1}{8})^{1/2}$	1	1	−1	−1	1	−1	1	−1
		i	$(\frac{1}{8})^{1/2}$	i	$-i$	i	$-i$	i	i	$-i$	$-i$

7. Δ is an internal point, hence its representation group $\mathbf{G}'_\Delta$ is isomorphic to the little co-group, the point group C_{4v}. From Table 8.3-6, we know that the CSCO-I of $C_{4\nu}$ is $C = (2C_{4z}, 2\sigma)$. We choose $C(s) = \sigma_y$; this corresponds to choosing (x, y) as the basis vectors of the two-dimensional irrep of C_{4v}. Due to the isomorphism between $\mathbf{G}'_\Delta$ and C_{4v}, the CSCO-I of $\mathbf{G}'_\Delta$ is

$$C = (C_3, C_4), \quad C_3 = \{C_{4z}^+|\tau\}' + \{C_{4z}^-|\tau'\} , \quad C_4 = \{\sigma_x|\tau\}' + \{\sigma_y|\tau\}' , \tag{10-184}$$

and the operator $C(s)$ is

$$C(s) = \{\sigma_y|\tau\}' . \tag{10-185}$$

By means of (10-163) and Table 10.19-1 and 10.15-5 we can construct the following submatrices:

$$\begin{aligned}
&\left[D^{(*X)(1)}(\{C_{4z}^+|\tau\})\right] = \begin{matrix} & \begin{matrix} 2 & 3 \end{matrix} \\ \begin{matrix} 2 \\ 3 \end{matrix} & \begin{pmatrix} 0 & \kappa \\ \lambda & 0 \end{pmatrix} \end{matrix} (-i) , \\
&\left[D^{(*X)(1)}(\{C_{4z}^-|\tau\})\right] = -i \begin{pmatrix} 0 & \lambda \\ \kappa & 0 \end{pmatrix} . \\
&\left[D^{(*X)(1)}(\{\sigma_x|\tau\})\right] = -i \begin{pmatrix} \kappa & 0 \\ 0 & \lambda \end{pmatrix} , \quad \left[D^{(*X)(1)}(\{\sigma_y|\tau\})\right] = -i \begin{pmatrix} \lambda & 0 \\ 0 & \kappa \end{pmatrix} ,
\end{aligned} \tag{10-186}$$

$$\begin{aligned}
&\left[D^{(*W)(1)}(\{C_{4z}^+|\tau\})\right] = \begin{matrix} & \begin{matrix} 5 & 6 \end{matrix} \\ \begin{matrix} 5 \\ 6 \end{matrix} & \begin{pmatrix} 0 & -\mu i \\ \mu^* & 0 \end{pmatrix} \end{matrix} e^{-\pi i/4} , \\
&\left[D^{(*W)(1)}(\{C_{4z}^-|\tau\})\right] = e^{-\pi i/4} \begin{pmatrix} 0 & -\mu^* i \\ \mu & 0 \end{pmatrix} , \\
&\left[D^{(*W)(1)}(\{\sigma_x|\tau\})\right] = e^{-\pi i/4} \begin{pmatrix} \kappa i & 0 \\ 0 & -\kappa \end{pmatrix} , \quad \left[D^{(*W)(1)}(\{\sigma_y|\tau\})\right] = e^{-\pi i/4} \begin{pmatrix} -\varphi i & 0 \\ 0 & \varphi \end{pmatrix} .
\end{aligned}$$

The 2×2 matrices $\kappa, \lambda, \ldots$, are defined in Eq. (10-172).

8. Using (10-164) and (10-186), we obtain

$$\begin{aligned}
&M(\{C_{4z}^+|\tau\}') = \begin{matrix} & \begin{matrix} 25 & & 36 \end{matrix} \\ \begin{matrix} 25 \\ 36 \end{matrix} & \begin{pmatrix} & \begin{pmatrix} 0 & -1 \\ 1 & 0 \end{pmatrix} \mu \\ \begin{pmatrix} -i & 0 \\ 0 & i \end{pmatrix} \mu^* & \end{pmatrix} \end{matrix} , \\
&M(\{C_{4z}^-|\tau\}') = \begin{pmatrix} & \begin{pmatrix} -1 & 0 \\ 0 & 1 \end{pmatrix} \mu^* \\ \begin{pmatrix} 0 & -i \\ i & 0 \end{pmatrix} \mu & \end{pmatrix} . \\
&M(\{\sigma_x|\tau\}') = \begin{pmatrix} \begin{pmatrix} 0 & 1 \\ -1 & 0 \end{pmatrix} \kappa & \\ & \begin{pmatrix} i & 0 \\ 0 & -i \end{pmatrix} \kappa \end{pmatrix} , \\
&M(\{\sigma_y|\tau\}') = \begin{pmatrix} \begin{pmatrix} -1 & 0 \\ 0 & 1 \end{pmatrix} \varphi & \\ & \begin{pmatrix} 0 & -i \\ i & 0 \end{pmatrix} \varphi \end{pmatrix} .
\end{aligned} \tag{10-187}$$

By setting $(\tau\tau') = (25)$, $(\sigma\sigma') = (36)$, and $h_\nu = h_{\nu'} = 2$ in Eq. (10-165), we get the ordering for the eight product basis vectors:

$$|2151\rangle\ ,|2152\rangle\ ,|2251\rangle\ ,|2252\rangle\ ,$$
$$|3161\rangle\ ,|3162\rangle\ ,|3261\rangle\ ,|3262\rangle\ . \qquad (10\text{-}188)$$

9. The eigenvalues of the matrices $M(C_3) = M(\{C_{4z}^+|\tau\}') + M(\{C_{4z}^-|\tau\}')$ and $M(C_4) = M(\{\sigma_x|\tau\}') + M(\{\sigma_y|\tau\}')$ are found to be

$$\begin{aligned} \text{Single roots: } & (2,2),(2,-2),(-2,2),(-2,-2)\ , \\ \text{Quartet root: } & (0,0)\ , \end{aligned} \qquad (10\text{-}189)$$

which corresponds, according to Table 8.3-6, to the point group reps A_1, A_2, B_1, B_2, and E, respectively. The first four are one dimensional, and the last one is two dimensional. We use $\Delta(1) - \Delta(5)$ to denote the corresponding reps for the group $\mathbf{G}'_\Delta$. The degeneracy 4 divided by the dimensionality 2 gives the number of times that the irrep $\Delta(5)$ occurs in the rep $*X(1) \otimes *W(1)$. Therefore, we have the CG series:

$$*X(1) \otimes *W(1) = *\Delta(1) \oplus *\Delta(2) \oplus *\Delta(3) \oplus *\Delta(4) \oplus 2 * \Delta(5)\ . \qquad (10\text{-}190)$$

For the first four single roots, we can immediately find the corresponding eigenvectors, as listed in the first four rows of Table 10.19-4. For the quartet root (0, 0), a further diagonalization of the matrix $M(C(s))$ is required. Its eigenvalues $a'' = 1$ and -1 correspond to the first and second components of the rep $\Delta(5)$, respectively. When we substitute the eigenvalues (0, 0, 1) of $(C_3, C_4, C(s))$ into their eigenequations, a degeneracy of 2 remains. Consequently there exist two linearly independent solutions. The two orthogonal eigenvectors can be chosen as

$$\begin{aligned} U_{\mathbf{k}'',1}^{(\theta=1)} &= \frac{1}{2}(1,-1,-1,-1,0,0,0,0)\ , \\ U_{\mathbf{k}'',1}^{(\theta=2)} &= \frac{1}{2}(0,0,0,0,1,i,-1,i)\ . \end{aligned} \qquad (10\text{-}191)$$

10. It is known that the matrix for the element C_{4z} in the rep E of C_{4v} is

$$D^{(E)}(C_{4z}^+) = \begin{bmatrix} 0 & -1 \\ 1 & 0 \end{bmatrix}\ . \qquad (10\text{-}192)$$

Hence (10-154) now reads

$$U_{\mathbf{k}''2}^{(\theta)} = \frac{1}{D_{21}^{(E)}(C_{4z}^+)} M(\{C_{4z}^+|\tau\}') U_{\mathbf{k}''1}^{(\theta)}\ . \qquad (10\text{-}193)$$

From (10-187) and (10-191)-(10-193) we can calculate $U_{\mathbf{k}''2}^{(\theta)}$; the result is listed in Table 10.19-4.

11. Using (10-158), we can get the CG coefficients with $\sigma \neq 1$ which are not listed here.

10.20. The Double Space Groups

So far we have only considered the single-valued reps of space groups. We now turn to the double-valued reps of space groups.

Corresponding to the space group $\mathbf{G}$ of (10-19b), we have the double space group $\mathbf{G}^\dagger$,

$$\mathbf{G}^\dagger = \{\{\alpha|\mathbf{V}(\alpha) + \mathbf{R}_n\}:\quad \alpha \in \mathbf{G}_0^\dagger,\quad \mathbf{R}_n \in L\}\ , \qquad (10\text{-}194)$$

Table 10.19-3. The coset representatives $\{\beta_{\sigma'}|\mathbf{V}_{\sigma'}\}$ and $\mathbf{k}'_{\sigma'}$ for star W.

σ'	1	2	3	4	5	6
$\{\beta_{\sigma'}\|\mathbf{V}_{\sigma'}\}$	$\{\varepsilon\|0\}$	$\{C_{2y}\|0\}$	$\{C_{31}^{+}\|0\}$	$\{C_{33}^{+}\|0\}$	$\{C_{31}^{-}\|0\}$	$\{C_{32}^{-}\|0\}$
$\mathbf{k}'_{\sigma'}$	$(\frac{1}{2}\frac{1}{4}\frac{3}{4})$	$(\frac{1}{2}-\frac{1}{4}\frac{1}{4})$	$(\frac{3}{4}\frac{1}{2}\frac{1}{4})$	$(\frac{1}{4}\frac{1}{2}-\frac{1}{4})$	$(\frac{1}{4}\frac{3}{4}\frac{1}{2})$	$(-\frac{1}{4}\frac{1}{4}\frac{1}{2})$

Table 10.19-4. The CG coefficients $\left(\begin{smallmatrix}X(1) & W(1) \\ \sigma a & \sigma' a'\end{smallmatrix}\middle|\begin{smallmatrix}\Delta(\nu'')\theta \\ 1a''\end{smallmatrix}\right)$ of O_h^7 for $*X(1) \otimes *W(1)$ and $\sigma'' = 1$.

						$\sigma a \sigma' a'$							
		(C_3, C_4)	$C(s)$	θ	N	$\|2151\rangle$	$\|2152\rangle$	$\|2251\rangle$	$\|2252\rangle$	$\|3161\rangle$	$\|3162\rangle$	$\|3261\rangle$	$\|3262\rangle$
$\Delta(1)$	A_1	$(2, 2)$			$(\frac{1}{8})^{1/2}$	1	-1	1	1	1	$-i$	1	i
$\Delta(2)$	A_2	$(2, -2)$			$(\frac{1}{8})^{1/2}$	1	1	1	-1	-1	$-i$	-1	i
$\Delta(3)$	B_1	$(-2, 2)$			$(\frac{1}{2})^{1/2}$	1	-1	1	1	-1	i	-1	$-i$
$\Delta(4)$	B_2	$(-2, -2)$			$(\frac{1}{8})^{1/2}$	1	1	1	-1	1	i	1	$-i$
$\Delta(5)$	E	$(0, 0)$	1	1	$\frac{1}{2}$	1	-1	-1	-1				
			-1	1	$\frac{1}{2}$					1	$-i$	-1	$-i$
$\Delta(5)$	E	$(0, 0)$	1	2	$\frac{1}{2}$					1	i	-1	i
			-1	2	$\frac{1}{2}$	1	1	-1	1				

where $\mathbf{G}_0^\dagger$ is the double point group defined by (8-105b). Note that

$$\mathbf{V}(\tilde{\alpha}) = \mathbf{V}(\alpha) , \quad \tilde{\alpha}\mathbf{V}(\beta) = \alpha\mathbf{V}(\beta) . \tag{10-195}$$

In the vector (single-valued) rep space, the representation of $\mathbf{G}^\dagger$ forms a group which is isomorphic to $\mathbf{G}$. In the spinor (double-valued) rep space, the representation of $\mathbf{G}^\dagger$ forms a rep group, denoted by

$$G^\dagger = \{\{\alpha|\mathbf{V}(\alpha) + \mathbf{R}_n\} : \alpha \in G_0^\dagger, \mathbf{R}_n \in L) , \tag{10-196}$$

where $G_0^\dagger$ is the rep group of the double point group $\mathbf{G}_0^\dagger$. Among the elements of the rep group $G^\dagger$, only half the elements are linearly independent, and they obey the multiplication rule

$$\{\alpha_i|\mathbf{V}(\alpha_i)\}\{\alpha_j|\mathbf{V}(\alpha_j)\} = \theta(i,j)\{\alpha_{ij}|\alpha_i\mathbf{V}(\alpha_j) + \mathbf{V}(\alpha_i)\} , \tag{10-197}$$

where the factor system $\theta(i,j)$ is defined by (8-108c), and can be found in Table 10-.21-1 for the group O and its subgroups.

The double little group is

$$\mathbf{G}(\mathbf{k})^\dagger = \{\{\gamma_i|\mathbf{V}(\gamma_i) + \mathbf{R}_n\} : \gamma_i \in \mathbf{G}_0(\mathbf{k})^\dagger, \quad \mathbf{R}_n \in L\} . \tag{10-198}$$

In spinor rep space and the eigenspace $\mathcal{L}_\mathbf{k}$ defined by (10-54b), the representation of the double little group form the "double rep group" $G'_\mathbf{k}$ of order $2m|G_0(\mathbf{k})|$ [cf. Eq. (10-70)],

$$\begin{aligned} G'_\mathbf{k} &= \{R_i : i = 1, 2, \dots, |\mathbf{G}_0(\mathbf{k})|\}_{2m} , \\ R_i &= \{\gamma_i|\mathbf{V}(\gamma_i)\}' = \exp[i\mathbf{k}\cdot\mathbf{V}(\gamma_i)]\,\{\gamma_i|\mathbf{V}(\gamma_i)\} . \end{aligned} \tag{10-199}$$

The multiplication relation of the "double rep group" is

$$R_i R_j = \theta(i,j)\eta(i,j)R_{ij} , \tag{10-200}$$

where the factor system $\eta(i,j)$ is still given by Eq. (10-67d) or (10-80a).

Equation (10-200) has the same form as (10-67a). Consequently, the problem of finding the double-valued irreps of space groups is also as easy as that of finding irreps of point groups with maximal order equal to 48, while with the conventional method (Bradley 1972), one needs to deal with a group of order $2m|\mathbf{G}_0(\mathbf{k})|$, which can be as high as $(2)(4)(48) = 384$. Therefore, the rep group approach to the double-valued reps of space groups drastically simplifies the problem.

The procedure for constructing the full double-valued rep matrices $D^{(*\mathbf{k})(\nu)}$ of a space group is the same as that for the single-valued rep matrices with the following modifications: In constructing the matrice $\mathcal{D}(\{\alpha|\mathbf{a}\})$ of (10-118), the product $\beta_\tau^{-1}\alpha\beta_\sigma$ should be calculated with the *double-point group multiplication rule*, and the $(\tau\sigma)$ block in (10-118) is again equal to $\Delta(\gamma)$ if $\beta_\tau^{-1}\alpha\beta_\sigma = \gamma$ belongs to the *double point group* $\mathbf{G}_0(\mathbf{k})^\dagger$, and $\Delta(-\gamma) = -\Delta(\gamma)$.

The technique for computing the CG coefficients for the double space group follows the discussion of Sec. 10.18. The only point worth mentioning is that the product of two spinor reps is a vector rep, while the product of a spinor rep and a vector rep is a spinor rep.

Example: Find the double-valued irreps of D_{4h}^{14} at the point Y. The group elements of D_{4h}^{14} are given in Sec. 10.6. The wave vector for the point Y and its symmetric group can be found from Table 3.6 in Bradley (1972, p. 116): $\mathbf{k}_y = (\alpha\frac{1}{2}0)$, and $\mathbf{G}_0(Y) = C_{2v} = (\varepsilon, C_{2x}, \sigma_y, \sigma_z)$.

The group table for the rep group G'_Y with the factor system $\eta(i,j)$ is given in Table 10.20-1 [where $\mathbf{v} = (\frac{1}{2}\frac{1}{2}\frac{1}{2})$]. For the double point group $C_{2v}^\dagger$, the group table can be found from Table 10.21-1, and is given in Table 10.20-2. Combining Tables 10.20-1 and 10.20-2, we obtain the group table for the "double rep group" $G_Y^{\dagger}$, shown in Table 10.20-3. Choosing $C(s) = R_2$, we can easily find the double-valued irreps $\Delta^{(Y)}$ of D_{4h}^{14}, listed in Table 10.20-3.

Table 10.20-1. Group table of G_Y'.

$\{\varepsilon\|0\}$	$\{C_{2x}\|\mathbf{v}\}'$	$\{\sigma_y\|\mathbf{v}\}'$	$\{\sigma_z\|0\}$
1	2	3	4
2	−1	−4	3
3	−4	−1	2
4	3	2	1

Table 10.20-2. Group table of $C_{2v}^{\dagger}$.

ε	C_{2x}	σ_y	σ_z
1	2	3	4
2	−1	4	−3
3	−4	−1	2
4	3	−2	−1

Table 10.20-3. Group table of $G_Y'^{\dagger}$ and double-valued rep of D_{4h}^{14}.

	R_1	R_2	R_3	R_4
	1	2	3	4
	2	1	−4	−3
	3	4	1	2
	4	3	−2	−1
$\Delta^{(Y)}$	$\begin{pmatrix}1&0\\0&1\end{pmatrix}$	$\begin{pmatrix}1&0\\0&-1\end{pmatrix}$	$\begin{pmatrix}0&1\\1&0\end{pmatrix}$	$\begin{pmatrix}0&-1\\1&0\end{pmatrix}$

Based on the rep group approach, a program package (Ping 1988) has been written which has the following multifunctions:

1. Computing the subgroup-symmetry adapted single-valued and double-valued irreps of space groups. The subgroup can be chosen either according to one's requirements or by computer.

2. Computing the wave vector selection rules, the CG series and subgroup-symmetry adapted CG coefficients of space groups for Kronecker products of any two reps, i.e., vector ⊗ vector, vector ⊗ spinor and spinor ⊗ spinor.

The program was written in Fortran-77 and is implemented in the personal computer IBM PC/XT.

Ex. 10.1. From the generators of D_{4h}^{19},

$$\{C_{4z}|0\tfrac{1}{2}0\}, \quad \{C_{2x}|000\}\ , \quad \{I|000\}\ ,$$

find the remaining group elements.

Ex. 10.2. Find the single-valued and double-valued irreps of D_{4h}^{14} at the point $T, \mathbf{p} = (\alpha, \frac{1}{2}\frac{1}{2})$, $\mathrm{G}_0(T) = (\varepsilon, C_{2x}, \sigma_y, \sigma_z)$. The group table of D_{4h} is given in Table 10.20-4.

Table 10.20-4. Group table for D_{4h}.

e	C_{2x}	C_{2y}	C_{2z}	C_{4z}^-	C_{4z}^+	C_{2a}	C_{2b}
1	2	3	4	5	6	7	8
2	1	4	3	7	8	5	6
3	4	1	2	8	7	6	5
4	3	2	1	6	5	8	7
5	8	7	6	4	1	2	3
6	7	8	5	1	4	3	2
7	6	5	8	3	2	1	4
8	5	6	7	2	3	4	1

Ex. 10.3. Find the irreducible bases for the single-valued irreps of D_{4h}^{14} at the point T.

Ex. 10.4. Find the single-valued and double-valued irreps of D_{4h}^{14} at the point $X, \mathbf{p} = (0, \frac{1}{2}, 0)$, $\mathbf{G}_0(X) = D_{2h} = (e, C_{2x}, C_{2y}, C_{2z}, I, \sigma_x, \sigma_y, \sigma_z)$.

Ex. 10.5. Find the single-valued and double-valued irreps of D_{4h}^{19} at the points $U[\mathbf{p} = (\frac{1}{2}, \frac{1}{2}, -\frac{1}{2} + \alpha)]$ and $Y[\mathbf{p} = (-\alpha, \alpha, \frac{1}{2})]$. $\mathbf{G}_0(U) = C_{2v} = (e, C_{2a}, \sigma_z, \sigma_{db})$, $\mathbf{G}_0(Y) = C_{2v} = (e, C_{2b}, \sigma_z, \sigma_{da})$. [The space group D_{4h}^{19} belongs to the tetragonal crystal system with body-centered Bravais lattice Γ_q^v. The operations of the group elements of D_{4h} on the primitive translations $\mathbf{t}_i$ can be read out from Table 10.21-2 under the heading Γ_c^v (the body-centered cubic).]

Ex. 10.6. Find the double-valued irreps for the little group of O_h^7 at the point W.

10.21. Appendices

Table 10.21-1. Group table for the point group O and the double point group $O^{\dagger}$, and the correspondence of notations for the group elements.

E	ε	1	2	3	4	5	6	7	8	9	10	11	12	13	14	15	16	17	18	19	20	21	22	23	24
C_2^x	C_{2x}	2	$\tilde{1}$	4	$\tilde{3}$	8	23	$\widetilde{19}$	$\tilde{5}$	$\widetilde{21}$	20	12	$\widetilde{11}$	14	$\widetilde{13}$	$\widetilde{18}$	17	$\widetilde{16}$	15	7	$\widetilde{10}$	9	$\widetilde{24}$	$\tilde{6}$	22
C_2^y	C_{2y}	3	$\tilde{4}$	$\tilde{1}$	2	22	$\tilde{9}$	20	24	6	19	$\widetilde{13}$	14	11	$\widetilde{12}$	16	$\widetilde{15}$	$\widetilde{18}$	17	$\widetilde{10}$	$\tilde{7}$	23	$\tilde{5}$	$\widetilde{21}$	$\tilde{8}$
C_2^z	C_{2z}	4	3	$\tilde{2}$	$\tilde{1}$	$\widetilde{24}$	21	$\widetilde{10}$	22	23	7	$\widetilde{14}$	$\widetilde{13}$	12	11	17	18	$\widetilde{15}$	$\widetilde{16}$	$\widetilde{20}$	19	$\tilde{6}$	$\tilde{8}$	$\tilde{9}$	5
$\overline{C}_4^x$	C_{4x}^-	5	8	24	22	$\tilde{2}$	16	11	1	18	13	19	7	$\widetilde{20}$	10	9	$\widetilde{23}$	6	21	$\widetilde{12}$	14	$\widetilde{15}$	3	17	$\tilde{4}$
$\overline{C}_4^y$	C_{4y}^-	6	$\widetilde{21}$	$\tilde{9}$	23	11	3	17	14	1	16	22	8	5	24	10	19	20	7	$\widetilde{15}$	$\widetilde{18}$	4	$\widetilde{13}$	2	$\widetilde{12}$
$\overline{C}_4^z$	C_{4z}^-	7	20	19	$\widetilde{10}$	18	11	4	17	12	1	21	23	9	6	8	5	22	$\widetilde{24}$	$\tilde{2}$	3	$\widetilde{14}$	$\widetilde{15}$	$\widetilde{13}$	16
C_4^x	C_{4x}^+	8	$\tilde{5}$	22	$\widetilde{24}$	1	17	12	2	15	14	7	$\widetilde{19}$	10	20	$\widetilde{21}$	6	23	9	11	$\widetilde{13}$	18	4	$\widetilde{16}$	3
C_4^y	C_{4y}^+	9	23	6	21	13	1	18	12	$\tilde{3}$	15	5	$\widetilde{24}$	$\widetilde{22}$	8	$\widetilde{19}$	10	7	$\widetilde{20}$	16	17	$\tilde{2}$	11	4	14
C_4^z	C_{4z}^+	10	$\widetilde{19}$	20	7	16	14	1	15	13	$\tilde{4}$	6	9	$\widetilde{23}$	$\widetilde{21}$	$\widetilde{22}$	24	8	5	3	2	11	17	12	$\widetilde{18}$
$\overline{C}_3^{xyz}$	C_{31}^-	11	14	$\widetilde{12}$	$\widetilde{13}$	21	19	22	6	7	5	$\widetilde{15}$	17	18	16	1	$\tilde{2}$	3	4	$\tilde{8}$	24	$\widetilde{10}$	$\tilde{9}$	20	$\widetilde{23}$
$\overline{C}_3^{\bar{x}\bar{y}z}$	C_{32}^-	12	$\widetilde{13}$	11	$\widetilde{14}$	9	7	$\widetilde{24}$	23	$\widetilde{19}$	8	18	$\widetilde{16}$	15	17	2	1	4	$\tilde{3}$	5	22	$\widetilde{20}$	21	$\widetilde{10}$	6
$\overline{C}_3^{x\bar{y}\bar{z}}$	C_{33}^-	13	12	14	11	$\widetilde{23}$	10	5	9	$\widetilde{20}$	$\widetilde{22}$	16	18	$\widetilde{17}$	15	$\tilde{3}$	$\tilde{4}$	1	$\tilde{2}$	24	8	19	6	7	$\widetilde{21}$
$\overline{C}_3^{\bar{x}y\bar{z}}$	C_{34}^-	14	$\widetilde{11}$	$\widetilde{13}$	12	6	20	8	$\widetilde{21}$	10	24	17	15	16	$\widetilde{18}$	$\tilde{4}$	3	2	1	22	$\tilde{5}$	7	23	$\widetilde{19}$	$\tilde{9}$
C_3^{xyz}	C_{31}^+	15	$\widetilde{16}$	17	18	10	8	9	$\widetilde{19}$	$\widetilde{22}$	$\widetilde{21}$	1	$\tilde{3}$	$\tilde{4}$	2	$\widetilde{11}$	14	12	13	6	23	5	7	$\widetilde{24}$	20
$C_3^{\bar{x}\bar{y}z}$	C_{32}^+	16	15	$\widetilde{18}$	17	19	24	6	10	5	$\widetilde{23}$	3	1	$\tilde{2}$	$\tilde{4}$	13	$\widetilde{12}$	14	11	$\tilde{9}$	$\widetilde{21}$	22	20	8	$\tilde{7}$
$C_3^{x\bar{y}\bar{z}}$	C_{33}^+	17	$\widetilde{18}$	$\widetilde{15}$	$\widetilde{16}$	7	22	23	20	8	6	4	2	1	3	14	11	$\widetilde{13}$	12	21	$\tilde{9}$	$\widetilde{24}$	$\widetilde{10}$	$\tilde{5}$	19
$C_3^{\bar{x}y\bar{z}}$	C_{34}^+	18	17	16	$\widetilde{15}$	$\widetilde{20}$	5	21	7	$\widetilde{24}$	9	$\tilde{2}$	4	$\tilde{3}$	1	12	13	11	$\widetilde{14}$	$\widetilde{23}$	6	$\tilde{8}$	19	22	10
C_2^{xy}	C_{2a}	19	10	$\tilde{7}$	20	$\widetilde{15}$	$\widetilde{12}$	3	16	11	$\tilde{2}$	$\tilde{9}$	6	21	$\widetilde{23}$	5	$\tilde{8}$	24	22	$\tilde{1}$	$\tilde{4}$	$\widetilde{13}$	$\widetilde{18}$	14	$\widetilde{17}$
$C_2^{x\bar{y}}$	C_{2b}	20	$\tilde{7}$	$\widetilde{10}$	$\widetilde{19}$	17	$\widetilde{13}$	2	$\widetilde{18}$	14	3	23	$\widetilde{21}$	6	$\tilde{9}$	24	22	$\tilde{5}$	8	4	$\tilde{1}$	12	$\widetilde{16}$	$\widetilde{11}$	$\widetilde{15}$
C_2^{xz}	C_{2c}	21	6	$\widetilde{23}$	$\tilde{9}$	$\widetilde{14}$	$\tilde{2}$	$\widetilde{15}$	11	4	18	$\tilde{8}$	22	$\widetilde{24}$	5	7	$\widetilde{20}$	19	$\widetilde{10}$	$\widetilde{17}$	16	$\tilde{1}$	$\widetilde{12}$	3	13
C_2^{yz}	C_{2d}	22	24	$\tilde{8}$	$\tilde{5}$	4	$\widetilde{15}$	$\widetilde{13}$	3	17	11	$\widetilde{10}$	20	7	19	6	21	$\tilde{9}$	23	$\widetilde{14}$	$\widetilde{12}$	$\widetilde{16}$	$\tilde{1}$	$\widetilde{18}$	$\tilde{2}$
$C_2^{x\bar{z}}$	C_{2e}	23	$\tilde{9}$	21	$\tilde{6}$	12	4	$\widetilde{16}$	$\widetilde{13}$	2	17	$\widetilde{24}$	$\tilde{5}$	8	22	20	7	$\widetilde{10}$	$\widetilde{19}$	18	$\widetilde{15}$	$\tilde{3}$	$\widetilde{14}$	$\tilde{1}$	11
$C_2^{y\bar{z}}$	C_{2f}	24	$\widetilde{22}$	$\tilde{5}$	8	3	$\widetilde{18}$	14	$\tilde{4}$	16	$\widetilde{12}$	20	10	19	$\tilde{7}$	$\widetilde{23}$	$\tilde{9}$	$\widetilde{21}$	6	$\widetilde{13}$	$\widetilde{11}$	17	2	15	$\tilde{1}$

Table 10.21-2. The operations of the point group elements on the primitive translation $\mathbf{t}_i$.

	simple cubic (Γ_c)			face-centered cubic (Γ_c^f)			body-centered cubic (Γ_c^v) body-centered tetragonal (Γ_g^v)		
ϵ	t_1	t_2	t_3	t_1	t_2	t_3	t_1	t_2	t_3
C_{2x}	t_1	$-t_2$	$-t_3$	$-t_1$	$-t_1+t_3$	$-t_1+t_2$	$-t_1-t_2-t_3$	t_3	t_2
C_{2y}	$-t_1$	t_2	$-t_3$	$-t_2+t_3$	$-t_2$	t_1-t_2	t_3	$-t_1-t_2-t_3$	t_1
C_{2z}	$-t_1$	$-t_2$	t_3	t_2-t_3	t_1-t_3	$-t_3$	t_2	t_1	$-t_1-t_2-t_3$
C_{31}^+	t_2	t_3	t_1	t_2	t_3	t_1	t_2	t_3	t_1
C_{32}^+	t_2	$-t_3$	$-t_1$	$-t_2$	t_1-t_2	$-t_2+t_3$	$-t_1-t_2-t_3$	t_1	t_3
C_{33}^+	$-t_2$	t_3	$-t_1$	t_1-t_3	$-t_3$	t_2-t_3	t_1	$-t_1-t_2-t_3$	t_2
C_{34}^+	$-t_2$	$-t_3$	t_1	$-t_1+t_3$	$-t_1+t_2$	$-t_1$	t_3	t_2	$-t_1-t_2-t_3$
C_{31}^-	t_3	t_1	t_2	t_3	t_1	t_2	t_3	t_1	t_2
C_{32}^-	$-t_3$	t_1	$-t_2$	$-t_1+t_2$	$-t_1$	$-t_1+t_3$	t_2	$-t_1-t_2-t_3$	t_3
C_{33}^-	$-t_3$	$-t_1$	t_2	t_1-t_2	$-t_2+t_3$	$-t_2$	t_1	t_3	$-t_1-t_2-t_3$
C_{34}^-	t_3	$-t_1$	$-t_2$	$-t_3$	t_2-t_3	t_1-t_3	$-t_1-t_2-t_3$	t_2	t_1
C_{4x}^+	t_1	t_3	$-t_2$	t_2-t_3	$-t_1+t_2$	t_2	$-t_3$	$-t_1$	$t_1+t_2+t_3$
C_{4y}^+	$-t_3$	t_2	t_1	t_3	$-t_1+t_3$	$-t_2+t_3$	$t_1+t_2+t_3$	$-t_1$	$-t_2$
C_{4z}^+	t_2	$-t_1$	t_3	t_1-t_3	t_1	t_1-t_2	$-t_3$	$t_1+t_2+t_3$	$-t_2$
C_{4x}^-	t_1	$-t_3$	t_2	$-t_2+t_3$	t_3	$-t_1+t_3$	$-t_2$	$t_1+t_2+t_3$	$-t_1$
C_{4y}^-	t_3	t_2	$-t_1$	t_1-t_2	t_1-t_3	t_1	$-t_2$	$-t_3$	$t_1+t_2+t_3$
C_{4z}^-	$-t_2$	t_1	t_3	t_2	t_2-t_3	$-t_1+t_2$	$t_1+t_2+t_3$	$-t_3$	$-t_1$
C_{2a}	t_2	t_1	$-t_3$	$-t_1+t_3$	$-t_2+t_3$	t_3	$-t_1$	$-t_2$	$t_1+t_2+t_3$
C_{2b}	$-t_2$	$-t_1$	$-t_3$	$-t_2$	$-t_1$	$-t_3$	$-t_2$	$-t_1$	$-t_3$
C_{2c}	t_3	$-t_2$	t_1	$-t_1+t_2$	t_2	t_2-t_3	$-t_1$	$t_1+t_2+t_3$	$-t_3$
C_{2d}	$-t_1$	t_3	t_2	t_1	t_1-t_2	t_1-t_3	$t_1+t_2+t_3$	$-t_2$	$-t_3$
C_{2e}	$-t_3$	$-t_2$	$-t_1$	$-t_3$	$-t_2$	$-t_1$	$-t_3$	$-t_2$	$-t_1$
C_{2f}	$-t_1$	$-t_3$	$-t_2$	$-t_1$	$-t_3$	$-t_2$	$-t_1$	$-t_3$	$-t_2$

[1]For example: for Γ_c^f, $C_{2x}\,\mathbf{t}_2 = -\mathbf{t}_1 + \mathbf{t}_3$, $C_{31}^+\,\mathbf{t}_1 = \mathbf{t}_2$.

Appendix

Table A1. Dimensions of irreps of the permutation group S_f ($f \leq 6$) and unitary groups SU_n ($n \leq 6$).

	[1]	[2]	[11]	[3]	[21]	$[1^3]$	[4]	[31]	[22]	[211]	$[1^4]$	[5]	[41]	[32]	[311]	[221]	$[21^3]$	$[1^5]$
S_f	1	1	1	1	2	1	1	3	2	3	1	1	4	5	6	5	4	1
SU_3	3	6	3	10	8	1	15	15	6	3	*	21	24	15	6	3	*	*
SU_4	4	10	6	20	20	4	35	45	20	15	1	56	84	60	36	20	4	*
SU_5	5	15	10	35	40	10	70	105	50	45	5	126	224	175	126	75	24	1
SU_6	6	21	15	56	70	20	126	210	105	105	15	252	504	420	336	210	84	6

	[6]	[51]	[42]	[411]	[33]	[321]	$[31^3]$	$[2^3]$	$[2^21^2]$	$[21^4]$	$[1^6]$
S	1	5	9	10	5	16	10	5	9	5	1
SU_3	28	35	27	10	10	8	*	1	*	*	*
SU_4	84	140	126	70	50	64	10	10	6	*	*
SU_5	210	420	420	280	175	280	70	50	45	5	*
SU_6	462	1050	1134	840	490	896	280	175	189	35	1

Table A2. Phase factors $\epsilon_1(\nu_1\nu_2\nu)$ for the permutation group IDC and SU_n CG coefficients [see Eq. (4-153) and (7-101)].

$[\nu_1]$	[1]		[2]		[11]		[3]		$[1^3]$		[2]		
$[\nu_2]$	[1]		[1]		[1]		[1]		[1]		[2]		
$[\nu]$	[2]	[11]	[3]	[21]	$[1^3]$	[21]	[4]	[31]	$[1^4]$	[211]	[4]	[31]	[22]
ϵ_1	1	−1	1	−1	1	1	1	−1	−1	1	1	−1	1

$[\nu_1]$	[11]			[2]		[21]			[4]		$[1^4]$		[31]		
$[\nu_2]$	[11]			[11]		[1]			[1]		[1]		[1]		
$[\nu]$	$[1^4]$	[211]	[22]	[31]	[211]	[31]	[211]	[22]	[5]	[41]	$[21^3]$	$[1^5]$	[41]	[32]	[311]
ϵ_1	1	−1	1	1	1	1	1	−1	1	−1	1	1	1	−1	1

$[\nu_1]$	[211]			[22]		[3]			$[1^3]$			[3]	
$[\nu_2]$	[1]			[1]		[2]			[11]			[11]	
$[\nu]$	[311]	[221]	$[21^3]$	[32]	[221]	[5]	[41]	[32]	[221]	$[21^3]$	$[1^5]$	[41]	[311]
ϵ_1	1	−1	−1	1	1	1	−1	1	1	1	1	1	1

$[\nu_1]$	$[1^3]$		[21]				[21]			
$[\nu_2]$	[2]		[2]				[11]			
$[\nu]$	[311]	$[21^3]$	[41]	[32]	[311]	[221]	[32]	[311]	[221]	$[21^3]$
ϵ_1	1	−1	1	−1	1	−1	1	−1	−1	1

References

Akiyama, Y.,

[1] *Nucl. Data* **A2** (1966) 403.

Akiyama, Y. and J.P. Draayer,

[1] *Computer Phys. Commun.* **5** (1973) 405.

Akyeampong, D.A. and M.A. Rashid,

[1] "On the finite transformation of $SU(3)$", *J. Math. Phys.* **13** (1972) 1218.

Altman, S.L.,

[1] *Induced Representation in Crystals and Molecules: Point, Space and Non-rigid Molecule Groups*, (Academic Press, New York, 1977).

Arima, A.J. Ginocchio and N. Yoshida,

[1] "The structure of Samaritan isotopes in a schematic monopole and quadrupole pairing model", *Nucl. Phys.* **A384** (1982) 112.

Arima, A. and F. Iachello,

Interacting boson model of collective states:

[1] (I) "The vibrational limit", *Ann. Phys.* **99** (1976) 253.

[2] (II) "The rotational limit", *Ann. Phys.* **111** (1978) 201.

[3] (III) "The transition from $SU(5)$ to $SU(3)$", *Ann. Phys.* **115** (1978) 325.

[4] (IV) "The $O(6)$ limit", *Ann. Phys.* **123** (1979) 468.

Arima, A. and F. Iachello,

[5] "The interacting boson model", *Ann. Rev. Nucl. Part. Sci.* **31** (1981) 75.

Bacry H.,

[1] *Lectures on Group Theory and Particle Theory* (Gordon and Breach, New York, 1977).

Baird, G.E. and L.C. Biedenharn,

On the representations of the semisimple Lie groups:

[1] (II) *J. Math. Phys.* **4** (1963) 1449.

[2] (III) "The explicit conjugation operation for $SU(n)$", *J. Math. Phys.* **5** (1964) 1723.

[3] (IV) "A canonical classification for tensor operators in $SU(3)$", *J. Math. Phys.* **5** (1964) 1730.

[4] (V) *J. Math. Phys.* **6** (1965) 1847.

Bayman, B.F.,

[1] *Some Lectures on Groups and Their Applications to Spectroscopy*, Nordita, 1960.

Bayman, B.F. and A. Lande,

[1] "Tables of identical-particle fractional parentage coefficients", *Nucl. Phys.* **77** (1966) 1.

Berenson, R. and J.L. Birman,

[1] "Clebsch-Gordan coefficients for crystal space groups", *J. Math. Phys.* **16** (1975) 227.

Berenson, R., I. Itzkan and J.L. Birman,

[1] "Clebsch-Gordan coefficients for $^*X \otimes ^*X$ in diamond O_h^7-Fd3m and rocksalt O_h^5-Fm3m", *J. Math. Phys.* **16** (1975) 236.

Bickerstaff, R.P.,

[1] "Simultaneity and reality $U(n)$ and $SU(n)$ $3jm$ and $6j$ symbols", *J. Math. Phys.* **25** (1984) 2808.

Biedenharn, L.C.,

[1] "On the representations of the semisimple Lie groups (I). The explicit construction of invariants for the unimodular unitary group in N dimensions", *J. Math. Phys.* **4** (1963) 436.

Biedenharn, L.C., A. Giovannini and J.D. Louck,

[1] "Canonical definition of Wigner coefficients in $U(n)$", *J. Math. Phys.* **11** (1970) 2368.

Biedenharn, L.C. and Louck, J.D.,

[1] *The Racah-Wigner Algebra in Quantum Theory* (Addison-Wesley, Reading, Mass., 1981).

Birman, J.L.,

[1] "Theory of Crystal Space Groups and Infra-Red and Raman Lattice Processes of Insulating Crystal", *Encyclopedia of Physics*, Vol. 25/2b, ed. L.Genzel (Springer-Verlag, Berlin, 1974).

[2] "Space group selection rules: diamond and zinc blend", *Phys. Rev.* **127** (1962) 1093.

Boerner, H.,

[1] *Representations of Group* (North-Holland, Amsterdam, 1963).

Bohr, A. and B.R. Mottelson,

[1] *Nuclear Structure*, Vol. I (Benjamin, New York, 1969).

[2] *Nuclear Structure*, Vol. II (Benjamin, New York, 1975).

Bradley, C.J. and A.P. Cracknell,

[1] *The Mathematical Theory of Symmetry in Solids* (Clarendon Press, Oxford, 1972).

Braunschweign, D.,

[1] "Reduced $SU(3)$ CFP's", *Computer Phys. Commun.* **14** (1978) 109.

Bremner, B.B., R.V. Moody and J. Patera,

[1] *Tables of Dominant Weight Multiplicities for Representations of Simple Lie Algebras* (Marcel Dekker, New York, 1985).

Burnside, W.,

[1] *Theory of Finite Order* (Dover, New York, 1955).

Butler, P.H.,

[1] "Calculation of j and jm symbols for arbitrary compact groups. (II) An alternative procedure for angular momentum", *Int. Quantum Chemistry* **10** (1976) 599.

[2] "Coupling coefficients and tensor operators for chains of groups", *Phil. Trans. Roy. Soc.* **A277** (1975) 545.

[3] *Point Group Symmetry Applications: Methods and Table* (Plenum, New York, 1981).

Butler, P.H. and B.G. Wybourne,

[1] "Calculation of j and jm symbols for arbitrary compact groups. (I) Methodology", *Int. Quantum Chemistry* **10** (1976) 581.

[2] (III) "Application to $SO_3 \supset T \supset C_3 \supset C_1$", *Int. Quantum Chemistry* **10** (1976) 615.

Castanos, O., E. Chacon and M. Moshinky,

[1] "Analytic expression for the matrix elements of generators of $SP(6)$ in an $SP(6) \supset U(3)$ basis", *J. Math. Phys.* **25** (1984) 1211.

Casten, R.F., C.L. Wu, D.H. Feng, J.N. Ginocchio and X.L. Han,

[1] "Empirical evidence for an $SO(7)$ fermion dynamical symmetry in nuclei", *Phys. Rev. Lett.* **56** (1986) 2578.

Chen, Jin-Quan,

[1] "On the extension of some sum rules in the Racah algebra", *Acta Phys. Sinica* **21** (1965) 1817.

[2] "$SU(mn) \supset SU(m) \times SU(n)$ isoscalar factors and $S(f_1 + f_2) \supset S(f_1) \otimes S(f_2)$ isoscalar factors", *J. Math. Phys.* **22** (1981) 1.

[3] "The $SU(m+n) \supset SU(m) \otimes SU(n)$ isoscalar factors and $S(f_1 + f_2) \supset S(f_1) \otimes S(f_2)$ outer-product isoscalar factors", *Acta Math. Scientia* **5** (1985) 19.

Chen, Jin-Quan, Xuan-Gen Chen and Mei-Juan Gao,

[4] "The Clebsch-Gordan coefficients and isoscalar factors of the graded unitary group $SU(m/n)$", *J. Phys. A* **16** (1983) L47.

[5] "The Casimir invariants and Gel'fand basis of the graded unitary group $U(m/n)$", *J. Phys. A* **16** (1983) 1361.

[6] "The Clebsch-Gordan coefficients of $SU(m/n)$ Gel'fand basis", *J. Phys. A* **17** (1984) 481.

[7] "The CFP for $U(m+p/m+q) \supset U(m/n) \otimes U(p/q)$ and $U(m/n) \supset U(m) \times U(n)$", *J. Phys. A* **17** (1984) 1941.

[8] "The CG coefficients for $U(m/n)$ Gel'fand and basis and isoscalar factors for $U(m/n) \supset U(m) \times U(n)$ group chain", *Nucl. Phys. A* **421** (1984) 387c.

Chen, Jin-Quan, D. Collinson and Mei-Juan Gao,

[9] "Transformation coefficients of permutation groups", *J. Math. Phys.* **24** (1983) 2695.

Chen, Jin-Quan and Mei-Juan Gao,

[10] *Reduction Coefficients of the Permutation Group and Their Applications* (Beijing Science Press, Beijing, 1981).

[11] "A new approach to permutation group representation theory", *J. Math. Phys.* **23** (1982) 928.

[12] "Some advances on group representation theory", in *Proceedings of XIVth International Colloquium on Group Theoretical Method in Physics*, Seoul, 1985, ed. Y.M. Cho (World Scientific, Singapore, 1986).

Chen, Jin-Quan, Mei-Juan Gao and Xuan-Gen Chen,

[13] "The Clebsch-Gordan coefficients for $SU(m/n)$ Gel'fand basis", *J. Phys. A* **17** (1984) 481.

Chen, Jin-Quan, Mei-Juan Gao and Guan-Qun Ma,

[14] "The eigenfunction method for the space group CG coefficients", *Commun. Theor. Phys.* (Beijing) **2** (1984) 1613.

[15] "The representation group and its application to space groups", *Rev. Mod. Phys.* **57** (1985) 211.

Chen, Jin-Quan, Mei-Juan Gao, Yi-Jin Shi, M. Vallieres and Da Hsuan Feng,

[16] "Tables of one-body CFP for the group chains $SU(mn) \supset SU(m) \times SU(n)$ and $SU(mp+nq/mq+np) \supset SU(m/n) \otimes SU(p/q)$", *Nucl. Phys.* **A419** (1984) 77.

Chen, Jin-Quan, Mei-Juan Gao and Fan Wang,

[17] "Physical method of group representation theory (V). The irreducible bases of $SU(3) \supset SO(3)$ and $SU(4) \supset SU(2) \times SU(2)$", *Acta Phys. Sinica* **27** (1978) 237.

[18] "On the phase problem and representation transformation of the $SU(n)$ wave functions for baryons and mesons", *Phys. Energ. Fortis Phys. Nucl.* **3** (1979) 408.

Chen, Jin-Quan, Mei-Juan Gao, Fan Wang and Tzu-Rong Yu,

[19] "A new approach to representation theory of finite groups. (III) Application of the eigenfunction method to point groups", *Nanjing Daxue Xuebao* (J. Nanjing Univ.) No. **4** (1979) 37.

Chen, Jin-Quan, Yi-Jin Shi, Da Hsuan Feng and M. Vallieres,

[20] "Symmetry and application of the $SU(mn) \supset SU(m) \times SU(n)$ coefficients of fractional parentage", *Nucl. Phys.* **A 23** (1983) 122.

Chen, Jin-Quan, Fan Wang and Mei-Juan Gao,

[21] "A new approach to representation theory of finite groups. (I) The eigenfunction method for characters and irreducible bases", *Nanjing Daxue Xuebao* (J. Nanjing Univ.) No. **2** (1977) 148.

[22] "A new approach to representation theory of finite groups. (II) The intrinsic group and resolution of regular representation", *Nanjing Daxue Xuebao* (J. Nanjing Univ.) No. **2** (1978) 110.

[23] "Physical method of group representation theory (I). A new approach to the theory of finite group representation", *Acta Phys. Sinica* **26** (1977) 307 (transl. *Chinese Phys.* **1** (1981) 533).

[24] "Physical method of group representation theory (II). The quasi-standard basis of permutation groups and the Gel'fand basis of unitary groups", *Acta Phys. Sinica* **26** (1977) 427 (transl. *Chinese Phys.* **1** (1981) 542).

[25] "Physical method of group representation theory (III). The outer-product reduction coefficients of the permutation group and the Clebsch-Gordan coefficients of $SU(n)$ group", *Acta. Phys. Sinica* **27** (1978) 31.

[26] "Physical method of group representation theory (IV). The irreducible basis of $SU(mn) \supset SU(m) \times SU(n)$ and $SU(m+n) \supset SU(m) \otimes SU(n)$", *Acta. Phys. Sinica* **27** (1978) 203.

[27] "Intrinsic Lie group and nuclear collective rotation about intrinsic axes", *J. Phys. A* **16** (1983) 1347.

Chen, Jin-Quan, Fan Wang, Mei-Juan Gao and Tzu-Rong Yu,

[28] "A new method of calculation for molecular shell model", *Scientia Sinica* **9** (1980) 1116.

Chen, Jin-Quan, Fan Wang, Tsu-Rong Yu and Mei-Juan Gao,

[29] "On the relation between permutation group and unitary group", *Kexu Tongbao* **23** (1978) 291.

Chen, Jin-Quan, Fan Wang and Tsu-Rong Yu,

[30] "A recursive formula for the Clebsch-Gordan coefficients of the group SU(n)", *Phys. Energ. Fortis et Phys. Nucl.* **3** (1979) 216.

Chen, Jin-Quan, Pei-Ning Wang, Zi-Ming Lu and Xiong-Biao Wu,

[31] *Tables of the Clebsch-Gordan, Racah and Subduction Coefficients of SU (n) Groups* (World Scientific, Singapore, 1987).

Chen, Jin-Quan, Da Hsuan Feng and Cheng-Li Wu,

[32] "Correspondence between the interacting boson model and the fermion dynamical symmetry model of nuclei", *Phys. Rev.* **C34** (1986) 2269.

Chen, Jin-Quan, A. Novoselsky, M. Vallieres and R. Gilmore,

[1] "A new approach to multi-shell calculation in multiple angular momentum coupling schemes" (1988, Drexel Univ. preprint).

Chen, Hsiao-Shen and Hsi-Teh Hsieh,

[1] "Reductions of the symmetrized and antisymmetrized space group representations", *Acta Phys. Sinica* **21** (1965) 519.

Chen, Xuan-Gen,

[1] "A recursive method for constructing the generalized projection operator", 1979 (unpublished).

Chihonov, A.N.,

[1] *Equations in Mathematical Physics*, Vol. I (Moscow, 1953) Chap. 2 (in Russian).

Cook, C.L. and G. Murtaza,

[1] "Clebsch-Gordan coefficients for the group $SU(6)$", *Nuovo Cimento* **39** (1965) 7095.

Cotton, F.A.,

[1] *Chemical Applications of Group Theory* (Wiley, New York, 1971).

Cracknell, A.P., B.L. Davies, S.C. Miller and W.F. Love,

[1] *Kronecker Product Tables*, Vol. I, *General Introduction and Tables of Irreducible Representations of Space Groups* (Plenum, New York, 1979).

Cracknell, A.P. and B.L. Davies,

[1] *Kronecker Product Tables*, Vol. III, *Wave Selection Rules and Reductions of Kronecker Products for Irreps of Triclinic, Monoclinic, Tetragonal and Hexagonal Space Groups* (Plenum, New York, 1979).

Dalitz, R.H. and A. Gal,

[1] "Strangeness and analogue states in $^{9}_{\Lambda}Be^{*}$", *Ann. Phys.* **131** (1981) 314.

Davies, B.L. and A.P. Cracknell,

[1] *Kronecker Product Tables*, Vol. II *Wave Vector Selection Rules and Reductions of Kronecker Products for Irreps of Orthorhombic and Cubic Space Groups* (Plenum, New York, 1979).

Davies, B.L. and A.P. Cracknell,

[1] *Kronecker Product Tables*, Vol. IV, *Symmetrized Powers of Irreps of Space Groups* (Plenum, New York, 1980).

Davies, B.L.,

[1] "Computational group theory in crystal physics", *Physica* **A114** (1982) 507.

Davies, B.L. and R. Dirl,

[1] in *Proceedings of XIIth International Colloquium on Group Theoretical Methods*, 1983 (Trieste).

DeGrand, T.A. and R.L. Jaffe,

[1] "Excited states of confined quarks", *Ann. Phys.* **100** (1976) 425.

Deonarine, S. and J.L. Birman,

[1] "Group-subgroup phase transitions, Hermann's space-group decomposition theorem, and chain subduction criterion in crystals", *Phys. Rev.* **B27** (1983) 4261.

Derome, J.R.,

[1] "Symmetry properties of 3j-symbols", *J. Math. Phys.* **7** (1966) 612.

Derome, J.R. and W.T. Sharp,

[1] "Racah algebra for arbitrary group", *J. Math. Phys.* **6** (1965) 1584.

DeShalit, A. and I. Talmi,

[1] *Nuclear Shell Theory* (Academic Press, New York, 1963).

de Swart, J.J.,

[1] "The octet model and its Clebsch-Gordan coefficients", *Rev. Mod. Phys.* **35** (1963) 916.

Dirac, P.A.M.,

[1] *The Principles of Quantum Mechanics* (Clarendon Press, Oxford, 1958).

Dirl, R.,

[1] "Clebsch-Gordan coefficients for corepresentation", *J. Math. Phys.* **21** (1980) 961; *ibid* 968; *ibid* 975; *ibid* 983; *ibid* 989; *ibid* 997.

[2] "Induced projective representations", *J. Math. Phys.* **18** (1977) 2065.

Dixson, J.D.,

[1] *Numer. Math.* **10** (1967) 446.

Draayer, J.P. and Y. Akiyama,

[1] "Wigner and Racah coefficients for $SU(3)$", *J. Math. Phys.* **14** (1973) 1904.

Edmonds, A.R.,

[1] *Angular Momentum in Quantum Mechanics*, (Princeton University Press, Princeton, N.J., 1957).

Eisenberg, J.M. and W. Greiner,

[1] *Nuclear Theory*, Vol. I (North Holland, Amsterdam, 1970).

[2] *Microscopic Theory of the Nucleus*, Vol. III (North Holland, Amsterdam, 1976).

Eisenhart, L.P.,

[1] *Continuous Groups of the Transformations* (Princeton University Press, Princeton, N.J., 1933).

Elliott, J.P.,

[1] "Collective motion in the nuclear shell model (I). Classification schemes for states of mixed configurations", *Proc. Roy. Soc.* (London) **A245** (1958) 128.

[2] "Collective motion in the nuclear shell model (II). The introduction of intrinsic wave functions", *Proc. Roy. Soc.* (London) **A245** (1958) 562.

Elliott, J.P. and A.M. Lane,

[1] "The nuclear shell model", in *Encyclopedia of Physics,* Vol. 39, p. 16, ed. Flügge, (Springer-Verlag, Berlin, 1957).

Elliott, J.P. and M. Harvey,

[1] "Collective motion in the nuclear shell model. (III) The calculation of spectra", *Proc. Roy. Soc.* (London) **A272** (1963) 557.

Elliott, J.P. and P.G. Dawber,

[1] *Symmetry in Physics*, Vols. 1, 2 (McMillan Press, London, 1979).

Elliott, J.P., J. Hope and H.A. Jahn,

[1] "Theoretical studies in nuclear structure IV B", *Phil. Trans. Roy. Soc.* (London) **A246** (1953) 241.

Fano, U. and G. Racah,

[1] *Irreducible Tensorial Sets* (Academic Press, New York, 1959).

Feld, B.T.,

[1] *Models of Elementary Particles* (Blaisdell, Boston, 1969).

Feynman, R.P.,

[1] *The Feynman Lectures on Physics*, Vol. III (Addison Wesley, Reading Mass, 1964).

Fieck, G.,

[1] "Symmetry adaption reduced to tabulated quantities", *Theor. Chim. Acta* **44** (1977) 279.

Flodmark, S. and Jannson P.-O.,

[1] "The 1984 edition of programm IRREP", May 1984.

[2] "Irreducible representations of finite groups", *Physica* **114A** (1982) 485.

Freudenthal, H. and H. de Vries,

[1] *Linear Lie Algebra* (Academic Press, New York, 1969).

Gao, Mei-Juan, Chong-Guo Yao, Yi-Mei Chen and Jin-Quan Chen,

[1] "The calculation of the characters of the permutation group S_{11} by the eigenfunction method", (1976) (unpublished).

Gao, Mei-Juan, Jin-Quan Chen and J. Paldus,

[1] "Point group symmetry adaptation in Clifford algebra unitary group approach", *Int. J. Quantum Chem.* **32** (1987) 133.

Gamba, A.,

[1] "Lie-like approach to the theory of representation of finite groups", *J. Math. Phys.* **10** (1969) 872.

Gel'fand, I.M.,

[1] "The center of an infinitesimal group algebra", *Mat. Sb.* **26** (1950) 103.

Gel'fand, I.M., R.A. Minlos and Z. Ya. Shapiro,

[1] *Representations of the Rotation and Lorentz and Their Applications* (Pergamon Press, New York, 1963).

Gel'fand, I.M. and M.L. Zetlin,

[1] "Matrix elements for the unitary groups", *Dokl. Akad. Nauk.* **71** (1950) 825.

Gell-Mann, M. and Y. Ne'eman,

[1] *The Eightfold Way* (Benjamin, New York, 1964).

Gilmore, R.,

[1] *Lie groups, Lie Algebras and Some of Their Applications* (Wiley, New York, 1974).

Ginocchio, J.N.,

[1] "An exact fermion model with monopole pairing and quadrupole pairing", *Phys. Lett.* **79B** (1978) 173;

[2] "On a generalization of quasispin to monopole and quadrupole pairing", *Phys. Lett.* **85B** (1978) 9;

[3] "A schematic model for monopole and quadrupole pairing in nuclei", *Ann. Phys.* **126** (1980) 234. Ginocchio, J.N., T Otsuka, R.D. Amado and D.A. Sparrow,

[4] "Medium energy probes and the interacting boson model of nuclei", 1987 (preprint).

Griffith, J.S.,

[1] *The Irreducible Tensor Method for Molecular Symmetry Groups* (Prentice-Hall, Englewood Cliffs, N.J., 1962).

[2] *The Theory of Transitional Ions* (Cambridge Univ. Press, London, 1961).

Guidry, M., C.L. Wu, Z.P. Li, D.H. Feng and J.N. Ginocchio,

[1] "An algebraic fermion description of band termination and loss of collectivity in heavy nuclei", *Phys. Lett.* **187B** (1987) 210.

Haacke, E.M., W. Moffat and P. Savaria,

[1] "A calculation of $SU(4)$ Clebsch-Gordan coefficients", *J. Math. Phys.* **17** (1976) 2041.

Hamermesh, M.,

[1] *Group Theory and Its Application to Physical Problems* (Addison-Wesley, Reading Mass, 1962).

Han, Qi-Zhi and Hong-Zhou,

[1] *Group Theory* (Beijing Univ. Press, Beijing, 1987).

Han, X.L., M. Guidry, D.H. Feng, K.X. Wang and C.L. Wu,

[1] "Evidence for nuclear shell symmetries", *Phys. Lett.* **128B** (1987) 253.

Harter, W.G. and C.W. Patterson,

[1] "Alternative basis for the theory of complex spectra II", *Phys. Rev.* **A13** (1976) 1067.

Harvey, M.,

[1] "On the fractional parentage expansion on color-singlet six quark state in a cluster model", *Nucl. Phys.* **352** (1981) 301.

Hecht, K.T.,

[1] "$SU(3)$ recoupling and fractional parentage in the $2s$–$1d$ shell", *Nucl. Phys.* **62** (1965) 1.

[2] "Coherent state theory for the LST quasispin group", *Nucl. Phys.* **A444** (1985) 189;

[3] "Vector coherent state theory for the S, D fermion pair algebra", *Nucl. Phys.* **A475** (1987) 276.

[4] *The vector coherent state method and its application to problems of higher symmetries*, Lecture Notes in Physics **290**, (Springer-Verlag, Berlin, 1987).

[5] "Vector coherent state theory for the SP(6)$\supset$U(3) branch of the S.D. fermion pair algebra 1987", (preprint).

[6] "Wigner coefficients for the proton-neutron quasispin group. An application of vector coherent state techniques", (preprint, 1988).

Hecht, K.T. and D. Braunschweign,

[1] "Few nucleon $SU(3)$ parentage coefficients and α-particle spectroscopic amplitudes for core excited states in s–d shell nuclei", *Nucl. Phys.* **A244** (1975) 365.

Hecht, K.T. and J.P. Elliott,

[1] "Coherent state theory for the proton-neutron quasispin group", *Nucl. Phys.* **A438** (1985) 29.

Hecht, K.T. and S.C. Pang,

[1] "On the Wigner supermultiplet scheme", *J. Math. Phys.* **10** (1969) 1571.

Horie, H.,

[1] "Representations of the symmetric group and the fractional parentage coefficients", *J. Phys. Soc.* Japan **19** (1964) 1783.

Hsieh Hsi-Teh and Hsiao-Shen Chen,

[1] "Space group selection rules", *Acta. Phys. Sinica* **20** (1964) 970.

Itzykson, C. and M. Nauenberg,

[1] "Unitary groups: Representations and decompositions", *Rev. Mod. Phys.* **38** (1966) 95.

Jahn, H.A.,

[1] "Theoretical studies in nuclear structure I. Enumeration and classification of the states arising from the filling of the nuclear d-shell", *Proc. Roy. Soc.* (London) **A201** (1950) 516.

[2] "Theoretical studies in nuclear structure II. Nuclear d^2, d^3 and d^4 configuration fractional parentage coefficients and central force matrix elements", *Proc. Roy. Soc.* (London) **A205** (1951) 192.

[3] "Direct evaluation of fractional parentage coefficients using Young operators. General theory and $\langle 4 | 2, 2\rangle$ coefficients", *Phys. Rev.* **96** (1954) 989.

Jahn. H.A. and H. van Wieringen,

[1] "Theoretical studies in nuclear structure IV. Wave functions for the nuclear p-shell, Part A. $\langle p^n | p^{n-1}, p\rangle$ fractional parentage coefficients", *Proc. Roy. Soc.* (London) **A209** (1951) 502.

Jartkina, M.E.,

[1] *Rudiments of Molecular Orbital Theory* (Moscow, 1975) Section 3.7 (in Russian).

Johnston, D.F.,

[1] "Group theory in solid state physics", *Rep. Prog. Phys.* **23** (1960) 66.

Jones, H.,

[1] *The Theory of Brillouin Zones and Electronic States* (North-Holland, Amsterdam, 1975) p.87.

Judd, B.R.,

[1] *Operator Techniques in Atomic Spectroscopy*, (New York: McGraw-Hill, 1963).

Kaplan, I.G.,

[1] "The transformation matrix of permutation group and construction of the orbital wave function of a multishell configuration", *Zh. Eksp. Theor. Fiz.* **41** (1961) 560.

[2] "Orbital fractional parentage coefficients for a configuration consisting of several shells", *Zh. Eksp. Theor. Fiz.* **41** (1961) 790.

Kerman, A.K.,

[1] *Ann. Phys.* (N.Y) **12** (1961) 300.

Killingbeck, J.,

[1] "Commutating operator approach to group representation theory", *J. Math. Phys.* **11** (1970) 2268.

[2] "The class sum operator approach for the point groups O and D_4", *J. Math. Phys.* **14** (1973) 185.

King, R.C. and A.H.A. Al-Qubanchi,

[1] "Natural labeling schemes for simple roots and irreps of exceptional Lie algebras", *J. Phys.* **A14** (1981) 15.

Koster, G.F.,

[1] "Space groups and their representations", *Solid State Physics*, Vol. 5, p. 173, ed. F. Seitz and D. Turnbull (Academic Press, New York, 1957).

Koster, G.F., J.O. Dimmock, R.G. Wheeler and H. Statz,

[1] *Properties of the Thirty-Two Point Groups* (MIT Press, Cambridge, Mass., 1963).

Kovalev, O.V.,

[1] *Irreducible Representations of the Space Groups* (Gordon-Breach, New York, 1965).

Kovalev, O.V. and G.Ya. Lyubarskii,

[1] "The contact of energy bands in crystals", *Zh. Tekh. Fiz.* **28** (1958) 1151.

Kramer, P.,

[1] "Orbital fractional parentage coefficients for the harmonic oscillator shell model", *Z. Phys.* **205** (1967) 181.

[2] "Recoupling coefficients of the symmetric group for shell and cluster model configuration", *Z. Phys.* **216** (1968) 68.

Kramer, P. and T.H. Seligman,

[1] "Studies in the nuclear cluster model II", *Nucl. Phys.* **A136** (1969) 545.

Kukulin, V.I., Yu. F. Smirnov and L. Majling,

[1] "Wave functions for mixed configurations and the Racah algebra of $SU(4)$", *Nucl. Phys.* **A103** (1967) 681.

Laskar, W.,

[1] "Highest weights of semisimple Lie algebras", *J. Math. Phys.* **18** (1977) 1162.

Le Blanc, R. and K.T. Hecht,

[1] "Wigner-Racah Calculus II", *J. Phys. A: Math. Gen.* **20** (1987) 4613. Lezuo, K.J.

Lezuo, K.J.,

[1] "The symmetric group and the Gel'fand basis of $U(3)$", *J. Math. Phys.* **13** (1972) 1389.

Lichtenberg, D.B.,

[1] *Unitary Symmetry and Elementary Particles* (Academic Press, New York, 1978).

Lipkin, H.J.,

[1] *Lie Groups for Pedestrians* (North-Holland, Amsterdam, 1966).

Littlewood, D.E.,

[1] *The Theory of Group Characters* (Oxford University Press, Oxford, 1958).

Lobel, E.M.,

[1] *Group Theory and its Applications*, Vol. I (Academic Press, New York, 1968).

[2] *Group Theory and its Applications*, Vol. II (Academic Press, New York, 1971).

[3] *Group Theory and its Applications*, Vol. III (Academic Press, New York, 1975).

Louck, J.,

[1] "Group theory of harmonic oscillators in n-dimensional space", *J. Math. Phys.* **6** (1965) 1786.

[2] "Recent progress toward a theory of tensor operators in the unitary groups", *Am. J. Phys.* **38** (1970) 3.

Louck, J.D. and L.C. Biedenharn,

[1] "Canonical unit adjoint tensor operator in $U(n)$", *J. Math. Phys.* **11** (1970) 2368.

Louck, J.D. and H.W. Galbraith,

[1] "Eckart vectors, Eckart frames and polyatomic molecules", *Rev. Mod. Phys.* **48** (1976) 69.

Löwdin, P.O.,

[1] "Group algebra, convolution algebra and applications to quantum mechanics", *Rev. Mod. Phys.* **39** (1967) 259.

Lü, Z.M., X.W. Pan, J.Q. Chen, X.G. Chen, D.H. Feng,

[1] "Correspondence between fermion dynamic symmetry model and interacting model in wave functions", *Phys. Rev.* **37C** (1988) 2789.

Machacek, M. and Y. Tomozawa,

[1] "$SU(6)$ isoscalar factor for the product $405\times 56 \to 56, 70^*$", *J. Math. Phys.* **17** (1976) 458.

Matsen, F.A. and C.J. Nein,

[1] "Spin-free quantum chemistry XXII", *Int. Quantum Chem.* **20** (1981) 861 and the references therein.

Matveev, V.A. and P. Sorba,

[1] "Quark analysis of multibaryonic system", *Nuovo Cimento* **45** (1978) 257.

McName, P.S.J. and F. Chilton,

[1] "Tables of Clebsch-Gordan coefficients of $SU(3)$", *Rev. Mod. Phys.* **36** (1964) 1005.

McWeeny, R.,

[1] *Symmetry-An Introduction to Group Theory and Its Applications* (Pergamon Press, New York, 1963).

Miller, S.C. and W.F. Love,

[1] "Tables of Irreducible Representations of Space Groups and Corepresentations of Magnetic Groups Colorado", (Pruett, Boulder, 1967).

Miller, W. Jr.,

[1] *Symmetry Groups and Their Applications* (Academic Press, New York, 1973).

Morse, P.M. and H. Feshbach,

[1] *Methods of Theoretical Physics*, Part 1 (McGraw-Hill, New York, 1953) p. 884.

Moshinsky, M.,

[1] "The harmonic oscillator and supermultiplet theory, I", *Nucl. Phys.* **31** (1962) 384.

[2] "Bases for the irreducible representations of the unitary groups and some applications", *J. Math. Phys.* **4** (1963) 1128.

[3] "Gel'fand states and the irreducible representations of the symmetric group", *J. Math. Phys.* **7** (1966) 691.

[4] *Group Theory and the Many Body Problem* (Gordon Breach, New York, 1968).

Neto, N.,

[1] *Comput. Phys. Commun.* **9** (1975) 231.

Novoselsky, A., J. Kartriel and R. Gilmore,

[1] *J. Math. Phys.* **29** (1988) 1363.

Novoselsky, A., M. Vallieres, R. Gilmore and J. Kartriel,

[1] "Coefficients of fractional parentage for states of arbitrary symmetry", (Drexel University, preprint, 1988).

Novoselsky, A., M. Vallieres, R. Gilmore and J.Q. Chen,

[2] "A recursive calculation of the inner- and outer-product isoscalar factors of permutation groups", (Drexel University, preprint, 1988).

Paldus, J.,

[1] "Group theoretical approach to the configuration interaction and perturbation theory calculations for atomic and molecular system", *J. Chem. Phys.* **61** (1974) 5321.

[2] "Unitary group approach to the many-electron correlation problem: Relation of Gel'fand and Weyl tableau formulations", *Phys. Rev.* **A14** (1976) 1620.

[3] "Many-electron correlation problem: A group theoretical approach", in *Theoretical Chemistry*, Vol. 2 Advances and Perspective (Academic Press, New York, 1976).

Paldus, J., Mei-Juan, Gao and Jin-Quan, Chen,

[1] "Clifford algebra unitary-group approach to many-electron system partitioning", *Phys. Rev. A* **35** (1987) 3197.

Paldus, J. and P.E.S. Wormer,

[1] "Configuration interaction matrix element II. Graphical approach to the relationship between unitary group generators and permutations", *Int. Quantum Chemistry* **16** (1979) 1307.

Parikh, J.C.,

[1] "The role of isospin in pair correlations for configurations of the type $(j)^N$", *Nucl. Phys.* **63** (1965) 214.

Parikh, J.C.,

[1] *Group Symmetries in Nuclear Structure* (Plenum, New York, 1978).

Partensky, A.,

[1] "On the eigenvalues of the invariant operators of the unitary unimodular group $SU(n)$", *J. Math. Phys.* **13** (1972) 621.

[2] "On the generalized exchange operators for $SU(n)$", *J. Math. Phys.* **13** (1972) 1503.

Patterson, C.W. and W.G. Harter,

[1] "Canonical symmetrization for unitary bases I. Canonical Weyl bases", *J. Math. Phys.* **17** (1976) 1125.

[2] "Canonical symmetrization for unitary bases II. Boson and fermion bases", *J. Math. Phys.* **17** (1976) 1137.

Ping, Jia-Lun, Qing-Rong Zheng, Bing-Qing Chen and Jin-Quan Chen,

[1] "Computer generated subgroup-symmetry adapted irreducible representations and CG coefficients of space groups by the eigenfunction method", 1988 (to be published in *Computer Phys. Commun.*)

Rabl, V., G. Campbell, Jr. and K.C. Wali,

[1] "$SU(4)$ Clebsch-Gordan Coefficients", *J. Math. Phys.* **16** (1975) 249.

Racah, G.,

[1] *Group Theory and Spectroscopy*, Lecture notes in Princeton, 1951.

[2] *Group Theoretical Concepts and Methods in Elementary Particle Physics*, ed. F. Gursey, (Gordon and Breach, New York, 1964).

Robinson, G.de B.,

[1] *Representation Theory of the Symmetric Group* (Edinburgh University Press, Edinburgh, 1961).

Rose, M.E.,

[1] *Elementary Theory of Angular Momentum* (Wiley, New York, 1957).

Rotenberg, R., R. Bivins, N. Metropolis and J.K. Wooten,

[1] *The 3-j and 6-j Symbols* (MIT Press, Cambridge, Mass., 1959).

Rowe, D.J.,

[1] "Some recent advances in coherent state theory and its application to nuclear collective motion", in *Proc. of Topical Meeting on Phase Space Approach to Nuclear Dynamics*, ed. M. diToro, (World Scientific, Singapore, 1986).

Rowe, J., G. Rosensteel, and R. Gilmore,

[1] "Vector coherent state representation theory", *J. Math. Phys.* **26** (1985) 2787.

Rutherford, D.E.,

[1] *Substitutional Analysis* (Edinburgh University Press, Edinburgh, 1948).

Salam, A.,

[1] "The formalism of Lie groups", in *Theoretical Physics* (International Atomic Energy Agency, Vienna, 1963) p. 173.

Saulevich, L.K., D.T. Sviridov and Yu. F. Smirnov,

[1] *Sov. Phys. — Crystall.* **15** (1970) 355.

Shelepin, L.A.,

[1] "Group theoretical method in physics", in *Proc. of the P.N. Lebdev Phys. Institute*, Vol. 70, ed. S.Y. Skolbe'tsyn.

Shindler, S. and R. Mirman,

[1] "The decomposition of the tensor product of representations of the symmetric group", *J. Math. Phys.* **18** (1977) 1678.

[2] "Generation of the Clebsch-Gordan coefficients for S_n", *Computer Phys. Commun.* **15** (1978) 131.

Slater, J.C.,

[1] *Quantum Theory of Molecules and Solids*, Vol. I (McGraw Hill, New York, 1960).

[2] *Quantum Theory of Molecules and Solids*, Vol. II (McGraw Hill, New York, 1962).

[3] "Space groups and wave function symmetry in crystals", *Rev. Mod. Phys.* **37** (1965) 68.

Slater, J.C., G.F. Koster and J.H. Wood,

[1] "Symmetry and free electron properties of the Gallium energy bands", *Phys. Rev.* **126** (1962) 1307.

Smirnov, V.E.,

[1] *Advanced Mathematics*, Vol. III, Part A (Moscow, 1951) in Russian.

So, S.I. and D. Strottman,

[1] "Wigner coefficients for $SU(6)\supset SU(3)\times SU(2)$", *J. Math. Phys.* **20** (1979) 153.

Sokolov, A.V. and V.P. Shirokovskii,

[1] "Group theoretical method in quantum solid state physics", *Usp. Fiz. Nauk* **60** (1956) 617.

Strottman, D.,

[1] "Evaluation of $SU(6)\supset SU(3)\times SU(2)$ Wigner coefficients", *J. Math. Phys.* **20** (1979) 1643.

[2] "Multiquark baryons and the MIT bag model", *Phys. Rev.* **D20** (1979) 748.

Sullivan, J.J.,

[1] "Permutation symmetry and the *N*-electron problem", *Phys. Rev.* **A5** (1972) 29.

Sun, Hong-Zhou,

[1] "A study of the $SU(3)$ wave functions", *Scientia Sinica* **14** (1965) 480.

[2] "On the irreducible representations of the compact simple Lie groups of rank two I", *Phys. Energ. Fortis. Phys. Nucl.* **4** (1980) 73.

[3] "On the irreducible representations of the compact simple Lie groups of rank two II", *Phys. Energ. Fortis. Phys. Nucl.* **4** (1980) 137.

[4] "On the irreducible representations of the compact simple Lie groups of rank two III", *Phys. Energ. Fortis. Phys. Nucl.* **4** (1980) 271.

Sun, Hong-Zhou and Qi-Zhi Han,

[1] "On the irreducible representations of the simple Lie groups I. The tensor basis for the infinitesimal generators of the classical Lie groups", *Phys. Energ. Fortis. Phys. Nucl.* **4** (1980) 588.

Suzuki, Y. and K.T. Hecht,

[1] "Sympletic and cluster excitations in nuclei: Evaluation of interaction matrix elements", *Nucl. Phys.* **A455** (1986) 315.

Szydick, P. and C. Werntz,

[1] "Alpha particle continuum states", *Phys. Rev.* **B138** (1965) 866.

Tang, Au-Chin *et al.*,

[1] "Studies on the ligand field theory I", *Journal Jilin University*, No. **3** (1975) 57.

[2] "Studies on the ligand field theory II", *Journal Jilin University* No. **4** (1975) 73.

[3] "Studies on the ligand field theory", *Scientia Sinica* **15** (1966) 610.

Tang, Y.C., M. LeMere and D.R. Thompson,

[1] "Resonating-group method for nuclear many body problems", *Phys. Reports* **47** (1978) 168.

Tao, Rui-Bao,

[1] *Group Theory in Physics* (Science and Technology Press, Shanghai, 1986).

Vanagas, V.,

[1] *Algebraic Method in Nuclear Theory* (Mintis, Vilnius, 1972) in Russian.

Vergados, J.D.,

[1] "$SU(3)\supset R_3$ Wigner coefficients in the $2s$–$1d$ shell", *Nucl. Phys.* **A111** (1968) 681.

Wen, Chen-Yi,

[1] "A unitary group treatment for strong crystal field in octehedral symmetry", *Acta. Phys. Sinica* **28** (1979) 88.

Weyl, H.,

[1] *The Classical Groups* (Princeton University Press, Princeton, N.J., 1946).

Wigner, E.P.,

[1] *Group Theory and its Applications to the Quantum Mechanics of Atomic Spectra* (Academic Press, New York, 1959).

Willians, A.S. and D.L. Pursey,

[1] "Particle permutation symmetry of multishell states I. Two shells", *J. Math. Phys.* **17** (1976) 1383.

Worlton, T.G.,

[1] *Comput. Phys. Commun.* **6** (1973) 149.

Wormer, P.E.S. and J. Paldus,

[1] "Configuration interaction matrix elements I. Algebraic approach to the relationship between unitary group generators and permutations", *Int. Quantum Chemistry* **116** (1979) 1307.

Wu, C.L., D.H. Feng, X.G. Chen, J.Q. Chen and M. Guidry,

[1] "Fermion dynamical symmetries and nuclear shell model", *Phys. Lett.* **168B** (1986) 313;

[2] "A fermion dynamical symmetry model of nuclei (I) basis, Hamiltonian and symmetries", *Phys. Rev.* **C36** (1987) 1157.

Wu, C.L., X.L. Han, Z.P. Li, M. Guidry and D.H. Feng,

[1] "A microscopic formula for actinides masses", *Phys. Lett.* **194B** (1987) (in press).

Wu, H., D.H. Feng, C.L. Wu, Z.P. Li and M. Guidry,

[1] "Fermion dynamical symmetry for odd-mass nuclei", **193B** (1987) 163.

Wu, Xiong-Biao, Mei-Juan Gao and Jin Quan Chen,

[1] *Tables of the $SU(mn)\supset SU(m)\times SU(n)$ Coefficients of Fractional Parentage*, (World Scientific, Singapore, to be published).

Wu, H.C. and J.Q. Chen,

[1] "The calculation of the $SU(15)$ $SU(3)$ isoscalar factors", 1983 (unpublished).

Wybourne, B.G.,

[1] *Classical Groups for Physicists* (Wiley, New York, 1974).

Xu, G.O., S.J. Wang and Y.T. Yang,

[1] "Generator coordinate method as a representation theory of the dynamic group and related topics", preprint, 1985.

Yamanouchi, T.,

[1] "On the construction of unitary irreducible representation of the symmetric group", *Proc. Phys. — Math. Soc.* Japan **19** (1937) 436.

Yamanouchi, T. *et al.*,

[1] *The Application of the Rotational Group and Symmetric Group* (Beijing Science Press, Beijing, 1966).

Ye Chong-Yuan and Yi-Qin Wu,

[1] "The finding of irreducible bases for the point M of the space group O_h^3 by the eigenfunction method of group representation theory", *Anhui Daxue Xuebao* No. **2** (1980) 98.

Yilmaz, H.,

[1] *Introduction to the Theory of Relativity* (Blaisdell, Boston, 1965).

Young, A.,

[1] 1900~1935 *Proc. London Math. Soc.,* see Rutherford for references.

Zhang, Zong-Ye and Guang-Lie Li,

[1] "Symmetry classification for excited states of hyper nuclei", *Acta. Phys. Sinica* **26** (1977) 467.

Zhu, Chen-Jiu,

[1] "The complete operator of finite group", *J. Jilin Univ.* No. **4**, 1980, 87.

Zhu, Chen-Jun and Jin-Quan Chen,

[1] "A new approach to permutation group representation II", *J. Math. Phys.* **24** (1983) 2266.

Index

Inner multiplicity, 105
Inner product, 152
Interacting boson model, 113, 439
Intrinsic group, 73, 420
Intrinsic Lie group, 237
Intrinsic permutation group, 78, 98
Intrinsic rotation group, 283, 284
Intrinsic state, 77, 98, 292
Invariant integration, 232
Invariant subgroup, 16
Invariant subspace, 26
Invariant, 107, 236
Inverse element, 4
Inversion (space), 6
Irreducible basis,
  $O_3 \supset G \supset G(s)$ basis, 399
  $S_n \supset S_{n1} \otimes S_{n2}$ basis, 182
  $SU_{mn} \supset SU_m \times SU_n$ basis, 330
  $SU_{m+n} \supset SU_m \otimes SU_n$ basis, 338
Irreducible space, 36
Irreducible representation (irrep), 35
Irreducible tensor, 106
  SO_3 and $\overline{SO}_3$, 295
  of intrinsic Lie group, 242
Irreducibility,
  Criterion for, 40
Isomorphism, 10, 230
Isoscalar factor (*ISF*), 106
  $G \supset G(s)$, 106
  $S_n \supset S_{n-1}$, 191, 196-201
  $S_n \supset S_{n1} \otimes S_{n2}$, 201
  $S_f \supset S_{f-1}$ outer-product, 366
  $S_f \supset S_{f_{12}} \times S_{34}$ outer-product *ISF*, $S_f \supset S_{f_{12}} \times S_{f_{34}}$ *ISF*, 369
  $SU_{mn} \supset SU_m \times SU_n$ *ISF*, 341, 358
  $SU_{m+n} \supset SU_m \times SU_n$ *ISF*, 341

Kernel (of a homomorphism), 11
Kronecker product (direct product), 101

Labelling of irrep, 54, 60
Lattice,
  empty, 462
  reciprocal, 470
Lattice vector, 462, 470
Layer, 260
Left coset, 15, 486
Level (power) of a weight, 261
Lie algebra, 212
Lie group, 208
  SO_3, 210, 282
  SO_n, 217-224
  $O(m, n)$, 217-226
  SU_2, 209